9780852642153

THE ADVANCED THEORY
OF STATISTICS

VOLUME 2

INFERENCE AND RELATIONSHIP

Other books on theoretical statistics

Scientific truth and statistical method	M. Boldrini
	(trans. Ruth Kendall)
Time-series	M. G. Kendall
Rank correlation methods	M. G. Kendall
An introduction to the theory of statistics	G. U. Yule and M. G. Kendall
Statistical papers of George Udny Yule	ed. A. Stuart and M. G. Kendall
Studies in the history of statistics and probability	
	ed. E. S. Pearson and M. G. Kendall
Exercises in theoretical statistics	M. G. Kendall
Exercises in probability and statistics (for mathematics undergraduates)	
	N. A. Rahman
Practical exercises in probability and statistics	N. A. Rahman
A course in theoretical statistics	N. A. Rahman
Rapid statistical calculations	M. H. Quenouille
Characteristic functions	E. Lukacs
Experimental design: selected papers	F. Yates
Biomathematics	Cedric A. B. Smith
Estimation of animal abundance and related parameters	G. A. F. Seber
Statistical method in biological assay	D. J. Finney

From the Series **"Griffin's Statistical Monographs and Courses"**—

Fundamentals of statistical reasoning	M. H. Quenouille
The analysis of multiple time-series	M. H. Quenouille
A course in multivariate analysis	M. G. Kendall
Families of frequency distributions	J. K. Ord
Families of bivariate distributions	K. V. Mardia
The analysis of variance	A. Huitson
Inequalities on distribution functions	H. J. Godwin
The method of paired comparisons	H. A. David
Statistical models and their experimental application	P. Ottestad
Statistical tolerance regions: classical and Bayesian	I. Guttman
Applications of characteristic functions	E. Lukacs and R. G. Laha
The logit transformation, with special reference to its uses in Bioassay	W. D. Ashton
Regression estimation from grouped observations	Y. Haitovsky

Descriptive catalogue from Charles Griffin & Co. Ltd.

THE ADVANCED THEORY OF STATISTICS

MAURICE G. KENDALL, M.A., Sc.D.
Formerly Chairman, Scientific Control Systems Ltd
Formerly Professor of Statistics, University of London
President of the Royal Statistical Society, 1960–1962

and

ALAN STUART, D.Sc. (ECON.)
Professor of Statistics in the University of London

IN THREE VOLUMES

VOLUME 2
INFERENCE AND RELATIONSHIP
THIRD EDITION

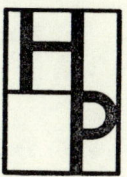

HAFNER PUBLISHING COMPANY
NEW YORK

Copyright © 1973
CHARLES GRIFFIN & COMPANY LIMITED
42 DRURY LANE, LONDON WC2B 5RX

All rights reserved

No part of this publication may be reproduced, stored in a retrieval system, or transmitted, in any form or by any means, electronic, mechanical, photocopying, recording or otherwise, without the prior permission of the Publishers, as above named.

THE ADVANCED THEORY OF STATISTICS

Volume 1			Volume 2		
First edition	..	1943	First edition	..	1946
Second ,,	..	1945	Second ,,	..	1947
Third ,,	..	1947	Third ,,	..	1951
Fourth ,,	..	1948	Second impression		1955
Fifth ,,	..	1952	Third ,,		1959
			Fourth ,,		1960

Three-volume edition

Volume 1: *Distribution Theory*
First edition .. 1958
Second ,, .. 1963
Third ,, .. 1969

Volume 2: *Inference and Relationship*
First edition .. 1961
Second ,, .. 1967
Third ,, .. 1973

Volume 3: *Design and Analysis, and Time-Series*
First edition .. 1966
Second ,, .. 1968

Made and printed in Great Britain by
Butler & Tanner Ltd., Frome and London

"You haven't told me yet," said Lady Nuttal, "what it is your fiancé does for a living."

"He's a statistician," replied Lamia, with an annoying sense of being on the defensive.

Lady Nuttal was obviously taken aback. It had not occurred to her that statisticians entered into normal social relationships. The species, she would have surmised, was perpetuated in some collateral manner, like mules.

"But Aunt Sara, it's a very interesting profession," said Lamia warmly.

"I don't doubt it," said her aunt, who obviously doubted it very much. "To express anything important in mere figures is so plainly impossible that there must be endless scope for well-paid advice on how to do it. But don't you think that life with a statistician would be rather, shall we say, humdrum?"

Lamia was silent. She felt reluctant to discuss the surprising depth of emotional possibility which she had discovered below Edward's numerical veneer.

"It's not the figures themselves," she said finally, "it's what you do with them that matters."

K. A. C. MANDERVILLE, *The Undoing of Lamia Gurdleneck*

PREFACE TO FIRST EDITION OF VOLUME TWO

We present herewith the second volume of this treatise on the advanced theory of statistics. It covers the theory of estimation and testing hypotheses, statistical relationship, distribution-free methods and sequential analysis. The third and concluding volume will comprise the design and analysis of sample surveys and experiments, including variance analysis, and the theory of multivariate analysis and time-series.

This volume bears very little resemblance to the original Volume 2 of Kendall's *Advanced Theory*. It has had to be planned and written practically *ab initio*, owing to the rapid development of the subject over the past fifteen years. A glance at the references will show how many of them have been published within that period.

As with the first volume, we have tried to make this volume self-contained in three respects: it lists its own references, it repeats the relevant tables given in the Appendices to Volume 1, and it has its own index. The necessity for taking up a lot of space with an extensive bibliography is being removed by the separate publication of Kendall and Doig's comprehensive *Bibliography of Statistical Literature*. We have made a special effort to provide a good set of exercises: there are about 400 in this volume.

For permission to quote some of the tables at the end of the book we are indebted to Professor Sir Ronald Fisher, Dr. Frank Yates, Messrs. Oliver and Boyd, and the editors of *Biometrika*. Mr. E. V. Burke of Charles Griffin and Company Limited has given his usual invaluable help in seeing the book through the press. We are also indebted to Mr. K. A. C. Manderville for permission to quote from an unpublished story the extract given on page v.

As always, we shall be glad to be notified of any errors, misprints or obscurities.

M. G. K.
A. S.

LONDON,
March, 1961

PREFACE TO THIRD EDITION

This volume has again been generally revised, as it was for the second edition. As a result, there are many minor and a few major changes in the text, a considerable number of new exercises, and many references to research published since the second edition appeared.

M. G. K.
A. S.

LONDON,
June 1972

CONTENTS

Chapter		Page
17.	Estimation	1
18.	Estimation: Maximum Likelihood	37
19.	Estimation: Least Squares and other Methods	78
20.	Interval Estimation: Confidence Intervals	103
21.	Interval Estimation: Other Methods	141
22.	Tests of Hypotheses: Simple Hypotheses	169
23.	Tests of Hypotheses: Composite Hypotheses	195
24.	Likelihood Ratio Tests and the General Linear Hypothesis	234
25.	The Comparison of Tests	273
26.	Statistical Relationship: Linear Regression and Correlation	290
27.	Partial and Multiple Correlation and Regression	330
28.	The General Theory of Regression	361
29.	Functional and Structural Relationship	391
30.	Tests of Fit	436
31.	Robust and Distribution-free Procedures	483
32.	Some Uses of Order-statistics	532
33.	Categorized Data	556
34.	Sequential Methods	612
	Appendix Tables	647
	References	659
	Index	696

GLOSSARY OF ABBREVIATIONS

The following abbreviations are sometimes used:

ARE	Asymptotic Relative Efficiency
ASN	Average Sample Number
AV	Analysis of Variance
BAN	Best Asymptotically Normal
BCR	Best Critical Region
BIB	Balanced Incomplete Blocks
c.f.	characteristic function
c.g.f.	cumulant-generating function
d.f.	distribution function
d.fr.	degrees of freedom
f.f.	frequency function
f.m.g.f.	factorial moment-generating function
LF	Likelihood Function
LR	Likelihood Ratio
LS	Least Squares
MCS	Minimum Chi-Square
m.g.f.	moment-generating function
ML	Maximum Likelihood
MS	Mean Square
m.s.e.	mean-square-error
MV	Minimum Variance
MVB	Minimum Variance Bound
$N(a, b)$	(multi-)normal with mean (-vector) a and variance (dispersion matrix) b
OC	Operating Characteristic
PBIB	Partially Balanced Incomplete Blocks
p.p.s.	probability proportional to size
s.e.	standard error
SPR	Sequential Probability Ratio
SS	Sum of Squares
UMP	Uniformly Most Powerful
UMPU	Uniformly Most Powerful Unbiassed
USF	Uniform Sampling Fraction

CHAPTER 17

ESTIMATION

The problem

17.1 On several occasions in previous chapters we have encountered the problem of estimating from a sample the values of the parameters of the parent population. We have hitherto dealt on somewhat intuitive lines with such questions as arose—for example, in the theory of large samples we have taken the means and moments of the sample to be satisfactory estimates of the corresponding means and moments in the parent.

We now proceed to study this branch of the subject in more detail. In the present chapter, we shall examine the sort of criteria which we require a " good " estimate to satisfy, and discuss the question whether there exist " best " estimates in an acceptable sense of the term. In the next few chapters, we shall consider methods of obtaining estimates with the required properties.

17.2 It will be evident that if a sample is not random and nothing precise is known about the nature of the bias operating when it was chosen, very little can be inferred from it about the parent population. Certain conclusions of a trivial kind are sometimes possible—for instance, if we take ten turnips from a pile of 100 and find that they weigh ten pounds altogether, the mean weight of turnips in the pile must be greater than one-tenth of a pound ; but such information is rarely of value, and estimation based on biassed samples remains very much a matter of individual opinion and cannot be reduced to exact and objective terms. We shall therefore confine our attention to random samples only. Our general problem, in its simplest terms, is then to estimate the value of a parameter in the parent from the information given by the sample. In the first instance we consider the case when only one parameter is to be estimated. The case of several parameters will be discussed later.

17.3 Let us in the first place consider what we mean by " estimation." We know, or assume as a working hypothesis, that the parent population is distributed in a form which is completely determinate but for the value of some parameter[*]θ. We are given a sample of observations $x_1, \ldots, x_n$. We require to determine, with the aid of observations, a number which can be taken to be the value of θ, or a range of numbers which can be taken to include that value.

Now the observations are random variables, and any function of the observations will also be a random variable. A function of the observations alone is called a *statistic*. If we use a statistic to estimate θ, it may on occasion differ considerably from the true value of θ. It appears, therefore, that we cannot expect to find any method of estimation which can be guaranteed to give us a close estimate of θ on every occasion and for every sample. We must content ourselves with formulating a rule which will give

---

[*] The term " parameter " is discussed in **22.3** below.

good results " in the long run " or " on the average," or which has " a high probability of success "—phrases which express the fundamental fact that we have to regard our method of estimation as generating a distribution of estimates and to assess its merits according to the properties of this distribution.

17.4 It will clarify our ideas if we draw a distinction between the method or rule of estimation, which we shall call an estimator, and the value to which it gives rise in particular cases, the estimate. The distinction is the same as that between a function $f(x)$, regarded as defined for a range of the variable x, and the particular value which the function assumes, say $f(a)$, for a specified value of x equal to a. Our problem is not to find estimates, but to find estimators. We do not reject an estimator because it may give a bad result in a particular case (in the sense that the estimate differs materially from the true value). We should only reject it if it gave bad results in the long run, that is to say, if the distribution of possible values of the estimator were seriously discrepant with the true value of θ. The merit of the estimator is judged by the distribution of estimates to which it gives rise, i.e. by the properties of its sampling distribution.

17.5 In the theory of large samples, we have often taken as an estimator of a parameter θ a statistic t calculated from the sample in exactly the same way as θ is calculated from the population : e.g. the sample mean is taken as an estimator of the parent mean. Let us examine how this procedure can be justified. Consider the case when the parent population is

$$dF(x) = (2\pi)^{-\frac{1}{2}} \exp\{-\tfrac{1}{2}(x-\theta)^2\}\, dx, \qquad -\infty \leqslant x \leqslant \infty. \tag{17.1}$$

Requiring an estimator for the parameter θ, which is the population mean, we take the sample mean

$$t = \sum_{j=1}^{n} x_j/n. \tag{17.2}$$

The distribution of t is (Example 11.12)

$$dF(t) = \{n/(2\pi)\}^{\frac{1}{2}} \exp\{-\tfrac{1}{2}n(t-\theta)^2\}\, dt, \tag{17.3}$$

that is to say, t is distributed normally about θ with variance $1/n$. We notice two things about this distribution : (a) it has a mean (and median and mode) at the true value θ, and (b) as n increases, the scatter of possible values of t about θ becomes smaller, so that the probability that a given t differs by more than a fixed amount from θ decreases. We may say that the accuracy of the estimator increases as n increases, i.e. with n.

17.6 Generally, it will be clear that the phrase " accuracy increasing with n " has a definite meaning whenever the sampling distribution of t has a variance which decreases with $1/n$ and a central value which is either identical with θ or differs from it by a quantity which also decreases with $1/n$. Many of the estimators with which we are commonly concerned are of this type, but there are exceptions. Consider, for example, the Cauchy population

$$dF(x) = \frac{1}{\pi} \frac{dx}{\{1+(x-\theta)^2\}}, \qquad -\infty \leqslant x \leqslant \infty. \tag{17.4}$$

If we estimate θ by the mean-statistic t we have, for the distribution of t,
$$dF(t) = \frac{1}{\pi} \frac{dt}{\{1+(t-\theta)^2\}}, \qquad (17.5)$$
(cf. Example 11.1). In this case the distribution of t is the same as that of any single value of the sample, and does not increase in accuracy as n increases.

Consistency

17.7 The possession of the property of increasing accuracy is evidently a very desirable one; and indeed, if the variance of the sampling distribution of an estimator decreases with increasing n, it is necessary that its central value should tend to θ, for otherwise the estimator would have values differing systematically from the true value. We therefore formulate our first criterion for a suitable estimator as follows:—

An estimator t_n, computed from a sample of n values, will be said to be a consistent estimator of θ if, for any positive ε and η, however small, there is some N such that the probability that
$$|t_n - \theta| < \varepsilon \qquad (17.6)$$
is greater than $1-\eta$ for all $n > N$. In the notation of the theory of probability,
$$P\{|t_n - \theta| < \varepsilon\} > 1-\eta, \qquad n > N. \qquad (17.7)$$

The definition bears an obvious analogy to the definition of convergence in the mathematical sense. Given any fixed small quantity ε, we can find a large enough sample number such that, for all samples over that size, the probability that t differs from the true value by more than ε is as near zero as we please. t_n is said to *converge in probability*, or to *converge stochastically*, to θ. Thus t is a consistent estimator of θ if it converges to θ in probability.

Example 17.1

The sample mean is a consistent estimator of the parameter θ in the population (17.1). This we have already established in general argument, but more formally the proof would proceed as follows:—

Suppose we are given ε. From (17.3) we see that $(t-\theta)n^{\frac{1}{2}}$ is distributed normally about zero with unit variance. Thus the probability that $|(t-\theta)n^{\frac{1}{2}}| \leq \varepsilon n^{\frac{1}{2}}$ is the value of the normal integral between limits $\pm \varepsilon n^{\frac{1}{2}}$. Given any positive η, we can always take n large enough for this quantity to be greater than $1-\eta$ and it will continue to be so for any larger n. N may therefore be determined and the inequality (17.7) is satisfied.

Example 17.2

Suppose we have a statistic t_n whose mean value differs from θ by terms of order n^{-1}, whose variance v_n is of order n^{-1} and which tends to normality as n increases. Clearly, as in Example 17.1, $(t_n-\theta)/v_n^{\frac{1}{2}}$ will then tend to zero in probability and t_n will be consistent. This covers a great many statistics encountered in practice.

Even if the limiting distribution of t_n is unspecified, the result will still hold, as can be seen from a direct application of the Bienaymé–Tchebycheff inequality (3.94). In fact, if $E(t_n) = \theta + k_n$, and $\operatorname{var} t_n = v_n$, where $\lim_{n\to\infty} k_n = \lim_{n\to\infty} v_n = 0$, we have at once

$$P\{|t_n - (\theta + k_n)| < \varepsilon\} \geqslant 1 - \frac{v_n}{\varepsilon^2} \xrightarrow[n\to\infty]{} 1,$$

so that (17.7) will be satisfied.

Unbiassed estimators

17.8 The property of consistency is a limiting property, that is to say, it concerns the behaviour of an estimator as the sample number tends to infinity. It requires nothing of the estimator's behaviour for finite n, and if there exists one consistent estimator t_n we may construct infinitely many others: e.g. for fixed a and b, $\frac{n-a}{n-b} t_n$ is also consistent. We have seen that in some circumstances a consistent estimator of the population mean is the sample mean $\bar{x} = \Sigma x_j / n$. But so is $\bar{x}' = \Sigma x_j / (n-1)$. Why do we prefer one to the other? Intuitively it seems absurd to divide the sum of n quantities by anything other than their number n. We shall see in a moment, however, that intuition is not a very reliable guide in such matters. There are reasons for preferring

$$\frac{1}{(n-1)} \sum_{j=1}^{n} (x_j - \bar{x})^2$$

to

$$\frac{1}{n} \sum_{j=1}^{n} (x_j - \bar{x})^2$$

as an estimator of the parent variance, notwithstanding the fact that the latter is the sample variance.

17.9 Consider the sampling distribution of an estimator t. If the estimator is consistent, its distribution must, for large samples, have a central value in the neighbourhood of θ. We may choose among the class of consistent estimators by requiring that θ shall be equated to this central value not merely for large, but for all samples.

If we require that for all n and θ the *mean* value of t shall be θ, we define what is known as an *unbiassed* estimator by the relation

$$E(t) = \theta. \tag{17.8}$$

This is an unfortunate word, like so many in statistics. There is nothing except convenience to exalt the arithmetic mean above other measures of location as a criterion of bias. We might equally well have chosen the median of the distribution of t or its mode as determining the "unbiassed" estimator. The mean value is used, as always, for its mathematical convenience. This is perfectly legitimate, and it is only necessary to remark that the term "unbiassed" should not be allowed to convey overtones of a non-technical nature.

Example 17.3

Since
$$E\left\{\frac{1}{n}\Sigma x\right\} = \frac{1}{n}\Sigma E(x) = \mu_1',$$

the mean-statistic is an unbiassed estimator of the parent mean whenever the latter exists. But the sample variance is not an unbiassed estimator of the parent variance, for
$$E\{\Sigma(x_j-\bar{x})^2\} = E\{\Sigma[x_j-\Sigma x_j/n]^2\} = E\left\{\frac{n-1}{n}\Sigma_j x_j^2 - \frac{1}{n}\Sigma\Sigma_{j\neq k} x_j x_k\right\}$$
$$= (n-1)\mu_2' - (n-1)\mu_1'^2 = (n-1)\mu_2.$$

Thus $\frac{1}{n}\Sigma(x-\bar{x})^2$ has a mean value $\frac{n-1}{n}\mu_2$. It follows that an unbiassed estimator is given by
$$s^2 = \frac{1}{n-1}\Sigma(x-\bar{x})^2,$$

and for this reason it is usually preferred to the sample variance.

Our discussion shows that consistent estimators are not necessarily unbiassed. We have already (Example 14.5) encountered an unbiassed estimator which is not consistent. Thus neither property implies the other. But a consistent estimator whose asymptotic distribution has finite mean value must be asymptotically unbiassed.

In certain circumstances, there may be no unbiassed estimator (cf. Exercise 17.12). Even if there is one, it may occur that it necessarily gives absurd results at times, or even always—cf. Exercise 17.26, due to E. L. Lehmann. For example, in estimating a parameter θ, $0 \leq \theta \leq 1$, no statistic distributed in the range $(0, 1)$ can be unbiassed, for if $\theta = 0$ its expectation must (except in trivial degenerate cases) exceed θ. We shall meet an important example of this in **27.34** below. Thus "impossible" estimates must sometimes occur if we insist on unbiassedness. See also Exercises 10.21–22 (Vol. 1).

Again, the sequential binomial sampling scheme discussed in Example 9.13 (Vol. 1) and in Example 34.1 below defines an unbiassed estimator of a probability which (when $m = 1$ in (34.1)) can only take the values 0 and 1.

Corrections for bias

17.10 If we have a biassed estimator t, and wish to remove its bias, this may be possible by direct evaluation of its expected value and the application of a simple adjustment, as in Example 17.3. But sometimes the expected value is a rather complicated function of the parameter, θ, being estimated, and it is not obvious what the correction should be. Quenouille (1956) has proposed an ingenious method of overcoming this difficulty in a fairly general class of situations.

We denote our biassed estimator by t_n, its suffix being the number of observations from which t_n is calculated. Now suppose that

$$E(t_n) - \theta = \sum_{r=1}^{\infty} a_r/n^r, \qquad (17.9)$$

where the a_r may be functions of θ but not of n. t_{n-1} may be calculated from each of the n possible subsets of $(n-1)$ observations. Let $\bar{t}_{n-1}$ denote the average of these n values, and consider the new statistic

$$t'_n = n t_n - (n-1)\bar{t}_{n-1} = t_n + (n-1)(t_n - \bar{t}_{n-1}). \tag{17.10}$$

It follows at once from (17.9) that

$$E(t'_n) - \theta = a_2\left(\frac{1}{n} - \frac{1}{n-1}\right) + a_3\left(\frac{1}{n^2} - \frac{1}{(n-1)^2}\right) + \cdots$$

$$= -\frac{a_2}{n^2} - O(n^{-3}). \tag{17.11}$$

Thus t'_n is only biassed to order $1/n^2$. Similarly,

$$t''_n = \{n^2 t'_n - (n-1)^2 \bar{t}'_{n-1}\}/\{n^2 - (n-1)^2\}$$

will be biassed only to order $1/n^3$, and so on. This method, which is usually called *the jack-knife*, provides a very direct means of removing bias to any required degree.

Because t'_n differs from t_n by a quantity of order n^{-1}, we see that if t_n has variance of order n^{-1} the variance of t'_n is asymptotically the same as that of t_n, so that reduction of bias carries no penalty in the variance—cf. Exercise 17.18; but this is not generally true for further-order corrections—cf. Robson and Whitlock (1964).

R. G. Miller (1964) discusses the asymptotic normality of t'_n and the estimation of its variance. See also Arvesen (1969), Schucany *et al.* (1971), Adams *et al.* (1971) and Gray *et al.* (1972).

Example 17.4

To find an unbiassed estimator of θ^2 in the binomial distribution

$$P\{x = r\} = \binom{n}{r}\theta^r (1-\theta)^{n-r}, \qquad r = 0, 1, 2, \ldots, n.$$

The intuitive estimator is

$$t_n = \left(\frac{r}{n}\right)^2,$$

since r/n is an unbiassed estimator of θ, but

$$E(t_n) = \operatorname{var}(r/n) + \{E(r/n)\}^2 = \theta^2 + \theta(1-\theta)/n.$$

Now t_{n-1} can only take the values

$$\left(\frac{r-1}{n-1}\right)^2 \quad \text{or} \quad \left(\frac{r}{n-1}\right)^2$$

according to whether a "success" or a "failure" is omitted from the sample. Thus

$$\bar{t}_{n-1} = \frac{1}{n}\left\{r\left(\frac{r-1}{n-1}\right)^2 + (n-r)\left(\frac{r}{n-1}\right)^2\right\} = \frac{r^2(n-2)+r}{n(n-1)^2}.$$

Hence, from (17.10),

$$t'_n = n t_n - (n-1)\bar{t}_{n-1}$$

$$= \frac{r^2}{n} - \frac{r^2(n-2)+r}{n(n-1)}$$

$$= \frac{r(r-1)}{n(n-1)}, \tag{17.12}$$

which, it may be directly verified, is exactly unbiassed for θ^2.

Exercises 17.17 and 17.13 give further applications of the method.

17.11 In general there will exist more than one consistent estimator of a parameter, even if we confine ourselves only to unbiassed estimators. Consider once again the estimation of the mean of a normal population with variance σ^2. The sample mean is consistent and unbiassed. We will now prove that the same is true of the median.

Consideration of symmetry is enough to show that the median is an unbiassed estimator of the population mean, which is, of course, the same as the population median. For large n the distribution of the median tends to the normal form (cf. **14.12**)

$$dF(x) \propto \exp\{-2nf_1^2(x-\theta)^2\}\,dx, \tag{17.13}$$

where f_1 is the population median ordinate, in our case equal to $(2\pi\sigma^2)^{-\frac{1}{2}}$. The variance of the sample median is therefore, from (17.13), equal to $\pi\sigma^2/(2n)$ and tends to zero for large n. Hence the estimator is consistent.

17.12 We must therefore seek further criteria to choose between estimators with the common property of consistency. Such a criterion arises naturally if we consider the sampling variances of the estimators. Generally speaking, the estimator with the smaller variance will be distributed more closely round the value θ; this will certainly be so for distributions of the normal type. An unbiassed consistent estimator with a smaller variance will therefore deviate less, on the average, from the true value than one with a larger variance. Hence we may reasonably regard it as better.

In the case of the mean and median of normal samples we have, for any n, from (17.3),

$$\text{var}(\text{mean}) = \sigma^2/n, \tag{17.14}$$

whereas (14.18) showed that for $n = 2r+1$, the variance of the median is, putting $\sigma^2 = 1$,

$$\text{var}(x_{(r+1)}) = \frac{\pi}{2(n+2)} + \frac{\pi^2}{4(n+2)(n+4)} + o(n^{-2}),$$

whence

$$\text{var}(\text{mean})/\text{var}(\text{median}) = \frac{2}{\pi} + \frac{1}{n}\left(\frac{4}{\pi} - 1\right) + o(n^{-1}). \tag{17.15}$$

Hodges (1967) gives the following exact values:

n:	1	3	5	7	9	11	13	15	17	19	∞
$\dfrac{\text{var}(\text{mean})}{\text{var}(\text{median})}$:	1	0·743	0·697	0·679	0·669	0·663	0·659	0·656	0·653	0·651	0·637

The approximation (17.15) is accurate to two decimal places for $n \geq 7$. The values for even n are always higher than the contiguous values for odd n—the definition of the median is different in this case—but tend monotonically to the same limit from the value 1 at $n = 2$. (The approximation (17.15) must now have the figure " 4 " replaced by " 6 ".) A few values of $\{\text{var}(\text{median})/\text{var}(\text{mean})\}^{\frac{1}{2}}$ for even n were given in **14.6**. Thus the mean always has smaller variance than the median in the normal case.

Example 17.5

For the Cauchy distribution
$$dF(x) = \frac{1}{\pi} \frac{dx}{\{1+(x-\theta)^2\}}, \quad -\infty \leqslant x \leqslant \infty,$$
we have already seen (**17.6**) that the sample mean is not a consistent estimator of θ, the population median. However, for the sample median, t, we have, since the median ordinate is $1/\pi$, the large-sample variance
$$\operatorname{var} t = \pi^2/4n$$
from (17.13). It is seen that the median is consistent, and although direct comparison with the mean is not possible because the latter does not possess a sampling variance, the median is evidently a better estimator of θ than the mean. This provides an interesting contrast with the case of the normal distribution, particularly in view of the similarity of the distributions.

> Pitman (1937c) defined t to be "closer" to θ than u is if $P\{|t-\theta|<|u-\theta|\} > \frac{1}{2}$, but this attractive concept is intransitive, as he pointed out. See also Geary (1944) and Johnson (1950).

Minimum variance estimators

17.13 It seems natural, then, to use the sampling variance of an estimator as a criterion of its acceptability, and it has, in fact, been so used since the days of Laplace and Gauss. But only in relatively recent times has it been established that, under fairly general conditions, there exists a bound below which the variance of an unbiassed estimator cannot fall. In order to establish this bound, we first derive some preliminary results, which will also be useful in other connexions later.

17.14 If the frequency function of the continuous or discrete population is $f(x|\theta)$, we define the *Likelihood Function*[*] of a sample of n independent observations by
$$L(x_1, x_2, \ldots, x_n | \theta) = f(x_1|\theta)f(x_2|\theta)\ldots f(x_n|\theta). \tag{17.16}$$
We shall often write this simply as L. Evidently, since L is the joint frequency function of the observations,
$$\int \ldots \int L \, dx_1 \ldots dx_n = 1. \tag{17.17}$$
Now suppose that the first two derivatives of L with respect to θ exist for all θ. If we differentiate both sides of (17.17) with respect to θ, and if we may interchange the operations of differentiation and integration on its left-hand side, we obtain
$$\int \ldots \int \frac{\partial L}{\partial \theta} dx_1 \ldots dx_n = 0,$$

---

[*] R. A. Fisher calls L the likelihood when regarded as a function of θ and the probability of the sample when it is regarded as a function of x for fixed θ. While appreciating this distinction, we use the term likelihood and the symbol L in both cases to preserve a single notation.

which we may rewrite

$$E\left(\frac{\partial \log L}{\partial \theta}\right) = \int \cdots \int \left(\frac{1}{L}\frac{\partial L}{\partial \theta}\right) L\, dx_1 \ldots dx_n = 0. \tag{17.18}$$

If we differentiate (17.18) again, we obtain, if we may again interchange operations,

$$\int \cdots \int \left\{\left(\frac{1}{L}\frac{\partial L}{\partial \theta}\right)\frac{\partial L}{\partial \theta} + L\frac{\partial}{\partial \theta}\left(\frac{1}{L}\frac{\partial L}{\partial \theta}\right)\right\} dx_1 \ldots dx_n = 0,$$

which becomes

$$\int \cdots \int \left\{\left(\frac{1}{L}\frac{\partial L}{\partial \theta}\right)^2 + \frac{\partial^2 \log L}{\partial \theta^2}\right\} L\, dx_1 \ldots dx_n = 0$$

or

$$E\left[\left(\frac{\partial \log L}{\partial \theta}\right)^2\right] = -E\left(\frac{\partial^2 \log L}{\partial \theta^2}\right). \tag{17.19}$$

17.15 Now consider an unbiassed estimator, t, of some function of θ, say $\tau(\theta)$. This formulation allows us to consider unbiassed and biassed estimators of θ itself, and also permits us to consider, for example, the estimation of the standard deviation when the parameter is equal to the variance. We thus have

$$E(t) = \int \cdots \int t L\, dx_1 \ldots dx_n = \tau(\theta). \tag{17.20}$$

We now differentiate (17.20), the result being

$$\int \cdots \int t \frac{\partial \log L}{\partial \theta} L\, dx_1 \ldots dx_n = \tau'(\theta),$$

which we may re-write, using (17.18), as

$$\tau'(\theta) = \int \cdots \int \{t - \tau(\theta)\} \frac{\partial \log L}{\partial \theta} L\, dx_1 \ldots dx_n. \tag{17.21}$$

By the Cauchy–Schwarz inequality, we have from (17.21)

$$\{\tau'(\theta)\}^2 \leq \int \cdots \int \{t - \tau(\theta)\}^2 L\, dx_1 \ldots dx_n \cdot \int \cdots \int \left(\frac{\partial \log L}{\partial \theta}\right)^2 L\, dx_1 \ldots dx_n,$$

which, on rearrangement, becomes

$$\operatorname{var} t = E\{t - \tau(\theta)\}^2 \geq \{\tau'(\theta)\}^2 / E\left[\left(\frac{\partial \log L}{\partial \theta}\right)^2\right]. \tag{17.22}$$

This is the fundamental inequality for the variance of an estimator, often known as the Cramér–Rao inequality, after two of its several discoverers (C. R. Rao (1945); Cramér (1946)); it was first given implicitly by Aitken and Silverstone (1942). Using (17.19), it may be written in what is often, in practice, the more convenient form

$$\operatorname{var} t \geq -\{\tau'(\theta)\}^2 / E\left(\frac{\partial^2 \log L}{\partial \theta^2}\right). \tag{17.23}$$

We shall call (17.22) and (17.23) the minimum variance bound (abbreviated to MVB) for the estimation of $\tau(\theta)$. An estimator which attains this bound for all θ will be called a MVB estimator.

It is only necessary that (17.18) hold for the MVB (17.22) to follow from (17.20). If (17.19) also holds, we may write the MVB in the form (17.23).

The interchange of the operations of differentiation and integration leading to (17.18) or (17.19) is permissible if, e.g., the limits of integration (i.e. the limits of variation of x) are finite and independent of θ, and also if these limits are infinite, provided that the integral resulting from the interchange is uniformly convergent for all θ and its integrand is a continuous function of x and θ. These are sufficient sets of conditions—Exercises 17.21–22 show that they are not necessary.

17.16 In the case where $\tau(\theta) \equiv \theta$, we have $\tau'(\theta) = 1$ in (17.22), so for an *unbiassed* estimator of θ

$$\operatorname{var} t \geqslant 1/E\left[\left(\frac{\partial \log L}{\partial \theta}\right)^2\right] = -1/E\left(\frac{\partial^2 \log L}{\partial \theta^2}\right). \tag{17.24}$$

In this case the quantity I defined as

$$I = E\left[\left(\frac{\partial \log L}{\partial \theta}\right)^2\right] \tag{17.25}$$

is sometimes called the *amount of information* in the sample, although this is not a universal usage.

17.17 It is very easy to establish the condition under which the MVB is attained. The inequality in (17.22) arose purely from the use of the Cauchy–Schwarz inequality, and the necessary and sufficient condition that the Cauchy–Schwarz inequality becomes an equality is (cf. **2.7**) that $\{t - \tau(\theta)\}$ is proportional to $\dfrac{\partial \log L}{\partial \theta}$ for all sets of observations. We may write this condition

$$\frac{\partial \log L}{\partial \theta} = A \cdot \{t - \tau(\theta)\}, \tag{17.26}$$

where A is independent of the observations but may be a function of θ. Thus (17.26) becomes

$$\frac{\partial \log L}{\partial \theta} = A(\theta)\{t - \tau(\theta)\}. \tag{17.27}$$

Multiplying both sides of (17.27) by $\{t - \tau(\theta)\}$ and taking expectations, we find, using (17.21),

$$\tau'(\theta) = A(\theta)\operatorname{var} t. \tag{17.28}$$

Thus $A(\theta)$ has the same sign as $\tau'(\theta)$ and

$$\operatorname{var} t = \tau'(\theta)/A(\theta). \tag{17.29}$$

We thus conclude that if (17.27) is satisfied, t is a MVB estimator of $\tau(\theta)$, with variance (17.29), which is then equal to the right-hand side of (17.23). If $\tau(\theta) \equiv \theta$, $\operatorname{var} t$ is just $1/A(\theta)$, which is then equal to the right-hand side of (17.24).

Example 17.6

To estimate θ in the normal population

$$dF(x) = \frac{1}{\sigma(2\pi)^{\frac{1}{2}}} \exp\left\{-\frac{1}{2}\left(\frac{x-\theta}{\sigma}\right)^2\right\} dx, \quad -\infty \leqslant x \leqslant \infty,$$

where σ is known.

We have

$$\frac{\partial \log L}{\partial \theta} = \frac{n}{\sigma^2}(\bar{x} - \theta).$$

This is of the form (17.27) with
$$t = \bar{x}, \quad A(\theta) = n/\sigma^2 \quad \text{and} \quad \tau(\theta) = \theta.$$
Thus $\bar{x}$ is the MVB estimator of θ, with variance σ^2/n.

Example 17.7

To estimate θ in the Cauchy distribution
$$dF(x) = \frac{1}{\pi} \frac{dx}{\{1+(x-\theta)^2\}}, \quad -\infty \leqslant x \leqslant \infty.$$
We have
$$\frac{\partial \log L}{\partial \theta} = 2\Sigma \frac{x-\theta}{\{1+(x-\theta)^2\}}.$$
This cannot be put in the form (17.27). Thus there is no MVB estimator in this case.

Example 17.8

To estimate θ in the Poisson distribution
$$f(x|\theta) = e^{-\theta}\theta^x/x!, \quad x = 0, 1, 2, \ldots, \infty.$$
We have
$$\frac{\partial \log L}{\partial \theta} = \frac{n}{\theta}(\bar{x}-\theta).$$
Thus $\bar{x}$ is the MVB estimator of θ, with variance θ/n.

Example 17.9

To estimate θ in the binomial distribution, for which
$$L(r|\theta) = \binom{n}{r}\theta^r(1-\theta)^{n-r}, \quad r = 0, 1, 2, \ldots, n.$$
We find
$$\frac{\partial \log L}{\partial \theta} = \frac{n}{\theta(1-\theta)}\left(\frac{r}{n}-\theta\right).$$
Hence r/n is the MVB estimator of θ, with variance $\theta(1-\theta)/n$.

17.18 It follows from our discussion that, where a MVB estimator exists, it will exist for one specific function $\tau(\theta)$ of the parameter θ, and for no other function of θ. The following example makes the point clear.

Example 17.10

To estimate θ in the normal distribution
$$dF(x) = \frac{1}{\theta(2\pi)^{\frac{1}{2}}} \exp\left(-\frac{x^2}{2\theta^2}\right)dx, \quad -\infty \leqslant x \leqslant \infty.$$
We find
$$\frac{\partial \log L}{\partial \theta} = -\frac{n}{\theta} + \frac{\Sigma x^2}{\theta^3} = \frac{n}{\theta^3}\left(\frac{1}{n}\Sigma x^2 - \theta^2\right).$$

We see at once that $\frac{1}{n}\Sigma x^2$ is a MVB estimator of θ^2 (the variance of the population) with sampling variance $\frac{\theta^3}{n}\cdot\frac{d}{d\theta}(\theta^2) = \frac{2\theta^4}{n}$, by (17.29). But there is no MVB estimator of θ itself.

Equation (17.27) determines a condition on the frequency function under which a MVB estimator of some function of θ, $\tau(\theta)$, exists. If the frequency is not of this form, there may still be an estimator of $\tau(\theta)$ which has, uniformly in θ, smaller variance than any other estimator; we then call it a minimum variance (MV) estimator. In other words, the least *attainable* variance may be greater than the MVB. Further, if the regularity conditions leading to the MVB do not hold, the least attainable variance may be less than the (in this case inapplicable) MVB. In any case, (17.27) demonstrates that there can only be one function of θ for which the MVB is attainable, namely, that function (if any) which is the expectation of a statistic t in terms of which $\partial \log L/\partial\theta$ may be linearly expressed.

17.19 From (17.27) we have on integration the necessary form for the Likelihood Function (continuing to write $A(\theta)$ for the integral of the arbitrary function $A(\theta)$ in (17.27))

$$\log L = tA(\theta) + P(\theta) + R(x_1, x_2, \ldots, x_n),$$

which we may re-write in the frequency-function form

$$f(x|\theta) = \exp\{A(\theta)B(x) + C(x) + D(\theta)\}, \qquad (17.30)$$

where $t = \sum_{i=1}^{n} B(x_i)$, $R(x_1, \ldots, x_n) = \sum_{i=1}^{n} C(x_i)$ and $P(\theta) = nD(\theta)$. (17.30) is often called the exponential family of distributions. We shall return to it in **17.36**.

17.20 We can find better (i.e. greater) lower bounds than the MVB (17.22) for the variance of an estimator in cases where the MVB is not attainable. The essential condition for (17.22) to be an attainable bound is that there be an estimator t for which $t - \tau(\theta)$ is a linear function of $\frac{\partial \log L}{\partial \theta} = \frac{1}{L}\frac{\partial L}{\partial \theta}$. But even if no such estimator exists, there may still be one for which $t - \tau(\theta)$ is a linear function of $\frac{1}{L}\frac{\partial L}{\partial \theta}$ and $\frac{1}{L}\frac{\partial^2 L}{\partial \theta^2}$ or, in general, of the higher derivatives of the Likelihood Function. It is this fact which leads to the following considerations, due to Bhattacharyya (1946).

Estimating $\tau(\theta)$ by a statistic t as before, we write

$$L^{(r)} = \frac{\partial^r L}{\partial \theta^r}$$

and

$$\tau^{(r)} = \frac{\partial^r \tau(\theta)}{\partial \theta^r}.$$

ESTIMATION

We now construct the function
$$D_s = t - \tau(\theta) - \sum_{r=1}^{s} a_r L^{(r)}/L, \tag{17.31}$$
where the a_r are constants to be determined. Since, on differentiating (17.17) r times, we obtain
$$E(L^{(r)}/L) = 0 \tag{17.32}$$
under the same conditions as before, we have from (17.31) and (17.32)
$$E(D_s) = 0. \tag{17.33}$$
The variance of D_s is therefore
$$E(D_s^2) = \int \ldots \int \{(t-\tau(\theta)) - \sum_r a_r L^{(r)}/L\}^2 L\, dx_1 \ldots dx_n. \tag{17.34}$$
We minimize (17.34) for variation of the a_r by putting
$$\int \ldots \int \{(t-\tau(\theta)) - \sum_r a_r L^{(r)}/L\} \frac{L^{(p)}}{L} \cdot L\, dx_1 \ldots dx_n = 0 \tag{17.35}$$
for $p = 1, 2, \ldots, s$. This gives
$$\int \ldots \int (t-\tau(\theta)) L^{(p)}\, dx_1 \ldots dx_n = \sum_r a_r \int \ldots \int \frac{L^{(r)}}{L} \cdot \frac{L^{(p)}}{L} L\, dx_1 \ldots dx_n. \tag{17.36}$$
The left-hand side of (17.36) is, from (17.32), equal to
$$\int \ldots \int t L^{(p)}\, dx_1 \ldots dx_n = \tau^{(p)}$$
on comparison with (17.20). The right-hand side of (17.36) is simply
$$\sum a_r E\left(\frac{L^{(r)}}{L} \cdot \frac{L^{(p)}}{L}\right).$$
On insertion of these values in (17.36) it becomes
$$\tau^{(p)} = \sum_{r=1}^{s} a_r E\left(\frac{L^{(r)}}{L} \cdot \frac{L^{(p)}}{L}\right), \qquad p = 1, 2, \ldots, s. \tag{17.37}$$
We may invert this set of linear equations, provided that the matrix of coefficients $J_{rp} = E\left(\frac{L^{(r)}}{L} \cdot \frac{L^{(p)}}{L}\right)$ is non-singular, to obtain
$$a_r = \sum_{p=1}^{s} \tau^{(p)} J_{rp}^{-1}, \qquad r = 1, 2, \ldots, s. \tag{17.38}$$
Thus, at the minimum of (17.34), (17.31) takes the value
$$D_s = t - \tau(\theta) - \sum_{r=1}^{s} \sum_{p=1}^{s} \tau^{(p)} J_{rp}^{-1} L^{(r)}/L, \tag{17.39}$$
and (17.34) itself has the value, from (17.39),
$$E(D_s^2) = \int \ldots \int \{(t-\tau(\theta)) - \sum_r \sum_p \tau^{(p)} J_{rp}^{-1} L^{(r)}/L\}^2 L\, dx_1 \ldots dx_n,$$
which, on using (17.35), becomes
$$= \int \ldots \int \{t-\tau(\theta)\}^2 L\, dx_1 \ldots dx_n - \sum_r \sum_p \tau^{(p)} J_{rp}^{-1} \int \ldots \int t L^{(r)}\, dx_1 \ldots dx_n,$$

and finally we have
$$E(D_s^2) = \operatorname{var} t - \sum_r \sum_p \tau^{(p)} J_{rp}^{-1} \tau^{(r)}. \tag{17.40}$$
Since its left-hand side is non-negative, (17.40) gives the required inequality
$$\operatorname{var} t \geqslant \sum_{r=1}^{s} \sum_{p=1}^{s} \tau^{(p)} J_{rp}^{-1} \tau^{(r)}. \tag{17.41}$$
In the particular case $s = 1$, (17.41) reduces to the MVB (17.22).

17.21 The condition for the bound in (17.41) to be attained is simply that $E(D_s^2) = 0$, which with (17.33) leads to $D_s = 0$ or, from (17.39),
$$t - \tau(\theta) = \sum_{r=1}^{s} \sum_{p=1}^{s} \tau^{(p)} J_{rp}^{-1} L^{(r)}/L, \tag{17.42}$$
which is the generalization of (17.27). (17.42) requires $t - \tau(\theta)$ to be a linear function of the quantities $L^{(r)}/L$. If it is a linear function of the first s such quantities, there is clearly nothing to be gained by adding further terms. On the other hand, the right-hand side of (17.41) is a non-decreasing function of s, as is easily seen by the consideration that the variance of D_s cannot be increased by allowing the optimum choice of a further coefficient a_r in (17.34). Thus we may expect, in some cases where the MVB (17.22) is not attained for the estimation of a particular function $\tau(\theta)$, to find the greater bound in (17.41) attained for some value of $s > 1$.

> Fend (1959) showed that if (17.27) holds, any polynomial of degree s in t attains the sth but not the $(s-1)$th variance bound (17.41), and is therefore the MV unbiassed estimator of its expectation.

17.22 We now investigate the improvement in the bound arising from taking $s = 2$ instead of $s = 1$ in (17.41). Remembering that we have defined
$$J_{rp} = E\left(\frac{L^{(r)}}{L} \cdot \frac{L^{(p)}}{L}\right), \tag{17.43}$$
we find that we may rewrite (17.41) in this case as
$$\begin{vmatrix} \operatorname{var} t & \tau' & \tau'' \\ \tau' & J_{11} & J_{12} \\ \tau'' & J_{12} & J_{22} \end{vmatrix} \geqslant 0, \tag{17.44}$$
which becomes, on expansion,
$$\operatorname{var} t \geqslant \frac{(\tau')^2}{J_{11}} + \frac{(\tau' J_{12} - \tau'' J_{11})^2}{J_{11}(J_{11} J_{22} - J_{12}^2)}. \tag{17.45}$$
The second item on the right of (17.45) is the improvement in the bound to the variance of t. It may easily be confirmed that, writing $J_{rp}(n)$ as a function of sample size,
$$\left.\begin{array}{l} J_{11}(n) = n J_{11}(1), \\ J_{12}(n) = n J_{12}(1), \\ J_{22}(n) = n J_{22}(1) + 2n(n-1)\{J_{11}(1)\}^2, \end{array}\right\} \tag{17.46}$$
and using (17.46), (17.45) becomes
$$\operatorname{var} t \geqslant \frac{(\tau')^2}{n J_{11}(1)} + \frac{\{\tau'' - \tau' J_{12}(1)/J_{11}(1)\}^2}{2n^2 \{J_{11}(1)\}^2} + o\left(\frac{1}{n^2}\right), \tag{17.47}$$
which makes it clear that the improvement in the bound is of order $1/n^2$ compared

with the leading term of order $1/n$, and is only of importance when $\tau' = 0$ and the first term disappears. In the case where t is estimating θ itself, (17.45–7) give

$$\operatorname{var} t \geqslant \frac{1}{J_{11}} + \frac{J_{12}^2}{J_{11}(J_{11}J_{22}-J_{12}^2)}$$
$$= \frac{1}{J_{11}} + \frac{J_{12}^2}{2J_{11}^4} + o\left(\frac{1}{n^2}\right). \tag{17.48}$$

If the bound with $s = 2$ equals the MVB ($s = 1$), this does not generally imply that the latter is attainable. For the exponential family (17.30), however, this equality does imply attainability—cf. Patil and Shorrock (1965).

Example 17.11

To estimate $\theta(1-\theta)$ in the binomial distribution of Example 17.9, it is natural to take as our estimator an unbiassed function of $r/n = p$, which is the MVB estimator of θ. We have seen in Example 17.4 that $r(r-1)/\{n(n-1)\}$ is an unbiassed estimator of θ^2. Hence

$$t = \frac{r}{n} - \frac{r(r-1)}{n(n-1)} = \left(\frac{n}{n-1}\right)p(1-p)$$

is an unbiassed estimator of $\theta(1-\theta)$. The exact variance of t is obtained by rewriting it as

$$\frac{n-1}{n}t \equiv \tfrac{1}{4} - (p-\tfrac{1}{2})^2$$

so that

$$\left(\frac{n-1}{n}\right)^2 \operatorname{var} t = \operatorname{var}\{(p-\tfrac{1}{2})^2\} = E\{(p-\tfrac{1}{2})^4\} - [E\{(p-\tfrac{1}{2})^2\}]^2$$
$$= \lambda_4' - (\lambda_2')^2,$$

where λ_r' is the rth moment of p about the value $\tfrac{1}{2}$. Using (3.8), Vol. 1, we express the λ_r' in terms of the central moments λ_r about $E(p) = \theta$:

$$\lambda_4' - (\lambda_2')^2 = \lambda_4 + 4(\theta - \tfrac{1}{2})\lambda_3 + 4(\theta - \tfrac{1}{2})^2 \lambda_2 - \lambda_2^2.$$

Now λ_r is $n^{-r}\mu_r$, where μ_r is the moment of the binomial variable np, so

$$\left(\frac{n-1}{n}\right)^2 \operatorname{var} t = \frac{\mu_4 - \mu_2^2}{n^4} - \frac{2(1-2\theta)\mu_3}{n^3} + (1-2\theta)^2 \frac{\mu_2}{n^2}$$

and we may substitute for the μ_r from (5.4) (where p is written for our θ), obtaining finally

$$\operatorname{var} t = \frac{\theta(1-\theta)}{n}\left\{(1-2\theta)^2 + \frac{2\theta(1-\theta)}{n-1}\right\}.$$

We cannot attain the MVB for $\tau = \theta(1-\theta)$, since it is attained for θ itself (cf. Example 17.9) and (cf. **17.18**) only one function can attain its MVB. However, when $\theta \neq \tfrac{1}{2}$ the term in n^{-1} in $\operatorname{var} t$ is $(\tau')^2/J_{11}$, the MVB, which is therefore *asymptotically* attained—cf. **17.23** below. For any θ, $\operatorname{var} t$ exactly attains the improved bound (17.45), as may be seen from substituting the results $J_{12} = 0$, $J_{22} = 2n(n-1)/\{\theta^2(1-\theta)^2\}$

(which are left to the reader in Exercise 17.3) in (17.45). Alternatively, this result follows from the fact that t is a linear function of L'/L and L''/L—cf. Exercise 17.3.

17.23 Example 17.11 brings out one point of some importance. If we have a MVB estimator t for $\tau(\theta)$, i.e.
$$\operatorname{var} t = \{\tau'(\theta)\}^2 / J_{11}, \qquad (17.49)$$
and we require to estimate some other function of θ, say $\psi\{\tau(\theta)\}$, we know from (10.14), Vol. 1, that in large samples
$$\operatorname{var}\{\psi(t)\} \sim \left(\frac{\partial \psi}{\partial \tau}\right)^2 \operatorname{var} t$$
$$\sim \left(\frac{\partial \psi}{\partial \tau}\right)^2 \left(\frac{\partial \tau}{\partial \theta}\right)^2 \Big/ J_{11}$$
from (17.49), provided that ψ' is non-zero. But this may be rewritten
$$\operatorname{var}\{\psi(t)\} \sim \left(\frac{\partial \psi}{\partial \theta}\right)^2 \Big/ J_{11},$$
so that *any* such function of θ has a MVB estimator in large samples if some function of θ has such an estimator for all n. Further, the estimator is always the corresponding function of t. We shall, in **17.35**, be able to supplement this result by a more exact one concerning functions of the MVB estimator.

17.24 There is no difficulty in extending the results of this chapter on MVB estimation to the case when the distribution has more than one parameter, and we are interested in estimating a single function of them all, say $\tau(\theta_1, \theta_2, \ldots, \theta_k)$. In this case, the analogue of the simplest result, (17.22), is
$$\operatorname{var} t \geqslant \sum_{i=1}^{k} \sum_{j=1}^{k} \frac{\partial \tau}{\partial \theta_i} \frac{\partial \tau}{\partial \theta_j} I_{ij}^{-1}, \qquad (17.50)$$
where the matrix which has to be inverted to obtain the terms in (17.50) is
$$\{I_{ij}\} = \left\{ E\left(\frac{1}{L}\frac{\partial L}{\partial \theta_i} \cdot \frac{1}{L}\frac{\partial L}{\partial \theta_j}\right) \right\}. \qquad (17.51)$$
As before, (17.50) only takes account of terms of order $1/n$, and a more complicated inequality is required if we are to take account of lower-order terms.

17.25 Even better lower bounds for the sampling variance of estimators than those imposing regularity conditions. See Kiefer (1952), Barankin (1949) and Blischke *et al.* (1969). Chapman and Robbins (1951) established that
$$\operatorname{var} t \geqslant \frac{1}{\inf\limits_{h} \frac{1}{h^2} \left\{ \int \left[\left(L(x|\theta+h) \right)^2 \big/ L(x|\theta) \right] dx - 1 \right\}}, \qquad (17.52)$$
the infimum being for all $h \neq 0$ for which $L(x|\theta) = 0$ implies $L(x|\theta+h) = 0$. (17.52) is in general at least as good a bound as (17.24), though more generally valid. For the denominator of the right-hand side of (17.24) is
$$E\left[\left(\frac{\partial \log L}{\partial \theta}\right)^2\right] = \int \lim_{h \to 0} \left[\frac{L(x|\theta+h) - L(x|\theta)}{h L(x|\theta)}\right]^2 L(x|\theta)\, dx,$$

and provided that we may interchange the integral and the limiting operation, this becomes
$$= \lim_{h \to 0} \frac{1}{h^2} \left[\int \frac{\{L(x\mid\theta+h)\}^2}{L(x\mid\theta)} dx - 1 \right]. \qquad (17.53)$$

The denominator on the right of (17.52) is the infimum of this quantity over all permissible values of h, and is consequently no greater than the limit as h tends to zero. Thus (17.52) is at least as good a bound as (17.24).

17.26 So far, we have been largely concerned with the uniform attainment of the MVB by a single estimator. In many cases, however, it will not be so attained, even when the conditions under which it is derived are satisfied. When this is so, there may still be a MV estimator for all values of θ. The question of the uniqueness of a MV estimator now arises. We can easily show that if a MV estimator exists, it is always unique, irrespective of whether any bound is attained.

Let t_1 and t_2 be MV unbiassed estimators of $\tau(\theta)$, each with variance V. Consider the new estimator
$$t_3 = \tfrac{1}{2}(t_1 + t_2)$$
which evidently also estimates $\tau(\theta)$ with variance
$$\operatorname{var} t_3 = \tfrac{1}{4} \{\operatorname{var} t_1 + \operatorname{var} t_2 + 2 \operatorname{cov}(t_1, t_2)\}. \qquad (17.54)$$
Now by the Cauchy–Schwarz inequality
$$\operatorname{cov}(t_1, t_2) = \int \ldots \int (t_1 - \tau)(t_2 - \tau) L \, dx_1 \ldots dx_n$$
$$\leqslant \left\{ \int \ldots \int (t_1-\tau)^2 L \, dx_1 \ldots dx_n \cdot \int \ldots \int (t_2-\tau)^2 L \, dx_1 \ldots dx_n \right\}^{\frac{1}{2}}$$
$$\leqslant (\operatorname{var} t_1 \, \operatorname{var} t_2)^{\frac{1}{2}} = V. \qquad (17.55)$$
Thus (17.54) and (17.55) give
$$\operatorname{var} t_3 \leqslant V,$$
which contradicts the assumption that t_1 and t_2 have MV unless the equality holds. This implies that the equality sign holds in (17.55). This can only be so if
$$(t_1 - \tau) = k(\theta)(t_2 - \tau), \qquad (17.56)$$
i.e. the variables are proportional. But (17.56) implies that
$$\operatorname{cov}(t_1, t_2) = k(\theta) \operatorname{var} t_2 = k(\theta) V$$
and this equals V since the equality sign holds in (17.55). Thus
$$k(\theta) = 1$$
and hence, from (17.56), $t_1 = t_2$ identically. Thus a MV estimator is unique.

17.27 The argument of the preceding section can be generalized to give an interesting inequality for the correlation ρ between any two estimators, which, it will be remembered from **16.23**, is defined as the ratio of their covariance to the square root of the product of their variances. By the argument leading to (17.55), we see that $\rho^2 \leqslant 1$ always.

Suppose that t is the MV unbiassed estimator of $\tau(\theta)$, with variance V, and that

t_1 and t_2 are any two unbiassed estimators of $\tau(\theta)$ with finite variances. Consider a new estimator
$$t_3 = at_1 + (1-a)t_2.$$
This will also estimate $\tau(\theta)$, with variance
$$\operatorname{var} t_3 = a^2 \operatorname{var} t_1 + (1-a)^2 \operatorname{var} t_2 + 2a(1-a)\operatorname{cov}(t_1, t_2). \tag{17.57}$$
If we now write
$$\operatorname{var} t_1 = k_1 V, \quad \operatorname{var} t_2 = k_2 V, \quad k_1, k_2 \geq 1,$$
we obtain, from (17.57),
$$\operatorname{var} t_3 / V = a^2 k_1 + (1-a)^2 k_2 + 2a(1-a)\operatorname{cov}(t_1, t_2)/V, \tag{17.58}$$
and writing
$$\rho = \frac{\operatorname{cov}(t_1, t_2)}{(\operatorname{var} t_1 \operatorname{var} t_2)^{\frac{1}{2}}} = \frac{\operatorname{cov}(t_1, t_2)}{(k_1 k_2)^{\frac{1}{2}} V},$$
we obtain from (17.58)
$$\operatorname{var} t_3 / V = a^2 k_1 + (1-a)^2 k_2 + 2a(1-a)\rho (k_1 k_2)^{\frac{1}{2}},$$
and since $\operatorname{var} t_3 \geq V$, this becomes
$$a^2 k_1 + (1-a)^2 k_2 + 2a(1-a)\rho (k_1 k_2)^{\frac{1}{2}} \geq 1. \tag{17.59}$$

(17.59) may be rearranged as a quadratic inequality in a,
$$a^2 \{k_1 + k_2 - 2\rho(k_1 k_2)^{\frac{1}{2}}\} + 2a\{\rho(k_1 k_2)^{\frac{1}{2}} - k_2\} + (k_2 - 1) \geq 0,$$
and the discriminant of the left-hand side cannot be positive, since the roots of the equation are complex or equal. Thus
$$\{\rho(k_1 k_2)^{\frac{1}{2}} - k_2\}^2 \leq \{k_1 + k_2 - 2\rho(k_1 k_2)^{\frac{1}{2}}\}(k_2 - 1)$$
which yields
$$\{\rho(k_1 k_2)^{\frac{1}{2}} - 1\}^2 \leq (k_1 - 1)(k_2 - 1).$$
Hence
$$\frac{1}{(k_1 k_2)^{\frac{1}{2}}} + \left\{\frac{(k_1 - 1)(k_2 - 1)}{k_1 k_2}\right\}^{\frac{1}{2}} \geq \rho \geq \frac{1}{(k_1 k_2)^{\frac{1}{2}}} - \left\{\frac{(k_1 - 1)(k_2 - 1)}{k_1 k_2}\right\}^{\frac{1}{2}}$$
or, finally, writing $E_1 = 1/k_1$, $E_2 = 1/k_2$,
$$(E_1 E_2)^{\frac{1}{2}} - \{(1-E_1)(1-E_2)\}^{\frac{1}{2}} \leq \rho \leq (E_1 E_2)^{\frac{1}{2}} + \{(1-E_1)(1-E_2)\}^{\frac{1}{2}}. \tag{17.60}$$
If either E_1 or $E_2 = 1$, i.e. either t_1 or t_2 is the MV estimator of $\tau(\theta)$, (17.60) collapses into the equality
$$\rho = E^{\frac{1}{2}}, \tag{17.61}$$
where E is the reciprocal of the relative variance of the other estimator, and in large samples is usually called its efficiency, for a reason to be discussed in **17.28–9**.

From the definition of ρ, it follows at once using (17.61) that the covariance of any unbiassed finite-variance estimator u with a MV unbiassed estimator t is exactly $\operatorname{var} t$. Since $u \equiv t + (u-t)$ this implies $\operatorname{cov}(t, u-t) = 0$—cf. Exercise 17.11. Hence t has zero covariance with *every* finite-variance function having zero expectation. Conversely, if $\operatorname{cov}(t, u-t) = 0$ for all u with finite variance and $E(u) = E(t)$, t must be the MV estimator of its expectation.

Efficiency

17.28 So far, our discussion of MV estimation has been exact, in the sense that it has not restricted sample size in any way. We now turn to consideration of large-sample properties. Even if there is no MV estimator for each value of n, there will often be one as n tends to infinity. Since most of the estimators we deal with are asymptotically normally distributed in virtue of the Central Limit theorem, the distribution of such an estimator will depend for large samples on only two parameters—its mean value and its variance. If it is a consistent estimator it will commonly be asymptotically unbiassed—cf. **17.9**. This leaves the variance as the means of discriminating between consistent, asymptotically normal estimators of the same parametric function.

Among such estimators, one with MV in large samples[*] is called an *efficient* estimator, or simply *efficient*, the term being due to Fisher (1921a).

17.29 If we compare asymptotically normal estimators in large samples, we may reasonably set up a *measure* of efficiency, something we have not attempted to do for small n and arbitrarily-distributed estimators. We shall define the efficiency of any other estimator, relative to an efficient estimator, as the reciprocal of the ratio of sample numbers required to give the estimators equal sampling variances, i.e. to make them equally precise.

If t_1 is an efficient estimator, and t_2 is another estimator, and (as is commonly the case) the variances are, in large samples, simple inverse functions of sample size, we may translate our definition of efficiency into a simple form. Let us suppose that

$$\left. \begin{array}{l} V_1 = \mathrm{var}\,(t_1 | n_1) \sim a_1/n_1^r \quad (r > 0), \\ V_2 = \mathrm{var}\,(t_2 | n_2) \sim a_2/n_2^s \quad (s > 0), \end{array} \right\} \quad (17.62)$$

where a_1, a_2 are constants independent of n, and we have shown sample size as an argument in the variances. If we are to have $V_1 = V_2$, we must have

$$1 = \frac{V_1}{V_2} = \lim_{n_1, n_2 \to \infty} \frac{a_1}{a_2} \frac{n_2^s}{n_1^r}.$$

Thus

$$\frac{a_2}{a_1} = \lim \frac{n_2^s}{n_1^r} = \lim \left(\frac{n_2}{n_1}\right)^s \cdot \frac{1}{n_1^{r-s}}. \quad (17.63)$$

Since t_1 is efficient, we must have $r \geqslant s$. If $r > s$, the last factor on the right of (17.63) will tend to zero, and hence, if the product is to remain equal to a_2/a_1, we must have

$$\frac{n_2}{n_1} \to \infty, \quad r > s, \quad (17.64)$$

and we would thus say that t_2 has zero efficiency. If, in (17.63), $r = s$, we have at once

$$\lim \frac{n_2}{n_1} = \left(\frac{a_2}{a_1}\right)^{1/r},$$

which from (17.62) may be written

$$\lim \frac{n_2}{n_1} = \lim \left(\frac{V_2}{V_1}\right)^{1/r},$$

and the efficiency of t_2 is the reciprocal of this, namely

$$E = \lim \left(\frac{V_1}{V_2}\right)^{1/r}. \quad (17.65)$$

---

[*] But see the discussion of "superefficiency" in **18.17**.

Note that if $r > s$, (17.65) gives the same result as (17.64). If $r = 1$, which is the most common case, (17.65) reduces to the inverse variance-ratio encountered at the end of **17.27**. Thus, when we are comparing estimators with variances of order $1/n$, we measure efficiency relative to an efficient estimator by the inverse of the variance ratio.

If the variance of an efficient estimator is not of the simple form (17.62)—cf. Exercise 18.21—the measurement of relative efficiency is not so simple.

> Although it follows from the result of **17.26** that efficient estimators tend asymptotically to equivalence, there will in general be a multiplicity of efficient estimators, for if t_1 is efficient so is any $t_2 = t_1 + cn^{-p}$, where p is chosen large enough. If (17.62) holds for two efficient estimators t_1, t_2, we must have $r = s$ and $a_1 = a_2 = a$, say, so that $E = 1$ at (17.65). Now take (17.62) to a further term, so that
>
> $$V_j = \frac{a}{n_j^r} + \frac{b_j}{n_j^{r+q}} + o(n_j^{-(r+q)}), \qquad j = 1, 2,$$
>
> where $q > 0$ and we assume $b_1 \leq b_2$. If we now equate V_1 to V_2 we obtain, instead of (17.63),
>
> $$\left(\frac{n_2}{n_1}\right)^r = \left(1 + \frac{b_2}{an_2^q}\right) \Big/ \left(1 + \frac{b_1}{an_1^q}\right)$$
>
> and writing $d_n = n_2 - n_1$, this gives, on expanding the right-hand side,
>
> $$1 + \frac{d_n}{n_1} = 1 + \frac{b_2}{ran_2^q} - \frac{b_1}{ran_1^q}.$$
>
> Since $E = 1$, $n_1/n_2 \to 1$ and we obtain
>
> $$d_n \to \frac{b_2 - b_1}{ra} \qquad \text{for } q = 1,$$
> $$\to \infty \qquad \text{for } q < 1,$$
> $$\to 0 \qquad \text{for } q > 1.$$
>
> The limiting value of d is called the *deficiency* of t_2 with respect to t_1 by Hodges and Lehmann (1970). It is the number of additional observations that t_2 requires asymptotically to attain the second-order performance of t_1. The commonest case is $r = q = 1$, when $d = (b_2 - b_1)/a$. Exercise 17.8 applies the deficiency concept to the estimation of a variance.

Example 17.12

We saw in Example 17.6 that the sample mean is a MVB estimator of the mean μ of a normal population, with variance σ^2/n. We saw in Example 11.12 that it is exactly normally distributed. *A fortiori*, it is an efficient estimator. In **17.11–12**, we saw that the sample median is asymptotically normal with mean μ and variance $\pi\sigma^2/(2n)$. Thus, from (17.65) with $r = 1$, the efficiency of the sample median is $2/\pi = 0.637$.

Example 17.13

Consider the estimation of the standard deviation of a normal population with variance σ^2 and unknown mean. Two possible estimators are the standard deviation of the sample, say t_1, and the mean deviation of the sample multiplied by $(\pi/2)^{\frac{1}{2}}$ (cf. **5.26**), say t_2. The latter is easier to calculate, as a rule, and if we have plenty of observations (as, for example, if we are using existing records and increasing sample size is merely a matter of turning up more records) it may be worth while using t_2 instead of t_1. Both estimators are asymptotically normally distributed.

In large samples the variance of the mean deviation is (cf. (10.39)) $\sigma^2(1-2/\pi)/n$. The variance of t_2 is then asymptotically $V_2 = \sigma^2(\pi-2)/2n$. The asymptotic variance of the standard deviation (cf. **10.8** (d)) is $V_1 = \sigma^2/(2n)$, and we shall see later that it is an efficient estimator. Thus, using (17.65) with $r = 1$, the efficiency of t_2 is

$$E = \lim V_1/V_2 = 1/(\pi-2) = 0{\cdot}876.$$

The precision of the estimator from the mean deviation of a sample of 1000 is then about the same as that from the standard deviation of a sample of 876. If it is easier to calculate the m.d. of 1000 observations than the s.d. of 876 and there is no shortage of observations, it may be more convenient to use the former.

It has to be remembered, nevertheless, that in adopting such a procedure we are deliberately wasting information. By taking greater pains we could improve the efficiency of our estimator from $0{\cdot}876$ to unity.

Minimum mean-square-error estimation

17.30 Our discussions of unbiassedness and the minimization of sampling variance have been conducted more or less independently. Sometimes, however, it is relevant to investigate both questions simultaneously. It is reasonable to argue that the presence of bias should not necessarily outweigh small sampling variance in an estimator. What we are really demanding of an estimator t is that it should be " close " to the true value θ. Let us, therefore, consider its mean-square-error (m.s.e.) about that true value, instead of about its own expected value. We have at once

$$E(t-\theta)^2 = E\{(t-E(t))+(E(t)-\theta)\}^2 = \operatorname{var} t + \{E(t)-\theta\}^2,$$

the cross-product term on the right being equal to zero. The last term on the right is simply the square of the bias of t in estimating θ. If t is unbiassed, this last term is zero, and m.s.e. becomes variance. In general, however, the minimization of m.s.e. gives different results.

> Exercise 17.15 shows that truncation of an estimator t to the known range of the parameter always improves m.s.e., whether or not t is unbiassed—if it is, variance is also always improved.

Example 17.14

What multiple a of an estimator t estimates $E(t) = \theta$ with smallest m.s.e.? We have, from **17.30**,

$$M(a) = E(at-\theta)^2 = a^2 \operatorname{var} t + \theta^2(a-1)^2,$$

and this can only be less than $\operatorname{var} t$ if $a < 1$. If V is the coefficient of variation of t, defined at (2.28), $M(a) < \operatorname{var} t$, if and only if $a > (1-V)/(1+V)$, and is minimized for variation in a when

$$a = \theta^2/(\theta^2 + \operatorname{var} t) = 1/(1+V^2).$$

The reduction in m.s.e. is then

$$\operatorname{var} t - E(at-\theta)^2 = \operatorname{var} t - \theta^2 \operatorname{var} t/(\theta^2 + \operatorname{var} t) = \operatorname{var} t/(1+1/V^2),$$

which is a monotone function of V, increasing from 0 to $\operatorname{var} t$ as V increases from 0 to ∞. Large reductions occur when V is large.

In general, V (and therefore a) is a function of θ, so at is not a statistic usable for estimation. Thus, e.g., if θ is the population mean and t is the sample mean, $\operatorname{var} t = \sigma^2/n$, where σ^2 is the population variance, and $V^2 = \sigma^2/(n\theta^2)$. For the exponential distribution $\theta^{-1}\exp(-x/\theta)$, $x \geq 0$, $\theta > 0$, we have $\sigma^2 = \theta^2$ and $a = n/(n+1)$, independent of θ; but for the Poisson distribution, $\sigma^2 = \theta$.

J. R. Thompson (1968) estimates a, replacing θ by t and σ^2 by its unbiased estimator s^2. Thus modified, at can have larger m.s.e. than t itself for moderate V, as he shows for the normal, binomial, Poisson and Gamma populations.

Exercise 17.16 deals with the estimation of powers of a normal standard deviation.

Pitman (1938) showed that among estimators $t(\mathbf{x})$ of the location parameter θ in $f(x-\theta)$ that satisfy

$$t(\mathbf{x}+a) = t(\mathbf{x})+a, \tag{A}$$

minimum m.s.e. is attained by

$$t_L(\mathbf{x}) = \int_{-\infty}^{\infty} \theta L(x\,|\,\theta)\,d\theta \bigg/ \int_{-\infty}^{\infty} L(x\,|\,\theta)\,d\theta,$$

which is always unbiased and is therefore MV unbiased subject to (A). For the scale parameter θ in $\theta^{-1}f(x\,|\,\theta)$, $\theta > 0$, $t(\mathbf{x})$ satisfying

$$t(c\mathbf{x}) = ct(\mathbf{x}), \quad c > 0 \tag{B}$$

minimizes m.s.e. when

$$t_S(\mathbf{x}) = \int_0^{\infty} \theta^{-2} L(x\,|\,\theta)\,d\theta \bigg/ \int_0^{\infty} \theta^{-3} L(x\,|\,\theta)\,d\theta,$$

and need not be unbiased—any multiple of t will also satisfy (B). Example 17.14 shows that we can sometimes improve on the unbiased estimator, the exponential distribution there being a case in point. For the location parameter, no multiple of t satisfies (A).

Minimum mean-square-error estimators are not much used, but it is as well to recognize that the objection to them is a practical, rather than a theoretical one. (Cf. the comparative study by Johnson (1950).) MV unbiased estimators are more tractable because they assume away the difficulty by insisting on unbiassedness.

Sufficient statistics

17.31 The criteria of estimation which we have so far discussed, namely consistency, unbiassedness, minimum variance and efficiency, are reasonable guides in assessing the properties of an estimator. To permit a more fundamental discussion, we now introduce the concept of *sufficiency*, which is due to Fisher (1921a, 1925).

Consider first the estimation of a single parameter θ. There is an unlimited number of possible estimators of θ, from among which we must choose. With a sample of $n \geq 2$ observations as before, consider the joint distribution of a set of r functionally independent statistics, $f_r(t, t_1, t_2, \ldots, t_{r-1}\,|\,\theta)$, $r = 2, 3, \ldots, n$, where we have selected the statistic t for special consideration. Using the multiplication theorem of probability (7.9), we may write this as the product of the marginal distribution of t and the conditional distribution of the other statistics given t, i.e.

$$f_r(t, t_1, \ldots, t_{r-1}\,|\,\theta) = g(t\,|\,\theta) h_{r-1}(t_1, \ldots, t_{r-1}\,|\,t, \theta). \tag{17.66}$$

Now if the last factor on the right of (17.66) is *independent* of θ, we clearly have a situation in which, given t, the set $t_1, \ldots, t_{r-1}$ contribute nothing further to our knowledge of θ. If, further, this is true for every r and *any* set of $(r-1)$ statistics t_i, we

may fairly say that t contains all the information in the sample about θ, and we therefore call it a *sufficient* statistic for θ. We thus formally define t as sufficient for θ if and only if
$$f_r(t, t_1, \ldots, t_{r-1} | \theta) = g(t | \theta) h_{r-1}(t_1, \ldots, t_{r-1} | t), \qquad (17.67)$$
where h_{r-1} is independent of θ, for $r = 2, 3, \ldots, n$ and any choice of $t_1, \ldots, t_{r-1}$.[*]

17.32 As it stands, the definition of (17.67) does not enable us to see whether, in any given situation, a sufficient statistic exists. However, we may reduce it to a condition on the Likelihood Function. For if the latter may be written
$$L(x_1, \ldots, x_n | \theta) = g(t | \theta) k(x_1, \ldots, x_n), \qquad (17.68)$$
where $g(t | \theta)$ is a function of t and θ alone[†] and k is independent of θ, it is easy to see that (17.67) is deducible from (17.68). For any fixed r, and any set of $t_1, \ldots, t_{r-1}$ insert the differential elements $dx_1 \ldots dx_n$ on both sides of (17.68) and make the transformation
$$\begin{cases} t = t(x_1, \ldots, x_n), \\ t_i = t_i(x_1, \ldots, x_n), & i = 1, 2, \ldots, r-1, \\ t_i = x_i, & i = r, \ldots, n-1. \end{cases}$$
The Jacobian of the transformation will not involve θ, and (17.68) will be transformed to
$$g(t|\theta) \, dt \, l(t, t_1, \ldots, t_{r-1}, t_r, \ldots, t_{n-1}) \prod_{i=1}^{n-1} dt_i, \qquad (17.69)$$
and if we now integrate out the redundant variables $t_r, \ldots, t_{n-1}$, we obtain, for the joint distribution of $t, t_1, \ldots, t_{r-1}$, precisely the form (17.67).

It should be noted that in performing the integration with respect to $t_r, \ldots, t_{n-1}$, i.e. $x_r, \ldots, x_{n-1}$, we have assumed that no factor in θ was thereby introduced. This is clearly so when the range of the distribution of the underlying variable is independent of θ; we shall see later (**17.40-1**) that these integrations remain independent of θ when one terminal of the range depends on θ, and also when both terminals depend on θ.

The converse result is also easily established. In (17.67) with $r = n$, put $t_i = x_i$ ($i = 1, 2, \ldots, n-1$). We then have
$$f_n(t, x_1, x_2, \ldots, x_{n-1} | \theta) = g(t | \theta) h_{n-1}(x_1, \ldots, x_{n-1} | t). \qquad (17.70)$$
On inserting the differential elements $dt \, dx_1 \ldots dx_{n-1}$ on either side of (17.70), the transformation
$$\begin{cases} x_n = x_n(t, x_1, \ldots, x_{n-1}), \\ x_i = x_i, & i = 1, 2, \ldots, n-1 \end{cases}$$
applied to (17.70) yields (17.68) at once. Thus (17.67) is necessary and sufficient for (17.68). This proof deals only with the case when the variates are continuous. In the discrete case the argument simplifies, as the reader will find on retracing the steps of the proof. A very general proof of the equivalence of (17.67) and (17.68) is given by Halmos and Savage (1949).

---

[*] This definition is usually given only for $r = 2$, but the definition for all r seems to us more natural. It adds no further restriction to the concept of sufficiency.

[†] We retain the notation $g(t | \theta)$, since the function of t and θ may always be expressed as the marginal distribution of t.

24 THE ADVANCED THEORY OF STATISTICS

We have discussed only the case $n \geqslant 2$. For $n = 1$ we take (17.68) as the definition of sufficiency.

Sufficiency and minimum variance

17.33 The necessary and sufficient condition for sufficiency at (17.68) has one immediate consequence of interest. On taking logarithms of both sides and differentiating, we have

$$\frac{\partial \log L}{\partial \theta} = \frac{\partial \log g(t|\theta)}{\partial \theta}. \tag{17.71}$$

On comparing (17.71) with (17.27), the condition that a MVB estimator of $\tau(\theta)$ exist, we see that such an estimator can only exist if there is a sufficient statistic. In fact, (17.27) is simply the special case of (17.71) when

$$\frac{\partial \log g(t|\theta)}{\partial \theta} = A(\theta)\{t - \tau(\theta)\}. \tag{17.72}$$

Thus sufficiency, which perhaps at first sight seems a more restrictive criterion than the attainment of the MVB, is in reality a less restrictive one. For whenever (17.27) holds, (17.71) holds also, while even if (17.27) does not hold we may still have a sufficient statistic.

Example 17.15

The argument of **17.33** implies that in all the cases (Examples 17.6, 17.8, 17.9, 17.10) where we have found MVB estimators to exist, they are also sufficient statistics. The reader should verify this in each case by direct factorization of L as at (17.68).

Example 17.16

Consider the estimation of θ in
$$dF(x) = dx/\theta, \quad 0 \leqslant x \leqslant \theta.$$
The Likelihood Function (LF) is
$$\begin{aligned} L(x|\theta) &= \theta^{-n} \quad \text{if} \quad x_{(n)} \leqslant \theta, \\ &= 0 \quad \text{otherwise,} \end{aligned}$$
since we know that all the observations, including the largest of them, $x_{(n)}$, cannot exceed θ.

We may write this in the form
$$L(x|\theta) = \theta^{-n} u(\theta - x_{(n)}), \tag{17.73}$$
where
$$\begin{aligned} u(z) &= 1, \quad z \geqslant 0 \\ &= 0, \quad z < 0 \end{aligned}.$$
(17.73) makes it clear at once that the LF can be factorized into a function of $x_{(n)}$ and θ alone,
$$g(x_{(n)}|\theta) = \theta^{-n} u(\theta - x_{(n)}),$$
and a second factor
$$k(x_1, \ldots, x_n) = 1.$$
Thus, from (17.68), $x_{(n)}$ is a sufficient statistic for θ.

17.34 If t is a sufficient statistic for θ, any one-to-one function of t, say u, will also be sufficient. For if $t(u)$ is one-to-one, we may write (17.68) as

$$L(x \mid \theta) = g(t(u) \mid \theta) k(x)$$
$$= g_1(u \mid \theta) k_1(x), \tag{17.74}$$

where k_1 is independent of θ, so that u is also sufficient for θ. To resolve the estimation problem, we choose a function of t that is a consistent, and usually also an unbiassed, estimator of θ.

Even if $t(u)$ is not one-to-one, the factorization (17.74) making u sufficient may still be possible in particular cases—Exercise 23.31 below treats one of these.

Apart from such functional relationships, the sufficient statistic is unique. If there were two distinct sufficient statistics, t_1 and t_2, (17.67) with $r = 2$ would give

$$f_2(t_1, t_2 \mid \theta) = g_1(t_1 \mid \theta) h_1(t_2 \mid t_1) = g_2(t_2 \mid \theta) h_2(t_1 \mid t_2)$$

so that, writing $h_3(t_1, t_2) = h_2(t_1 \mid t_2)/h_1(t_2 \mid t_1)$, we have

$$g_1(t_1 \mid \theta) \equiv h_3(t_1, t_2) g_2(t_2 \mid \theta) \tag{17.75}$$

identically in θ. This cannot hold unless t_1 and t_2 are functionally related.

17.35 We have seen in **17.33** that a sufficient statistic provides the MVB estimator, where there is one. We now prove a more general result, due to C. R. Rao (1945) and Blackwell (1947), that irrespective of the attainability of any variance bound, the MV unbiassed estimator of $\tau(\theta)$, if one exists, is always a function of the sufficient statistic.

Let t be sufficient for θ, and t_1 another statistic with finite variance and

$$E(t_1) = \tau(\theta). \tag{17.76}$$

Let us carry out the expectation operation in (17.76) in two stages, first holding the sufficient statistic t constant, and then allowing t to vary over its distribution. Symbolically, we have

$$E(t_1) = E_t\{E(t_1 \mid t)\}. \tag{17.77}$$

Now (17.67) ensures that $E(t_1 \mid t)$ does not depend upon θ, so that it is purely a function of t, say $p(t)$. (17.76–7) imply that

$$E_t\{p(t)\} = \tau(\theta), \tag{17.78}$$

i.e., $p(t)$ is also unbiassed for $\tau(\theta)$. Moreover,

$$\operatorname{var} t_1 = E\{t_1 - \tau(\theta)\}^2 = E\{[t_1 - p(t)] + [p(t) - \tau(\theta)]\}^2$$
$$= E\{t_1 - p(t)\}^2 + E\{p(t) - \tau(\theta)\}^2,$$

since $E\{t_1 - p(t)\}\{p(t) - \tau(\theta)\} = 0$ on taking the conditional expectation given t. Thus

$$\operatorname{var} t_1 = E\{t_1 - p(t)\}^2 + \operatorname{var}\{p(t)\} \geqslant \operatorname{var}\{p(t)\},$$

and the equality holds if and only if $t_1 \equiv p(t)$. Thus $p(t) = E(t_1 \mid t)$ has smaller variance than t_1. We shall see in **23.9–11** that usually there is only one function of a sufficient statistic with any given expectation. Then, whatever t_1 we start from, $E(t_1 \mid t)$ must be the same, and it is the *unique* MV unbiassed estimator of $\tau(\theta)$. It is not always easy to use this result constructively, since the conditional expectation $E(t_1 \mid t)$ may be difficult to evaluate—Exercise 17.24 gives an important class of cases where explicit results can be obtained—but it does assure us that an unbiassed estimator with finite variance that is a function of a sufficient statistic is the unique MV estimator.

Since $p(t)$ has the same expectation as, and smaller variance than, t_1, it follows from **17.30** that its m.s.e. is also smaller than that of t_1 no matter which function of θ is being estimated. We may be able to improve m.s.e. further by using a function of t with a different expectation. Example 17.14 implies that in estimating $\tau(\theta)$, even a constant multiple $ap(t)$, where $a < 1$, can have better m.s.e. than $p(t)$ itself—in the case given there

$$dF = \theta^{-1}\exp(-x/\theta), \qquad \theta > 0, \quad x \geqslant 0,$$

it is easily seen that $t = \bar{x}$ is sufficient for θ and unbiassed, while $nt/(n+1)$ has smaller m.s.e.

Distributions possessing sufficient statistics

17.36 We now seek to define the class of distributions in which a sufficient statistic exists for a parameter. We first consider the case where the range of the variate does not depend on the parameter θ. From (17.71) we have, if t is sufficient for θ in a sample of n independent observations,

$$\frac{\partial \log L}{\partial \theta} = \sum_{j=1}^{n} \frac{\partial \log f(x_j|\theta)}{\partial \theta} = K(t, \theta), \tag{17.79}$$

where K is some function of t and θ. Regarding this as an equation in t, we see that it remains true for any particular value of θ, say zero. It is then evident that t must be expressible in the form

$$t = M\left\{ \sum_{j=1}^{n} k(x_j) \right\}, \tag{17.80}$$

where M and k are arbitrary functions. If $w = \Sigma k(x_j)$, then K is a function of θ and w only, say $N(\theta, w)$. We have then, from (17.79), if the derivatives exist,

$$\frac{\partial^2 \log L}{\partial \theta \, \partial x_j} = \frac{\partial N}{\partial w} \frac{\partial w}{\partial x_j}. \tag{17.81}$$

Now the left-hand side of (17.81) is a function of θ and x_j only and $\partial w/\partial x_j$ is a function of x_j only. Hence $\partial N/\partial w$ is a function of θ and x_j only. But it must be symmetrical in the x's and hence is a function of θ only. Hence, integrating it with respect to w, we have

$$N(\theta, w) = wp(\theta) + q(\theta),$$

where p and q are arbitrary functions of θ. Thus (17.79) becomes

$$\frac{\partial}{\partial \theta} \log L = \frac{\partial}{\partial \theta} \sum_j \log f(x_j|\theta) = p(\theta) \Sigma k(x_j) + q(\theta), \tag{17.82}$$

whence

$$\frac{\partial}{\partial \theta} \log f(x|\theta) = p(\theta) k(x) + q(\theta)/n,$$

giving the necessary condition for a sufficient statistic to exist,

$$f(x|\theta) = \exp\{A(\theta)B(x) + C(x) + D(\theta)\}. \tag{17.83}$$

This result, which is due to Darmois (1935), Pitman (1936) and Koopman (1936), is precisely the form of the exponential family of distributions, obtained at (17.30) as a condition for the existence of a MVB estimator for some function of θ.

L. Brown (1964) gives a rigorous treatment of the regularity conditions sufficient for this result to hold, with references to related work. See also Denny (1967); E. B. Andersen (1970) treats the discrete case.

If (17.83) holds, it is easily verified that if the range of $f(x|\theta)$ is independent of θ, the Likelihood Function yields a sufficient statistic for θ. Thus, under this condition, (17.83) is sufficient for the distribution to possess a sufficient statistic.

All the parent distributions of Example 17.15 are of the form (17.83).

17.37 Under regularity conditions, there is therefore a one-to-one correspondence between the existence of a sufficient statistic for θ and the existence of a MVB estimator of some function of θ. If (17.83) holds, a sufficient statistic exists for θ, and there will be just one function, t, of that statistic (itself sufficient) which will satisfy (17.27) and so estimate some function $\tau(\theta)$ with variance equal to the MVB. In large samples, moreover (cf. **17.23**), *any* function of the sufficient statistic will estimate its expected value with MVB accuracy. Finally, for any n (cf. **17.35**), any function of the sufficient statistic will have the minimum *attainable* variance in estimating its expected value.

Sufficient statistics for several parameters

17.38 All the ideas of the previous sections generalize immediately to the case where the distribution is dependent upon several parameters $\theta_1, \ldots, \theta_k$. It also makes no difference to the essentials if we have a multivariate distribution, rather than a univariate one. Thus if we define each x_i as a vector variate with $p (\geqslant 1)$ components, **t** and **t**$_i$ as vectors of statistics, and **θ** as a vector of parameters with k components, **17.31** and **17.32** remain substantially unchanged. If we can write

$$L(\mathbf{x}|\boldsymbol{\theta}) = g(\mathbf{t}|\boldsymbol{\theta})h(\mathbf{x}) \qquad (17.84)$$

we call the components of **t** a set of (jointly) sufficient statistics for **θ**. The property (17.67) follows as before.

If **t** has s components, we may have s greater than, equal to, or less than k. If $s = 1$, we may call **t** a *single* sufficient statistic. (Thus the term "sufficient" used earlier in this chapter should now be read "single sufficient".) If we put $\mathbf{t} = \mathbf{x}$ we see that the observations themselves *always* constitute a set of sufficient statistics for **θ** with $s = n$. In order to reduce the problem of analysing the data as far as possible, we naturally desire s to be as small as possible. Even this is not quite restrictive enough —see Exercise 18.13. In Chapter 23 we shall define the concept of a *minimal* set of sufficient statistics for **θ**, which is a function of all other sets of sufficient statistics.

It evidently does not follow from the joint sufficiency of **t** for **θ** that any particular component of **t**, say $t^{(1)}$, is individually sufficient for θ_1. This will only be so if $g(\mathbf{t}|\boldsymbol{\theta})$ factorizes with $g_1(t^{(1)}|\theta_1)$ as one factor. Nor is the converse true always: individual sufficiency of all the $t^{(i)}$ when the others are known does not imply joint sufficiency.

If $k = 1$, the result of **17.35** holds unchanged if t is a vector with $s < n$ components (if $s = n$, the result is an empty one); the result is most easily applied with s as small as possible.

Example 17.17

Consider the estimation of the parameters μ and σ^2 in

$$dF(x) = \frac{1}{\sigma(2\pi)^{\frac{1}{2}}} \exp\left\{-\frac{1}{2}\left(\frac{x-\mu}{\sigma}\right)^2\right\} dx, \quad -\infty \leqslant x \leqslant \infty.$$

We have

$$L(x|\mu,\sigma^2) = \frac{1}{\sigma^n (2\pi)^{\frac{1}{2}n}} \exp\left\{-\frac{1}{2}\sum_{i=1}^{n}\left(\frac{x_i-\mu}{\sigma}\right)^2\right\} \qquad (17.85)$$

and we have seen (Example 11.7) that the joint distribution of $\bar{x}$ and s^2 in normal samples is

$$g(\bar{x},s^2|\mu,\sigma^2) \propto \frac{1}{\sigma}\exp\left\{-\frac{n}{2\sigma^2}(\bar{x}-\mu)^2\right\}\cdot\frac{1}{\sigma^{n-1}}s^{n-3}\exp\left\{-\frac{ns^2}{2\sigma^2}\right\},$$

so that, remembering that $\Sigma(x-\mu)^2 = n\{s^2+(\bar{x}-\mu)^2\}$, we have

$$L(x|\mu,\sigma^2) = g(\bar{x},s^2|\mu,\sigma^2)k(x)$$

and therefore $\bar{x}$ and s^2 are jointly sufficient for μ and σ^2. We have already seen (Examples 17.6, 17.15) that $\bar{x}$ is sufficient for μ when σ^2 is known and (Examples 17.10, 17.15) that $\frac{1}{n}\Sigma(x-\mu)^2$ is sufficient for σ^2 when μ is known. It is easily seen directly from (17.85) that s^2 is not sufficient for σ^2 alone.

17.39 The principal results for sufficient statistics generalize to the k-parameter case in a natural way. The condition (generalizing (17.83)) for a distribution to possess a set of k jointly sufficient statistics for its k parameters becomes, under similar conditions of continuity and the existence of derivatives,

$$f(x) = \exp\left\{\sum_{j=1}^{k} A_j(\boldsymbol{\theta})B_j(x)+C(x)+D(\boldsymbol{\theta})\right\}, \qquad (17.86)$$

a result due to Darmois (1935), Koopman (1936) and Pitman (1936). More general results of this kind are given by Barankin and Maitra (1963). The result of **17.35** on the unique MV properties of functions of a sufficient statistic finds its generalization in a theorem due to C. R. Rao (1947): for the simultaneous estimation of $r(\leqslant k)$ functions τ_j of the k parameters θ_s, the unbiased functions of a minimal set of k sufficient statistics, say t_i, have the minimum attainable variances, and (if the range is independent of the θ_s) the (not necessarily attainable) lower bounds to their variances are given by

$$\operatorname{var} t_i \geqslant \sum_{j=1}^{k}\sum_{l=1}^{k}\frac{\partial \tau_i}{\partial \theta_j}\frac{\partial \tau_i}{\partial \theta_l}I_{jl}^{-1}, \quad i=1,2,\ldots,r, \qquad (17.87)$$

where the *information matrix*

$$(I_{jl}) = \left\{E\left(\frac{1}{L}\frac{\partial L}{\partial \theta_j}\cdot\frac{1}{L}\frac{\partial L}{\partial \theta_l}\right)\right\} \qquad (17.88)$$

is to be inverted. (17.87), in fact, is a further generalization of (17.22) and (17.50), its simplest cases. Like (17.22), (17.87) takes account only of terms of order $1/n$ in the variance.

Sufficiency when the range depends on the parameter

17.40 Now consider the situation when the range of the variable does depend on θ. We omit the trivial case $n = 1$. First, we take the case when only one terminal of the range, say the lower terminal, depends on θ. We then have the frequency function $f(x \mid \theta)$, $a(\theta) \leqslant x \leqslant b$, where $a(\theta)$ is a monotone function of θ, and θ is in some non-degenerate interval.

Just as in Example 17.16, we have

$$L(x \mid \theta) = \prod_{i=1}^{n} f(x_i \mid \theta) \, u(x_{(1)} - a(\theta))$$

where

$$u(z) = 1 \quad \text{if} \quad z \geqslant 0,$$
$$ = 0 \quad \text{otherwise.}$$

It is at once obvious from the Likelihood Function that the smallest observation $x_{(1)}$ cannot be factored away from θ in the u-function; thus, if there is a single sufficient statistic, it must be $x_{(1)}$. But $x_{(1)}$ can only be sufficient if $f(x_i \mid \theta)$ can be factored into a function of x_i alone and a function of θ alone, i.e. if

$$f(x \mid \theta) = g(x)/h(\theta). \tag{17.89}$$

Then and only then is

$$L(x \mid \theta) = \frac{u(x_{(1)} - a(\theta))}{\{h(\theta)\}^n} \cdot \prod_{i=1}^{n} g(x_i)$$

of the form required at (17.68) for $x_{(1)}$ to be sufficient.

Evidently the same result will hold, with $x_{(n)}$ instead of $x_{(1)}$, if the range of the variate is $a \leqslant x \leqslant b(\theta)$, where $b(\theta)$ is monotone in θ; if and only if (17.89) holds, $x_{(n)}$ is singly sufficient for θ.

In these situations, Exercise 17.24 uses the result of **17.35** to establish an explicit form for the unique MV unbiassed estimator of any function $\tau(\theta)$. The reader should notice that there the problem is reparametrized so that the affected terminal is at θ itself.

17.41 If both terminals of the range depend on θ, we have $a(\theta) \leqslant x \leqslant b(\theta)$ and

$$L(x \mid \theta) = \prod_{i=1}^{n} f(x_i \mid \theta) \, u(x_{(1)} - a(\theta)) \, u(b(\theta) - x_{(n)}).$$

We see by exactly the argument in **17.40** that the extreme observations $x_{(1)}$ and $x_{(n)}$ are a pair of sufficient statistics for θ if and only if (17.89) holds. We now consider whether there can be a *single* sufficient statistic in this case; if there is it must clearly be a function of $x_{(1)}$ and $x_{(n)}$.

Essentially, we are asking whether we can find a single statistic which will tell us whether the product $u(x_{(1)} - a(\theta)) \, u(b(\theta) - x_{(n)})$ in $L(x \mid \theta)$ is equal to 1 or 0. It is only equal to 1 if both

$$x_{(1)} \geqslant a(\theta), \quad b(\theta) \geqslant x_{(n)}. \tag{17.90}$$

There are four possibilities:

(i) $a(\theta)$, $b(\theta)$ are both increasing functions of θ. (17.90) becomes

$$a^{-1}(x_{(1)}) \geqslant \theta \geqslant b^{-1}(x_{(n)}).$$

(ii) They are both decreasing functions of θ. (17.90) becomes
$$b^{-1}(x_{(n)}) \geqslant \theta \geqslant a^{-1}(x_{(1)}).$$
(iii) $a(\theta)$ is increasing, $b(\theta)$ decreasing. (17.90) becomes
$$a^{-1}(x_{(1)}) \geqslant \theta, \quad b^{-1}(x_{(n)}) \geqslant \theta. \tag{17.91}$$
(iv) $a(\theta)$ is decreasing, $b(\theta)$ increasing. (17.90) becomes
$$a^{-1}(x_{(1)}) \leqslant \theta, \quad b^{-1}(x_{(n)}) \leqslant \theta. \tag{17.92}$$
In cases (i) and (ii), we need both $x_{(1)}$ and $x_{(n)}$, and no single sufficient statistic exists. But (17.91) shows that in case (iii), $u(t_1 - \theta)$ is equivalent to the product of the original u-functions, where
$$t_1 = \min \{a^{-1}(x_{(1)}), b^{-1}(x_{(n)})\}, \tag{17.93}$$
so t_1 is singly sufficient for θ. Similarly, in case (iv),
$$t_2 = \max \{a^{-1}(x_{(1)}), b^{-1}(x_{(n)})\} \tag{17.94}$$
is singly sufficient since $u(\theta - t_2)$ is enough. We may summarize by saying that if the upper terminal is a monotone decreasing function of the lower terminal and (17.89) holds, there is a single sufficient statistic, given by t_1 or by t_2 according as the lower terminal is an increasing or a decreasing function of θ. Exercise 20.13 gives the distribution of t_1, from which that of t_2 is immediately obtainable.

These results were originally due to Pitman (1936) and R. C. Davis (1951). The condition that θ lies in a non-degenerate interval is important—cf. Example 17.23.

Example 17.18

For the rectangular distribution
$$dF = dx/(2\theta), \quad -\theta \leqslant x \leqslant \theta.$$
we are in case (iv) of **17.41**. The single sufficient statistic (17.92) is
$$t_2 = \max \{-x_{(1)}, x_{(n)}\}$$
and since
$$x_{(1)} \leqslant x_{(n)},$$
this is the same as
$$t_2 = \max \{|x_{(1)}|, |x_{(n)}|\}$$
which is intuitively acceptable.

Example 17.19

The distribution $dF(x) = \exp\{-(x-\alpha)\} dx$, $\alpha \leqslant x \leqslant \infty$, is of the form (17.89), since it may be written
$$f(x) = \exp(-x)/\exp(-\alpha).$$
Here the smallest observation, $x_{(1)}$, is sufficient for the lower terminal α.

Example 17.20

The distribution
$$dF(x) \propto \exp(-x\theta) dx, \quad 0 \leqslant x \leqslant \theta,$$
evidently cannot be put in the form (17.89). Thus there is no single sufficient statistic for θ when $n \geqslant 2$.

Example 17.21

In the two-parameter distribution
$$dF(x) = dx/(\beta - \alpha), \qquad \alpha \leqslant x \leqslant \beta,$$
it is clear that, given β, $x^{(1)}$ is sufficient for α; and given α, $x^{(n)}$ is sufficient for β. Also $x_{(1)}$ and $x_{(n)}$ are a set of jointly sufficient statistics for α and β. This is confirmed by observing that the joint distribution of $x_{(1)}$ and $x_{(n)}$ is, by (14.2) with $r = 1$, $s = n$,
$$g(x_{(1)}, x_{(n)}) = n(n-1)(x_{(n)} - x_{(1)})^{n-2}/(\beta - \alpha)^n,$$
so that we may write
$$L(x \mid \alpha, \beta) = (\beta - \alpha)^{-n} = g(x_{(1)}, x_{(n)}) k(x).$$

Example 17.22

The rectangular distribution $dF(x) = dx/\theta$, $k\theta \leqslant x \leqslant (k+1)\theta$; $k > 0$ comes under case (i) of **17.41**. No single sufficient statistic exists, but $(x_{(1)}, x_{(n)})$ are a sufficient pair since (17.89) holds.

Example 17.23

The rectangular distribution $dF = dx$, $\theta \leqslant x < \theta + 1$; $\theta = 0, 1, 2, \ldots$, in which θ is confined to integer values, does not satisfy the condition that the upper terminal be a monotone decreasing function of the lower. But, evidently, *any* single observation, x_i, in a sample of n is a single sufficient statistic for θ. In fact, $[x_i]$ estimates θ with zero variance. If the integer restriction on θ is removed, no single sufficient statistic exists, in accordance with **17.41**.

17.42 We have now concluded our discussion of the basic ideas of the theory of estimation. In Chapter 23 we shall be developing the theory of sufficient statistics further. Meanwhile, we continue our study of estimation from another point of view, by studying the properties of the estimators given by the Method of Maximum Likelihood.

EXERCISES

17.1 Show that in samples from
$$dF = \frac{1}{\Gamma(p)\theta^p} x^{p-1} e^{-x/\theta} dx, \qquad p > 0; \ 0 \leq x \leq \infty,$$
the MVB estimator of θ for known p is $\bar{x}/p$, with variance $\theta^2/(np)$, while if $\theta = 1$ that of $\frac{\partial}{\partial p} \log \Gamma(p)$ is $\frac{1}{n} \sum_{i=1}^{n} \log x_i$ with variance $\left\{ \frac{\partial^2 \log \Gamma(p)}{\partial p^2} \right\} / n$.

17.2 A random variable x has f.f.
$$f(x \mid \theta) = \theta f_1(x) + (1-\theta) f_2(x)$$
where $0 < \theta < 1$ and f_1, f_2 are completely specified f.f.'s whose ranges of variation do not involve θ. Show that the MVB in estimating θ from a sample of n observations is not generally attainable, but is equal to
$$\frac{\theta(1-\theta)}{n} \left\{ 1 - \int_{-\infty}^{\infty} \frac{f_1(x) f_2(x)}{f(x \mid \theta)} dx \right\}^{-1},$$
reducing to the binomial result (Example 17.9) when the ranges of f_1 and f_2 do not overlap.
(Hill, 1963b)

17.3 In Example 17.11, show that
$$L'/L = (r - n\theta)/\{\theta(1-\theta)\}$$
and
$$L''/L = \frac{\partial^2 \log L}{\partial \theta^2} + \left(\frac{\partial \log L}{\partial \theta} \right)^2$$
$$= \frac{1}{\theta^2(1-\theta)^2} [\{(r-n\theta) - \tfrac{1}{2}(1-2\theta)\}^2 - \{\tfrac{1}{4} + (n-1)\theta(1-\theta)\}].$$

Hence show that $J_{12} = 0$. Using (17.19), show that $E(L''/L) = 0$ and that $J_{22} = \text{var}(L''/L) = 2n(n-1)/\{\theta^2(1-\theta)^2\}$. Use these results to verify that var t coincides with (17.45). Show that t is a linear function of the uncorrelated variables L'/L and L''/L.

17.4 Writing (17.27) as
$$(t - \tau) = \frac{\tau'(\theta)}{J_{11}} \cdot \frac{L'}{L},$$
and the characteristic function of t about its mean as $\phi(z)$ $(z = u)$, show that for a MVB estimator
$$\frac{\partial \phi(z)}{\partial z} = \frac{\tau'(\theta)}{J_{11}} \left\{ \frac{\partial \phi(z)}{\partial \theta} + z \tau'(\theta) \phi(z) \right\}$$
and that its cumulants are given by
$$\varkappa_{r+1} = \frac{\partial \tau}{\partial \theta} \cdot \frac{\partial \varkappa_r}{\partial \theta} \Big/ J_{11}, \qquad r = 2, 3, \ldots \tag{A}$$

Hence show that the covariance between t and an unbiased estimator of its rth cumulant is equal to its $(r+1)$th cumulant.

Establish the inequality (17.50) for an estimated function of several parameters, and show that (A) holds in this case also when the bound is attained.
(Bhattacharyya, 1946–7)

17.5 Show that for the estimation of θ in the logistic distribution
$$f(x) = e^{-(x-\theta)}\{1+e^{-(x-\theta)}\}^{-2},$$
the MVB is exactly $3/n$, whereas the sample mean has exact variance $\pi^2/(3n)$ and the sample median asymptotic variance $4/n$.

17.6 Show that in estimating σ in the distribution
$$dF = \frac{1}{\sigma(2\pi)^{\frac{1}{2}}}\exp\left(-\frac{1}{2}\frac{x^2}{\sigma^2}\right)dx, \quad -\infty \leqslant x \leqslant \infty,$$
$$s_1 = \left(\frac{1}{2}\sum_{i=1}^{n}x_i^2\right)^{\frac{1}{2}}\Gamma\left(\frac{1}{2}n\right) \bigg/ \Gamma\left\{\frac{1}{2}(n+1)\right\}$$
and
$$s_2 = \left\{\frac{1}{2}\sum_{i=1}^{n}(x_i-\bar{x})^2\right\}^{\frac{1}{2}}\Gamma\left\{\frac{1}{2}(n-1)\right\} \bigg/ \Gamma\left(\frac{1}{2}n\right)$$
are both unbiassed. Hence show that s in Example 17.3 has expectation $\sigma\left(1-\frac{1}{4n}\right)$ approximately, agreeing with the result of Exercise 10.20, Vol. 1.

Show that (17.52) generally gives a greater bound than (17.24), which gives $\sigma^2/(2n)$, but, by considering the case $n = 2$, that even this greater bound is not attained for small n by s_1.

(cf. Chapman and Robbins, 1951)

17.7 In estimating μ^2 in
$$dF = \frac{1}{\sqrt{(2\pi)}}\exp\{-\tfrac{1}{2}(x-\mu)^2\}dx,$$
show that $(\bar{x}^2 - 1/n - \mu^2)$ is a linear function of $\frac{1}{L}\frac{\partial L}{\partial \mu}$ and $\frac{1}{L}\frac{\partial^2 L}{\partial \mu^2}$, and hence, from (17.42), that $\bar{x}^2 - 1/n$ is an unbiassed estimator of μ^2 with minimum attainable variance.

17.8 In estimating the variance σ^2 of a distribution with known mean μ from a sample of n observations x_i, consider the two unbiassed estimators
$$t_1 = \frac{1}{n}\sum_{i=1}^{n}(x_i-\mu)^2, \quad t_2 = \frac{1}{n-1}\sum_{i=1}^{n}(x_i-\bar{x})^2.$$
Using (10.5) and (12.35), show that in the terminology of **17.29** they have the same efficiency, and that the deficiency of t_2 with respect to t_1 is $d = 2/(2+\gamma_2)$, where $\gamma_2 = \kappa_4/\kappa_2^2$ is the kurtosis coefficient of the distribution. From Exercise 3.19, $\gamma_2 \geqslant -2$, so that $0 \leqslant d \leqslant \infty$, with $d = 1$ in the normal case, reflecting the single d.fr. lost in passing from t_1 to t_2 (cf. Example 11.7).

(Hodges and Lehmann, 1970)

17.9 For the three-parameter distribution
$$dF(x) = \frac{1}{\Gamma(p)}\exp\left\{-\left(\frac{x-\alpha}{\sigma}\right)\right\}\left(\frac{x-\alpha}{\sigma}\right)^{p-1}\frac{dx}{\sigma}, \quad p, \sigma > 0;\ \alpha \leqslant x \leqslant \infty,$$
show that there are single sufficient statistics for p and σ individually when the other two parameters are known; that there is a pair of sufficient statistics for p and σ jointly if α is known; and that if σ is known and $p = 1$, there is a single sufficient statistic for α.

17.10 In samples of size n from a normal distribution with cumulants κ_1, κ_2, show that the sample mean k_1 has moments
$$E(k_1^r) = \sum_{j=0}^{r}\binom{r}{j}\kappa_1^{r-j}E(k_1-\kappa_1)^j = \sum_{i=0}^{[\frac{1}{2}r]}\frac{r!}{i!(r-2i)!}\kappa_1^{r-2i}\left(\frac{\kappa_2}{2n}\right)^i, \quad r = 1, 2, \ldots$$
Hence show that
$$\kappa_1^r = \sum_{i=0}^{[\frac{1}{2}r]}(-1)^i\frac{r!}{i!(r-2i)!}E(k_1^{r-2i})\left(\frac{\kappa_2}{2n}\right)^i,$$

and finally, using (16.7), that for $n \geq 2$ and $m > -\tfrac{1}{2}(n-1)$ the MV unbiassed estimator of $\kappa_1^r \kappa_2^m$ is

$$\sum_{i=0}^{[\frac{1}{2}r]} (-1)^i \frac{r!}{i!(r-2i)!} \frac{\Gamma\{\tfrac{1}{2}(n-1)\}}{\Gamma\{\tfrac{1}{2}(n-1)+m+i\}} \frac{k^{r-2i}}{(2n)^i} \left\{\frac{(n-1)k_2}{2}\right\}^{m+i}$$

where k_2 is the second k-statistic. (Cf. Hoyle, 1968)

17.11 Show that if t is the MV unbiassed estimator, and u another unbiassed estimator, of θ, the covariance of t and $(u-t)$ is zero. Hence show that we may regard the variation of $(u-\theta)$ as composed of two orthogonal parts, one being the variation of $(t-\theta)$ and the other a component due to inefficiency of estimation.

(cf. Fisher, 1925)

17.12 In the binomial distribution of Example 17.9, let a and b be integers or zero and define $\tau_{ab} = \theta^a(1-\theta)^b$. Show that if $a, b \geq 0$ and $a+b \leq n$, τ_{ab} has the unbiassed estimator $r^{(a)}(n-r)^{(b)}/n^{(a+b)}$, but that otherwise no unbiassed estimator of τ_{ab} exists.

17.13 For a sample of n observations from a distribution of form (17.89), with $a \leq x \leq \theta$, show that the sufficient statistic $x_{(n)}$ is not an unbiassed estimator of θ. Using the method of **17.10**, show that $t' = 2x_{(n)} - x_{(n-1)}$ is unbiassed to order n^{-1} and has the same m.s.e. as $x_{(n)}$ to order n^{-2}, namely $\dfrac{2}{n^2}\left\{\dfrac{h(\theta)}{h'(\theta)}\right\}^2$. Show further that $t'' = 3(x_{(n)} - x_{(n-1)}) + x_{(n-2)}$.

(cf. Robson and Whitlock, 1964)

17.14 For a sample of n observations from a distribution with frequency function (17.86), and the range of the variates independent of the parameters, show that the statistics

$$t_j = \sum_{i=1}^{n} B_j(x_i), \qquad j = 1, 2, \ldots, k,$$

are jointly sufficient for the k parameters $\theta_1, \ldots, \theta_k$, and that their joint f.f. is

$$g(t_1, t_2, \ldots, t_k | \theta) = \exp\{nD(\theta)\} h(t_1, t_2, \ldots, t_k) \exp\left\{\sum_{j=1}^{k} A_j(\theta) t_j\right\}$$

which is itself of the form (17.86). Use this result to derive the distribution of $\bar{x}$ in Example 17.6.

17.15 t is an estimator of a parameter θ known to lie in the interval (a, b). u is defined equal to t if $a \leq t \leq b$, but equal to a if $t < a$ and to b if $t > b$. Show that u has better m.s.e. than t in estimating θ. If $E(t) = \theta$, show that u also has better variance than t.

17.16 In a sample of size $n = \nu+1$ from a normal population with variance σ^2, show that σ^p is unbiassedly estimated if $\nu+p > 0$ by

$$t_p = S^p 2^{-\frac{1}{2}p} \Gamma(\tfrac{1}{2}\nu)/\Gamma\{\tfrac{1}{2}(\nu+p)\},$$

where $S^2 = \Sigma(x_i - \bar{x})^2$. Verify that $p = 1$ gives the second estimator of σ in Exercise 17.6 and $p = 2$ gives the estimator of σ^2 in Example 17.3.

Using Example 17.14, show that the multiple of t_p with smallest m.s.e. in estimating σ^p is exactly t_p with ν replaced by $(\nu+p)$. Verify that when $p = 2$, this gives $(n+1)$ as the divisor of S^2.

17.17 Use the method of **17.10** to correct the bias of the sample variance in estimating the population variance and to correct the bias in using the square of the sample mean to estimate the square of the population mean.

(cf. Quenouille, 1956)

ESTIMATION

17.18 Show that if the method of **17.10** is used to correct for bias, and $\operatorname{var} t_n \sim c/n$, $\operatorname{var} t'_n \sim \operatorname{var} t_n$. Show that the m.s.e. of t'_n is consequently no greater than that of t_n.

(cf. Quenouille, 1956)

17.19 In a sample of n observations x_i from a Poisson distribution with parameter λ, we write $X = \sum_{i=1}^{n} x_i$, $\bar{x} = X/n$ and k_p for the pth k-statistic of the sample, with expectation $\varkappa_p$, the pth cumulant. Show using **17.35** that $E(k_p \mid X) = \bar{x}$ and that $\operatorname{var}(k_p - \bar{x}) = \sum_{r=1}^{p} c_r \lambda^r$ is unbiassedly estimated by

$$\hat{V}_p = \operatorname{var}(k_p \mid X) = \sum_{r=1}^{p} c_{rp} X^{(r)} n^{-r}$$

(cf. Example 11.17 and Exercise 5.26, Vol. 1, 3rd edn). Hence, using the results in Chapter 12, show that

$$\operatorname{var}(k_2 \mid X) = \frac{2}{n} \frac{X^{(2)}}{n^{(2)}}, \quad \operatorname{var}(k_3 \mid X) = \frac{6}{n}\left\{\frac{3 X^{(2)}}{n^{(2)}} + \frac{X^{(3)}}{n^{(3)}}\right\}$$

and similarly that

$$\mu_3(k_2 \mid X) = \frac{4}{n^3(n-1)^2}\{(n-2) X^{(2)} + 2 X^{(3)}\}.$$

(Gart and Pettigrew, 1970)

17.20 For a (positive definite) matrix of variances and covariances, the product of any diagonal element with the corresponding diagonal element of the reciprocal matrix cannot be less than unity. Hence show that if, in (17.87), we have $r = k$ and $\tau_i = \theta_i$ (all i), the resulting bound for an estimator of θ_i is not less than the bound given by (17.24). Give a reason for this result.

(C. R. Rao, 1952)

17.21 The MVB (17.22) holds good for a distribution $f(x \mid \theta)$ whose range (a, b) depends on θ, provided that (17.18) remains true. Show that this is so if

$$f(a \mid \theta) = f(b \mid \theta) = 0,$$

and that if in addition

$$\left[\frac{\partial f(x \mid \theta)}{\partial \theta}\right]_{x=a} = \left[\frac{\partial f(x \mid \theta)}{\partial \theta}\right]_{x=b} = 0,$$

(17.19) also remains true and we may write the MVB in the form (17.23).

17.22 Apply the result of Exercise 17.21 to show that the MVB holds for the estimation of θ in

$$dF(x) = \frac{1}{\Gamma(p)}(x-\theta)^{p-1} \exp\{-(x-\theta)\} dx, \quad \theta \leq x \leq \infty; \; p > 2,$$

and equals $(p-2)/n$, but is not attainable since there is no single sufficient statistic for θ.

17.23 x is a random variable in the range (a, b) (which may depend on θ) whose distribution is $f(x \mid \theta)$. If $E\{t(x, \theta)\} = \tau(\theta)$ and f and $\partial f/\partial x$ vanish at a and at b, show that

$$\operatorname{var} t \geq \left[E\left(\frac{\partial t}{\partial x}\right)\right]^2 \Big/ E\left\{\left(\frac{\partial \log f}{\partial x}\right)^2\right\} = \left[E\left(\frac{\partial t}{\partial x}\right)\right]^2 \Big/ -E\left(\frac{\partial^2 \log f}{\partial x^2}\right).$$

(Cf. Exercise 17.21 to establish (17.22–3).)

(B. R. Rao (1958b). The analogue of (17.45) also holds, $\tau^{(r)}$ being replaced by $E\left(\dfrac{\partial^r t}{\partial x^r}\right)$ and $\dfrac{L^{(r)}}{L}$ by $\dfrac{1}{f}\dfrac{\partial^r f}{\partial x^r}$ in J_{rv}—see Sankaran (1964).)

17.24 For a distribution of form
$$f(x) = g(x)/h(\theta), \quad \theta \leq x \leq b,$$
show that if a function $t(x)$ of a single observation is an unbiassed estimator of $\tau(\theta)$, then
$$-t(x) = \{\tau(x) h'(x) + \tau'(x) h(x)\}/g(x).$$
Hence show, using **17.35**, that the unique MV unbiassed estimator of $\tau(\theta)$ in samples of size n is
$$p(x_{(1)}) = \tau(x_{(1)}) - \frac{\tau'(x_{(1)})}{n} \frac{h(x_{(1)})}{g(x_{(1)})},$$
and similarly if $a \leq x \leq \theta$ that it is
$$p(x_{(n)}) = \tau(x_{(n)}) + \frac{\tau'(x_{(n)})}{n} \frac{h(x_{(n)})}{g(x_{(n)})}.$$
Show that for $f(x) = 1/\theta$, $0 \leq x \leq \theta$, $\frac{n+1}{n} x_{(n)}$ is the estimator of θ,

while for $f(x) = \exp -\{(x-\theta)\}$, $\theta \leq x \leq \infty$, $x_{(1)} - \frac{1}{n}$ estimates θ.

(Tate, 1959)

17.25 If the pair of statistics (t_1, t_2) is jointly sufficient for two parameters (τ_1, τ_2), and t_1 is sufficient for τ_1 when τ_2 is known, show that the conditional distribution of t_2, given t_1, is independent of τ_1. As an illustration, consider c independent binomial distributions with sample sizes $n_i (i = 1, 2, \ldots, c)$ and parameters θ_i connected by the relation
$$\lambda_i = \log\left(\frac{\theta_i}{1-\theta_i}\right) = \alpha + \beta x_i,$$
where the x_i are known constants, and show that if y_i is the number of " successes " in the ith sample, the conditional distribution of $\sum_i x_i y_i$, given $\sum_i y_i$, is independent of α.

(D. R. Cox, 1958a)

17.26 If the zero frequency of a Poisson distribution cannot be observed, it is called a truncated Poisson distribution. Show that from a single observation x ($x = 1, 2, \ldots$), on a truncated Poisson $e^{-\theta} \theta^x / x!$, the only unbiassed estimator of $1 - e^{-\theta}$ takes the values 0 when x is odd, 2 when x is even.

17.27 As in Example 17.13, show that the efficiency of the estimator of σ based on the mean difference discussed in **10.14** is 0·978 in normal samples.

CHAPTER 18

ESTIMATION: MAXIMUM LIKELIHOOD

18.1 We have already (**8.6–10**) encountered the Maximum Likelihood (abbreviated ML) principle in its general form. In this chapter we shall be concerned with its application to the problems of estimation, and its properties when used as a method of estimation. We shall confine our discussion for the most part to the case of samples of n independent observations from the same distribution. The joint probability of the observations, regarded as a function of a single unknown parameter θ, is called the Likelihood Function (abbreviated LF) of the sample, and is written

$$L(x\,|\,\theta) = f(x_1\,|\,\theta)f(x_2\,|\,\theta)\ldots f(x_n\,|\,\theta), \tag{18.1}$$

where we write $f(x\,|\,\theta)$ indifferently for a univariate or multivariate, continuous or discrete distribution.

The ML principle, whose extensive use in statistical theory dates from the work of Fisher (1921a), directs us to take as our estimator of θ that value (say, $\hat{\theta}$) within the admissible range of θ which makes the LF as large as possible. That is, we choose $\hat{\theta}$ so that for any admissible value θ

$$L(x\,|\,\hat{\theta}) \geqslant L(x\,|\,\theta). \tag{18.2}$$

We assume that θ may take any real value in an interval (which may be infinite in either or both directions).

18.2 The determination of the form of the ML estimator becomes relatively simple in one general situation. If the LF is a twice-differentiable function of θ throughout its range, stationary values of the LF within the admissible range of θ will, if they exist, be given by roots of

$$L'(x\,|\,\theta) = \frac{\partial L(x\,|\,\theta)}{\partial \theta} = 0. \tag{18.3}$$

A sufficient (though not a necessary) condition that any of these stationary values (say, $\tilde{\theta}$) be a local maximum is that

$$L''(x\,|\,\tilde{\theta}) < 0. \tag{18.4}$$

If we find all the local maxima of the LF in this way (and, if there are more than one, choose the largest of them) we shall have found the solution(s) of (18.2), provided that there is no terminal maximum of the LF at the extreme permissible values of θ.

18.3 In practice, it is often simpler to work with the logarithm of the LF than with the function itself. Under the conditions of the last section, they will have maxima together, since

$$\frac{\partial}{\partial \theta}\log L = L'/L$$

and $L > 0$. We therefore seek solutions of
$$(\log L)' = 0 \qquad (18.5)$$
for which
$$(\log L)'' < 0, \qquad (18.6)$$
if these are simpler to solve than (18.3) and (18.4). (18.5) is often called the likelihood equation.

Maximum Likelihood and sufficiency

18.4 If a single sufficient statistic exists for θ, we see at once that the ML estimator of θ must be a function of it. For sufficiency of t for θ implies the factorization of the LF (17.84). That is,
$$L(x|\theta) = g(t|\theta)h(x), \qquad (18.7)$$
the second factor on the right of (18.7) being independent of θ. Thus choice of $\hat{\theta}$ to maximize $L(x|\theta)$ is equivalent to choosing $\hat{\theta}$ to maximize $g(t|\theta)$, and hence $\hat{\theta}$ will be a function of t alone.

18.5 If a MVB estimator t exists for $\tau(\theta)$, and the likelihood equation (18.5) has a solution $\hat{\theta}$, then $t = \tau(\hat{\theta})$ and the solution $\hat{\theta}$ is unique, occurring at a maximum of the LF. For we have seen (**17.33**) that, when there is a single sufficient statistic, the LF is of the form in which MVB estimation of some function of θ is possible. Thus, as at (17.27), the LF is of the form
$$(\log L)' = A(\theta)\{t - \tau(\theta)\}, \qquad (18.8)$$
so that the solutions of (18.5) are of form
$$t = \tau(\hat{\theta}). \qquad (18.9)$$
Differentiating (18.8) again, we have
$$(\log L)'' = A'(\theta)\{t - \tau(\theta)\} - A(\theta)\tau'(\theta). \qquad (18.10)$$
But since, from (17.29), $\tau'(\theta)/A(\theta) = \operatorname{var} t$, the last term in (18.10) may be written
$$-A(\theta)\tau'(\theta) = -\{A(\theta)\}^2 \operatorname{var} t. \qquad (18.11)$$
Moreover, at $\hat{\theta}$ the first term on the right of (18.10) is zero in virtue of (18.9). Hence (18.10) becomes, on using (18.11),
$$(\log L)''_{\hat{\theta}} = -\{A(\theta)\}^2 \operatorname{var} t < 0. \qquad (18.12)$$
By (18.12), every solution of (18.5) is a maximum of the LF. But under regularity conditions there must be a minimum between successive maxima. Since there is no minimum, it follows that there cannot be more than one maximum. This is otherwise obvious from the uniqueness of the MVB estimator t.

(18.9) shows that where a MVB (unbiassed) estimator exists, it is given by the ML method. This result does not extend to the more general variance bound (17.41)—cf. Exercise 18.37.

18.6 The uniqueness of the ML estimator where a single sufficient statistic exists extends to the case where the range of $f(x|\theta)$ depends upon θ, but the argument is somewhat different in this case. We have seen **(17.40–1)** that a single sufficient statistic can only exist if

$$f(x|\theta) = g(x)/h(\theta). \tag{18.13}$$

The LF is thus also of form

$$L(x|\theta) = \prod_{i=1}^{n} g(x_i)/\{h(\theta)\}^n, \tag{18.14}$$

and (18.14) is as large as possible if $h(\theta)$ is as small as possible. Now from (18.13)

$$1 = \int f(x|\theta)\,dx = \int g(x)\,dx/h(\theta),$$

where integration is over the whole range of x. Hence

$$h(\theta) = \int g(x)\,dx. \tag{18.15}$$

From (18.15) it follows that to make $h(\theta)$ as small as possible, we must choose $\hat{\theta}$ so that the value of the integral on the right (one or both of whose limits of integration depend on θ) is minimized.

Now a single sufficient statistic for θ exists **(17.40–1)** only if one terminal of the range is independent of θ or if the upper terminal is a monotone decreasing function of the lower terminal. In either of these situations, the value of (18.15) is a monotone function of the range of integration on the right-hand side, reaching a unique terminal minimum when that range is as small as is possible, consistent with the observations. The ML estimator $\hat{\theta}$ obtained by minimizing this range is thus unique, and the LF (18.14) has a terminal maximum at $L(x|\hat{\theta})$.

The results of this and the previous section were originally obtained by Huzurbazar (1948), who used a different method in the " regular " case of **18.5**.

18.7 Thus we have seen that where a single sufficient statistic t exists for a parameter θ, the ML estimator $\hat{\theta}$ of θ is a function of t alone. Further, $\hat{\theta}$ is unique, the LF having a single maximum in this case. The maximum is a stationary value (under regularity conditions) or a terminal maximum according to whether the range is independent of, or dependent upon, θ.

18.8 It follows from our results that all the optimum properties of single sufficient statistics are conferred upon ML estimators which are one-to-one functions of them. For example, we need only obtain the solution of the likelihood equation, and find the function of it which is unbiassed for the parameter. It then follows from the results of **17.35** that this will be the unique MV estimator of the parameter, attaining the MVB (17.22) if this is possible.

The sufficient statistics derived in Examples 17.8, 17.9, 17.10, 17.16, 17.18 and 17.19 are all easily obtained by the ML method.

Example 18.1

To estimate θ in
$$dF(x) = dx/\theta, \qquad 0 \leqslant x \leqslant \theta,$$
we see at once from the LF in Example 17.16 that $\hat{\theta} = x_{(n)}$, the sufficient statistic, the LF having a sharp (non-differentiable) maximum there.

Obviously, $\hat{\theta}$ is not an unbiassed estimator of θ. A modified unbiassed estimator is easily seen to be
$$t = (n+1)\, x_{(n)}/n.$$

Example 18.2

To estimate the mean θ of a normal distribution with known variance. We have seen (Example 17.6) that
$$(\log L)' = \frac{n}{\sigma^2}(\bar{x} - \theta).$$
We obtain the ML estimator by equating this to zero, and find
$$\hat{\theta} = \bar{x}.$$
In this case, $\hat{\theta}$ is unbiassed for θ.

The general case

18.9 If no single sufficient statistic for θ exists, the LF no longer necessarily has a unique maximum value (cf. Exercises 18.17, 18.33), and we choose the ML estimator to satisfy (18.2). We now have to consider the properties of the estimators obtained by this method. We shall see that, under very broad conditions, the ML estimator is consistent; and that under regularity conditions, the most important of which is that the range of $f(x \mid \theta)$ does not depend on θ, the ML estimator is asymptotically normally distributed and is an efficient estimator. These, however, are large-sample properties and, important as they are, it should be borne in mind that they are not such powerful recommendations of the ML method as the properties, inherited from sufficient statistics, which we have discussed in sections **18.4** onwards. Perhaps it would be unreasonable to expect any method of estimation to produce "best" results under all circumstances and for all sample sizes. However that may be, the fact remains that, outside the field of sufficient statistics, the optimum properties of ML estimators are asymptotic ones.

Example 18.3

As an example of the general situation, consider the estimation of the correlation parameter ρ in samples of n from the standardized bivariate normal distribution
$$dF = \frac{1}{2\pi(1-\rho^2)^{\frac{1}{2}}} \exp\left\{-\frac{1}{2(1-\rho^2)}(x^2 - 2\rho xy + y^2)\right\} dx\, dy,$$
$$-\infty \leqslant x, y \leqslant \infty;\ |\rho| < 1.$$
We find
$$\log L = -n\log(2\pi) - \tfrac{1}{2}n\log(1-\rho^2) - \frac{1}{2(1-\rho^2)}(\Sigma x^2 - 2\rho \Sigma xy + \Sigma y^2),$$

whence, for $\frac{\partial \log L}{\partial \rho} = 0$ we have

$$\frac{n\rho}{1-\rho^2} - \frac{\rho}{(1-\rho^2)^2}(\Sigma x^2 - 2\rho \Sigma xy + \Sigma y^2) + \frac{1}{1-\rho^2}\Sigma xy = 0,$$

reducing to the cubic equation

$$g(\rho) = \rho(1-\rho^2) + (1+\rho^2)\frac{1}{n}\Sigma xy - \rho\left(\frac{1}{n}\Sigma x^2 + \frac{1}{n}\Sigma y^2\right) = 0.$$

This has three roots, two of which may be complex. There is always at least one real root in the admissible interval $-1 < \rho < 1$, for

$$g(-1) = \frac{1}{n}\Sigma(x+y)^2 > 0$$

and

$$g(+1) = -\frac{1}{n}\Sigma(x-y)^2 < 0.$$

Since

$$g(0) = \frac{1}{n}\Sigma xy,$$

this root will have the sign of Σxy. If more than one real root lies in this interval, then in accordance with (18.2) we choose as the ML estimator that which corresponds to the largest value of the LF.

There are three real roots of $g(\rho) = 0$ if and only if $g(\rho)$ has two real turning points satisfying

$$g'(\rho) = 1 - 3\rho^2 + 2\rho\frac{1}{n}\Sigma xy - \left(\frac{1}{n}\Sigma x^2 + \frac{1}{n}\Sigma y^2\right) = 0,$$

a quadratic equation that has distinct real roots if and only if its discriminant

$$\left(\frac{1}{n}\Sigma xy\right)^2 - 3\left(\frac{1}{n}\Sigma x^2 + \frac{1}{n}\Sigma y^2 - 1\right) > 0, \tag{18.16}$$

one root of $g'(\rho) = 0$ then lying on either side of $\rho_0 = \frac{1}{3}\frac{1}{n}\Sigma xy$. It follows that at least one root of $g(\rho) = 0$ lies on each side of ρ_0, and since we may have $|\rho_0| > 1$, that there may be one or two real roots of $g(\rho) = 0$ outside the interval $(-1, 1)$. Such real roots, and complex roots, are inadmissible.

Since, by the results of **10.3** and **10.9**, the sample moments on the left in (18.16) are consistent estimators of the corresponding population moments, that left-hand side will converge in probability to $\rho^2 - 3(1+1-1) < 0$. Thus, as $n \to \infty$, there will tend to be only one real root of the likelihood equation.

The consistency of Maximum Likelihood estimators

18.10 We now show that, under very general conditions, ML estimators are consistent.

As at (18.2), we consider the case of n independent observations from a distribution $f(x|\theta)$, and for each n we choose the ML estimator $\hat{\theta}$ so that, if θ is any admissible value of the parameter, we have[*]
$$\log L(x|\hat{\theta}) \geqslant \log L(x|\theta). \tag{18.17}$$
We denote the true value of θ by θ_0, and let E_0 represent the operation of taking expectations when the true value θ_0 holds. Consider the random variable $L(x|\theta)/L(x|\theta_0)$. In virtue of the fact that the geometric mean of a non-degenerate distribution cannot exceed its arithmetic mean, we have, for all $\theta^* \neq \theta_0$,
$$E_0\left\{\log \frac{L(x|\theta^*)}{L(x|\theta_0)}\right\} < \log E_0\left\{\frac{L(x|\theta^*)}{L(x|\theta_0)}\right\}. \tag{18.18}$$
Now the expectation on the right-hand side of (18.18) is
$$\int \ldots \int \frac{L(x|\theta^*)}{L(x|\theta_0)} L(x|\theta_0) \, dx_1 \ldots dx_n = 1.$$
Thus (18.18) becomes
$$E_0\left\{\log \frac{L(x|\theta^*)}{L(x|\theta_0)}\right\} < 0$$
or, inserting a factor $1/n$,
$$E_0\left\{\frac{1}{n}\log L(x|\theta^*)\right\} < E_0\left\{\frac{1}{n}\log L(x|\theta_0)\right\} \tag{18.19}$$
provided that the expectation on the right exists, as it does very generally.

Now for any value of θ
$$\frac{1}{n}\log L(x|\theta) = \frac{1}{n}\sum_{i=1}^{n} \log f(x_i|\theta)$$
is the mean of a set of n independent identical variates with expectation
$$E_0\{\log f(x|\theta)\} = E_0\left\{\frac{1}{n}\log L(x|\theta)\right\}.$$
By the Strong Law of Large Numbers of **7.25**, therefore, $\frac{1}{n}\log L(x|\theta)$ converges with probability unity to its expectation, as n increases. Thus as $n \to \infty$ we have, from (18.19), with probability unity
$$\frac{1}{n}\log L(x|\theta^*) < \frac{1}{n}\log L(x|\theta_0)$$
or
$$\lim_{n\to\infty}\text{prob}\{\log L(x|\theta^*) < \log L(x|\theta_0)\} = 1, \quad \theta^* \neq \theta_0. \tag{18.20}$$
On the other hand, (18.17) with $\theta = \theta_0$ gives
$$\log L(x|\hat{\theta}) \geqslant \log L(x|\theta_0). \tag{18.21}$$
(18.20) and (18.21) imply that, as $n \to \infty$, $L(x|\hat{\theta})$ cannot take any other value than $L(x|\theta_0)$. If $L(x|\theta)$ is a one-to-one function of θ, this implies that
$$\text{prob}\{\lim_{n\to\infty} \hat{\theta} = \theta_0\} = 1. \tag{18.22}$$

---

[*] Because of the equality sign in (18.17), the sequence of values of $\hat{\theta}$ may be determinable in more than one way. See **18.11** and **18.13** below.

This is a heuristic form of Wald's (1949) rigorous proof of the consistency of ML estimators, which requires further conditions. See also an extension by Huber (1967).

18.11 We have shown that any sequence of estimators $\hat{\theta}$ obtained by use of (18.2) is consistent. This result is strengthened by the fact that Huzurbazar (1948) has shown under regularity conditions that ultimately, as n increases, there is a *unique* consistent ML estimator.

Suppose that the LF possesses two derivatives. It follows from the convergence in probability of $\hat{\theta}$ to θ_0 that

$$\frac{1}{n}\left[\frac{\partial^2}{\partial \theta^2}\log L(x\mid\theta)\right]_{\theta=\hat{\theta}} \xrightarrow[n\to\infty]{} \frac{1}{n}\left[\frac{\partial^2}{\partial \theta^2}\log L(x\mid\theta)\right]_{\theta=\theta_0} \tag{18.23}$$

Now by the Strong Law of Large Numbers, once more,

$$\frac{1}{n}\frac{\partial^2}{\partial \theta^2}\log L(x\mid\theta) = \frac{1}{n}\sum_{i=1}^{n}\frac{\partial^2}{\partial \theta^2}\log f(x_i\mid\theta)$$

is the mean of n independent identical variates and converges with probability unity to its mean value. Thus we may write (18.23) as

$$\lim_{n\to\infty}\text{prob}\left\{\left[\frac{\partial^2}{\partial \theta^2}\log L(x\mid\theta)\right]_{\theta=\hat{\theta}} = E_0\left[\frac{\partial^2}{\partial \theta^2}\log L(x\mid\theta)\right]_{\theta=\theta_0}\right\} = 1. \tag{18.24}$$

But we have seen at (17.19) that under regularity conditions

$$E\left[\frac{\partial^2}{\partial \theta^2}\log L(x\mid\theta)\right] = -E\left\{\left(\frac{\partial \log L(x\mid\theta)}{\partial \theta}\right)^2\right\} < 0. \tag{18.25}$$

Thus (18.24) becomes

$$\lim_{n\to\infty}\text{prob}\left\{\left[\frac{\partial^2}{\partial \theta^2}\log L(x\mid\theta)\right]_{\theta=\hat{\theta}} < 0\right\} = 1. \tag{18.26}$$

18.12 Now suppose that the conditions of **18.2** hold, and that two local maxima of the LF, at $\hat{\theta}_1$ and $\hat{\theta}_2$ are roots of (18.5) satisfying (18.6). If $\log L(x\mid\theta)$ has a second derivative everywhere, as we have assumed in the last section, there must be a minimum between the maxima at $\hat{\theta}_1$ and $\hat{\theta}_2$. If this is at $\hat{\theta}_3$, we must have

$$\left[\frac{\partial^2 \log L(x\mid\theta)}{\partial \theta^2}\right]_{\theta=\hat{\theta}_3} \geqslant 0. \tag{18.27}$$

But since $\hat{\theta}_1$ and $\hat{\theta}_2$ are consistent estimators, $\hat{\theta}_3$, which lies between them in value, must also be consistent and must satisfy (18.26). Since (18.26) and (18.27) directly contradict each other, it follows that we can only have one consistent estimator $\hat{\theta}$ obtained as a root of the likelihood equation (18.5).

18.13 A point which should be discussed in connexion with the consistency of ML estimators is that, for particular samples, there is the possibility that the LF has two (or more) equal suprema, i.e. that the equality sign holds in (18.2). How can we choose between the values $\hat{\theta}_1$, $\hat{\theta}_2$, etc., at which they occur? There seems to be an essential indeterminacy here. Fortunately, however, it is not an important one, since the difficulty in general only arises when particular configurations of sample values

Example 18.4

In Example 18.3 put
$$\cos \theta = \rho.$$
To each real solution of the cubic likelihood equation, say $\hat{\rho}$, there will now correspond an infinity of estimators of θ, of form
$$\hat{\theta}_r = \arccos \hat{\rho} + 2r\pi$$
where r is any integer. The parameter θ is essentially incapable of estimation. Considered as a function of θ, the LF is periodic, with an infinite number of equal maxima at $\hat{\theta}_r$, and the $\hat{\theta}_r$ differ by multiples of 2π. There can be only one consistent estimator of θ_0, the true value of θ, but we have no means of deciding which $\hat{\theta}_r$ is consistent. In such a case, we must recognize that only $\cos \theta$ is directly estimable. We say that θ is *unidentifiable*.

Consistency and bias of ML estimators

18.14 Although, under the conditions of **18.10**, the ML estimator is consistent, it is not unbiassed generally. We have already seen in Example 18.1 that there may be bias even when the ML estimator is a function of a single sufficient statistic. In general, we must expect bias, for if the ML estimator is $\hat{\theta}$ and we seek to estimate a function $\tau(\theta)$, we have seen in **8.9** that the ML estimator of $\tau(\theta)$ is $\tau(\hat{\theta})$. But in general
$$E\{\tau(\hat{\theta})\} \neq \tau\{E(\hat{\theta})\}, \qquad (18.28)$$
so that if $\hat{\theta}$ is unbiassed for θ, $\tau(\hat{\theta})$ cannot be unbiassed for $\tau(\theta)$. If the ML estimator is consistent, the paragraph below Example 17.3 may apply.

> Brillinger (1964) shows that if ML bias is removed by the method of **17.10**, t_n' at (17.10) is asymptotically normal under regularity conditions, and he gives expansions for its bias and mean-square-error.

The efficiency and asymptotic normality of ML estimators

18.15 When we turn to the discussion of the efficiency of ML estimators, we cannot obtain a result as clear-cut as that of **18.10**. The following example is enough to show that we must make restrictions before we can obtain optimum results on efficiency.

Example 18.5

We saw in Example 17.22 that in the distribution
$$dF(x) = dx/\theta, \qquad k\theta \leqslant x \leqslant (k+1)\theta; \quad k > 0,$$
there is no single sufficient statistic for θ, but that the extreme observations $x_{(1)}$ and $x_{(n)}$ are a pair of jointly sufficient statistics for θ. Let us now find the ML estimator of θ. We maximize the LF as in Example 18.1. Since
$$L(x \mid \theta) = \theta^{-n} u(x_{(1)} - k\theta) u((k+1)\theta - x_{(n)})$$
we have
$$\hat{\theta} = x_{(n)}/(k+1),$$

which is accordingly the ML estimator. We see at once that $\hat{\theta}$ is a function of $x_{(n)}$ only, although $x_{(1)}$ and $x_{(n)}$ are both required for sufficiency.

Now by symmetry, $x_{(1)}$ and $x_{(n)}$ have the same variance, say V. The ML estimator has variance
$$\text{var } \hat{\theta} = V/(k+1)^2,$$
and the estimator
$$\theta^* = x_{(1)}/k$$
has variance
$$\text{var } \theta^* = V/k^2.$$
Since $x_{(1)}$ and $x_{(n)}$ are asymptotically independently distributed (**14.23**), the function
$$\bar{\theta} = a\hat{\theta} + (1-a)\theta^*$$
will, like $\hat{\theta}$ and θ^*, be a consistent estimator of θ, and its variance is
$$\text{var } \bar{\theta} = V\left\{\frac{a^2}{(k+1)^2} + \frac{(1-a)^2}{k^2}\right\}$$
which is minimized for variation in a when $a = \dfrac{(k+1)^2}{k^2+(k+1)^2}$. Then
$$\text{var } \bar{\theta} = V/\{k^2+(k+1)^2\}.$$
Thus, for all $k > 0$,
$$\frac{\text{var } \bar{\theta}}{\text{var } \hat{\theta}} = \frac{(k+1)^2}{k^2+(k+1)^2} < 1$$
and the ML estimator has larger variance. If k is large, the variance of $\hat{\theta}$ is nearly twice that of the other estimator.

18.16 To prove the asymptotic normality and efficiency of $\hat{\theta}$, we shall assume that the first two derivatives of $\log L(x \mid \theta)$ exist, and that (17.18–19) hold, i.e. that
$$E\left(\frac{\partial \log L}{\partial \theta}\right) = 0 \tag{18.29}$$
and
$$R^2(\theta) = -E\left(\frac{\partial^2 \log L}{\partial \theta^2}\right) = E\left\{\left(\frac{\partial \log L}{\partial \theta}\right)^2\right\}, \tag{18.30}$$
where $R^2(\theta) > 0$.

Using Taylor's theorem, we have
$$\left(\frac{\partial \log L}{\partial \theta}\right)_{\hat{\theta}} = \left(\frac{\partial \log L}{\partial \theta}\right)_{\theta_0} + (\hat{\theta} - \theta_0)\left(\frac{\partial^2 \log L}{\partial \theta^2}\right)_{\theta^*}, \tag{18.31}$$
where θ^* is some value between $\hat{\theta}$ and θ_0. As we pointed out in **18.2**, our differentiability assumptions imply that $\hat{\theta}$ is a root of the likelihood equation $\partial \log L/\partial \theta = 0$. Thus the left-hand side of (18.31) is zero. On its right-hand side, both $\partial \log L/\partial \theta$ and $\partial^2 \log L/\partial \theta^2$ are sums of independent identical variates, and as $n \to \infty$ each therefore converges to its expectation by the Strong Law of Large Numbers, as in the argument of **18.10**. The first of these expectations is zero by (18.29) and the second

non-zero by (18.30). Since the right-hand side of (18.31) as a whole must converge to zero, to remain equal to the left, we see that we must have $(\hat{\theta}-\theta_0)$ converging to zero as $n \to \infty$, so that $\hat{\theta}$ is a consistent estimator under our assumptions.

We now re-write (18.31) in the form

$$(\hat{\theta}-\theta_0)R(\theta_0) = \frac{\left(\frac{\partial \log L}{\partial \theta}\right)_{\theta_0} \bigg/ R(\theta_0)}{\left(\frac{\partial^2 \log L}{\partial \theta^2}\right)_{\theta^*} \bigg/ \{-R^2(\theta_0)\}}. \qquad (18.32)$$

In the denominator on the right of (18.32) we have, since $\hat{\theta}$ is consistent for θ_0 and θ^* lies between them, from (18.23-4) and (18.30),

$$\lim_{n \to \infty} \text{prob} \left\{ \left[\frac{\partial^2 \log L}{\partial \theta^2}\right]_{\theta^*} = -R^2(\theta_0) \right\} = 1, \qquad (18.33)$$

so that the denominator converges to unity. The numerator on the right of (18.32) is the ratio to $R(\theta_0)$ of the sum of the n independent identical variates $\partial \log f(x_i | \theta_0)/\partial \theta$. This sum has zero mean by (18.29) and variance defined at (18.30) to be $R^2(\theta_0)$. The Central Limit Theorem (**7.26**) therefore applies, and the numerator is asymptotically a standardized normal variate; the same is therefore true of the right-hand side as a whole. Thus the left-hand side of (18.32) is asymptotically standard normal or, in other words, the ML estimator $\hat{\theta}$ is asymptotically normally distributed with mean θ_0 and variance $1/R^2(\theta_0)$.

A more rigorous proof on these lines is given by Cramér (1946). Daniels (1961) relaxes the conditions for the asymptotic normality and efficiency of ML estimators. See also Huber (1967) and Lecam (1970).

18.17 This result, which gives the ML estimator an asymptotic variance equal to the MVB (17.24), implies that under these regularity conditions the ML estimator is efficient. Since the MVB can only be attained in the presence of a sufficient statistic (cf. **17.33**) we are also justified in saying that the ML estimator is "asymptotically sufficient."

Lecam (1953) has objected to the use of the term "efficient" because it implies absolute minimization of variance in large samples, and in the strict sense this is not achieved by the ML (or any other) estimator. For example, consider a consistent estimator t of θ, asymptotically normally distributed with variance of order n^{-1}. Define a new statistic

$$t' = \begin{cases} t & \text{if } |t| \geq n^{-\frac{1}{4}}, \\ kt & \text{if } |t| < n^{-\frac{1}{4}}. \end{cases} \qquad (18.34)$$

We have

$$\lim_{n \to \infty} \text{var } t'/\text{var } t = \begin{cases} 1 & \text{if } \theta \neq 0, \\ k^2 & \text{if } \theta = 0, \end{cases}$$

and k may be taken very small, so that at one point t' is more efficient than t, and nowhere is it worse. Lecam has shown (cf. also Bahadur (1964)) that such "superefficiency" can arise only for a set of θ-values of measure zero. In view of this, we shall retain the term "efficiency" in its ordinary use. However, C. R. Rao (1962b) shows that even this limited paradox can be avoided by redefining the efficiency of an estimator in terms of its correlation with $\frac{\partial \log L}{\partial \theta}$—cf. **17.15–17** and (17.61). Walker (1963) gives sufficient

Example 18.6

In Example 18.3 we found that the ML estimator $\hat{\rho}$ of the correlation parameter in a standardized bivariate normal distribution is a root of the cubic equation
$$\frac{\partial \log L}{\partial \rho} = \frac{n\rho}{1-\rho^2} - \frac{\rho}{(1-\rho^2)^2}\{\Sigma x^2 - 2\rho \Sigma xy + \Sigma y^2\} + \frac{1}{1-\rho^2}\Sigma xy = 0.$$
If we differentiate again, we have
$$\frac{\partial^2 \log L}{\partial \rho^2} = \frac{n(1+\rho^2)}{(1-\rho^2)^2} - \frac{(1+3\rho^2)}{(1-\rho^2)^3}(\Sigma x^2 - 2\rho \Sigma xy + \Sigma y^2) + \frac{4\rho}{(1-\rho^2)^2}\Sigma xy$$
so that, since $E(x^2) = E(y^2) = 1$ and $E(xy) = \rho$,
$$E\left(\frac{\partial^2 \log L}{\partial \rho^2}\right) = \frac{n(1+\rho^2)}{(1-\rho^2)^2} - \frac{2n(1+3\rho^2)}{(1-\rho^2)^2} + \frac{4n\rho^2}{(1-\rho^2)^2}$$
$$= -\frac{n(1+\rho^2)}{(1-\rho^2)^2}.$$
Hence, from **18.16**, we have asymptotically
$$\operatorname{var}\hat{\rho} = \frac{(1-\rho^2)^2}{n(1+\rho^2)}.$$

Example 18.7

The distribution
$$dF(x) = \tfrac{1}{2}\exp\{-|x-\theta|\}dx, \qquad -\infty \leqslant x \leqslant \infty,$$
yields the log likelihood
$$\log L(x\mid \theta) = -n\log 2 - \sum_{i=1}^{n}|x_i - \theta|.$$
This is maximized when $\sum_i |x_i - \theta|$ is minimized, and by the result of Exercise 2.1 this occurs when θ is the median of the n values of x. (If n is odd, the value of the middle observation is the median; if n is even, any value in the interval including the two middle observations is a median.) Thus the ML estimator is $\hat{\theta} = \tilde{x}$, the sample median. It is easily seen from (14.20) that its asymptotic variance in this case is
$$\operatorname{var}\hat{\theta} = 1/n.$$
We cannot use the result of **18.16** to check the efficiency of $\hat{\theta}$, since the differentiability conditions there imposed do not hold for this distribution. But since
$$\frac{\partial \log f(x\mid \theta)}{\partial \theta} = \begin{cases} +1 & \text{if } x > \theta, \\ -1 & \text{if } x < \theta, \end{cases}$$
only fails to exist at $x = \theta$, we have
$$\left(\frac{\partial \log f(x\mid \theta)}{\partial \theta}\right)^2 = 1, \qquad x \neq \theta.$$
For $\varepsilon > 0$, we now interpret $E\left[\left(\dfrac{\partial \log f(x\mid \theta)}{\partial \theta}\right)^2\right]$ as
$$\lim_{\varepsilon \to 0}\left\{\int_{-\infty}^{\theta-\varepsilon} + \int_{\theta+\varepsilon}^{\infty}\right\}\left(\frac{\partial \log f(x\mid \theta)}{\partial \theta}\right)^2 dF(x) = 1.$$

Thus we have
$$E\left\{\left(\frac{\partial \log L}{\partial \theta}\right)^2\right\} = nE\left\{\left(\frac{\partial \log f(x|\theta)}{\partial \theta}\right)^2\right\} = n,$$
so that the MVB for an estimator of θ is
$$\operatorname{var} t \geqslant 1/n,$$
which is attained asymptotically by $\hat{\theta}$.

18.18 The result of **18.16** simplifies for a distribution admitting a single sufficient statistic for the parameter. For in that case, from (18.10), (18.11) and (18.12),
$$E\left(\frac{\partial^2 \log L}{\partial \theta^2}\right) = -A(\theta)\tau'(\theta) = \left(\frac{\partial^2 \log L}{\partial \theta^2}\right)_{\hat{\theta}=\theta}, \tag{18.35}$$
so that there is no need to evaluate the expectation in this case: the MVB becomes simply $-1\bigg/\left(\dfrac{\partial^2 \log L}{\partial \theta^2}\right)_{\hat{\theta}=\theta}$, is attained exactly when $\hat{\theta}$ is unbiassed for θ, and asymptotically in any case under the conditions of **18.16**.

If there is no single sufficient statistic, the asymptotic variance of $\hat{\theta}$ may be estimated in the usual way from the sample, an unbiassed estimator commonly being sought.

Example 18.8

To estimate the standard deviation σ of a normal distribution
$$dF(x) = \frac{1}{\sigma(2\pi)^{\frac{1}{2}}}\exp\left(-\frac{x^2}{2\sigma^2}\right)dx, \quad -\infty \leqslant x \leqslant \infty.$$
We have
$$\log L(x|\sigma) = -n\log\sigma - \frac{\Sigma x^2}{2\sigma^2},$$
$$(\log L)' = -n/\sigma + \Sigma x^2/\sigma^3,$$
so that the sufficient ML estimator is $\hat{\sigma} = \sqrt{\left(\dfrac{1}{n}\Sigma x^2\right)}$ and
$$(\log L)'' = n/\sigma^2 - 3\Sigma x^2/\sigma^4 = \frac{n}{\sigma^2}\left(1 - \frac{3\hat{\sigma}^2}{\sigma^2}\right).$$
Thus, using (18.35), we have as n increases
$$\operatorname{var}\hat{\sigma} \to -1/(\log L)''_{\hat{\sigma}=\sigma} = \sigma^2/(2n).$$

18.19 Even when the regularity conditions for its efficiency do not hold, the ML estimator may have minimum variance properties conferred upon it by the sufficient statistic(s) of which it is always a function. Thus, e.g., in Example 18.1 $\hat{\theta} = x_{(n)}$ is a multiple $n/(n+1)$ of the unbiassed estimator t with minimum variance (cf. Exercise 17.24), and consequently $\hat{\theta}$ will be asymptotically unbiassed with the same asymptotic variance as t. However, we saw in Example 11.4 (Vol. 1, 3rd edn—cf. also (19.85) below) that here $\operatorname{var} x_{(n)} = \dfrac{n\theta^2}{(n+1)^2(n+2)}$. Thus when we consider the m.s.e. of $x_{(n)}$ as in **17.30**, the square of its bias, $\theta^2/(n+1)^2$, is of the same order of magnitude as its variance. If we apply the result of Example 17.14, we find that $\dfrac{n(n+2)}{(n+1)^2}t = \left(\dfrac{n+2}{n+1}\right)x_{(n)}$ has m.s.e.

$\theta^2/(n+1)^2$, against $\theta^2/n(n+2)$ for the unbiassed t and $2\theta^2/(n+1)(n+2)$ for $\hat{\theta}$. Thus $\hat{\theta}$ in this case has asymptotically, and indeed almost exactly, twice as large a m.s.e. as its best multiple has.

The cumulants of a ML estimator

18.20 Haldane and Smith (1956) obtain expressions for the first four cumulants of a ML estimator. Suppose that the distribution sampled is grouped into k classes, and that the probability of an observation falling into the rth class is π_r ($r = 1, 2, \ldots, k$). We thus reduce any distribution to a multinomial distribution (5.30), and if the range of the original distribution is independent of the unknown parameter θ, we seek solutions of the likelihood equation (18.5). Since the probabilities π_r are functions of θ, we write, from (5.78),

$$L(x|\theta) \propto \prod_r \pi_r^{n_r}, \qquad (18.36)$$

where n_r is the number of observations in the rth class and $\sum_r n_r = n$, the sample size. (18.36) gives the likelihood equation as

$$\frac{\partial \log L}{\partial \theta} = \sum_r n_r \frac{\pi_r'}{\pi_r} = 0, \qquad (18.37)$$

where a prime denotes differentiation with respect to θ. Of course, the ML estimator $\hat{\theta}$ will now differ from what it would be for the ungrouped observations, when more information is available. Exercises 18.24–5 discuss the difference, which is important in some contexts—cf. **30.15-19** below. However, as $k \to \infty$ the difference disappears.

Now, using Taylor's theorem, we expand π_r and π_r' about the true value θ_0, and obtain

$$\left.\begin{array}{l}\pi_r(\hat{\theta}) = \pi_r(\theta_0) + (\hat{\theta}-\theta_0)\pi_r'(\theta_0) + \tfrac{1}{2}(\hat{\theta}-\theta_0)^2 \pi_r''(\theta_0) + \ldots,\\ \pi_r'(\hat{\theta}) = \pi_r'(\theta_0) + (\hat{\theta}-\theta_0)\pi_r''(\theta_0) + \tfrac{1}{2}(\hat{\theta}-\theta_0)^2 \pi_r'''(\theta_0) + \ldots.\end{array}\right\} \qquad (18.38)$$

If we insert (18.38) into (18.37), expand binomially, and sum the series, we have, writing

$$A_i = \sum_r \{\pi_r'(\theta_0)\}^{i+1} / \{\pi_r(\theta_0)\}^i,$$

$$B_i = \sum_r \{\pi_r'(\theta_0)\}^i \pi_r''(\theta_0) / \{\pi_r(\theta_0)\}^i,$$

$$C_i = \sum_r \{\pi_r'(\theta_0)\}^{i-1} \{\pi_r''(\theta_0)\}^2 / \{\pi_r(\theta_0)\}^i,$$

$$D_i = \sum_r \{\pi_r'(\theta_0)\}^i \pi_r'''(\theta_0) / \{\pi_r(\theta_0)\}^i,$$

$$\alpha_i = \sum_r \{\pi_r'(\theta_0)\}^i \left\{\frac{n_r}{n} - \pi_r(\theta_0)\right\} / \{\pi_r(\theta_0)\}^i,$$

$$\beta_i = \sum_r \{\pi_r'(\theta_0)\}^{i-1} \pi_r''(\theta_0) \left\{\frac{n_r}{n} - \pi_r(\theta_0)\right\} / \{\pi_r(\theta_0)\}^i,$$

$$\delta_i = \sum_r \{\pi_r'(\theta_0)\}^{i-1} \pi_r'''(\theta_0) \left\{\frac{n_r}{n} - \pi_r(\theta_0)\right\} / \{\pi_r(\theta_0)\}^i,$$

the expansion

$$\alpha_1 - (A_1 + \alpha_2 - \beta_1)(\hat{\theta}-\theta_0) + \tfrac{1}{2}(2A_2 - 3B_1 + 2\alpha_3 - 3\beta_2 + \delta_1)(\hat{\theta}-\theta_0)^2$$
$$- \tfrac{1}{6}(6A_3 - 12B_2 + 3C_1 + 4D_1)(\hat{\theta}-\theta_0)^3 + \ldots = 0. \qquad (18.39)$$

For large n, (18.39) may be inverted by Lagrange's theorem to give
$$(\hat{\theta}-\theta_0) = A_1^{-1}\alpha_1 + A_1^{-3}\alpha_1[(A_2-\tfrac{3}{2}B_1)\alpha_1 - A_1(\alpha_2-\beta_1)]$$
$$+ A_1^{-5}\alpha_1[\{2(A_2-\tfrac{3}{2}B_1)^2 - A_1(A_3-2B_2+\tfrac{1}{2}C_1+\tfrac{2}{3}D_1)\}\alpha_1^2$$
$$- 3A_1(A_2-\tfrac{3}{2}B_1)\alpha_1(\alpha_2-\beta_1) + \tfrac{1}{2}A_1^2\alpha_1(2\alpha_3-3\beta_2+\delta_1)$$
$$+ A_1^2(\alpha_2-\beta_1)^2] + O(n^{-3}). \tag{18.40}$$

(18.40) enables us to obtain the moments of $\hat{\theta}$ as series in powers of n^{-1}.

Consider the sampling distribution of the sum
$$W = \sum_r h_r \left\{ \frac{n_r}{n} - \pi_r(\theta_0) \right\},$$
where the h_r are any constant weights. From the moments of the multinomial distribution (cf. (5.80)), we obtain for the moments of W, writing $S_i = \sum_r h_r^i \pi_r(\theta_0)$,

$$\left.\begin{aligned}
\mu_1'(W) &= 0, \\
\mu_2(W) &= n^{-1}(S_2 - S_1^2), \\
\mu_3(W) &= n^{-2}(S_3 - S_1 S_2 + 2S_1^3), \\
\mu_4(W) &= 3n^{-2}(S_2 - S_1^2)^2 + n^{-3}(S_4 - 4S_1 S_3 - 3S_2^2 + 12S_1^2 S_2 - 6S_1^4), \\
\mu_5(W) &= 10n^{-3}(S_2 - S_1^2)(S_3 - 3S_1 S_2 + 2S_1^3) + O(n^{-4}), \\
\mu_6(W) &= 15n^{-3}(S_2 - S_1^2)^3 + O(n^{-4}).
\end{aligned}\right\} \tag{18.41}$$

From (18.41) we can derive the moments and product-moments of the random variables α_i, β_i and δ_i appearing in (18.40), for all of these are functions of form W. Finally, we substitute these moments into the powers of (18.40) to obtain the moments of $\hat{\theta}$. Expressed as cumulants, these are

$$\left.\begin{aligned}
\kappa_1 &= \theta_0 - \tfrac{1}{2}n^{-1} A_1^{-2} B_1 + O(n^{-2}), \\
\kappa_2 &= n^{-1} A_1^{-1} + n^{-2} A_1^{-4}[-A_2^2 + \tfrac{7}{2}B_1^2 + A_1(A_3 - B_2 - D_1) - A_1^3] + O(n^{-3}), \\
\kappa_3 &= n^{-2} A_1^{-3}(A_2 - 3B_1) + O(n^{-3}), \\
\kappa_4 &= n^{-3} A_1^{-5}[-12B_1(A_2 - 2B_1) + A_1(A_3 - 4D_1) - 3A_1^3] + O(n^{-4}),
\end{aligned}\right\} \tag{18.42}$$

whence
$$\left.\begin{aligned}
\gamma_1 &= \kappa_3/\kappa_2^{3/2} = n^{-\frac{1}{2}} A_1^{-\frac{3}{2}}(A_2 - 3B_1) + o(n^{-\frac{1}{2}}), \\
\gamma_2 &= \kappa_4/\kappa_2^2 = n^{-1} A_1^{-3}[-12B_1(A_2 - 2B_1) + A_1(A_3 - 4D_1) - 3A_1^3] + o(n^{-1}).
\end{aligned}\right\} \tag{18.43}$$

The first cumulant in (18.42) shows that the bias in $\hat{\theta}$ is of the order of magnitude n^{-1} unless $B_1 = 0$, when it is of order n^{-2}, as may be confirmed by calculating a further term in the first cumulant. The leading term in the second cumulant corresponds to the asymptotic variance previously established in **18.16**. (18.43) illustrates the rapidity of the tendency to normality, established in **18.16** for ungrouped observations.

If the terms in (18.42) were all evaluated, and unbiassed estimates made of each of the first four moments of $\hat{\theta}$, a Pearson distribution (cf. **6.2–12**) could be fitted and an estimate of the small-sample distribution of $\hat{\theta}$ obtained which would provide a better approximation than the ultimate normal approximation of **18.16**.

The next higher-order terms in (18.42–3) are derived by Shenton and Bowman (1963).

Successive approximation to ML estimators

18.21 In most of the examples we have considered, the ML estimator has been obtained in explicit form. The exception was in Example 18.3, where we were left with a cubic equation to solve for the estimator, and even this can be done without much trouble when the values of x are given. Sometimes, however, the likelihood equation is so complicated that iterative methods must be used to find a root, starting from some trial value t.

As at (18.31), we expand $\partial \log L / \partial \theta$ in a Taylor series, but this time about its value at t, obtaining

$$0 = \left(\frac{\partial \log L}{\partial \theta}\right)_{\hat\theta} = \left(\frac{\partial \log L}{\partial \theta}\right)_t + (\hat\theta - t)\left(\frac{\partial^2 \log L}{\partial \theta^2}\right)_{\theta^*},$$

where θ^* lies between $\hat\theta$ and t. Thus

$$\hat\theta = t - \left(\frac{\partial \log L}{\partial \theta}\right)_t \bigg/ \left(\frac{\partial^2 \log L}{\partial \theta^2}\right)_{\theta^*} \tag{18.44}$$

If we can choose t so that it is likely to be in the neighbourhood of $\hat\theta$, we can replace θ^* in (18.44) by t and obtain

$$\hat\theta = t - \left(\frac{\partial \log L}{\partial \theta}\right)_t \bigg/ \left(\frac{\partial^2 \log L}{\partial \theta^2}\right)_t, \tag{18.45}$$

which will give a closer approximation to $\hat\theta$. The process can be repeated until no further correction is achieved to the desired degree of accuracy.

The most common method for choice of t is to take it as the value of some (preferably simply-calculated) consistent estimator of θ. Then, as $n \to \infty$, we shall have the two consistent estimators t and $\hat\theta$ converging to θ_0, and θ^* consequently also doing so. The three random variables $\left(\frac{\partial^2 \log L}{\partial \theta^2}\right)_{\theta^*}$, $\left(\frac{\partial^2 \log L}{\partial \theta^2}\right)_t$ and $\left[E\left(\frac{\partial^2 \log L}{\partial \theta^2}\right)\right]_t$ will all converge to $\left[E\left(\frac{\partial^2 \log L}{\partial \theta^2}\right)\right]_{\theta_0}$. Use of the second of these variables, instead of the first, in (18.44) gives (18.45) above: use of the third instead of the first gives the alternative iterative procedure

$$\hat\theta = t - \left(\frac{\partial \log L}{\partial \theta}\right)_t \bigg/ \left[E\left(\frac{\partial^2 \log L}{\partial \theta^2}\right)\right]_t = t + \left(\frac{\partial \log L}{\partial \theta}\right)_t (\text{var } \hat\theta)_t, \tag{18.46}$$

var $\hat\theta$ being the asymptotic variance obtained in **18.16**. (18.45) is the Newton–Raphson iterative process: (18.46) is sometimes known as "the method of scoring for parameters," and is due to Fisher (1925). Kale (1961) shows that (18.46) will usually be the quicker process for large n unless extremely high accuracy is ultimately required. It is usually less laborious.

Both (18.45) and (18.46) may fail to converge in particular cases. Even when they do converge, if the likelihood equation has multiple roots there is no guarantee that they will converge to the root corresponding to the absolute maximum of the LF; this should be verified by examining the changes in sign of $\partial \log L / \partial \theta$ from positive to negative and searching the intervals in which these changes occur to locate, evaluate and compare the maxima. V. D. Barnett (1966a) discusses a systematic method of doing this, using "the method of false positions."

Example 18.9

To estimate the location parameter θ in the Cauchy distribution

$$dF(x) = \frac{dx}{\pi\{1+(x-\theta)^2\}}, \qquad -\infty \leqslant x \leqslant \infty.$$

The likelihood equation is

$$\frac{\partial \log L}{\partial \theta} = 2 \sum_{i=1}^{n} \frac{(x_i - \theta)}{\{1+(x_i-\theta)^2\}} = 0,$$

an equation of degree $(2n-1)$ in θ. From **18.16** the asymptotic variance of $\hat{\theta}$ is given by

$$-\frac{1}{\operatorname{var} \hat{\theta}} \sim E\left(\frac{\partial^2 \log L}{\partial \theta^2}\right) = nE\left(\frac{\partial^2 \log f}{\partial \theta^2}\right)$$

$$= \frac{n}{\pi} \int_{-\infty}^{\infty} \frac{2(x-\theta)^2 - 2}{\{1+(x-\theta)^2\}^3} dx$$

$$= \frac{4n}{\pi} \int_{0}^{\infty} \frac{(x^2-1)\,dx}{(1+x^2)^3}$$

$$= -n/2.$$

Hence
$$\operatorname{var} \hat{\theta} = 2/n.$$

The equation has multiple roots in general, and for small n, (18.45) or (18.46) may not converge—cf. V. D. Barnett (1966a). For $n \geqslant 9$, however, $\hat{\theta}$ is almost always—cf. Haas *et al.* (1970)—the nearest maximum to the sample median t, which has large-sample variance (Example 17.5) $\operatorname{var} t = \pi^2/(4n)$ and thus has efficiency $8/\pi^2 = 0{\cdot}81$ approximately. For $n \geqslant 15$, we may confidently use the median as our starting-point in seeking the value of $\hat{\theta}$, and solve (18.46), which here becomes

$$\hat{\theta} = t + \frac{4}{n} \sum_i \frac{(x_i - t)}{\{1+(x_i-t)^2\}}.$$

This is our first approximation to $\hat{\theta}$, which we may improve by further iterations of the process.

Bloch (1966) gives an estimator with efficiency $>0{\cdot}95$, namely the linear combination of 5 order-statistics $x_{(r)}$ ($r = 0{\cdot}13n, 0{\cdot}4n, 0{\cdot}5n, 0{\cdot}6n, 0{\cdot}87n$) with weights $-0{\cdot}052$, $0{\cdot}3485$, $0{\cdot}407$, $0{\cdot}3485$, $-0{\cdot}052$. This would therefore be an excellent starting-point for ML iteration. Rothenberg *et al.* (1964) show that the mean of the central 24 per cent of a Cauchy sample has asymptotic variance $2{\cdot}28/n$, its efficiency therefore being $0{\cdot}88$. See also V. D. Barnett (1966b).

> Haas *et al.* (1970) and Chan (1970) also discuss ML estimation of a scale parameter for the Cauchy distribution, with or without a location parameter.

Example 18.10

We now examine the iterative method of solution in more detail, and for this purpose we use some data due to Fisher (1925–, Chapter 9).

Consider a multinomial distribution (cf. **5.30**) with four classes, their probabilities being

$$p_1 = (2+\theta)/4,$$
$$p_2 = p_3 = (1-\theta)/4,$$
$$p_4 = \theta/4.$$

ESTIMATION : MAXIMUM LIKELIHOOD

The parameter θ, which lies in the range $(0, 1)$, is to be estimated from the observed frequencies (a, b, c, d) falling into the classes, the sample size n being equal to $a+b+c+d$. We have
$$L(a, b, c, d \mid \theta) \propto (2+\theta)^a (1-\theta)^{b+c} \theta^d,$$
so that
$$\frac{\partial \log L}{\partial \theta} = \frac{a}{2+\theta} - \frac{(b+c)}{1-\theta} + \frac{d}{\theta},$$
and if this is equated to zero, we obtain the quadratic equation in θ
$$n\theta^2 + \{2(b+c)+d-a\}\theta - 2d = 0.$$
Since the product of the coefficient of θ^2 and the constant term is negative, the product of the roots of the quadratic must also be negative, and only one root can be positive. Only this positive root falls into the permissible range for θ. Its value $\hat{\theta}$ is given by
$$2n\hat{\theta} = \{a-d-2(b+c)\} + [\{a+2(b+c)+3d\}^2 - 8a(b+c)]^{\frac{1}{2}}.$$

The ML estimator $\hat{\theta}$ can very simply be evaluated from this formula. For Fisher's (genetical) example, where the observed frequencies are
$$a = 1997, \; b = 906, \; c = 904, \; d = 32, \; n = 3839$$
the value of $\hat{\theta}$ is 0·0357.

It is easily verified from a further differentiation that
$$\operatorname{var}\hat{\theta} \sim -\frac{1}{E\left(\dfrac{\partial^2 \log L}{\partial \theta^2}\right)} = \frac{2\theta(1-\theta)(2+\theta)}{n(1+2\theta)},$$
the value being 0·0000336 in this case, when $\hat{\theta}$ is substituted for θ in $\operatorname{var}\hat{\theta}$.

For illustrative purposes, we now suppose that we wish to find $\hat{\theta}$ iteratively in this case, starting from the value of an inefficient estimator. A simple inefficient estimator which was proposed by Fisher is
$$t = \{a+d-(b+c)\}/n,$$
which is easily seen to be consistent and has variance
$$\operatorname{var} t = (1-\theta^2)/n.$$
The value of t for the genetical data is
$$t = \{1997+32-(906+904)\}/3839 = 0{\cdot}0570.$$
This is a long way from the value of $\hat{\theta}$, 0·0357, which we seek. Using (18.46) we have, for our first approximation to $\hat{\theta}$,
$$\hat{\theta}_1 = 0{\cdot}0570 + \left(\frac{\partial \log L}{\partial \theta}\right)_{\theta=t} (\operatorname{var}\hat{\theta})_{\theta=t}.$$
Now
$$\left(\frac{\partial \log L}{\partial \theta}\right)_{\theta=0{\cdot}0570} = \frac{1997}{2{\cdot}0570} - \frac{1810}{0{\cdot}9430} + \frac{32}{0{\cdot}0570} = -387{\cdot}1713,$$
$$(\operatorname{var}\hat{\theta})_{\theta=0{\cdot}0570} = \frac{2 \times 0{\cdot}057 \times 0{\cdot}943 \times 2{\cdot}057}{3839 \times 1{\cdot}114} = 0{\cdot}00005170678,$$
so that our improved estimator is
$$\hat{\theta}_1 = 0{\cdot}0570 - 387{\cdot}1713 \times 0{\cdot}00005170678 = 0{\cdot}0570 - 0{\cdot}0200$$
$$= 0{\cdot}0370,$$

which is in fairly close agreement with the value sought, 0·0357. A second iteration gives

$$\left(\frac{\partial \log L}{\partial \theta}\right)_{\theta=0\cdot0370} = \frac{1997}{2\cdot037} - \frac{1810}{0\cdot963} + \frac{32}{0\cdot037} = -34\cdot31495,$$

$$(\operatorname{var} \hat{\theta})_{\theta=0\cdot0370} = \frac{2 \times 0\cdot037 \times 0\cdot963 \times 2\cdot037}{3839 \times 1\cdot074} = 0\cdot00003520681,$$

and hence
$$\hat{\theta}_2 = 0\cdot0370 - 34\cdot31495 \times 0\cdot00003520681 = 0\cdot0370 - 0\cdot0012$$
$$= 0\cdot0358.$$

This is very close to the value sought. At least one further iteration would be required to bring the value to 0·0357 correct to 4 d.p., and a further iteration to confirm that the value of $\hat{\theta}$ arrived at was stable to a sufficient number of decimal places to make further iterations unnecessary. The reader should carry through these further iterations to satisfy himself that he can use the method.

This example makes it clear that care must be taken to carry the iteration process far enough for practical purposes. It is a somewhat unfavourable example, in that t has an efficiency of $\frac{2\theta(2+\theta)}{(1+\theta)(1+2\theta)}$, which takes the value of 0·13, or 13 per cent, when $\hat{\theta} = 0\cdot0357$ is substituted for θ. One would usually seek to start from the value of an estimator with greater efficiency than this.

Exercise 18.35 shows that a single iteration is always enough if there is a MVB estimator of θ. See also Exercise 18.36.

ML estimators for several parameters

18.22 We now turn to discussion of the general case, in which more than one parameter are to be estimated simultaneously, whether in a univariate or multivariate distribution. If we interpret θ, and possibly also x, as a vector, the formulation of the ML principle at (18.2) holds good: we have to choose the *set* of admissible values of the parameters $\theta_1, \ldots, \theta_k$ which makes the LF an absolute maximum. Under the regularity conditions of **18.2–3**, the necessary condition for a local turning-point in the LF is that

$$\frac{\partial}{\partial \theta_r} \log L(x \mid \theta_1, \ldots, \theta_k) = 0, \qquad r = 1, 2, \ldots, k, \tag{18.47}$$

and a sufficient condition that this be a maximum is that the matrix

$$\left(\frac{\partial^2 \log L}{\partial \theta_r \partial \theta_s}\right) \tag{18.48}$$

be negative definite. The k equations (18.47) are to be solved for the k ML estimators $\hat{\theta}_1, \ldots, \hat{\theta}_k$.

The case of joint sufficiency

18.23 Just as in **18.4**, we see that if there exists a set of s statistics $t_1, \ldots, t_s$ which are jointly sufficient for the parameters $\theta_1, \ldots, \theta_k$, the ML estimators $\hat{\theta}_1, \ldots, \hat{\theta}_k$ must be functions of the sufficient statistics. As before, this follows immediately from the *forms* of the distributions of the errors: we make unbiassed estimators of the parameters and, further, unbiassed estimators of the sampling variances and covariances of these estimators, without distributional assumptions. However, if we wish to test hypotheses concerning the parameters, distributional assumptions are necessary. We shall be discussing the problems of testing hypotheses in the linear model in Chapter 24; here we shall only point out some fundamental features of the situation.

19.11 If we postulate that the ε_i are normally distributed, the fact that they are uncorrelated implies their *independence* (cf. **15.3**), and we may use the result of **15.11** to the effect that an idempotent quadratic form in independent standardized normal variates is a chi-squared variate with degrees of freedom given by the rank of the quadratic form. Applying this to the sum of squared residuals (19.37), we have, in the notation of (19.41), the result that $(n-k)s^2/\sigma^2$ is a chi-squared variate with $(n-k)$ degrees of freedom.

Further, we have the identity
$$\mathbf{y}'\mathbf{y} \equiv (\mathbf{y}-\mathbf{X}\hat{\boldsymbol{\theta}})'(\mathbf{y}-\mathbf{X}\hat{\boldsymbol{\theta}})+(\mathbf{X}\hat{\boldsymbol{\theta}})'(\mathbf{X}\hat{\boldsymbol{\theta}}), \qquad (19.42)$$
which is easily verified using (19.12). The second term on the right of (19.42) is
$$\hat{\boldsymbol{\theta}}'\mathbf{X}'\mathbf{X}\hat{\boldsymbol{\theta}} = \mathbf{y}'\mathbf{X}(\mathbf{X}'\mathbf{X})^{-1}\mathbf{X}'\mathbf{y} = (\boldsymbol{\varepsilon}'+\boldsymbol{\theta}'\mathbf{X}')\mathbf{X}(\mathbf{X}'\mathbf{X})^{-1}\mathbf{X}'(\mathbf{X}\boldsymbol{\theta}+\boldsymbol{\varepsilon}). \qquad (19.43)$$
From (19.43) it follows that if $\boldsymbol{\theta} = \mathbf{0}$,
$$\hat{\boldsymbol{\theta}}'\mathbf{X}'\mathbf{X}\hat{\boldsymbol{\theta}} = \boldsymbol{\varepsilon}'\mathbf{X}(\mathbf{X}'\mathbf{X})^{-1}\mathbf{X}'\boldsymbol{\varepsilon}, \qquad (19.44)$$
and (19.42) may then be rewritten, using (19.37) and (19.44),
$$\boldsymbol{\varepsilon}'\boldsymbol{\varepsilon} = \boldsymbol{\varepsilon}'\{\mathbf{I}_n-\mathbf{X}(\mathbf{X}'\mathbf{X})^{-1}\mathbf{X}'\}\boldsymbol{\varepsilon}+\boldsymbol{\varepsilon}'\{\mathbf{X}(\mathbf{X}'\mathbf{X})^{-1}\mathbf{X}'\}\boldsymbol{\varepsilon}. \qquad (19.45)$$

We have already seen in **19.9** that the rank of the first matrix in braces on the right of (19.45) is $(n-k)$, and we also established there that the rank of the second matrix in braces in (19.45) is k. Thus the ranks on the right-hand side add to n, the rank on the left, and Cochran's theorem (**15.16**) applies. Thus the two quadratic forms on the right of (19.45) are *independently* distributed (after division by σ^2 in each case to adjust the scale) like chi-squared with $(n-k)$ and k degrees of freedom.

19.12 It will have been noticed that, in **19.11**, the chi-squared distribution of $(\mathbf{y}-\mathbf{X}\hat{\boldsymbol{\theta}})'(\mathbf{y}-\mathbf{X}\hat{\boldsymbol{\theta}})$ holds whatever the true value of $\boldsymbol{\theta}$, while the second term in (19.42), $(\mathbf{X}\hat{\boldsymbol{\theta}})'(\mathbf{X}\hat{\boldsymbol{\theta}})$, is only so distributed if the true value of $\boldsymbol{\theta}$ is $\mathbf{0}$. Whether or not this is so, we have from (19.43), using (19.9),
$$E\{(\mathbf{X}\hat{\boldsymbol{\theta}})'(\mathbf{X}\hat{\boldsymbol{\theta}})\} = E\{\boldsymbol{\varepsilon}'\mathbf{X}(\mathbf{X}'\mathbf{X})^{-1}\mathbf{X}'\boldsymbol{\varepsilon}\}+\boldsymbol{\theta}'\mathbf{X}'\mathbf{X}\boldsymbol{\theta}. \qquad (19.46)$$
We saw in **19.9** that the first term on the right has the value $k\sigma^2$. Thus
$$E\{(\mathbf{X}\hat{\boldsymbol{\theta}})'(\mathbf{X}\hat{\boldsymbol{\theta}})\} = k\sigma^2+(\mathbf{X}\boldsymbol{\theta})'(\mathbf{X}\boldsymbol{\theta}), \qquad (19.47)$$
which exceeds $k\sigma^2$ unless $\mathbf{X}\boldsymbol{\theta} = \mathbf{0}$, which requires $\boldsymbol{\theta} = \mathbf{0}$ unless $\mathbf{X}$ takes special values. Thus it is intuitively reasonable to use the ratio $(\mathbf{X}\hat{\boldsymbol{\theta}})'(\mathbf{X}\hat{\boldsymbol{\theta}})/(ks^2)$ (where s^2, defined at (19.41), *always* has expected value σ^2) to test the hypothesis that $\boldsymbol{\theta} = \mathbf{0}$. We shall be returning to the justification of this and similar procedures from a less intuitive point of view in Chapter 24.

The singular case

19.13 In **19.4** we assumed $\mathbf{X'X}$ to be non-singular, so that (19.12) was valid, and $n > k$, so that (19.41) could be valid. If $n = k$, (19.12) still holds if $(\mathbf{X'X})^{-1}$ exists, but (19.41) is useless since the sum of squared residuals is identically zero, as (19.37) shows. If $n < k$, the rank of $\mathbf{X}$ (and that of $\mathbf{X'X}$, which is the same) is less than k, so $\mathbf{X'X}$ has no inverse.

We now let $\mathbf{X}$ (and $\mathbf{X'X}$) have rank $r < k$ and suppose that $n \geq r$. The LS estimation problem must be discussed afresh, since $\mathbf{X'X}$ has no inverse and (19.12) is invalid. The treatment follows Plackett (1950).

The condition (19.19) is still necessary and sufficient for $\mathbf{C\theta}$ to be unbiassedly estimable by $\mathbf{Ty}$. In this singular case, it cannot be satisfied if we wish to estimate $\mathbf{\theta}$ itself, when it becomes

$$\mathbf{TX} = \mathbf{I}. \tag{19.48}$$

For, remembering that $\mathbf{X}$ is of rank r, we partition it into

$$\mathbf{X} = \begin{pmatrix} \mathbf{X}_{r,r} & \vdots & \mathbf{X}_{r,k-r} \\ \cdots & \vdots & \cdots \\ \mathbf{X}_{n-r,r} & \vdots & \mathbf{X}_{n-r,k-r} \end{pmatrix}, \tag{19.49}$$

the suffixes of the matrix elements of (19.49) indicating the numbers of rows and columns. We assume, without loss of generality, that $\mathbf{X}_{r,r}$ is non-singular, and therefore has inverse $\mathbf{X}_{r,r}^{-1}$. The last $n-r$ rows of $\mathbf{X}$ are linearly dependent upon the first r rows, so that $\mathbf{X}_{n-r,r} = \mathbf{CX}_{r,r}$ and $\mathbf{X}_{n-r,k-r} = \mathbf{CX}_{r,k-r}$ for some $(n-r) \times r$ matrix $\mathbf{C}$. Define a new matrix, of order $k \times (k-r)$,

$$\mathbf{D} = \begin{pmatrix} \mathbf{X}_{r,r}^{-1} \cdot \mathbf{X}_{r,k-r} \\ \cdots \\ -\mathbf{I}_{k-r} \end{pmatrix}, \tag{19.50}$$

where $\mathbf{I}_{k-r}$ is the identity matrix of that order. Evidently, $\mathbf{D}$ is of rank $k-r$. If we form the product $\mathbf{XD}$, we see at once that

$$\mathbf{XD} = \mathbf{0}. \tag{19.51}$$

If we postmultiply (19.48) by $\mathbf{D}$, we obtain, using (19.51),

$$\mathbf{D} = \mathbf{TXD} = \mathbf{0}. \tag{19.52}$$

This contradicts the fact that $\mathbf{D}$ has rank $k-r$. Hence (19.48) cannot hold.

Of course, as (19.19) implies, some linear functions $\mathbf{C\theta}$ may still be estimated unbiassedly, for (19.52) is then generalized to $\mathbf{CD} = \mathbf{0}$, which does not contradict the rank $(k-r)$ of $\mathbf{D}$ if $\mathbf{C} \neq \mathbf{I}$.

19.14 We may proceed as in the non-singular case if we first introduce a set of $(k-r)$ linear constraints upon the parameters

$$\mathbf{a} \quad \mathbf{B\theta}, \tag{19.53}$$

where $\mathbf{a}$ is a $(k-r) \times 1$ vector of constants and $\mathbf{B}$ is a known $(k-r) \times k$ matrix, of rank $(k-r)$. We now seek an estimator of the form $\mathbf{t} = \mathbf{Ly} + \mathbf{Na}$. The condition (19.48) now becomes

$$\mathbf{I} = \mathbf{LX} + \mathbf{NB}. \tag{19.54}$$

ESTIMATION: LEAST SQUARES AND OTHER METHODS

Provided that
$$|BD| \neq 0. \tag{19.55}$$
the matrix **B**, of rank $(k-r)$, makes up the deficiency in rank of **X**. In fact, we treat **a** as a vector of dummy observations and solve (19.8) and (19.53) together, in the augmented model
$$\begin{pmatrix} y \\ a \end{pmatrix} = \begin{pmatrix} X \\ B \end{pmatrix} \theta + \begin{pmatrix} \epsilon \\ 0 \end{pmatrix}. \tag{19.56}$$
The matrix $\begin{pmatrix} X \\ B \end{pmatrix}' \begin{pmatrix} X \\ B \end{pmatrix} = X'X + B'B$ is positive definite, for a non-null vector **d** makes $d'X'Xd = (Xd)'Xd$ equal to zero only if $Xd = 0$, whence **d** must be a column of **D**. But (19.55) ensures that $Bd \neq 0$, so that $d'B'Bd > 0$. Thus $X'X + B'B$ is strictly positive definite and may be inverted.

(19.56) therefore yields, as at (19.22), the solution
$$C\hat{\theta} = C(X'X + B'B)^{-1}(X'y + B'a), \tag{19.57}$$
which as before is the MV linear unbiassed estimator of θ. Using (19.21), its dispersion matrix is, since **a** is constant,
$$V(C\hat{\theta}) = \sigma^2 C(X'X + B'B)^{-1} X'X (X'X + B'B)^{-1} C'. \tag{19.58}$$
The matrix **B** in (19.53) is arbitrary, subject to (19.55). In fact, if for **B** we substitute **UB**, where **U** is any non-singular $(k-r) \times (k-r)$ matrix, (19.57) and (19.58) are unaltered in value. Thus we may choose **B** for convenience in computation in any particular case.

19.15 Exercise 19.8 shows that σ^2 is estimated unbiassedly by the sum of squared residuals divided by $(n-r)$ if $n > r$.

Chipman (1964) gives a detailed discussion of LS theory in the singular case. See also T. O. Lewis and Odell (1966).

Example 19.9

As a simple example of a singular situation suppose that we have
$$\theta = \begin{pmatrix} \theta_1 \\ \theta_2 \\ \theta_3 \end{pmatrix} \quad X = \begin{pmatrix} 1 & 1 & 0 \\ 1 & 0 & 1 \\ 1 & 1 & 0 \\ 1 & 0 & 1 \end{pmatrix}.$$
Here $n = 4$, $k = 3$ and **X** has rank $2 < k$ because of the linear relation between its column vectors
$$x_1 - x_2 - x_3 = 0.$$
We first verify that θ cannot be unbiassedly estimated, as we saw in **19.13**.

The matrix **D** at (19.50) is of order 3×1, being
$$D = \begin{pmatrix} \begin{pmatrix} 1 & 1 \\ 1 & 0 \end{pmatrix}^{-1} \cdot \begin{pmatrix} 0 \\ 1 \end{pmatrix} \\ \cdots\cdots\cdots \\ -1 \end{pmatrix} = \begin{pmatrix} 1 \\ -1 \\ -1 \end{pmatrix},$$

expressing the linear relation. We now introduce the matrix of order 1×3
$$\mathbf{B} = (1 \ 0 \ 0),$$
which satisfies (19.55) since $\mathbf{BD} = 1$, a scalar in this case of a single linear relation. Hence (19.53) is
$$c = (1 \ 0 \ 0)\begin{pmatrix}\theta_1\\ \theta_2\\ \theta_3\end{pmatrix} = \theta_1,$$
again a scalar in this simple case. From (19.57), the LS estimator is

$$\begin{pmatrix}\hat{\theta}_1\\ \hat{\theta}_2\\ \hat{\theta}_3\end{pmatrix} = \left[\begin{pmatrix}1 & 1 & 1 & 1\\ 1 & 0 & 1 & 0\\ 0 & 1 & 0 & 1\end{pmatrix}\begin{pmatrix}1 & 1 & 0\\ 1 & 0 & 1\\ 1 & 1 & 0\\ 1 & 0 & 1\end{pmatrix} + \begin{pmatrix}1\\ 0\\ 0\end{pmatrix}(1 \ 0 \ 0)\right]^{-1}\left[\begin{pmatrix}1 & 1 & 1 & 1\\ 1 & 0 & 1 & 0\\ 0 & 1 & 0 & 1\end{pmatrix}\begin{pmatrix}y_1\\ y_2\\ y_3\\ y_4\end{pmatrix} + \begin{pmatrix}1\\ 0\\ 0\end{pmatrix}c\right]$$

$$= \begin{pmatrix}5 & 2 & 2\\ 2 & 2 & 0\\ 2 & 0 & 2\end{pmatrix}^{-1}\begin{pmatrix}y_1+y_2+y_3+y_4+c\\ y_1+y_3\\ y_2+y_4\end{pmatrix}$$

$$= \begin{pmatrix}1 & -1 & -1\\ -1 & \tfrac{3}{2} & 1\\ -1 & 1 & \tfrac{3}{2}\end{pmatrix}\begin{pmatrix}\Sigma y + c\\ y_1+y_3\\ y_2+y_4\end{pmatrix} = \begin{pmatrix}c\\ \tfrac{1}{2}(y_1+y_3)-c\\ \tfrac{1}{2}(y_2+y_4)-c\end{pmatrix}.$$

Since we chose $\mathbf{B}$ so that $c = \theta_1$, we can obviously make no progress in estimating $\boldsymbol{\theta}$ itself. However, by (19.19), any set of linear functions $\mathbf{C}\boldsymbol{\theta}$ are unbiasedly estimated by $\mathbf{Ty}$ if $\mathbf{TX} = \mathbf{C}$. Thus $(\theta_1+\theta_2)$ and $(\theta_1+\theta_3)$ are estimable, since $\mathbf{C} = \begin{pmatrix}1 & 1 & 0\\ 1 & 0 & 1\end{pmatrix}$ satisfies (19.19) with $\mathbf{T} = \begin{pmatrix}\tfrac{1}{2} & 0 & \tfrac{1}{2} & 0\\ 0 & \tfrac{1}{2} & 0 & \tfrac{1}{2}\end{pmatrix}$. The estimator of $(\theta_1+\theta_2)$ is therefore $\tfrac{1}{2}(y_1+y_3)$ and that of $(\theta_1+\theta_3)$ is $\tfrac{1}{2}(y_2+y_4)$. It may be verified that $\mathbf{CD} = 0$ as stated in **19.13**.

From (19.58),
$$\mathbf{V}(\hat{\boldsymbol{\theta}}) = \sigma^2\begin{pmatrix}1 & -1 & -1\\ -1 & \tfrac{3}{2} & 1\\ -1 & 1 & \tfrac{3}{2}\end{pmatrix}\begin{pmatrix}4 & 2 & 2\\ 2 & 2 & 0\\ 2 & 0 & 2\end{pmatrix}\begin{pmatrix}1 & -1 & -1\\ -1 & \tfrac{3}{2} & 1\\ -1 & 1 & \tfrac{3}{2}\end{pmatrix}$$

$$= \sigma^2\begin{pmatrix}0 & 0 & 0\\ 0 & \tfrac{1}{2} & 0\\ 0 & 0 & \tfrac{1}{2}\end{pmatrix}$$

so that
$$\text{var}(\hat{\theta}_1+\hat{\theta}_2) = \text{var}\,\hat{\theta}_2 = \text{var}(\hat{\theta}_1+\hat{\theta}_3) = \text{var}\,\hat{\theta}_3 = \sigma^2/2,$$
as is evident from the fact that each estimator is a mean of two observations with variance σ^2. Also
$$\text{cov}(\hat{\theta}_1+\hat{\theta}_2, \hat{\theta}_1+\hat{\theta}_3) = 0,$$
a useful property which is due to the orthogonality of the second and third columns of $\mathbf{X}$. When we come to discuss the application of LS theory to the Analysis of Variance in Volume 3, we shall be returning to this subject.

Least Squares with known linear constraints

19.16 Suppose now that, whether or not $X'X$ is singular, (19.53) represents a set of p (instead of $(k-r)$) linear relations among the k parameters θ which are known *a priori* to hold. Provided that the augmented model (19.56) is of full rank k, (19.57-8) follow as before, while θ itself may now be unbiassedly estimable. In the non-singular case, Exercise 19.17 expresses $\hat{\theta}$ as an adjustment to the unconstrained LS estimator (19.12).

A more general linear model

19.17 The LS theory which we have been developing assumes throughout that (19.10) holds, i.e. that the errors are uncorrelated and have constant variance. There is no difficulty in generalizing the linear model to the situation where the dispersion matrix of errors is $\sigma^2 V$, V being positive definite (it is always non-negative definite by (18.59)), and we find (the details we left to the reader in Exercises 19.2 and 19.5) that (19.22) generalizes to

$$t = C(X'V^{-1}X)^{-1}X'V^{-1}y, \qquad (19.59)$$

and that this is the MV unbiassed linear estimator of $C\theta$. Further, (19.26) becomes

$$V(t) = \sigma^2 C(X'V^{-1}X)^{-1}C'. \qquad (19.60)$$

In particular, if V is diagonal but not equal to I, so that the ε_i are uncorrelated but with unequal variances, (19.59) provides the required set of estimators.

> (19.59) coincides with the simple form (19.22) whenever $y = Xz$, i.e. y is in the space spanned by the columns of X. Thus a necessary and sufficient condition that (19.59) and (19.22) coincide is that they do so for y orthogonal to X, i.e. that $X'y = 0$ implies $X'V^{-1}y = 0$. McElroy (1967) shows that if X contains a column of limits (so that one parameter is a constant term, as in Exercise 19.1), (19.59) coincides with (19.22) if and only if the errors have equal variances and are all equally non-negatively correlated. See also Mitra and Rao (1969) and Zyskind (1969).
>
> It should be observed that (19.22) will remain unbiassed in the present more general model, since (19.9) alone is sufficient to establish (19.14). However, the variance estimator (19.41) is generally biassed—Swindel (1968) gives attainable bounds for the bias.
>
> Watson (1967) gives a general exposition of LS theory for arbitrary V.

To use (19.59-60), of course, we need to know V. In practical cases this is usually unknown; if an estimate of V is available from past experience, or from replicated observations on the model, it can be substituted for V, but the optimum properties of (19.59) no longer necessarily hold—cf. C. R. Rao (1967). Bement and Williams (1969) discuss the effect of such estimation on the variances in (19.60) in the case when V is diagonal. C. R. Rao (1970) and Chew (1970) discuss the estimation of the elements of V in the diagonal and other structured cases.

Ordered Least Squares estimation of location and scale parameters

19.18 A particular situation in which (19.59) and (19.60) are of value is in the estimation of location and scale parameters from the *order-statistics*, i.e. the sample observations ordered according to magnitude. The results are due to Lloyd (1952) and Downton (1953).

We denote the order-statistics, as previously (14.1), by $y_{(1)}, y_{(2)}, \ldots, y_{(n)}$. As usual, we write μ and σ for the location and scale parameters to be estimated (which are not necessarily the mean and standard deviation of the distribution), and

$$z_{(r)} = (y_{(r)} - \mu)/\sigma, \quad r = 1, 2, \ldots, n. \tag{19.61}$$

Let
$$\begin{aligned} E(\mathbf{z}) &= \boldsymbol{\alpha}, \\ \mathbf{V}(\mathbf{z}) &= \mathbf{V}, \end{aligned} \tag{19.62}$$

where $\mathbf{z}$ is the $(n \times 1)$ vector of the $z_{(r)}$. Since $\mathbf{z}$ has been standardized by (19.61) $\boldsymbol{\alpha}$ and $\mathbf{V}$ are independent of μ and σ.

Now, from (19.61) and (19.62),

$$E(\mathbf{y}) = \mu \mathbf{1} + \sigma \boldsymbol{\alpha}, \tag{19.63}$$

where $\mathbf{y}$ is the vector of $y_{(r)}$ and $\mathbf{1}$ is a vector of units, while

$$\mathbf{V}(\mathbf{y}) = \sigma^2 \mathbf{V}. \tag{19.64}$$

We may now apply (19.59) and (19.60) to find the LS estimators of μ and σ. We have

$$\begin{pmatrix} \hat{\mu} \\ \hat{\sigma} \end{pmatrix} = \{(\mathbf{1} \vdots \boldsymbol{\alpha})' \mathbf{V}^{-1} (\mathbf{1} \vdots \boldsymbol{\alpha})\}^{-1} (\mathbf{1} \vdots \boldsymbol{\alpha})' \mathbf{V}^{-1} \mathbf{y} \tag{19.65}$$

and

$$\mathbf{V}\begin{pmatrix} \hat{\mu} \\ \hat{\sigma} \end{pmatrix} = \sigma^2 \{(\mathbf{1} \vdots \boldsymbol{\alpha})' \mathbf{V}^{-1} (\mathbf{1} \vdots \boldsymbol{\alpha})\}^{-1}. \tag{19.66}$$

Now

$$\{(\mathbf{1} \vdots \boldsymbol{\alpha})' \mathbf{V}^{-1} (\mathbf{1} \vdots \boldsymbol{\alpha})\}^{-1} = \begin{pmatrix} \mathbf{1}' \mathbf{V}^{-1} \mathbf{1} & \mathbf{1}' \mathbf{V}^{-1} \boldsymbol{\alpha} \\ \mathbf{1}' \mathbf{V}^{-1} \boldsymbol{\alpha} & \boldsymbol{\alpha}' \mathbf{V}^{-1} \boldsymbol{\alpha} \end{pmatrix}^{-1}$$

$$= \frac{1}{\Delta} \begin{pmatrix} \boldsymbol{\alpha}' \mathbf{V}^{-1} \boldsymbol{\alpha} & -\mathbf{1}' \mathbf{V}^{-1} \boldsymbol{\alpha} \\ -\mathbf{1}' \mathbf{V}^{-1} \boldsymbol{\alpha} & \mathbf{1}' \mathbf{V}^{-1} \mathbf{1} \end{pmatrix} \tag{19.67}$$

where
$$\Delta = \{(\mathbf{1}' \mathbf{V}^{-1} \mathbf{1})(\boldsymbol{\alpha}' \mathbf{V}^{-1} \boldsymbol{\alpha}) - (\mathbf{1}' \mathbf{V}^{-1} \boldsymbol{\alpha})^2\}. \tag{19.68}$$

From (19.65) and (19.67),

$$\begin{aligned} \hat{\mu} &= -\boldsymbol{\alpha}' \mathbf{V}^{-1} (\mathbf{1} \boldsymbol{\alpha}' - \boldsymbol{\alpha} \mathbf{1}') \mathbf{V}^{-1} \mathbf{y} / \Delta, \\ \hat{\sigma} &= \mathbf{1}' \mathbf{V}^{-1} (\mathbf{1} \boldsymbol{\alpha}' - \boldsymbol{\alpha} \mathbf{1}') \mathbf{V}^{-1} \mathbf{y} / \Delta. \end{aligned} \tag{19.69}$$

From (19.66) and (19.67)

$$\begin{aligned} \operatorname{var} \hat{\mu} &= \sigma^2 \boldsymbol{\alpha}' \mathbf{V}^{-1} \boldsymbol{\alpha} / \Delta, \\ \operatorname{var} \hat{\sigma} &= \sigma^2 \mathbf{1}' \mathbf{V}^{-1} \mathbf{1} / \Delta, \\ \operatorname{cov}(\hat{\mu}, \hat{\sigma}) &= -\sigma^2 \mathbf{1}' \mathbf{V}^{-1} \boldsymbol{\alpha} / \Delta. \end{aligned} \tag{19.70}$$

19.19 Now since $\mathbf{V}$ and $\mathbf{V}^{-1}$ are positive definite, we may write

$$\begin{aligned} \mathbf{V} &= \mathbf{TT}', \\ \mathbf{V}^{-1} &= (\mathbf{T}^{-1})' \mathbf{T}^{-1}, \end{aligned} \tag{19.71}$$

so that for an arbitrary vector $\mathbf{b}$

$$\mathbf{b}' \mathbf{V} \mathbf{b} = \mathbf{b}' \mathbf{T} \mathbf{T}' \mathbf{b} = (\mathbf{T}' \mathbf{b})' (\mathbf{T}' \mathbf{b}) = \sum_{i=1}^{n} h_i^2,$$

where h_i is the ith row element of the vector $\mathbf{T}' \mathbf{b}$.

Similarly, for a vector **c**,
$$\mathbf{c'V^{-1}c} = (\mathbf{T^{-1}c})'(\mathbf{T^{-1}c}) = \sum_{i=1}^{n} k_i^2,$$
k_i being the element of $\mathbf{T^{-1}c}$. Now by the Cauchy inequality,
$$\sum h_i^2 \sum k_i^2 = \mathbf{b'Vb} \cdot \mathbf{c'V^{-1}c} \geq (\sum h_i k_i)^2 = \{(\mathbf{T'b})'(\mathbf{T^{-1}c})\}^2 = (\mathbf{b'c})^2. \qquad (19.72)$$
In (19.72), put
$$\left. \begin{array}{l} \mathbf{b} = (\mathbf{V^{-1}-I})\mathbf{1}, \\ \mathbf{c} = \boldsymbol{\alpha}. \end{array} \right\} \qquad (19.73)$$
We obtain
$$\mathbf{1'(V^{-1}-I)V(V^{-1}-I)1} \cdot \boldsymbol{\alpha'V^{-1}\alpha} \geq \{\mathbf{1'(V^{-1}-I)\alpha}\}^2. \qquad (19.74)$$
If μ and σ^2 are the mean and variance, we find that
$$\left. \begin{array}{l} E(\mathbf{1'z}) = \mathbf{1'\alpha} = 0, \\ V(\mathbf{1'z}) = \mathbf{1'V1} = n = \mathbf{1'1}. \end{array} \right\} \qquad (19.75)$$
Using (19.75) in (19.74), it becomes
$$(\mathbf{1'V^{-1}1} - n)\boldsymbol{\alpha'V^{-1}\alpha} \geq (\mathbf{1'V^{-1}\alpha})^2,$$
which we may rewrite, using (19.70) and (19.68),
$$\operatorname{var} \hat{\mu} \leq \sigma^2/n = \operatorname{var} \bar{y}. \qquad (19.76)$$
(19.76) is obvious enough, since $\bar{y}$, the sample mean, is a linear estimator and therefore cannot have variance less than the MV estimator $\hat{\mu}$. But the point of the above argument is that it enables us to determine when (19.76) becomes a strict equality. This happens when the Cauchy inequality (19.72) becomes an equality, i.e. when $h_i = \lambda k_i$ for some constant λ, or
$$\mathbf{T'b} = \lambda \mathbf{T^{-1}c}.$$
From (19.73) this is, in our case, the condition
$$\mathbf{T'(V^{-1}-I)1} = \lambda \mathbf{T^{-1}\alpha},$$
or
$$\mathbf{TT'(V^{-1}-I)1} = \lambda \boldsymbol{\alpha}. \qquad (19.77)$$
Using (19.71), (19.77) finally becomes
$$(\mathbf{I-V})\mathbf{1} = \lambda \boldsymbol{\alpha}, \qquad (19.78)$$
the condition that $\operatorname{var} \hat{\mu} = \operatorname{var} \bar{y} = \sigma^2/n$. If (19.78) holds, we must also have, by the uniqueness of the LS solution,
$$\hat{\mu} = \bar{y}, \qquad (19.79)$$
and this may be verified by using (19.78) on $\hat{\mu}$ in (19.69).

19.20 If the parent distribution is symmetrical, the situation simplifies. For then the vector of expectations
$$\boldsymbol{\alpha} = \begin{pmatrix} E(z_{(1)}) \\ \vdots \\ E(z_{(n)}) \end{pmatrix} = \begin{pmatrix} \alpha_1 \\ \vdots \\ \alpha_n \end{pmatrix}$$
has
$$\alpha_i = -\alpha_{n+1-i}, \quad \text{all } i \qquad (19.80)$$

as follows immediately from (14.2). Hence
$$\alpha' V^{-1} 1 = 1' V^{-1} \alpha = 0 \tag{19.81}$$
and thus (19.69) becomes
$$\left. \begin{array}{l} \hat{\mu} = 1' V^{-1} y / 1' V^{-1} 1, \\ \hat{\sigma} = \alpha' V^{-1} y / \alpha' V^{-1} \alpha, \end{array} \right\} \tag{19.82}$$
while (19.70) simplifies to
$$\left. \begin{array}{l} \operatorname{var} \hat{\mu} = \sigma^2 / 1' V^{-1} 1, \\ \operatorname{var} \hat{\sigma} = \sigma^2 / \alpha' V^{-1} \alpha, \\ \operatorname{cov}(\hat{\mu}, \hat{\sigma}) = 0. \end{array} \right\} \tag{19.83}$$

Thus the ordered LS estimators $\hat{\mu}$ and $\hat{\sigma}$ are uncorrelated if the parent distribution is symmetrical, an analogous result to that for ML estimators obtained in **18.34**.

Example 19.10

To estimate the midrange (or mean) μ and range σ of the rectangular distribution
$$dF(y) = dy/\sigma, \qquad \mu - \tfrac{1}{2}\sigma \leqslant y \leqslant \mu + \tfrac{1}{2}\sigma.$$
Using (14.1), it is easy to show that, standardizing as in (19.61),
$$\alpha_r = E(z_{(r)}) = \{r/(n+1)\} - \tfrac{1}{2}, \tag{19.84}$$
and from (14.2) that the elements of the dispersion matrix V of the $z_{(r)}$ are (cf. Example 11.4 for the variances)
$$V_{rs} = r(n-s+1)/\{(n+1)^2(n+2)\}, \qquad r \leqslant s. \tag{19.85}$$
The inverse of V is
$$V^{-1} = (n+1)(n+2) \begin{pmatrix} 2 & -1 & & & & \\ -1 & 2 & & & 0 & \\ & & \ddots & & & \\ & 0 & & & 2 & -1 \\ & & & & -1 & 2 \end{pmatrix}. \tag{19.86}$$

From (19.86),
$$1' V^{-1} = (n+1)(n+2) \begin{pmatrix} 1 \\ 0 \\ 0 \\ \vdots \\ 0 \\ 0 \\ 1 \end{pmatrix}'. \tag{19.87}$$

and, from (19.84) and (19.86),
$$\alpha' V^{-1} = \tfrac{1}{2}(n+1)(n+2) \begin{pmatrix} -1 \\ 0 \\ 0 \\ \vdots \\ 0 \\ 0 \\ 1 \end{pmatrix}. \tag{19.88}$$

Using (19.87) and (19.88), (19.82) and (19.83) give

$$\left.\begin{aligned}
\hat{\mu} &= \tfrac{1}{2}(y_{(1)}+y_{(n)}), \\
\hat{\sigma} &= (n+1)(y_{(n)}-y_{(1)})/(n-1), \\
\operatorname{var}\hat{\mu} &= \sigma^2/\{2(n+1)(n+2)\}, \\
\operatorname{var}\hat{\sigma} &= 2\sigma^2/\{(n-1)(n+2)\}, \\
\operatorname{cov}(\hat{\mu},\hat{\sigma}) &= 0.
\end{aligned}\right\} \quad (19.89)$$

Apart from the bias correction to $\hat{\sigma}$, these are essentially the results we obtained by the ML method in Example 18.12. The agreement is to be expected, since $y_{(1)}$ and $y_{(n)}$ are a pair of jointly sufficient statistics for μ and σ, as we saw in effect in Example 17.21.

19.21 As will have been made clear by Example 19.10, in order to use the theory in **19.18–20**, we must determine the dispersion matrix **V** of the standardized order-statistics, and this is a function of the form of the parent distribution. This is in direct contrast with the general LS theory using unordered observations, discussed earlier in this chapter, which does not presuppose knowledge of the parent form. In Chapter 32 we shall return to the properties of order-statistics in the estimation of parameters.

The general LS theory developed in this chapter is fundamental in many branches of statistical theory, and we shall use it repeatedly in later chapters.

Other methods of estimation

19.22 We saw in the preceding chapter that, apart from the fact that they are functions of sufficient statistics for the parameters being estimated, the desirable properties of the ML estimators are all asymptotic ones, namely:

(i) consistency;

(ii) asymptotic normality; and

(iii) efficiency.

As we saw in **17.29**, the ML estimator, $\hat{\theta}$, cannot be unique in the possession of these properties. For example, the addition to $\hat{\theta}$ of an arbitrary constant C/n^r will make no difference to its first-order properties if r is large enough. It is thus natural to inquire, as Neyman (1949) did, concerning the class of estimators which share the asymptotic properties of $\hat{\theta}$. Added interest is lent to the inquiry by the numerical tedium sometimes involved (cf. Examples 18.3, 18.9, 18.10) in evaluating the ML estimator.

19.23 Suppose that we have $s(\geqslant 1)$ samples, with n_i observations in the ith sample. As at **18.19**, we simplify the problem by supposing that each observation in the ith sample is classified into one of k_i mutually exclusive and exhaustive classes. If π_{ij} is the probability of an observation in the ith sample falling into the jth class, we therefore have

$$\sum_{j=1}^{k_i} \pi_{ij} = 1, \quad (19.90)$$

and we have reduced the problem to one concerning a set of s multinomial distributions.

Let n_{ij} be the number of ith sample observations actually falling into the jth class, and $p_{ij} = n_{ij}/n_i$ the corresponding relative frequency. The probabilities π_{ij} are functions of a set of unknown parameters $(\theta_1, \ldots, \theta_r)$.

A function T of the random variables p_{ij} is called a Best Asymptotically Normal estimator (abbreviated BAN estimator) of θ_1, one of the unknown parameters, if

(i) $T(\{p_{ij}\})$ is consistent for θ_1;

(ii) T is asymptotically normal as $N = \sum_{i=1}^{s} n_i \to \infty$;

(iii) T is efficient; and

(iv) $\partial T/\partial p_{ij}$ exists and is continuous in p_{ij} for all i,j.

The first three of these conditions are precisely those we have already proved for the ML estimator in Chapter 18. It is easily verified that the ML estimator also possesses the fourth property in this multinomial situation. Thus the class of BAN estimators contains the ML estimator as a special case.

19.24 Neyman showed that a set of necessary and sufficient conditions for an estimator to be BAN is that

(i) $T(\{\pi_{ij}\}) \equiv \theta_1$;

(ii) condition (iv) of **19.23** holds; and

(iii) $\sum_{i=1}^{s} \frac{1}{n_i} \sum_{j=1}^{k_i} \left[\left(\frac{\partial T}{\partial p_{ij}} \right)_{p_{ij} = \pi_{ij}} \right]^2 \pi_{ij}$ be minimized for variation in $\partial T/\partial p_{ij}$.

Condition (i) is enough to ensure consistency: it is, in general, a stronger condition than consistency.[*] In this case, since the statistic T is a continuous function of the p_{ij}, and the p_{ij} converge in probability to the π_{ij}, T converges in probability to $T(\{\pi_{ij}\})$, i.e. to θ_1.

Condition (iii) is simply the efficiency condition, for the function there to be minimized is simply the variance of T subject to the necessary condition for a minimum $\sum_j \left(\frac{\partial T}{\partial p_{ij}} \right)_{p_{ij} = \pi_{ij}} \pi_{ij} = 0$.

As they stand, these three conditions are not of much practical value. However, Neyman also showed that a sufficient set of conditions is obtainable by replacing (iii) by a direct condition on $\partial T/\partial p_{ij}$, which we shall not give here. From this he deduced that

(a) the ML estimator is a BAN estimator, as we have already seen;

(b) that the class of estimators known as *Minimum Chi-Square* estimators are also BAN estimators.

We now proceed to examine this second class of estimators.

Minimum Chi-Square estimators

19.25 Referring to the situation described in **19.23**, a statistic T is called a Mini-

---

[*] In fact, (i) is the form in which consistency was originally defined by Fisher (1921a).

mum Chi-Square (abbreviated MCS) estimator of θ_1, if it is obtained by minimizing, with respect to θ_1, the expression

$$\chi^2 = \sum_{i=1}^{s} n_i \sum_{j=1}^{k_i} \frac{(p_{ij}-\pi_{ij})^2}{\pi_{ij}} = \sum_{i=1}^{s} n_i \left(\sum_{j=1}^{k_i} \frac{p_{ij}^2}{\pi_{ij}} - 1 \right), \tag{19.91}$$

where the π_{ij} are functions of $\theta_1, \ldots, \theta_r$. To minimize (19.91), we put

$$\frac{\partial \chi^2}{\partial \theta_1} = -\sum_i n_i \sum_j \left(\frac{p_{ij}}{\pi_{ij}}\right)^2 \frac{\partial \pi_{ij}}{\partial \theta_1} = 0, \tag{19.92}$$

and a root of (19.92), regarded as an equation in θ_1, is the MCS estimator of θ_1. Evidently, we may generalize (19.92) to a set of r equations to be solved together to find the MCS estimators of $\theta_1, \ldots, \theta_r$.

The procedure for finding MCS estimators is quite analogous to that for finding ML estimators, discussed in Chapter 18. Moreover, the (asymptotic) properties of MCS estimators are similar to those of ML estimators. In fact, there is, with probability 1, a unique consistent root of the MCS equations, and this corresponds to the absolute minimum (infimum) value of (19.91). The proofs are given, for the commonest case $s = 1$, by C. R. Rao (1957).

19.26 A modified form of MCS estimator is obtained by minimizing

$$(\chi')^2 = \sum_{i=1}^{s} n_i \sum_{j=1}^{k_i} \frac{(p_{ij}-\pi_{ij})^2}{p_{ij}} = \sum_i n_i \left(\sum_j \frac{\pi_{ij}^2}{p_{ij}} - 1 \right) \tag{19.93}$$

instead of (19.91). In (19.93), we assume that no $p_{ij} = 0$. To minimize it for variation in θ_1, we put

$$\frac{\partial (\chi')^2}{\partial \theta_1} = 2 \sum_i n_i \sum_j \left(\frac{\pi_{ij}}{p_{ij}}\right) \frac{\partial \pi_{ij}}{\partial \theta_1} = 0 \tag{19.94}$$

and solve for the estimator of θ_1. These modified MCS estimators have also been shown to be BAN estimators by Neyman (1949).

19.27 Since the ML, the MCS and the modified MCS methods all have the same asymptotic properties, the choice between them must rest, in any particular case, either on the grounds of computational convenience, or on those of superior sampling properties in small samples, or on both. As to the first ground, there is little that can be said in general. Sometimes the ML, and sometimes the MCS, equation is the more difficult to solve. But when dealing with a continuous distribution, the observations *must* be grouped in order to make use of the MCS method, and it seems rather wasteful to impose an otherwise unnecessary grouping for estimation purposes. Furthermore, there is, especially for continuous distributions, preliminary inconvenience in having to determine the π_{ij} in terms of the parameters to be estimated. Our own view is therefore that the now traditional leaning towards ML estimation is fairly generally justifiable on computational grounds. The following example illustrates the MCS computational procedure in a relatively simple case.

Example 19.11

Consider the estimation, from a single sample of n observations, of the parameter θ of a Poisson distribution. We have seen (Examples 17.8, 17.15) that the sample mean $\bar{x}$

is a MVB sufficient estimator of θ, and it follows from **18.5** that $\bar{x}$ is also the ML estimator.

The MCS estimator of θ, however, is not equal to $\bar{x}$, illustrating the point that MCS methods do not necessarily yield a single sufficient statistic if one exists.

The theoretical probabilities here are

$$\pi_j = e^{-\theta}\theta^j/j!, \quad j = 0, 1, 2, \ldots,$$

so that

$$\frac{\partial \pi_j}{\partial \theta} = \pi_j\left(\frac{j}{\theta}-1\right).$$

The minimizing equation (19.92) is therefore, dropping the factor n,

$$-\Sigma\left(\frac{p_j}{\pi_j}\right)^2 \pi_j\left(\frac{j}{\theta}-1\right) = \Sigma_j \frac{p_j^2}{\pi_j}\left(1-\frac{j}{\theta}\right) = 0. \tag{19.95}$$

This is the equation to be solved for θ, and we use an iterative method of solution similar to that used for the ML estimator at **18.21**. We expand the left-hand side of (19.95) in a Taylor series as a function of θ about the sample mean $\bar{x}$, regarded as a trial value. We obtain to the first order of approximation

$$\Sigma_j \frac{p_j^2}{\pi_j}\left(1-\frac{j}{\theta}\right) = \Sigma_j \frac{p_j^2}{m_j}\left(1-\frac{j}{\bar{x}}\right) + (\theta-\bar{x})\Sigma_j \frac{p_j^2}{m_j}\left\{\frac{j}{\bar{x}^2}+\left(1-\frac{j}{\bar{x}}\right)^2\right\}, \tag{19.96}$$

where we have written $m_j = e^{-\bar{x}}\bar{x}^j/j!$. If (19.96) is equated to zero, by (19.95), we find

$$(\theta-\bar{x}) = \bar{x} \cdot \frac{\Sigma_j \dfrac{p_j^2}{m_j}(j-\bar{x})}{\Sigma_j \dfrac{p_j^2}{m_j}\{j+(j-\bar{x})^2\}}. \tag{19.97}$$

We use (19.97) to find an improved estimate of θ from $\bar{x}$, and repeat the process as necessary—cf. (18.45) for the ML estimator.

As a numerical example, we use Whitaker's (1914) data on the number of deaths of women over 85 years old reported in *The Times* newspaper for each day of 1910–1912, 1,096 days in all. The distribution is given in the first two columns of the table on page 95.

The mean number of deaths reported is found to be $\bar{x} = 1295/1096 = 1\cdot181569$. This is therefore the ML estimator, and we use it as our first trial value for the MCS estimator. The third column of the table gives the expected frequencies in a Poisson distribution with parameter equal to $\bar{x}$, and the necessary calculations for (19.97) are set out in the remaining five columns.

Thus, from (19.97), we have

$$\theta = 1\cdot 1816\left\{1+\frac{42\cdot 2}{3242\cdot 4}\right\}$$

$$= 1\cdot 198$$

as our improved value. K. Smith (1916) reported a value of $1\cdot 196903$ when working to greater accuracy, with more than one iteration of this procedure.

No. of deaths (j)	Frequency reported (np_j)	nm_j	$\dfrac{(np_j)^2}{nm_j} = \dfrac{np_j^2}{m_j}$	$(j-\bar{x})$	$\dfrac{np_j^2}{m_j}(j-\bar{x})$	$\{j+(j-\bar{x})^2\}$	$\dfrac{np_j^2}{m_j}\{j+(j-\bar{x})^2\}$
0	364	336·25	394·1	−1·1816	−465·7	1·396	551·1
1	376	397·30	355·8	−0·1816	− 64·6	1·033	365·9
2	218	234·72	202·5	0·8184	165·8	2·670	540·6
3	89	92·45	85·69	1·8184	155·8	6·307	540·4
4	33	27·31	39·87	2·8184	112·4	11·943	476·1
5	13	6·45	26·20	3·8184	100·0	19·580	512·9
6	2	1·27	3·15	4·8184	15·2	29·217	92·0
7	1	0·25	4·00	5·8184	23·3	40·854	163·4
Total	$n = 1096$	1096·00			+42·2		3242·4

Smith also gives details of the computational procedure when we are estimating the parameters of a continuous distribution. This is considerably more laborious.

19.28 Small-sample properties, the second ground for choice between the ML and MCS methods, are more amenable to general inquiry. C. R. Rao (1961, 1962a) defines a concept of second-order efficiency—cf. also **17.29** above—and shows that in the multinomial model of **18.19**, the ML is the only BAN estimator with optimum second-order efficiency, under regularity conditions. Berkson (1955, 1956) has carried out sampling experiments which show that, in a situation arising in a bio-assay problem, the ML estimator presents difficulties, being sometimes infinite, while another BAN estimator has smaller mean-square-error. These papers should be read in the light of another by Silverstone (1957), which points out some errors in them.

EXERCISES

19.1 In the linear model (19.8), suppose that a further parameter θ_0 is introduced, so that we have the new model
$$\mathbf{y} = \mathbf{X}\boldsymbol{\theta} + \mathbf{1}\theta_0 + \boldsymbol{\epsilon},$$
where $\mathbf{1}$ is an $(n \times 1)$ vector of units. Show that the LS estimator in the new model of $\boldsymbol{\theta}$, the original vector of k parameters, remains of exactly the same form (19.12) as in the original model, with y_i replaced by $(y_i - \bar{y})$ and x_{ij} by $(x_{ij} - \bar{x}_j)$ for $i = 1, 2, \ldots, n$, and $j = 1, 2, \ldots, k$.

19.2 If, in the linear model (19.8), we replace the simple dispersion matrix (19.10) by a non-singular dispersion matrix $\sigma^2 \mathbf{V}$ which allows correlations and unequal variances among the ε_i, show by putting $\mathbf{w} = \mathbf{T}'\mathbf{y}$, where $\mathbf{TT}' = \mathbf{V}^{-1}$, that the LS and MV unbiassed estimator of $\mathbf{C}\boldsymbol{\theta}$ is
$$\mathbf{C}\hat{\boldsymbol{\theta}} = \mathbf{C}(\mathbf{X}'\mathbf{V}^{-1}\mathbf{X})^{-1}\mathbf{X}'\mathbf{V}^{-1}\mathbf{y}.$$
(cf. Aitken, 1935; Plackett, 1949)

19.3 Generalizing (19.38), show that if $E(\boldsymbol{\epsilon}\boldsymbol{\epsilon}') = \sigma^2 \mathbf{V}$, $E(\boldsymbol{\epsilon}'\mathbf{B}\boldsymbol{\epsilon}) = \sigma^2 \operatorname{tr}(\mathbf{BV})$. Show further that $\operatorname{var}(\boldsymbol{\epsilon}'\mathbf{B}\boldsymbol{\epsilon}) = 2\sigma^4 \operatorname{tr}(\mathbf{BVBV})$ if $\boldsymbol{\epsilon}$ is multinormal.

19.4 Show that in **19.12** the ratio $(\mathbf{X}\hat{\boldsymbol{\theta}})'(\mathbf{X}\hat{\boldsymbol{\theta}})/(ks^2)$ is distributed in Fisher's F distribution with k and $(n-k)$ degrees of freedom if $\boldsymbol{\theta} = \mathbf{0}$.

19.5 In Exercise 19.2, show that, generalizing (19.26),
$$\mathbf{V}(\mathbf{C}\hat{\boldsymbol{\theta}}) = \sigma^2 \cdot \mathbf{C}(\mathbf{X}'\mathbf{V}^{-1}\mathbf{X})^{-1}\mathbf{C}',$$
and that the generalization of (19.40) is, using Exercise 19.3,
$$E\{(\mathbf{y}-\mathbf{X}\hat{\boldsymbol{\theta}})'\mathbf{V}^{-1}(\mathbf{y}-\mathbf{X}\hat{\boldsymbol{\theta}})\} = E\{\boldsymbol{\epsilon}'[\mathbf{V}^{-1}-\mathbf{V}^{-1}\mathbf{X}(\mathbf{X}'\mathbf{V}^{-1}\mathbf{X})^{-1}\mathbf{X}'\mathbf{V}^{-1}]\boldsymbol{\epsilon}\} = (n-k)\sigma^2.$$

19.6 Prove the statement in **19.14** to the effect that $\mathbf{C}\hat{\boldsymbol{\theta}}$ and $\mathbf{V}(\mathbf{C}\hat{\boldsymbol{\theta}})$ in the singular case are unaffected by replacing $\mathbf{B}$ by $\mathbf{UB}$, where $\mathbf{U}$ is non-singular.

19.7 In **19.14**, show using (19.51) that $(\mathbf{X}'\mathbf{X}+\mathbf{B}'\mathbf{B})^{-1}\mathbf{B}'\mathbf{B} = \mathbf{D}(\mathbf{BD})^{-1}\mathbf{B}$ and hence that (19.58) gives
$$\mathbf{V}(\hat{\boldsymbol{\theta}})(\mathbf{X}'\mathbf{X}) = \sigma^2\{\mathbf{I}_k - \mathbf{D}(\mathbf{BD})^{-1}\mathbf{B}\}.$$
(Plackett, 1950)

19.8 Using the result of Exercise 19.7,
$$(\mathbf{X}'\mathbf{X}+\mathbf{B}'\mathbf{B})^{-1}\mathbf{B}'\mathbf{B} = \mathbf{D}(\mathbf{BD})^{-1}\mathbf{B},$$
modify the argument of **19.9** to show that the unbiassed estimator of σ^2 in the singular case is $\dfrac{1}{(n-r)}(\mathbf{y}-\mathbf{X}\hat{\boldsymbol{\theta}})'(\mathbf{y}-\mathbf{X}\hat{\boldsymbol{\theta}})$.
(Plackett, 1950)

19.9 For the linear model (19.8-10), show using **19.9** that the quadratic form $\boldsymbol{\theta}'\mathbf{A}\boldsymbol{\theta}$ is unbiassedly estimated by $\hat{\boldsymbol{\theta}}'\mathbf{A}\hat{\boldsymbol{\theta}} - s^2 \operatorname{tr}\{(\mathbf{X}'\mathbf{X})^{-1}\mathbf{A}\}$.

19.10 Show that in the case of a symmetrical parent distribution, the condition that the ordered LS estimator $\hat{\mu}$ in (19.82) is equal to the sample mean $\bar{y} = \mathbf{1}'\mathbf{y}/\mathbf{1}'\mathbf{1}$ is that
$$\mathbf{V}\mathbf{1} = \mathbf{1},$$

i.e. that the sum of each row of the dispersion matrix be unity. Show that this property holds for the univariate normal distribution. (Lloyd, 1952)

19.11 For the exponential distribution
$$dF(y) = \exp\left\{-\left(\frac{y-\mu}{\sigma}\right)\right\} dy/\sigma, \qquad \sigma > 0;\ \mu \leqslant y \leqslant \infty,$$
show that in (19.62) the elements of $\boldsymbol{\alpha}$ are
$$\alpha_r = \sum_{i=1}^{r} (n-i+1)^{-1}$$
and that those of $\mathbf{V}$ are
$$V_{rs} = \sum_{i=1}^{m} (n-i+1)^{-2} \qquad \text{where} \qquad m = \min(r, s).$$
Hence verify that $\mathbf{V}^{-1}$ has elements all zero except
$$V^{-1}_{r,r+1} = V^{-1}_{r+1,r} = -(n-r)^2,$$
$$V^{-1}_{r,r} = (n-r)^2 + (n-r+1)^2.$$

19.12 In Exercise 19.11, show from (19.69) that the MV unbiased estimators are
$$\hat{\mu} = y_{(1)} - (\bar{y} - y_{(1)})/(n-1) \text{ and } \hat{\sigma} = n(\bar{y} - y_{(1)})/(n-1).$$
Compare these with the ML estimators of the same parameters.
(Sarhan, 1954)

19.13 Show that when all the n_i are large the minimization of the chi-squared expression (19.91) or (19.93) gives the same estimator as the ML method.

19.14 In the case $s = 1$, show that the first two moments of the statistic (19.91) are given by
$$n_1^2 E(\chi^2) = k_1 - 1,$$
$$n_1^4 \operatorname{var}(\chi^2) = 2(k_1-1)\left(1 - \frac{1}{n_1}\right) + \frac{1}{n_1}\sum_{j=1}^{k_1} \frac{1}{\pi_{1j}} - \frac{k_1^2}{n_1},$$
and that for any $c > 0$, the generalization of (19.93) has expectation
$$E\left\{\sum_{j=1}^{k_1} \frac{(n_{1j}-n_1)^2 + b}{n_{1j}+c}\right\} = k_1 - 1 + \frac{1}{n_1}\left[(b-c+2)\sum_{j=1}^{k_1}\frac{1}{\pi_{1j}} - (3-c)k_1 + 1\right] + O\left(\frac{1}{n_1^2}\right).$$
Thus, to the second order at least, the π_{1j} disappear from the expectation if $b = c-2$. If $b = 0$, $c = 2$, it is $(k_1-1)\left(1 - \frac{1}{n_1}\right)$ and if $b = 1$, $c = 3$, it is $(k_1-1) + \frac{1}{n_1}$, which for $k_1 > 2$ is even closer to the expectation of (19.91).
(F. N. David, 1950; Haldane, 1955a)

19.15 For a binomial distribution with probability of success equal to π, show that the MCS estimator of π obtained from (19.91) is identical with the ML estimator for any n; and that if the number of successes is not 0 or n, the modified MCS estimator obtained from (19.93) is also identical with the ML estimator.

19.16 y_i is a Poisson variable with parameter θx_i, where x_i is a constant observed with y_i, $(i = 1, 2, \ldots, n)$. Show that the ML estimator of θ is $\Sigma y_i/\Sigma x_i$, with asymptotic variance $\theta/\Sigma x_i$, and that the LS estimator $\Sigma y_i x_i/\Sigma x_i^2$ has exact variance $\theta \Sigma x_i^3/(\Sigma x_i^2)^2$. Hence show that the LS estimator is inefficient unless the x_i are all equal. Explain the result.

19.17 Show by using a vector of Lagrangian undetermined multipliers that the minimization of (19.11) subject to p linear constraints (19.53) yields the LS estimator
$$\hat{\boldsymbol{\theta}} = (\mathbf{X'X})^{-1}[\mathbf{X'y} - \mathbf{B'}\{\mathbf{B(X'X)}^{-1}\mathbf{B'}\}^{-1}\{\mathbf{B(X'X)}^{-1}\mathbf{X'y} - \mathbf{a}\}]$$
and verify that this satisfies (19.57).

19.18 In **19.9**, show that the LS residuals $\mathbf{y} - \mathbf{X}\hat{\boldsymbol{\theta}}$ have dispersion matrix $\{\mathbf{I}_n - \mathbf{X(X'X)}^{-1}\mathbf{X'}\}\sigma^2$, so that they are correlated. Show that if the linear model (19.8) is fitted to the first m of N observations only, the m residuals thus obtained are uncorrelated with the $(N-m)$ residuals obtained from the other observations when (19.8) is fitted to all N observations, and that this result holds also if the model is singular. Hence, that if we successively put $m = 1, 2, \ldots, n-1$ and $N = m+1$, we generate a set of n uncorrelated residuals (independent if the ε_i are normal).

(cf. Hedayat and Robson, 1970)

CHAPTER 20

INTERVAL ESTIMATION: CONFIDENCE INTERVALS

20.1 In the previous three chapters we have been concerned with methods which will provide an estimate of the value of one or more unknown parameters; and the methods gave functions of the sample values—the estimators—which, for any given sample, provided a unique estimate. It was, of course, fully recognized that the estimate might differ from the parameter in any particular case, and hence that there was a margin of uncertainty. The extent of this uncertainty was expressed in terms of the sampling variance of the estimator. With the somewhat intuitive approach which has served our purpose up to this point, we might say that it is probable that θ lies in the range $t \pm \sqrt{(\operatorname{var} t)}$, very probable that it lies in the range $t \pm 2\sqrt{(\operatorname{var} t)}$, and so on. In short, what we might do is, in effect, to locate θ in a range and not at a particular point, although regarding one point in the range, viz. t itself, as having a claim to be considered as the "best" estimate of θ.

20.2 In the present chapter we shall examine this procedure more closely and look at the problem of estimation from a different point of view. We now abandon attempts to estimate θ by a function which, for a specified sample, gives a unique estimate. Instead, we shall consider the specification of a range in which θ lies. Three methods, of which two are similar but not identical, arise for discussion. The first, known as the method of Confidence Intervals, relies only on the frequency theory of probability without importing any new principle of inference. The second, which we shall call the method of Fiducial Intervals, explicitly requires something beyond a frequency theory. The third relies on Bayes' theorem and some form of Bayes' postulate (8.4). In the present chapter we shall attempt to explain the basic ideas and methods of Confidence Interval estimation, which are due to Neyman—the memoir of 1937 should be particularly mentioned (see Neyman (1937*b*)). In Chapter 21 we shall be concerned with the same aspects of Fiducial Intervals and Bayes' estimation.

Confidence statements

20.3 Consider first a distribution dependent on a single unknown parameter θ and suppose that we are given a random sample of n values $x_1, \ldots, x_n$ from the population. Let z be a variable dependent on the x's and on θ, whose sampling distribution is independent of θ. (The examples given below will show that in some cases at least such a function may be found.) Then, given any probability $1-\alpha$, we can find a value z_1 such that

$$\int_{z_1}^{\infty} dF(z) = 1-\alpha,$$

and this is true whatever the value of θ. In the notation of the theory of probability we shall then have

$$P(z \geqslant z_1) = 1-\alpha. \tag{20.1}$$

Now it may happen that the inequality $z \geqslant z_1$ can be written in the form $\theta \leqslant t_1$ or $\theta \geqslant t_1$, where t_1 is some function depending on the value z_1 and the x's but not on θ. For instance, if $z = \bar{x} - \theta$ we shall have
$$\bar{x} - \theta \geqslant z_1$$
and hence
$$\theta \leqslant \bar{x} - z_1.$$
If we can rewrite this inequality in this way, we have, from (20.1),
$$P(\theta \leqslant t_1) = 1 - \alpha. \tag{20.2}$$

More generally, whether or not the distribution of z is independent of θ, suppose that we can find a statistic t_1, depending on $1 - \alpha$ and the x's but not on θ, such that 20.2) is true for all θ. Then we may use this equation in probability to make certain statements about θ.

20.4 Note, in the first place, that we cannot assert that the probability is $1 - \alpha$ that θ does not exceed a constant t_1. This statement (in the frequency theory of probability) can only relate to the variation of θ in a population of θ's, and in general we do not know that θ varies at all. If it is merely an unknown constant, then the probability that $\theta \leqslant t_1$ is either unity or zero. We do not know which of these values is correct, but we do know that one of them is correct.

We therefore look at the matter in another way. Although θ is not a random variable, t_1 is, and will vary from sample to sample. Consequently, if we *assert* that $\theta \leqslant t_1$ in each case presented for decision, we shall be right in a proportion $1 - \alpha$ of the cases in the long run. The statement that the probability of θ is less than or equal to some assigned value has no meaning except in the trivial sense already mentioned; but the statement that a statistic t_1 is greater than or equal to θ (whatever θ happens to be) has a definite probability of being correct. If therefore we make it a rule to assert the inequality $\theta \leqslant t_1$ for any sample values which arise, we have the assurance of being right in a proportion $1 - \alpha$ of the cases " on the average " or " in the long run."

This idea is basic to the theory of confidence intervals which we proceed to develop, and the reader should satisfy himself that he has grasped it. In particular, we stress that the confidence statement holds *whatever the value of θ*: we are not concerned with repeated sampling from the same population, but just with repeated sampling.

20.5 To simplify the exposition we have considered only a single quantity t_1 and the statement that $\theta \leqslant t_1$. In practice, however, we usually seek two quantities t_0 and t_1, such that for all θ
$$P(t_0 \leqslant \theta \leqslant t_1) = 1 - \alpha,^{(*)} \tag{20.3}$$
and make the assertion that θ lies in the interval t_0 to t_1, which is called a Confidence Interval for θ. t_0 and t_1 are known as the Lower and Upper Confidence Limits respectively. They depend only on $1 - \alpha$ and the sample values. For any fixed $1 - \alpha$, the totality of Confidence Intervals for different samples determines a field within which θ

(*) We shall almost always write $1 - \alpha$ for the probability of the interval covering the parameter, but practice in the literature varies, and α is often written instead. Our convention is nowadays the more common.

INTERVAL ESTIMATION: CONFIDENCE INTERVALS

is asserted to lie. This field is called the Confidence Belt. We shall give a graphical representation of the idea below. The fraction $1-\alpha$ is called the Confidence Coefficient.

Example 20.1

Suppose we have a sample of size n from the normal population with known variance (taken without loss of generality to be unity)

$$dF = \frac{1}{\sqrt{(2\pi)}} \exp\{-\tfrac{1}{2}(x-\mu)^2\} dx, \qquad -\infty \leq x \leq \infty.$$

The distribution of the sample mean $\bar{x}$ is

$$dF = \sqrt{\left(\frac{n}{2\pi}\right)} \exp\left\{-\frac{n}{2}(\bar{x}-\mu)^2\right\} d\bar{x}, \qquad -\infty \leq \bar{x} \leq \infty.$$

From the tables of the normal integral we know that the probability of a positive deviation from the mean not greater than twice the standard deviation is 0·97725. We have then

$$P(\bar{x}-\mu \leq 2/\sqrt{n}) = 0 \cdot 97725,$$

which is equivalent to

$$P(\bar{x}-2/\sqrt{n} \leq \mu) = 0 \cdot 97725.$$

Thus, if we assert that μ is greater than or equal to $\bar{x}-2/\sqrt{n}$ we shall be right in about 97·725 per cent of cases in the long run.

Similarly we have

$$P(\bar{x}-\mu \geq -2/\sqrt{n}) = P(\mu \leq \bar{x}+2/\sqrt{n}) = 0 \cdot 97725.$$

Thus, combining the two results,

$$P(\bar{x}-2/\sqrt{n} \leq \mu \leq \bar{x}+2/\sqrt{n}) = 1 - 2(1-0 \cdot 97725)$$
$$= 0 \cdot 9545. \qquad (20.4)$$

Hence, if we assert that μ lies in the range $\bar{x} \pm 2/\sqrt{n}$, we shall be right in about 95·45 per cent of cases in the long run.

Conversely, given the confidence coefficient, we can easily find from the tables of the normal integral the deviation d such that

$$P(\bar{x}-d/\sqrt{n} \leq \mu \leq \bar{x}+d/\sqrt{n}) = 1-\alpha.$$

For instance, if $1-\alpha = 0 \cdot 8$, $d = 1 \cdot 28$, so that if we assert that μ lies in the range $\bar{x} \pm 1 \cdot 28/\sqrt{n}$ the odds are 4 to 1 that we shall be right.

The reader to whom this approach is new will probably ask: but is this not a roundabout method of using the standard error to set limits to an estimate of the mean? In a way, it is. Effectively, what we have done in this example is to show how the use of the standard error of the mean in normal samples may be justified on logical grounds without appeal to new principles of inference other than those incorporated in the theory of probability itself. In particular we make no use of Bayes' postulate (8.4).

Another point of interest in this example is that the upper and lower confidence limits derived above are equidistant from the mean $\bar{x}$. This is not by any means necessary, and it is easy to see that we can derive any number of alternative limits for the same confidence coefficient $1-\alpha$. Suppose, for instance, we take $1-\alpha = 0 \cdot 9545$, and select two numbers α_0 and α_1 which obey the condition

$$\alpha_0 + \alpha_1 = \alpha = 0 \cdot 0455$$

say $\alpha_0 = 0.01$ and $\alpha_1 = 0.0355$. From the tables of the normal integral we have
$$P(\bar{x}-\mu \leqslant 2.326/\sqrt{n}) = 0.99$$
$$P(\bar{x}-\mu \geqslant -1.806/\sqrt{n}) = 0.9645,$$
and hence
$$P\left(\bar{x}-\frac{2.326}{\sqrt{n}} \leqslant \mu \leqslant \bar{x}+\frac{1.806}{\sqrt{n}}\right) = 0.9545. \tag{20.5}$$

Thus, with the same confidence coefficient we can assert that μ lies in the interval $\bar{x}-2/\sqrt{n}$ to $\bar{x}+2/\sqrt{n}$, or in the interval $\bar{x}-2.326/\sqrt{n}$ to $\bar{x}+1.806/\sqrt{n}$. In either case we shall be right in about 95.45 per cent of cases in the long run.

We note that in the first case the interval has length $4/\sqrt{n}$, while in the second case its length is $4.132/\sqrt{n}$. Other things being equal, we should choose the first set of limits since they locate the parameter in a narrower range. We shall consider this point in more detail in **20.16** below. It does not always happen that there is an infinity of possible confidence limits or that the choice between them can be made on such clear-cut grounds as in this example.

Graphical representation

20.6 In a number of simple cases, including that of Example 20.1, the confidence limits can be represented in a useful graphical form. We take two orthogonal axes, OX relating to the observed $\bar{x}$ and OY to μ (see Fig. 20.1).

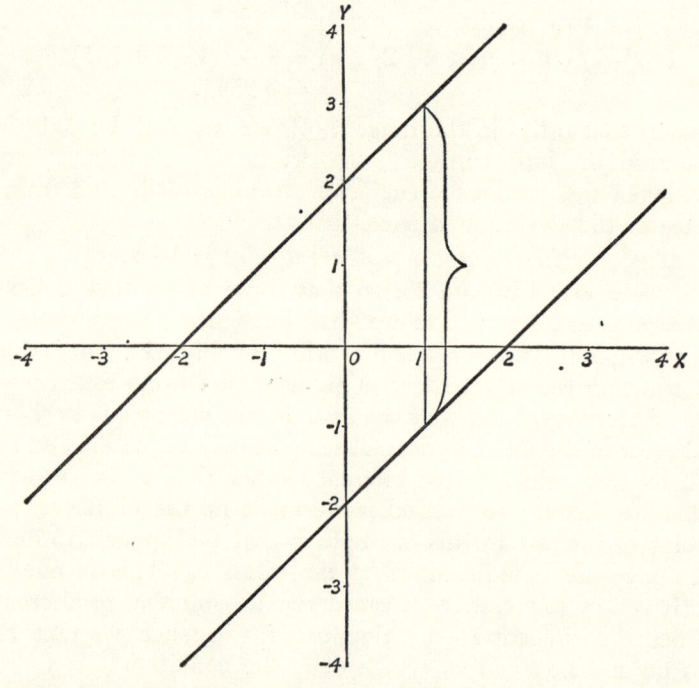

Fig. 20.1—Confidence limits in Example 20.1 for $n = 1$

The two straight lines shown have as their equations
$$\mu = \bar{x}+2, \qquad \mu = \bar{x}-2.$$
Consequently, for any point between the lines,
$$\bar{x}-2 \leqslant \mu \leqslant \bar{x}+2.$$
Hence, if for any observed $\bar{x}$ we read off the two ordinates on the lines corresponding to that value, we obtain the two confidence limits. The vertical interval between the limits is the confidence interval (shown in the diagram for $\bar{x} = 1$), and the total zone between the lines is the confidence belt. We may refer to the two lines as upper and lower confidence lines respectively.

This example relates to the case $n = 1$ in Example 20.1. For different values of n, there will be different confidence lines, all parallel to $\mu = \bar{x}$, and getting closer to each other as n increases. They may be shown on a single diagram for selected values of n, and a figure so constructed provides a useful method of reading off confidence limits in practical work.

Alternatively, we may wish to vary the confidence coefficient $1-\alpha$, which in our example is $0\cdot9545$. Again, we may show a series of pairs of confidence lines, each pair corresponding to a selected value of $1-\alpha$, on a single diagram relating to some fixed value of n. In this case, of course, the lines become farther apart with increasing $1-\alpha$. In fact, in many practical situations, we are interested in the variation of the confidence interval with $1-\alpha$, and we may validly make assertions of the form (20.3) simultaneously for a number of values of α: each will be true in the corresponding proportion of cases in the long run. Indeed, this procedure may be taken to its extreme form, when we consider *all* values of $1-\alpha$ in $(0, 1)$ simultaneously, and thus generate "a confidence distribution" of the parameter—the term is due to D. R. Cox (e.g. 1958b): we then have an infinite sequence of simultaneous confidence statements, each contained within the preceding one, with increasing values of $1-\alpha$.

Central and non-central intervals

20.7 In Example 20.1 the sampling distribution on which the confidence intervals were based was symmetrical, and hence, by taking equal deviations from the mean we obtained equal values of
$$1-\alpha_0 = P(t_0 \leqslant \theta)$$
and
$$1-\alpha_1 = P(\theta \leqslant t_1).$$
In general, we cannot achieve this result with equal deviations, but subject always to the condition $\alpha_0+\alpha_1 = \alpha$, α_0 and α_1 may be chosen arbitrarily.

If α_0 and α_1 are taken to be equal, we shall say that the intervals are *central*. In such a case we have
$$P(t_0 > \theta) = P(\theta > t_1) = \alpha/2. \tag{20.6}$$
In the contrary case the intervals will be called *non-central*. It should be observed that centrality in this sense does not mean that the confidence limits are equidistant from the sample statistic, unless the sampling distribution is symmetrical.

20.8 In the absence of other considerations it is usually convenient to employ central intervals, but circumstances sometimes arise in which non-central intervals are more serviceable. Suppose, for instance, we are estimating the proportion of some drug in a medicinal preparation and the drug is toxic in large doses. We must then clearly err on the safe side, an excess of the true value over our estimate being more serious than a deficiency. In such a case we might like to take α_1 equal to zero, so that

$$P(\theta \leqslant t_1) = 1$$
$$P(t_0 \leqslant \theta) = 1-\alpha,$$

in order to be certain that θ is not greater than t_1. But if our statistic has a sampling distribution with infinite range, this is only possible with t_1 infinite, so we must content ourselves with making α_1 very close to zero.

Again, if we are estimating the proportion of viable seed in a sample of material that is to be placed on the market, we are more concerned with the accuracy of the lower limit than that of the upper limit, for a deficiency of germination is more serious than an excess from the grower's point of view. In such circumstances we should probably take α_0 as small as conveniently possible so as to be near to certainty about the minimum value of viability. This kind of situation often arises in the specification of the quality of a manufactured product, the seller wishing to guarantee a minimum standard but being much less concerned with whether his product exceeds expectation.

20.9 On a somewhat similar point, it may be remarked that in certain circumstances it is enough to know that $P(t_0 \leqslant \theta \leqslant t_1) \geqslant 1-\alpha$. We then know that in asserting θ to lie in the range t_0 to t_1 we shall be right in *at least* a proportion $1-\alpha$ of the cases. Mathematical difficulties in ascertaining confidence limits exactly for given $1-\alpha$, or theoretical difficulties when the distribution is discontinuous may, for example, lead us to be content with this inequality rather than the equality of (20.3).

Example 20.2

To find confidence intervals for the probability ϖ of "success" in sampling for attributes.

In samples of size n the distribution of successes is arrayed by the binomial $(\chi+\varpi)^n$, where $\chi = 1-\varpi$. We will determine the limits for the case $n = 20$ and confidence coefficient 0·95.

We require in the first instance the distribution function of the binomial. The table overleaf shows the functions for certain specified values up to $\varpi = 0·5$ (the remainder being obtainable by symmetry). For the accurate construction of the confidence belt we require more detailed information such as is obtainable from the comprehensive tables of the binomial function referred to in **5.7**. These, however, will serve for purposes of illustration.

The final figures may be a unit or two in error owing to rounding up, but that need not bother us to the degree of approximation here considered.

We note in the first place that the variate p is discontinuous. On the other hand, we are prepared to consider any value of ϖ in the range 0 to 1. For given ϖ we cannot in general find limits to p for which $1 - \alpha$ is exactly 0·95; but we will take p to be

Proportion of Successes p	$\varpi = 0.1$	$\varpi = 0.2$	$\varpi = 0.3$	$\varpi = 0.4$	$\varpi = 0.5$
0·00	0·1216	0·0115	0·0008	—	—
0·05	0·3918	0·0691	0·0076	0·0005	—
0·10	0·6770	0·2060	0·0354	0·0036	0·0002
0·15	0·8671	0·4114	0·1070	0·0159	0·0013
0·20	0·9569	0·6296	0·2374	0·0509	0·0059
0·25	0·9888	0·8042	0·4163	0·1255	0·0207
0·30	0·9977	0·9133	0·6079	0·2499	0·0577
0·35	0·9997	0·9678	0·7722	0·4158	0·1316
0·40	1·0001	0·9900	0·8866	0·5955	0·2517
0·45	1·0002	0·9974	0·9520	0·7552	0·4119
0·50	—	0·9994	0·9828	0·8723	0·5881
0·55	—	0·9999	0·9948	0·9433	0·7483
0·60	—	1·0000	0·9987	0·9788	0·8684
0·65	—	—	0·9997	0·9934	0·9423
0·70	—	—	0·9999	0·9983	0·9793
0·75	—	—	—	0·9996	0·9941
0·80	—	—	—	0·9999	0·9987
0·85	—	—	—	—	0·9998
0·90	—	—	—	—	1·0000
0·95	—	—	—	—	—

the sample proportion which gives a confidence coefficient at least equal to 0·95, so as to be on the safe side. We will consider only central intervals, so that for given ϖ we have to find ϖ_0 and ϖ_1 such that

$$P(p \geqslant \varpi_0) \geqslant 0.975$$
$$P(p \leqslant \varpi_1) \geqslant 0.975,$$

the inequalities for P being as near to equality as we can make them.

Consider the diagrammatic representation of the type shown in Fig. 20.2.

From the table we can find, for any assigned ϖ, the values ϖ_0 and ϖ_1 such that $P(p \geqslant \varpi_0) \geqslant 0.975$ and $P(p \leqslant \varpi_1) \geqslant 0.975$. Note that in determining ϖ_1 the distribution function gives the probability of obtaining a proportion p or less of successes, so that the complement of the function gives the probability of a proportion strictly greater than p. Here, for example, on the horizontal through $\varpi = 0.1$ we find $\varpi_0 = 0$ and $\varpi_1 = 0.25$ from our table; and for $\varpi = 0.4$ we have $\varpi_0 = 0.15$ and $\varpi_1 = 0.60$. The points so obtained lie on stepped curves which have been drawn in. For example, when $\varpi = 0.3$ the greatest value of ϖ_0 such that $P(p \geqslant \varpi_0) \geqslant 0.975$ is 0·1. By the time ϖ has increased to 0·4 the value of ϖ_0 has increased to 0·20. Somewhere between is the marginal value of ϖ such that $P(p \geqslant 0.1)$ is exactly 0·975. If we tabulated the probabilities for finer intervals of ϖ these step curves would be altered slightly; and in the limit, if we calculate values of ϖ such that $P(p \geqslant \varpi_0) = 0.975$ exactly we obtain points lying inside our present step curves. These points have been joined by dotted lines in Fig. 20.2.

The zone between the stepped lines is the confidence belt. For any p the probability that we shall be wrong in locating ϖ inside the belt is at the most 0·05. We determine p_0 and p_1 by drawing a vertical at the observed value of p on the abscissa and

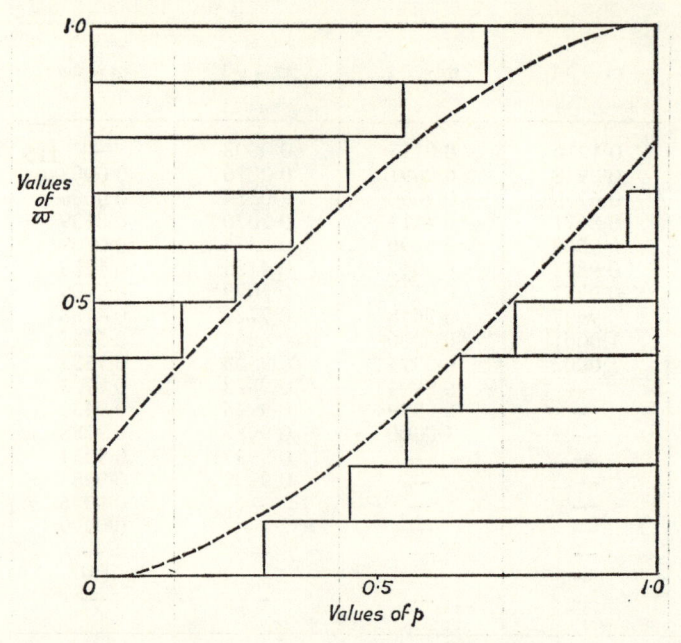

Fig. 20.2—Confidence limits for a binomial parameter

reading off the values where it intersects the appropriate lines giving ϖ_0 and ϖ_1. That these are, in fact, the required limits will be shown in a moment.

We consider a more sophisticated method of dealing with discontinuities below **(20.22)**.

It is, perhaps, worth noticing that the points on the curves of Fig. 20.2 were constructed by selecting an ordinate ϖ and then finding the corresponding abscissae ϖ_0 and ϖ_1. The diagram is, so to speak, constructed *horizontally*. In applying it, however, we read it *vertically*, that is to say, with observed abscissa p we read off two values p_0 and p_1 and assert that $p_0 \leq \varpi \leq p_1$. It is instructive to observe how this change of viewpoint can be justified without reference to Bayes' postulate.

Considering the diagram horizontally we see that, for any given ϖ, an observation falls in the confidence belt with probability $\geq 1-\alpha$. If and only if the observation is in the belt, the pair of values (p_0, p_1) will contain between them the true value of ϖ. Thus the latter event has probability $\geq 1-\alpha$, whatever the true value of ϖ.

Confidence intervals for large samples

20.10 We have seen **(18.16)** that the first derivative of the logarithm of the Likelihood Function is, under regularity conditions, asymptotically normally distributed with zero mean and

$$\mathrm{var}\left(\frac{\partial \log L}{\partial \theta}\right) = E\left\{\left(\frac{\partial \log L}{\partial \theta}\right)^2\right\} = -E\left\{\frac{\partial^2 \log L}{\partial \theta^2}\right\}. \qquad (20.7)$$

We may use this fact to set confidence intervals for θ in large samples. Writing

$$\psi = \frac{\partial \log L}{\partial \theta} \bigg/ \left[E\left\{\left(\frac{\partial \log L}{\partial \theta}\right)^2\right\}\right]^{\frac{1}{2}}, \qquad (20.8)$$

so that ψ is a standardized normal variate in large samples, we may from the normal integral determine confidence limits for θ in large samples if ψ is a monotonic function of θ, so that inequalities in one may be transformed to inequalities in the other. The following examples illustrate the procedure.

Example 20.3

Consider again the problem of Example 20.1. We have seen in Example 17.6 that in this case

$$\frac{\partial \log L}{\partial \mu} = n(\bar{x} - \mu), \qquad (20.9)$$

so that

$$-\frac{\partial^2 \log L}{\partial \mu^2} = n \qquad (20.10)$$

and, from (20.7) and (20.8),

$$\psi = (\bar{x} - \mu)\sqrt{n} \qquad (20.11)$$

is normally distributed with unit variance for large n. (We know, of course, that this is true for any n in this particular case.) Confidence limits for μ may then be set as in Example 20.1.

Example 20.4

Consider the Poisson distribution whose general term is

$$f(x, \lambda) = \frac{e^{-\lambda} \lambda^x}{x!}, \qquad x = 0, 1, \ldots$$

We have seen in Example 17.8 that

$$\frac{\partial \log L}{\partial \lambda} = \frac{n}{\lambda}(\bar{x} - \lambda). \qquad (20.12)$$

Hence

$$-\frac{\partial^2 \log L}{\partial \lambda^2} = \frac{n\bar{x}}{\lambda^2}$$

and

$$E\left(-\frac{\partial^2 \log L}{\partial \lambda^2}\right) = \frac{n}{\lambda}. \qquad (20.13)$$

Hence, from (20.7),

$$\psi = (\bar{x} - \lambda)\sqrt{(n/\lambda)}. \qquad (20.14)$$

For example, with $1-\alpha = 0.95$, corresponding to a normal deviate ± 1.96, we have, for the central confidence limits,

$$(\bar{x} - \lambda)\sqrt{(n/\lambda)} = \pm 1.96,$$

giving, on squaring,

$$\lambda^2 - \left(2\bar{x} + \frac{3.84}{n}\right)\lambda + \bar{x}^2 = 0$$

or
$$\lambda = \bar{x} + \frac{1 \cdot 92}{n} \pm \sqrt{\left(\frac{3 \cdot 84 \bar{x}}{n} + \frac{3 \cdot 69}{n^2}\right)},$$
the ambiguity in the square root giving upper and lower limits respectively.

To order $n^{-\frac{1}{2}}$ this is equivalent to
$$\lambda = \bar{x} \pm 1 \cdot 96 \sqrt{(\bar{x}/n)}, \tag{20.15}$$
from which the upper and lower limits are seen to be equidistant from the mean $\bar{x}$, as we should expect.

20.11 The procedure we have used in arriving at (20.15) requires some further examination. If we have probabilities like
$$P(\theta \leq t) \quad \text{or} \quad P(\theta^{-1} \leq t), \quad \theta > 0$$
they can be immediately "inverted" so as to give
$$P(t \geq \theta) \quad \text{or} \quad P(t^{-1} \leq \theta).$$
But we may encounter more complicated forms such as
$$P\{g(t,\theta) \leq 0\}$$
where g is, say, a polynomial in t or θ or both, of degree greater than unity. The problem of translating such an expression into terms of intervals for θ may be far from straightforward.

Let us reconsider (20.14) in the form
$$\psi = n^{\frac{1}{2}}(\bar{x} - \lambda)/\lambda^{\frac{1}{2}}. \tag{20.16}$$
Take a confidence coefficient $1 - \alpha$ and let the corresponding values of ψ be ψ_0 and ψ_1, i.e.
$$P\{\psi_0 \leq \psi \leq \psi_1\} = 1 - \alpha. \tag{20.17}$$
Equation (20.16) may be written
$$\lambda^2 - (2\bar{x} + n^{-1}\psi^2)\lambda + \bar{x}^2 = 0 \tag{20.18}$$
and if the intervals of ψ are central, that is to say, if $\psi_0 = -\psi_1$, the roots in λ of (20.18) are the same whether we put $\psi = \psi_0$ or $\psi = \psi_1$. Moreover, the roots are always real. Let λ_0, λ_1 be the roots of the equation with $\psi = \psi_0$ (or ψ_1), and let λ_1 be the larger. Then, as ψ goes from $-\infty$ to ψ_0, λ is seen from (20.16) to go from $+\infty$ to λ_1; as ψ goes from ψ_0 to ψ_1, λ goes (downwards) from λ_1 to λ_0; and as ψ goes from ψ_1 to $+\infty$, λ goes from λ_0 to $-\infty$. Thus
$$P(\psi_0 \leq \psi \leq \psi_1) = 1 - \alpha$$
is equivalent to
$$P(\lambda_0 \leq \lambda \leq \lambda_1) = 1 - \alpha,$$
and our confidence intervals are of the kind required.

It is instructive to consider this diagrammatically, as in Fig. 20.3.

From (20.16) we see that, for given n and ψ, the curves relating λ (ordinate) and $\bar{x}$ (abscissa) may be represented as
$$(\lambda - \bar{x})^2 = k\lambda, \tag{20.19}$$
where k is a positive constant. For varying k, these are parabolas with $\lambda = \bar{x}$ as the major axis, passing through the origin. The line $\lambda = \bar{x}$ corresponds to $k = 0$ or

INTERVAL ESTIMATION: CONFIDENCE INTERVALS

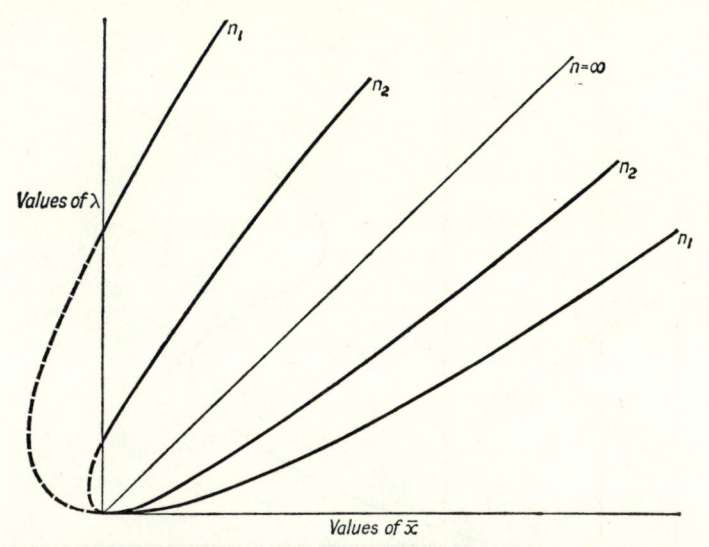

Fig. 20.3—Confidence parabolas (20.19) for varying k or n

$n = \infty$ and we have shown two other curves (not to scale) for values n_1 and n_2 ($n_1 < n_2$). From our previous discussion it follows that, for given n, the values of λ corresponding to values of ψ *inside* the range ψ_0 to ψ_1 lie *inside* the appropriate parabolas. It is also evident that the parabola for n_1 lies wholly inside the parabola for any smaller n.

Thus, given any $\bar{x}$, we may read off ordinate-wise two corresponding values of λ and assert that the unknown λ lies between them. The confidence lines in Fig. 20.3 have similar properties of convexity and nestedness to those for the binomial distribution in Example 20.2.

20.12 Let us now consider a more complicated case. Suppose we have a statistic t from which we can set limits t_0 and t_1, independent of θ, with assigned confidence coefficient $1-\alpha$. And suppose that

$$t = a\theta^3 + b\theta^2 + c\theta + d, \tag{20.20}$$

where a, b, c, d are constants. Sometimes, but not always, there will be three real values of θ corresponding to a value of t. How do we use them to set limits to θ?

Again the position is probably clearest from a diagram. In Fig. 20.4 we graph θ as ordinate against t as abscissa, again not to scale.

We have supposed the constants to be such that the cubic has a real maximum and minimum, as shown. For various values of t, the cubic of equation (20.20) is translated along the t-axis. To avoid confusing the diagram we will suppose that only the lines for one value of n are shown. We also take $a > 0$.

Now for a given value of t, say t_0, there will be a cubic, as shown in the diagram, such that for the area on the right $a\theta^3 + b\theta^2 + c\theta + d > t_0$ and for the area on the left that cubic is $< t_0$. Similarly for t_1. With the appropriate confidence coefficient we may then say that for an observed t, the limits to θ are given by reading vertically along the ordinate at t.

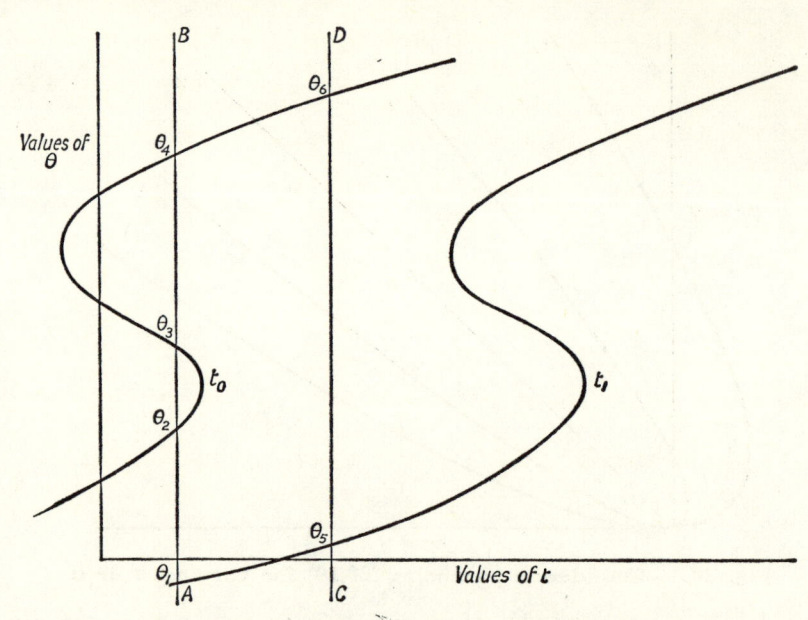

Fig. 20.4—Confidence cubics (20.20) (see text)

We now begin to encounter difficulties. If we take a value such as that along AB in the diagram, we shall have to assert that θ lies in the broken range $\theta_1 \leqslant \theta \leqslant \theta_2$ and $\theta_3 \leqslant \theta \leqslant \theta_4$. On the other hand, at CD we have the unbroken range $\theta_5 \leqslant \theta \leqslant \theta_6$.

Devotees of pathological mathematics will have no difficulty in constructing further examples in which the intervals are broken up even more, or in which we have to assert that the parameter θ lies *outside* a closed interval. A case of some consequence is discussed in Example 20.7 below. Cf. also Exercise 28.21.

20.13 The point to observe in such cases is that the statements concerning the intervals may still be made with exactitude. The question is, are they still useful and do they solve the problem with which we began, that of specifying a neighbourhood of the parameter value? Shall we, in fact, admit them as confidence intervals or shall we deny this name to them?

No simple answer to such questions has been given, but we may record our own opinion on the subject.

(a) The most satisfactory situation is that in which the confidence lines are monotonic in the sense that an ordinate meets each line only once, the parameter then being asserted to lie inside a connected interval. Further desiderata are that, for fixed α, the confidence belt for any n should lie inside the belt for any smaller n; and that for fixed n, the belt for any $(1-\alpha)$ should lie inside that for larger $(1-\alpha)$. These conditions are obeyed in Examples 20.1 to 20.4.

(b) Where such conditions are not obeyed, the case should be considered on its merits. Instances may arise where a disconnected interval such as that of Fig. 20.4 occurs and is acceptable. Where possible, the confidence regions should be graphed.

The automatic " inversion " of probability statements without attention to such points must be avoided.

20.14 We may, at this stage, notice another point of a rather different kind which sometimes leads to difficulty. When considering the quadratic (20.18) we remarked that, under the conditions of the problem, the roots were always real. It may happen, however, that for some confidence coefficients we set ourselves an impossible task in the construction of real intervals. The following example will illustrate the point.

Example 20.5

If $x_1, \ldots, x_n$ are a sample of n observations from a normal population with unit variance and mean μ, the statistic $\chi^2 = \Sigma(x-\mu)^2$ is distributed in the chi-squared form with n degrees of freedom. For assigned confidence coefficient $1-\alpha$ we can determine χ_0^2, χ_1^2 (say as a central interval, to simplify the exposition) such that

$$P\{\chi_0^2 \leqslant \chi^2 \leqslant \chi_1^2\} = 1-\alpha. \tag{20.21}$$

Now if $s^2 = \Sigma(x-\bar{x})^2/n$ we have the identity

$$\chi^2 = \Sigma(x-\mu)^2 = n\{s^2 + (\bar{x}-\mu)^2\},$$

and hence the limits for $(\bar{x}-\mu)^2$ are given by

$$\frac{\chi_0^2}{n} - s^2 \leqslant (\bar{x}-\mu)^2 \leqslant \frac{\chi_1^2}{n} - s^2. \tag{20.22}$$

Now it may happen that s^2 is greater than χ_1^2/n, in which case (since $\chi_0^2 < \chi_1^2$) the inequality (20.22) asserts that $(\bar{x}-\mu)^2$ lies between two negative quantities. What are we to make of such an assertion?

The matter becomes clearer if we again consider a geometrical argument. Since χ^2 now depends on two statistics, s and $\bar{x}$ (which are, incidentally, independent), we require three dimensions to represent the position, one for μ and one each for s and $\bar{x}$. Fig. 20.5 attempts the representation.

The quantity χ^2 is constant on the surfaces

$$(\bar{x}-\mu)^2 + s^2 = \text{constant}.$$

For fixed μ (i.e. planes perpendicular to the μ-axis) these surfaces intersect the plane $\mu = $ constant in a circle centred at $(\mu, 0)$. These centres all lie on the line in the $(\mu, \bar{x})$ plane, with equation $\mu = \bar{x}$; and the surfaces of constant χ^2 are cylinders with this line as axis. (They are not right circular cylinders; only the sections perpendicular to the μ-axis are circles.)

Moreover, the cylinder for χ_1^2 completely encloses that for χ_0^2, as illustrated in the diagram. Given now an observed $\bar{x}, s$ we draw a line parallel to the μ-axis. If this meets each cylinder in two points, μ_{00}, μ_{01} for χ_0^2 and μ_{10}, μ_{11} for χ_1^2, we assert that $\mu_{00} \leqslant \mu \leqslant \mu_{10}$ and $\mu_{01} \leqslant \mu \leqslant \mu_{11}$. (There are two intervals corresponding to the ambiguity of sign when we take square roots in (20.22).)

The point of the present example is that the line may not meet the cylinders at all. The roots for μ of (20.22) are then imaginary. Such a situation cannot arise in, for example, the binomial case of Example 20.2, where every line parallel to the ϖ axis in the range $0 \leqslant p \leqslant 1$ must cross the confidence belt. Apart from complications

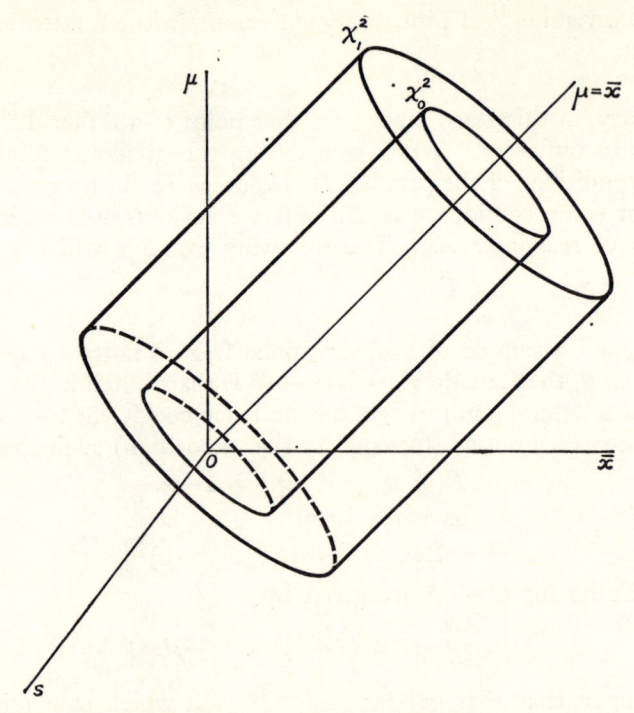

Fig. 20.5—**Confidence cylinders (20.22)** (see text)

due to inverting inequalities such as we considered in **20.11** to **20.13,** this usually happens whenever the parameter θ has a single sufficient statistic t which is used to set the intervals. But it can fail to happen as soon as we use more than one statistic and go into more than two dimensions, as in the present example.

In such cases, it seems to us, we must recognize that we are being set an impossible task.[*] We require to make an assertion with confidence coefficient $1-\alpha$, *using these particular statistics*, which is to be valid for all observed $\bar{x}$ and s. This cannot be done. It can only be done for certain sets of values of $\bar{x}$ and s, those for which the limits of (20.22) are positive. For some specified $\bar{x}$ and s we may be able to lower our confidence level, increase the radii of the cylinders and ensure that the line through $\bar{x}, s$ does meet the cylinders. But however low we make it, there may always arise a sample, however infrequently, for which we cannot set bounds to μ by this method.

In our present example, the remedy is clear. We have chosen the wrong method of setting confidence intervals; in fact, if we use the method of Example 20.1 and set bounds to $\bar{x}-\mu$ from the normal curve, no difficulty arises. $\bar{x}$ is then sufficient for μ. In general, where no single sufficient statistic exists, the difficulty may be unavoidable

---

[*] As we understand him, Neyman would say that such intervals are not confidence intervals in his sense. The conditions of **20.28** below are violated. Other writers have used the expression for intervals obtained by inverting a probability statement without regard to these conditions.

and must be faced squarely, if not after our own suggested manner, then by some equally explicit interpretation.

20.15 We revert to the approximation to confidence intervals for large samples discussed in **20.10**. If it is considered to be too inaccurate to assume that ψ is normally distributed for the sample size n with which we are concerned, a closer approximation may be derived. In fact, we find the higher moments of ψ and use an expansion of the Cornish–Fisher type (**6.25–6**). Write, using (17.19),

$$I = E\left(\frac{\partial \log L}{\partial \theta}\right)^2 = -E\left(\frac{\partial^2 \log L}{\partial \theta^2}\right), \tag{20.23}$$

$$J = \frac{\partial \log L}{\partial \theta}. \tag{20.24}$$

From (17.18), under regularity conditions,

$$\kappa_1(J) = 0 \tag{20.25}$$

whence
$$\kappa_2(J) = I. \tag{20.26}$$

We now prove that

$$\kappa_3(J) = 3\frac{\partial I}{\partial \theta} + 2E\left(\frac{\partial^3 \log L}{\partial \theta^3}\right), \tag{20.27}$$

$$\kappa_4(J) = 6\frac{\partial^2 I}{\partial \theta^2} + 8\frac{\partial}{\partial \theta} E\left(\frac{\partial^3 \log L}{\partial \theta^3}\right) - 3E\left(\frac{\partial^4 \log L}{\partial \theta^4}\right) + 3 \operatorname{var}\left(\frac{\partial^2 L}{\partial \theta^2}\right). \tag{20.28}$$

In fact, differentiating I, we have

$$\frac{\partial I}{\partial \theta} = 2E\left(\frac{\partial^2 \log L}{\partial \theta^2}\frac{\partial \log L}{\partial \theta}\right) + E\left(\frac{\partial \log L}{\partial \theta}\right)^3, \tag{20.29}$$

and differentiating

$$0 = I + E\left(\frac{\partial^2 \log L}{\partial \theta^2}\right)$$

we have

$$0 = \frac{\partial I}{\partial \theta} + E\left(\frac{\partial^3 \log L}{\partial \theta^3}\right) + E\left(\frac{\partial^2 \log L}{\partial \theta^2}\frac{\partial \log L}{\partial \theta}\right). \tag{20.30}$$

Eliminating $E\{(\partial^2 \log L/\partial \theta^2)(\partial \log L/\partial \theta)\}$ from (20.29) and (20.30) we arrive at (20.27).

Differentiating twice both relations for I given by (20.23) and eliminating $E\{(\partial^2 \log L/\partial \theta^2)(\partial \log L/\partial \theta)^2\}$ we find

$$6\frac{\partial^2 I}{\partial \theta^2} = E\left(\frac{\partial \log L}{\partial \theta}\right)^4 - 8E\left(\frac{\partial^3 \log L}{\partial \theta^3}\frac{\partial \log L}{\partial \theta}\right) - 5E\left(\frac{\partial^4 \log L}{\partial \theta^4}\right) - 3E\left(\frac{\partial^2 \log L}{\partial \theta^2}\right)^2.$$

Using the relation

$$\frac{\partial}{\partial \theta} E\left(\frac{\partial^3 \log L}{\partial \theta^3}\right) = E\left(\frac{\partial^3 \log L}{\partial \theta^3}\frac{\partial \log L}{\partial \theta}\right) + E\left(\frac{\partial^4 \log L}{\partial \theta^4}\right)$$

and transferring to cumulants we arrive at (20.28). The formulae are due to Bartlett (1953).

Using the first four terms in (6.54) with $l_1 = l_2 = 0$, we then have the statistic

$$T(\theta) = \frac{1}{\sqrt{I}} \left[\frac{\partial \log L}{\partial \theta} - \frac{1}{6} \frac{\kappa_3(J)}{I^2} \left\{ \left(\frac{\partial \log L}{\partial \theta} \right)^2 - I \right\} \right.$$
$$\left. - \frac{1}{24} \frac{\kappa_4(J)}{I^3} \left\{ \left(\frac{\partial \log L}{\partial \theta} \right)^3 - 3I \frac{\partial \log L}{\partial \theta} \right\} \right], \quad (20.31)$$

which is, to the next order of approximation, normal with zero mean and unit variance. The first term is the quantity we have called ψ. The corrective terms involve the standardized cumulants of J which are equivalent to the cumulants of ψ.

Example 20.6

Let us consider the problem of setting confidence limits to the variance of a normal population. The distribution of the sample variance is known to be skew, and we can compare the exact results with those given by the foregoing approximation.

Defining

$$s^2 = \frac{1}{n} \Sigma (x - \bar{x})^2,$$

we know that in samples of n the quantity s^2/σ^2 is distributed in the Type III form, or alternatively that ns^2/σ^2 is distributed as χ^2 with $n-1$ degrees of freedom.

Thus, for a sample of size 10 we have (since the upper and lower 5 per cent points of χ^2 are 3·3251 and 16·9190)

$$P\left\{ 3 \cdot 3251 \leqslant \frac{10 s^2}{\sigma^2} \leqslant 16 \cdot 9190 \right\} = 0 \cdot 90.$$

The inequalities may be inverted to give

$$P\{0 \cdot 5911 s^2 \leqslant \sigma^2 \leqslant 3 \cdot 001 s^2\} = 0 \cdot 90. \quad (20.32)$$

For example, with $s^2 = 1$ the limits are 0·5911 and 3·001.

We find, taking $\theta = \sigma^2$,

$$\frac{\partial \log L}{\partial \theta} = \frac{n}{2\theta^2} \left\{ \frac{1}{n} \Sigma (x - \mu)^2 - \theta \right\}, \quad (20.33)$$

whence we confirm that

$$E\left(\frac{\partial \log L}{\partial \theta} \right) = 0, \quad (20.34)$$

as required. It follows from Example 17.10 that

$$I = \frac{n}{2\theta^2}, \quad (20.35)$$

whence

$$\frac{\partial I}{\partial \theta} = -\frac{n}{\theta^3}. \quad (20.36)$$

Differentiating (20.33) twice, we find on taking expectations

$$E\left(\frac{\partial^3 \log L}{\partial \theta^3} \right) = \frac{2n}{\theta^3}. \quad (20.37)$$

INTERVAL ESTIMATION: CONFIDENCE INTERVALS

Hence, from (20.27), (20.36) and (20.37) we have

$$\kappa_3(J) = \frac{n}{\theta^3}. \tag{20.38}$$

We will take the expansion (20.31) as far as κ_3 only, obtaining

$$T(\theta) = \sqrt{\left(\frac{2\theta^2}{n}\right)} \left[\frac{n}{2\theta^2}\left\{\frac{1}{n}\Sigma(x-\mu)^2 - \theta\right\} - \frac{n}{6\theta^3} \frac{4\theta^4}{n}\left(\frac{n}{2\theta^2}\left\{\frac{n}{2\theta^2}\left(\frac{1}{n}\Sigma(x-\mu)^2 - \theta\right)^2\right\} - \frac{n}{2\theta^2}\right)\right]$$

$$= \sqrt{\frac{2}{n}} \left[\frac{n}{2\theta}\left\{\frac{1}{n}\Sigma(x-\mu)^2 - \theta\right\} - \frac{n}{6\theta^2}\left\{\frac{1}{n}\Sigma(x-\mu)^2 - \theta\right\}^2 + \frac{1}{3}\right]. \tag{20.39}$$

We shall replace $\Sigma(x-\mu)^2/n$ by $ns^2/(n-1)$, which has the same mean value, and obtain

$$T = \sqrt{\frac{2}{n}} \left[\frac{n}{2}\left(\frac{ns^2}{(n-1)\theta} - 1\right) - \frac{n}{6}\left(\frac{ns^2}{(n-1)\theta} - 1\right)^2 + \frac{1}{3}\right]. \tag{20.40}$$

The first term gives us the confidence limits for θ based on ψ alone. The other terms will be corrective terms of lower order in n. We then have approximately from (20.40)

$$T = \sqrt{\frac{2}{n}} \left[\frac{n}{2}\left(\frac{ns^2}{(n-1)\theta} - 1\right) - \frac{n}{6} \cdot \frac{2}{n} T^2 + \frac{1}{3}\right],$$

giving

$$\frac{ns^2}{(n-1)\theta} = 1 + T\sqrt{\frac{2}{n}} + \frac{2}{3n} T^2 - \frac{2}{3n}. \tag{20.41}$$

For example, with $n = 10$, $1-\alpha = 0.90$, $s^2 = 1$, and $T = \pm 1.6449$ (the 5 per cent points of the standardized normal distribution), we find for the limits of s^2/θ the values 0·3403 to 1·6644 and hence limits for θ of 0·6008 and 2·939. The true values, as we saw at (20.32) above, are 0·5911 and 3·001. For so low a value as $n = 10$ the approximation seems very fair.

Shortest sets of confidence intervals

20.16 It has been seen in Example 20.1 that in some circumstances at least there exist more than one set of confidence intervals, and it is now necessary to consider whether any particular set can be regarded as better than the others in some useful sense. The problem is analogous to that of estimation, where we found that in general there are many different estimators for a parameter, but that we could sometimes find one (such as that with minimum variance) which was superior to the rest.

In Example 20.1 the problem presented itself in rather a specialized form. We found that for the intervals based on the mean $\bar{x}$ there were infinitely many sets of intervals according to the way in which we selected α_0 and α_1 (subject to the condition that $\alpha_0 + \alpha_1 = \alpha$). Among these the central intervals are obviously the shortest, for a given range will include the greatest area of the normal distribution if it is centred at the mean. We might reasonably say that the central intervals are the best among those determined by $\bar{x}$.

But it does not follow that they are the shortest of all possible intervals, or even that such a shortest set exists. In general, for two sets of intervals c_1 and c_2, those of c_1 may be shorter than those of c_2 for some samples and longer in others.

Exercise 20.10 shows that the simplicity of Example 20.1 does not persist even in the case of the normal variance, discussed in Example 20.6; essentially this is because s^2, unlike $\bar{x}$, is asymmetrically distributed.

20.17 We will therefore consider sets of intervals which are shortest on the average. That is to say, if
$$\delta = t_1 - t_0 \tag{20.42}$$
we require to minimize $\int \delta \, dF$, where the integral is taken over all x's and is therefore equivalent to
$$\int_{-\infty}^{\infty} \ldots \int_{-\infty}^{\infty} L\delta \, dx_1 \ldots dx_n. \tag{20.43}$$

We now prove a theorem, due to Wilks (1938b), which is very similar to the result of **18.16** that Maximum Likelihood estimators in the limit have minimum variance, namely that in a certain class of intervals the method of **20.10** gives those which are shortest on the average in large samples.

Let $h(x, \theta)$ be a statistic which has a zero mean value and is such that the sum of a number of similar functions obeys the Central Limit Theorem. That is to say,
$$\zeta = \frac{\sum_{j=1}^{n} h(x_j, \theta)}{\sqrt{(n \operatorname{var} h)}} \tag{20.44}$$
is normally distributed in the limit with zero mean and unit variance. ψ of equation (20.8) is a member of the class ζ, having $h = \partial \log f(x, \theta)/\partial \theta$. We first show that the absolute average rate of change of ψ with respect to θ, for each fixed θ, is greater than that of any ζ except in the trivial case
$$h = k \frac{\partial \log f}{\partial \theta}.$$

Writing $\quad g(x, \theta) = \dfrac{\partial \log f}{\partial \theta}$, we have

$$\frac{\partial \psi}{\partial \theta} = \frac{1}{\sqrt{(n \operatorname{var} g)}} \left\{ \Sigma \frac{\partial g}{\partial \theta} - \frac{1}{2 \operatorname{var} g} \Sigma g \frac{\partial \operatorname{var} g}{\partial \theta} \right\},$$
$$\frac{\partial \zeta}{\partial \theta} = \frac{1}{\sqrt{(n \operatorname{var} h)}} \left\{ \Sigma \frac{\partial h}{\partial \theta} - \frac{1}{2 \operatorname{var} h} \Sigma h \frac{\partial \operatorname{var} h}{\partial \theta} \right\}. \tag{20.45}$$

Hence $\quad E\left(\dfrac{\partial \psi}{\partial \theta}\right) = \dfrac{1}{\sqrt{(n \operatorname{var} g)}} \left\{ \Sigma E\left(\dfrac{\partial g}{\partial \theta}\right) - \dfrac{1}{2 \operatorname{var} g} \Sigma E(g) \dfrac{\partial \operatorname{var} g}{\partial \theta} \right\}. \tag{20.46}$

Now $E(g) = 0$ by (20.25) and by (20.26)
$$E\left(\frac{\partial g}{\partial \theta}\right) = E\left(\frac{\partial^2 \log f}{\partial \theta^2}\right) = -E\left(\frac{\partial \log f}{\partial \theta}\right)^2 = -\operatorname{var} g.$$

Thus (20.46) becomes $\quad E\left(\dfrac{\partial \psi}{\partial \theta}\right) = -\dfrac{n \operatorname{var} g}{\sqrt{(n \operatorname{var} g)}} = -\sqrt{(n \operatorname{var} g)}. \tag{20.47}$

Similarly, since $E(h) = 0$, (20.45) gives
$$E\left(\frac{\partial \zeta}{\partial \theta}\right) = \sqrt{\left(\frac{n}{\operatorname{var} h}\right)} E\left(\frac{\partial h}{\partial \theta}\right). \tag{20.48}$$

Since $\quad 0 = E(h) = \int h f \, dx$

we have, differentiating under the integral sign,
$$0 = \frac{\partial}{\partial \theta} E(h) = \int \frac{\partial h}{\partial \theta} f \, dx + \int h \frac{\partial f}{\partial \theta} \, dx,$$
whence
$$E\left(\frac{\partial h}{\partial \theta}\right) = \int \frac{\partial h}{\partial \theta} f \, dx = -\int h \frac{\partial f}{\partial \theta} \, dx = -\operatorname{cov}(h, g). \qquad (20.49)$$
Hence, from (20.47–20.49)
$$\left| E\left(\frac{\partial \psi}{\partial \theta}\right) \right| - \left| E\left(\frac{\partial \zeta}{\partial \theta}\right) \right| = \sqrt{\frac{n}{\operatorname{var} h}} \{ \sqrt{(\operatorname{var} h \operatorname{var} g)} - |\operatorname{cov}(h,g)| \}. \qquad (20.50)$$
By the Cauchy–Schwarz inequality, the factor in braces in (20.50) is positive unless h is a constant multiple of g, when the factor is zero. Excluding this case, we have
$$\left| E\left(\frac{\partial \psi}{\partial \theta}\right) \right| > \left| E\left(\frac{\partial \zeta}{\partial \theta}\right) \right| \qquad (20.51)$$
which is our preliminary result. Now if λ_α is defined by
$$(2\pi)^{-\frac{1}{2}} \int_0^{\lambda_\alpha} \exp(-\tfrac{1}{2}x^2) \, dx = \tfrac{1}{2}(1-\alpha),$$
the confidence limits for θ, say t_0 and t_1, obtained from ψ satisfy
$$\Sigma g(x, \theta)/\sqrt{(n \operatorname{var} g)} = \pm \lambda_\alpha,$$
which we may write $\psi(t) = \pm \lambda_\alpha$. Similarly those obtained from ζ, say u_0 and u_1, satisfy
$$\Sigma h(x, \theta)/\sqrt{(n \operatorname{var} h)} = \pm \lambda_\alpha,$$
which we write
$$\zeta(u) = \pm \lambda_\alpha.$$
Taylor expansions about the true value θ_0 give
$$\pm \lambda_\alpha = \psi(\theta_0) + (t - \theta_0)\left(\frac{\partial \psi}{\partial \theta}\right)_{\theta'} = \zeta(\theta_0) + (u - \theta_0)\left(\frac{\partial \zeta}{\partial \theta}\right)_{\theta''}, \qquad (20.52)$$
where θ', θ'' are values in the neighbourhood of θ_0 which converge in probability to θ_0 as n increases. Putting $t = u = \theta_0$ in (20.52), we find $\psi(\theta_0) = \zeta(\theta_0)$.

Hence
$$(t - \theta_0)\left(\frac{\partial \psi}{\partial \theta}\right)_{\theta'} = (u - \theta_0)\left(\frac{\partial \zeta}{\partial \theta}\right)_{\theta''} \qquad (20.53)$$
Now the derivatives in (20.53) will converge in probability to their expectations. Hence, from (20.51) and (20.53), we have for large n
$$|t - \theta_0| < |u - \theta_0|,$$
so that the confidence limits t_0, t_1 are closer together, on the average, than any other limits u_0, u_1 given by a member of the class (20.44).

20.18 The result of **20.17** illustrates the close relation between the theory of confidence intervals and that of point estimation developed in Chapter 17. In **17.15–17** we showed that the MVB for the estimation of θ was equal to $1/E\{(\partial \log L/\partial \theta)^2\}$ and could be attained only by an estimator which was a linear function of $\partial \log L/\partial \theta$. It is natural to expect that interval estimates of θ based on $\partial \log L/\partial \theta$ should have the corresponding property of being shortest on the average. We have now seen that this is so in large samples.

20.19 Neyman (1937b) proposed to apply the phrase "shortest confidence intervals" to sets of intervals defined in quite a different way. As it is clear that such intervals are not necessarily the shortest in the sense of possessing the least length, even on the average, we shall attempt to avoid confusion by calling them "most selective."

Consider a set of intervals s_0, typified by δ_0, obeying the condition that
$$P\{\delta_0 \text{ c } \theta \mid \theta\} = 1-\alpha, \tag{20.54}$$
where we write $\delta_0 \text{ c } \theta$—that is, δ_0 "contains" θ—for the more usual $t_0 \leqslant \theta \leqslant t_1$. Let s_1 be some other set, typified by δ_1, such that
$$P\{\delta_1 \text{ c } \theta \mid \theta\} = 1-\alpha. \tag{20.55}$$
Either set is a permissible set of intervals, as the probability is $1-\alpha$ in both cases that the interval δ contains θ.

If now for every s_1 we have, for any value θ' other than the true value,
$$P\{\delta_0 \text{ c } \theta' \mid \theta\} \leqslant P\{\delta_1 \text{ c } \theta' \mid \theta\}, \tag{20.56}$$
s_0 is said to be *most selective*.

20.20 The ideas underlying this definition will be clearer from a reading of Chapters 22 and 23 dealing with the theory of tests. We anticipate them here to the extent of remarking that the object of most selective intervals is to cover the true value with assigned probability $1-\alpha$, but to cover other values as little as possible. We may say of both s_0 and s_1 that the assertion $\delta \text{ c } \theta$ is true in proportion $1-\alpha$ of the cases. What marks out s_0 for choice as the most selective set is that it covers false values less frequently than the remaining sets.

The difference between this approach and the one leading to shortest intervals is that the latter is concerned only with the physical length of the confidence interval, whereas the former gives weight to the frequency with which alternative values of θ are covered. The one concentrates on locating the true value θ with the smallest margin of error; the other takes into account the desirability of excluding so far as possible false values of θ from the interval, so that mistakes of taking the wrong value are minimized. It turns out that the "selectivity" approach is easier to handle mathematically, so that much more attention has been given to it. See Exercises 20.17–18 below for a relationship between the two approaches. Madansky (1962) gives an example which illuminates the difference between them. Harter (1964) discusses other criteria for intervals, preferring one based on the mean squared deviation of the confidence limits from the true parameter value.

Neyman himself has shown that most selective sets do not usually exist (for instance, if the distribution is continuous) and has proposed two alternative systems:

(a) most selective one-sided systems (Neyman's "shortest one-sided" sets) which obey (20.56) only for values of $\theta'-\theta$ which are always positive or always negative;

(b) selective unbiased systems (Neyman's "short unbiased" sets) which obey the further relation
$$P\{\delta \text{ c } \theta \mid \theta\} = 1-\alpha \geqslant P\{\delta \text{ c } \theta \mid \theta'\}. \tag{20.57}$$

These definitions, also, amount to a translation into terms of confidence intervals of certain ideas in the theory of tests, and we may defer consideration of them until

Chapter 23. We therefore need make no systematic study of " optimum " confidence intervals in this chapter.

Tables and charts of confidence intervals

20.21 (1) *Binomial distribution*—Clopper and Pearson (1934) give two central confidence interval charts for the parameter, for $\alpha = 0.01$ and 0.05; each gives contours for $n = 8$ (2) 12 (4) 24, 30, 40, 60, 100, 200, 400 and 1000. The charts are reproduced in the *Biometrika Tables*. Upper or lower one-sided intervals can be obtained for half these values of α. Incomplete B-function tables may also be used—see **5.7** and the *Biometrika Tables*.

Pachares (1960) gives central limits for $\alpha = 0.01, 0.02, 0.05, 0.10$ and $n = 55$ (5) 100, and references to other tables, including those of Clark (1953) for the same values of α and $n = 10$ (1) 50.

> Sterne (1954) has proposed an alternative method of setting confidence limits for a proportion. Instead of being central, the belt contains the values of p with the largest probabilities of occurrence. Since the distribution of p is skew in general, we clearly shorten the interval in this way. Walton (1970) extended Sterne's tables. Crow (1956) has shown that these intervals constitute a confidence belt with minimum total area, and has tabulated a slightly modified set of intervals for sample sizes up to 30 and confidence coefficients $0.90, 0.95$ and 0.99. See also **20.23** below.

(2) *Poisson distribution*—(a) The *Biometrika Tables*, using the work of Garwood (1936), give central confidence intervals for the parameter, for observed values $x = 0$ (1) 30 (5) 50 and $\alpha = 0.002, 0.01, 0.02, 0.05, 0.10$. As in (1), one-sided intervals are available for $\alpha/2$. (b) Woodcock and Eames (1970) give similar tables for $x = 0$ (1) 100 (2) 200 (5) 500 (10) 1200 and α, 10α, 100α or 1000α equal to $0.1, 0.2$ and 0.5. (c) Przyborowski and Wilénski (1935) give upper confidence limits only for $x = 0$ (1) 50, $\alpha = 0.001, 0.005, 0.01, 0.02, 0.05, 0.10$. (d) Walton (1970) tabulates Sterne-type intervals and Crow and Gardner (1959) tabulate modified intervals of the Sterne–Crow binomial type for $x = 0$ (1) 300 and $\alpha = \cdot 001, \cdot 01, \cdot 05, \cdot 10, \cdot 20$. See also **20.23** below.

(3) *Variance of a normal distribution*—Tate and Klett (1959) give the most selective unbiassed confidence intervals, and the physically shortest intervals, based on multiples of the sufficient statistic for $\alpha = 0.001, 0.005, 0.01, 0.05, 0.10$ and $n = 3$ (1) 30. The former are also given by Pachares (1961) for $\alpha = \cdot 01, \cdot 05, \cdot 10$ and $n - 1 = 1$ (1) 20, 24, 30, 40, 60, 120; and by Lindley *et al.* (1960) for $\alpha = \cdot 001, \cdot 01, \cdot 05$ and $n - 1 = 1$ (1) 100.

(4) *Ratio of normal variances*—Ramachandran (1958) gives the most selective unbiassed intervals for $\alpha = 0.05$ and $n_1 - 1, n_2 - 1 = 2$ (1) 4 (2) 12 (4) 24, 30, 40, 60.

(5) *Correlation parameter*—F. N. David (1938) gives four central confidence interval charts for the correlation parameter ρ of a bivariate normal population, for $\alpha = 0.01, 0.02, 0.05, 0.10$; each gives contours for $n = 3$ (1) 8, 10, 12, 15, 20, 25, 50, 100, 200 and 400. The *Biometrika Tables* reproduce the $\alpha = 0.01$ and $\alpha = 0.05$ charts. One-sided intervals may be obtained as in (1).

Discontinuities

20.22 In discussing the binomial distribution in Example 20.2, we remarked on the fact that, as the number of successes (say c) is necessarily integral, and the proportion of successes $p\,(=c/n)$ therefore discontinuous, the confidence belt obtained is not exact, but provides confidence statements of form $P \geq 1-\alpha$ instead of $P = 1-\alpha$. By a rather peculiar device, we can always make exact statements of form $P = 1-\alpha$ even in the presence of discontinuity. The method was given by Stevens (1950).

In fact, after we have drawn our sample and observed c successes, let us from elsewhere draw a random number x from the rectangular distribution $dF = dx, 0 \leq x \leq 1$, e.g. by selecting a random number of four digits from the usual tables and putting a decimal point in front. Then the variate

$$y = c + x \qquad (20.58)$$

can take all values in the range 0 to $n+1$ (assuming that four decimal places is enough to specify a continuous variate). If y_0 is some given value $c_0 + x_0$, we have, writing ϖ for the probability to be estimated,

$$P(y \geq y_0) = P(c > c_0) + P(c = c_0)P(x \geq x_0)$$

$$= \sum_{j=c_0+1}^{n} \binom{n}{j} \varpi^j (1-\varpi)^{n-j} + \binom{n}{c_0} \varpi^{c_0} (1-\varpi)^{n-c_0} (1-x_0)$$

$$= x_0 \sum_{j=c_0+1}^{n} \binom{n}{j} \varpi^j (1-\varpi)^{n-j} + (1-x_0) \sum_{j=c_0}^{n} \binom{n}{j} \varpi^j (1-\varpi)^{n-j}. \qquad (20.59)$$

This defines a continuous probability distribution for y. It is clearly continuous as x_0 moves from $0+$ to $1-$, for c_0 is then constant. And at the points where $x_0 = 0$ the probability approaches the same value from the left and from the right. We can therefore use this distribution to set confidence limits for ϖ and our confidence statements based upon them will be exact statements of the form $P = 1-\alpha$.

The confidence intervals are of the type exhibited in Fig. 20.6. The upper limit is now shifted to the right by amounts which, in effect, join up the discontinuities by a series of arcs. The lower limit also has a series of arcs, but there is no displacement to the right, and we have therefore shown on the diagram only the (dotted) approximate

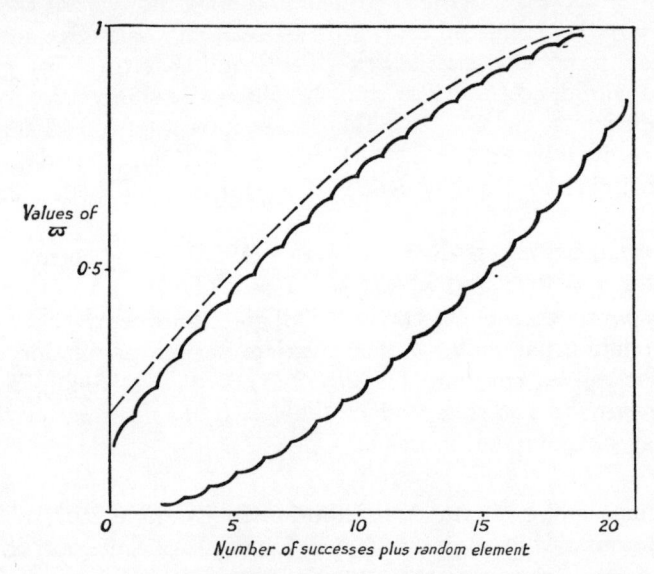

Fig. 20.6—Randomized confidence intervals for a binomial parameter

upper limit of Fig. 20.2. On our scale the lower approximate limit would almost coincide with the lower series of arcs. The general effect is to shorten the confidence interval

20.23 It is at first sight surprising that the intervals set up in this way lie inside the approximate step-intervals of Fig. 20.2, and are therefore no less accurate; for by taking an additional random number x we have imported additional uncertainty into the situation. A little reflection will show, however, that we have not got something for nothing. We have removed one uncertainty, associated with the inequality in $P \geqslant 1-\alpha$, by bringing in another so as to make statements of the kind $P=1-\alpha$; and what we lose on the second we more than offset by removing the first.

Central intervals for the binomial parameter may easily be derived by use of (20.59), but they are not the most selective unbiassed randomized intervals. The latter are tabulated by Blyth and Hutchinson (1960) for $\alpha = \cdot 01, \cdot 05$, and $n = 2\,(1)\,24\,(2)\,50$. The same authors (1961) give intervals with the same property for the Poisson parameter, for observed x ranging to 250 and $\alpha = \cdot 01, \cdot 05$.

Generalization to the case of several parameters

20.24 We now proceed to generalize the foregoing theory to the case of a distribution dependent upon several parameters. Although, to simplify the exposition, we shall deal in detail only with a single variate, the theory is quite general. We begin by extending our notation and introducing a geometrical terminology which may be regarded as an elaboration of the diagrams of Fig. 20.1 and 20.2.

Suppose we have a frequency function of known form depending on l unknown parameters, $\theta_1, \ldots, \theta_l$, and denoted by $f(x, \theta_1, \ldots, \theta_l)$. We may require to estimate either θ_1 only or several of the θ's simultaneously. In the first place we consider only the estimation of a single parameter. To determine confidence limits we require to find two functions, u_0 and u_1, dependent on the sample values but not on the θ's, such that

$$P\{u_0 \leqslant \theta_1 \leqslant u_1\} = 1-\alpha, \tag{20.60}$$

where $1-\alpha$ is the confidence coefficient chosen in advance.

With a sample of n values, $x_1, \ldots, x_n$, we can associate a point in an n-dimensional Euclidean space, and the frequency distribution will determine a density function for each such point. The quantities u_0 and u_1, being functions of the x's, are determined in this space, and for any given $1-\alpha$ will lie on two hypersurfaces (the natural extension of the confidence lines of Fig. 20.1). Between them will lie a Confidence Zone.

In general we also have to consider a range of values of θ which are *a priori* possible. There will thus be an l-dimensional space of θ's subjoined to the n-space, the total region of variation having $(l+n)$ dimensions; but if we are considering only the estimation of θ_1, this reduces to an $(n+1)$-space, the other $(l-1)$ parameters not appearing.

We shall call the sample-space W and denote a point whose co-ordinates are $x_1, \ldots, x_n$ by E. We may then write $u_0(E), u_1(E)$ to show that the confidence functions depend on E. The interval $u_1(E)-u_0(E)$ we denote by $\delta(E)$ or δ, and as above we write $\delta \, \mathrm{c} \, \theta_1$ to denote $u_0 \leqslant \theta_1 \leqslant u_1$. The confidence zone we denote by A, and may write $E \in \delta$ or $E \in A$ to indicate that the sample-point lies in the interval δ or the region A.

20.25 In Fig. 20.7 we have shown two axes, x_1 and x_2, and a third axis corresponding to the variation of θ_1. The sample-space W is thus two-dimensional. For any given θ_1, say θ_1', the space W is a hyperplane (or part of it), one such being shown.

Take any given pair of values (x_1, x_2) and draw through the point so defined a line parallel to the θ_1-axis, such as PQ in the figure, cutting the hyperplane at R. The two values of u_0 and u_1 will give two limits to θ_1 corresponding to two points on this

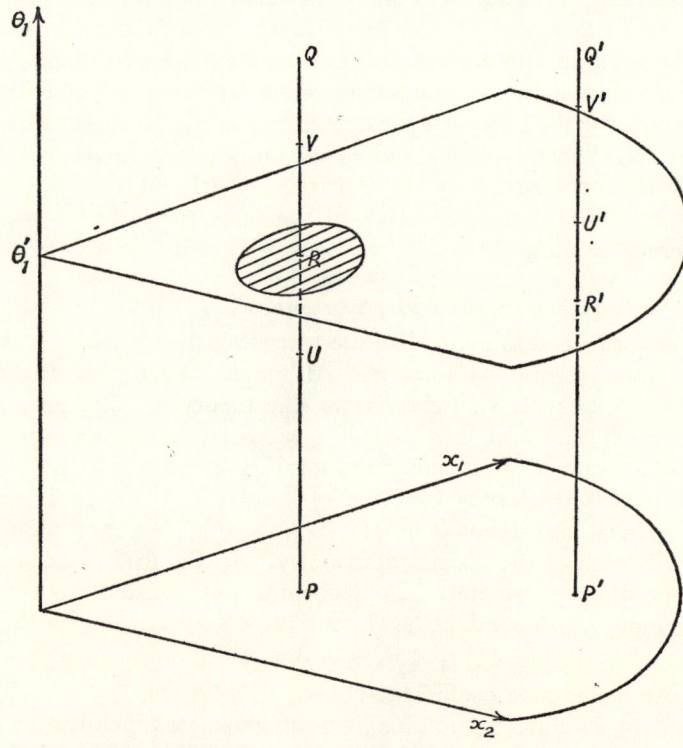

Fig. 20.7—Confidence intervals for $n = 2$ (see text)

line, say U, V. Consider now the lines PQ as x_1, x_2 vary. In some cases U, V will lie on opposite sides of R, and θ_1 lies inside the interval UV. In other cases (as for instance in $U'V'$ shown in the figure), the contrary is true. The totality of points in the former category determines the region in A, shaded in the figure. If for any point in A we assert δ c θ', we shall be right; if we assert it for points outside A we shall be wrong.

20.26 Evidently, if the sample-point E falls in the region A, the corresponding θ_1 lies in the confidence interval, and conversely. It follows that the probability of any fixed θ_1 being covered by the confidence interval is the probability that E lies in $A(\theta_1)$; or in symbols—

$$P\{\delta \text{ c } \theta_1' \mid \theta_1, \ldots, \theta_i\} = P\{u_0 \leqslant \theta_1' \leqslant u_1 \mid \theta_1, \ldots, \theta_i\}$$
$$= P\{E \in A(\theta_1') \mid \theta_1, \ldots, \theta_i\}. \qquad (20.61)$$

From this it follows that if the confidence functions are determined so that
$$P\{u_0 \leqslant \theta_1 \leqslant u_1\} = 1-\alpha$$
we shall have, for all θ_1,
$$P\{E \in A(\theta_1) \mid \theta_1, \ldots, \theta_i\} = 1-\alpha. \qquad (20.62)$$
It follows also that for no θ_1 can the region A be empty, for if it were the probability in (20.62) would be zero.

20.27 If the functions u_0 and u_1 are single-valued and determined for all E, then any sample-point will fall into at least one region $A(\theta_1')$. For on the line PQ corresponding to the given E we take an R between U and V, and this will define a value of θ_1, say θ_1', such that $E \in A(\theta_1')$.

More importantly, if a sample-point falls in the regions $A(\theta_1')$ and $A(\theta_1'')$ corresponding to two values of θ_1, θ_1' and θ_1'', it will fall in the region $A(\theta_1''')$, where θ_1''' is any value between θ_1' and θ_1''. For we have
$$u_0 \leqslant \theta_1' \leqslant u_1, \qquad u_0 \leqslant \theta_1'' \leqslant u_1,$$
and hence
$$u_0 \leqslant \theta_1' \leqslant \theta_1''' \leqslant \theta_1'' \leqslant u_1,$$
if θ_1'' is the greater of θ_1' and θ_1''.

Further, if a sample-point falls in any of the regions $A(\theta_1)$ for the range of θ-values $\theta_1' < \theta_1 < \theta_1''$ it must also fall within $A(\theta_1')$ and $A(\theta_1'')$.

20.28 The conditions referred to in the two previous sections are necessary. We now prove that they are sufficient, that is to say: if for each value of θ_1 there is defined in the sample-space W a region A such that
 (1) $P\{E \in A(\theta_1) \mid \theta_1\} = 1-\alpha$, whatever the value of the θ's;
 (2) for any E there is at least one θ_1, say θ_1', such that $E \in A(\theta_1')$;
 (3) if $E \in A(\theta_1')$ and $E \in A(\theta_1'')$, then $E \in A(\theta_1''')$ for any θ_1''' between θ_1' and θ_1'';
 (4) if $E \in A(\theta_1)$ for any θ_1 satisfying $\theta_1' \leqslant \theta_1 \leqslant \theta_1''$, $E \in A(\theta_1')$ and $E \in A(\theta_1'')$;
then confidence limits for θ, u_0 and u_1 are given by taking the lower and upper bounds of values of θ_1 for which a fixed sample-point falls within $A(\theta_1)$. They are determinate and single-valued for all E, $u_0 \leqslant u_1$, and $P\{u_0 \leqslant \theta_1 \leqslant u_1 \mid \theta_1\} = 1-\alpha$ for all θ_1.

The lower and upper bounds exist in virtue of condition (2), and the lower is not greater than the upper. We have then merely to show that $P\{u_0 \leqslant \theta_1 \leqslant u_1 \mid \theta_1\} = 1-\alpha$ and for this it is sufficient, in virtue of condition (1), to show that
$$P\{u_0 \leqslant \theta_1 \leqslant u_1 \mid \theta_1\} = P\{E \in A(\theta_1) \mid \theta_1\}. \qquad (20.63)$$
We already know that if $E \in A(\theta_1)$ then $u_0 \leqslant \theta_1 \leqslant u_1$; and our result will be established if we demonstrate the converse.

Suppose it is not true that when $u_0 \leqslant \theta_1 \leqslant u_1$, $E \in A(\theta_1)$. Let E' be a point outside $A(\theta_1)$ for which $u_0 \leqslant \theta_1 \leqslant u_1$. Then either $u_0 = \theta_1$, or $u_1 = \theta_1$, or both: for otherwise, u_0 and u_1 being the bounds of the values of θ_1 for which E lies in $A(\theta_1)$, there would exist values θ_1' and θ_1'', such that $E \in A(\theta_1')$ and $E \in A(\theta_1'')$ and
$$u_0 \leqslant \theta_1' \leqslant \theta_1 \leqslant \theta_1'' \leqslant u_1,$$
so that, from condition (3), $E \in A(\theta_1)$, which is contrary to assumption.

Thus $u_0 = \theta_1$ or $u_1 = \theta_1$ or both. If both, then E must fall in $A(\theta_1')$, for u_0 and u_1 are the bounds of θ-values for which this is so. Finally, if $u_0 = \theta_1 < u_1$ (and similarly if $u_0 < \theta_1 = u_1$) we see that for $u_0 < \theta_1 < u_1$, E must fall in $A(\theta_1)$ from condition (3), and hence, from condition (4), E must fall in $A(\theta_1')$ and $A(\theta_1'')$ where $\theta_1' = u_0$ and $\theta_1'' = u_1$. Hence it falls in $A(\theta_1)$.

Choice of statistic

20.29 The foregoing theorem gives us a formal solution of the problem of finding confidence intervals for a single parameter in the general case, but it does not provide a method of finding the intervals in particular instances. In practice we have four lines of approach : (1) to use a single sufficient statistic if one exists ; (2) to adopt the process known as " studentization " (cf. **20.31**) ; (3) to " guess " a set of intervals in the light of general knowledge and experience and to verify that they do or do not satisfy the required conditions ; and (4) to approximate by an extension of the method of **20.15**.

20.30 Consider the use of a single sufficient statistic in the general case. If t_1 is sufficient for θ_1, we have

$$L = g(t_1 \mid \theta_1) L_2(x_1, \ldots, x_n, \theta_2, \ldots, \theta_l). \qquad (20.64)$$

The locus $t_1 = $ constant determines a series of hypersurfaces in the sample-space W. If we regard these hypersurfaces as determining regions in W, then $t_1 \leqslant k$, say, determines a fixed region K. The probability that E falls in K is then clearly dependent only on t_1 and θ_1. By appropriate choice of k we can determine K so that

$$P\{E \in K \mid \theta_1\} = 1 - \alpha$$

and hence set up regions based on values of t_1. We can do so, moreover, in an infinity of ways, according to the values selected for α_0 and α_1. We shall see in **23.3**, when discussing this problem in terms of testing hypotheses, that the most selective intervals (equivalent to the most powerful test of $\theta_1 = \theta_1^0$) are always obtainable from the sufficient statistics.

Studentization

20.31 In Example 20.1 we considered a simplified problem of estimating the mean in samples from a normal population with known variance. Suppose now that we require to determine confidence limits for the mean μ in samples from

$$dF = \frac{1}{\sigma\sqrt{(2\pi)}} \exp\left\{-\frac{1}{2}\left(\frac{x-\mu}{\sigma}\right)^2\right\} dx,$$

when σ is unknown.

Consider the distribution of $z = (\bar{x}-\mu)/s$, where s^2 is the sample variance. This is known to be the " Student " form

$$dF = \frac{k\, dz}{(1+z^2)^{\frac{1}{2}n}}, \qquad (20.65)$$

(cf. Example 11.8). Given α, we can now find z_0 and z_1, such that

$$\int_{-\infty}^{-z_1} dF = \int_{z_0}^{\infty} dF = \frac{\alpha}{2},$$

INTERVAL ESTIMATION : CONFIDENCE INTERVALS

and hence
$$P(-z_1 \leqslant z \leqslant z_0) = 1-\alpha,$$
which is equivalent to
$$P(\bar{x}-sz_0 \leqslant \mu \leqslant \bar{x}+sz_1) = 1-\alpha. \qquad (20.66)$$
Hence we may say that μ lies in the range $\bar{x}-sz_0$ to $\bar{x}+sz_1$ with confidence coefficient $1-\alpha$, the range now being independent of either μ or σ. In fact, owing to the symmetry of " Student's " distribution, $z_0 = z_1$, but this is an accidental circumstance not necessary to the argument.

It should be noted that (20.66), like (20.4), is linear in the statistic $\bar{x}$; the confidence lines in this case also are parallel straight lines as in Fig. 20.1. The difference is that whereas, with σ known, the vertical distance between the confidence lines is fixed as a function of σ, in the present case the distance is a random variable, being a function of s. Thus we cannot here fix the length of the confidence interval in advance of taking the observations.

20.32 The possibility of finding confidence intervals in this case arose from our being able to find a statistic z, depending only on the parameter under estimate, whose distribution did not contain σ. A scale parameter can often be eliminated in this way, although the resulting distributions are not always easy to handle. If, for instance, we have a statistic t which is of degree p in the variables, then t/s^p is of degree zero, and its distribution must be independent of the scale parameter. When a statistic is reduced to independence of the scale in this way it is said to be " studentized," after " Student " (W. S. Gosset), who was the first to perceive the significance of the process.

20.33 It is interesting to consider the relation between the studentized mean-statistic and confidence zones based on sufficient statistics in the normal case. The joint distribution of mean and variance in normal samples is (Example 11.7)
$$dF = \left(\frac{n}{2\pi\sigma^2}\right)^{\frac{1}{2}} \exp\left\{-\frac{n}{2\sigma^2}(\bar{x}-\mu)^2\right\} d\bar{x} \frac{k}{\sigma^{n-1}} s^{n-3} \exp\left\{-\frac{ns^2}{2\sigma^2}\right\} ds^2 \qquad (20.67)$$
and $\bar{x}, s$ are jointly sufficient (Example 17.17). In the sample space W the regions of constant $\bar{x}$ are hyperplanes and those of constant s are hyperspheres. If we fix $\bar{x}$ and s the sample-point E lies on a hypersphere of $(n-2)$ dimensions (Example 11.7). Choose a region on this hypersphere of content $1-\alpha$. Then the confidence zone A will be obtained by combining all such areas for all $\bar{x}$ and s.

One such region is seen to be the " slice " of the sample-space obtained by rotating the hyperplane passing through the origin and the point $(1, 1, \ldots, 1)$ through an angle $\pi(1-\alpha)$ (not $2\pi(1-\alpha)$ because a half-turn of the plane covers the whole space). The situation is illustrated for $n = 2$ in Fig. 20.8.

For any given μ' the axis of rotation meets the hyperplane $\mu = \mu'$ in the point $x_1 = x_2 = \mu'$, and the hypercones $(\bar{x}-\mu)/s = $ constant in the W space become the plane areas between two straight lines (shaded in the figure). A set of regions A is obtained by rotating a plane about the line $x_1 = x_2 = \mu$ through an angle so as to cut off in any plane $\mu = \mu'$ an angle $\frac{1}{2}\pi(1-\alpha)$ on each side of
$$x_1-\mu' = x_2-\mu'.$$

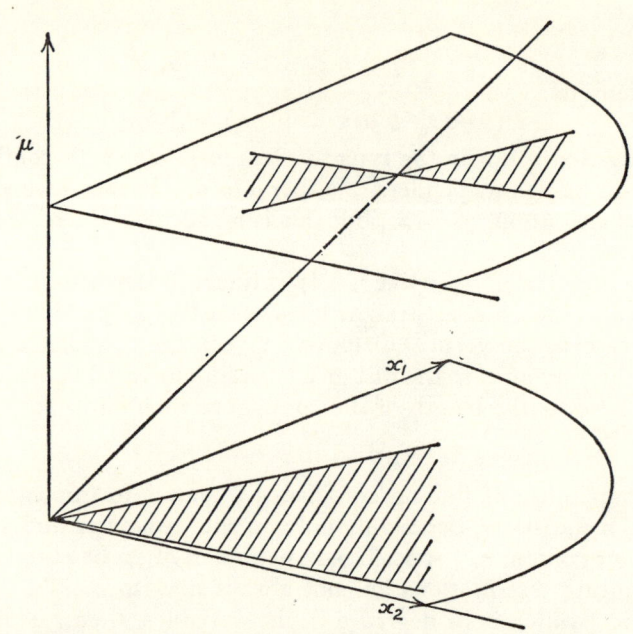

Fig. 20.8—Confidence intervals based on "Student's" t for $n = 2$ (see text)

The boundary planes are given by
$$x_1 - \mu = (x_2 - \mu) \tan(\tfrac{1}{4}\pi - \tfrac{1}{2}\beta),$$
$$x_1 - \mu = (x_2 - \mu) \tan(\tfrac{1}{4}\pi + \tfrac{1}{2}\beta),$$
where $\beta = \pi\alpha$; or, after a little reduction,
$$\mu = \tfrac{1}{2}(x_1 + x_2) + \tfrac{1}{2}(x_1 - x_2) \cot \tfrac{1}{2}\beta,$$
$$\mu = \tfrac{1}{2}(x_1 + x_2) - \tfrac{1}{2}(x_1 - x_2) \cot \tfrac{1}{2}\beta.$$

μ then lies in the region of acceptance if
$$\tfrac{1}{2}(x_1 + x_2) - \tfrac{1}{2}|x_1 - x_2| \cot \tfrac{1}{2}\beta \leqslant \mu \leqslant \tfrac{1}{2}(x_1 + x_2) + \tfrac{1}{2}|x_1 - x_2| \cot \tfrac{1}{2}\beta.$$

These are, in fact, the limits given by "Student's" distribution for $n = 2$, since the sample standard deviation then becomes $\tfrac{1}{2}|x_1 - x_2|$ and
$$\frac{1}{\pi}\int_{z_0}^{\infty} \frac{dz}{1 + z^2} = \frac{1}{\pi}(\tfrac{1}{2}\pi - \tan^{-1} z_0) = \alpha/2 = \beta/(2\pi)$$
so that
$$z_0 = \tan(\tfrac{1}{2}\pi - \tfrac{1}{2}\beta) = \cot \tfrac{1}{2}\beta.$$

20.34 As a further example of the use of studentization in setting confidence intervals, and the results to which it may lead, we consider Example 20.7.

Example 20.7—Estimation of the ratio of means of two normal variables

Let x, y be bivariate normally distributed as in Example 18.14, all five parameters of the distribution being unknown, as in case (c) of that Example. We make a convenient re-parametrization and write $\rho\sigma_1\sigma_2 = \sigma_{12}$ in order to use the covariance σ_{12} as the fifth parameter instead of the correlation coefficient ρ.

From a sample of n observations on (x, y), we now wish to estimate the ratio of means $\theta = \mu_2/\mu_1$, where we assume $\mu_1 \neq 0$. From Example 18.14(c) it follows from the fact that ML estimators are unaffected by re-parametrization (cf. **8.9** and **18.14**) that the ML estimator of σ_{12} is

$$\hat{\sigma}_{12} = \hat{\sigma}_1 \hat{\sigma}_2 \hat{\rho} = \frac{1}{n} \Sigma (x - \bar{x})(y - \bar{y}) = s_{12},$$

say. Similarly, we now write s_1^2, s_2^2 for the ML estimators of σ_1^2, σ_2^2 found at (18.78). The ML estimator of θ, which is our primary interest here, is $\hat{\theta} = \hat{\mu}_2/\hat{\mu}_1 = \bar{y}/\bar{x}$. It is intuitively clear that if $|\bar{x}|$ is small, we must expect poor precision in estimating θ. We now turn to the problem of finding confidence intervals for θ, where we shall see that if $\bar{x}^2/s_1^2$ is small, difficulties of the kind discussed in **20.12–14** arise.

Consider the new variable $z_i = y_i - \theta x_i$ $(i = 1, 2, \ldots, n)$, which is normally distributed (since it is a linear function of bivariate normal variables—cf. **15.4**) with zero expectation. The sample mean of the z_i is

$$\bar{z} = \bar{y} - \theta \bar{x},$$

and also has zero mean. For fixed θ, the variance of $\bar{z}$ is

$$V_\theta(\bar{z}) = \sigma_z^2/n = \frac{1}{n}(\sigma_2^2 - 2\theta\sigma_{12} + \theta^2 \sigma_1^2)$$

with unbiased estimator

$$\hat{V}_\theta = \frac{1}{n(n-1)} \sum_{i=1}^{n} (z_i - \bar{z})^2 = \frac{1}{n-1}(s_2^2 - 2\theta s_{12} + \theta^2 s_1^2)$$

and $\hat{V}_\theta/V_\theta(\bar{z})$ is distributed, independently of $\bar{z}$, as a $\chi^2_{n-1}/(n-1)$ variable. Thus

$$\bar{z}/\hat{V}_\theta^{\frac{1}{2}} = (\bar{y} - \theta\bar{x}) \bigg/ \left\{\frac{1}{n-1}(s_2^2 - 2\theta s_{12} + \theta^2 s_1^2)\right\}^{\frac{1}{2}} \tag{20.68}$$

has a Student's t-distribution with $\nu = n-1$ degrees of freedom, a result due to Fieller (1940). Equivalently, $\bar{z}^2/\hat{V}_\theta$ has a t^2 distribution, i.e. an F-distribution with $(1, n-1)$ degrees of freedom (cf. **16.15**). From the tables of Student's t, we may now find critical values t_α such that, for any θ,

$$P\left\{\frac{\bar{z}^2}{\hat{V}_\theta} \leq t_\alpha^2\right\} = 1 - \alpha. \tag{20.69}$$

We now examine the nature of the confidence intervals for θ which emerge from (20.69).

The event $\bar{z}^2/\hat{V}_\theta \leq t_\alpha^2$ in (20.69) can be re-written in the form of a quadratic inequality in θ

$$\left(\bar{x}^2 - \frac{t_\alpha^2}{n-1} s_1^2\right)\theta^2 - 2\left(\bar{x}\bar{y} - \frac{t_\alpha^2}{n-1} s_{12}\right)\theta + \left(\bar{y}^2 - \frac{t_\alpha^2}{n-1} s_2^2\right) \leq 0 \tag{20.70}$$

and the problem is to determine the values of θ that satisfy this inequality—these will constitute the $100(1-\alpha)$ per cent confidence interval for θ.

The left-hand side of (20.70) is a parabola in θ, of standard type $a\theta^2 - 2b\theta + c$. We require the values of θ for which this parabola lies on or below the θ-axis, and we shall see that not only the confidence interval, but also its *form*, will change with the relationships between the randomly varying coefficients a, b, c. We ignore the degenerate case $a = 0$, which occurs with probability zero.

The parabola has its unique turning-point at $\theta = b/a$, where the value of the left-hand side of (20.70) is seen to be $(ac-b^2)/a$. If $a > 0$, the turning-point is a minimum, and if the minimum value $(ac-b^2)/a$ were positive, there would be no value of θ at all which satisfied (20.70), the parabola having no real roots since $b^2 < ac$. This absurd situation cannot arise, however, for we see that the value $\theta = \hat{\theta} = \bar{y}/\bar{x}$ always satisfies (20.70), whose left-hand side upon this substitution becomes equal to $-t_\alpha^2 \hat{V}_{\hat{\theta}}$, which cannot be positive. There are thus always some values of θ satisfying (20.70), and if $a > 0$ we must also have $b^2 \geqslant ac$, with real roots of the parabola at

$$\theta = \{b \pm (b^2 - ac)^{\frac{1}{2}}\}/a. \qquad (20.71)$$

(20.70) will then be satisfied by all values of θ lying between these roots, in the finite interval determined by (20.71)—the reader should draw the parabola roughly to help him see this. If $b^2 = ac$, the interval contains only the point $\theta = b/a$, which then must be equal to $\hat{\theta} = \bar{y}/\bar{x}$.

On the other hand, if $a < 0$, the turning-point of the parabola is a maximum. If also $b^2 \geqslant ac$, so that (20.71) gives the real roots as before, we see that (20.70) is satisfied by all values of θ lying *outside* the finite interval defined by (20.71), i.e. by its infinite complement on the real line—again, a drawing may help. Further, if $a < 0$ and $b^2 < ac$, the whole of the parabola lies below the θ-axis, and (20.71) is satisfied by any real value of θ. These results, given by Fieller (1954), are less startling than they first seem. If the confidence interval for θ lies outside a finite interval, this can be looked on loosely as saying that θ^{-1} lies inside one. (Nevertheless, this does not imply that we only need to work in terms of θ^{-1} to avoid the problem—the coefficient c in (20.70) is the same function of the y's as a there is of the $\bar{x}$'s, while b is symmetric in x and y. We could thus find $a < 0$ and $c < 0$ together, with both θ and θ^{-1} poorly determined by confidence intervals.) Similarly, it is not so difficult to accept a confidence interval consisting of all possible values of θ if we reflect that this simply means that the observations are completely uninformative with respect to θ.

Finally, we should observe that the condition $a < 0$ which led to our difficulties is, from (20.70), simply $\dfrac{\bar{x}^2}{s_1^2} < \dfrac{t_\alpha^2}{n-1}$, so that, as we anticipated at the beginning of this Example, it is the relative smallness of $|\bar{x}|$ that produces the trouble.

> Scheffé (1970a) avoids the whole-real-line confidence interval by reformulating the problem slightly. In (20.70), he replaces $t_\alpha^2 = F_\alpha(1, n-1)$ by a positive monotone function of an F-statistic which decreases from $2F$ at $F = F_\alpha(2, n-1)$ to $F_\alpha(1, n-1)$ as $F \to \infty$. As a result, one obtains *either* a finite interval or its complement for θ or an elliptical region for (μ_1, μ_2), with preassigned overall probability.

Simultaneous confidence intervals for several parameters

20.35 Cases fairly frequently arise in which we wish to estimate more than one parameter of a population, for example the mean and variance. The extension of the theory of confidence intervals for one parameter to this case of two or more parameters is a matter of very considerable difficulty. What we should like to be able to do, given,

say, two parameters θ_1 and θ_2 and two statistics t and u, is to make simultaneous interval assertions of the type
$$P\{t_0 \leq \theta_1 \leq t_1 \text{ and } u_0 \leq \theta_2 \leq u_1\} = 1-\alpha. \qquad (20.72)$$
This, however, is rarely possible. Sometimes we can make a statement giving a *confidence region* for the two parameters together, e.g. such as
$$P\{w_0 \leq \theta_1^2 + \theta_2^2 \leq w\} = 1-\alpha. \qquad (20.73)$$
But this is not entirely satisfactory; we do not know, so to speak, how much of the uncertainty of the region to assign to each parameter. It may be that, unless we are prepared to lay down some new rule on this point, the problem of locating the parameters in separate intervals is insoluble.

Even for large samples the problems are severe. We may then find that we can determine intervals of the type
$$P\{t_0(\theta_2) \leq \theta_1 \leq t_1(\theta_2)\} = 1-\alpha$$
and substitute a (large sample) estimate of θ_2 in the limits $t_0(\theta_2)$ and $t_1(\theta_2)$. This is very like the familiar procedure in the theory of standard errors, where we replace parameters occurring in the error variances by estimates obtained from the samples.

20.36 We shall not attempt to develop the theory of simultaneous confidence intervals any further here. The reader who is interested may consult papers by S. N. Roy and Bose (1953) and S. N. Roy (1954) on the theoretical aspect. Bartlett (1953, 1955) discussed the generalization of the method of **20.15** to the case of two or more unknown parameters. Beale (1960), Halperin and Mantel (1963) and Halperin (1964) consider intervals for non-linear functions of parameters, especially in large samples.

The theorem of **20.17** concerning shortest intervals was generalized by Wilks and Daly (1939). Under fairly general conditions the large-sample regions for l parameters which are smallest on the average are given by
$$\sum_{i=1}^{l} \sum_{j=1}^{l} \left\{ I_{ij}^{-1} \frac{\partial \log L}{\partial \theta_i} \frac{\partial \log L}{\partial \theta_j} \right\} \leq \chi_\alpha^2 \qquad (20.74)$$
where $\mathbf{I}^{-1}$ is the inverse matrix to the information matrix whose general element is
$$I_{ij} = E\left(\frac{\partial \log L}{\partial \theta_i} \frac{\partial \log L}{\partial \theta_j}\right)$$
and χ_α^2 is such that $P(\chi^2 \leq \chi_\alpha^2) = 1-\alpha$, the probability being calculated from the χ^2 distribution with l degrees of freedom. This is clearly related to the result of **17.39** giving the minimum attainable variances (and, by a simple extension, covariances) of a set of unbiassed estimators of several parametric functions.

In Volume 3, when we discuss the Analysis of Variance, we shall meet the problem of simultaneously setting confidence intervals for a number of means.

Tolerance intervals

20.37 Throughout this chapter we have been discussing the setting of confidence intervals for the parameters entering explicitly into the specification of a distribution. But the technique of confidence intervals can be used for other problems. We shall see in later chapters that intervals can be found for the quantiles of a parent distribution and also for the entire distribution function itself, without any assumption on the form of the distribution beyond its continuity. There is another type of problem, commonly met with in practical sampling, which may be solved by these methods. Suppose that, on the basis of a sample of n independent observations from a distribution, we wish to find two limits, L_1 and L_2, between which at least a given proportion γ of the distribution may be asserted to lie. Clearly, we can only make such an assertion in probabilistic form, i.e. we assert that, with given probability β, at least a proportion γ of the distribution lies between L_1 and L_2. L_1 and L_2 are called *tolerance limits* for the distribution; we shall call them the (β, γ) tolerance limits. The interval (L_1, L_2) is called a *tolerance interval*. In **32.11-13** we shall see that tolerance limits, also, may be set without assumptions (except continuity) on the form of the parent distribution. In this chapter, however, we shall discuss the derivation of tolerance limits for a normal distribution, due to Wald and Wolfowitz (1946).

20.38 Since the sample mean and variance are a pair of sufficient statistics for the parameters of a normal distribution (Example 17.17), it is natural to base tolerance limits for the distribution upon them. In a sample of size n, we work with the unbiassed statistics

$$\bar{x} = \Sigma x/n, \quad s'^2 = \Sigma(x-\bar{x})^2/(n-1),$$

and define
$$A(\bar{x}, s', \lambda) = \int_{\bar{x}-\lambda s'}^{\bar{x}+\lambda s'} f(t)\,dt, \qquad (20.75)$$

where $f(t)$ is the normal frequency function. We now seek to determine the value λ so that

$$P\{A(\bar{x}, s', \lambda) > \gamma\} = \beta. \qquad (20.76)$$

$L_1 = \bar{x} - \lambda s'$ and $L_2 = \bar{x} + \lambda s'$ will then be a pair of central (β, γ) tolerance limits for the parent distribution. Since we are concerned only with the proportion of that distribution covered by the interval (L_1, L_2), we may without any loss of generality standardize the population mean at 0 and its variance at 1. Thus

$$f(t) = (2\pi)^{-\frac{1}{2}} \exp\left(-\tfrac{1}{2}t^2\right). \qquad (20.77)$$

20.39 Consider first the conditional probability, given $\bar{x}$, that $A(\bar{x}, s', \lambda)$ exceeds γ. We denote this by $P\{A > \gamma \mid \bar{x}\}$. Now A is a monotone increasing function of s', and the equation in s'

$$A(\bar{x}, s', \lambda) = \gamma \qquad (20.78)$$

has just one root, which we denote by $s'(\bar{x}, \gamma, \lambda)$. Let

$$\lambda s'(\bar{x}, \gamma, \lambda) = r(\bar{x}, \gamma). \qquad (20.79)$$

Given $\bar{x}$ and γ, $r = r(\bar{x}, \gamma)$ is obtainable from a table of the normal distribution, since

$$\int_{\bar{x}-r}^{\bar{x}+r} f(t)\,dt = \gamma. \qquad (20.80)$$

From (20.80) it is clear that r does not depend upon λ. Moreover, since A is monotone increasing in s', the inequality $A > \gamma$ is equivalent to

$$s' > s'(\bar{x}, \gamma, \lambda) = r(\bar{x}, \gamma)/\lambda.$$

Thus we may write
$$P\{A > \gamma \mid \bar{x}\} = P\left\{s' > \frac{r}{\lambda} \mid \bar{x}\right\}, \qquad (20.81)$$
and since $\bar{x}$ and s' are independently distributed, (20.81) becomes
$$P\{A > \gamma \mid \bar{x}\} = P\{(n-1)s'^2 > (n-1)r^2/\lambda^2\}. \qquad (20.82)$$
Since $(n-1)s'^2 = \Sigma(x-\bar{x})^2$ is distributed as χ^2 with $(n-1)$ degrees of freedom, we have
$$P\{A > \gamma \mid \bar{x}\} = P\{\chi^2_{n-1} > (n-1)r^2/\lambda^2\}. \qquad (20.83)$$

20.40 To obtain the unconditional probability $P(A > \gamma) = \beta$ from (20.83), we must integrate it over the distribution of $\bar{x}$, which is normal with zero mean and variance $1/n$. This is a tedious numerical operation, but fortunately an excellent approximation is available. We expand $P(A > \gamma \mid \bar{x})$ in a Taylor series about $\bar{x} = \mu = 0$, and since it is an even function of $\bar{x}$, the odd powers in the expansion vanish,[*] leaving
$$P(A > \gamma \mid \bar{x}) = P(A > \gamma \mid 0) + \frac{\bar{x}^2}{2!} P''(A > \gamma \mid 0) + \ldots \qquad (20.84)$$
Integrating over the distribution of $\bar{x}$, we have, using the moments of $\bar{x}$,
$$P(A > \gamma) = P(A > \gamma \mid 0) + \frac{1}{2n} P''(A > \gamma \mid 0) + O(n^{-2}) \ldots \qquad (20.85)$$
But from (20.84) with $\bar{x} = 1/\sqrt{n}$, we also have
$$P\left(A > \gamma \mid \frac{1}{\sqrt{n}}\right) = P\{A > \gamma \mid 0\} + \frac{1}{2n} P''(A > \gamma \mid 0) + O(n^{-2}). \qquad (20.86)$$
(20.85) and (20.86) give the approximation
$$P(A > \gamma) \doteqdot P\left(A > \gamma \mid \frac{1}{\sqrt{n}}\right) \qquad (20.87)$$
and (20.76), (20.87) and (20.83) yield finally
$$\beta = P\{A > \gamma\} \doteqdot P\{\chi^2_{n-1} > (n-1)r^2/\lambda^2\}, \qquad (20.88)$$
where r is defined from (20.80) by $\int_{n^{-\frac{1}{2}}-r}^{n^{-\frac{1}{2}}+r} f(t)\,d(t) = \gamma$. Given γ, β, n and $\bar{x}$, we can determine λ approximately from (20.88), and hence the tolerance limits $\bar{x} \pm \lambda s'$. Wald and Wolfowitz (1946) showed that the approximation is extremely good even for values of n as low as 2 if β and γ are ≥ 0.95, as they usually are in practice. Bowker (1947) gives tables of λ (his k) for β (his γ) = 0.75, 0.90, 0.95, 0.99 and γ (his P) = 0.75, 0.90, 0.99 and 0.999, for sample sizes n = 2 (1) 102 (2) 180 (5) 300 (10) 400 (25) 750 (50) 1000.

On examination of the argument above, it will be seen to hold if $\bar{x}$ is replaced by any estimator $\hat{\mu}$ of the mean, and s'^2 by any independent estimator $\hat{\sigma}^2$ of the variance, of a normal population, as pointed out by Wallis (1951). If the mean is estimated from n observations and the variance estimate has ν degrees of freedom, Ellison (1964) has shown that the approximation corresponding to (20.88) is valid only to order ν/n^2 (reducing

---

[*] This is because the interval is symmetric about $\bar{x}$; it could not happen otherwise.

to $1/n$ for (20.88) itself, where $v = n-1$), and Howe (1969) gives better tolerance limits for the case where v/n^2 is large. In the contrary case, there are some useful tables. Taguti (1958) gives tables of λ (his k) for β (his $1-\alpha$) and γ (his P) = 0·90, 0·95 and 0·99; and n = 0·5 (0·5) 2 (1) 10 (2) 20 (5) 30 (10) 60 (20) 100, 200, 500, 1000, ∞; v = 1 (1) 20 (2) 30 (5) 100 (100) 1000, ∞. The small fractional values of n are useful in some applications discussed by Taguti. Weissberg and Beatty (1960) give tables of λ/r (their u) for v (their f) = 1 (1) 150 (2) 250 (5) 500 (10) 1000 (1000) 10,000, ∞ and β (their γ) = 0·90, 0·95, 0·99; and of r for n = 1 (1) 100 (5) 200 (10) 300 (20) 500 (100) 1000 (1000) 10,000, ∞ and γ (their P) = 0·5, 0·75, 0·9, 0·95, 0·99, 0·999.

Fraser and Guttman (1956) and Guttman (1957) consider tolerance intervals which cover a given proportion of a normal parent distribution *on the average*. Sharpe (1970) studies the robustness of both kinds of tolerance intervals.

> Zacks (1970) considers tolerance limits, and their relationship with confidence limits, for a large class of discrete distributions including the binomial, the negative binomial, and the Poisson.

EXERCISES

20.1 For a sample of n from the distribution
$$dF = \frac{x^{p-1}e^{-x/\theta}}{\Gamma(p)\theta^p} dx, \qquad 0 \leq x \leq \infty, \; p > 0,$$
we have seen (Exercise 17.1) that, for known p, a sufficient statistic for θ is $\bar{x}/p$. Hence derive confidence intervals for θ.

20.2 Show that for the rectangular population
$$dF = dx/\theta, \qquad 0 \leq x \leq \theta$$
and confidence coefficient $1-\alpha$, confidence limits for θ are t and t/ψ, where t is the sample range and ψ is given by
$$\psi^{n-1}\{n-(n-1)\psi\} = \alpha.$$
(Wilks, 1938c)

20.3 Show that, for the distribution of the previous exercise, confidence limits for θ from samples of two, x_1 and x_2, are
$$(x_1+x_2)/[1 \pm \{1-(1-\alpha)^{\frac{1}{2}}\}].$$
(Neyman, 1937b)

20.4 In Exercise 20.2, show also that if L is the larger of a sample of size two, confidence limits for θ are
$$L, \quad L/\sqrt{\alpha}$$
and that if M is the largest of samples of size four, limits are
$$M, \quad M/\alpha^{\frac{1}{4}}.$$
(Neyman, 1937b)

20.5 Using the asymptotic multivariate normal distribution of Maximum Likelihood estimators (**18.26**) and the χ^2 distribution of the exponent of a multivariate normal distribution (**15.10**), show that (20.74) gives a large-sample confidence region for a set of parameters. From it, derive a confidence region for the mean and variance of a univariate normal distribution.

20.6 In setting confidence limits to the variance of a normal population by the use of the distribution of the sample variance (Example 20.6), sketch the confidence belts for some value of the confidence coefficient, and show graphically that they always provide a connected range within which σ^2 is located.

20.7 Show how to set confidence limits to the ratio of variances σ_1^2/σ_2^2 in two normal populations, based on independent samples of n_1 observations from the first and n_2 observations from the second. (Use the distribution of the ratio of sample variances at (16.24).)

20.8 Use the method of **20.10** to show that large-sample 95 per cent confidence limits for ϖ in the binomial distribution of Example 20.2 are given by
$$\frac{1}{1+(1\cdot 96)^2/n}\left\{p+\frac{(1\cdot 96)^2}{2n} \pm 1\cdot 96\sqrt{\left(\frac{p(1-p)}{n}+\frac{(1\cdot 96)^2}{4n^2}\right)}\right\}.$$

20.9 Using Geary's theorem (Exercise 11.11), show that large-sample 95 per cent. confidence limits for the ratio ϖ_2/ϖ_1 of the parameters of two binomial distributions based on independent samples of size n_2, n_1 respectively, are given by

$$\frac{p_2/p_1}{1+(1\cdot 96)^2/n_2}\left\{1+\frac{(1\cdot 96)^2}{2n_2 p_2}+1\cdot 96\sqrt{\left[\frac{1-p_1}{n_1 p_1}+\frac{1-p_2}{n_2 p_2}+\frac{(1\cdot 96)^2}{4}\left(\frac{1}{n_2^2 p_2^2}+\frac{4(1-p_1)}{n_1 n_2 p_1}\right)\right]}\right\}.$$

(Noether, 1957)

20.10 In Example 20.6, show that the confidence interval based on

$$P\left\{\frac{ns^2}{\chi_0^2}\leqslant\sigma^2\leqslant\frac{ns^2}{\chi_1^2}\right\}=1-\alpha$$

(where χ_0^2 and χ_1^2 are the upper and lower $\tfrac{1}{2}\alpha$ points of the χ^2 distribution with $(n-1)$ d.f.) is not the physically shortest interval for σ^2 based on the χ^2 distribution of ns^2/σ^2.

(cf. Tate and Klett, 1959)

20.11 From two normal populations with means μ_1 and μ_2 and variances $\sigma_1^2=\sigma_2^2=\sigma^2$, independent samples of sizes n_1 and n_2 respectively are drawn. Show that

$$t=\{(\bar{x}_1-\mu_1)-(\bar{x}_2-\mu_2)\}\bigg/\left\{\frac{n_1 s_1^2+n_2 s_2^2}{n_1+n_2-2}\left(\frac{1}{n_1}+\frac{1}{n_2}\right)\right\}^{\frac{1}{2}}$$

(where $\bar{x}_1$, $\bar{x}_2$ and s_1^2, s_2^2 are the sample means and variances) is distributed in "Student's" distribution with (n_1+n_2-2) d.f., and hence set confidence limits for $(\mu_1-\mu_2)$.

20.12 In Exercise 20.11, if $\sigma_1^2\neq\sigma_2^2$, show that the ratio distributed in "Student's" distribution is no longer t, but

$$t'=\frac{(\bar{x}_1-\mu_1)-(\bar{x}_2-\mu_2)}{(\sigma_1^2/n_1+\sigma_2^2/n_2)^{\frac{1}{2}}}\bigg/\left\{\left(\frac{n_1 s_1^2}{\sigma_1^2}+\frac{n_2 s_2^2}{\sigma_2^2}\right)\bigg/(n_1+n_2-2)\right\}^{\frac{1}{2}}.$$

20.13 If $f(x\,|\,\theta)=g(x)/h(\theta)$, $(a(\theta)\leqslant x\leqslant b(\theta))$, and $b(\theta)$ is a monotone decreasing function of $a(\theta)$, show (cf. 17.40-1) that the extreme observations $x_{(1)}$ and $x_{(n)}$ are a pair of jointly sufficient statistics for θ. From their joint distribution, show that the single sufficient statistic for θ,

$$\hat{\theta}=\min\{a^{-1}(x_{(1)}),\,b^{-1}(x_{(n)})\},$$

has distribution

$$dF=\frac{n\{h(\hat{\theta})\}^{n-1}}{\{h(\theta)\}^n}\{-h'(\hat{\theta})\}\,d\hat{\theta},\qquad \theta\leqslant\hat{\theta}\leqslant\theta^*,$$

where θ^* is defined by $a(\theta^*)=b(\theta^*)$.

20.14 In Exercise 20.13, show that $\psi=h(\hat{\theta})/h(\theta)$ has distribution

$$dF=n\psi^{n-1}\,d\psi,\qquad 0\leqslant\psi\leqslant 1.$$

Show that

$$P\{\alpha^{1/n}\leqslant\psi\leqslant 1\}=1-\alpha,$$

and hence set a confidence interval for θ. Show that this is shorter than any other interval based on the distribution of ψ.

(Huzurbazar, 1955)

20.15 Apply the result of Exercise 20.14 to show that a confidence interval for θ in

$$dF=dx/\theta,\qquad 0\leqslant x\leqslant\theta,$$

is obtainable from
$$P\{x_{(n)} \leqslant \theta \leqslant x_{(n)}\alpha^{-1/n}\} = 1-\alpha$$
and that this is shorter than the interval in Exercise 20.2.

20.16 Use the result of Exercise 20.14 to show that a confidence interval for θ in
$$dF = e^{-(x-\theta)}\,dx, \qquad \theta \leqslant x \leqslant \infty$$
is obtainable from
$$P\left\{x_{(1)} + \frac{1}{n}\log\alpha \leqslant \theta \leqslant x_{(1)}\right\} = 1-\alpha.$$

(Huzurbazar, 1955)

20.17 If $I(x)$ is a confidence interval for θ calculated from the distribution of a sample, $f(x\mid\theta)$, and θ_0 is the true value of θ, show that the expected length of $I(x)$ may be written as
$$E(L) = \int\left\{\int_{\theta\in I(x)} d\theta\right\} dF(x\mid\theta_0)$$
and that
$$E(L) = \int_{\theta\neq\theta_0} \text{prob}\,\{\theta\in I(x)\mid\theta_0\}\,d\theta,$$
the integral over all false values of θ of the probability of inclusion in the confidence interval.

(Pratt, 1961, 1963)

20.18 If $x \in A(\theta)$ if and only if $\theta \in I(x)$ in Exercise 20.17, show that $E(L)$ is minimized by choosing $A(\theta)$ for each θ so that prob $\{x \in A(\theta) \mid \theta_0\}$ is minimized. (This is equivalent to choosing the most powerful test for each θ against the alternative value θ_0—cf. **23.26** below.)

(Pratt, 1961, 1963)

20.19 x and y have a bivariate normal distribution with variances σ_1^2, σ_2^2, and correlation parameter ρ. Show that the variables
$$u = \frac{x}{\sigma_1} + \frac{y}{\sigma_2}, \qquad v = \frac{x}{\sigma_1} - \frac{y}{\sigma_2},$$
are independently normally distributed. In a sample of n observations with sample variances s_x^2 and s_y^2 and correlation coefficient r_{xy}, show that the sample correlation coefficient of u and v may be written
$$r_{uv}^2 = \frac{(l-\lambda)^2}{(l+\lambda)^2 - 4r_{xy}^2 l\lambda},$$
where $l = s_1^2/s_2^2$ and $\lambda = \sigma_1^2/\sigma_2^2$. Hence show that, whatever the value of ρ, confidence limits for λ are given by
$$l\{K-(K^2-1)^{\frac{1}{2}}\}, \quad l\{K+(K^2-1)^{\frac{1}{2}}\}$$
where
$$K = 1 + \frac{2(1-r_{xy}^2)}{n-2} t_\alpha^2$$
and t_α^2 is the 100α per cent point of "Student's" t^2 distribution.

(Pitman, 1939a)

20.20 In **20.39**, show that $r(\bar{x}, y)$ defined at (20.80) is, asymptotically in n,
$$r(\bar{x}, y) \sim r(0, y)\left(1 + \frac{1}{2n}\right).$$

(Bowker, 1946)

20.21 Using the method of Example 6.4, show that for a χ^2 distribution with ν degrees of freedom, the value above which 100β per cent of the distribution lies is χ_β^2 where

$$\frac{\chi_\beta^2}{\nu} \sim 1 + \left(\frac{2}{\nu}\right)^{\frac{1}{2}} d_{1-\beta} + \frac{2}{3\nu}(d_{1-\beta}^2 - 1) + o\left(\frac{1}{\nu}\right),$$

where

$$\int_{-\infty}^{d_\alpha} (2\pi)^{-\frac{1}{2}} \exp\left(-\tfrac{1}{2}t^2\right) dt = \alpha.$$

20.22 Combine the results of Exercises 20.20–20.21 to show that, from (20.83),

$$\lambda \sim r(0, \gamma) \left\{ 1 + \frac{d_\beta}{(2n)^{\frac{1}{2}}} + \frac{5(d_\beta^2 + 2)}{12n} \right\}.$$

(Bowker, 1946)

CHAPTER 21

INTERVAL ESTIMATION: OTHER METHODS

21.1 At the outset of this chapter it is desirable to make a few remarks on matters of terminology. Problems of interval estimation in the precise sense began to engage the attention of statisticians round about the period 1925–1930. The approach from confidence intervals, as we have defined them in the previous chapter, and that from fiducial intervals, which we shall now expound and compare with other methods, were presented respectively by J. Neyman and by R. A. Fisher; and since they seemed to give identical results there was at first a very natural belief that the two methods were only saying the same things in different terms. In consequence, the earlier literature of the subject often contains references to "fiducial" intervals in the sense of our "confidence" intervals; and (less frequently) to "confidence" intervals in some sense more nearly related to the "fiducial" line of argument.

Although this confusion of nomenclature has never been adequately cleared up, it is now generally recognized that fiducial intervals are different in kind from confidence intervals. But their devotees have, so it seems to us, not always made it quite clear where the difference lies; nor have they always used the term "fiducial" in strict conformity with the usage of Fisher, who, having invented it, may be allowed the right of precedence by way of definition. We shall present what we believe to be the basic ideas of the fiducial approach, but the reader who goes to the original literature may expect to find considerable variation in terminology.

21.2 To fix the ideas, consider a sample of size n from a normal population of unknown mean, μ, and unit variance. The sample mean $\bar{x}$ is a sufficient statistic for μ, and its distribution is

$$dF = \sqrt{\left(\frac{n}{2\pi}\right)} \exp\{-\tfrac{1}{2}n(\bar{x}-\mu)^2\}\, d\bar{x}. \tag{21.1}$$

(21.1), of course, expresses the distribution of different values of $\bar{x}$ for a fixed unknown value of μ. Now suppose that we have a single sample of n observations, yielding a sample mean $\bar{x}_1$. We recall from (17.68) that the Likelihood Function of the sample, $L(\bar{x}_1|\mu)$, will (since $\bar{x}$ is sufficient for μ) depend on μ only through the distribution of $\bar{x}$ at (21.1), which may therefore be taken to represent the Likelihood Function. Thus

$$L(\bar{x}_1|\mu) \propto \sqrt{\left(\frac{n}{2\pi}\right)} \exp\{-\tfrac{1}{2}n(\bar{x}_1-\mu)^2\}. \tag{21.2}$$

If we are prepared, perhaps somewhat intuitively, to use the Likelihood Function (21.2) as measuring the intensity of our credence in a particular value of μ, we finally write

$$dF = \left(\sqrt{\frac{n}{2\pi}}\right) \exp\{-\tfrac{1}{2}n(\bar{x}_1-\mu)^2\}\, d\mu, \tag{21.3}$$

which we shall call the *fiducial distribution* of the parameter μ. We note that the integral of (21.3) over the range $(-\infty, \infty)$ for μ is 1, so that no constant adjustment is necessary.

21.3 This fiducial distribution is not a frequency distribution in the sense in which we have used the expression hitherto. It is a new concept, expressing the intensity of our belief in the various possible values of a parameter. It so happens, in this case, that the non-differential element in (21.3) is the same as that in (21.1). This is not essential, though it is not infrequent.

Nor is the fiducial distribution a probability distribution in the sense of the frequency theory of probability. It may be regarded as a distribution of probability in the sense of degrees of belief; the consequent link with interval estimation based on the use of Bayes' theorem will be discussed below. Or it may be regarded as a new concept, giving formal expression to our somewhat intuitive ideas about the extent to which we place credence in various values of μ.

21.4 The fiducial distribution can now be used to determine intervals within which μ is located. We select some arbitrary numbers, say 0·02275 and 0·97725, and decide to regard those values as critical in the sense that any acceptable value of μ must not give to the observed $\bar{x}_1$ a (cumulative) probability less than 0·02275 or greater than 0·97725. Then, since these values correspond to deviations of $\pm 2\sigma$ from the mean of a normal distribution, and $\sigma = 1/\sqrt{n}$, we have

$$-2 \leq (\bar{x}_1 - \mu)\sqrt{n} \leq 2,$$

which is equivalent to

$$\bar{x}_1 - 2/\sqrt{n} \leq \mu \leq \bar{x}_1 + 2/\sqrt{n}. \tag{21.4}$$

This, as it happens, is the same inequality as that to which we were led by central confidence intervals based on (21.1) in Example 20.1. But it is essential to note that it is not reached by the same line of thought. The confidence approach says that if we assert (21.4) we shall be right in about 95·45 per cent of the cases *in the long run*. Under the fiducial approach the assertion of (21.4) is equivalent to saying that (in some sense not defined) we are 95·45 per cent sure of being right *in this particular case*. The shift of emphasis is evidently the one we encountered in considering the Likelihood Function itself, where the function $L(x|\theta)$ can be considered as an elementary probability in which θ is fixed and x varies, or as a likelihood in which x is fixed and θ varies. So here, we can make an inference about the range of θ either by regarding it as a constant and setting up containing intervals which are random variables, or by regarding the observations as fixed and setting up intervals based on some undefined intensity of belief in the values of the parameter generating those observations.

21.5 There is one further fundamental distinction between the two methods. We have seen in the previous chapter that in confidence theory it is possible to have different sets of intervals for the same parameter based on different statistics (although we naturally discriminate between the different sets, and chose the shortest or most selective set). This is explicitly ruled out in fiducial theory (even in the sense that we may choose central or non-central intervals for the same distribution when using both its tails). We must, in fact, use all the information about the parameter which the Likelihood Function contains. This implies that if we are to set limits to θ by a single statistic t, the latter must be sufficient for θ. (We also reached this conclusion from the standpoint of most selective confidence intervals in **20.30**.)

As we pointed out in **17.38**, there is always a *set* of jointly sufficient statistics for an unknown parameter, namely the n observations themselves. But this tautology offers little consolation: even a sufficient set of two statistics would be difficult enough to handle; a larger set is almost certainly practically useless. As to what should be done to construct an interval for a single parameter θ where a single sufficient statistic does not exist, writers on fiducial theory are for the most part silent.

21.6 Let $f(t, \theta)$ be a continuous frequency function and $F(t, \theta)$ the distribution function of a statistic t which is sufficient for θ. Consider the behaviour of f for some fixed t, as θ varies. Suppose also that we know beforehand that θ must lie in a certain range, which may in particular be $(-\infty, \infty)$. Take some critical probability $1-\alpha$ (analogous to a confidence coefficient) and let θ_α be the value of θ for which $F(t, \theta) = 1-\alpha$.

Now suppose also that over the permissible range of θ, $f(t_1, \theta)$ is a monotonic non-increasing function of θ for any t_1. Then for all $\theta \leqslant \theta_\alpha$ the observed t_1 has at least as high a probability density as $f(t_1, \theta_\alpha)$, and for $\theta > \theta_\alpha$ it has a lower probability density. We then choose $\theta \leqslant \theta_\alpha$ as our fiducial interval. It includes all those values of the parameter which give to the probability density a value greater than or equal to $f(t_1, \theta_\alpha)$.

21.7 If we require a fiducial interval of type

$$\theta_{\alpha_1} \leqslant \theta \leqslant \theta_{\alpha_2}$$

we look for two values of θ such that $f(t_1, \theta_{\alpha_1}) = f(t_1, \theta_{\alpha_2})$ and $F(t_1, \theta_{\alpha_2}) - F(t_1, \theta_{\alpha_1}) = 1-\alpha$. If, between these values, $f(t_1, \theta)$ is greater than the extreme values $f(t_1, \theta_{\alpha_1})$ or $f(t_1, \theta_{\alpha_2})$, and is less than those values outside it, the interval again comprises values for which the probability density is at least as great as the density at the critical points.

If the distribution of t is symmetrical this involves taking a range which cuts off equal tail areas on it. For a non-symmetrical distribution the tails are to be such that their total probability content is α; but the contents of the two tails are not equal. It is the extreme ordinates of the interval which must be equal. Similar considerations have already been discussed in connexion with central confidence intervals in **20.7**.

21.8 On this understanding, if our fiducial interval is increased by an element $d\theta$ at each end, the probability ordinate at the end decreases by $(\partial F(t_1, \theta)/\partial \theta) d\theta$. For the fiducial distribution we then have

$$dF = -\frac{\partial F(t_1, \theta)}{\partial \theta} d\theta. \tag{21.5}$$

This formula, however, requires that $f(t_1, \theta)$ shall be a non-decreasing function of θ at the lower end and a non-increasing function of θ at the upper end of the interval.

Example 21.1

Consider again the normal distribution of (21.1). For any fixed $\bar{x}_1$, as μ varies from $-\infty$ through $\bar{x}_1$ to $+\infty$, the probability density varies from zero monotonically to a maximum at $\bar{x}_1$ and then monotonically to zero. Thus for any value in the range $\bar{x}_1 - k$ to $\bar{x}_1 + k$ the density is greater than at the points $\bar{x}_1 - k$ or $\bar{x}_1 + k$. We can therefore set a fiducial interval

$$\bar{x}_1 - k \leqslant \mu \leqslant \bar{x}_1 + k,$$

for any convenient value of $k > 0$. In (21.4) we took k to be $2/\sqrt{n}$.

Example 21.2

As an example of a non-symmetrical sampling distribution, consider the distribution

$$dF = \frac{x^{p-1} e^{-x/\theta}}{\theta^p \, \Gamma(p)} dx, \qquad p > 0; \ 0 \leqslant x \leqslant \infty. \tag{21.6}$$

If p is known, $t \equiv \bar{x}/p$ is sufficient for θ (cf. Exercise 17.1) and its sampling distribution is easily seen to be

$$dF = \left(\frac{\beta}{\theta}\right)^{\beta} \frac{t^{\beta-1} e^{-\beta t/\theta}}{\Gamma(\beta)} dt, \tag{21.7}$$

where $\beta = np$. Now in this case θ may vary only from 0 to ∞. As it does so the ordinate of (21.7) for fixed t rises monotonically from zero to a maximum and then falls again to zero, being in fact an inversion of a Type III distribution. Thus, if we determine θ_{α_1} and θ_{α_2} such that the ordinates at those two values are equal and the integral of (21.7) between them has the assigned value $1 - \alpha$, the fiducial range is $\theta_{\alpha_1} \leqslant \theta \leqslant \theta_{\alpha_2}$.

We may write (21.7) in the form

$$dF = \left(\frac{\beta t}{\theta}\right)^{\beta-1} \frac{e^{-\beta t/\theta}}{\Gamma(\beta)} d\left(\frac{\beta t}{\theta}\right), \tag{21.8}$$

and hence

$$F(t, \theta) = \int_0^{\beta t/\theta} \frac{u^{\beta-1} e^{-u}}{\Gamma(\beta)} du. \tag{21.9}$$

Thus

$$-\frac{\partial F}{\partial \theta} = -\left[\frac{u^{\beta-1} e^{-u}}{\Gamma(\beta)}\right]_{u = \beta t/\theta} \frac{\partial}{\partial \theta}\left(\frac{\beta t}{\theta}\right)$$

$$= \left(\frac{\beta t}{\theta}\right)^{\beta-1} \frac{e^{-\beta t/\theta}}{\Gamma(\beta)} \frac{\beta t}{\theta^2}.$$

Thus the fiducial distribution of θ is

$$\left(\frac{\beta t}{\theta}\right)^{\beta} \frac{e^{-\beta t/\theta}}{\Gamma(\beta)} \frac{d\theta}{\theta}. \tag{21.10}$$

The integral of this from $\theta = 0$ to $\theta = \infty$ is unity.

In comparing (21.7) with (21.10) it should be noticed that we have replaced dt, not by $d\theta$, but by $t\, d\theta/\theta$; or, putting it slightly differently, we have replaced dt/t by $d\theta/\theta$. It is worth while considering why this should be so, and to restate in specific form the argument of **21.8**.

We determine our fiducial interval by reference to the probability $F(t, \theta)$. Looking at (21.9) we see that this is an integral whose upper limit is, apart from a constant, t/θ.

INTERVAL ESTIMATION: OTHER METHODS

Thus for variation in θ we have the ordinate of the frequency function (the integrand) multiplied by $d_\theta(t/\theta) = -t\,d\theta/\theta^2$, while for variation in t the multiplying factor is $d_t(t/\theta) = dt/\theta$. Thus, from (21.5), $-(\partial F/\partial \theta)\,d\theta = t\,d\theta/\theta^2$, while $(\partial F/\partial t)\,dt = dt/\theta$. It is by equating these expressions that we obtain $d\theta/\theta = dt/t$.

21.9 When we try to extend our theory to cover the case where two or more parameters are involved, we begin to meet difficulties. In point of fact, practical examples in this field are so rare that any general theory is apt to be left in the air for want of exemplification. We shall therefore concentrate the exposition on two important standard cases, the estimation of the mean in normal samples where the variance is unknown, and the estimation of the difference of two means in samples from two normal populations with unequal variances.

Fiducial inference in "Student's" distribution

21.10 It is known that in normal samples the sample mean $\bar{x}$ and the sample variance $s^2 (= \Sigma(x-\bar{x})^2/n)$ are jointly sufficient for the parent mean μ and variance σ^2. Their distribution may be written as

$$dF \propto \frac{1}{\sigma}\exp\left\{-\frac{n}{2\sigma^2}(\bar{x}-\mu)^2\right\}d\bar{x}\left(\frac{s}{\sigma}\right)^{n-2}\exp\left\{-\frac{ns^2}{2\sigma^2}\right\}\frac{ds}{\sigma}. \quad (21.11)$$

If we were considering fiducial limits for μ with known σ, we should use the first factor on the right of (21.11); but if we were considering limits for σ with known μ we should *not* use the second factor, the reason being that σ itself enters into the first factor. In fact (cf. Example 17.10), the sufficient statistic in this case is not s^2 but $\Sigma(x-\mu)^2/n$, whose distribution is obtained by merging the two factors in (21.11).

For known σ, we should, as in Example 21.1, replace $d\bar{x}$ by $d\mu$ to obtain the fiducial distribution of μ. For known μ, we should use the fact that $\Sigma(x-\mu)^2 = s'^2$ is distributed like x in (21.6) with $p = n$ and $\theta = \sigma^2$, and hence, as in Example 21.2, replace ds'/s' by $d\sigma/\sigma$. In (21.11), s is distributed as s', but with $p = n-1$. The question is, can we here replace $d\bar{x}\,ds/s$ in (21.11) by $d\mu\,d\sigma/\sigma$ to obtain the joint fiducial distribution of μ and σ?

Fiducialists assume that this is so. The question appears to us to be very debatable.(*) However, let us make the assumption and see where it leads us. For the fiducial distribution we shall then have

$$dF \propto \frac{1}{\sigma}\exp\left\{-\frac{n}{2\sigma^2}(\bar{x}-\mu)^2\right\}d\mu\left(\frac{s}{\sigma}\right)^{n-1}\exp\left\{-\frac{ns^2}{2\sigma^2}\right\}\frac{d\sigma}{\sigma}. \quad (21.12)$$

We now integrate for σ to obtain the fiducial distribution of μ.

We arrive at
$$dF \propto \frac{d\mu/s}{\left\{1+\frac{(\bar{x}-\mu)^2}{s^2}\right\}^{\frac{1}{2}n}}. \quad (21.13)$$

(*) Although $\bar{x}$ and s are statistically independent, μ and σ are not independent in any fiducial sense. The laws of transformation from the frequency to the fiducial distribution have not been elucidated to any extent for the multi-parameter case. In the above case some support for the process can be derived *a posteriori* from the reflexion that it leads to "Student's" distribution, but if fiducial theory is to be accepted on its own merits, something more is required.

This is a form of "Student's" distribution, with $\frac{(\mu-\bar{x})}{s}\sqrt{(n-1)}$ in place of the usual t, and $n-1$ degrees of freedom. Thus, given α, we can find two values of t, t_0 and t_1, such that
$$P\{-t_0 \leqslant t \leqslant t_1\} = 1-\alpha$$
and this is equivalent to locating μ in the range
$$\{\bar{x}-st_0/\sqrt{(n-1)},\ \bar{x}+st_1/\sqrt{(n-1)}\}. \qquad (21.14)$$
This may be interpreted, as in **20.31**, in the sense of confidence intervals, i.e. as implying that if we assert μ to lie in the range (21.14) we shall be right in a proportion $1-\alpha$ of the cases. But this is by no means essential to the fiducial argument, as we shall see later.

The problem of two means

21.11 We now turn to the problem of finding an interval estimate for the difference between the means of two normal distributions, which was left undiscussed in the previous chapter in order to facilitate a unified exposition here. We shall first discuss several confidence-interval approaches to the problem, and then proceed to the fiducial-interval solution. Finally, we shall examine the problem from the standpoint of Bayes' theorem.

21.12 Suppose, then, that we have two normal distributions, the first with mean and variance parameters μ_1, σ_1^2 and the second with parameters μ_2, σ_2^2. Samples of size n_1, n_2 respectively are taken, and the sample means and variances observed are $\bar{x}_1, s_1^2$ and $\bar{x}_2, s_2^2$. Without loss of generality, we assume $n_1 \leqslant n_2$.

Now if $\sigma_1^2 = \sigma_2^2 = \sigma^2$, the problem of finding an interval for $\mu_1 - \mu_2 = \delta$ is simple. For in this case $d = \bar{x}_1 - \bar{x}_2$ is normally distributed with
$$\left.\begin{array}{l} E(d) = \delta, \\ \operatorname{var} d = \sigma^2\left(\dfrac{1}{n_1}+\dfrac{1}{n_2}\right), \end{array}\right\} \qquad (21.15)$$
and $n_1 s_1^2/\sigma_1^2$, $n_2 s_2^2/\sigma_2^2$ are each distributed like χ^2 with n_1-1, n_2-1 d.f. respectively. Since the two samples are independent, $(n_1 s_1^2 + n_2 s_2^2)/\sigma^2$ will be distributed like χ^2 with n_1+n_2-2 d.f., and hence, writing
$$s^2 = (n_1 s_1^2 + n_2 s_2^2)/(n_1+n_2-2)$$
we have
$$E(s^2) = \sigma^2. \qquad (21.16)$$
Now
$$y = \frac{d-\delta}{\left\{\sigma^2\left(\dfrac{1}{n_1}+\dfrac{1}{n_2}\right)\right\}^{\frac{1}{2}}} \bigg/ \left(\frac{s^2}{\sigma^2}\right)^{\frac{1}{2}} \qquad (21.17)$$
$$= \frac{d-\delta}{\left\{s^2\left(\dfrac{1}{n_1}+\dfrac{1}{n_2}\right)\right\}^{\frac{1}{2}}} \qquad (21.18)$$
is a ratio of a standardized normal variate to the square root of an unbiassed estimator of its sampling variance, which is distributed independently of it (since s_1^2 and s_2^2 are

independent of $\bar{x}_1$ and $\bar{x}_2$). Moreover, $(n_1+n_2-2)s^2/\sigma^2$ is a χ^2 variable with n_1+n_2-2 d.f. Thus y is of exactly the same form as the one-sample ratio

$$\frac{\bar{x}_1-\mu_1}{\{s_1^2/(n_1-1)\}^{\frac{1}{2}}} = \frac{\bar{x}_1-\mu_1}{(\sigma^2/n_1)^{\frac{1}{2}}} \bigg/ \left\{\frac{n_1 s_1^2/(n_1-1)}{\sigma^2}\right\}^{\frac{1}{2}}$$

which we have on several occasions (e.g. Example 11.8) seen to be distributed in "Student's" distribution with n_1-1 d.f. Hence (21.18) is also a "Student's" variable, but with n_1+n_2-2 d.f., a result which may easily be proved directly.

There is therefore no difficulty in setting confidence intervals or fiducial intervals for δ in this case: we simply use the method of **20.31** or **21.10**, and of course, as in the one-sample case, we obtain identical results, quite incidentally.

21.13 When we leave the case $\sigma_1^2 = \sigma_2^2$, complications arise. The variate distributed in "Student's" form, with n_1+n_2-2 d.f., by analogy with (21.17), is now

$$t = \frac{d-\delta}{\left\{\frac{\sigma_1^2}{n_1}+\frac{\sigma_2^2}{n_2}\right\}^{\frac{1}{2}}} \bigg/ \left\{\frac{\frac{n_1 s_1^2}{\sigma_1^2}+\frac{n_2 s_2^2}{\sigma_2^2}}{n_1+n_2-2}\right\}^{\frac{1}{2}}. \tag{21.19}$$

The numerator of (21.19) is a standardized normal variate, and its denominator is the square root of an independently distributed χ^2 variate divided by its degrees of freedom, as for (21.17). The difficulty is that (21.19) involves the unknown ratio of variances $\theta = \sigma_1^2/\sigma_2^2$. If we also define $u = s_1^2/s_2^2$, $N = n_1/n_2$, we may rewrite (21.19) as

$$t = \frac{(d-\delta)(n_1+n_2-2)^{\frac{1}{2}}}{s_2\left\{\left(1+\frac{\theta}{N}\right)\left(1+\frac{Nu}{\theta}\right)\right\}^{\frac{1}{2}}}, \tag{21.20}$$

which clearly displays its dependence upon the unknown θ. If $\theta = 1$, of course, (21.20) reduces to (21.18).

21.14 We now have to consider methods by which the "nuisance parameter," θ, can be eliminated from interval statements concerning δ. We must clearly seek some statistic other than t of (21.20). One possibility suggests itself immediately from inspection of the alternative form, (21.18), to which (21.20) reduces when $\theta = 1$. The statistic

$$z = \frac{d-\delta}{\left(\frac{s_1^2}{n_1-1}+\frac{s_2^2}{n_2-1}\right)^{\frac{1}{2}}} \tag{21.21}$$

is, like (21.18), the ratio of a normal variate with zero mean to the square root of an independently distributed unbiassed estimator of its sampling variance. However, that estimator is not a multiple of a χ^2 variate, and hence z is not distributed in "Student's" form. The statistic z is the basis of the fiducial approach and one approximate confidence interval approach to this problem, as we shall see below.

An alternative possibility is to investigate the distribution of (21.18) itself, i.e. to see how far the statistic appropriate to the case $\theta = 1$ retains its properties when $\theta \neq 1$. This, too, has been investigated from the confidence interval standpoint.

However, before proceeding to discuss the approaches outlined in this section, we

THE ADVANCED THEORY OF STATISTICS

examine at some length an exact confidence interval solution to this problem, based on "Student's" distribution, and its properties. The results are due to Scheffé (1943a, 1944).

Exact confidence intervals based on "Student's" distribution

21.15 If we desire an exact confidence interval for δ based on the "Student" distribution, it will be sufficient if we can find a linear function of the observations, L, and a quadratic function of them, Q, such that, for all values of σ_1^2, σ_2^2,

(i) L and Q are independently distributed;
(ii) $E(L) = \delta$ and $\operatorname{var} L = V$; and
(iii) Q/V has a χ^2 distribution with k d.f.

Then
$$t = \frac{L-\delta}{(Q/k)^{\frac{1}{2}}} \tag{21.22}$$

has "Student's" distribution with k d.f. We now prove a remarkable result due to Scheffé (1944), to the effect that no statistic of the form (21.22) can be a symmetric function of the observations in each sample; that is to say, t cannot be invariant under permutation of the first sample members x_{1i} ($i = 1, 2, \ldots, n_1$) among themselves and of the second sample members x_{2i} ($i = 1, 2, \ldots, n_2$) among themselves.

21.16 Suppose that t is symmetric in the sense indicated. Then we must have

$$\left. \begin{array}{l} L = c_1 \sum\limits_i x_{1i} + c_2 \sum\limits_i x_{2i}, \\ Q = c_3 \sum x_{1i}^2 + c_4 \sum\limits_{i \neq j} x_{1i} x_{1j} + c_5 \sum x_{2i}^2 + c_6 \sum\limits_{i \neq j} x_{2i} x_{2j} + c_7 \sum\limits_{i,j} x_{1i} x_{2j}, \end{array} \right\} \tag{21.23}$$

where the c's are constants independent of the parameters.

Now from (ii) in **21.15**

$$E(L) = \delta = \mu_1 - \mu_2, \tag{21.24}$$

while from (21.23)

$$E(L) = c_1 n_1 \mu_1 + c_2 n_2 \mu_2. \tag{21.25}$$

(21.24) and (21.25) are identities in μ_1 and μ_2; hence

$$c_1 n_1 \mu_1 = \mu_1, \quad c_2 n_2 \mu_2 = -\mu_2$$

so that

$$c_1 = 1/n_1, \quad c_2 = -1/n_2. \tag{21.26}$$

From (21.26) and (21.23),

$$L = \bar{x}_1 - \bar{x}_2 = d, \tag{21.27}$$

and hence

$$\operatorname{var} L = V = \sigma_1^2/n_1 + \sigma_2^2/n_2. \tag{21.28}$$

Since Q/V has a χ^2 distribution with k d.f.,

$$E(Q/V) = k,$$

so that, using (21.28),

$$E(Q) = k(\sigma_1^2/n_1 + \sigma_2^2/n_2), \tag{21.29}$$

while, from (21.23),

$$E(Q) = c_3 n_1 (\sigma_1^2 + \mu_1^2) + c_4 n_1 (n_1 - 1) \mu_1^2 + c_5 n_2 (\sigma_2^2 + \mu_2^2)$$
$$+ c_6 n_2 (n_2 - 1) \mu_2^2 + c_7 n_1 n_2 \mu_1 \mu_2. \tag{21.30}$$

Equating (21.29) and (21.30), we obtain expression for the c's, and thence, from (21.23),

$$Q = k\left\{\frac{s_1^2}{n_1-1}+\frac{s_2^2}{n_2-1}\right\}. \qquad (21.31)$$

(21.27) and (21.31) reduce (21.22) to (21.21). Now a linear function of two independent χ^2 variates can only itself have a χ^2 distribution if it is a simple sum of them, and $n_1 s_1^2/\sigma_1^2$ and $n_2 s_2^2/\sigma_2^2$ are independent χ^2 variates. Thus, from (21.31), Q will only be a χ^2 variate if

$$\frac{k\sigma_1^2}{n_1(n_1-1)} = \frac{k\sigma_2^2}{n_2(n_2-1)} = 1$$

or

$$\theta = \frac{\sigma_1^2}{\sigma_2^2} = \frac{n_1(n_1-1)}{n_2(n_2-1)}. \qquad (21.32)$$

Given n_1, n_2, this is only true for special values of σ_1^2, σ_2^2. Since we require it to be true for *all* values of σ_1^2, σ_2^2 we have established a contradiction. Thus t cannot be a symmetric function in the sense stated.

21.17 Since we cannot find a symmetric function of the desired type having " Student's " distribution, we now consider others. We specialize (21.22) to the situation where

$$\left.\begin{array}{l} L = \sum\limits_{i=1}^{n_1} d_i/n_1, \\[2mm] Q = \dfrac{1}{n_1}\sum\limits_{i=1}^{n_1} (d_i-L)^2, \end{array}\right\} \qquad (21.33)$$

and the d_i are independent identical normal variates with

$$E(d_i) = \delta, \quad \text{var } d_i = \sigma^2, \text{ all } i. \qquad (21.34)$$

It will be remembered that we have taken $n_1 \leqslant n_2$. (21.22) now becomes

$$t = \frac{L-\delta}{\{Q/(n_1-1)\}^{\frac{1}{2}}} = (L-\delta)\left\{\frac{n_1(n_1-1)}{\Sigma(d_i-L)^2}\right\}^{\frac{1}{2}}, \qquad (21.35)$$

which is a " Student " variate with (n_1-1) d.f.

Suppose now that in terms of the original observations

$$d_i = x_{1i} - \sum_{j=1}^{n_2} c_{ij} x_{2j}. \qquad (21.36)$$

The d_i are multinormally distributed, since they are linear functions of normal variates (cf. **15.4**). Necessary and sufficient conditions that (21.34) holds are

$$\left.\begin{array}{l} \sum\limits_{j} c_{ij} = 1, \\[2mm] \sum\limits_{j} c_{ij}^2 = c^2, \\[2mm] \sum\limits_{j} c_{ij} c_{kj} = 0, \quad i \neq k. \end{array}\right\} \qquad (21.37)$$

Thus, from (21.36) and (21.37)

$$\text{var } d_i = \sigma^2 = \sigma_1^2 + c^2 \sigma_2^2. \qquad (21.38)$$

21.18 The central confidence interval, with confidence coefficient $1-\alpha$, derived from (21.35) is

$$|L-\delta| \leq t_{n_1-1,\alpha}\{Q/(n_1-1)\}^{\frac{1}{2}}, \tag{21.39}$$

where $t_{n_1-1,\alpha}$ is the appropriate deviate for n_1-1 d.f. The interval-length l has expected value, from (21.39),

$$E(l) = 2t_{n_1-1,\alpha}\frac{\sigma}{\{n_1(n_1-1)\}^{\frac{1}{2}}}E\left\{\left(\frac{n_1 Q}{\sigma^2}\right)^{\frac{1}{2}}\right\}, \tag{21.40}$$

the last factor on the right being found, from the fact that $n_1 Q/\sigma^2$ has a χ^2 distribution with n_1-1 d.f., to be

$$E\left\{\left(\frac{n_1 Q}{\sigma^2}\right)^{\frac{1}{2}}\right\} = \frac{\sqrt{2}\,\Gamma(\frac{1}{2}n_1)}{\Gamma\{\frac{1}{2}(n_1-1)\}}. \tag{21.41}$$

To minimize the expected length (21.40), we must minimize σ, or equivalently, minimize c^2 in (21.38), subject to (21.37). The problem may be visualized geometrically as follows: consider a space of n_2 dimensions, with one axis for each *second* suffix of the c_{ij}. Then $\sum_j c_{ij} = 1$ is a hyperplane, and $\sum_j c_{ij}^2 = c^2$ is an n_2-dimensional hypersphere which is intersected by the plane in an (n_2-1)-dimensional hypersphere. We require to locate $n_1 \leq n_2$ vectors through the origin which touch this latter hypersphere and (to satisfy the last condition of (21.37)) are mutually orthogonal, in such a way that the radius of the n_2-dimensional hypersphere is minimized. This can be done by making our vectors coincide with n_1 axes, and then $c^2 = 1$. But if $n_1 < n_2$, we can improve upon this procedure, for we can, while keeping the vectors orthogonal, space them symmetrically about the equiangular vector, and reduce c^2 from 1 to its minimum value n_1/n_2, as we shall now show.

21.19 Written in vector form, the conditions (21.37) are

$$\left.\begin{array}{ll}\mathbf{c}_i\mathbf{u}' = 1 & \\ \mathbf{c}_i\mathbf{c}_k' = c^2 & i=k, \\ \quad\quad = 0 & i \neq k,\end{array}\right\} \tag{21.42}$$

where $\mathbf{c}_i$ is the ith row vector of the matrix $\{c_{ij}\}$ and $\mathbf{u}$ is a row vector of units.

If the n_1 vectors $\mathbf{c}_i$ satisfy (21.42), we can add another (n_2-n_1) vectors, satisfying the second (normalizing and orthogonalizing) condition of (21.42), so that the augmented set forms a basis for an n_2-space. We may therefore express $\mathbf{u}$ as a linear function of the n_2 $\mathbf{c}$-vectors,

$$\mathbf{u} = \sum_{k=1}^{n_2} g_k \mathbf{c}_k, \tag{21.43}$$

where the g_k are scalars. Now, using (21.42) and (21.43),

$$1 = \mathbf{c}_i\mathbf{u}' = \mathbf{c}_i\sum_{k=1}^{n_2} g_k \mathbf{c}_k' = \sum g_k \mathbf{c}_i \mathbf{c}_k'$$
$$= g_i c^2.$$

Thus

$$g_i = 1/c^2, \quad i = 1, 2, \ldots, n_1. \tag{21.44}$$

Also, since **u** is a row vector of units,
$$n_2 = \mathbf{u}\mathbf{u}' = \left(\sum_k g_k \mathbf{c}_k\right)\left(\sum_k g_k \mathbf{c}'_k\right),$$
which, on using (21.42), becomes
$$n_2 = \sum_{k=1}^{n_2} g_k^2 \mathbf{c}_k \mathbf{c}'_k$$
$$= c^2 \left(\sum_{k=1}^{n_1} + \sum_{n_1+1}^{n_2}\right) g_k^2. \tag{21.45}$$
Use of (21.44) gives, from (21.45),
$$n_2 = c^2 \left\{ n_1/c^4 + \sum_{n_1+1}^{n_2} g_k^2 \right\}$$
or
$$n_2 \geq \frac{n_1}{c^2}.$$
Hence
$$c^2 \geq n_1/n_2, \tag{21.46}$$
the required result.

21.20 The equality sign holds in (21.46) whenever $g_k = 0$ for $k = n_1+1, \ldots, n_2$. Then the equiangular vector **u** lies entirely in the space spanned by the original n_1 **c**-vectors. From (21.44), these will be symmetrically disposed around it. Evidently, there is an infinite number of ways of determining c_{ij}, merely by rotating the set of n_1 vectors. Scheffé (1943a) obtained the particularly appealing solution

$$\left.\begin{array}{ll} c_{ii} = (n_1/n_2)^{\frac{1}{2}} - (n_1 n_2)^{-\frac{1}{2}} + 1/n_2, & j = 1, 2, \ldots, n_1, \\ c_{ij} = \phantom{(n_1/n_2)^{\frac{1}{2}}} -(n_1 n_2)^{-\frac{1}{2}} + 1/n_2, & j(\neq i) = 1, 2, \ldots, n_1, \\ c_{ij} = \phantom{(n_1/n_2)^{\frac{1}{2}} -(n_1 n_2)^{-\frac{1}{2}} +} 1/n_2, & j = n_1+1, \ldots, n_2. \end{array}\right\} \tag{21.47}$$

It may easily be confirmed that (21.47) satisfies the conditions (21.37) with $c^2 = n_1/n_2$. Substituted into (21.36), (21.47) gives
$$d_i = x_{1i} - (n_1/n_2)^{\frac{1}{2}} x_{2i} + (n_1 n_2)^{-\frac{1}{2}} \sum_{j=1}^{n_1} x_{2j} + (1/n_2) \sum_{j=1}^{n_2} x_{2j}, \tag{21.48}$$
which yields in (21.33)
$$\left.\begin{array}{l} L = \bar{x}_1 - \bar{x}_2, \\ Q = \dfrac{1}{n_1} \sum_{i=1}^{n_1} (u_i - \bar{u})^2, \end{array}\right\} \tag{21.49}$$
where
$$\left.\begin{array}{l} u_i = x_{1i} - (n_1/n_2)^{\frac{1}{2}} x_{2i}, \\ \bar{u} = \sum_{i=1}^{n_1} u_i/n_1. \end{array}\right\} \tag{21.50}$$
Hence, from (21.35) and (21.48–21.50),
$$\{\bar{x}_1 - \bar{x}_2 - \delta\} \left\{\frac{n_1(n_1-1)}{\sum(u_i - \bar{u})^2}\right\}^{\frac{1}{2}} \tag{21.51}$$
is a "Student's" variate with $n_1 - 1$ d.f., and we may proceed to set confidence limits for $\delta = \mu_1 - \mu_2$.

Bain (1967) derives this procedure by a purely algebraic method, and also applies the method to other problems.

21.21 It is rather remarkable that we have been able to find an exact solution of the confidence interval problem in this case only by abandoning the seemingly natural requirement of symmetry. (21.51) holds for *any* randomly selected subset of n_1 of the n_2 variates in the second sample. Just as, in **20.22**, we resorted to randomization to remove the difficulty in making exact confidence interval statements about a discrete variable, so we find here that randomization alone allows us to bypass the nuisance parameter θ. But the extent of the randomization should not be exaggerated. The numerator of (21.51) uses the sample means of both samples, complete; only the denominator varies with different random selections of the subset in the second sample. It is impossible to assess intuitively how much efficiency is lost by this procedure. We now proceed to examine the length of the confidence intervals it provides.

21.22 From (21.38) and (21.46), we have for the optimum solution (21.48),

$$\operatorname{var} d_i = \sigma^2 = \sigma_1^2 + (n_1/n_2)\sigma_2^2. \tag{21.52}$$

Putting (21.52) into (21.40), and using (21.41), we have for the expected length of the confidence interval

$$E(l) = 2t_{n_1-1,\alpha}\left\{\frac{\sigma_1^2+(n_1/n_2)\sigma_2^2}{n_1(n_1-1)}\right\}^{\frac{1}{2}} \frac{\sqrt{2}\,\Gamma(\tfrac{1}{2}n_1)}{\Gamma\{\tfrac{1}{2}(n_1-1)\}}. \tag{21.53}$$

We now compare this interval l with the interval L obtained from (21.19) if $\theta = \sigma_1^2/\sigma_2^2$ is known. The latter has expected length

$$E(L) = 2t_{n_1+n_2-2,\alpha}\left\{\frac{\sigma_1^2/n_1+\sigma_2^2/n_2}{n_1+n_2-2}\right\}^{\frac{1}{2}} E\left\{\frac{n_1 s_1^2}{\sigma_1^2}+\frac{n_2 s_2^2}{\sigma_2^2}\right\}^{\frac{1}{2}}, \tag{21.54}$$

the last factor being evaluated from the χ^2 distribution with (n_1+n_2-2) d.f. as

$$\frac{\sqrt{2}\,\Gamma\{\tfrac{1}{2}(n_1+n_2-1)\}}{\Gamma\{\tfrac{1}{2}(n_1+n_2-2)\}}. \tag{21.55}$$

(21.53–55) give for the ratio of expected lengths

$$E(l)/E(L) = \frac{t_{n_1-1,\alpha}}{t_{n_1+n_2-2,\alpha}} \cdot \left(\frac{n_1+n_2-2}{n_1-1}\right)^{\frac{1}{2}} \cdot \frac{\Gamma(\tfrac{1}{2}n_1)\,\Gamma\{\tfrac{1}{2}(n_1+n_2-2)\}}{\Gamma\{\tfrac{1}{2}(n_1-1)\}\,\Gamma\{\tfrac{1}{2}(n_1+n_2-1)\}}. \tag{21.56}$$

As $n_1 \to \infty$, with n_2 fixed, each of the three factors of (21.56) tends to 1, and therefore the ratio of expected interval length does so, as is intuitively reasonable. For small n_1, the first two factors exceed 1, but the last is less than 1. The following table gives the exact values of (21.56) for $1-\alpha = 0.95, 0.99$ and a few sample sizes.

Table of $E(l)/E(L)$ *(from Scheffé, 1943a)*

n_1-1 \ n_2-1	\multicolumn{5}{c}{$1-\alpha = 0.95$}	\multicolumn{5}{c}{$1-\alpha = 0.99$}
	5	10	20	40	∞	5	10	20	40	∞
5	1·15	1·20	1·23	1·25	1·28	1·27	1·36	1·42	1·47	1·52
10		1·05	1·07	1·09	1·11		1·10	1·13	1·16	1·20
20			1·03	1·03	1·05			1·05	1·06	1·09
40				1·01	1·02				1·02	1·04
∞					1					1

Evidently, l is a very efficient interval even for moderate sample sizes, having an expected length no greater than 11 per cent in excess of that of L for $n_1-1 \geqslant 10$ at $1-\alpha = 0.95$, and no greater than 9 per cent in excess for $n_1-1 \geqslant 20$ at $1-\alpha = 0.99$. Furthermore, we are comparing it with an interval *based on knowledge of θ*. Taking this into account, we may fairly say that l puts up a very good performance indeed: the element of randomization cannot have resulted in very much loss of efficiency.

In addition to this solution to the two-means problem there are also approximate confidence-interval solutions, which we shall now summarize.

Approximate confidence-interval solutions

21.23 Welch (1938) has investigated the approximate distribution of the statistic (21.18), which is a "Student's" variate when $\sigma_1^2 = \sigma_2^2$, in the case $\sigma_1^2 \neq \sigma_2^2$. In this case, the sampling variance of its numerator is

$$\operatorname{var}(d-\delta) = \sigma_1^2/n_1 + \sigma_2^2/n_2,$$

so that, writing

$$\left.\begin{aligned} u &= (d-\delta)/(\sigma_1^2/n_1 + \sigma_2^2/n_2)^{\frac{1}{2}}, \\ w^2 &= s^2\left(\frac{1}{n_1}+\frac{1}{n_2}\right)\Big/\left(\frac{\sigma_1^2}{n_1}+\frac{\sigma_2^2}{n_2}\right), \end{aligned}\right\} \quad (21.57)$$

(21.18) may be written

$$y = u/w. \quad (21.58)$$

The difficulty now is that w^2, although distributed independently of u, is not a multiple of a χ^2 variate when $\theta \neq 1$. However, by equating its first two moments to those of a χ^2 variate, we can determine a number of degrees of freedom, ν, for which it is *approximately* a χ^2 variate. Its mean and variance are, from (21.57),

$$\left.\begin{aligned} E(w^2) &= b(\nu_1\theta+\nu_2), \\ \operatorname{var}(w^2) &= 2b^2(\nu_1\theta^2+\nu_2), \end{aligned}\right\} \quad (21.59)$$

where we have written

$$\left.\begin{aligned} \nu_1 &= n_1-1, \quad \nu_2 = n_2-1, \\ b &= (n_1+n_2)\sigma_2^2/\{(n_1+n_2-2)(n_2\sigma_1^2+n_1\sigma_2^2)\}. \end{aligned}\right\} \quad (21.60)$$

If we identify (21.59) with the moments of a multiple g of a χ^2 variate with ν d.f.,

$$\mu_1' = g\nu, \quad \mu_2 = 2g^2\nu, \quad (21.61)$$

we find

$$\left.\begin{aligned} g &= b(\theta^2\nu_1+\nu_2)/(\theta\nu_1+\nu_2), \\ \nu &= (\theta\nu_1+\nu_2)^2/(\theta^2\nu_1+\nu_2). \end{aligned}\right\} \quad (21.62)$$

With these values of g and ν, w^2/g is approximately a χ^2 variate with ν degrees of freedom, and hence, from (21.57),

$$t = u\Big/\left\{\frac{w^2}{g\nu}\right\}^{\frac{1}{2}} \quad (21.63)$$

is a "Student's" variate with ν d.f. If $\theta = 1$, $\nu = \nu_1+\nu_2 = n_1+n_2-2$, $g = b = 1/\nu$, and (21.63) reduces to (21.18), as it should. But in general, g and ν depend upon θ.

21.24 Welch (1938) investigated the extent to which the assumption that $\theta = 1$ in (21.63), when in reality it takes some other value, leads to erroneous conclusions. His discussion was couched in terms of testing hypotheses rather than of interval estimation, which is our present concern, but his conclusion should be briefly mentioned. He found that, so long as $n_1 = n_2$, no great harm was done by ignorance of the true value of θ, but that if $n_1 \neq n_2$, serious errors could arise. To overcome this difficulty, he used exactly the technique of **21.23** to approximate the distribution of the statistic z of (21.21). In this case he found that, whatever the values of n_1 and n_2, z itself was approximately distributed in " Student's " form with

$$\nu = \left(\frac{\theta}{n_1}+\frac{1}{n_2}\right)^2 \bigg/ \left(\frac{\theta^2}{n_1^2(n_1-1)}+\frac{1}{n_2^2(n_2-1)}\right) \tag{21.64}$$

degrees of freedom, and that the influence of a wrongly assumed value of θ was now very much smaller. This is what we should expect, since the denominator of z at (21.21) estimates the variances σ_1^2, σ_2^2 separately, while that of (21.58) uses a " pooled " estimate s^2 which is clearly not appropriate when $\sigma_1^2 \neq \sigma_2^2$.

Lawton (1965) extends earlier work by J. Hájek to obtain close bounds upon the size and power of the " equal-tails " test based on z.

Mickey and Brown (1966) show that the exact distribution of z is bounded by Student's t-distributions with n_1+n_2-2 and $\min(n_1-1, n_2-1)$ d.fr.—cf. Exercise 21.12.

21.25 Welch (1947) has refined the approximate approach of the last section. His argument is a general one, but for the present problem may be summarized as follows. Defining s_1^2, s_2^2 with n_1-1, n_2-1 as divisors respectively, so that they are unbiassed estimators of variances, we seek a statistic $h(s_1^2, s_2^2, P)$ such that

$$P\{(d-\delta) < h(s_1^2, s_2^2, P)\} = P \tag{21.65}$$

whatever the value of θ. Now since $(d-\delta)$ is normally distributed independently of s_1^2, s_2^2, with zero mean and variance $\sigma_1^2/n_1+\sigma_2^2/n_2 = D^2$, we have

$$P\{(d-\delta) \leqslant h(s_1^2, s_2^2, P) | s_1^2, s_2^2\} = I\left(\frac{h}{D}\right) \tag{21.66}$$

where $I(x) = \int_{-\infty}^{x} (2\pi)^{-\frac{1}{2}} \exp(-\frac{1}{2}t^2) dt$. Thus, from (21.65) and (21.66),

$$P = \int \int I(h/D) f(s_1^2) f(s_2^2) ds_1^2 ds_2^2. \tag{21.67}$$

Now we may expand $I(h/D)$, which is a function of s_1^2, s_2^2, in a Taylor series about the true values σ_1^2, σ_2^2. We write this symbolically

$$I\left\{\frac{h(s_1^2, s_2^2, P)}{D}\right\} = \exp\left\{\sum_{i=1}^{2}(s_i^2-\sigma_i^2)\partial_i\right\} I\left\{\frac{h(s_1^2, s_2^2, P)}{s}\right\}, \tag{21.68}$$

where the operator ∂_i represents differentiation with respect to s_i^2, and then putting $s_i^2 = \sigma_i^2$, and $s^2 = s_1^2/n_1+s_2^2/n_2$. We may put (21.68) into (21.67) to obtain

$$P = \prod_{i=1}^{2}\left[\int \exp\{(s_i^2-\sigma_i^2)\partial_i\}f(s_i^2)d(s_i^2)\right] \times I\left\{\frac{h(s_1^2, s_2^2, P)}{s}\right\}. \tag{21.69}$$

Now since we have

$$f(s_i^2) ds_i^2 = \frac{1}{\Gamma(\frac{1}{2}\nu_i)}\left(\frac{\nu_i s_i^2}{2\sigma_i^2}\right)^{\frac{1}{2}\nu_i-1} \exp\left(-\frac{\nu_i s_i^2}{2\sigma_i^2}\right) d\left(\frac{\nu_i s_i^2}{\sigma_i^2}\right),$$

on carrying out each integration in the symbolic expression (21.69) we find
$$\int \exp\{(s_i^2 - \sigma_i^2)\partial_i\} f(s_i^2) \, ds_i^2 = \left(1 - \frac{2\sigma_i^2 \partial_i}{\nu_i}\right)^{-\frac{1}{2}\nu_i} \exp(-\sigma_i^2 \partial_i)$$
which, put into (21.69), gives
$$P = \prod_{i=1}^{2}\left[\left(1 - \frac{2\sigma_i^2 \partial_i}{\nu_i}\right)^{-\frac{1}{2}\nu_i} \exp(-\sigma_i^2 \partial_i)\right] I\left\{\frac{h(s_1^2, s_2^2, P)}{s}\right\}. \qquad (21.70)$$

We can solve (21.70) to obtain the form of the function h, and hence find $h(s_1^2, s_2^2, P)$, for any known P.

Welch gave a series expansion for h, which in our special case becomes
$$\frac{h(s_1^2, s_2^2, P)}{s} = \xi\left[1 + \frac{(1+\xi)^2}{4}\sum_{i=1}^{2} c_i^2/\nu_i - \frac{(1+\xi^2)}{2}\sum_{i=1}^{2} c_i^2/\nu_i^2 + \ldots\right], \qquad (21.71)$$
where $c_i = \frac{s_i^2}{n_i}\left(\frac{s_1^2}{n_1} + \frac{s_2^2}{n_2}\right)^{-1}$, $\nu_i = n_i - 1$ and ξ is defined by $I(\xi) = P$.

Since $(d-\delta)/s = z$ of (21.21), (21.71) gives the distribution function of z.

Following further work by Welch, Aspin (1948, 1949) and Trickett *et al.* (1956), (21.71) has now been tabled as a function of ν_1, ν_2 and c_1, for $P = 0.95, 0.975, 0.99$ and 0.995. These tables enable us to set central confidence limits for δ with $1-\alpha = 0.90, 0.95, 0.98$ and 0.99. Some of the tables are reproduced as Table 11 of the *Biometrika Tables*.

Asymptotic expressions of the type (21.71) have been justified by Chernoff (1949) and D. L. Wallace (1958). (21.71) is asymptotic in the sense that each succeeding term on the right is an order lower in ν_i.

Wald (1955) carried the Welch approach much further for the case $n_1 = n_2$.

Press (1966) shows that for $1-\alpha = 0.90$, $n_1 \leq n_2 \leq 30$, Welch's intervals have smaller expected length than (21.53) if θ is small (when (21.51) discards information about the more variable population) but not if θ is large. The two sets of intervals are shown to be asymptotically equivalent, and never differ by more than 10 per cent in expected length when $n_1 > 10$ if $0.01 \leq \theta \leq 100$. See also some comparisons by Mehta and Srinivasan (1970) and Scheffé (1970b).

The fiducial solution

21.26 The fiducial solution of the two-means problem starts from the joint distribution of sample means and variances, which may be written
$$dF \propto \frac{1}{\sigma_1 \sigma_2} \exp\left\{-\frac{n_1}{2\sigma_1^2}(\bar{x}_1 - \mu_1)^2 - \frac{n_2}{2\sigma_2^2}(\bar{x}_2 - \mu_2)^2\right\} d\bar{x}_1 d\bar{x}_2 \times$$
$$\frac{s_1^{n_1-2} s_2^{n_2-2}}{\sigma_1^{n_1-2} \sigma_2^{n_2-2}} \exp\left\{-\frac{n_1}{2}\frac{s_1^2}{\sigma_1^2} - \frac{n_2}{2}\frac{s_2^2}{\sigma_2^2}\right\} ds_1 ds_2. \qquad (21.72)$$

In accordance with the fiducial argument, we replace $d\bar{x}_1, d\bar{x}_2$ by $d\mu_1, d\mu_2$ and ds_1/s_1, ds_2/s_2 by $d\sigma_1/\sigma_1$, $d\sigma_2/\sigma_2$, as in **21.10**. Then for the fiducial distribution (omitting powers of s_1 and s_2, which are now constants) we have
$$dF \propto \frac{1}{\sigma_1^{n_1+1} \sigma_2^{n_2+1}} \exp\left\{-\frac{n_1}{2\sigma_1^2}(\bar{x}_1 - \mu_1)^2 - \frac{n_2}{2\sigma_2^2}(\bar{x}_2 - \mu_2)^2\right\} d\mu_1 d\mu_2 \times$$
$$\exp\left\{-\frac{n_1 s_1^2}{2\sigma_1^2} - \frac{n_2 s_2^2}{2\sigma_2^2}\right\} d\sigma_1 d\sigma_2. \qquad (21.73)$$

Writing
$$t_1 = \frac{(\mu_1-\bar{x}_1)\sqrt{(n_1-1)}}{s_1}, \quad t_2 = \frac{(\mu_2-\bar{x}_2)\sqrt{(n_2-1)}}{s_2}, \tag{21.74}$$
we find, as in (21.10), the joint distribution of μ_1 and μ_2
$$dF \propto \frac{d_\mu t_1}{\{1+t_1^2/(n_1-1)\}^{\frac{1}{2}n_1}} \frac{d_\mu t_2}{\{1+t_2^2/(n_2-1)\}^{\frac{1}{2}n_2}}, \tag{21.75}$$
where we write $d_\mu t_1$ to remind ourselves that the differential element is $\sqrt{(n_1-1)}\,d\mu_1/s_1$ and similarly for the second sample.

We cannot proceed at once to find an interval for $\delta = \mu_1-\mu_2$. In fact, from (21.74) we have
$$(\mu_1-\bar{x}_1)-(\mu_2-\bar{x}_2) = \delta-d = t_1 s_1/\sqrt{(n_1-1)}-t_2 s_2/\sqrt{(n_2-1)}, \tag{21.76}$$
and to set limits to δ we require the fiducial distribution of the right-hand side of (21.76) or some convenient function of it. This is a linear function of t_1 and t_2, whose fiducial distribution is given by (21.75). In actual fact Fisher (1935b, 1939), following Behrens, chose the statistic (21.21)
$$z = \frac{d-\delta}{\left(\dfrac{s_1^2}{n_1-1}+\dfrac{s_2^2}{n_2-1}\right)^{\frac{1}{2}}}$$
as the most convenient function. We have
$$z_1 = t_1 \cos\psi - t_2 \sin\psi, \tag{21.77}$$
where
$$\tan^2\psi = \frac{s_2^2}{n_2-1} \bigg/ \frac{s_1^2}{n_1-1}. \tag{21.78}$$
For given ψ the distribution of z (usually known as the Fisher–Behrens distribution) can be found from (21.76). It has no simple form, but Fisher and Yates' *Statistical Tables* give tables of significance points for z with assigned values of n_1, n_2, ψ, and the probability $1-\alpha$. In using these tables (and in consulting Fisher's papers generally) the reader should note that our $s^2/(n-1)$ is written by him as s'^2.

21.27 In this case, the most important yet noticed, the fiducial argument does not give the same result as the approach from confidence intervals. That is to say, if we determine from a probability $1-\alpha$ the corresponding points of z, say z_0 and z_1, and then assert
$$\bar{x}_1-\bar{x}_2-z_0\sqrt{\left(\frac{s_1^2}{n_1-1}+\frac{s_2^2}{n_2-1}\right)} \leqslant \mu_1-\mu_2 \leqslant \bar{x}_1-\bar{x}_2+z_1\sqrt{\left(\frac{s_1^2}{n_1-1}+\frac{s_2^2}{n_2-1}\right)}, \tag{21.79}$$
we shall not be correct in a proportion $1-\alpha$ of cases in the long run, as is obvious from the fact that z may be expressed as
$$z = t\left\{\frac{s_2^2(1+\theta/N)(1+Nu/\theta)}{(n_2-1)s_1^2+(n_1-1)s_2^2}\right\}^{\frac{1}{2}} \left\{\frac{(n_1-1)(n_2-1)}{n_1+n_2-2}\right\}^{\frac{1}{2}}$$
where t, defined by (21.20), has an exact "Student's" distribution. Since t is distributed independently of θ, z cannot be.

This fact has been made the ground of criticism by adherents of the confidence-interval approach. The fiducialist reply is that there is no particular reason why such statements should be correct in a proportion of cases in the long run; and that to impose such a desideratum is to miss the point of the fiducial approach. We return to this point later.

A. W. Davis and Scott (1971) use the symbolic method of **21.25** to obtain an asymptotic expansion of the confidence coefficient corresponding to (21.79), which always exceeds $1-\alpha$.

Bayesian intervals

21.28 We proceed now to consider the relation between fiducial theory and interval estimation based on Bayes' theorem, as developed by Jeffreys (1948). The theorem (8.2–3) states that the probability of q_r on data p and H is proportional to the product of the probability of q_r on H and the probability of p on q_r and H. Symbolically

$$P\{q_r|p,H\} \propto P(q_r|H)P(p|q_r,H). \tag{21.80}$$

From the Bayesian viewpoint, we take q_r to be a value of the parameter θ under estimate and $P(q_r|H)$ as its prior probability distribution. $P\{q_r|p,H\}$ then becomes the posterior probability distribution of θ and we can use it to set limits within which θ lies, to assigned degrees of probability *in this sense*.

The major problem, as we have noted earlier, is to assign values to the prior distribution $P(q_r|H)$. Jeffreys has extended Bayes' postulate (which stated that if nothing is known about θ and its range is finite, the prior distribution should be proportional to $d\theta$) to take account of various situations. In particular, (1) if the range of θ is infinite in both directions the prior probability is still taken as proportional to $d\theta$; (2) if θ ranges from 0 to ∞ the prior distribution is taken as proportional to $d\theta/\theta$.

Example 21.3

In the case of the normal distribution considered in **21.2** we have, with $\bar{x}$ sufficient for μ,

$$P(\bar{x}|\mu,H) = \frac{n^{\frac{1}{2}}}{(2\pi)^{\frac{1}{2}}} \exp\left\{-\frac{n}{2}(\bar{x}-\mu)^2\right\}, \tag{21.81}$$

and if μ can lie anywhere in $(-\infty, +\infty)$, the prior distribution is taken as

$$P(d\mu|H) = d\mu. \tag{21.82}$$

Hence, for the posterior distribution of μ,

$$P(d\mu|\bar{x},H) \propto \frac{n^{\frac{1}{2}}}{(2\pi)^{\frac{1}{2}}} \exp\left\{-\frac{n}{2}(\bar{x}-\mu)^2\right\} d\mu. \tag{21.83}$$

Integration over the range of μ from $-\infty$ to ∞ shows that the proportionality is in fact an equality. Thus we may, for any given level of probability, determine the range of μ. This is, in fact, the same as that given by confidence-interval theory or fiducial theory.

On the other hand, for the distribution (21.6) of Example 21.2 we take the prior distribution of θ, which is in $(0, \infty)$, to be

$$P(d\theta|H) = d\theta/\theta. \tag{21.84}$$

The essential similarity to the fiducial procedure in Example 21.2 will be evident. We also have

$$P(d\theta \mid t, H) \propto \left(\frac{\beta}{\theta}\right)^\beta \frac{t^{\beta-1} e^{-\beta t/\theta}}{\Gamma(\beta)} \frac{d\theta}{\theta}. \tag{21.85}$$

Evaluation of the constant, required to make the integral from $\theta = 0$ to ∞ equal to unity, gives

$$P(d\theta \mid t, H) = \left(\frac{\beta}{\theta}\right)^\beta \frac{t^\beta e^{-\beta t/\theta}}{\theta \, \Gamma(\beta)} d\theta, \tag{21.86}$$

which again is the same distribution as the one obtained by the confidence-interval and the fiducial approaches.

21.29 Let us now consider the case of setting limits to the mean in normal samples when the variance is unknown. For the "Student" distribution we have

$$P(dt \mid \mu, \sigma, H) = \frac{k \, dt}{(1 + t^2/\nu)^{\frac{1}{2}(\nu+1)}}, \tag{21.87}$$

where k is some constant and $\nu = n-1$. The parameters μ and σ do not appear on the right and hence are irrelevant to $P(dt \mid H)$ and may be suppressed. Thus

$$P(dt \mid H) = \frac{k \, dt}{(1 + t^2/\nu)^{\frac{1}{2}(\nu+1)}}. \tag{21.88}$$

Suppose now that we *assume* that

$$P(dt \mid \bar{x}, s, H) = f(t) \, dt. \tag{21.89}$$

Then, as before, $\bar{x}$ and s may be suppressed, and we have

$$P(dt \mid H) = f(t) \, dt, \tag{21.90}$$

and hence, by comparison with (21.88),

$$P(dt \mid \bar{x}, s, H) = \frac{k \, dt}{(1 + t^2/\nu)^{\frac{1}{2}(\nu+1)}}. \tag{21.91}$$

We can then proceed to find limits to t, given $\bar{x}$ and s, in the usual way. Jeffreys emphasizes, however, that this depends on a new postulate expressed by (21.89) which, though natural, is not trivial. It amounts to an assumption that if we are comparing different distributions, samples from which give different $\bar{x}$'s and s's, the scale of the distribution of μ must be taken proportional to s and its mean displaced by the difference of sample means.

21.30 In a similar way it will be found that to arrive at the Fisher–Behrens distribution it is necessary to postulate that

$$P\{dt_1, dt_2 \mid \bar{x}_1, \bar{x}_2, s_1, s_2, H\} = f_1(t_1) f_2(t_2) \, dt_1 \, dt_2. \tag{21.92}$$

Jeffreys' derivation of the Fisher–Behrens form from Bayes' theorem would be as follows:

The prior probability of $d\mu_1 \, d\mu_2 \, d\sigma_1 \, d\sigma_2 \mid H$ is

$$P\{d\mu_1 \, d\mu_2 \, d\sigma_1 \, d\sigma_2 \mid H\} \propto \frac{d\mu_1 \, d\mu_2 \, d\sigma_1 \, d\sigma_2}{\sigma_1 \sigma_2}.$$

The likelihood (denoting the data by D) is

$$P\{D \mid \mu_1, \mu_2, \sigma_1, \sigma_2, H\} \propto \frac{1}{\sigma_1^{n_1} \sigma_2^{n_2}} \exp\left[-\frac{n_1}{2\sigma_1^2}\{(\mu_1 - \bar{x}_1)^2 + s_1^2\} - \frac{n_2}{2\sigma_2^2}\{(\mu_2 - \bar{x}_2)^2 + s_2^2\}\right].$$

INTERVAL ESTIMATION: OTHER METHODS

Hence, by Bayes' theorem,

$$P\{d\mu_1 d\mu_2 d\sigma_1 d\sigma_2 | D, H\} = \frac{1}{\sigma_1^{n_1+1} \sigma_2^{n_2+1}}$$
$$\exp\left[-\frac{n_1}{2\sigma_1^2}\{(\mu_1 - \bar{x}_1)^2 + s_1^2\} - \frac{n_2}{2\sigma_2^2}\{(\mu_2 - \bar{x}_2)^2 + s_2^2\}\right].$$

Integrating out the values of σ_1 and σ_2, we find for the posterior distribution of μ_1 and μ_2 a form which is easily reducible to (21.75).

Discussion

21.31 There has been so much controversy about the various methods of estimation we have described that, at this point, we shall have to leave our customary objective standpoint and descend into the arena ourselves. The remainder of this chapter is an expression of personal views. We think that it is the correct viewpoint; and it represents the result of many years' silent reflexion on the issues involved, a serious attempt to understand what the protagonists say, and an even more serious attempt to divine what they mean. Whether it will command their approval is more than we can conjecture.

21.32 We have, then, to examine three methods of approach: confidence intervals, fiducial intervals and Bayesian intervals. We must not be misled by, though we may derive some comfort from, the similarity of the results to which they lead in certain simple cases. We shall, however, develop the thesis that, where they differ, the basic reason is not that one or more are wrong, but that they are consciously or unconsciously either answering different questions or resting on different postulates.

21.33 It will be simplest if we begin with the Bayesian approach. If it be granted that probability is a measure of belief, or an undefined idea obeying the usual postulates, the use of prior probabilities is unexceptionable. We have to recognize, however, that by abandoning an attempt to base our probability theory on frequencies of events, we have lost something in objectivity.

The second hurdle to be taken is the acceptance of rules expressing prior probability distributions. Jeffreys has very persuasively argued for the rules referred to above and nothing better has been proposed. At the same time there seems to be something arbitrary, for example, in requiring the prior distribution of a parameter which may range from $-\infty$ to $+\infty$ to be $d\mu$, whereas if it varies only over the range 0 to ∞ it should have a prior distribution proportional to $d\mu/\mu$. Sophisticated arguments concerning the distinction somehow fail to impress us as touching the root of the problem.

21.34 It should also be noticed that we have applied the Bayes argument to cases where a small set of sufficient statistics exists. This is not essential. If L is the Likelihood Function, we can always write

$$P\{\theta | x_1, \ldots x_n, H\} \propto P(\theta | H) L(x_1, \ldots, x_n | \theta, H) \qquad (21.93)$$

and, given $P(\theta | H)$, determine the posterior distribution of θ. From this viewpoint, the only advantage of a small set of sufficient statistics is that it summarizes all the relevant information in the Likelihood Function in fewer statistics than the n sample values. As we have remarked previously, these sample values themselves *always* constitute a set of sufficient statistics, though in practice this may be only a comforting tautology.

21.35 Confidence-interval theory is also general in the sense that the existence of a single sufficient statistic for the unknown parameter is a convenience, not a necessity. We have, however, noted that where no single sufficient statistic exists there may be imaginary or otherwise nugatory intervals in some cases at least, and we know of no case in which these difficulties appear in the presence of single sufficiency, so that confidence-interval theory is possibly not so free from the need for sufficiency as might appear; but perhaps it would be better to say that where nested and simply connected intervals cannot be obtained, there are special difficulties of interpretation.

The principal argument in favour of confidence intervals, however, is that they can be derived in terms of a frequency theory of probability without any assumptions concerning prior distributions such as are essential to the Bayes approach. This, in our opinion, is undeniable. But it is fair to ask whether they achieve this economy of basic assumption without losing something which the Bayes theory possesses. Our view is that they do, in fact, lose something on occasion, and that this something may be important for the purposes of estimation.

21.36 Consider the case where we are estimating the mean μ of a normal population with known variance. And let us suppose that we *know* that μ lies between 0 and 1. According to Bayes' postulate, we should have

$$P(d\mu \mid \bar{x}) = \frac{\exp\left\{-\frac{n}{2}(\mu-\bar{x})^2\right\} d\mu}{\int_0^1 \exp\left\{-\frac{n}{2}(\mu-\bar{x})^2\right\} d\mu}, \qquad (21.94)$$

and the problem of setting limits to μ, though not free from mathematical complexity, is determinate. What has confidence-interval theory to say on this point? It can do no more than reiterate statements like

$$P\{\bar{x}-1\cdot 96/\sqrt{n} \leqslant \mu \leqslant \bar{x}+1\cdot 96/\sqrt{n}\} = 0\cdot 95.$$

These are still true in the required proportion of cases, but the statements take no account of our prior knowledge about the range of μ and may occasionally be idle. It may be true, but would be absurd, to assert $-1 \leqslant \mu \leqslant 2$ if we know already that $0 \leqslant \mu \leqslant 1$. Of course, we may truncate our interval to accord with the prior information. In our example, we could assert only that $0 \leqslant \mu \leqslant 1$: the observations would have added nothing to our knowledge.

In fact, so it seems to us, confidence-interval theory has the defect of its principal virtue: it attains its generality at the price of being unable to incorporate prior knowledge into its statements. When we make our final judgment about μ, we have to synthesize the information obtained from the observations with our prior knowledge.

Bayes' theorem attempts this synthesis at the outset. Confidence theory leaves it until the end (and, we feel bound to remark, in most current expositions ignores the point completely).

21.37 Fiducial theory, as we have remarked, has been confined by Fisher to the case where sufficient statistics are used, or, quite generally, to cases where all the information in the Likelihood Function can be utilized. No systematic exposition has been given of the procedure to be followed when prior information is available, but there seems no reason why a similar method to that exemplified by equation (21.94) should not be used. That is to say, if we derive the fiducial distribution $f(\mu)$ over a general range but have the supplementary information that the parameter must lie in the range μ_0 to μ_1 (within that general range), we modify the fiducial distribution by truncation to

$$f(\mu) \bigg/ \int_{\mu_0}^{\mu_1} f(\mu)\, d\mu.$$

21.38 One critical difficulty of fiducial theory is exemplified by the derivation of "Student's" distribution in fiducial form given in **21.10**. It appears to us that this particular matter has been widely misunderstood, except by Jeffreys. Since the "Student" distribution gives the same result for fiducial theory as for confidence theory, whereas the two methods differ on the problem of two means, both sides seem to have sought for their basic differences in the second, not in the first. But in our view *c'est le premier test qui coûte*. If the logic of this is agreed, the more general Fisher-Behrens result follows by a very simple extension. This is also evident from the Bayes-Jeffreys approach, in which (21.92) is an obvious extension of (21.90) for two independent samples.

The question, as noted in **21.10**, is whether, given the joint distribution of $\bar{x}$ and s (which are independent in the ordinary sense), we can replace $d\bar{x}\,ds$ by $d\mu\,d\sigma/\sigma$. It appears to us that this is not obvious and, indeed, requires a new postulate, just as (21.90) requires a new postulate. On this point, the paper by Yates (1939a) is explicit. A penetrating general discussion of the fiducial argument is given by Dempster (1964) and in the Symposium published in the Bulletin of the International Statistical Institute (1964, **40(2)**, pp. 833–939).

Paradoxes and restrictions in fiducial theory

21.39 If (t_1, t_2) are jointly sufficient for (θ_1, θ_2), we may write the alternative factorizations

$$L(x \mid \theta_1, \theta_2) \propto g(t_1, t_2 \mid \theta_1, \theta_2) = g_1(t_1 \mid t_2, \theta_1, \theta_2) g_2(t_2 \mid \theta_1, \theta_2)$$
$$= g_3(t_2 \mid t_1, \theta_1, \theta_2) g_4(t_1 \mid \theta_1, \theta_2).$$

If t_1 and t_2 each depend on only one of the parameters, there is no difficulty, each statistic being singly sufficient for its parameter.

More generally, we may distinguish two special structures of the sufficient statistics:

(a) One of the statistics depends on only one parameter. The factorization becomes either
$$L \propto g_1(t_1 \mid t_2, \theta_1, \theta_2)\, g_2(t_2 \mid \theta_2)$$
or
$$L \propto g_3(t_2 \mid t_1, \theta_1, \theta_2)\, g_4(t_1 \mid \theta_1)$$

(21.95)

(b) One of the conditional distributions depends on only one parameter, giving either
$$L \propto g_1(t_1 \mid t_2, \theta_1)\, g_2(t_2 \mid \theta_1, \theta_2)$$
or
$$L \propto g_3(t_2 \mid t_1, \theta_2)\, g_4(t_1 \mid \theta_1, \theta_2).$$
(21.96)

If the first line of (21.96) holds, t_2 is singly sufficient for θ_2 when θ_1 is known; is the second line holds, t_1 is singly sufficient for θ_1 when θ_2 is known.

Either line of (21.95) or of (21.96) permits a joint fiducial distribution to be constructed by first obtaining the fiducial distribution of one parameter from the factor in which it appears alone, and then obtaining the conditional fiducial distribution of the other parameter (the value of the first parameter being fixed) from the factor in which both parameters appear. The product of these distributions is taken as the joint fiducial distribution, on the analogy of the multiplication theorem for probabilities. (21.95) and (21.96) were used in this way by Fisher (1956) and Quenouille (1958) respectively.

It will be seen that in **21.10** and **21.26**, both (21.95) and (21.96) held—this was possible because the sample mean (or difference of means) t_1 was distributed independently of the sample variance(s) t_2. In general, however, even these special sufficiency structures are not enough to guarantee the uniqueness of the joint fiducial distribution, as Tukey (1957) and Brillinger (1962) showed by counter-examples. The non-uniqueness arises precisely because *both* lines of (21.95) (or of (21.96)) can hold simultaneously, and the joint fiducial distribution may depend on which line we use to construct it. See also Mauldon (1955) and Dempster (1963).

We shall see in **23.37-9** that either of the sufficiency structures (21.95-6) ensures optimum properties for conditional tests under a certain condition.

Fraser (1961a, b) discusses the relationship of fiducial inference and some invariance properties. See also Hora and Buehler (1966, 1967).

21.40 Lindley (1958a) has obtained a simple yet far-reaching result which not only illuminates the relationship between fiducial and Bayesian arguments, but also limits the claims of fiducial theory to provide a general method of inference, consistent with and combinable with Bayesian methods. In fact, Lindley shows that the fiducial argument is consistent with Bayesian methods if and only if it is applied to a random variable x and a parameter θ which may be (separately) transformed to u and τ respectively so that τ is a location parameter for u; and in this case, it is equivalent to a Bayesian argument with a uniform prior distribution for τ. The criticism applies equally to "confidence distributions" so defined at the end of **20.6** above, in so far as they coincide with fiducial distributions.

21.41 Using (21.5), we write for the fiducial distribution of θ (without confusion with the usual notation for the characteristic function)
$$\phi_x(\theta) = -\frac{\partial}{\partial \theta} F(x \mid \theta), \tag{21.97}$$
while the posterior distribution for θ given a prior distribution $p(\theta)$ is, by Bayes' theorem,
$$\pi_x(\theta) = p(\theta) f(x \mid \theta) \Big/ \int p(\theta) f(x \mid \theta)\, d\theta, \tag{21.98}$$

where $f(x|\theta) = \partial F(x|\theta)/\partial x$, the frequency function. Writing $r(x)$ for the denominator on the right of (21.98), we thus have

$$\pi_x(\theta) = \frac{p(\theta)}{r(x)} \frac{\partial F(x|\theta)}{\partial x}. \tag{21.99}$$

If there is some prior distribution $p(\theta)$ for which the fiducial distribution is equivalent to a Bayes posterior distribution, (21.97) and (21.99) will be equal, or

$$\frac{-\frac{\partial}{\partial \theta} F(x|\theta)}{\frac{\partial}{\partial x} F(x|\theta)} = \frac{p(\theta)}{r(x)}. \tag{21.100}$$

(21.100) shows that the ratio on its left-hand side must be a product of a function of θ and a function of x. We rewrite it

$$\frac{1}{r(x)} \frac{\partial F}{\partial x} + \frac{1}{p(\theta)} \frac{\partial F}{\partial \theta} = 0. \tag{21.101}$$

For given $p(\theta)$ and $r(x)$, we solve (21.101) for F. The only non-constant solution is

$$F = G\{R(x) - P(\theta)\}, \tag{21.102}$$

where G is an arbitrary function and R, P are respectively the integrals of r, p with respect to their arguments. If we write $u = R(x)$, $\tau = P(\theta)$, (21.102) becomes

$$F = G\{u - \tau\}, \tag{21.103}$$

so that τ is a location parameter for u. Conversely, if (21.103) holds, (21.100) is satisfied with u and τ for x and θ and $p(\tau)$ a uniform distribution. Thus (21.103) is a necessary and sufficient condition for (21.100) to hold, i.e. for the fiducial distribution to be equivalent to some Bayes posterior distribution.

21.42 Now consider the situation where we have two independent samples, summarized by sufficient statistics x, y, from which to make an inference about θ. We can do this in two ways:

(a) we may consider the combined evidence of the two samples simultaneously, and derive the fiducial distribution $\phi_{x,y}(\theta)$;

(b) we may derive the fiducial distribution $\phi_x(\theta)$ from the first sample above, and use this as the prior distribution in a Bayesian argument on the second sample, to produce a posterior distribution $\pi_{x,y}(\theta)$.

Now if the fiducial argument is consistent with Bayesian arguments, (a) and (b) are logically equivalent and we should have $\phi_{x,y}(\theta) = \pi_{x,y}(\theta)$.

Take the simplest case, where x and y have the same distribution. Since it admits a single sufficient statistic for θ, the parent frequency function is of the form (17.83), from which we may assume (cf. Exercise 17.14) that the distribution of x itself is of form

$$f(x|\theta) = f(x) g(\theta) \exp(x\theta), \tag{21.104}$$

and similarly for y in the other sample. Moreover, in the combined samples, $x+y$ is evidently sufficient for θ, and thus the combined fiducial distribution $\phi_{x,y}(\theta)$ is a function of $(x+y)$ and θ only. We now ask for the conditions under which $\pi_{x,y}(\theta)$ is also a function of $(x+y)$ and θ only. Since by Bayes' theorem

$$\pi_{x,y}(\theta) = \frac{\phi_x(\theta)f(y|\theta)}{\int \phi_x(\theta)f(y|\theta)\,d\theta},$$

if $\pi_{x,y}(\theta)$ is a function of $(x+y)$ and θ only, so also will be the ratio for two different values of θ

$$\frac{\pi_{x,y}(\theta)}{\pi_{x,y}(\theta')} = \frac{\phi_x(\theta)f(y|\theta)}{\phi_x(\theta')f(y|\theta')}. \tag{21.105}$$

Thus (21.105) must be invariant under interchange of x and y. Using (21.104) in (21.105), we therefore have

$$\frac{\phi_x(\theta)g(\theta)}{\phi_x(\theta')g(\theta')}\exp\{y(\theta-\theta')\} = \frac{\phi_y(\theta)g(\theta)}{\phi_y(\theta')g(\theta')}\exp\{x(\theta-\theta')\},$$

so that

$$\frac{\phi_x(\theta)}{\phi_x(\theta')}\cdot\frac{\phi_y(\theta')}{\phi_y(\theta)} = \exp\{(x-y)(\theta-\theta')\}$$

or

$$\phi_x(\theta) = \frac{\phi_x(\theta')e^{-x\theta'}\cdot\phi_y(\theta)e^{-y\theta}}{\phi_y(\theta')e^{-y\theta'}}e^{x\theta}, \tag{21.106}$$

and if we regard θ' and y as constants, we may write (21.106) as

$$\phi_x(\theta) = A(x).B(\theta)e^{x\theta}, \tag{21.107}$$

where A and B are arbitrary functions. Using (21.97), (21.104) and (21.107), we have

$$\frac{-\frac{\partial}{\partial\theta}F(x|\theta)}{\frac{\partial}{\partial x}F(x|\theta)} = \frac{\phi_x(\theta)}{f(x|\theta)} = \frac{A(x)B(\theta)}{f(x)g(\theta)}. \tag{21.108}$$

But (21.108) is precisely the condition (21.100), for which we saw (21.103) to be necessary and sufficient. Thus we can have $\phi_{x,y}(\theta) = \pi_{x,y}(\theta)$ if and only if x and θ are transformable to (21.103) with τ a location parameter for u, and $p(\tau)$ a uniform distribution. Thus the fiducial argument is consistent with Bayes' theorem if and only if the problem is transformable into a location parameter problem, the prior distribution of the parameter then being uniform. An example where this is not so is given as Exercise 21.11.

Lindley goes on to show that in the exponential family of distributions (17.83), the normal and the Gamma distributions are the only ones obeying the condition of transformability to (21.103): this explains the identity of the results obtained by fiducial and Bayesian methods in these cases (cf. Example 21.3). Sprott (1960, 1961) shows that these remain the only such distributions if x and y are differently distributed.

Welch and Peers (1963), Welch (1965), and Peers (1965) examine the problem of correspondence of Bayesian and confidence intervals with special reference to asymptotic solutions. Thatcher (1964) examines this correspondence for binomial predictions. Geisser and Cornfield (1963) and Fraser (1964) display further difficulties with fiducial distributions in the multivariate case. See also the I.S.I. Symposium cited at the end of **21.38**.

Fraser (1962) proposes a modification of the fiducial method which extends the range of its consistency with Bayesian methods.

21.43 Still another objection to fiducial theory is one which has already been mentioned in respect of the Bayes approach. It abandons a strict frequency approach to the problem of interval estimation. It is possible, indeed, as Barnard (1950) has shown, to justify the Fisher–Behrens solution of the two-means problem from a different frequency standpoint, but as he himself goes on to argue, the idea of a fixed " reference set," in terms of which frequencies are to be interpreted, is really foreign to the fiducial approach. And it is at this point that the statistician must be left to choose between confidence intervals, which make precise frequency-interpretable statements which may on exceptional occasions be trivial, and the other methods, which forgo frequency interpretations in the interests of what are, perhaps intuitively, felt to be more relevant inferences.

EXERCISES

21.1 If $\bar{x}$ is the mean of a sample of n values from

$$dF = \frac{1}{\sigma\sqrt{(2\pi)}} \exp\left\{-\frac{(x-\mu)^2}{2\sigma^2}\right\} dx,$$

s'^2 is equal to $\Sigma(x-\bar{x})^2/(n-1)$, and x is a further independent sample value, show that

$$t = \frac{x-\bar{x}}{s'}\sqrt{\frac{n}{n+1}}$$

is distributed in " Student's " form with $n-1$ d.f. Hence show that fiducial limits for x are

$$\bar{x} \pm s' t_1 \sqrt{\frac{n+1}{n}},$$

where t_1 is chosen so that the integral of " Student's " form between $-t_1$ and t_1 is an assigned probability $1-\alpha$.

> (Fisher, 1935b. This gives an estimate of the next value when n values have already been chosen, and extends the idea of fiducial limits from parameters to variates dependent on them.)

21.2 Show similarly that if a sample of n_1 values gives mean $\bar{x}_1$ and estimated variance s'^2_1, the fiducial distribution of mean $\bar{x}_2$ and estimated variance s'^2_2 in a second sample of n_2 is

$$dF \propto \frac{s_1'^{(n_1-1)} s_2'^{(n_2-2)} d\bar{x}_2 ds_2'}{\left\{(n_1-1)s_1'^2 + (n_2-1)s_2'^2 + (\bar{x}_1-\bar{x}_2)^2 \frac{n_1 n_2}{n_1+n_2}\right\}^{\frac{1}{2}(n_1+n_2-1)}}.$$

Hence, allowing n_2 to tend to infinity, derive the simultaneous fiducial distribution of μ and σ.

> (Fisher, 1935b)

21.3 If the cumulative binomial distribution is given by

$$G(f, \pi) = \sum_{j=f}^{n} \binom{n}{j} \pi^j (1-\pi)^{n-j}$$

show that f/n is sufficient for π and that

$$h_0(\pi) d\pi \equiv \frac{\partial G(f, \pi)}{\partial \pi} d\pi = \binom{n}{f-1} \pi^{f-1}(1-\pi)^{n-f} d\pi$$

is an admissible fiducial distribution of π. Show that

$$h_1(\pi) d\pi \equiv \frac{\partial G(f+1, \pi)}{\partial \pi} d\pi = \binom{n}{f} \pi^f (1-\pi)^{n-f-1} d\pi$$

is also admissible.

Hence show how to determine π_0 from h_0 and π_1 from h_1, such that the fiducial interval $\pi_0 \leq \pi \leq \pi_1$ has *at least* the associated probability $1-\alpha$.

> (Stevens, 1950. The use of two fiducial distributions is necessitated by discontinuity in the observed f. Compare 20.22 on the analogous difficulty in confidence intervals.)

21.4 Let $l_{11}, l_{12}, \ldots, l_{1,n-1}$ be $(n-1)$ linear functions of the observations which are orthogonal to one another and to $\bar{x}_1$, and let them have zero mean and variance σ_1^2. Similarly define $l_{21}, l_{22}, \ldots, l_{2,n-1}$.

Then, in two samples of size n from normal populations with equal means and variances σ_1^2 and σ_2^2, the function

$$\frac{(\bar{x}_1 - \bar{x}_2) n^{\frac{1}{2}}}{\{\sum (l_{1j} + l_{2j})^2/(n-1)\}^{\frac{1}{2}}}$$

will be distributed as " Student's " t with $n-1$ degrees of freedom. Show how to set confidence intervals to the difference of two means by this result, and show that the solution (21.51) is a member of this class of statistics when $n_1 = n_2$.

21.5 Given two samples of n_1, n_2 members from normal populations with unequal variances, show that by picking n_1 members at random from the n_2 (where $n_1 \leqslant n_2$) and pairing them at random with the members of the first sample, confidence intervals for the difference of means can be based on " Student's " distribution independently of the variance ratio in the populations. Show that this is equivalent to putting $c_{ij} = 0 \, (i \neq j); = 1 \, (i = j)$ in (21.36), and hence that this is an inefficient solution of the two-means problem.

21.6 Use the method of **21.23** to show that the statistic z of (21.21) is distributed approximately in " Student's " form with degrees of freedom given by (21.64).

21.7 From Fisher's F distribution (16.24), find the fiducial distribution of $\theta = \sigma_1^2/\sigma_2^2$, and show that if we regard the " Student's " distribution of the statistic (21.20) as the joint fiducial distribution of δ and θ, and integrate out θ over its fiducial distribution, we arrive at the result of **21.26** for the distribution of z.

(Fisher, 1939)

21.8 Prove the statement in **21.16** to the effect that if $ax + by = z$, where x and y are independent random variables and x, y, z are all χ^2 variates, the constants $a = b = 1$.

(Scheffé, 1944)

21.9 Show that if we take the first two terms in the expansion on the right of (21.71), (21.65) is, to order $1/n$, the approximation of (21.21) given in **21.24**, i.e. a " Student's " distribution with degrees of freedom (21.64).

21.10 Show that for $n_1 = n_2 = n$, the conditional distribution of the statistic z of (21.21) *for fixed* s_1/s_2 is obtainable from the fact that

$$\left\{ \frac{2}{\left(1 - \frac{s_1^2}{s_2^2}\right)\left(\frac{s_1^2}{s_2^2}\right)} + \frac{\left(\frac{s_1^2}{s_2^2}\right)}{\left(1 - \frac{\sigma_1^2}{\sigma_2^2}\right)\left(\frac{\sigma_1^2}{\sigma_2^2}\right)} \right\}^{\frac{1}{2}} z$$

is distributed like " Student's " t with $2(n-1)$ degrees of freedom.

(Bartlett, 1936)

21.11 Show that if the distribution of a sufficient statistic x is

$$f(x \mid \theta) = \frac{\theta^2}{\theta + 1}(x+1)e^{-x\theta}, \qquad x > 0, \; \theta \geqslant 0,$$

the fiducial distribution of θ for combined samples with sufficient statistics $x, y,$ is

$$\phi_{x,y}(\theta) = \frac{e^{-z\theta}}{(\theta+1)^3}[\theta^3(2z^2+\tfrac{4}{3}z^3+\tfrac{1}{6}z^4)+\theta^4(z^2+z^3+\tfrac{1}{6}z^4)]$$

(where $z = x+y$), while that for a single sample is

$$\phi_x(\theta) = \frac{\theta\,x\,e^{-x\theta}}{(\theta+1)^2}[1+(1+\theta)(1+x)].$$

(Note that the minus sign in (21.5) is unnecessary here, since $F(x|\theta)$ is an increasing function of θ.) Hence show that the Bayes posterior distribution from the second sample, using $\phi_x(\theta)$ as prior distribution, is

$$\pi_{x,y}(\theta) \propto e^{-z\theta}\left(\frac{\theta}{\theta+1}\right)^3 x(1+y)[1+(1+\theta)(1+x)],$$

so that $\pi_{x,y}(\theta) \neq \phi_{x,y}(\theta)$. Note that $\pi_{x,y}(\theta) \neq \pi_{y,x}(\theta)$ also.

(Lindley, 1958a)

21.12 Show that ν, the approximate degrees of freedom of z given at (21.64), increases steadily from the value n_2-1 at $\theta = 0$ to its unique maximum value n_1+n_2-2 at $\theta = n_1(n_1-1)/\{(n_2n_2-1)\}$ and then decreases steadily to (n_1-1) as $\theta \to \infty$.

CHAPTER 22

TESTS OF HYPOTHESES: SIMPLE HYPOTHESES

22.1 We now pass from the problems of estimating parameters to those of testing hypotheses concerning parameters. Instead of seeking the best (unique or interval) estimator of an unknown parameter, we shall now be concerned with deciding whether some pre-designated value is acceptable in the light of the observations.

In a sense, the testing problem is logically prior to that of estimation. If, for example, we are examining the difference between the means of two normal populations, our first question is whether the observations indicate that there is *any* true difference between the means. In other words, we have to compare the observed differences between the two samples with what might be expected on the hypothesis that there is no true difference at all, but only random sampling variation. If this hypothesis is not sustained, we proceed to the second step of estimating the *magnitude* of the difference between the population means.

Quite obviously, the problems of testing hypotheses and of estimation are closely related, but it is nevertheless useful to preserve a distinction between them, if only for expository purposes. Many of the ideas expounded in this and the following chapters are due to Neyman and E. S. Pearson, whose remarkable series of papers (1928, 1933a, b, 1936a, b, 1938) is fundamental. See also the monograph by Lehmann (1959).

22.2 The kind of hypothesis which we test in statistics is more restricted than the general scientific hypothesis. It is a scientific hypothesis that every particle of matter in the universe attracts every other particle, or that life exists on Mars; but these are not hypotheses such as arise for testing from the statistical viewpoint. Statistical hypotheses concern the behaviour of observable random variables. More precisely, suppose that we have a set of random variables $x_1, \ldots, x_n$. As before, we may represent them as the co-ordinates of a point (**x**, say) in the n-dimensional sample space, one of whose axes corresponds to each variable. Since **x** is a random variable, it has a probability distribution, and if we select any region, say w, in the sample space W, we may (at least in principle) calculate the probability that the sample point **x** falls in w, say $P(\mathbf{x} \in w)$. We shall say that any hypothesis concerning $P(\mathbf{x} \in w)$ is a statistical hypothesis. In other words, any hypothesis concerning the behaviour of observable random variables is a statistical hypothesis.

For example, the hypothesis (a) that a normal distribution has a specified mean and variance is statistical; so is the hypothesis (b) that it has a given mean but unspecified variance; so is the hypothesis (c) that a distribution is of normal form, both mean and variance remaining unspecified; and so, finally, is the hypothesis (d) that two unspecified continuous distributions are identical. Each of these four examples implies certain properties of the sample space. Each of them is therefore translatable into statements concerning the sample space, which may be tested by comparison with observation.

Parametric and non-parametric hypotheses

22.3 It will have been noticed that in the examples (a) and (b) in the last paragraph, the distribution underlying the observations was taken to be of a certain form (the normal) and the hypothesis was concerned entirely with value of one or both of its parameters. Such a hypothesis, for obvious reasons, is called *parametric*.

Hypothesis (c) was of a different nature. It may be expressed in an alternative way, since it is equivalent to the hypothesis that the distribution has all cumulants finite, and all cumulants above the second equal to zero (cf. Example 3.11). Now the term " parameter " is often used to denote a cumulant or moment of the population, in order to distinguish it from the corresponding sample quantity. This is an understandable, but rather imprecise use of the term. The normal distribution

$$dF(x) = (2\pi)^{-\frac{1}{2}} \exp\left\{-\frac{1}{2}\left(\frac{x-\mu}{\sigma}\right)^2\right\} dx/\sigma$$

has just two parameters, μ and σ. (Sometimes it is more convenient to regard μ and σ^2 as the parameters, this being a matter of convention. We cannot affect the *number* of parameters by minor considerations of this kind.) We know that the mean of the distribution is equal to μ, and the variance to σ^2, but the mean and variance are no more parameters of the distribution than are, say, the median (also equal to μ), the mean deviation about the mean ($= \sigma(2/\pi)^{\frac{1}{2}}$), or any other of the infinite set of constants, including all the moments and cumulants, which we may be interested in. By "parameters", then, we refer to a finite number of constants appearing in the specified probability distribution of our random variable.

With this understanding, hypothesis (c), and also (d), of **22.2** are *non-parametric* hypotheses. We shall be discussing non-parametric hypotheses at length in Chapters 30 onwards, but most of the theoretical discussion in this and the next chapter is equally applicable to the parametric and the non-parametric case. However, our particularized discussions will mostly be of parametric hypotheses.

Simple and composite hypotheses

22.4 There is a distinction between the hypotheses (a) and (b) in **22.2**. In (a), the values of *all* the parameters of the distribution were specified by the hypothesis; in (b) only a subset of the parameters was specified by the hypothesis. This distinction is important for the theory. To formulate it generally, if we have a distribution depending upon l parameters, and a hypothesis specifies unique values for k of these parameters, we call the hypothesis *simple* if $k = l$ and we call it composite if $k < l$. In geometrical terms, we can represent the possible values of the parameters as a region in a space of l dimensions, one for each parameter. If the hypothesis considered selects a unique point in this parameter space, it is a simple hypothesis; if the hypothesis selects a sub-region of the parameter space which contains more than one point, it is composite.

$l-k$ is known as the number of *degrees of freedom* of the hypothesis, and k as the number of *constraints* imposed by the hypothesis. This terminology is obviously related to the geometrical picture in the last paragraph.

Critical regions and alternative hypotheses

22.5 To test any hypothesis on the basis of a (random) sample of observations, we must divide the sample space (i.e. all possible sets of observations) into two regions. If the observed sample point $\mathbf{x}$ falls into one of these regions, say w, we shall reject the hypothesis; if $\mathbf{x}$ falls into the complementary region, $W-w$, we shall accept the hypothesis. w is known as the *critical region* of the test, and $W-w$ is called the *acceptance region*.

It is necessary to make it clear at the outset that the rather peremptory terms "reject" and "accept" which we have used of a hypothesis under test are now conventional usage, to which we shall adhere, and are not intended to imply that any hypothesis is ever finally accepted or rejected in science. If the reader cannot overcome his philosophical dislike of these admittedly inapposite expressions, he will perhaps agree to regard them as code words, "reject" standing for "decide that the observations are unfavourable to" and "accept" for the opposite. We are concerned to investigate procedures which make such decisions with calculable probabilities of error, in a sense to be explained.

22.6 Now if we know the probability distribution of the observations under the hypothesis being tested, which we shall call H_0, we can determine w so that, given H_0, the probability of rejecting H_0 (i.e. the probability that $\mathbf{x}$ falls in w) is equal to a pre-assigned value α, i.e.

$$\text{Prob}\{\mathbf{x} \in w \mid H_0\} = \alpha. \tag{22.1}$$

If we are dealing with a discontinuous distribution, it may not be possible to satisfy (22.1) for every α in the interval $(0, 1)$. The value α is called the *size* of the test.(*) For the moment, we shall regard α as determined in some way. We shall discuss the choice of α later.

Evidently, we can in general find many, and often even an infinity, of sub-regions w of the sample space, all obeying (22.1). Which of them should we prefer to the others? This is the problem of the theory of testing hypotheses. To put it in everyday terms, which sets of observations are we to regard as favouring, and which as disfavouring, a given hypothesis?

22.7 Once the question is put in this way, we are directed to the heart of the problem. For it is of no use whatever to know merely what properties a critical region will have when H_0 holds. What happens when some other hypothesis holds? In other words, we cannot say whether a given body of observations favours a given hypothesis unless we know to what alternative(s) this hypothesis is being compared.

(*) The hypothesis under test is often called "the null hypothesis," and the size of the test "the level of significance." We shall not use these terms, since the words "null" and "significance" can be misleading.

It is perfectly possible for a sample of observations to be a rather " unlikely " one if the original hypothesis were true ; but it may be much more " unlikely " on another hypothesis. If the situation is such that we are forced to choose one hypothesis or the other, we shall obviously choose the first, notwithstanding the " unlikeliness " of the observations. The problem of testing a hypothesis is essentially one of choice between it and some other or others. It follows immediately that whether or not we accept the original hypothesis depends crucially upon the alternatives against which it is being tested.

The power of a test

22.8 The discussion of **22.7** leads us to the recognition that a critical region (or, synonymously, a test) must be judged by its properties both when the hypothesis tested is true and when it is false. Thus we may say that the errors made in testing a statistical hypothesis are of two types:

(I) We may wrongly reject it, when it is true ;
(II) We may wrongly accept it, when it is false.

These are known as Type I and Type II errors respectively. The probability of a Type I error is equal to the size of the critical region used, α. The probability of a Type II error is, of course, a function of the alternative hypothesis (say, H_1) considered, and is usually denoted by β. Thus

$$\text{Prob } \{\mathbf{x} \in W-w \mid H_1\} = \beta$$

or

$$\text{Prob } \{\mathbf{x} \in w \mid H_1\} = 1-\beta. \qquad (22.2)$$

This complementary probability, $1-\beta$, is called the *power* of the test of the hypothesis H_0 against the alternative hypothesis H_1. The specification of H_1 in the last sentence is essential, since power is a function of H_1.

Example 22.1

Consider the problem of testing a hypothetical value for the mean of a normal distribution with unit variance. Formally, in

$$dF(x) = (2\pi)^{-\frac{1}{2}} \exp\{-\tfrac{1}{2}(x-\mu)^2\} dx, \qquad -\infty \leqslant x \leqslant \infty,$$

we test the hypothesis

$$H_0 : \mu = \mu_0.$$

This is a simple hypothesis, since it specifies $F(x)$ completely. The alternative hypothesis will also be taken as the simple

$$H_1 : \mu = \mu_1 > \mu_0.$$

Thus, essentially, we are to choose between a smaller given value (μ_0) and a larger (μ_1) for the mean of our distribution.

We may represent the situation diagrammatically for a sample of $n = 2$ observations. In Fig. 22.1 we show the scatters of sample points which would arise, the lower cluster being that arising when H_0 is true, and the higher when H_1 is true.

In this case, of course, the sampling distributions are continuous, but the dots indicate roughly the condensations of sample densities around the true means.

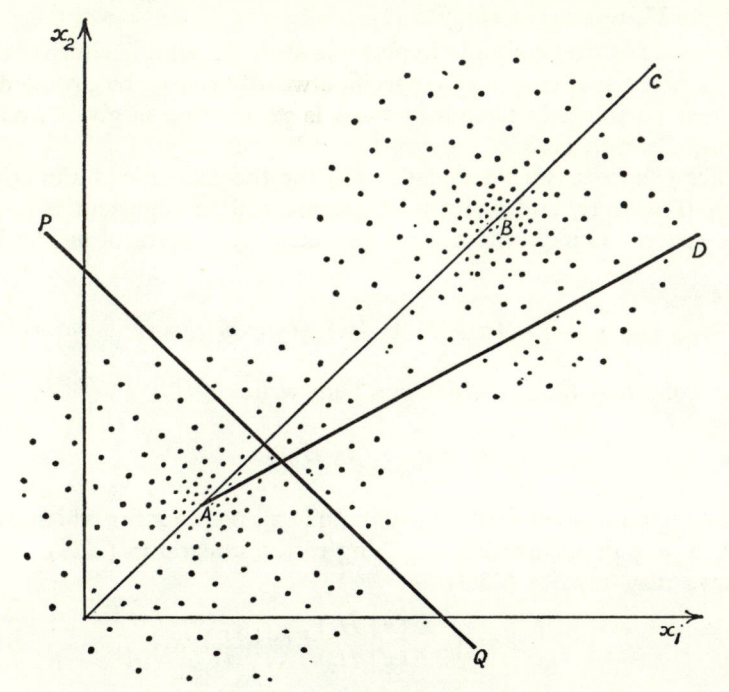

Fig. 22.1—Critical regions for $n = 2$ (see text)

To choose a critical region, we need, in accordance with (22.1), to choose a region in the plane containing a proportion α of the distribution on H_0. One such region is represented by the area above the line PQ, which is perpendicular to the line AB connecting the hypothetical means. (A is the point (μ_0, μ_0), and B the point (μ_1, μ_1).) Another possible critical region of size α is the region CAD.

We see at once from the circular symmetry of the clusters that the first of these critical regions contains a very much larger proportion of the H_1 cluster than does the CAD region. The first region will reject H_0 rightly, when H_1 is true, in a higher proportion of cases than will the second region. Consequently, its value of $1 - \beta$ in (22.2), or in other words its power, will be the greater.

22.9 Example 22.1 directs us to an obvious criterion for choosing among critical regions, all satisfying (22.1). We seek a critical region w such that its power, defined at (22.2), is as large as possible. Then, in addition to having controlled the probability of Type I errors at α, we shall have minimized the probability of a Type II error, β. This is the fundamental idea, first expressed explicitly by J. Neyman and E. S. Pearson, which underlies the theory of this and following chapters.

A critical region, whose power is no smaller than that of any other region of the same size for testing a hypothesis H_0 against the alternative H_1, is called a best critical region (abbreviated BCR), and a test based on a BCR is called a most powerful (abbreviated MP) test.

Testing a simple H_0 against a simple H_1

22.10 If we are testing a simple hypothesis against a simple alternative hypothesis, i.e. choosing between two completely specified distributions, the problem of finding a BCR of size α is particularly straightforward. Its solution is given by a lemma due to Neyman and Pearson (1933b), which we now prove.

As in earlier chapters, we write $L(x \mid H_i)$ for the Likelihood Function given the hypothesis H_i ($i = 0, 1$), and write a single integral to represent n-fold integration in the sample space. Our problem is to maximize, for choice of w, the integral form of (22.2),

$$1 - \beta = \int_w L(x \mid H_1)\,dx, \tag{22.3}$$

subject to the condition (22.1), which we here write

$$\alpha = \int_w L(x \mid H_0)\,dx. \tag{22.4}$$

The critical region w should obviously include all points $\mathbf{x}$ for which $L(x \mid H_0) = 0$, $L(x \mid H_1) > 0$; these points contribute nothing to the integral in (22.4). For the other points in w, we may rewrite (22.3) as

$$1 - \beta = \int_w \frac{L(x \mid H_1)}{L(x \mid H_0)} L(x \mid H_0)\,dx. \tag{22.5}$$

From (22.4–5), we see that $(1-\beta)/\alpha$ is the average within w, when H_0 holds, of $\dfrac{L(x \mid H_1)}{L(x \mid H_0)}$. Clearly this will be maximized if and only if w consists of that fraction α of the sample space containing the largest values of $\dfrac{L(x \mid H_1)}{L(x \mid H_0)}$. Thus the BCR consists of the points in W satisfying

$$\frac{L(x \mid H_0)}{L(x \mid H_1)} \leq k_\alpha \tag{22.6}$$

when H_0 holds. To any constant k_α in (22.6) there corresponds a value α for the size (22.4). If the x's are continuously distributed, we can also find a k_α for any α.

22.11 If the distribution of the x's is not continuous, we may effectively render it so by a randomization device (cf. **20.22**). In this case,

$$\frac{L(x \mid H_0)}{L(x \mid H_1)} = k_\alpha \tag{22.7}$$

with some non-zero probability p, while in general, owing to discreteness, we can only choose k_α in (22.6) to make the size of the test equal to $\alpha - q$ ($0 < q < p$). To convert the test into one of exact size α, we simply arrange that, whenever (22.7) holds, we use a random device (e.g. a table of random sampling numbers) so that with probability q/p we reject H_0, while with probability $1 - (q/p)$ we accept H_0. The overall probability of rejection will then be $(\alpha - q) + p \cdot q/p = \alpha$, as required, whatever the value of α desired. In this case, the BCR is clearly not unique, being subject to random sampling fluctuation.

Example 22.2

Consider again the normal distribution of Example 22.1,
$$dF(x) = (2\pi)^{-\frac{1}{2}} \exp\left[-\tfrac{1}{2}(x-\mu)^2\right] dx, \qquad -\infty \leq x \leq \infty, \tag{22.8}$$
where we are now to test $H_0 : \mu = \mu_0$ against the alternative $H_1 : \mu = \mu_1 (\neq \mu_0)$. We have
$$L(x \mid H_i) = (2\pi)^{-\frac{1}{2}n} \exp\left[-\tfrac{1}{2} \sum_{j=1}^n (x_j - \mu_i)^2\right], \qquad i = 0, 1,$$
$$= (2\pi)^{-\frac{1}{2}n} \exp\left[-\frac{n}{2}\{s^2 + (\bar{x}-\mu_i)^2\}\right] \tag{22.9}$$
where $\bar{x}$, s^2 are the sample mean and variance respectively. Thus, for the BCR, we have from (22.6)
$$\frac{L(x \mid H_0)}{L(x \mid H_1)} = \exp\left[\frac{n}{2}\{(\bar{x}-\mu_1)^2 - (\bar{x}-\mu_0)^2\}\right]$$
$$= \exp\left[\frac{n}{2}\{(\mu_0 - \mu_1) 2\bar{x} + (\mu_1^2 - \mu_0^2)\}\right] \leq k_\alpha, \tag{22.10}$$
or
$$(\mu_0 - \mu_1) \bar{x} \leq \tfrac{1}{2}(\mu_0^2 - \mu_1^2) + \frac{1}{n} \log k_\alpha. \tag{22.11}$$

Thus, given μ_0, μ_1 and α, the BCR is determined by the value of the sample mean $\bar{x}$ alone. This is what we might have expected from the fact (cf. Examples 17.6, 17.15) that $\bar{x}$ is a MVB sufficient statistic for μ. Further, from (22.11), we see that if $\mu_0 > \mu_1$ the BCR is
$$\bar{x} \leq \tfrac{1}{2}(\mu_0 + \mu_1) + \log k_\alpha / \{n(\mu_0 - \mu_1)\}, \tag{22.12}$$
while if $\mu_0 < \mu_1$ it is
$$\bar{x} \geq \tfrac{1}{2}(\mu_0 + \mu_1) - \log k_\alpha / \{n(\mu_1 - \mu_0)\}, \tag{22.13}$$
which is again intuitively reasonable: in testing the hypothetical value μ_0 against a smaller value μ_1, we reject μ_0 if the sample mean falls below a certain value, which depends on α, the size of the test; in testing μ_0 against a larger value μ_1, we reject μ_0 if the sample mean exceeds a certain value.

22.12 A feature of Example 22.2 which is worth remarking, since it occurs in a number of problems, is that the BCR turns out to be determined by a single statistic, rather than by the whole configuration of sample values. This simplification permits us to carry on our discussion entirely in terms of the sampling distribution of that statistic, called a "test statistic," and to avoid the complexities of n-dimensional distributions.

Example 22.3

In Example 22.2, we know (cf. Example 11.12) that whatever the value of μ, $\bar{x}$ is itself exactly normally distributed with mean μ and variance $1/n$. Thus, to obtain the BCR (22.13) of size α for testing μ_0 against $\mu_1 > \mu_0$, we determine $\bar{x}_\alpha$ so that

$$\int_{\bar{x}_\alpha}^{\infty} \left(\frac{n}{2\pi}\right)^{\frac{1}{2}} \exp\left\{-\frac{n}{2}(\bar{x}-\mu_0)^2\right\} d\bar{x} = \alpha.$$

Writing

$$G(x) = \int_{-\infty}^{x} (2\pi)^{-\frac{1}{2}} \exp\left(-\tfrac{1}{2}y^2\right) dy \qquad (22.14)$$

for the normal d.f., we see by substituting $y = n^{\frac{1}{2}}(\bar{x}-\mu_0)$ that

$$\alpha = 1 - G\{n^{\frac{1}{2}}(\bar{x}_\alpha - \mu_0)\} = G\{-n^{\frac{1}{2}}(\bar{x}_\alpha - \mu_0)\}$$

since $G(x) = 1 - G(-x)$ by symmetry. If we now write

$$d_\alpha = n^{\frac{1}{2}}(\bar{x}_\alpha - \mu_0) \qquad (22.15)$$

we have

$$G(-d_\alpha) = \alpha. \qquad (22.16)$$

For example, from a table of the normal d.f., we find for $\alpha = 0.05$ that $d_{0.05} = 1.6449$, so that when $\mu_0 = 2$ and $n = 25$, we have from (22.15)

$$\bar{x}_\alpha = 2 + 1.6449/5 = 2.3290.$$

In this normal case, the power of the test may be written down explicitly. It is

$$\int_{\bar{x}_\alpha}^{\infty} \left(\frac{n}{2\pi}\right)^{\frac{1}{2}} \exp\left\{-\frac{n}{2}(\bar{x}-\mu_1)^2\right\} d\bar{x} = 1 - \beta. \qquad (22.17)$$

Substituting $y = n^{\frac{1}{2}}(\bar{x}-\mu_1)$, this becomes

$$1 - \beta = 1 - G\{n^{\frac{1}{2}}(\mu_0 - \mu_1) + d_\alpha\} = G\{n^{\frac{1}{2}}(\mu_1 - \mu_0) - d_\alpha\}, \qquad (22.18)$$

again using the symmetry. From (22.18) it is clear that the power is a monotone increasing function both of n, the sample size, and of $(\mu_1 - \mu_0)$, the difference between the hypothetical values between which the test has to choose. It should be observed that although (22.13) is only the BCR for $\mu_1 > \mu_0$, the expression for its power at (22.18) remains valid for any μ_1 whatever. As $\mu_1 \to -\infty$, the power $\to 0$, and it remains less than α for all $\mu_1 < \mu_0$.

Example 22.4

As a contrast, consider the Cauchy distribution

$$dF(x) = \frac{dx}{\pi\{1+(x-\theta)^2\}}, \qquad -\infty \leqslant x \leqslant \infty,$$

and suppose that we wish to test

$$H_0 : \theta = 0$$

against

$$H_1 : \theta = \theta_1.$$

For simplicity, we shall confine ourselves to the case $n = 1$. According to (22.6), the BCR is given by

$$\frac{L(x \mid H_0)}{L(x \mid H_1)} = \frac{1+(x-\theta_1)^2}{1+x^2} \leqslant k_\alpha.$$

This is equivalent to
$$x^2(k_\alpha-1)+2\theta_1 x+(k_\alpha-1-\theta_1^2) \geqslant 0. \qquad (22.19)$$

The form of the BCR thus defined depends upon the value of α chosen. If $k_\alpha = 1$, (22.19) reduces to $\theta_1 x \geqslant \tfrac{1}{2}\theta_1^2$, i.e. to $x \geqslant \tfrac{1}{2}\theta_1$ if $\theta_1 > 0$ and to $x \leqslant \tfrac{1}{2}\theta_1$ if $\theta_1 < 0$ (cf. Example 22.2). But if $k_\alpha < 1$, the quadratic on the left of (22.19) will be non-negative only within a finite interval for x, while for $k_\alpha > 1$ it will be so only outside a finite interval for x. Thus the BCR changes its form with α.

Since the Cauchy distribution is a Student's distribution with one degree of freedom, and accordingly $F(x) = \tfrac{1}{2}+\dfrac{1}{\pi}\,\text{arc tan } \theta$, we may calculate the size of the test for any k_α and θ_1. Thus, for $\theta_1 = 1$ and $k_\alpha = 1$, the size is
$$\text{prob}\,(x \geqslant \tfrac{1}{2}) = 0\cdot 352,$$
while for $\theta_1 = 1$, $k_\alpha = 0\cdot 5$, (22.19) holds when $1 \leqslant x \leqslant 3$ and
$$\text{prob}\,(1 \leqslant x \leqslant 3) = 0\cdot 148.$$

This method may also be used to determine the powers of these tests. We leave this to the reader as Exercise 22.4 at the end of this chapter.

22.13 The examples we have given so far of the use of the Neyman–Pearson lemma have related to the testing of a parametric hypothesis for some given form of distribution. But, as will be seen on inspection of the proof in **22.10–22.11**, (22.6) gives the BCR for *any* test of a simple hypothesis against a simple alternative. For instance, we might be concerned to test the *form* of a distribution with known location parameter, as in the following example.

Example 22.5

Suppose that we know that a distribution is standardized, but wish to investigate its form. We wish to choose between the alternative forms
$$\left.\begin{array}{l} H_0: dF = (2\pi)^{-\frac{1}{2}}\exp(-\tfrac{1}{2}x^2)\,dx, \\ H_1: dF = 2^{-\frac{1}{2}}\exp(-2^{\frac{1}{2}}|x|)\,dx. \end{array}\right\} \quad -\infty \leqslant x \leqslant \infty,$$
For simplicity, we again take sample size $n = 1$.

Using (22.6), the BCR is given by
$$\frac{L(x\mid H_0)}{L(x\mid H_1)} = \pi^{-\frac{1}{2}}\exp(2^{\frac{1}{2}}|x|-\tfrac{1}{2}x^2) \leqslant k_\alpha.$$
Thus we reject H_0 when
$$2^{\frac{1}{2}}|x|-\tfrac{1}{2}x^2 \leqslant \log(k_\alpha \pi^{\frac{1}{2}}) = c_\alpha.$$
The BCR therefore consists of extreme positive and negative values of the observation, supplemented, if $k_\alpha > \pi^{-\frac{1}{2}}$ (i.e. $c_\alpha > 0$), by values in the neighbourhood of $x = 0$. Just as in Example 22.4, the form of the BCR depends upon α. The reader should verify this by drawing a diagram.

BCR and sufficient statistics

22.14 If both hypotheses being compared refer to the value of a parameter θ, and there is a sufficient statistic t for θ, it follows from the factorization of the Likelihood

Function at (17.68) that (22.6) becomes

$$\frac{L(x\mid\theta_0)}{L(x\mid\theta_1)} = \frac{g(t\mid\theta_0)}{g(t\mid\theta_1)} \leq k_\alpha, \qquad (22.20)$$

so that the BCR is a function of the value of the sufficient statistic t, as might be expected. We have already encountered an instance of this in Example 22.2. (The same result evidently holds if θ is a set of parameters for which t is a jointly sufficient set of statistics.) Exercise 22.13 shows that the ratio of likelihoods on the left of (22.20) is itself a sufficient statistic, so that the BCR is a function of its value.

However, it will not always be the case that the BCR will, as in Example 22.2, be of the form $t \geq a_\alpha$ or $t \leq b_\alpha$: Example 22.4, in which the single observation x is a sufficient statistic for θ, is a counter-example. Inspection of (22.20) makes it clear that the BCR will be of this particularly simple form if $g(t\mid\theta_0)/g(t\mid\theta_1)$ is a non-decreasing function of t for $\theta_0 > \theta_1$. This will certainly be true if

$$\frac{\partial^2}{\partial\theta\,\partial t}\log g(t\mid\theta) \geq 0, \qquad (22.21)$$

a condition which is satisfied by nearly all the distributions met with in statistics.

Example 22.6

For the distribution

$$dF(x) = \begin{cases} \exp\{-(x-\theta)\}\,dx, & \theta \leq x \leq \infty, \\ 0 & \text{elsewhere,} \end{cases}$$

the smallest sample observation $x_{(1)}$ is sufficient for θ (cf. Example 17.19). For a sample of n observations, we have, for testing θ_0 against $\theta_1 > \theta_0$,

$$\frac{L(x\mid\theta_0)}{L(x\mid\theta_1)} = \begin{cases} \infty & \text{if } x_{(1)} < \theta_1 \\ \exp\{n(\theta_0-\theta_1)\} & \text{otherwise.} \end{cases}$$

Thus we require for a BCR

$$\exp\{n(\theta_0-\theta_1)\} \leq k_\alpha. \qquad (22.22)$$

Now the left-hand side of (22.22) does not depend on the observations at all, being a constant, and (22.22) will therefore be satisfied by *every* critical region of size α with $x_{(1)} \geq \theta_1$. Thus every such critical region is of equal power, and is therefore a BCR.

If we allow θ_1 to be greater or less than θ_0, we find

$$\frac{L(x\mid\theta_0)}{L(x\mid\theta_1)} = \begin{cases} \infty & \text{if } \theta_0 \leq x_{(1)} < \theta_1, \\ \exp\{n(\theta_0-\theta_1)\} > 1 & \text{if } x_{(1)} \geq \theta_0 > \theta_1, \\ \exp\{n(\theta_0-\theta_1)\} < 1 & \text{if } x_{(1)} \geq \theta_1 > \theta_0, \\ 0 & \text{if } \theta_1 \leq x_{(1)} < \theta_0. \end{cases}$$

Thus the BCR is given by

$$(x_{(1)}-\theta_0) < 0, \quad (x_{(1)}-\theta_0) > c_\alpha.$$

The first of these events has probability zero on H_0. The value of c_α is determined to give probability α that the second event occurs when H_0 is true.

Estimating efficiency and power

22.15 The use of a statistic which is efficient in estimation (cf. **17.28–9**) does not imply that a more powerful test will be obtained than if a less efficient estimator had been used for testing purposes. This result is due to Sundrum (1954).

Let t_1 and t_2 be two asymptotically normally distributed estimators of a parameter θ, and suppose that, at least asymptotically,

$$\left. \begin{array}{l} E(t_1) = E(t_2) = \theta, \\ \operatorname{var}(t_i | \theta = \theta_0) = \sigma^2_{i0}, \\ \operatorname{var}(t_i | \theta = \theta_1) = \sigma^2_{i1}. \end{array} \right\} \quad i = 1, 2,$$

We now test $H_0 : \theta = \theta_0$ against $H_1 : \theta = \theta_1 > \theta_0$. Exactly as at (22.15) in Example 22.3, we have the critical regions, one for each test,

$$t_i \geqslant \theta_0 + d_\alpha \sigma_{i0}, \quad i = 1, 2, \tag{22.23}$$

where d_α is the normal deviate defined by (22.14) and (22.16). The powers of the tests are (generalizing (22.18) which dealt with a case where $\sigma_{i0} = \sigma_{i1}$)

$$1 - \beta(t_i) = G\left\{ \frac{(\theta_1 - \theta_0) - d_\alpha \sigma_{i0}}{\sigma_{i1}} \right\}. \tag{22.24}$$

Since $G(x)$ is a monotone increasing function of its argument, t_1 will provide a more powerful test than t_2 if and only if, from (22.24),

$$\frac{(\theta_1 - \theta_0) - d_\alpha \sigma_{10}}{\sigma_{11}} > \frac{(\theta_1 - \theta_0) - d_\alpha \sigma_{20}}{\sigma_{21}},$$

i.e. if

$$\theta_1 - \theta_0 > d_\alpha \left(\frac{\sigma_{10} \sigma_{21} - \sigma_{20} \sigma_{11}}{\sigma_{21} - \sigma_{11}} \right). \tag{22.25}$$

If we put $E_j = \sigma_{2j}/\sigma_{1j}$ ($j = 0, 1$), (22.25) becomes

$$\theta_1 - \theta_0 > d_\alpha \left(\frac{E_1 - E_0}{E_1 - 1} \right) \sigma_{10}. \tag{22.26}$$

E_0, E_1 are simply powers (usually square roots) of the estimating efficiency of t_1 relative to t_2 when H_0 and H_1 respectively hold. Now if

$$E_0 = E_1 > 1, \tag{22.27}$$

the right-hand side of (22.25) is zero, and (22.26) always holds. Thus if the estimating efficiency of t_1 exceeds that of t_2 *by the same amount* on both hypotheses, the more efficient statistic t_1 always provides a more powerful test, whatever value α or $\theta_1 - \theta_0$ takes. But if

$$E_1 > E_0 \geqslant 1 \tag{22.28}$$

we can always find a test size α small enough for (22.26) to be falsified. Hence, the less efficient estimator t_2 will provide a more powerful test if (22.28) holds, i.e. if its relative efficiency is greater on H_0 than on H_1. Alternatively if $E_0 > E_1 > 1$, we can find α *large* enough to falsify (22.26). If E_1 is continuous in θ, $E_1 \to E_0$ as $\theta_1 \to \theta_0$, so that (22.26) is not falsified in the immediate neighbourhood of θ_0.

This result, though a restrictive one, is enough to show that the relation between estimating efficiency and test power is rather loose. In Chapter 25 we shall again consider this relationship when we discuss the measurement of test efficiency.

Example 22.7

In Examples 18.3 and 18.6 we saw that in estimating the parameter ρ of a standardized bivariate normal distribution, the ML estimator $\hat{\rho}$ is a root of a cubic equation, with large-sample variance equal to $(1-\rho^2)^2/\{n(1+\rho^2)\}$, while the sample correlation coefficient r has large-sample variance $(1-\rho^2)^2/n$. Both estimators are consistent and asymptotically normal, and the ML estimator is efficient. In the notation of **22.15**,

$$E = (1+\rho^2)^{\frac{1}{2}}.$$

If we test $H_0 : \rho = 0$ against $H_1 : \rho = 0\cdot 1$, we have $E_0 = 1$, and (22.26) simplifies to

$$0\cdot 1 > d_\alpha \sigma_{10} = d_\alpha \left(\frac{1}{n}\right)^{\frac{1}{2}}. \tag{22.29}$$

If we choose n to be, say, 400, so that the normal approximations are adequate, we require

$$d_\alpha > 2$$

to falsify (22.29). This corresponds to $\alpha < 0\cdot 023$, so that for tests of size $< 0\cdot 023$, the inefficient estimator r has greater power asymptotically in this case than the efficient $\hat{\rho}$. Since tests of size $0\cdot 01$, $0\cdot 05$ are quite commonly used, this is not merely a theoretical example : it cannot be assumed in practice that " good " estimators are " good " test statistics.

Testing a simple H_0 against a class of alternatives

22.16 So far we have been discussing the most elementary problem, where in effect we have only to choose between two completely specified competitive hypotheses. For such a problem, there is a certain symmetry about the situation—it is only a matter of convention or convenience which of the two hypotheses we regard as being " under test " and which as " the alternative." As soon as we proceed to the generalization of the testing situation, this symmetry disappears.

Consider now the case where H_0 is simple, but H_1 is composite and consists of a class of simple alternatives. The most frequently occurring case is the one in which we have a class Ω of simple parametric hypotheses of which H_0 is one and H_1 comprises the remainder ; for example, the hypothesis H_0 may be that the mean of a certain distribution has some value μ_0 and the hypothesis H_1 that it has some other value unspecified.

For each of these other values we may apply the foregoing results and find, for given α, corresponding to any particular member of H_1 (say H_t) a BCR w_t. But this region in general will vary from one H_t to another. We obviously cannot determine a different region for all the unspecified possibilities and are therefore led to inquire whether there exists one BCR which is the best for all H_t in H_1. Such a region is called Uniformly Most Powerful (UMP) and the test based on it a UMP test.

22.17 Unfortunately, as we shall find below, a UMP test does not usually exist unless we restrict our alternative class Ω in certain ways. Consider, for instance, the case dealt with in Example 22.2. We found there that for $\mu_1 < \mu_0$ the BCR for a simple alternative was defined by

$$\bar{x} \leqslant a_\alpha. \tag{22.30}$$

Now so long as $\mu_1 < \mu_0$, the regions determined by (22.30) do not depend on μ_1 and can be found directly from the sampling distribution of $\bar{x}$ when the test size, α, is given. Consequently the test based on (22.30) is UMP for the class of hypotheses that $\mu_1 < \mu_0$.

However, from Example 22.2, if $\mu_1 > \mu_0$, the BCR is defined by $\bar{x} \geq b_\alpha$. Here again, if our class Ω is confined to the values of μ_1 greater than μ_0 the test is UMP. But if μ_1 can be either greater or less than μ_0, no UMP test is possible, for one or other of the two UMP regions we have just discussed will be better than any compromise region against this class of alternatives.

22.18 We now prove that for a simple $H_0 : \theta = \theta_0$ concerning a parameter θ defining a class of hypotheses, no UMP test exists in general against an interval including positive and negative values of $\theta - \theta_0$, under regularity conditions, in particular that the derivative of the likelihood with respect to θ is continuous in θ.

We expand the Likelihood Function in a Taylor series about θ_0, getting

$$L(x|\theta_1) = L(x|\theta_0) + (\theta_1 - \theta_0) L'(x|\theta^*) \tag{22.31}$$

where θ^* is some value in the interval (θ_1, θ_0). For the BCR, if any, we must have, from (22.6) and (22.31),

$$\frac{L(x|\theta_1)}{L(x|\theta_0)} = 1 + \frac{(\theta_1 - \theta_0) L'(x|\theta^*)}{L(x|\theta_0)} \geq k_\alpha(\theta_1). \tag{22.32}$$

Thus the BCR is defined by

$$\frac{L'(x|\theta^*)}{L(x|\theta_0)} \geq a_\alpha, \quad \theta_1 > \theta_0, \tag{22.33}$$

$$\leq b_\alpha, \quad \theta_1 < \theta_0. \tag{22.34}$$

Now consider what happens as θ_1 approaches θ_0. θ^* necessarily does the same, and in the immediate neighbourhood of θ_0, (22.33-4) become, in virtue of the continuity of L' in θ,

$$\frac{L'(x|\theta_0)}{L(x|\theta_0)} = \left[\frac{\partial \log L}{\partial \theta}\right]_{\theta=\theta_0} \geq a_\alpha, \quad \theta > \theta_0, \tag{22.35}$$

$$\leq b_\alpha, \quad \theta < \theta_0. \tag{22.36}$$

We thus establish, incidentally, that in the immediate neighbourhood of θ_0, one-sided tests based on $\left[\dfrac{\partial \log L}{\partial \theta}\right]_{\theta=\theta_0}$ are UMP. This is a testing analogue of the confidence interval result obtained in **20.17**.

Our main result now follows at once. If we are considering an interval of alternatives including positive and negative values of $(\theta_1 - \theta_0)$, (22.35) and (22.36) cannot both hold (and there can therefore be no BCR) unless

$$\left[\frac{\partial \log L}{\partial \theta}\right]_{\theta=\theta_0} = \text{constant}. \tag{22.37}$$

(22.37) is the essential condition for the existence of a two-sided BCR. It cannot be satisfied if (17.18) holds (e.g. for distributions with range independent of θ) unless the constant is zero, for the condition $E\left(\dfrac{\partial \log L}{\partial \theta}\right) = 0$ with (22.37) implies $\left[\dfrac{\partial \log L}{\partial \theta}\right]_{\theta=\theta_0} = 0$.

In Example 22.6, we have already encountered an instance where a two-sided BCR exists. The reader should verify that for that distribution $\left[\dfrac{\partial \log L}{\partial \theta}\right]_{\theta=\theta_0} = n$ exactly, so that (22.37) is satisfied.

UMP tests with more than one parameter

22.19 If the distribution considered has more than one parameter, and we are testing a simple hypothesis, it remains possible that a common BCR exists for a class of alternatives varying with these parameters. The following two examples discuss the case of the two-parameter normal distribution, where we might expect to find such a BCR, but where none exists, and the two-parameter exponential distribution, where a BCR does exist.

Example 22.8

Consider the normal distribution with mean μ and variance σ^2. The hypothesis to be tested is

$$H_0 : \mu = \mu_0, \ \sigma = \sigma_0,$$

and the alternative, H_1, is restricted only in that it must differ from H_0. For any such

$$H_1 : \mu = \mu_1, \ \sigma = \sigma_1,$$

the BCR is, from (22.6), given by

$$\frac{L(x|H_0)}{L(x|H_1)} = \left(\frac{\sigma_1}{\sigma_0}\right)^n \exp\left[-\tfrac{1}{2}\left\{\Sigma\left(\frac{x-\mu_0}{\sigma_0}\right)^2 - \Sigma\left(\frac{x-\mu_1}{\sigma_1}\right)^2\right\}\right] \leqslant k_\alpha.$$

This may be written in the form

$$s^2\left(\frac{1}{\sigma_1^2} - \frac{1}{\sigma_0^2}\right) + \frac{(\bar{x}-\mu_1)^2}{\sigma_1^2} - \frac{(\bar{x}-\mu_0)^2}{\sigma_0^2} \leqslant \frac{2}{n}\log\left\{\left(\frac{\sigma_0}{\sigma_1}\right)^n k_\alpha\right\}$$

where $\bar{x}$, s^2 are sample mean and variance respectively. If $\sigma_0 \neq \sigma_1$, we may further simplify this to

$$\left(\frac{\sigma_1^2 - \sigma_0^2}{\sigma_0^2 \sigma_1^2}\right)\Sigma(x-\rho)^2 \geqslant c_\alpha, \tag{22.38}$$

where c_α is independent of the observations, and

$$\rho = \frac{\mu_0 \sigma_1^2 - \mu_1 \sigma_0^2}{\sigma_1^2 - \sigma_0^2}.$$

We have already dealt with the case $\sigma_0 = \sigma_1$ in Example 22.2, where we took them both equal to 1.

(22.38), when a strict equality, is the equation of a hypersphere, centred at

$x_1 = x_2 = \ldots = x_n = \rho$. Thus the BCR is always bounded by a hypersphere. When $\sigma_1 > \sigma_0$, (22.38) yields
$$\Sigma (x-\rho)^2 \geq a_\alpha,$$
so that the BCR lies outside the sphere; when $\sigma_1 < \sigma_0$, we find from (22.38)
$$\Sigma (x-\rho)^2 \leq b_\alpha,$$
and the BCR is inside the sphere.

Since ρ is a function of μ_1 and σ_1, it is clear that there will not generally be a common BCR for different members of H_1, even if we limit ourselves by $\sigma_1 < \sigma_0$ and $\mu_1 < \mu_0$ or similar restrictions. We may illustrate the situation by a diagram of the $(\bar{x}, s)$ plane, for
$$\Sigma (x-\rho)^2 = \Sigma (x-\bar{x})^2 + n(\bar{x}-\rho)^2$$
$$= n \{s^2 + (\bar{x}-\rho)^2\}, \qquad (22.39)$$
and for (22.39) constant, we obtain a circle with centre $(\rho, 0)$ and fixed radius a function of α.

Fig. 22.2 (adapted from Neyman and Pearson, 1933b) illustrates some of the contours

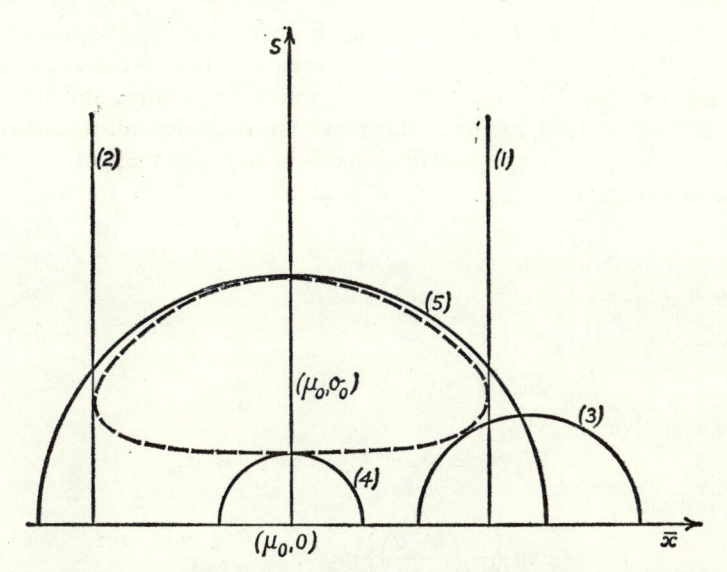

Fig. 22.2—Contours of constant likelihood ratio k (see text)

for particular cases. A single curve, corresponding to a fixed value of k in (22.37), is shown in each case.

Cases (1) and (2): $\sigma_1 = \sigma_0$ and $\rho = \pm \infty$. The BCR lies on the right of the line (1) if $\mu_1 > \mu_0$ and on the left of (2) if $\mu_1 < \mu_0$. This is the case discussed in Example 22.2.

Case (3): $\sigma_1 < \sigma_0$, say $\sigma_1 = \frac{1}{2}\sigma_0$. Then $\rho = \mu_0 + \frac{4}{3}(\mu_1 - \mu_0)$ and the BCR lies inside the semicircle marked (3).

Case (4): $\sigma_1 < \sigma_0$ and $\mu_1 = \mu_0$. The BCR is inside the semicircle (4).

Case (5): $\sigma_1 > \sigma_0$ and $\mu_1 = \mu_0$. The BCR is outside the semicircle (5).

There is evidently no common BCR for these cases. The regions of acceptance, however, may have a common part, centred round the value (μ_0, σ_0), and we should expect them to do so. Let us find the envelope of the BCR, which is, of course, the same as that of the regions of acceptance. The likelihood ratio is differentiated with respect to μ_1 and to σ_1, and these derivatives equated to zero. This gives precisely the ML solutions (cf. Example 18.11)

$$\hat{\mu}_1 = \bar{x},$$
$$\hat{\sigma}_1 = s.$$

Substituting in the likelihood ratio, we find for the envelope

$$1 - \frac{s^2}{\sigma_0^2} - \left(\frac{\bar{x}-\mu_0}{\sigma_0}\right)^2 = \frac{2}{n}\log\left\{\left(\frac{\sigma_0}{s}\right)^n k_\alpha\right\}$$

or

$$\left(\frac{\bar{x}-\mu_0}{\sigma_0}\right)^2 - \log\left(\frac{s^2}{\sigma_0^2}\right) + \frac{s^2}{\sigma_0^2} = 1 - \frac{2}{n}\log k_\alpha. \tag{22.40}$$

The dotted curve in Fig. 22.2 shows one such envelope. It touches the boundaries of all the BCR which have the same k (and hence are not of the same size α). The space inside may be regarded as a "good" region of acceptance and the space outside according as a good critical region. There is no BCR for all alternatives, but the regions determined by envelopes of likelihood-ratio regions effect a compromise by picking out and amalgamating parts of critical regions which are best for individual alternatives.

Example 22.9

To test the simple hypothesis

$$H_0: \theta = \theta_0, \; \sigma = \sigma_0$$

against the alternative

$$H_1: \theta = \theta_1 \leqslant \theta_0, \; \sigma = \sigma_1 < \sigma_0,$$

for the distribution

$$dF = \exp\left\{-\left(\frac{x-\theta}{\sigma}\right)\right\} dx/\sigma, \quad \begin{matrix} \theta \leqslant x \leqslant \infty; \\ \sigma > 0. \end{matrix} \tag{22.41}$$

From (22.6), the BCR is given by

$$\frac{L_0}{L_1} = \left(\frac{\sigma_1}{\sigma_0}\right)^n \exp\left\{-\frac{n(\bar{x}-\theta_0)}{\sigma_0} + \frac{n(\bar{x}-\theta_1)}{\sigma_1}\right\} \leqslant k_\alpha,$$

so that whatever the values of θ_1, σ_1 in H_1, the BCR is of form

$$x_{(1)} \leqslant \theta_0, \quad \bar{x} \leqslant \frac{\frac{1}{n}\log\left\{k_\alpha\left(\frac{\sigma_0}{\sigma_1}\right)^n\right\} + \left(\frac{\theta_1}{\sigma_1} - \frac{\theta_0}{\sigma_0}\right)}{\left(\frac{1}{\sigma_1} - \frac{1}{\sigma_0}\right)} \tag{22.42}$$

The first of these events has probability zero when H_0 holds. There is therefore a common BCR for the whole class of alternatives H_1, on which a UMP test may be based.

We have already effectively dealt with the case $\sigma_1 = \sigma_0$ in Example 22.6.

UMP tests and sufficient statistics

22.20 In **22.14** we saw that in testing a simple parametric hypothesis against a simple alternative, the BCR is necessarily a function of the value of the (jointly) sufficient statistic for the parameter(s), if one exists. In testing a simple H_0 against a composite H_1 consisting of a class of simple parametric alternatives, it evidently follows from the argument of **22.14** that if a common BCR exists, providing a UMP test against H_1, and if t is a sufficient statistic for the parameter(s), then the BCR will be a function of t. But, since a UMP test does not always exist, new questions now arise. Does the existence of a UMP test imply the existence of a corresponding sufficient statistic? And, conversely, does the existence of a sufficient statistic guarantee the existence of a corresponding UMP test?

22.21 The first of these questions may be affirmatively answered if an additional condition is imposed. In fact, as Neyman and Pearson (1936a) showed, if (1) there is a common BCR for, and therefore a UMP test of, H_0 against H_1 for every size α in an interval $0 < \alpha \leqslant \alpha_0$ (where α_0 is not necessarily equal to 1); and (2) if every point in the sample space W (save possibly a set of measure zero) forms part of the boundary of the BCR for at least one value of α, and then corresponds to a value of $L(x|H_0) > 0$; then a single sufficient statistic exists for the parameter(s) whose variation provides the class of admissible alternatives H_1.

To establish this result, we first note that, if a common BCR exists for H_0 against H_1 for two test sizes α_1 and $\alpha_2 < \alpha_1$, a common BCR of size α_2 can always be formed as a sub-region of that of size α_1. This follows from the fact that any common BCR satisfies (22.6). We may therefore, without loss of generality, take it that as α decreases, the BCR is adjusted simply by exclusion of some of its points.(*)

Now, suppose that conditions (1) and (2) are satisfied. If a point (say, x) of w forms part of the boundary of the BCR for only one value of α, we define the statistic

$$t(x) = \alpha. \tag{22.43}$$

If a point x forms part of the BCR boundary for more than one value of α, we define

$$t(x) = \tfrac{1}{2}(\alpha_1 + \alpha_2), \tag{22.44}$$

where α_1 and α_2 are the smallest and largest values of α for which it does so: it follows from the remark of the last paragraph that x will also be part of the BCR boundary for all α in the interval (α_1, α_2). The statistic t is thus defined by (22.43) and (22.44) for all points in W (except possibly a zero-measure set). Further, if t has the same value at two points, they must lie on the same boundary. Thus, from (22.6), we have

$$\frac{L(x|\theta_0)}{L(x|\theta_1)} = k(t, \theta),$$

(*) This is not true of critical regions in general—see, e.g., Chernoff (1951).

where k does not contain the observations except in the statistic t. Thus we must have

$$L(x|\theta) = g(t|\theta)h(x) \tag{22.45}$$

so that the single statistic t is sufficient for θ, the set of parameters concerned.

22.22 We have already considered in Example 22.2 a situation where single sufficiency and a UMP test exist together. Exercises 22.1 to 22.3 give further instances. But condition (2) of **22.21** is not always fulfilled, and then the existence of a single sufficient statistic may not follow from that of a UMP test. The following example illustrates the point.

Example 22.10

In Example 22.9, we showed that the distribution (22.41) admits a UMP test of the H_0 against the H_1 there described. The UMP test is based on the BCR (22.42). depending on $x_{(1)}$ and $\bar{x}$.

We have already seen (cf. **17.36**, Example 17.19 and Exercise 17.9) that the smallest observation $x_{(1)}$ is sufficient for θ if σ is known, and that $\bar{x}$ is sufficient for σ if θ is knows. The pair of statistics $x_{(1)}$ and $\bar{x}$ are jointly sufficient, but there is no *single* sufficient statistic for θ and σ.

22.23 On the other hand, the possibility that a single sufficient statistic exist without a one-sided UMP test, even where only a single parameter is involved, is made clear by Example 22.11.

Example 22.11

Consider the multinormal distribution of n variates $x_1, \ldots, x_n$, with

$$E(x_1) = n\theta, \quad \theta > 0,$$
$$E(x_r) = 0, \quad r > 1;$$

and dispersion matrix

$$\mathbf{V} = \begin{pmatrix} n-1+\theta^2, & -1, & \ldots & -1 \\ -1 & 1 & & \\ \vdots & & \ddots & 0 \\ \vdots & & & \ddots \\ -1 & & 0 & & 1 \end{pmatrix}. \tag{22.46}$$

The determinant of this matrix is easily seen to be

$$|\mathbf{V}| = \theta^2$$

and its inverse matrix is

$$\mathbf{V}^{-1} = \frac{1}{\theta^2} \begin{pmatrix} 1 & 1 & \ldots & 1 \\ 1, & 1+\theta^2, & & \\ \vdots & & \ddots & 1 \\ \vdots & & 1 & \ddots \\ 1 & & & & 1+\theta^2 \end{pmatrix}. \tag{22.47}$$

Thus, from **15.3**, the joint distribution is

$$dF = \frac{1}{\theta(2\pi)^{\frac{1}{2}n}} \exp\left\{-\frac{1}{2}\left[\frac{n^2}{\theta^2}(\bar{x}-\theta)^2 + \sum_{i=2}^{n} x_i^2\right]\right\} dx_1 \ldots dx_n. \tag{22.48}$$

Consider now the testing of the hypothesis $H_0: \theta = \theta_0 > 0$ against $H_1: \theta = \theta_1 > 0$ on the basis of a single observation. From (22.6), the BCR is given by

$$\frac{L(x|\theta_0)}{L(x|\theta_1)} = \left(\frac{\theta_1}{\theta_0}\right) \exp\left\{-\frac{n^2}{2}\left[\frac{(\bar{x}-\theta_0)^2}{\theta_0^2} - \frac{(\bar{x}-\theta_1)^2}{\theta_1^2}\right]\right\} \leqslant k_\alpha,$$

which reduces to

$$\frac{(\bar{x}-\theta_1)^2}{\theta_1^2} - \frac{(\bar{x}-\theta_0)^2}{\theta_0^2} \leqslant \frac{2}{n^2} \log(k_\alpha \theta_0/\theta_1)$$

or

$$\bar{x}^2(\theta_0^2 - \theta_1^2) - 2\bar{x}\,\theta_0\theta_1(\theta_0 - \theta_1) \leqslant \frac{2\theta_0^2\theta_1^2}{n^2} \log(k_\alpha \theta_0/\theta_1).$$

If $\theta_0 > \theta_1$, this is of form

$$\bar{x}^2(\theta_0+\theta_1) - 2\bar{x}\,\theta_0\theta_1 \leqslant a_\alpha, \tag{22.49}$$

which implies

$$b_\alpha \leqslant \bar{x} \leqslant c_\alpha. \tag{22.50}$$

If $\theta_0 < \theta_1$, the BCR is of form

$$\bar{x}^2(\theta_0+\theta_1) - 2\bar{x}\,\theta_0\theta_1 \geqslant d_\alpha, \tag{22.51}$$

implying

$$\bar{x} \leqslant e_\alpha \quad \text{or} \quad \bar{x} \geqslant f_\alpha. \tag{22.52}$$

In both (22.50) and (22.52), the limits between which (or outside which) $\bar{x}$ has to lie are functions of the exact value of θ_1. This difficulty, which arises from the fact that θ_1 appears in the coefficient of $\bar{x}^2$ in the quadratics (22.49) and (22.51), means that there is no BCR even for a one-sided set of alternatives, and therefore no UMP test.

It is easily verified that $\bar{x}$ is a single sufficient statistic for θ, and this completes the demonstration that single sufficiency does not imply the existence of a UMP test.

The power function

22.24 Now that we are considering the testing of a simple H_0 against a composite H_1, we generalize the idea of the power of a test defined at (22.2). As we stated there, the power is an explicit function of H_1. If, as is usual, H_1 is formed by the variation of a set of parameters θ, the power of a test of $H_0: \theta = \theta_0$ against the simple alternative $H_1: \theta = \theta_1 > \theta_0$ will be a function of the value of θ_1. For instance, we saw in Example 22.3 that the power of the most powerful test of the hypothesis that the mean μ of a normal population is μ_0 against the alternative value $\mu_1 > \mu_0$, is given by (22.18), a monotone increasing function of μ_1. (22.18) is called the *power function* of this test of H_0 against the class of alternatives $H_1: \mu > \mu_0$. We indicate the compositeness of H_1 by writing it thus, instead of the form used for a simple $H_1: \mu = \mu_1 > \mu_0$.

The evaluation of a power function is rarely so easy as in Example 22.3, since even if the sampling distribution of the test statistic is known exactly for both H_0 and the class of alternatives H_1 (and more commonly only approximations are available, especially for H_1), there is still the problem of evaluating (22.2) for each value of θ in H_1, which usually is a matter of numerical mathematics : only rarely is the power function exactly obtainable from a tabulated integral, as at (22.18). Asymptotically, however, the Central Limit theorem comes to our aid : the distributions of many test statistics tend to normality, given either H_0 or H_1, as sample size increases, and then the asymptotic power function will be of the form (22.18), as we shall see later.

Example 22.12

The general shape of the power function (22.18) in Example 22.3 is simply that of the normal distribution function. It increases from the value

$$G\{-d_\alpha\} = \alpha$$

at $\mu = \mu_0$ (in accordance with the size requirement) to the value

$$G\{0\} = 0.5$$

at $\mu = \mu_0 + \dfrac{d_\alpha}{n^{\frac{1}{2}}}$, the first derivative G' increasing up to this point ; as μ increases beyond it, G' declines to its limiting value of zero as G increases to its asymptote 1.

22.25 Once the power function of a test has been determined, it is of obvious value in determining how large the sample should be in order to test H_0 with given size and power. The procedure is illustrated in the next example.

Example 22.13

How many observations should be taken in Example 22.3 so that we may test $H_0: \mu = 3$ with $\alpha = 0.05$ (i.e. $d_\alpha = 1.6449$) and power of at least 0.75 against the alternatives that $\mu \geq 3.5$? Put otherwise, how large should n be to ensure that the probability of a Type I error is 0.05, and that of a Type II error at most 0.25 for $\mu \geq 3.5$?

From (22.18), we require n large enough to make

$$G\{n^{\frac{1}{2}}(3.5-3)-1.6449\} = 0.75, \qquad (22.53)$$

it being obvious that the power will be greater than this for $\mu > 3.5$. Now, from a table of the normal distribution

$$G\{0.6745\} = 0.75, \qquad (22.54)$$

and hence, from (22.53) and (22.54),

$$0.5n^{\frac{1}{2}} - 1.6449 = 0.6745,$$

whence

$$n = (4.6388)^2 = 21.5 \text{ approx.},$$

so that $n = 22$ will suffice to give the test the required power property.

One- and two-sided tests

22.26 We have seen in **22.18** that in general no UMP test exists when we test a parametric hypothesis $H_0: \theta = \theta_0$ against a two-sided alternative hypothesis, i.e. one in which $\theta - \theta_0$ changes sign. Nevertheless, situations often occur in which such an alternative hypothesis is appropriate; in particular, when we have no prior knowledge about the values of θ likely to occur. In such circumstances, it is tempting to continue to use as our test statistic one which is known to give a UMP test against one-sided alternatives ($\theta > \theta_0$ or $\theta < \theta_0$) but to modify the critical region in the distribution of the statistic by compromising between the BCR for $\theta > \theta_0$ and the BCR for $\theta < \theta_0$.

22.27 For instance, in Example 22.2 and in **22.17** we saw that the mean $\bar{x}$, used to test $H_0 : \mu = \mu_0$ for the mean μ of a normal population, gives a UMP test against $\mu_1 < \mu_0$ with common BCR $\bar{x} \leq a_\alpha$, and a UMP test for $\mu_1 > \mu_0$ with common BCR $\bar{x} \geq b_\alpha$. Suppose, then, that for the alternative $H_1 : \mu \neq \mu_0$, which is two-sided, we construct a compromise critical region defined by

$$\left. \begin{array}{l} \bar{x} \leq a_{\alpha/2}, \\ \bar{x} \geq b_{\alpha/2}, \end{array} \right\} \qquad (22.55)$$

in other words, combining the one-sided critical regions and making each of them of size $\tfrac{1}{2}\alpha$, so that the critical region as a whole remains one of size α.

We know that the critical region defined by (22.55) will always be less powerful than one or other of the one-sided BCR, but we also know that it will always be more powerful than the other. For its power will be, exactly as in Example 22.3,

$$G\{n^{\frac{1}{2}}(\mu - \mu_0) - d_{\alpha/2}\} + G\{n^{\frac{1}{2}}(\mu_0 - \mu) - d_{\alpha/2}\}. \qquad (22.56)$$

(22.56) is an even function of $(\mu - \mu_0)$, with a minimum at $\mu = \mu_0$. Hence it is always intermediate in value between $G\{n^{\frac{1}{2}}(\mu - \mu_0) - d_\alpha\}$ and $G\{n^{\frac{1}{2}}(\mu_0 - \mu) - d_\alpha\}$, which are the power functions of the one-sided BCR, except when $\mu = \mu_0$, when all three expressions are equal. The comparison is worth making diagrammatically, in Fig. 22.3, where a single fixed value of n and of α is illustrated.

22.28 We shall see later that other, less intuitive, justifications can be given for splitting the critical region in this way between the tails of the distribution of the test statistic. For the moment, the procedure is to be regarded as simply a common-sense way of insuring against the risk of extreme loss of power which, as Fig. 22.3 makes

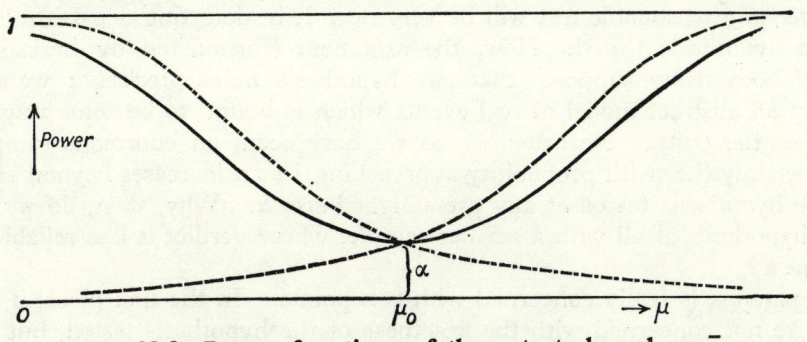

Fig. 22.3—Power functions of three tests based on $\bar{x}$
——— Critical region in both tails equally.
- - - - Critical region in upper tail.
—·—·— Critical region in lower tail.

Choice of test size

22.29 Throughout our exposition so far we have assumed that the test size α has been fixed in some way, and all our results are valid however that choice was made. We now turn to the question of how α is to be determined.

In the first place, it is natural to suggest that α should be made " small " according to some acceptable criterion, and indeed it is customary to use certain conventional values of α, such as 0·05, 0·01 or 0·001. But we must take care not to go too far in this direction. We can only fix two of the quantities n, α and β, even in testing a simple H_0 against a simple H_1. If n is fixed, we can only in general decrease the value of α, the probability of Type I error, by increasing the value of β, the probability of Type II error. In other words, reduction in the size of a test decreases its power.

This point is well illustrated in Example 22.3 by the expression (22.18) for the power of the BCR in a one-sided test for a normal population mean. We see there that as $\alpha \to 0$, by (22.16) $d_\alpha \to \infty$, and consequently the power (22.18) $\to 0$.

Thus, for fixed sample size, we have essentially to reconcile the size and power of the test. If the practical risks attaching to a Type I error are great, while those attaching to a Type II error are small, there is a case for reducing α, at the expense of increasing β, if n is fixed. If, however, sample size is at our disposal, we may, as in Example 22.13, ensure that n is large enough to reduce both α and β to any pre-assigned levels. These levels have still to be fixed, but unless we have supplementary information in the form of the *costs* (in money or other common terms) of the two types of error, and the costs of making observations, we cannot obtain an " optimum " combination of α, β and n for any given problem. It is sufficient for us to note that, however α is determined, we shall obtain a *valid* test.

22.30 The point discussed in **22.29** is reflected in another, which has sometimes been made the basis of criticism of the theory of testing hypotheses.

Suppose that we carry out a test with α fixed, no matter how, and n extremely large. The power of a reasonable test will be very near 1, in detecting departure of any sort from the hypothesis tested. Now, the argument (formulated by Berkson (1938)) runs : Nobody really supposes that any hypothesis holds precisely : we are simply setting up an abstract model of real events which is bound to be some way, if only a little, from the truth. Nevertheless, as we have seen, an enormous sample would almost certainly (i.e. with probability approaching 1 as n increases beyond any bound) reject the hypothesis tested at any pre-assigned size α. Why, then, do we bother to test the hypothesis at all with a smaller sample, whose verdict is less reliable than the larger one's ?

This paradox is really concerned with two points. In the first place, if n is fixed, and we are not concerned with the exactness of the hypothesis tested, but only with its approximate validity, our alternative hypothesis would embody this fact by being sufficiently distant from the hypothesis tested to make the difference of practical interest. This in itself would tend to increase the power of the test. But if we had no wish to reject the hypothesis tested on the evidence of small deviations from it, we should want the power of the test to be very low against these small deviations, and this would imply a small α and a correspondingly high β and low power.

But the crux of the paradox is the argument from increasing sample size. The hypothesis tested will only be rejected with probability near 1 if we keep α fixed as n increases. There is no reason why we should do this: we can determine α in any way we please, and it is rational, in the light of the discussion of **22.29**, to apply the gain in sensitivity arising from increased sample size to the reduction of α as well as of β. It is only the habit of fixing α at certain conventional levels that leads to the paradox. If we allow α to decline as n increases, it is no longer certain that a very small departure from H_0 will cause H_0 to be rejected: this now depends on the rate at which α declines.

22.31 There is a converse to the paradox discussed in **22.30**. Just as, for large n, inflexible use of conventional values of α will lead to very high power, which may possibly be too high for the problem in hand, so for very small fixed n their use will lead to very low power, perhaps too low. Again, the situation can be remedied by allowing α to rise and consequently reducing β. It is always incumbent upon the statistician to satisfy himself that, for the conditions of his problem, he is not sacrificing sensitivity in one direction to sensitivity in another.

Example 22.14

E. S. Pearson (discussion on Lindley (1953a)) has calculated a few values of the power function (22.56) of the two-sided test for a normal mean, which we reproduce to illustrate our discussion.

Table 22.1—Power function calculated from (22.56)

The entries in the first row of the table give the sizes of the tests.

Value of $\|\mu-\mu_0\|$	Sample size (n)						
	10			20	100		
0	0·050	0·072	0·111	0·050	0·050	0·019	0·0056
0·1				0·073	0·170	0·088	0·038
0·2	0·097	0·129	0·181	0·145	0·516	0·362	0·221
0·3				0·269	0·851	0·741	0·592
0·4	0·244	0·298	0·373	0·432	0·979	0·950	0·891
0·5				0·609	0·999	0·996	0·987
0·6	0·475	0·539	0·619	0·765			

It will be seen from the table that when sample size is increased from 20 to 100, the reductions of α from 0·050 to 0·019 and 0·0056 successively reduce the power of the test for each value of $|\mu-\mu_0|$. In fact, for $\alpha = 0.0056$ and $|\mu-\mu_0| = 0.1$, the power actually falls below the value attained at $n = 20$ with $\alpha = 0.05$. Conversely, on reduction of sample size from 20 to 10, the increase in α to 0·072 and 0·111 increases the power correspondingly, though only in the case $\alpha = 0·111$, $|\mu-\mu_0| = 0·2$, does it exceed the power at $n = 20$, $\alpha = 0·05$.

22.32 Bartholomew (1967) discusses the choice of α and β when n is a random variable, distributed free of θ. (22.6) then remains valid, but the distribution of n enters into the determination of k_α, which remains the same whatever the value of n observed. See also the discussion following Bartholomew's paper.

EXERCISES

22.1 Show directly by use of (22.6) that the BCR for testing a simple hypothesis $H_0: \mu = \mu_0$ concerning the mean μ of a Poisson distribution against a simple alternative $H_1: \mu = \mu_1$ is of the form

$$\bar{x} \leqslant a_\alpha \quad \text{if} \quad \mu_0 > \mu_1,$$
$$\bar{x} \geqslant b_\alpha \quad \text{if} \quad \mu_0 < \mu_1,$$

where $\bar{x}$ is the sample mean and a_α, b_α are constants.

22.2 Show similarly that for the parameter π of a binomial distribution, the BCR is of the form

$$x \leqslant a_\alpha \quad \text{if} \quad \pi_0 > \pi_1,$$
$$x \geqslant b_\alpha \quad \text{if} \quad \pi_0 < \pi_1,$$

where x is the observed number of " successes " in the sample.

22.3 Show that for the normal distribution with zero mean and variance σ^2, the BCR for $H_0: \sigma = \sigma_0$ against the alternative $H_1: \sigma = \sigma_1$ is of form

$$\sum_{i=1}^n x_i^2 \leqslant a_\alpha \quad \text{if} \quad \sigma_0 > \sigma_1,$$

$$\sum_{i=1}^n x_i^2 \geqslant b_\alpha \quad \text{if} \quad \sigma_0 < \sigma_1.$$

Show that the power of the BCR when $\sigma_0 > \sigma_1$ is $F\left\{\frac{\sigma_0^2}{\sigma_1^2} \chi^2_{\alpha, n}\right\}$, where $\chi^2_{\alpha, n}$ is the lower 100α per cent point and F is the d.f. of the χ^2 distribution with n degrees of freedom.

22.4 In Example 22.4, show that the power of the test when $\theta_1 = 1$ is 0·648 when $k_\alpha = 1$ and 0·352 when $k_\alpha = 0·5$. Draw a diagram of the two Cauchy distributions to illustrate the power and size of each test. Show that when $\theta_1 = 1$, $k_\alpha = 1·5$, the BCR consists of values of x outside the interval $(-2-\sqrt{5}, -2+\sqrt{5})$.

22.5 In Exercise 22.3, verify that the power is a monotone increasing function of σ_0^2/σ_1^2, and also verify numerically from a table of the χ^2 distribution that the power is a monotone increasing function of n.

22.6 Confirm that (22.21) holds for the sufficient statistics on which the BCR of Example 22.2, and Exercises 22.1–22.3 are based.

22.7 In **22.15** show that the more efficient estimator always gives the more powerful test if its test power exceeds 0·5.

(Sundrum, 1954)

22.8 Show that for testing $H_0: \mu = \mu_0$ in samples from the distribution

$$dF = dx, \quad \mu \leqslant x \leqslant \mu+1,$$

there is a pair of UMP one-sided tests, and hence no UMP test for all alternatives.

22.9 In Example 22.11, show that $\bar{x}$ is normally distributed with mean θ and variance θ^2/n^2, and that it is a sufficient statistic for θ.

22.10 Verify that the distribution of Example 22.10 does not satisfy condition (2) of **22.21**.

22.11 In Example 22.9, let σ be any positive increasing function of θ. Show that to test $H_0: \theta = \theta_0$ against $H_1: \theta = \theta_1 < \theta_0$, there is still a common BCR.

(Neyman & Pearson, 1936a)

22.12 Generalizing the discussion of **22.27**, write down the power function of any test based on the distribution of $\bar{x}$ with its critical region of form

$$\bar{x} \leqslant a_{\alpha_1},$$
$$\bar{x} \geqslant b_{\alpha_2},$$

where $\alpha_1 + \alpha_2 = \alpha$ (α_1 and α_2 not necessarily being equal). Show that the power function of any such test lies completely between those for the cases $\alpha_1 = 0$, $\alpha_2 = 0$ illustrated in Fig. 22.3.

22.13 Referring to the discussion of **22.14**, show that the likelihood ratio (for testing a simple $H_0: \theta = \theta_0$ against a simple $H_1: \theta = \theta_1$) is a sufficient statistic for θ on either hypothesis by writing the Likelihood Function as

$$L(x|\theta) = L(x|\theta_1) \left[\frac{L(x|\theta_0)}{L(x|\theta_1)} \right]^{(\theta-\theta_1)/(\theta_0-\theta_1)}$$

(Pitman, 1957)

22.14 For the most powerful test based on the BCR (22.6), show from (22.5) that $(1-\beta)/\alpha \geqslant 1/k_\alpha$ and hence by interchanging H_0 and H_1 that $\beta/(1-\alpha) \leqslant 1/k_\alpha \leqslant (1-\beta)/\alpha$.

22.15 From Exercise 22.14, show that as $n \to \infty$ with α fixed, the inequality $\beta \leqslant (1-\alpha)/k_\alpha$ becomes an equality and $\log \beta \sim E\{\log(L_1/L_0)\}$.

(cf. C. R. Rao (1962a). Efron (1967) gets a fuller result.)

CHAPTER 23

TESTS OF HYPOTHESES: COMPOSITE HYPOTHESES

23.1 We have seen in Chapter 22 that, when the hypothesis tested is simple (specifying the distribution completely), there is *always* a BCR, providing a most powerful test, against a simple alternative hypothesis; that there *may* be a UMP test against a class of simple hypotheses constituting a composite parametric alternative hypothesis; and that there will not, in general, be a UMP test if the parameter whose variation generates the alternative hypothesis is free to vary in both directions from the value tested.

If the hypothesis tested is composite, leaving at least one parameter value unspecified. it is to be expected that UMP tests will be even rarer than for simple hypotheses, but we shall find that progress can be made if we are prepared to restrict the class of tests considered in certain ways.

Composite hypotheses

23.2 First, we formally define the problem. We suppose that the n observations have a distribution dependent upon the values of $l(\leqslant n)$ parameters which we shall write

$$L(x|\theta_1, \ldots, \theta_l)$$

as before. The hypothesis to be tested is

$$H_0: \theta_1 = \theta_{10};\ \theta_2 = \theta_{20};\ \ldots;\ \theta_k = \theta_{k0}, \qquad (23.1)$$

where $k \leqslant l$, and the second suffix 0 denotes the value specified by the hypothesis. We lose no generality by thus labelling the k parameters whose values are specified by H_0 as the first k of the set of l parameters. H_0 as defined at (23.1) is said to impose k constraints, or alternatively to have $l-k$ degrees of freedom, though this latter (older) usage requires some care since we already use the term " degrees of freedom " in another sense.

Hypotheses of the form

$$H_0: \theta_1 = \theta_2;\ \theta_3 = \theta_4;\ \ldots,$$

which do not specify the values of parameters whose equality we are testing, may be transformable into the form (23.1) by reparametrizing the problem in terms of $\theta_1-\theta_2$, $\theta_3-\theta_4$, etc., and testing the hypothesis that these new parameters have zero values. Thus (23.1) is a more general composite hypothesis than at first appears.

To keep our notation simple, we shall write $L(x|\theta_r, \theta_s)$ and

$$H_0: \theta_r = \theta_{r0}, \qquad (23.2)$$

where it is to be understood that θ_r, θ_s may each consist of more than one parameter, the " nuisance parameter " θ_s being left unspecified by the hypothesis tested.

An optimum property of sufficient statistics

23.3 This is a convenient place to prove an optimum test property of sufficient statistics analogous to the result proved in **17.35**. There we saw that if t_1 is an unbiassed estimator of θ and t is a sufficient statistic for θ, then the statistic $E(t_1|t)$ is unbiassed for θ with variance no greater than that of t_1. We now prove a result due to Lehmann (1950): if w is a critical region for testing H_0, a hypothesis concerning θ in $L(x|\theta)$, against some alternative H_1, and t is a sufficient statistic, both on H_0 and on H_1, for θ, then there is a test of the same size, based on a function of t, which has the same power as w.

We first define a function (*)

$$c(w) \begin{cases} = 1 & \text{if the sample point is in } w, \\ = 0 & \text{otherwise.} \end{cases} \qquad (23.3)$$

Then the integral
$$\int c(w) L(x|\theta) dx = E\{c(w)\} \qquad (23.4)$$

gives the probability that the sample point falls into w, and is therefore equal to the size (α) of the test when H_0 is true and to the power of the test when H_1 is true. Using the factorization property (17.68) of the Likelihood Function in the presence of a sufficient statistic, (23.4) becomes

$$E\{c(w)\} = \int c(w) h(x|t) g(t|\theta) dx$$
$$= E\{E(c(w)|t)\}, \qquad (23.5)$$

the expectation operation outside the braces being with respect to the distribution of t. Thus the particular function of t, $E(c(w)|t)$, not dependent upon θ since t is sufficient, has the same expectation as $c(w)$. There is therefore a test based on the sufficient statistic t which has the same size and power as the original region w. We may therefore without loss of power confine the discussion of any test problem to functions of a sufficient statistic.

This result is quite general, and therefore also covers the case of a simple H_0 discussed in Chapter 22.

Test size for composite hypotheses: similar regions

23.4 Since a composite hypothesis leaves some parameter values unspecified, a new problem immediately arises, for the size of the test of H_0 will obviously be a function, in general, of these unspecified parameter values, θ_s.

If we wish to keep Type I errors down to some preassigned level, we must seek critical regions whose size can be kept down to that level for all possible values of θ_s. Thus we require

$$\alpha(\theta_s) \leqslant \alpha. \qquad (23.6)$$

If a critical region has
$$\alpha(\theta_s) = \alpha \qquad (23.7)$$

(*) $c(w)$ is known in measure theory as the characteristic function of the set of points w. We shall avoid this terminology, since there is some possibility of confusion with the use of "characteristic function" for the Fourier transform of a distribution function, with which we have been familiar since Chapter 4.

as a strict equality for all θ_s, it is called a (critical) region *similar to the sample space*(*) with respect to θ_s, or, more briefly, a similar (critical) region. The test based on a similar critical region is called a similar size-α test.

23.5 It is not obvious that similar regions exist at all generally, but, in one sense, as Feller (1938) pointed out, they exist whenever we are dealing with a set of n independent identically distributed observations on a continuous variate x. For no matter what the distribution or its parameters, we have
$$P\{x_1 < x_2 < x_3 < \ldots < x_n\} = 1/n! \tag{23.8}$$
(cf. 11.4), since any of the $n!$ permutations of the x_i is equally likely. Thus, for α an integral multiple of $1/n!$, there are similar regions based on the $n!$ hypersimplices in the sample space obtained by permuting the n suffixes in (23.8).

23.6 If we confine ourselves to regions defined by symmetric functions of the observations (so that similar regions based on (23.8) are excluded) it is easy to see that, where similar regions do exist, they need not exist for all sample sizes. For example, for a sample of n observations from the normal distribution
$$dF(x) = (2\pi)^{-\frac{1}{2}} \exp\{-\tfrac{1}{2}(x-\theta)^2\} dx,$$
there is no similar region with respect to θ for $n = 1$, but for $n \geqslant 2$ the fact that $ns^2 = \sum_{i=1}^{n}(x_i - \bar{x})^2$ has a chi-squared distribution with $(n-1)$ degrees of freedom, whatever the value of θ, ensures that similar regions of any size can be found from the distribution of s^2. This is because $\bar{x}$ is a single sufficient statistic for θ, and to find a similar region we must, by Exercise 23.3, find a statistic uncorrelated with
$$\frac{\partial \log L}{\partial \theta} = n(\bar{x} - \theta).$$
This is impossible when $n = 1$, since $\bar{x} = x$ is then the whole sample, but for $n \geqslant 2$, $\Sigma(x - \bar{x})^2$ is distributed independently of $\bar{x}$ and thus gives similar regions. The same argument holds in Exercise 23.1, where there is a pair of sufficient statistics for two parameters, and at least three observations are required so that we may have a statistic independent of both sufficient statistics.

23.7 Even if n is large, symmetric similar regions will not exist if each observation bring a new parameter with it, as in the following example, due to Feller (1938).

Example 23.1
Consider a sample of n observations, where the ith observation has distribution
$$dF(x_i) = (2\pi)^{-\frac{1}{2}} \exp\{-\tfrac{1}{2}(x_i - \theta_i)^2\} dx_i,$$
so that
$$L(x|\theta) = (2\pi)^{-\frac{1}{2}n} \exp\{-\tfrac{1}{2}\sum_i (x_i - \theta_i)^2\}.$$
For a similar region w of size α, we require, identically in θ,
$$\int_w L(x|\theta) dx = \alpha.$$

(*) The term arose because, trivially, the entire sample space is a similar region with $\alpha = 1$.

Using (23.3), we may re-write this size condition as

$$\int_W L(x|\theta)\frac{c(w)}{\alpha}dx = 1, \tag{23.9}$$

where W is the whole sample space. Differentiating (23.9) with respect to θ_i, we find

$$\int_W L(x|\theta)\frac{c(w)}{\alpha}(x_i-\theta_i)dx = 0. \tag{23.10}$$

A second differentiation with respect to θ_i gives

$$\int_W L(x|\theta)\frac{c(w)}{\alpha}\{(x_i-\theta_i)^2-1\}dx = 0. \tag{23.11}$$

Now from the definition of $c(w)$,

$$g(x|\theta) = L(x|\theta)\frac{c(w)}{\alpha} \tag{23.12}$$

is a (joint) frequency function. (23.10) and (23.11) express the facts that the marginal distribution of x_i in $g(x|\theta)$ has

$$E(x_i) = \theta_i, \quad \text{var } x_i = 1,$$

just as it has in the initial distribution of x_i.

If we examine the form of $g(x|\theta)$, we see that if we were to proceed with further differentiations, we should find that *all* the moments and product-moments of $g(x|\theta)$ are identical with those of $L(x|\theta)$, which is uniquely determined by its moments. Thus, from (23.12), $c(w)/\alpha = 1$ identically. But since $c(w)$ is either 0 or 1, we see finally that the trivial values $\alpha = 0$ or 1 are the only values for which similar regions can exist. The difficulty here is that all n observations are required to form a sufficient set for the n parameters, and we can find no statistic independent of them all.

23.8 It is nevertheless true that for many problems of testing composite hypotheses, similar regions exist for any size α and any sample size n. We now have to consider how they are to be found.

Let t be a sufficient statistic for the parameter θ_s unspecified by the hypothesis H_0, and suppose that we have a critical region w such that for all values of t, when H_0 is true,

$$E\{c(w)|t\} = \alpha. \tag{23.13}$$

Then, on taking expectations with respect to t, we have, as at (23.5),

$$E\{c(w)\} = E\{E(c(w)|t)\} = \alpha \tag{23.14}$$

so that the original critical region w is similar of size α, as Neyman (1937b) and Bartlett (1937) pointed out. Thus w is composed of a fraction α of the probability content of each contour of constant t.

It should be noticed that here t need be sufficient only for the unspecified parameter θ_s, and only when H_0 is true. This should be contrasted with the more demanding requirements of **23.3**.

Our argument has shown that (23.13) is a sufficient condition that w be similar. We shall show in **23.19** that it is necessary and sufficient, provided that a further condition is fulfilled, and in order to state that condition we must now introduce, following

TESTS OF HYPOTHESES : COMPOSITE HYPOTHESES

Lehmann and Scheffé (1950), the concept of the *completeness* of a parametric family of distributions, a concept which also permits us to supplement the discussion of sufficient statistics in Chapter 17.

Complete parametric families and complete statistics

23.9 Consider a parametric family of (univariate or multivariate) distributions, $f(x|\theta)$, depending on the value of a vector of parameters θ. Let $h(x)$ be any statistic, independent of θ. If

$$E\{h(x)\} = \int h(x) f(x|\theta) dx = 0 \qquad (23.15)$$

for all θ implies that
$$h(x) = 0 \qquad (23.16)$$

identically (save possibly on a zero-measure set), then the family $f(x|\theta)$ is called *complete*. The term is apt, since no non-zero function can be found that is orthogonal to all members of the family. If (23.15) implies (23.16) only for all bounded $h(x)$, $f(x|\theta)$ is called *boundedly complete*.

In the statistical applications of the concept of completeness, the family of distributions we are interested in is often the sampling distribution of a (possibly vector-) statistic t, say $g(t|\theta)$. We then call t a complete (or boundedly complete) statistic if, for all θ, $E\{h(t)\} = 0$ implies $h(t) = 0$ identically, for all functions (or bounded functions) $h(t)$. In other words, we label the statistic t with the completeness property of its distribution.

An evident immediate consequence of the completeness of a statistic t is that only one function of that statistic can have a given expected value. Thus if one function of t is an unbiased estimator of a certain function of θ, no other function of t will be. Completeness confers a uniqueness property upon an estimator.

J. K. Ghosh and Singh (1966) use a theorem by Wiener to show that if θ is a location parameter (i.e., $f = f(x-\theta)$), bounded completeness is equivalent to the c.f. $\phi(t)$ being non-zero for all t. Thus, e.g., the Cauchy distribution of Example 17.7 is boundedly complete—the c.f. is given in Example 4.2, Vol. 1.

The completeness of sufficient statistics

23.10 The special case of the exponential family (17.83) with $A(\theta) = \theta$, $B(x) = x$ has
$$f(x|\theta) = \exp\{\theta x + C(x) + D(\theta)\}, \qquad -\infty \leqslant x \leqslant \infty. \qquad (23.17)$$

If, for all θ,
$$\int h(x) f(x|\theta) dx = 0,$$

we must have
$$\int [h(x) \exp\{C(x)\}] \exp(\theta x) dx = 0. \qquad (23.18)$$

The integral in (23.18) is the two-sided Laplace transform (*) of the function in

---

(*) The two-sided Laplace transform of a function $g(x)$ is defined by
$$\lambda(\theta) = \int_{-\infty}^{\infty} \exp(\theta x) g(x) dx.$$
The integral converges in a strip of the complex plane $\alpha < R(\theta) < \beta$, where one or both of α, β may be infinite. (The strip may degenerate to a line.) Except possibly for a zero-measure set, there is a one-to-one correspondence between $g(x)$ and $\lambda(\theta)$. See, e.g., D. V. Widder (1941), *The Laplace Transform*, Princeton U.P., and compare also the Inversion Theorem for c.f.'s in 4.3.

square brackets in the integrand. By the uniqueness property of the transform, the only function having a transform of zero value is zero itself; i.e.,
$$h(x)\exp\{C(x)\} = 0$$
identically, whence
$$h(x) = 0$$
identically. Thus $f(x|\theta)$ is complete.

This result generalizes to the multi-parameter case, as shown by Lehmann and Scheffé (1955): the k-parameter, k-variate exponential family

$$f(\mathbf{x}|\boldsymbol{\theta}) = \exp\left\{\sum_{j=1}^{k} \theta_j x_j + C(\mathbf{x}) + D(\boldsymbol{\theta})\right\} \tag{23.19}$$

is a complete family. We have seen (Exercise 17.14) that the joint distribution of the set of k sufficient statistics for the k parameters of the general univariate exponential form (17.86) takes a form of which (23.19) is the special case, with $A_j(\theta) = \theta_j$. (We have replaced nD and Q of the Exercise by D and $\exp(C)$ respectively.) By **23.3**, we may confine ourselves, in testing hypotheses about the parent parameters, to the sufficient statistics.

Example 23.2

Consider the family of normal distributions

$$f(x|\theta_1, \theta_2) = (2\pi\theta_2)^{-\frac{1}{2}} \exp\left\{-\frac{1}{2\theta_2}(x-\theta_1)^2\right\}, \quad -\infty \leqslant x \leqslant \infty; \; \theta_2 > 0.$$

(a) If θ_2 is known (say $= 1$), the family is complete with respect to θ_1, for we are then considering a special case of (23.17) with
$$\theta = \theta_1, \quad \exp\{C(x)\} = (2\pi)^{-\frac{1}{2}}\exp(-\tfrac{1}{2}x^2)$$
and
$$D(\theta) = -\tfrac{1}{2}\theta_1^2.$$

(b) If, on the other hand, θ_1 is known (say $= 0$), the family is not even boundedly complete with respect to θ_2, for $f(x|0, \theta_2)$ is an even function of x, so that any odd function $h(x)$ will have zero expectation without being identically zero. However, if we transform to $y = x^2$, we see that the distribution of y is complete, since $g(y|\theta_2) = (2\pi\theta_2 y)^{-\frac{1}{2}}\exp\{-y/(2\theta_2)\}$ is again a special case of (23.17). Cf. the remark in **23.16** below.

23.11 In **23.10** we discussed the completeness of the characteristic form of the joint distribution of sufficient statistics in samples from a parent distribution with range independent of the parameters. Hogg and Craig (1956) have established the completeness of the sufficient statistic for parent distributions whose range is a function of a single parameter θ and which possess a single sufficient statistic for θ. We recall from **17.40–1** that the parent must then be of form

$$f(x|\theta) = g(x)/h(\theta) \tag{23.20}$$

and that

(i) if a single terminal of $f(x|\theta)$ is a function of θ (which may be taken to be θ itself without loss of generality), the corresponding extreme order-statistic is sufficient;

TESTS OF HYPOTHESES: COMPOSITE HYPOTHESES

(ii) if both terminals are functions of θ, the upper terminal ($b(\theta)$) must be a monotone decreasing function of the lower terminal (θ) for a single sufficient statistic to exist, and that statistic is then

$$\min\{x_{(1)}, b^{-1}(x_{(n)})\}. \tag{23.21}$$

We consider the cases (i) and (ii) in turn.

23.12 In case (i), take the upper terminal equal to θ, the lower equal to a constant a. $x_{(n)}$ is then sufficient for θ. Its distribution is, from (11.34) and (23.20),

$$dG(x_{(n)}) = n\{F(x_{(n)})\}^{n-1} f(x_{(n)}) dx_{(n)}$$

$$= \frac{n\left\{\int_a^{x_{(n)}} g(x)\,dx\right\}^{n-1} g(x_{(n)})}{\{h(\theta)\}^n} dx_{(n)}, \qquad a \leqslant x_{(n)} \leqslant \theta. \tag{23.22}$$

Now suppose that for a statistic $u(x_{(n)})$ we have

$$\int_a^\theta u(x_{(n)})\,dG(x_{(n)}) = 0,$$

or, substituting from (23.22), and dropping the factor in $h(\theta)$,

$$\int_a^\theta u(x_{(n)}) \left\{\int_a^{x_{(n)}} g(x)\,dx\right\}^{n-1} g(x_{(n)})\,dx_{(n)} = 0. \tag{23.23}$$

If we differentiate (23.23) with respect to θ, we find

$$u(\theta) \left\{\int_a^\theta g(x)\,dx\right\}^{n-1} g(\theta) = 0, \tag{23.24}$$

and since the integral in braces equals $h(\theta)$, while $g(\theta) \neq 0 \neq h(\theta)$, (23.24) implies

$$u(\theta) = 0$$

for any θ. Hence the function $u(x_{(n)})$ is identically zero, and the distribution of $x_{(n)}$, given at (23.22), is complete. Exactly the same argument holds for the lower terminal and $x_{(1)}$.

23.13 In case (ii), the distribution function of the sufficient statistic (23.21) is

$$G(t) = P\{x_{(1)}, b^{-1}(x_{(n)}) \leqslant t\}$$
$$= P\{x_{(1)} \leqslant t, x_{(n)} \leqslant b(t)\}$$
$$= \left\{\int_t^{b(t)} \frac{g(x)}{h(\theta)}\,dx\right\}^n. \tag{23.25}$$

Differentiating (23.25) with respect to t, we obtain the frequency function of the sufficient statistic,

$$g(t) = \frac{n}{\{h(\theta)\}^n}\left\{\int_t^{b(t)} g(x)\,dx\right\}^{n-1} [g\{b(t)\}b'(t) - g(t)],$$

$$\theta \leqslant t \leqslant c(\theta). \tag{23.26}$$

If there is a statistic $u(t)$ for which

$$\int_\theta^{c(\theta)} u(t)g(t)\,dt = 0, \tag{23.27}$$

we find, on differentiating (23.27) with respect to θ and by following through the argument of **23.12**, that $u(\theta) = 0$ for any θ, as before. Thus $u(t) = 0$ identically and $g(t)$ at (23.26) is complete.

23.14 The following example is of a non-complete single sufficient statistic.

Example 23.3

Consider a sample of a single observation x from the rectangular distribution
$$dF = dx, \qquad \theta \leqslant x \leqslant \theta+1.$$
x is evidently a sufficient statistic. (There would be no single sufficient statistic for $n \geqslant 2$, since the condition (ii) of **23.11** is not satisfied.)

Any bounded periodic function $h(x)$ of period 1 which satisfies
$$\int_0^1 h(x)\,dx = 0$$
will give us
$$\int_\theta^{\theta+1} h(x)\,dF = \int_\theta^{\theta+1} h(x)\,dx = \int_0^1 h(x)\,dx = 0,$$
so that the distribution is not even boundedly complete, since $h(x)$ is not identically zero.

Minimal sufficiency

23.15 We recall from **17.38** that, when we consider the problem of sufficient statistics in general (i.e. without restricting ourselves, as we did earlier in Chapter 17, to the case of a single sufficient statistic), we have to consider the choice between alternative sets of sufficient statistics. In a sample of n observations we *always* have a set of n sufficient statistics (namely, the observations themselves) for the $k\ (\geqslant 1)$ parameters of the distribution we are sampling from. Sometimes, though not always, there will be a set of $s\ (<n)$ statistics sufficient for the parameters. Often, $s = k$; e.g. all the cases of sufficiency discussed in Examples 17.15–16 have $s = k = 1$, while in Example 17.17 we have $s = k = 2$. By contrast, the following is an example in which $s < k$.[*]

Example 23.4

Consider again the problem of Example 22.11, with the alteration that
$$E(x_1) = n\mu$$
instead of $n\theta$ as previously. Exactly as before, we find for the joint distribution
$$dF = \frac{1}{\theta(2\pi)^{\frac{1}{2}n}} \exp\left\{-\tfrac{1}{2}\left[\frac{n^2}{\theta^2}(\bar{x}-\mu)^2 + \sum_{i=2}^n x_i^2\right]\right\} dx_1 \ldots dx_n.$$
Here it is clear that the single statistic $\bar{x}$ is sufficient for the parameters μ,θ.

---

[*] Fisher (e.g., 1956) called a sufficient set of statistics "sufficient" only if $s = k$ and "exhaustive" if $s > k$.

23.16 We thus have to ask ourselves: what is the *smallest* number s of statistics which constitute a sufficient set in any problem? With this in mind, Lehmann and Scheffé (1950) define a vector of statistics as *minimal sufficient* if it is a single-valued function of all other vectors of statistics which are sufficient for the parameters of the distribution.[*] The problems which now raise themselves are: how can we be sure, in any particular situation, that a sufficient vector is the minimal sufficient vector? And can we find a construction which yields the minimal sufficient vector?

A partial answer to the first of these questions is supplied by the following result: if the vector **t** is a boundedly complete sufficient statistic for **θ**, and the vector **u** is a minimal sufficient statistic for **θ**, then **t** is equivalent to **u**, i.e. they are identical, except possibly for a zero-measure set.

The proof is simple. Let w be a region in the sample space for which
$$D = E(c(w)\,|\,\mathbf{t}) - E(c(w)\,|\,\mathbf{u}) \neq 0, \tag{23.28}$$
where the function $c(w)$ is defined at (23.3). From (23.28), we find, on taking expectations over the entire sample space,
$$E(D) = 0. \tag{23.29}$$
Now since **u** is minimal sufficient, it is a function of **t**, another sufficient statistic, by definition. Hence we may write (23.28)
$$D = h(\mathbf{t}) \neq 0. \tag{23.30}$$
Since D is a bounded function, (23.29) and (23.30) contradict the assumed bounded completeness of **t**, and thus there can be no region w for which (23.28) holds. Hence **t** and **u** are equivalent statistics, i.e. **t** is minimal sufficient.

The converse does not hold: while bounded completeness implies minimal sufficiency, we can have minimal sufficiency without bounded completeness. An important instance is discussed in Example 23.10 below.

A consequence of the result of this section is that there cannot be more than one boundedly complete sufficient statistic for a parameter. E.g., in Example 23.2(b) x^2 is minimal sufficient and complete, while x is sufficient and not complete—cf. also Exercises 18.13 and 23.31 for other instances of minimal and non-minimal single sufficient statistics.

An alternative formulation of the problem of minimal sufficiency is given by Dynkin (1951).

23.17 In view of the results of **23.10–13** concerning the completeness of sufficient statistics, a consequence of **23.16** is that all the examples of sufficient statistics we have discussed in earlier chapters are minimal sufficient, as one would expect on intuitive grounds.

23.18 The result of section **23.16**, though useful, is less direct than the following procedure for finding a minimal sufficient statistic, given by Lehmann and Scheffé (1950).

---

[*] That this is for practical purposes equivalent to a sufficient statistic with minimum number of components is shown by Barankin and Katz (1959). See also Barankin (1960a, 1960b, 1961) and Fraser (1963).

We have seen in **22.14** and **22.20** that in testing a simple hypothesis, the ratio of likelihoods is a function of the sufficient (set of) statistic(s). We may now, so to speak, put this result into reverse, and use it to find the minimal sufficient set. Writing $L(x|\theta)$ for the LF as before, where x and θ may be vectors, consider a particular set of values x_0 and select all those values of x within the permissible range for which $L(x|\theta)$ is non-zero and

$$\frac{L(x|\theta)}{L(x_0|\theta)} = k(x, x_0) \qquad (23.31)$$

is independent of θ. Now any sufficient statistic t (possibly a vector) will satisfy (17.68), whence

$$\frac{L(x|\theta)}{L(x_0|\theta)} = \frac{g(t|\theta)}{g(t_0|\theta)} \cdot \frac{h(x)}{h(x_0)}, \qquad (23.32)$$

so that if $t = t_0$, (23.32) reduces to the form (23.31). Conversely, if (23.31) holds for all θ, this implies the constancy of the sufficient statistic t at the value t_0. This may be used to identify sufficient statistics, and to select the minimal set, in the manner of the following examples.

Example 23.5

We saw in Example 17.17 that the set of statistics $(\bar{x}, s^2)$ is jointly sufficient for the parameters (μ, σ^2) of a normal distribution. For this distribution, $L(x|\theta)$ is non-zero for all $\sigma^2 > 0$, and the condition (23.31) is, on taking logarithms, that

$$-\frac{1}{2\sigma^2}\left\{\left(\sum_i x_i^2 - \sum_i x_{0i}^2\right) - 2\mu n(\bar{x} - \bar{x}_0)\right\} \qquad (23.33)$$

be independent of (μ, σ^2), i.e. that the term in braces be equal to zero. This will be so, for example, if every x_i is equal to the corresponding x_{0i}, confirming that the set of n observations is a jointly sufficient set, as we have remarked that they always are.

It will also be so if the x_i are any rearrangement (permutation) of the x_{0i}: thus the set of *order-statistics* is sufficient, as it is again obvious that they always are. Finally, the condition in (23.33) will be satisfied if

$$\bar{x} = \bar{x}_0, \qquad \sum_i x_i^2 = \sum_i x_{0i}^2, \qquad (23.34)$$

and clearly, from inspection, nothing less than this will do. Thus the pair $(\bar{x}, \Sigma x^2)$ is minimal sufficient: equivalently, since $ns^2 = \Sigma x^2 - n\bar{x}^2$, $(\bar{x}, s^2)$ is minimal sufficient.

Example 23.6

As a contrast, consider the Cauchy distribution of Example 17.7. $L(x|\theta)$ is everywhere non-zero and (23.31) requires that

$$\prod_{i=1}^{n}\{1+(x_{0i}-\theta)^2\}\bigg/\prod_{i=1}^{n}\{1+(x_i-\theta)^2\} \qquad (23.35)$$

be independent of θ. As in the previous example, the set of order-statistics is sufficient, but nothing less will do here, for (23.35) is the ratio of two polynomials, each of degree $2n$ in θ. If the ratio is to be independent of θ, each polynomial must have the same set of roots, possibly permuted. Thus we are thrown back on the order-statistics as the minimal sufficient set.

Completeness and similar regions

23.19 After our lengthy excursus on completeness, we return to the discussion of similar regions in **23.8**. We may now show that if, given H_0, the sufficient statistic t is boundedly complete, *all* size-α similar regions must satisfy (23.13). For any such region, (23.14) holds and may be re-written

$$E\{E(c(w)\,|\,t) - \alpha\} = 0. \tag{23.36}$$

The expression in braces in (23.36) is bounded. Thus if t is boundedly complete, (23.36) implies that $E(c(w)\,|\,t) - \alpha = 0$ identically, i.e. that (23.13) holds.

The converse result also holds: if all similar regions satisfy (23.13), then Lehmann and Scheffé (1950) proved that t must be boundedly complete. Thus the bounded completeness of a sufficient statistic is equivalent to the condition that all similar regions w satisfy (23.13).

The choice of most powerful similar regions

23.20 The importance of the result of **23.19** is that it permits us to reduce the problem of finding most powerful similar regions for a composite hypothesis to the familiar problem of finding a BCR for a simple hypothesis.

By **23.19**, the bounded completeness of the statistic t, sufficient for θ_s on H_0, implies that all similar regions w satisfy (23.13), i.e. every similar region is composed of a fraction α of the probability content of each contour of constant t. We therefore may conduct our discussion with t held constant. Constancy of the sufficient statistic, t, for θ_s implies from (17.68) that the conditional distribution of the observations in the sample space will be independent of θ_s. Thus the composite H_0 with θ_s unspecified is reduced to a simple H_0 with t held constant. If t is also sufficient for θ_s when H_1 holds, the composite H_1 is also reduced to a simple H_1 with t constant (and, incidentally, the power of any critical region with t constant, as well as its size, will be independent of θ_s). If, however, t is not sufficient for θ_s when H_1 holds, we consider H_1 as a class of simple alternatives to the simple H_0, in just the manner of the previous chapter.

Thus, by keeping t constant, we reduce the problem to that of testing a simple H_0 concerning θ_r against a simple H_1 (or a class of simple alternatives constituting H_1). We use the methods of the last chapter, based on the Neyman–Pearson lemma (22.6), to seek a BCR (or common BCR) for H_0 against H_1. If there is such a BCR for each fixed value of t, it will evidently be an unconditional BCR, and gives the most powerful similar test of H_0 against H_1. Just as previously, if this test remains most powerful against a class of alternative values of θ_r, it is a UMP similar test.

Example 23.7

To test $H_0 : \mu = \mu_0$ against $H_1 : \mu = \mu_1$ for the normal distribution

$$dF = \frac{1}{\sigma(2\pi)^{\frac{1}{2}}} \exp\left\{-\tfrac{1}{2}\left(\frac{x-\mu}{\sigma}\right)^2\right\} dx, \quad -\infty \leqslant x \leqslant \infty.$$

H_0 and H_1 are composite with one degree of freedom, σ^2 being unspecified.

From Examples 17.10 and 17.15, the statistic (calculated from a sample of n independent observations) $u = \sum_{i=1}^{n}(x_i - \mu_0)^2$ is sufficient for σ^2 when H_0 holds, but not otherwise. From **23.10**, u is a complete statistic. All similar regions for H_0 therefore consist of fractions α of each contour of constant u.

Holding u fixed, we now test
$$H_0 : \mu = \mu_0 \quad \text{against} \quad H_1' : \mu = \mu_1, \; \sigma = \sigma_1,$$
both hypotheses being simple. The BCR obtained from (22.6) is that for which
$$\frac{L(x|H_0)}{L(x|H_1)} \leqslant k_\alpha.$$
This reduces, on simplification, to the condition
$$\bar{x}(\mu_1 - \mu_0) \geqslant C(\mu_0, \mu_1, \sigma^2, \sigma_1^2, k_\alpha, u) \tag{23.37}$$
where C is a constant containing no function of x except u. Thus the BCR consists of large values of $\bar{x}$ if $\mu_1 - \mu_0 > 0$ and of small values of $\bar{x}$ if $\mu_1 - \mu_0 < 0$, and this is true whatever the values of σ^2 and σ_1^2, and whatever the magnitude of $|\mu_1 - \mu_0|$. Thus we have a common BCR for the class of alternatives $H_1 : \mu = \mu_1$ for each one-sided situation $\mu_1 > \mu_0$ and $\mu_1 < \mu_0$.

We have been holding u fixed. Now
$$u = \Sigma(x - \mu_0)^2 = \Sigma(x - \bar{x})^2 + n(\bar{x} - \mu_0)^2 \tag{23.38}$$
$$= \Sigma(x - \bar{x})^2 \left\{ 1 + \frac{n(\bar{x} - \mu_0)^2}{\Sigma(x - \bar{x})^2} \right\}. \tag{23.39}$$

Since the BCR for fixed u consists of extreme values of $\bar{x}$, (23.38) implies that the BCR consists of small values of $\Sigma(x - \bar{x})^2$, which by (23.39) implies large values of
$$\frac{t^2}{n-1} = \frac{n(\bar{x} - \mu_0)^2}{\Sigma(x - \bar{x})^2}. \tag{23.40}$$

t^2 as defined by (23.40) is the square of the "Student's" t statistic whose distribution was derived in Example 11.8. By Exercise 23.7, t, which is distributed free of σ^2, is consequently distributed independently of the complete sufficient statistic, u, for σ^2. Remembering the necessary sign of $\bar{x}$, we have finally that the unconditional UMP similar test of H_0 against H_1 is to reject the largest or smallest 100α per cent of the distribution of t according to whether $\mu_1 > \mu_0$ or $\mu_1 < \mu_0$.

As we have seen, the distribution of t does not depend on σ^2. The power of the UMP similar test, however, does depend on σ^2, for u is not sufficient for σ^2 when H_0 does not hold. Since every similar region for H_0 consists of fractions α of each contour of constant u, and the distribution on any such contour is a function of σ^2 when H_1 holds, there can be no similar region for H_0 with power independent of σ^2, a result first established by Dantzig (1940).

Example 23.8

For two normal distributions with means μ, $\mu + \theta$ and common variance σ^2, to test
$$H_0 : \theta = \theta_0 \; (= 0, \text{ without loss of generality})$$
against
$$H_1 : \theta = \theta_1$$
on the basis of independent samples of size n_1, n_2 with means $\bar{x}_1$, $\bar{x}_2$.

Write
$$n = n_1+n_2,$$
$$n\bar{x} = n_1\bar{x}_1+n_2\bar{x}_2, \qquad (23.41)$$
$$s^2 = \sum_{i=1}^{2}\sum_{j=1}^{n_i}(x_{ij}-\bar{x})^2 = \sum\sum x_{ij}^2 - n\bar{x}^2.$$

The hypotheses are composite with two degrees of freedom. When H_0 holds, but not otherwise, the pair of statistics $(\bar{x}, s^2)$ is sufficient for the unspecified parameters (μ, σ^2), and it follows from **23.10** that $(\bar{x}, s^2)$ is complete. Thus all similar regions for H_0 satisfy (23.13), and we hold $(\bar{x}, s^2)$ fixed, and test the simple

$$H_0 : \theta = 0$$

against $\qquad H_1' : \theta = \theta_1, \ \mu = \mu_1, \ \sigma = \sigma_1.$

Our original H_1 consists of the class of H_1' for all μ_1, σ_1.

The BCR obtained from (22.6) reduces, on simplification, to

$$\bar{x}_2 \theta_1 \leqslant g_\alpha$$

where g_α is a constant function of all the parameters, and of $\bar{x}$ and s^2, but not otherwise of the observations. For fixed $\bar{x}, s^2$, the BCR is therefore characterized by extreme values of $\bar{x}_2$ of opposite sign to θ_1, and this is true whatever the values of the other parameters. (23.41) then implies that for each fixed $(\bar{x}, s^2)$, the BCR will consist of large values of $\frac{(\bar{x}_1-\bar{x}_2)^2}{s^2}$, and hence of the equivalent monotone increasing function

$$\frac{(\bar{x}_1-\bar{x}_2)^2}{\Sigma(x_1-\bar{x}_1)^2+\Sigma(x_2-\bar{x}_2)^2} = \frac{t^2}{(n-2)}\cdot\frac{n}{n_1 n_2}. \qquad (23.42)$$

(23.42) is the definition of the usual "Student's" t^2 statistic for this problem, which we have encountered as an interval estimation problem in **21.12**. By Exercise 23.7, t^2, which is distributed free of μ and σ^2, is distributed independently of the complete sufficient statistic $(\bar{x}, s^2)$ for (μ, σ^2). Thus, unconditionally, the UMP similar test or H_0 against H_1 is given by rejecting the 100α per cent largest or smallest values in the distribution of t, according to whether θ_1 (or, more generally, $\theta_1-\theta_0$) is positive of negative.

Here, as in the previous example, the power of the BCR depends on (μ, σ^2), since $(\bar{x}, s^2)$ is not sufficient when H_0 does not hold.

Example 23.9

To test the composite $H_0 : \sigma = \sigma_0$ against $H_1 : \sigma = \sigma_1$ for the distribution

$$dF = \exp\left\{-\left(\frac{x-\theta}{\sigma}\right)\right\}dx/\sigma, \qquad \theta \leqslant x \leqslant \infty; \ \sigma > 0.$$

We have seen (Example 17.19) that $x_{(1)}$, the smallest of a sample of n independent observations, is sufficient for the unspecified parameter θ, whether H_0 or H_1 holds. By **23.12** it is also complete. Thus all similar regions consist of fractions α of each contour of constant $x_{(1)}$.

The comprehensive sufficiency of $x_{(1)}$ renders both H_0 and H_1 simple when $x_{(1)}$ is fixed. The BCR obtained from (22.6) consists of points satisfying

$$\sum_{i=1}^{n} x_i \left(\frac{1}{\sigma_1} - \frac{1}{\sigma_0} \right) \leqslant g_\alpha,$$

where g_α is a constant, a function of σ_0, σ_1. For each fixed $x_{(1)}$, we therefore have the BCR defined by

$$\left. \begin{array}{rl} \sum_{i=1}^{n} x_i \leqslant a_\alpha & \text{if } \sigma_1 < \sigma_0, \\ \geqslant b_\alpha & \text{if } \sigma_1 > \sigma_0. \end{array} \right\} \qquad (23.43)$$

The statistic in (23.43), Σx_i, is not distributed independently of $x_{(1)}$. To put (23.43) in a form of more practical value, we observe that the statistic

$$z = \sum_{i=1}^{n} (x_{(i)} - x_{(1)})$$

is distributed independently of $x_{(1)}$. (This is a consequence of the completeness and sufficiency of $x_{(1)}$—see Exercise 23.7 below.) Thus if we rewrite (23.43) for fixed $x_{(1)}$ as

$$\left. \begin{array}{rl} z \leqslant c_\alpha, & \sigma_1 < \sigma_0, \\ \geqslant d_\alpha, & \sigma_1 > \sigma_0, \end{array} \right\} \qquad (23.44)$$

where $c_\alpha = a_\alpha - n x_{(1)}$, $d_\alpha = b_\alpha - n x_{(1)}$, we have on the left of (23.44) a statistic which for every fixed $x_{(1)}$ determines the BCR by its extreme values and whose distribution does not depend on $x_{(1)}$. Thus (23.44) gives an unconditional BCR for each of the one-sided situations $\sigma_1 < \sigma_0$, $\sigma_1 > \sigma_0$, and we have the usual pair of UMP tests.

Note that in this example, the comprehensive sufficiency of $x_{(1)}$ makes the power of the UMP tests independent of θ (which is only a location parameter).

23.21 Examples 23.7 and 23.8 afford a sophisticated justification for two of the standard normal distribution test procedures for means. Exercises 23.13 and 23.14 at the end of this chapter, by following through the same argument, similarly justify two other standard procedures for variances, arriving in each case at a pair of UMP similar one-sided tests. Unfortunately, not all the problems of normal test theory are so tractable: the thorniest of them, the problem of two means which we discussed at length in Chapter 21, does not yield to the present approach, as the next example shows.

Example 23.10

For two normal distributions with means and variances (θ, σ_1^2), $(\theta + \mu, \sigma_2^2)$, to test $H_0: \mu = 0$ on the basis of independent samples of n_1 and n_2 observations.

Given H_0, the sample means and variances $(\bar{x}_1, \bar{x}_2, s_1^2, s_2^2) = \mathbf{t}$ form a set of four jointly sufficient statistics for the three parameters θ, σ_1^2, σ_2^2 left unspecified by H_0. They may be seen to be minimal sufficient by use of (23.31)—cf. Lehmann and Scheffé (1950). But $\mathbf{t}$ is not boundedly complete, since $\bar{x}_1, \bar{x}_2$ are normally distributed independently of s_1^2, s_2^2 and of each other, so that any bounded odd function of $(\bar{x}_1 - \bar{x}_2)$ alone will have zero expectation. We therefore cannot rely on (23.13) to find all similar regions, though regions satisfying (23.13) would certainly be similar, by **23.8**. But it is easy to see, from the fact that the Likelihood Function contains the four components of **t** and no other functions of the observations, that any region consisting entirely of a fraction α of each surface of constant **t** will have the same probability content in the sample space *whatever the value of μ*, and will therefore be an ineffective critical region with power exactly equal to its size. This disconcerting aspect of a familiar and useful property of normal distributions was pointed out by Watson (1957a).

No useful exact unrandomized similar regions exist for this problem—see Linnik (1964, 1967). If we are prepared to use asymptotically similar regions, we may use Welch's method expounded in **21.25** as an interval estimation technique; similarly, if we are prepared to introduce an element of randomization, Scheffé's method of **21.15–22** is available. The relation between the terminology of confidence intervals and that of the theory of tests is discussed in **23.26** below.

23.22 The discussion of **23.20** and Examples 23.8–10 make it clear that, if there is a complete sufficient statistic for the unspecified parameter, the problem of selecting a most powerful test for a composite hypothesis is considerably reduced if we restrict our choice to similar regions. But something may be lost by this—for specific alternatives there may be a non-similar test, satisfying (23.6), with power greater than the most powerful similar test.

Lehmann and Stein (1948) considered this problem for the composite hypotheses considered in Example 23.7 and Exercise 23.13. In the former, where we are testing the mean of a normal distribution, they found that if $\alpha \geqslant \tfrac{1}{2}$ there is no non-similar test more powerful than " Student's " t, whatever the true values μ_1, σ_1, but that for $\alpha < \tfrac{1}{2}$ (as in practice it always is) there is a more powerful critical region, which is of form

$$\sum_i \{x_i - c_\alpha(\mu_1, \sigma_1)\}^2 \leqslant k_\alpha(\mu_1, \sigma_1). \tag{23.45}$$

Similarly, for the variance of a normal distribution (Exercise 23.13 below), they found that if $\sigma_1 > \sigma_0$ no more powerful non-similar test exists, but if $\sigma_1 < \sigma_0$ the region

$$\sum_i (x_i - \mu_1)^2 \leqslant k_\alpha \tag{23.46}$$

is more powerful than the best similar critical region.

Thus if we restrict the alternative class H_1 sufficiently, we can sometimes improve the power of the test, while reducing the average value of the Type I error below the size α, by abandoning the requirement of similarity. In practice, this is not a very strong argument against using similar regions, precisely because we are not usually in a position to be very restrictive about the alternatives to a composite hypothesis.

Bias in tests

23.23 In the previous chapter (**22.26–8**) we briefly discussed the problem of testing a simple H_0 against a two-sided class of alternatives, where no UMP test generally exists. We now return to this subject from another viewpoint, although the two-sided nature of the alternative hypothesis is not essential to our discussion, as we shall see.

Example 23.11

Consider again the problem of Examples 22.2–3 and of **22.27**, that of testing the mean μ of a normal population with known variance, taken as unity for convenience. Suppose that we restrict ourselves to tests based on the distribution of the sample mean $\bar{x}$, as we may do by **23.3** since $\bar{x}$ is sufficient. Generalizing (22.55), consider the size-α region defined by

$$\bar{x} \leqslant a_{\alpha_1}, \quad \bar{x} \geqslant b_{\alpha_2}, \tag{23.47}$$

where $\alpha_1 + \alpha_2 = \alpha$, and α_1 is not now necessarily equal to α_2. a and b are defined, as at (22.15), by

$$a_\alpha = \mu_0 - d_\alpha/n^{\frac{1}{2}}, \quad b_\alpha = \mu_0 + d_\alpha/n^{\frac{1}{2}},$$

and

$$G(-d_\alpha) = \int_{-\infty}^{-d_\alpha} (2\pi)^{-\frac{1}{2}} \exp(-\tfrac{1}{2} y^2) \, dy = \alpha.$$

We take $d_\alpha > 0$ without loss of generality.

Exactly as at (22.56), the power of the critical region (23.47) is seen to be

$$P = G\{n^{\frac{1}{2}}\Delta - d_{\alpha_2}\} + G\{-n^{\frac{1}{2}}\Delta - d_{\alpha_1}\}, \tag{23.48}$$

where $\Delta = \mu_1 - \mu_0$.

We consider the power (23.48) as a function of Δ. Its first two derivatives are

$$P' = \left(\frac{n}{2\pi}\right)^{\frac{1}{2}} [\exp\{-\tfrac{1}{2}(n^{\frac{1}{2}}\Delta - d_{\alpha_2})^2\} - \exp\{-\tfrac{1}{2}(n^{\frac{1}{2}}\Delta + d_{\alpha_1})^2\}] \tag{23.49}$$

and

$$P'' = \frac{n}{(2\pi)^{\frac{1}{2}}}[(d_{\alpha_2} - n^{\frac{1}{2}}\Delta)\exp\{-\tfrac{1}{2}(n^{\frac{1}{2}}\Delta - d_{\alpha_2})^2\} \\ + (n^{\frac{1}{2}}\Delta + d_{\alpha_1})\exp\{-\tfrac{1}{2}(n^{\frac{1}{2}}\Delta + d_{\alpha_1})^2\}]. \tag{23.50}$$

From (23.49), we can only have $P' = 0$ if

$$\Delta = (d_{\alpha_2} - d_{\alpha_1})/(2n^{\frac{1}{2}}). \tag{23.51}$$

When (23.51) holds, we have from (23.50)

$$P'' = \frac{n}{(2\pi)^{\frac{1}{2}}}(d_{\alpha_1} + d_{\alpha_2})\exp\{-\tfrac{1}{2}(n^{\frac{1}{2}}\Delta + d_{\alpha_1})^2\}. \tag{23.52}$$

Since we have taken d_α always positive, we therefore have $P'' > 0$ at the stationary value, which is therefore a minimum. From (23.51), it occurs at $\Delta = 0$ only when $\alpha_1 = \alpha_2$, the case discussed in **22.27**. Otherwise, the unique minimum occurs at some value μ_m where

$$\mu_m > \mu_0 \text{ if } \alpha_1 > \alpha_2, \qquad \mu_m < \mu_0 \text{ if } \alpha_1 < \alpha_2.$$

23.24 The implication of Example 23.11 is that, except when $\alpha_1 = \alpha_2$, there exist values of μ in the alternative class H_1 for which the probability of rejecting H_0 is actually smaller when H_0 is false than when it is true. (Note that if we were considering a one-sided class of alternatives (say, $\mu_1 > \mu_0$), the same situation would arise if we used the critical region located in the wrong tail of the distribution of $\bar{x}$ (say, $\bar{x} \leqslant a_\alpha$).) It is clearly undesirable to use a test which is more likely to reject the hypothesis when it is true than when it is false. In fact, we can improve on such a test by using a table of random numbers to reject the hypothesis with probability α—the power of this procedure will always be α.

We may now generalize our discussion. If a size-α critical region w for $H_0: \theta = \theta_0$ against the simple $H_1: \theta = \theta_1$ is such that its power

$$P\{\mathbf{x} \in w \,|\, \theta_1\} \geqslant \alpha, \tag{23.53}$$

it is said to give an *unbiassed*(*) test of H_0 against H_1; in the contrary case, the region w, and the test it provides, are said to be *biassed*.(*) If H_1 is composite, and (23.53) holds for every member of H_1, w is said to be an unbiassed critical region against H_1. It should be noted that unbiassedness does not require that the power function should actually have a regular minimum at θ_0, as we found to be the case in Example 23.11 when $\alpha_1 = \alpha_2$, although this is often found to be so in practice. Fig. 22.3 on page 189 illustrates the appearance of the power function for an unbiassed test (the full line) and two biassed tests.

If no unbiassed test exists, there may be a " locally unbiassed Type M " test (Krishnan, 1966) which has average power $\geqslant \alpha$ in a neighbourhood of H_0.

The criterion of unbiassedness for tests has such strong intuitive appeal that it is natural to restrict oneself to the class of unbiassed tests when investigating a problem, and to seek UMP unbiassed (UMPU) tests, which may exist even against two-sided alternative hypotheses, for which we cannot generally hope to find UMP tests without some restriction on the class of tests considered. Thus, in Example 23.11, the " equal-tails " test based on $\bar{x}$ is at once seen to be UMPU in the class of tests there considered. That it is actually UMPU among all tests of H_0 will be seen in **23.33**.

Example 23.12

We have left over to Exercise 23.13 the result that, for a normal distribution with mean μ and variance σ^2, the statistic $z = \sum_{i=1}^{n} (x_i - \bar{x})^2$ gives a pair of one-sided UMP similar tests of the hypothesis $H_0: \sigma^2 = \sigma_0^2$, the BCR being

$$z \geqslant a_\alpha \text{ if } \sigma_1 > \sigma_0, \qquad z \leqslant b_\alpha \text{ if } \sigma_1 < \sigma_0.$$

Now consider the two-sided alternative hypothesis

$$H_1: \sigma^2 \neq \sigma_0^2.$$

By **22.18** there is no UMP test of H_0 against H_1, but we are intuitively tempted to use the statistic z, splitting the critical region equally between its tails in the hope of achieving unbiassedness, as in Example 23.11. Thus we reject H_0 if

$$z \geqslant a_{\frac{1}{2}\alpha} \quad \text{or} \quad z \leqslant b_{\frac{1}{2}\alpha}.$$

This critical region is certainly similar, for the distribution of z is not dependent on μ, the nuisance parameter. Since z/σ^2 has a chi-square distribution with $(n-1)$ d.f., whether H_0 or H_1 holds, we have

$$a_{\frac{1}{2}\alpha} = \sigma_0^2 \chi^2_{1-\frac{1}{2}\alpha}, \quad b_{\frac{1}{2}\alpha} = \sigma_0^2 \chi^2_{\frac{1}{2}\alpha},$$

(*) This use of " bias " is unconnected with that of the theory of estimation, and is only prevented from being confusing by the fortunate fact that the context rarely makes confusion possible.

where χ_α^2 is the 100α per cent point of that chi-square distribution. When H_1 holds, it is z/σ_1^2 which has the distribution, and H_0 will then be rejected when

$$\frac{z}{\sigma_1^2} \geq \frac{\sigma_0^2}{\sigma_1^2}\chi_{1-\frac{1}{2}\alpha}^2 \quad \text{or} \quad \frac{z}{\sigma_1^2} \leq \frac{\sigma_0^2}{\sigma_1^2}\chi_{\frac{1}{2}\alpha}^2.$$

The power of the test against any alternative value σ_1^2 is the sum of the probabilities of these two events. We thus require the probability that a chi-square variable will fall outside its $100(\frac{1}{2}\alpha)$ per cent and $100(1-\frac{1}{2}\alpha)$ per cent points each multiplied by a constant σ_0^2/σ_1^2. For each value of α and $(n-1)$, the degrees of freedom, this probability can be calculated from a table of the distribution for each value of σ_0^2/σ_1^2. Fig. 23.1 shows the power function resulting from such calculations by Neyman and Pearson

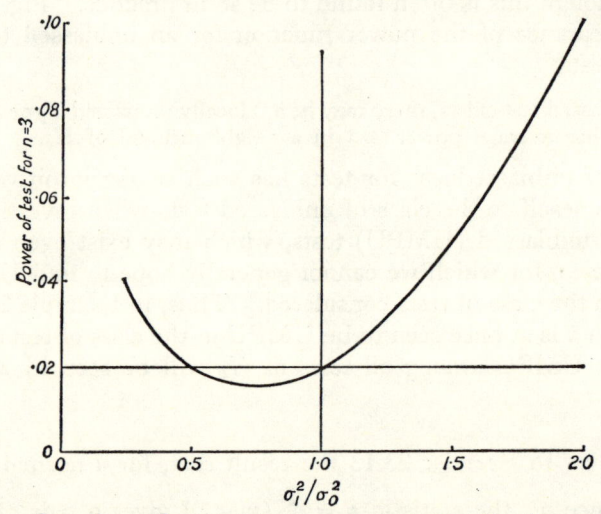

Fig. 23.1—Power function of a test for a normal distribution variance (see text)

(1936b) for the case $n = 3$, $\alpha = 0.02$. The power is less than α in this case when $0.5 < \sigma_1^2/\sigma_0^2 < 1$, and the test is therefore biassed.

We now enquire whether, by modifying the apportionment of the critical region between the tails of the distribution of z, we can remove the bias. Suppose that the critical region is

$$z \geq a_{1-\alpha_1} \quad \text{or} \quad z \leq b_{\alpha_2},$$

where $\alpha_1 + \alpha_2 = \alpha$. As before, the power of the test is the probability that a chi-square variable with $(n-1)$ degrees of freedom, say y_{n-1}, falls outside the range of its $100\alpha_2$ per cent and $100(1-\alpha_1)$ per cent points, each multiplied by the constant $\theta = \sigma_0^2/\sigma_1^2$. Writing F for the distribution function of y_{n-1}, we have

$$P = F(\theta\chi_{\alpha_2}^2) + 1 - F(\theta\chi_{1-\alpha_1}^2). \tag{23.54}$$

Regarded as a function of θ, this is the power function. We now choose α_1 and α_2 so that this power function has a regular minimum at $\theta = 1$, where it equals the size of the test. Differentiating (23.54), we have

$$P' = \chi_{\alpha_2}^2 f(\theta\chi_{\alpha_2}^2) - \chi_{1-\alpha_1}^2 f(\theta\chi_{1-\alpha_1}^2), \tag{23.55}$$

where f is the frequency function of y_{n-1}. If this is to be zero when $\theta = 1$, we require
$$\chi^2_{\alpha_2} f(\chi^2_{\alpha_2}) = \chi^2_{1-\alpha_1} f(\chi^2_{1-\alpha_1}). \tag{23.56}$$
Substituting for the frequency function
$$f(y) \propto e^{-\frac{1}{2}y} y^{\frac{1}{2}(n-3)} dy, \tag{23.57}$$
we have finally from (23.56) the condition for unbiassedness
$$\left\{ \frac{\chi^2_{1-\alpha_1}}{\chi^2_{\alpha_2}} \right\}^{\frac{1}{2}(n-1)} = \exp\left\{ \tfrac{1}{2}(\chi^2_{1-\alpha_1} - \chi^2_{\alpha_2}) \right\}. \tag{23.58}$$

Values of α_1 and α_2 satisfying (23.58) will give a test whose power function has zero derivative at the origin. To investigate whether it is strictly unbiassed, we write (23.55), using (23.57) and (23.58), as

$$P' = c\theta^{\frac{1}{2}(n-3)} \chi^{n-1}_{\alpha_2} \exp(-\tfrac{1}{2}\chi^2_{\alpha_2}) [\exp\{\tfrac{1}{2}\chi^2_{\alpha_2}(1-\theta)\} - \exp\{\tfrac{1}{2}\chi^2_{1-\alpha_1}(1-\theta)\}], \tag{23.59}$$

where c is a positive constant. Since $\chi^2_{1-\alpha_1} > \chi^2_{\alpha_2}$, we have from (23.59)
$$P' \begin{cases} < 0, & \theta < 1, \\ = 0, & \theta = 1, \\ > 0, & \theta > 1. \end{cases} \tag{23.60}$$

(23.60) shows that the test with α_1, α_2 determined by (23.58) is unbiassed in the strict sense, for the power function is monotonic decreasing as θ increases from 0 to 1 and monotonic increasing as θ increases from 1 to ∞.

Tables of the values of $\chi^2_{\alpha_2}$ and $\chi^2_{1-\alpha_1}$ satisfying (23.58) are given by Ramachandran (1958) for $\alpha = 0.05$ and $n-1 = 2(1)8(2)24, 30, 40$ and 60; other tables are described in **20.21**(3), where the terminology of confidence intervals is used—cf. **23.26** for the correspondence with tests. Table 23.1 compares some of Ramachandran's values with the corresponding limits for the biassed "equal-tails" test which we have considered, obtained from the *Biometrika Tables*.

Table 23.1—Limits outside which the chi-square variable $\Sigma(x-\bar{x})^2/\sigma_0^2$ must fall for $H_0 : \sigma^2 = \sigma_0^2$ to be rejected ($\alpha = 0.05$)

Degrees of freedom ($n-1$)	Unbiassed test limits	"Equal-tails" test limits	Differences
2	(0·08, 9·53)	(0·05, 7·38)	(0·03, 2·15)
5	(0·99, 14·37)	(0·83, 12·83)	(0·16, 1·54)
10	(3·52, 21·73)	(3·25, 20·48)	(0·27, 1·25)
20	(9·96, 35·23)	(9·59, 34·17)	(0·37, 1·06)
30	(17·21, 47·96)	(16·79, 46·98)	(0·42, 0·98)
40	(24·86, 60·32)	(24·43, 59·34)	(0·43, 0·98)
60	(40·93, 84·23)	(40·48, 83·30)	(0·45, 0·93)

It will be seen that the differences in both limits are proportionately large for small n, that the lower limit difference increases steadily with n, and the larger limit difference decreases steadily with n. At $n-1 = 60$, both differences are just over 1 per cent of the values of the limits.

We defer the question whether the unbiassed test is UMPU to Example 23.14 below.

Unbiassed tests and similar tests

23.25 There is a close connection between unbiassedness and similarity, which often leads to the best unbiassed test emerging directly from an analysis of the similar regions for a problem.

We consider a more general form of hypothesis than (23.2), namely

$$H'_0 : \theta_r \leq \theta_{r0}, \qquad (23.61)$$

which is to be tested against

$$H_1 : \theta_r > \theta_{r0}. \qquad (23.62)$$

If we can find a critical region w satisfying (23.6) for all θ_r in H'_0 as well as for all values of the unspecified parameters θ_s, i.e.

$$P(H'_0, \theta_s) \leq \alpha, \qquad (23.63)$$

(where P is the power function whose value is the probability of rejecting H_0), the test based on w will be of size α as before. If it also unbiassed, we have from (23.53)

$$P(H_1, \theta_s) \geq \alpha. \qquad (23.64)$$

Now if the power function P is a continuous function of θ_r, (23.63) and (23.64) imply, in view of the form of H'_0 and H_1,

$$P(\theta_{r0}, \theta_s) = \alpha, \qquad (23.65)$$

i.e. that w is a similar critical region for the "boundary" hypothesis

$$H_0 : \theta_r = \theta_{r0}.$$

All unbiassed tests of H'_0 are similar tests of H_0. If we confine our discussions to similar tests of H_0, using the methods we have encountered, and find a test with optimum properties—e.g., a UMP similar test—then *provided that this test is unbiassed* it will retain the optimum properties in the class of unbiassed tests of H'_0—e.g. it will be a UMPU test.

Exactly the same argument holds if H'_0 specifies that the parameter point θ_r lies within a certain region R (which may consist of a number of subregions) in the parameter space, and H_1 that the θ_r lies in the remainder of that space : if the power function is continuous in θ_r, then if a critical region w is unbiassed for testing H'_0, it is a similar region for testing the hypothesis H_0 that θ_r lies on the boundary of R. If w gives an unbiassed test of H'_0, it will carry over into the class of unbiassed tests of H'_0 any optimum properties it may have as a similar test of H_0. There will not always be a UMP similar test of H_0 if the alternatives are two-sided : a UMPU test may exist against such alternatives, but it must be found by other methods.

Example 23.13

We return to the hypothesis of Example 23.12. One-sided critical regions based on the statistic $z \geq a_\alpha$, $z \leq b_\alpha$, give UMP similar tests against one-sided alternatives. Each of them is easily seen to be unbiassed in testing one of

$$H'_0 : \sigma^2 \leq \sigma_0^2, \quad H''_0 : \sigma^2 \geq \sigma_0^2$$

respectively against

$$H'_1 : \sigma^2 > \sigma_0^2, \quad H''_1 : \sigma^2 < \sigma_0^2.$$

Thus they are, by the argument of **23.25**, UMPU tests for these one-sided situations.

For the two-sided alternative $H_1 : \sigma^2 \neq \sigma_0^2$, the unbiassed test based on (23.58) cannot be shown to be UMPU by this method, since we have not shown it to be UMP similar.

Tests and confidence intervals

23.26 The early work on unbiassed tests was largely carried out by Neyman and Pearson in the series of papers mentioned in **22.1**, and by Neyman (1935, 1938b), Scheffé (1942a) and Lehmann (1947). Much of the detail of their work has now been superseded, as pioneering work usually is, but their terminology is still commonly used, and it is desirable to provide a " dictionary " connecting it with the present terminology, where it differs. We take the opportunity of translating the ideas of the theory of hypothesis-testing into those of the theory of confidence intervals, as promised in **20.20**.

If a sample is observed, we may ask the question: for which values of θ does the sample point **x** form part of the acceptance region A complementary to the size-α critical region for a certain test on the parameter θ? If we aggregate these " acceptable " values of θ, we obtain the level-$(1-\alpha)$ confidence interval C for θ corresponding to that test, for θ is in C if and only if **x** is in A, i.e. with probability $1-\alpha$. We used this method of constructing confidence intervals in **20.3**, and indeed throughout Chapter 20. There is thus no need to derive optimum properties separately for tests and for intervals: there is a one-to-one correspondence between the problems.

Present terminology	Property of test — Older terminology	Property of corresponding confidence interval
UMP		" Shortest " (= most selective)
Unbiassed		Unbiassed
UMPU " locally " (i.e. near H_0)	Type A(*) (simple H_0, one parameter) Type B(*) (composite H_0) Type C(*) (simple H_0, two or more parameters)	} " Short " unbiassed
UMPU	Type A_1(*) (simple H_0, one parameter) Type B_1(*) (composite H_0)	} " Shortest " unbiassed
Unbiassed similar	Bisimilar	

(*) Subject to regularity conditions.

For example, in **20.31**, we noticed that in setting confidence intervals for the mean μ of a normal distribution with unspecified variance, using " Student's " t distribution, the length of the interval was a random variable, being a multiple of the sample standard deviation. In Example 23.7, on the other hand, we remarked that the power of the similar test based on " Student's " t was a function of the unknown variance. Now the power of the test is the probability of rejecting the hypothesis when false, i.e. in confidence interval terms, is the probability of not covering another value of μ than the true one, μ_0. If this probability is a function of the unknown variance, for all values of μ, we evidently cannot pre-assign the length of the interval as well as the confidence coefficient. Our earlier statement was a consequence of the later one.

UMPU tests for the exponential family

23.27 We now give an account of some remarkably comprehensive results, due to Lehmann and Scheffé (1955), which establish the existence of, and give a construction for, UMPU tests for a variety of parametric hypotheses in distributions belonging to the exponential family (17.86). We write the joint distribution of n independent observations from such a distribution as

$$f(\mathbf{x}) = D(\mathbf{\tau}) h(\mathbf{x}) \exp\left\{\sum_{j=1}^{r+1} b_j(\mathbf{\tau}) u_j(\mathbf{x})\right\}, \qquad (23.66)$$

where $\mathbf{x}$ is the column vector $(x_1, \ldots, x_n)$ and $\mathbf{\tau}$ is a vector of $(r+1)$ parameters $(\tau_1, \ldots, \tau_{r+1})$. In matrix notation, the exponent in (23.66) may be concisely written $\mathbf{u}'\mathbf{b}$, where $\mathbf{u}$ and $\mathbf{b}$ are column vectors.

Suppose now that we are interested in the particular linear function of the parameters

$$\theta = \sum_{j=1}^{r+1} a_{j1} b_j(\mathbf{\tau}), \qquad (23.67)$$

where $\sum_{j=1}^{r+1} a_{j1}^2 = 1$. Write $\mathbf{A}$ for an orthogonal matrix (a_{uv}) whose first column contains the coefficients in (23.67), and transform to a new vector of $(r+1)$ parameters $(\theta, \mathbf{\psi})$, where $\mathbf{\psi}$ is the column vector $(\psi_1, \ldots, \psi_r)$, by the equation

$$\begin{pmatrix} \theta \\ \mathbf{\psi} \end{pmatrix} = \mathbf{A}'\mathbf{b}. \qquad (23.68)$$

The first row of (23.68) is (23.67). We now suppose that there is a column vector of statistics $\mathbf{T} = (s, t_1, \ldots, t_r)$ defined by the relation

$$\mathbf{T}' \begin{pmatrix} \theta \\ \mathbf{\psi} \end{pmatrix} = \mathbf{u}'\mathbf{b}, \qquad (23.69)$$

i.e. we suppose that the exponent in (23.66) may be expressed as $\theta s(\mathbf{x}) + \sum_{j=1}^{r} \psi_j t_j(\mathbf{x})$. Using (23.68), (23.69) becomes

$$\mathbf{T}' \begin{pmatrix} \theta \\ \mathbf{\psi} \end{pmatrix} = \mathbf{u}'\mathbf{A} \begin{pmatrix} \theta \\ \mathbf{\psi} \end{pmatrix}. \qquad (23.70)$$

(23.70) is an identity in $(\theta, \mathbf{\psi})$, so we have $\mathbf{T}' = \mathbf{u}'\mathbf{A}$ or

$$\mathbf{T} = \mathbf{A}'\mathbf{u}. \qquad (23.71)$$

Comparing (23.71) with (23.68), we see that each component of $\mathbf{T}$ is the same function of the $u_j(\mathbf{x})$ as the corresponding component of $(\theta, \mathbf{\psi})$ is of the $b_j(\mathbf{\tau})$. In particular, the first component is, from (23.67),

$$s(x) = \sum_{j=1}^{r+1} a_{j1} u_j(\mathbf{x}) \qquad (23.72)$$

while the $t_j(\mathbf{x})$, $j = 1, 2, \ldots, r$, are orthogonal to $s(\mathbf{x})$.

Note that the orthogonality condition $\sum_{j=1}^{r+1} a_{j1}^2 = 1$ does not hamper us in testing hypotheses about θ defined by (23.67), since only a constant factor need be changed and the hypothesis adjusted accordingly.

23.28 If, therefore, we can reduce a hypothesis-testing problem (usually through its sufficient statistics) to the standard form of one concerning θ in

$$f(\mathbf{x}|\theta, \boldsymbol{\psi}) = C(\theta, \boldsymbol{\psi}) h(\mathbf{x}) \exp\left\{\theta s(\mathbf{x}) + \sum_{i=1}^{r} \psi_i t_i(\mathbf{x})\right\}, \qquad (23.73)$$

by the device of the previous section, we can avail ourselves of the results summarized in **23.10**: given a hypothesis value for θ, the r-component vector $\mathbf{t} = (t_1, \ldots, t_r)$ will be a complete sufficient statistic for the r-component parameter $\boldsymbol{\psi} = (\psi_1, \ldots, \psi_r)$, and we now consider the problem of using s and $\mathbf{t}$ to test various composite hypotheses concerning θ, $\boldsymbol{\psi}$ being an unspecified ("nuisance") parameter. Simple hypotheses are the special case when $r = 0$, with no nuisance parameter.

23.29 For this purpose we shall need an extended form of the Neyman–Pearson lemma of **22.10**. Let $f(\mathbf{x}|\boldsymbol{\theta})$ be a frequency function, and $\boldsymbol{\theta}_i$ a subset of admissible values of the vector of parameters $\boldsymbol{\theta}$, $(i = 1, 2, \ldots, k)$. A specific element of $\boldsymbol{\theta}_i$ is written $\boldsymbol{\theta}_i^0$. $\boldsymbol{\theta}^*$ is a particular value of $\boldsymbol{\theta}$. The vector $\mathbf{u}_i(\mathbf{x})$ is sufficient for $\boldsymbol{\theta}$ when $\boldsymbol{\theta}$ is in $\boldsymbol{\theta}_i$ and its distribution is $g_i(\mathbf{u}_i|\boldsymbol{\theta}_i)$. Since the Likelihood Function factorizes in the presence of sufficiency, the conditional value of $f(\mathbf{x}|\boldsymbol{\theta}_i)$, given $\mathbf{u}_i$, will be independent of $\boldsymbol{\theta}_i$, and we write it $f(\mathbf{x}|\mathbf{u}_i)$. Finally, we define $l_i(\mathbf{x})$, $m_i(\mathbf{u}_i)$ to be non-negative functions, of the observations and of $\mathbf{u}_i$ respectively.

Now suppose we have a critical region w for which

$$\int_w \{l_i(\mathbf{x}) f(\mathbf{x}|\mathbf{u}_i)\} d\mathbf{x} = \alpha_i. \qquad (23.74)$$

Since the product in braces in (23.74) is non-negative, it may be regarded as a frequency function, and we may say that the conditional size of w, given $\mathbf{u}_i$, is α_i with respect to this distribution. We now write

$$\beta_i = \alpha_i \int m_i(\mathbf{u}_i) g_i(\mathbf{u}_i|\boldsymbol{\theta}_i^0) d\mathbf{u}_i$$

$$= \int_w l_i(\mathbf{x}) m_i(\mathbf{u}_i) \left\{\int f(\mathbf{x}|\mathbf{u}_i) g_i(\mathbf{u}_i|\boldsymbol{\theta}_i^0) d\mathbf{u}_i\right\} d\mathbf{x}$$

$$= \int_w \{l_i(\mathbf{x}) m_i(\mathbf{u}_i) f(\mathbf{x}|\boldsymbol{\theta}_i^0)\} d\mathbf{x}. \qquad (23.75)$$

The product in braces is again essentially a frequency function, say $p(\mathbf{x}|\boldsymbol{\theta}_i^0)$. To test the simple hypothesis that $p(\mathbf{x}|\boldsymbol{\theta}_i^0)$ holds against the simple alternative that $f(\mathbf{x}|\boldsymbol{\theta}^*)$ holds, we use (22.6) and find that the BCR w of size β_i consists of points satisfying

$$[f(\mathbf{x}|\boldsymbol{\theta}^*)]/[p(\mathbf{x}|\boldsymbol{\theta}_i^0)] \geq c_i(\beta_i), \qquad (23.76)$$

where c_i is a non-negative constant. (23.76) will hold for every value of i. Thus for testing the composite hypothesis that *any* of $p(\mathbf{x}|\boldsymbol{\theta}_i^0)$ holds $(i = 1, 2, \ldots, k)$, we require all k of the inequalities (23.76) to be satisfied by w. If we now write $k m_i(\mathbf{u}_i)/c_i(\beta_i)$ for $m_i(\mathbf{u}_i)$ in $p(\mathbf{x}|\boldsymbol{\theta}_i^0)$, as we may since $m_i(\mathbf{u}_i)$ is arbitrary, we have from (23.76), adding the inequalities for $i = 1, 2, \ldots, k$, the necessary and sufficient condition for a BCR

$$f(\mathbf{x}|\boldsymbol{\theta}^*) \geq \sum_{i=1}^{k} l_i(\mathbf{x}) m_i(\mathbf{u}_i) f(\mathbf{x}|\boldsymbol{\theta}_i^0). \qquad (23.77)$$

This is the required generalization. (22.6) is its special case with $k = 1$, $l_1(\mathbf{x}) = k_\alpha$ (constant), $m_1(u_1) \equiv 1$. (23.77) will play a role for composite hypotheses similar to that of (22.6) for simple hypotheses.

One-sided alternatives

23.30 Reverting to (23.73), we now investigate the problem of testing
$$H_0^{(1)} : \theta \leq \theta_0$$
against
$$H_1^{(1)} : \theta > \theta_0,$$
which we discussed in general terms in **23.25**. Now that we are dealing with the exponential form (23.73), we can show that there is always a UMPU test of $H_0^{(1)}$ against $H_1^{(1)}$. By our discussion in **23.25**, if a size-α critical region is unbiased for $H_0^{(1)}$ against $H_1^{(1)}$, it is a similar region for testing $\theta = \theta_0$.

Consider testing the simple
$$H_0 : \theta = \theta_0, \quad \boldsymbol{\psi} = \boldsymbol{\psi}^0$$
against the simple
$$H_1 : \theta = \theta^* > \theta_0, \quad \boldsymbol{\psi} = \boldsymbol{\psi}^*.$$
We now apply the result of **23.29**. Putting $k = 1$, $l_1(\mathbf{x}) \equiv 1$, $\alpha_1 = \alpha$, $\boldsymbol{\theta} = (\theta, \boldsymbol{\psi})$, $\boldsymbol{\theta}_1 = (\theta_0, \boldsymbol{\psi})$, $\boldsymbol{\theta}^* = (\theta^*, \boldsymbol{\psi}^*)$, $\boldsymbol{\theta}_1^0 = (\theta_0, \boldsymbol{\psi}^0)$, $u_1 = \mathbf{t}$, we have the result that the BCR for testing H_0 against H_1 is defined from (23.77) and (23.73) as

$$\frac{C(\theta^*, \boldsymbol{\psi}^*) \exp\left\{\theta^* s(\mathbf{x}) + \sum_{i=1}^{r} \psi_i^* t_i(\mathbf{x})\right\}}{C(\theta_0, \boldsymbol{\psi}^0) \exp\left\{\theta_0 s(\mathbf{x}) + \sum_{i=1}^{r} \psi_i^0 t_i(\mathbf{x})\right\}} \geq m_1(\mathbf{t}). \tag{23.78}$$

This may be rewritten
$$s(\mathbf{x})(\theta^* - \theta_0) \geq c_\alpha(\mathbf{t}, \theta^*, \theta_0, \boldsymbol{\psi}^*, \boldsymbol{\psi}^0). \tag{23.79}$$

We now see that c_α is not a function of $\boldsymbol{\psi}$, for since, by **23.28**, $\mathbf{t}$ is a sufficient statistic for $\boldsymbol{\psi}$ when H_0 holds, the value of c_α for given $\mathbf{t}$ will be independent of $\boldsymbol{\psi}^0, \boldsymbol{\psi}^*$. Further, from (23.79) we see that so long as the sign of $(\theta^* - \theta_0)$ does not change, the BCR will consist of the largest 100α per cent of the distribution of $s(\mathbf{x})$ given θ_0. We thus have a BCR for $\theta = \theta_0$ against $\theta > \theta_0$, giving a UMP test. This UMP test cannot have smaller power than a randomized test against $\theta > \theta_0$ which ignores the observations. The latter test has power equal to its size α, so the UMP test is unbiased against $\theta > \theta_0$, i.e. by **23.25** it is UMPU. Its size for $\theta < \theta_0$ will not exceed its size at θ_0, as is evident from the consideration that the critical region (23.79) has *minimum* power against $\theta < \theta_0$ and therefore its power (size) there is less than α. Thus finally we have shown that the largest 100α per cent of the conditional distribution of $s(\mathbf{x})$, given $\mathbf{t}$, gives a UMPU size-α test of $H_0^{(1)}$ against $H_1^{(1)}$.

Two-sided alternatives

23.31 We now consider the problem of testing
$$H_0^{(2)} : \theta = \theta_0$$
against
$$H_1^{(2)} : \theta \neq \theta_0.$$

Our earlier examples stopped short of establishing UMPU tests for two-sided hypotheses of this kind (cf. Examples 23.12 and 23.13). Nevertheless a UMPU test does exist for the linear exponential form (23.73).

From **23.25** we have that if the power function of a critical region is continuous in θ, and unbiassed, it is similar for $H_0^{(2)}$. Now for any region w, the power function is

$$P(w|\theta) = \int_w f(\mathbf{x}|\theta, \boldsymbol{\psi}) \, d\mathbf{x}, \qquad (23.80)$$

where f is defined by (23.73). (23.80) is continuous and differentiable under the integral sign with respect to θ. For the test based on the critical region w to be unbiassed we must therefore have, for each value of $\boldsymbol{\psi}$, the necessary condition

$$P'(w|\theta_0) = 0. \qquad (23.81)$$

Differentiating (23.80) under the integral sign and using (23.73) and (23.81), we find the condition for unbiassedness

$$0 = \int_w \left[s(\mathbf{x}) + \frac{C'(\theta_0, \boldsymbol{\psi})}{C(\theta_0, \boldsymbol{\psi})} \right] f(\mathbf{x}|\theta_0, \boldsymbol{\psi}) \, d\mathbf{x}$$

or

$$E\{s(\mathbf{x})c(w)\} = -\alpha C'(\theta_0, \boldsymbol{\psi})/C(\theta_0, \boldsymbol{\psi}). \qquad (23.82)$$

Since, from (23.73),

$$1/C(\theta, \boldsymbol{\psi}) = \int h(\mathbf{x}) \exp\left\{\theta s(\mathbf{x}) + \sum_i \psi_i t_i(\mathbf{x})\right\} d\mathbf{x}$$

we have

$$\frac{C'(\theta, \boldsymbol{\psi})}{C(\theta, \boldsymbol{\psi})} = -E\{s(\mathbf{x})\}, \qquad (23.83)$$

and putting (23.83) into (23.82) gives

$$E\{s(\mathbf{x})c(w)\} = \alpha E\{s(\mathbf{x})\}. \qquad (23.84)$$

Taking the expectation first conditionally upon the value of $\mathbf{t}$, and then unconditionally, (23.84) gives

$$E_t[E\{s(\mathbf{x})c(w) - \alpha s(\mathbf{x})|\mathbf{t}\}] = 0. \qquad (23.85)$$

Since $\mathbf{t}$ is complete, (23.85) implies

$$E\{s(\mathbf{x})c(w) - \alpha s(\mathbf{x})|\mathbf{t}\} = 0 \qquad (23.86)$$

and since all similar regions for H_0 satisfy

$$E\{c(w)|\mathbf{t}\} = \alpha, \qquad (23.87)$$

(23.86) and (23.87) combine into

$$E\{s^{i-1}(\mathbf{x})c(w)|\mathbf{t}\} = \alpha E\{s^{i-1}(\mathbf{x})|\mathbf{t}\} = \alpha_i, \quad i = 1, 2. \qquad (23.88)$$

All our expectations are taken when θ_0 holds.

Now consider a simple

$$H_0: \theta = \theta_0, \quad \boldsymbol{\psi} = \boldsymbol{\psi}^0$$

against the simple

$$H_1: \theta = \theta^* \neq \theta_0, \quad \boldsymbol{\psi} = \boldsymbol{\psi}^*,$$

and apply the result of **23.29** with $k = 2$, α_i as in (23.88), $\theta = (\theta, \psi)$, $\theta_1 = \theta_2 = (\theta_0, \psi)$, $\theta^* = (\theta^*, \psi^*)$, $\theta_1^0 = \theta_2^0 = (\theta_0, \psi^0)$, $l_i(\mathbf{x}) = s^{i-1}(\mathbf{x})$, $u_1 = u_2 = \mathbf{t}$. We find that the BCR w for testing H_0 against H_1 is given by (23.77) and (23.73) as

$$\frac{C(\theta^*, \psi^*) \exp\left\{\theta^* s(\mathbf{x}) + \sum_{i=1}^{r} \psi_i^* t_i(\mathbf{x})\right\}}{C(\theta_0, \psi^0) \exp\left\{\theta_0 s(\mathbf{x}) + \sum_{i=1}^{r} \psi_i^0 t_i(\mathbf{x})\right\}} \geq m_1(\mathbf{t}) + s(\mathbf{x}) m_2(\mathbf{t}). \tag{23.89}$$

(23.89) reduces to

$$\exp\{s(\mathbf{x})(\theta^* - \theta_0)\} \geq c_1(\mathbf{t}, \theta^*, \theta_0, \psi^*, \psi^0) + s(\mathbf{x}) c_2(\mathbf{t}, \theta^*, \theta_0, \psi^*, \psi^0)$$

or

$$\exp\{s(\mathbf{x})(\theta^* - \theta_0)\} - s(\mathbf{x}) c_2 \geq c_1. \tag{23.90}$$

(23.90) is equivalent to $s(\mathbf{x})$ lying outside an interval, i.e.

$$s(\mathbf{x}) \leq v(\mathbf{t}), \quad s(\mathbf{x}) \geq w(\mathbf{t}), \tag{23.91}$$

where $v(\mathbf{t}) < w(\mathbf{t})$ are possibly functions also of the parameters. We now show that they are not dependent on the parameters other than θ_0. As before, the sufficiency of $\mathbf{t}$ for ψ rules out the dependence of v and w on ψ when $\mathbf{t}$ is given. That they do not depend on θ^* follows at once from (23.86), which states that when H_0 holds

$$\int_w \{s(\mathbf{x}) | \mathbf{t}\} f d\mathbf{x} = \alpha \int_W \{s(\mathbf{x}) | \mathbf{t}\} f d\mathbf{x}. \tag{23.92}$$

The right-hand side of (23.92), which is integrated over the whole sample space, clearly does not depend on θ^* at all. Hence the left-hand side is also independent of θ^*, so that the BCR w defined by (23.91) depends only on θ_0, as it must. The BCR therefore gives a UMP test of $H_0^{(2)}$ against $H_1^{(2)}$. Its unbiassedness follows by precisely the argument at the end of **23.30**. Thus, finally, we have established that the BCR defined by (23.91) gives a UMPU test of $H_0^{(2)}$ against $H_1^{(2)}$. If we determine from the conditional distribution of $s(\mathbf{x})$, given $\mathbf{t}$, an interval which excludes 100α per cent of the distribution when $H_0^{(2)}$ holds, and take the excluded values as our critical region, then if the region is unbiassed it gives the UMPU size-α test.

Finite-interval hypotheses

23.32 We may also consider the hypothesis

$$H_0^{(3)} : \theta_0 \leq \theta \leq \theta_1$$

against

$$H_1^{(3)} : \theta < \theta_0 \quad \text{or} \quad \theta > \theta_1,$$

or the complementary

$$H_0^{(4)} : \theta \leq \theta_0 \quad \text{or} \quad \theta \geq \theta_1$$

against

$$H_1^{(4)} : \theta_0 < \theta < \theta_1.$$

We now set up two hypotheses

$$H_0' : \theta = \theta_0, \quad \psi = \psi^0, \qquad H_0'' : \theta = \theta_1, \quad \psi = \psi^1,$$

to be tested against

$$H_1 : \theta = \theta^*, \quad \psi = \psi^*, \quad \text{where} \quad \theta_0 \neq \theta^* \neq \theta_1.$$

We use the result of **23.29** again, this time with $k = 2$, $\alpha_1 = \alpha_2 = \alpha$, $\theta = (\theta, \psi)$,

$\theta_1 = (\theta_0, \psi)$, $\theta_2 = (\theta_1, \psi)$, $\theta^* = (\theta^*, \psi^*)$, $\theta_1^0 = (\theta_0, \psi^0)$, $\theta_2^0 = (\theta_1, \psi^1)$, $l_i(\mathbf{x}) \equiv 1$, $u_1 = u_2 = \mathbf{t}$. We find that the BCR w for testing H_0' or H_0'' against H_1 is defined by

$$f(\mathbf{x}|\theta^*, \psi^*) \geq m_1(\mathbf{t}) f(\mathbf{x}|\theta_0, \psi^0) + m_2(\mathbf{t}) f(\mathbf{x}|\theta_1, \psi^1). \tag{23.93}$$

On substituting $f(\mathbf{x})$ from (23.73), (23.93) is equivalent to

$$H(s) = c_1 \exp\{(\theta_0 - \theta^*) s(\mathbf{x})\} + c_2 \exp\{(\theta_1 - \theta^*) s(\mathbf{x})\} < 1, \tag{23.94}$$

where c_1, c_2 may be functions of all the parameters and of $\mathbf{t}$. If $\theta_0 < \theta^* < \theta_1$, (23.94) requires that $s(\mathbf{x})$ lie inside an interval, i.e.

$$v(\mathbf{t}) \leq s(\mathbf{x}) \leq w(\mathbf{t}). \tag{23.95}$$

On the other hand, if $\theta^* < \theta_0$ or $\theta^* > \theta_1$, (23.94) requires that $s(\mathbf{x})$ lie *outside* the interval $(v(\mathbf{t}), w(\mathbf{t}))$. The proof that the end-points of the interval are not dependent on the values of the parameters, other than θ_0 and θ_1, follows the same lines as before, as does the proof of unbiassedness. Thus we have a UMPU test for $H_0^{(3)}$ and another for $H_0^{(4)}$. The test is similar at values θ_0 and θ_1, as follows from **23.25**. To obtain a UMPU test for $H_0^{(3)}$ (or $H_0^{(4)}$), we determine an interval in the distribution of $s(\mathbf{x})$ for given $\mathbf{t}$ which excludes (for $H_0^{(4)}$ includes) 100α per cent of the distribution both when $\theta = \theta_0$ and $\theta = \theta_1$. The excluded (or included) region, if unbiassed, will give a UMPU test of $H_0^{(3)}$ (or $H_0^{(4)}$).

23.33 We now turn to some applications of the fundamental results of **23.30–2** concerning UMPU tests for the exponential family of distributions. We first mention briefly that in Example 23.11 and Exercises 22.1–3 above, UMPU tests for all four types of hypothesis are obtained directly from the distribution of the single sufficient statistic, no conditional distribution being involved since there is no nuisance parameter.

Example 23.14

For n independent observations from a normal distribution, the statistics $(\bar{x}, s^2)$ are jointly sufficient for (μ, σ^2), with joint distribution (cf. Example 17.17)

$$g(\bar{x}, s^2 | \mu, \sigma^2) \propto \frac{s^{n-3}}{\sigma^n} \exp\left\{-\frac{\Sigma(x-\mu)^2}{2\sigma^2}\right\}. \tag{23.96}$$

(23.96) may be written

$$g \propto C(\mu, \sigma^2) \exp\left\{(-\tfrac{1}{2}\Sigma x^2)\left(\frac{1}{\sigma^2}\right) + (\Sigma x)\left(\frac{\mu}{\sigma^2}\right)\right\}, \tag{23.97}$$

which is of form (23.73). Remembering the discussion of **23.27**, we now consider a linear form in the parameters of (23.97). We put

$$\theta = A\left(\frac{1}{\sigma^2}\right) + B\left(\frac{\mu}{\sigma^2}\right), \tag{23.98}$$

where A and B are arbitrary known constants. We specialize A and B to obtain from the results of **23.30–2** UMPU tests for the following hypotheses:

(1) Put $A = 1$, $B = 0$ and test hypotheses concerning $\theta = \frac{1}{\sigma^2}$, with $\psi = \frac{\mu}{\sigma^2}$ as nuisance parameter. Here $s(\mathbf{x}) = -\tfrac{1}{2}\Sigma x^2$ and $t(\mathbf{x}) = \Sigma x$. From (23.97) there is

a UMPU test of $H_0^{(1)}$, $H_0^{(2)}$, $H_0^{(3)}$ and $H_0^{(4)}$ concerning $1/\sigma^2$, and hence concerning σ^2, based on the conditional distribution of Σx^2 given Σx, i.e. of $\Sigma (x-\bar{x})^2$ given Σx. Since these two statistics are independently distributed, we may use the unconditional distribution of $\Sigma(x-\bar{x})^2$, or of $\Sigma(x-\bar{x})^2/\sigma^2$, which is a χ^2 distribution with $(n-1)$ degrees of freedom. $H_0^{(2)}$ was discussed in Examples 23.12-13, where the UMP similar test was given for $\theta = \theta_0$ against one-sided alternatives and an unbiassed test based on $\Sigma(x-\bar{x})^2$ given for $H_0^{(2)}$; it now follows that this is a UMPU test for $H_0^{(2)}$, while the one-sided test is UMPU for $H_0^{(1)}$.

Graphs of the critical values of the UMPU tests of $H_0^{(3)}$ and $H_0^{(4)}$ for $\alpha = 0.05, 0.10$ are given by Guenther and Whitcomb (1966).

(2) To test hypotheses concerning μ, invert (23.98) into $\mu = (\theta \sigma^2 - A)/B$.

If we specify a value μ_0 for μ, we cannot choose A and B to make this correspond uniquely to a value θ_0 for θ (without knowledge of σ^2) if $\theta_0 \neq 0$. But if $\theta_0 = 0$ we have $\mu_0 = -A/B$. Thus from our UMPU tests for $H_0^{(1)} : \theta \leq 0$, $H_0^{(2)} : \theta = 0$, we get UMPU tests of $\mu \leq \mu_0$ and of $\mu = \mu_0$. We use (23.71) to see that the test statistic $s(\mathbf{x})\,|\,t$ is here $(-\tfrac{1}{2}\Sigma x^2)A + (\Sigma x)B$ given an orthogonal function, say $(-\tfrac{1}{2}\Sigma x^2)B - (\Sigma x)A$. This reduces to the conditional distribution of Σx given Σx^2. Clearly we cannot get tests of $H_0^{(3)}$ or $H_0^{(4)}$ for μ in this case.

The test of $\mu = \mu_0$ against one-sided alternatives has been discussed in Example 23.7, where we saw that the "Student's" t test to which it reduces is the UMP similar test of $\mu = \mu_0$ against one-sided alternatives. This test is now seen to be UMPU for $H_0^{(1)}$. It also follows that the two-sided "equal-tails" "Student's" t-test, which is unbiassed for $H_0^{(2)}$ against $H_1^{(2)}$, is the UMPU test of $H_0^{(2)}$.

Example 23.15

Consider k independent samples of n_i ($i = 1, 2, \ldots, k$) observations from normal distributions with means μ_i and common variance σ^2. Write $n = \sum_{i=1}^{k} n_i$. It is easily confirmed that the k sample means $\bar{x}_i$ and the pooled sum of squares $S^2 = \sum_{i=1}^{k} \sum_{j=1}^{n_i} (x_{ij} - \bar{x}_i)^2$ are jointly sufficient for the $(k+1)$ parameters. The joint distribution of the sufficient statistics is

$$g(\bar{x}_1, \ldots, \bar{x}_k, S^2) \propto \frac{S^{n-k-2}}{\sigma^n} \exp\left\{-\frac{1}{2\sigma^2} \sum_i \sum_j (x_{ij} - \mu_i)^2\right\}. \tag{23.99}$$

(23.99) is a simple generalization of (23.96), obtained by using the independence of the $\bar{x}_i$ of each other and of S^2, and the fact that S^2/σ^2 has a χ^2 distribution with $(n-k)$ degrees of freedom. (23.99) may be written

$$g \propto C(\mu_i, \sigma^2) \exp\left\{\left(-\tfrac{1}{2}\sum_i \sum_j x_{ij}^2\right)\left(\frac{1}{\sigma^2}\right) + \sum_i \left(\sum_j x_{ij}\right)\left(\frac{\mu_i}{\sigma^2}\right)\right\}, \tag{23.100}$$

in the form (23.73). We now consider the linear function

$$\theta = A\left(\frac{1}{\sigma^2}\right) + \sum_{i=1}^{k} B_i \left(\frac{\mu_i}{\sigma^2}\right). \tag{23.101}$$

(1) Put $A = 1$, $B_i = 0$ (all i). Then $\theta = \dfrac{1}{\sigma^2}$, and $\psi_i = \dfrac{\mu_i}{\sigma^2}$ ($i = 1, \ldots, k$) is the set of nuisance parameters. There is a UMPU test of each of the four $H_0^{(r)}$ discussed in 23.30-2 for $\dfrac{1}{\sigma^2}$ and therefore for σ^2. The tests are based on the conditional distribution of $\sum\sum_{j} x_{ij}^2$ given the vector $(\sum_j x_{1j}, \sum_j x_{2j}, \ldots, \sum_j x_{kj})$, i.e. of $S^2 = \sum_i\sum_j (x_{ij} - \bar{x}_i)^2$ given that vector. Just as in Example 23.14, this leads to the use of the unconditional distribution of S^2 to obtain the UMPU tests.

(2) Exactly analogous considerations to those of Example 23.14 (2) show that by putting $\theta_0 = 0$, we obtain UMPU tests of $\sum_{i=1}^{k} c_i \mu_i \leq c_0$, $\sum c_i \mu_i = c_0$, where c_0 is any constant. (Cf. Exercise 23.19.) Just as before, no "interval" hypothesis can be tested, using this method, concerning the linear form $\sum c_i \mu_i$.

(3) The substitution $k = 2$, $c_1 = 1$, $c_2 = -1$, $c_0 = 0$, reduces (2) to testing $H_0^{(1)}: \mu_1 - \mu_2 \leq 0$, $H_0^{(2)}: \mu_1 - \mu_2 = 0$. The test of $\mu_1 - \mu_2 = 0$ has been discussed in Example 23.8, where it was shown to reduce to a "Student's" t-test and to be UMP similar. It is now seen to be UMPU for $H_0^{(1)}$. The "equal-tails" two-sided "Student's" t-test, which is unbiassed, is also seen to be UMPU for $H_0^{(2)}$.

Example 23.16

We generalize the situation in Example 23.15 by allowing the variances of the k normal distributions to differ. We now have a set of $2k$ sufficient statistics for the $2k$ parameters, which are the sample sums and sums of squares $\sum_{j=1}^{n_i} x_{ij}, \sum_{j=1}^{n_i} x_{ij}^2$, $i = 1, 2, \ldots, k$. We now write

$$\theta = \sum_{i=1}^{k} A_i \left(\frac{1}{\sigma_i^2}\right) + \sum_{i=1}^{k} B_i \left(\frac{\mu_i}{\sigma_i^2}\right). \tag{23.102}$$

(1) Put $B_i = 0$ (all i). We get UMPU tests for all four hypotheses concerning

$$\theta = \sum_i A_i \left(\frac{1}{\sigma_i^2}\right),$$

a weighted sum of the reciprocals of the population variances. The case $k = 2$ reduces this to

$$\theta = \frac{A_1}{\sigma_1^2} + \frac{A_2}{\sigma_2^2}.$$

If we want to test hypotheses concerning the variance ratio σ_2^2/σ_1^2, then just as in (2) of Examples 23.14-15, we have to put $\theta = 0$ to make any progress. If we do this, the UMPU tests of $\theta = 0$, ≤ 0 reduce to those of

$$\frac{\sigma_2^2}{\sigma_1^2} = -\frac{A_2}{A_1}, \leq -\frac{A_2}{A_1},$$

and we therefore have UMPU tests of $H_0^{(1)}$ and $H_0^{(2)}$ concerning the variance ratio. The joint distribution of the four sufficient statistics may be written

$$g(\sum x_{1j}, \sum x_{2j}, \sum x_{1j}^2, \sum x_{2j}^2) \propto C(\mu_i, \sigma_i^2) \exp\left\{-\tfrac{1}{2}\left(\frac{1}{\sigma_1^2}\sum x_{1j}^2 + \frac{1}{\sigma_2^2}\sum x_{2j}^2\right) + \frac{\mu_1}{\sigma_1^2}\sum x_{1j} + \frac{\mu_2}{\sigma_2^2}\sum x_{2j}\right\}.$$

By **23.27**, the coefficient $s(\mathbf{x})$ of θ when (23.103) is transformed to make θ one of its parameters, will be the same function of $-\frac{1}{2}\Sigma x_{1j}^2$, $-\frac{1}{2}\Sigma x_{2j}^2$ as θ is of $1/\sigma_1^2$, $1/\sigma_2^2$, i.e.
$$-2s(\mathbf{x}) = A_1\Sigma x_{1j}^2 + A_2\Sigma x_{2j}^2,$$
and the UMPU tests will be based on the conditional distribution of $s(\mathbf{x})$ given any three functions of the sufficient statistics, orthogonal to $s(\mathbf{x})$ and to each other, say
$$\Sigma x_{1j}, \Sigma x_{2j}, \text{ and } A_2\Sigma x_{1j}^2 - A_1\Sigma x_{2j}^2.$$
This is equivalent to holding $\bar{x}_1, \bar{x}_2$ and $t = \Sigma(x_{1j}-\bar{x}_1)^2 - \dfrac{A_1}{A_2}\Sigma(x_{2j}-\bar{x}_2)^2$ fixed, so that $s(\mathbf{x})$ is equivalent to $\Sigma(x_{1j}-\bar{x}_1)^2 + \dfrac{A_2}{A_1}\Sigma(x_{2j}-\bar{x}_2)^2$ for fixed t. In turn, this is equivalent to considering the distribution of the ratio $\Sigma(x_{1j}-\bar{x}_1)^2/\Sigma(x_{2j}-\bar{x}_2)^2$, so that the UMPU tests of $H_0^{(1)}$, $H_0^{(2)}$ are based on the distribution of the sample variance ratio—cf. Exercises 23.14 and 23.17.

(2) We cannot get UMPU tests concerning functions of the μ_i free of the σ_i^2, as is obvious from (23.102). In the case $k = 2$, this precludes us from finding a solution to the problem of two means by this method.

23.34 The results of **23.27–33** may be better appreciated with the help of a partly geometrical explanation. From (23.73), the characteristic function of $s(\mathbf{x})$ is
$$\phi(u) = E\{\exp(ius)\} = \frac{C(\theta)}{C(\theta+iu)}, \tag{23.103}$$
so that its cumulant-generating function is
$$\psi(u) = \log\phi(u) = \log C(\theta) - \log C(\theta+iu). \tag{23.104}$$
From the form of (23.104), it is clear that the rth cumulant of $s(\mathbf{x})$ is
$$\kappa_r = \left[\frac{\partial^r}{\partial(iu)^r}\psi(u)\right]_{u=0} = -\frac{\partial^r}{\partial\theta^s}\log C(\theta), \tag{23.105}$$
whence
$$E(s) = \kappa_1 = -\frac{\partial}{\partial\theta}\log C(\theta) \tag{23.106}$$
and
$$\kappa_r = \frac{\partial^{r-1}}{\partial\theta^{r-1}}E(s), \quad r \geq 2. \tag{23.107}$$
Consider the derivative
$$\mathbf{D}^q f \equiv \frac{\partial^q}{\partial\theta^q}f(x|\theta,\boldsymbol{\psi}).$$
From (23.73) and (23.106),
$$\mathbf{D}f = \left\{s + \frac{C'(\theta)}{C(\theta)}\right\}f = \{s - E(s)\}f. \tag{23.108}$$
By Leibniz's rule, we have from (23.108)
$$\mathbf{D}^q f = \mathbf{D}^{q-1}[\{s-E(s)\}f]$$
$$= \{s-E(s)\}\mathbf{D}^{q-1}f + \sum_{i=1}^{q-1}\binom{q-1}{i}[\mathbf{D}^i\{s-E(s)\}][\mathbf{D}^{q-1}f], \tag{23.109}$$

which, using (23.107), may be written
$$\mathbf{D}^q f = \{s - E(s)\} \mathbf{D}^{q-1} f - \sum_{i=1}^{q-1} \binom{q-1}{i} \kappa_{i+1} \mathbf{D}^{q-1-i} f. \tag{23.110}$$

23.35 Now consider any critical region w of size α. Its power function is defined at (23.80), and we may alternatively express this as an integral in the sample space of the sufficient statistics $(s, \mathbf{t})$ by
$$P(w | \theta) = \int_w f \, ds \, d\mathbf{t}, \tag{23.111}$$
where f now stands for the joint frequency function of $(s, \mathbf{t})$, which is of the form (23.73) as we have seen. The derivatives of the power function (23.111) are
$$P^{(q)}(w | \theta) = \int_w \mathbf{D}^q f \, ds \, d\mathbf{t}, \tag{23.112}$$
since we may differentiate under the integral sign in (23.111). Using (23.108) and (23.110), (23.111) gives
$$P'(w | \theta) = \int_w \{s - E(s)\} f \, ds \, d\mathbf{t} = \text{cov}\{s, c(w)\}, \tag{23.113}$$
and
$$P^{(q)}(w | \theta) = \int_w \{s - E(s)\} \mathbf{D}^{q-1} f \, ds \, d\mathbf{t} - \sum_{i=1}^{q-1} \binom{q-1}{i} \kappa_{i+1} P^{(q-1-i)}(w | \theta), \quad q \geq 2, \tag{23.114}$$
a recurrence relation which enables us to build up the value of any derivative from lower derivatives. In particular, (23.114) gives
$$P''(w | \theta) = \text{cov}\{[s - E(s)]^2, c(w)\}, \tag{23.115}$$
$$\left. \begin{array}{l} P'''(w | \theta) = \text{cov}\{[s - E(s)]^3, c(w)\} - 3\kappa_2 P'(w | \theta), \\ P^{(iv)}(w | \theta) = \text{cov}\{[s - E(s)]^4, c(w)\} - 6\kappa_2 P''(w | \theta) - 4\kappa_3 P'(w | \theta). \end{array} \right\} \tag{23.116}$$
(23.113) and (23.115) show that the first two derivatives are simply the covariances of $c(w)$ with s, and with the squared deviation of s from its mean, respectively. The third and fourth derivatives given by (23.116) are more complicated functions of covariances and of the cumulants of s, as are the higher derivatives.

23.36 We are now in a position to interpret geometrically some of the results of **23.27–33**. To maximize the power we must choose w to maximize, for all admissible alternatives, the covariance of $c(w)$ with s, or some function of $s - E(s)$, in accordance with (23.113) and (23.114). In the $(r+1)$-dimensional space of the sufficient statistics, $(s, \mathbf{t})$, it is obvious that this will be done by confining ourselves to the subspace orthogonal to the r co-ordinates corresponding to the components of $\mathbf{t}$, i.e. by confining ourselves to the conditional distribution of s given $\mathbf{t}$.

If we are testing $\theta = \theta_0$ against $\theta > \theta_0$ we maximize $P(w | \theta)$ for all $\theta > \theta_0$ by maximizing $P'(w | \theta)$, i.e. by maximizing $\text{cov}(s, c(w))$ for all $\theta > \theta_0$. This is easily seen to be done if w consists of the 100α per cent *largest* values of the distribution of s given $\mathbf{t}$. Similarly for testing $\theta = \theta_0$ against $\theta < \theta_0$, we maximize P by *minimizing P'*, and this is done if w consists of the 100α per cent *smallest* values of the distribution of s given $\mathbf{t}$. Since $P'(w | \theta)$ is always of the same sign, the one-sided tests are unbiassed.

For the two-sided $H_0^{(2)}$ of **23.31**, (23.81) and (23.115) require us to maximize $P''(w|\theta)$, i.e. $\text{cov}\{[s-E(s)]^2, c(w)\}$. By exactly the same argument as in the one-sided case, we choose w to include the 100α per cent largest values of $\{s-E(s)\}^2$, so that we obtain a two-sided test, which is only an "equal-tails" test if the distribution of s given $\mathbf{t}$ is symmetrical. It follows that the boundaries of the UMPU critical region are equidistant from $E(s|\mathbf{t})$.

> Spjøtvoll (1968) gives results on most powerful tests for some non-exponential families of distributions.

Ancillary statistics: a conditionality principle

23.37 We have seen that there is always a set of $r+s$ ($r \geqslant 1$, $s \geqslant 0$) statistics, written (T_r, T_s), which are minimal sufficient for $k+l$ ($k \geqslant 1$, $l \geqslant 0$) parameters, which we shall write (θ_k, θ_l). Suppose now that the subset T_s has a distribution free of θ_k. (This is only possible if the distribution of (T_r, T_s) is not complete—cf. Exercise 23.7.) We then have the factorization of the Likelihood Function into

$$L(\mathbf{x}|\theta_k, \theta_l) = g(T_r, T_s|\theta_k, \theta_l)h(\mathbf{x})$$
$$= g_1(T_r|T_s, \theta_k, \theta_l)g_2(T_s|\theta_l)h(\mathbf{x}). \qquad (23.117)$$

This is (21.95) in different notation.

Fisher (e.g., 1956) calls T_s an *ancillary statistic*, while Bartlett (e.g., 1939) calls the conditional statistic $(T_r|T_s)$ a *quasi-sufficient statistic* for θ_k, the term arising from the resemblance of (23.117) when θ_l is known to the factorization (17.84) which characterizes a sufficient statistic.

Fisher has suggested a *Conditionality Principle* for statistical inference in general, and testing hypotheses in particular: if T_s is distributed free of θ_k, as in (23.117), the conditional distribution of $T_r|T_s$ is all that we need to consider in making inferences about θ_k. Now if T_s is sufficient for θ_l when θ_k is known, it immediately follows that (23.117) becomes

$$L(\mathbf{x}|\theta_k, \theta_l) = g_1(T_r|T_s, \theta_k)g_2(T_s|\theta_l)h(\mathbf{x}), \qquad (23.118)$$

and the two distributions of $(T_r|T_s)$ and T_s are separated off, each depending on separate parameters and each sufficient for its parameters. There is then no doubt that, in accordance with the general principle of **23.3**, we may confine ourselves to functions of $(T_r|T_s)$ in testing θ_k.

However, the real question is whether we should confine ourselves to the conditional statistic when T_s is *not* sufficient for θ_l. The difficulty is essentially that only the *marginal* distribution of T_s is free of θ_k; T_s remains a component of the set of minimal sufficient statistics for all the parameters, whose *joint* distribution depends on θ_k. Welch (1939) gave an example (Exercise 23.23) which showed that the conditional test based on $(T_r|T_s)$ may be uniformly less powerful than an alternative (unconditional) test.

We must now consider the relationship of the Conditionality Principle to the *Likelihood Principle*, which states (cf. **18.32**) that *only* the LF need be regarded in making any statistical inference from observations. In particular, this has the consequence that the details of the sampling procedure which produced the observations (and the LF) are strictly irrelevant to subsequent statistical inference. Many, perhaps most, statisticians will find it intuitively unacceptable to eliminate the sample space from consideration in making inferences from observations. If so, they must reject the Likelihood Principle.

> Birnbaum (1962) showed that the Conditionality Principle implies, as well as obviously being implied by, the Likelihood Principle; Durbin (1970)—see also the papers by Savage (1970) and Birnbaum (1970)—has shown, however, that this result does not hold if, as above, we first reduce the problem to consideration of the *minimal* sufficient statistics.

Example 23.17

We have seen (Example 17.17) that in normal samples the pair $(\bar{x}, s^2)$ is jointly sufficient for (μ, σ^2), and we know that the distribution of s^2 does not depend on μ. Thus we have
$$L(x|\mu, \sigma^2) = g_1(\bar{x}|s^2, \mu, \sigma^2) g_2(s^2|\sigma^2) h(x),$$
a case of (23.117) with $k = l = r = s = 1$. The conditionality principle states that the statistic $\bar{x}|s^2$ is to be used in testing hypotheses about μ. (It happens that $\bar{x}$ is actually independent of s^2 in this case, but this is merely a simplification irrelevant to the general argument.) But s^2 is not a sufficient statistic for the nuisance parameter σ^2, so that the distribution of $\bar{x}|s^2$ is not free of σ^2. If we have no prior distribution given for σ^2 we can only make progress by integrating out σ^2 in some more or less arbitrary way. If we are prepared to use its fiducial distribution and integrate over that, we arrive back at the discussion of **21.10**, where we found that this gives the same result as that obtained from the standpoint of maximizing power in Examples 23.7 and 23.14, namely that "Student's" t-distribution should be used.

Another conditional test principle

23.38 Another principle of test construction may be invoked (cf. D. R. Cox (1958a)) to suggest the use of $(T_r | T_s)$ whenever T_s is sufficient for θ_l when θ_k is known, irrespective of whether its distribution depends on θ_k, for then we have
$$L(\mathbf{x}|\theta_k, \theta_l) = g_1(T_r|T_s, \theta_k) g_2(T_s|\theta_k, \theta_l) h(\mathbf{x}), \qquad (23.119)$$
so that the conditional statistic is distributed independently of the nuisance parameter θ_l. Here again, we have no obvious reason to suppose that the test is optimum in any sense. (23.119) is (21.96) in different notation.

The justification of conditional tests

23.39 The results of **23.30–2** enable us to see that, if the distribution of the sufficient statistics (T_r, T_s) is of the exponential form (23.73), then the use of the conditional distribution of T_r for given T_s will give UMPU tests, for in our previous notation the statistic T_r is $s(\mathbf{x})$ and T_s is $\mathbf{t}(\mathbf{x})$, and we have seen that the UMPU tests are always based on the distribution of T_r for given T_s. If the sufficient statistics are not distributed in the form (23.73) (e.g. in the case of a distribution with range depending on the parameters) this justification is no longer valid. However, following Lindley (1958b), we may derive a further justification of the conditional statistic $T_r | T_s$, provided only that the distribution of T_s, $g_2(T_s | \theta_k, \theta_l)$, is boundedly complete when H_0 holds and that T_s is then sufficient for θ_l. For then, by **23.19**, every size-α critical region similar with respect to θ_l will consist of a fraction α of all surfaces of constant

T_s. Thus any similar test of H_0 will be a conditional test based on $T_r \mid T_s$, and any optimum conditional test will be an optimum similar test.

Welch's (1939) counter-example, which is given in Exercise 23.23, falls within the scope of neither of our justifications of the use of conditional test statistics, for there the two-component minimal sufficient statistic for the single parameter is not complete, so that Exercise 23.7 does not preclude the existence of an ancillary statistic (the sample range R).

EXERCISES

23.1 Show that for samples of n observations from a normal distribution with mean θ and variance σ^2, no symmetric similar region with respect to θ and σ^2 exists for $n \leq 2$, but that such regions do exist for $n \geq 3$.

(Feller, 1938)

23.2 Show, as in Example 23.1, that for a sample of n observations, the ith of which has distribution

$$\frac{1}{\Gamma(\theta_i)} e^{-x_i} x_i^{\theta_i - 1} dx_i, \qquad 0 \leq x_i \leq \infty \,;\; \theta_i > 0,$$

no similar size-α region exists for $0 < \alpha < 1$.

(Feller, 1938)

23.3 If $L(x|\theta)$ is a Likelihood Function and $E\left(\frac{\partial \log L}{\partial \theta}\right) = 0$, show that if the distribution of a statistic z does not depend on θ then $\operatorname{cov}\left(z, \frac{\partial \log L}{\partial \theta}\right) = 0$. As a corollary, show that no similar region with respect to θ exists if no statistic exists which is uncorrelated with $\frac{\partial \log L}{\partial \theta}$.

(Neyman, 1938a)

23.4 Show, using the c.f. of z, that the converses of the result and the corollary of Exercise 23.3 are true.

Together, this exercise and the last state that $\operatorname{cov}\left(z, \frac{\partial \log L}{\partial \theta}\right) = 0$ is a necessary and sufficient condition for $\operatorname{cov}\left(e^{iuz}, \frac{\partial \log L}{\partial \theta}\right) = 0$, where u is a dummy variable.

(Neyman, 1938a)

23.5 Show that the Cauchy family of distributions

$$dF = \frac{dx}{\left\{\pi \theta^{\frac{1}{2}}\left(1 + \frac{x^2}{\theta}\right)\right\}}, \qquad -\infty \leq x \leq \infty,$$

is not complete.

(Lehmann and Scheffé, 1950)

23.6 Show that if a statistic z is distributed independently of t, a sufficient statistic for θ, then the distribution of z does not depend on θ.

23.7 In Exercise 23.6, write $H_1(z)$ for the d.f. of z, $H_2(z|t)$ for its conditional d.f. given t, and $g(t|\theta)$ for the frequency function of t. Show that

$$\int \{H_1(z) - H_2(z|t)\} g(t|\theta) dt = 0$$

for all θ. Hence show that if t is a *complete* sufficient statistic for θ, the converse of the result of Exercise 23.6 holds, namely, if the distribution of z does not depend upon θ, z is distributed independently of t.

(cf. Basu, 1955)

23.8 Use the result of Exercise 23.7 to show directly that, in univariate normal samples:

(a) any moment about the sample mean $\bar{x}$ is distributed independently of $\bar{x}$;
(b) the quadratic form $\mathbf{x}'\mathbf{A}\mathbf{x}$ is distributed independently of $\bar{x}$ if and only if the elements of each row of the matrix $\mathbf{A}$ add to zero (cf. 15.15);
(c) the sample range is distributed independently of $\bar{x}$;
(d) $(x_{(n)} - \bar{x})/(x_{(n)} - x_{(1)})$ is distributed independently both of $\bar{x}$ and of s^2, the sample variance.

(Hogg and Craig, 1956)

23.9 Use Exercise 23.7 to show that:

(a) in samples from a bivariate normal distribution with $\rho = 0$, the sample correlation coefficient is distributed independently of the sample means and variances (cf. 16.28);
(b) in independent samples from two univariate normal populations with the same variance σ^2, the statistic

$$F = \frac{\sum\limits_{j}(x_{1j} - \bar{x}_1)^2/(n_1 - 1)}{\sum\limits_{j}(x_{2j} - \bar{x}_2)^2/(n_2 - 1)}$$

is distributed independently of the set of three jointly sufficient statistics

$$\bar{x}_1, \bar{x}_2, \sum_{j}(x_{1j} - \bar{x}_1)^2 + \sum_{j}(x_{2j} - \bar{x}_2)^2$$

and therefore of the statistic

$$t^2 = \frac{(\bar{x}_1 - \bar{x}_2)^2}{\sum(x_{1j} - \bar{x}_1)^2 + \sum(x_{2j} - \bar{x}_2)^2} \left\{ \frac{n_1 n_2 (n_1 + n_2 - 2)}{n_1 + n_2} \right\}$$

which is a function of the sufficient statistics. This holds whether or not the population means are equal.

(Hogg and Craig, 1956)

23.10 In samples of size n from the distribution

$$dF = \exp\{-(x - \theta)\}\, dx, \qquad \theta \leqslant x \leqslant \infty,$$

show that $x_{(1)}$ is distributed independently of

$$z = \sum_{i=1}^{r}(x_{(i)} - x_{(1)}) + (n - r)(x_{(r)} - x_{(1)}), \qquad r \leqslant n.$$

(Epstein and Sobel, 1954)

23.11 Show that for the binomial distribution with parameter π, the sample proportion p is minimal sufficient for π.

(Lehmann and Scheffé, 1950)

23.12 For the rectangular distribution

$$dF = dx, \qquad \theta - \tfrac{1}{2} \leqslant x \leqslant \theta + \tfrac{1}{2},$$

show that the pair of statistics $(x_{(1)}, x_{(n)})$ is minimal sufficient for θ.

(Lehmann and Scheffé, 1950)

23.13 For a normal distribution with variance σ^2 and unspecified mean μ, show by the method of **23.20** that the UMP similar test of $H_0 : \sigma^2 = \sigma_0^2$ against $H_1 : \sigma^2 = \sigma_1^2$ takes the form

$$\sum (x - \bar{x})^2 \geq a_\alpha \quad \text{if} \quad \sigma_1^2 > \sigma_0^2,$$
$$\sum (x - \bar{x})^2 \leq b_\alpha \quad \text{if} \quad \sigma_1^2 < \sigma_0^2.$$

23.14 Two normal distributions have unspecified means and variances $\sigma^2, \theta\sigma^2$. From independent samples of sizes n_1, n_2, show by the method of **23.20** that the UMP similar test of $H_0 : \theta = 1$ against $H_1 : \theta = \theta_1$ takes the form

$$s_1^2 / s_2^2 \geq a_\alpha \quad \text{if} \quad \theta_1 > 1,$$
$$s_1^2 / s_2^2 \leq b_\alpha \quad \text{if} \quad \theta_1 < 1,$$

where s_1^2, s_2^2 are the sample variances.

(Harter (1963) shows that a test based on the ratio of sample ranges is almost as powerful as the UMP similar test.)

23.15 Independent samples, each of size n, are taken from the distributions

$$dF = \exp\left(-\frac{x}{\theta_1}\right) dx / \theta_1, \quad \theta_1, \theta_2 > 0,$$
$$dG = \exp(-y\theta_2) \theta_2 \, dy, \quad 0 \leq x, y \leq \infty.$$

Show that $t = (\sum x, \sum y) = (X, Y)$ is minimal sufficient for (θ_1, θ_2) and remains so if $H_0 : \theta_1 = \theta_2 = \theta$ holds. By considering the function $XY - E(XY)$ show that the distribution of t is not boundedly complete given H_0, so that not all similar regions satisfy (23.13). Finally, show that the statistic XY is then distributed independently of θ, so that H_0 may be tested by similar regions from it.

(Watson, 1957a)

23.16 In Example 23.14, show from (23.98) that there is a UMPU test of the hypothesis that the parameter point (μ, σ) lies between the two parabolas

$$\mu = \mu_0 + c_1 \sigma^2, \quad \mu = \mu_0 + c_2 \sigma^2,$$

tangent to each other at $(\mu_0, 0)$.

(Lehmann and Scheffé, 1955)

23.17 In Exercise 23.14, show that the critical region

$$s_1^2 / s_2^2 \geq a_{\frac{1}{2}\alpha}, \leq b_{\frac{1}{2}\alpha},$$

is biassed against the two-sided alternative $H_1 : \theta \neq 1$ unless $n_1 = n_2$. By exactly the same argument as in Example 23.12, show that an unbiassed critical region

$$t = s_1^2 / s_2^2 \geq a_{1-\alpha_1}, \leq b_{\alpha_2}, \quad \alpha_1 + \alpha_2 = \alpha,$$

is determined by the condition (cf. (23.56))

$$V_{\alpha_2} f(V_{\alpha_2}) = V_{1-\alpha_1} f(V_{1-\alpha_1}),$$

where f is the frequency function of the variance-ratio statistic t and V_α its 100α per cent point. Show that the power function of the unbiassed test is monotone increasing for $\theta > 1$, monotone decreasing for $\theta < 1$.

(Ramachandran (1958) gives values of $V_{1-\alpha_1}, V_{\alpha_2}$ for $\alpha = 0{\cdot}05$, $n_1 - 1$ and $n_2 - 1 = 2\,(1)\,4\,(2)\,12\,(4)\,24\,;\ 30, 40, 60$.)

23.18 In Exercise 23.17, show that the unbiassed confidence interval for θ given by $\left(\dfrac{t}{V_{\alpha_2}}, \dfrac{t}{V_{1-\alpha_1}}\right)$ minimizes the expectation of $(\log U - \log L)$ for confidence intervals (L, U) based on the tails of the distribution of t.

(Scheffé, 1942b)

23.19 In Example 23.15, use **23.27** to show that the UMPU tests for $\sum_i c_i \mu_i$ are based on the distribution of
$$t = \sum c_i(\bar{x}_i - \mu_i) \Big/ \left\{\frac{S^2}{n-k} \sum \frac{c_i^2}{n_i}\right\}^{\frac{1}{2}},$$
which is a "Student's" t with $(n-k)$ degrees of freedom.

23.20 In Example 23.16, show that there is a UMPU test of the hypothesis
$$\frac{\mu_i}{\mu_j} = a\frac{\sigma_i^2}{\sigma_j^2}, \qquad i \neq j,\ \mu_j \neq 0.$$

23.21 For independent samples from two Poisson distributions with parameters μ_1, μ_2, show that there are UMPU tests for all four hypotheses considered in **23.30-2** concerning μ_1/μ_2, and that the test of $\mu_1/\mu_2 = 1$ consists of testing whether the sum of the observations is binomially distributed between the samples with equal probabilities.

(Lehmann and Scheffé, 1955)

23.22 For independent binomial distributions with parameters θ_1, θ_2, find the UMPU tests for all four hypotheses in **23.30-2** concerning the "odds ratio" $\left(\dfrac{\theta_1}{1-\theta_1}\right) \Big/ \left(\dfrac{\theta_2}{1-\theta_2}\right)$, and the UMPU tests for $\theta_1 = \theta_2$, $\theta_1 \leqslant \theta_2$.

(Lehmann and Scheffé, 1955)

23.23 For the rectangular distribution
$$dF = dx, \qquad \theta - \tfrac{1}{2} \leqslant x \leqslant \theta + \tfrac{1}{2},$$
the conditional distribution of the midrange M given the range R, and the marginal distribution of M, are given by the results of Exercise 14.12 (Vol. 1, 3rd edition). For testing $H_0: \theta = \theta_0$ against the two-sided alternative $H_1: \theta \neq \theta_0$ show that the "equal-tails" test based on M given R, when integrated over all values of R, gives uniformly less power than the "equal-tails" test based on the marginal distribution of M; use the value $\alpha = 0.08$ for convenience.

(Welch, 1939)

23.24 In Example 23.9, show that the UMPU test of $H_0: \sigma = \sigma_0$ against $H_1: \sigma \neq \sigma_0$ is of the form
$$\sum_{i=1}^{n} x_i \geqslant a_{\alpha_1},\ \leqslant b_{\alpha_2}.$$

(Lehmann, 1947)

23.25 For the distribution of Example 23.9, show that the UMP similar test of $H_0: \theta = \theta_0$ against $H_1: \theta \neq \theta_0$ is of the form
$$\frac{x_{(1)} - \theta_0}{\bar{x}} < 0,\ \geqslant c_\alpha.$$
(Lehmann, 1947; see also Takeuchi (1969) for the $H_1: \theta < \theta_0$).

23.26 For the rectangular distribution
$$dF = dx/\theta, \quad \mu \leqslant x \leqslant \mu+\theta,$$
show that the UMP similar test of $H_0: \mu = \mu_0$ against $H_1: \mu \neq \mu_0$ is of the form
$$\frac{x_{(1)} - \mu_0}{x_{(n)} - x_{(1)}} < 0, \geqslant c_\alpha.$$
Cf. the simple hypothesis with $\theta = 1$, where it was seen in Exercise 22.8 that no UMP test exists.

(Lehmann, 1947)

23.27 If $x_1, \ldots, x_n$ are independent observations from the distribution
$$dF = \frac{1}{\theta^p \Gamma(p)} \exp(-x/\theta) x^{p-1} dx, \quad p > 0, \, 0 \leqslant x \leqslant \infty,$$
use Exercises 23.6 and 23.7 to show that a necessary and sufficient condition that a statistic $h(x_1, \ldots, x_n)$ be independent of $S = \sum_{i=1}^{n} x_i$ is that $h(x_1, \ldots, x_n)$ be homogeneous of degree zero in x. (Cf. refs. to Exercise 15.22.)

23.28 From (23.113) and (23.114), show that if the first non-zero derivative of the power function is the mth, then
$$P^{(m)}(w|\theta) = \text{cov}\{[s - E(s)]^m, c(w)\}$$
and
$$\frac{\{P^{(m)}(w|\theta)\}^2}{\mu_{2m}} \leqslant \frac{1}{4},$$
where μ_r is the rth central moment of s. In particular,
$$|P'(w|\theta)| \leqslant \tfrac{1}{2}\mu_2^{\frac{1}{2}}.$$

23.29 From **23.35**, show that w is a similar region for a hypothesis for which θ is a nuisance parameter if and only if
$$\text{cov}\{s, c(w)\} = 0$$
identically in θ. Cf. Exercises 23.3–4.

23.30 Generalize the argument of the last paragraph of Example 23.7 to show that for any distribution of form
$$dF = f\left(\frac{x-\mu}{\sigma}\right) \frac{dx}{\sigma},$$
admitting a complete sufficient statistic for σ when μ is known, there can be no similar critical region for $H_0: \mu = \mu_0$ against $H_1: \mu = \mu_1$ with power independent of σ.

23.31 For a normal distribution with mean and variance both equal to θ, show that for a single observation, x and x^2 are each singly sufficient for θ, x^2 being minimal. Hence it follows that single sufficiency does not imply minimal sufficiency. (Exercise 18.13 gives a bivariate instance of the same phenomenon.)

23.32 In the problem of the ratio of two normal means (Example 20.7), assume that $\sigma_{xy} = 0$, $\sigma_x^2 = \sigma_y^2 = 1$. To test the composite $H_0: \theta = \theta_0$ against $H_1: \theta = \theta_1$, show that when H_0 holds, the nuisance parameter μ_1 has a complete sufficient statistic $u = \theta_0 \bar{y} + \bar{x}$. Hence show, using **23.20**, that the UMP similar test of H_0 against H_1 is based on large (small) values of $\bar{y} - \theta_0 \bar{x}$ if $(\theta_1 - \theta_0)\mu_1 > 0 \, [<0)$.

(cf. D. R. Cox (1967).)

CHAPTER 24

LIKELIHOOD RATIO TESTS AND THE GENERAL LINEAR HYPOTHESIS

24.1 The ML method discussed in Chapter 18 is a constructive method of obtaining estimators which, under certain conditions, have desirable properties. A method of test construction closely allied to it is the Likelihood Ratio (LR) method, proposed by Neyman and Pearson (1928). It has played a role in the theory of tests analogous to that of the ML method in the theory of estimation.

As before, we have a LF

$$L(x|\theta) = \prod_{i=1}^{n} f(x_i|\theta),$$

where $\theta = (\theta_r, \theta_s)$ is a vector of $r+s = k$ parameters ($r \geq 1$, $s \geq 0$) and x may also be a vector. We wish to test the hypothesis

$$H_0 : \theta_r = \theta_{r0}, \tag{24.1}$$

which is composite unless $s = 0$, against

$$H_1 : \theta_r \neq \theta_{r0}.$$

We know that there is generally no UMP test in this situation, but that there may be a UMPU test—cf. **23.31**.

The LR method first requires us to find the ML estimators of (θ_r, θ_s), giving the unconditional maximum of the LF

$$L(x|\hat{\theta}_r, \hat{\theta}_s), \tag{24.2}$$

and also to find the ML estimators of θ_s, when H_0 holds,(*) giving the conditional maximum of the LF

$$L(x|\theta_{r0}, \hat{\hat{\theta}}_s). \tag{24.3}$$

$\hat{\hat{\theta}}_s$ in (24.3) has been given a double circumflex to emphasize that it does not in general coincide with $\hat{\theta}_s$ in (24.2). Now consider the likelihood ratio(†)

$$l = \frac{L(x|\theta_{r0}, \hat{\hat{\theta}}_s)}{L(x|\hat{\theta}_r, \hat{\theta}_s)}. \tag{24.4}$$

Since (24.4) is the ratio of a conditional maximum of the LF to its unconditional maximum, we clearly have

$$0 \leq l \leq 1. \tag{24.5}$$

Intuitively, l is a reasonable test statistic for H_0: it is the maximum likelihood under

---

(*) When $s = 0$, H_0 being simple, no maximization process is needed, for L is uniquely determined.

(†) The ratio is usually denoted by λ, and the LR statistic is sometimes called "the lambda criterion," but we use the Roman letter in accordance with the convention that Greek symbols are reserved for parameters.

H_0 as a fraction of its largest possible value, and large values of l signify that H_0 is reasonably acceptable. The critical region for the test statistic is therefore

$$l \leqslant c_\alpha, \qquad (24.6)$$

where c_α is determined from the distribution $g(l)$ of l to give a size-α test, i.e.

$$\int_0^{c_\alpha} g(l)\,dl = \alpha. \qquad (24.7)$$

24.2 For the LR method to be useful in the construction of similar tests, i.e. tests based on similar critical regions, the distribution of l should be free of nuisance parameters, and it is a fact that for many common statistical problems it is so. The next two examples illustrate the method in cases where it does and does not lead to a similar test.

Example 24.1

For the normal distribution

$$dF(x) = (2\pi\sigma^2)^{-\frac{1}{2}} \exp\left\{-\tfrac{1}{2}\left(\frac{x-\mu}{\sigma}\right)^2\right\} dx,$$

we wish to test

$$H_0: \mu = \mu_0.$$

Here

$$L(x\,|\,\mu,\sigma^2) = (2\pi\sigma^2)^{-\frac{1}{2}n} \exp\left\{-\tfrac{1}{2}\Sigma\left(\frac{x-\mu}{\sigma}\right)^2\right\}.$$

Using Example 18.11, we have for the unconditional ML estimators

$$\hat{\mu} = \bar{x},$$
$$\hat{\sigma}^2 = \frac{1}{n}\Sigma(x-\bar{x})^2 = s^2,$$

so that

$$L(x\,|\,\hat{\mu},\hat{\sigma}^2) = (2\pi s^2)^{-\frac{1}{2}n} \exp(-\tfrac{1}{2}n). \qquad (24.8)$$

When H_0 holds, the ML estimator is (cf. Example 18.8)

$$\hat{\hat{\sigma}}^2 = \frac{1}{n}\Sigma(x-\mu_0)^2 = s^2 + (\bar{x}-\mu_0)^2,$$

so that

$$L(x\,|\,\mu_0,\hat{\hat{\sigma}}^2) = [2\pi\{s^2 + (\bar{x}-\mu_0)^2\}]^{-\frac{1}{2}n} \exp(-\tfrac{1}{2}n). \qquad (24.9)$$

From (24.4), (24.8) and (24.9), we find

$$l = \left\{\frac{s^2}{s^2 + (\bar{x}-\mu_0)^2}\right\}^{\frac{1}{2}n}$$

or

$$l^{2/n} = \frac{1}{1 + \dfrac{t^2}{n-1}},$$

where t is "Student's" t-statistic with $(n-1)$ degrees of freedom. Thus l is a monotone decreasing function of t^2. Hence we may use the known exact distribution of t^2

as equivalent to that of l, rejecting the 100α per cent largest values of t^2, which correspond to the 100α per cent smallest values of l. We thus obtain an "equal-tails" test based on the distribution of "Student's" t, half of the critical region consisting of extreme positive values, and half of extreme negative values, of t. This is a very reasonable test: we have seen that it is UMPU for H_0 in Example 23.14.

Example 24.2

Consider again the problem of two means, extensively discussed in Chapters 21 and 23. We have samples of sizes n_1, n_2 from normal distributions with means and variances $(\mu_1, \sigma_1^2), (\mu_2, \sigma_2^2)$ and wish to test $H_0: \mu_1 = \mu_2$, which we may re-parametrize (cf. **23.2**) as $H_0: \theta \equiv \mu_1 - \mu_2 = 0$. We call the common unknown value of the means μ. We have

$$L(x|\mu_1, \mu_2, \sigma_1^2, \sigma_2^2) = (2\pi)^{-\frac{1}{2}(n_1+n_2)} \sigma_1^{-n_1} \sigma_2^{-n_2} \exp\left\{-\frac{1}{2}\left(\sum_{j=1}^{n_1} \frac{(x_{1j}-\mu_1)^2}{\sigma_1^2} + \sum_{j=1}^{n_2} \frac{(x_{2j}-\mu_2)^2}{\sigma_2^2}\right)\right\}.$$

The unconditional ML estimators are

$$\hat{\mu}_1 = \bar{x}_1, \quad \hat{\mu}_2 = \bar{x}_2, \quad \hat{\sigma}_1^2 = s_1^2, \quad \hat{\sigma}_2^2 = s_2^2,$$

so that

$$L(x|\hat{\mu}_1, \hat{\mu}_2, \hat{\sigma}_1^2, \hat{\sigma}_2^2) = (2\pi)^{-\frac{1}{2}(n_1+n_2)} s_1^{-n_1} s_2^{-n_2} \exp\left\{-\frac{1}{2}(n_1+n_2)\right\}.$$

When H_0 holds, the ML estimators are roots of the set of three equations

$$\left.\begin{aligned}\frac{n_1(\bar{x}_1-\mu)}{\sigma_1^2} + \frac{n_2(\bar{x}_2-\mu)}{\sigma_2^2} &= 0, \\ \sigma_1^2 = \frac{1}{n_1}\sum_{j=1}^{n_1}(x_{1j}-\mu)^2 &= s_1^2 + (\bar{x}_1-\mu)^2, \\ \sigma_2^2 = \frac{1}{n_2}\sum_{j=1}^{n_2}(x_{2j}-\mu)^2 &= s_2^2 + (\bar{x}_2-\mu)^2.\end{aligned}\right\} \quad (24.10)$$

When the solutions of (24.10) are substituted into the LF, we get

$$L(x|\hat{\hat{\mu}}, \hat{\hat{\sigma}}_1^2, \hat{\hat{\sigma}}_2^2) = (2\pi)^{-\frac{1}{2}(n_1+n_2)} \hat{\hat{\sigma}}_1^{-n_1} \hat{\hat{\sigma}}_2^{-n_2} \exp\left\{-\frac{1}{2}(n_1+n_2)\right\},$$

and the likelihood ratio is

$$l = \left(\frac{s_1}{\hat{\hat{\sigma}}_1}\right)^{n_1}\left(\frac{s_2}{\hat{\hat{\sigma}}_2}\right)^{n_2} = \left\{\frac{s_1^2}{s_1^2+(\bar{x}_1-\hat{\hat{\mu}})^2}\right\}^{\frac{1}{2}n_1}\left\{\frac{s_2^2}{s_2^2+(\bar{x}_2-\hat{\hat{\mu}})^2}\right\}^{\frac{1}{2}n_2}. \quad (24.11)$$

We need then only to determine $\hat{\hat{\mu}}$ to be able to use (24.11). Now by (24.10), we see that $\hat{\hat{\mu}}$ is a solution of a cubic equation in μ whose coefficients are functions of the n_i and of the sums and sums of squares of the two sets of observations. We cannot therefore write down $\hat{\hat{\mu}}$ as an explicit function, though we can solve for it numerically in any given case. Its distribution is, in any case, not independent of the ratio σ_1^2/σ_2^2, for $\hat{\hat{\mu}}$ is a function of both s_1^2 and s_2^2 and l is therefore of the form

$$l = g(s_1^2, s_2^2) h(s_1^2, s_2^2).$$

Thus the LR method fails in this case to give us a similar test.

24.3 If, as in Example 24.1, we find that the LR test statistic is a one-to-one function of some statistic whose distribution is either known exactly (as in that Example)

LIKELIHOOD RATIO TESTS

or can be found, there is no difficulty in constructing a valid test of H_0, though we shall have shortly to consider what desirable properties LR tests as a class possess. However, it frequently occurs that the LR method is not so convenient, when the test statistic is a more or less complicated function of the observations whose exact distribution cannot be obtained, as in Example 24.2. In such a case, we have to resort to approximations to its distribution.

Since l is distributed on the interval $(0, 1)$, we see that for any fixed constant $c > 0$, $w = -2c \log l$ will be distributed on the interval $(0, \infty)$. It is therefore natural to seek an approximation to its distribution by means of a χ^2 variate, which is also on the interval $(0, \infty)$, adjusting c to make the approximation as close as possible. The inclination to use such an approximation is increased by the fact, to be proved in **24.7**, that as n increases, the distribution of $-2 \log l$ when H_0 holds tends to a χ^2 distribution with r degrees of freedom. In fact, we shall be able to find the asymptotic distribution of $-2 \log l$ when H_1 holds also, but in order to do this we must introduce a generalization of the χ^2 distribution.

The non-central χ^2 distribution

24.4 We have seen in **16.2–3** that the sum of squares of n independent standardized normal variates is distributed in the χ^2 form with n degrees of freedom, (16.1), and c.f. given by (16.3). We now consider the distribution of the statistic

$$z = \sum_{i=1}^{n} x_i^2$$

where the x_i are still independent normal variates with unit variance, but where their means can differ from zero and

$$E(x_i) = \mu_i, \quad \Sigma \mu_i^2 = \lambda. \tag{24.12}$$

We write the joint distribution of the x_i as

$$dF \propto \exp\{-\tfrac{1}{2}(\mathbf{x}-\boldsymbol{\mu})'(\mathbf{x}-\boldsymbol{\mu})\} \prod_i dx_i,$$

and make the orthogonal transformation to a new set of independent normal variates with variances unity,

$$\mathbf{y} = \mathbf{Bx}.$$

Since
$$E(\mathbf{x}) = \boldsymbol{\mu},$$
we have
$$\boldsymbol{\theta} = E(\mathbf{y}) = \mathbf{B}\boldsymbol{\mu},$$
so that
$$\boldsymbol{\theta}'\boldsymbol{\theta} = \boldsymbol{\mu}'\boldsymbol{\mu} = \lambda, \tag{24.13}$$

since $\mathbf{B}'\mathbf{B} = \mathbf{I}$. We now choose the first $(n-1)$ components of $\boldsymbol{\theta}$ equal to zero. Then by (24.13),
$$\theta_n^2 = \lambda.$$

Thus
$$z = \mathbf{x}'\mathbf{x} = \mathbf{y}'\mathbf{y}$$

is a sum of squares of n independent normal variates, the first $(n-1)$ of which are standardized, and the last of which has mean $\lambda^{\frac{1}{2}}$ and variance 1. We write
$$u = \sum_{i=1}^{n-1} y_i^2, \quad v = y_n^2$$
and we know that u is distributed like χ^2 with $(n-1)$ degrees of freedom. The distribution of y_n is
$$dF \propto \exp\{-\tfrac{1}{2}(y_n - \lambda^{\frac{1}{2}})^2\} dy_n,$$
so that the distribution of v is
$$f_1(v)\, dv \propto \frac{dv}{2v^{\frac{1}{2}}} [\exp\{-\tfrac{1}{2}(v^{\frac{1}{2}} - \lambda^{\frac{1}{2}})^2\} + \exp\{-\tfrac{1}{2}(-v^{\frac{1}{2}} - \lambda^{\frac{1}{2}})^2\}]$$
$$\propto v^{-\frac{1}{2}} \exp\{-\tfrac{1}{2}(v+\lambda)\} \sum_{r=0}^{\infty} \frac{(v\lambda)^r}{(2r)!}\, dv. \tag{24.14}$$

The joint distribution of v and u is
$$dG \propto f_1(v) f_2(u)\, dv\, du, \tag{24.15}$$
where f_2 is the χ^2 distribution with $(n-1)$ degrees of freedom
$$f_2(u)\, du \propto e^{-\frac{1}{2}u} u^{\frac{1}{2}(n-3)}\, du. \tag{24.16}$$

We put (24.14) and (24.16) into (24.15) and make the transformation
$$\left. \begin{array}{l} z = u+v, \\ w = \dfrac{u}{u+v}, \end{array} \right\}$$
with Jacobian equal to z. We find for the joint distribution of z and w
$$dG(z,w) \propto e^{-\frac{1}{2}(z+\lambda)} z^{\frac{1}{2}(n-2)} w^{\frac{1}{2}(n-3)} (1-w)^{-\frac{1}{2}} \sum_{r=0}^{\infty} \frac{\lambda^r z^r}{(2r)!} (1-w)^r\, dw\, dz.$$

We now integrate out w over its range from 0 to 1, getting for the marginal distribution of z
$$dH(z) \propto e^{-\frac{1}{2}(z+\lambda)} z^{\frac{1}{2}(n-2)} \sum_{r=0}^{\infty} \frac{\lambda^r z^r}{(2r)!} B\{\tfrac{1}{2}(n-1), \tfrac{1}{2}+r\}\, dz. \tag{24.17}$$

To obtain the constant in (24.17), we recall that it does not depend on λ, and put $\lambda = 0$. (24.17) should then reduce to (16.1), which is the ordinary χ^2 distribution with n degrees of freedom. The non-constant factors agree, but whereas (16.1) has a constant term $\dfrac{1}{2^{\frac{1}{2}n} \Gamma(\frac{1}{2}n)}$, (24.17) has $B\{\tfrac{1}{2}(n-1), \tfrac{1}{2}\} = \dfrac{\Gamma\{\tfrac{1}{2}(n-1)\} \Gamma(\tfrac{1}{2})}{\Gamma(\tfrac{1}{2}n)}$. We must therefore divide (24.17) by the factor $2^{\frac{1}{2}n} \Gamma\{\tfrac{1}{2}(n-1)\} \Gamma(\tfrac{1}{2})$ and finally, writing ν for n, we have for any λ
$$dH(z) = \frac{e^{-\frac{1}{2}(z+\lambda)} z^{\frac{1}{2}(\nu-2)}}{2^{\frac{1}{2}\nu} \Gamma\{\tfrac{1}{2}(\nu-1)\} \Gamma(\tfrac{1}{2})} \sum_{r=0}^{\infty} \frac{\lambda^r z^r}{(2r)!} B\{\tfrac{1}{2}(\nu-1), \tfrac{1}{2}+r\}\, dz. \tag{24.18}$$

Guenther (1964) gives a simple geometric derivation of (24.18) with references to other geometric proofs. McNolty (1962) obtains it by inverting its c.f., which the reader is asked to find in Exercise 24.1.

24.5 The distribution (24.18) is called the non-central χ^2 distribution with ν degrees of freedom and non-central parameter[*] λ, and sometimes written $\chi'^2(\nu, \lambda)$. It was first

---

[*] In the literature, $\tfrac{1}{2}\lambda$ is sometimes used as the parameter, and occasionally λ^2 written for λ, but our notation is now the standard one.

given by Fisher (1928a), and has been studied by Wishart (1932), Patnaik (1949) and by Tiku (1965a).

> Johnson and Pearson (1969) give the 0·5, 1, 2·5, 5, 95, 97·5, 99 and 99·5 percentage points of the distribution of $\sqrt{z}$ for $\sqrt{\lambda} = 0\cdot2$ $(0\cdot2)$ $6\cdot0$ and $\nu = 1$ (1) 12, 15, 20 to 4 significant figures, and also approximations for $\sqrt{\lambda} = 8, 10$.

Since the first two cumulants of χ'^2 are (cf. Exercise 24.1)

$$\left.\begin{array}{l}\kappa_1 = \nu + \lambda, \\ \kappa_2 = 2(\nu + 2\lambda),\end{array}\right\} \quad (24.19)$$

it can be approximated by a (central) χ^2 distribution as follows. The first two cumulants of a χ^2 with ν^* degrees of freedom are (putting $\lambda = 0$, $\nu = \nu^*$ in (24.19))

$$\kappa_1 = \nu^*, \quad \kappa_2 = 2\nu^*. \quad (24.20)$$

If we equate the first two cumulants of χ'^2 with those of $\rho\chi^2$, where ρ is a constant to be determined, we have, from (24.19) and (24.20),

$$\left.\begin{array}{l}\nu + \lambda = \rho\nu^*, \\ 2(\nu + 2\lambda) = 2\rho^2\nu^*,\end{array}\right\}$$

so that χ'^2/ρ is approximately a central χ^2 variate with

$$\left.\begin{array}{l}\rho = \dfrac{\nu + 2\lambda}{\nu + \lambda} = 1 + \dfrac{\lambda}{\nu + \lambda}, \\ \nu^* = \dfrac{(\nu + \lambda)^2}{\nu + 2\lambda} = \nu + \dfrac{\lambda^2}{\nu + 2\lambda},\end{array}\right\} \quad (24.21)$$

ν^* in general being fractional. If $\nu \to \infty$, $\rho \to 1$ and $\nu^* \sim \nu$; but if $\lambda \to \infty$, $\rho \to 2$ and $\nu^* \sim \frac{1}{2}\lambda$.

Patnaik (1949) shows that this approximation to the d.f. of χ'^2 is adequate for many purposes, but he also gives better approximations obtained from Edgeworth series expansions.

If ν^* is large, we may make the approximation simpler by approximating the χ^2 approximating distribution itself, for (cf. **16.6**) $(2\chi'^2/\rho)^{\frac{1}{2}}$ tends to normality with mean $(2\nu^* - 1)^{\frac{1}{2}}$ and variance 1, while, more slowly, χ'^2/ρ becomes normal with mean ν^* and variance $2\nu^*$.

> E. S. Pearson (1959) gives a more accurate central χ^2 approximation by equating three moments. Johnson and Pearson (1969) fit a Pearson Type I distribution using four moments.

24.6 We may now generalize our derivation of **24.4**. Suppose that $\mathbf{x}$ is a vector of n multinormal variates with mean $\boldsymbol{\mu}$ and non-singular dispersion matrix $\mathbf{V}$. We can find an orthogonal transformation $\mathbf{x} = \mathbf{By}$ which reduces the quadratic form $\mathbf{x'V^{-1}x}$ to the diagonal form $\mathbf{y'B'V^{-1}By} = \mathbf{y'Cy}$, the elements of the diagonal of $\mathbf{C}$ being the latent roots of $\mathbf{V^{-1}}$. To $\mathbf{y'Cy}$ we apply a further scaling transformation $\mathbf{y} = \mathbf{Dz}$, where the leading diagonal elements of the diagonal matrix $\mathbf{D}$ are the reciprocals of the square roots of the corresponding elements of $\mathbf{C}$, so that $\mathbf{D^2} = \mathbf{C^{-1}}$. Thus $\mathbf{x'V^{-1}x} = \mathbf{y'Cy} = \mathbf{z'z}$, and $\mathbf{z}$ is a vector of n independent normal variates with unit variances and mean vector $\boldsymbol{\theta}$ satisfying $\boldsymbol{\mu} = \mathbf{BD\theta}$. Thus $\lambda = \boldsymbol{\theta'\theta} = \boldsymbol{\mu'V^{-1}\mu}$.

We have now reduced our problem to that considered in **24.4**. We see that the distribution of $\mathbf{x}'\mathbf{V}^{-1}\mathbf{x}$, where $\mathbf{x}$ is a multinormal vector with dispersion matrix $\mathbf{V}$ and mean vector $\boldsymbol{\mu}$, is a non-central χ^2 distribution with n degrees of freedom and non-central parameter $\boldsymbol{\mu}'\mathbf{V}^{-1}\boldsymbol{\mu}$. This generalizes the result of **15.10** for multinormal variates with zero means.

> Graybill and Marsaglia (1957) have generalized the theorems on the distribution of quadratic forms in normal variates, discussed in **15.10–21** and Exercises 15.13–17, to the case where $\mathbf{x}$ has mean $\boldsymbol{\mu} \neq 0$. Idempotency of a matrix is then a necessary and sufficient condition that its quadratic form is distributed in a non-central χ^2 distribution, and all the theorems of Chapter 15 hold with this modification.

The asymptotic distribution of the LR statistic

24.7 We saw in **18.17** that under regularity conditions the ML estimator (temporarily written t) of a single parameter θ attains the MVB asymptotically. It follows from **17.17** that the LF is asymptotically of the form

$$\frac{\partial \log L}{\partial \theta} = -E\left(\frac{\partial^2 \log L}{\partial \theta^2}\right)(t-\theta), \tag{24.22}$$

which is the leading term (of order $n^{\frac{1}{2}}$) obtained by differentiating the logarithm of

$$L \propto \exp\left\{\tfrac{1}{2} E\left(\frac{\partial^2 \log L}{\partial \theta^2}\right)(t-\theta)^2\right\}, \tag{24.23}$$

showing that the LF reduces to the normal distribution of the "asymptotically sufficient" statistic t.

For a k-component vector of parameters $\boldsymbol{\theta}$, the matrix analogue of (24.22) is

$$\frac{\partial \log L}{\partial \boldsymbol{\theta}} = (\mathbf{t}-\boldsymbol{\theta})'\mathbf{V}^{-1}, \tag{24.24}$$

where $\mathbf{V}^{-1}$ is defined by (cf. **18.26**)

$$V_{ij}^{-1} = -E\left(\frac{\partial^2 \log L}{\partial \theta_i \partial \theta_j}\right).$$

When integrated, (24.24) gives the analogue of (24.23)

$$L \propto \exp\left\{-\tfrac{1}{2}(\mathbf{t}-\boldsymbol{\theta})'\mathbf{V}^{-1}(\mathbf{t}-\boldsymbol{\theta})\right\}. \tag{24.25}$$

We saw in **18.26** that under regularity conditions the vector of ML estimators $\mathbf{t}$ is asymptotically multinormally distributed with the dispersion matrix $\mathbf{V}$. Thus the LF reduces to the multinormal distribution of $\mathbf{t}$. This result was rigorously proved by Wald (1943a).

We may now easily establish the asymptotic distribution of the LR statistic l defined at (24.4). In virtue of (24.25), we may reduce the problem to considering the ratio of the maximum of the right-hand side of (24.25) given H_0 to its maximum given H_1. When H_1 holds, the maximum of (24.25) is when $\boldsymbol{\theta} = \hat{\boldsymbol{\theta}} = \mathbf{t}$, so that every component of $(\mathbf{t}-\boldsymbol{\theta})$ is equal to zero and we have

$$L(x|\hat{\boldsymbol{\theta}}_r,\hat{\boldsymbol{\theta}}_s) \propto 1. \tag{24.26}$$

When H_0 holds, the s components of $(\mathbf{t}-\boldsymbol{\theta})$ corresponding to $\boldsymbol{\theta}_s$ will still be zero, for the maximum of (24.25) occurs when $\boldsymbol{\theta}_s = \hat{\hat{\boldsymbol{\theta}}}_s = \mathbf{t}_s$, say. (24.25) may now be written

$$L(x|\boldsymbol{\theta}_{r0},\hat{\hat{\boldsymbol{\theta}}}_s) \propto \exp\left\{-\tfrac{1}{2}(\mathbf{t}_r - \boldsymbol{\theta}_{r0})'\mathbf{V}_r^{-1}(\mathbf{t}_r - \boldsymbol{\theta}_{r0})\right\}, \tag{24.27}$$

the suffix r signifying that we are now confined to an r-dimensional distribution. Thus, from (24.26) and (24.27),

$$l = \frac{L(x|\boldsymbol{\theta}_{r0},\hat{\boldsymbol{\theta}}_s)}{L(x|\boldsymbol{\theta}_r,\hat{\boldsymbol{\theta}}_s)} = \exp\{-\tfrac{1}{2}(\mathbf{t}_r - \boldsymbol{\theta}_{r0})'\mathbf{V}_r^{-1}(\mathbf{t}_r - \boldsymbol{\theta}_{r0})\}.$$

Thus
$$-2\log l = (\mathbf{t}_r - \boldsymbol{\theta}_{r0})'\mathbf{V}_r^{-1}(\mathbf{t}_r - \boldsymbol{\theta}_{r0}).$$

Now we have seen that $\mathbf{t}_r$ is multinormal with dispersion matrix $\mathbf{V}_r$ and mean vector $\boldsymbol{\theta}_{r0}$. Thus, by the result of **24.6**, $-2\log l$ for a hypothesis imposing r constraints is asymptotically distributed in the non-central χ^2 distribution with r degrees of freedom and non-central parameter

$$\lambda = (\boldsymbol{\theta}_r - \boldsymbol{\theta}_{r0})'\mathbf{V}_r^{-1}(\boldsymbol{\theta}_r - \boldsymbol{\theta}_{r0}), \qquad (24.28)$$

a result due to Wald (1943a). If λ is to be bounded away from infinity, we must, since $\mathbf{V}_r^{-1}$ is of order n, restrict $(\boldsymbol{\theta}_r - \boldsymbol{\theta}_{r0})$ to be of order $n^{-\frac{1}{2}}$. When H_0 holds, $\lambda = 0$ and this reduces to a central χ^2 distribution with r degrees of freedom, a result originally due to Wilks (1938a). A simple rigorous proof of the H_0 result is given by K. P. Roy (1957). Using the LR test is therefore asymptotically equivalent to basing a test on the ML estimators of the parameters tested. It should be emphasized that these results only hold if the conditions for the asymptotic normality and efficiency of the ML estimators are satisfied.

See also the generalizations by Chernoff (1954) and Feder (1968).

The asymptotic power of LR tests

24.8 The result of **24.7** makes it possible to calculate the asymptotic power function of the LR test in any case satisfying the conditions for that result to be valid. We have first to evaluate the matrix $\mathbf{V}_r^{-1}$, and then to evaluate the integral

$$P = \int_{\chi_\alpha'^2(\nu,0)}^{\infty} d\chi'^2(\nu,\lambda) \qquad (24.29)$$

where $\chi_\alpha'^2(\nu,0)$ is the $100(1-\alpha)$ per cent point of the central χ^2 distribution. P is the power of the test, and its size when $\lambda = 0$.

Patnaik (1949) gives a table of P for $\alpha = 0.05$, degrees of freedom 2 (1) 12 (2) 20 and $\lambda = 2$ (2) 20. For 1 degree of freedom, we may use the normal d.f. to evaluate P as in Example 24.3 below. Fix (1949b) gives inverse tables of λ for degrees of freedom 1 (1) 20 (2) 40 (5) 60 (10) 100, $\alpha = 0.05$ and $\alpha = 0.01$ and P (her β) $= 0.1$ (0.1) 0.9.

If we use the approximation of **24.5** for the non-central distribution, (24.29) becomes, using (24.21) with $\nu = r$,

$$P = \int_{\left(\frac{r+\lambda}{r+2\lambda}\right)\chi_\alpha^2(r)} d\chi^2\left(r + \frac{\lambda^2}{r+2\lambda}\right), \qquad (24.30)$$

where $\chi^2(r)$ is the central χ^2 distribution with r degrees of freedom and $\chi_\alpha^2(r)$ its $100(1-\alpha)$ per cent point. Putting $\lambda = 0$ in (24.30) gives the size of the test.

The degrees of freedom in (24.30) are usually fractional, and interpolation in the tables of χ^2 is necessary.

From the fact that the non-central parameter λ defined by (24.28) is, under the regularity conditions assumed, a quadratic form with the elements of $\mathbf{V}_r^{-1}$ as coefficients,

Example 24.3

To test $H_0: \sigma^2 = \sigma_0^2$ for the normal distribution of Example 24.1. The unconditional ML estimators are as given there, so that (24.8) remains the unconditional maximum of the LF. Given our present H_0, the ML estimator of μ is $\hat{\hat{\mu}} = \bar{x}$ (Example 18.2). Thus

$$L(x \mid \hat{\hat{\mu}}, \sigma_0^2) = (2\pi \sigma_0^2)^{-\frac{1}{2}n} \exp\left\{-\tfrac{1}{2}\frac{\Sigma(x-\bar{x})^2}{\sigma_0^2}\right\}. \tag{24.31}$$

The ratio of (24.31) to (24.8) gives

$$l = \left(\frac{s^2}{\sigma_0^2}\right)^{n/2} \exp\left[-\tfrac{1}{2}n\left\{\frac{s^2}{\sigma_0^2}-1\right\}\right],$$

so that

$$z = e^{-1} l^{2/n} = \frac{t}{n} e^{-t/n}, \tag{24.32}$$

where $t = ns^2/\sigma_0^2$. z is a monotone function of l, but is not a monotone function of t/n its derivative being

$$\frac{dz}{dt} = \frac{1}{n}\left(1-\frac{t}{n}\right)e^{-t/n},$$

so that z increases steadily for $t < n$ to a maximum at $t = n$ and then decreases steadily. Putting $l \leqslant c_\alpha$ is therefore equivalent to putting

$$t \leqslant a_\alpha, \quad t \geqslant b_\alpha,$$

where a_α, b_α are determined, using (24.32), by

$$\left.\begin{array}{l} P\{t \leqslant a_\alpha\}+P\{t \geqslant b_\alpha\} = \alpha, \\ a_\alpha e^{-a_\alpha/n} = b_\alpha e^{-b_\alpha/n}. \end{array}\right\} \tag{24.33}$$

Since the statistic t has a χ^2 distribution with $(n-1)$ d.f. when H_0 holds, we can use tables of that distribution to satisfy (24.33).

Now consider the approximate distribution of

$$-2\log l = (t-n) - n\log(t/n).$$

Since $E(t) = n-1$, $\text{var } t = 2(n-1)$, we may write

$$\begin{aligned} -2\log l &= (t-n) - n\log\left\{1+\frac{t-n}{n}\right\} \\ &= (t-n) - n\sum_{r=1}^{\infty}(-1)^{r-1}\left(\frac{t-n}{n}\right)^r\!/r \\ &= (t-n) - n\left\{\frac{t-n}{n} - \frac{(t-n)^2}{2n^2} + o(n^{-2})\right\} \\ &= \frac{(t-n)^2}{2n} + o(n^{-1}). \end{aligned} \tag{24.34}$$

We have seen (**16.6**) that, as $n \to \infty$, a χ^2 distribution with $(n-1)$ degrees of freedom is asymptotically normally distributed with mean $(n-1)$ and variance $2(n-1)$; or equivalently, that $(t-n)/(2n)^{\frac{1}{2}}$ tends to a standardized normal variate. Its square, the first term on the right of (24.34), is therefore a χ^2 variate with 1 degree of freedom.

This is precisely the distribution of $-2\log l$ given by the general result of **24.7** when H_0 holds. This result also tells us that when H_0 is false, $-2\log l$ has a non-central χ^2 distribution with 1 degree of freedom and non-central parameter, by (24.28),

$$\lambda = -E\left\{\frac{\partial^2 \log L}{\partial (\sigma^2)^2}\right\}(\sigma^2-\sigma_0^2)^2 = \frac{n}{2\sigma^4}(\sigma^2-\sigma_0^2)^2 = \frac{n}{2}\left(1-\frac{\sigma_0^2}{\sigma^2}\right)^2.$$

Thus the expression (24.30) for the approximate power of the LR test in this case, where $r = 1$, is

$$P = \int_{\left(\frac{1+\lambda}{1+2\lambda}\right)\chi_\alpha^2(1)}^\infty d\chi^2\left(1+\frac{\lambda^2}{1+2\lambda}\right). \tag{24.35}$$

For illustrative purposes we shall evaluate P for one value of λ and of n conveniently chosen. Choose $\chi_\alpha^2(1) = 3.84$ to give a test of size 0.05. Consider the alternative $\sigma^2 = \sigma_1^2 = 1.25\,\sigma_0^2$. We then have $\lambda = 0.02n$, and we choose $n = 50$ to give $\lambda = 1$. (24.35) is then

$$P = \int_{2\cdot 56}^\infty d\chi^2\left(\frac{4}{3}\right),$$

and from the *Biometrika Tables* we find by simple interpolation between 1 and 2 degrees of freedom that $P = 0.166$ approximately. The exact power may be obtained from the normal d.f.: it is the power of an equal-tails size-α test against an alternative with mean $\lambda^{\frac{1}{2}} = 1$ standard deviations distant from the mean on H_0, i.e. the proportion of the alternative distribution lying outside the interval $(-2\cdot96, +0\cdot96)$ standard deviations from its mean. The normal tables give the value $P = 0.170$. The approximation to the power function is thus quite accurate enough.

Closer approximations to the distribution of the LR statistic

24.9 Confining ourselves now to the distribution of l when H_0 holds, we may seek closer approximations than the asymptotic result of **24.7**. As indicated in **24.3**, if we wish to find χ^2 approximations to the distribution of a function of l, we can gain some flexibility by considering the distribution of $w = -2c\log l$ and adjusting c to improve the approximation.

The simplest way of doing this would be to find the expected value of w and adjust c so that

$$E(w) = r,$$

the expectation of a χ^2 variate with r degrees of freedom. An approximation of this kind was first given by Bartlett (1937), and a general method for deriving the value of c has been given by Lawley (1956), who uses essentially the methods of **20.15** to investigate the moments of $-2\log l$. If

$$E(-2\log l) = r\left\{1+\frac{a}{n}+O\!\left(\frac{1}{n^2}\right)\right\}, \tag{24.36}$$

Lawley shows that by putting either

$$\left.\begin{array}{c} w_1 = -2\left(\dfrac{1}{1+\dfrac{a}{n}}\right)\log l \\[2ex] w_2 = -2\left(1-\dfrac{a}{n}\right)\log l \end{array}\right\} \tag{24.37}$$

or

we not only have
$$E(w) = r + O\left(\frac{1}{n^2}\right),$$
which follows immediately from (24.36) and (24.37), but also that *all* the cumulants of w conform, to order n^{-1}, with those of a χ^2 distribution with r degrees of freedom. The simple scaling correction which adjusts the means of w to the correct value is therefore an unequivocal improvement.

If even closer approximations are required, they can be obtained in a large class of situations by methods due to G. E. P. Box (1949), who gives improved χ^2 approximations (cf. **42.11**, Vol. 3), shows how to derive a function of $-2 \log l$ which is distributed in the variance-ratio distribution, and also derives an asymptotic expansion for its distribution function in terms of Incomplete Gamma-functions.

Example 24.4

k independent samples of sizes n_i ($i = 1, 2, \ldots, k$; $n_i \geq 2$) are taken from different normal populations with means μ_i and variances σ_i^2. To test
$$H_0 : \sigma_1^2 = \sigma_2^2 = \ldots = \sigma_k^2,$$
a composite hypothesis imposing the $r = k-1$ constraints
$$\frac{\sigma_2^2}{\sigma_1^2} = \frac{\sigma_3^2}{\sigma_1^2} = \ldots = \frac{\sigma_k^2}{\sigma_1^2} = 1,$$
and having $s = k+1$ degrees of freedom. Call the common unknown value of the variances σ^2.

The unconditional maximum of the LF is obtained, just as in Example 24.1, by putting
$$\left.\begin{array}{l}\hat{\mu}_i = \bar{x}_i, \\ \hat{\sigma}_i^2 = \dfrac{1}{n_i} \sum\limits_{j=1}^{n_i} (x_{ij} - \bar{x}_i)^2 = s_i^2,\end{array}\right\}$$
giving
$$L(x \mid \hat{\mu}_1, \ldots, \hat{\mu}_k, \hat{\sigma}_1^2, \ldots, \hat{\sigma}_k^2) = (2\pi)^{-n/2} \prod_{i=1}^{k} (s_i^2)^{-n_i/2} e^{-n/2}, \qquad (24.38)$$
where
$$n = \sum_{i=1}^{k} n_i.$$
Given H_0, the ML estimators of the means and the common variance σ^2 are
$$\left.\begin{array}{l}\hat{\hat{\mu}}_i = \bar{x}_i, \\ \hat{\hat{\sigma}}^2 = \dfrac{1}{n} \sum\limits_{i=1}^{k} n_i s_i^2 = s^2,\end{array}\right\}$$
so that
$$L(x \mid \hat{\hat{\mu}}_1, \ldots, \hat{\hat{\mu}}_k, \hat{\hat{\sigma}}^2) = (2\pi)^{-n/2} (s^2)^{-n/2} e^{-n/2}. \qquad (24.39)$$
From (24.4), (24.38) and (24.39),
$$l = \prod_{i=1}^{k} \left(\frac{s_i^2}{s^2}\right)^{n_i/2}, \qquad (24.40)$$
so that
$$-2 \log l = n \log(s^2) - \sum_{i=1}^{k} n_i \log(s_i^2). \qquad (24.41)$$

Now when H_0 holds, each of the statistics $\left(\dfrac{n_i s_i^2}{2\sigma^2}\right)$ has a Gamma distribution with parameter $\frac{1}{2}(n_i-1)$, and their sum $\dfrac{n s^2}{2\sigma^2}$ has the same distribution with parameter $\sum_{i=1}^{k} \frac{1}{2}(n_i-1) = \frac{1}{2}(n-k)$. For a Gamma variate x with parameter p, we have

$$E\{\log(ax)\} = \frac{1}{\Gamma(p)} \int_0^\infty \log(ax) e^{-x} x^{p-1} dx$$
$$= \log a + \frac{d}{dp} \log \Gamma(p),$$

which, using Stirling's series (3.63), becomes

$$E\{\log(ax)\} = \log a + \log p - \frac{1}{2p} - \frac{1}{12p^2} + O\left(\frac{1}{p^3}\right). \qquad (24.42)$$

Using (24.42) in (24.41), we have

$$E\{-2\log l\} = n\left\{\log\left(\frac{2\sigma^2}{n}\right) + \log\{\tfrac{1}{2}(n-k)\} - \frac{1}{(n-k)} - \frac{1}{3(n-k)^2} + O\left(\frac{1}{n^3}\right)\right\}$$
$$- \sum_{i=1}^{k} n_i \left\{\log\left(\frac{2\sigma^2}{n_i}\right) + \log\{\tfrac{1}{2}(n_i-1)\} - \frac{1}{(n_i-1)} - \frac{1}{3(n_i-1)^2} + O\left(\frac{1}{n_i^3}\right)\right\}$$
$$= n\left\{\log\left(1 - \frac{k}{n}\right) - \frac{1}{(n-k)} - \frac{1}{3(n-k)^2} + O\left(\frac{1}{n^3}\right)\right\}$$
$$- \sum_{i=1}^{k} n_i \left\{\log\left(1 - \frac{1}{n_i}\right) - \frac{1}{(n_i-1)} - \frac{1}{3(n_i-1)^2} + O\left(\frac{1}{n_i^3}\right)\right\}$$
$$= (k-1) + \left[\left(\sum_{i=1}^{k} \frac{1}{n_i-1} - \frac{k}{n-k}\right) + \tfrac{1}{2}\left(\sum_{i=1}^{k} \frac{1}{n_i} - \frac{k^2}{n}\right)\right.$$
$$\left. + \tfrac{1}{3}\left\{\sum_{i=1}^{k} \frac{n_i}{(n_i-1)^2} - \frac{n}{(n-k)^2}\right\}\right] + O\left(\frac{1}{N^3}\right), \qquad (24.43)$$

where we now write N indifferently for n_i and n. We could now improve the χ^2 approximation, in accordance with (24.37), with the expression in square brackets in (24.43) as $(k-1)\dfrac{a}{n}$.

Now consider Bartlett's (1937) modification of the LR statistic (24.40) in which n_i is replaced throughout by the "degrees of freedom" $n_i - 1 = \nu_i$, so that n is replaced by $\nu = \sum_{i=1}^{k}(n_i-1) = n-k$. We write this

$$l^* = \prod_{i=1}^{k} \left(\frac{s_i^2}{s^2}\right)^{\frac{1}{2}\nu_i},$$

where now

$$\left.\begin{aligned} s_i^2 &= \frac{1}{\nu_i} \sum_{j=1}^{n_i} (x_{ij} - \bar{x}_i)^2, \\ s^2 &= \frac{1}{\nu} \sum_{i=1}^{k} \nu_i s_i^2. \end{aligned}\right\}$$

Thus
$$-2\log l^* = \nu \log s^2 - \sum_{i=1}^{k} \nu_i \log s_i^2. \tag{24.44}$$

We shall see in Example 24.6 that l^* has the advantage over l that it gives an unbiassed test for any values of the n_i. If we retrace the passage from (24.42) to (24.43), we find that

$$E(-2\log l^*) = -\nu\left\{\frac{1}{\nu}+\frac{1}{3\nu^2}+O\left(\frac{1}{\nu^3}\right)\right\} + \sum_{i=1}^{k}\nu_i\left\{\frac{1}{\nu_i}+\frac{1}{3\nu_i^2}+O\left(\frac{1}{\nu_i^3}\right)\right\}$$
$$= (k-1)+\tfrac{1}{3}\left(\sum_{i=1}^{k}\frac{1}{\nu_i}-\frac{1}{\nu}\right)+O\left(\frac{1}{\nu_i^3}\right). \tag{24.45}$$

From (24.37) and (24.45) it follows that $-2\log l^*$ defined at (24.44) should be divided by the scaling constant

$$1+\frac{1}{3(k-1)}\left(\sum_{i=1}^{k}\frac{1}{\nu_i}-\frac{1}{\nu}\right)$$

to give a closer approximation to a χ^2 distribution with $(k-1)$ degrees of freedom.

LR tests when the range depends upon the parameter

24.10 The asymptotic distribution of the LR statistic, given in **24.7**, depends essentially on the regularity conditions necessary to establish the asymptotic normality of ML estimators. We have seen in Example 18.5 that these conditions break down where the range of the parent distribution is a function of the parameter θ. What can be said about the distribution of the LR statistic in such cases? It is a remarkable fact that, as Hogg (1956) showed, for certain hypotheses concerning such distributions the statistic $-2\log l$ is distributed *exactly* as χ^2, but with $2r$ degrees of freedom, i.e. twice as many as there are constraints imposed by the hypothesis.

24.11 We first derive some preliminary results concerning rectangular distributions. If k variables z_i are independently distributed as

$$dF = dz_i, \qquad 0 \leqslant z_i \leqslant 1, \tag{24.46}$$

the distribution of

$$t_i = -2\log z_i$$

is at once seen to be

$$dG = \tfrac{1}{2}\exp(-\tfrac{1}{2}t_i)\,dt_i, \qquad 0 \leqslant t_i \leqslant \infty,$$

a χ^2 distribution with 2 degrees of freedom, so that the sum of k such independent variates

$$t = \sum_{i=1}^{k} t_i = -2\sum_{i=1}^{k}\log z_i = -2\log \prod_{i=1}^{k} z_i$$

has a χ^2 distribution with $2k$ degrees of freedom.

It follows from (14.1) that the distribution of $y_{(n_i)}$, the largest among n_i independent observations from a rectangular distribution on the interval $(0, 1)$, is

$$dH = n_i\, y_{(n_i)}^{n_i-1}\, dy_{(n_i)}, \qquad 0 \leqslant y_{(n_i)} \leqslant 1, \tag{24.47}$$

and hence that $y_{(n_i)}^{n_i} = z_i$ is uniformly distributed as in (24.46). Hence for k independent samples of size n_i, $t = -2\log \prod_{i=1}^{k} y_{(n_i)}^{n_i}$ has a χ^2 distribution with $2k$ degrees of freedom.

Now consider the distribution of the largest of the k largest values $y_{(n_i)}$. Since all the observations are independent, this is simply the largest of $n = \sum_{i=1}^{k} n_i$ observations from the original rectangular distribution. If we denote this largest value by $y_{(n)}$ the distribution of $-2\log y_{(n)}^n$ will, by the argument above, be a χ^2 distribution with 2 degrees of freedom. We now show that the statistics $y_{(n)}$ and $u = \prod_{i=1}^{k} y_{(n_i)}^{n_i}/y_{(n)}^n$ are independently distributed. Introduce the parameter θ, so that the original rectangular distribution is on the interval $(0, \theta)$. The joint frequency function of the $y_{(n_i)}$ then becomes, from (24.47),

$$f = \prod_{i=1}^{k} \{n_i y_{(n_i)}^{n_i-1}/\theta^{n_i}\} = \frac{1}{\theta^n} \prod_{i=1}^{k} n_i y_{(n_i)}^{n_i}.$$

By **17.40**, $y_{(n)}$ is sufficient for θ, and by **23.12** its distribution is complete. Thus by Exercise 23.7 we need only observe that the distribution of u is free of the parameter θ to establish the result that u is distributed independently of the complete sufficient statistic $y_{(n)}$.

Since $y_{(n)}$ and u are independent, $y_{(n)}^n$ and u are likewise. If we write $\phi_1(t)$ for the c.f. of $-2\log y_{(n)}^n$, $\phi_2(t)$ for the c.f. of $-2\log \prod_{i=1}^{k} y_{(n_i)}^{n_i}$, and $\phi(t)$ for the c.f. of $-2\log u$, we then have

$$(-2\log u) + (-2\log y_{(n)}^n) = -2\log \prod_{i=1}^{k} y_{(n_i)}^{n_i},$$

and using our previous results concerning χ^2 distributions, and the fact that the c.f. of a sum of independent variates is the product of their c.f.'s (cf. **7.18**), we have

$$\phi(t).(1-2it)^{-1} = (1-2it)^{-k},$$

whence

$$\phi(t) = (1-2it)^{-(k-1)},$$

so that $-2\log u$ has a χ^2 distribution with $2(k-1)$ degrees of freedom.

Collecting our results finally, we have established that if we have k variates $y_{(n_i)}$ independently distributed as in (24.47), then $-2\log \prod_{i=1}^{k} y_{(n_i)}^{n_i}$ has a χ^2 distribution with $2k$ degrees of freedom, while, if $y_{(n)}$ is the largest of the $y_{(n_i)}$, $-2\log \left\{ \prod_{i=1}^{k} y_{(n_i)}^{n_i}/y_{(n)}^n \right\}$ has a χ^2 distribution with $2(k-1)$ degrees of freedom.

24.12 We now consider in turn the two classes of situation in which a single sufficient statistic exists for θ when the range is a function of θ, taking first the case when only one terminal (say the upper) depends on θ. We then have, from **17.40**, the necessary form for the frequency function

$$f(x|\theta) = g(x)/h(\theta), \quad a \leqslant x \leqslant \theta. \tag{24.48}$$

Now suppose we have $k (\geqslant 1)$ separate populations of this form, $f(x_i|\theta_i)$, and wish to test
$$H_0 : \theta_1 = \theta_2 = \ldots = \theta_k = \theta_0,$$
a simple hypothesis imposing k constraints, on the basis of samples of sizes n_i ($i = 1, 2, \ldots, k$). We now find the LR criterion for H_0. The unconditional ML estimator of θ_i is the largest observation $x_{(n_i)}$ (cf. Example 18.1). Thus

$$L(x|\hat{\theta}_1, \ldots, \hat{\theta}_k) = \prod_{i=1}^{k} \prod_{j=1}^{n_i} \{g(x_{ij})/[h(x_{(n_i)})]^{n_i}\}. \tag{24.49}$$

Since H_0 is simple, the LF when it holds is determined and no ML estimator is necessary. We have

$$L(x|\theta_0, \theta_0, \ldots, \theta_0) = \prod_{i=1}^{k} \prod_{j=1}^{n_i} \{g(x_{ij})\}/[h(\theta_0)]^n.$$

Hence the LR statistic is

$$l = \frac{L(x|\theta_0, \ldots, \theta_0)}{L(x|\hat{\theta}_1, \ldots, \hat{\theta}_k)} = \prod_{i=1}^{k} \left[\frac{h(x_{(n_i)})}{h(\theta_0)}\right]^{n_i}. \tag{24.50}$$

When H_0 holds, $y_i = h(x_{(n_i)})/h(\theta_0)$ is the probability that an observation falls below or at $x_{(n_i)}$ and is itself a random variable with distribution obtained from that of $x_{(n_i)}$ as

$$dF = n_i y_i^{n_i-1} dy_i, \quad 0 \leqslant y_i \leqslant 1,$$

of the form (24.47). Thus, from the result of the last section,

$$-2\log \prod_{i=1}^{k} y_i^{n_i} = -2\log l$$

has a χ^2 distribution with $2k$ degrees of freedom.

24.13 We now investigate the composite hypothesis, for $k \geqslant 2$ populations,
$$H_0 : \theta_1 = \theta_2 = \ldots = \theta_k$$
which imposes $(k-1)$ constraints, leaving the common value of θ unspecified. The unconditional maximum of the LF is given by (24.49) as before. The maximum under our present H_0 is $L(x|\hat{\hat{\theta}}, \hat{\hat{\theta}}, \ldots, \hat{\hat{\theta}})$, where $\hat{\hat{\theta}}$ is the ML estimator for the pooled samples, which is $x_{(n)}$. Thus we have the LR statistic

$$l = \frac{L(x|\hat{\hat{\theta}}, \ldots, \hat{\hat{\theta}})}{L(x|\hat{\theta}_1, \ldots, \hat{\theta}_k)} = \prod_{i=1}^{k} \frac{[h(x_{(n_i)})]^{n_i}}{[h(x_{(n)})]^n}. \tag{24.51}$$

By writing this as

$$l = \prod_{i=1}^{k} \left[\frac{h(x_{(n_i)})}{h(\theta)}\right]^{n_i} \bigg/ \left[\frac{h(x_{(n)})}{h(\theta)}\right]^n,$$

where θ is the common unspecified value of the θ_i, we see that in the notation of the last section,

$$l = \left[\prod_{i=1}^{k} y_{(n_i)}^{n_i}\right] \bigg/ y_{(n)}^n,$$

so that by **24.11** we have that in this case $-2\log l$ is distributed like χ^2 with $2(k-1)$ degrees of freedom.

24.14 When both terminals of the range are functions of θ, we have from **17.41** that if there is a single sufficient statistic for θ, then

$$f(x\mid\theta) = g(x)/h(\theta), \qquad \theta \leqslant x \leqslant b(\theta), \tag{24.52}$$

where $b(\theta)$ must be a monotone decreasing function of θ. For $k \geqslant 1$ such populations $f(x_i\mid\theta_i)$, we again test the simple

$$H_0: \theta_1 = \theta_2 = \ldots = \theta_k = \theta_0$$

on the basis of samples of sizes n_i. The unconditional ML estimator of θ_i is the sufficient statistic

$$t_i = \min\{x_{(1i)}, b^{-1}(x_{(n_i)})\},$$

where $x_{(1i)}, x_{(n_i)}$ are respectively the smallest and largest observations in the ith sample. When H_0 holds, the LF is specified by $L(x\mid\theta_0,\ldots,\theta_0)$. Thus the LR statistic

$$l = \frac{L(x\mid\theta_0,\ldots,\theta_0)}{L(x\mid\hat{\theta}_1,\ldots,\hat{\theta}_k)} = \prod_{i=1}^{k}\left[\frac{h(t_i)}{h(\theta_0)}\right]^{n_i}. \tag{24.53}$$

Just as for (24.50), we see that

$$l = \prod_{i=1}^{k} y_i^{n_i},$$

where the y_i are distributed in the form (24.47), and hence $-2\log l$ again has a χ^2 distribution with $2k$ degrees of freedom.

Similarly for the composite hypothesis with $(k-1)$ constraints $(k \geqslant 2)$

$$H_0: \theta_1 = \theta_2 = \ldots = \theta_k,$$

we find, just as in **24.13**, that the LR statistic is

$$l = \frac{L(x\mid\hat{\theta},\ldots,\hat{\theta})}{L(x\mid\hat{\theta}_1,\ldots,\hat{\theta}_{k,})} = \prod_{i=1}^{k} [h(t_i)]^{n_i}/[h(t)]^n$$

where $t = \min\{t_i\}$ is the combined ML estimator $\hat{\theta}$, so that by writing

$$l = \prod_{i=1}^{k}\left[\frac{h(t_i)}{h(\theta)}\right]^{n_i} \bigg/ \left[\frac{h(t)}{h(\theta)}\right]^{n}$$

we again reduce l to the form required in **24.11** for $-2\log l$ to be distributed like χ^2 with $2(k-1)$ degrees of freedom.

24.15 We have thus obtained exact χ^2 distributions for two classes of hypotheses concerning distributions whose terminals depend upon the parameter being tested. Exercises 24.8 and 24.9 give further examples, one exact and one asymptotic, of LR tests for which $-2\log l$ has a χ^2 distribution with twice as many degrees of freedom as there are constraints imposed by the hypothesis tested. It will have been noted that these χ^2 forms spring not from any tendency to multinormality on the part of the ML estimators, as did the limiting results of **24.7** for "regular" situations, but from the intimate connexion between the rectangular and χ^2 distributions explored in **24.11**. One effect of this difference is that the power functions of the tests take a quite different form from that obtained by use of the non-central χ^2 distribution in **24.8**.

Barr (1966) finds the power function of the test of the simple hypothesis based on the LR statistic (24.50), and shows that the test is unbiassed. When $k = 1$, the test is shown to be UMP (cf. Example 22.6 and the condition (22.37)), but there is no UMP test for $k > 1$, and the LR test is not even UMPU. For the composite hypothesis of **24.13** with $k = 2$, the power function of the LR test statistic (24.51) is used to show that the LR test is UMPU.

The properties of LR tests

24.16 So far, we have been concerned entirely with the problems of determining the distribution of the LR statistic, or a function of it. We now have to inquire into the properties of LR tests, in particular the question of their unbiassedness and whether they are optimum tests in any sense. First, however, we turn to consider a weaker property, that of consistency, which we now define for the first time.

Test consistency

24.17 A test of a hypothesis H_0 against a class of alternatives H_1 is said to be consistent if, when any member of H_1 holds, the probability of rejecting H_0 tends to 1 as sample size(s) tend to infinity. If w is the critical region, and $\mathbf{x}$ the sample point, we write this

$$\lim_{n \to \infty} P\{\mathbf{x} \in w \mid H_1\} = 1. \qquad (24.54)$$

The idea of test consistency, which is a simple and natural one, was first introduced by Wald and Wolfowitz (1940). It seems perfectly reasonable to require that, as the number of observations increases, any test worth considering should reject a false hypothesis with increasing certainty, and in the limit with complete certainty. Test consistency is as intrinsically acceptable as is consistency in estimation (**17.7**), of which it is in one sense a generalization. For if a test concerning the value of θ is based on a statistic which is a consistent estimator of θ, it is immediately obvious that the test will be consistent too. But an inconsistent estimator may still provide a consistent test. For example, if t tends in probability to $a\theta$, t will give a consistent test of hypotheses about θ. In general, it is clear that it is sufficient for test consistency that the test statistic, when regarded as an estimator, should tend in probability to some one-to-one function of θ.

Since the condition that a size-α test be unbiassed is (cf. (23.53)) that

$$P\{\mathbf{x} \in w \mid H_1\} \geq \alpha, \qquad (24.55)$$

it is clear from (24.54) and (24.55) that a consistent test will lose its bias, if any, as $n \to \infty$. However, an unbiassed test need not be consistent.[*]

The consistency and unbiassedness of LR tests

24.18 We saw in **18.10** and **18.22** that under a very generally satisfied condition, the ML estimator $\hat{\theta}$ of a parameter-vector θ is consistent, though in other circumstances

---

[*] Cf. the remark in **17.9** on consistent and asymptotically unbiassed estimators.

it need not be. If we take it that we are dealing with a situation in which all the ML estimators are consistent, we see from the definition of the LR statistic at (24.4) that, as sample sizes increase,

$$l \to \frac{L(x \mid \theta_{r_0}, \theta_s)}{L(x \mid \theta_r, \theta_s)}, \qquad (24.56)$$

where θ_r, θ_s are the true values of those parameters, and θ_{r_0} is the hypothetical value of θ_r being tested. Thus, when H_0 holds, $l \to 1$ in probability, and the critical region (24.6) will therefore have its boundary c_α approaching 1. When H_0 does not hold, the limiting value of l in (24.56) will be some constant k satisfying (cf. (18.20))

$$0 \leqslant k < 1$$

and thus we have

$$P\{l \leqslant c_\alpha\} \to 1 \qquad (24.57)$$

and the LR test is consistent.

In **24.8** we confirmed from the approximate power function that LR tests are consistent under regularity conditions, and in **24.15** we deduced consistency in another case, not covered by **24.8**. Both of these examples are special cases of our present discussion.

24.19 When we turn to the question of unbiassedness, we recall the penultimate sentence of **24.17** which, coupled with the result of **24.18**, ensures that most LR estimators are asymptotically unbiassed. Of itself, this is not very comforting (though it would be so if it could be shown under reasonable restrictions that the maximum extent of the bias is always small), for the criterion of unbiassedness in tests is intuitively attractive enough to impose itself as a necessity for all sample sizes. Example 24.5 shows that an important LR test is biassed.

Example 24.5

Consider again the hypothesis H_0 of Example 24.3. The LR test uses as its critical region the tails of the χ^2_{n-1} distribution of $t = ns^2/\sigma_0^2$ determined by (24.33). Now in Examples 23.12 and 23.14 we saw that the unbiassed (actually UMPU) test of H- was determined from the distribution of t by the relations

$$\left.\begin{array}{l} P\{t \leqslant a_\alpha\} + P\{t \geqslant b_\alpha\} = \alpha, \\ a_\alpha^{(n-1)/2} \exp(-a_\alpha/2) = b_\alpha^{(n-1)/2} \exp(-b_\alpha/2). \end{array}\right\} \qquad (24.58)$$

It is clear on comparison of (24.58) with (24.33) that they would only give the same result if $a_\alpha = b_\alpha$, which cannot hold except in the trivial case $a_\alpha - b_\alpha = 0, \alpha = 1$. In all other cases, the tests have different critical regions, the LR test having higher values of a_α and b_α than the unbiassed test, i.e. a larger fraction of α concentrated in the lower tail. It is easy to see that for alternative values of σ^2 just larger than σ_0^2, for which the distribution of t is slightly displaced towards higher values of t compared to its H_0 distribution, the probability content lost to the critical region of the LR test through its larger value of b_α will exceed the gain due to the larger value of a_α; and thus the LR test will be biassed.

It will be seen by reference to Example 23.12 that whereas the LR test has values of a_α, b_α too large for unbiassedness, the " equal-tails " test there discussed has a_α, b_α too small for unbiassedness. Thus the two more or less intuitively acceptable critical regions " bracket " the unbiassed critical region.

If in (24.33) we replace n by $n-1$, it becomes precisely equivalent to the unbiassed (24.58), confirming the general fact that the LR test loses its bias asymptotically. It is suggestive to trace this bias to its source. If, in constructing the LR statistic in Example 24.3, we had adjusted the unconditional ML estimator of σ^2 to be unbiassed, s^2 would have been replaced by $\left(\dfrac{n}{n-1}\right) s^2$, and the adjusted LR test would have been unbiassed: the estimation bias of the ML estimator lies behind the test bias of the LR test.

Unbiassed invariant tests for location and scale parameters

24.20 Example 24.5 suggests that a good principle in constructing a LR test is to adjust all the ML estimators used in the process so that they are unbiassed. A further confirmation of this principle is contained in Example 24.4, where we stated that the adjusted LR statistic l^* gives an unbiassed test. We now prove this, developing a method due to Pitman (1939b) for this purpose.

If the hypothesis being tested concerns a set of k location parameters θ_i ($i = 1, 2, \ldots, k$), we write the joint distribution of the variates as

$$dF = f(x_1-\theta_1, x_2-\theta_2, \ldots, x_k-\theta_k)\, dx_1 \ldots dx_k. \qquad (24.59)$$

We wish to test

$$H_0 : \theta_1 = \theta_2 = \ldots = \theta_k. \qquad (24.60)$$

Any test statistic t, to be satisfactory intuitively, must satisfy the invariance condition

$$t(x_1, x_2, \ldots, x_k) = t(x_1-\lambda, x_2-\lambda, \ldots, x_k-\lambda), \qquad (24.61)$$

for a change in the origin of measurement should not affect the test. We may therefore without loss of generality take the common value of the θ_i in (24.60) to be zero. We suppose that $t > 0$, and that w_0, the size-α critical region based on the distribution of t, is defined by

$$t \leqslant c_\alpha; \qquad (24.62)$$

if either or both of these statements were not true, we could transform to a function of t for which they were.

Because of its invariance property (24.61), t must be constant in the k-dimensional sample space W on any line L parallel to the equiangular vector V defined by $x_1 = x_2 = \ldots = x_k$. When H_0 holds, the content of w_0 is its size

$$\alpha = \int_{w_0} dF(x_1, x_2, \ldots, x_k), \qquad (24.63)$$

and when H_0 is not true the content of w_0 is its power

$$1-\beta = \int_{w_0} dF(x_1-\theta_1, x_2-\theta_2, \ldots, x_k-\theta_k) = \int_{w_1} dF(x_1, x_2, \ldots, x_k), \quad (24.64)$$

where w_1 is derived from w_0 by translation in W without rotation. We define the integral, on any line L parallel to V,

$$P(L) = \int_L f(x_1, x_2, \ldots, x_k) \, d\bar{x}; \qquad (24.65)$$

the variation along any L being the same for each co-ordinate x_j, it can be summarized in the variation of the mean co-ordinate $\bar{x}$. Since the aggregate of lines L is the whole of W, we have

$$\int P(L) \, dL = \int \left\{ \int_L f \, d\bar{x} \right\} dL = \int \cdots \int f \, dx_1 \ldots dx_k = 1. \qquad (24.66)$$

We now determine w_0 as the aggregate of all lines L for which the statistic $P(L) \leq$ some constant h. Then $P(L)$ will exceed h on any L which is in w_1 but not in w_0. Hence, from (24.63), (24.64) and (24.66),

$$\alpha = \int_{w_0} dF \leq \int_{w_1} dF = 1 - \beta,$$

so that the test is unbiassed. We therefore define the test statistic t so that at any point on a line L, parallel to V, it is equal to $P(L)$. Now using the invariance property (24.61) with $\lambda = \bar{x}$ we have from (24.65)

$$t(x) = P(L) = \int_L f(x_1 - \bar{x}, x_2 - \bar{x}, \ldots, x_k - \bar{x}) \, d\bar{x},$$

and replacing $\bar{x}$ by u, this is

$$t(x) = \int_{-\infty}^{\infty} f(x_1 - u, x_2 - u, \ldots, x_k - u) \, du, \qquad (24.67)$$

the unbiassed size-α region being defined by (24.62). It will be seen that the unbiassed test thus obtained is unique. An example of the use of (24.67) is given in Exercise 24.15.

24.21 Turning now to tests concerning scale parameters, which are more to our present purpose, suppose that the joint distribution of k variates is

$$dG = g\left(\frac{y_1}{\phi_1}, \frac{y_2}{\phi_2}, \ldots, \frac{y_k}{\phi_k}\right) \frac{dy_1}{\phi_1} \cdot \frac{dy_2}{\phi_2} \cdots \frac{dy_k}{\phi_k}, \qquad (24.68)$$

where all the scale parameters ϕ_i are positive. We make the transformation

$$x_i = \log|y_i|, \qquad \theta_i = \log \phi_i$$

and find for the distribution of the x_i

$$dF = g\{\exp(x_1 - \theta_1), \exp(x_2 - \theta_2), \ldots, \exp(x_k - \theta_k)\}$$
$$\exp\left\{\sum_{i=1}^{k} (x_i - \theta_i)\right\} dx_1 \ldots dx_k. \qquad (24.69)$$

(24.69) is of the form (24.59) which we have already discussed. To test

$$H_0' : \phi_1 = \phi_2 = \ldots = \phi_k \qquad (24.70)$$

is the same as to test H_0 of (24.60). The statistic (24.67) becomes

$$t(x) = \int_{-\infty}^{\infty} g\{\exp(x_1-u), \exp(x_2-u), \ldots, \exp(x_k-u)\} \exp\left\{\sum_{i=1}^{k} x_i - ku\right\} du,$$

which when expressed in terms of the y_i becomes

$$t(y) = \prod_{i=1}^{k} |y_i| \int_0^{\infty} g\left(\frac{y_1}{v}, \frac{y_2}{v}, \ldots, \frac{y_k}{v}\right) \frac{dv}{v^{k+1}}. \tag{24.71}$$

24.22 Now consider the special case of k independently distributed Gamma variates $\frac{y_i}{\phi_i}$ with parameters m_i. Their joint distribution is

$$dG = \prod_{i=1}^{k} \left\{\frac{1}{\Gamma(m_i)} \left(\frac{y_i}{\phi_i}\right)^{m_i-1}\right\} \exp\left(-\sum_{i=1}^{k} \frac{y_i}{\phi_i}\right) \prod_{i=1}^{k} \frac{dy_i}{\phi_i}. \tag{24.72}$$

To test H'_0 of (24.70), we use (24.71) and obtain

$$t(y) = \prod_{i=1}^{k} \left\{\frac{y_i^m}{\Gamma(m_i)}\right\} \int_0^{\infty} \exp\left(-\sum_{i=1}^{k} y_i/v\right) \frac{dv}{v^{m+1}}, \tag{24.73}$$

where $m = \sum_{i=1}^{k} m_i$. On substituting $u = \sum_{i=1}^{k} y_i/v$ in (24.73), we find

$$t(y) = \left[\frac{\Gamma(m)}{\prod_i \Gamma(m_i)}\right] \frac{\prod_i y_i^{m_i}}{(\sum_i y_i)^m}. \tag{24.74}$$

We now neglect the constant factor in square brackets in (24.74). From the remainder, T, the maximum attainable value of t, occurs when $y_i/\sum_i y_i = m_i/m$, when

$$T = \prod m_i^{m_i}/m^m. \tag{24.75}$$

We now write

$$t^* = -\log\left(\frac{t}{T}\right) = m \log\left(\frac{\sum_i y_i}{m}\right) - \sum_i m_i \log\left(\frac{y_i}{m_i}\right). \tag{24.76}$$

t^* will be unbiassed for H'_0, and will range from 0 to ∞, large values being rejected.

Example 24.6

We may now apply (24.76) to the problem of testing the equality of k normal variances, discussed in Example 24.4. For each of the quantities $\sum_j (x_{ij} - \bar{x}_i)^2/(2\sigma^2)$ is, when H_0 holds, a Gamma variate with parameter $\frac{1}{2}(n_i - 1)$. We thus have to substitute in (24.76)

$$\left.\begin{array}{l} y_i = \sum (x_{ij} - \bar{x}_i)^2 = \nu_i s_i^2, \\ m_i = \frac{1}{2}(n_i - 1) = \frac{1}{2}\nu_i, \\ m = \sum_i m_i = \frac{1}{2}(n-k) = \frac{1}{2}\nu, \end{array}\right\} \tag{24.77}$$

and we find for the unbiassed test statistic

$$2t^* = \nu \log\left(\frac{\sum_i \nu_i s_i^2}{\nu}\right) - \Sigma \nu_i \log(s_i^2). \qquad (24.78)$$

(24.78) is identical with (24.44), so that $2t^*$ is simply $-2\log l^*$ which we discussed there. Thus the l^* test is unbiassed, as stated in Example 24.4. From this, it is fairly evident that the unadjusted LR test statistic l of (24.40), which employs another weighting system, cannot also be unbiassed in general. When all sample sizes are equal, the two tests are equivalent, as Exercise 24.7 shows. Even in the case $k = 2$, the unadjusted LR test is biassed when $n_1 \neq n_2$: this is left to the reader to prove in Exercise 24.14.

A quite different proof is given by A. Cohen and Strawderman (1971).

24.23 Before leaving the question of the unbiassedness of LR tests, it should be mentioned that Paulson (1941) investigated the bias of a number of LR tests for exponential distributions—some of his results are given in Exercises 24.16 and 24.18. We shall discuss LR tests in the multinormal case in Chapter 42, Volume 3.

Other properties of LR tests

24.24 Apart from questions of consistency and unbiassedness, what can be said in general concerning the properties of LR tests? In the first place, we know that ML estimators are functions of the sufficient statistics (cf. **18.4**) so that the LR statistic (24.4) may be re-written

$$l = \frac{L(x \mid \theta_{r0}, t_s)}{L(x \mid T_{r+s})} \qquad (24.79)$$

where t_s is the vector minimal sufficient for θ_s when H_0 holds and T_{r+s} is the statistic sufficient for all the parameters when H_0 does not hold. As we have seen in **17.38**, it is not true in general that the components of T_{r+s} include the components of t_s—the sufficient statistic for θ_s when H_0 holds may no longer form part of the sufficient set when H_0 does not hold, and even when it does may not then be separately sufficient for θ_s, merely forming part of T_{r+s} which is sufficient for (θ_r, θ_s). Thus all that we can say of l is that it is *some* function of the two sets of sufficient statistics involved. There is, in general, no reason to suppose that it will be the right function of them.

It is easily seen that the LR method does not necessarily produce a UMP test when one exists, by observing that even in the case of testing a simple H_0 against a simple H_1, it does not yield the BCR (**22.6**).

If we are seeking a UMPU test, the LR method is handicapped by its own general biassedness, but we have seen that a simple bias adjustment will sometimes remove this difficulty. The adjustment takes the form of a " reweighting " of the test statistic by substituting unbiassed estimators for the ordinary ML estimators (Examples 24.4 and 24.6, Example 24.5), or sometimes equivalently of an adjustment of the critical region of the statistic to which the LR method leads (Exercise 24.14). Exercise 24.16 shows that two UMP tests derived in Exercises 23.25–26 for an exponential and a rectangular distribution are produced by the LR method, while the UMPU test for

256   THE ADVANCED THEORY OF STATISTICS

an exponential distribution given in Exercise 23.24 is not equivalent to the LR test, which is biassed.

Wald (1943a) shows that the LR test *asymptotically* has a number of optimum power properties under regularity conditions—but see 25.4 and Example 25.1. Hoeffding (1965) derives an optimum property of LR tests for multinomial distributions when test size $\alpha \to 0$ as sample size $\to \infty$.

The LR principle is an intuitively appealing one when there is no "optimum" test. It is of particular value in tests of linear hypotheses (which we shall discuss in the second part of this chapter) for which, in general, no UMPU test exists. But it is as well to be reminded of the possible fallibility of the LR method in exceptional circumstances, and the following example, adapted from one due to C. Stein and given by Lehmann (1950), is a salutary warning against using the method without investigation of its properties in the particular situation concerned.

Example 24.7

A discrete random variable x is defined at the values $0, \pm 1, \pm 2$, and the probabilities at these points given a hypothesis H_1 are:

$$x: \quad 0 \qquad \pm 1 \qquad +2 \qquad -2$$
$$P|H_1: \quad \alpha\left(\frac{1-\theta_1}{1-\alpha}\right) \quad (\tfrac{1}{2}-\alpha)\left(\frac{1-\theta_1}{1-\alpha}\right) \quad \theta_1\theta_2 \quad \theta_1(1-\theta_2) \tag{24.80}$$

The parameters θ_1, θ_2, are restricted by the inequalities

$$0 \leqslant \theta_1 \leqslant \alpha < \tfrac{1}{2}, \quad 0 \leqslant \theta_2 \leqslant 1,$$

where α is a known constant. We wish to test the simple

$$H_0: \theta_1 = \alpha, \quad \theta_2 = \tfrac{1}{2},$$

H_1 being the general alternative (24.80), on the evidence of a single observation. The probabilities on H_0 are:

$$x: \quad 0 \qquad \pm 1 \qquad +2 \qquad -2$$
$$P|H_0: \quad \alpha \quad \tfrac{1}{2}-\alpha \quad \tfrac{1}{2}\alpha \quad \tfrac{1}{2}\alpha \tag{24.81}$$

The LF is independent of θ_2 when $x = 0, \pm 1$, and is maximized unconditionally by making θ_1 as small as possible, i.e. putting $\hat{\theta}_1 = 0$. The LR statistic is therefore

$$l = \frac{L(x|H_0)}{L(x|\hat{\theta}_1,\hat{\theta}_2)} = 1-\alpha, \quad x = 0, \pm 1. \tag{24.82}$$

When $x = +2$ or -2, the LF is maximized unconditionally by choosing θ_2 respectively as large or as small as possible, i.e. $\hat{\theta}_2 = 1, 0$, respectively; and by choosing θ_1 as large as possible, i.e. $\hat{\theta}_1 = \alpha$. The maximum value of the LF is therefore α and the LR statistic is

$$l = \tfrac{1}{2}, \quad x = \pm 2. \tag{24.83}$$

Since $\alpha < \tfrac{1}{2}$, it follows from (24.82) and (24.83) that the LR test consists of rejecting H_0 when $x = \pm 2$. From (24.81) this test is seen to be of size α. But from (24.80) its power is seen to be θ_1 exactly, so for any value of θ_1 in

$$0 \leqslant \theta_1 < \alpha \tag{24.84}$$

the LR test will be biassed for all θ_2, while for $\theta_1 = \alpha$ the test will have power equal to its size α for all θ_2. In this latter extreme case the test is useless, but in the former case it is worse than useless, for we can get a test of size and power α by using a table of random numbers as the basis for our decision concerning H_0. Furthermore, a useful test exists, for if we reject H_0 when $x = 0$, we still have a size-α test by (24.81) and its power, from (24.80), is $\alpha \left(\dfrac{1-\theta_1}{1-\alpha} \right)$ which exceeds α when (24.84) holds and equals α when $\theta_1 = \alpha$.

Apart from the fact that the random variable is discrete, the noteworthy feature of this cautionary example is that the range of one of the parameters is determined by α, the size of the test.

> D. R. Cox (1961, 1962) considers the distribution of LR statistics when H_0 and H_1 are entirely separate families of composite hypotheses (so that (24.5) no longer holds) and obtains some large-sample results. See also Feder (1968) and Atkinson (1970).

The general linear hypothesis and its canonical form

24.25 We are now in a position to discuss the problem of testing hypotheses in the general linear model of Chapter 19. As at (19.8), we write

$$\mathbf{y} = \mathbf{X}\,\boldsymbol{\theta} + \boldsymbol{\epsilon}, \tag{24.85}$$

where the ε_i have zero means, equal variances σ^2 and are uncorrelated. For the moment, we make no further assumptions about the form of their distribution. We take $\mathbf{X'X}$ to be non-singular: if it were not, we could make it so by augmentation, as in **19.13–16**.

Suppose that we wish to test the hypothesis

$$H_0 : \mathbf{A}\,\boldsymbol{\theta} = \mathbf{c}_0, \tag{24.86}$$

where $\mathbf{A}$ is a $(r \times k)$ matrix and $\mathbf{c}_0$ a $(r \times 1)$ vector, each of known constants. (24.86) imposes $r (\leqslant k)$ constraints, which we take to be functionally independent, so that $\mathbf{A}$ is of rank r. H_1 is simply the negation of H_0. When $r = k$, $\mathbf{A'A}$ is non-singular and (24.86) is equivalent to $H_0 : \boldsymbol{\theta} = (\mathbf{A'A})^{-1} \mathbf{A'} \mathbf{c}_0$. If $\mathbf{A}$ is the first r rows of the $(n \times k)$ matrix $\mathbf{X}$, we also have a particularly direct H_0, in which we are testing the means of the first r y_i.

Consider the $(n \times 1)$ vector

$$\mathbf{z} = \mathbf{C}(\mathbf{X'X})^{-1} \mathbf{X'y} = \mathbf{C}\hat{\boldsymbol{\theta}}, \tag{24.87}$$

where $\mathbf{C}$ is a $(n \times k)$ matrix and $\hat{\boldsymbol{\theta}}$ is the Least Squares (LS) estimator of $\boldsymbol{\theta}$ given at (19.12). Then, from (24.87) and (24.85),

$$\mathbf{z} = \mathbf{C}\boldsymbol{\theta} + \mathbf{C}(\mathbf{X'X})^{-1} \mathbf{X'} \boldsymbol{\epsilon},$$

so that

$$\boldsymbol{\mu} = E(\mathbf{z}) = \mathbf{C}\boldsymbol{\theta} \tag{24.88}$$

and the dispersion matrix of $\mathbf{z}$ is, as in **19.6**,

$$\mathbf{V} = \sigma^2 \mathbf{C}(\mathbf{X'X})^{-1} \mathbf{C'}. \tag{24.89}$$

Let us now choose $\mathbf{C}$ so that the components of $\mathbf{z}$ are all uncorrelated, i.e. so that
$$\mathbf{V} = \sigma^2 \mathbf{I}.$$
From (24.89), this requires that
$$\mathbf{C}(\mathbf{X}'\mathbf{X})^{-1}\mathbf{C}' = \mathbf{I}$$
or, if $\mathbf{C}'\mathbf{C}$ is non-singular, that
$$\mathbf{C}'\mathbf{C} = \mathbf{X}'\mathbf{X}. \tag{24.90}$$
(24.90) is the condition that the z_i be uncorrelated, and implies, with (24.87) and (24.88), that
$$(\mathbf{z}-\boldsymbol{\mu})'(\mathbf{z}-\boldsymbol{\mu}) = \boldsymbol{\epsilon}'\mathbf{X}(\mathbf{X}'\mathbf{X})^{-1}\mathbf{X}'\boldsymbol{\epsilon} = \{\mathbf{X}(\hat{\boldsymbol{\theta}}-\boldsymbol{\theta})\}'\{\mathbf{X}(\hat{\boldsymbol{\theta}}-\boldsymbol{\theta})\}. \tag{24.91}$$

24.26 We now write
$$\mathbf{C} = \begin{pmatrix} \mathbf{A} \\ \mathbf{D} \\ \mathbf{F} \end{pmatrix},$$
where $\mathbf{A}$ is the $(r \times k)$ matrix in (24.86), $\mathbf{D}$ is a $((k-r) \times k)$ matrix and $\mathbf{F}$ is a $((n-k) \times k)$ matrix satisfying
$$\mathbf{F}\boldsymbol{\theta} = \mathbf{0}. \tag{24.92}$$
Since $\mathbf{A}$ is of rank r, we can choose $\mathbf{D}$ so that the $(k \times k)$ matrix $\begin{pmatrix} \mathbf{A} \\ \mathbf{D} \end{pmatrix}$ is non-singular, and thus $\mathbf{C}$ is of rank k. $\mathbf{C}'\mathbf{C}$ is then also of rank k, and hence non-singular as required above (24.90).

From (24.88), we have
$$\boldsymbol{\mu} = E(\mathbf{z}) = \begin{pmatrix} \mathbf{A} \\ \mathbf{D} \\ \mathbf{F} \end{pmatrix} \boldsymbol{\theta}. \tag{24.93}$$
Thus the means of the first r z_i are precisely the left-hand side of (24.86), so that H_0 is equivalent to testing
$$H_0 : \mu_i = E(z_i) = c_{0i}, \quad i = 1, 2, \ldots, r, \tag{24.94}$$
a composite hypothesis imposing r constraints upon the parameters. Since, by (24.92), the last $(n-k)$ of the μ_i are zero, there are k non-zero parameters μ_i, which together with σ^2 make up the total of $(k+1)$ parameters.

24.27 Thus we have reduced our problem to the following terms: we have a set of n mutually uncorrelated variates z_i with equal variances σ^2. $(n-k)$ of the z_i have zero means, and the others non-zero means. The hypothesis to be tested is that r of these k variates have specified means. This is called the canonical form of the general linear hypothesis.

In order to make progress with the hypothesis-testing problem, we need to make assumptions about the distribution of the errors in the linear model (24.85): specifically, we take each ε_i to be normal and hence, since they are uncorrelated, independent. The z_i, being linear functions of them, will also be normally distributed and, being uncorrelated, independently normally distributed. Their joint distribution therefore gives the LF

$$L(z\mid \mu, \sigma^2) = (2\pi\sigma^2)^{-n/2}\exp\left\{-\frac{1}{2\sigma^2}(z-\mu)'(z-\mu)\right\}$$

$$= (2\pi\sigma^2)^{-n/2}\exp\left[-\frac{1}{2\sigma^2}\left\{(z_r-\mu_r)'(z_r-\mu_r)\right.\right.$$

$$\left.\left.+(z_{k-r}-\mu_{k-r})'(z_{k-r}-\mu_{k-r})+z'_{n-k}z_{n-k}\right\}\right], \quad (24.95)$$

where suffixes to vectors denote the number of components in the vector. Our hypothesis is

$$H_0 : \mu_r = c_0 \quad (24.96)$$

and H_1 is its negation.

We saw in Example 23.14 that if we have only one constraint ($r = 1$), there is a UMPU test of H_0 against H_1, as is otherwise obvious in our present application from the fact that we are then testing the mean of a single normal population with unknown variance: the UMPU test is, as we saw in Example 23.14, the ordinary "equal-tails" "Student's" t-test for this hypothesis.

Kolodzieczyk (1935), to whom the first general results concerning the linear hypothesis are due, demonstrated the impossibility of a UMP test with more than one constraint, and showed that there is a pair of one-sided UMP similar tests when $r = 1$: these are the one-sided "Student's" t-tests (cf. Example 23.7). We have just seen that there is a two-sided "Student's" t-test which is UMPU for $r = 1$, but the critical region of this test is different according to which of the μ_i is being tested: thus there is no common UMPU critical region for $r > 1$.

Since there is no "optimum" test in any sense we have so far discussed, we are tempted to use the LR method to give an intuitively reasonable test.

24.28 The derivation of the LR statistic is simple enough. The unconditional maximum of (24.95) is obtained by solving the set of equations

$$\left.\begin{array}{l}\dfrac{\partial \log L}{\partial \mu_i} = 0, \quad i = 1, 2, \ldots, k,\\[6pt] \dfrac{\partial \log L}{\partial(\sigma^2)} = 0.\end{array}\right\}$$

The ML estimators thus obtained are

$$\left.\begin{array}{l}\hat{\mu}_i = z_i, \quad i = 1, 2, \ldots, k,\\[4pt] \hat{\sigma}^2 = \dfrac{1}{n}\sum\limits_{i=k+1}^{n} z_i^2\end{array}\right\}$$

whence

$$(z-\hat{\mu})'(z-\hat{\mu}) = n\hat{\sigma}^2.$$

Thus the unconditional maximum of the LF is

$$L(z\mid \hat{\mu}, \hat{\sigma}^2) = (2\pi\hat{\sigma}^2 e)^{-n/2} = \left(\frac{2\pi e}{n}\sum_{i=k+1}^{n} z_i^2\right)^{-n/2}. \quad (24.97)$$

When the hypothesis (24.96) holds, the ML estimators of the unspecified parameters are

$$\hat{\hat{\mu}}_i = z_i, \quad i = r+1, r+2, \ldots, k,$$
$$\hat{\hat{\sigma}}^2 = \frac{1}{n}\left\{\sum_{i=k+1}^{n} z_i^2 + \sum_{i=1}^{r} (z - c_{0i})^2\right\},$$

whence

$$(\mathbf{z}-\hat{\hat{\mu}})'(\mathbf{z}-\hat{\hat{\mu}}) = n\hat{\hat{\sigma}}^2,$$

so that the conditional maximum of the LF is

$$L(\mathbf{z}\,|\,\mathbf{c}_0, \hat{\hat{\mu}}, \hat{\hat{\sigma}}^2) = (2\pi\hat{\hat{\sigma}}^2 e)^{-n/2} = \left[\frac{2\pi e}{n}\left\{\sum_{i=k+1}^{n} z_i^2 + \sum_{i=1}^{r}(z_i - c_{0i})^2\right\}\right]^{-n/2} \quad (24.98)$$

From (24.97) and (24.98) the LR statistic l is given by

$$l^{2/n} = \frac{\hat{\sigma}^2}{\hat{\hat{\sigma}}^2} = \frac{1}{1+W}, \quad (24.99)$$

where

$$W = (\mathbf{z}_r - \mathbf{c}_0)'(\mathbf{z}_r - \mathbf{c}_0)/\mathbf{z}'_{n-k}\mathbf{z}_{n-k}$$

$$= \frac{\sum_{i=1}^{r}(z_i - c_{0i})^2}{\sum_{i=k+1}^{n} z_i^2} = \frac{\hat{\hat{\sigma}}^2 - \hat{\sigma}^2}{\hat{\sigma}^2}. \quad (24.100)$$

It will be observed from above that $n\hat{\hat{\sigma}}^2$, $n\hat{\sigma}^2$ are respectively the minima of $(\mathbf{z}-\mu)'(\mathbf{z}-\mu)$ with respect to μ under H_0 and H_1. By (24.91), these are the same as the minima of

$$R = \{\mathbf{X}(\hat{\theta}-\theta)\}'\{\mathbf{X}(\hat{\theta}-\theta)\}$$

with respect to θ. The identity in θ

$$S = \epsilon'\epsilon = (\mathbf{y}-\mathbf{X}\theta)'(\mathbf{y}-\mathbf{X}\theta) \equiv (\mathbf{y}-\mathbf{X}\hat{\theta})'(\mathbf{y}-\mathbf{X}\hat{\theta}) + R$$

is easily verified by direct expansion, and the term on its right

$$(\mathbf{y}-\mathbf{X}\hat{\theta})'(\mathbf{y}-\mathbf{X}\hat{\theta})$$

does not depend on θ. Minimization of R with respect to θ is therefore equivalent to minimization of S. But the process of minimizing S for θ is precisely the means by which we arrived at the Least Squares solution in 19.4. To obtain $\hat{\hat{\sigma}}^2$, $\hat{\sigma}^2$ in (24.100), therefore, we minimize S in the original model under H_0 and H_1 respectively.

Since l is a monotone decreasing function of W, the LR test is equivalent to rejecting H_0 when W is large. If we divide the numerator and denominator of W by σ^2, we see that when H_0 holds, W is the ratio of the sum of squares of r independent normal variates to an independent sum of squares of $(n-k)$ such variates, i.e. is the ratio of two independent χ^2 variates with r and $(n-k)$ degrees of freedom. Thus, when H_0 holds, $F = \frac{(n-k)}{r} W$ is distributed in the variance-ratio distribution (cf. 16.15) with $(r, n-k)$ degrees of freedom and the LR test is carried out in terms of F, large values forming the critical region.

Many of the standard tests in statistics may be reduced to tests of a linear hypothesis, and we shall be encountering them frequently in later chapters.

Example 24.8

As a special case of particular importance, consider the hypothesis
$$H_0 : \theta_r = 0,$$
where θ_r is a $(r \times 1)$ subvector of θ in (24.85). We may therefore rewrite (24.85) as
$$\mathbf{y} = (\mathbf{X}_1 \ \mathbf{X}_2) \begin{pmatrix} \theta_r \\ \theta_{k-r} \end{pmatrix} + \boldsymbol{\epsilon}$$
where $\mathbf{X}_1$ is of order $(n \times r)$ and $\mathbf{X}_2$ is of order $(n \times (k-r))$. Then H_0 becomes equivalent to specifying
$$\mathbf{y} = \mathbf{X}_2 \theta_{k-r} + \boldsymbol{\epsilon}.$$
In accordance with **24.28**, we find the minima of $S = \boldsymbol{\epsilon}' \boldsymbol{\epsilon}$. Since we are here estimating all the parameters of a linear model both on H_0 and on H_1, we may use the result of **19.9**. We have at once that the minimum under H_0 is
$$n\hat{\hat{\sigma}}^2 = \mathbf{y}'\{\mathbf{I} - \mathbf{X}_2(\mathbf{X}_2' \mathbf{X}_2)^{-1} \mathbf{X}_2'\}\mathbf{y}$$
while under H_1 it is
$$n\hat{\sigma}^2 = \mathbf{y}'\{\mathbf{I} - \mathbf{X}(\mathbf{X}'\mathbf{X})^{-1}\mathbf{X}'\}\mathbf{y}$$
where $\mathbf{X} = (\mathbf{X}_1 \ \mathbf{X}_2)$. The statistic
$$F = \frac{n-k}{r} \left(\frac{\hat{\hat{\sigma}}^2 - \hat{\sigma}^2}{\hat{\sigma}^2} \right)$$
is distributed in the variance-ratio distribution with $(r, n-k)$ degrees of freedom, the critical region for H_0 being the 100α per cent largest values of F.

24.29 In **24.28** we saw that the LR test is based on the statistic (24.100) which may be rewritten
$$W = \frac{\sum_{i=1}^{r} (z_i - c_{0i})^2 / \sigma^2}{\sum_{i=k+1}^{n} z_i^2 / \sigma^2}.$$

Whether H_0 holds or not, the denominator of W is distributed like χ^2 with $(n-k)$ degrees of freedom. When H_0 holds, as we saw, the numerator is also a χ^2 variate with r degrees of freedom, but when H_0 does not hold this is no longer so: in fact, the numerator will always be a non-central χ^2 variate (cf. **24.4**) with r degrees of freedom and non-central parameter
$$\lambda = \sum_{i=1}^{r} (c_{0i} - \mu_i)^2 / \sigma^2 = (\boldsymbol{\mu}_r - \mathbf{c}_0)'(\boldsymbol{\mu}_r - \mathbf{c}_0)/\sigma^2, \tag{24.101}$$
where μ_i is the true mean of z_i. Only when H_0 holds is λ equal to zero, giving the central χ^2 distribution of **24.28**. Since we wish to investigate the distribution of W (or equivalently of F) when H_0 is not true, so that we can evaluate the power of the LR test, we are led to the study of the ratio of a non-central to a central χ^2 variate.

The non-central F distribution

24.30 Consider first the ratio of two variates z_1, z_2, independently distributed in the non-central χ^2 form (24.18) with degrees of freedom ν_1, ν_2 and non-central parameters λ_1, λ_2 respectively. Using (11.74) the distribution of $u = z_1/z_2$ is given by

$$dH(u) = du \int_0^\infty \frac{e^{-\frac{1}{2}(uv+\lambda_1)}(uv)^{\frac{1}{2}\nu_1 - 1}}{2^{\nu_1/2}\Gamma\{\frac{1}{2}(\nu_1-1)\}\Gamma(\frac{1}{2})} \sum_{r=0}^\infty \frac{\lambda_1^r(uv)^r}{(2r)!} B\{\frac{1}{2}(\nu_1-1), \frac{1}{2}+r\}$$

$$\times \frac{e^{-\frac{1}{2}(v+\lambda_2)}v^{\frac{1}{2}\nu_2 - 1}}{2^{\nu_2/2}\Gamma\{\frac{1}{2}(\nu_2-1)\}\Gamma(\frac{1}{2})} \sum_{s=0}^\infty \frac{\lambda_2^s v^s}{(2s)!} B\{\frac{1}{2}(\nu_2-1), \frac{1}{2}+s\} v\, dv.$$

If we write $\lambda = \lambda_1 + \lambda_2$, $\nu = \nu_1 + \nu_2$, and simplify, this becomes

$$dH(u) = \frac{e^{-\frac{1}{2}\lambda}}{2^{\nu/2}} \sum_{r=0}^\infty \sum_{s=0}^\infty \frac{\lambda_1^r}{(2r)!} \frac{\lambda_2^s}{(2s)!} \frac{\Gamma(\frac{1}{2}+r)}{\Gamma(\frac{1}{2})\Gamma(\frac{1}{2}\nu_1+r)} \cdot \frac{\Gamma(\frac{1}{2}+s)}{\Gamma(\frac{1}{2})\Gamma(\frac{1}{2}\nu_2+s)}$$

$$\times \left\{ \int_0^\infty e^{-\frac{1}{2}v(1+u)} v^{\frac{1}{2}\nu+r+s-1} dv \right\} u^{\frac{1}{2}\nu_1 + r - 1} du. \qquad (24.102)$$

The integral in (24.102) is equal to $\Gamma(\frac{1}{2}\nu + r + s)\big/\left(\frac{1+u}{2}\right)^{\frac{1}{2}\nu + r + s}$. Thus

$$dH(u) = e^{-\frac{1}{2}\lambda} \sum_{r=0}^\infty \sum_{s=0}^\infty \frac{\lambda_1^r \lambda_2^s}{(2r)!(2s)!} \frac{\Gamma(\frac{1}{2}+r)\Gamma(\frac{1}{2}+s) 2^{r+s}}{B(\frac{1}{2}\nu_1+r, \frac{1}{2}\nu_2+s)\{\Gamma(\frac{1}{2})\}^2}$$

$$\times u^{\frac{1}{2}\nu_1 + r - 1} \left(\frac{1}{1+u}\right)^{\frac{1}{2}\nu + r + s} du. \qquad (24.103)$$

Since

$$\Gamma(x)\Gamma(x+\tfrac{1}{2}) = \Gamma(2x)\Gamma(\tfrac{1}{2})/2^{2x-1},$$

$$\frac{\Gamma(\frac{1}{2}+r)\Gamma(\frac{1}{2}+s) 2^{r+s}}{(2r)!(2s)!\{\Gamma(\frac{1}{2})\}^2} = \frac{1}{2^{r+s}\, r!\, s!},$$

and (24.103) may finally be simplified to

$$dH(u) = e^{-\frac{1}{2}\lambda} \sum_{r=0}^\infty \sum_{s=0}^\infty \frac{(\frac{1}{2}\lambda_1)^r}{r!} \frac{(\frac{1}{2}\lambda_2)^s}{s!} u^{\frac{1}{2}\nu_1 + r - 1} \left(\frac{1}{1+u}\right)^{\frac{1}{2}\nu + r + s} \frac{du}{B(\frac{1}{2}\nu_1+r, \frac{1}{2}\nu_2+s)}, \qquad (24.104)$$

a result obtained by Tang (1938) and studied by Price (1964). If we put

$$F'' = \frac{z_1/\nu_1}{z_2/\nu_2} = \frac{\nu_2}{\nu_1} u$$

in (24.104), we obtain the doubly non-central F-distribution, the computation of whose d.f. is considered by Bulgren (1971).

24.31 If now we put $\lambda_2 = 0$ and $\lambda_1 = \lambda$, F'' is only singly non-central, and we write it F'. From (24.104), its distribution is

$$dG(F') = e^{-\frac{1}{2}\lambda} \sum_{r=0}^\infty \frac{(\frac{1}{2}\lambda)^r}{r!} \left(\frac{\nu_1}{\nu_2}\right)^{\frac{1}{2}\nu_1 + r} \frac{(F')^{\frac{1}{2}\nu_1 + r - 1}}{B(\frac{1}{2}\nu_1 + r, \frac{1}{2}\nu_2)\left(1+\frac{\nu_1}{\nu_2}F'\right)^{\frac{1}{2}(\nu_1+\nu_2)+r}} dF'. \qquad (24.105)$$

(24.105) is a generalization of the variance-ratio (F) distribution (16.24), to which it reduces when $\lambda = 0$. It is called the non-central F distribution with degrees of freedom ν_1, ν_2 and non-central parameter λ. We sometimes write it $F'(\nu_1, \nu_2, \lambda)$. Like

(24.18), it was first discussed by Fisher (1928a), and it has been studied by Wishart (1932), Tang (1938), Patnaik (1949), Price (1964), and Tiku (1965a). Formulae for its moments are given in Exercise 24.21.

The 50, 75, 90 and 95 percentage points of F' are tabulated by Wallace and Toro-Vizcarrondo (1969) for v_1 and $v_2 = 1\,(1)\,30, 40, 60, 120, 200, 400, 1000$ and $\lambda = 1$ only (they use $\tfrac{1}{2}\lambda$ as parameter).

As $v_2 \to \infty$, $v_1 F'(v_1, v_2, \lambda) \to \chi'^2(v_1, \lambda)$ defined at (24.18)—this result when $\lambda = 0$ has already been noted at **16.22** (7). Cf. Exercise 24.20.

The power function of the LR test of the linear hypothesis

24.32 It follows at once from **24.28–9** and **24.31** that the power function of the LR test of the general linear hypothesis is

$$P = \int_{F'_\alpha(v_1, v_2, 0)}^{\infty} dG\{F'(v_1, v_2, \lambda)\}, \tag{24.106}$$

where F'_α is the $100\,(1-\alpha)$ per cent point of the distribution, $v_1 = r$, $v_2 = n-k$, and λ is defined at (24.101).

Several tables and charts of P have been constructed:

(1) Tang (1938) gives $1-P$ (i.e. the Type II error β) to 3 d.p. for test sizes $\alpha = 0.01, 0.05$; v_1 (his f_1) $= 1\,(1)\,8$; v_2 (his f_2) $= 2\,(2)\,6\,(1)\,30, 60, \infty$; and $\phi = \{\lambda/(v_1+1)\}^{\frac{1}{2}} = 1\,(0.5)\,3\,(1)\,8$. These tables are reproduced in Mann (1949) and in Kempthorne (1952).

(2) Tiku (1967a) extends Tang's tables (1) to 4 d.p. for $\alpha = 0.005, 0.01, 0.025, 0.05$; $v_1 = 1\,(1)\,10, 12$; $v_2 = 2\,(2)\,30, 40, 60, 120, \infty$; and $\phi = 0.5, 1.0\,(0.2)\,2.2\,(0.4)\,3.0$.

(3) Lehmer (1944) gives inverse tables of ϕ for $\alpha = 0.01, 0.05$; $v_1 = 1\,(1)\,10, 12, 15, 20, 24, 30, 40, 60, 80, 120, \infty$; $v_2 = 2\,(2)\,20, 24, 30, 40, 60, 80, 120, 240, \infty$; and P (her β) $= 0.7, 0.8$.

(4) Dasgupta (1968) gives tables of $\tfrac{1}{2}\lambda$ (his δ) for v_1 (his M) $= 1\,(1)\,10$; v_2 (his N) $= 10\,(5)\,50\,(10), 100, \infty$; $\alpha = 0.01, 0.05$ and P (his β) $= 0.1\,(0.1)\,0.9$.

(5) Kastenbaum *et al.* (1970a) give tables of $\tau = \{2\lambda(v_1+1)/(v_1+v_2+1)\}^{\frac{1}{2}}$ to 3 d.p. for $\alpha = 0.01, 0.05, 0.1, 0.2$; $\beta = 1-P = 0.005, 0.01, 0.05, 0.1, 0.2, 0.3$; v_1 (their $k-1$) $= 1\,(1)\,5$; and $(v_1+v_2+1)/(v_1+1)$ (their N) $= 2\,(1)\,8\,(2)\,30, 40\,(20)\,100, 200, 500, 1000$. See also their related smaller tables (1970b) and Bowman (1972).

(6) E. S. Pearson and Hartley (1951) give eight charts of the power function, one for each value of v_1 from 1 to 8. Each chart shows the power for $v_2 = 6\,(1)\,10, 12, 15, 20, 30, 60, \infty$; $\alpha = 0.01, 0.05$; and ϕ ranging from 1 (except when $v_1 = 1$, when ϕ ranges from 2 for $\alpha = 0.01$, 1.2 for $\alpha = 0.05$) to a value large enough for the power to be at least 0.98. The table for $v_1 = 1$ is reproduced in the *Biometrika Tables*.

(7) M. Fox (1956) gives inverse charts, one for each of the combinations of $\alpha = 0.01, 0.05$, with power P (his β) $= 0.5, 0.7, 0.8, 0.9$. Each chart shows, for
$v_1 = 3\,(1)\,10\,(2)\,20\,(20)\,100, 200, \infty$; $v_2 = 4\,(1)\,10\,(2)\,20\,(20)\,100, 200, \infty$,
the contours of constant ϕ. He also gives a nomogram for each α to facilitate interpolation in β.

(8) A. J. Duncan (1957) gives two charts, one for $\alpha = 0.01$ and one for $\alpha = 0.05$. Each shows, for $v_2 = 6\,(1)\,10, 12, 15, 20\,(10)\,40, 60, \infty$, the values of v_1 (ranging from 1 to 8) and ϕ required to attain power $P = 1-\beta = 0.50$ and 0.90.

Approximation to the power function of the LR test

24.33 As will be seen from the form of (24.105), the computation of the exact power function (24.106) is a laborious matter, and even now its tabulation is far from complete.

However, we may obtain a simple approximation to the power function in the manner of, and using the results of, our approximation to the non-central χ^2 distribution in **24.5**. If z_1 is a non-central χ^2 variate with ν_1 degrees of freedom and non-central parameter λ, we have from (24.21) that $z_1 \Big/ \left(\dfrac{\nu_1+2\lambda}{\nu_1+\lambda}\right)$ is approximately a central χ^2 variate with degrees of freedom $\nu^* = (\nu_1+\lambda)^2/(\nu_1+2\lambda)$. Thus

$$z_1 \Big/ \left\{\nu^*\left(\frac{\nu_1+2\lambda}{\nu_1+\lambda}\right)\right\} = z_1/(\nu_1+\lambda)$$

is approximately a central χ^2 variate divided by its degrees of freedom ν^*. Hence z_1/ν_1 is approximately a multiple $(\nu_1+\lambda)/\nu_1$ of such a variate. If we now define the non-central F'-variate

$$F' = \frac{z_1/\nu_1}{z_2/\nu_2},$$

where z_2 is a central χ^2 variate with ν_2 degrees of freedom, it follows at once that approximately

$$F' = \frac{\nu_1+\lambda}{\nu_1}F, \qquad (24.107)$$

where F is a central F-variate with degrees of freedom $\nu^* = (\nu_1+\lambda)^2/(\nu_1+2\lambda)$ and ν_2.

The simple approximation (24.107) is surprisingly effective. By making comparisons with Tang's (1938) exact tables, Patnaik (1949) shows that the power function calculated by use of (24.107) is generally accurate to two significant figures; it will therefore suffice for all practical purposes.

To calculate the power of the LR test of the linear hypothesis, we therefore replace (24.106) by the approximate central F-integral

$$P = \int_{\left(\frac{\nu_1}{\nu_1+\lambda}\right)F_\alpha(\nu_1,\nu_2)}^{\infty} dG\left\{F\left(\frac{(\nu_1+\lambda)^2}{(\nu_1+2\lambda)}, \nu_2\right)\right\}, \qquad (24.108)$$

the size of the test being determined by putting $\lambda = 0$. $(\nu_1+\lambda)^2/(\nu_1+2\lambda)$ is generally fractional, and interpolation is necessary. Even the central F distribution, however, is not yet so very well tabulated as to make the accurate evaluation of (24.108) easy—see the list of tables in **16.19**.

> A central F-approximation due to Tiku (1965a, 1966) obtained by equating three moments is even more accurate—see also Pearson and Tiku (1970).
>
> Dar (1962) gives a simple normal approximation to the distribution of the ratio of two independent identical non-central F-variables.

The non-central t-distribution

24.34 When $\nu_1 = 1$, the non-central F distribution (24.105) reduces to the non-central t^2 distribution, just as for the central distributions (cf. **16.15**). If we transform from t^2 to t, we obtain the non-central t-distribution, which we call the t'-distribution. Evidently, from the derivation of non-central χ^2 as the sum of non-central squared normal variates, we may write

$$t' = (z+\delta)/w^{\frac{1}{2}}, \qquad (24.109)$$

where z is a normal variate with zero mean and w is independently distributed like χ^2/f with f degrees of freedom (we write f instead of ν_2 in (24.105), and $\delta^2 = \lambda$, in this case and sometimes write the variate as $t'(f, \delta)$). Our discussion of the F' distribution covers the t'^2 distribution, but the t' distribution has received special attention because of its importance in applications.

Johnson and Welch (1939) studied the distribution and gave tables for finding $100(1-\alpha)$ per cent points of the distribution of t' for α or $1-\alpha = 0.005, 0.01, 0.025, 0.05, 0.1 (0.1) 0.5$, $f = 4(1) 9, 16, 36, 144, \infty$, and any δ; and conversely for finding δ for given values of t'. Resnikoff (1962) gives additional tables.

Resnikoff and Lieberman (1957) have given tables of the frequency function and the distribution function of t' to 4 d.p., at intervals of 0.05 for $t/f^{\frac{1}{2}}$, for $f = 2(1)24(5)49$, and for the values of δ defined by

$$\int_{\delta/(f+1)^{\frac{1}{2}}}^{\infty} (2\pi)^{-\frac{1}{2}} \exp(-\tfrac{1}{2}x^2) dx = \alpha,$$

$\alpha = 0.001, 0.0025, 0.004, 0.01, 0.025, 0.04, 0.065, 0.10, 0.15, 0.25$. They, and also Scheuer and Spurgeon (1963), give some percentage points of the distributions. Locks *et al.* (1963) give similar tables at intervals of 0.2 for t', with $f = 1(1)20(5)40$ and δ defined by $\delta(f+1)^{-\frac{1}{2}}$ or $\delta(f+2)^{-\frac{1}{2}} = 0(0.25) 3$. Owen (1963) gives very extensive tables of percentage points. Hogben *et al.* (1961) give a method of obtaining the moments, with tables for the first four—explicit formulae are given in Exercise 24.22. Amos (1964) studies series approximations of the distribution.

Krishnan (1967, 1968) and Bulgren and Amos (1968) study and tabulate the doubly non-central t-distribution obtained from F'' of **24.30** by putting $\nu_1 = 1$ to give t''^2.

24.35 A particular important application of the t' distribution is in evaluating the power of a "Student's" t-test for which the critical region is in one tail only (the "equal-tails" case, of course, corresponds to the t'^2 distribution). The test is that $\delta = 0$ in (24.109), the critical region being determined from the central t-distribution. Its power is evidently just the integral of the non-central t-distribution over the critical region. It has been specially tabulated by Neyman *et al.* (1935), who give, for $\alpha = 0.05, 0.01$, f (their n) $= 1(1) 30, \infty$ and δ (their ρ) $= 1(1)10$, tables and charts of the complement $1-P$ of the power of the test, together with the values of δ for which $P = 1-\alpha$. Neyman and Tokarska (1936) give inverse tables of δ for the same values of α and f and $1-P = 0.05, 0.10, (0.10) 0.90$. Owen (1965) gives 5 d.p. tables of δ for $\alpha = 0.05, 0.025, 0.01$ and 0.005; $n-1 = 1(1) 30(5) 100 (10) 200$, ∞ and $1-P = 0.01, 0.05, 0.10 (0.10) 0.90$. Hodges and Lehmann (1967) give an asymptotic series for the power and use it (1968) to construct an extremely compact table.

Optimum properties of the LR test of the general linear hypothesis

24.36 We saw in **24.27** that, apart from the case $r = 1$, there is no UMPU test of the general linear hypothesis. Nevertheless, the LR test of that hypothesis has certain optimum properties which we now proceed to develop, making use of simplified proofs due to Wolfowitz (1949) and Lehmann (1950).

In **24.28** we derived the ML estimators of the $(k-r+1)$ unspecified parameters when H_0 holds. They are the components of

$$\mathbf{t} = (\hat{\hat{\mu}}, \hat{\hat{\sigma}}^2),$$

which are defined above (24.98). When H_0 holds, the components of $\mathbf{t}$ are a set of $(k-r+1)$ sufficient statistics for the unspecified parameters. By **23.10**, their distribution is complete. Thus, by **23.19**, every similar size-α critical region w for H_0 will consist of a fraction α of every surface $\mathbf{t} = $ constant. Here every component of $\mathbf{t}$ is to be constant, and in particular the component $\hat{\sigma}^2$. Let

$$n\hat{\sigma}^2 = \sum_{=k+1}^{n} z_i^2 + \sum_{=1}^{r} (z_i - c_{0i})^2 = a^2, \qquad (24.110)$$

where a is a constant.

Now consider a fixed value of λ, defined at (24.101), say $\lambda = d^2 > 0$. The power of any similar region on this surface will consist of the aggregate of its power on (24.110) for all a. For fixed a, the power on the surface $\lambda = d^2$ is

$$P(w \mid \lambda, a) = \int_{\lambda = d^2} L(\mathbf{z} \mid \mathbf{\mu}, \sigma^2) \, d\mathbf{z}, \qquad (24.111)$$

where L is the LF defined at (24.95). We may write this out fully as

$$P(w \mid \lambda, a) = (2\pi\sigma^2)^{-n/2} \int_{\lambda = d^2} \exp\left\{-\frac{1}{2\sigma^2}[\{(\mathbf{z}_r - \mathbf{c}_0) - (\mathbf{\mu}_r - \mathbf{c}_0)\}'\{(\mathbf{z}_r - \mathbf{c}_0) - (\mathbf{\mu}_r - \mathbf{c}_0)\} \right.$$
$$\left. + (\mathbf{z}_{k-r} - \mathbf{\mu}_{k-r})'(\mathbf{z}_{k-r} - \mathbf{\mu}_{k-r}) + \mathbf{z}'_{n-k}\mathbf{z}_{n-k}]\right\} d\mathbf{z}. \qquad (24.112)$$

Using (24.110) and (24.101), (24.112) becomes

$$P(w \mid \lambda, a) = (2\pi\sigma^2)^{-\frac{1}{2}(n-k+r)} \exp\left\{-\frac{1}{2}\left(d^2 + \frac{a^2}{\sigma^2}\right)\right\} \int_{\lambda = d^2} \exp\{(\mathbf{z}_r - \mathbf{c}_0)'(\mathbf{\mu}_r - \mathbf{c}_0)\} d\mathbf{z}_r, \quad (24.113)$$

the vector $\mathbf{z}_{k-r}$ having been integrated out over its whole range since its distribution is free of λ and independent of a. The only non-constant factor in (24.113) is the integral, which is to be maximized to obtain the critical region w with maximum P. The integral is over the surface $\lambda = d^2$ or $(\mathbf{\mu}_r - \mathbf{c}_0)'(\mathbf{\mu}_r - \mathbf{c}_0) = $ constant. It is clearly a monotone increasing function of $|\mathbf{z}_r - \mathbf{c}_0|$ i.e. of $(\mathbf{z}_r - \mathbf{c}_0)'(\mathbf{z}_r - \mathbf{c}_0) = \sum_{i=1}^{r}(z_i - c_{0i})^2$.

Now if $\sum_{i=1}^{r}(z_i - c_{0i})^2$ is maximized for fixed a in (24.110), W defined at (24.100) is also maximized. Thus for any fixed λ and a, the maximum value of $P(w \mid \lambda, a)$ is attained when w consists of large values of W. Since this holds for each a, it holds when the restriction that a be fixed is removed. We have therefore established that on any surface $\lambda = d^2 > 0$, the LR test, which consists of rejecting large values of W, has maximum power, a result due to Wald (1942).

An immediate consequence is P. L. Hsu's (1941) result, that the LR test is UMP among all tests whose power is a function of λ only.

Invariant tests

24.37 In developing unbiassed tests for location parameters in **24.20**, we found it quite natural to introduce the invariance condition (24.61) as a necessary condition which any acceptable test must satisfy. Similarly for scale parameters in **24.21**, the logarithmic transformation from (24.68) to (24.69) requires implicitly that the test statistic t satisfies

$$t(y_1, y_2, \ldots, y_n) = t(cy_1, cy_2, \ldots, cy_n), \qquad c > 0. \qquad (24.114)$$

Frequently, it is reasonable to restrict the class of tests considered to those which are invariant under transformations which leave the hypothesis to be tested invariant;

if this is not done, e.g. in the problem of testing the equality of location (or scale) parameters, it would mean that a change of origin (or unit) of measurement would affect the conclusions reached by the test. The relationship between invariance and sufficiency principles in general is discussed by Hall *et al.* (1965), with a theorem due to C. Stein which gives conditions under which it does not matter in which order the principles are applied.

If we examine the canonical form of the general linear hypothesis in **24.27** from this point of view, we see at once that the problem is invariant under:

(a) any orthogonal transformation of $(\mathbf{z}_r - \mathbf{c}_0)$ (this leaves $(\mathbf{z}_r - \mathbf{c}_0)'(\mathbf{z}_r - \mathbf{c}_0)$ unchanged);

(b) any orthogonal transformation of $\mathbf{z}_{n-k}$ (this leaves $\mathbf{z}'_{n-k} \mathbf{z}_{n-k}$ unchanged);

(c) the addition of any constant a to each component of $\mathbf{z}_{k-r}$ (the mean vector of which is arbitrary);

(d) the multiplication of all the variables by $c > 0$ (which affects only the common variance σ^2).

It is easily seen that a statistic t is invariant under all the operations (a) to (d) if, and only if, it is a function of $W = (\mathbf{z}_r - \mathbf{c}_0)'(\mathbf{z}_r - \mathbf{c}_0)/\mathbf{z}'_{n-k} \mathbf{z}_{n-k}$ alone. Clearly if t is a function of W alone, its power function, like that of W, will depend only on λ. By the last sentence of **24.36**, therefore, the LR test, rejecting large values of W, is UMP among invariant tests of the general linear hypothesis.

EXERCISES

24.1 Show that the c.f. of the non-central χ^2 distribution (24.18) is
$$\phi(t) = (1-2it)^{-\nu/2} \exp\left\{\frac{\lambda it}{1-2it}\right\},$$
giving cumulants $\kappa_r = (\nu+r\lambda)2^{r-1}(r-1)!$. In particular,
$$\kappa_1 = \nu+\lambda, \qquad \kappa_2 = 2(\nu+2\lambda),$$
$$\kappa_3 = 8(\nu+3\lambda), \qquad \kappa_4 = 48(\nu+4\lambda).$$
Hence show that the sum of two independent non-central χ^2 variates is another such, with both degrees of freedom and non-central parameter equal to the sum of those of the component distributions.

(Wishart, 1932; Tang, 1938)

24.2 Show that if the non-central normal variates x_i of **24.4** are subjected to k orthogonal linear constraints
$$\sum_{i=1}^{n} a_{ij} x_i = b_j \qquad j = 1, 2, \ldots, k,$$
where
$$\sum_{i=1}^{n} a_{ij}^2 = 1, \quad \sum_{i=1}^{n} a_{ij} a_{il} = 0, \qquad j \neq l,$$
then
$$y^2 = \sum_{i=1}^{n} x_i^2 - \sum_{j=1}^{k} b_j^2$$
has the non-central χ^2 distribution with $(n-k)$ degrees of freedom and non-central parameter $\lambda = \sum_{i=1}^{n} \mu_i^2 - \sum_{j=1}^{k}\left(\sum_{i=1}^{n} a_{ij}\mu_i\right)^2$.

(Patnaik (1949). Cf. also Bateman (1949).)

24.3 Show that for any fixed r, the first r moments of a non-central χ^2 distribution with fixed λ tend, as degrees of freedom increase, to the corresponding moments of the central χ^2 distribution with the same degrees of freedom. Hence show that, in testing a hypothesis H_0 distinguished from the alternative hypothesis by the value of a parameter θ, if the test statistic has a non-central χ^2 distribution with degrees of freedom an increasing function of sample size n, and non-central parameter λ a non-increasing function of n such that $\lambda = 0$ when H_0 holds, then the test will become ineffective as $n \to \infty$, i.e. its power will tend to its size α.

24.4 Show that the LR statistic l defined by (24.40) for testing the equality of k normal variances has moments about zero
$$\mu'_r = \frac{n^{\frac{1}{2}rn}\,\Gamma\{\frac{1}{2}(n-k)\}}{\Gamma\{\frac{1}{2}[(r+1)n-k]\}} \prod_{i=1}^{k} \frac{\Gamma\{\frac{1}{2}[(r+1)n_i-1]\}}{n_i^{\frac{1}{2}rn_i}\,\Gamma\{\frac{1}{2}(n_i-1)\}}$$

(Neyman and Pearson, 1931)

24.5 For testing the hypothesis H_0 that k normal distributions are identical in mean and variance, show that the LR statistic is, for sample sizes $n_i \geq 2$,
$$l_0 = \prod_{i=1}^{k} \left(\frac{s_i^2}{s_0^2}\right)^{n_i/2}$$

where
$$s_i^2 = \frac{1}{n_i} \sum_{j=1}^{n_i} (x_{ij}-\bar{x}_i)^2, \quad \bar{x} = \frac{1}{n} \sum_{i=1}^{k} n_i \bar{x}_i$$

and
$$s_0^2 = \frac{1}{n} \sum_{i=1}^{k} n_i \{s_i^2 + (\bar{x}_i-\bar{x})^2\},$$

and that its moments about zero are

$$\mu'_r = \frac{n^{\frac{1}{2}rn}\Gamma\{\frac{1}{2}(n-1)\}}{\Gamma\{\frac{1}{2}[(r+1)n-1]\}} \prod_{i=1}^{k} \frac{\Gamma\{\frac{1}{2}[(r+1)n_i-1]\}}{n_i^{\frac{1}{2}rn_i}\Gamma\{\frac{1}{2}(n_i-1)\}}.$$

(Neyman and Pearson, 1931)

24.6 For testing the hypothesis H_2 that k normal distributions with the same variance have equal means, show that the LR statistic (with sample sizes $n_i \geqslant 2$) is

$$l_2 = l_0/l$$

where l and l_0 are as defined for Exercises 24.4 and 24.5, and that the exact distribution of $z = 1 - l_2^{2/n}$ when H_2 holds is

$$dF \propto z^{\frac{1}{2}(k-3)}(1-z)^{\frac{1}{2}(n-k-2)}dz, \quad 0 \leqslant z \leqslant 1.$$

Find the moments of l_2 and hence show that when the hypothesis H_0 of Exercise 24.5 holds, l and l_2 are independently distributed.

(Neyman and Pearson, 1931; Hogg (1961). See also Hogg (1962) for a test of H_2.)

24.7 Show that when all the sample sizes n_i are equal, the LR statistic l of (24.40) and its modified form l^* of (24.44) are connected by the relation

$$n \log l^* = (n-k) \log l,$$

so that in this case the tests based on l and l^* are equivalent.

24.8 For samples from k distributions of form (24.48) or (24.52), show that if l is the LR statistic for testing the hypothesis

$$H_0: \theta_1 = \theta_2 = \ldots = \theta_{p_1}; \; \theta_{p_1+1} = \theta_{p_1+2} = \ldots = \theta_{p_2}; \; \theta_{p_2+1} = \ldots = \theta_{p_3};$$
$$\ldots; \theta_{p_{r-1}+1} = \ldots = \theta_{p_r}$$

that the θ_i fall into r distinct groups (not necessarily of equal size) within which they are equal, then $-2 \log l$ is distributed exactly like χ^2 with $2(n-r)$ degrees of freedom.

(Hogg, 1956)

24.9 In a sample of n observations from
$$dF = dx/2\theta, \quad \mu-\theta \leqslant x \leqslant \mu+\theta,$$
show that the LR statistic for testing $H_0: \mu = 0$ is

$$l = \left(\frac{x_{(n)}-x_{(1)}}{2z}\right)^n = \left(\frac{R}{2z}\right)^n$$

where $z = \max\{-x_{(1)}, x_{(n)}\}$. Using Exercise 23.7, show that l and z are independently distributed, so that we have the factorization of c.f's

$$E[\exp\{(-2\log R^n)it\}] = E[\exp\{(-2\log l)it\}] E[\exp\{[-2\log(2z)^n]it\}].$$

Hence show that the c.f. of $-2 \log l$ is $\phi(t) = \dfrac{(n-1)}{n(1-2it)-1}$ so that, as $n \to \infty$, $-2 \log l$ is distributed as χ^2 with 2 degrees of freedom.

(Hogg and Craig, 1956)

24.10 In **24.6**, show that a quadratic form $\mathbf{x'Ax}$ has a non-central χ^2 distribution if and only if $\mathbf{AV}$ is idempotent, and that if the distribution has n degrees of freedom this implies $\mathbf{A} = \mathbf{V}^{-1}$.

24.11 k independent samples, of sizes $n_i \geqslant 2$, $\sum_{i=1}^{k} n_i = n$, are taken from exponential populations

$$dF_i(x) = \exp\left\{-\left(\frac{x-\theta_i}{\sigma_i}\right)\right\} dx/\sigma_i, \qquad \theta_i \leqslant x \leqslant \infty.$$

Show that the LR statistic for testing

$$H_0: \theta_1 = \theta_2 = \ldots = \theta_k; \quad \sigma_1 = \sigma_2 = \ldots = \sigma_k$$

is

$$l_0 = \prod_{i=1}^{k} d_i^{n_i}/d^n$$

where $d_i = \bar{x}_i - (x_{1i,})$ the difference between the mean and smallest observation in the ith sample, and d is the same function of the combined samples, i.e.

$$d = \bar{x} - x_{(1)}.$$

Show that the moments of $l_0^{1/n}$ are

$$\mu'_p = \frac{n^p \Gamma(n-1)}{\Gamma(n+p-1)} \prod_{i=1}^{k} \frac{\Gamma\{(n_i-1)+pn_i/n\}}{n_i^{pn_i/n} \Gamma(n_i-1)},$$

(P. V. Sukhatme, 1936)

24.12 In Exercise 24.11, show that for testing

$$H_1: \sigma_1 = \sigma_2 = \ldots = \sigma_k,$$

the θ_i being unspecified, the LR statistic is

$$l_1 = \frac{\prod_{i=1}^{k} d_i^{n_i}}{\left(\frac{1}{n}\sum_{i=1}^{k} n_i d_i\right)^n},$$

and that the moments of $l_1^{1/n}$ are

$$\mu'_p = \frac{n^p \Gamma(n-k)}{\Gamma(n-k+p)} \prod_{i=1}^{k} \frac{\Gamma\{(n_i-1)+pn_i/n\}}{n_i^{pn_i/n} \Gamma(n_i-1)}.$$

(P. V. Sukhatme, 1936)

24.13 In Exercise 24.11, show that if it is known that the σ_i are all equal, the LR statistic for testing

$$H_2: \theta_1 = \theta_2 = \ldots = \theta_k$$

is

$$l_2 = l_0/l_1,$$

where l_0 and l_1 are defined in Exercises 24.11–12. Show that the exact distribution of $l_2^{1/n} = u$ is

$$dF = \frac{1}{B(n-k, k-1)} u^{n-k-1}(1-u)^{k-2} du, \qquad 0 \leqslant u \leqslant 1,$$

and find the moments of u. Show that when H_0 of Exercise 24.11 holds, l_1 and l_2 are independently distributed.

(P. V. Sukhatme, 1936; Hogg (1961). Cf. also Hogg and Tanis (1963) for other tests of these hypotheses.)

24.14 Show by comparison with the unbiassed test of Exercise 23.17 that the LR test for the hypothesis that two normal populations have equal variances is biassed for unequal sample sizes n_1, n_2.

24.15 Show by using (24.67) that an unbiassed similar size-α test of the hypothesis H_0 that k independent observations x_i ($i = 1, 2, \ldots, k$) from normal populations with unit variance have equal means is given by the critical region

$$\sum_{i=1}^{k} (x_i - \bar{x})^2 \geqslant c_\alpha,$$

where c_α is the $100(1-\alpha)$ per cent point of the distribution of χ^2 with $(n-1)$ degrees of freedom. Show that this is also the LR test.

24.16 Show that the three test statistics of Exercises 23.24–26 are equivalent to the LR statistics in the situations given; that the critical region of the LR test in Exercise 23.24 is not the UMPU region and is in fact biassed; but that in the other two Exercises the LR test coincides with the UMP similar test.

24.17 Extending the results of **23.10–13**, show that if a distribution is of form

$$f(x \mid \theta_1, \theta_2, \ldots, \theta_k) = Q(\theta) M(x) \exp\{\sum_{j=3}^{k} B_j(x) A_j(\theta_3, \theta_4, \ldots, \theta_k)\},$$

$$a(\theta_1, \theta_2) \leqslant x \leqslant b(\theta_1, \theta_2)$$

(the terminals of the distribution depending only on the two parameters not entering into the exponential term), the statistics $t_1 = x_{(1)}$, $t_2 = x_{(n)}$, $t_j = \sum_{i=1}^{n} B_j(x_i)$ are jointly sufficient for θ in a sample of n observations, and that their distribution is complete.

(Hogg and Craig, 1956)

24.18 Using the result of Exercise 24.17, show that in independent samples of sizes n_1, n_2 from two distributions

$$dF = \exp\left\{-\frac{(x_i - \theta_i)}{\sigma}\right\} \frac{dx_i}{\sigma}, \quad \sigma > 0; \; x_i \geqslant \theta_i; \; i = 1, 2,$$

the statistics

$$z_1 = \min\{x_{i(1)}\},$$

$$z_2 = \sum_{j=1}^{n_1} x_{1j} + \sum_{j=1}^{n_2} x_{2j},$$

are sufficient for θ_1 and θ_2 and complete.

Show that the LR statistic for $H_0 : \theta_1 = \theta_2$ is

$$l = \left\{\frac{z_2 - (n_1 x_{1(1)} + n_2 x_{2(1)})}{z_2 - (n_1 + n_2) z_1}\right\}^{n_1 + n}$$

and that l is distributed independently of z_1, z_2 and hence of its denominator. Show that l gives an unbiassed test of H_0.

(Paulson, 1941)

24.19 Show that (24.18) may be written

$$h(z) = \sum_{r=0}^{\infty} e^{-\frac{1}{2}\lambda} \frac{(\frac{1}{2}\lambda)^r}{r!} \cdot \frac{e^{-\frac{1}{2}z} z^{\frac{1}{2}\nu + r - 1}}{2^{\frac{1}{2}\nu + r} \Gamma(\frac{1}{2}\nu + r)},$$

displaying it as a mixture (cf. **5.13**) of central χ^2 distributions, with Poisson frequencies as the mixing distribution. By representing these $\chi^2(\nu+2r)$ distributions as the sum of a $\chi^2(\nu)$ and r $\chi^2(2)$ distributions, all independent, use Exercise 5.22 to establish the c.f. of (24.18) given in Exercise 24.1.

Using the second result in Exercise 16.7, show that the d.f. of (24.18) is given, for even ν, by
$$H(z) = \text{Prob}\{u - v \geqslant \tfrac{1}{2}\nu\},$$
where u and v are independent Poisson variates with parameters $\tfrac{1}{2}z$ and $\tfrac{1}{2}\lambda$ respectively.

(c.f. Johnson, 1959a)

24.20 For the LR test of the general linear hypothesis based on (24.99), show that the asymptotic non-central χ^2 distribution of $-2\log l$, given in **24.7**, agrees with the asymptotic result for $\nu_1 F'(\nu_1, \nu_2, \lambda)$ at the end of **24.31**.

24.21 As in **24.31**, write the non-central F variate in the form $F'(\nu_1, \nu_2, \lambda) = \dfrac{z_1/\nu_1}{z_2/\nu_2}$, where z_1 is a non-central $\chi'^2(\nu_1, \lambda)$ variate and z_2 an independent central $\chi^2(\nu_2)$ variate. Let z_3 be a central $\chi^2(1)$ variate, independent of z_1 and z_2. Show that
$$E\{(F')^r\} = E\left\{\left(\frac{z_3}{z_2/\nu_2}\right)^r\right\} E(z_1^r)/\{\nu_1^r E(z_3^r)\}$$
whence symbolically
$$\mu_r'\{F'(\nu_1, \nu_2, \lambda)\} = \frac{\mu_r'\{F(1, \nu_2)\}\,\mu_r'\{\chi'^2(\nu_1, \lambda)\}}{\nu_1^r \mu_r'\{\chi^2(1)\}}$$
if $2r < \nu_2$, so that the central F moments exist by Exercise 16.1. Hence, using (16.4–5), (16.28) and (24.19), show that $F'(\nu_1, \nu_2, \lambda)$ has mean and variance given by
$$\mu_1' = (\nu_1 + \lambda)\nu_2/\{\nu_1(\nu_2 - 2)\}, \quad \nu_2 > 2,$$
$$\mu_2 = \frac{2\nu_2^2}{\nu_1^3(\nu_2-2)^2(\nu_2-4)}\{(\nu_1+2\lambda)(\nu_1+\nu_2-2)+\lambda^2\}, \quad \nu_2 > 4.$$

(cf. Bain, 1969; Pearson and Tiku (1970) give μ_3 and μ_4.)

24.22 Apply the method of Exercise 24.21 to the non-central t' variate defined at (24.109) by multiplying and dividing $E\{(t')^r\}$ by $E\{z_3^{\frac{1}{2}r}\}$, and show that
$$\mu_r'\{t'(f, \delta)\} = \frac{\mu_r'\{N(\delta, 1)\}\,\mu_{\frac{1}{2}r}'\{F(1, f)\}}{\mu_{\frac{1}{2}r}'\{\chi^2(1)\}}, \quad r < f,$$
where $N(\delta, 1)$ is a normal variate with mean δ and variance 1. If r is even, say $= 2s$, show that this may be written
$$\mu_{2s}'\{t'(f, \delta)\} = \mu_{2s}'\{N(\delta, 1)\}\,\mu_{2s}\{t(f)\}/\mu_{2s}\{N(0, 1)\},$$
where $t(f)$ is the central t-distribution with f degrees of freedom.

(cf. Bain, 1969)

CHAPTER 25

THE COMPARISON OF TESTS

25.1 In Chapters 22–24 we have been concerned with the problems of finding "optimum" tests, i.e. of selecting the test with the "best" properties in a given situation, where "best" means the possession by the test of some desirable property such as being UMP, UMPU, etc. We have not so far considered the question of comparing two or more tests for a given situation with the aim of evaluating their relative efficiencies. Some investigation of this subject is necessary to permit us to evaluate the loss of efficiency incurred in using any other test than the optimum one. It may happen, for example, that a UMP test is only very slightly more powerful than another test, which is perhaps much simpler to compute; in such circumstances we might well decide to use the less efficient test in routine testing. Before we can decide an issue such as this, we must make some quantitative comparison between the tests.

We discussed the analogous problem in the theory of estimation in **17.29**, where we derived a measure of estimating efficiency. The reader will perhaps ask how it comes about that, whereas in the theory of estimation the measurement of efficiency was discussed almost as soon as the concept of efficiency had been defined, we have left over the question of measuring test efficiency to the end of our general discussion of the theory of tests. The answer is partly that the concept of test efficiency turns out to be more complicated than that of estimating efficiency, and therefore could not be so shortly treated. For the most part, however, we are simply following the historical development of the subject: it was not until, from about 1935 onwards, the attention of statisticians turned to the computationally simple tests to be discussed in Chapters 31 and 32 that the need arose to measure test efficiency. Even the idea of test consistency, which we encountered in **24.17**, was not developed by Wald and Wolfowitz (1940) until nearly twenty years after the first definition of a consistent estimator by Fisher (1921a); only when "inefficient" tests became of practical interest was it necessary to investigate the weaker properties of tests.

The comparison of power functions

25.2 In testing a given hypothesis against a given alternative for fixed sample size, the simplest way of comparing two tests is by direct examination of their power functions. If sample size is at our disposal (e.g. in the planning of a series of observations), it is natural to seek a definition of test efficiency of the same form as that used for estimating efficiency in **17.29**. If an "efficient" test (i.e. the most powerful in the class considered) of size α requires to be based on n_1 observations to attain a certain power, and a second size-α test requires n_2 observations to attain the same power against the same alternative, we may define the *relative efficiency* of the second test in attaining that power against that alternative as n_1/n_2. This measure is, as in the case of estimation, the reciprocal of the ratio of sample sizes required for a given performance, but it will be noticed that our definition of relative efficiency is not asymptotic, and that it imposes no restriction upon the forms of the sampling distributions of the test statistics being compared. We can compare any two tests in this way because the power functions of the tests, from which the relative efficiency is calculated, take comprehensive account of the distributions of the test statistics; the power functions contain all the information relevant to our comparison.

Asymptotic comparisons

25.3 The concept of relative efficiency, although comprehensive, is not concise. Like the power functions on which it is based, it is a function of three arguments—the size α of the tests, the " distance " (in terms of some parameter θ) between the hypothesis tested and the alternative, and the sample size (n_1) required by the efficient test. Even if we may confine ourselves to a few typical values of α, a table of double entry is still required for the comparison of tests by this measure. It would be much more convenient if we could find a single summary measure of efficiency, and it is clear that we can only hope to achieve this by imposing some limiting process. We have thus been brought back to the necessity for restriction to asymptotic results.

25.4 A different approach which suggests itself is that we let sample sizes tend to infinity, as in **17.29**, and take the ratio of the powers of the tests as our measure of test efficiency. If we consider this suggestion we immediately encounter a difficulty. If the tests we are considering are both size-α consistent tests against the class of alternative hypotheses in the problem (and henceforth we shall always assume this to be so), it follows by definition that the power function of each tends to 1 as sample size increases. If we compare the tests against some fixed alternative value of θ, it follows that the efficiency thus defined will always tend to 1 as sample size increases. The suggested measure of test efficiency is therefore quite useless.

More generally, it is easy to see that consideration of the power functions of consistent tests asymptotically in n is of limited value. For instance, Wald (1941) defined an asymptotically most powerful test as one whose power function cannot be bettered as sample size tends to infinity, i.e. which is UMP asymptotically. The following example, due to Lehmann (1949), shows that one asymptotically UMP test may in fact be decidedly inferior to another such test, even asymptotically.

Example 25.1

Consider again the problem, discussed in Examples 22.1 and 22.2, of testing the mean θ of a normal distribution with known variance, taken to be equal to 1. We wish to test $H_0: \theta = \theta_0$ against the one-sided alternative $H_1: \theta = \theta_1 > \theta_0$. In **22.17**, we saw that a UMP test of H_0 against H_1 is given by the critical region $\bar{x} \geqslant \theta_0 + d_\alpha/n^{\frac{1}{2}}$, and in Example 22.3 that its power function is

$$P_1 = G\{\Delta n^{\frac{1}{2}} - d_\alpha\} = 1 - G\{d_\alpha - \Delta n^{\frac{1}{2}}\}, \tag{25.1}$$

where $\Delta = \theta_1 - \theta_0$ and the fixed value d_α defines the size α of the test as at (22.16).

THE COMPARISON OF TESTS

We now construct a two-tailed size-α test, rejecting H_0 when

$$\bar{x} \geq \theta_0 + d_{\alpha_2}/n^{\frac{1}{2}} \quad \text{or} \quad \bar{x} \leq \theta_0 - d_{\alpha_1}/n^{\frac{1}{2}},$$

where d_{α_1} and d_{α_2}, functions of n, may be chosen arbitrarily subject to the condition $\alpha_1 + \alpha_2 = \alpha$, which implies that d_{α_1} and d_{α_2} both exceed d_α. (23.48) shows that the power function of this second test is

$$P_2 = G\{\Delta n^{\frac{1}{2}} - d_{\alpha_2}\} + G\{-\Delta n^{\frac{1}{2}} - d_{\alpha_1}\}, \tag{25.2}$$

and since G is always positive, it follows that

$$P_2 > G\{\Delta n^{\frac{1}{2}} - d_{\alpha_2}\} = 1 - G\{d_{\alpha_2} - \Delta n^{\frac{1}{2}}\}. \tag{25.3}$$

Since the first test is UMP, we have, from (25.1) and (25.3),

$$G\{d_{\alpha_2} - \Delta n^{\frac{1}{2}}\} - G\{d_\alpha - \Delta n^{\frac{1}{2}}\} > P_1 - P_2 \geq 0. \tag{25.4}$$

It is easily seen that the difference between $G\{x\}$ and $G\{y\}$ for fixed $(x-y)$ is maximized when x and y are symmetrically placed about zero, i.e. when $x = -y$, i.e. that

$$G\{\tfrac{1}{2}(x-y)\} - G\{-\tfrac{1}{2}(x-y)\} \geq G\{x\} - G\{y\}. \tag{25.5}$$

Applying (25.5) to (25.4), we have

$$G\{\tfrac{1}{2}(d_{\alpha_2} - d_\alpha)\} - G\{-\tfrac{1}{2}(d_{\alpha_2} - d_\alpha)\} > P_1 - P_2 \geq 0. \tag{25.6}$$

Thus if we choose d_{α_2} for each n so that

$$\lim_{n \to \infty} d_{\alpha_2} = d_\alpha, \tag{25.7}$$

the left-hand side of (25.6) will tend to zero, whence $P_1 - P_2$ will tend to zero uniformly in Δ. The two-tailed test will therefore be asymptotically UMP.

Now consider the ratio of Type II errors of the tests. From (25.1) and (25.2), we have

$$\frac{1-P_2}{1-P_1} = \frac{G\{d_{\alpha_2} - \Delta n^{\frac{1}{2}}\} - G\{-d_{\alpha_1} - \Delta n^{\frac{1}{2}}\}}{G\{d_\alpha - \Delta n^{\frac{1}{2}}\}}. \tag{25.8}$$

As $n^{\frac{1}{2}} \to \infty$, numerator and denominator of (25.8) tend to zero. Using L'Hôpital's rule, we find, using a prime to denote differentiation with respect to $n^{\frac{1}{2}}$ and writing g for the normal f.f.,

$$\lim_{n^{\frac{1}{2}} \to \infty} \frac{1-P_2}{1-P_1} = \lim_{n^{\frac{1}{2}} \to \infty} \left[\frac{(d'_{\alpha_2} - \Delta)g\{d_{\alpha_2} - \Delta n^{\frac{1}{2}}\}}{-\Delta g\{d_\alpha - \Delta n^{\frac{1}{2}}\}} + \frac{(d'_{\alpha_1} + \Delta)g\{-d_{\alpha_1} - \Delta n^{\frac{1}{2}}\}}{-\Delta g\{d_\alpha - \Delta n^{\frac{1}{2}}\}} \right]. \tag{25.9}$$

Now (25.7) implies that $d_{\alpha_1} \to \infty$ with n, and therefore that the second term on the right of (25.9) tends to zero: (25.7) also implies that the first term on the right of (25.9) will tend to infinity if

$$\lim_{n \to \infty} \frac{-d'_{\alpha_2} g\{d_{\alpha_2} - \Delta n^{\frac{1}{2}}\}}{g\{d_\alpha - \Delta n^{\frac{1}{2}}\}} = \lim_{n \to \infty} -d'_{\alpha_2} \frac{\exp\{-\tfrac{1}{2}(d_{\alpha_2} - \Delta n^{\frac{1}{2}})^2\}}{\exp\{-\tfrac{1}{2}(d_\alpha - \Delta n^{\frac{1}{2}})^2\}}$$

$$= \lim_{n \to \infty} -d'_{\alpha_2} \exp\{-\tfrac{1}{2}(d_{\alpha_2}^2 - d_\alpha^2) + \Delta n^{\frac{1}{2}}(d_{\alpha_2} - d_\alpha)\} \tag{25.10}$$

does so. By (25.7), the first term in the exponent on the right of (25.10) tends to zero. If we put

$$d_{\alpha_2} = d_\alpha + n^{-\delta}, \quad 0 < \delta < \tfrac{1}{2}, \tag{25.11}$$

(25.7) is satisfied and (25.10) tends to infinity with n. Thus, although both tests are asymptotically UMP, the ratio of Type II errors (25.8) tends to infinity with n. It is clear, therefore, that the criterion of being asymptotically UMP is not a very selective one.

Asymptotic relative efficiency

25.5 In order to obtain a useful asymptotic measure of test efficiency from the relative efficiency, we consider the limiting relative efficiency of tests against a sequence of alternative hypotheses in which θ approaches the value tested, θ_0, as n increases. We do this in order to avoid forcing the powers of tests to be nearly 1, as in **25.4**. This type of alternative was first investigated by Pitman (1948), whose work was generalized by Noether (1955). Other types of limiting process on relative efficiency are considered by Dixon (1953b) and Hodges and Lehmann (1956).

Let t_1 and t_2 be consistent test statistics for the hypothesis $H_0 : \theta = \theta_0$ against the one-sided alternative $H_1 : \theta > \theta_0$. We assume for the moment that t_1 and t_2 are asymptotically normally distributed whatever the value of θ—we shall relax this restriction in **25.14–15**. For brevity, we shall write

$$\left. \begin{aligned} E(t_i | H_j) &= E_{ij}, \\ \mathrm{var}(t_i | H_j) &= D_{ij}^2, \\ E_i^{(r)}(\theta) &= \frac{\partial^r}{\partial \theta^r} E_{i1}, \\ D_{i1}^{(r)} &= \frac{\partial^r}{\partial \theta^r} D_{i1}, \quad D_{i0}^{(r)} = D_{i1}^{(r)}(\theta_0), \end{aligned} \right\} \quad i = 1, 2; \; j = 0, 1.$$

Large-sample size-α tests are defined by the critical regions

$$t_i > E_{i0} + \lambda_\alpha D_{i0} \tag{25.12}$$

the sign of t_i being changed if necessary to make the region of this form), where λ_α is the normal deviate defined by $G\{-\lambda_\alpha\} = \alpha$, G being the standardized normal d.f. as before. Just as in Example 22.3, the asymptotic power function of t_i is

$$P_i(\theta) = G\{[E_{i1} - (E_{i0} + \lambda_\alpha D_{i0})]/D_{i1}\}. \tag{25.13}$$

Writing $u_i(\theta, \lambda_\alpha)$ for the argument of G in (25.13), we expand $(E_{i1} - E_{i0})$ in a Taylor series, obtaining

$$u_i(\theta, \lambda_\alpha) = \left[E_i^{(m_i)}(\theta_i^*) \frac{(\theta - \theta_0)^{m_i}}{m_i!} - \lambda_\alpha D_{i0} \right] \Big/ D_{i1}, \tag{25.14}$$

where $\theta_0 < \theta_i^* < \theta$ and m_i is the first non-zero derivative at θ_0, i.e., m_i is defined by

$$\left. \begin{aligned} E_i^{(r)}(\theta_0) &= 0, \quad r = 1, 2, \ldots, m_i - 1, \\ E_i^{(m_i)}(\theta_0) &\neq 0. \end{aligned} \right\} \tag{25.15}$$

In order to define the alternative hypothesis, we assume that, as $n \to \infty$,

$$R_i = [E_i^{(m_i)}(\theta_0)/D_{i0}] \sim c_i n^{m_i \delta_i}. \tag{25.16}$$

(25.16) defines the constants $\delta_i > 0$ and c_i. Now consider the sequences of alternatives, approaching θ_0 as $n \to \infty$,

$$\theta = \theta_0 + \frac{k_i}{n^{\delta_i}}, \tag{25.17}$$

where k_i is an arbitrary positive constant. If the regularity conditions

$$\lim_{n \to \infty} \frac{E_i^{(m_i)}(\theta)}{E_i^{(m_i)}(\theta_0)} = 1, \quad \lim_{n \to \infty} \frac{D_{i1}}{D_{i0}} = 1, \tag{25.18}$$

are satisfied, (25.16), (25.17) and (25.18) reduce (25.14) to

$$u_i(\theta, \lambda_\alpha) = \frac{c_i k_i^{m_i}}{m_i!} - \lambda_\alpha, \qquad (25.19)$$

and the asymptotic powers of the tests are $G\{u_i\}$ from (25.13).

25.6 If the two tests are to have equal power against the same sequence of alternatives for any fixed α, we must have, from (25.17) and (25.19),

$$\frac{k_1}{n_1^{\delta_1}} = \frac{k_2}{n_2^{\delta_2}} \qquad (25.20)$$

and

$$\frac{c_1 k_1^{m_1}}{m_1!} = \frac{c_2 k_2^{m_2}}{m_2!}, \qquad (25.21)$$

where n_1 and n_2 are the sample sizes upon which t_1 and t_2 are based. We combine (25.20) and (25.21) into

$$\frac{n_1^{\delta_1}}{n_2^{\delta_2}} = \left(\frac{c_2}{c_1} \frac{m_1!}{m_2!} k_2^{m_2 - m_1}\right)^{\frac{1}{m_1}}. \qquad (25.22)$$

The right-hand side of (25.22) is a positive constant. Thus if we let $n_1, n_2 \to \infty$, the ratio n_1/n_2 will tend to a constant if and only if $\delta_1 = \delta_2$. If $\delta_1 > \delta_2$, we must have $n_1/n_2 \to 0$, while if $\delta_1 < \delta_2$ we have $n_1/n_2 \to \infty$. If we define the *asymptotic relative efficiency* (ARE) of t_2 compared to t_1 as

$$A_{21} = \lim \frac{n_1}{n_2}, \qquad (25.23)$$

we therefore have the result

$$A_{21} = 0, \qquad \delta_1 > \delta_2. \qquad (25.24)$$

Thus to compare two tests by the criterion of ARE, we first compare their values of δ: if one has a smaller δ than the other, it has ARE of zero compared to the other. The value of δ plays the same role here as the order of magnitude of the variance plays in measuring efficiency of estimation (cf. **17.29**).

We may now confine ourselves to the case $\delta_1 = \delta_2 = \delta$. (25.22) and (25.23) then give

$$A_{21} = \lim \frac{n_1}{n_2} = \left(\frac{c_2}{c_1} \frac{m_1!}{m_2!} k_2^{m_2 - m_1}\right)^{1/(m_1 \delta)} \qquad (25.25)$$

If, in addition,

$$m_1 = m_2 = m, \qquad (25.26)$$

(25.25) reduces to

$$A_{21} = \left(\frac{c_2}{c_1}\right)^{1/(m\delta)}$$

which on using (25.16) becomes

$$A_{21} = \lim_{n \to \infty} \left\{\frac{E_2^{(m)}(\theta_0)/D_{20}}{E_1^{(m)}(\theta_0)/D_{10}}\right\}^{1/(m\delta)}. \qquad (25.27)$$

(25.27) is simple to evaluate in most cases, and we shall be using it extensively in later chapters to evaluate the ARE of particular tests. Most commonly, $\delta = \frac{1}{2}$ (corresponding to an estimation variance of order n^{-1}) and $m = 1$. For an interpretation of the value of m, see **25.10** below.

In passing, we may note that if $m_2 \neq m_1$, (25.25) is indeterminate, depending as it does on the arbitrary constant k_2. We therefore see that tests with equal values of δ do not have the same ARE against all sequences of alternatives (25.17) unless they also have equal values of m. We shall be commenting on the reasons for this in **25.10**.

25.7 If we wish to test H_0 against the two-sided $H_1: \theta \neq \theta_0$, our results for the ARE are unaffected if we use "equal-tails" critical regions of the form

$$t_i > E_{i0} + \lambda_{\frac{1}{2}\alpha} D_{i0} \quad \text{or} \quad t_i < E_{i0} - \lambda_{\frac{1}{2}\alpha} D_{i0},$$

for the asymptotic power functions (25.13) are replaced by

$$Q_i(\theta) = G\{u_i(\theta, \lambda_{\frac{1}{2}\alpha})\} + 1 - G\{u_i(\theta, -\lambda_{\frac{1}{2}\alpha})\}, \tag{25.28}$$

and $Q_1 = Q_2$ against the alternative (25.17) (where k_i need no longer be positive) is (25.20) and (25.21) hold, as before. Konijn (1956) gives a more general treatment of two-sided tests, which need not necessarily be "equal-tails" tests.

Example 25.2

Let us compare the sample median $\tilde{x}$ with the UMP sample mean $\bar{x}$ in testing the mean θ of a normal distribution with known variance σ^2. Both statistics are asymptotically normally distributed. We know that

$$E(\bar{x}) = \theta, \quad D^2(\bar{x}|\theta) = \sigma^2/n$$

and $\tilde{x}$ is a consistent estimator of θ, symmetrically distributed about θ, with

$$E(\tilde{x}) = \theta, \quad D^2(\tilde{x}|\theta) \sim \pi\sigma^2/(2n)$$

(cf. Example 10.7). Thus we have

$$E'(\theta_0) = 1$$

for both tests, so that $m_1 = m_2 = 1$, while from (25.16), $\delta_1 = \delta_2 = \frac{1}{2}$. Thus, from (25.27),

$$A_{\tilde{x},\bar{x}} = \lim_{n \to \infty} \left\{ \frac{1/(\pi\sigma^2/2n)^{\frac{1}{2}}}{1/(\sigma^2/n)^{\frac{1}{2}}} \right\}^2 = \frac{2}{\pi}.$$

This is precisely the result we obtained in Example 17.12 for the efficiency of $\tilde{x}$ in estimating θ. We shall see in **25.13** that this is a special case of a general relationship between estimating efficiency and ARE for tests.

ARE and the derivatives of the power functions

25.8 The nature of the sequence of alternative hypotheses (25.17), which approaches θ_0 as $n \to \infty$, makes it clear that the ARE is in some way related to the behaviour, near θ_0, of the power functions of the tests being compared. We shall make this relationship more precise by showing that, under certain conditions, the ARE is a simple function of the ratio of derivatives of the power functions.

We first treat the case of the one-sided H_1 discussed in **25.5–6**, where the power

functions of the tests are asymptotically given by (25.13), which we write, as before,
$$P_i(\theta) = G\{u_i(\theta, \lambda_\alpha)\}. \tag{25.29}$$
Differentiating with respect to θ, we have
$$P_i'(\theta) = g\{u_i\} u_i'(\theta, \lambda_\alpha), \tag{25.30}$$
where g is the normal frequency function. From (25.13) we find
$$u_i'(\theta, \lambda_\alpha) = \frac{E_{i1}'}{D_{i1}} - \frac{D_{i1}'}{D_{i1}^2}(E_{i1} - E_{i0} - \lambda_\alpha D_{i0}). \tag{25.31}$$
As $n \to \infty$, we find, using (25.18), and the further regularity conditions
$$\lim_{n \to \infty} \frac{D_{i1}'}{D_{i0}'} = 1, \quad \lim_{n \to \infty} \frac{E_{i1}}{E_{i0}} = 1,$$
that (25.31) becomes
$$u_i'(\theta, \lambda_\alpha) = \frac{E_i'(\theta_0)}{D_{i0}} + \frac{D_{i0}'}{D_{i0}} \lambda_\alpha, \tag{25.32}$$
so that if $m_i = 1$ in (25.15) and if
$$\lim_{n \to \infty} \frac{D_{i0}'}{E_i'(\theta_0)} = 0, \tag{25.33}$$
(25.32) reduces at θ_0 to
$$u_i'(\theta_0, \lambda_\alpha) \sim E_i'(\theta_0)/D_{i0}. \tag{25.34}$$
Since, from (25.13),
$$g\{u_i(\theta_0, \lambda_\alpha)\} = g\{-\lambda_\alpha\}, \tag{25.35}$$
(25.30) becomes, on substituting (25.34) and (25.35),
$$P_i'(\theta_0) = P_i'(\theta_0, \lambda_\alpha) \sim g\{-\lambda_\alpha\} E_i'(\theta_0)/D_{i0}. \tag{25.36}$$
Remembering that $m_i = 1$, we therefore have from (25.36) and (25.27)
$$\lim_{n \to \infty} \frac{P_2'(\theta_0)}{P_1'(\theta_0)} = A_{21}^\delta, \quad m_1 = m_2 = 1, \tag{25.37}$$
so that the asymptotic ratio of the first derivatives of the power functions of the tests at θ_0 is simply the ARE raised to the power δ (commonly $\tfrac{1}{2}$). Thus if we were to use this ratio as a criterion of asymptotic efficiency of tests, we should get precisely the same results as by using the ARE. This criterion was, in fact, proposed (under the name " asymptotic local efficiency ") by Blomqvist (1950).

25.9 If $m_i > 1$, i.e. $E_i'(\theta_0) = 0$, (25.36) is zero to our order of approximation and the result of **25.8** is of no use. The differentiation process has to be taken further to yield useful results.

From (25.30), we obtain
$$P_i''(\theta) = \frac{\partial g\{u_i\}}{\partial u_i}[u_i'(\theta, \lambda_\alpha)]^2 + g\{u_i\} u_i''(\theta, \lambda_\alpha). \tag{25.38}$$
From (25.31),
$$u_i''(\theta, \lambda_\alpha) = \frac{E_{i1}''}{D_{i1}} - \frac{2 E_{i1}' D_{i1}'}{D_{i1}^2} - (E_{i1} - E_{i0} - \lambda_\alpha D_{i0})\left[\frac{D_{i1}''}{D_{i1}^2} - \frac{2(D_{i1}')^2}{D_{i1}^3}\right]. \tag{25.39}$$

If (25.18) holds with $m_i = 2$ and also the regularity conditions below (25.31) and

$$\lim_{n \to \infty} E'_{i1} = E'_i(\theta_0) = 0, \quad \lim_{n \to \infty} \frac{D'_{i1}}{D'_{i0}} = 1, \quad \lim_{n \to \infty} \frac{D''_{i1}}{D''_{i0}} = 1, \quad \lim_{n \to \infty} \frac{E_{i1}}{E_{i0}} = 1, \qquad (25.40)$$

(25.39) gives

$$u''_i(\theta_0, \lambda_\alpha) = \frac{E''_i(\theta_0)}{D_{i0}} + \lambda_\alpha \left[\frac{D''_{i0}}{D_{i0}} - 2\left(\frac{D'_{i0}}{D_{i0}}\right)^2 \right]. \qquad (25.41)$$

Instead of (25.33), we now assume the conditions

$$\lim_{n \to \infty} \frac{D''_{i0}}{E''_i(\theta_0)} = 0, \quad \lim_{n \to \infty} \frac{(D'_{i0})^2}{D_{i0} E''_i(\theta_0)} = 0. \qquad (25.42)$$

(25.42) reduces (25.41) to

$$u''_i(\theta_0, \lambda_\alpha) \sim E''_i(\theta_0)/D_{i0}. \qquad (25.43)$$

Returning now to (25.38), we see that since

$$\frac{\partial g\{u_i\}}{\partial u_i} = -u_i g\{u_i\},$$

we have, using (25.32), (25.35) and (25.43) in (25.38),

$$P''_i(\theta_0) \sim g\{-\lambda_\alpha\} \left\{ \lambda_\alpha \left[\frac{E'_i(\theta_0)}{D_{i0}} + \frac{D'_{i0}}{D_{i0}} \lambda_\alpha \right]^2 + \frac{E''_i(\theta_0)}{D_{i0}} \right\}. \qquad (25.44)$$

Since we are considering the case $m_i = 2$ here, the term in $E'_i(\theta_0)$ is zero, and from the second condition of (25.42), (25.44) may finally be written

$$P''_i(\theta_0) \sim g\{-\lambda_\alpha\} E''_i(\theta_0)/D_{i0}, \qquad (25.45)$$

whence, with $m = 2$, (25.27) gives

$$\lim_{n \to \infty} \frac{P''_2(\theta_0)}{P''_1(\theta_0)} = A_{21}^{2\delta} \qquad (25.46)$$

for the limiting ratio of the second derivatives.

(25.37) and (25.46) may be expressed concisely by the statement that for $m = 1, 2$, the ratio of the mth derivatives of the power functions of one-sided tests is asymptotically equal to the ARE raised to the power $m\delta$.

If, instead of (25.33) and (25.42), we had imposed the stronger conditions

$$\lim_{n \to \infty} D'_{i0}/D_{i0} = 0, \quad \lim_{n \to \infty} D''_{i0}/D_{i0} = 0, \qquad (25.47)$$

which with (25.16) imply (25.33) and (25.42), (25.34) and (25.43) would have followed from (25.32) and (25.41) as before. (25.47) may be easier to verify in particular cases.

The interpretation of the value of m

25.10 We now discuss the general conditions under which m will take the value 1 or 2. Consider again the asymptotic power function (25.13) for a one-sided alternative $H_1: \theta > \theta_0$ and a one-tailed test (25.12). For brevity, we drop the suffix "i" in this section. If $\theta \to \theta_0$, and $D_1 \to D_0$ by (25.18), it becomes

$$P(\theta) = G\left\{ \frac{E_1 - E_0}{D_0} - \lambda_\alpha \right\},$$

a monotone increasing function of $(E_1 - E_0)$.

THE COMPARISON OF TESTS

If (E_1-E_0) is an increasing function of $(\theta-\theta_0)$, $P(\theta) \to 0$ as $\theta \to -\infty$ (which implies that the other "tail" of the distribution of the test statistic would be used as a critical region if $\theta < \theta_0$). If $E'(\theta_0)$ exists, it is non-zero and $m = 1$, and $P'(\theta_0) \neq 0$ also, by (25.36).

If, on the other hand, (E_1-E_0) is an even function of $(\theta-\theta_0)$, (which implies that the same "tail" would be used as critical region whatever the sign of $(\theta-\theta_0)$), and an increasing function of $|\theta-\theta_0|$, and $E'(\theta_0)$ exists, it must under regularity conditions equal zero, and $m > 1$—in practice, we find $m = 2$. By (25.36), $P'(\theta_0) = 0$ also to this order of approximation.

We are now in a position to see why, as remarked at the end of **25.6**, the ARE is not useful in comparing tests with differing values of m, which in practice are 1 and 2. For we are then comparing tests whose power functions behave essentially differently at θ_0, one having a regular minimum there and the other not. The indeterminacy of (25.25) in such circumstances is not really surprising. It should be added that this indeterminacy is, at the time of writing, of purely theoretical interest, since no case seems to be known to which it applies.

Example 25.3

Consider the problem of testing $H_0: \theta = \theta_0$ for a normal distribution with mean θ and variance 1. The pair of one-tailed tests based on the sample mean $\bar{x}$ are UMP (cf. **22.17**), the upper or lower tail being selected according to whether H_1 is $\theta > \theta_0$ or $\theta < \theta_0$. From Example 25.2, $\delta = \frac{1}{2}$ and $m = 1$ for $\bar{x}$.

We could also use as a test statistic

$$S = \sum_{i=1}^{n}(x_i-\theta_0)^2.$$

S has a non-central chi-squared distribution with n degrees of freedom and non-central parameter $n(\theta-\theta_0)^2$, so that (cf. Exercise 24.1)

$$E(S|\theta) = n\{1+(\theta-\theta_0)^2\},$$
$$D^2(S|\theta_0) = 2n,$$

and as $n \to \infty$, S is asymptotically normally distributed. We have $E'(\theta) = 2n(\theta-\theta_0)$, $E'(\theta_0) = 0$, $E''(\theta) = 2n = E''(\theta_0)$, so that $m = 2$ and

$$\frac{E''(\theta_0)}{D_0} = \frac{2n}{(2n)^{\frac{1}{2}}} = (2n)^{\frac{1}{2}}.$$

From (25.16), since $m = 2$, $\delta = \frac{1}{4}$. Since $\delta = \frac{1}{2}$ for $\bar{x}$, the ARE of S compared to $\bar{x}$ is zero by (25.24). The critical region for S consists of the upper tail, whatever the value of θ.

25.11 We now turn to the case of the two-sided alternative $H_1: \theta \neq \theta_0$. The power function of the "equal-tails" test is given asymptotically by (25.28). Its derivative at θ_0 is

$$Q'_i(\theta_0) = P'_i(\theta_0, \lambda_{\frac{1}{2}\alpha}) - P'_i(\theta_0, -\lambda_{\frac{1}{2}\alpha}), \qquad (25.48)$$

where P'_i is given by (25.36) if $m_i = 1$ and (25.33) or (25.47) holds. Since $g\{-\lambda_\alpha\}$ in (25.36) is an even function of λ_α, (25.48) immediately gives the asymptotic result

$$Q'_i(\theta_0) \sim 0$$

so that the slope of the power function at θ_0 is asymptotically zero. This result is also implied (under regularity conditions) by the remark in **24.17** concerning the asymptotic unbiassedness of consistent tests.

The second derivative of the power function (25.28) is

$$Q''_i(\theta_0) = P''_i(\theta_0, \lambda_{\frac{1}{2}\alpha}) - P''_i(\theta_0, -\lambda_{\frac{1}{2}\alpha}). \tag{25.49}$$

We have evaluated P''_i at (25.44) where we had $m_i = 2$. (25.44) still holds for $m_i = 1$ if we strengthen the first condition in (25.47) to

$$D'_{i0}/D_{i0} = o(n^{-\delta}), \tag{25.50}$$

for then by (25.16) the second term on the right of (25.39) may be neglected and we obtain (25.44) as before. Substituted into (25.49), it gives

$$Q''_i(\theta_0) \sim 2\lambda_{\frac{1}{2}\alpha} g\{-\lambda_{\frac{1}{2}\alpha}\} \left\{ \left(\frac{E'_i(\theta_0)}{D_{i0}}\right)^2 + \left(\frac{D'_{i0}}{D_{i0}}\lambda_\alpha\right)^2 \right\},$$

and (25.50) reduces this to

$$Q''_i(\theta_0) \sim 2\lambda_{\frac{1}{2}\alpha} g\{-\lambda_{\frac{1}{2}\alpha}\} \left(\frac{E'_i(\theta_0)}{D_{i0}}\right)^2. \tag{25.51}$$

In this case, therefore, (25.27) and (25.51) give

$$\frac{Q''_2(\theta_0)}{Q''_1(\theta_0)} = A^{2\delta}_{21}. \tag{25.52}$$

Thus for $m = 1$, the asymptotic ratio of second derivatives of the power functions of two-sided tests is exactly that given by (25.46) for one-sided tests when $m = 2$, and exactly the square of the one-sided test result for $m = 1$ at (25.37).

The case $m = 2$ does not seem of much importance for two-tailed tests: the remarks in **25.10** suggest that where $m = 2$ a one-tailed test would often be used even against a two-sided H_1.

Example 25.4

Reverting to Example 25.2, we saw that both tests have $\delta = \frac{1}{2}$, $m = 1$ and $E'(\theta_0) = 1$. Since the variance of each statistic is independent of θ, at least asymptotically, we see that (25.33) and (25.50) are satisfied and, the regularity conditions being satisfied, it follows from (25.37) that for one-sided tests

$$\lim_{n \to \infty} \frac{P'_{\tilde{x}}(\theta_0)}{P'_{\bar{x}}(\theta_0)} = A^{\frac{1}{2}}_{\tilde{x}, \bar{x}} = \left(\frac{2}{\pi}\right)^{\frac{1}{2}},$$

while for two-sided tests, from (25.52),

$$\lim_{n \to \infty} \frac{Q''_{\tilde{x}}(\theta_0)}{Q''_{\bar{x}}(\theta_0)} = A_{\tilde{x}, \bar{x}} = \frac{2}{\pi}.$$

The maximum power loss and the ARE

25.12 Although the ARE of tests essentially reflects their power properties in the neighbourhood of θ_0, it does have some implications for the asymptotic power function as a whole, at least for the case $m = 1$, to which we now confine ourselves.

The power function $P_i(\theta)$ of a one-sided test is $G\{u_i(\theta)\}$, where $u_i(\theta)$, given at (25.14), is asymptotically equal, under regularity conditions (25.18), to

$$u_i(\theta) = \frac{E_i'(\theta_0)}{D_{i0}}(\theta - \theta_0) - \lambda_\alpha, \qquad (25.53)$$

when $m_i = 1$. Thus $u_i(\theta)$ is asymptotically linear in θ. If we write $R_i = E_i'(\theta_0)/D_{i0}$ as at (25.16), we may write the difference between two such power functions as

$$d(\theta) = P_2(\theta) - P_1(\theta) = G\{(\theta - \theta_0)R_2 - \lambda_\alpha\} - G\left\{(\theta - \theta_0)R_2\frac{R_1}{R_2} - \lambda_\alpha\right\}, \qquad (25.54)$$

where we assume $R_2 > R_1$ without loss of generality. Consider the behaviour of $d(\theta)$ as a function of θ. When $\theta = \theta_0$, $d = 0$, and again as θ tends to infinity P_1 and P_2 both tend to 1 and d to zero. The maximum value of $d(\theta)$ depends only on the ratio R_1/R_2, for although R_2 appears in the right-hand side of (25.54) it is always the coefficient of $(\theta - \theta_0)$, which is being varied from 0 to ∞, so that $R_2(\theta - \theta_0)$ also goes from 0 to ∞ whatever the value of R_2. We therefore write $\Delta = R_2(\theta - \theta_0)$ in (25.54), obtaining

$$d(\Delta) = G\{\Delta - \lambda_\alpha\} - G\left\{\Delta\frac{R_1}{R_2} - \lambda_\alpha\right\}. \qquad (25.55)$$

The first derivative of (25.55) with respect to Δ is

$$d'(\Delta) = g\{\Delta - \lambda_\alpha\} - \frac{R_1}{R_2}g\left\{\Delta\frac{R_1}{R_2} - \lambda_\alpha\right\},$$

and if this is equated to zero, we have

$$\frac{R_1}{R_2} = \frac{g\{\Delta - \lambda_\alpha\}}{g\left\{\Delta\frac{R_1}{R_2} - \lambda_\alpha\right\}} = \exp\left\{-\tfrac{1}{2}(\Delta - \lambda_\alpha)^2 + \tfrac{1}{2}\left(\Delta\frac{R_1}{R_2} - \lambda_\alpha\right)^2\right\}$$

$$= \exp\left\{-\tfrac{1}{2}\Delta^2\left(1 - \frac{R_1^2}{R_2^2}\right) + \lambda_\alpha\Delta\left(1 - \frac{R_1}{R_2}\right)\right\}. \qquad (25.56)$$

(25.56) is a quadratic equation in Δ, whose only positive root is

$$\Delta = \frac{\lambda_\alpha + \left\{\lambda_\alpha^2 + 2\left(\dfrac{1 + \dfrac{R_1}{R_2}}{1 - \dfrac{R_1}{R_2}}\right)\log\dfrac{R_2}{R_1}\right\}^{\frac{1}{2}}}{\left(1 + \dfrac{R_1}{R_2}\right)}. \qquad (25.57)$$

This is the value at which (25.55) is maximized. Consider, for example, the case $\alpha = 0.05$ ($\lambda_\alpha = 1.645$) and $R_1/R_2 = 0.5$. (25.57) gives

$$\Delta = \frac{1.645 + \{1.645^2 + 6\log_e 2\}^{\frac{1}{2}}}{1.5} = 2.85.$$

(25.55) then gives, using tables of the normal d.f.,
$$P_2 = G\{2.85-1.64\} = G\{1.21\} = 0.89,$$
$$P_1 = G\{1.42-1.64\} = G\{-0.22\} = 0.41.$$

D. R. Cox and Stuart (1955) gave values of P_2 and P_1 at the point of maximum difference, obtained by the graphical equivalent of the above method, for a range of values of α and R_1/R_2. Their table is reproduced below, our worked example above being one of the entries.

Asymptotic powers per cent. of tests at the point of greatest difference
(*D. R. Cox and Stuart, 1955*)

α	0.10		0.05		0.01		0.001	
R_1/R_2	P_1	P_2	P_1	P_2	P_1	P_2	P_1	P_2
0.9	67	73	63	71	49	60	54	67
0.8	61	74	56	72	49	71	43	72
0.7	59	80	51	77	42	77	39	83
0.6	54	84	47	84	39	86	29	87
0.5	48	88	41	89	30	90	20	93
0.3	35	96	27	96	14	97	7	99

It will be seen from the table that as α decreases for fixed R_1/R_2, the maximum difference between the asymptotic power functions increases steadily—it can, in fact, be made as near to 1 as desired by taking α small enough. Similarly, for fixed α, the maximum difference increases steadily as R_1/R_2 falls.

The practical consequence of the table is that if R_1/R_2 is 0.9 or more, the loss of power *along the whole course* of the asymptotic power function will not exceed 0.08 for $\alpha = 0.05$, 0.11 for $\alpha = 0.01$, and 0.13 for $\alpha = 0.001$, the most commonly used test sizes. Since R_1/R_2 is, from (25.36), the ratio of first derivatives of the power functions, we have from (25.37) that $(R_1/R_2)^{1/\delta} = A_{12}$, where δ is commonly $\frac{1}{2}$, and thus the ARE needs to be $(0.9)^{1/\delta}$ for the statements above to be true.

ARE and estimating efficiency

25.13 There is a simple connexion between the ARE and estimating efficiency. If we have two consistent test statistics t_i as before, we define functions f_i, independent of n, such that the statistics

$$T_i = f_i(t_i) \tag{25.58}$$

are consistent *estimators* of θ. If we write

$$\theta = f_i(\tau_i), \tag{25.59}$$

it follows from (25.58) that since $T_i \to \theta$ in probability, $t_i \to \tau_i$ and $E(t_i)$ if it exists also tends to τ_i. Expanding (25.58) by Taylor's theorem about τ_i, we have, using (25.59),

$$T_i = \theta + (t_i - \tau_i)\left[\frac{\partial f(t_i)}{\partial \tau_i}\right]_{t_i = \tau_i} \tag{25.60}$$

where t_i^*, intermediate in value between t_i and τ_i, tends to τ_i as n increases. Thus (25.60) may be written

$$T_i - \theta \sim (t_i - \tau_i) \left[\frac{\partial \theta}{\partial E(t_i)} \right]$$

whence
$$\operatorname{var} T_i \sim \operatorname{var} t_i \bigg/ \left(\frac{\partial E(t_i)}{\partial \theta} \right)^2. \tag{25.61}$$

If 2δ is the order of magnitude in n of the variances of the T_i, the estimating efficiency of T_2 compared to T_1 is, by (17.65) and (25.61),

$$\lim_{n \to \infty} \left(\frac{\operatorname{var} T_1}{\operatorname{var} T_2} \right)^{1/(2\delta)} = \left[\frac{\{\partial E(t_2)/\partial \theta\}^2/\operatorname{var} t_2}{\{\partial E(t_1)/\partial \theta\}^2/\operatorname{var} t_1} \right]^{1/(2\delta)} \tag{25.62}$$

At θ_0, (25.62) is precisely equal to the ARE (25.27) when $m_i = 1$. Thus the ARE essentially gives the relative estimating efficiencies of transformations of the test statistics which are consistent estimators of the parameter concerned. But this correspondence is a local one: in **22.15** we saw that the connexion between estimating efficiency and power is not strong in general. It follows at once that tests based upon efficient estimators have maximum ARE and (from **25.8–11**) that the derivatives of their power functions at θ_0 are maximized. (25.62) and (17.61) also imply that if T_1 is efficient, $A_{21} = \{\rho(T_1, T_2)\}^{1/\delta}$. A more general result in terms of $\rho(t_1, t_2)$ is given in Exercise 25.9.

Example 25.5

The result we have just obtained explains the fact, noted in Example 25.2, that the ARE of the sample median, compared to the sample mean, in testing the mean of a normal distribution has exactly the same value as its estimating efficiency for that parameter.

Non-normal cases

25.14 From **25.5** onwards, we have confined ourselves to the case of asymptotically normally distributed test statistics. However, examination of **25.5–7** will show that in deriving the ARE we made no specific use of the normality assumption. We were concerned to establish the conditions under which the arguments u_i of the power functions $G\{u_i\}$ in (25.19) would be equal against the sequence of alternatives (25.17). G played no role in the discussion other than of ensuring that the asymptotic power functions were of the same form, and we need only require that G is a regularly behaved d.f.

It follows that if two tests have asymptotic power functions of any two-parameter form G, only one of whose parameters is a function of θ, the results of **25.5–7** will hold, for (25.17) will fix this parameter and u_i in (25.19) then determines the other. Given the form G, the critical region for one-tailed tests can always be put in the form (25.12), where λ_α is more generally interpreted as the multiple of D_{i0} required to make (25.12) a size-α critical region.

25.15 The only important limiting distributions other than the normal are the non-central χ^2 distributions whose properties were discussed in **24.4–5**. Suppose that for the hypothesis $H_0 : \theta = \theta_0$ we have two test statistics t_i with such distributions, the degrees of freedom being ν_i (independent of θ) and the non-central parameters $\lambda_i(\theta)$, where $\lambda_i(\theta_0) = 0$, so that the χ^2 distributions are central when H_0 holds. We have (cf. Exercise 24.1)

$$E_{i1} = \nu_i + \lambda_i(\theta), \quad D_{i0}^2 = 2\nu_i. \tag{25.63}$$

All the results of **25.5–6** for one-sided tests therefore hold for the comparison of test statistics distributed in the non-central χ^2 form (central when H_0 holds) with degrees of freedom independent of θ. In particular, when $\delta_1 = \delta_2 = \delta$ and $m_1 = m_2 = m$, (25.63) substituted into (25.27) gives

$$A_{21} = \lim_{n \to \infty} \left\{ \frac{\lambda_2^{(m)}(\theta_0)/\nu_2^{\frac{1}{2}}}{\lambda_1^{(m)}(\theta_0)/\nu_1^{\frac{1}{2}}} \right\}^{1/(m\delta)} \tag{25.64}$$

A different derivation of this result is given by E. J. Hannan (1956).

Other measures of test efficiency

25.16 Although in later chapters we shall use only the relative efficiency and the ARE as measures of test efficiency, we conclude this chapter by discussing two alternative methods which have been proposed.

Walsh (1946) proposed the comparison of two tests for fixed size α by a measure which takes into account the performance of the tests for all alternative hypothesis values of the parameter θ. If the tests t_i are based on sample sizes n_i and have power functions $P_i(\theta, n_i)$, the efficiency of t_2 compared to t_1 is $n_1/n_2 = e_{12}$ where

$$\int [P_1(\theta, n_1) - P_2(\theta, n_2)] d\theta = 0. \tag{25.65}$$

Thus, given one of the sample sizes (say, n_2), we choose n_1 so that the algebraic sum of the areas between the power functions is zero, and measure efficiency by n_1/n_2.

This measure removes the effect of θ from the table of triple entry required to compare two power functions, and does so in a reasonable way. However, e_{12} is still a function of α and, more important, of n_2. Moreover, the calculation of n_1/n_2 so that (25.65) is satisfied is inevitably tedious and probably accounts for the fact that this measure has rarely been used. As an asymptotic measure, however, it is equivalent to the use of the ARE, at least for asymptotically normally distributed test statistics with $m_i = 1$ in (25.15). For we then have, as in **25.12**,

$$P_i(\theta, n_i) = G\{(\theta - \theta_0)R_i - \lambda_\alpha\},$$

where $R_i = E'_i(\theta_0)/D_{i0}$ as at (25.16), and (25.65) then becomes

$$\int [G\{(\theta - \theta_0)R_1 - \lambda_\alpha\} - G\{(\theta - \theta_0)R_2 - \lambda_\alpha\}] d\theta = 0. \tag{25.66}$$

Clearly, (25.66) holds asymptotically only when $R_1 = R_2$, or, from (25.16),

$$\frac{R_1}{R_2} \sim \frac{c_1}{c_2}\left(\frac{n_1}{n_2}\right)^\delta = 1$$

whence

$$\lim \frac{n_1}{n_2} = \left(\frac{c_2}{c_1}\right)^{1/\delta} = A_{21},$$

exactly as at (25.27) with $m = 1$.

25.17 Finally, we summarize a quite different approach to the problem of measuring asymptotic efficiency for tests, due to Chernoff (1952). For a variate x with moment-generating function $M_x(t) = E(e^{xt})$, we define

$$m(a) = \inf_t M_{x-a}(t), \tag{25.67}$$

the absolute minimum value of the m.g.f. of $(x-a)$. If $E(x \mid H_i) = \mu_i$ for simple hypotheses H_0, H_1, we further define

$$\rho = \inf_{\mu_0 < a < \mu_1} \max\{m_0(a), m_1(a)\}, \tag{25.68}$$

where the suffix to m, defined at (25.67), indicates the hypothesis. For a one-sided test of H_0 against H_1 based on a sum of n identically distributed x_i, with size α and power $1-\beta$, Chernoff shows that if *any* linear function $l(\alpha,\beta)$ of the probabilities of error α and β is minimized, its minimum value behaves as $n \to \infty$ like ρ^n, where ρ is defined by (25.68). Consider two such tests t_i, based on samples of size n_i. If they have equal minima for $l(\alpha,\beta)$, we therefore have

$$\lim \frac{\rho_1^{n_1}}{\rho_2^{n_2}} = 1$$

or

$$\lim \frac{n_1}{n_2} = \frac{\log \rho_2}{\log \rho_1}. \tag{25.69}$$

Thus the right-hand side of (25.69) is a measure of the asymptotic efficiency of t_2 compared to t_1. Its use is restricted to test statistics based on sums of independent observations, and the computation required may be considerable.

25.18 Hoeffding (1965) develops a method of comparison of tests when $\alpha \to 0$ as $n \to \infty$ (as distinct from the approach of **25.5** onwards, where α is held fixed and $H_1 \to H_0$ as $n \to \infty$) and shows using this method that LR tests have an optimum property for tests on multinomial distributions.

25.19 Bahadur (1967) reviews results (largely his own) on the comparison of tests by means of the rate of convergence to zero of the maximum size of critical region that includes the observed value of the test statistic. Dempster and Schatzoff (1965) use the expected value of this maximum size as a criterion for choosing between tests. See also Joiner (1969), who compares these with other methods including that of Exercise 25.10, and Sievers (1969). L. D. Brown (1971) shows that appropriate LR tests are asymptotically optimum in Bahadur's sense.

EXERCISES

25.1 The Sign Test for the hypothesis H_0 that a population median takes a specified value θ_0 consists of counting the number of sample observations exceeding θ_0 and rejecting H_0 when this number is too large. Show that for a normal population this test has ARE $2/\pi$ compared to the "Student's" t-test for H_0, and connect this with the result of Example 25.2.

(Cochran, 1937)

25.2 Generalizing the result of Exercise 25.1, show that for any continuous frequency function f with variance σ^2, the ARE of the Sign Test compared to the t-test is $4\sigma^2\{f(\theta_0)\}^2$.

(Pitman, 1948)

25.3 The difference between the means of two normal populations with equal variances is tested from two independent samples by comparing every observation y_j in the second sample with every observation x_i in the first sample, and counting the number of times a y_j exceeds an x_i. Show that the ARE of this S-test compared to the two-sample "Student's" t-test is $3/\pi$.

(Pitman, 1948)

25.4 Generalizing Exercise 25.3, show that if any two continuous frequency functions $f(x), f(x-\theta)$, differ only by a location parameter θ, and have variance σ^2, the ARE of the S-test compared to the t-test is

$$12\sigma^2 \left\{ \int_{-\infty}^{\infty} \{f(x)\}^2 dx \right\}^2.$$

(Pitman, 1948)

25.5 If x is normally distributed with mean μ_i and variance σ_i^2, given H_i ($i = 0, 1$; $\mu_0 < \mu_1$), show that (25.68) has the value

$$\rho = \exp\{-\tfrac{1}{2}[(\mu_1-\mu_0)/(\sigma_1+\sigma_0)]^2\}.$$

(Chernoff, 1952)

25.6 If x/σ_i^2 has a chi-squared distribution with r degrees of freedom, given H_i ($i = 0, 1$; $\sigma_0^2/\sigma_1^2 = \tau < 1$), show that ρ in (25.68) satisfies

$$\log \rho = -\tfrac{1}{2}r(\delta - 1 - \log \delta)$$

where

$$\delta = (\log \tau)/(\tau - 1).$$

(Chernoff, 1952)

25.7 t_1 and t_2 are unbiassed estimators of θ, jointly normally distributed in large samples with variances σ^2, σ^2/e respectively ($0 < e \leq 1$). Using the results of **16.23** and **17.29**, show that

$$E(t_1 | t_2) = \theta(1-e) + t_2 e,$$

and hence that if t_2 is observed to differ from θ by a multiple d of its standard deviation we expect t_1 to differ from θ by a multiple $de^{\frac{1}{2}}$ of its standard deviation.

(D. R. Cox, 1956)

25.8 Using Exercise 25.7, show that if t_2 is used to test $H_0 : \theta = \theta_0$, we may calculate the "expected result" of a test based on the more efficient statistic t_1. In particular, show that if a one-tail test of size 0·01, using t_2, rejects H_0, we should expect a one-tail test of size 0·05, using t_1, to do so if $e > 0·50$; while if an "equal-tails" size-0·01 test on t_2 rejects H_0, we should expect an "equal-tails" size-0·05 test on t_1 to do so if $e > 0·58$.

25.9 Let t_1 be a statistic with maximum ARE and t_2 any other test statistic for the same problem with δ and m as for t_1. By considering
$$t_3 = a\frac{t_1}{D_{10}} + (1-a)\frac{t_2}{D_{20}},$$
show that, in (25.16),
$$R_3 = \{aR_1 + (1-a)R_2\}/\{a^2 + (1-a)^2 + 2a(1-a)\rho\}^{\frac{1}{2}},$$
where ρ is the asymptotic correlation coefficient of t_1 and t_2. Hence show that $\rho = R_2/R_1$, the $(m\delta)$th power of the ARE (25.27). Cf. (17.61) for estimators.

(Cf. van Eeden, 1963)

25.10 In **25.5**, let $\delta_1 = \delta_2 = \delta$ and $m_1 = m_2 = m$. It is proposed to measure the efficiency of the tests t_1, t_2 by the reciprocal of the ratio of the distances $(\theta - \theta_0)$ which they require for $E(t|\theta)$ to fall on the boundary of the critical region (25.12). Show that this is approximately the same as using the δth power of the ARE.

(This *average critical value* method is due to R. C. Geary—cf. Stuart (1967).)

CHAPTER 26

STATISTICAL RELATIONSHIP: LINEAR REGRESSION AND CORRELATION

26.1 For this and the next three chapters we shall be concerned with one or another aspect of the relationships between two or more variables. We have already, at various points in our exposition, discussed bivariate and multivariate distributions, their moments and cumulants; in particular, we have discussed the properties of bivariate and multivariate normal distributions. However, a systematic discussion of the relationships between variables was deferred until the theory of estimation and testing hypotheses had been explored. Even in this group of four chapters, we shall not be able to address ourselves to the whole problem, the more complicated distributional problems of three or more variables being deferred until we discuss Multivariate Analysis in Volume 3.

26.2 Even so, the area which we are about to study is a very large one, and it will be helpful if we begin by reviewing it in a general way.

Most of our work stems from an interest in the joint distribution of a pair of random variables: we may describe this as the problem of *statistical relationship*. There is a quite distinct field of interest concerning relationships of a strictly functional kind between variables, such as those of classical physics; this subject is of statistical interest because the functionally related variables are subject to observational or instrumental errors. We call this the problem of *functional relationship*, and discuss it in Chapter 29 below. Before we reach that chapter, we shall be concerned with the problem of statistical relationship alone, where the variables are not (except in degenerate cases) functionally related, although they may also be subject to observational or instrumental errors; we regard them simply as members of a distributional complex.

26.3 Within the field of statistical relationship there is a further useful distinction to be made. We may be interested either in the *interdependence* between a number (not necessarily all) of our variables or in the *dependence* of one or more variables upon others. For example, we may be interested in whether there is a relationship between length of arm and length of leg in men; put this way, it is a problem of interdependence. But if we are interested in using leg-length measurements to convey information about arm-length, we are considering the dependence of the latter upon the former. This is a case in which either interdependence or dependence may be of interest. On the other hand, there are situations when only dependence is of interest. The relationship of crop-yields and rainfall is an example in which non-statistical considerations make it clear that there is an essential asymmetry in the situation: we say, loosely, that rainfall " causes " crop-yield to vary, and we are quite certain that crops do not affect the rainfall, so we measure the dependence of yield upon rainfall.

There is no clear-cut distinction in statistical terminology for the techniques appropriate to these essentially different types of problem. For example, we shall see in Chapter 27 that if we are interested in the interdependence of two variables with the effects of other variables eliminated, we use the method called " partial correlation," while if we are interested in the dependence of a single variable upon a group of others, we use " multiple correlation." Nevertheless, it is true in the main that the study of *interdependence* leads to the theory of correlation dealt with in Chapters 26–27, while the study of *dependence* leads to the theory of regression discussed in these chapters and in Chapter 28.

26.4 Before proceeding to the exposition of the theory of correlation (largely developed around the beginning of this century by Karl Pearson and by Yule), which will occupy most of this chapter, we make one final general point. A statistical relationship, however strong and however suggestive, can never *establish* a causal connexion : our ideas on causation must come from outside statistics, ultimately from some theory or other. Even in the simple example of crop-yield and rainfall discussed in **26.3**, we had no *statistical* reason for dismissing the idea of dependence of rainfall upon crop-yield : the dismissal is based on quite different considerations. Even if rainfall and crop-yields were in perfect functional correspondence, we should not dream of reversing the " obvious " causal connexion. We need not enter into the philosophical implications of this ; for our purposes, we need only reiterate that statistical relationship, of whatever kind, cannot logically imply causation.

G. B. Shaw made this point brilliantly in his Preface to *The Doctor's Dilemma* (1906) : " Even trained statisticians often fail to appreciate the extent to which statistics are vitiated by the unrecorded assumptions of their interpreters . . . It is easy to prove that the wearing of tall hats and the carrying of umbrellas enlarges the chest, prolongs life, and confers comparative immunity from disease. . . . A university degree, a daily bath, the owning of thirty pairs of trousers, a knowledge of Wagner's music, a pew in church, anything, in short, that implies more means and better nurture . . . can be statistically palmed off as a magic-spell conferring all sorts of privileges. . . . The mathematician whose correlations would fill a Newton with admiration, may, in collecting and accepting data and drawing conclusions from them, fall into quite crude errors by just such popular oversights as I have been describing."

Although Shaw was on this occasion supporting a characteristically doubtful cause, his logic was valid. In the first flush of enthusiasm for correlation techniques, it was easy for early followers of Karl Pearson and Yule to be incautious. It was not until twenty years after Shaw wrote that Yule (1926) frightened statisticians by adducing cases of very high correlations which were obviously not causal : e.g. the annual suicide rate was highly correlated with the membership of the Church of England. Most of these " nonsense " correlations operate through concomitant variation in time, and they had the salutary effect of bringing home to the statistician that causation cannot be deduced from any observed co-variation, however close. Now, half a century later, the reaction has perhaps gone too far : correlation analysis is very unfashionable among statisticians. Yet there are large fields of application (the social sciences and psychology, for example) where patterns of causation are not yet sufficiently well understood for correlation analysis to be replaced by more specifically structured statistical methods, and also large areas of multivariate analysis where the computation of what is in effect a matrix of correlation coefficients is a necessary prelude to the detailed statistical analysis; on both these accounts, some study of the subject is necessary.

26.5 In Chapter 1 (Tables 1.15, 1.23 and 1.24) we gave a few examples of bivariate distributions arising in practice. Tables 26.1 and 26.2 give two further examples which will be used for illustrative purposes.

Table 26.1—Distribution of weight and stature for 4,995 women in Great Britain, 1951
Reproduced, by permission, from *Women's Measurements and Sizes*, London, H.M.S.O., 1957

Weight (y): central values of groups, in pounds	\multicolumn{11}{c	}{Stature (x): central values of groups in inches}	Total
	54	56	58	60	62	64	66	68	70	72	74	
278·5						1						1
272·5												—
266·5						1						1
260·5							1					1
254·5												—
248·5					1	1						2
242·5							1					1
236·5							1					1
230·5					2				1			3
224·5					1	2	1					4
218·5			1		2	1		1				5
212·5			2	1	6		1	1				11
206·5			2	2	3	2		1				10
200·5		4	2	6	2							14
194·5			1	3	7	7	4	1				23
188·5		1	5	14	8	12	3	1	2			46
182·5		1	7	12	26	9	5		1	2		63
176·5		5	8	18	21	15	11	7		2		87
170·5		2	11	17	44	21	13	3	1			112
164·5	1	3	12	35	48	30	15	5	3			152
158·5		8	17	52	42	36	21	9				185
152·5	1	7	30	81	71	58	21	2	2			273
146·5	2	13	36	76	91	82	36	8	1			345
140·5	1	6	55	101	138	89	50	8				448
134·5		15	64	95	175	122	45	5				521
128·5	1	19	73	155	207	101	25	3				584
122·5	3	34	91	168	200	81	12	1	1			591
116·5	3	24	108	184	184	50	8					561
110·5	5	33	119	165	124	22	4					472
104·5	1	3	33	87	95	35	6					260
98·5	2	5	29	59	45	16	3					159
92·5		6	10	21	9							46
86·5		1	5	3								9
80·5	2	1	1									4
Total	5	33	254	813	1340	1454	750	275	56	11	4	4995

Table 26.2—Distribution of bust girth and stature for 4,995 women in Great Britain, 1951

Data from same source as those of Table 26.1

Bust girth (y) central values of groups, in inches	\multicolumn{11}{c	}{Stature (x): central values of groups, in inches}	Total
	54	56	58	60	62	64	66	68	70	72	74	
56					1							1
54					1	2						3
52			1		3	4	1		1			10
50			1	3	5	4	1					14
48		1	3	9	7	6	3	1				30
46			4	11	17	17	7		1			57
44		2	11	26	50	45	17	10	1			162
42		2	11	42	85	73	31	12	3	2		261
40		2	20	76	132	131	71	31	9	4	3	479
38		2	36	98	158	203	126	65	17	3	1	709
36		6	48	188	317	410	263	89	15	1		1337
34	1	9	67	210	376	427	196	59	8			1353
32	3	5	39	131	163	122	31	8	1	1		504
30	1	4	11	18	25	10	2					71
28				2	1		1					4
Total	5	33	254	813	1340	1454	750	275	56	11	4	4995

For the moment, we treat these data as populations, leaving aside the question of sampling until later in the chapter.

Just as, for univariate distributions, we constructed summarizing constants such as the mean, variance, etc., we should like to summarize the relationship between the variables, and in particular their interdependence. Summarizing constants for a bivariate distribution arise naturally from the following considerations.

We call the two variables x, y. For any given value of x, say X, the distribution of y is called a y-array. The y-array is, of course, the conditional distribution of y given that $x = X$. This conditional distribution has a mean which we write

$$\bar{y}_X = E(y \mid X), \tag{26.1}$$

which will be a function of X, and vary with it. Similarly, by considering the x-array for $y = Y$, we have

$$\bar{x}_Y = E(x \mid Y). \tag{26.2}$$

(26.1) and (26.2) are called the *regression curves* (or, more shortly, the *regressions*) of y on x and of x on y respectively.

Although we have done so here for explicitness, we shall not use a capital letter to denote the variable being held constant where the context makes the notation $E(y \mid x)$, $E(x \mid y)$ clear.

Fig. 26.1 and 26.2 show, for the data of Tables 26.1 and 26.2, the means of y-arrays (marked by crosses) and of x-arrays (marked by circles). Lines CC' and RR' have been drawn to fit the array-means as closely as possible for straight lines, in the sense of Least Squares—cf. **26.8**. These diagrams summarize the properties of a bivariate distribution in the same way that a mean summarizes a univariate distribution.

294   THE ADVANCED THEORY OF STATISTICS

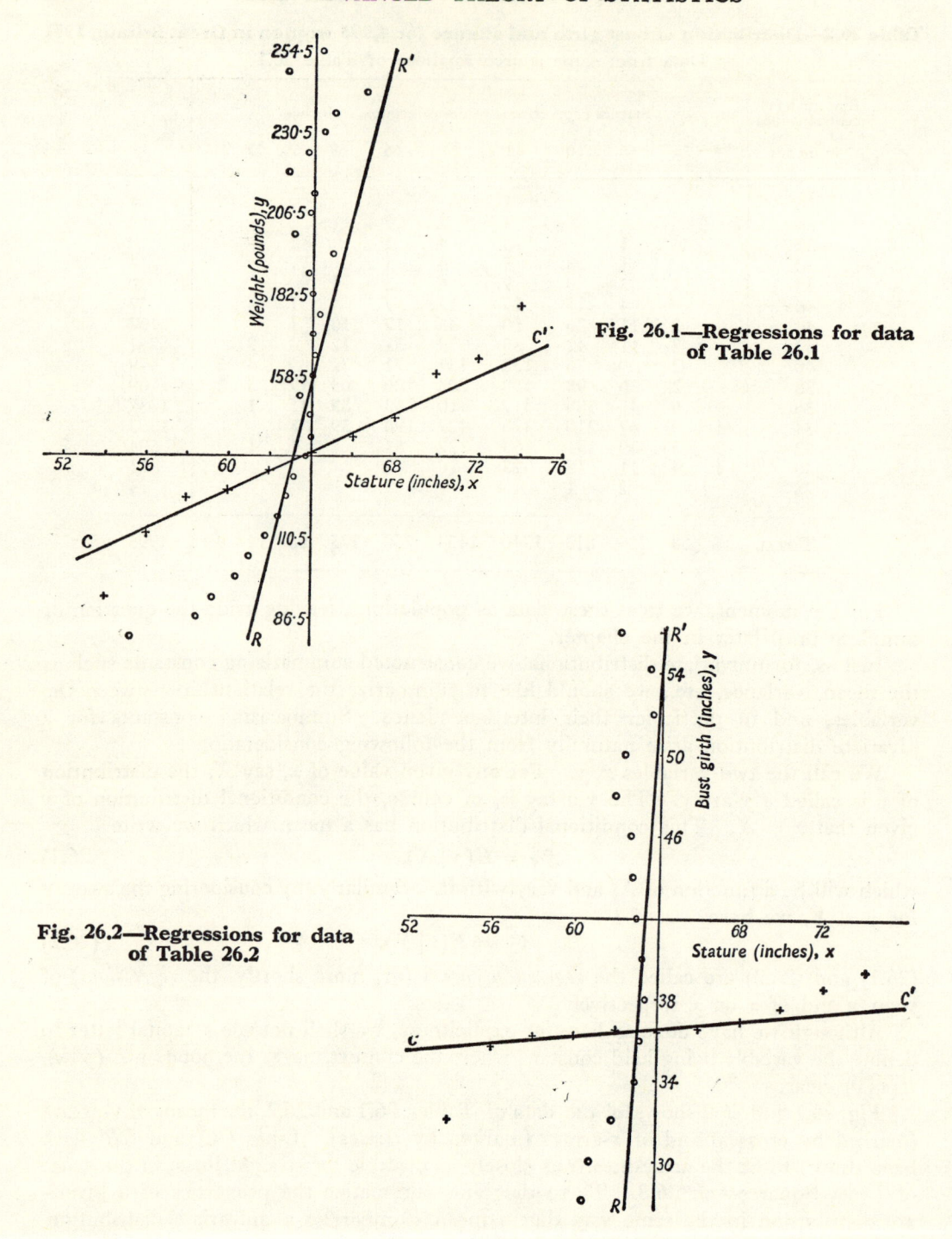

Fig. 26.1—Regressions for data of Table 26.1

Fig. 26.2—Regressions for data of Table 26.2

Since these are grouped data, the only arrays for which we have information are those corresponding to the grouped values of x and y. Conventionally, the y-arrays are taken to refer to the central value of the x-group within which they are observed, and similarly for the x-arrays.

We shall study regression in its own right in Chapter 28—here we are using the ideas of linear regression mainly as an introduction to a measure of interdependence, the coefficient of (product-moment) correlation, though we shall also take the opportunity to complete our study of the bivariate normal distribution from the standpoint of regression and correlation.

Covariance and regression

26.6 It is natural to consider using as the basis of a measure of dependence the product-moment μ_{11}, which we have encountered several times already in earlier chapters. μ_{11}, which is known as the *covariance* of x and y, is defined for a discrete population by

$$\mu_{11} = \sum_{i=1}^{n}(x_i-\mu_x)(y_i-\mu_y)/n \equiv \sum_{i=1}^{n} x_i y_i/n - \mu_x \mu_y, \qquad (26.3)$$

where n is the number of pairs of values x, y, and μ_x, μ_y are the means of x, y. For a continuous population defined by

$$dF(x,y) = f(x,y)\,dx\,dy, \qquad (26.4)$$

the corresponding expression is (cf. **3.27**)

$$\mu_{11} = \int_{-\infty}^{\infty}\int_{-\infty}^{\infty}(x-\mu_x)(y-\mu_y)\,dF(x,y)$$

$$= E\{(x-\mu_x)(y-\mu_y)\} \equiv E(xy) - E(x)E(y). \qquad (26.5)$$

If the variates x, y are independent,

$$\mu_{11} = \kappa_{11} = 0, \qquad (26.6)$$

as we saw in Example 12.7. By that Example, too, the converse is not generally true: (26.6) does not generally imply independence, which requires

$$\kappa_{rs} = 0 \quad \text{for all } r \times s \neq 0. \qquad (26.7)$$

For a bivariate *normal* distribution, however, we know that $\kappa_{rs} = 0$ for all $r+s > 2$, so that κ_{11} is the only non-zero product-cumulant. Thus (26.6) implies (26.7) and independence for normal variables. It may also do so for other specified distributions, but it does not in general do so. Example 26.1 gives a non-normal distribution for which $\kappa_{11} = 0$ implies independence; Example 26.2 gives one where it does not.

Example 26.1

If x and y are bivariate normally distributed standardized variables, the joint characteristic function of x^2 and y^2 is

$$\phi(t,u) = \frac{1}{2\pi(1-\rho^2)^{\frac{1}{2}}}\int_{-\infty}^{\infty}\int_{-\infty}^{\infty} \exp\left\{-\frac{1}{2(1-\rho^2)}[x^2\{1-2(1-\rho^2)it\}-2\rho xy\right.$$

$$\left.+y^2\{1-2(1-\rho^2)iu\}]\right\}dx\,dy.$$

The integral is, by Exercise 1.5, equal to

$$\pi \, 2(1-\rho^2) \begin{vmatrix} 1-2(1-\rho^2)it & \rho \\ \rho & 1-2(1-\rho^2)iu \end{vmatrix}^{-\frac{1}{2}}$$

so that

$$\phi(t,u) = (1-\rho^2)^{\frac{1}{2}}[\{1-2(1-\rho^2)it\}\{1-2(1-\rho^2)iu\}-\rho^2]^{-\frac{1}{2}}$$
$$= [(1-2it)(1-2iu)+4\rho^2 tu]^{-\frac{1}{2}}.$$

We see that

$$\phi(t,0) = (1-2it)^{-\frac{1}{2}},$$
$$\phi(0,u) = (1-2iu)^{-\frac{1}{2}},$$

so that the marginal distributions are chi-squares with one degree of freedom, as we know. By differentiating the logarithm of $\phi(t,u)$ we find

$$\mu_{11} = \kappa_{11} = \left[\frac{\partial^2 \log \phi(t,u)}{\partial(it)\partial(iu)}\right]_{t=u=0} = 2\rho^2.$$

Now when $\rho = 0$, we see that

$$\phi(t,u) = \phi(t,0)\phi(0,u),$$

a necessary and sufficient condition for independence of x^2 and y^2 by **4.16–17**. Thus $\mu_{11} = 0$ implies independence in this case.

Example 26.2

Consider a bivariate distribution with uniform probability over a unit circle centred at the means of x and y. We have

$$dF(x,y) = dx\,dy/\pi, \qquad 0 \leqslant x^2+y^2 \leqslant 1,$$

whence

$$\mu_{11} = \iint xy\,dF = \frac{1}{\pi}\iint xy\,dx\,dy$$
$$= \frac{1}{\pi}\int x\left[\int y\,dy\right]dx$$
$$= \frac{1}{\pi}\int x\left[\frac{1}{2}y^2\right]_{-(1-x^2)^{1/2}}^{+(1-x^2)^{1/2}} dx = 0,$$

as is otherwise obvious. But clearly x and y are not independent, since the range of variation of each depends on the value of the other.

Linear regression

26.7 If the regression of x on y, defined at (26.2), is exactly linear, we have the equation

$$E(x|y) = \alpha_1 + \beta_1 y, \qquad (26.8)$$

in which we now determine α_1 and β_1. Taking expectations on both sides of (26.8) with respect to y, we find

$$\mu_x = \alpha_1 + \beta_1 \mu_y. \qquad (26.9)$$

If we subtract (26.9) from (26.8), multiply both sides by $(y-\mu_y)$ and take expectations again, we find
$$E\{(x-\mu_x)(y-\mu_y)\} = \beta_1 E\{(y-\mu_y)^2\},$$
or
$$\beta_1 = \mu_{11}/\sigma_2^2, \qquad (26.10)$$
where σ_2^2 is the variance of y. Similarly, we obtain from (26.1)
$$\beta_2 = \mu_{11}/\sigma_1^2 \qquad (26.11)$$
for the coefficient in an exactly linear regression of y on x. (26.10) and (26.11) define the (linear) *regression coefficients*[*] of x on y (β_1) and of y on x (β_2). Using (26.8), (26.9) and (26.10), we have
$$E(x|y)-\mu_x = \beta_1(y-\mu_y) \qquad (26.12)$$
and similarly
$$E(y|x)-\mu_y = \beta_2(x-\mu_x). \qquad (26.13)$$
(26.12) and (26.13) are the *linear regression equations*.

We have already encountered a case of exact linear regressions in our discussion of the bivariate normal distribution in **16.23**.

Example 26.3

The regressions of x^2 and y^2 on each other in Example 26.1 are strictly linear. For, from **16.23**, putting $\sigma_1 = \sigma_2 = 1$ in (16.46), we have
$$E(x|y) = \rho y,$$
$$\text{var}(x|y) = 1-\rho^2.$$
Thus
$$E(x^2|y) \equiv \text{var}(x|y) + \{E(x|y)\}^2$$
$$= 1-\rho^2+\rho^2 y^2.$$
To each value of y^2 there correspond values $+y$ and $-y$ which occur with equal probability. Thus, since $E(x^2|y)$ is a function of y^2 only,
$$E(x^2|y^2) = \tfrac{1}{2}\{E(x^2|y)+E(x^2|-y)\} = E(x^2|y) = 1-\rho^2+\rho^2 y^2,$$
which we may rewrite, in the form (26.12),
$$E(x^2|y^2)-1 = \rho^2(y^2-1),$$
and the regression of y^2 on x^2 is strictly linear. Similarly
$$E(y^2|x^2)-1 = \rho^2(x^2-1).$$
Since we saw in Example 26.1 that $\mu_{11} = 2\rho^2$, and we know that the variances $= 2$, since these are chi-squared distributions with one degree of freedom, we may confirm from (26.10) and (26.11) that ρ^2 is the regression coefficient in each of the linear regression equations.

---

[*] The notation β_1, β_2 is unconnected with the symbolism for skewness and kurtosis in **3.31-2**; they are unlikely to be confused, since they arise in different contexts.

Example 26.4

In Example 26.2 it is easily seen that $E(x|y) = E(y|x) = 0$, so that we have linear regressions here, too, the coefficients being zero.

Example 26.5

Consider the variables x and y^2 in Example 26.3. We saw there that the regression of y^2 is linear on x^2, with coefficient ρ^2, and it is therefore not linear on x when $\rho \neq 0$. However, since $E(x|y) = \rho y$ we have

$$E(x|y^2) = \tfrac{1}{2}\{E(x|y)+E(x|-y)\} = 0,$$

so that the regression of x on y^2 is linear with regression coefficient zero.

Approximate linear regression: Least Squares

26.8 Examples 26.3–5 give instances where one or both regressions are exactly linear. When the population is an observed and not a theoretical one, however (and *a fortiori* when we have to take sampling fluctuation into account), it is very rare to find an exactly linear regression. Nevertheless, as in Fig. 26.1, the regression may be near enough to the linear form for us to wish to use a linear regression as an approximation. We are therefore led to the problem of " fitting " a straight line to the regression curve of y on x.

When there are no sampling considerations involved, the choice of a method of fitting is essentially arbitrary, in exactly the same way that, from the point of view of the description of data, the choice between mean and median as a measure of location is arbitrary. If we are fitting the regression of y on x, it is clearly desirable that in some sense the deviations of the points (y, x) from the fitted line should be small if the line is to represent them adequately. We might consider choosing the line to minimize the sum of the absolute deviations of the points from the line, but this gives rise to the usual mathematical difficulties accompanying an expression involving a modulus sign. Just as these difficulties lead us to prefer the standard deviation to the mean deviation as a measure of dispersion, they lead us here to propose that the sum of *squares* of the deviations of the points should be minimized.

We have still to determine how the deviations are to be taken: in the y-direction, the x-direction, or as " normal " deviations obtained by dropping a perpendicular from each point to the line. As we are considering the dependence of y on x, it seems natural to minimize the sum of squared deviations in the y-direction. Thus we are led back to the Method of Least Squares: we choose the " best-fitting " regression line of y on x,

$$y = \alpha_2 + \beta_2 x, \qquad (26.14)$$

so that the sum of squared deviations of the n observations from the fitted regression line, i.e.

$$S = \sum_{i=1}^{n} \{y_i - (\alpha_2 + \beta_2 x_i)\}^2 \qquad (26.15)$$

is minimized. The problem is to determine α_2 and β_2. We have already considered a much more general form of this problem in **19.4**. In the matrix notation we used there, (26.14) is

$$\underset{(n\times 1)}{\mathbf{y}} = \underset{(n\times 2)}{(\mathbf{1} \vdots \mathbf{x})} \underset{(2\times 1)}{\begin{pmatrix}\alpha_2\\ \beta_2\end{pmatrix}}$$

where **1** is a $(n\times 1)$ vector of units. The solution is, from (19.12),

$$\begin{pmatrix}\alpha_2\\ \beta_2\end{pmatrix} = \{(\mathbf{1} \vdots \mathbf{x})'(\mathbf{1} \vdots \mathbf{x})\}^{-1}(\mathbf{1} \vdots \mathbf{x})'\mathbf{y}$$

$$= \begin{pmatrix}n & \Sigma x\\ \Sigma x & \Sigma x^2\end{pmatrix}^{-1}\begin{pmatrix}\Sigma y\\ \Sigma xy\end{pmatrix}$$

$$= \frac{1}{n\Sigma x^2-(\Sigma x)^2}\begin{pmatrix}\Sigma x^2\Sigma y-\Sigma x\Sigma xy\\ n\Sigma xy-\Sigma x\Sigma y\end{pmatrix}.$$

Thus

$$\beta_2 = \frac{n\Sigma xy-\Sigma x\Sigma y}{n\Sigma x^2-(\Sigma x)^2} = \frac{\mu_{11}}{\sigma_1^2},$$

just as at (26.11) for the case of exact linearity of regression; while

$$\alpha_2 = \frac{\Sigma x^2\Sigma y-\Sigma x\Sigma xy}{n\Sigma x^2-(\Sigma x)^2} = \mu_y - \beta_2\mu_x,$$

the equivalent of (26.9). Thus (26.14) becomes

$$y-\mu_y = \beta_2(x-\mu_x),$$

the analogue of (26.13).

We have thus reached the conclusion that the calculation of an approximate regression line by the Method of Least Squares gives results which are the same as the correct ones in the case of exact linearity of regression.

The correlation coefficient

26.9 In view of the result of **26.8**, we now make our discussion cover the general case where regression is not exactly linear. The linear regression coefficients, (26.10) and (26.11), are generally the coefficients in approximate regression lines, though on occasions these lines may be exact.

We now define the product-moment *correlation coefficient* ρ by

$$\rho = \mu_{11}/(\sigma_1\sigma_2), \tag{26.16}$$

whence, from (26.10), (26.11) and (26.16),

$$\rho^2 = \beta_1\beta_2. \tag{26.17}$$

ρ is a symmetric function of x and y, as any coefficient of interdependence should be. Since it is a homogeneous function of moments about the means, it is invariant under changes of origin and scale. ρ has the same sign as β_1 and β_2, since all three have μ_{11} as numerator and a positive denominator; when $\mu_{11} = 0$, $\rho = 0$. If $\sigma_1 = \sigma_2$, $\rho = \beta_1 = \beta_2$. From (26.17) we see that $|\rho|$ is the geometric mean of $|\beta_1|$ and $|\beta_2|$.

By the Cauchy–Schwarz inequality

$$\mu_{11}^2 = \left\{\iint (x-\mu_x)(y-\mu_y)\,dF\right\}^2 \leq \left\{\iint (x-\mu_x)^2\,dF\right\}\left\{\iint (y-\mu_y)^2\,dF\right\} = \sigma_1^2 \sigma_2^2$$

so that

$$0 \leq \rho^2 \leq 1, \qquad (26.18)$$

the upper equality in (26.18) holding (cf. **2.7**) if and only if $(x-\mu_x)$ and $(y-\mu_y)$ are strictly proportional, i.e. x and y are in strict linear functional relationship. Essentially, therefore, ρ is the covariance μ_{11} divided by a factor which ensures that ρ will lie in the interval $(-1, +1)$.

It may easily be shown that the angle between the two regression lines (26.12) and (26.13) is

$$\theta = \arctan\left\{\frac{\sigma_1\sigma_2}{\sigma_1^2+\sigma_2^2}\left(\frac{1}{\rho}-\rho\right)\right\}, \qquad (26.19)$$

so that as ρ varies over its range from -1 to $+1$, θ increases steadily from 0 to its maximum of $\tfrac{1}{2}\pi$ when $\rho = 0$, and then decreases steadily to 0 again. Thus, if and only if x and y are in strict linear functional relationship, the two regression lines coincide ($\rho^2 = 1$). If and only if $\rho = 0$, when x and y are said to be *uncorrelated*, the regression lines are at right angles to each other.

It may be shown that

$$\rho^2 = \operatorname{var}(\alpha_2+\beta_2 x)/\sigma_2^2 = \operatorname{var}(\alpha_1+\beta_1 y)/\sigma_1^2, \qquad (26.20)$$

where "var" here means simply the calculated variance. The proof of (26.20) is left to the reader as Exercise 26.13.

> The terms "regression," "lines of regression" and "correlation" were first used by Galton in 1886–8 (his 1877 synonym for "regression" was "reversion"); "coefficient of correlation" was first used by Edgeworth in 1892. The term "correlation" arose naturally from "co-relation," but "regression" requires some explanation. Galton found that the average stature of adult offspring increased with parents' stature, but not by as much, and he called this a "regression to mediocrity." The term has stuck firmly, but the apparently dramatic result is theoretically trivial, for as we have seen, if $\sigma_1 = \sigma_2$ (as we may reasonably suppose here), then $\beta_1 = \beta_2 = \rho \leq 1$. Thus the same phenomenon would be observed if we interchanged the roles of parents and offspring, although, as in **26.3**, there is an essential asymmetry in the variables for the geneticist.

ρ as a coefficient of interdependence

26.10 From **26.6** and Example 26.2 we see that while independence of x and y implies $\mu_{11} = \rho = 0$, the converse does not generally apply. It does apply for jointly normal variables, and sometimes for others (Example 26.1). In this lies the difficulty of interpreting ρ as a coefficient of interdependence in general. In fact, we have seen that ρ is essentially a coefficient of *linear* interdependence, and more complex forms of interdependence lie outside its vocabulary. In general, the problem of joint variation is too complex to be comprehended in a single coefficient.

To *express* a quality, moreover, is not the same as to *measure* it. If $\rho = 0$ implies independence, we know from **26.9** that as $|\rho|$ increases, the interdependence also increases until when $|\rho| = 1$ we have the limiting case of linear functional relationship. Even so, it remains an open question which function of ρ should be used as a *measure* of interdependence: we see from (26.20) that ρ^2 is more directly interpretable than ρ itself, being the ratio of the variance of the fitted line to the overall variance. Leaving this point aside, ρ gives us a measure in such cases, though there may be better measures. On the other hand, if $\rho = 0$ does not imply independence, it is difficult to interpret ρ as a measure of interdependence, and perhaps wiser to use it as an indicator rather than as a precise measure. In practical work, we would recommend the use of ρ as a *measure* of interdependence only in cases of normal or near-normal variation.

Computation of coefficients

26.11 From the definitions at (26.10), (26.11) and (26.16) we see that the linear regression coefficients and the correlation coefficient require for their computation the two variances and the covariance μ_{11}. The calculation of variances was discussed in **2.19** and Example 2.7. The covariance is calculated similarly, using the identity (26.3),

$$\mu_{11} = \sum_{i=1}^{n} (x_i - \mu_x)(y_i - \mu_y)/n \equiv \sum_{i=1}^{n} x_i y_i / n - \mu_x \mu_y$$
$$\equiv \sum_{i=1}^{n} x_i y_i / n - \left(\sum_{i=1}^{n} x_i\right)\left(\sum_{i=1}^{n} y_i\right)\bigg/ n^2.$$

For convenience, we often take arbitrary origins a, b, for x and y respectively. Then

$$\mu_{11} = \Sigma(x-a)(y-b)/n - \{\Sigma(x-a)\}\{\Sigma(y-b)\}/n^2 \qquad (26.21)$$

identically in a and b. In other words, μ_{11} is invariant under changes of origin, as it must be since it is a product-moment about the means. (26.21) holds if we put $(x-a) \equiv (y-b)$, when it reduces to (2.21) for the calculation of variances. We usually also find it convenient to take an arbitrary unit u_x for x and another arbitrary unit u_y for y. It is easy to see that the effect of this is to divide μ_{11} by $u_x u_y$, σ_1^2 by u_x^2, and σ_2^2 by u_y^2. Thus β_1 is multiplied by a factor u_y/u_x, β_2 by a factor u_x/u_y, and ρ is quite unaffected by a change of scales.

To summarize, μ_{11}, σ_1^2 and σ_2^2 are invariant under changes of origin, so β_1, β_2 and ρ are. If *different* arbitrary scale factors are introduced, for computational purposes, β_1 and β_2 require adjustment by the appropriate ratio; if the *same* scale factor is used for each variable, β_1 and β_2 are unaffected. ρ is unaffected by any scale change.

Example 26.6. Computation of coefficients for grouped data

For grouped data, such as those in Table 26.1, we choose the group-width of each variable as the working unit for that variable (if the groups of a variable are of unequal width, we usually take the *smallest* group width). We also choose a working origin for each variable somewhere near the mean, estimated by eye. Thus we take the x-origin in Table 26.1 at 64, the centre of the modal frequency-group, the marginal distribution of x being near symmetry; the y-origin is placed at 134·5, since the mean is likely to lie appreciably above the modal frequency group (122·5) for a very skew distribution like the marginal distribution of y. The group widths (2 and 6) are taken as working units. The sum of products, Σxy, is calculated by multiplying each frequency in turn by its " co-ordinates " in the table in the arbitrary units. Thus the extreme " south-eastern " entry in the table, the frequency 4, for which $x = 68$,

$y = 110.5$, is multiplied by $(+2)(-4) = -8$, contributing -32 to the sum. We find

$$\Sigma x = -2,353, \qquad \Sigma y = -1,400,$$
$$\Sigma x^2 = 10,161, \qquad \Sigma y^2 = 70,802,$$
$$\Sigma xy = +8,786,$$

giving

$$\mu_x = -\frac{2353}{4995} \cdot 2 + 64 = 63.06,$$

$$\mu_y = -\frac{1400}{4995} \cdot 6 + 134.5 = 132.82,$$

$$\sigma_1^2 = \left\{\frac{10,161}{4995} - \left(\frac{2353}{4995}\right)^2\right\} \times 2^2 = 7.25,$$

$$\sigma_2^2 = \left\{\frac{70802}{4995} - \left(\frac{1400}{4995}\right)^2\right\} \times 6^2 = 507.46,$$

$$\mu_{11} = \left\{\frac{8786}{4995} - \left(-\frac{2353}{4995}\right)\left(-\frac{1400}{4995}\right)\right\} \times 2 \times 6 = +19.52,$$

whence

$$\rho = \frac{\mu_{11}}{\sigma_1 \sigma_2} = 0.322,$$
$$\beta_1 = \mu_{11}/\sigma_2^2 = 0.0385,$$
$$\beta_2 = \mu_{11}/\sigma_1^2 = 2.692.$$

The (approximate) linear regression equations are:

x on y: $\qquad x - 63.06 = 0.0385 (y - 132.82)$ or
$$x = 0.0385y + 57.95,$$
y on x: $\qquad y - 132.82 = 2.692 (x - 63.06)$ or
$$y = 2.692x - 36.96.$$

These lines are drawn in on Fig. 26.1 (page 294) as RR' and CC' respectively.

Example 26.7. Computation of coefficients for ungrouped data

Table 26.3 shows the yields of wheat and of potatoes in 48 counties of England in 1936. For ungrouped data such as these, it is rarely worth taking an arbitrary origin and unit for calculations. Using the natural origins and units, we find

$\Sigma x = \quad 758.0, \quad \mu_x = 15.792, \quad \beta_1 = \mu_{11}/\sigma_2^2 = 0.612,$
$\Sigma y = \quad 291.1, \quad \mu_y = 6.065, \quad \beta_2 = \mu_{11}/\sigma_1^2 = 0.078,$
$\Sigma x^2 = 12,170.48, \quad \sigma_1^2 = 4.174, \quad \rho = \mu_{11}/(\sigma_1 \sigma_2) = 0.219.$
$\Sigma y^2 = 1,791.03, \quad \sigma_2^2 = 0.534,$
$\Sigma xy = 4,612.64, \quad \mu_{11} = 0.327,$

The (approximate) linear regression equations are therefore:

Regression of x on y: $\quad x - 15.792 = 0.612 (y - 6.065)$
Regression of y on x: $\quad y - 6.065 = 0.078 (x - 15.792)$

The data and the regression lines are shown diagrammatically in Fig. 26.3, one point corresponding to each pair of values (x, y). A diagram such as this, on which all points

LINEAR REGRESSION AND CORRELATION

Table 26.3—Yields of wheat and potatoes in 48 counties in England in 1936

County	Wheat (cwt per acre) x	Potatoes (tons per acre) y	County	Wheat (cwt per acre) x	Potatoes (tons per acre) y
Bedfordshire	16·0	5·3	Northamptonshire	14·3	4·9
Huntingdonshire	16·0	6·6	Peterborough	14·4	5·6
Cambridgeshire	16·4	6·1	Buckinghamshire	15·2	6·4
Ely	20·5	5·5	Oxfordshire	14·1	6·9
Suffolk, West	18·2	6·9	Warwickshire	15·4	5·6
Suffolk, East	16·3	6·1	Shropshire	16·5	6·1
Essex	17·7	6·4	Worcestershire	14·2	5·7
Hertfordshire	15·3	6·3	Gloucestershire	13·2	5·0
Middlesex	16·5	7·8	Wiltshire	13·8	6·5
Norfolk	16·9	8·3	Herefordshire	14·4	6·2
Lincs (Holland)	21·8	5·7	Somersetshire	13·4	5·2
" (Kesteven)	15·5	6·2	Dorsetshire	11·2	6·6
" (Lindsey)	15·8	6·0	Devonshire	14·4	5·8
Yorkshire (East Riding)	16·1	6·1	Cornwall	15·4	6·3
Kent	18·5	6·6	Northumberland	18·5	6·3
Surrey	12·7	4·8	Durham	16·4	5·8
Sussex, East	15·7	4·9	Yorkshire (North Riding)	17·0	5·9
Sussex, West	14·3	5·1	" (West Riding)	16·9	6·5
Berkshire	13·8	5·5	Cumberland	17·5	5·8
Hampshire	12·8	6·7	Westmorland	15·8	5·7
Isle of Wight	12·0	6·5	Lancashire	19·2	7·2
Nottinghamshire	15·6	5·2	Cheshire	17·7	6·5
Leicestershire	15·8	5·2	Derbyshire	15·2	5·4
Rutland	16·6	7·1	Staffordshire	17·1	6·3

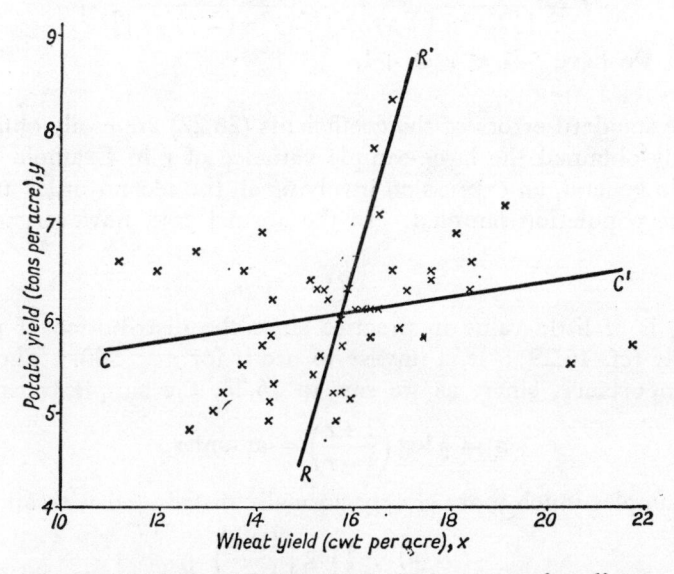

Fig. 26.3—Data of Table 26.3, with regression lines

304   THE ADVANCED THEORY OF STATISTICS

are plotted, is called a *scatter diagram*: its use is strongly recommended, since it conveys quickly and simply an idea of the adequacy of the fitted regression lines (not very good in our example). Indeed, a scatter diagram, plotted in advance of the analysis, will often make it clear whether the fitting of regression lines is worth while.

Sample coefficients: standard errors

26.12 We now turn to the consideration of sampling problems for correlation and regression coefficients. As usual, we observe the convention that a Roman letter (actually italic) represents a sample statistic, the Greek letter being the population equivalent. Thus we write

$$\left.\begin{array}{l} b_1 = m_{11}/s_2^2 = \dfrac{1}{n}\Sigma(x-\bar{x})(y-\bar{y}) \Big/ \dfrac{1}{n}\Sigma(y-\bar{y})^2, \\[6pt] b_2 = m_{11}/s_1^2 = \dfrac{1}{n}\Sigma(x-\bar{x})(y-\bar{y}) \Big/ \dfrac{1}{n}\Sigma(x-\bar{x})^2, \\[6pt] r = m_{11}/(s_1 s_2) = \dfrac{1}{n}\Sigma(x-\bar{x})(y-\bar{y}) \Big/ \left\{\dfrac{1}{n}\Sigma(x-\bar{x})^2 \cdot \dfrac{1}{n}\Sigma(y-\bar{y})^2\right\}^{\frac{1}{2}}, \end{array}\right\} \quad (26.22)$$

for the sample regression coefficients and correlation coefficient, the summations now being over sample values. Just as for the population coefficients β_1, β_2, ρ, we may simplify these expressions for computational purposes to

$$\left.\begin{array}{l} b_1 = \dfrac{\Sigma xy - (\Sigma x)(\Sigma y)/n}{\Sigma y^2 - (\Sigma y)^2/n}, \\[6pt] b_2 = \dfrac{\Sigma xy - (\Sigma x)(\Sigma y)/n}{\Sigma x^2 - (\Sigma x)^2/n}, \\[6pt] r = \dfrac{\Sigma xy - (\Sigma x)(\Sigma y)/n}{[\{\Sigma x^2 - (\Sigma x)^2/n\}\{\Sigma y^2 - (\Sigma y)^2/n\}]^{\frac{1}{2}}}. \end{array}\right\} \quad (26.23)$$

Just as before, we have $-1 \leqslant r \leqslant +1$.

26.13 The standard errors of the coefficients (26.22) are easily obtained. In fact, we have already obtained the large-sample variance of r in Example 10.6, where we saw that it is, in general, an expression involving all the second-order and fourth-order moments of the population sampled. In the normal case, however, we found that it simplified to

$$\operatorname{var} r = (1-\rho^2)^2/n, \quad (26.24)$$

though (26.24) is of little value in practice since the distribution of r tends to normality so slowly (cf. **16.29**): it is unwise to use it for $n < 500$. The difficulty is of no practical importance, since, as we saw in **16.33**, the simple transformation of r,

$$z = \tfrac{1}{2}\log\left(\dfrac{1+r}{1-r}\right) = \operatorname{ar\,tanh} r, \quad (26.25)$$

is for normal samples much more closely normally distributed with approximate mean

$$E(z) \doteqdot \tfrac{1}{2}\log\left(\dfrac{1+\rho}{1-\rho}\right) \quad (26.26)$$

LINEAR REGRESSION AND CORRELATION

and variance approximately

$$\operatorname{var} z \doteqdot \frac{1}{n-3}, \qquad (26.27)$$

independent of ρ. For $n > 50$, the use of this standard error for z is adequate; closer approximations are given in **16.33**.

For the sample regression coefficient of y on x,

$$b_2 = m_{11}/s_1^2,$$

the use of (10.17) gives, just as in Example 10.6 for r,

$$\operatorname{var} b_2 = \left(\frac{\mu_{11}}{\sigma_1^2}\right)^2 \left\{ \frac{\operatorname{var} m_{11}}{\mu_{11}^2} + \frac{\operatorname{var}(s_1^2)}{\sigma_1^4} - \frac{2 \operatorname{cov}(m_{11}, s_1^2)}{\mu_{11} \sigma_1^2} \right\}.$$

Substituting for the variances and covariance from (10.23) and (10.24), this becomes

$$\operatorname{var} b_2 = \frac{1}{n}\left(\frac{\mu_{11}}{\sigma_1^2}\right)^2 \left\{ \frac{\mu_{22}}{\mu_{11}^2} + \frac{\mu_{40}}{\sigma_1^4} - \frac{2\mu_{31}}{\mu_{11} \sigma_1^2} \right\}. \qquad (26.28)$$

For a normal parent population, we substitute the relations of Example 3.17 and obtain

$$\operatorname{var} b_2 = \frac{1}{n}\left(\frac{\mu_{11}}{\sigma_1^2}\right)^2 \left\{ \frac{(1+2\rho^2)\sigma_1^2 \sigma_2^2}{\mu_{11}^2} + \frac{3\sigma_1^2}{\sigma_1^4} - \frac{6\rho \sigma_1^3 \sigma_2}{\mu_{11} \sigma_1^2} \right\}$$

$$= \frac{1}{n}\left(\frac{\mu_{11}}{\sigma_1^2}\right)^2 \left\{ \frac{\sigma_1^2 \sigma_2^2}{\mu_{11}^2} - 1 \right\}$$

$$= \frac{1}{n}\frac{\sigma_2^2}{\sigma_1^2}(1-\rho^2). \qquad (26.29)$$

Similarly, for the regression coefficient of x on y,

$$\operatorname{var} b_1 = \frac{1}{n}\frac{\sigma_1^2}{\sigma_2^2}(1-\rho^2). \qquad (26.30)$$

The expressions (26.29) and (26.30) are rather more useful for standard error purposes (when, of course, we substitute s_1^2, s_2^2 and r for σ_1^2, σ_2^2 and ρ in them) than (26.24), for we saw at **16.35** that the exact distribution of b_2 is symmetrical about β_2: it is left to the reader as Exercise 26.9 to show from (16.86) that (26.29) is exact when multiplied by a factor $n/(n-3)$, and that the distribution of b_2 tends to normality rapidly, its measure of kurtosis being of order $1/n$.

The estimation of ρ in normal samples

26.14 The sampling theory of the bivariate normal distribution was developed in **16.23–36**: we may now discuss, in particular, the problem of estimating ρ from a sample, in the light of our results in the theory of estimation (Chapters 17–18).

In **16.24** we saw in effect that the Likelihood Function is given by (16.52), which contains the observations only in the form of the five statistics $\bar{x}$, $\bar{y}$, s_1^2, s_2^2, r. These are therefore a set of sufficient statistics for the five parameters μ_1, μ_2, σ_1^2, σ_2^2, ρ, and their distribution is complete by **23.10**. Further, (16.52) makes it clear that even if all four other parameters are known, we still require this five-component sufficient statistic for ρ alone.

In Chapter 18 we saw that the Maximum Likelihood estimator of ρ takes a different form according to which, if any, other parameters are being simultaneously estimated: the ML estimator is always a function of the set of sufficient statistics, but it is a *different* function in different situations. When ρ alone is being estimated, the ML estimator is the root of a cubic equation (Example 18.3); when all five parameters are being estimated, the ML estimator is the sample correlation coefficient r (Example 18.14). In practice, the latter is by far the most common case, and we therefore now consider the estimation of ρ by r or functions of it.

26.15 The exact distribution of r, which depends only upon ρ, is given by (16.61) or, more conveniently, by (16.66). Its mean value is given by the hypergeometric series (16.73). Expanding the gamma functions in (16.73) by Stirling's series (3.64) and taking the two leading terms of the hypergeometric function, we find

$$E(r) = \rho \left\{ 1 - \frac{(1-\rho^2)}{2n} + O(n^{-2}) \right\}. \tag{26.31}$$

Thus r is a slightly biassed estimator of ρ when $0 \neq \rho^2 \neq 1$. The bias is generally small, but it is interesting to inquire whether it can be removed.

26.16 We may approach the problem in two ways. First, we may ask: is there a function $g(r)$ such that

$$E\{g(r)\} = g(\rho) \tag{26.32}$$

holds identically in ρ? Hotelling (1953) showed that if g is not dependent on n, $g(r)$ could only be a linear function of $\arcsin r$, and Harley (1956-7) showed that in fact

$$E(\arcsin r) = \arcsin \rho, \tag{26.33}$$

a simple proof of Harley's result being given by Daniels and Kendall (1958).

26.17 The second, more direct, approach, is to seek a function of r unbiassed for ρ itself. By Hotelling's result in **26.16**, this function must involve n. Since r is a function of a set of complete sufficient statistics, the unbiassed function of r must be unique (cf. **23.9**). Olkin and Pratt (1958) found the unbiassed estimator of ρ, say r_u, to be the hypergeometric function

$$r_u = r F[\tfrac{1}{2}, \tfrac{1}{2}, \tfrac{1}{2}(n-2), (1-r^2)] \tag{26.34}$$

which, expanded into series, gives

$$r_u = r \left\{ 1 + \frac{1-r^2}{2(n-2)} + \frac{9(1-r^2)^2}{8n(n-2)} + O(n^{-3}) \right\}. \tag{26.35}$$

No term in the series is negative, so that

$$|r_u| \geq |r|,$$

the equality holding only if $r^2 = 0$ or 1. Since $F(\tfrac{1}{2}, \tfrac{1}{2}, \tfrac{1}{2}(n-2), 0) = 1$ and r_u is an increasing function of r, we have $0 \leq r^2 \leq r_u^2 \leq 1$.

Evidently, the first correction term in (26.35) is counteracting the downward bias of the term in $1/n$ in (26.31). Olkin and Pratt recommend the use of the two-term expansion

$$r_u^* = r \left\{ 1 + \frac{1-r^2}{2(n-4)} \right\}. \tag{26.36}$$

The term in braces in (26.36) gives r_u/r within 0·01 for $n \geqslant 8$ and within 0·001 for $n \geqslant 18$, uniformly in r.

Olkin and Pratt give exact tables of r_u for $n = 2(2)30$ and $|r| = 0(0\cdot1)1$ which show that for $n \geqslant 14$, $|r_u|$ never exceeds $|r|$ by more than 5 per cent.

Finally, we note that as n appears only in the denominators of the hypergeometric series, $r_u \to r$ as $n \to \infty$, so that it has the same limiting distribution as r, namely a normal distribution with mean ρ and variance $(1-\rho^2)^2/n$.

Confidence limits and tests for ρ

26.18 For testing that $\rho = 0$, the tests based on r are UMPU (cf. Exercise 31.21); this is not so when we are testing a non-zero value of ρ. However, if we confine ourselves to test statistics which are invariant under changes in location and scale, one-sided tests based on r are UMP invariant, as Lehmann (1959) shows.

For interval estimation purposes, we may use F. N. David's (1938) charts, described in **20.21**. By the duality between confidence intervals and tests remarked in **23.26**, these charts may also be used to read off the values of ρ to be rejected by a size-α test, i.e. all values of ρ not covered by the confidence interval for that α. F. N. David (1937) has shown that this test is slightly biassed (and the confidence intervals correspondingly so). This may most easily be seen from the standpoint of the z-transformation standard-error test given in **26.13**: if the latter were exact, and z were *exactly* normal with variance independent of ρ, the test of ρ would simply be a test of the value of the mean of a normal distribution with known variance, and we know from Example 23.11 that if we use an "equal-tails" test for this hypothesis, it is unbiassed. Since z is a one-to-one function of r, the "equal-tails" test on r would then also be unbiassed. Thus the slight bias in the "equal-tails" r-test may be regarded as a reflection of the approximate nature of the z-transformation.

Exercise 26.15 shows that the LR test is based on r, but is not "equal-tails" except when testing $\rho = 0$.

26.19 Alternatively, we may make an approximate test using Fisher's z-transformation, the simplest results for which are given in **26.13**: to test a hypothetical value of ρ we compute (26.25) and test that it is normally distributed about (26.26) with variance (26.27). A one- or two-tailed test is appropriate according to whether the alternative to this simple hypothesis is one- or two-sided.

In the same way, we may use the z-transformation to test the composite hypothesis that the correlation parameters of two independently sampled bivariate normal populations are the same. For if so, the two transformed statistics z_1, z_2 will each be distributed as in **26.13**, and $(z_1 - z_2)$ will have zero mean and variance $1/(n_1-3)+1/(n_2-3)$, where n_1 and n_2 are the sample sizes. Exercises 26.19–21 show that $(z_1 - z_2)$ is exactly the Likelihood Ratio statistic when $n_1 = n_2$, and approximately so when $n_1 \neq n_2$. In either case, however, the test is approximate, being a standard-error test.

The more general composite hypothesis, that the two correlation parameters ρ_1, ρ_2 differ by an amount Δ, cannot be tested in this way. For then

$$E(z_1 - z_2) = \tfrac{1}{2}\log\left(\frac{1+\rho_1}{1-\rho_1}\right) - \tfrac{1}{2}\log\left(\frac{1+\rho_2}{1-\rho_2}\right) = \tfrac{1}{2}\log\left\{\left(\frac{1+\rho_1}{1-\rho_1}\right)\left(\frac{1-\rho_2}{1+\rho_2}\right)\right\}$$

is not a function of $|\rho_1-\rho_2|$ alone. The z-transformation could be used to test

$$H_0: \frac{1+\rho_1}{1-\rho_1} = a\left(\frac{1+\rho_2}{1-\rho_2}\right)$$

for any constant a, but this is not a hypothesis of interest. Except in the very large sample case, when we may use standard errors, there seems to be no way of testing $H_0: \rho_1-\rho_2 = \Delta$, for the exact distribution of the difference r_1-r_2 has not been investigated.

Dunn and Clark (1969, 1971) examine methods of testing the equality of correlation parameters when the sample coefficients are not independent.

Tests of independence and regression tests

26.20 In the particular case when we wish to test $\rho = 0$, i.e. the *independence* of the normal variables, we may use the exact result of **16.28**, that

$$t = \{(n-2)r^2/(1-r^2)\}^{\frac{1}{2}} \tag{26.37}$$

is distributed in Student's t-distribution with $(n-2)$ degrees of freedom. t^2 is essentially the LR test statistic for $\rho = 0$—cf. Exercise 26.15—and this is equivalent to an "equal-tails" test on t.

Essentially, we are testing here that μ_{11} of the population is zero, and clearly this implies that the population regression coefficients β_1, β_2 are zero. Now in **16.36**, we showed that

$$t = (b_2-\beta_2)\left\{\frac{s_1^2(n-2)}{s_2^2(1-r^2)}\right\}^{\frac{1}{2}} \tag{26.38}$$

has Student's distribution with $(n-2)$ degrees of freedom. When $\beta_2 = 0$, (26.38) is seen to be identical with (26.37). Thus the test of independence may be regarded as a test that a regression coefficient is zero, a special case of the general test (26.38) which we use for hypotheses concerning β_2. It will be noted that the exact test for any hypothetical value of β_2 is of a much simpler form than that for ρ.

We shall see in Chapter 31 that tests of independence can be made without any assumption of normality in the parent distribution.

Correlation ratios and linearity of regression

26.21 In **26.8** we discussed the fitting of approximate regression lines in cases where regressions are not exactly linear. We can make further progress in analysing the linearity of regression. Consider first the case of exact linear regression of x on y, when (26.12) holds. Squaring (26.12) and taking expectations with respect to y, we have

$$\text{var}\{E(x|y)\} = \underset{y}{E}\{(E(x|y)-E(x))^2\} = \beta_1^2\sigma_2^2 = \mu_{11}^2/\sigma_2^2. \tag{26.39}$$

We now define the *correlation ratio* of x on y, η_1, by

$$\eta_1^2 = \text{var}\{E(x|y)\}/\sigma_1^2, \tag{26.40}$$

the ratio of the variance of x-array means to the variance of x. Unlike ρ, η_1^2 is evidently not symmetric in x and y. (26.40) implies that η_1^2 is invariant under permutation of the x-arrays, since var$\{E(x|y)\}$ does not depend on the order in which the values of

$E(x|y)$ occur. This is in sharp contrast to ρ, which is sensitive to any change in the order of arrays.

From (26.39) we see that if the regression is exactly linear,
$$\eta_1^2 = \mu_{11}^2/(\sigma_1^2 \sigma_2^2) = \rho^2.$$

Now consider the general case where the regression is not necessarily exactly linear. We have
$$\begin{aligned}\sigma_1^2 &= E[\{x-E(x)\}]^2 = E[(\{x-E(x|y)\}+\{E(x|y)-E(x)\})^2]\\ &= E[\{x-E(x|y)\}^2]+E[\{E(x|y)-E(x)\}^2]\\ &= E[\{x-E(x|y)\}^2]+\mathrm{var}\{E(x|y)\},\end{aligned} \qquad (26.41)$$
since the cross-product term
$$2E[\{x-E(x|y)\}\{E(x|y)-E(x)\}] = \underset{y}{E}[\{E(x|y)-E(x)\}\underset{x|y}{E}\{x-E(x|y)\}] = 0.$$
Thus, from (26.40) and (26.41), we have
$$0 \leqslant \eta_1^2 \leqslant 1, \qquad (26.42)$$
and $\eta_1^2 = 0$ if and only if $\mathrm{var}\{E(x|y)\} = 0$, i.e. if all x-array means are equal, while $\eta_1^2 = 1$ if and only if $E[\{x-E(x|y)\}^2] = 0$, i.e. every observation lies on the regression curve, so that x and y are strictly functionally related. Further,
$$\rho^2/\eta_1^2 = \mu_{11}^2/[\sigma_2^2 \mathrm{var}\{E(x|y)\}]. \qquad (26.43)$$
By the Cauchy–Schwarz inequality,
$$\begin{aligned}\mu_{11}^2 &= \left(\underset{x,y}{E}[\{y-E(y)\}\{x-E(x)\}]\right)^2 = \left(\underset{y}{E}[\{y-E(y)\}\{E(x|y)-E(x)\}]\right)^2 \\ &\leqslant \underset{y}{E}[\{y-E(y)\}^2]\underset{y}{E}[\{E(x|y)-E(x)\}^2] = \sigma_2^2 \mathrm{var}\{E(x|y)\},\end{aligned} \qquad (26.44)$$
the equality holding if and only if $\{y-E(y)\}$ is proportional to $\{E(x|y)-E(x)\}$, i.e. if $E(x|y)$ is a strict linear function of y. Thus, from (26.43) and (26.44),
$$\rho^2/\eta_1^2 \leqslant 1, \qquad (26.45)$$
the equality holding only when the regression of x on y is exactly linear. Hence, from (26.42), (26.45) and (26.18), we finally have the inequalities
$$0 \leqslant \rho^2 \leqslant \eta_1^2 \leqslant 1. \qquad (26.46)$$
We may summarize our results on the attainment of the inequalities in (26.46), given in **26.9–10** and in this section, as follows:

(a) $\rho^2 = \eta_1^2 = 0$ if, but not only if, x and y are independent—all array means are equal;
(b) $\rho^2 = \eta_1^2 = 1$ if, and only if, x and y are in strict linear functional relationship;
(c) $\rho^2 < \eta_1^2 = 1$ if, and only if, x and y are in strict non-linear functional relationship;
(d) $\rho^2 = \eta_1^2 < 1$ if, and only if, the regression of x on y is exactly linear, but there is no strict linear or non-linear functional relationship;
(e) $\rho^2 < \eta_1^2 < 1$ implies that there is no strict functional relationship, and *some* non-linear regression curve is a better " fit " than the " best " straight line, for (26.20) and (26.40) then imply that $\mathrm{var}\{E(x|y)\} > \mathrm{var}(\alpha_1+\beta_1 y)$, so that the array means are more dispersed than in the straight-line regression most nearly " fitting " them. (Of course, there may be no better-fitting *simple* regression curve.)

Since η_1 takes no account of the order of the x-arrays, it does not measure any particular type of dependence of x on y, but the value of $\eta_1^2 - \rho^2$ is an indicator of non-linearity of regression: it is important to remember that it is an indicator, not a measure, and in order to assess its importance the number of arrays (and the number of observations) must also be taken into account. We discuss this matter in **26.24**.

26.22 Similarly, we define, for the regression of y on x, the correlation ratio

$$\eta_2^2 = \text{var}\{E(y|x)\}/\sigma_2^2, \tag{26.47}$$

and again

$$0 \leqslant \rho^2 \leqslant \eta_2^2 \leqslant 1.$$

Since $\eta_1^2 = 1$ if and only if there is a strict functional relationship, $\eta_1^2 = 1$ implies $\eta_2^2 = 1$ and conversely. In general, both squared correlation ratios exceed ρ^2, but we shall have $\eta_1^2 = \rho^2 < \eta_2^2$ if the regression of x on y is linear while that of y on x is not, as in the following Example.

Example 26.8

Consider again the situation in Example 26.3. The regression of x on y^2 was linear with regression coefficient 0, so that the correlation between x and y^2 is zero also. Since we found $E(x|y^2) = 0$, it follows that $\text{var}\{E(x|y^2)\} = 0$ also, so that the correlation ratio of x on y^2 is 0, as it must be, since the correlation coefficient is zero and the regression linear.

The regression of y^2 on x was not linear: we found in Example 26.3 that

$$E(y^2|x) = 1 + \rho^2(x^2 - 1)$$

so that

$$\text{var}\{E(y^2|x)\} = E[\{\rho^2(x^2-1)\}^2] = \rho^4 E[\{x^2-1\}^2] = 2\rho^4$$

and $\sigma_2^2 = 2$, so that the correlation ratio of y^2 on x is ρ^4, which always exceeds the correlation coefficient between x and y^2, which is zero, when $\rho \neq 0$.

When correlation ratios are being calculated from sample data, we use the observed variance of array means and the observed variance in (26.40) and (26.41), properly weighted, obtaining for the observed correlation ratio of x on y

$$e_1^2 = \frac{\sum\limits_{i=1}^{k} n_i(\bar{x}_i - \bar{x})^2}{\sum\limits_{i=1}^{k}\sum\limits_{j=1}^{n_i}(x_{ij}-\bar{x})^2} \equiv \frac{\sum\limits_{i} n_i \bar{x}_i^2 - n\bar{x}^2}{\sum\limits_{i}\sum\limits_{j} x_{ij}^2 - n\bar{x}^2}, \tag{26.48}$$

where $\bar{x}_i$ is the mean of the ith x-array, and n_i the number of observations in the array, there being k arrays. A similar expression holds for e_2^2, the observed correlation ratio of y on x. As for populations,

$$0 \leqslant r^2 \leqslant e_i^2 \leqslant 1, \quad i = 1, 2. \tag{26.49}$$

Example 26.9. Computation of the correlation ratio

Let us calculate the correlation ratio of y on x for the data of Table 26.1, which we now treat as a sample. The computation is set out in Table 26.4.

Table 26.4

Stature (x)	Mean weight in array ($\bar{y}_i$)	$\bar{y}_i^2$	n_i	$n_i\bar{y}_i^2$
54	92·50	8,556·25	5	42,781·25
56	111·41	12,412·19	33	409,602·27
58	122·05	14,896·20	254	3,783,634·80
60	124·43	15,482·82	813	12,587,532·66
62	130·22	16,957·25	1340	22,722,715·00
64	134·59	18,114·47	1454	26,338,439·38
66	140·48	19,734·63	750	14,800,972·50
68	146·37	21,424·18	275	5,891,649·50
70	157·32	24,749·58	56	1,385,976·48
72	163·41	26,702·83	11	293,731·13
74	179·50	32,220·25	4	128,881·00
			$n = 4995$	88,385,915·97

In Example 26.6 we found the mean of y to be

$$\bar{y} = 132 \cdot 82$$

and the variance of y to be 507·46. Thus, from (26.48), the correlation ratio of y on x is

$$e_2^2 = \frac{88{,}385{,}915 \cdot 97 - 4{,}995 \, (132 \cdot 82)^2}{4{,}995 \times 507 \cdot 46}$$

$$= \frac{88{,}385{,}915 \cdot 97 - 88{,}117{,}544 \cdot 25}{2{,}534{,}762 \cdot 70}$$

$$= \frac{268{,}359 \cdot 73}{2{,}534{,}762 \cdot 70} = 0 \cdot 106.$$

This is only slightly greater than the squared correlation coefficient

$$r^2 = (0 \cdot 322)^2 = 0 \cdot 104.$$

Fig. 26.1 shows that the linear approximation CC' to the regression is indeed rather good.

Testing correlation ratios and linearity of regression

26.23 We saw in **26.21** that $\eta_1^2 = \rho^2$ indicates that no better regression curve than a straight line can be found, and hence that a positive value of $\eta_1^2 - \rho^2$ is an indicator of non-linearity of regression. Now that we have defined the sample correlation ratios, e_i^2, it is natural to ask whether the statistic $(e_1^2 - r^2)$ will provide a test of the linearity of regression of x on y. In the following discussion, we take the opportunity to give also a test for the hypothesis that $\eta_1^2 = 0$ and also to bring these tests into relation with the test of $\rho = 0$ given at (26.37). These problems were first solved by R. A. Fisher.

The identity

$$n s_1^2 \equiv n s_1^2 r^2 + n s_1^2 (e_1^2 - r^2) + n s_1^2 (1 - e_1^2), \tag{26.50}$$

has all terms on the right non-negative, by (26.49). Since

$$\sum_{i=1}^{k} \sum_{j=1}^{n_i} \{(\bar{x}_i - \bar{x}) - b_1(\bar{y}_i - \bar{y})\}^2 \equiv \sum_i \sum_j (\bar{x}_i - \bar{x})^2 - r^2 \sum_i \sum_j (x_{ij} - \bar{x})^2,$$

(26.50) may be rewritten in x as

$$\sum_i \sum_j (x_{ij}-\bar{x})^2 \equiv r^2 \sum_i \sum_j (x_{ij}-\bar{x})^2 + \sum_i \sum_j \{(\bar{x}_i-\bar{x})-b_1(\bar{y}_i-\bar{y})\}^2 + \sum_i \sum_j (x_{ij}-\bar{x}_i)^2. \quad (26.51)$$

Now (26.51) is a decomposition of a quadratic form in the x_{ij} into three other such forms. We now assume that the y_i are fixed and that all the x_{ij} are normally distributed, independently of each other, with the same variance (taken to be unity without loss of generality). We leave open for the moment the question of the means of the x_{ij}.

On the hypothesis H_0 that every x_{ij} has the same mean, i.e. that the regression curve is a line parallel to the y-axis, we know that the left-hand side of (26.51) is distributed in the chi-squared form with $(n-1)$ degrees of freedom. It is a straightforward, though tedious, task to show that the quadratic forms on the right of (26.51) have ranks 1, $(k-2)$ and $(n-k)$ respectively. Since these add to $(n-1)$, it follows from Cochran's theorem (**15.16**) that the three terms on the right are independently distributed in the chi-squared form with degrees of freedom equal to their ranks. By **16.15**, it follows that the ratio of any two of them (divided by their ranks) has an F distribution, with the appropriate degrees of freedom. We may use this fact in two ways to test H_0:

(a) The ratio of the first to the sum of the second and third terms, divided by their ranks,

$$\frac{r^2/1}{(1-r^2)/(n-2)} \quad \text{is } F_{1,\,n-2}, \quad (26.52)$$

suffixes denoting degrees of freedom.

This, it will be seen, is identical with the test of (26.37), since $t_{n-2}^2 \equiv F_{1,\,n-2}$ by **16.15**. We derived it at **16.28** for a bivariate normal population. Here we are taking the y's as fixed and the distribution within each x-array as normal.

(b) The ratio of the sum of the first and second terms to the third, divided by their ranks,

$$\frac{e_1^2/(k-1)}{(1-e_1^2)/(n-k)} \quad \text{is } F_{k-1,\,n-k}. \quad (26.53)$$

For both tests, large values of the test statistic lead to the rejection of H_0.

The tests based on (26.52) and (26.53) are quite distinct and are both valid tests of H_0, but (26.52) essentially tests $\rho^2 = 0$ while (26.53) tests $\eta_1^2 = 0$. If the alternative hypothesis is that the regression of x on y is linear, the test (26.52) will have higher power; but if the alternative is that the regression may be of any form other than that specified by H_0, (26.53) is evidently more powerful. It is almost universal practice to use (26.52) in the form of a linear regression test (**26.20**), but there certainly are situations to which (26.53) is more appropriate. We discuss the tests further in **26.24**, but first discuss the test of linearity of regression.

26.24 If the x_{ij} do not all have the same mean, the left-hand side of (26.51) is no longer a χ_{n-1}^2. However, whatever the means of the x_{ij} are, if we take the first term on the right over to the left, we get

$$n s_1^2 (1-r^2) \equiv n s_1^2 (e_1^2 - r^2) + n s_1^2 (1 - e_1^2). \quad (26.54)$$

Since
$$ns_1^2(1-r^2) \equiv \sum_i \sum_j \{x_{ij} - (a_1 + b_1 y_i)\}^2,$$

the sum of squared residuals from the fitted linear regression, we see that on the hypothesis H_0' that the regression of x on y is exactly linear, and distributions within arrays are normal as before, $ns_1^2(1-r^2)$ is distributed in the chi-squared form with $(n-2)$ degrees of freedom, one degree of freedom being lost for each parameter fitted in the regression line (cf. **19.9**). The ranks of the quadratic forms on the right of (26.54) are $(k-2)$ and $(n-k)$ as before, and they are therefore independently distributed in the chi-squared form with those degrees of freedom. Hence their ratio, after division by their ranks,

$$\frac{(e_1^2 - r^2)/(k-2)}{(1-e_1^2)/(n-k)} \text{ is } F_{k-2,\,n-k}. \tag{26.55}$$

(26.55) may be used to test H_0', the hypothesis of linearity of regression, H_0' being rejected for large values of the test statistic.

Thus our intuitive notion that $(e_1^2 - r^2)$ must afford a test of linearity of regression is correct, but (26.55) shows that the test result will be a function of $(1-e_1^2)$, k and n, so that a value of $(e_1^2 - r^2)$ alone means little.

All three tests which we have discussed in this and the last section are LR tests of linear hypotheses, of the type discussed in the second part of Chapter 24. For example, the hypothesis that all the variables x_{ij} have the same mean may be regarded in two ways: we may regard them as lying on a straight line which has two parameters, and test the hypothesis that the line has zero slope, which imposes one constraint on the two parameters. In the notation of **24.27-8**, $k=2$ and $r=1$, so that we get an F-test with $(1, n-2)$ degrees of freedom: this is (26.52). Alternatively, we may consider that the k array means are on a k-parameter curve (a polynomial of degree $(k-1)$, say), and test the hypothesis that all the polynomial's coefficients except the constant are zero, imposing $(k-1)$ constraints. We then get an F-test with $(k-1, n-k)$ degrees of freedom: this is (26.53). Finally, if in this second formulation we test the hypothesis that all the polynomial coefficients except the constant and the linear one are zero, so that the array means lie on a straight line, we impose $(k-2)$ constraints and get an F-test with $(k-2, n-k)$ degrees of freedom: this is (26.55).

It follows that for fixed values of y_i the results of Chapter 24 concerning the power of the LR test, based on the non-central F-distribution, are applicable to these tests, which are UMP invariant tests by **24.37**. However, the distributions in the bivariate normal case, which allow the y_i to vary, will *not* coincide with those derived by holding the y_i fixed as above, except when the hypothesis tested is true, when the variation of the y_i is irrelevant (as we shall see in **27.29**). For example, the distribution of r^2 obtained from the non-central F-distribution for (26.52) does not coincide with the bivariate normal result obtainable from (16.61) or (16.66). The power functions of the test of $\rho = 0$ are therefore different in the two cases, even though the same test is valid in each case. For large n, however, the results do coincide: we shall observe this more generally in connexion with the multiple correlation coefficient (of which r^2 is a special case) in **27.29** and **27.31**.

Intra-class correlation

26.25 There sometimes occur, mainly in biological work, cases in which we require the correlation between members of one or more families. We might, for example, wish to examine the correlation between heights of brothers. The question then arises, which is the first variate and which the second? In the simplest case we might have a number of families each containing two brothers. Our correlation table has two variates, both height, but in order to complete it we must decide which brother is to be related to which variate. One way of doing so would be to take the elder brother first, or the taller brother; but this would provide us with the correlation between elder and younger brothers, or between taller and shorter brothers, and not the correlation between brothers in general, which is what we require.

The problem is met by entering in the correlation table both possible pairs, i.e. those obtained by taking each brother first. If the family, or, more generally, the class, contains k members, there will be $k(k-1)$ entries, each member being taken first in association with each other member second. If there are p classes with $k_1, k_2, \ldots, k_p$ members there will be $\sum_{i=1}^{p} k_i(k_i - 1) = N$ entries in the correlation table.

As a simple illustration consider five families of three brothers with heights in inches respectively: 69, 70, 72; 70, 71, 72; 71, 72, 72; 68, 70, 70; 71, 72, 73. There will be 30 entries in the table, which will be as follows:

Table 26.5

Height (inches)

	68	69	70	71	72	73	TOTALS
68	–	–	2	–	–	–	2
69	–	–	1	–	1	–	2
70	2	1	2	1	2	–	8
71	–	–	1	–	4	1	6
72	–	1	2	4	2	1	10
73	–	–	–	1	1	–	2
TOTALS	2	2	8	6	10	2	30

Here, for example, the pair 69, 70 in the first family is entered as (69, 70) and (70, 69) and the pair 72, 72 in the third family *twice* as (72, 72).

The table is symmetrical about its leading diagonal, as it evidently must be. We may calculate the product-moment correlation coefficient in the usual way. We find $\sigma_1^2 = \sigma_2^2 = 1\cdot716$, $\mu_{11} = 0\cdot516$ and hence $\rho = \dfrac{0\cdot516}{1\cdot716} = 0\cdot301$.

A correlation coefficient of this kind is called an *intra-class* correlation coefficient. It can be found more directly as follows:

Suppose there are p classes with variate-values $x_{11}, \ldots, x_{1k_1}$; $x_{21}, \ldots, x_{2k_2}$; $\ldots$; $x_{p1}, \ldots, x_{pk_p}$. In the correlation table, each member of the ith class will appear $k_i - 1$ times (once in association with each other member of its class), and thus the mean of each variate is given by

$$\mu = \frac{1}{N} \sum_{i=1}^{p} (k_i - 1) \sum_{j=1}^{k_i} x_{ij},$$

and the variance of each variate by

$$\sigma^2 = \frac{1}{N} \sum_{i=1}^{p} (k_i - 1) \sum_{j=1}^{k_i} (x_{ij} - \mu)^2.$$

The covariance is

$$\mu_{11} = \frac{1}{N} \sum_{i=1}^{p} \sum_{\substack{j,l=1 \\ j \neq l}}^{k_i} (x_{ij} - \mu)(x_{il} - \mu)$$

$$= \frac{1}{N} \left\{ \sum_{i=1}^{p} \sum_{j,l=1}^{k_i} (x_{ij} - \mu)(x_{il} - \mu) - \sum_{i=1}^{p} \sum_{j=1}^{k_i} (x_{ij} - \mu)^2 \right\}$$

$$= \frac{1}{N} \left\{ \sum_{i=1}^{p} \left[\sum_{j=1}^{k_i} (x_{ij} - \mu) \right]^2 - \sum_i \sum_j (x_{ij} - \mu)^2 \right\}$$

$$= \frac{1}{N} \left\{ \sum_i k_i^2 (\mu_i - \mu)^2 - \sum_i \sum_j (x_{ij} - \mu)^2 \right\},$$

where μ_i is the mean of the ith class. Thus we have for the correlation coefficient

$$\rho = \frac{\sum\limits_i k_i^2 (\mu_i - \mu)^2 - \sum\limits_i \sum\limits_j (x_{ij} - \mu)^2}{\sum\limits_i (k_i - 1) \sum\limits_j (x_{ij} - \mu)^2}, \tag{26.56}$$

If $k_i = k$ for all i, (26.56) simplifies to

$$\rho = \frac{k^2 p \sigma_{\mu_i}^2 - k p \sigma^2}{(k-1) k p \sigma^2} = \frac{1}{k-1} \left(\frac{k \sigma_{\mu_i}^2}{\sigma^2} - 1 \right), \tag{26.57}$$

where $\sigma_{\mu_i}^2$ is the variance of class means, $\dfrac{1}{p} \sum\limits_{i=1}^{p} (\mu_i - \mu)^2$.

To distinguish the intra-class coefficient from the ordinary product-moment correlation coefficient ρ, we shall denote it by ρ_i and sample values of it by r_i.

Example 26.10

Let us use (26.57) to find the intra-class coefficient for the data of Table 26.5. With a working mean at 70 inches, the values of the variates are $-1, 0, 2$; $0, 1, 2$; $1, 2, 2$; $-2, 0, 0$; $1, 2, 3$.

Hence $\mu = \dfrac{13}{15}$, $\mu_2' = \dfrac{1}{15}\{(-1)^2 + 0^2 + \ldots\} = \dfrac{37}{15}$, and $\sigma^2 = \dfrac{386}{225}$.

The means of families, μ_i, are

$$\frac{5}{15}, \frac{15}{15}, \frac{25}{15}, \frac{-10}{15}, \frac{30}{15},$$

and their deviations from μ are

$$\frac{-8}{15}, \frac{2}{15}, \frac{12}{15}, \frac{-23}{15}, \frac{17}{15}.$$

Thus

$$\sigma_{\mu_i}^2 = \frac{1}{5}\left\{\left(\frac{-8}{15}\right)^2 + \ldots\right\} = \frac{1030}{1125}.$$

Hence, from (26.57),

$$\rho_i = \frac{1}{2}\left\{\frac{3.1030.225}{1125.386} - 1\right\} = 0\cdot 301,$$

a result we have already found directly in **26.25**.

26.26 Caution is necessary in the interpretation of the intra-class correlation coefficient. From (26.57) it is seen that ρ_i cannot be less than $\frac{-1}{k-1}$, though it may attain $+1$ when $\sigma_{\mu_i}^2 = \sigma^2$. It is thus a skew coefficient in the sense that a negative value has not the same significance (as a departure from independence) as the equivalent positive value.

In point of fact, the intra-class coefficient is, from most points of view, more con- and within classes in the Analysis of Variance. Fisher (1921c) thus derived the distribution of intra-class r_i in normal samples when families are of the same size k. When $k = 2$, he found, as for the product-moment coefficient r, that the transformation

$$z = \operatorname{ar\,tanh} r_i$$

gives a statistic (z) very nearly normally distributed with mean $\zeta = \operatorname{ar\,tanh} \rho_i$ and variance independent of ρ_i. For $k > 2$, a more complicated transformation is necessary. His results are given in Exercise 26.14.

Tetrachoric correlation

26.27 We now discuss the estimation of ρ in a bivariate normal population when the data are not given in full detail. We take first of all an extreme case exemplified by Table 26.6. This is based on the distribution of cows according to age and milk-yield given in Table 1.24, Exercise 1.4. Suppose that, instead of being given that table we had only

Table 26.6—Cows by age and milk-yield

	Age 6 and over	Age 3–5 years	TOTAL
Yield 8–18 galls.	1078	1407	2485
Yield 19 galls. and over	1546	881	2427
TOTAL	2624	2288	4912

LINEAR REGRESSION AND CORRELATION

This is a highly condensed version of the original. Suppose we assume that the underlying distribution is bivariate normal. How can we estimate ρ from this table? In general, for a table of this "2×2" type with frequencies

$$\begin{array}{cc|c} a & b & a+b \\ c & d & c+d \\ \hline a+c & b+d & a+b+c+d = n \end{array} \qquad (26.58)$$

we require to estimate ρ. In (26.58) we shall always take d to be a frequency such that neither of its marginal frequencies contain the median value of the variate.

If this table is derived by a double dichotomy of the bivariate normal distribution $f(x, y)$ defined at the beginning of Example 18.14, we can find h' such that

$$\int_{-\infty}^{h'} \int_{-\infty}^{\infty} f(x,y) \, dx \, dy = \frac{a+c}{n}. \qquad (26.59)$$

Putting $h = (h' - \mu_1)/\sigma_1$, we find this is

$$F(h) = (2\pi)^{-\frac{1}{2}} \int_{-\infty}^{h} \exp\left(-\tfrac{1}{2}x^2\right) dx = \frac{a+c}{n}, \qquad (26.60)$$

and thus h is determinable from tables of the univariate normal distribution function. Likewise there is a k such that

$$F(k) = (2\pi)^{-\frac{1}{2}} \int_{-\infty}^{k} \exp\left(-\tfrac{1}{2}y^2\right) dy = \frac{a+b}{n}. \qquad (26.61)$$

On our convention as to the arrangement of table (26.58), h and k are never negative.

Having fitted univariate normal distributions to the marginal frequencies of the table in this way, we now solve for ρ the equation

$$\frac{d}{n} = \int_{h}^{\infty} \int_{k}^{\infty} \frac{1}{2\pi(1-\rho^2)^{\frac{1}{2}}} \exp\left\{\frac{-1}{2(1-\rho^2)}(x^2 - 2\rho xy + y^2)\right\} dx \, dy. \qquad (26.62)$$

The choice of d, rather than a, b or c, to determine the estimator of ρ is convenient but loses no generality—we shall see that the estimator is a function of $(ad - bc)$ only.

The characteristic function of the integrand is (Example 15.1)

$$\phi(t, u) = \exp\{-\tfrac{1}{2}(t^2 + 2\rho t u + u^2)\}.$$

Thus, using the bivariate form of the Inversion Theorem (4.17), (26.62) becomes

$$\frac{d}{n} = \int_{h}^{\infty} \int_{k}^{\infty} \left\{\frac{1}{4\pi^2} \int_{-\infty}^{\infty} \int_{-\infty}^{\infty} \phi(t,u) \exp(-itx - iuy) \, dt \, du\right\} dx \, dy$$

$$= \int_{h}^{\infty} \int_{k}^{\infty} \left\{\frac{1}{4\pi^2} \int_{-\infty}^{\infty} \int_{-\infty}^{\infty} \exp\{-\tfrac{1}{2}(t^2 + u^2) - itx - iuy\} \sum_{j=0}^{\infty} \frac{(-\rho)^j t^j u^j}{j!} \, dt \, du\right\} dx \, dy. \qquad (26.63)$$

The coefficient of $(-\rho)^j/j!$ is the product of two integrals, of which the first is

$$I(x, h, t) = \int_{h}^{\infty} \left\{\frac{1}{2\pi} \int_{-\infty}^{\infty} t^j \exp\left(-\tfrac{1}{2}t^2 - itx\right) dt\right\} dx \qquad (26.64)$$

and the second is $I(y, k, u)$. Now from **6.18** the integral in braces in (26.64) is equal to
$$(-i)^j H_j(x) \alpha(x)$$
where
$$\alpha(x) = (2\pi)^{-\frac{1}{2}} \exp(-\tfrac{1}{2}x^2).$$
For $j = 0$, this is simply $\alpha(x)$. For $j \geq 1$, we have from (6.21),
$$-\frac{d}{dx}\{H_{j-1}(x)\alpha(x)\} = H_j(x)\alpha(x).$$
Hence the double integral in (26.64) is, for $j \geq 1$,
$$I(x, h, t) = \left[(-1)^{j-1} i^j H_{j-1}(x)\alpha(x)\right]_h^\infty = (-i)^j H_{j-1}(h)\alpha(h). \tag{26.65}$$
For $j = 0$, the double integral is $\{1-F(h)\}\{1-F(k)\} = \dfrac{(b+d)}{n} \cdot \dfrac{(c+d)}{n}$ by (26.60–1). Substituting this and (26.65) for $I(x, h, t)$, $I(y, k, u)$ in (26.63), we have
$$\frac{d}{n} - \frac{(b+d)(c+d)}{n^2} = \frac{ad-bc}{n^2} = \alpha(h)\alpha(k) \sum_{j=1}^\infty \frac{\rho^j}{j!} H_{j-1}(h) H_{j-1}(k). \tag{26.66}$$
In terms of the tetrachoric functions which were defined at (6.44) for the purpose,
$$\frac{d}{n} = \sum_{j=0}^\infty \rho^j \tau_j(h) \tau_j(k). \tag{26.67}$$

26.28 Formally, (26.67) provides a soluble equation for ρ, but in practice the solution by successive approximation can be very tedious. (The series (26.67) always converges, but may do so slowly.) It is simpler to interpolate in tables which have been prepared giving the integral d/n in terms of ρ for various values of h and k (*Tables for Statisticians and Biometricians*, Vol. 2).

The estimate of ρ derived from a sample of n in this way is known as *tetrachoric r*. and is due to K. Pearson. We shall denote it by r_t.

Example 26.11

For the data of Table 26.6 we find the normal deviate corresponding to $2624/4912 = 0\cdot5342$ as $h = 0\cdot086$, and similarly for $2485/4912 = 0\cdot5059$ we find $k = 0\cdot015$. We have also for d/n the value $881/4912 = 0\cdot1794$.

From the tables, we find for varying values of h, k and ρ the following values of d:

		$h = 0$	$h = 0\cdot1$			$h = 0$	$h = 0\cdot1$
$\rho = -0\cdot30$	$k = 0$	$0\cdot2015$	$0\cdot1818$	$\rho = -0\cdot35$	$k = 0$	$0\cdot1931$	$0\cdot1735$
	$k = 0\cdot1$	$0\cdot1818$	$0\cdot1639$		$k = 0\cdot1$	$0\cdot1735$	$0\cdot1555$

Linear interpolation gives us for $h = 0\cdot086$, $k = 0\cdot015$, the result $\rho = -0\cdot32$ approximately. In the table, we have inverted the order of columns, and taking account of this gives us an estimate of $\rho = +0\cdot32$. We therefore write $r_t = +0\cdot32$. (The product-moment coefficient for Table 1.24 is $r = 0\cdot22$.)

26.29 Tetrachoric r_t has been used mainly by psychologists, whose material is often of the 2×2 type. Karl Pearson (1913) gave a complicated asymptotic expression for its standard error. There are, however, simpler methods of calculation based on nomograms (Hayes (1946); Hamilton (1948); Jenkins (1955)) and tables for the standard error in approximate form (Guilford and Lyons (1942); Hayes (1943); Goheen and Kavruck (1948)).

Since the estimation procedure equates observed frequencies with the corresponding probabilities (say, θ) in the bivariate normal distribution, it is using the ML estimator $\hat{\theta}$. Now the ML estimator of ρ, which is a function $\rho(\theta)$, is $\rho(\hat{\theta})$. Thus r_t is the ML estimator of ρ from a 2×2 table. Its approximate large-sample variance, obtained from **18.18** by Hamdan (1970), is $\left\{n^2\left(\frac{1}{a}+\frac{1}{b}+\frac{1}{c}+\frac{1}{d}\right)f^2(h,k\,|\,r_t)\right\}^{-1}$, where $f(x,\,y\,|\rho)$ is the standardized bivariate normal distribution.

For a generalization to polychoric estimation in $r \times c$ tables, see **33.35** below.

Biserial correlation

26.30 Suppose now that we have a $(2 \times q)$-fold table, the dichotomy being according to some qualitative factor and the other classification either to a numerical variate or to a qualitative one, which may or may not be ordered.

Table 26.7 will illustrate the type of material under discussion. The data relate to 1426 criminals classified according to whether they were alcoholic or not and according

Table 26.7—Showing 1426 criminals classified according to alcoholism and type of crime

(C. Goring's data, quoted by K. Pearson, 1909)

	Arson	Rape	Violence	Stealing	Coining	Fraud	TOTALS
Alcoholic . . .	50	88	155	379	18	63	753
Non-alcoholic .	43	62	110	300	14	144	673
TOTALS . .	93	150	265	679	32	207	1426

to the crime for which they were imprisoned. Even though the columns of the table are not unambiguously ordered (they are shown arranged in order of an association of the crimes with intelligence, but this ordering is somewhat arbitrary), we may still derive an estimate of ρ on the assumption that there is an underlying bivariate normal distribution. For in such a distribution, $\rho^2 = \eta^2$, the regressions both being linear, and we remarked in **26.21** that η^2 is invariant under permutation of arrays. We therefore proceed to estimate $\eta^2 (= \rho^2)$ as follows.

Consider each column of Table 26.7 as a y-array, and let n_p be the number of observations in the pth array, $n = \Sigma n_p$, μ_p the mean of y in that array, μ_y the mean and σ_y^2 the variance of y, and σ_p^2 the variance of y in the pth array. We suppose all measurements in y to be made from the value k which is the point of dichotomy; this involves no loss of generality, since ρ^2 and η^2 are invariant under a change of origin. Then the correlation ratio of y on x (cf. (26.40)) is estimated by

$$\frac{\frac{1}{n}\sum_{p=1}^{q} n_p \mu_p^2 - \mu_y^2}{\sigma_y^2} = \frac{1}{n}\sum_{p=1}^{q} \frac{n_p \mu_p^2}{\sigma_p^2} \cdot \frac{\sigma_p^2}{\sigma_y^2} - \frac{\mu_y^2}{\sigma_y^2}. \qquad (26.68)$$

But for the bivariate normal distribution $\eta^2 = \rho^2$ and (cf. **16.23**)

$$\sigma_p^2/\sigma_y^2 = \mathrm{var}\,(y\,|\,x)/\sigma_y^2 = (1-\rho^2),$$

so we replace σ_p^2/σ_y^2 by $(1-\rho^2)$ in (26.68), obtaining

$$\rho^2 \doteq \frac{1-\rho^2}{n}\sum_{p=1}^{q} \frac{n_p \mu_p^2}{\sigma_p^2} - \frac{\mu_y^2}{\sigma_y^2}, \qquad (26.69)$$

which we solve for ρ^2 to obtain the estimator

$$r_\eta^2 = \frac{\frac{1}{n}\sum_{p=1}^{q} n_p \left(\frac{\mu_p}{\sigma_p}\right)^2 - \left(\frac{\mu_y}{\sigma_y}\right)^2}{1 + \frac{1}{n}\sum_{p=1}^{q} n_p \left(\frac{\mu_p}{\sigma_p}\right)^2}. \qquad (26.70)$$

This estimator is known as *biserial* η because of the analogy with the correlation ratio. We shall write it as r_η when estimating from a sample, to maintain our convention about the use of Roman letters for statistics.

The use of the expression (26.70) lies in the fact that the quantities in it can be estimated from the data. Our assumption that there is an underlying bivariate normal distribution implies that the quantity according to which dichotomy has been made (in our example, alcoholism) is capable of representation by a variate which is normally distributed, and that each y-array is a dichotomy of a univariate normal distribution. Thus the ratios (μ_p/σ_p) and (μ_y/σ_y) can be estimated from the tables of the normal integral. For example, in Table 26.7, the two frequencies " alcoholic " and " non-alcoholic " are, for arson, 50 and 43. Thus the proportional frequency in the alcoholic group is $50/93 = 0\cdot5376$ and the normal deviate corresponding to this frequency is seen from the tables to be $0\cdot0944$, which is thus an estimate of $|\mu_p/\sigma_p|$ for this array.

Example 26.12

For the data of Table 26.7, the proportional frequencies, the estimated values of $|\mu_p/\sigma_p|$ and $|\mu_y/\sigma_y|$, and the n_p are:

	Arson	Rape	Violence	Stealing	Coining	Fraud	TOTALS
Alcoholic	0·5376	0·5867	0·5849	0·5582	0·5625	0·3043	0·5281
$\|\mu_p/\sigma_p\|$	0·0944	0·2190	0·2144	0·1463	0·1573	0·5119	$0\cdot0704 = \|\mu_y/\sigma_y\|$
n_p	93	150	265	679	32	207	$1426 = n$

Then from (26.70) we have

$$r_\eta^2 = \frac{\frac{1}{1426}\{93\,(0\cdot0944)^2 + \ldots\} - (0\cdot0704)^2}{1 + \frac{1}{1426}\{93\,(0\cdot0944)^2 + \ldots\}} = 0\cdot05456$$

or
$$|r_\eta| = 0.234,$$
which, on our assumptions, may be taken as estimating the supposed product-moment correlation coefficient.

26.31 As for the tetrachoric r_t, the sampling distribution of biserial r_η is unknown. An asymptotic expression for its sampling variance was derived by K. Pearson (1917), but it is not known how large n must be for this to be valid.

Neither r_t nor r_η can be expected to estimate ρ very efficiently, since they are based on so little information about the variables, and it should be (though it has not always been) remembered that the assumption of underlying bivariate normality is crucial to both methods. In the absence of the normality assumption, we do not know what r_t and r_η are estimating in general.

26.32 If in the $(2 \times q)$-fold table the q-fold classification, instead of being defined by an unordered classification as in Table 26.7, is actually given by variate-value, we may proceed directly to estimate ρ instead of η. For we may now use the extra information to estimate the variance σ_x^2 of this measured variate and its means, μ_1, μ_2, in the two halves of the dichotomy according to y. Since the regression of x on y is linear we have (cf. (26.12))

$$E(x|y) - \mu_x = \rho \frac{\sigma_x}{\sigma_y}(y - \mu_y). \tag{26.71}$$

We can, as in **26.27**, find k such that

$$1 - F(k) = (2\pi)^{-\frac{1}{2}} \int_k^\infty \exp(-\tfrac{1}{2}u^2)\, du = \frac{n_1}{n_1 + n_2}, \tag{26.72}$$

where n_1 is the total number of individuals bearing one attribute of the y-class ("higher" values of y) and n_2 is the number bearing the other. k is the point of dichotomy of the normal distribution of y.

From (26.71), the means (y_i, μ_i), $(i = 1, 2)$ of each part of the dichotomy will be on the regression line (26.71). Thus, for the part of the dichotomy with the "higher" value of y, say y_1,

$$\rho = \left(\frac{E(x|y_1) - \mu_x}{\sigma_x}\right) \Big/ \left(\frac{y_1 - \mu_y}{\sigma_y}\right).$$

Thus we may estimate ρ by

$$\left(\frac{\bar{x}_1 - \bar{x}}{s_x}\right) \Big/ \left(\frac{y_1 - \mu_y}{\sigma_y}\right), \tag{26.73}$$

where $\bar{x}_1, \bar{x}$ are the means of x in the "high-y" observations and the whole table respectively, while s_x^2 is the observed variance of x in the whole table. The denominator of (26.73) is given by

$$\frac{y_1 - \mu_y}{\sigma_y} = (2\pi)^{-\frac{1}{2}} \int_k^\infty u \exp(-\tfrac{1}{2}u^2)\, du \Big/ (2\pi)^{-\frac{1}{2}} \int_k^\infty \exp(-\tfrac{1}{2}u^2)\, du$$
$$= (2\pi)^{-\frac{1}{2}} \exp(-\tfrac{1}{2}k^2) \Big/ \left(\frac{n_1}{n_1 + n_2}\right) \tag{26.74}$$

by (26.72).

If, then, we denote the ordinate of the normal distribution at k by z_k, we have the estimator of ρ

$$r_b = \left(\frac{\bar{x}_1 - \bar{x}}{s_x}\right)\frac{n_1}{(n_1+n_2)}\frac{1}{z_k}.$$

We write the estimator based on this equation as r_b, the suffix denoting "biserial": r_b is called "biserial r."

The equation is usually put in a more symmetrical form. Since
$$\bar{x} = (n_1\bar{x}_1 + n_2\bar{x}_2)/(n_1+n_2),$$
$\bar{x}_1 - \bar{x}$ is equal to $n_2(\bar{x}_1 - \bar{x}_2)/(n_1+n_2)$. Writing p for the proportion $n_1/(n_1+n_2)$ and $q = 1-p$, we have the alternative expression for (26.74)

$$r_b = \frac{\bar{x}_1 - \bar{x}_2}{s_x}\frac{p\,q}{z_k}. \tag{26.75}$$

Example 26.13 (from K. Pearson, 1909)

Table 26.8 shows the returns for 6156 candidates for the London University Matriculation Examination for 1908/9. The average ages for the two higher age-groups have been estimated.

Table 26.8

Age of candidate	Passed	Failed	TOTALS
16	583	563	1146
17	666	980	1646
18	525	868	1393
19–21	383	814	1197
22–30 (mean 25)	214	439	653
over 30 (mean 33)	40	81	121
TOTALS	2411	3745	6156

Taking the suffix "1" as relating to successful candidates, we have
$$\bar{x}_1 = 18\cdot 4280.$$
For all candidates together
$$\bar{x} = 18\cdot 7685, \quad s_x^2 = (3\cdot 2850)^2.$$
The value of p is $2411/6156 = 0\cdot 3917$.

(26.72) gives $1 - F(k) = 0\cdot 3917$, and we find $k = 0\cdot 275$ and $z_k = 0\cdot 384$. Hence, from (26.74),
$$r_b = -\frac{0\cdot 3405}{3\cdot 2850}\cdot\frac{0\cdot 3917}{0\cdot 384} = -0\cdot 11.$$
The estimated correlation between age and success is small.

26.33 As for r_t and r_n, the assumption of underlying normality is crucial to r_b. The distribution of biserial r_b is not known, but Soper (1914) derived the expression for its variance in normal samples

$$\operatorname{var} r_b \sim \frac{1}{n}\left[\rho^4 + \rho^2\left\{\frac{pqk^2}{z_k^2} + (2p-1)\frac{k}{z_k} - \frac{5}{2}\right\} + \frac{pq}{z_k^2}\right], \tag{26.76}$$

and showed that (26.76) is generally well approximated by
$$\operatorname{var} r_b \sim \frac{1}{n}\left[r_b^2 - \frac{(pq)^{\frac{1}{2}}}{z_k}\right]^2.$$

More recently r_b has been extensively studied by Maritz (1953) and by Tate (1955), who showed that in normal samples it is asymptotically normally distributed with mean ρ and variance (26.76), and considered the Maximum Likelihood estimation of ρ in biserial data. It appears, as might be expected, that the variance of r_b is least, for fixed ρ, when the dichotomy is at the middle of the dichotomized variate's range ($y = 0$). When $\rho = 0$, r_b is an efficient estimator of ρ, but when $\rho^2 \to 1$ the efficiency of r_b tends to zero. Tate also tables Soper's formula (26.76) for $\operatorname{var} r_b$. Cf. Exercises 26.10–12.

Point-biserial correlation

26.34 This is a convenient place at which to mention another coefficient, the *point-biserial correlation*, which we shall denote by ρ_{pb}, and by r_{pb} for a sample. Suppose that the dichotomy according to y is regarded, not as a section of a normal distribution, but as defined by a variable taking two values only. So far as correlations are concerned, we can take these values to be 1 and 0. For example, in Table 26.8 it is not implausible to suppose that success in the examination is a dichotomy of a normal distribution of ability to pass it. But if the y-dichotomy were according, say, to sex, this is no longer a reasonable assumption and a different approach is necessary.

Such a situation is, in fact, fundamentally different from the one we have so far considered, for we are now no longer estimating ρ in a bivariate normal population: we consider instead the product-moment of a $0-1$ variable y and the variable x. If P is the true proportion of values of y with $y = 1$, $Q = 1-P$, we have from binomial distribution theory
$$E(y) = P, \quad \sigma_y^2 = PQ$$
and thus, by definition,
$$\rho_{pb} = \frac{\mu_{11}}{\sigma_x \sigma_y} = \frac{E(xy) - PE(x)}{\sigma_x (PQ)^{\frac{1}{2}}}.$$

We estimate $E(xy)$ by $m_{11} = \frac{1}{n_1 + n_2} \sum_{i=1}^{n_1} x_i$, $E(x)$ by $\bar{x}$, σ_x by s_x, and P by $p = \frac{n_1}{n_1 + n_2}$, obtaining
$$\begin{aligned}r_{pb} &= \frac{p\bar{x}_1 - p(p\bar{x}_1 + q\bar{x}_2)}{s_x (pq)^{\frac{1}{2}}} \\ &= \frac{(\bar{x}_1 - \bar{x}_2)(pq)^{\frac{1}{2}}}{s_x}.\end{aligned} \qquad (26.77)$$

26.35 r_{pb} in (26.77) may be compared with the biserial r_b defined at (26.75). We have
$$\frac{r_{pb}}{r_b} = \frac{z_k}{(pq)^{\frac{1}{2}}}. \qquad (26.78)$$

It has been shown by Tate (1953) by a consideration of Mills' ratio (cf. **5.22**) that the expression on the right of (26.78) is $\leq (2/\pi)^{\frac{1}{2}}$ and the values of the coefficients will thus, in general, be appreciably different.

Tate (1954) shows that r_{pb} is asymptotically normally distributed with mean ρ_{pb} and variance

$$\operatorname{var} r_{pb} \sim \frac{(1-\rho_{pb}^2)^2}{n}\left(1-\frac{3}{2}\rho_{pb}^2+\frac{\rho_{pb}^2}{4pq}\right), \qquad (26.79)$$

which is a minimum when $p = q = \frac{1}{2}$.

Apart from the measurement of correlation, it is clear from (26.77) that, in effect, for a point-biserial situation, we are simply comparing the means of two samples of a variate x, the y-classification being no more than a labelling of the samples. In fact

$$\frac{r_{pb}^2}{1-r_{pb}^2} = \frac{(\bar{x}_1-\bar{x}_2)\dfrac{n_1 n_2}{n_1+n_2}}{\Sigma(x_{1i}-\bar{x}_1)^2+\Sigma(x_{2i}-\bar{x}_2)^2} = \frac{t^2}{n_1+n_2-2}, \qquad (26.80)$$

where t is the usual "Student's" t-test used for comparing the means of two normal populations with equal variance (cf. Example 23.8). Thus if the distribution of x is normal for $y = 0, 1$, the point-biserial coefficient is a simple transformation of the t^2 statistic, which may be used to test it.

26.36 The above account does not exhaust the possible estimators of bivariate normal ρ from data which are classified in a two-way table. In Chapter 33 we shall discuss some estimators based on rank-order statistics.

EXERCISES

26.1 Show that the correlation coefficient for the data of Table 26.2 is $+0.072$. Show that the regression lines in Fig. 26.2 are:
$$CC' : y = 0.0938x + 30.56; \quad RR' : x = 0.0547y + 61.06.$$

26.2 If x_i/σ^2 ($i = 1, 2, 3$) are independent χ^2 variates with ν_i degrees of freedom, $y_1 = x_1/x_3$ and $y_2 = x_2/x_3$, show that the joint distribution of y_1 and y_2 is
$$g(y_1, y_2) = \frac{\Gamma(\frac{1}{2}\nu)}{\prod_{i=1}^{3} \Gamma(\frac{1}{2}\nu_i)} \frac{y_1^{\frac{1}{2}\nu_1 - 1} y_2^{\frac{1}{2}\nu_2 - 1}}{(1 + y_1 + y_2)^{\frac{1}{2}\nu}}, \quad 0 \leq y_1, y_2 \leq \infty,$$
where $\nu = \sum_{i=1}^{3} \nu_i$. Show that the regression of y_1 on y_2, and of y_2 on y_1, is linear. If $\nu_3 > 4$, show that their correlation coefficient is
$$\rho = [\nu_1 \nu_2 / \{(\nu_1 + \nu_3 - 2)(\nu_2 + \nu_3 - 2)\}]^{\frac{1}{2}}.$$

26.3 A bivariate normal distribution is dichotomized at some value of y. The variance of x for the whole distribution is known to be σ_x^2 and that for one part of the dichotomy is σ_1^2. The correlation between x and y for the latter is c. Show that the correlation of x and y in the whole distribution may be estimated by r, where
$$r^2 = 1 - \frac{\sigma_1^2}{\sigma_x^2}(1 - c^2).$$

26.4 In the previous exercise, if σ_y^2 is the variance of y in the whole distribution and σ_2^2 is its variance in the part of the dichotomy, show that ρ for the whole distribution may be estimated by
$$r^2 = \frac{c^2 \sigma_y^2}{\sigma_2^2 + c^2(\sigma_y^2 - \sigma_2^2)}.$$

26.5 Show that whereas tetrachoric r_t, biserial r_η, and point-biserial r_{pb} can never exceed unity in absolute value, biserial r_b may do so.

26.6 Prove that the tetrachoric series (26.67) always converges for $|\rho| < 1$.

26.7 A set of variables $x_1, x_2, \ldots, x_n$ are distributed so that the product-moment correlation of x_i and x_j is ρ_{ij}. They all have the same variance. Show that the average value of ρ_{ij} defined by
$$\bar{\rho} = \frac{1}{n(n-1)} \sum_{i=1}^{n} \sum_{j=1}^{n} \rho_{ij}, \quad i \neq j,$$
must be not less than $-1/(n-1)$.

26.8 In the previous exercise show that $|\rho_{ij}|$, the determinant of the array of correlation coefficients, is non-negative. Hence show that
$$\rho_{12}^2 + \rho_{13}^2 + \rho_{23}^2 \leq 1 + 2\rho_{12}\rho_{13}\rho_{23}.$$

26.9 Show from (16.86) that in samples from a bivariate normal population the sampling distribution of b_2, the regression coefficient of y on x, has exact variance
$$\operatorname{var} b_2 = \frac{1}{n-3} \frac{\sigma_2^2}{\sigma_1^2}(1 - \rho^2), \quad n \geq 4,$$

and that its skewness and kurtosis coefficients are
$$\gamma_1 = 0, \quad n \geqslant 5,$$
$$\gamma_2 = \frac{6}{n-5}, \quad n \geqslant 6.$$

Show that when $\rho = 0$, var b_2 is the expectation of the variance given for fixed x's in Example 19.6, which is in our present notation var $(b_2 | x) = \sigma_2^2/(ns_1^2)$.

26.10 Let $\psi\{(x-\mu)/\sigma, y\}$ denote the bivariate normal frequency with means of x and y equal to μ and 0 respectively, variances equal to σ^2 and 1 respectively, and correlation ρ. Define
$$\xi(x, \omega) = \int_\omega^\infty \psi\,dy, \quad \eta(x, \omega) = \int_{-\infty}^\omega \psi\,dy.$$

If z_i is a random variable taking the values 0, 1 according as $y < \omega$ or $y \geqslant \omega$, show that in a biserial table the Likelihood Function may be written
$$L(x, y | \omega, \rho, \mu, \sigma) = \prod_{i=1}^n \left\{ z_i \xi\left(\frac{x_i - \mu}{\sigma}, \omega\right) + (1 - z_i) \eta\left(\frac{x_i - \mu}{\sigma}, \omega\right) \right\}.$$

If ∂^2 represents a partial differential of the second order with respect to any pair of parameters, show that
$$E(\partial^2 \log L) = n\{1 - p(x)\} E_0(\partial^2 \log \eta) + np(x) E_1(\partial^2 \log \xi)$$
where
$$p(x) = \int_x^\infty (2\pi)^{-\frac{1}{2}} \exp(-\tfrac{1}{2}t^2)\,dt,$$
and E_0, E_1 are conditional expectations with respect to x for $y < \omega$, $y \geqslant \omega$ respectively. Hence derive the inverse of the dispersion matrix for the Maximum Likelihood estimators of the four parameters (the order of rows and columns being the same as the order of the parameters in the LF):

$$\mathbf{V}^{-1} = \frac{n}{(1-\rho^2)} \begin{pmatrix} a_0 & \dfrac{\rho\omega a_0 - a_1}{1-\rho^2} & \dfrac{\rho a_0}{\sigma} & \dfrac{\rho a_1}{\sigma} \\ & \dfrac{a_2 - 2\rho\omega a_1 + \rho^2 \omega^2 a_0}{(1-\rho^2)^2} & \dfrac{\rho^2 \omega a_0 - \rho a_1}{\sigma(1-\rho^2)} & \dfrac{\rho^2 \omega a_1 - \rho a_2}{\sigma(1-\rho^2)} \\ & & \dfrac{1-\rho^2 + \rho^2 a_0}{\sigma^2} & \dfrac{\rho^2 a_1}{\sigma^2} \\ & & & \dfrac{2(1-\rho^2) + \rho^2 a_2}{\sigma^2} \end{pmatrix}$$

where
$$a_s = \int_{-\infty}^\infty x^s g(x, \omega, \rho)\,dx,$$
$$g(x, \omega, \rho) = (2\pi)^{-\frac{1}{2}} \exp(-\tfrac{1}{2}x^2)\,\phi\left(\frac{\omega - \rho x}{(1-\rho^2)^{\frac{1}{2}}}\right) \phi\left(\frac{\rho x - \omega}{(1-\rho^2)^{\frac{1}{2}}}\right),$$
and
$$\phi(x) = (2\pi)^{-\frac{1}{2}} \exp(-\tfrac{1}{2}x^2)/\{1 - p(x)\}.$$

By inverting this matrix, derive the asymptotic variance of the Maximum Likelihood estimator $\hat{\rho}_b$ in the form
$$\operatorname{var} \hat{\rho}_b = \frac{(1-\rho^2)^3}{n} \left\{ \frac{\int_{-\infty}^\infty g\,dx}{\int_{-\infty}^\infty g\,dx \int_{-\infty}^\infty x^2 g\,dx - \left(\int_{-\infty}^\infty xg\,dx\right)^2} \right\} + \frac{\rho^2(1-\rho^2)}{n}.$$

(Tate, 1955)

26.11 In Exercise 26.10, show that when $\rho = 0$,
$$\operatorname{var} \hat{\rho}_b = \frac{2\pi p(\omega)\{1-p(\omega)\}}{n \exp(-k^2)}.$$

By comparing this with the large-sample formula (26.76), show that when $\rho = 0$, r_b is a fully efficient estimator.

(Tate, 1955)

26.12 In Exercise 26.10, show that $n \operatorname{var} \hat{\rho}_b$ tends to zero as $|\rho|$ tends to unity, and from (26.76) that $n \operatorname{var} r_b$ does not, and hence that r_b is of zero efficiency near $|\rho| = 1$.

(Tate, 1955; the results of Exercises 26.10–12 are extended to the multinormal distribution by J. F. Hannan and Tate (1965).)

26.13 Establish equations (26.19) and (26.20).

26.14 Writing l for the sample intra-class correlation coefficient (26.57) and λ for the parent value, show that the exact distribution of l is given by
$$dF \propto \frac{(1-l)^{\frac{1}{2}p(k-1)-1}\{1+(k-1)l\}^{\frac{1}{2}(p-3)}\,dl}{\{1-\lambda+\lambda(k-1)(1-l)\}^{\frac{1}{2}(kp-1)}},$$
reducing in the case $k = 2$ to
$$dF = \frac{\Gamma(p-\tfrac{1}{2})}{\Gamma(p-1)(2\pi)^{\frac{1}{2}}} \operatorname{sech}^{p-\frac{1}{2}}(z-\xi) \exp\{-\tfrac{1}{2}(z-\xi)\}$$
where $l = \tanh z$, $\lambda = \tanh \xi$. Hence show that, for $k = 2$, $z - \xi$ is nearly normal with mean zero and variance $\dfrac{1}{n-3/2}$.

(Fisher, 1921c)

26.15 Show that for testing $\rho = \rho_0$ in a bivariate normal population, the Likelihood Ratio statistic is given by
$$l^{1/n} = \frac{(1-r^2)^{\frac{1}{2}}(1-\rho_0^2)^{\frac{1}{2}}}{(1-r\rho_0)},$$
so that $l^{1/n} = (1-r^2)^{\frac{1}{2}}$ when $\rho_0 = 0$, and when $\rho_0 \neq 0$ we have
$$(1-\rho_0^2)^{-\frac{1}{2}} l^{1/n} = 1 + r\rho_0 + r^2(\rho_0^2 - \tfrac{1}{2}) + \dots$$

26.16 Show that the effect of applying Sheppard's corrections to the moments is always to increase the value of the correlation coefficient.

26.17 Show that if x and y are respectively subject to errors of observation u, v, where u and v are uncorrelated with x, y and each other, the correlation coefficient is reduced ("attenuated") by a factor
$$\left\{\left(1+\frac{\sigma_u^2}{\sigma_x^2}\right)\left(1+\frac{\sigma_v^2}{\sigma_y^2}\right)\right\}^{\frac{1}{2}}.$$

26.18 If x_i ($i = 1, 2, 3$) are mutually independent variates with means μ_i, variances σ_i^2 and coefficients of variation $V_i = \sigma_i/\mu_i$, show that the correlation between x_1/x_3 and x_2/x_3 is
$$\rho = \frac{\mu_1 \mu_2}{\left\{\sigma_1^2\left(1+\dfrac{1}{V_4^2}\right)+\mu_1^2\right\}^{\frac{1}{2}} \left\{\sigma_2^2\left(1+\dfrac{1}{V_4^2}\right)+\mu_2^2\right\}^{\frac{1}{2}}}$$

exactly, where V_4 is the coefficient of variation of $1/x_3$. Thus $\rho = 0$ if either μ_1 or $\mu_2 = 0$; if neither is, ρ takes the sign of their product, and

$$|\rho| = \left\{1 + V_1^2\left(1 + \frac{1}{V_4^2}\right)\right\}^{-\frac{1}{2}} \left\{1 + V_2^2\left(1 + \frac{1}{V_4^2}\right)\right\}^{-\frac{1}{2}}.$$

Show that if V_4^2 is small, we obtain approximately

$$|\rho| = V_3^2 / \{(V_1^2 + V_3^2)(V_2^2 + V_3^2)\}^{\frac{1}{2}},$$

since $V_3 \doteqdot V_4$ by the argument of **10.7** (Vol. 1).

(The approximation was differently obtained by K. Pearson (1897) who dealt with the case where μ_1, μ_2 and therefore ρ are positive; he also treated the case where x_1, x_2 and x_3 are correlated. He called ρ a " spurious " correlation because the original x_i are uncorrelated, but the term is inapt if one is fundamentally interested in the ratios.)

26.19 If two bivariate normal populations have $\rho_1 = \rho_2 = \rho$, the other parameters being unspecified, show that the Maximum Likelihood estimator of ρ is

$$\hat{\rho} = \frac{n(1 + r_1 r_2) - \{n^2(1 - r_1 r_2)^2 - 4n_1 n_2 (r_1 - r_2)^2\}^{\frac{1}{2}}}{2(n_1 r_2 + n_2 r_1)},$$

where n_i, r_i are the sample sizes and correlation coefficients ($i = 1, 2$) and $n = n_1 + n_2$. If $n_1 = n_2 = \frac{1}{2}n$, show that if z_1, z_2 are defined by (26.25), and

$$\zeta = \tfrac{1}{2} \log\left(\frac{1 + \hat{\rho}}{1 - \hat{\rho}}\right)$$

then

$$\zeta = \tfrac{1}{2}(z_1 + z_2)$$

exactly.

26.20 Using the result of the previous exercise, show that the Likelihood Ratio test of $\rho_1 = \rho_2$ when $n_1 = n_2$ uses the statistic

$$l^{1/n} = \operatorname{sech}\{\tfrac{1}{2}(z_1 - z_2)\},$$

so that it is a one-to-one function of $z_1 - z_2$, the statistic suggested in **26.19**.

(Brandner, 1933)

26.21 In Exercise 26.19, show that if $n_1 \neq n_2$, we have *approximately* for the ML estimator of ζ

$$\zeta = \frac{1}{n}(n_1 z_1 + n_2 z_2),$$

and hence that the LR test of $\rho_1 = \rho_2$ uses the statistic

$$l = \left[\operatorname{sech}\left\{\frac{n_1}{n}(z_1 - z_2)\right\}\right]^{n_2} \left[\operatorname{sech}\left\{\frac{n_2}{n}(z_1 - z_2)\right\}\right]^{n_1}$$

approximately, again a one-to-one function of $(z_1 - z_2)$.

(Brandner, 1933)

26.22 To estimate a common value of ρ for two bivariate normal populations, show that

$$\zeta^* = \frac{(n_1 - 3) z_1 + (n_2 - 3) z_2}{n_1 + n_2 - 6}$$

is the linear combination of z_1 and z_2 with minimum variance as an estimator of ζ, but that when $n_1 \neq n_2$ this does not give the Maximum Likelihood estimator of ρ given in Exercise 26.19.

LINEAR REGRESSION AND CORRELATION

26.23 Show that the correlation coefficient between x and y, ρ_{xy}, satisfies
$$\rho_{xy}^2 = \frac{\operatorname{var}(\alpha_2 + \beta_2 x)}{\sigma_2^2} \equiv 1 - \frac{E\{[y-(\alpha_2+\beta_2 x)]^2\}}{\sigma_2^2}$$
and hence establish (26.18).

26.24 Writing $z = E(x|y)$, show that
$$\rho_{xz}^2 = \eta_1^2$$
and that
$$\rho_{yz}^2 = \rho_{xy}^2/\eta_1^2.$$
Hence show that (26.18) implies (26.46) and establish the conditions under which the various equalities in (26.46) hold.

(M. Fréchet published these relations in 1933–1935; see Kruskal (1958))

26.25 Show from (26.5) that the covariance
$$\mu_{11} = \int_{-\infty}^{\infty} \int_{-\infty}^{\infty} \{F(x,y) - F(x,\infty) F(\infty, y)\}\, dx\, dy.$$

CHAPTER 27

PARTIAL AND MULTIPLE CORRELATION AND REGRESSION

27.1 In normal or nearly-normal variation, the correlation parameter ρ between two variables can, as we saw in **26.10**, be used as a measure of interdependence. When we come to interpret "interdependence" in practice, however, we often meet difficulties of the kind discussed in **26.4**: if a variable is correlated with a second variable, this may be merely incidental to the fact that both are correlated with another variable or set of variables. This consideration leads us to examine the correlations between variables when other variables are held constant, i.e. conditionally upon those other variables taking certain fixed values. These are the so-called *partial correlations*.

If we find that holding another variable fixed reduces the correlation between two variables, we infer that their interdependence arises in part through the agency of that other variable; and, if the partial correlation is zero or very small, we infer that their interdependence is entirely attributable to that agency. Conversely, if the partial correlation is larger than the original correlation between the variables we infer that the other variable was obscuring the stronger connection or, as we may say, "masking" the correlation. But it must be remembered that even in the latter case we still have no warrant to presume a causal connection: by the argument of **26.4**, some quite different variable, overlooked in our analysis, may be at work to produce the correlation. As with ordinary product-moment correlations, so with partial correlations: the presumption of causality must always be extra-statistical.

27.2 In this branch of the subject, it is difficult at times to arrive at a notation which is unambiguous and flexible without being impossibly cumbrous. Basing ourselves on Yule's (1907) system of notation, we shall do our best to steer a middle course, but we shall at times have to make considerable demands on the reader's tolerance of suffixes.

As in Chapter 26, we shall discuss linear regression incidentally, but we leave over the main discussion of regression problems to Chapter 28.

Partial correlation

27.3 Consider three multinormally distributed variables. We exclude the singular case (cf. **15.2**), and lose no generality, so far as correlations are concerned, if we standardize the variables. Their dispersion matrix then becomes the matrix of their correlations, which we shall call the *correlation matrix* and denote by **C**. Thus if the correlation between x_i and x_j is ρ_{ij}, the frequency function becomes, from (15.19),

$$f(x_1, x_2, x_3) = (2\pi)^{-3/2} |C|^{-\frac{1}{2}} \exp\left\{-\frac{1}{2|C|} \sum_{i,j=1}^{3} C_{ij} x_i x_j\right\}, \qquad (27.1)$$

where C_{ij} is the cofactor of the (i,j)th element in the symmetric correlation determinant

$$|C| = \begin{vmatrix} 1 & \rho_{12} & \rho_{13} \\ & 1 & \rho_{23} \\ & & 1 \end{vmatrix}. \tag{27.2}$$

$C_{ij}/|C| = C^{ij}$ is the element of the reciprocal of C. We shall sometimes write the determinant or matrix of correlations in this way, leaving the entries below the leading diagonal to be filled in by symmetry.

The c.f. of the distribution is, by (15.20),

$$\phi(t_1, t_2, t_3) = \exp\left\{-\tfrac{1}{2} \sum_{i,j=1}^{3} \rho_{ij} t_i t_j\right\}. \tag{27.3}$$

27.4 Consider the correlation between x_1 and x_2 for a fixed value of x_3. The conditional distribution of x_1 and x_2, given x_3, is

$$g(x_1, x_2 | x_3) \propto \exp\{-\tfrac{1}{2}(C^{11}x_1^2 + 2C^{12}x_1x_2 + C^{22}x_2^2 + 2C^{13}x_1x_3 + 2C^{23}x_2x_3)\}$$
$$\propto \exp\{-\tfrac{1}{2}[C^{11}(x_1-\xi_1)^2 + 2C^{12}(x_1-\xi_1)(x_2-\xi_2) + C^{22}(x_2-\xi_2)^2]\}, \tag{27.4}$$

where

$$C^{11}\xi_1 + C^{12}\xi_2 = -C^{13}x_3,$$
$$C^{12}\xi_1 + C^{22}\xi_2 = -C^{23}x_3.$$

From (27.4) we see that, given x_3, x_1 and x_2 are bivariate-normally distributed, with correlation coefficient which we write

$$\rho_{12.3} = -\frac{C^{12}}{(C^{11}C^{22})^{\frac{1}{2}}}.$$

Clearly, $\rho_{12.3}$ does not depend on the actual value at which x_3 is fixed. Furthermore, cancelling the factor in $|C|$, we have

$$\left.\begin{array}{l} \rho_{12.3} = -\dfrac{C_{12}}{(C_{11}C_{22})^{\frac{1}{2}}} \\ \\ \phantom{\rho_{12.3}} = \dfrac{\rho_{12} - \rho_{13}\rho_{23}}{\{(1-\rho_{13}^2)(1-\rho_{23}^2)\}^{\frac{1}{2}}} \end{array}\right\} \tag{27.5}$$

from (27.2). $\rho_{12.3}$ is called the partial correlation coefficient of x_1 and x_2 with x_3 fixed. It is symmetric in its *primary subscripts* 1, 2. Its *secondary subscript*, 3, refers to the variable held fixed.

Although (27.5) has been derived under the assumption of normality, we now *define* the partial correlation coefficient by (27.5) for any parent distribution.

27.5 Similarly, if we have a p-variate non-singular multinormal distribution and fix $(p-2)$ of the variates, the resulting partial correlation of the other two (say x_1, x_2) is

$$\rho_{12.34\ldots p} = \frac{-C_{12}}{(C_{11}C_{22})^{\frac{1}{2}}}, \tag{27.6}$$

where C_{ij} is the cofactor of ρ_{ij} in

$$|C| = \begin{vmatrix} 1 & \rho_{12} & \rho_{13} & \cdots & \rho_{1p} \\ & 1 & \rho_{23} & \cdots & \rho_{2p} \\ & & 1 & \cdots & \rho_{3p} \\ & & & \ddots & \vdots \\ & & & & 1 \end{vmatrix}. \qquad (27.7)$$

Like (27.5), (27.6) is to be regarded as a general definition of the partial correlation coefficient between x_1 and x_2 with $x_3, \ldots, x_p$ fixed.

27.6 It is instructive to consider the same problem from another angle. Write $f(x_1, \ldots, x_k | x_{k+1}, \ldots, x_p)$ for the conditional joint frequency function of $x_1, \ldots, x_k$ when $x_{k+1}, \ldots, x_p$ are fixed, and $g(x_{k+1}, \ldots, x_p)$ for the marginal joint distribution of $x_{k+1}, \ldots, x_k$.

The joint c.f. of the p variables is

$$\phi(t_1, \ldots, t_p)$$
$$= \int \cdots \int f(x_1, \ldots, x_k | x_{k+1}, \ldots, x_p) g(x_{k+1}, \ldots, x_p) \exp\left(\sum_{j=1}^{p} it_j x_j\right) dx_1 \ldots dx_p$$
$$= \int \cdots \int \phi_k(t_1, \ldots, t_k | x_{k+1}, \ldots, x_p) g(x_{k+1}, \ldots, x_p) \exp\left(\sum_{j=k+1}^{p} it_j x_j\right) dx_{k+1} \ldots dx_p,$$

where $\phi_k(t_1, \ldots, t_k | x_{k+1}, \ldots, x_p)$ is the conditional joint c.f. of $x_1, \ldots, x_k$. It follows from the multivariate Inversion Theorem (4.17) that

$$\phi_k g = \frac{1}{(2\pi)^{p-k}} \int \cdots \int \phi(t_1, \ldots, t_p) \exp\left(-\sum_{j=k+1}^{p} it_j x_j\right) dt_{k+1} \ldots dt_p. \qquad (27.8)$$

If we put $t_1 = t_2 = \ldots = t_k = 0$ in (27.8), we obtain, since ϕ_k then becomes equal to unity,

$$g = \frac{1}{(2\pi)^{p-k}} \int \cdots \int \phi(0, \ldots, 0, t_{k+1}, \ldots, t_p) \exp\left(-\sum_{j=k+1}^{p} it_j x_j\right) dt_{k+1} \ldots dt_p. \qquad (27.9)$$

Hence, dividing (27.8) by (27.9),

$$\phi_k = \frac{\int \cdots \int \phi(t_1, \ldots, t_p) \exp\left(-\sum_{j=k+1}^{p} it_j x_j\right) dt_{k+1} \ldots dt_p}{\int \cdots \int \phi(0, \ldots, 0, t_{k+1}, \ldots, t_p) \exp\left(-\sum_{j=k+1}^{p} it_j x_j\right) dt_{k+1} \ldots dt_p}. \qquad (27.10)$$

This is a general result, suggested by a theorem of Bartlett (1938).

If we now assume that the p variables are multinormal, the integrand of the numerator in (27.10) becomes, using the c.f. (15.20),

$$\exp\left(-\tfrac{1}{2} \sum_{l,j=1}^{p} \rho_{lj} t_l t_j - \sum_{j=k+1}^{p} it_j x_j\right)$$
$$= \exp\left(-\tfrac{1}{2} \sum_{l,j=1}^{k} \rho_{lj} t_l t_j\right) \exp\left(-\tfrac{1}{2} \sum_{l,j=k+1}^{p} \rho_{lj} t_l t_j\right) \exp\left(-\sum_{l=1}^{k} \sum_{j=k+1}^{p} \rho_{lj} t_l t_j\right) \exp\left(-\sum_{j=k+1}^{p} it_j x_j\right)$$
$$= \exp\left(-\tfrac{1}{2} \sum_{l,j=1}^{k} \rho_{lj} t_l t_j\right) \exp\left(-\tfrac{1}{2} \sum_{l,j=k+1}^{p} \rho_{lj} t_l t_j\right) \exp\left\{-\sum_{j=k+1}^{p} it_j \left(x_j - i \sum_{l=1}^{k} \rho_{lj} t_l\right)\right\}. \qquad (27.11)$$

Now the integral with respect to $t_{k+1}, \ldots, t_p$ of the last two factors on the right of (27.11) is the inversion of the multinormal c.f. of $x_{k+1}, \ldots, x_p$ with x_j measured from the value $i \sum_{m=1}^{k} \rho_{jm} t_m$. This change of origins does not affect correlations. If we write **D** for the correlation matrix of $x_{k+1}, \ldots, x_p$ alone, this gives for the integral of (27.11) a constant times

$$\exp\left(-\tfrac{1}{2} \sum_{l,j=1}^{k} \rho_{lj} t_l t_j\right) \exp\left\{-\tfrac{1}{2} \sum_{l,j=k+1}^{p} D^{lj}\left(x_l - i \sum_{m=1}^{k} \rho_{lm} t_m\right)\left(x_j - i \sum_{m=1}^{k} \rho_{jm} t_m\right)\right\}.$$

From (27.10) we then find

$$\phi_k(t_1, \ldots, t_k \mid x_{k+1}, \ldots, x_p) = \exp\left(-\tfrac{1}{2} \sum_{l,j=1}^{k} \rho_{lj} t_l t_j\right) \times$$

$$\exp\left\{-\tfrac{1}{2} \sum_{l,j=k+1}^{p} D^{lj}\left(x_l - i \sum_{m=1}^{k} \rho_{lm} t_m\right)\left(x_j - i \sum_{m=1}^{k} \rho_{jm} t_m\right) + \tfrac{1}{2} \sum_{l,j=k+1}^{p} D^{lj} x_l x_j\right\}. \quad (27.12)$$

Thus if σ'_{uv} denotes the covariance of x_u and x_v in the conditional distribution of $x_1, \ldots, x_k$, and σ_{uv} their covariance unconditionally, we find, on identifying coefficients of $t_u t_v$ in (27.12),

$$\sigma'_{uv} = \sigma_{uv} - \sum_{l,j=k+1}^{p} D^{lj} \rho_{lu} \rho_{jv}. \quad (27.13)$$

This is in terms of standardized initial variables. If we now destandardize, the variance of x_i being σ_i^2, each ρ is replaced by its corresponding σ, D^{ij} is replaced by the dispersion matrix elements $D^{ij}/(\sigma_i \sigma_j)$ and we have the more general form of (27.13)

$$\sigma'_{uv} = \sigma_{uv} - \sum_{l,j=k+1}^{p} D^{lj} \sigma_{lu} \sigma_{jv}/(\sigma_l \sigma_j). \quad (27.14)$$

(27.14) does not depend on the values at which $x_{k+1}, \ldots, x_p$ are fixed.

If we write **A** for the $(k \times k)$ unconditional dispersion matrix $\{\sigma_{uv}\}$, **B**′ for the $(k \times (p-k))$ matrix $\{\sigma_{lu}\}$ and **E** for the $((p-k) \times (p-k))$ dispersion matrix of which **D** is the standardized form, (27.14) states that the conditional dispersion matrix

$$\{\sigma'_{uv}\} = \mathbf{A} - \mathbf{B}' \mathbf{E}^{-1} \mathbf{B}.$$

27.7 In particular, if we fix only one variable, say x_p, we have $D^{pp} = 1$ and the conditional covariance (27.14) becomes simply

$$\sigma'_{uv} = \sigma_{uv} - \sigma_{pu} \sigma_{pv}/\sigma_p^2 = \sigma_u \sigma_v (\rho_{uv} - \rho_{up} \rho_{vp}), \quad (27.15)$$

and if $u = v$ we have from (27.15) the conditional variance of u

$$\sigma_u'^2 = \sigma_u^2 (1 - \rho_{up}^2),$$

and the last two formulae give the conditional correlation coefficient

$$\rho_{uv \cdot p} = \frac{\rho_{uv} - \rho_{up} \rho_{vp}}{\{(1 - \rho_{up}^2)(1 - \rho_{vp}^2)\}^{\frac{1}{2}}},$$

another form of (27.5).

334   THE ADVANCED THEORY OF STATISTICS

If we fix all but two variables, say x_1 and x_2, we have from (27.14)

$$\rho_{12.34\ldots p} = \frac{\rho_{12} - \sum_{l,j=3}^{p} D^{lj} \rho_{l1} \rho_{j2}}{\left\{\left(1 - \sum_{l,j=3}^{p} D^{lj} \rho_{l1} \rho_{j1}\right)\left(1 - \sum_{l,j=3}^{p} D^{lj} \rho_{l2} \rho_{j2}\right)\right\}^{\frac{1}{2}}}. \tag{27.16}$$

Inspection of (27.7) shows that the minor of ρ_{12}, namely,

$$\begin{vmatrix} \rho_{21} & \rho_{23} & \rho_{24} & \cdots & \rho_{2p} \\ \rho_{31} & 1 & \rho_{34} & \cdots & \rho_{3p} \\ \vdots & & & \ddots & \\ \rho_{p1} & \rho_{p3} & \rho_{p4} & \cdots & 1 \end{vmatrix} = \begin{vmatrix} \rho_{21} & \rho_{32} & \cdots & \rho_{p2} \\ \rho_{31} & & & \\ \vdots & & D & \\ \rho_{p1} & & & \end{vmatrix}$$

may be expanded by its first row and column as

$$\rho_{21}|D| - \sum_{l,j=3}^{p} D_{lj} \rho_{l1} \rho_{j2},$$

and similarly for the minors of ρ_{11}, ρ_{22}. Thus (27.16) may be written

$$\rho_{12.34\ldots p}^2 = \frac{C_{12}^2}{C_{11} C_{22}},$$

which is (27.6) again.

Linear partial regressions

27.8 We now consider the extension of the linear regression relations of **26.7** to p variates. For p multinormal variates x_i with zero means and variances σ_i^2, the mean of x_1 if $x_2, \ldots, x_p$ are fixed is seen from the exponent of the distribution to be

$$\frac{E(x_1 | x_2, \ldots, x_p)}{\sigma_1} = -\sum_{j=2}^{p} \frac{C_{1j}}{C_{11}} \frac{x_j}{\sigma_j}. \tag{27.17}$$

We shall denote the regression coefficient of x_1 on x_j with the other $(p-2)$ variables held fixed by $\beta_{1j.23\ldots,j-1,j+1,\ldots p}$ or, for brevity, by $\beta_{1j.q_j}$, where q stands for "the other variables than those in the primary subscripts," and the suffix to q is to distinguish different q's. The $\beta_{1j.q}$ are the *partial regression coefficients*.

We have, therefore,

$$E(x_1 | x_2, \ldots, x_p) = \beta_{12.q_2} x_2 + \beta_{13.q_3} x_3 + \ldots + \beta_{1p.q_p} x_p. \tag{27.18}$$

Comparison of (27.18) with (27.17) gives, in the multinormal case,

$$\beta_{1j.q_j} = -\frac{\sigma_1}{\sigma_j} \frac{C_{1j}}{C_{11}}. \tag{27.19}$$

Similarly, the regression coefficient of x_j upon x_1 with the other variables fixed is

$$\beta_{j1.q_j} = -\frac{\sigma_j}{\sigma_1} \frac{C_{j1}}{C_{jj}}, \tag{27.20}$$

and thus, since $C_{1j} = C_{j1}$, (27.6), (27.19) and (27.20) give

$$\rho_{1j.q_j}^2 = \frac{C_{1j}^2}{C_{11} C_{jj}} = \beta_{1j.q_j} \beta_{j1.q_j}, \quad (27.21)$$

an obvious generalization of (26.17). (27.19) and (27.20) make it obvious that $\beta_{1j.q_j}$ is not symmetric in x_1 and x_j, which is what we should expect from a coefficient of dependence. Like (27.5) and (27.6), (27.19) and (27.20) are *definitions* of the partial coefficients in the general case.

Errors from linear regression

27.9 We define the *error*[*] of order $(p-1)$

$$x_{1.2\ldots p} = x_1 - E(x_1 \,|\, x_2, \ldots, x_p).$$

It has zero mean and its variance is

$$\sigma_{1.2\ldots p}^2 = E(x_{1.2\ldots p}^2) = E[\{x_1 - E(x_1 \,|\, x_2, \ldots, x_p)\}^2],$$

so that $\sigma_{1.2\ldots p}^2$ is the *error variance* of x_1 about the regression. We have at once, from (27.18),

$$\sigma_{1.2\ldots p}^2 = E\left[\left\{x_1 - \sum_{j=2}^{p} \beta_{1j.q_j} x_j\right\}^2\right] \quad (27.22)$$

$$= E\left[x_1\left(x_1 - \sum_{j=2}^{p} \beta_{1j.q_j} x_j\right) - \sum_{j=2}^{p} \beta_{1j.q_j} x_j \left(x_1 - \sum_{j=2}^{p} \beta_{1j.q_j} x_j\right)\right]. \quad (27.23)$$

If we take expectations in two stages, first keeping $x_2, \ldots, x_p$ fixed, we find that the conditional expectation of the second product in the right of (27.23) is zero by (27.18), so that

$$\sigma_{1.2\ldots p}^2 = E\left\{x_1^2 - \sum_{j=2}^{p} \beta_{1j.q_j} x_1 x_j\right\} = \sigma_1^2 - \sum_{j=2}^{p} \beta_{1j.q_j} \sigma_{1j}. \quad (27.24)$$

The error variance (27.24) is independent of the values fixed for $x_3, \ldots, x_p$ if the $\beta_{1j.q_j}$ are independent of these values. The distribution of x_1 in arrays is then said to be *homoscedastic* (or *heteroscedastic* in the contrary case). This constancy of error variance makes the interpretation of regressions and correlations easier. For example, in the normal case, the conditional variances and covariances obtained by fixing a set of variates does not depend on the values at which they are fixed (cf. (27.14)). In other cases, we must make due allowance for observed heteroscedasticity in our interpretations: the partial regression coefficients are then, perhaps, best regarded as *average* relationships over all possible values of the fixed variates.

Relations between variances, regressions and correlations of different orders

27.10 Given p variables, we may examine the correlation between any pair when any subset of the others is fixed, and similarly we may be interested in the regression of any one upon any subset of the others. The number of possible coefficients becomes very large as p increases. When a coefficient contains k secondary subscripts, it is said to be of order k. Thus $\rho_{12.34}$ is of order 2, $\rho_{12.3}$ of order 1 and ρ_{12} of order zero,

---

(*) This is often called a "residual" in the literature, but as in Chapter 19 we distinguish between *errors* from population regressions and *residuals* from regressions fitted to sample data.

while $\beta_{12.678}$ is of order 3 and $\sigma^2_{1.2678}$ is of order 4. In our present notation, the linear regression coefficients of the last chapter, β_1 and β_2, would be written β_{12} and β_{21} respectively and are of order zero, as is an ordinary variance σ^2.

We have already seen in **27.4** and **27.7** how any correlation coefficient of order 1 can be expressed in terms of those of order zero. We will now obtain more general results of this kind for all types of coefficient.

27.11 From (27.24) and (27.19) we have

$$\sigma^2_{1.2\ldots p} = \sigma^2_1 + \sum_{j=2}^{p} \frac{\sigma_1}{\sigma_j} \frac{C_{1j}}{C_{11}} \sigma_{1j}, \qquad (27.25)$$

whence

$$\sigma^2_{1.2\ldots p}/\sigma^2_1 = 1 + \frac{1}{C_{11}} \sum_{j=2}^{p} C_{1j} \rho_{1j}$$

$$= 1 + \frac{1}{C_{11}}(|C| - C_{11}) = \frac{|C|}{C_{11}},$$

or, using the definition of q given in **27.8**,

$$\sigma^2_{1.q} = \sigma^2_1 \frac{|C|}{C_{11}}, \qquad (27.26)$$

and similarly if 1 is replaced by any other suffix. More generally, it may be seen in the same way that

$$\operatorname{cov}(x_{l.q_l}, x_{m.q_m}) = \sigma_l \sigma_m \frac{|C|}{C_{lm}}, \qquad (27.27)$$

which reduces to (27.26) when $l = m$. (27.27) applies to the case where the secondary subscripts of each variable include the primary subscript of the other. If, on the other hand, both sets of secondary subscripts exclude l and m, we denote a common set of secondary subscripts by r. The covariance of two errors $x_{l.r} x_{m.r}$ is related to their correlation and variances by the natural extension of the definitions (26.10), (26.11) and (26.17), namely

$$\left.\begin{array}{l} \operatorname{cov}(x_{l.r}\, x_{m.r})/\sigma^2_{m.r} = \beta_{lm.r}, \\ \operatorname{cov}(x_{l.r}\, x_{m.r})/\sigma^2_{l.r} = \beta_{ml.r}, \\ \operatorname{cov}(x_{l.r}\, x_{m.r})/(\sigma_{l.r}\, \sigma_{m.r}) = \rho_{lm.r} \end{array}\right\} \qquad (27.28)$$

agreeing with the relationship (27.21) already found. By adjoining a set of suffixes, r, to both variables x_l, x_m we simply do the same to all their coefficients.

27.12 We may now use (27.26) to obtain the relation between error variances of different orders. Writing $|D|$ for the correlation determinant of all the variables except x_2, we have, from (27.26),

$$\sigma^2_{1.q-2} = \sigma^2_1 \frac{|D|}{D_{11}},$$

(where the suffix $q-2$ denotes the set q excluding x_2) and

$$\sigma^2_{1.q} = \sigma^2_1 \frac{|C|}{C_{11}},$$

whence
$$\frac{\sigma_{1.q}^2}{\sigma_{1.q-2}^2} = \frac{D_{11}}{C_{11}} \cdot \frac{|C|}{|D|}. \qquad (27.29)$$

Now $|D| = C_{22}$ by definition, and by Jacobi's generalized theorem on determinants

$$\begin{vmatrix} C_{11} & C_{12} \\ C_{12} & C_{22} \end{vmatrix} = |C|D_{11}, \qquad (27.30)$$

since D_{11} is the complementary minor of

$$\begin{vmatrix} \rho_{11} & \rho_{12} \\ \rho_{12} & \rho_{22} \end{vmatrix}$$

in C. Thus, using (27.30), (27.29) becomes

$$\frac{\sigma_{1.q}^2}{\sigma_{1.q-2}^2} = \frac{\begin{vmatrix} C_{11} & C_{12} \\ C_{12} & C_{22} \end{vmatrix}}{C_{11} C_{22}} = 1 - \frac{C_{12}^2}{C_{11} C_{22}} \qquad (27.31)$$

or, using (27.6),

$$\sigma_{1.q}^2 = \sigma_{1.q-2}^2 (1 - \rho_{12.q}^2). \qquad (27.32)$$

(27.32) is a generalization of the bivariate result given in Exercise 26.23, which in our present notation would be written

$$\sigma_{2.1}^2 = \sigma_2^2 (1 - \rho_{12}^2).$$

We have also met this result in the special context of the bivariate normal distribution at (16.46).

27.13 (27.32) enables us to express the error variance of order $(p-1)$ in terms of the error variance and a correlation coefficient of order $(p-2)$. If we now again use (27.32) to express $\sigma_{1.q-2}^2$, we find in exactly the same way

$$\sigma_{1.q-2}^2 = \sigma_{1.q-2-3}^2 (1 - \rho_{13.q-2}^2).$$

We may thus apply (27.32) successively to obtain, writing subscripts more fully,

$$\sigma_{1.2...p}^2 = \sigma_1^2 (1 - \rho_{1p}^2)(1 - \rho_{1(p-1).p}^2)(1 - \rho_{1(p-2).(p-1)p}^2) \cdots (1 - \rho_{12.3...p}^2). \qquad (27.33)$$

In (27.33), the order in which the secondary subscripts of $\sigma_{1.23...p}^2$ are taken is evidently immaterial; we may permute them as desired. In particular, we may write for simplicity

$$\frac{\sigma_{1.2...p}^2}{\sigma_1^2} = (1-\rho_{12}^2)(1-\rho_{13.2}^2)(1-\rho_{14.23}^2) \cdots (1-\rho_{1p.23...(p-1)}^2)$$

$$= \frac{|C|}{C_{11}} \qquad (27.34)$$

by (27.26), the subscripts other than 1 in (27.34) being permutable. (27.34) enables us to express any error variance of order s in terms of the error variance of zero order and s correlation coefficients, one of each order from zero up to $(s-1)$.

27.14 We now turn to the regression coefficients. (27.15) may be written, for the covariance of x_1 and x_2 with x_p fixed,

$$\sigma_{12.p} = \sigma_{12} - \sigma_{1p}\sigma_{p2}/\sigma_p^2,$$

and if we adjoin the suffixes $3, \ldots, (p-1)$ throughout, we have

$$\sigma_{12.3\ldots p} = \sigma_{12.3\ldots(p-1)} - \frac{\sigma_{1p.3\ldots(p-1)} \sigma_{p2.3\ldots(p-1)}}{\sigma^2_{p.3\ldots(p-1)}}. \tag{27.35}$$

Using the definition (27.28) of a regression coefficient as the ratio of a covariance to a variance, i.e.

$$\beta_{ij.k} = \sigma_{ij.k}/\sigma^2_{j.k},$$

we have from (27.35), writing r for the set $3, \ldots, (p-1)$,

$$\beta_{12.pr} \sigma^2_{2.pr} = \beta_{12.r} \sigma^2_{2.r} - \beta_{1p.r} \beta_{p2.r} \sigma^2_{2.r}$$

or

$$\beta_{12.pr} = \frac{\sigma^2_{2.r}}{\sigma^2_{2.pr}} (\beta_{12.r} - \beta_{1p.r} \beta_{p2.r}). \tag{27.36}$$

If we put $x_1 \equiv x_2$ in (27.36), we obtain

$$\sigma^2_{2.pr} = \sigma^2_{2.r}(1 - \beta_{2p.r} \beta_{p2.r}) = \sigma^2_{2.r}(1 - \rho^2_{2p.r}), \tag{27.37}$$

another form of (27.32). Thus, from (27.36) and (27.37),

$$\beta_{12.pr} = \frac{\beta_{12.r} - \beta_{1p.r} \beta_{p2.r}}{1 - \beta_{2p.r} \beta_{p2.r}}, \tag{27.38}$$

the required formula for expressing a regression coefficient in terms of some of those of next lower order. Repeated applications of (27.38) give any regression coefficient in terms of those of zero order.

Finally, using (27.21), we find from (27.38)

$$\rho_{12.pr} = (\beta_{12.pr} \beta_{21.pr})^{\frac{1}{2}} = \frac{\rho_{12.r} - \rho_{1p.r} \rho_{2p.r}}{\{(1-\rho^2_{1p.r})(1-\rho^2_{2p.r})\}^{\frac{1}{2}}}, \tag{27.39}$$

which is (27.5) generalized by adjoining the set of suffixes r.

Approximate linear partial regressions

27.15 In our discussion from **27.8** onwards we have taken the regression relationships to be exactly linear, of type (27.18). Just as in **26.8**, we now consider the question of fitting regression relationships of this type to observed populations, whose regressions are almost never exactly linear, and by the same reasoning as there, we are led to the Method of Least Squares. We therefore choose the β_{ij} to minimize the sum of squared deviations of the n observations from the fitted regression,

$$\sum_{i=1}^{n} (x_{1i} - \sum_{j=2}^{p} \beta_{1j.q_j} x_{ji})^2, \tag{27.40}$$

where we measure from the means of the x's, and assume $n > p$. The solution is, from (19.12),

$$\boldsymbol{\beta} = (\mathbf{X}'\mathbf{X})^{-1} \mathbf{X}' \mathbf{x}_1, \tag{27.41}$$

where the matrix $\mathbf{X}$ refers to the observations on the $(p-1)$ variables $x_2, \ldots, x_p$, and $\mathbf{x}_1$ is the vector of observations on that variable. (27.41) may be written

$$\boldsymbol{\beta} = (n\mathbf{V}_{p-1})^{-1}(n\mathbf{M}) = \mathbf{V}^{-1}_{p-1} \mathbf{M}, \tag{27.42}$$

where $\mathbf{V}_{p-1}$ is the dispersion matrix of $x_2, \ldots, x_p$ and $\mathbf{M}$ is the vector of covariances of x_1 with x_j ($j = 2, \ldots, p$). Thus

$$\beta_{1j.q_j} = \frac{1}{|V_{p-1}|} \sum_{l=2}^{p} (V_{p-1})_{jl} \sigma_{1l}. \tag{27.43}$$

Since $|V_{p-1}|$ is the minor V_{11} of the dispersion matrix $\mathbf{V}$ of all p variables, $(V_{p-1})_{jl}$ is the complementary minor of

$$\begin{vmatrix} \sigma_1^2 & \sigma_{1l} \\ \sigma_{1j} & \sigma_{jl} \end{vmatrix}$$

in $\mathbf{V}$, so that the sum on the right of (27.43) is the cofactor of $(-\sigma_{1j})$ in $\mathbf{V}$. Thus (27.43) becomes

$$\beta_{1j.q_j} = -\frac{V_{1j}}{V_{11}} = -\frac{\sigma_1}{\sigma_j}\frac{C_{1j}}{C_{11}}. \tag{27.44}$$

(27.44) is identical with (27.19). Thus, as in **26.8**, we reach the conclusion that the Least Squares approximation gives us the same regression coefficients as in the case of exact linearity of regression.

It follows that all the results of this chapter are valid when we fit regressions by Least Squares to observed populations.

Sample coefficients

27.16 If we are using a sample of n observations and fit regressions by Least Squares, all the relationships we have discussed will hold between the sample coefficients. Following our usual convention, we shall use r instead of ρ, b instead of β, and s^2 instead of σ^2 to distinguish the sample coefficients from their population equivalents. The b's are determined by minimizing the analogue of (27.40)

$$n s_{1.23\ldots p}^2 = \sum_{i=1}^{n} \left(x_{1i} - \sum_{j=2}^{p} b_{1j.q_j} x_{ji} \right)^2, \tag{27.45}$$

and we have as at (27.21)

$$r_{ij.k}^2 = b_{ij.k} b_{ji.k},$$

while the analogues of (27.34), (27.38) and (27.39) also hold.

If we equate to zero the derivatives of (27.45) with respect to the b_{1j} (which is the method by which we determine the b_{1j}), we have the $(p-1)$ equations

$$\sum_{i=1}^{n} x_{ji} \left(x_{1i} - \sum_{j=2}^{p} b_{1j.q_j} x_{ji} \right) = 0, \quad i = 2, 3, \ldots, p,$$

which we may write

$$\sum x_j x_{1.23\ldots p} = 0, \quad j = 2, 3, \ldots, p, \tag{27.46}$$

the summation being over the n observations. $x_{1.23\ldots p}$ is the *residual* from the fitted regression—cf. **27.9**. From (27.46) it follows that

$$\sum x_{1.q}^2 = \sum x_{1.q}(x_1 - \sum b_{1j.q_j} x_j) = \sum x_{1.q} x_1, \tag{27.47}$$

and similarly

$$\sum x_{1.r} x_{2.r} = \sum x_{1.r} x_2 = \sum x_1 x_{2.r}, \tag{27.48}$$

where r is any common set of secondary subscripts. Relations like (27.47) and (27.48) hold for the population errors as well as the sample residuals, but we shall find them of use mainly in sampling problems, which is why we have expressed them in terms of residuals. Exercise 27.5 gives the most general rule for the omission of common secondary subscripts in summing products of residuals or of errors.

Estimation of population coefficients

27.17 As for the zero-order correlations and regressions of the previous chapter, we may use the sample coefficients as estimators of their population equivalents. If the regression concerned is linear, we know from the Least Squares theory in Chapter 19 that any b is an unbiassed estimator of the corresponding β and that $\dfrac{n}{n-(p-1)} s^2_{1.23\ldots p}$ is an unbiassed estimator of $\sigma^2_{1.23\ldots p}$. However, no r is an unbiassed estimator of its ρ: we saw in **26.15–17** that even for a zero-order coefficient in the normal case, r is not unbiassed for ρ, but that the modification (26.34) or (26.35) is an unbiassed estimator. A result to be obtained in **27.22** will enable us to estimate any partial correlation coefficient analogously in the normal case.

Geometrical interpretation of partial correlation

27.18 From our results, it is clear that the whole complex of partial regressions, correlations and variances or covariances of errors or residuals is completely determined by the variances and correlations, or by the variances and regressions, of zero order. It is interesting to consider this result from the geometrical point of view.

Suppose in fact that we have n observations on $p\ (< n)$ variates

$$x_{11}, \ldots, x_{1p};\ x_{21}, \ldots, x_{2p};\ \ldots;\ x_{n1}, \ldots, x_{np}.$$

Consider a (Euclidean) sample space of n dimensions. To the observations $x_{1k}, \ldots, x_{nk}$ on the kth variate, there will correspond one point in this space, and there are p such points, one for each variate. Call these points $Q_1, Q_2, \ldots, Q_p$. We will assume that the x's are measured about their means, and take the origin to be P.

The quantity $n\sigma_l^2$ may then be interpreted as the square of the length of the vector joining the point Q_l (with co-ordinates $x_{1l}, \ldots, x_{nl}$) to P. Similarly ρ_{lm} may be interpreted as the cosine of the angle $Q_l P Q_m$, for

$$\rho_{lm} = \frac{\sum\limits_{j=1}^{n} x_{jl} x_{jm}}{\left(\sum\limits_{j=1}^{n} x_{jl}^2 \sum\limits_{j=1}^{n} x_{jm}^2 \right)^{\frac{1}{2}}},$$

which is the formula for the cosine of the angle between PQ_l and PQ_m.

Our result may then be expressed by saying that all the relations connecting the p points in the n-space are expressible in terms of the lengths of the vectors PQ_i and of the angles between them; and the theory of partial correlation and regression is thus exhibited as formally identical with the trigonometry of an n-dimensional constellation of points.

27.19 The reader who prefers the geometrical way of looking at this branch of the subject will have no difficulty in translating the foregoing equations into trigonometrical terminology. We will indicate only the more important results required for later sampling investigations.

Note in the first place that the p points Q_i and the point P determine (except perhaps in degenerate cases) a sub-space of p dimensions in the n-space. Consider the point $Q_{1.2\ldots p}$ whose co-ordinates are the n residuals $x_{1.2\ldots p}$. In virtue of (27.46) the vector $PQ_{1.2\ldots p}$ is orthogonal to each of the vectors $PQ_2, \ldots, PQ_p$ and hence to the space of $(p-1)$ dimensions spanned by $P, Q_2, \ldots, Q_p$.

Consider now the residual vectors $Q_{1.r}, Q_{2.r}$, where r represents the secondary subscripts $3, 4, \ldots, (p-1)$. The cosine of the angle between them, say θ, is $\rho_{12.r}$ and each is orthogonal to the space spanned by $P, Q_3, \ldots, Q_{(p-1)}$. In Fig. 27.1, let M be the foot of the perpendicular from $Q_{1.r}$ on to PQ_p and $Q'_{2.r}$ a point on $PQ_{2.r}$ such that $Q'_{2.r}M$ is also perpendicular to PQ_p. Then $MQ_{1.r}$ and $MQ'_{2.r}$ are orthogonal

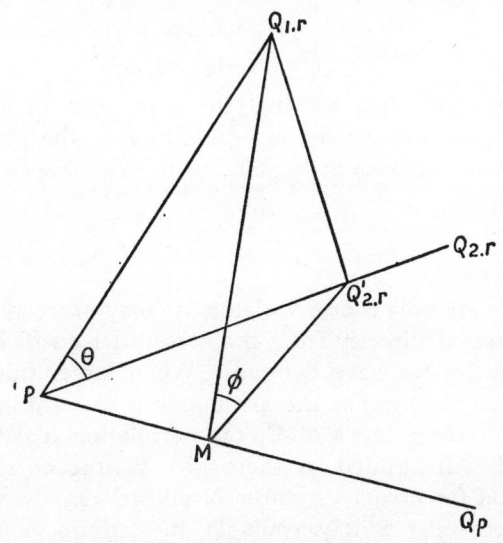

Fig. 27.1—The geometry of partial correlation

to the space spanned by $P, Q_3, \ldots, Q_p$, and the cosine of the angle between them, say ϕ, is $\rho_{12.rp}$. Thus, to express $\rho_{12.rp}$ in terms of $\rho_{12.r}$ we have to express ϕ in terms of θ, or the angle between the vectors $PQ_{1.r}$ and $PQ'_{2.r}$ in terms of that between their projections on the hyperplane perpendicular to PQ_p. We now drop the prime in $Q'_{2.r}$ for convenience. By Pythagoras' theorem,

$$(Q_{1.r}Q_{2.r})^2 = PQ_{1.r}^2 + PQ_{2.r}^2 - 2PQ_{1.r}.PQ_{2.r}\cos\theta$$
$$= MQ_{1.r}^2 + MQ_{2.r}^2 - 2MQ_{1.r}.MQ_{2.r}\cos\phi.$$

Further,
$$PQ_{1.r}^2 = PM^2 + MQ_{1.r}^2$$

and
$$PQ_{2.r}^2 = PM^2 + MQ_{2.r}^2,$$

and hence we find
$$MQ_{1.r} MQ_{2.r} \cos\phi = PQ_{1.r} PQ_{2.r} \cos\theta - PM^2$$
or
$$\frac{MQ_{1.r}}{PQ_{1.r}} \cdot \frac{MQ_{2.r}}{PQ_{2.r}} \cos\phi = \cos\theta - \frac{PM}{PQ_{1.r}} \cdot \frac{PM}{PQ_{2.r}}. \tag{27.49}$$

Now $\frac{MQ_{1.r}}{PQ_{1.r}}$ and $\frac{PM}{PQ_{1.r}}$ are the sine and cosine of the angle between PQ_p and $PQ_{1.r}$. Since $PQ_{1.r}$ is orthogonal to the space spanned by $P, Q_3, \ldots, Q_{p-1}$, its angle with PQ_p is unchanged if the latter is projected orthogonally to that space, i.e. if we replace PQ_p by $PQ_{p.r}$. The cosine of the angle between $PQ_{1.r}$ and $PQ_{p.r}$ is $\rho_{1p.r}$, and hence $\frac{PM}{PQ_{1.r}} = \rho_{1p.r}$, $\frac{MQ_{1.r}}{PQ_{1.r}} = (1 - \rho_{1p.r}^2)^{\frac{1}{2}}$. The same result holds with the suffix 2 replacing 1. Thus, substituting in (27.49),

$$\rho_{12.rp} = \frac{\rho_{12.r} - \rho_{1p.r} \rho_{2p.r}}{\{(1 - \rho_{1p.r}^2)(1 - \rho_{2p.r}^2)\}^{\frac{1}{2}}}, \tag{27.50}$$

which is (27.39) again. We thus see that the expression of a partial correlation in terms of that of next lower order may be represented as the projection of an angle in the sample space on to a subspace orthogonal to the variable held fixed in the higher-order coefficient alone.

Computation of coefficients

27.20 Where there are only 3 or 4 variates, we may proceed to calculate the partial correlations and regressions directly from the zero-order coefficients, using the appropriate one of the formulae we have derived. When larger numbers of variables are involved, it is as well to systematize the arithmetic in determinantal form. In effect, we need to evaluate all the minors of **C**, the correlation matrix, and then formulae (27.6), (27.19) and (27.26) applied to them give us the correlation and regression coefficients and residual (or error) variances of all orders. Now that electronic computing facilities are becoming widely available, the tedium of manual calculation can be avoided.

For p small, tables of quantities such as $1 - \rho^2$, $(1-\rho^2)^{\frac{1}{2}}$ and $\{(1-\rho_1^2)(1-\rho_2^2)\}^{-\frac{1}{2}}$ are useful. Trigonometrical tables are also useful; for instance, given ρ we can find $\theta = \arccos\rho$ and hence $\sin\theta = (1-\rho^2)^{\frac{1}{2}}$, $\csc\theta = (1-\rho^2)^{-\frac{1}{2}}$, and so on.

The Kelley Statistical Tables (Harvard U.P., 1948) give $(1-\rho^2)^{\frac{1}{2}}$ for
$$\rho = 0.0001 (0.0001) 0.9999.$$

The two examples which follow are of interpretational, as well as computational, interest.

Example 27.1

In an investigation into the relationship between weather and crops, Hooker (1907) found the following means, standard deviations and correlations between the yields of " seeds' hay " (x_1) in cwt per acre, the spring rainfall (x_2) in inches, and the accumulated temperature above 42° F in the spring (x_3) for an English area over 20 years :—

$$\mu_1 = 28\cdot02, \quad \sigma_1 = 4\cdot42, \quad \rho_{12} = +0\cdot80,$$
$$\mu_2 = 4\cdot91, \quad \sigma_2 = 1\cdot10, \quad \rho_{13} = -0\cdot40,$$
$$\mu_3 = 594, \quad \sigma_3 = 85, \quad \rho_{23} = -0\cdot56.$$

The question of primary interest here is the influence of weather on crop yields, and we consider only the regression of x_1 on the other two variates. From the correlations of zero order, it appears that yield and rainfall are positively correlated but that yield and accumulated spring temperature are negatively correlated. The question is, what interpretation is to be placed on this latter result? Does high temperature adversely affect yields or may the negative correlation be due to the fact that high temperature involves less rain, so that the beneficial effect of warmth is more than offset by the harmful effect of drought?

To throw some light on this question, let us calculate the partial correlations. From (27.5) we have

$$\rho_{12.3} = \frac{\rho_{12} - \rho_{13}\rho_{23}}{\{(1-\rho_{13}^2)(1-\rho_{23}^2)\}^{\frac{1}{2}}}$$
$$= \frac{0\cdot80 - (-0\cdot40)(-0\cdot56)}{\{(1-0\cdot40^2)(1-0\cdot56^2)\}^{\frac{1}{2}}}$$
$$= 0\cdot759.$$

Similarly

$$\rho_{13.2} = 0\cdot097, \quad \rho_{23.1} = -0\cdot436.$$

We next require the regressions and the error variances. We have

$$\beta_{12.3} = \frac{\operatorname{cov}(x_{1.3}, x_{2.3})}{\operatorname{var} x_{2.3}}$$
$$= \rho_{12.3}\frac{\sigma_{1.3}}{\sigma_{2.3}}.$$

This, however, involves the calculation of $\sigma_{1.3}$ and $\sigma_{2.3}$, which are not in themselves of interest. We can avoid these calculations by noting from (27.33) that

$$\left.\begin{array}{l}\sigma_{1.23} = \sigma_{1.3}(1-\rho_{12.3}^2)^{\frac{1}{2}},\\ \sigma_{2.13} = \sigma_{2.3}(1-\rho_{12.3}^2)^{\frac{1}{2}},\end{array}\right\} \qquad (27.51)$$

so that

$$\beta_{12.3} = \rho_{12.3}\frac{\sigma_{1.23}}{\sigma_{2.13}}. \qquad (27.52)$$

The standard deviations $\sigma_{1.23}$ and $\sigma_{2.13}$ are of some interest and may be calculated from (27.33). We have

$$\sigma_{1.23} = \sigma_1\{(1-\rho_{12}^2)(1-\rho_{13.2}^2)\}^{\frac{1}{2}}$$
$$= \sigma_1\{(1-\rho_{13}^2)(1-\rho_{12.3}^2)\}^{\frac{1}{2}},$$

the two forms offering a check on each other. From the first we have
$$\sigma_{1.23} = 4\cdot 42\{(1-0\cdot 8^2)(1-0\cdot 097^2)\}^{\frac{1}{2}}$$
$$= 2\cdot 64.$$

Similarly
$$\sigma_{2.13} = 0\cdot 594, \quad \sigma_{3.12} = 70\cdot 1.$$

Thus
$$\beta_{12.3} = 0\cdot 759 \frac{2\cdot 64}{0\cdot 594} = 3\cdot 37,$$

and we also find
$$\beta_{13.2} = 0\cdot 00364.$$

The regression equation of x_1 on x_2 and x_3 is then
$$x_1 - 28\cdot 02 = 3\cdot 37(x_2 - 4\cdot 91) + 0\cdot 00364(x_3 - 594).$$

This equation shows that for increasing rainfall the yield increases, and that for increasing temperature the yield also increases, *other things being equal*. It enables us to isolate the effects of rainfall from those of temperature and to study each separately. The fact that $\beta_{13.2}$ is positive means that there is a positive relation between yield and temperature when the effect of rainfall is eliminated. The partial correlations tell the same story. Although ρ_{13} is negative, $\rho_{13.2}$ is positive (though small), indicating that the negative value of ρ_{13} is due to complications introduced by the rainfall factor.

The foregoing procedure avoids the use of determinantal arithmetic, but the latter may be used if preferred. (27.2) is
$$|C| = \begin{vmatrix} 1 & 0\cdot 80 & -0\cdot 40 \\ 0\cdot 80 & 1 & -0\cdot 56 \\ -0\cdot 40 & -0\cdot 56 & 1 \end{vmatrix} = 0\cdot 2448,$$
$$C_{11} = \begin{vmatrix} 1 & -0\cdot 56 \\ -0\cdot 56 & 1 \end{vmatrix} = 0\cdot 6864,$$

from which, for example, by (27.34),
$$\sigma_{1.23} = \sigma_1 \left(\frac{|C|}{C_{11}}\right)^{\frac{1}{2}} = 2\cdot 64, \text{ as before.}$$

Example 27.2

In some investigations into the variation of crime in 16 large cities in the U.S.A., Ogburn (1935) found a correlation of $-0\cdot 14$ between crime rate (x_1) as measured by the number of known offences per thousand inhabitants and church membership (x_5) as measured by the number of church members of 13 years of age or over per 100 of total population of 13 years of age or over. The obvious inference is that religious belief acts as a deterrent to crime. Let us consider this more closely.

If x_2 = percentage of male inhabitants,

x_3 = percentage of total inhabitants who are foreign-born males, and

x_4 = number of children under 5 years old per 1000 married women between 15 and 44 years old,

Ogburn finds the values

$$\rho_{12} = +0.44, \quad \rho_{24} = -0.19,$$
$$\rho_{13} = -0.34, \quad \rho_{25} = -0.35,$$
$$\rho_{14} = -0.31, \quad \rho_{34} = +0.44,$$
$$\rho_{15} = -0.14, \quad \rho_{35} = +0.33,$$
$$\rho_{23} = +0.25, \quad \rho_{45} = +0.85.$$

From these and other data given in his paper it may be shown that we have, for the regression of x_1 on the other four variates,

$$x_1 - 19.9 = 4.51(x_2 - 49.2) - 0.88(x_3 - 30.2) - 0.072(x_4 - 4814) + 0.63(x_5 - 41.6),$$

and for certain partial correlations

$$\rho_{15.3} = -0.03,$$
$$\rho_{15.4} = +0.25,$$
$$\rho_{15.34} = +0.23.$$

Now we note from the regression equation that when the other factors are constant x_1 and x_5 are positively related, i.e. church membership appears to be positively associated with crime. How does this effect come to be masked so as to give a negative correlation in the coefficient of zero order ρ_{15}?

We note in the first place that the correlation between crime and church membership when the effect of x_3, the percentage of foreigners, is eliminated, is near zero. The correlation when x_4, the number of young children, is eliminated, is positive; and the correlation when both x_3 and x_4 are eliminated is again positive. It appears, in fact, from the regression equation that a high percentage of foreigners and a high proportion of children are negatively associated with the crime-rate. Now both these factors are positively correlated with church membership (foreign immigrants being mainly Catholic and more fecund). These correlations submerge the positive association with crime of church membership among other members of the population. The apparently negative association of church membership with crime appears to be due to the more law-abiding spirit of the foreign immigrants and the fact that they are also more zealous churchmen.

The reader may care to refer to Ogburn's paper for a more complete discussion.

Sampling distributions of partial correlation and regression coefficients in the normal case

27.21 We now consider the sampling distributions of the partial correlation and regression coefficients in the normal case.

For large samples, the standard errors appropriate to zero-order coefficients (cf. **26.13**) may be used with obvious adjustments. Writing m for a set of secondary subscripts, we have, from (26.24),

$$\operatorname{var} r_{12.m} = \frac{1}{n}(1 - \rho_{12.m}^2)^2, \tag{27.53}$$

and from (26.30)

$$\operatorname{var} b_{12.m} = \frac{1}{n}\frac{\sigma_{1.m}^2}{\sigma_{2.m}^2}(1 - \rho_{12.m}^2) = \frac{1}{n}\frac{\sigma_{1.2m}^2}{\sigma_{2.m}^2}, \tag{27.54}$$

by (27.32). The proof of (27.53) and (27.54) by the direct methods of Chapter 10 is very tedious. They follow directly, however, from noting that the joint distribution of any two errors $x_{1.m}$ and $x_{2.m}$ is bivariate normal with correlation coefficient $\rho_{12.m}$. It follows, as Yule (1907) pointed out, that the sample correlation and regressions between the corresponding residuals have at least the large-sample distribution of a zero-order coefficient. We shall see in **27.22** that in a sample of size n, the exact distribution of $r_{12.m}$ is that of a zero-order correlation based on $(n-d)$ observations, where d is the number of secondary subscripts in m. However, since (27.53) and (27.54) are correct only to order n^{-1}, we need not adjust them by this small factor.

27.22 Consider now the geometrical representation of **27.18–19**. Suppose that we have three vectors PQ_1, PQ_2, PQ_3, representing n observations on x_1, x_2, x_3. As we saw in **27.19**, the partial correlation $r_{12.3}$ is the cosine of the angle between PQ_1 and PQ_2 projected on to the subspace orthogonal to PQ_3, which is of dimension $(n-1)$. If we make an orthogonal transformation (i.e. a rotation of the co-ordinate axes), the correlations, being cosines of angles, are unaffected; moreover, if the n original observations on the three variables are independent of each other, the n observations on the orthogonally transformed variables will also be. (This is a generalization of the result of Examples 11.2 and 11.3 and of **15.27** for independent x_1, x_2, x_3, and its proof is left for the reader as Exercise 27.7; it is geometrically obvious from the radial symmetry of the standardized multinormal distribution.) If PQ_3 is taken as one of the new co-ordinate axes in the orthogonal transformation, the distribution of $r_{12.3}$ is at once seen to be the same as that of a zero-order coefficient based on $(n-1)$ independent observations. By repeated application of this argument, it follows that the distribution of a correlation coefficient of order d based on n observations is that of a zero-order coefficient based on $(n-d)$ observations: each secondary subscript involves a projection in the sample space orthogonal to that variable and a loss of one degree of freedom. The result is due to Fisher (1924a).

The results of the previous chapter are thus immediately applicable to partial correlations, with this adjustment. If d is small compared with n, the distribution of partial correlations as n increases is effectively the same as that of zero-order coefficients, confirming the approximation (27.53) to the standard error.

It also follows for partial regression coefficients that the zero-order coefficient distribution (16.86) persists when the set m of secondary subscripts is adjoined throughout, with n replaced by $(n-d)$. In particular, the "Student's" distribution of (26.38) becomes, for the regression of x_1 on x_2, that of

$$t = (b_{12.m} - \beta_{12.m}) \left\{ \frac{s_{2.m}^2 (n-d-2)}{s_{1.m}^2 (1 - r_{12.m}^2)} \right\}^{\frac{1}{2}} \qquad (27.55)$$

with $(n-d-2)$ degrees of freedom. If the set m consists of all $(p-2)$ other variates, there are $(n-p)$ degrees of freedom. Since the regression coefficients are functions of distances (variances) as well as angles in the sample space, the distribution of b_{12} itself, unlike that of r, is not directly preserved under projection with only degrees of freedom being reduced; the statistics $s_{1.m}^2$, $s_{2.m}^2$ in (27.55) make the necessary "distance" adjustments for the projections.

The multiple correlation coefficient

27.23 The variance in the population of x_1 about its regression on the other variates (27.18) is $\sigma^2_{1.2\ldots p}$, defined in **27.9**. We now define the *multiple correlation coefficient*[*] $R_{1(2\ldots p)}$ between x_1 and $x_2, \ldots, x_p$ by

$$1 - R^2_{1(2\ldots p)} = \sigma^2_{1.2\ldots p}/\sigma^2_1. \tag{27.56}$$

From (27.56) and (27.34),

$$0 \leqslant R^2 \leqslant 1.$$

We shall define R as the positive square root of R^2: it is always non-negative. R is evidently not symmetric in its subscripts, and it is, indeed, a measure of the *dependence* of x_1 upon $x_2, \ldots, x_p$.

To justify its name, we have to show that it is in fact a correlation coefficient. We have, from **27.9**,

$$\sigma^2_{1.2\ldots p} = E(x^2_{1.2\ldots p}), \tag{27.57}$$

and by the population analogue of (27.47),

$$E(x^2_{1.2\ldots p}) = E(x_1 x_{1.2\ldots p}). \tag{27.58}$$

(27.57) and (27.58) give, since $E(x_{1.2\ldots p}) = 0$,

$$\sigma^2_{1.2\ldots p} = \operatorname{var}(x_{1.2\ldots p}) = \operatorname{cov}(x_1, x_{1.2\ldots p}). \tag{27.59}$$

If we now consider the correlation between x_1 and its conditional expectation

$$E(x_1 | x_2, \ldots, x_p) = x_1 - x_{1.2\ldots p},$$

we find that this is

$$\frac{\operatorname{cov}(x_1, x_1 - x_{1.2\ldots p})}{\{\operatorname{var} x_1 \operatorname{var}(x_1 - x_{1.2\ldots p})\}^{\frac{1}{2}}} = \frac{\operatorname{var} x_1 - \operatorname{cov}(x_1, x_{1.2\ldots p})}{[\operatorname{var} x_1 \{\operatorname{var} x_1 + \operatorname{var} x_{1.2\ldots p} - 2\operatorname{cov}(x_1, x_{1.2\ldots p})\}]^{\frac{1}{2}}},$$

and using (27.59) this is

$$\frac{\sigma^2_1 - \sigma^2_{1.2\ldots p}}{\{\sigma^2_1(\sigma^2_1 - \sigma^2_{1.2\ldots p})\}^{\frac{1}{2}}} = \left\{\frac{\sigma^2_1 - \sigma^2_{1.2\ldots p}}{\sigma^2_1}\right\}^{\frac{1}{2}} = R_{1(2\ldots p)}, \tag{27.60}$$

by (27.56). Thus $R_{1(2\ldots p)}$ is the ordinary product-moment correlation coefficient between x_1 and the conditional expectation $E(x_1 | x_2, \ldots, x_p)$. Since the sum of squared errors (and therefore their mean $\sigma^2_{1.2\ldots p}$) is minimized in finding the Least Squares regression, which is identical with $E(x_1 | x_2, \ldots, x_p)$ (cf. **27.15**), it follows from (27.60) that $R_{1(2\ldots p)}$ is the correlation between x_1 and the "best-fitting" linear combination of $x_2, \ldots, x_p$. No other linear function of $x_2, \ldots, x_p$ will have greater correlation with x_1.

27.24 From (27.56) and (27.34), we have

$$1 - R^2_{1(2\ldots p)} = \frac{|C|}{C_{11}} = (1 - \rho^2_{12})(1 - \rho^2_{13.2}) \cdots (1 - \rho^2_{1p.23\ldots(p-1)}), \tag{27.61}$$

---

[*] We use a bold-face R for the population coefficient, and will later use an ordinary capital R for the corresponding sample coefficient: we are reluctant to use the Greek capital for the population coefficient, in accordance with our usual convention, because it resembles a capital P, which might be confusing.

expressing the multiple correlation coefficient in terms of the correlation determinant or of the partial correlations. Since permutation of the subscripts other than 1 is allowed in (27.34), it follows at once from (27.61) that, since each factor on the right is in the interval (0, 1),

$$1 - R^2_{1(2\ldots p)} \leqslant 1 - \rho^2_{1j.s}$$

where $\rho_{1j.s}$ is any partial or zero-order coefficient having 1 as a primary subscript. Thus

$$R_{1(2\ldots p)} \geqslant |\rho_{1j.s}|; \qquad (27.62)$$

the multiple correlation coefficient is no less in value than the absolute value of any correlation coefficient with a common primary subscript. It follows that if $R_{1(2\ldots p)} = 0$, all the corresponding $\rho_{1j.s} = 0$ also, so that x_1 is completely uncorrelated with all the other variables. On the other hand, if $R_{1(2\ldots p)} = 1$, at least one $\rho_{1j.s}$ must be 1 also to make the right-hand side of (27.61) equal to zero (Exercise 27.22 shows that all zero-order ρ_{1j} may nevertheless be arbitrarily small). In this case, (27.56) shows that $\sigma^2_{1.2\ldots p} = 0$, so that all points in the distribution of x_1 lie on the regression line, and x_1 is a strict linear function of $x_2, \ldots, x_p$.

Thus $R_{1(2\ldots p)}$ is a measure of the linear dependence of x_1 upon $x_2, \ldots, x_p$.

27.25 So far, we have considered the multiple correlation coefficient between x_1 and all the other variates, but we may evidently also consider the multiple correlation of x_1 and any subset. Thus we define

$$R^2_{1(s)} = 1 - \frac{\sigma^2_{1.s}}{\sigma^2} \qquad (27.63)$$

for any set of subscripts s. It now follows immediately from (27.34) that

$$\sigma^2_{1.s} \leqslant \sigma^2_{1.r}, \qquad (27.64)$$

where r is any subset of s: the error variance cannot be increased by the addition of a further variate. We thus have, from (27.63) and (27.64), relations of the type

$$R^2_{1(2)} \leqslant R^2_{1(23)} \leqslant R^2_{1(234)} \leqslant \ldots \leqslant R^2_{1(2\ldots p)}, \qquad (27.65)$$

expressing the fact that the multiple correlation coefficient can never be reduced by adding to the set of variables upon which the dependence of x_1 is to be measured.

In the particular case $p = 2$, we have from (27.61)

$$R^2_{1(2)} = \rho^2_{12}, \qquad (27.66)$$

so that $R_{1(2)}$ is the absolute value of the ordinary correlation coefficient between x_1 and x_2.

Geometrical interpretation of multiple correlation

27.26 We may interpret $R_{1(2\ldots p)}$ in the geometrical terms of **27.18–19**. Consider first the interpretation of the Least Squares regression (27.18): by **27.23**, it is that linear function of the variables $x_2, \ldots, x_p$ which minimizes the sum of squares (27.40). Thus we choose the vector PV in the $(p-1)$-dimensional sub-space spanned by P, $Q_2, \ldots, Q_p$, which minimizes the distance Q_1V, i.e. which minimizes the angle between PQ_1 and PV. By (27.60), $R_{1(2\ldots p)}$ is the cosine of this minimized angle. But this means that $R_{1(2\ldots p)}$ is the cosine of the angle between PQ_1 and the $(p-1)$-dimensional subspace itself, for otherwise the angle would not be minimized.

If $R_{1(2...p)} = 0$, PQ_1 is orthogonal to the $(p-1)$-subspace so that x_1 is uncorrelated with $x_2, \ldots, x_p$ and with any linear function of them. If, on the other hand, $R_{1(2...p)} = 1$, PQ_1 lies in the $(p-1)$-subspace, so that x_1 is a strict linear function of $x_2, \ldots, x_p$. These are the results we obtained in **27.24**.

We shall find this geometrical interpretation helpful in deriving the distribution of the sample coefficient $R_{1(2...p)}$ in the normal case. It is a direct generalization of the representation used in **16.24** to obtain the distribution of the ordinary product-moment correlation coefficient r which, as we observed at (27.66), is essentially the signed value of $R_{1(2)}$.

The discarding of variables

27.27 In investigatory work, there is often a large number of variables which are known or suspected to influence the value of the dependent variable x_1. Despite the rapid computing facilities now commonly available through the development of, electronic machines, it is still often necessary to reduce this number to a smaller one, which we shall call $(p-1)$ in accordance with our earlier notation. This may be required to reduce costs, or computation time, or to achieve economy in the intellectual effort required to comprehend and use the relationship between x_1 and $x_2, x_3, \ldots, x_p$.

We wish to choose from among a larger set of candidate variables a set $x_2, x_3, \ldots, x_p$ to be used in an approximate linear regression model, as in **27.15**. We can write the model explicitly

$$x_1 = \beta_2 x_2 + \beta_3 x_3 + \ldots + \beta_p x_p + \varepsilon. \tag{27.67}$$

In (27.67) we have written β_j for $\beta_{1j.qj}$ and have made the error term explicit; otherwise it is identical with (27.18).

Our problem may be considered in two parts:

(a) How many variables should we use? i.e., how large should p be?
(b) Given p, which of the candidate variables should be used?

(a) We recall from (27.65) that R^2 can never be reduced by increasing p. Thus if we wish to determine p by using the value of R^2, we must use a rule of some such type as (i) " increase p only if R^2 is increased *by more than a certain value* " or (ii) " increase p until R^2 reaches a certain value ".

A fairly widely-used method of effecting (i), made possible by electronic computing facilities, is the so-called *Stepwise Regression* method (cf. Anscombe, 1967). Variables are introduced into the regression equation one by one, the order of their introduction being determined so that the increase in R^2 is maximized at each step. When this maximum possible increase in R^2 falls below a certain critical level, the process stops and p is determined.

This Stepwise Regression method can also be carried out in reverse: all candidate variables are used initially, and they are then rejected one by one so that at each step the reduction in R^2 is minimized, the process stopping when the minimum possible decrease exceeds a critical level.

Unfortunately there is no guarantee that these Stepwise Forward and Stepwise Backward procedures will give the same set of $(p-1)$ variables, or even the same value of

p, notwithstanding the fact that they may use the same sets of critical levels. Nor will either of them necessarily give the highest value of R^2 possible for the value of p arrived at: maximization of the gain in R^2 at each step does not necessarily produce the maximum R^2 for a fixed number of steps, since the value of R^2, as we have seen, depends on the whole complex of partial correlations. Further, the critical levels used in present practice are purely conventional, being based on the sampling distribution of the sample coefficient R^2 (to be discussed in **27.28**) when a *fixed* set of variables are included in the regression equation; R^2 as determined by the Stepwise Regression process stands in relation to this distribution as does a rather extreme order-statistic to that of a random observation from the same population. Until more is known of the distribution of R^2 under selection, this arbitrary element cannot be removed (cf. **28.23** below), but even without it, Stepwise Regression leaves something to be desired. Fortunately for human impatience, life has a habit of being less complicated than it need be, and we usually escape the worst possible consequences of simplifying procedures for the selection of "predictor" variables; we usually have enough background knowledge, even in new fields, to help us to avoid the more egregious oversights, but the theoretical objection remains.

(b) If p is determined by other means, the subset of $(p-1)$ variables giving the maximum R^2 can be obtained by a computerized search method due to Beale *et al.* (1967) which uses the theory of multiple correlation to reduce search time. Suppose there are s candidate variables from which we wish to choose our best $(p-1)$, and denote by $\bar{j}$ the set of $(s-1)$ variables obtained by excluding x_j from the candidates.

Define the *unconditional threshold* of x_j,
$$T_j = R_{1(\bar{j})}. \qquad (27.68)$$
By (27.65), this maximizes the multiple correlation obtainable without using x_j; now if, during the search for the best subset of $(p-1)$ variables, we achieve for some $(p-1)$-subset u a value of R^2 satisfying
$$R^2_{1(u)} > T_j, \qquad (27.69)$$
we can obviously only obtain a larger R^2 by including x_j in the subset to be considered. Similarly, we may define *conditional thresholds*
$$T_{jk} = R^2_{1(\overline{jk})}, \qquad (27.70)$$
obtained by excluding the pair (x_j, x_k). If we find an R^2 exceeding T_{jk}, we can only improve it by including one or both of x_j, x_k. By calculating such thresholds, the best value of R^2 is found much more quickly than by complete enumeration.

The sample multiple correlation coefficient and its conditional distribution

27.28 We now define the sample analogue of $R^2_{1(2\ldots p)}$ by
$$1 - R^2_{1(2\ldots p)} = \frac{s^2_{1.2\ldots p}}{s^2_1}, \qquad (27.71)$$
and all the relations of **27.23-6** hold with the appropriate substitutions of r for ρ, and s for σ. We proceed to discuss the sampling distribution of R^2 in detail. Since, by **27.23**, it is a correlation coefficient, whose value is independent of location and scale, its distribution will be free of location and scale parameters.

First, consider the conditional distribution of R^2 when the values of $x_2, \ldots, x_p$ are fixed. As at (26.50) we write the identity

$$ns_1^2 \equiv ns_1^2 R_{1(2\ldots p)}^2 + ns_1^2(1 - R_{1(2\ldots p)}^2),$$
$$\equiv n(s_1^2 - s_{1.2\ldots p}^2) + ns_{1.2\ldots p}^2, \qquad (27.72)$$

by (27.71). If the observations on x_1 are independent normal variates, so that $R_{1(2\ldots p)}^2 = 0$, and we standardize them, the left-hand side of (27.72) is distributed in the chi-squared form with $(n-1)$ degrees of freedom, and the quadratic forms in x_1 on the right of (27.72) may be shown to have ranks $(p-1)$ and $(n-p)$ respectively. If follows by Cochran's theorem (15.16) that they are independently distributed in the chi-squared form with these degrees of freedom and that the ratio

$$F = \frac{R_{1(2\ldots p)}^2/(p-1)}{(1-R_{1(2\ldots p)}^2)/(n-p)} \qquad (27.73)$$

has the F distribution with $(p-1, n-p)$ degrees of freedom, a result first given by Fisher (1924b). (26.52) is the special case of (27.73) for $p = 2$, when $R_{1(2)}^2 = r_{12}^2$ (cf. (27.66)).

This is another example of a LR test of a linear hypothesis. We postulate that the mean of the observations of x_1 is a linear function of $(p-1)$ other variables, with $(p-1)$ coefficients and a constant term, p parameters in all. We test the hypothesis that all $(p-1)$ coefficients are zero, i.e. $H_0: R^2 = 0$. In the notation of **24.27-8**, we have $k = p, r = p-1$ so that the F-test (27.73) has $(p-1, n-p)$ degrees of freedom, as we have seen. It follows immediately from **24.29-31** that when H_0 is not true, F at (27.73) has a non-central F-distribution with degrees of freedom $p-1$ and $n-p$. Its non-central parameter λ is now derived. (27.72) reflects the identity

$$x_1 \equiv (x_1 - x_{1.2\ldots p}) + x_{1.2\ldots p}.$$

With $x_2, \ldots, x_p$ fixed, the variance of x_1 is $\sigma_{1.(2\ldots p)}^2 = \sigma_1^2(1 - R_{1(2\ldots p)}^2)$ by (27.56), and $x_{1.2\ldots p}$ has zero mean, but $(x_1 - x_{1.2\ldots p})$ has mean $\sum_{j=2}^{p} \beta_{1j.q_j} x_j$. Thus $\dfrac{ns_{1.2\ldots p}^2}{\sigma_1^2(1-R^2)}$ has a central χ^2 distribution with $(n-p)$ degrees of freedom, but $\dfrac{n(s_1^2 - s_{1.2\ldots p}^2)}{\sigma_1^2(1-R^2)}$ has a non-central χ^2 distribution with $(p-1)$ degrees of freedom, and, from **24.4**,

$$\lambda = \sum_{i=1}^{n} \left(\sum_{j=2}^{p} \beta_{1j.q_j} x_{ji} \right)^2 \Big/ \{\sigma_1^2(1-R^2)\}$$
$$= \sum_i \boldsymbol{\beta}_1' \mathbf{x}_i . \mathbf{x}_i' \boldsymbol{\beta} / \{\sigma_1^2(1-R^2)\}$$
$$= \boldsymbol{\beta}' \mathbf{V} \boldsymbol{\beta} / \{\sigma_1^2(1-R^2)\},$$

where $\boldsymbol{\beta}$ is the $(p-1)\times 1$ vector of partial regression coefficients and $\mathbf{V}$ is the $(p-1)\times(p-1)$ observed dispersion matrix of $x_2, \ldots, x_p$. From **24.31**, λ is also the parameter of the non-central F-distribution of (27.73), reducing to zero when $R^2 = 0$, i.e. $\boldsymbol{\beta} = 0$.

The multinormal (unconditional) case

27.29 If we now allow the values of $x_2, \ldots, x_p$ to vary also, and suppose that we are sampling from a multinormal population, we find that the distribution of R^2 is unchanged if $R^2 = 0$, but quite different otherwise from that of R^2 with $x_2, \ldots, x_p$ fixed. Thus the power function of the test of $R^2 = 0$ is different in the two cases, although the same test is valid in each case. As $n \to \infty$, however, the results are identical in the two cases.

We derive the multinormal result for $R^2 = 0$ geometrically, and proceed to generalize it in **27.30**.

Consider the geometrical representation of **27.26**. R is the cosine of the angle, say θ, between PQ_1 (the x_1-vector) and the vector PV, in the $(p-1)$-dimensional space S_{p-1} of the other variables, which makes the minimum angle with PQ_1. If the parent $R = 0$, x_1 is, since the population is multinormal, independent of $x_2, \ldots, x_p$, and the vector PQ_1 will then, because of the radial symmetry of the normal distribution, be randomly directed with respect to S_{p-1}, which we may therefore regard as fixed in the subsequent argument. (We therefore see how it is that the conditional and unconditional results coincide when $R^2 = 0$.)

We have to consider the relative probabilities with which different values of θ may arise. For fixed variance s_1^2, the probability density of the sample of n observations is constant upon the $(n-2)$-dimensional surface of an $(n-1)$-dimensional hypersphere. If θ and PV are fixed, PQ_1 is constrained to lie upon a hypersphere of $(n-2)-(p-1) = (n-p-1)$ dimensions, whose content is proportional to $(\sin\theta)^{n-p-1}$ (cf. **16.24**). Now consider what happens when PV varies. PV is free to vary within S_{p-1}, where by radial symmetry it will be equiprobable on the $(p-2)$-dimensional surface of a $(p-1)$-sphere. This surface has content proportional to $(\cos\theta)^{p-2}$. For fixed θ, therefore, we have the probability element $(\sin\theta)^{n-p-1}(\cos\theta)^{p-2}d\theta$. Putting $R = \cos\theta$, and $d\theta \propto d(R^2)/\{R(1-R^2)^{\frac{1}{2}}\}$, we find for the distribution of R^2 the Beta distribution

$$dF \propto (R^2)^{\frac{1}{2}(p-3)}(1-R^2)^{\frac{1}{2}(n-p-2)}d(R^2), \qquad 0 \leq R^2 \leq 1. \tag{27.74}$$

The constant of integration is easily seen to be $\dfrac{1}{B\{\frac{1}{2}(p-1), \frac{1}{2}(n-p)\}}$. The transformation (27.73) applied to (27.74) then gives us exactly the same F-distribution as that derived for $x_2, \ldots, x_p$ fixed in **27.28**. When $p = 2$, (27.74) reduces to (16.62), which is expressed in terms of dR rather than $d(R^2)$.

27.30 We now turn to the case when $R \neq 0$. The distribution of R in this case was first given by Fisher (1928a) by a considerable development of the geometrical argument of **27.29**. We give a much simpler derivation due to Moran (1950).

We may write (27.61) for the sample coefficient as

$$1 - R_{1(2\ldots p)}^2 = (1-r_{12}^2)(1-T^2), \tag{27.75}$$

say, where T is the multiple correlation coefficient between $x_{1.2}$ and $x_{3.2}, x_{4.2}, \ldots, x_{p.2}$. Now $R_{1(2\ldots p)}$ and the distribution of $R_{1(2\ldots p)}$ are unaffected if we make an orthogonal transformation of $x_2, \ldots, x_p$ so that x_2 itself is the linear function of $x_2, \ldots, x_p$ which has maximum correlation with x_1 in the population, i.e. $\rho_{12} = R_{1(2\ldots p)}$. It then follows from (27.61) that

$$\rho_{13.2}^2 = \rho_{14.23}^2 = \ldots = \rho_{1p.23\ldots(p-1)}^2 = 0, \tag{27.76}$$

PARTIAL AND MULTIPLE CORRELATION AND REGRESSION

and since subscripts other than 1 may be permuted in (27.61), it follows that *all* partial coefficients of form $\rho_{1j.2s} = 0$. Thus $x_{1.2}$ is uncorrelated with (and since the variation is normal, independent of) $x_{3.2}, x_{4.2}, \ldots, x_{p.2}$, and T in (27.75) is distributed as a multiple correlation coefficient, based on $(n-1)$ observations (since we lose one dimension by projection for the residuals), between one variate and $(p-2)$ others, with the parent $R = 0$. Moreover, T is distributed independently of r_{12}, for all the variates $x_{j.2}$ are orthogonal to x_2 by (27.46). Thus the two factors on the right of (27.75) are independently distributed. The distribution of r_{12}, say $f_1(r)$, is (16.60) with $\rho = R_{1(2\ldots p)}$, integrated for β over its range, while that of T^2, say $f_2(R^2)$, is (27.74) with n and p each reduced by 1. We therefore have from (27.75) the distribution of R^2

$$dF = \int_{-R}^{R} \left\{ f_2\left(\frac{R^2 - r^2}{1 - r^2}\right) \right\} dF_1(r) \tag{27.77}$$

which, dropping all suffixes for convenience, is

$$= \frac{(n-2)}{\pi}(1-R^2)^{\frac{1}{2}(n-1)} \int_{-R}^{R} (1-r^2)^{\frac{1}{2}(n-4)} \int_0^\infty \frac{d\beta}{(\cosh\beta - Rr)^{n-1}}$$

$$\times \left\{ \frac{1}{B\{\frac{1}{2}(p-2), \frac{1}{2}(n-p)\}} \left(\frac{R^2 - r^2}{1-r^2}\right)^{\frac{1}{2}(p-4)} \left(\frac{1-R^2}{1-r^2}\right)^{\frac{1}{2}(n-p-2)} \frac{dR^2}{(1-r^2)} \right\} dr$$

$$= \frac{(n-2)}{\pi} \frac{(1-R^2)^{\frac{1}{2}(n-1)}(1-R^2)^{\frac{1}{2}(n-p-2)} d(R^2)}{B\{\frac{1}{2}(p-2), \frac{1}{2}(n-p)\}} \int_{-R}^{R} (R^2 - r^2)^{\frac{1}{2}(p-4)} \left[\int_0^\infty \frac{d\beta}{(\cosh\beta - Rr)^{n-1}} \right] dr. \tag{27.78}$$

We can substitute for the inner integral as at (16.64–5). If in (27.78) we put $r = R \cos \psi$ and write the integral with respect to β from $-\infty$ to ∞, dividing by 2 to compensate for this, we obtain Fisher's form of the distribution,

$$dF = \frac{\Gamma(\frac{1}{2}n)(1-R^2)^{\frac{1}{2}(n-1)}}{\pi\Gamma\{\frac{1}{2}(p-2)\}\Gamma\{\frac{1}{2}(n-p)\}} (R^2)^{\frac{1}{2}(p-3)}(1-R^2)^{\frac{1}{2}(n-p-2)} d(R^2)$$

$$\int_0^\pi \sin^{p-3}\psi \left\{ \int_{-\infty}^{\infty} \frac{d\beta}{(\cosh\beta - R R \cos\psi)^{n-1}} \right\} d\psi. \tag{27.79}$$

27.31 The distribution (27.79) may be expressed as a hypergeometric function. Expanding the integrand in a uniformly convergent series of powers of $\cos \psi$, it becomes, since odd powers of $\cos \psi$ will vanish on integration from 0 to π,

$$\sum_{j=0}^{\infty} \binom{n+2j-2}{2j} \frac{\sin^{p-3}\psi \cos^{2j}\psi}{(\cosh\beta)^{n-1+2j}} (R R)^{2j}$$

and since

$$\int_0^\pi \cos^{2j}\psi \sin^{p-3}\psi\, d\psi = B\{\frac{1}{2}(p-2), \frac{1}{2}(2j+1)\}$$

and

$$\int_{-\infty}^{\infty} \frac{d\beta}{(\cosh\beta)^{n-1+2j}} = B\{\frac{1}{2}, \frac{1}{2}(n+2j-1)\},$$

the integral in (27.79) becomes

$$\sum_{j=0}^{\infty} \binom{n+2j-2}{2j} B\{\tfrac{1}{2}(p-2), \tfrac{1}{2}(2j+1)\} B\{\tfrac{1}{2}, \tfrac{1}{2}(n+2j-1)\}(R\,R)^{2j},$$

and on writing this out in terms of Gamma functions and simplifying, it becomes

$$= \frac{\pi \Gamma\{\tfrac{1}{2}(p-2)\} \Gamma\{\tfrac{1}{2}(n-1)\}}{\Gamma(\tfrac{1}{2}n) \Gamma\{\tfrac{1}{2}(p-1)\}} F\{\tfrac{1}{2}(n-1), \tfrac{1}{2}(n-1), \tfrac{1}{2}(p-1), R^2 R^2\}. \tag{27.80}$$

Substituting (27.80) for the integrand in (27.79), we obtain

$$dF = \frac{(R^2)^{\tfrac{1}{2}(p-3)}(1-R^2)^{\tfrac{1}{2}(n-p-2)} d(R^2)}{B\{\tfrac{1}{2}(p-1), \tfrac{1}{2}(n-p)\}} \cdot (1-R^2)^{\tfrac{1}{2}(n-1)} F\{\tfrac{1}{2}(n-1),$$

$$\tfrac{1}{2}(n-1), \tfrac{1}{2}(p-1), R^2 R^2\}. \tag{27.81}$$

This unconditional distribution should be compared with the conditional distribution of R^2 given in Exercise 27.13. Exercise 27.14 shows that as $n \to \infty$, both yield the same non-central χ^2 distribution for nR^2.

The first factor on the right of (27.81) is the distribution (27.74) when $R = 0$, the second factor then being unity. (27.81) generally converges slowly, for the first two arguments in the hypergeometric function are $\tfrac{1}{2}(n-1)$. Lee (1971) gives recurrence relations for the f.f. and d.f. of R^2 (cf. Exercise 16.14 for r) which make the computation of its distribution straightforward when p is even.

Exercises 27.23–5 derive results for the distribution of $R^2/(1-R^2)$ in the multinormal case from the conditional distribution in **27.28**.

The moments and limiting distributions of R^2

27.32 It may be shown (cf. Wishart (1931)) that the mean value of R^2 in the multinormal case is

$$E(R^2) = 1 - \frac{n-p}{n-1}(1-R^2) F\{1, 1, \tfrac{1}{2}(n+1), R^2\}, \tag{27.82}$$

$$= R^2 + \frac{p-1}{n-1}(1-R^2) - \frac{2(n-p)}{(n^2-1)} R^2 (1-R^2) + O\left(\frac{1}{n^2}\right). \tag{27.83}$$

In particular, when $R^2 = 0$, (27.83) reduces to

$$E(R^2 \mid R^2 = 0) = \frac{p-1}{n-1}, \tag{27.84}$$

an exact result also obtainable directly from (27.74)

Similarly, the variance may be shown to be

$$\mathrm{var}(R^2) = \frac{(n-p)(n-p+2)}{(n^2-1)}(1-R^2)^2 F\{2, 2, \tfrac{1}{2}(n+3), R^2\} - \{E(R^2)-1\}^2 \tag{27.85}$$

$$= \frac{(n-p)}{(n^2-1)(n-1)}(1-R^2)^2 \left[2(p-1) + \frac{4R^2\{(n-p)(n-1)+4(p-1)\}}{(n+3)} + O\left\{\left(\frac{R^2}{n}\right)^2\right\} \right]. \tag{27.86}$$

(27.86) may be written

$$\text{var}(R^2) = \frac{4R^2(1-R^2)^2(n-p)^2}{(n^2-1)(n+3)} + O\left(\frac{1}{n^2}\right), \qquad (27.87)$$

so that if $R^2 \neq 0$

$$\text{var}(R^2) \sim 4R^2(1-R^2)^2/n. \qquad (27.88)$$

But if $R^2 = 0$, (27.87) is of no use, and we return to (27.86), finding

$$\text{var}(R^2) = \frac{2(n-p)(p-1)}{(n^2-1)(n-1)} \sim 2(p-1)/n^2, \qquad (27.89)$$

the exact result in (27.89) being obtainable from (27.74).

27.33 The different orders of magnitude of the asymptotic variances (27.88) and (27.89) when $R \neq 0$ and $R = 0$ reflect the fundamentally different behaviour of the distribution of R^2 in the two circumstances. Although (27.84) shows that R^2 is a biased estimator of R^2, it is clearly consistent; for large n, $E(R^2) \to R^2$ and $\text{var}(R^2) \to 0$. When $R \neq 0$, the distribution of R^2 is asymptotically normal with mean R^2 and variance given by (27.88) (cf. Exercise 27.15). When $R = 0$, however, R, which is confined to the interval $(0, 1)$, is converging to the value 0 at the lower extreme of its range, and this alone is enough to show that its distribution is not normal in this case (cf. Exercises 27.14–15). It is no surprise in these circumstances that its variance is of order n^{-2}: the situation is analogous to the estimation of a terminal of a distribution, where we saw in Example 19.10 and Exercise 19.11 that variances of order n^{-2} occur.

The distribution of R behaves similarly in respect of its limiting normality to that of R^2, though we shall see that its variance is always of order $1/n$.

One direct consequence of the singularity in the distribution of R at $R^2 = 0$ should be mentioned. It follows from (27.88) that

$$\text{var } R \sim (1-R^2)^2/n, \qquad (27.90)$$

which is the same as the asymptotic expression for the variance of the product-moment correlation coefficient (cf. (26.24))

$$\text{var } r \sim (1-\rho^2)^2/n.$$

It is natural to apply the variance-stabilizing z-transformation of **16.33** (cf. also Exercise 16.18) to R also, obtaining a transformed variable $z = \text{ar tanh } R$ with variance close to $1/n$, independent of the value of R. But this will not do near $R = 0$, as Hotelling (1953) pointed out, since (27.90) breaks down there; its asymptotic variance then will be given by (27.84) as

$$\text{var } R = E(R^2) - \{E(R)\}^2 \sim (p-1)/n, \qquad (27.91)$$

agreeing with the value $1/n$ obtained from (27.90) only for $p = 2$, when $R = |r|$. Lee (1971) investigates the approximation numerically, finds it inadequate, and proposes better ones.

Unbiassed estimation of R^2 in the multinormal case

27.34 Since, by (27.83), R^2 is a biassed estimator of R^2, we may wish to adjust it for the bias. Olkin and Pratt (1958) show that an unbiassed estimator of $R^2_{1(2...p)}$ is

$$t = 1 - \frac{n-3}{n-p}(1 - R^2_{1(2...p)}) F(1, 1, \tfrac{1}{2}(n-p+2), 1 - R^2_{1(2...p)}), \qquad (27.92)$$

where $n > p \geqslant 3$. t is the unique unbiassed function of R^2 since it is a function of the complete sufficient statistics. (27.92) may be expanded into series as

$$t = R^2 - \frac{p-3}{n-p}(1 - R^2) - \left\{ \frac{2(n-3)}{(n-p)(n-p+2)}(1-R^2)^2 + O\left(\frac{1}{n^2}\right) \right\}, \qquad (27.93)$$

whence it follows that $t \leqslant R^2$. If $R^2 = 1$, $t = 1$ also. When R^2 is zero or small, on the other hand, t is negative, as we might expect. We cannot (cf. **17.9**) find an unbiassed estimator of R^2 (i.e. an estimator whose expectation is R^2 *whatever the true value of R^2*) which takes only non-negative values, even though we know that R^2 is non-negative. We may remove the absurdity of negative estimates by using as our estimator

$$t' = \max(t, 0) \qquad (27.94)$$

but (27.94) is no longer unbiassed.

27.35 Lehmann (1959) shows that for testing $R^2 = 0$ in the multinormal case, tests rejecting large values of R^2 are UMP among test statistics which are invariant under location and scale changes.

Ezekiel and Fox (1959) and Kramer (1963) give charts and tables for constructing confidence intervals for R^2 from the value of R^2. Lee (1972) gives tables of the upper 5 per cent. and 1 per cent. points of the d.f. of R for $R = 0$ (0·1) 0·9; $p-1 = 1$ (1) 10, 12, 15, 20, 24, 30, 40; and $(n-p)^{\frac{1}{2}} = 60/\nu$ where $\nu = 1$ (1) 6 (2) 20.

PARTIAL AND MULTIPLE CORRELATION AND REGRESSION

EXERCISES

27.1 Show that
$$\beta_{12.34\ldots(p-1)} = \frac{\beta_{12.34\ldots p} + \beta_{1p.23\ldots(p-1)}\beta_{p2.13\ldots(p-1)}}{1 - \beta_{1p.23\ldots(p-1)}\beta_{p1.23\ldots(p-1)}},$$

and that

$$\rho_{12.34\ldots(p-1)} = \frac{\rho_{12.34\ldots p} + \rho_{1p.23\ldots(p-1)}\rho_{2p.13\ldots(p-1)}}{\{(1-\rho^2_{1p.23\ldots(p-1)})(1-\rho^2_{2p.13\ldots(p-1)})\}^{\frac{1}{2}}}.$$

(Yule, 1907)

27.2 Show that for p variates there are $\binom{p}{2}$ correlation coefficients of order zero and $\binom{p-2}{s}\binom{p}{2}$ of order s. Show further that there are $\binom{p}{2}2^{p-2}$ correlation coefficients altogether and $\binom{p}{2}2^{p-1}$ regression coefficients.

27.3 If the correlations of zero order among a set of variables are all equal to ρ, show that every partial correlation of the sth order is equal to $\frac{\rho}{(1+s\rho)}$.

27.4 Prove equation (27.27), and show that it implies that the coefficient of $x_l x_m$ in the exponent of the multinormal distribution of $x_1, x_2, \ldots, x_p$ is $1/\mathrm{cov}(x_{l.q_l}, x_{m.q_m})$.

27.5 Show from (27.46) that in summing the product of two residuals, any or all of the secondary subscripts may be omitted from a residual *all* of whose secondary subscripts are included among those of the other residual, i.e. that

$$\Sigma x_{1.stu} x_{2.st} = \Sigma x_{1.stu} x_{2.s} = \Sigma x_{1.stu} x_2,$$

but that

$$\Sigma x_{1.stu} x_{2.st} \neq \Sigma x_{1.su} x_{2.st},$$

where s, t, u are sets of subscripts. (The same result holds for products of errors.)

(Chandler, 1950)

27.6 By the transformation

$$y_1 = x_1,$$
$$y_2 = x_{2.1},$$
$$y_3 = x_{3.21},$$
etc.,

show that the multivariate normal distribution may be written

$$dF = \frac{1}{(2\pi)^{\frac{1}{2}p}\sigma_1 \sigma_{2.1}\sigma_{3.12}\ldots}\exp\left\{-\frac{1}{2}\left(\frac{x_1^2}{\sigma_1^2}+\frac{x_{2.1}^2}{\sigma_{2.1}^2}+\frac{x_{3.12}^2}{\sigma_{3.12}^2}+\ldots\right)\right\}dx_1\,dx_{2.1}\ldots$$

so that the residuals $x_1, x_{2.1}, \ldots$ are independent of each other. Hence show that any two residuals $x_{j.r}$ and $x_{k.r}$ (where r is a set of common subscripts) are distributed in the bivariate normal form with correlation $\rho_{jk.r}$.

27.7 Show that if an orthogonal transformation is applied to a set of n independent observations on p multinormal variates, the transformed set of n observations will also be independent.

27.8 For the data of Tables 26.1 and 26.2, we saw in Example 26.6 and Exercise 26.1 that
$$r_{12} = 0.34, \quad r_{13} = 0.07,$$
where subscripts 1, 2, 3 refer to Stature, Weight and Bust Girth respectively. Given also that
$$r_{23} = 0.86$$
show that
$$R^2_{3(12)} = 0.80,$$
indicating that Bust Girth is fairly well determined by a linear function of Stature and Weight.

27.9 Show directly that no linear function of $x_2, \ldots, x_p$ has a higher correlation with x_1 than the Least Squares estimate of x_1.

27.10 Establish (27.83), the expression for $E(R^2)$.

(Wishart, 1931)

27.11 Establish (27.85), the expression for var (R^2).

(Wishart, 1931)

27.12 Verify that (27.92) is an unbiassed estimator of R^2.

27.13 Show from the non-central F-distribution of F at (27.73) when $R^2 \neq 0$, that the distribution of R^2 in this case, when $x_2, \ldots, x_p$ are fixed, is
$$dF = \frac{1}{B\{\tfrac{1}{2}(p-1), \tfrac{1}{2}(n-p)\}} (R^2)^{\tfrac{1}{2}(p-3)} (1-R^2)^{\tfrac{1}{2}(n-p-2)} dR^2 \cdot \exp\{-\tfrac{1}{2}(n-p)R^2\}$$
$$\times \sum_{j=0}^{\infty} \frac{\Gamma\{\tfrac{1}{2}(n-1+2j)\} \Gamma\{\tfrac{1}{2}(p-1)\}}{\Gamma\{\tfrac{1}{2}(n-1)\} \Gamma\{\tfrac{1}{2}(p-1+2j)\}} \frac{\{\tfrac{1}{2}(n-p)R^2 R^2\}^j}{j!}.$$

(Fisher, 1928a)

27.14 Show from (27.81) that for $n \to \infty$, p fixed, the distribution of $nR^2 = B^2$ is
$$dF = \frac{(B^2)^{\tfrac{1}{2}(p-3)}}{2^{\tfrac{1}{2}(p-1)} \Gamma\{\tfrac{1}{2}(p-1)\}} \exp(-\tfrac{1}{2}\beta^2 - \tfrac{1}{2}B^2)$$
$$\times \left\{1 + \frac{\beta^2 B^2}{(p-1).2} + \frac{(\beta^2 B^2)^2}{(p-1)(p+1).2.4} + \ldots\right\} d(B^2),$$
where $\beta^2 = nR^2$, and hence that nR^2 is a non-central χ^2 variate of form (24.18) with $\nu = p-1$, $\lambda = nR^2$. Show that the same result holds for the conditional distribution of nR^2, from Exercise 27.13.

(Fisher, 1928a)

27.15 In Exercise 27.14, use the c.f. of a non-central χ^2 variate given in Exercise 24.1 to show that as $n \to \infty$ for fixed p, R^2 is asymptotically normally distributed when $R \neq 0$, but not when $R = 0$. Extend the result to R.

27.16 Show that the distribution function of R^2 in multinormal samples may be written, if $n-p$ is even, in the form
$$(1-R^2)^{\tfrac{1}{2}(n-1)} R^{p-1} \sum_{j=0}^{\tfrac{1}{2}(n-p-2)} \frac{\Gamma\{\tfrac{1}{2}(p-1+2j)\}}{\Gamma\{\tfrac{1}{2}(p-1)\}} \frac{(1-R^2)^j}{(1-R^2 R^2)^{\tfrac{1}{2}(n-1+2j)}}$$
$$\times F\{-j, -\tfrac{1}{2}(n-p), \tfrac{1}{2}(p-1), R^2 R^2\}.$$

(Fisher, 1928a)

27.17 Show that in a sample $(x_1, \ldots, x_n)$ of one observation from an n-variate multinormal population with all means μ, all variances σ^2 and all correlations equal to ρ, the statistic

$$t^2 = \frac{(\bar{x}-\mu)^2}{\sum_{i=1}^{n}(x_i-\bar{x})^2/\{n(n-1)\}} \cdot \left\{\frac{1-\rho}{1+(n-1)\rho}\right\}$$

has a "Student's" t^2-distribution with $(n-1)$ degrees of freedom. When $\rho = 0$, this reduces to the ordinary test of a mean of n independent normal variates.

(Walsh, 1947)

27.18 If $x_0, x_1, \ldots, x_n$ are normal variates with common variance α^2, $x_1, \ldots, x_n$ being independent of each other and x_0 having zero mean and correlation λ with each of the others, show that the n variates

$$y_i = x_i - a x_0, \quad i = 1, 2, \ldots, n,$$

are multinormally distributed with all correlations equal to

$$\rho = (a^2 - 2a\lambda)/(1 + a^2 - 2a\lambda)$$

and all variances equal to

$$\sigma^2 = \alpha^2/(1-\rho).$$

(Stuart, 1958)

27.19 Use the result of Exercise 27.18 to establish that of Exercise 27.17.

(Stuart, 1958)

27.20 Show that if each pair from $x_2, \ldots, x_p$ is uncorrelated,

$$R^2_{1(2\ldots p)} = \sum_{s=2}^{p} \rho^2_{1s} = \sum_{s=2}^{p} (\beta_{1s.q}\sigma_s)^2/\sigma^2_1.$$

27.21 Generalizing (27.17), show in the matrix notation given at the end of **27.6** that the conditional mean of the vector $(x_1, \ldots, x_k)'$, when the vector $(x_{k+1}, \ldots, x_p)'$ is fixed at $\mathbf{x}_0'$, is $\mathbf{B}'\mathbf{E}^{-1}\mathbf{x}_0'$.

(Marsaglia (1964) shows that this result and (27.14) hold even for singular multinormal distributions if $\mathbf{E}^{-1}$ is replaced by the *pseudo-inverse* $\mathbf{E}^+ = \mathbf{T}'(\mathbf{T}\mathbf{T}')^{-2}\mathbf{T}$, where $\mathbf{E} = \mathbf{T}'\mathbf{T}$.)

27.22 Consider variables x_1, x_2, x_3 for which

$$\rho_{13} = 0, \quad \rho_{12} = \cos\theta, \quad \rho_{23} = \sin\theta, \quad 0 < \theta < \pi/2.$$

Show that $\rho_{12.3} = 1$ and hence that $R^2_{1(23)} = 1$. By letting $\theta \to \pi/2$, show that $R^2_{1(23)} = 1$ is consistent with $\rho_{13} = 0$, $\rho_{12} = \varepsilon$ for any $\varepsilon > 0$. Interpret the result geometrically.

27.23 In **27.28**, show that when $x_2, \ldots, x_p$ are allowed to vary, $\lambda(1-R^2)/R^2$ is distributed as a χ^2 with $(n-1)$ d.fr., and hence, using the c.f. of χ'^2 in Exercise 24.1, that the distribution of $R^2/(1-R^2)$ in the multinormal case is that of the ratio of independent variables y, z, with c.f.'s

$$\phi_y(t) = (1-2it)^{\frac{1}{2}(n-p)}\{1-2it(1+\theta)\}^{-\frac{1}{2}(n-1)}$$
$$\phi_z(t) = (1-2it)^{\frac{1}{2}(n-p)},$$

where $\theta = R^2/(1-R^2)$, and hence that y may be represented as

$$y = \chi^2_{-2} + (\chi_1 + \theta^{\frac{1}{2}}\chi_{n-1})^2,$$

all the variables being independent. (Cf. Exercise 16.6 when $p = 2$.)

(Gurland, 1968; Lee, 1971)

27.24 In Exercise **27.23**, show that if $n-p = 2k$ is even,

$$\phi_y(t) = \{1-2(1+\theta)it\}^{-\frac{1}{2}(p-1)}(1+\theta)^{-k} \sum_{j=0}^{k} \binom{k}{j}\left(\frac{\theta}{1-2(1+\theta)it}\right)^j,$$

and hence that $R^2/(1-R^2)$ has the d.f.

$$G(x) = \sum_{j=0}^{k} \binom{k}{j}\left(\frac{\theta}{1+\theta}\right)^j (1+\theta)^{j-k} H_{p-1+2j,\,2k}\left(\frac{x}{1+\theta}\right),$$

where $H_{a,b}(x)$ is the d.f. of a ratio of independent χ^2 variates with a and b degrees of freedom

(Gurland, 1968, who also gives two infinite series for $G(x)$, one for odd $(n-p)$, $R^2 < \frac{1}{2}$, and one for all n, p and R. A general class of series is given by Gurland and Milton (1970).)

27.25 In Exercise **27.23**, approximate the distribution of $R^2/(1-R^2)$ by assuming that it is a multiple g of the ratio of independent central χ^2 variates with ν and $n-p$ degrees of freedom, and equate the first two moments with the exact ones, as in **21.23**. Show that $g = B/A$ and $\nu = A^2/B$, where

$$A = (n-1)\theta + p - 1 \quad \text{and} \quad B = (n-1)\theta(\theta+2) + p - 1.$$

(Gurland, 1968; the result is exact when $R^2 = \theta = 0$, and the approximation seems generally good. See also Gurland and Milton (1970).)

CHAPTER 28

THE GENERAL THEORY OF REGRESSION

28.1 In the two previous chapters we have developed the theory of linear regression of one variable upon one or more others, but our main preoccupation there was with the theory of correlation. We now, so to speak, bring the theory of regression to the centre of the stage. In this chapter we shall generalize and draw together the results of Chapters 26 and 27, and we shall also make use of the theory of Least Squares developed in Chapter 19.

When discussing the regression of y upon one or more variables x, it has been customary to call y a "dependent" variable and x the "independent" variables. This usage, taken over from ordinary algebra, is a bad one, for the x-variables are not in general independent of each other in the probability sense; indeed, we shall see that they need not be random variables at all. Further, since the whole purpose of a regression analysis is to investigate the dependence of y upon x, it is particularly confusing to call the x-variables "independent." Notwithstanding common usage, therefore, we shall follow some more recent writers, e.g. Hannan (1956), and call x the *regressor* variables (or *regressors*, for short).

We first consider the extension of the analytical theory of regression from the linear situations discussed in Chapters 26 and 27. The distinguishing feature of the analytical theory is that knowledge of the joint distribution of the variables, or equivalently of their joint characteristic function, is assumed.

The analytical theory of regression

28.2 Let $f(x, y)$ be the joint frequency function of the variables x, y. Then, for any fixed value of x, say X, the mean value of y is defined by

$$E(y|X) = \int_{-\infty}^{\infty} y f(X, y) \, dy \bigg/ \int_{-\infty}^{\infty} f(X, y) \, dy. \tag{28.1}$$

(28.1) is the *regression (curve)* discussed in **26.5**; it gives the relation between X and the mean value of y for that value of X, which is a mathematical relationship, not a probabilistic one.

We may also consider the more general regression (curve) of order r, defined by

$$\mu'_{rX} = E(y^r|X) = \int_{-\infty}^{\infty} y^r f(X, y) \, dy \bigg/ \int_{-\infty}^{\infty} f(X, y) \, dy, \tag{28.2}$$

which expresses the dependence of the rth moment of y, for fixed X, upon X. Similarly

$$\mu_{rX} = E[\{y - E(y|X)\}^r | X]$$
$$= \int_{-\infty}^{\infty} \{y - E(y|X)\}^r f(X, y) \, dy \bigg/ \int_{-\infty}^{\infty} f(X, y) \, dy \tag{28.3}$$

gives the dependence of the central moments of y, for fixed X, upon X.

If $r = 2$ in (28.3), it is called the *scedastic* curve, giving the dependence of the variance of y for fixed X upon X. If the skewness coefficient $\beta_{1X} = \frac{\mu_{3X}^2}{\mu_{2X}^3}$ is plotted against X, we obtain the *clitic* curve, and if $\beta_{2X} = \frac{\mu_{4X}}{\mu_{2X}^2}$ is plotted, we have the *kurtic* curve.[*] These are not, in fact, in common use. The regression curve of outstanding importance is that for $r = 1$, which is (28.1); so much so, that whenever "regression" is mentioned without qualification, the regression of the mean, (28.1), is to be understood.

As we saw in **26.5**, we are sometimes interested in the regression of x upon y as well as that of y upon x. We then have the obvious analogues of (28.2) and (28.3), and in particular that of (28.1).

$$\mu_{1Y}' = E(x|Y) = \int_{-\infty}^{\infty} x f(x, Y) dx \bigg/ \int_{-\infty}^{\infty} f(x, Y) dx. \qquad (28.4)$$

28.3 Just as we can obtain the moments from a c.f. without explicitly evaluating the frequency function, so we can find the regression of any order from the joint c.f. of x and y without explicitly determining their joint f.f., $f(x, y)$. Write

$$f(x, y) = g(x) \cdot h_x(y), \qquad (28.5)$$

where $g(x)$ is the marginal distribution of x and $h_x(y)$ the conditional distribution of y for given x.[†] The joint c.f. of x and y is

$$\phi(t_1, t_2) = \int_{-\infty}^{\infty} \int_{-\infty}^{\infty} \exp(it_1 x + it_2 y) g(x) h_x(y) dx \, dy \qquad (28.6)$$

$$= \int_{-\infty}^{\infty} \exp(it_1 x) g(x) \phi_x(t_2) dx, \qquad (28.7)$$

where $\qquad \phi_x(t_2) = \int_{-\infty}^{\infty} \exp(it_2 y) h_x(y) dy$

is the conditional c.f. of y for given x. If the rth moment of y for given x is μ_{rx}', as in **28.2**, we have

$$i^r \mu_{rx}' = \left[\frac{\partial^r}{\partial t_2^r} \phi_x(t_2) \right]_{t_2 = 0} \qquad (28.8)$$

and hence, from (28.7) and (28.8),

$$\left[\frac{\partial^r}{\partial t_2^r} \phi(t_1, t_2) \right]_{t_2 = 0} = i^r \int_{-\infty}^{\infty} \exp(it_1 x) g(x) \mu_{rx}' dx. \qquad (28.9)$$

Hence, by the Inversion Theorem (**4.3**),

$$g(x) \mu_{rx}' = \frac{(-i)^r}{2\pi} \int_{-\infty}^{\infty} \exp(-it_1 x) \left[\frac{\partial^r}{\partial t_2^r} \phi(t_1, t_2) \right]_{t_2 = 0} dt_1. \qquad (28.10)$$

(28.10) is the required expression, from which the regression of any order may be written down.

---

[*] Although, so far as we know, such a thing has never been done, it might be more advantageous to plot the cumulants of y, rather than its moments, against X.

[†] We now no longer use X for the fixed value of x.

From (28.10) with $r = 1$, we have

$$g(x)\mu'_{1x} = \frac{-i}{2\pi} \int_{-\infty}^{\infty} \exp(-it_1 x) \left[\frac{\partial}{\partial t_2} \phi(t_1, t_2)\right]_{t_2=0} dt_1. \quad (28.11)$$

28.4 If all cumulants exist, we have the definition of bivariate cumulants at (3.74)

$$\phi(t_1, t_2) = \exp\left\{\sum_{r,s=0}^{\infty} \kappa_{rs} \frac{(it_1)^r (it_2)^s}{r! \, s!}\right\},$$

where κ_{00} is defined to be equal to zero. Hence

$$\left[\frac{\partial \phi(t_1, t_2)}{\partial t_2}\right]_{t_2=0} = \left[i\phi(t_1, t_2) \sum_{r=0}^{\infty} \sum_{s=1}^{\infty} \kappa_{rs} \frac{(it_1)^r (it_2)^{s-1}}{r! \, (s-1)!}\right]_{t_2=0}$$

$$= i\phi(t_1, 0) \sum_{r=0}^{\infty} \kappa_{r1} \frac{(it_1)^r}{r!}. \quad (28.12)$$

In virtue of (28.12), (28.11) becomes

$$g(x)\mu'_{1x} = \frac{1}{2\pi} \int_{-\infty}^{\infty} \exp(-it_1 x) \phi(t_1, 0) \sum_{r=0}^{\infty} \kappa_{r1} \frac{(it_1)^r}{r!} dt_1, \quad (28.13)$$

and if the interchange of integration and summation operations is permissible, (28.13) becomes

$$g(x)\mu'_{1x} = \frac{1}{2\pi} \sum_{r=0}^{\infty} \frac{\kappa_{r1}}{r!} i^r \int_{-\infty}^{\infty} t_1^r \exp(-it_1 x) \phi(t_1, 0) dt_1. \quad (28.14)$$

Since, by the Inversion Theorem,

$$g(x) = \frac{1}{2\pi} \int_{-\infty}^{\infty} \exp(-it_1 x) \phi(t_1, 0) dt_1,$$

we have, subject to existence conditions,

$$(-D)^j g(x) \equiv (-1)^j \frac{d^j}{dx^j} g(x) = \frac{i^j}{2\pi} \int_{-\infty}^{\infty} t_1^j \exp(-it_1 x) \phi(t_1, 0) dt_1. \quad (28.15)$$

Using (28.15), (28.14) becomes

$$g(x)\mu'_{1x} = \sum_{r=0}^{\infty} \frac{\kappa_{r1}}{r!} (-D)^r g(x). \quad (28.16)$$

Thus, for the regression of the mean of y on x, we have

$$\mu'_{1x} = \sum_{r=0}^{\infty} \frac{\kappa_{r1}}{r!} \frac{(-D)^r g(x)}{g(x)}, \quad (28.17)$$

a result due to Wicksell (1934). (28.17) is valid if cumulants of all orders exist and if the interchange of integration and summation in (28.13) is legitimate; this will be so, in particular, if $g(x)$ and all its derivatives are continuous within the range of x and zero at its extremes.

If $g(x)$ is normal and standardized, we have the particular case of (28.17)

$$\mu'_{1x} = \sum_{r=0}^{\infty} \frac{\kappa_{r1}}{r!} H_r(x), \quad (28.18)$$

where $H_r(x)$ is the Tchebycheff–Hermite polynomial of order r, defined at (6.21).

Example 28.1

For the bivariate normal distribution
$$f(x,y) = (2\pi\sigma_1\sigma_2)^{-1}(1-\rho^2)^{-\frac{1}{2}}\exp\left[-\frac{1}{2(1-\rho^2)}\left\{\left(\frac{x-\mu_1}{\sigma_1}\right)^2 - 2\rho\left(\frac{x-\mu_1}{\sigma_1}\right)\left(\frac{y-\mu_2}{\sigma_2}\right)+\left(\frac{y-\mu_2}{\sigma_2}\right)^2\right\}\right],$$

the joint c.f. of $\frac{x-\mu_1}{\sigma_1}$ and $\frac{y-\mu_2}{\sigma_2}$ is (cf. Example 15.1)
$$\phi(t_1,t_2) = \exp\{-\tfrac{1}{2}(t_1^2+t_2^2+2\rho t_1 t_2)\},$$
whence
$$\kappa_{01} = 0,$$
$$\kappa_{r1} = 0, \quad r > 1,$$
so that
$$\kappa_{11} = \rho$$
is the only non-zero cumulant in (28.17). The marginal distribution $g(x)$ is standard normal, so that (28.17) becomes (28.18) and we have, using (6.23),
$$\mu'_{1x} = \kappa_{11}H_1(x) = \rho x.$$
This is the regression of $(y-\mu_2)/\sigma_2$ on $(x-\mu_1)/\sigma_1$. If we now de-standardize, we find for the regression of y on x,
$$E(y\,|\,x) - \mu_2 = \frac{\rho\sigma_2}{\sigma_1}(x-\mu_1),$$
a more general form of the first equation in (16.46), which has x and y interchanged and $\mu_1 = \mu_2 = 0$.

Example 28.2

In a sample of n observations from the bivariate normal distribution of the previous example, consider the joint distribution of
$$u = \tfrac{1}{2}\sum_{i=1}^n (x_i-\mu_1)^2/\sigma_1^2 \quad \text{and} \quad v = \tfrac{1}{2}\sum_{i=1}^n (y_i-\mu_2)^2/\sigma_2^2.$$
The joint c.f. of u and v is easily found from Example 26.1 to be
$$\phi(t_1,t_2) = \{(1-\theta_1)(1-\theta_2)-\rho^2\theta_1\theta_2\}^{-\frac{1}{2}n}, \tag{28.19}$$
where $\theta_1 = it_1$, $\theta_2 = it_2$. The joint f.f. of u and v cannot be expressed in a simple form, but we may determine the regressions without it. From (28.19),
$$\left[\frac{\partial^r \phi(t_1,t_2)}{\partial t_2^r}\right]_{t_2=0} = i^r\left[\frac{\partial^r \phi}{\partial \theta_2^r}\right]_{\theta_2=0} = i^r(\tfrac{1}{2}n+r-1)^{(r)}\frac{\{1-(1-\rho^2)\theta_1\}^r}{(1-\theta_1)^{\frac{1}{2}n+r}}. \tag{28.20}$$
Thus, from (28.10) and (28.20),
$$g(u)\mu'_{rv} = (\tfrac{1}{2}n+r-1)^{(r)}\frac{1}{2\pi}\int_{-\infty}^{\infty}\exp(-\theta_1 u)\frac{\{\rho^2+(1-\rho^2)(1-\theta_1)\}^r}{(1-\theta_1)^{\frac{1}{2}n+r}}dt_1. \tag{28.21}$$

Now, from the inversion of the c.f. in Example 4.4,
$$\frac{1}{2\pi}\int_{-\infty}^{\infty}\frac{\exp(-\theta_1 u)}{(1-\theta_1)^k}dt_1 = \frac{1}{\Gamma(k)}e^{-u}u^{k-1},$$
while the marginal distribution $g(u)$ is (cf. (11.8))
$$g(u) = \frac{1}{\Gamma(\frac{1}{2}n)}e^{-u}u^{\frac{1}{2}n-1}.$$
Substituting into (28.21), we find, putting $r = 1, 2$ successively,
$$\mu'_{1v} = \tfrac{1}{2}n\left\{\rho^2 \cdot \frac{v}{(\tfrac{1}{2}n)} + (1-\rho^2)\right\} = \rho^2 v + \tfrac{1}{2}n(1-\rho^2) \tag{28.22}$$
and
$$\mu'_{2v} = \tfrac{1}{2}n(\tfrac{1}{2}n+1)\left\{\rho^4 \cdot \frac{v^2}{(\tfrac{1}{2}n)(\tfrac{1}{2}n+1)} + 2\rho^2(1-\rho^2)\frac{v}{(\tfrac{1}{2}n)} + (1-\rho^2)^2\right\}$$
$$= \rho^4 v^2 + 2\rho^2(1-\rho^2)v(\tfrac{1}{2}n+1) + (1-\rho^2)^2 \tfrac{1}{2}n(\tfrac{1}{2}n+1),$$
so that
$$\mu_{2v} = \mu'_{2v} - (\mu'_{1v})^2 = (1-\rho^2)\{2\rho^2 v + \tfrac{1}{2}n(1-\rho^2)^2\}. \tag{28.23}$$
(28.22) and (28.23) indicate that the regressions upon v of both the mean and variance of u are linear.

Criteria for linearity of regression

28.5 Let $\psi(t_1, t_2) = \log\phi(t_1, t_2)$ be the joint c.g.f. of x and y. We now prove: if the regression of y upon x is linear, so that
$$\mu'_{1x} = E(y|x) = \beta_0 + \beta_1 x, \tag{28.24}$$
then
$$\left[\frac{\partial\psi(t_1, t_2)}{\partial t_2}\right]_{t_2=0} = i\beta_0 + \beta_1\frac{\partial\psi(t_1, 0)}{\partial t_1}; \tag{28.25}$$
and conversely, if a completeness condition is satisfied, (28.25) is sufficient as well as necessary for (28.24).

From (28.9) with $r = 1$, we have, using (28.24),
$$\left[\frac{\partial\phi(t_1, t_2)}{\partial t_2}\right]_{t_2=0} = i\int_{-\infty}^{\infty}\exp(it_1 x)g(x)(\beta_0+\beta_1 x)dx \tag{28.26}$$
$$= i\beta_0\phi(t_1, 0) + \beta_1\frac{\partial}{\partial t_1}\phi(t_1, 0). \tag{28.27}$$

Putting $\psi = \log\phi$ in (28.27), and dividing through by $\phi(t_1, 0)$, we obtain (28.25).

Conversely, if (28.25) holds, we rewrite it, using (28.9), in the form
$$i\int_{-\infty}^{\infty}\exp(it_1 x)(\beta_0+\beta_1 x-\mu'_{1x})g(x)dx = 0. \tag{28.28}$$
We now see that (28.28) implies
$$\beta_0 + \beta_1 x - \mu'_{1x} = 0 \tag{28.29}$$
identically in x if $\exp(it_1 x)g(x)$ is complete, and hence (28.24) follows.

28.6 If all cumulants exist, (28.25) gives, on using (28.12)

$$\sum_{r=0}^{\infty} \kappa_{r1} \frac{(it_1)^r}{r!} = \beta_0 + \beta_1 \sum_{r=0}^{\infty} \kappa_{r0} \frac{(it_1)^{r-1}}{(r-1)!}. \tag{28.30}$$

Identifying coefficients of t^r in (28.30) gives

$$(r = 0) \quad \kappa_{01} = \beta_0 + \beta_1 \kappa_{10}, \tag{28.31}$$

as is obvious from (28.24);

$$(r \geqslant 1) \quad \kappa_{r1} = \beta_1 \kappa_{r+1, 0}. \tag{28.32}$$

The condition (28.32) for linearity of regression is also due to Wicksell (1934). (28.31) and (28.32) together are sufficient, as well as necessary, for (28.25) and thence (given the completeness of $g(x)$, as before) for the linearity condition (28.24).

If we express (28.25) in terms of the c.f. ϕ, instead of its logarithm ψ, as in (28.27), and carry through the process leading to (28.32), we find the analogue of (28.32) for the central moments,

$$\mu_{r1} = \beta_1 \mu_{r+1, 0}. \tag{28.33}$$

If the regression of x on y is also linear, of form

$$x = \beta_0' + \beta_1' y,$$

we shall also have

$$\kappa_{1r} = \beta_1' \kappa_{0, r+1}, \quad r \geqslant 1. \tag{28.34}$$

When $r = 1$, (28.32) and (28.34) give

$$\kappa_{11} = \beta_1 \kappa_{20} = \beta_1' \kappa_{02},$$

whence

$$\beta_1 \beta_1' = \kappa_{11}^2 / (\kappa_{20} \kappa_{02}) = \rho^2, \tag{28.35}$$

which is (26.17) again, ρ being the correlation coefficient between x and y.

28.7 We now impose a further restriction on our variables: we suppose that the conditional distribution of y about its mean value (which, as before, is a function of the fixed value of x) is the same for any x, i.e. that only the mean of y changes with x. We shall refer to this restriction by saying that y "has identical errors." There is thus a variate ε such that

$$y = \mu_{1x}' + \varepsilon. \tag{28.36}$$

In particular, if the regression is linear (28.36) is

$$y = \beta_0 + \beta_1 x + \varepsilon. \tag{28.37}$$

If y has identical errors, (28.5) becomes

$$f(x, y) = g(x) h(\varepsilon) \tag{28.38}$$

where h is now the conditional distribution of ε. Conversely, (28.38) implies identical errors for y.

The corresponding result for c.f.s is not quite so obvious: if the regression of y on x is linear with identical errors, then the joint c.f. of x and y factorizes into

$$\phi(t_1, t_2) = \phi_g(t_1 + t_2 \beta_1) \phi_h(t_2) \exp(it_2 \beta_0), \tag{28.39}$$

$$\left[\frac{\partial^r}{\partial u^r}\phi(u,t_1,\ldots,t_p)\right]_{u=0} = i^r \int \exp(i\Sigma t_j x_j) g(\mathbf{x}) \mu'_{r\mathbf{x}} d\mathbf{x}, \qquad (28.53)$$

giving the generalization of (28.10)

$$g(\mathbf{x})\mu'_{r\mathbf{x}} = \frac{(-i)^r}{2\pi} \int \exp(-i\Sigma t_j x_j) \left[\frac{\partial^r}{\partial u^r}\phi(u,t_1,\ldots,t_p)\right]_{u=0} dt. \qquad (28.54)$$

28.10 The reader should have no difficulty in extending the criterion of **28.5** for linearity of regression: if (28.51) is to hold, we must have

$$\left[\frac{\partial \psi(u,t_1,\ldots,t_p)}{\partial u}\right]_{u=0} = i\beta_0 + \sum_{j=1}^p \beta_j \frac{\partial}{\partial t_j} \psi(0,t_1,\ldots,t_p), \qquad (28.55)$$

generalizing (28.25). Similarly, the extension of the criterion of (28.32) is

$$\kappa_{1,r_1,r_2,\ldots,r_p} = \beta_1 \kappa_{0,r_1+1,r_2,\ldots,r_p} + \beta_2 \kappa_{0,r_1,r_2+1,r_3,\ldots,r_p} + \ldots + \beta_p \kappa_{0,r_1,\ldots,r_{p-1},r_p+1} \qquad (28.56)$$

The condition (28.38) for identical errors generalizes to

$$f(y,x_1,\ldots,x_p) = g(\mathbf{x}) h(t)$$

and (28.39) generalizes to

$$\phi(u,t_1,\ldots,t_p) = \phi_g(t_1+u\beta_1, t_2+u\beta_2, \ldots, t_p+u\beta_p) \phi_h(u) \exp(iu\beta_0). \qquad (28.57)$$

Finally, generalizing **28.8**, if each of the linear regressions of a set of p variables has identical errors, the variables are multinormally distributed unless they are mutually completely independent or they are functionally related.

28.11 If the regression of y on x is a polynomial, of type

$$E(y|x) = \beta_0 + \beta_1 x + \beta_2 x^2 + \ldots + \beta_p x^p, \qquad (28.58)$$

we may obtain similar results. However, as we shall see later in this chapter, this is best treated as a particular case of the p-regressor situation where the regressors are functionally related, so that any results we require for the polynomial regression situation may be obtained by specializing the results of **28.9–10**. For example, a condition that (28.58) holds is

$$\left[\frac{\partial \psi(u,t)}{\partial u}\right]_{u=0} = i\left\{\beta_0 + \beta_1 \frac{\partial \psi(0,t)}{\partial(it)} + \beta_2 \frac{\partial^2 \psi(0,t)}{\partial(it)^2} + \ldots + \beta_p \frac{\partial^p \psi(0,t)}{\partial(it)^p}\right\},$$

which reduces to (28.25) (in a slightly different notation) when $p = 1$, and is easily obtained as a special case of (28.55) by noting that the c.f. of x^r is $E\{\exp(itx^r)\}$, whose derivative with respect to t is

$$E\{ix^r \exp(itx^r)\} = \frac{1}{i^{r-1}} \frac{\partial^r}{\partial t^r} E\{\exp(itx)\}.$$

The general linear regression model

28.12 The analytical theory of regression, which we have so far discussed, is of interest in statistical theory but not in the practice of experimental statistics, precisely because it requires a detailed knowledge of the form of the underlying distribution. We now turn to the discussion of the general linear regression model, which is extensively used in practice because of the simplified (but nevertheless reasonably realistic) assumptions which it embodies. This is, in fact, simply the general linear model of **19.4**, with the parameters $\boldsymbol{\theta}$ as regression coefficients. (19.8) is thus rewritten

$$\mathbf{y} = \mathbf{X}\boldsymbol{\beta} + \boldsymbol{\epsilon}, \qquad (28.59)$$

where $\boldsymbol{\beta}$ is a $(k \times 1)$ vector of regression coefficients, $\mathbf{X}$ is an $(n \times k)$ matrix of known coefficients, and $\boldsymbol{\epsilon}$ an $(n \times 1)$ vector of "error" random variables (not necessarily normally distributed) with means and dispersion matrix

$$\left. \begin{array}{l} E(\boldsymbol{\epsilon}) = \mathbf{0}, \\ V(\boldsymbol{\epsilon}) = \sigma^2 \mathbf{I}. \end{array} \right\} \qquad (28.60)$$

We assume $n \geqslant k$ and $|\mathbf{X}'\mathbf{X}| \neq 0$. The results of Chapter 19 now apply. From (19.12),

$$\hat{\boldsymbol{\beta}} = (\mathbf{X}'\mathbf{X})^{-1}\mathbf{X}'\mathbf{y} \qquad (28.61)$$

is the vector of LS estimators of $\boldsymbol{\beta}$; from (19.16), its dispersion matrix is

$$\mathbf{V}(\hat{\boldsymbol{\beta}}) = \sigma^2 (\mathbf{X}'\mathbf{X})^{-1} \qquad (28.62)$$

and from **19.6** it is the MV unbiassed linear estimator of $\boldsymbol{\beta}$. Finally, from (19.41), an unbiassed estimator of σ^2 is s^2, where

$$(n-k)s^2 = (\mathbf{y} - \mathbf{X}\hat{\boldsymbol{\beta}})'(\mathbf{y} - \mathbf{X}\hat{\boldsymbol{\beta}}) \equiv \mathbf{y}'\mathbf{y} - \hat{\boldsymbol{\beta}}'\mathbf{X}'\mathbf{y}. \qquad (28.63)$$

s^2 is the sum of squared residuals divided by the number of observations minus the number of parameters estimated.

We have already applied this model to regression situations in **26.8** and **27.15**.

The meaning of "linear"

28.13 Before proceeding further, it is as well to emphasize the meaning of the adjective "linear" in the general regression model (28.59): it is *linear in the parameters* β_i, not necessarily in the x's. In fact, as we have remarked, the elements of $\mathbf{X}$ can be any set of known constants, related to each other in any desired manner. Up to **28.11**, on the other hand, we understood by "linear regression" that the conditional mean value of y is a linear function of the regressors $x_1, \ldots, x_p$. From the point of view of our present (Least Squares) analysis, the latter (perhaps more "natural") definition of linearity is irrelevant; it is linearity in the parameters that is essential. Thus the linear regression model includes all manner of "polynomial" or "curvilinear" forms of dependence of y upon $x_1, \ldots, x_p$. For example, the straightforward polynomial relationship

$$y_j = \beta_0 + \beta_1 x_{1j} + \beta_2 x_{1j}^2 + \ldots + \beta_k x_{1j}^k + \varepsilon_j, \qquad j = 1, 2, \ldots, n \qquad (28.64)$$

is linear in the β's, and thus is a special case of (28.59) (cf. the remarks in **28.11**). Similarly, the "multiple curvilinear" case

$$y_j = \beta_0 + \beta_1 x_{1j} + \beta_2 x_{1j}^2 + \beta_3 x_{2j} + \beta_4 x_{2j}^2 + \beta_5 x_{1j} x_{2j} + \varepsilon_j, \qquad j = 1, 2, \ldots, n, \qquad (28.65)$$

is a linear regression model. However,

$$y_j = \beta_0 + \beta_1 x_{1j} + \beta_2 x_{2j} + \beta_1^2 x_{3j} + \varepsilon_j, \qquad j = 1, 2, \ldots, n$$

is not, since β_1 and β_1^2 both appear.

Other functions than polynomials may also appear in a linear model. Thus $y_j = \beta_0 + \beta_1 x_{1j} + \beta_2 x_{1j}^2 + \beta_3 \sin x_{1j} \cos x_{2j} + \varepsilon_j$ is a linear model.

With this understanding, we see that any linear regression analysis reduces to the mathematical problem of inverting the matrix of sums of squares and products of the regressors, $\mathbf{X'X}$. The inverse is required both for estimation in (28.61) and for estimating the dispersion matrix of the estimators from (28.62) and (28.63).

Cochran (1938) and Kabe (1963) give formulae for adjusting an analysis when one or two of the original x-variables are omitted, or one or two new ones added.

Hudson (1966, 1969) discusses generally the fitting by LS of segmented curves whose join points must be estimated and of polynomials constrained in various ways—see also Bellman and Roth (1969) and Hinkley (1969).

Conditional and unconditional inferences

28.14 If $\mathbf{X}$ is the observed value of a set of random variables, the use of the linear model, in which $\mathbf{X}$ is a matrix of known coefficients, is conditional upon the observed $\mathbf{X}$. It is easy to see that conditionally unbiassed estimators remain unbiassed unconditionally, while conditional confidence intervals and tests will remain valid unconditionally, since the value of any fixed conditional probability is unaffected by integrating it over the distribution of $\mathbf{X}$—the identity of (26.52) and (26.37) is an example. However, the unconditional efficiency (i.e. selectivity and power) of tests and intervals will generally differ from the conditional (linear model) values—cf. 27.28–31 for the multiple correlation coefficient—because the underlying statistics will have different distributions. Similarly, other properties of estimators (e.g. their variances) will not generally persist unconditionally—cf. Exercise 26.9 for the regression coefficient.

Orthogonal regression analyses

28.15 It is evidently a convenience in carrying out a regression analysis if the estimators $\hat{\beta}_i$ are uncorrelated: in fact, if the ε_j are normally distributed, so will the $\hat{\beta}_i$ be, since they are linear functions of $\mathbf{y}$, and lack of correlation will then imply independence. A regression analysis with uncorrelated estimators is called *orthogonal*. If the regressors are not random variables, but constants which are at choice in experimental work, we may now ask a new type of question: can, and if so how should, the elements of $\mathbf{X}$ be chosen so that the estimators $\hat{\beta}_i$ are uncorrelated?

This is a question arising in the theory of experimental design, and we defer a detailed discussion of design problems to Volume 3. However, we observe from (28.62) that if, and only if, $(\mathbf{X'X})^{-1}$ is diagonal, the analysis is orthogonal; and $(\mathbf{X'X})^{-1}$ is diagonal only if $\mathbf{X'X}$ is. Thus, to obtain an orthogonal analysis, we must choose the elements of $\mathbf{X}$ so that $\mathbf{X'X}$ is diagonal. It follows at once that we must have

$$\sum_{j=1}^{n} x_{ij} x_{hj} = 0, \qquad i \neq h. \tag{28.66}$$

The diagonal elements of $\mathbf{X'X}$ are, of course, simply

$$(\mathbf{X'X})_{ii} = \sum_{j=1}^{n} x_{ij}^2,$$

whence the corresponding inverse element is

$$[(\mathbf{X'X})^{-1}]_{ii} = 1 \bigg/ \sum_{j=1}^{n} x_{ij}^2. \tag{28.67}$$

(28.61) and (28.62) are then particularly simple.

Polynomial regression: orthogonal polynomials

28.16 For a polynomial dependence of y upon x, as in (28.64), $\mathbf{X}'\mathbf{X}$ cannot be diagonal, since the off-diagonal elements will be sums of powers of a single variable x. However, we can choose polynomials of degree i in x, say $\phi_i(x)$, $(i = 0, 1, 2, \ldots, k)$, which are mutually orthogonal. Then (28.64) is replaced by

$$y_j = \alpha_0 \phi_0(x_j) + \alpha_1 \phi_1(x_j) + \ldots + \alpha_k \phi_k(x_j) + \varepsilon_j, \qquad j = 1, 2, \ldots, n, \quad (28.68)$$

which we may write in matrix form $\mathbf{y} = \mathbf{\Phi}\boldsymbol{\alpha} + \boldsymbol{\epsilon}$. The α's are a new set of parameters (functions of the original β's), in terms of which it is more convenient to work, for we now have, from (28.67),

$$[(\mathbf{\Phi}'\mathbf{\Phi})^{-1}]_{ii} = 1 \bigg/ \sum_{j=1}^{n} \phi_i^2(x_j), \quad (28.69)$$

which we may use in (28.62). Furthermore, (28.61) becomes, using (28.69),

$$\hat{\alpha}_i = [(\mathbf{\Phi}'\mathbf{\Phi})^{-1}]_{ii} (\mathbf{\Phi}'\mathbf{y})_i = \sum_j y_j \phi_i(x_j) \bigg/ \sum_j \phi_i^2(x_j). \quad (28.70)$$

Thus each estimator $\hat{\alpha}_i$ depends only on the corresponding polynomial. This is extremely convenient in "fitting" a polynomial regression whose degree is not determined in advance: we increase the value of k, step by step, until a sufficiently good "fit" is obtained. If we had used the non-orthogonal regression model (28.64), the whole set of estimators $\hat{\beta}_0, \hat{\beta}_1, \ldots, \hat{\beta}_{k-1}$ would have had to be recalculated when a further term, in x^k, was added to raise the degree of the polynomial. Of course, if we reassemble the estimated regression $y = \sum_{i=0}^{k} \hat{\alpha}_i \phi_i(x)$ as $y = \sum_{i=0}^{k} \hat{\beta}_i x^i$, the $\hat{\beta}_i$ are precisely those we should have obtained directly, though less conveniently, from (28.64). This follows from the fact that both methods minimize the same sum of squared residuals.

Using (28.70), (28.63) becomes in the orthogonal case

$$(n-k)s^2 = \mathbf{y}'\mathbf{y} - \hat{\boldsymbol{\alpha}}' \mathbf{\Phi}' \mathbf{y}$$

$$= \sum_{j=1}^{n} y_j^2 - \sum_{i=0}^{k} \left[\left\{ \sum_{j=1}^{n} y_j \phi_i(x_j) \right\}^2 \bigg/ \sum_{j=1}^{n} \phi_i^2(x_j) \right] \quad (28.71)$$

$$= \sum_{j=1}^{n} y_j^2 - \sum_{i=0}^{k} \hat{\alpha}_i^2 \sum_{j=1}^{n} \phi_i^2(x_j). \quad (28.72)$$

These very simple expressions for the sum of squared residuals from the fitted regression permit the rapid calculation of the additional reduction in residual variance brought about by increasing the degree k of the fitted polynomial.

28.17 We now have to see how the orthogonal polynomials $\phi_i(x)$ are to be evaluated. We require

$$\sum_{j=1}^{n} \phi_i(x_j) \phi_h(x_j) = 0, \qquad i, h = 0, 1, 2, \ldots, k; \; i \neq h, \quad (28.73)$$

where

$$\phi_i(x) = \sum_{r=0}^{i} c_{ir} x^r. \quad (28.74)$$

There are $(i+1)$ coefficients c_{ir} in (28.74), and hence in all the polynomials ϕ_i there are $\sum_{i=0}^{k}(i+1) = \frac{1}{2}(k+1)(k+2)$ coefficients to be determined. On these, (28.73) imposes only $\frac{1}{2}k(k+1)$ constraints. We determine the excess $(k+1)$ constants by requiring, as is convenient, that $c_{ii} = 1$, all i. We then have at once from (28.74)
$$\phi_0(x) = c_{00} \equiv 1 \tag{28.75}$$
identically in x. (28.73) is now just sufficient to determine the c_{ir}, apart from an arbitrary constant multiplier, say λ_{in}, for each $\phi_i(x)$, $i > 0$. (28.73) and (28.74) give, with $h = k$,
$$\sum_{j=1}^{n} \sum_{r=0}^{i} c_{ir} x_j^r \sum_{s=0}^{k} c_{ks} x_j^s = 0, \qquad i \neq k,$$
or
$$\sum_{r=0}^{i} c_{ir} \sum_{s=0}^{k} c_{ks} \mu'_{r+s} = 0, \qquad i \neq k, \tag{28.76}$$
where μ'_p is the pth moment of the set of x's. Since (28.76) holds for all $i = 0, 1, 2, \ldots, k-1$, we must have
$$\sum_{s=0}^{k} c_{ks} \mu'_{r+s} = 0, \qquad r = 0, 1, \ldots, k-1. \tag{28.77}$$
Writing the determinant
$$|M_k| = \begin{vmatrix} \mu'_0 & \mu'_1 & \cdots & \mu'_{k-1} & \mu'_k \\ \mu'_1 & \mu'_2 & \cdots & \mu'_k & \mu'_{k+1} \\ \vdots & \vdots & & \vdots & \vdots \\ \mu'_{k-1} & \mu'_k & \cdots & \mu'_{2k-2} & \mu'_{2k-1} \\ \mu'_k & \mu'_{k+1} & \cdots & \mu'_{2k-1} & \mu'_{2k} \end{vmatrix},$$
and $|M_k^{u,v}|$ for the cofactor of the element in the uth row and vth column of $|M_k|$, the solution of (28.77) is (remembering that $c_{kk} = 1$)
$$c_{ks} = |M_k^{k+1,s+1}|/|M_{k-1}|, \qquad s = 0, 1, 2, \ldots, k-1. \tag{28.78}$$
Thus, from (28.74) and (28.78),
$$\phi_k(x) = \frac{1}{|M_{k-1}|} \begin{vmatrix} \mu'_0 & \mu'_1 & \cdots & \mu'_k \\ \mu'_1 & \mu'_2 & \cdots & \mu'_{k+1} \\ \vdots & \vdots & & \vdots \\ \mu'_{k-1} & \mu'_k & \cdots & \mu'_{2k-1} \\ 1 & x & \cdots & x^k \end{vmatrix}. \tag{28.79}$$

(28.79) is used to evaluate the polynomial for any k. Of course, $\mu'_0 \equiv 1$, and we simplify by measuring from the mean of the x's, so that $\mu'_1 = 0$ and we may drop the primes. It will be observed that the determinant in the denominator of (28.79) is simply that in the numerator with its last row and column deleted.

We find, for example,
$$\phi_1(x) = \frac{\begin{vmatrix} 1 & 0 \\ 1 & x \end{vmatrix}}{|1|} = x, \tag{28.80}$$

$$\phi_2(x) = \frac{\begin{vmatrix} 1 & 0 & \mu_2 \\ 0 & \mu_2 & \mu_3 \\ 1 & x & x^2 \end{vmatrix}}{\begin{vmatrix} 1 & 0 \\ 0 & \mu_2 \end{vmatrix}} = x^2 - \frac{\mu_3}{\mu_2}x - \mu_2, \tag{28.81}$$

and so on. A simpler recursive method of obtaining the polynomials is given in Exercise 28.23.

The case of equally-spaced x-values

28.18 The most important applications of orthogonal polynomials in regression analysis are to situations where the regressor variable, x, takes values at equal intervals. This is often the case with observations taken at successive times, and with data grouped into classes of equal width. If we have n such equally-spaced values of x, we measure from their mean and take the natural interval as unit, thus obtaining as working values of x: $-\frac{1}{2}(n-1), -\frac{1}{2}(n-3), -\frac{1}{2}(n-5), \ldots, \frac{1}{2}(n-3), \frac{1}{2}(n-1)$. For this simple case, the values of the moments in (28.79) can be explicitly calculated: in fact, apart from the mean which has been taken as origin, these are the moments of the first n natural numbers, obtainable from the cumulants given in Exercise 3.23. The odd moments are zero by symmetry; the even moments are

$$\mu_2 = (n^2-1)/12,$$
$$\mu_4 = \mu_2(3n^2-7)/20,$$
$$\mu_6 = \mu_2(3n^4-18n^2+31)/112,$$

and so on. Substituting these and higher moments into (28.79), we obtain for the first six polynomials

$$\left.\begin{aligned}
\phi_0(x) &\equiv 1, \\
\phi_1(x) &= \lambda_{1n} x, \\
\phi_2(x) &= \lambda_{2n}\{x^2 - \tfrac{1}{12}(n^2-1)\}, \\
\phi_3(x) &= \lambda_{3n}\{x^3 - \tfrac{1}{20}(3n^2-7)x\}, \\
\phi_4(x) &= \lambda_{4n}\{x^4 - \tfrac{1}{14}(3n^2-13)x^2 + \tfrac{3}{560}(n^2-1)(n^2-9)\}, \\
\phi_5(x) &= \lambda_{5n}\{x^5 - \tfrac{5}{18}(n^2-7)x^3 + \tfrac{1}{1008}(15n^4-230n^2+407)x\}, \\
\phi_6(x) &= \lambda_{6n}\{x^6 - \tfrac{5}{44}(3n^2-31)x^4 + \tfrac{1}{176}(5n^4-110n^2+329)x^2 \\
&\qquad - \tfrac{5}{14874}(n^2-1)(n^2-9)(n^2-25)\}.
\end{aligned}\right\} \tag{28.82}$$

Allan (1930) also gives $\phi_i(x)$ for $i = 7, 8, 9, 10$. Following Fisher (1921b), the arbitrary constants λ_{in} in (28.82), referred to below (28.75), are determined conveniently so that $\phi_i(x_j)$ is an integer for all $j = 1, 2, \ldots, n$. It will be observed that

$$\phi_{2i}(x) = \phi_{2i}(-x) \quad \text{and} \quad \phi_{2i-1}(x) = -\phi_{2i-1}(-x);$$

even-degree polynomials are even functions and odd-degree polynomials odd functions.

Tables of orthogonal polynomials

28.19 The *Biometrika Tables* give $\phi_i(x_j)$ for all j, $n = 3(1)52$ and $i = 1(1)$ min $(6, n-1)$, together with the values of λ_{in} and $\sum_{j=1}^{n} \phi_i^2(x_j)$.

Fisher and Yates' *Tables* give $\phi_i(x_j)$ (their ξ'_i), λ_{in} and $\sum_{j=1}^{n} \phi_i^2(x_j)$ for all j, $n = 3(1)75$ and $i = 1(1)\min(5, n-1)$.

The *Biometrika Tables* give references to more extensive tabulations, ranging to $i = 9$, $n = 52$, by van der Reyden, and to $i = 5$, $n = 104$, by Anderson and Houseman.

28.20 There is a large literature on orthogonal polynomials. For theoretical details, the reader should refer to the paper by Fisher (1921b) which first applied them to polynomial regression, to a paper by Allan (1930), and three papers by Aitken (1933). More recently, Rushton (1951) discussed the case of unequally-spaced x-values, and C. P. Cox (1958) gave a concise determinantal derivation of general orthogonal polynomials, while Guest (1954, 1956) has considered grouping problems.

We shall content ourselves here with a single example of fitting orthogonal polynomials in the equally-spaced case.

Example 28.3

The first two columns of Table 28.1 show the human population of England and Wales at the decennial Censuses from 1811 to 1931. These observations are clearly not uncorrelated, so that the regression model (28.64) is not strictly appropriate, but we carry through the fitting process for purely illustrative purposes.

Table 28.1

Year	Population (millions) y	$\frac{\text{Year}-1871}{10} = x =$	$\phi_1(x)$	$\phi_2(x)$	$\phi_3(x)$	$\phi_4(x)$
1811	10·16		−6	22	−11	99
1821	12·00		−5	11	0	−66
1831	13·90		−4	2	6	−96
1841	15·91		−3	−5	8	−54
1851	17·93		−2	−10	7	11
1861	20·07		−1	−13	4	64
1871	22·71		0	−14	0	84
1881	25·97		1	−13	−4	64
1891	29·00		2	−10	−7	11
1901	32·53		3	−5	−8	−54
1911	36·07		4	2	−6	−96
1921	37·89		5	11	0	−66
1931	39·95		6	22	11	99
Σy :	314·09					
		$\lambda_{1, 13}$: 1	1	1/6	7/12	
		$\sum_{j=1}^{13} \phi_i^2(x_j)$: 182	2002	572	68,068	

Here $n = 13$, and from the *Biometrika Tables*, Table 47, we read off the values in the last four columns of Table 28.1. From that Table, we have

$$\Sigma y_j \phi_0(x_j) = \Sigma y_j = 314·09,$$
$$\Sigma y_j \phi_1(x_j) = 474·77,$$
$$\Sigma y_j \phi_2(x_j) = 123·19,$$
$$\Sigma y_j \phi_3(x_j) = -39·38,$$
$$\Sigma y_j \phi_4(x_j) = -374·30.$$

Hence, using (28.70),

$$\hat{\alpha}_0 = 314 \cdot 09/13 = 24 \cdot 160, 8,$$
$$\hat{\alpha}_1 = 474 \cdot 77/182 = 2 \cdot 608, 63,$$
$$\hat{\alpha}_2 = 123 \cdot 19/2,002 = 0 \cdot 061, 533, 5,$$
$$\hat{\alpha}_3 = -39 \cdot 38/572 = -0 \cdot 068, 846, 2,$$
$$\hat{\alpha}_4 = -374 \cdot 30/68,068 = -0 \cdot 005, 498, 91.$$

For the estimated fourth-degree orthogonal polynomial regression of y on x, we then have, using (28.68) and (28.82),

$$y = 24 \cdot 160,8 + 2 \cdot 608, 63\, x + 0 \cdot 061, 533, 5\, (x^2 - 14)$$
$$- 0 \cdot 068, 846, 2 \{\tfrac{1}{6}(x^3 - 25\,x)\} - 0 \cdot 005, 498, 91\{\tfrac{7}{12}(x^4 - \tfrac{247}{7}x^2 + 144)\}.$$

If we collected the terms on the right so that we had

$$y = \hat{\beta}_0 + \hat{\beta}_1 x + \hat{\beta}_2 x^2 + \hat{\beta}_3 x^3 + \hat{\beta}_4 x^4,$$

the coefficients $\hat{\beta}_i$ would be exactly those we should have obtained if we had used (28.64) instead of the orthogonal form (28.68). The advantage of the latter, apart from its computational simplicity, is that we can simply examine the improvement in "fit" of the regression equation as its degree increases. We require only the calculation of

$$\sum_j y_j^2 = 8,839 \cdot 939,$$

and we may substitute the quantities already calculated into (28.72) for this purpose. Thus we have:

Total sum of squares		8,839·939
Reduction due to $\hat{\alpha}_0 = \hat{\alpha}_0^2 \Sigma \phi_0^2 = (24 \cdot 160, 8)^2 \cdot 13$	$=$	7,588·656
	Residual:	1,251·283
,, ,, ,, $\hat{\alpha}_1 = \hat{\alpha}_1^2 \Sigma \phi_1^2 = (2 \cdot 608, 63)^2 \cdot 182$	$=$	1,238·497
	Residual:	12·786
,, ,, ,, $\hat{\alpha}_2 = \hat{\alpha}_2^2 \Sigma \phi_2^2 = (0 \cdot 061, 533, 5)^2 \cdot 2,002$	$=$	7·580
	Residual:	5·206
,, ,, ,, $\hat{\alpha}_3 = \hat{\alpha}_3^2 \Sigma \phi_3^2 = (0 \cdot 068, 846, 2)^2 \cdot 572$	$=$	2·711
	Residual:	2·495
,, ,, ,, $\hat{\alpha}_4 = \hat{\alpha}_4^2 \Sigma \phi_4^2 = (0 \cdot 005, 498, 91)^2 \cdot 68,068$	$=$	2·058
	Residual:	0·437

Evidently, the cubic and quartic expressions are good "fits": they are displayed in Fig. 28.1.

The reader should not need to be warned against the dangers of extrapolating from a fitted regression, however close, which has no theoretical basis. In this case, for example, he can satisfy himself visually that the value "predicted" by the quartic regression for 1951 ($x = 8$) is a good deal less than the Census population of 43·7 millions actually found in that year.

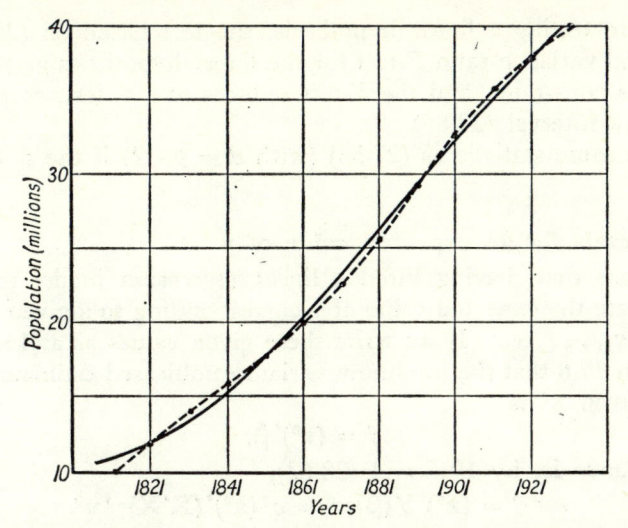

Fig. 28.1—Cubic (full line) and quartic (broken line) polynomials fitted to the data of Table 28.1

Confidence intervals and tests for the parameters of the linear model

28.21 In **28.12** we discussed the point estimation of the parameters $\boldsymbol{\beta}, \sigma^2$ of the general linear regression model (28.59). If we now assume $\boldsymbol{\epsilon}$ to be a vector of *normal* error variables, as we shall do for the remainder of this chapter, we may set confidence intervals for (and correspondingly test hypotheses concerning) any component of the parameter vector $\boldsymbol{\beta}$. These are all linear hypotheses in the sense of Chapter 24 and the tests are all LR tests.

Any estimator $\hat{\beta}_i$ is a linear function of the y_j and is therefore normally distributed with mean β_i and variance, from (28.62),

$$\operatorname{var}(\hat{\beta}_i) = \sigma^2 [(\mathbf{X}'\mathbf{X})^{-1}]_{ii}. \tag{28.83}$$

(If the analysis is orthogonal, (28.67) is used in (28.83).) From **19.11**, s^2, the estimator of σ^2 defined at (28.63), is distributed independently of $\hat{\boldsymbol{\beta}}$ (and hence of any component of $\hat{\boldsymbol{\beta}}$), the distribution of $(n-k)s^2/\sigma^2$ being of the χ^2 form with $\nu = (n-k)$ degrees of freedom. It follows immediately that the statistic

$$t = (\hat{\beta}_i - \beta)/\{s^2[(\mathbf{X}'\mathbf{X})^{-1}]_{ii}\}^{\frac{1}{2}}, \tag{28.84}$$

being the ratio of a standardized normal variate to the square root of an independent χ^2/ν variate, has a "Student's" t-distribution with $\nu = (n-k)$ degrees of freedom. This enables us to set confidence intervals for β_i or to test hypotheses concerning its value. The central confidence interval with coefficient $(1-\alpha)$ is simply

$$\hat{\beta}_i \pm t_{1-\frac{1}{2}\alpha}\{s^2[(\mathbf{X}'\mathbf{X})^{-1}]_{ii}\}^{\frac{1}{2}} \tag{28.85}$$

where $t_{1-\frac{1}{2}\alpha}$ is the value of "Student's" t for ν degrees of freedom for which its distribution function

$$F(t_{1-\frac{1}{2}\alpha}) = 1 - \frac{1}{2}\alpha.$$

Since we are here testing a linear hypothesis, the test based on (28.84) is a special case of the general variance-ratio F-test for the linear hypothesis given in **24.28**: here we have only one constraint, and the F-test reduces to a t^2 test, corresponding to the central confidence interval (28.85).

(28.84) is the same statistic as (27.55) (with $d = p-2$) if the notation is changed suitably.

Confidence intervals for an expected value of y

28.22 Suppose that, having fitted a linear regression model to n observations, we wish to estimate the expected value of y corresponding to a given value for each of the k regressors $x_1, \ldots, x_k$. If we write these given values as a $(1 \times k)$ vector $\mathbf{x}^0$, we have at once from **19.6** that the minimum variance unbiassed estimator of the expected value of y for given $\mathbf{x}^0$ is

$$\hat{y} = (\mathbf{x}^0)' \hat{\boldsymbol{\beta}}, \tag{28.86}$$

and that its variance is, by **19.6** and (28.62),

$$\operatorname{var} \hat{y} = (\mathbf{x}^0)' \mathbf{V}(\hat{\boldsymbol{\beta}}) \mathbf{x}^0 = \sigma^2 (\mathbf{x}^0)' (\mathbf{X}'\mathbf{X})^{-1} \mathbf{x}^0. \tag{28.87}$$

Just as in **28.21**, we estimate the sampling variance (28.87) by inserting s^2 for σ^2, and set confidence limits from "Student's" t-distribution, which here applies to the statistic

$$t = \{\hat{y} - E(y|\mathbf{x}^0)\} / \{s^2 (\mathbf{x}^0)' (\mathbf{X}'\mathbf{X})^{-1} \mathbf{x}^0\}^{\frac{1}{2}} \tag{28.88}$$

with $\nu = (n-k)$ as before.

Confidence intervals for the expectation of a further value of y: prediction intervals

28.23 The results of **28.22** may be applied to obtain a confidence interval for the expectation of a further $((n+1)$th$)$ value of y, y_{n+1}, not taken into account in fitting the regression model. If $\mathbf{x}^0$ represents the given values of the regressors for which y_{n+1} is to be observed, (28.86) gives us the unbiassed estimator

$$\hat{y}_{n+1} = (\mathbf{x}^0)' \hat{\boldsymbol{\beta}} \tag{28.89}$$

just as before, but the fact that y_{n+1} will have variance σ^2 about its expectation increases its sampling variance over (28.87) by that amount, giving us

$$\operatorname{var} \hat{y}_{n+1} = \sigma^2 \{(\mathbf{x}^0)' (\mathbf{X}'\mathbf{X})^{-1} \mathbf{x}^0 + 1\} \tag{28.90}$$

which we estimate, putting s^2 for σ^2 as before, to obtain the "Student's" variate

$$t = \{\hat{y}_{n+1} - E(y_{n+1}|\mathbf{x}^0)\} / [s^2 \{(\mathbf{x}^0)' (\mathbf{X}'\mathbf{X})^{-1} \mathbf{x}^0 + 1\}]^{\frac{1}{2}} \tag{28.91}$$

again with $\nu = (n-k)$, from which to set our confidence intervals.

Similarly, if a set of N further observations are to be made on y at the same $\mathbf{x}_0$, (28.89)–(28.91) hold for the estimation of the mean $\bar{y}_N$ to be observed, with the obvious adjustment that the unit in the braces in (28.90) and (28.91) is replaced by $1/N$, the additional variance now being σ^2/N.

Confidence intervals for further values, such as those discussed in this section, are sometimes called *prediction intervals*; it must always be borne in mind that these "predictions" are conditional upon the assumption that the linear model fitted to the previous n observations is valid for the further observations too, i.e. that there is no structural change in the model.

If the regressors have been selected from a larger set of candidate variables by some procedure of the Stepwise Regression type (cf. **27.27**), neither the interval estimation results of **28.21–3** nor the unbiassed estimation results of **28.12** remain valid. Larson and Bancroft (1963a, b) and Kennedy and Bancroft (1971) investigate the bias and m.s.e. of the predictor (**28.89**). Draper *et al.* (1971) show that, roughly, the true size of the standard size-α test is $k\alpha$ when selecting one of k candidate variables.

Example 28.4

In the simple case
$$y_j = \beta_1 + \beta_2 x_j + \varepsilon_j, \qquad j = 1, 2, \ldots, n, \tag{28.92}$$
we have seen in Examples 19.3, 19.6, that
$$\hat{\beta}_2 = \sum_j (y_j - \bar{y})(x_j - \bar{x}) / \sum_j (x_j - \bar{x})^2,$$
$$\hat{\beta}_1 = \bar{y} - \hat{\beta}_2 \bar{x},$$
$$s^2 = \frac{1}{n-2} \sum_j \{y_j - (\hat{\beta}_1 + \hat{\beta}_2 x_j)\}^2,$$
and
$$(\mathbf{X}'\mathbf{X})^{-1} = \frac{1}{\sum_j (x_j - \bar{x})^2} \begin{pmatrix} \sum x^2 / n & -\bar{x} \\ -\bar{x} & 1 \end{pmatrix}.$$

Here $\mathbf{x}^0$ is the two-component vector $\begin{pmatrix} 1 \\ x^0 \end{pmatrix}$, and we may proceed to set confidence intervals for β_1, β_2, $E(y|\mathbf{x}^0)$ and $E(y_{n+1}|\mathbf{x}^0)$, using (28.84), (28.88) and (28.91); in each case we have a " Student's " variate with $(n-2)$ degrees of freedom.

(a) It will be noticed that the analysis is orthogonal if and only if $\bar{x} = 0$, so that in this case we need only make a change of origin in x to obtain orthogonality. Also, the variances of the estimators (the diagonal elements of their dispersion matrix) are minimized when $\bar{x} = 0$ and $\sum x^2$ is as large as possible. Both orthogonality and minimized sampling variances are therefore achieved if we choose the x_j so that (assuming n to be even)
$$x_1, x_2, \ldots, x_{\frac{1}{2}n} = +a,$$
$$x_{\frac{1}{2}n+1}, x_{\frac{1}{2}n+2}, \ldots, x_n = -a,$$
and a is as large as possible. This corresponds to the intuitively obvious fact that if we are certain that the dependence of y upon x is linear with constant variance, we can most efficiently " fix " the line at its end-points. However, if the dependence were non-linear, we should be unable to detect this if all our observations had been made at two values of x only, and it is therefore usual to spread the x-values more evenly over its range; it is always as well to be able to check the structural assumptions of our model in the course of the analysis.

(b) Our confidence interval in this case for $E(y|\mathbf{x}^0)$ is, from (28.88)
$$(\mathbf{x}^0)'\hat{\boldsymbol{\beta}} \pm t_{1-\frac{1}{2}\alpha} \left\{ \frac{s^2}{\sum (x-\bar{x})^2} \begin{pmatrix} 1 \\ x^0 \end{pmatrix}' \begin{pmatrix} \sum x^2/n & -\bar{x} \\ -\bar{x} & 1 \end{pmatrix} \begin{pmatrix} 1 \\ x^0 \end{pmatrix} \right\}^{\frac{1}{2}}$$
$$= (\hat{\beta}_1 + \hat{\beta}_2 x^0) \pm t_{1-\frac{1}{2}\alpha} \left\{ s^2 \left(\frac{1}{n} + \frac{(x^0 - \bar{x})^2}{\sum (x-\bar{x})^2} \right) \right\}^{\frac{1}{2}}. \tag{28.93}$$

If we consider this as a function of the value x^0, we see that (28.93) defines the two branches of a hyperbola of which the fitted regression $(\hat{\beta}_1 + \hat{\beta}_2 x^0)$ is a diameter. The confidence interval obviously has minimum length when $x^0 = \bar{x}$, the observed mean, and its length increases steadily as $|x^0 - \bar{x}|$ increases, confirming the intuitive notion that we can estimate most accurately near the "centre" of the observed values of x. Fig. 28.2 illustrates the loci of the confidence limits given by (28.93).

Robison (1964) gives ML estimates and confidence intervals for the intersection abscissa of two polynomial regressions, and a bibliography of related work.

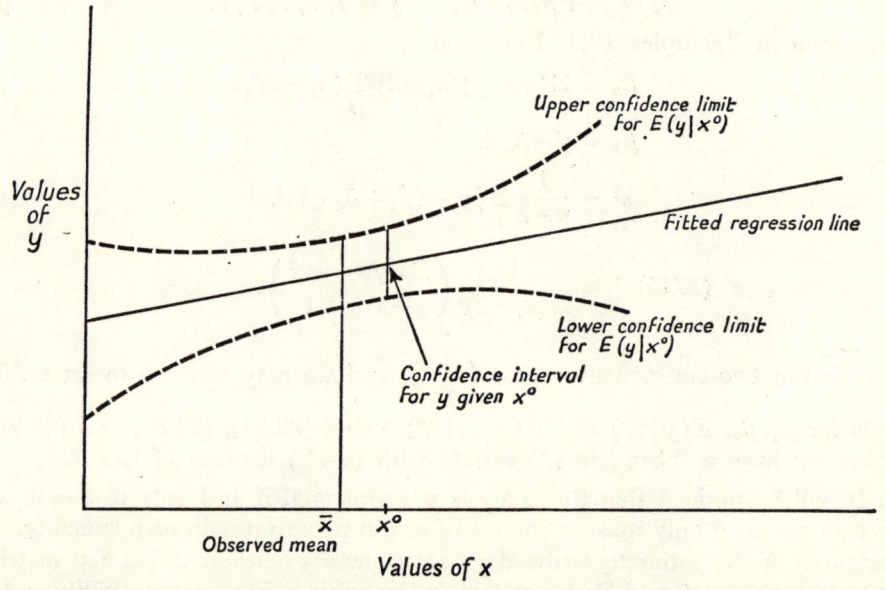

Fig. 28.2—Hyperbolic loci of confidence limits (28.93) for an expected value of y in simple linear regression

28.24 The confidence limits for an expected value of y discussed in Example 28.4(b), and more generally in **28.22**, refer to the value of y corresponding to a particular $\mathbf{x}^0$; in Fig. 28.2, any particular confidence interval is given by that part of the vertical line through x_0 lying between the branches of the hyperbola. Suppose now that we require a *confidence region for an entire regression line*, i.e. a region R in the (x, y) plane (or, more generally, in the $(\mathbf{x}, y)$ space) such that there is probability $1-\alpha$ that the true regression line $y = \mathbf{x}\boldsymbol{\beta}$ is contained in R. This, it will be seen, is a quite distinct problem from that just discussed; we are now seeking a confidence region, not an interval, and it covers the whole line, not one point on the line. We now consider this problem, first solved in the simplest case by Working and Hotelling (1929) in a remarkable paper; our discussion follows that of P.G. Hoel (1951).

Confidence regions for a regression line

28.25 We first treat the simple case of Example 28.4 and assume σ^2 known, restrictions to be relaxed in **28.31-2**. For convenience, we measure the x_j from their mean, so that $\bar{x} = 0$ and, from Example 28.4(a), the analysis is orthogonal. We then

have, from the dispersion matrix, $\operatorname{var} \hat{\beta}_1 = \sigma^2/n$, $\operatorname{var} \hat{\beta}_2 = \sigma^2/\Sigma x^2$, and $\hat{\beta}_1$ and $\hat{\beta}_2$ are normally and independently distributed. Thus
$$u = n^{\frac{1}{2}}(\hat{\beta}_1 - \beta_1)/\sigma, \quad v = (\Sigma x^2)^{\frac{1}{2}}(\hat{\beta}_2 - \beta_2)/\sigma, \tag{28.94}$$
are independent standardized normal variates.

Let $g(u^2, v^2)$ be a single-valued even function of u and v, and let
$$g(u^2, v^2) = g_{1-\alpha}, \quad 0 < \alpha < 1, \tag{28.95}$$
define a family of closed curves in the (u, v) plane such that (a) whenever $g_{1-\alpha}$ decreases, the new curve is contained inside that corresponding to the larger value of $1-\alpha$; and (b) every interior point of a curve lies on some other curve. To the implicit relation (28.95) between u and v, we assume that there corresponds an explicit relation
$$u^2 = p(v^2)$$
or
$$u = \pm h(v). \tag{28.96}$$
We further assume that $h'(v) = dh(v)/dv$ exists for all v and is a monotone decreasing function of v taking all real values.

28.26 We see from (28.94) that for any given set of observations to which a regression has been fitted, there will correspond to the true regression line,
$$y = \beta_1 + \beta_2 x, \tag{28.97}$$
values of u and v such that
$$\beta_1 + \beta_2 x = \left(\hat{\beta}_1 + \frac{\sigma}{n^{\frac{1}{2}}} u\right) + \left(\hat{\beta}_2 + \frac{\sigma}{(\Sigma x^2)^{\frac{1}{2}}} v\right) x. \tag{28.98}$$
Substituting (28.96) into (28.98), we have two families of regression lines, with v as parameter,
$$\left(\hat{\beta}_1 \pm \frac{\sigma}{n^{\frac{1}{2}}} h(v)\right) + \left(\hat{\beta}_2 + \frac{\sigma}{(\Sigma x^2)^{\frac{1}{2}}} v\right) x, \tag{28.99}$$
one family corresponding to each sign in (28.96). We now find the envelopes of these families.

Differentiating (28.99) with respect to v and equating the derivative to zero, we obtain
$$x = \mp \left(\frac{\Sigma x^2}{n}\right)^{\frac{1}{2}} h'(v). \tag{28.100}$$
Substituted into (28.99), (28.100) gives the required envelopes:
$$(\hat{\beta}_1 + \hat{\beta}_2 x) \pm \frac{\sigma}{n^{\frac{1}{2}}} \{h(v) - v h'(v)\}, \tag{28.101}$$
where the functions of v are to be substituted for in terms of x from (28.100). The restrictions placed on $h'(v)$ below (28.96) ensure that the two envelopes in (28.101) exist for all x, are single-valued, and that all members of each family lie on one side only of its envelope. In fact, the curve given taking the upper signs in (28.101) always lies above the curve obtained by taking the lower signs in (28.101), and all members of the two families (28.99) lie between them.

28.27 Any pair of values (u, v) for which
$$g(u^2, v^2) < g_{1-\alpha} \tag{28.102}$$
will correspond to a regression line lying between the pair of envelopes (28.101), because

for any fixed v, $u^2 = \{h(v)\}^2$ will be reduced, so that the constant term in (28.99) will be reduced in magnitude as a function of v, while the coefficient of x is unchanged. Thus if u and v satisfy (28.102), the true regression line will lie between the pair of envelopes (28.101). Now choose $g_{1-\alpha}$ so that the continuous random variable $g(u^2, v^2)$ satisfies

$$P\{g(u^2, v^2) < g_{1-\alpha}\} = 1-\alpha. \tag{28.103}$$

Then we have probability $1-\alpha$ that (28.102) holds, and the region R between the pair of envelopes (28.101) is a confidence region for the true regression line with confidence coefficient $1-\alpha$.

28.28 We now have to consider how to choose the function $g(u^2, v^2)$ so that, for fixed $1-\alpha$, the confidence region R is in some sense as small as possible. We cannot simply minimize the area of R, since its area is always infinite. We therefore introduce a weight function $w(x)$ and choose R to minimize the integral

$$I = \int_{-\infty}^{\infty} (y_2 - y_1) w(x) \, dx, \tag{28.104}$$

where y_1, y_2 are respectively the lower and upper envelopes (28.101), the boundaries of R, and $\int_{-\infty}^{\infty} w(x) \, dx = 1$. We may rewrite (28.104)

$$I = E(y_2) - E(y_1), \tag{28.105}$$

expectations being with respect to $w(x)$.

Obviously, the optimum R resulting from the minimization will depend on the weight function chosen. Putting $S^2 = \Sigma x^2/n$, consider the normal weight-function

$$w(x) = (2\pi S^2)^{-\frac{1}{2}} \exp\left(-\frac{x^2}{2S^2}\right), \tag{28.106}$$

which is particularly appropriate if the values of x, here regarded as fixed, are in fact sampled from a normal distribution, e.g. if x and y are bivariate normally distributed. Putting (28.101) and (28.106) into (28.105), it becomes

$$I = \frac{2\sigma}{n^{\frac{1}{2}}} [E\{h(v)\} - E\{v h'(v)\}]. \tag{28.107}$$

From (28.100) we have, since $h'(v)$ is decreasing,

$$dx = -S h''(v) \, dv, \tag{28.108}$$

so that if we transform the integrals in (28.107) to the variable v, we find

$$\left.\begin{array}{l} E\{h\} = -(2\pi)^{-\frac{1}{2}} \int h h'' \exp\{-\frac{1}{2}(h')^2\} \, dv, \\[4pt] E\{v h'\} = -(2\pi)^{-\frac{1}{2}} \int v h' h'' \exp\{-\frac{1}{2}(h')^2\} \, dv, \end{array}\right\} \tag{28.109}$$

the integration in each case being over the whole range of v. Since $h(v)$ is an even function, both the integrals need be taken for positive v only, and (28.109) gives, in (28.107),

$$I = -\frac{4\sigma}{(2\pi n)^{\frac{1}{2}}} \int_0^{v_{\max}} h''(h - v h') \exp\{-\frac{1}{2}(h')^2\} \, dv. \tag{28.110}$$

This is to be minimized, subject to (28.103), which by the independence, normality and symmetry of the distributions of u and v is equivalent to the condition that

$$(2\pi)^{-1} \int_0^{v_{\max}} \left\{ \int_0^{h(v)} \exp(-\tfrac{1}{2}u^2)\,du \right\} \exp(-\tfrac{1}{2}v^2)\,dv = \tfrac{1}{4}(1-\alpha). \qquad (28.111)$$

It must also be remembered that we have required $h'(v)$ to be a monotone decreasing function of v taking all real values.

28.29 So that we can proceed effectively to the minimization of I, we choose a general form for $g(u^2, v^2)$, and here, too, there is a "natural" choice, the family of ellipses, which we write

$$h(v) = b(a^2 - v^2)^{\frac{1}{2}}, \qquad (28.112)$$

and we now have to minimize (28.110) for variation in a, subject to (28.111). Since (28.112) gives

$$\left. \begin{array}{l} h'(v) = -b^2 v/h(v), \\ h''(v) = -b^2 [\{h(v)\}^2 + b^2 v^2]/\{h(v)\}^3, \end{array} \right\} \qquad (28.113)$$

we therefore have to minimize (dropping the constant)

$$J = b^2 \int_0^a \frac{[\{h(v)\}^2 + b^2 v^2]^2}{\{h(v)\}^5} \exp[-\tfrac{1}{2}\{b^2 v/h(v)\}^2]\,dv \qquad (28.114)$$

for a choice of a, subject to

$$\int_0^a \left\{ \int_0^{b(a^2-v^2)^{1/2}} \exp(-\tfrac{1}{2}u^2)\,du \right\} \exp(-\tfrac{1}{2}v^2)\,dv = k. \qquad (28.115)$$

By differentiating (28.114) with respect to a, and replacing db/da in that derivative by its value obtained by differentiating (28.115), we find, after some reduction,

$$\frac{dJ}{da} = b^2 \exp(\tfrac{1}{2}b^2) \left[\int_0^1 \frac{\exp\{-\tfrac{1}{2}b^2/(1-t^2)\}\,dt}{(1-t^2)^2} \right. $$
$$\left. - \frac{\int_0^1 \frac{\exp\{-\tfrac{1}{2}b^2/(1-t^2)\}\,dt}{(1-t^2)} \int_0^1 \frac{\exp\{-\tfrac{1}{2}a^2(1-b^2)t^2\}\,dt}{(1-t^2)^{\frac{1}{2}}}}{\int_0^1 (1-t^2)^{\frac{1}{2}} \exp\{-\tfrac{1}{2}a^2(1-b^2)t^2\}\,dt} \right] \qquad (28.116)$$

and for a minimum, we put $\dfrac{dJ}{da} = 0$ and solve for a, and thence b.

P. G. Hoel (1951) has carried this through for $1 - \alpha = 0.95$, and found the ellipse (28.112) to have semi-axes of 2·62 and 2·32, not very far from equal. If we specialize the ellipse to a circle by putting $b = 1$ in (28.112), we find the radius a to be 2·45 for $1 - \alpha = 0.95$. Hoel found in this case that the value of J was less than one per cent larger than the minimum.

28.30 The choice of a circle for $g(u^2, v^2)$ corresponds to the original solution of this problem by Working and Hotelling (1929), who derived it simply by observing that, since u and v in (28.94) are independent standardized normal variates, u^2+v^2 is a χ^2 variate with 2 degrees of freedom, and a^2 ($= g_{1-\alpha}$ in (28.103)) is simply the $100(1-\alpha)$ per cent point obtained from the tables of that distribution. The boundaries of the confidence region are, putting (28.112) and (28.113) with $b = 1$ into (28.101),

$$(\hat{\beta}_1+\hat{\beta}_2 x) \pm \frac{\sigma}{n^{\frac{1}{2}}}\left\{h(v)+\frac{v^2}{h(v)}\right\}$$

$$= (\hat{\beta}_1+\hat{\beta}_2 x) \pm \frac{\sigma}{n^{\frac{1}{2}}}[\{h(v)\}^2+v^2]^{\frac{1}{2}}\left[1+\frac{v^2}{\{h(v)\}^2}\right]^{\frac{1}{2}}$$

$$= (\hat{\beta}_1+\hat{\beta}_2 x) \pm \frac{\sigma}{n^{\frac{1}{2}}}(g_{1-\alpha})^{\frac{1}{2}}[1+\{h'(v)\}^2]. \tag{28.117}$$

Using (28.100), (28.117) becomes

$$(\hat{\beta}_1+\hat{\beta}_2 x) \pm (g_{1-\alpha})^{\frac{1}{2}}\left\{\frac{\sigma^2}{n}+x^2\frac{\sigma^2}{\Sigma x^2}\right\}^{\frac{1}{2}}, \tag{28.118}$$

the terms in the braces being $\{\operatorname{var} \hat{\beta}_1 + x^2 \operatorname{var} \hat{\beta}_2\}$.

If (28.118) is compared with the confidence limits (28.93) for $E(y|\mathbf{x}^0)$ derived in Example 28.4 (where we now put $\bar{x} = 0$, as we have done here), we see that apart from the replacement of s^2 by σ^2, and of the $t_{1-\frac{1}{2}\alpha}$ multiple by the χ multiple $(g_{1-\alpha})^{\frac{1}{2}}$, the equations are of exactly the same form. Thus the confidence region (28.118) will look exactly like the loci of the confidence limits (28.93) plotted in Fig. 28.2, being a hyperbola with the fitted line as diameter. As might be expected, for given α the branches of the hyperbola (28.118) are farther apart than those of (28.93), for we are now setting a region for the whole line where previously we had loci of limits for a single value on the line. For example, with $\alpha = 0.05$, $t_{1-\frac{1}{2}\alpha}$ (with infinite degrees of freedom, corresponding to σ^2 known) $= 1.96$, while $g_{1-\alpha}$ for a χ^2 distribution with two degrees of freedom $= 5.99$, the value 2.45 given for a at the end of **28.29** being the square root of this.

28.31 If σ^2 is unknown, only slight modifications of the argument of **28.25-30** are required. Define the variable

$$w^2 = (n-2)s^2/\sigma^2, \tag{28.119}$$

so that w^2 is the ratio of the sum of squared residuals from the fitted regression to the true error variance, which (cf. **28.21**) has a χ^2 distribution with $n-2$ degrees of freedom. From (28.84) and (28.119), we see that the statistics

$$u^* = (n-2)^{\frac{1}{2}}u/w = n^{\frac{1}{2}}(\hat{\beta}_1-\beta_1)/s \quad \text{and} \quad v^* = (n-2)^{\frac{1}{2}}v/w = n^{\frac{1}{2}}(\hat{\beta}_2-\beta_2)/s$$

each have a "Student's" distribution with $n-2$ degrees of freedom. If we now re-trace the argument of **28.25-30** using u^* and v^* in place of u and v, we find that $g(u^{*2}, v^{*2})$ is distributed independently of the parameters $\beta_1, \beta_2, \sigma^2$. The solution of the weighted area minimization problem of **28.28-9** now becomes too difficult in practice, and we proceed directly to the classical solution given in **28.30**.

Using Working and Hotelling's direct argument, we see that since, from **28.21**, u^2, v^2 and w^2 are distributed independently of one another as χ^2 with 1, 1, and $(n-2)$

degrees of freedom respectively, the ratio $\left(\dfrac{u^2+v^2}{2}\right) \Big/ \left(\dfrac{w^2}{n-2}\right) = \tfrac{1}{2}(u^{*2}+v^{*2})$ has a variance-ratio (F) distribution with 2 and $n-2$ degrees of freedom. Thus if we replace σ by s in **28.30**, and put $g_{1-\alpha}$ equal to twice the $100(1-\alpha)$ per cent point of this F-distribution, we obtain the required confidence region from (28.118). As in **28.30**, we find that the boundaries of the region are always farther apart than the loci of the confidence limits for $E(y \mid \mathbf{x}^0)$.

28.32 There is no difficulty in extending our results to the case of more than one regressor, a sketch of such a generalization having been given by Hoel (1951). With k regressors we find, generalizing **28.31**, that $(u^{*2}+\sum_{i=1}^{k} v_i^{*2})/(k+1)$ has a variance-ratio distribution with $(k+1, n-k-1)$ d.fr.

Wynn and Bloomfield (1971) give tables of the fractional multiplying factors to be applied to the Working–Hotelling region when only part of the line is to be covered for the same α. They also give tables extending the Working–Hotelling region to the quadratic regression case. See also related work by Gafarian (1964), Graybill and Bowden (1967), Halperin et al. (1967), Folks and Antle (1967), Halperin and Gurian (1968), Dunn (1968), Bowden (1970), and Bohrer and Francis (1972).

EXERCISES

28.1 The bivariate distribution of x and y is uniform over the region in the (x, y) plane bounded by the ellipse
$$ax^2 + 2hxy + by^2 = c, \quad h \neq 0; \quad h^2 < ab; \quad a, b > 0.$$
Show that the regression of each variable on the other is linear and that the scedastic curves are quadratic parabolas.

28.2 The bivariate distribution of x and y is uniform over the parallelogram bounded by the lines $x = 3(y-1)$, $x = 3(y+1)$, $x = y+1$, $x = y-1$. Show that the regression of y on x is linear, but that the regression of x on y consists of sections of three straight lines joined together.

28.3 Show that if (28.59–60) holds, but $\mathbf{X}$ has elements which are non-linear functions of r further parameters $\gamma_1, \ldots, \gamma_r$, making $(k+r)$ in all, the regression model can be augmented (cf. **19.13–16**) to
$$\mathbf{y} = (\mathbf{X}, \mathbf{D})\binom{\boldsymbol{\beta}}{\mathbf{0}} + \boldsymbol{\epsilon},$$
where $\mathbf{D}$ is any $(n \times r)$ matrix chosen so that $(\mathbf{X}, \mathbf{D})$ is of full rank $(k+r)$, and $\mathbf{0}$ is a $r \times 1$ vector of zeros. Hence obtain confidence regions for (a) the complete set of $(k+r)$ parameters, (b) the r further parameters alone.

(Halperin (1963); cf. also Hartley (1964))

28.4 From (28.17), show that if the marginal distribution of a bivariate distribution is of the Gram–Charlier form
$$f = \alpha(x)\{1 + a_3 H_3 + a_4 H_4 + \ldots\},$$
then the regression of y on x is
$$\mu'_{1x} = \frac{\sum\limits_{r=0}^{\infty} \sum\limits_{s=0}^{\infty} \frac{\varkappa_{r1}}{r!} a_s H_{r+s}(x)}{1 + \sum\limits_{r=3}^{\infty} a_r H_r(x)}.$$

(Wicksell, 1917)

28.5 x_1, x_2, x_3 are trivariate normally distributed. Use **28.9** to show that the regression of each variate on the other two is linear.

28.6 Verify equation (28.33).

28.7 If, for each fixed y, the conditional distribution of x is normal, show that their bivariate distribution must be of the form
$$f(x, y) = \exp\{-(a_1 x^2 + a_2 x + a_3)\}$$
where the a_i are functions of y. Show that if, in addition, the equiprobable contours of $f(x, y)$ are similar concentric ellipses, f must be bivariate normal.

(Bhattacharyya, 1943)

28.8 Show that if the regression of x on y is linear, if the conditional distribution of x for each fixed y is normal and homoscedastic and if the marginal distribution of y is normal, then $f(x, y)$ must be bivariate normal.

(Bhattacharyya, 1943)

28.9 If the conditional distributions of x for each fixed y, and of y for each fixed x, are normal, and one of these conditional distributions is homoscedastic, show that $f(x, y)$ is bivariate normal. (Bhattacharyya, 1943)

28.10 Show that if every non-degenerate linear function of x and y is normal, then $f(x, y)$ is bivariate normal. (Bhattacharyya, 1943)

28.11 If the regressions of x on y and of y on x are both linear, and the conditional distribution of each for every fixed value of the other is normal, show that $f(x, y)$ is either bivariate normal or may (with a suitable choice of origin and scale) be written in the form
$$f = \exp\{-(x^2+a^2)(y^2+b^2)\}.$$ (Bhattacharyya, 1943)

28.12 Show that for interval estimation of β in the linear regression model $y_i = \beta x_i + \varepsilon_i$, the interval based on the "Student's" variate
$$t = (b-\beta)/(s^2/\sum x_i^2)^{\frac{1}{2}}$$
is physically shorter *for every sample* than that based on
$$u = (\bar{y} - \beta\bar{x})/(s^2/n)^{\frac{1}{2}}.$$

28.13 In setting confidence regions for a regression line in **28.28**, show that if the weight function used is
$$w(x) = \left(1 + \frac{x}{S^2}\right)^{-3/2}$$
instead of (28.106), the Working–Hotelling solution of **28.30** is strictly optimum (area-minimizing) in the family of ellipses (28.112). (P.G. Hoel, 1951)

28.14 Show that if there are two different vectors $\mathbf{y}_1$, $\mathbf{y}_2$ each related to the same set of regressors $\mathbf{x}$ in a linear model, the difference between any pair of corresponding parameters in the models may be tested by applying the method of **28.21** to the differences $(y_{1i} - y_{2i})$.

(Yates (1939b) also considers the case where the regressors are different and the **y**-vectors correlated.)

28.15 Independent samples of sizes n_i are taken from two regression models
$$y = \alpha_i + \beta_i x + \varepsilon, \quad i = 1, 2,$$
with independently normally distributed errors. The error variance σ^2 is the same in both models. If b_1, b_2 are the separate Least Squares estimators of β_1, β_2, show that $(b_1 - b_2)$ is normally distributed with mean $(\beta_1 - \beta_2)$ and variance
$$\sigma^2\left\{\left(\sum_{i=1}^{n_1}(x_i-\bar{x}_1)^2\right)^{-1} + \left(\sum_{j=1}^{n_2}(x_j-\bar{x}_2)^2\right)^{-1}\right\},$$
and that
$$t = \{(b_1-b_2)-(\beta_1-\beta_2)\} \bigg/ \left\{s^2\left(\frac{1}{\sum_i(x_i-\bar{x}_1)^2} + \frac{1}{\sum_j(x_j-\bar{x}_2)^2}\right)\right\}^{\frac{1}{2}}$$
has a "Student's" t-distribution with n_1+n_2-4 degrees of freedom, where
$$s^2 = \frac{(n_1-2)s_1^2 + (n_2-2)s_2^2}{n_1+n_2-4}$$
and s_1^2, s_2^2 are the separate estimators of σ^2 in the two models. Hence show that t may be used to test the hypothesis that $\beta_1 = \beta_2$ against $\beta_1 \neq \beta_2$. (cf. Fisher, 1922b)

28.16 For the simple linear model $y = \beta_0 + \beta_1 x + \varepsilon$, two independent samples, of sizes m and n, have means $(\bar{y}_m, \bar{x}_m)$ and $(\bar{y}_n, \bar{x}_n)$. Show that $b_1 = (\bar{y}_m - \bar{y}_n)/(\bar{x}_m - \bar{x}_n)$ is an unbiassed estimator of β_1, with variance $\sigma^2 \left(\dfrac{1}{m} + \dfrac{1}{n}\right)/(\bar{x}_m - \bar{x}_n)^2$. Show that b_1 is not consistent (as $m, n \to \infty$ with m/n fixed) if the two samples were formed by random subdivision of an original sample of $(m+n)$ observations.

28.17 We are given n observations on the model
$$y = \beta_1 x_1 + \beta_2 x_2 + \varepsilon$$
with error variance σ^2, and, in addition, an extraneous unbiassed estimator b_1 of β_1 together with an unbiassed estimator s_1^2 of its sampling variance σ_1^2. To estimate β_2, consider the regression of $(y - b_1 x_1)$ on x_2. Show that the estimator
$$b_2 = \Sigma (y - b_1 x_1) x_2 / \Sigma x_2^2$$
is unbiassed, with variance
$$\operatorname{var} b_2 = (\sigma^2 + \sigma_1^2 r^2 \Sigma x_1^2)/\Sigma x_2^2,$$
where r is the observed correlation between x_1 and x_2. If b_1 is ignored, show that the ordinary Least Squares estimator of β_2 has variance $\sigma^2/\{\Sigma x_2^2 (1 - r^2)\}$ and hence that the use of the extraneous information about β_1 increases efficiency in estimating β_2 if and only if
$$\sigma_1^2 < \frac{\sigma^2}{\Sigma x_1^2 (1 - r^2)},$$
i.e. if the variance of b_1 is less than that of the ordinary Least Squares estimator of β_1. Show that an unbiassed estimator of $\operatorname{var} b_2$ is given by
$$\hat{V} = \frac{1}{(n-2)\Sigma x_2^2} [\Sigma (y - b_1 x_1 - b_2 x_2)^2 + s_1^2 \Sigma x_1^2 \{(n-1) r^2 - 1\}],$$
but that if the errors are normally distributed this is not distributed as a multiple of a χ^2 variate.

(Durbin, 1953)

28.18 In generalization of the situation of Exercise 28.17, let $\mathbf{b}_1$ be a vector of unbiassed estimators of the h parameters $(\beta_1, \beta_2, \ldots, \beta_h)$, with dispersion matrix $\mathbf{V}_1$; and let $\mathbf{b}_2$ be an independently distributed vector of unbiassed estimators of the $k(>h)$ parameters $(\beta_1, \beta_2, \ldots, \beta_h, \beta_{h+1}, \ldots, \beta_k)$, with dispersion matrix $\mathbf{V}_2$. Using Aitken's generalization of Gauss's Least Squares Theorem (**19.17**), show that the minimum variance unbiassed estimators of $(\beta_1, \ldots, \beta_k)$ which are linear in the elements of $\mathbf{b}_1$ and $\mathbf{b}_2$ are the components of the vector
$$\mathbf{b} = \{(\mathbf{V}_1^{-1})^* + \mathbf{V}_2^{-1}\}^{-1} \{(\mathbf{V}_1^{-1})^* \mathbf{b}_1^* + \mathbf{V}_2^{-1} \mathbf{b}_2\},$$
with dispersion matrix
$$\mathbf{V}(\mathbf{b}) = \{(\mathbf{V}_1^{-1})^* + \mathbf{V}_2^{-1}\}^{-1},$$
where an asterisk denotes the conversion of an $(h \times 1)$ vector into a $(k \times 1)$ vector or an $(h \times h)$ matrix into a $(k \times k)$ matrix by putting it into the leading position and augmenting it with zeros.

Show that $\mathbf{V}(\mathbf{b})$ reduces, in the particular case $h=1$, to
$$\mathbf{V}(\mathbf{b}) = \sigma^2 \begin{pmatrix} \Sigma x_1^2 + \dfrac{\sigma^2}{\sigma_1^2} & \Sigma x_1 x_2 & \cdots & \Sigma x_1 x_k \\ \Sigma x_1 x_2 & \Sigma x_2^2 & \cdots & \Sigma x_2 x_k \\ \vdots & & \ddots & \vdots \\ \Sigma x_1 x_k & \Sigma x_2 x_k & \cdots & \Sigma x_k^2 \end{pmatrix}^{-1},$$

differing only in its leading term from the usual Least Squares dispersion matrix $\sigma^2 (\mathbf{X'X})^{-1}$.
(Durbin, 1953)

28.19 A simple graphical procedure may be used to fit an ordinary Least Squares regression of y on x *without computations* when the x-values are equally spaced, say at intervals of s. Let the n observed points on the scatter diagram of (y, x) be $P_1, P_2, \ldots, P_n$ in increasing order of x. Find the point Q_2 on $P_1 P_2$ with x-coordinate $\tfrac{2}{3}s$ above that of P_1; find Q_3 on $Q_2 P_3$ with x-coordinate $\tfrac{2}{3}s$ above that of Q_2; and so on by equal steps, joining each Q-point to the next P-point and finding the next Q-point $\tfrac{2}{3}s$ above, until finally $Q_{n-1} P_n$ gives the last point, Q_n. Carry out the same procedure backwards, starting from $P_n P_{n-1}$ and determining Q'_2, say, $\tfrac{2}{3}s$ below P_n in x-coordinate, and so on until Q'_n on $Q'_{n-1} P_1$ is reached, $\tfrac{2}{3}s$ below Q'_{n-1}. Then $Q_n Q'_n$ is the Least Squares line. Prove this.
(Askovitz, 1957)

28.20 A matrix $\mathbf{A}$ of sums of squares and cross-products of n observations on p variables is inverted. A vector $\mathbf{x}$ containing one further observation on each variable becomes available. Show that the inverse of $\mathbf{B} = \mathbf{A} + \mathbf{xx'}$ is

$$\mathbf{B}^{-1} = \mathbf{A}^{-1} - (\mathbf{A}^{-1} \mathbf{xx'} \mathbf{A}^{-1})/(1 + \mathbf{x'} \mathbf{A}^{-1} \mathbf{x}).$$

Hence show that a quadratic form $\mathbf{x'} \mathbf{A}^{-1} \mathbf{x}$ may be evaluated by

$$1 + \mathbf{x'} \mathbf{A}^{-1} \mathbf{x} = |\mathbf{A} + \mathbf{xx'}|/|\mathbf{A}|. \qquad \text{(Cf. Bartlett, 1951)}$$

28.21 In the regression model

$$y_i = \alpha + \beta x_i + \varepsilon_i, \qquad i = 1, 2, \ldots, n,$$

suppose that the observed mean $\bar{x} = 0$ and let x_0 satisfy $\alpha + \beta x_0 = 0$. Use the random variable $\hat{\alpha} + \hat{\beta} x_0$ to set up a confidence statement for a quadratic function, of form

$$P\{Q(x_0) \geqslant 0\} = 1 - \alpha.$$

Hence derive a confidence statement for x_0 itself, and show that, depending on the coefficients in the quadratic function, this may place x_0:

(i) in a finite interval;
(ii) outside a finite interval;
(iii) in the infinite interval consisting of the whole real line.
(cf. Lehmann, 1959)

28.22 To determine which of the models
$$y = \beta'_0 + \beta'_1 x_1 + \varepsilon', \qquad y = \beta''_0 + \beta'_2 x_2 + \varepsilon'',$$
is more effective in predicting y, consider the model
$$y_i = \beta_0 + \beta_1 x_{1i} + \beta_2 x_{2i} + \varepsilon_i, \qquad i = 1, 2, \ldots, n,$$
with independent normal errors of variance σ^2, estimated by s^2 with $(n-2)$ degrees of freedom. Show that the statistics
$$z_s = \sum_i (y_i - \bar{y})(x_{si} - \bar{x}_s) / \{\sum_i (x_{si} - \bar{x}_s)^2\}^{\frac{1}{2}}, \qquad s = 1, 2,$$
have $\operatorname{var} z_1 = \operatorname{var} z_2 = \sigma^2$, $\operatorname{cov}(z_1, z_2) = \sigma^2 r_{12}$,
where r_{12} is the observed correlation between x_1 and x_2. Hence show that $(z_1 - z_2)$ is exactly normally distributed with mean $\beta'_1 \{\sum_i (x_{1i} - \bar{x}_1)^2\}^{\frac{1}{2}} - \beta'_2 \{\sum_i (x_{2i} - \bar{x}_2)^2\}^{\frac{1}{2}}$ and variance $2\sigma^2 (1 - r_{12})$. Using the fact that $\sum_i (y_i - \bar{y})^2 - (\beta'_s)^2 \sum_i (x_{si} - \bar{x}_s)^2$ is the sum of squares of deviations from the regression of y on x_s alone, show that the hypothesis of equality of these

two sums of squares may be tested by the statistic $t = (z_1 - z_2)/\{2s^2(1-r_{12})\}^{\frac{1}{2}}$, distributed in "Student's" form with $(n-3)$ degrees of freedom.

(Hotelling (1940); Healy (1955). See also E. J. Williams (1959) and Dunn and Clark (1969, 1971).)

28.23 By consideration of the case when $y_j = x_j^k$, $j = 1, 2, \ldots, n$, exactly, show that if the orthogonal polynomials defined at (28.73) and (28.74) are *orthonormal* (i.e, $\sum_{j=1}^{n} \phi_i^2(x_j) = 1$, all i) then they satisfy the recurrence relation

$$\phi_k(x_j) = \frac{1}{b_k}\left\{x_j^k - \sum_{i=0}^{k-1} \phi_i(x_j) \sum_{j=1}^{n} x_j^k \phi_i(x_j)\right\},$$

where the normalizing constant b_k is defined by

$$b_k^2 = \sum_{j=1}^{n}\left\{x_j^k - \sum_{i=0}^{k-1} \phi_i(x_j) \sum_{j=1}^{n} x_j^k \phi_i(x_j)\right\}^2.$$

Hence verify (28.80) and (28.81), with appropriate adjustments. (Robson, 1959)

28.24 In the linear model $\mathbf{y} = \mathbf{X}_1\boldsymbol{\beta}_1 + \mathbf{X}_2\boldsymbol{\beta}_2 + \boldsymbol{\epsilon}$, show that the LS estimators may be written as
$$\hat{\boldsymbol{\beta}}_1 = (\mathbf{X}_1'\mathbf{X}_1)^{-1}\mathbf{X}_1'(\mathbf{y} - \mathbf{X}_2\hat{\boldsymbol{\beta}}_2), \qquad \hat{\boldsymbol{\beta}}_2 = (\mathbf{X}_2\mathbf{D}\mathbf{X}_2)^{-1}\mathbf{X}_2'\mathbf{D}\mathbf{y},$$
where
$$\mathbf{D} = \mathbf{I} - \mathbf{X}_1(\mathbf{X}_1'\mathbf{X}_1)^{-1}\mathbf{X}_1'.$$

If $\boldsymbol{\beta}_1$ is first estimated from

$$\mathbf{y} = \mathbf{X}_1\boldsymbol{\beta}_1 + \boldsymbol{\epsilon}^* \qquad\qquad\qquad (A)$$

and $\boldsymbol{\beta}_2$ is then estimated, using the residuals $\mathbf{y}_r$ in (A) as though they were uncorrelated, from
$$\mathbf{y}_r = \mathbf{X}_2\boldsymbol{\beta}_2 + \boldsymbol{\eta},$$
show that the estimators obtained are
$$\boldsymbol{\beta}_1^* = (\mathbf{X}_1'\mathbf{X}_1)^{-1}\mathbf{X}_1'\mathbf{y}, \qquad \boldsymbol{\beta}_2^* = (\mathbf{X}_2'\mathbf{X}_2)^{-1}\mathbf{X}_2'\mathbf{D}\mathbf{y},$$
and that $\boldsymbol{\beta}_1^*$ and $\boldsymbol{\beta}_2^*$ are biassed unless $\mathbf{X}_1'\mathbf{X}_2 = 0$ or $\boldsymbol{\beta}_2 = 0$. If β_2 is a scalar parameter, show that
$$\beta_2^* = (1 - R^2)\hat{\beta}_2,$$
where R is the multiple correlation coefficient of the single variable x_2 upon all the variables in $\mathbf{X}_1$.

In the case $y_j = \beta_1 x_{1j} + \beta_2 x_{2j} + \varepsilon_j$, show that the mean-square-errors of the biassed two-stage estimators β_1^*, β_2^* are less than the variances of the unbiassed LS estimators $\hat{\beta}_1$, $\hat{\beta}_2$ if $\beta_2^2/V(\hat{\beta}_2) < 1$.

(Cf. Freund *et al.*, (1961) Goldberger and Jochems (1961), Goldberger (1961), Zyskind (1963) and T. D. Wallace (1964))

CHAPTER 29

FUNCTIONAL AND STRUCTURAL RELATIONSHIP

Functional relations between mathematical variables

29.1 It is common in the natural sciences, and to some extent in the social sciences, to set up a model of a system in which certain mathematical (not random) variables are functionally related. A well-known example is Boyle's law, which states that, at constant temperature, the pressure (P) and the volume (V) of a given quantity of gas are related by the equation

$$PV = \text{constant.} \tag{29.1}$$

(29.1) may not hold near the liquefaction point of the gas, or possibly in other parts of the range of P and V. If we wish to discuss the pressure–volume relationship in the so-called adiabatic expansion, when internal heat does not have time to adjust itself to surrounding conditions, we may have to modify (29.1) to

$$PV^\gamma = \text{constant,} \tag{29.2}$$

where γ is an additional constant which may have to be estimated. Moreover, at some stage we may wish to take temperature (T) into account and extend (29.1) to the form

$$PVT^{-1} = \text{constant.}$$

In general, we have a set of variables $X_1, \ldots, X_k$ related in p functional forms

$$f_j(X_1, \ldots, X_k; \alpha_1, \ldots, \alpha_l) = 0, \qquad j = 1, 2, \ldots, p, \tag{29.3}$$

depending on l parameters α_r, $r = 1, 2, \ldots, l$. Our object is usually to estimate the α_r from a set of observations, and possibly also to determine the actual functional forms f_j, especially in cases where neither theoretical considerations nor previous experience provide a complete specification of these forms. If we were able to observe values of X without error, there would be no statistical problem here at all: we should simply have a set of values satisfying (29.3) and the problem would be merely the mathematical one of solving the set of equations. However, experimental or observational error usually affects our measurements. What we then observe is not a "true" value X, but X together with some random element. We thus have to estimate the parameters α_r (and possibly the forms f_j) from data which are, to some extent at least, composed of samples from frequency distributions of error. Our problem then immediately becomes statistical.

29.2 In our view, it is particularly important in this subject, which has suffered from confusion in the past, to use a clear terminology and notation. In this chapter, we shall denote mathematical variables by capital Roman letters (actually italic). As usual, we denote parameters by small Greek letters (here we shall particularly use α and β) and random variables generally by a small Roman letter or, in the case of Maximum Likelihood estimators, by the parameter covered by a circumflex, e.g. $\hat{\alpha}$. Error random variables will be symbolized by other small Greek letters, particularly

δ and ε, and the observed random variables corresponding to unobservable variables will be denoted by a "corresponding" [*] Greek letter, e.g., ξ for X. The only possible source of confusion in this system of notation is that Greek letters are performing three roles (parameters, error variables, observable variables) but distinct groups of letters are used throughout, and there is a simple way of expressing our notation which may serve as a rescuer: any Greek letter "corresponding" to a capital Roman letter is the observable random variable emanating from that mathematical variable; all other Greek letters are unobservables, being either parameters or error variables.

29.3 We begin with the simplest case. Two mathematical variables X and Y are known to be linearly related, so that we have
$$Y = \alpha_0 + \alpha_1 X, \qquad (29.4)$$
and we wish to estimate the parameters α_0, α_1. We are not able to observe X and Y; we observe only the values of two random variables ξ, η defined by
$$\left. \begin{array}{l} \xi_i = X_i + \delta_i, \\ \eta_i = Y_i + \varepsilon_i, \end{array} \right\} \quad i = 1, 2, \ldots, n. \qquad (29.5)$$
The suffixes in (29.5) are important. Observations about any "true" value are distributed in a frequency distribution of an "error" random variable, and the form of this distribution may depend on i. For example, errors may tend to be larger for large values of X than for small X, and this might be expressed by an increase in the variance of the error variable δ.

In this simplest case, however, we suppose the δ_i to be identically distributed, so that δ_i has the same mean (taken to be zero without loss of generality) and variance for all X_i; and thus also for ε and Y. We also suppose the errors δ, ε to be uncorrelated amongst themselves and with each other. For the present, we do not assume that δ and ε are normally distributed. Our model is thus (29.4) and (29.5) with
$$\left. \begin{array}{l} E(\delta_i) = E(\varepsilon_i) = 0, \quad \text{var } \delta_i = \sigma_\delta^2, \quad \text{var } \varepsilon_i = \sigma_\varepsilon^2, \quad \text{all } i, \\ \text{cov}(\delta_i, \delta_j) = \text{cov}(\varepsilon_i, \varepsilon_j) = 0, \quad i \neq j, \\ \text{cov}(\delta_i, \varepsilon_j) = 0, \quad \text{all } i, j. \end{array} \right\} \qquad (29.6)$$
The restrictive assumption on the means of the δ_i is only that they are all equal, and similarly for the ε_i—we may reduce their means μ_δ and μ_ε to zero by absorbing them into α_0, since we clearly could not distinguish α_0 from these biases in any case.

In view of (29.6) we may on occasion unambiguously write the model as
$$\left. \begin{array}{l} \xi = X + \delta, \\ \eta = Y + \varepsilon. \end{array} \right\} \qquad (29.7)$$

29.4 At first sight, the estimation of the parameters in (29.4) looks like a problem in regression analysis; and indeed, this resemblance has given rise to much confusion. In a regression situation, however, we are concerned with the dependence of the mean

---

[*] It will be seen that the Roman–Greek "correspondence" is not so much strictly alphabetical as aural and visual. In any case, it would be more logical to use the ordinary lower-case Roman letter, i.e. the observed x corresponding to the mathematical variable X, but there is danger of confusion in suffixes, and besides, we need x for another purpose—cf. **29.6**.

FUNCTIONAL AND STRUCTURAL RELATIONSHIP

value of η (which is Y) upon X, which is not subject to error; the error variable δ is identically zero in value, so that $\sigma_\delta^2 = 0$. Thus the regression situation is essentially a special case of our present model. In addition (though this is a difference of background, not of formal analysis), the variation of the dependent variable in a regression analysis is not necessarily, or even usually, due to error alone. It may be wholly or partly due to the inherent structure of the relationship between the variables. For example, body weight varies with height in an intrinsic way, quite unconnected with any errors of measurement.

We may easily convince ourselves that the existence of errors in both X and Y poses a problem quite distinct from that of regression. If we substitute for X and Y from (29.7) into (29.4), we obtain

$$\eta = \alpha_0 + \alpha_1 \xi + (\varepsilon - \alpha_1 \delta). \qquad (29.8)$$

This is not a simple regression situation: ξ is a random variable, and it is correlated with the error term $(\varepsilon - \alpha_1 \delta)$. For, from (29.6) and (29.7),

$$\operatorname{cov}(\xi, \varepsilon - \alpha_1 \delta) = E\{\xi(\varepsilon - \alpha_1 \delta)\} = E\{(X + \delta)(\varepsilon - \alpha_1 \delta)\}$$
$$= -\alpha_1 \sigma_\delta^2, \qquad (29.9)$$

which is only zero if $\sigma_\delta^2 = 0$, which is the regression situation, or in the trivial case $\alpha_1 = 0$.

The equation (29.8) is called a *structural relation* between the observable random variables ξ, η. This structural relation is a result of the *functional relation* between the mathematical variables X, Y.

29.5 In regression analysis, the values of the regressor variable X may be selected arbitrarily, e.g. at equal intervals along its effective range. But they may also emerge as the result of some random selection, i.e. n pairs of observations may be randomly chosen from a bivariate distribution and the regression of one variable upon the other examined. (We have already discussed these alternative regression models in **26.24**, **27.29**.) In our present model also, the values of X might appear as a result of some random process or as a result of deliberate measurement at particular points, but in either case X remains unobserved due to the errors of observation. We now discuss the situation where X, and hence Y, becomes a random variable, so that the functional relation (29.4) itself becomes a *structural relation* between the unobservables.

Structural relations between random variables

29.6 Suppose that X, Y are themselves random variables (in accordance with our conventions we shall therefore now write them as x, y) and that (29.4), (29.5) and (29.6) hold as before. (29.8) will once more follow, but (29.9) will no longer hold without further assumptions, for in it X was treated as a constant. The correct version of (29.9) is now

$$\operatorname{cov}(\xi, \varepsilon - \alpha_1 \delta) = E\{(x + \delta)(\varepsilon - \alpha_1 \delta)\} = E(x\varepsilon) - \alpha_1 E(x\delta) - \alpha_1 \sigma_\delta^2, \qquad (29.10)$$

and we now make the further assumptions (two for x and two for y)

$$\operatorname{cov}(x, \delta) = \operatorname{cov}(x, \varepsilon) = \operatorname{cov}(y, \delta) = \operatorname{cov}(y, \varepsilon) = 0. \qquad (29.11)$$

(29.11) reduces (29.10) to (29.9) as before.

The present model is therefore
$$\left.\begin{aligned}\xi_i &= x_i+\delta_i,\\ \eta_i &= y_i+\varepsilon_i,\end{aligned}\right\} \qquad (29.12)$$
$$y_i = \alpha_0+\alpha_1 x_i, \qquad (29.13)$$

subject to (29.6) and (29.11), leading to (29.8) as before. We have replaced the *functional relation* (29.4) between mathematical variables by the *structural relation* (29.13) expressing an exact linear relationship between two unobservable random variables x, y. The present model is a generalization of our previous one, which is simply the case where x_i degenerates to a constant, X_i. The relation (29.8) between the observables ξ, η is a structural one, as before, but we also have a structural relation at the heart of the situation, so to speak.

Structural relation models will be appropriate, for example, when a random variable can be measured by either of two instruments, each subject to its own measurement error, and we wish to use a hypothesized linear relation between the true values obtained by the two instruments in order to compare results obtained using different instruments. Similarly, the alcohol content of a unit volume of human blood is a random variable even if the blood is always taken from a single site and there is no measurement error. If two sites are available, it would not be unreasonable to suppose the two determinations of alcohol content to be linearly related, with measurement error added to each. The essential point is that there is both inherent variability in each fundamental quantity with which we are concerned *and* observational error in determining each.

29.7 One consequence of the distinctions we have been making has frequently puzzled scientists. The investigator who is looking for a unique linear relationship between variables cannot accept two different lines, but he was liable in the early days of the subject (and perhaps sometimes even today) to be presented with a pair of regression lines. Our discussion should have made it clear that a regression line does not purport to represent a functional relation between mathematical variables or a structural relation between random variables: it either exhibits a property of a bivariate distribution or, when the regressor variable is not subject to error, gives the relation between the mean of the dependent variable and the value of the regressor variable. The methods of this chapter, which our references will show to have been developed largely since 1940, permit the mathematical model to be more precisely fitted to the needs of the scientific situation.

29.8 It is interesting to consider how the approach from Least Squares regression analysis breaks down when applied to the estimation of α_0 and α_1 in (29.8). If we have n pairs of observed values (ξ_i, η_i), $i = 1, 2, \ldots, n$, we find on averaging (29.8) over these values

$$\bar{\eta} = \alpha_0+\alpha_1\bar{\xi}+\frac{1}{n}\Sigma(\varepsilon-\alpha_1\delta). \qquad (29.14)$$

The last term on the right of (29.14) has a zero expectation, and we therefore have the estimating equation

$$\bar{\eta} = \alpha_0+\alpha_1\bar{\xi}, \qquad (29.15)$$

which is unbiassed in the sense that both sides have the same expectation. If we measure from the sample means $\bar{\xi}$, $\bar{\eta}$, we therefore have, as an estimator of α_0,
$$a_0 = 0. \tag{29.16}$$
Similarly, multiplying (29.8) by ξ, we have on averaging
$$\frac{1}{n}\Sigma \eta \xi = \frac{a_1}{n}\Sigma \xi^2 + \frac{1}{n}\Sigma \xi(\varepsilon - \alpha_1 \delta), \tag{29.17}$$
where a_1 is the estimator of α_1. The last term on the right of (29.17) does not vanish, even as $n \to \infty$, for it tends to $\operatorname{cov}\{\xi, \varepsilon - \alpha_1 \delta\}$, a multiple of σ_δ^2 by (29.9). It seems, then, that we require knowledge of σ_δ^2 before we can estimate α_1, by this method at least. Indeed, we shall find that the error variances play an essential role in the estimation of α_1.

ML estimation of structural relationship

29.9 If we are prepared to make the further assumption that the pairs of observables ξ_i, η_i are jointly normally and identically distributed, we may use the Maximum Likelihood method to estimate the parameters of the structural relationship model specified by (29.6) and (29.11)–(29.13). (This joint normality would follow from the x_i being identically normally distributed, and similarly for the y_i, δ_i and ε_i; if x, y degenerate to constants X, Y, bivariate normality of δ, ε would be sufficient for the joint normality of ξ, η.) We then have, by (29.6) and (29.11)–(29.13), the moments

$$\left. \begin{array}{l} E(\xi) = E(x) = \mu, \\ E(\eta) = E(y) = \alpha_0 + \alpha_1 \mu, \\ \operatorname{var} \xi = \operatorname{var} x + \sigma_\delta^2 = \sigma_x^2 + \sigma_\delta^2, \\ \operatorname{var} \eta = \operatorname{var} y + \sigma_\varepsilon^2 = \alpha_1^2 \sigma_x^2 + \sigma_\varepsilon^2, \\ \operatorname{cov}(\xi, \eta) = \operatorname{cov}(x, y) = \alpha_1 \sigma_x^2. \end{array} \right\} \tag{29.18}$$

It should be particularly noted that in (29.18) all the structural variables x_i have the same mean, and hence all the y_i have the same mean. This is of importance in the ML process, as we shall see, and it also means that the results which we are about to obtain for structural relations are only of trivial value in the functional relation case, since they will apply only to the case where X_i (the constant to which x_i degenerates when $\sigma_x^2 = 0$) takes the same value (μ) for all i. See **29.13** below.

29.10 From (16.47) and Examples 18.14–15, the set of sample means, variances and covariance are sufficient statistics for the five parameters of a bivariate normal distribution, and are also the ML estimators of these parameters. Thus if s_η^2, s_ξ^2 (both >0) are the sample variances and $s_{\xi\eta}$ the sample covariance, the solutions of the equations

$$\left. \begin{array}{rl} (a) & \mu = \bar{\xi} \\ (b) & \alpha_0 + \alpha_1 \mu = \bar{\eta} \\ (c) & \sigma_x^2 + \sigma_\delta^2 = s_\xi^2 \\ (d) & \alpha_1^2 \sigma_x^2 + \sigma_\varepsilon^2 = s_\eta^2 \\ (e) & \alpha_1 \sigma_x^2 = s_{\xi\eta} \end{array} \right\} \tag{29.19}$$

for the unknowns among the six parameters μ, α_0, α_1, σ_x^2, σ_δ^2 and σ_ε^2 will be the ML estimators of these parameters also, provided that these solutions yield admissible values for all of them. Since μ, α_0 and α_1 are unrestricted *a priori*, we need only ensure that the solutions for σ_x^2, σ_δ^2 and σ_ε^2 are non-negative. From (29.19)(c)–(e), these give the restrictions:

$$\left.\begin{array}{ll} \text{For } \hat\sigma_\delta^2 \geqslant 0 & \text{(i) } \alpha_1 s_\xi^2 \geqslant s_{\xi\eta}, \\ \text{For } \hat\sigma_\varepsilon^2 \geqslant 0 & \text{(ii) } s_\eta^2 \geqslant \alpha_1 s_{\xi\eta}, \\ & \text{(iii) } s_\xi^2 - \sigma_\delta^2 \geqslant 0, \\ \text{For } \hat\sigma_x^2 \geqslant 0 & \text{(iv) } s_\eta^2 - \sigma_\varepsilon^2 \geqslant 0, \\ & \text{(v) if } \sigma_x^2 > 0, \alpha_1 \gtreqless 0 \text{ with } s_{\xi\eta}, \\ & \text{(vi) if } \sigma_x^2 = 0, \alpha_1 \text{ is indeterminate.} \end{array}\right\} \quad (29.20)$$

If the restrictions (29.20) are not satisfied, the solutions of (29.19) are not the ML estimators for our problem—they must instead be obtained by direct maximation of the LF. (29.20)(vi) will remain true in that case, as the moments (29.18) show.

(29.19)(c)–(e) give the equalities

$$\alpha_1(s_\xi^2 - \sigma_\delta^2) = s_{\xi\eta}; \quad s_\eta^2 - \sigma_\varepsilon^2 = \alpha_1 s_{\xi\eta}, \quad (29.21)$$

and making the coefficients of α_1 equal in these, we find

$$\alpha_1 s_{\xi\eta}(s_\xi^2 - \sigma_\delta^2) = s_{\xi\eta}^2 = (s_\eta^2 - \sigma_\varepsilon^2)(s_\xi^2 - \sigma_\delta^2). \quad (29.22)$$

(29.22) implies that

$$\frac{|s_{\xi\eta}|}{s_\xi^2} \leqslant |\alpha_1| \leqslant \frac{s_\eta^2}{|s_{\xi\eta}|}, \quad \text{if } \alpha_1, s_{\xi\eta} \neq 0, \quad (29.23)$$

so that a ML estimate of the slope of the structural line obtained from (29.19) is bounded in absolute value by the LS regression coefficient of η on ξ and by the reciprocal of the LS regression coefficient of ξ on η. (29.19)(a)–(b) then implies that the estimated structural line will lie between the two estimated regression lines, as is intuitively reasonable.

29.11 Whether the ML estimation is accomplished through (29.19) or by direct maximization of the LF, (29.19)(a)–(b) will always give the ML estimators $\hat\mu$ and $\hat\alpha_0$ once $\hat\alpha_1$ is determined. When (29.19) is used, equations (c)–(e) must be solved for α_1, but we cannot do this without some further assumptions, since there are four unknowns in these three equations.

The reason for this difficulty is not far to seek. Looking back at (29.18), we see that a change in the true value of α_1 need not change the values of the five moments given there. For example, suppose μ and α_1 are positive; then any increase in the value of α_1 may be offset (a) in $E(\eta)$ by a reduction in α_0, (b) in $\text{cov}(\xi, \eta)$ by a reduction in σ_x^2, and (c) in $\text{var } \eta$ by an appropriate adjustment of σ_ε^2. (The reader will, perhaps, like to try a numerical example.) What this means is that α_1 is intrinsically impossible to estimate, however large the sample; it is said to be *unidentifiable*. In fact, μ alone of the six parameters is identifiable. We do not wish to assume knowledge of α_0 and α_1, whose estimation is our primary objective, or of σ_x^2, since x is unobservable. μ is already identifiable, so we cannot improve matters there. Clearly, we must make an assumption about the error variances.

29.12 *Case 1: σ_δ^2 known*

(29.21) gives at once

$$\hat{\alpha}_1 = s_{\xi\eta}/(s_\xi^2 - \sigma_\delta^2) \tag{29.24}$$

if $s_\xi^2 > \sigma_\delta^2$, which ensures that (29.20 (iii)–(v)) hold, but (29.20)(ii) must be imposed as the condition $s_\eta^2 \geqslant s_{\xi\eta}^2/(s_\xi^2 - \sigma_\delta^2)$ for all the restrictions in (29.20) to be satisfied and (29.24) to be the ML estimator. If these conditions are not satisfied, (29.19) does not give the ML estimator, which is (cf. Exercise 29.18)

$$\hat{\alpha}_1 = s_\eta^2/s_{\xi\eta}. \tag{29.25}$$

Note that as $\sigma_\delta^2 \to 0$, $|\hat{\alpha}_1|$ in (29.24) tends to its lower bound in (29.23), while (29.25) is its upper bound there.

Case 2: σ_ε^2 known

(29.21) gives

$$\hat{\alpha}_1 = (s_\eta^2 - \sigma_\varepsilon^2)/s_{\xi\eta}, \quad s_{\xi\eta} \neq 0 \tag{29.26}$$

and, as in Case 1 above, the conditions $s_\eta^2 > \sigma_\varepsilon^2$, $s_\xi^2 \geqslant s_{\xi\eta}^2/(s_\eta^2 - \sigma_\varepsilon^2)$ ensure that all the restrictions in (29.20) are satisfied and (29.26) is the ML estimator. Failing these conditions, the ML estimator is the analogue of (29.25),

$$\hat{\alpha}_1 = s_{\xi\eta}/s_\xi^2.$$

The last sentence of Case 1 applies here, too, with obvious modifications.

Case 3: $\sigma_\varepsilon^2/\sigma_\delta^2$ known

This is the classical method of resolving the identifiability problem. Putting $\sigma_\varepsilon^2 = \lambda\sigma_\delta^2$, elimination of σ_δ^2 between the equations of (29.21) gives

$$\alpha_1^2 s_{\xi\eta} + \alpha_1(\lambda s_\xi^2 - s_\eta^2) - \lambda s_{\xi\eta} = 0. \tag{29.27}$$

Unless $s_{\xi\eta} = 0$ (in which case $\hat{\alpha}_1 = 0$ unless $s_\eta^2/s_\xi^2 = \lambda$, when $\hat{\alpha}_1$ is indeterminate and $\hat{\sigma}_x^2 = 0$—see (29.20)(vi)) this quadratic has necessarily non-zero roots

$$\frac{(s_\eta^2 - \lambda s_\xi^2) \pm \{(s_\eta^2 - \lambda s_\xi^2)^2 + 4\lambda s_{\xi\eta}^2\}^{\frac{1}{2}}}{2s_{\xi\eta}} = \frac{N}{2s_{\xi\eta}}, \tag{29.28}$$

say. By (29.19)(e), $\hat{\sigma}_x^2 = s_{\xi\eta}/\alpha_1 = 2s_{\xi\eta}^2/N$, so to satisfy $\hat{\sigma}_x^2 \geqslant 0$, N must be non-negative and therefore the positive square root must always be taken in it. Thus

$$\hat{\alpha}_1 = \frac{(s_\eta^2 - \lambda s_\xi^2) + \{(s_\eta^2 - \lambda s_\xi^2)^2 + 4\lambda s_{\xi\eta}^2\}^{\frac{1}{2}}}{2s_{\xi\eta}} \tag{29.29}$$

$= \dfrac{N_+}{2s_{\xi\eta}}$, say, and $\hat{\sigma}_x^2 > 0$. We need only check that (29.20) (i) or (ii) holds, since now $\hat{\sigma}_\varepsilon^2 = \lambda\hat{\sigma}_\delta^2$. (29.20)(ii) requires that $2s_\eta^2 \geqslant N_+$. Replacing $s_{\xi\eta}^2$ by its upper bound $s_\xi^2 s_\eta^2$ establishes this. (29.29) is therefore the ML estimator.

Case 4: σ_δ^2 and σ_ε^2 both known

Only two unknowns (α_1, σ_x^2) now remain in (29.19) (c)–(e), and we can deduce both (29.24) and (29.26), which are inconsistent with each other. (29.19) therefore cannot give the ML estimators in this case, and we must maximize the LF directly, following Birch (1964a).

Using the moments in the last three equations of (29.18),

$$-\frac{2}{n}\log L = \log|V| + |V|^{-1}\{s_\xi^2(\alpha_1^2\sigma_x^2+\sigma_\varepsilon^2) - 2s_{\xi\eta}\alpha_1\sigma_x^2 + s_\eta^2(\sigma_x^2+\sigma_\delta^2)\} \qquad (29.30)$$

where

$$|V| = (\sigma_x^2+\sigma_\delta^2)(\alpha_1^2\sigma_x^2+\sigma_\varepsilon^2) - (\alpha_1\sigma_x^2)^2 = \alpha_1^2\sigma_x^2\sigma_\delta^2 + \sigma_\varepsilon^2(\sigma_x^2+\sigma_\delta^2).$$

We standardize the known constants σ_δ and σ_ε out of (29.30) for simplicity, measuring x and ξ in units of σ_δ, y and η in units of σ_ε. (29.30) is then of the form

$$G(u,v) = -\frac{n}{2}\left\{\log(1+u^2+v^2) + \frac{s_\xi^2(1+v^2) - 2s_{\xi\eta}uv + s_\eta^2(1+u^2)}{1+u^2+v^2}\right\}$$

where $u^2 = \sigma_x^2$, $v^2 = \alpha_1^2 u^2$. $G(u,v)$ is differentiable and $\to -\infty$ when $(u^2+v^2) \to \infty$, so is maximized at a stationary value obtained by equating to zero the derivatives $\frac{\partial G}{\partial u}$, $\frac{\partial G}{\partial v}$. We thence obtain

$$\left.\begin{array}{l}(1+u^2)\dfrac{\partial G}{\partial u} + uv\dfrac{\partial G}{\partial v} = n\left(\dfrac{s_\xi^2 u + s_{\xi\eta}v}{1+u^2+v^2} - u\right) = 0,\\[2mm] uv\dfrac{\partial G}{\partial u} + (1+v^2)\dfrac{\partial G}{\partial v} = n\left(\dfrac{s_{\xi\eta}u + s_\eta^2 v}{1+u^2+v^2} - v\right) = 0.\end{array}\right\} \qquad (29.31)$$

Eliminating $(1+u^2+v^2)$ from these equations, we have

$$u^2\{\alpha_1^2 s_{\xi\eta} + \alpha_1(s_\xi^2 - s_\eta^2) - s_{\xi\eta}\} = 0.$$

Thus either we must have u^2 (and hence v^2 also) $= 0$ or the quadratic in braces must be equated to zero. In the latter case, we are back at (29.27) (remembering that we have made $\lambda = 1$ by our standardizations), and on destandardizing (29.29) is again the solution for α_1. From (29.31),

$$s_\xi^2 + s_{\xi\eta}\frac{v}{u} - (1+u^2+v^2) = 0$$

which gives, since $v^2 = \alpha_1^2 u^2$,

$$\sigma_x^2 = u^2 = (s_\xi^2 + \alpha_1 s_{\xi\eta} - 1)/(1+\alpha_1^2). \qquad (29.32)$$

The stationary values taken by G at the points $(\pm u, v)$ given by (29.29) and (29.32) will only be maxima if the stationary value at $u = v = 0$ is not. From the second-order derivatives of G, this is when one or more of the conditions $(s_\xi^2 - 1) > 0$, $(s_\eta^2 - 1) > 0$, $s_{\xi\eta}^2 > (1-s_\xi^2)(1-s_\eta^2)$ is satisfied. Destandardizing, we therefore have that (29.29) and (29.32) give the ML estimators $(\hat{\alpha}_1, \hat{\sigma}_x^2)$ when one or more of the conditions $s_\xi^2 > \sigma_\delta^2$, $s_\eta^2 > \sigma_\varepsilon^2$, $s_{\xi\eta}^2 > (\sigma_\delta^2 - s_\xi^2)(\sigma_\varepsilon^2 - s_\eta^2)$ holds. If none holds (which seems unlikely in practice) $u = v = 0$ is a maximum, so $\hat{\sigma}_x^2 = 0$ and $\hat{\alpha}_1$ is indeterminate—cf. (29.20) (vi) and below it.

The identifiability problem of **29.11** disappears if we have replicated observations, i.e., if there are r_i observations $\xi_{ij}(j = 1, 2, \ldots, r_i)$ corresponding to the true value x_i, and s_i observations $\eta_{ik}(k = 1, 2, \ldots, s_i)$ corresponding to y_i, with at least one r_i and one s_i exceeding unity. We write $\xi_{i.} = \sum\limits_{j=1}^{r_i}\xi_{ij}/r_i$, $\eta_{i.} = \sum\limits_{k=1}^{s_i}\eta_{ik}/s_i$, $R = \sum\limits_{i=1}^{n} r_i$, $S = \sum\limits_{i=1}^{n} s_i$,

$\xi.. = \sum_{i=1}^{n} r_i \xi_i./R$, $\eta.. = \sum_{i=1}^{n} s_i \eta_i./S$. Then it follows very simply that

$$\hat{\sigma}_\delta^2 = \frac{1}{R-n} \sum_i \sum_j (\xi_{ij} - \xi_i.)^2$$

and

$$\hat{\sigma}_\varepsilon^2 = \frac{1}{S-n} \sum_i \sum_k (\eta_{ik} - \eta_i.)^2$$

are unbiassed estimators of the error variances and that $\hat{\lambda} = \hat{\sigma}_\varepsilon^2/\hat{\sigma}_\delta^2$ is a consistent (though biased) estimator of their ratio. Any of the estimators of α_1 in Cases 1-4 may now be used with the appropriate estimator substituted, and it will be consistent. Madansky (1959) discusses various estimators derived by methods similar to these, which are essentially simple applications of the ideas of the Analysis of Variance (Volume 3). Dorff and Gurland (1961a) make asymptotic variance comparisons when s_i/r_i is constant which favour the use of the estimator obtained by using $\hat{\lambda}$ in (29.29).

Generalization of the structural relationship model

29.13 As we remarked below (29.18), the structural relationship model discussed in **29.9-12** is a restrictive one because of the condition that all x_i have the same mean, which implies the same for the y_i. We had

$$E(\xi_i) = E(x_i) = \mu, \quad \text{all } i, \tag{29.33}$$

$$E(\eta_i) = E(y_i) = \alpha_0 + \alpha_1\mu, \quad \text{all } i. \tag{29.34}$$

Suppose now that we relax (29.33) and postulate that

$$E(\xi_i) = E(x_i) = \mu_i, \quad i = 1, 2, \ldots, n. \tag{29.35}$$

(29.34) is then replaced by

$$E(\eta_i) = \alpha_0 + \alpha_1\mu_i. \tag{29.36}$$

This is a more comprehensive structural relationship model, which may be specialized to the functional relationship model without loss of generality by putting $\sigma_x^2 = \sigma_y^2 = 0$, so that $X_i = \mu_i$, $Y_i = \alpha_0 + \alpha_1 X_i$.

However, in taking this more general model, we have radically changed the estimation problem. For all the μ_i are unknown parameters, and thus instead of six parameters to estimate, as in (29.18), we have $(n+5)$ parameters. The essentially new feature is that every new observation brings with it a new parameter to be estimated, and it is not surprising that we discover new problems in this case. These parameters, specific to individual observations, were called "incidental" parameters by Neyman and Scott (1948); other parameters, common to sets of observations, were called "structural." We have already encountered a problem involving incidental parameters in Example 18.16.

We have now to consider the ML estimation process in the presence of incidental parameters, and we shall proceed directly to the case of functional relationship, which is what interests us here.

ML estimation of functional relationship

29.14 Let us, then, suppose that (29.4), (29.5) and (29.6) hold, and that the δ_i and ε_i are independent normal variables. Since the X_i are mathematical (not random) variables, $\sigma_x^2 = 0$ and there are $(n+4)$ parameters, namely α_0, α_1, σ_δ^2, σ_ε^2 and the n values X_i. Our Likelihood Function is

$$L \propto \sigma_\delta^{-n} \sigma_\varepsilon^{-n} \exp\left[-\frac{1}{2\sigma_\delta^2}\sum_i (\xi_i - X_i)^2 - \frac{1}{2\sigma_\varepsilon^2}\sum_i \{\eta_i - (\alpha_0 + \alpha_1 X_i)\}^2\right]. \tag{29.37}$$

We thus have

$$-\frac{2}{n}\log L = \log \sigma_\delta^2 + \log \sigma_\varepsilon^2 + \frac{S_1}{\sigma_\delta^2} + \frac{S_2}{\sigma_\varepsilon^2} \tag{29.38}$$

where $S_1 = \frac{1}{n}\Sigma(\xi_i - X_i)^2$

and $S_2 = \frac{1}{n}\Sigma\{\eta_i - (\alpha_0 + \alpha_1 X_i)\}^2$.

We may write (29.38) as

$$-\frac{2}{n}\log L = h(\sigma_\delta^2, S_1) + h(\sigma_\varepsilon^2, S_2), \tag{29.39}$$

where $h(a,b) = \log a + b/a$.

Now the function $h(a,b)$ has $h(a,0) = \log a$, so that as $a \to 0$, $h(a,0) \to -\infty$. Thus (29.39), the sum of two such functions, will also $\to -\infty$ if either of the functions on its right-hand side does so while the other remains bounded. Now consider the parameter values

$$\left.\begin{array}{l}X_i = \xi_i, \quad i = 1, 2, \ldots, n \\ \sigma_\delta^2 \to 0.\end{array}\right\} \tag{29.40}$$

These make $S_1 = 0$ and $-\frac{2}{n}\log L \to -\infty$, and L itself $\to +\infty$ irrespective of the value assigned to α_1. Thus no ML estimation of α_1 is possible here, as Solari (1969) first pointed out. The situation is similar to that at (29.20) (vi) above.

29.15 We therefore cannot obtain an ML estimator of α_1 in the functional relationship without a further assumption, and indeed this was so even in the structural relationship case of **29.9–12**. This need for a further assumption often seems strange to the user of statistical method, who has perhaps too much faith in its power to produce a simple and acceptable solution to any problem that can be posed simply. A geometrical illustration is therefore possibly useful.

Consider the points (ξ_i, η_i) plotted as in Fig. 29.1.

Any observed point (ξ_i, η_i) has emanated from a "true" point $(X_i, Y_i) = (\xi_i - \delta_i, \eta_i - \varepsilon_i)$ whose situation is unknown. Since, in our model, δ_i and ε_i are independent normal variates, (ξ_i, η_i) is equiprobable on any ellipse centred at (X_i, Y_i), whose axes are parallel to the co-ordinate axes. Conversely, since the frequency function of (ξ_i, η_i) is symmetric in (ξ_i, η_i) and (X_i, Y_i), there is an elliptical confidence region for (X_i, Y_i) at any given probability level, centred at (ξ_i, η_i). These are the regions shown in Fig. 29.1. Heuristically, our problem of estimating α_1 may be conceived as that of finding a straight line to intersect as many as possible of these confidence regions. The difficulty is now plain to see: the problem as specified does not tell us what the lengths

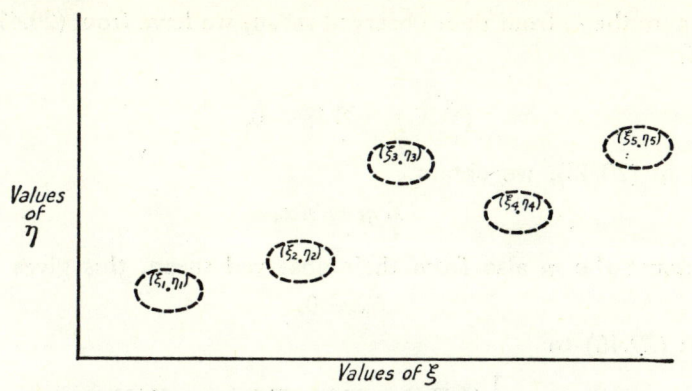

Fig. 29.1—Confidence regions for (X_i, Y_i)—see text

of the axes of the ellipses should be—these depend on the scale parameters σ_δ, σ_ε. It is clear that to make the problem definite we need only know the eccentricity of the ellipses, i.e. the ratio $\sigma_\varepsilon/\sigma_\delta$. It will be remembered that in the structural relationship problem of **29.9-10**, we found a knowledge of this ratio sufficient to solve the problem of estimating α_1.

29.16 Let us, then, suppose that $\sigma_\varepsilon^2/\sigma_\delta^2 = \lambda$ is known. If we substitute $\sigma_\varepsilon^2/\lambda$ for σ_δ^2 in (29.38), it becomes

$$-\frac{2}{n}\log L = -\log\lambda + 2\log\sigma_\varepsilon^2 + \frac{\lambda S_1 + S_2}{\sigma_\varepsilon^2}. \tag{29.41}$$

(29.40) no longer makes this $\to -\infty$, for S_2 in the numerator of the last term remains positive. Moreover, it is obvious from the definitions of S_1 and S_2 that we cannot choose parameter values that will make $S_1 = S_2 = 0$. Thus (29.41) is bounded away from $-\infty$ and L itself from $+\infty$. We proceed to the ML solution through the likelihood equations. We find

$$\frac{\partial \log L}{\partial X_i} = \frac{1}{\sigma_\varepsilon^2}[\lambda(\xi_1 - X_i) + \alpha_1\{\eta_i - (\alpha_0 + \alpha_1 X_i)\}] = 0, \quad i = 1, 2, \ldots, n, \tag{29.42}$$

$$\frac{\partial \log L}{\partial \alpha_0} = \frac{1}{\sigma_\varepsilon^2}\sum_i\{\eta_i - (\alpha_0 + \alpha_1 X_i)\} = 0, \tag{29.43}$$

$$\frac{\partial \log L}{\partial \alpha_1} = \frac{1}{\sigma_\varepsilon^2}\sum_i X_i\{\eta_i - (\alpha_0 + \alpha_1 X_i)\} = 0, \tag{29.44}$$

$$\frac{\partial \log L}{\partial \sigma_\varepsilon} = -\frac{2n}{\sigma_\varepsilon} + \frac{n}{\sigma_\varepsilon^3}(\lambda S_1 + S_2) = 0. \tag{29.45}$$

(29.45) at once gives

$$\sigma_\varepsilon^2 = \tfrac{1}{2}(\lambda S_1 + S_2). \tag{29.46}$$

Summing (29.42) over all values of i, and using (29.43), we obtain

$$\sum_i (\xi_i - X_i) = 0 \tag{29.47}$$

and if we measure the ξ_i from their observed mean, we have from (29.47) the ML estimator of $\sum\limits_i X_i$

$$(\widehat{\sum_i X_i}) = \sum_i \xi_i = 0. \qquad (29.48)$$

Using (29.48) in (29.43), we obtain

$$\sum_i \eta_i = n\alpha_0$$

and if we measure the η_i also from their observed mean, this gives

$$\hat{\alpha}_0 = 0, \qquad (29.49)$$

which reduces (29.46) to

$$\hat{\sigma}_\epsilon^2 = \frac{1}{2n}\{\lambda \sum_i (\xi_i - X_i)^2 + \sum_i (\eta_i - \alpha_1 X_i)^2\}. \qquad (29.50)$$

(29.42) now gives directly, using (29.49),

$$\lambda(\xi_i - X_i) + \hat{\alpha}_1(\eta_i - \hat{\alpha}_1 X_i) = 0$$

or

$$\hat{X}_i = \frac{\lambda \xi_i + \hat{\alpha}_1 \eta_i}{\lambda + \hat{\alpha}_1^2}. \qquad (29.51)$$

Using (29.51) and (29.49) in (29.44), we find

$$\hat{\alpha}_1 = \frac{(\lambda + \hat{\alpha}_1^2)\{\lambda \sum_i \xi_i \eta_i + \hat{\alpha}_1 \sum_i \eta_i^2\}}{\lambda^2 \sum_i \xi_i^2 + \hat{\alpha}_1^2 \sum_i \eta_i^2 + 2\lambda \hat{\alpha}_1 \sum_i \xi_i \eta_i},$$

which simplifies to

$$\hat{\alpha}_1^2 \sum_i \xi_i \eta_i + \hat{\alpha}_1 (\lambda \sum_i \xi_i^2 - \sum_i \eta_i^2) - \lambda \sum_i \xi_i \eta_i = 0. \qquad (29.52)$$

(29.52) is just (29.27) written in a slightly different notation. Thus the result of **29.12**, Case 3, holds good: (29.29) is the ML estimator of α_1 in the linear functional relationship, as well as in the simplest structural relationship, when the error variance ratio λ is known.

29.17 As we remarked at the end of **29.10** (in a discussion which applies here since we estimate α_1 as in Case 3 of **29.12**), the estimated regression lines "bracket" the estimated functional line. This also follows from the fact that $\hat{\alpha}_1$ defined at (29.29) is a monotone function of λ (the proof of this is left to the reader as Exercise 29.1). Thus the estimated regression lines set mathematical limits to the estimated functional line. However, these limits may be too far apart to be of much practical use. In any case, they are not, of course, probabilistic limits of any kind.

29.18 In the presence of incidental parameters, the ML estimators of structural parameters are not necessarily consistent, as Neyman and Scott (1948) showed. More recently, Kiefer and Wolfowitz (1956) have shown that *if the incidental parameters are themselves independent, identically distributed random variables,* and the structural parameters are identifiable, the ML estimators of structural parameters are consistent, under regularity conditions. The italicized condition evidently takes us back from our present functional relationship model to the structural relationship model considered in **29.9-12**, where we derived the ML estimators of α_1 under various assumptions. Neyman (1951) had previously proved the existence of consistent estimators of α_1 in the structural relationship.

29.19 We demonstrate the consistency of $\hat{\alpha}_1$ by observing that from the general results of Chapter 10, the sample variances and covariance in (29.29) converge in probability to their expectations. Thus, if we write the variance of the unobservable X_i as S_X^2, we have (cf. (29.18) for the structural relationship)

$$\left. \begin{array}{l} s_\xi^2 \to S_X^2 + \sigma_\delta^2 = S_X^2 + \dfrac{\sigma_\varepsilon^2}{\lambda}, \\ s_\eta^2 \to \alpha_1^2 S_X^2 + \sigma_\varepsilon^2 = \alpha_1^2 S_X^2 + \lambda \sigma_\delta^2, \\ s_{\xi\eta} \to \alpha_1 S_X^2. \end{array} \right\} \qquad (29.53)$$

Substituting (29.53) in (29.29), we see that

$$\hat{\alpha}_1 \to \left(\{\alpha_1^2 S_X^2 + \lambda \sigma_\delta^2 - \lambda (S_X^2 + \sigma_\delta^2)\} + [\{\alpha_1^2 S_X^2 + \lambda \sigma_\delta^2 - \lambda (S_X^2 + \sigma_\delta^2)\}^2 + 4\lambda (\alpha_1 S_X^2)^2]^{\frac{1}{2}} \right) / \{2\alpha_1 S_X^2\}$$
$$= \alpha_1, \qquad (29.54)$$

which establishes consistency. The same argument holds for the structural relationship with σ_x^2 replacing S_X^2 throughout.

But our troubles are not yet over, for although $\hat{\alpha}_1$ is a consistent estimator of α_1, $\hat{\sigma}_\varepsilon^2$ is not a consistent estimator of σ_ε^2, as Lindley (1947) showed.

Substituting (29.51) into (29.50), we have the alternative forms

$$\hat{\sigma}_\varepsilon^2 = \frac{\lambda}{2(\lambda + \hat{\alpha}_1^2)} \cdot \frac{1}{n} \Sigma_i (\eta_i - \hat{\alpha}_1 \xi_i)^2, \qquad (29.55)$$

$$= \frac{\lambda}{2(\lambda + \hat{\alpha}_1^2)} (s_\eta^2 - 2\hat{\alpha}_1 s_{\xi\eta} + \hat{\alpha}_1^2 s_\xi^2). \qquad (29.56)$$

Using (29.53) and (29.54) in (29.56), we have

$$\hat{\sigma}_\varepsilon^2 \to \frac{\lambda}{2(\lambda + \alpha_1^2)} \left\{ \alpha_1^2 S_X^2 + \sigma_\varepsilon^2 - 2\alpha_1^2 S_X^2 + \alpha_1^2 \left(S_X^2 + \frac{\sigma_\varepsilon^2}{\lambda} \right) \right\} = \frac{1}{2} \sigma_\varepsilon^2. \qquad (29.57)$$

This substantial inconsistency in the ML estimator reminds one of the inconsistency noticed in Example 18.16; the difficulty there was directly traceable to the use of samples of size 2 together with the characteristic bias of order $1/n$ in ML estimators. Here, too, we are essentially estimating σ_ε^2 from the pairs (ξ_i, η_i), as the form (29.55) for $\hat{\sigma}_\varepsilon^2$ makes clear. The inconsistency of the ML estimator is therefore a reflection of the small-sample bias of ML estimators in general. This particular inconsistent estimator causes no difficulty, a consistent estimator of σ_ε^2 being given by replacing the number of observations, $2n$, by the number of degrees of freedom, $2n - (n+2) = n-2$, in the divisor of $\hat{\sigma}_\varepsilon^2$. The consistent estimator is therefore $\dfrac{2n}{n-2} \hat{\sigma}_\varepsilon^2$.

We have thus seen that in the functional relationship, even knowledge of $\lambda = \sigma_\varepsilon^2/\sigma_\delta^2$ is not enough for ML estimators to estimate all structural parameters consistently.

Example 29.1

R. L. Brown (1957) gives 9 pairs of observations

| ξ: | 1·8 | 4·1 | 5·8 | 7·5 | 9·3 | 10·6 | 13·4 | 14·7 | 18·9 |
| η: | 6·9 | 12·5 | 20·0 | 15·7 | 24·9 | 23·4 | 30·2 | 35·6 | 39·1 |

which were generated from a true linear functional relationship $Y = \alpha_0 + \alpha_1 X$ with error variances $\sigma_\delta^2 = \sigma_\varepsilon^2$. Thus we have $\lambda = 1$, $n = 9$, and we compute

$$\Sigma\xi = 86\cdot1, \qquad \Sigma\eta = 208\cdot3$$
$$\bar{\xi} = 9\cdot57, \qquad \bar{\eta} = 23\cdot14,$$

and, rounding to three figures,

$$n s_\xi^2 = 238, \quad n s_\eta^2 = 906, \quad n s_{\xi\eta} = 451.$$

Thus (29.29) gives

$$\hat{\alpha}_1 = \frac{(906-238) + \{(906-238)^2 + 4(451)^2\}^{\frac{1}{2}}}{2 \times 451}$$

$$= \frac{668 + 1122}{902} = 1\cdot99.$$

If we measure from the observed means, therefore, we have $\hat{\alpha}_0 = 0$ by (29.43) and the estimated line is

$$Y - 23\cdot14 = 1\cdot99(X - 9\cdot57)$$

or

$$Y = 1\cdot99 X + 4\cdot01.$$

The consistent estimator of σ_ε^2 is, by **29.19**, $s_\varepsilon^2 = \dfrac{2n}{n-2}\hat{\sigma}_\varepsilon^2$, where $\hat{\sigma}_\varepsilon^2$ is defined at (29.56). We thus have as our estimator in this case

$$s_\varepsilon^2 = \frac{9}{7} \cdot \frac{1}{1+\hat{\alpha}_1^2}(s_\eta^2 - 2\hat{\alpha}_1 s_{\xi\eta} + \hat{\alpha}_1^2 s_\xi^2)$$

$$= \frac{1}{7(1+1\cdot99^2)}\{906 - (3\cdot98 \times 451) + (1\cdot99)^2 238\} = 1\cdot53.$$

In point of fact, the data were generated by adding to the linear functional relationship

$$(Y - \bar{\eta}) = 2(X - \bar{\xi})$$

random normal errors δ, ε with common variance $\sigma_\varepsilon^2 = 1$. Thus the estimators, particularly $\hat{\alpha}_1$, have performed rather well, even with n as low as 9.

Confidence interval estimation and tests

29.20 So far, we have only discussed the point estimation of the parameters. We now consider the question of interval estimation, and turn first to the problem of finding confidence intervals (and the corresponding tests of hypotheses) for α_1 alone, which has been solved by Creasy (1956) in the case where the ratio of error variances λ is known. We can always reduce this to the case $\lambda = 1$ by dividing the observed values of η by $\lambda^{\frac{1}{2}}$. Hence we may without loss of generality consider only the case where the error variances are known to be equal. In this case, the Likelihood Function is

$$L \propto \sigma_\varepsilon^{-2n} \exp\left\{-\frac{1}{2\sigma_\varepsilon^2}\left(\sum_{i=1}^n \delta_i^2 + \sum_{i=1}^n \varepsilon_i^2\right)\right\} \qquad (29.58)$$

whether the relationship is structural or functional. Maximizing (29.58) is the same as minimizing the term in parentheses, which may be rewritten as $\sum_{i=1}^n (\delta_i^2 + \varepsilon_i^2)$. We therefore see, by Pythagoras' theorem, that the ML estimation procedure minimizes the sum of squares of perpendicular distances from the observed points (ξ_i, η_i) to the estimated line. This is intuitively obvious from the equality of the error variances.

We now define

$$\left.\begin{array}{l}\alpha_1 = \tan\theta,\\ \hat{\alpha}_1 = \tan\hat{\theta},\end{array}\right\} \qquad (29.59)$$

and we have at once from (29.29) and the invariance of ML estimators under transformation that the ML estimator of $\tan 2\theta$ is

$$\tan 2\hat{\theta} = \frac{2\tan\hat{\theta}}{1-\tan^2\hat{\theta}} = \frac{2\hat{\alpha}_1}{1-\hat{\alpha}_1^2} = \frac{2s_{\xi\eta}}{|s_\xi^2 - s_\eta^2|}, \qquad (29.60)$$

the modulus in the denominator on the right of (29.60) ensuring that the sign of $\tan 2\hat{\theta}$ is that of $\hat{\alpha}_1$ and $s_{\xi\eta}$.

If and only if $\alpha_1 = 0$, ξ and η are uncorrelated, by (29.18), and since they are normal, this implies that their observed correlation coefficient

$$r = s_{\xi\eta}/(s_\xi s_\eta)$$

will be distributed in the form (16.62), or equivalently, by (16.63), that

$$t = \{(n-2)r^2/(1-r^2)\}^{\frac{1}{2}} \qquad (29.61)$$

is distributed in "Student's" form with $(n-2)$ degrees of freedom. Since

$$\sin^2 2\hat{\theta} = \tan^2 2\hat{\theta}/(1+\tan^2 2\hat{\theta}),$$

(29.61) may be rewritten, using (29.60), as

$$t = \left\{(n-2)\sin^2 2\hat{\theta} \left[\frac{\frac{1}{4}(s_\xi^2 - s_\eta^2)^2 + s_{\xi\eta}^2}{s_\xi^2 s_\eta^2 - s_{\xi\eta}^2}\right]\right\}^{\frac{1}{2}}. \qquad (29.62)$$

The statistic (29.61) or (29.62) may be used to test the hypothesis that $\alpha_1 = \theta = 0$.

29.21 If we wish to test the hypothesis that α_1 takes some non-zero value, a difficulty arises, for the correlation between ξ and η is seen from (29.18) to be, with $\sigma_\delta^2 = \sigma_\varepsilon^2$,

$$\rho = \frac{\alpha_1 \sigma_x^2}{\{(\alpha_1^2 \sigma_x^2 + \sigma_\varepsilon^2)(\sigma_x^2 + \sigma_\varepsilon^2)\}^{\frac{1}{2}}},$$

a function of the unknown σ_x^2 (for which we understand S_X^2 in the functional case, as previously). To remove this difficulty we observe that for any known value α_1, and therefore of θ, we can transform the observed values (ξ_i, η_i) to new values (ξ_i', η_i') by the orthogonal transformation

$$\xi' = \eta\sin\theta + \xi\cos\theta, \qquad \eta' = \eta\cos\theta - \xi\sin\theta,$$

which simply rotates the co-ordinate axes through the angle θ. Thus to test that α_1 takes any specified value, we simply test that $\alpha_1 = \theta = 0$ for the transformed variables (ξ', η'). Since variances and covariances are invariant under orthogonal transformation, this means that (29.62) remains our test statistic, except that in it $\hat{\theta}$ is replaced by $(\hat{\theta} - \theta)$.

There remains the difficulty that to each value of t in (29.62) thus modified, there correspond four values of θ, as a result of the periodicity of the sine function. If we may take it that the probability that $|\hat{\theta} - \theta|$ exceeds $\tfrac{1}{4}\pi$ is negligible, the problem disappears, and we may use (29.62), with $(\hat{\theta} - \theta)$ written for θ, to test any value of $\alpha_1 = \tan \theta$ or to set confidence limits for θ and thence α_1. The confidence limits for θ are, of course, simply

$$\hat{\theta} \pm \tfrac{1}{2} \arcsin \left[2t \left\{ \frac{s_\xi^2 s_\eta^2 - s_{\xi\eta}^2}{(n-2)[(s_\xi^2 - s_\eta^2)^2 + 4s_{\xi\eta}^2]} \right\}^{\tfrac{1}{2}} \right], \qquad (29.63)$$

where t is the appropriate " Student's " deviate for $(n-2)$ degrees of freedom and the confidence coefficient being used. Because of the condition that $|\hat{\theta} - \theta| < \tfrac{1}{4}\pi$, this is essentially an approximate method.

Example 29.2

For the data of Example 29.1, we find

$$\hat{\theta} = \arctan 1 \cdot 99 = 0 \cdot 35\pi$$

and for 7 degrees of freedom and a central confidence interval with coefficient 0·95, we have from a table of " Student's " distribution

$$t = 2 \cdot 36.$$

Thus (29.63) becomes, using the computations for $\hat{\alpha}_1$,

$$0 \cdot 35\pi \pm \tfrac{1}{2} \arcsin \left[4 \cdot 72 \frac{(238 \times 906 - 451^2)^{\tfrac{1}{2}}}{\sqrt{7} \times 1122} \right]$$

$$= 0 \cdot 35\pi \pm \tfrac{1}{2} \arcsin 0 \cdot 1742 = 0 \cdot 35\pi \pm 0 \cdot 03\pi.$$

The 95 per cent confidence limits for θ are therefore $0 \cdot 32\pi$ and $0 \cdot 38\pi$. Those for α_1 are simply the tangents of these angles, namely 1·58 and 2·53. The ML estimate 1·99 is not central between these limits precisely because we had to transform to θ to obtain them. The limits are rather wide apart as a result of the few degrees of freedom available.

29.22 As well as setting confidence limits for α_1 in the manner of the preceding section, we may, as R. L. Brown (1957) pointed out, find a confidence region for the whole line *if the error variances are both known*. For notwithstanding the fact proved at (29.9) that the error term $(\varepsilon - \alpha_1 \delta)$ is correlated with ξ, we may rewrite (29.8) as

$$\eta - (\alpha_0 + \alpha_1 \xi) = \varepsilon - \alpha_1 \delta. \qquad (29.64)$$

The right-hand side of (29.64) is a normally distributed random variable with zero mean and variance $\sigma_\varepsilon^2+\alpha_1^2\sigma_\delta^2$, and the left-hand side of (29.64) contains only the observables ξ, η and the parameters α_0, α_1. We thus have the fact that

$$\frac{\sum_{i=1}^n \{\eta_i-(\alpha_0+\alpha_1\xi_i)\}^2}{\sigma_\varepsilon^2+\alpha_1^2\sigma_\delta^2} \tag{29.65}$$

is distributed in the chi-squared form with n degrees of freedom (χ_n^2). If σ_δ^2 and σ_ε^2 are both known, we may without loss of generality take them both to be equal to unity, since we need only divide ξ_i by σ_δ and η_i by σ_ε to achieve this. (29.65) then becomes the χ_n^2 variate

$$\sum_{i=1}^n \frac{\{\eta_i-(\alpha_0+\alpha_1\xi_i)\}^2}{1+\alpha_1^2}. \tag{29.66}$$

We may use (29.66) to find a confidence region for the line. For if we determine c_γ from tables of χ^2 by

$$P\{\chi_n^2 > c_\gamma\} = 1-\gamma,$$

we have probability γ that

$$\sum_{i=1}^n \{\eta_i-(\alpha_0+\alpha_1\xi_i)\}^2 \leqslant c_\gamma(1+\alpha_1^2). \tag{29.67}$$

Measuring from the observed means $\bar{\xi}$, $\bar{\eta}$ as before, (29.67) becomes

$$s_\eta^2+\alpha_0^2+\alpha_1^2 s_\xi^2-2\alpha_1 s_{\xi\eta} \leqslant c_\gamma(1+\alpha_1^2)$$

or

$$\alpha_1^2(s_\xi^2-c_\gamma)-2\alpha_1 s_{\xi\eta}+\alpha_0^2 \leqslant c_\gamma-s_\eta^2. \tag{29.68}$$

If we take the equality sign in (29.68), it defines a conic in the (α_0, α_1) plane. If c_γ is increased (i.e. γ is increased), the new conic lies inside the previous one. The conic is a $100(1-\gamma)$ per cent confidence region for (α_0, α_1).

This confidence region is bounded if the conic is an ellipse, but unbounded if the conic is a hyperbola. There may, in fact, be no real values of (α_0, α_1) satisfying (29.68). We have already discussed this difficulty in another context in Example 20.5.

We may now, just as in **28.26**, treat (29.68) as a constraint which the true line must satisfy, and then find the envelope of the family of possible lines, which will again be a conic, by differentiation. The result, given by R. L. Brown (1957), is the region in the (X, Y) plane bounded by

$$\frac{(Y-\hat{\alpha}_1 X)^2}{c_\gamma-b_1}-\frac{(\hat{\alpha}_1 Y+X)^2}{b_2-c_\gamma} = 1+\hat{\alpha}_1^2 \tag{29.69}$$

where $\hat{\alpha}_1$ is defined by (29.29) and

$$b_1 = s_\xi^2-\frac{s_{\xi\eta}}{\hat{\alpha}_1}, \quad b_2 = s_\xi^2+\hat{\alpha}_1 s_{\xi\eta}.$$

$\hat{\alpha}_1$ is assumed positive, so that $b_2 > b_1$. (29.69) is the required confidence region, which is a hyperbola if $b_1 < c_\gamma < b_2$, an ellipse if $c_\gamma > b_2$. If $c_\gamma < b_1$, the conic is not real.

29.23 The result at (29.69) is expressed in terms of the estimator (29.29), which may not be the ML estimator when both error variances are known, as was assumed in **29.22**—see the discussion of Case 4 in **29.12**.

Quite apart from this point and the conceptual difficulties arising from confidence regions for the line in the cases stated at the end of **29.22**, another remark is worth making. It is not an efficient procedure to set confidence limits for the mean of a normal distribution with known variance by using the χ_n^2 distribution of the sum of squares about that mean—we remarked this in Example 20.5 and again in Example 25.3, where we showed that the ARE of the corresponding test procedure is zero. We thus should not expect the confidence region (29.69), which is based essentially on this inefficient procedure, to be efficient. It is given here only because better results are not available.

Linear functional relationship between several variables

29.24 We now consider the estimation of a linear relationship in k variables. To make the notation more symmetrical, we will consider the variables $X_1, X_2, \ldots, X_k$ and the dummy variable X_0 ($\equiv 1$) related by the equation

$$\sum_{j=0}^{k} \alpha_j X_j = 0. \tag{29.70}$$

Apart from X_0, the variables are subject to error, so that we observe ξ_j given by

$$\xi_{ji} = X_{ji} + \delta_{ji}, \qquad i = 1, 2, \ldots, n; \; j = 1, 2, \ldots, k. \tag{29.71}$$

Of course, $\xi_{0i} = X_{0i} = 1$ for all i. As before, we assume the δ's normally distributed, independently of X and of each other, with zero means; and we make the situation identifiable by postulating knowledge of the ratios of error variances. If we suppose that

$$\frac{\operatorname{var} \delta_1}{\lambda_1} = \frac{\operatorname{var} \delta_2}{\lambda_2} = \ldots = \frac{\operatorname{var} \delta_k}{\lambda_k}, \tag{29.72}$$

where the λ's are known, we may remove them from the analysis at the outset by dividing the observed ξ_j by $\sqrt{\lambda_j}$. These standardized variables will all have the same unknown variance, say σ_δ^2.

The logarithm of the Likelihood Function of the error variables is then

$$\log L = \text{constant} - nk \log \sigma_\delta - \frac{1}{2\sigma_\delta^2} \sum_{j=1}^{k} \sum_{i=1}^{n} (\xi_{ji} - X_{ji})^2. \tag{29.73}$$

If we regard our data as n points in k dimensions, the problem is to determine the hyperplane (29.70). Maximizing the likelihood is equivalent to minimizing the double sum in (29.73) and this is the sum of squares of distances from the observed points ξ to the estimated points X, as in (29.58). This sum is a minimum when the estimated X's are the feet of perpendiculars from the ξ's on to the hyperplane. Thus the ML estimator of the hyperplane is determined so that the sum of squares of perpendiculars on to it from the observed set of points is a minimum.

29.25 This is a problem of familiar type in many mathematical contexts. The distance of a point $\xi_{1i}, \xi_{2i}, \ldots, \xi_{ki}$ from the hyperplane (29.70) is

$$\sum_{j=0}^{k} \alpha_j \xi_{ji} \bigg/ \left(\sum_{j=0}^{k} \alpha_j^2 \right)^{\frac{1}{2}}.$$

The quantity to be minimized is then

$$S = \sum_{i=1}^{n} \left(\sum_{j=0}^{k} \alpha_j \xi_{ji} \right)^2 \bigg/ \sum_{j=0}^{k} \alpha_j^2.$$

It is convenient to regard this as a minimization of

$$S' = \sum_i \left(\sum \alpha_j \xi_{ji} \right)^2 \qquad (29.74)$$

subject to the constraint

$$\sum \alpha^2 = \text{constant},$$

and we may take the constant to be 1, without loss of generality, by (29.70). We have then to minimize unconditionally

$$\sum_i \left(\sum_j \alpha_j \xi_{ji} \right)^2 - \mu \sum_j \alpha_j^2,$$

where μ is a Lagrange undetermined multiplier. Differentiating with respect to α_l, we have

$$\sum_i \xi_{li} \left(\sum_j \alpha_j \xi_{ji} \right) = \mu \alpha_l, \qquad l = 0, 1, \ldots, k. \qquad (29.75)$$

The first of these equations, with $l = 0$, ($\xi_{0i} = 1$) may be removed if we take an origin at the mean of the ξ's, i.e.

$$\sum_i \xi_{ji} = 0.$$

Writing c_{ij} for the covariance of the ith and jth variate, we then have, from (29.75),

$$\sum_{j=1}^{k} c_{lj} \alpha_j = \frac{\mu}{n} \alpha_l, \qquad l = 1, 2, \ldots, k. \qquad (29.76)$$

Taking the right-hand terms over to the left and eliminating the α's between these equations, we find

$$\begin{vmatrix} c_{11} - \frac{\mu}{n} & c_{12} & c_{13} & \cdots & c_{1k} \\ c_{12} & c_{22} - \frac{\mu}{n} & c_{23} & \cdots & c_{2k} \\ c_{13} & c_{23} & c_{33} - \frac{\mu}{n} & \cdots & c_{3k} \\ \vdots & \vdots & \vdots & \ddots & \vdots \\ c_{1k} & c_{2k} & c_{3k} & \cdots & c_{kk} - \frac{\mu}{n} \end{vmatrix} = 0. \qquad (29.77)$$

If we divide row i by the observed standard deviation of ξ_i, S_i, and column j by S_j, and write r_{ij} for correlations, (29.77) becomes

$$\begin{vmatrix} 1-\theta_1 & r_{12} & r_{13} & \cdots & r_{1k} \\ r_{12} & 1-\theta_2 & r_{23} & \cdots & r_{2k} \\ r_{13} & r_{23} & 1-\theta_3 & \cdots & r_{3k} \\ \vdots & & & \ddots & \vdots \\ r_{1k} & r_{2k} & r_{3k} & \cdots & 1-\theta_k \end{vmatrix} = 0, \qquad (29.78)$$

where $\theta_j = \mu/(nS_j^2)$.

We can solve (29.78) for μ and hence find the α's from (29.76). In actual computational practice it is customary to follow an iterative process which evaluates the α's simultaneously. These solutions for the α's are, of course, the ML estimators of the true values.

Note that (29.78) is an equation of degree k in μ, with k roots (which, incidentally, are always real since the matrix, of which the left-hand side is the determinant, is non-negative definite). We require the smallest of these roots, for if we multiply the lth equation in (29.75) by α_l and add the last k of them, we find that the left-hand side sums to (29.74). Hence

$$S' = \mu \sum_{l=1}^{k} \alpha_l^2 = \mu \qquad (29.79)$$

is to be minimized.

29.26 The method of **29.22** for two variables can be extended to give a quadric surface as confidence region in k dimensions for $\alpha_1, \alpha_2, \ldots, \alpha_k$. Likewise a quadric can be found "within" which should lie the hyperplane (29.70) representing the functional relation. Such regions are, of course, difficult to visualize and impossible to draw for $k > 3$. Reference may be made to R. L. Brown and Fereday (1958) for details. (Some of their results are given in Exercises 29.6–8.) The remarks of **29.23** will apply here also.

> Villegas (1961) considers the case where the error variances are unknown and they are estimated from replicated observations (cf. **29.12**) by ML, and (1964) bases a confidence region for the linear relation on these results. His discussion covers the case of correlated errors. A. P. Basu (1969) extends the results to the case of several linear relations.
>
> Sprent (1966) gives a general method of estimating the coefficients when the errors are correlated.

29.27 So far, we have essentially been considering situations in which identifiability is assured by some knowledge or assumption concerning the error variances, or by replicated observations. The question now arises whether there is any other way of making progress in the problem of estimating a linear functional or structural relationship. Different approaches have been made to this question, which we now consider in turn.

Geary's method of using product-cumulants

29.28 The first method we consider was proposed by Geary (1942b, 1943) in the structural relationship context, but applies also to the functional relationship situation. We write the linear structural relationship in the homogeneous form

$$\alpha_1 x_1 + \alpha_2 x_2 + \ldots + \alpha_k x_k = 0. \tag{29.80}$$

Each of the x_j is subject to an error of observation, δ_j, which is a random variable independent of x_j and the observable is $\xi_j = x_j + \delta_j$. The δ_j are mutually independent. Consider the joint cumulant-generating function of the ξ_j. It will be the sum of the joint c.g.f. of the x_j and that of the δ_j. The product-cumulants of the latter are all zero, by Example 12.7. Thus the product-cumulants of the other two sets, the ξ_j and the x_j, must be identical. If we write κ_x for cumulants of the x's, κ_ξ for cumulants of the ξ's and write the multiple subscripts as arguments in parentheses we have

$$\kappa_x(p_1, p_2, \ldots, p_k) = \kappa_\xi(p_1, p_2, \ldots, p_k), \tag{29.81}$$

provided that at least two p_i exceed zero. Thus the product-cumulants of the x's can be estimated by estimating those of the ξ's.

29.29 The joint c.f. of the x's, measured from their true means, is

$$\phi(t_1, t_2, \ldots, t_k) = E\left\{\exp\left(\sum_{j=1}^{k} \theta_j x_j\right)\right\}, \tag{29.82}$$

where $\theta_j = it_j$. Differentiation of (29.82) with respect to each θ_j yields

$$\sum_{j=1}^{k} \alpha_j \frac{\partial \phi}{\partial \theta_j} = E\left\{\left(\sum_j \alpha_j x_j\right) \exp\left(\sum_j \theta_j x_j\right)\right\} = 0, \tag{29.83}$$

using (29.80). For the c.g.f. $\psi = \log \phi$ also, we have from (29.83)

$$\sum \alpha_j \frac{\partial \psi}{\partial \theta_j} = \frac{1}{\phi} \sum \alpha_j \frac{\partial \phi}{\partial \theta_j} = 0. \tag{29.84}$$

Since, by definition,

$$\psi = \Sigma \kappa(p_1, p_2, \ldots, p_k) \frac{\theta_1^{p_1} \theta_2^{p_2} \ldots \theta_k^{p_k}}{p_1! p_2! \ldots p_k!}$$

we have from (29.84) for all $p_i \geq 0$

$$\alpha_1 \kappa(p_1+1, p_2, \ldots, p_k) + \alpha_2 \kappa(p_1, p_2+1, \ldots, p_k) + \ldots + \alpha_p \kappa(p_1, p_2, \ldots, p_k+1) = 0. \tag{29.85}$$

The relations (29.85) will also be true for the product-cumulants of the observed ξ's, in virtue of (29.81), provided that at least two of the arguments in each cumulant exceed zero, i.e. if two or more $p_i > 0$. In the functional relationship situation, the same argument holds. The random variable x_j, on which n observations are made, is now replaced by a set of n fixed values $X_{j1}, \ldots, X_{jn}$. If this is regarded as a finite population which is exhaustively sampled, our argument remains intact.

29.30 Unfortunately, the method of estimating the α_j from (29.85) (with estimators substituted for the product-cumulants) is completely useless if the x's are jointly normally distributed, the most important case in practice. For the total order of each product-cumulant in (29.85) is $\sum_{i=1}^{k} p_i + 1 \geq 3$ since two or more $p_i > 0$. All cumulants of order ≥ 3 are zero in normal systems, as we have seen in **15.3**. Thus the equations (29.85) are nugatory in this case. This is not at all surprising, for we are dealing here with the unidentifiable situation of **29.9**, and we have made no further assumption to render the situation identifiable.

Even in non-normal cases, there remains the problem to decide which of the relations (29.85) should be used to estimate the k coefficients α_j. We need only k equations, but (assuming that all cumulants exist) have a choice of an infinite number. The obvious course is to use the lowest-order equations, taking the p_i as small as possible, for then the estimation of the product-cumulants in (29.85) will be less subject to sampling fluctuations (cf. **10.8**(e)). However, we must be careful, even in the simplest case, which we now discuss.

29.31 Consider the simplest case, with $k = 2$, which we specified by (29.13). We rewrite this in the form $\alpha_1 x - y = 0$, which is (29.80) with $x \equiv x_1$, $y \equiv x_2$, $\alpha_2 = -1$, $\alpha_0 = 0$ because we are measuring from the means of x and y. (29.85) gives in this case the relations

$$\alpha_1 \kappa(p_1+1, p_2) - \kappa(p_1, p_2+1) = 0$$

or, if $\kappa(p_1+1, p_2) \neq 0$,

$$\alpha_1 = \frac{\kappa(p_1, p_2+1)}{\kappa(p_1+1, p_2)}. \tag{29.86}$$

This holds for any $p_1, p_2 > 0$, and is therefore, as remarked in **29.30**, useless in the normal case. Even if the distribution of the observables (ξ, η) is not normal, its marginal distributions may be symmetrical, and if so all odd-order product-moments and hence product-cumulants will be zero. Thus even in the absence of normality, we must ensure that (p_1+p_2+1) is even in order to guard against the danger of symmetry. The lowest-order generally useful relations are therefore

$$\left. \begin{array}{l} p_1 = 1, \; p_2 = 2: \quad \alpha_1 = \kappa_{13}/\kappa_{22}, \\ p_1 = 2, \; p_2 = 1: \quad \alpha_1 = \kappa_{22}/\kappa_{31}, \end{array} \right\} \tag{29.87}$$

the cumulants being those of (ξ, η), which are to be estimated from the observations.

There remains the question of deciding which of the relations (29.87) to use, or more generally, which combination of them to use. Madansky (1959) suggests finding a minimum variance linear combination, but the algebra would be formidable and not necessarily conclusive in the absence of some assumptions on the parent (ξ, η) distribution.

Even in the absence of symmetry, we may still be unfortunate enough to be sampling a distribution for which the denominator product-cumulant used in (29.87) is equal to zero or nearly so; then we may expect large sampling fluctuations in the estimator.

Example 29.3

Let us reconsider the data of Example 29.1 from our present viewpoint. We find, with $n = 9$,

$$s_{21} = \Sigma(\xi-\bar{\xi})^2(\eta-\bar{\eta}) = 445 \cdot 853 = n\mu_{21},$$
$$s_{12} = \Sigma(\xi-\bar{\xi})(\eta-\bar{\eta})^2 = 542 \cdot 877 = n\mu_{12},$$
$$s_{31} = \Sigma(\xi-\bar{\xi})^3(\eta-\bar{\eta}) = 24{,}635 \cdot 041 = n\mu_{31},$$
$$s_{22} = \Sigma(\xi-\bar{\xi})^2(\eta-\bar{\eta})^2 = 46{,}677 \cdot 679 = n\mu_{22}.$$

Thus (3.81) gives the observed cumulants

$$\kappa_{21} = \mu_{21} = 49 \cdot 539; \quad \kappa_{12} = \mu_{12} = 60 \cdot 320;$$
$$\kappa_{31} = \mu_{31} - 3\mu_{20}\mu_{11} = -1232 \cdot 45;$$
$$\kappa_{22} = \mu_{22} - \mu_{20}\mu_{02} - 2\mu_{11}^2 = -2493 \cdot 613.$$

Using these values in equation (29.86) we find the estimate of α_1:

$$p_1 = 1, \; p_2 = 1: \; \frac{\kappa_{12}}{\kappa_{21}} = \frac{60 \cdot 320}{49 \cdot 539} = 1 \cdot 22, \tag{29.88}$$

while from the second equation in (29.87), we have the much closer estimate

$$p_1 = 2, \; p_2 = 1: \; \frac{\kappa_{22}}{\kappa_{31}} = \frac{-2493 \cdot 613}{-1232 \cdot 45} = 2 \cdot 02. \tag{29.89}$$

It might be considered preferable to use k-statistics instead of cumulants in these equations. From **13.2** we have, since we are using central moments,

$$k_{21} = \frac{ns_{21}}{(n-1)(n-2)},$$
$$k_{12} = \frac{ns_{12}}{(n-1)(n-2)},$$
$$k_{31} = \frac{n(n+1)s_{31} - 3(n-1)s_{11}s_{20}}{(n-1)(n-2)(n-3)},$$
$$k_{22} = \frac{n(n+1)s_{22} - 2(n-1)s_{11}^2 - (n-1)s_{20}s_{02}}{(n-1)(n-2)(n-3)}.$$

The use of k-statistics rather than sample cumulants as estimators therefore makes no difference to the estimate (29.88). We find

$$k_{31} = -1057 \cdot 19, \quad k_{22} = -2308 \cdot 79,$$

and the estimate (29.89) is now replaced by $\dfrac{-2308 \cdot 79}{-1057 \cdot 19} = 2 \cdot 18.$ \hfill (29.90)

It will be remembered that these data were actually generated from random normal deviates. It is not surprising, therefore, that the estimate (29.88) is so wide of the mark. (The ML estimator in Example 29.1 was 1·99.) The remarks in **29.30–1** would lead us to expect this estimator to behave very wildly in the normal case, since it is essentially estimating a ratio of zero quantities.

It will be noticed that (29.89) is slightly closer to the ML estimator than the apparently more refined (29.90). This "refinement" is illusory, for although the k-statistics are unbiassed estimators of the cumulants, we are here estimating a *ratio* of cumulants. Both estimators are biassed; (29.89) is slightly simpler to compute.

The reader may like to verify that if the first equation in (29.87) is used we find $\mu_{13} = 10,003$, $\kappa_{13} = -5131$ and thus the estimate $\kappa_{13}/\kappa_{22} = 2\cdot06$, very close to (29.89).

In large samples from a normal system, none of our estimators would be at all reliable.

29.32 We conclude that the product-cumulant method of estimating α_1, while it is free from additional assumptions, is vulnerable in a rather unexpected way. It always estimates α_1 by a ratio of cumulants, and if the denominator cumulant is zero, or near zero, we must expect sharp fluctuation in the estimator. This is not a phenomenon which disappears as sample size increases—indeed it may get worse.

The use of supplementary information: instrumental variables

29.33 Suppose now that, when we observe ξ and η, we also observe a further variable ζ, which is correlated with the unobservable true value x but not with the errors of observation. The observations on ζ clearly furnish us with supplementary information about x which we may turn to good account. ζ is called an *instrumental* variable, because it is used merely as an instrument in the estimation of the relationship between y and x. We measure ξ, η and ζ from their observed means.

Consider the estimator of α_1

$$a_1 = \sum_{i=1}^{n} \zeta_i \eta_i \bigg/ \sum_{i=1}^{n} \zeta_i \xi_i, \qquad (29.91)$$

which we write in the form

$$a_1 \sum_{i=1}^{n} \zeta_i \xi_i = \sum_{i=1}^{n} \zeta_i \eta_i,$$

or, on substitution for η and ξ,

$$a_1 \sum_i \zeta_i (x_i + \delta_i) = \sum_i \zeta_i (\alpha_0 + \alpha_1 x_i + \varepsilon_i). \qquad (29.92)$$

Each of the sample covariances in (29.92) will converge in probability to its expectation. Thus, since ζ is uncorrelated with δ and ε, we obtain from (29.92)

$$a_1 \operatorname{cov}(\zeta, x) \to \alpha_1 \operatorname{cov}(\zeta, x). \qquad (29.93)$$

If and only if

$$\lim_{n \to \infty} \operatorname{cov}(\zeta, x) \neq 0, \qquad (29.94)$$

(29.93) gives

$$a_1 \to \alpha_1, \qquad (29.95)$$

so that a_1 is a consistent estimator. It will be seen that nothing has been assumed about the instrumental variable ζ beyond its correlation with x and its lack of correlation with the errors. In particular, it may be a discrete variable. Exercise 29.17 gives an indication of how efficient a_1 is. See also Exercises 29.15–16.

29.34 Whatever the form of the instrumental variable, it not only enables us to estimate α_1 consistently by (29.91) but also to obtain confidence regions for (α_0, α_1), as Durbin (1954) pointed out.

The random variable $(\eta - \alpha_0 - \alpha_1 \xi) = \varepsilon - \alpha_1 \delta$ by (29.8). Since ζ is uncorrelated with δ and with ε, it is uncorrelated with $(\eta - \alpha_0 - \alpha_1 \xi)$. It follows (cf. **26.23**(a)) that, given α_0 and α_1, the observed correlation r between ζ and $(\eta - \alpha_0 - \alpha_1 \xi)$ is distributed so that

$$t^2 = (n-2)r^2/(1-r^2) \tag{29.96}$$

has a "Student's" t^2-distribution with $(n-2)$ degrees of freedom. If we denote by $t^2_{1-\gamma}$ the value of such a variate satisfying

$$P\{t^2 \leqslant t^2_{1-\gamma}\} = 1-\gamma,$$

we have, since $r^2 = t^2/\{t^2 + (n-2)\}$, a monotone increasing function of t^2,

$$P\left\{\frac{[\Sigma \zeta(\eta-\alpha_0-\alpha_1\xi)]^2}{\Sigma \zeta^2 \Sigma(\eta-\alpha_0-\alpha_1\xi)^2} \leqslant r^2_{1-\gamma}\right\} = 1-\gamma$$

or

$$P\left\{\frac{(\Sigma \zeta\eta)^2 - 2\alpha_1 \Sigma \zeta\eta \Sigma \zeta\xi + \alpha_1^2 (\Sigma \zeta\xi)^2}{\Sigma \zeta^2 (\Sigma \eta^2 + n\alpha_0^2 - 2\alpha_1 \Sigma \eta\xi + \alpha_1^2 \Sigma \xi^2)} \leqslant r^2_{1-\gamma}\right\} = 1-\gamma. \tag{29.97}$$

It will be seen that (29.97) depends only on α_0 and α_1, apart from the observables ζ, η, ξ. It defines a quadratic confidence region in the (α_0, α_1) plane, with confidence coefficient $1-\gamma$. If α_0 is known, (29.97) gives a confidence interval for α_1, but t^2 now has $(n-1)$ degrees of freedom, since only one parameter is now involved. We shall see later that for particular instrumental variables, confidence intervals for α_1 may be obtained even when α_0 is unknown.

29.35 The general difficulty in using instrumental variables is the practical one of finding a random variable known to be correlated with x and known not to be correlated with δ and with ε: we rarely know enough of a system to be sure that these conditions are satisfied. However, if we use as instrumental variable a discrete "grouping" variable (i.e. we classify the observations according to whether they fall into certain discrete groups, and treat this classification as a discrete-valued variable) we have more hope of satisfying the conditions. For we may know from the nature of the situation that the observations come from several distinct groups, which materially affect the true values of x; while the errors of observation have no connexion with this classification at all. For example, referring to the pressure-volume relationship discussed in **29.1**, suppose that (29.2) were believed to hold. If we take logarithms, the relationship becomes

$$\log P = C - \gamma \log V,$$

precisely the form we have been discussing, with $\alpha_0 = C$ and $\alpha_1 = \gamma$. But see **29.55**.

Suppose now that we knew that the determinations of volume had been made sometimes by one method, sometimes by another; and suppose it is known that Method 1 produces a slightly different result from Method 2. The Method 1–Method 2 classification will then be correlated with the volume determination. The errors in this determination, and certainly those in the pressure determination (which is supposed to be made in the same way for all observations), may be quite uncorrelated with the Method classification. Thus we have an instrumental variable of a special kind, essentially a grouping into two groups.

We now discuss instrumental variables of this grouping kind in some detail.

Two groups of equal size

29.36 Suppose that n, the number of observations, is even, and that we divide them into two equal groups of $\frac{1}{2}n = m$ observations each. (We shall discuss how the allocation to groups is to be made in a moment.) Let $\bar{\xi}$ be the mean observed ξ in the first group and $\bar{\xi}'$ that for the second group, and similarly define $\bar{\eta}$ and $\bar{\eta}'$. Then we may estimate α_1 by putting an instrumental variable ζ equal to $+1$ for each observation in the first group and -1 for each observation in the second group. (29.91) then becomes

$$a_1 = (\bar{\eta}' - \bar{\eta})/(\bar{\xi}' - \bar{\xi}), \tag{29.98}$$

and using (29.98) in (29.19) or (29.43), we estimate α_0 by

$$2a_0 = (\bar{\eta}' + \bar{\eta}) - a_1(\bar{\xi}' + \bar{\xi}). \tag{29.99}$$

Geometrically, this procedure means that in the (ξ, η) plane we divide the points into equal groups according to the value of ξ, and determine the centre of gravity of each group. The slope of the true linear relationship is then estimated by that of the join of these centres of gravity.

Wald (1940), to whom these estimators are due, showed that a_1 defined at (29.98) is a consistent estimator of α_1 if the grouping is independent of the errors and if the true x-values satisfy

$$\lim_{n \to \infty} \inf |\bar{x}' - \bar{x}| > 0, \tag{29.100}$$

which is (29.94) again, since here $\text{cov}(\zeta, x) = \bar{x}' - \bar{x}$. (29.100) clearly will not be satisfied if the observations are randomly allocated to the two groups, when $\lim_{n \to \infty} |\bar{x}' - \bar{x}| = 0$. (Cf. Exercise 28.16 for the simple linear model.) Nor is it satisfactory to allocate the m smallest observed ξ's to one group and the m largest to the other—Neyman and Scott (1951) show that in this case the estimator will not be consistent. (It is easy to see that the grouping is not now, in general, independent of the errors.) It follows that Wald's method is only of interest if we have prior information (like that mentioned in **29.35**) to validate (29.100).

29.37 We may use the estimator (29.98) to obtain estimators of the two error variances. For since, by (29.18),

$$\left. \begin{array}{l} \sigma_\delta^2 = \text{var}\,\xi - \dfrac{\text{cov}(\xi, \eta)}{\alpha_1}, \\[2mm] \sigma_\varepsilon^2 = \text{var}\,\eta - \alpha_1 \text{cov}(\xi, \eta), \end{array} \right\} \tag{29.101}$$

we need only substitute the consistent estimators s_ξ^2, s_η^2 and $s_{\xi\eta}$ for the variances and covariances (multiplying each by $\frac{n}{n-1}$ to remove bias), and a_1 for α_1, to obtain the estimators

$$\left.\begin{array}{l} s_\delta^2 = \dfrac{n}{n-1}\left(s_\xi^2 - \dfrac{s_{\xi\eta}}{a_1}\right), \\ s_\varepsilon^2 = \dfrac{n}{n-1}(s_\eta^2 - a_1 s_{\xi\eta}). \end{array}\right\} \quad (29.102)$$

Example 29.4

Let us apply this method purely illustratively to the data of Example 29.1. There are 9 observations, so we omit that with the median value of ξ, and take our two groups to be:

$$\xi: \quad 1\cdot 8 \quad 4\cdot 1 \quad 5\cdot 8 \quad 7\cdot 5\,; \quad 10\cdot 6 \quad 13\cdot 4 \quad 14\cdot 7 \quad 18\cdot 9$$
$$\eta: \quad 6\cdot 9 \quad 12\cdot 5 \quad 20\cdot 0 \quad 15\cdot 7\,; \quad 23\cdot 4 \quad 30\cdot 2 \quad 35\cdot 6 \quad 39\cdot 1.$$

We find

$$\bar{\xi} = 19\cdot 2/4 = 4\cdot 800\,; \quad \bar{\xi}' = 57\cdot 6/4 = 14\cdot 400$$
$$\bar{\eta} = 55\cdot 1/4 = 13\cdot 775\,; \quad \bar{\eta}' = 128\cdot 3/4 = 32\cdot 075.$$

The estimate is

$$a_1 = \frac{32\cdot 075 - 13\cdot 775}{14\cdot 400 - 4\cdot 800} = 1\cdot 91,$$

reasonably close to the true value 2.

For these 8 observations, we find

$$s_\xi^2 = 29\cdot 735, \quad s_\eta^2 = 112\cdot 709, \quad s_{\xi\eta} = 56\cdot 764.$$

Substituting in (29.102), we find the estimates

$$s_\delta^2 = -0\cdot 054,$$
$$s_\varepsilon^2 = 5\cdot 16.$$

These are very bad estimates, the true values being unity; s_δ^2 is actually negative and therefore " impossible."

Inaccurate estimates of the error variances are quite likely to appear with these estimators, as we may easily see. If the true values (x, y) are widely spaced relative to the errors of observation, the observed values (ξ, η) will be highly correlated, their two regression lines will be close to each other, and a_1 will then be close to the regression coefficient of η on ξ, $s_{\xi\eta}/s_\xi^2$, and to the reciprocal of the regression coefficient of ξ on η, $s_{\xi\eta}/s_\eta^2$. Thus, from (29.102), both s_δ^2 and s_ε^2 will be near zero, and quite small variations in a_1 will alter them substantially. In our present example, the correlation between ξ and η is 0·98, and even the small deviation of a_1 from the true value α_1 is enough to swing s_δ^2 violently downwards and s_ε^2 violently upwards.

29.38 A confidence interval for α_1 was also obtained by Wald (1940). For each of the two groups, we compute sums of squares and products about its own means, and define the pooled estimators, each therefore based on $(m-1)+(m-1) = n-2$ degrees of freedom,

$$\left.\begin{aligned} S_\xi^2 &= \frac{1}{n-2}\left\{\sum_{i=1}^{m}(\xi_i-\bar{\xi})^2 + \sum_{i=1}^{m}(\xi_i'-\bar{\xi}')^2\right\}, \\ S_\eta^2 &= \frac{1}{n-2}\left\{\sum_{i=1}^{m}(\eta_i-\bar{\eta})^2 + \sum_{i=1}^{m}(\eta_i'-\bar{\eta}')^2\right\}, \\ S_{\xi\eta} &= \frac{1}{n-2}\left\{\sum_{i=1}^{m}(\xi_i-\bar{\xi})(\eta_i-\bar{\eta}) + \sum_{i=1}^{m}(\xi_i'-\bar{\xi}')(\eta_i'-\bar{\eta}')\right\}. \end{aligned}\right\} \quad (29.103)$$

These three quantities, in normal variation, are distributed independently of the means $\bar{\xi}$, $\bar{\xi}'$, $\bar{\eta}$, $\bar{\eta}'$, and therefore of the estimator (29.98). In (29.101), we substitute (29.103) to obtain the random variables, still functions of α_1,

$$\left.\begin{aligned} S_\delta^2 &= S_\xi^2 - S_{\xi\eta}/\alpha_1, \\ S_\varepsilon^2 &= S_\eta^2 - \alpha_1 S_{\xi\eta}. \end{aligned}\right\} \quad (29.104)$$

Now consider
$$S^2 = S_\varepsilon^2 + \alpha_1^2 S_\delta^2 = S_\eta^2 + \alpha_1^2 S_\xi^2 - 2\alpha_1 S_{\xi\eta}$$
$$= \frac{1}{n-2}\left[\sum_{i=1}^{m}\{(\eta_i-\alpha_0-\alpha_1\xi_i)-(\bar{\eta}-\alpha_0-\alpha_1\bar{\xi})\}^2\right.$$
$$\left.+\sum_{i=1}^{m}\{(\eta_i'-\alpha_0-\alpha_1\xi_i')-(\bar{\eta}'-\alpha_0-\alpha_1\bar{\xi}')\}^2\right]. \quad (29.105)$$

$(n-2)S^2$ is seen to be the sum of two sums of squares; each of these is itself the sum of squares of m independent normal variables $(\eta_i-\alpha_0-\alpha_1\xi_i)$ about their mean, and from (29.8) we see that each of these has variance $\sigma_\varepsilon^2+\alpha_1^2\sigma_\delta^2$. Thus

$$\frac{(n-2)S^2}{\sigma_\varepsilon^2+\alpha_1^2\sigma_\delta^2}$$

has a χ^2 distribution with $(n-2)$ degrees of freedom. We also define

$$u = \tfrac{1}{2}(\bar{\xi}'-\bar{\xi})(a_1-\alpha_1) = \tfrac{1}{2}\{(\bar{\eta}'-\bar{\eta})-\alpha_1(\bar{\xi}'-\bar{\xi})\}$$
$$= \tfrac{1}{2}\{(\bar{\eta}'-\alpha_0-\alpha_1\bar{\xi}')-(\bar{\eta}-\alpha_0-\alpha_1\bar{\xi})\} = \tfrac{1}{2}\{(\bar{\varepsilon}'-\alpha_1\bar{\delta}')-(\bar{\varepsilon}-\alpha_1\bar{\delta})\}.$$

The two components on the extreme right, being functions of the error means in the separate groups, are independently distributed. We thus see that u is normally distributed with zero mean and variance $\tfrac{1}{4}\cdot 2\dfrac{(\sigma_\varepsilon^2+\alpha_1^2\sigma_\delta^2)}{m} = \dfrac{1}{n}(\sigma_\varepsilon^2+\alpha_1^2\sigma_\delta^2)$. Moreover, u is a function only of $\bar{\xi}'$, $\bar{\xi}$, $\bar{\eta}'$ and $\bar{\eta}$, and is therefore distributed independently of S^2. Thus

$$t = \frac{un^{\frac{1}{2}}}{S} = \frac{(\bar{\xi}'-\bar{\xi})(a_1-\alpha_1)n^{\frac{1}{2}}}{2(S_\eta^2-2\alpha_1 S_{\xi\eta}+\alpha_1^2 S_\xi^2)^{\frac{1}{2}}} \quad (29.106)$$

has a "Student's" distribution with $(n-2)$ degrees of freedom. For any given confidence coefficient $1-\gamma$, we have

$$P\{t^2 \leq t_{1-\gamma}^2\} = 1-\gamma. \quad (29.107)$$

The extreme values of α_1 for which (29.107) is satisfied are, from (29.106), the roots of

$$(\bar{\xi}'-\bar{\xi})^2(a_1-\alpha_1)^2 = \frac{4t_{1-\gamma}^2}{n}(S_\eta^2 - 2\alpha_1 S_{\xi\eta} + \alpha_1^2 S_\xi^2)$$

or

$$\alpha_1^2 \left\{ \frac{4t_{1-\gamma}^2}{n} S_\xi^2 - (\bar{\xi}'-\bar{\xi})^2 \right\} + 2\alpha_1 \left\{ a_1(\bar{\xi}'-\bar{\xi})^2 - \frac{4t_{1-\gamma}^2}{n} S_{\xi\eta} \right\}$$
$$+ \left\{ \frac{4t_{1-\gamma}^2}{n} S_\eta^2 - a_1^2(\bar{\xi}'-\bar{\xi})^2 \right\} = 0, \qquad (29.108)$$

a quadratic equation in α_1 of which the discriminant is

$$\left(\frac{4t_{1-\gamma}^2}{n}\right)^2 (S_{\xi\eta}^2 - S_\xi^2 S_\eta^2) + \frac{4t_{1-\gamma}^2}{n}(a_1^2 S_\xi^2 - 2a_1 S_{\xi\eta} + S_\eta^2). \qquad (29.109)$$

The first term in (29.109) is negative, by Cauchy's inequality, and the second term positive, since its factor in brackets is $\frac{1}{n-2}\sum_i(\eta_i - a_i\xi_i)^2$. If n is large enough, the positive term, which has a multiplier $\frac{4t_{1-\gamma}^2}{n}$, will be greater than the negative, with a multiplier $\left(\frac{4t_{1-\gamma}^2}{n}\right)^2$. Then the quadratic (29.108) will have two real roots, which are the confidence limits for α_1.

29.39 Similarly, we may derive a confidence region for (α_0, α_1). From (29.99), we estimate α_0 by a_0. Consider the variable

$$v = a_0 - \alpha_0 = (\bar{\eta}' + \bar{\eta}) - \alpha_0 - \alpha_1(\bar{\xi}' + \bar{\xi}). \qquad (29.110)$$

v is normally distributed, with zero mean and variance

$$\frac{1}{n}\operatorname{var}(\eta - \alpha_0 - \alpha_1 \xi) = \frac{1}{n}(\sigma_\varepsilon^2 + \alpha_1^2 \sigma_\delta^2),$$

i.e. its variance is the same as that of u in **29.38**. v, like u, is easily seen to be distributed independently of S^2, so that if we substitute v for u in (29.106), we still have a "Student's" t variable with $(n-2)$ degrees of freedom. If α_1 is known, we may use this variable to set confidence intervals for α_0, the process being simple in this case, since α_0 appears only in the numerator of t. However, this is of little practical importance, since we rarely know α_1 and not α_0.

But we may also see that u and v are independently distributed. To establish this we have, by the definitions of u and v, only to show that

$$2u = (\bar{\eta}' - \bar{\eta}) - \alpha_1(\bar{\xi}' - \bar{\xi})$$

is independent of

$$\alpha_0 + v = (\bar{\eta}' + \bar{\eta}) - \alpha_1(\bar{\xi}' + \bar{\xi}).$$

These two variables are normally distributed, the first of them with zero mean. Their covariance is

$$E(\bar{\eta}' + \bar{\eta})(\bar{\eta}' - \bar{\eta}) + \alpha_1^2 E(\bar{\xi}' + \bar{\xi})(\bar{\xi}' - \bar{\xi}) - 2\alpha_1 E(\bar{\eta}'\,\bar{\xi}' - \bar{\eta}\,\bar{\xi}).$$

Each of the first two expectations is that of a difference between identically distributed squares, and is therefore zero. The third expectation is a difference of identically distributed products, and is also zero. Thus the covariance is zero, and these variables are independent. Hence u and v are independent.

It now follows that $\dfrac{u^2+v^2}{\dfrac{1}{n}(\sigma_\varepsilon^2+\alpha_1^2\sigma_\delta^2)}$ is a χ^2 variate with 2 degrees of freedom and hence that

$$F = \frac{\frac{1}{2}(u^2+v^2)}{S^2/n} \qquad (29.111)$$

is distributed in the variance-ratio distribution with $2, n-2$ degrees of freedom. From this, we may obtain a confidence region for α_0 and α_1, which is (cf. Exercise 29.5) an ellipse, as we should expect from the independence and normality of u and v, which are linear functions respectively of α_1 and α_0.

This confidence region is not equivalent to that obtained by putting the instrumental variable $\zeta = \pm 1$ in (29.97). Our present region is based on the distribution of $F = \dfrac{n}{2} \cdot \dfrac{(u^2+v^2)}{S^2}$, but the random variable in (29.97) has a numerator depending only on u^2 and is not a monotone function of F. Intuitively, the latter seems likely to give a better interval, but we know of no result to this effect.

Three groups

29.40 It was pointed out by Nair and Shrivastava (1942) and by Bartlett (1949) that the efficiency of the grouping method may be increased by using three groups instead of two, and estimating α_1 by the slope of the line joining the centres of gravity of the two extreme groups. (We have already done this implicitly in Example 29.4, where we omitted the central observation in order to carry out a two-group analysis.) The three-group method may be formulated as follows.

We divide the n observations into three groups, the first group containing np_1 observations, and the third group np_2 observations. p_1 and p_2 are proportions, the choice of which is to be discussed below. The two-group method is a special case with $p_1 = p_2 = \frac{1}{2}$ when n is even (when the middle group is empty) and $p_1 = p_2 = \dfrac{\frac{1}{2}(n-1)}{n}$ when n is odd (as in Example 29.4). The grouping now corresponds to an instrumental variable ζ in (29.91) taking values $+1, 0$ and -1 for the third, second and first groups respectively. The estimator is

$$a_1 = (\bar{\eta}' - \bar{\eta})/(\bar{\xi}' - \bar{\xi}) \qquad (29.112)$$

as before, but the primed symbols now refer to the third group, and the unprimed symbols to the first group. The estimator is consistent under the same condition as before.

Nair and Shrivastava (1942) and Bartlett (1949) studied the case $p_1 = p_2 = \frac{1}{3}$. In this case, as in **29.38**, we define S_ξ^2, S_η^2 and $S_{\xi\eta}$ in (29.103) by pooling the observed variances and covariances within the three groups, but now dividing by $(n-3)$, the number of degrees of freedom in the present case. (29.104) then defines S_δ^2 and S_ε^2 as before and $S^2 = S_\varepsilon^2 + \alpha_1^2 S_\delta^2$. $(n-3)S^2/(\sigma_\varepsilon^2 + \alpha_1^2 \sigma_\delta^2)$ is a χ^2 variate with $(n-3)$ degrees of freedom. In this case

$$(\bar{\xi}' - \bar{\xi})(a_1 - \alpha_1)\left(\frac{n}{6}\right)^{\frac{1}{2}}$$

will be a normal variate distributed independently of S^2, with zero mean and variance $\sigma_\varepsilon^2 + \alpha_1^2 \sigma_\delta^2$. Thus the analogue of (29.106) is

$$t = \left(\frac{n}{6}\right)^{\frac{1}{2}} \frac{(\bar{\xi}' - \bar{\xi})(a_1 - \alpha_1)}{S}, \qquad (29.113)$$

distributed in "Student's" distribution with $(n-3)$ degrees of freedom, and we set confidence intervals from (29.113) as before.

The results of **29.39** extend similarly to the three-group case.

29.41 The optimum choice of p_1 and p_2 has been investigated for various distributions of x, *assumed free from error*. Bartlett's (1949) result in the rectangular case is given as Exercise 29.11. Other results are given by Theil and van Yzeren (1956) and by Gibson and Jowett (1957). Summarized, the results indicate that for a rather wide range of symmetrical distributions for x, we should take $p_1 = p_2 = \frac{1}{3}$ approximately, the efficiency achieved compared with the minimum variance LS estimator being of the order of 80 or 85 per cent.

The evidence of the relative efficiency of the two- and three-group methods in the presence of errors of observation is limited and indecisive. Nair and Banerjee (1942) found the three-group method more efficient in a sampling experiment. An example given by Madansky (1959) leads strongly to the opposite conclusion.

Richardson and Wu (1970) compare the grouping method generally with the LS estimator. Their overall recommendation is that it is safest to use at least 15 groups to avoid the possibility of a large increase in m.s.e.

Example 29.5

Applied to the data of Example 29.1, the method with $p_1 = p_2 = \frac{1}{3}$ gives 3 observations in each group. We find

$$\bar{\xi}' = 15.67, \quad \bar{\xi} = 3.90,$$
$$\bar{\eta}' = 34.97, \quad \bar{\eta} = 13.13,$$

whence

$$a_1 = \frac{34.97 - 13.13}{15.67 - 3.90} = 1.86,$$

close to the value 1.91 obtained by the two-group method in Example 29.4, but actually a little further from the true value, 2.

Halperin (1961b) develops a generalization of (29.108) and (29.111) which requires, instead of a grouping method, the guessing of the x_i individually to permit maximization of the probability of obtaining a closed confidence interval or region; this is particularly advantageous for small n.

The use of ranks

29.42 To conclude our discussion of grouping and instrumental variable methods, we discuss the use of ranks. Suppose that we can arrange the individual observations in their true order according to the observed value of one of the variables. We now suppose, not merely that two or three groups can be so arranged, but that the values of x are so far spread out compared with error variances that the series of observed ξ's is in the same order as the series of unobserved x's. We now take suffixes as referring to the ordered observations. Again we make the usual assumptions about the independence of the errors and consider an even number of values $2m = n$. To any pair ξ_i, η_i there is a corresponding pair ξ_{m+i}, η_{m+i} and we can form an estimator of α_1 from each of the m statistics

$$a(i) = \frac{\eta_{m+i} - \eta_i}{\xi_{m+i} - \xi_i}, \qquad i = 1, 2, \ldots, m, \qquad (29.114)$$

and we may choose either their mean or their median as an estimator of α_1.

Alternatively, we could consider all possible pairs of values

$$a(i,j) = \frac{\eta_i - \eta_j}{\xi_i - \xi_j}, \qquad i,j = 1, 2, \ldots, n. \qquad (29.115)$$

There are $\frac{1}{2}n(n-1)$ of these values, and again we could estimate α_1 from their mean or median.

These methods, due to Theil (1950), obviously use more information than the grouping methods discussed earlier. The advantage of using the median rather than the mean resides in the fact that from median estimators it is fairly easy to construct confidence intervals, as we shall see in **29.43**.

Example 29.6

Reverting once more to the data of Examples 29.4 with the middle value omitted, we find for the four values of $a(i)$ in (29.114),

$$\frac{23\cdot 4 - 6\cdot 9}{10\cdot 6 - 1\cdot 8} = 1\cdot 875, \qquad \frac{30\cdot 2 - 12\cdot 5}{13\cdot 4 - 4\cdot 1} = 1\cdot 903,$$

$$\frac{35\cdot 6 - 20\cdot 0}{14\cdot 7 - 5\cdot 8} = 1\cdot 753, \qquad \frac{39\cdot 1 - 15\cdot 7}{18\cdot 9 - 7\cdot 5} = 1\cdot 887.$$

The median (half-way between the two middle values) is 1·88. The mean is 1·85.

If we use (29.115), we can use all nine observations. There are 36 values of $a(i,j)$ which, in order of magnitude, are $-2\cdot 529$, $-1\cdot 154$, $0\cdot 708$, $0\cdot 833$, $0\cdot 941$, $1\cdot 293$, $1\cdot 342$, $1\cdot 400$, $1\cdot 458$, $1\cdot 479$, $1\cdot 544$, $1\cdot 618$, $1\cdot 677$, $1\cdot 753$, $1\cdot 797$, $1\cdot 875$, $1\cdot 883$, $1\cdot 892$, $1\cdot 903$, $1\cdot 981$, $2\cdot 009$, $2\cdot 053$, $2\cdot 179$, $2\cdot 225$, $2\cdot 385$, $2\cdot 400$, $2\cdot 429$, $2\cdot 435$, $2\cdot 458$, $2\cdot 484$, $2\cdot 764$, $2\cdot 976$, $3\cdot 275$, $4\cdot 154$, $4\cdot 412$, $5\cdot 111$. The median value (half-way between the 18th and 19th values) is 1·90. The mean is 1·93.

29.43 We now relax the normality assumptions on the errors and impose a milder condition on the term $(\varepsilon_i - \alpha_1 \delta_i)$, namely, that it shall have the same continuous distribution for all i. In the terminology of **28.7**, we have identical errors in $\eta - \alpha_0 - \alpha_1 \xi$, together with continuity. It then follows that the probability of one value, say $\varepsilon_i - \alpha_1 \delta_i$, exceeding another, $\varepsilon_{m+i} - \alpha_1 \delta_{m+i}$, is $\frac{1}{2}$.

Since from (29.114)

$$a(i) = \frac{\eta_{m+i} - \eta_i}{\xi_{m+i} - \xi_i} = \alpha_1 + \frac{(\eta_{m+i} - \alpha_1 \xi_{m+i}) - (\eta_i - \alpha_1 \xi_i)}{\xi_{m+i} - \xi_i},$$

we have

$$a(i) - \alpha_1 = \frac{(\varepsilon_{m+i} - \alpha_1 \delta_{m+i}) - (\varepsilon_i - \alpha_1 \delta_i)}{\xi_{m+1} - \xi_i}.$$

The denominator $\xi_{m+1} - \xi_i$ is positive, and consequently the probability that $a(i) - \alpha_1 > 0$ is $\frac{1}{2}$. Thus the probability that exactly j of the $(a(i) - \alpha_1)$ exceed zero, i.e. $\alpha_1 < a(i)$, is given binomially as $\binom{m}{j}\frac{1}{2^m}$, so that the probability that the r greatest $a(i)$ exceed α_1 and the r smallest $a(i)$ are less than α_1 is

$$P\{a(r) < \alpha_1 < a(m-r+1)\} = 1 - 2 \sum_{j=0}^{r} \binom{m}{j}\frac{1}{2^m}, \qquad (29.116)$$

which may be expressed in terms of the Incomplete Beta Function by 5.7 if desired. This is a confidence interval for α_1.

29.44 If, in addition, we assume that δ and ε have zero medians, we have

$$P\{\eta_i - \alpha_1 \xi_i > \alpha_0\} = \tfrac{1}{2}.$$

Given any α_1 we can arrange the quantities $\eta_i - \alpha_1 \xi_i$, say z_i, in order of magnitude, and in the same manner we have

$$P\{z_r < \alpha_0 < z_{m-r+1}\} = 1 - 2 \sum_{j=0}^{r} \binom{m}{j}\frac{1}{2^m}. \qquad (29.117)$$

It does not appear possible by this method to give joint confidence intervals for α_0 and α_1 together, except with an upper bound to the confidence coefficient (cf. Exercise 29.10). Exercise 29.9 indicates a test of linearity.

The use of (29.115), when all pairs are considered, is more complicated, the distributions no longer being binomial. They are, in fact, those required for the distribution of a rank correlation coefficient, t, which we discuss in Chapter 31. Given that distribution, confidence intervals may be set in a similar manner.

29.45 These methods may be generalized to deal with a linear relation in k variables. If we can divide the n observations into k groups whose order, according to one variable, is the same for the observed as for the unobserved variable, we may find the centre of gravity of each group and determine the hyperplane which passes through the k points. If, in addition, the order of observed and unobserved variable is the same for every point, we may calculate $[n/k] = l$ relations for the points $(\xi_1, \xi_{l+1}, \xi_{2l+1}, \ldots, \xi_{kl+1})$, $(\xi_2, \xi_{l+2}, \xi_{2l+2}, \ldots, \xi_{kl+2})$, etc., and average them. Theoretically the use of (29.115) may also be generalized, but in practice it would probably be too tedious to calculate all the $\binom{n}{k}$ possible relations. See Exercise 29.10.

29.46 A more thoroughgoing use of ranks is to use the rank values of the ξ's, i.e. the natural numbers from 1 to n, as an instrumental variable. This method ought to be superior in efficiency to grouping methods, since it uses more information. Dorff and Gurland (1961b) find it to be generally superior, for small samples, to the two- and three-group methods, with smaller bias and mean-square-error. We illustrate the method by an example.

Example 29.7

For the data of Example 29.1, we use the ranks of ξ from 1 to 9 as the values of the instrumental variables ζ. Since the ξ-values are already arranged in order, we simply number them from 1 to 9 across the page. Then

$$\sum_i \zeta_i \eta_i = (1 \times 6\cdot9) + (2 \times 12\cdot5) + \ldots + (9 \times 39\cdot1) = 1267\cdot7,$$

$$\sum_i \zeta_i \xi_i = (1 \times 1\cdot8) + (2 \times 4\cdot1) + \ldots + (9 \times 18\cdot9) = 549\cdot0.$$

From our earlier computations,

$$\bar{\eta} = 23\cdot14,$$
$$\bar{\xi} = 9\cdot57,$$

while $\sum_i \zeta_i = \tfrac{1}{2} n(n+1) = 45$,

so that *from the observed means* the covariances are

$$\sum_i \zeta_i \eta_i - \bar{\eta} \sum_i \zeta_i = 1267\cdot7 - 23\cdot14 \times 45 = 226\cdot40,$$

$$\sum_i \zeta_i \xi_i - \bar{\xi} \sum_i \zeta_i = 549\cdot0 - 9\cdot57 \times 45 = 118\cdot35.$$

Thus from (29.91) we have

$$a_1 = \frac{226\cdot40}{118\cdot55} = 1\cdot91,$$

the same value as we obtained for the two-group method in Example 29.4, closer to the true value 2 than the three-group method's estimate of 1·86 in Example 29.5.

Controlled variables

29.47 Berkson (1950) (cf. also Lindley (1953b)) has adduced an argument to show that in certain types of experiment the estimation of a linear relation in two variables may be reduced to a regression problem. We recall from **29.4** that the relationship

$$y = \alpha_0 + \alpha_1 \xi + (\varepsilon - \alpha_1 \delta)$$

cannot be regarded as an ordinary regression because ξ is correlated with $(\varepsilon - \alpha_1 \delta)$.

Suppose now that we are conducting an experiment to determine the relationship between y and x, in which we can adjust ξ to a certain series of values, and then measure the corresponding values of η. For example, in determining the relation between the extension and the tension in a spring, we might hang weights (ξ) of 10 grams, 20 grams, 30 grams, ..., 100 grams and measure the extensions (η) which are regarded as the result of a random error ε acting on a true value y. However, our weights may also be imperfect, and in attaching a nominal weight of $\xi = 50$ grams we may in fact be attaching a weight x with error $\delta = 50 - x$. Under repetitions of such experiments with different weights, each purporting to be 50 grams, we are really applying a series of true weights x_i with errors $\delta_i = 50 - x_i$. Thus the real weights applied are the values of a random variable x. ξ is called a *controlled* variable, for its values are fixed in advance, while the unknown true values x are fluctuating.

We suppose that the errors δ have zero mean. This implies that x has a mean of $50 = \xi$. We now have
$$\xi_i = x_i + \delta_i,$$
where x_i and δ_i are perfectly negatively correlated. If we suppose, as before, that δ_i has the same distribution for all ξ_i we may write
$$x = \xi - \delta$$
and, as before,
$$\eta = (\alpha_0 + \alpha_1 x) + \varepsilon. \qquad (29.118)$$
Putting the previous equation into (29.118) we find
$$\eta = \alpha_0 + \alpha_1 \xi + (\varepsilon - \alpha_1 \delta), \qquad (29.119)$$
which is of the same form as (29.8) but is radically different. For ξ is not now a random variable, and neither ε nor δ is correlated with it. Thus (29.119) is an ordinary regression equation, to which the ordinary Least Squares methods may be applied without modification, and α_0 and α_1 may be estimated and tested without difficulty.

29.48 Even if the values at which ξ is controlled are themselves random variables (i.e. determined by some process of random selection) the analysis above applies if the errors δ and ε are uncorrelated with ξ. The latter assumption is usually fulfilled, but the former may be more difficult. In terms of our previous example, suppose that we made a random selection of the weights available, and used these for the experiment. The requirement that δ be uncorrelated with ξ now implies, e.g. that the larger weights should not tend to have larger or smaller errors of determination in their nominal values than do the smaller weights. Whether this is so is a matter for empirical study.

There is no doubt that, in many experimental situations, the preceding analysis is valid. ξ is often an instrumental reading, and the experimenter often tries to hold ξ to certain preassigned values (not chosen at random, but to cover a specified range adequately). In doing so, he is well aware that the instrument is subject to error and will not read the true values x precisely. It is comforting, after our earlier discussions of the difficulties of situations involving errors of measurement, that the standard LS analysis may be made in this common experimental situation. This fact illustrates the point, which cannot be too heavily stressed, that a thorough analysis of the sources of error and of the nature of the observational process is essential to the use of correct inferential methods, and may, as in this case, lead to a simple solution of an apparently difficult problem.

29.49 The analysis of situations of a more complex kind, when some variables are controlled and some are not, or when replicated observations are obtained for certain values of the controlled variables, requires a careful specification of the model under discussion. We have not the space to go into the complications here. Reference may be made to T. W. Anderson (1955) and to Scheffé (1958) for some interesting work in this field.

Non-linear relations

29.50 Up to this point we have considered only linear relations between the variables. The extension of the methods to non-linear relationship is not as straightforward a matter as it is in the theory of regression; and indeed some of the problems which arise have successfully resisted attack hitherto. We proceed with an account of some work, mainly due to Geary (1942b, 1943, 1949, 1953), in what is, as yet, only a partially explored field.

It will illustrate the kind of difficulty with which we have to contend if we consider the quadratic functional relationship

$$Y = \alpha_0 + \alpha_1 X + \alpha_2 X^2. \qquad (29.120)$$

On the same assumptions concerning errors ε in Y and δ in X as we made in the linear case, and with the further simplification that their variances σ_δ^2, σ_ε^2 are in a known ratio which we take to be 1 without loss of generality, we have, for the Likelihood Function when the errors are normal,

$$\log L = \text{constant} - 2n \log \sigma_\varepsilon - \frac{1}{2\sigma_\varepsilon^2}\{\sum_i (\xi_i - X_i)^2 + \sum (\eta_i - \alpha_0 - \alpha_1 X_i - \alpha_2 X_i^2)^2\}. \qquad (29.121)$$

Differentiation of (29.121) gives

$$\xi_i - X_i + (\alpha_0 + \alpha_1 X_i + \alpha_2 X_i^2 - \eta_i)(\alpha_1 + 2\alpha_2 X_i) = 0, \quad i = 1, 2, \ldots, n, \qquad (29.122)$$

$$\sum_i (\alpha_0 + \alpha_1 X_i + \alpha_2 X_i^2 - \eta_i) = 0, \qquad (29.123)$$

$$\sum_i (\alpha_0 + \alpha_1 X_i + \alpha_2 X_i^2 - \eta_i) X_i = 0, \qquad (29.124)$$

$$\sum_i (\alpha_0 + \alpha_1 X_i + \alpha_2 X_i^2 - \eta_i) X_i^2 = 0, \qquad (29.125)$$

$$-\frac{2n}{\sigma_\varepsilon} + \frac{1}{\sigma_\varepsilon^3}\left[\sum_i (\xi_i - X_i)^2 + \sum_i \left(\eta_i - \alpha_0 - \alpha_1 X_i - \alpha_2 X_i^2\right)^2\right] = 0. \qquad (29.126)$$

Summing (29.122) over i, and using (29.123) and (29.124) we find, as before at (29.42), that if we measure the ξ's from their observed mean

$$(\sum_i \hat{X}_i) = \sum_i \xi_i = 0. \qquad (29.127)$$

If we also measure the η's from their mean, we find, from (29.123)–(29.125), and (29.127),

$$\left.\begin{array}{l} n\alpha_0 \qquad\qquad\qquad + \alpha_2 \sum X_i^2 = 0, \\ \alpha_1 \sum X_i^2 + \alpha_2 \sum X_i^3 = \sum \eta_i X_i, \\ \alpha_0 \sum X_i^2 + \alpha_1 \sum X_i^3 + \alpha_2 \sum X_i^4 = \sum \eta_i X_i^2. \end{array}\right\} \qquad (29.128)$$

(29.128) is of the pattern familiar in regression analysis; but the X's here are not observed quantities. To obtain the ML estimators we must solve the $(n+3)$ equations in (29.122) and (29.128) for the $(n+3)$ unknowns X_i, α_0, α_1, α_2. The estimator of σ_ε^2 then follows from (29.126). In practice we should probably solve these equations by iterative methods.

The complication is also obvious from the geometrical viewpoint. We are now,

FUNCTIONAL AND STRUCTURAL RELATIONSHIP

from (29.121), seeking to determine a quadratic curve such that the sum of squares of perpendicular distances of points from it is a minimum. The joins of the different points to the curve are not parallel and may even not be unique. A solution, though arithmetically attainable, is clearly not expressible in concise form.

29.51 The product-cumulant method of estimating the coefficients (cf. **29.28**) can be extended. Consider the cubic structural relationship

$$y = \alpha_0 + \alpha_1 x + \alpha_2 x^2 + \alpha_3 x^3. \tag{29.129}$$

We drop the assumption that the errors are normally distributed. The joint c.f. of y and x is

$$\phi(t_1, t_2) = \int \exp(\theta_1 y + \theta_2 x) \, dF(x, y),$$

where $\theta_j = it_j$. We then have

$$\left(\frac{\partial}{\partial \theta_1} - \alpha_0 - \alpha_1 \frac{\partial}{\partial \theta_2} - \alpha_2 \frac{\partial^2}{\partial \theta_2^2} - \alpha_3 \frac{\partial^3}{\partial \theta_2^3}\right)\phi$$

$$= \int (y - \alpha_0 - \alpha_1 x - \alpha_2 x^2 - \alpha_3 x^3) \exp(\theta_1 y + \theta_2 x) \, dF = 0 \tag{29.130}$$

by (29.129). Putting $\psi = \log \phi$ and using the relations

$$\left.\begin{aligned}\frac{\partial \phi}{\partial \theta} &= \phi \frac{\partial \psi}{\partial \theta}, \\ \frac{\partial^2 \phi}{\partial \theta^2} &= \phi \left\{\frac{\partial^2 \psi}{\partial \theta^2} + \left(\frac{\partial \psi}{\partial \theta}\right)^2\right\}, \\ \frac{\partial^3 \phi}{\partial \theta^3} &= \phi \left\{\frac{\partial^3 \psi}{\partial \theta^3} + 3 \frac{\partial^2 \psi}{\partial \theta^2} \cdot \frac{\partial \psi}{\partial \theta} + \left(\frac{\partial \psi}{\partial \theta}\right)^3\right\},\end{aligned}\right\} \tag{29.131}$$

we find, from (29.130)

$$\frac{\partial \psi}{\partial \theta_1} - \alpha_0 - \alpha_1 \frac{\partial \psi}{\partial \theta_2} - \alpha_2 \left\{\frac{\partial^2 \psi}{\partial \theta_2^2} + \left(\frac{\partial \psi}{\partial \theta_2}\right)^2\right\} - \alpha_3 \left\{\frac{\partial^3 \psi}{\partial \theta_2^3} + 3 \frac{\partial^2 \psi}{\partial \theta_2^2} \frac{\partial \psi}{\partial \theta_2} + \left(\frac{\partial \psi}{\partial \theta_2}\right)^3\right\} = 0. \tag{29.132}$$

The equating to zero of coefficients in (29.132) when ψ is expressed as a power series in cumulants gives us a set of equations for the determination of the α's. These equations are linear in the α's but not, in general, in the cumulants. By (29.81), the product-cumulants of x and y are the same as those of ξ and η. If x and y are normally distributed, the method breaks down as in **29.30**.

29.52 This process is not entirely straightforward. We will illustrate the point by considering the estimation of α_0, α_1 and α_2 in the quadratic case ($\alpha_3 = 0$ in (29.129).) We have

$$\psi = \Sigma \kappa_{rs} \frac{\theta_1^r \theta_2^s}{r! s!}.$$

Without loss of generality we take $\kappa_{10} = \kappa_{01} = 0$. This is equivalent to taking an origin at the mean of the x's, which is estimated by the mean of the ξ's. From (29.132) we then have

$$\Sigma \kappa_{rs} \frac{\theta_1^{r-1}\theta_2^s}{(r-1)!s!} - \alpha_0 - \alpha_1 \Sigma \kappa_{rs} \frac{\theta_1^r \theta_2^{s-1}}{r!(s-1)!} - \alpha_2 \left\{ \Sigma \kappa_{rs} \frac{\theta_1^r \theta_2^{s-2}}{r!(s-2)!} + \left(\Sigma \kappa_{rs} \frac{\theta_1^r \theta_2^{s-1}}{r!(s-1)!}\right)^2 \right\} = 0. \qquad (29.133)$$

Equating coefficients to zero in (29.133), in order, we find

$$\text{Constant terms:} \quad -\alpha_0 - \alpha_2 \kappa_{02} = 0. \qquad (29.134)$$

This is the only equation involving α_0. It is useless except for estimating α_0, which means that we must estimate not only α_2 but κ_{02} (which depends on the variance of the error term). We also have

$$\left. \begin{array}{ll} \text{Terms in } \theta_1: & \kappa_{20} - \alpha_1 \kappa_{11} - \alpha_2 \kappa_{12} = 0. \\ \text{Terms in } \theta_2: & \kappa_{11} - \alpha_1 \kappa_{02} - \alpha_2 \kappa_{03} = 0. \\ \text{Terms in } \theta_1^2: & \kappa_{30} - \alpha_1 \kappa_{21} - \alpha_2(\kappa_{22} + 2\kappa_{11}^2) = 0. \\ \text{Terms in } \theta_1\theta_2: & \kappa_{21} - \alpha_1 \kappa_{12} - \alpha_2(\kappa_{13} + 2\kappa_{11}\kappa_{02}) = 0. \\ \text{Terms in } \theta_2^2: & \kappa_{12} - \alpha_1 \kappa_{03} - \alpha_2(\kappa_{04} + 2\kappa_{02}^2) = 0. \end{array} \right\} \qquad (29.135)$$

The first equation in (29.135) involves κ_{20}, which does not occur again. This equation is thus also useless for estimating the α's. It will be plain from (29.135) that the coefficient of any power of θ_1, say θ_1^r, contains κ_{r0}, which will not occur in any other equation. Such equations are therefore useless for estimating the α's unless we assume that the errors ε are normally distributed (with cumulants above the second equal to zero), in which case the cumulants κ_{r0}, as well as the product-cumulants, may be estimated from the observables, and equations such as the third in (29.135) are usable. We then can eliminate κ_{02} between the second and fourth equation in (29.135). The result, in conjunction with the third equation, enables us to solve for α_1 and α_2.

If we do not assume the errors to be normal, we require further equations. We have from (29.133)

$$\left. \begin{array}{ll} \text{Terms in } \theta_1^2\theta_2: & \kappa_{31} - \alpha_1 \kappa_{22} - \alpha_2(\kappa_{23} + 2\kappa_{02}\kappa_{21} + 4\kappa_{11}\kappa_{12}) = 0, \\ \text{Terms in } \theta_1\theta_2^2: & \kappa_{22} - \alpha_1 \kappa_{13} - \alpha_2(\kappa_{14} + 2\kappa_{03}\kappa_{11} + 4\kappa_{02}\kappa_{12}) = 0, \\ \text{Terms in } \theta_2^3: & \kappa_{13} - \alpha_1 \kappa_{04} - \alpha_2(\kappa_{05} + 6\kappa_{02}\kappa_{03}) = 0. \end{array} \right\} \qquad (29.136)$$

The first two equations of (29.136) contain, apart from product-cumulants, κ_{02} and κ_{03}. We can eliminate these with the help of the second and fourth equations in (29.135), and solve for α_1 and α_2. We can thus estimate κ_{02} and hence, from (29.134), the value of α_0. Some of the eliminants may be non-linear, in which case we might get more than one set of estimators.

29.53 The two-group estimation method of **29.36** clearly generalizes to polynomials of order k in x if we can divide the observations into $(k+1)$ groups which are in the same order by ξ as by x; we then determine the centre of gravity of each group and fit the polynomial to the $(k+1)$ resulting points. Theil's method (**29.42**) also generalizes in a fairly obvious way. If we divide the observations into $(k+1)$ groups and fit $[n/(k+1)]$ parabolas to the points obtained by picking one observation from each group, we have only to average the resulting parabolas. This is not as simple as it sounds, however. It is not necessarily true that, if we get a set of parabolas $a_0 + a_1 x + a_2 x^2$, the best estimated parabola is $\bar{a}_0 + \bar{a}_1 x + \bar{a}_2 x^2$. Some heuristic amalgamation of the set seems to be indicated, such as fitting by Least Squares in the direction of the y-axis, or drawing the curves and selecting one which seems to represent the median position so far as possible along its length.

> Villegas (1969) shows that if replicated observations are available, LS estimators of the parameters of a general non-linear model are consistent and asymptotically efficient.

29.54 The extension of the analysis of controlled variables to the non-linear case involves one or two new points. Although the linear case can be reduced to regression analysis, the curvilinear case cannot. Consider the cubic functional relationship
$$Y = \alpha_0 + \alpha_1 X + \alpha_2 X^2 + \alpha_3 X^3.$$
If we put $\xi = X + \delta$, $\eta = Y + \varepsilon$ we find
$$\eta = \alpha_0 + \alpha_1(\xi - \delta) + \alpha_2(\xi - \delta)^2 + \alpha_3(\xi - \delta)^3, \tag{29.137}$$
where the ξ's, the controlled quantities, are fixed. Let us consider repetitions of the observations *over the same set of ξ's* and denote by E the expectation in such a reference set. Summing (29.137) over the observations, we have
$$\Sigma \eta = n\alpha_0 + \alpha_1 \Sigma \xi + \alpha_2 \Sigma \xi^2 + \alpha_2 \Sigma \delta^2 + \alpha_3 \Sigma \xi^3 + 3\alpha_3 \Sigma \xi \delta^2 + \text{terms of odd order in } \delta \text{ or } \varepsilon.$$
Taking expectations, we then have
$$E(\Sigma \eta) = n(\alpha_0 + \alpha_2 \sigma_\delta^2) + (\alpha_1 + 3\alpha_3 \sigma_\delta^2) \Sigma \xi + \alpha_2 \Sigma \xi^2 + \alpha_3 \Sigma \xi^3.$$
Likewise, multiplying (29.137) by ξ and summing, we get equations
$$E(\Sigma \eta \xi) = (\alpha_0 + \alpha_2 \sigma_\delta^2) \Sigma \xi + (\alpha_1 + 3\alpha_3 \sigma_\delta^2) \Sigma \xi^2 + \alpha_2 \Sigma \xi^3 + \alpha_3 \Sigma \xi^4, \tag{29.138}$$
and so on. These equations can be solved for the quantities
$$(\alpha_0 + \alpha_2 \sigma_\delta^2), (\alpha_1 + 3\alpha_3 \sigma_\delta^2), \alpha_2, \alpha_3.$$
It is rather remarkable that, although α_2 and α_3 are identifiable, α_0 and α_1 are not so without knowledge, or an estimate, of σ_δ^2. The only way round the difficulty seems to be to replicate the experiment with the same set of ξ's. The papers by Geary (1953) and Scheffé (1958) should be consulted for further details.

> Dolby and Lipton (1972) consider the general non-linear functional relationship with replicated observations.

29.55 We may be able to reduce a non-linear relationship to a linear one by a transformation. Consider, for example, a functional relationship of the type
$$Y^\beta X^\gamma = \text{constant}.$$
Here the obvious procedure is to take logarithms. In general, if we can transform data to linearity before the estimation begins we shall have gained a great deal. The theoretical drawback of this procedure is that if errors in Y and X are, say, normal and homoscedastic, those of the transforms $\log Y$ and $\log X$ will not be. The moral,

here as elsewhere, is that we should endeavour to obtain as much prior information as possible about the nature of the observational errors ; and that, when the errors are substantial and of unknown distribution, we should use methods of estimation which make as few assumptions about their nature as possible.

The effect of observational errors on regression analysis

29.56 It is convenient to conclude this chapter with a brief account of an allied but rather different subject, the effect of errors of observation on regression analysis. Suppose that x and y, a pair of random variables, are affected by errors of observation δ and ε, so that we observe
$$\xi = x + \delta,$$
$$\eta = y + \varepsilon.$$
As before, we suppose the δ's independent, the ε's independent, and δ and ε independent. Our question now is: suppose that we determine the regression of one observed variable on the other, say η on ξ; what relation does this bear to the regression of y on x, which is of primary interest?

The argument of **29.28** shows us that the product-cumulants of ξ and η are those of x and y. Thus $\mathrm{cov}(\xi,\eta) = \mathrm{cov}(x,y)$. But regression coefficients also depend on variances, which are not unchanged. The linear regression equation
$$y = \beta_1 x, \quad \text{with} \quad \beta_1 = \mathrm{cov}(x,y)/\sigma_x^2$$
is replaced by
$$\eta = \beta_1' \xi, \quad \text{with} \quad \beta_1' = \mathrm{cov}(\xi,\eta)/\sigma_\xi^2 = \mathrm{cov}(x,y)/\{\sigma_x^2 + \sigma_\delta^2\},$$
and clearly $\beta_1' < \beta_1$. The effect of the errors is thus to diminish the slope of the regression lines. It follows that the correlation between ξ, η will also be weaker than that between x, y.

29.57 However, this attenuation of the coefficients is not the whole story. Let us suppose that the true regression of y on x is exactly linear with identical errors (cf. **28.7**). Does it follow that the true regression of η on ξ is also exactly linear? The answer is, in general, no; only under certain quite stringent conditions will linearity be unimpaired. We will prove a theorem stated in an elegant form by Lindley (1947): a necessary and sufficient condition for the regression to continue to be linear is that the c.g.f. of the variable x is a multiple of the c.g.f. of the error δ. More precisely, in terms of the c.g.f.s of x and ξ, we must have
$$\beta_1 \psi_x = \beta_1' \psi_\xi. \tag{29.139}$$
We have seen in **28.7** that it is necessary and sufficient for the regression of y on x to be exactly linear, of the form
$$y = \beta_0 + \beta_1 x + \varepsilon$$
with identical errors, that the joint f.f. of x and y factorizes into
$$f(x,y) = g(x) h(y - \beta_0 - \beta_1 x), \tag{29.140}$$
where $g(x)$ is the marginal distribution of x; or equivalently that the c.g.f.s satisfy
$$\psi(t_1, t_2) = \psi_g(t_1 + t_2 \beta_1) + \psi_h(t_2) + i t_2 \beta_0. \tag{29.141}$$

FUNCTIONAL AND STRUCTURAL RELATIONSHIP

Now we know (28.5) that if $\zeta(t_1, t_2)$ is the joint c.g.f. of ξ, η and the regression of η on ξ is exactly linear of form $\eta = \beta_0' + \beta_1' \xi + \varepsilon$, then

$$\left[\frac{\partial}{\partial t_2}\zeta(t_1, t_2)\right]_{t_2=0} = i\beta_0' + \beta_1' \frac{\partial \zeta(t_1, 0)}{\partial t_1}. \tag{29.142}$$

But if δ, ε are independent of each other and of x, y we have

$$\zeta(t_1, t_2) = \psi(t_1, t_2) + \psi_\delta(t_1) + \psi_\varepsilon(t_2). \tag{29.143}$$

Substituting (29.143) and (29.141) into (29.142) we find

$$\left[\frac{\partial \psi_g(t_1+t_2\beta_1)}{\partial t_2}\right]_{t_2=0} + \left[\frac{\partial \psi_h(t_2)}{\partial t_2}\right]_{t_2=0} + i\beta_0 + \left[\frac{\partial \psi_\varepsilon(t_2)}{\partial t_2}\right]_{t_2=0} = i\beta_0' + \beta_1' \frac{\partial \zeta(t_1, 0)}{\partial t_1}. \tag{29.144}$$

Equating coefficients of t_1 in (29.144) we have

$$\beta_1 \frac{\partial \psi_g(t_1)}{\partial t_1} = \beta_1' \frac{\partial \zeta(t_1, 0)}{\partial t_1}. \tag{29.145}$$

Since $\zeta(t_1, 0) = \psi_\xi(t_1)$, (29.145) integrates at once to (29.139).

The other terms in (29.144) give, writing μ for means,

$$\mu_h + \beta_0 + \mu_\varepsilon = \beta_0'. \tag{29.146}$$

In particular, if δ and ε have zero means,

$$\beta_0 = \beta_0'. \tag{29.147}$$

This proves the necessity of (29.139). Its sufficiency follows easily.

29.58 If we also require identical errors in the regression of η on ξ, we obtain a much stronger result (Kendall, 1951–2). For then we have (29.141) holding as well as (29.143) for the c.g.f. $\zeta(t_1, t_2)$ of ξ, η. Thus, writing primes in (29.141),

$$\zeta(t_1, t_2) = \psi(t_1, t_2) + \psi_\delta(t_1) + \psi_\varepsilon(t_2) = \psi_{g'}(t_1+t_2\beta_1') + \psi_{h'}(t_2) + it_2\beta_0'. \tag{29.148}$$

Substituting for $\psi(t_1, t_2)$ from (29.141), (29.148) becomes

$$\psi_g(t_1+t_2\beta_1) + \psi_h(t_2) + it_2\beta_0 + \psi_\delta(t_1) + \psi_\varepsilon(t_2) = \psi_{g'}(t_1+t_2\beta_1') + \psi_{h'}(t_2) + it_2\beta_0'. \tag{29.149}$$

Putting $t_1 = 0$ in (29.149) and subtracting the resulting equation from (29.149), we find

$$\psi_g(t_1+t_2\beta_1) - \psi_g(t_2\beta_1) + \psi_\delta(t_1) = \psi_{g'}(t_1+t_2\beta_1') - \psi_{g'}(t_2\beta_1'). \tag{29.150}$$

Denoting cumulants of x, ξ by superfixes (not powers) g and g', we find from (29.150)

$$\Sigma \kappa_r^g \frac{(t_1+t_2\beta_1)^r}{r!} - \Sigma \kappa_r^g \frac{(t_2\beta_1)^r}{r!} + \Sigma \kappa_r^\delta \frac{t_1^r}{r!} = \Sigma \kappa_r^{g'} \frac{(t_1+t_2\beta_1')^r}{r!} - \Sigma \kappa_r^{g'} \frac{(t_2\beta_1')^r}{s!}. \tag{29.151}$$

Consider a term of order > 2, say $r = 3$. Identifying powers of the third degree we have

$$\left.\begin{aligned}
\kappa_3^g + \kappa_3^\delta &= \kappa_3^{g'}, \\
\beta_1 \kappa_3^g &= \beta_1' \kappa_3^{g'}, \\
\beta_1^2 \kappa_3^g &= (\beta_1')^2 \kappa_3^{g'}.
\end{aligned}\right\} \tag{29.152}$$

The second and third results are only possible if $\beta_1 = \beta_1'$ or if κ_3^g and $\kappa_3^{g'}$ vanish. It follows that all third cumulants, and similarly all cumulants higher than the second, vanish. The converse again follows. Thus, if and only if x, δ, and ξ are all normal will the exact linear regression with identical errors, $y = \beta_0 + \beta_1 x$ become the exact linear regression with identical errors $\eta = \beta_0' + \beta_1' \xi$.

29.59 Various other theorems on this subject have been proved. Apparently the first was given by Allen (1938), who proved under restrictive conditions that if

$$\left. \begin{array}{l} \xi = lx + \delta, \\ \eta = mx + \varepsilon, \end{array} \right\} \quad l, m \neq 0,$$

then the necessary and sufficient condition for the regression of η on ξ to be exactly linear for all l in a closed interval is that x and δ are normal. Some of her conditions have been relaxed by Fix (1949a), who requires only that x, δ and ε have finite means and that either x or δ has finite variance. Fix's result has been generalized further by Laha (1956) to the case where the error variables δ, ε are not independent.

Lindley (1947) proved a more general result than (29.139)—see Exercises, 29.12, 29.14. The result of **29.58** may also be extended to several variables—see Exercise 29.13.

EXERCISES

29.1 Show that the ML estimator $\hat{\alpha}_1$ defined at (29.29) is a monotone function of λ.

29.2 Referring to the method of **29.28**, show that: (a) if the errors δ are completely independent of the x's, but (b) if $(k-1)$ product-cumulants of the δ's can be found which vanish, then equations (29.85) can still be used to estimate the α's. In particular, this is so if the δ's are distributed in the multivariate normal form. (Geary, 1942b)

29.3 Show that equations (29.85) are equally true if there are substituted for the κ's the corresponding moments of the x's, but that it does not then hold for the observed ξ's.

29.4 Show directly that the estimator a_1 of (29.98) is a consistent estimator of α_1, and hence that the estimators (29.102) of the error variances are consistent.
(Wald, 1940)

29.5 Referring to equation (29.111), show that the confidence region for α_0 and α_1 consists of the interior of an ellipse. (Wald, 1940)

29.6 We have n observations, $x_1, x_2, \ldots, x_n$, on a vector variate, the components of which are distributed independently and normally with unit variances. It is desired to test the acceptability of the relationship

$$\alpha_0 + \boldsymbol{\alpha}' \mathbf{x} = 0 \qquad (A)$$

where α_0 is scalar and $\boldsymbol{\alpha}'$ the transpose of a column vector $\boldsymbol{\alpha}$. Show that

$$n\phi = \sum_{j=1}^{n} \frac{(\alpha_0 + \boldsymbol{\alpha}' \mathbf{x}_j)'(\alpha_0 + \boldsymbol{\alpha}' \mathbf{x}_j)}{\boldsymbol{\alpha}' \boldsymbol{\alpha}} \qquad (B)$$

is distributed as χ^2 with n degrees of freedom. If $\mathbf{V}$ is the dispersion matrix of the observations, show that the envelope of (A) subject to a constraint imposed by putting $n\phi$ at (B) equal to a constant χ_0^2 is given by

$$|\mathbf{V} - \phi\mathbf{I}| + V_{ij}^* x_i x_j = 0,$$

where $\mathbf{I}$ is the identity matrix and V_{ij}^* is the cofactor of the (i,j)th element in $|\mathbf{V} - \phi\mathbf{I}|$. Show that this may also be written

$$1 + \mathbf{x}'(\mathbf{V} - \phi\mathbf{I})^{-1}\mathbf{x} = 0.$$

(R. L. Brown and Fereday, 1958)

29.7 In the previous exercise, if the roots of $|\mathbf{V} - \phi\mathbf{I}| = 0$ are $\phi_1, \ldots, \phi_k$ and $\mathbf{V}$ is the diagonal matrix with elements ϕ_j; and if $\mathbf{L}$ is the orthogonal $(k \times k)$ matrix of vectors determined by $\mathbf{VL} = \mathbf{L}\boldsymbol{\Lambda}$, show that by transforming to new variables $\mathbf{y} = \mathbf{L}'\mathbf{x}$ the equation of the envelope is $1 + \mathbf{y}'(\boldsymbol{\Lambda} - \phi\mathbf{I})^{-1}\mathbf{y} = 0$ and hence is

$$\sum_{j=1}^{k} \frac{y_j^2}{\phi_j - \phi} + 1 = 0.$$

(R. L. Brown and Fereday, 1958)

29.8 In the previous two exercises, show that if, corresponding to (A) of Exercise 29.6, we have in Exercise 29.7

$$\beta_0 + \boldsymbol{\beta}' \mathbf{y} = 0,$$

then the joint confidence region for the β's may be written

$$\beta_0^2 + \sum_{j=1}^{k} (\phi_j - \phi_0)\beta_j^2 = 0,$$

where ϕ_0 is the critical value of ϕ at (B) in Exercise 29.6, obtained from the χ^2 tables.

(R. L. Brown and Fereday, 1958)

29.9 Show that the statistics $a(i)$ of equation (29.114) can be used to provide a test of linearity of relationship (as against a convex or concave relationship), by considering the correlation between $a(i)$ and i. (Theil, 1950)

29.10 A set of variables X_j are connected with Y by the relation

$$Y = \alpha_0 + \sum_{j=1}^{k} \alpha_j X_j.$$

The errors in the X's are such that the order of the observed $\xi_j\ (= X_j + \delta_j)$ is the same as that of the corresponding X_j for all $j = 1, 2, \ldots, k$. Show that if the δ's are such that

$$P\{z_i < z_j\} = \tfrac{1}{2},$$

where $z_i = \varepsilon_i - \sum_{j=1}^{k} \alpha_j \delta_j$, then a confidence interval in the manner of **29.43** can be set up for any α_j if the other α's are given. Hence show how to set up a conservative confidence region for all α's by taking the union of the individual intervals.

(Theil, 1950)

29.11 $n\,(= 2l+1)$ observations on $\eta\,(= Y+\varepsilon)$ are made, at equally-spaced unit intervals of X, which is not subject to error, so that $\xi = X$. The ε's have variance σ^2. If the parameters in $Y = \alpha_0 + \alpha_1 X$ are estimated by Least Squares, giving minimum variance unbiassed estimators, show that the estimator of α_1 has variance

$$3\sigma^2/\{l(l+1)(2l+1)\}.$$

Show further that if the observations are divided into three groups, consisting of nk, $n-2nk$, nk observations, the estimator a_1 of (29.112) has maximum efficiency when $k = \tfrac{1}{3}$, and is then $\geqslant \tfrac{8}{9}$, while the efficiency when $k = \tfrac{1}{2}$ is only $\tfrac{27}{32}$ of this.

(Bartlett, 1949)

29.12 The regression of y on $x_1, x_2, \ldots, x_k$ is given by

$$y = \sum_j \beta_j x_j,$$

and the errors are independent of the x's. If x_j is subject to error $\delta_j\,(\xi_j = x_j + \delta_j)$, and y to error $\varepsilon\,(\eta = y + \varepsilon)$, the δ's being independent of each other and of ε, show that the regression of η on the ξ's is exactly linear of the form

$$\eta = \sum_j \beta_j' \xi_j$$

if and only if

$$\sum_j (\beta_j - \beta_j') \frac{\partial \psi_{x_j}}{\partial t_j} = \sum_j \beta_j' \frac{\partial \psi_{\delta_j}}{\partial t_j},$$

where the ψ's are c.g.f.s of their suffix variables. This generalizes **29.57**.

(Lindley, 1947)

29.13 In Exercise **29.12**, show further that the errors in the second regression are independent of the ξ's if and only if the distribution of the x's, ξ's and δ's are all normal. This generalizes **29.58**.

(Kendall, 1951-2)

29.14 In Exercise 29.12 show that

$$\sum_j \beta_j \frac{\partial^2 \psi_x}{\partial t_i \, \partial t_j} = \sum \beta'_j \frac{\partial^2 \psi_\xi}{\partial t_i \, \partial t_j}.$$

Hence, if **S** is the dispersion matrix of the ξ's and Λ (a diagonal matrix) that of the δ's

$$\boldsymbol{\beta} - (\boldsymbol{\beta}') = \boldsymbol{\beta}\Lambda \mathbf{S}^{-1}$$

where $\boldsymbol{\beta}$, $(\boldsymbol{\beta}')$ are the row vectors of the β's and β'''s respectively.

(Lindley, 1947)

29.15 Show that if the unobservables x, y and the instrumental variable ζ are normally distributed with zero means, and a_1 is defined by (29.91), the variable

$$t^2 = (n-1) \bigg/ \left\{ \frac{\sigma_\zeta^2 (a_1^2 \sigma_x^2 - 2a_1 \sigma_{xy} + \sigma_y^2)}{(a_1 \sigma_{x\zeta} - \sigma_{y\zeta})^2} - 1 \right\}$$

is distributed as "Student's" t^2 with $(n-1)$ degree of freedom. (Geary, 1949)

29.16 From the result of the last exercise, show that approximately

$$2n^2 \operatorname{var} a_1 = \bigg[\frac{\sigma_{x\zeta}(\alpha_1 \sigma_{x\zeta} - \sigma_{y\zeta})}{\sigma_\zeta^2 (\alpha_1^2 \sigma_x^2 - 2\alpha_1 \sigma_{xy} + \sigma_y^2) - (\alpha_1 \sigma_{x\zeta} - \sigma_{y\zeta})^2} \\ - \frac{(\alpha_1 \sigma_{x\zeta} - \sigma_{y\zeta})^2 \{\alpha_1(\sigma_x^2 \sigma_\zeta^2 - \sigma_{x\zeta}^2) + (\sigma_{x\zeta} \sigma_{y\zeta} - \sigma_{xy} \sigma_\zeta^2)\}}{\{\sigma_\zeta^2 (\alpha_1^2 \sigma_x^2 - 2\alpha_1 \sigma_{xy} + \sigma_y^2) - (\alpha_1 \sigma_{x\zeta} - \sigma_{y\zeta})^2\}^2} \bigg]^{-2}$$

(Madansky, 1959)

29.17 Assuming that there are no errors of observation in x or y (i.e. $\xi = x$, $\eta = y$), show that, *for fixed x's and ζ's*, the estimating efficiency of the estimator a_1 of (29.91), compared to the LS estimator, is equal to the square of the correlation between x and ζ.

(Durbin, 1954)

29.18 When the conditions $s_\xi^2 > \sigma_\delta^2$, $s_\eta^2 \geqslant s_{\xi\eta}^2/(s_\xi^2 - \sigma_\delta^2)$ are not satisfied in Case 1 of **29.12**, show first that the LF is maximized either when $\sigma_\varepsilon^2 = 0$ or when $\sigma_x^2 = 0$, and by comparing the maxima that the former always gives the overall maximum. Hence show that

$$\hat{\alpha}_1 = s_\eta^2/s_{\xi\eta}, \quad \hat{\sigma}_\varepsilon^2 = 0, \quad \hat{\sigma}_x^2 = s_{\xi\eta}^2/s_\eta^2.$$

(Birch, 1964a)

CHAPTER 30

TESTS OF FIT

30.1 In our discussions of estimation and test procedures from Chapter 17 onwards, we have concentrated entirely on problems concerning the parameters of distributions of known form. In our classification of hypothesis-testing problems in **22.3** we did indeed define a non-parametric hypothesis, but we have not yet investigated non-parametric hypotheses or estimation problems. In the group of four chapters of which this is the first, we shall be pursuing these subjects systematically.

We shall find it convenient to defer a general discussion of non-parametric problems, and their special features, until Chapter 31. In the present chapter, we confine ourselves to a particular class of procedures which stand slightly apart from the others, and are of sufficient practical importance to justify this special treatment.

Tests of fit

30.2 Let $x_1, x_2, \ldots, x_n$ be independent observations on a random variable with distribution function $F(x)$ which is unknown. Suppose that we wish to test the hypothesis

$$H_0 : F(x) = F_0(x), \tag{30.1}$$

where $F_0(x)$ is some particular d.f., which may be continuous or discrete. The problem of testing (30.1) is called a *goodness-of-fit problem*. Any test of (30.1) is called a *test of fit*.

Hypotheses of fit, like parametric hypotheses, divide naturally into simple and composite hypotheses. (30.1) is a simple hypothesis if $F_0(x)$ is completely specified; e.g. the hypothesis (a) that the n observations have come from a normal distribution with specified mean and variance is a simple hypothesis. On the other hand, we may wish to test (b) whether the observations have come from a normal distribution whose parameters are unspecified, and this would be a composite hypothesis (in this case it would often be called a " test of normality "). Similarly, if (c) the normal distribution has its mean, but not its variance, specified, the hypothesis remains composite. This is precisely the distinction we discussed in the parametric case in **22.4**.

30.3 It is clear that (30.1) is no more than a restatement of the general problem of testing hypotheses; we have merely expressed the hypothesis in terms of the d.f. instead of the frequency function. What is the point of this? Shall we not merely be retracing our previous steps?

The reasons for the new formulation are several. The parametric hypothesis-testing methods developed earlier were necessarily concerned with hypotheses imposing one or more constraints (cf. **22.4**) in the parameter space; they afford no means whatever of testing a hypothesis like (b) in **30.2**, where no constraint is imposed upon parameters and we are testing the non-parametric hypothesis that the parent d.f. is a member of a specified (infinite) family of distributions. In such cases, and even in cases where the hypothesis does impose one or more parametric constraints, as in (a) or (c) of **30.2**, the reformulation of the hypothesis in the form (30.1) provides us with new methods. For we are led by intuition to expect the whole distribution of the sample observations to mimic closely that of the true d.f. $F(x)$. It is therefore natural to seek to use the whole observed distribution directly as a means of testing (30.1), and we shall find that the most important tests of fit do just this. Furthermore, the "optimum" tests we have devised for parametric hypotheses, H_0, have been recommended by the properties of their power functions against alternative hypotheses which differ from H_0 only in the values of the parameters specified by H_0. It seems at least likely that a test based on the whole distribution of the sample will have reasonable power properties against a wider (infinite) class of alternatives, even though it may not be optimum against any one of them.

The LR and Pearson tests of fit for simple H_0

30.4 Two well-known methods of testing goodness-of-fit depend on a very simple device. We consider it first in the case when $F_0(x)$ is completely specified, so that (30.1) is a simple hypothesis.

Suppose that the range of the variate x is arbitrarily divided into k mutually exclusive classes. (These need not be, though in practice they are usually taken as, successive intervals in the range of x.)(*) Then, since $F_0(x)$ is specified, we may calculate the probability of an observation falling in each class. If these are denoted by p_{0i}, $i = 1, 2, \ldots, k$, and the observed frequencies in the k classes by $n_i \left(\sum_{i=1}^{k} n_i = n \right)$, the n_i are multinomially distributed (cf. **5.30**), and from (5.78) we see that the LF is

$$L(n_1, n_2, \ldots, n_k | p_{01}, p_{02}, \ldots, p_{0k}) \propto \prod_{i=1}^{k} p_{0i}^{n_i}. \tag{30.2}$$

On the other hand, if the true distribution function is $F_1(x)$, where F_1 may be any d.f., we may denote the probabilities in the k classes by p_{1i}, $i = 1, 2, \ldots, k$, and the likelihood is

$$L(n_1, n_2, \ldots, n_k | p_{11}, p_{12}, \ldots, p_{1k}) \propto \prod_{i=1}^{k} p_{1i}^{n_i}. \tag{30.3}$$

We may now easily find the Likelihood Ratio test of the hypothesis (30.1), the composite alternative hypothesis being

$$H_1 : F(x) = F_1(x).$$

The likelihood (30.3) is maximized when we substitute the ML estimators for p_{1i}

$$\hat{p}_{1i} = n_i / n.$$

The LR statistic for testing H_0 against H_1 is therefore

$$l = \frac{L(n_1, n_2, \ldots, n_k | p_{01}, p_{02}, \ldots, p_{0k})}{L(n_1, n_2, \ldots, n_k | \hat{p}_{11}, \hat{p}_{12}, \ldots, \hat{p}_{1k})}$$

$$= n^n \prod_{i=1}^{k} (p_{0i} / n_i)^{n_i}. \tag{30.4}$$

H_0 is rejected when l is small enough.

(*) We discuss the choice of k and of the classes in **30.20-3, 30.28-30** below. For the present, we allow them to be arbitrary.

The exact distribution of (30.4) is unknown. However, we know from **24.7** that as $n \to \infty$ when H_0 holds, $-2\log l$ is asymptotically distributed in the χ^2 form, with $k-1$ degrees of freedom (since there are $r = k-1$ independent constraints p_{1i} because $\sum_{i=1}^{k} p_{1i} = 1$).

30.5 (30.4) is not, however, the classical test statistic put forward by Karl Pearson (1900) for this situation. This procedure, which has been derived already as Example 15.3, uses the asymptotic k-variate normality of the multinomial distribution of the n_i, and the fact that, given H_0, the quadratic form in the exponent of this distribution is distributed in the χ^2 form with degrees of freedom equal to its rank, $k-1$. In our present notation, this quadratic form was found in Example 15.3 to be (*)

$$X^2 = \sum_{i=1}^{k} \frac{(n_i - np_{0i})^2}{np_{0i}}. \tag{30.5}$$

From (30.4) we have

$$-2\log l = 2 \sum_{i=1}^{k} n_i \log(n_i / np_{0i}). \tag{30.6}$$

The two distinct statistics (30.5) and (30.6) thus have the same distribution asymptotically, given H_0. More than this, however, they are asymptotically equivalent statistics when H_0 holds, for if we write $\Delta_i = \frac{n_i - np_{0i}}{np_{0i}}$, we have

$$-2\log l = 2\sum_i n_i \log(1+\Delta_i)$$
$$= 2\sum_i \{(n_i - np_{0i}) + np_{0i}\}\{\Delta_i - \tfrac{1}{2}\Delta_i^2 + O(n^{-\frac{3}{2}})\}$$
$$= 2\sum_i \left\{(n_i - np_{0i})\Delta_i + np_{0i}\Delta_i - \frac{np_{0i}}{2}\Delta_i^2 + O(n^{-\frac{1}{2}})\right\},$$

and since $\sum p_{0i}\Delta_i = 0$, we have

$$-2\log l = \sum_i \{np_{0i}\Delta_i^2 + O(n^{-\frac{1}{2}})\} = X^2\{1 + O(n^{-\frac{1}{2}})\}. \tag{30.7}$$

For small n, the test statistics differ. Pearson's form (30.5) may alternatively be expressed as

$$X^2 = \frac{1}{n}\sum_i \frac{n_i^2}{p_{0i}} - n, \tag{30.8}$$

which is easier to compute; but (30.5) has the advantage over (30.8) of being a direct function of the differences between the observed frequencies n_i and their hypothetical expectations np_{0i}, differences which are themselves of obvious interest. The corresponding simplification of (30.6) is computationally inconvenient.

M. E. Wise (1963, 1964) examines the approximations involved in using (30.5) as a χ^2_{k-1} variable, and shows that the error is particularly small when the np_{0i} are equal or nearly so—the latter need not then be large (cf. **30.22**, **30.30**, which also cover composite H_0). Some small-sample tabulations of (30.5) and (30.6) in the case when each $p_{0i} = 1/k$ are given by Good et al. (1970). Zahn and Roberts (1971) tabulate the case $p_{0i} = 1/n$.

(*) Following recent practice, we write X^2 for the test statistic and reserve the symbol χ^2 for the distributional form we have so frequently discussed. Earlier writers confusingly wrote χ^2 for the statistic as well as the distribution.

Choice of critical region

30.6 Since H_0 is rejected for small values of l, (30.7) implies that when using (30.5) as test statistic, H_0 is to be rejected when X^2 is large. There has been some uncertainty in the literature about this, the older practice being to reject H_0 for small as well as large values of X^2, i.e. to use a two-tailed rather than an upper-tail test. For example, Cochran (1952) approves this practice on the grounds that extremely small X^2 values are likely to have resulted from numerical errors in computation, while on other occasions such values have apparently been due to the frequencies n_i having been biassed, perhaps inadvertently, to bring them closer to the hypothetical expectations np_{0i}.

Now there is no doubt that computations should be checked for accuracy, but there are likely to be more direct and efficient methods of doing this than by examining the value of X^2 reached. After all, we have no assurance that a moderate and acceptable value of X^2 has been any more accurately computed than a very small one. Cochran's second consideration is a more cogent one, but it is plain that in this case we are considering a different and rarer hypothesis (that there has been voluntary or involuntary irregularity in collecting the observations) which must be precisely formulated before we can determine the best critical region to use (cf. Stuart (1954a)). Leaving such irregularities aside, we use the upper tail of the distribution of X^2 as critical region. This will be justified from the point of view of its asymptotic power in **30.27**.

30.7 The essence of the LR and Pearson tests of fit is the reduction of the problem to one concerning the multinomial distribution. The need to group the data into classes clearly involves the sacrifice of a certain amount of information, especially if the underlying variable is continuous. However, this defect also carries with it a corresponding virtue: we do not need to know the values of the individual observations, so long as we have k classes for which the hypothetical p_{0i} can be computed. In fact, there need be no underlying variable at all—we may use either of these tests of fit even if the original data refer to a non-numerical classification. The point is illustrated by Example 30.1.

Example 30.1

In some classical experiments on pea-breeding, Mendel observed the frequencies of different kinds of seeds in crosses from plants with round yellow seeds and plants with wrinkled green seeds. They are given below, together with the theoretical probabilities on the Mendelian theory of inheritance.

Seeds	Observed frequency n_i	Theoretical probability p_{0i}
Round and yellow	315	9/16
Wrinkled and yellow	101	3/16
Round and green	108	3/16
Wrinkled and green	32	1/16
	$n = 556$	1

(30.8) gives

$$X^2 = \frac{1}{556} \cdot 16 \left\{ \frac{315^2}{9} + \frac{101^2}{3} + \frac{108^2}{3} + \frac{32^2}{1} \right\} - 556$$

$$= \frac{16}{556} \cdot 19{,}337 \cdot 3 - 556 = 0 \cdot 47.$$

For $(k-1) = 3$ degrees of freedom, tables of χ^2 show that the probability of a value exceeding 0·47 lies between 0·90 and 0·95, so that the fit of the observations to the theory is very good indeed : a test of any size $\alpha \leqslant 0 \cdot 90$ would not reject the hypothesis.

For the LR statistic, (30.6) gives, after considerably more computation, $-2 \log l = 0 \cdot 48$, very close to the value for X^2.

Composite H_0

30.8 Confining our attention now to Pearson's test statistic (30.5), we consider the situation which arises when the hypothesis tested is composite—the LR test remains asymptotically equivalent when H_0 holds—cf. Exercise 30.11. Suppose that $F_0(x)$ is specified as to its form, but that some (or perhaps all) of the parameters are left unspecified, as in (b) or (c) of **30.2**. In the multinomial formulation of **30.4**, the new feature is that the theoretical probabilities p_{0i} are not now immediately calculable, since they are functions of the s (assumed $< k-1$) unspecified parameters $\theta_1, \theta_2, \ldots, \theta_s$, which we may denote collectively by $\boldsymbol{\theta}$. Thus we must write them $p_{0i}(\boldsymbol{\theta})$. In order to make progress, we must estimate $\boldsymbol{\theta}$ by some vector of estimators $\mathbf{t}$, and use (30.5) in the form

$$X^2 = \sum_{i=1}^{k} \frac{\{n_i - n p_{0i}(\mathbf{t})\}^2}{n p_{0i}(\mathbf{t})}.$$

This clearly changes our distribution problem, for now the $p_{0i}(\mathbf{t})$ are themselves random variables, and it is not obvious that the asymptotic distribution of X^2 will be of the same form as in the case of a simple H_0. In fact, the term $n_i - n p_{0i}(\mathbf{t})$ does not necessarily have a zero expectation. We may write X^2 identically as

$$X^2 = \sum_{i=1}^{k} \frac{1}{n p_{0i}(\mathbf{t})} [\{n_i - n p_{0i}(\boldsymbol{\theta})\}^2 + n^2 \{p_{0i}(\mathbf{t}) - p_{0i}(\boldsymbol{\theta})\}^2$$
$$- 2n \{n_i - n p_{0i}(\boldsymbol{\theta})\} \{p_{0i}(\mathbf{t}) - p_{0i}(\boldsymbol{\theta})\}]. \tag{30.9}$$

Now we know from the theory of the multinomial distribution that asymptotically

$$n_i - n p_{0i}(\boldsymbol{\theta}) \sim c n^{\frac{1}{2}},$$

so that the first term in the square brackets in (30.9) is of order n. If we also have

$$p_{0i}(\mathbf{t}) - p_{0i}(\boldsymbol{\theta}) = o(n^{-\frac{1}{4}}), \tag{30.10}$$

the second and third term will be of order less than n, and relatively negligible, so that (30.9) asymptotically behaves like its first term. Even this, however, still has the random variable $n p_{0i}(\mathbf{t})$ as its denominator, but to the same order of approximation we may replace this by $n p_{0i}(\boldsymbol{\theta})$. We thus see that if (30.10) holds, (30.8) behaves asymptotically just as (30.5)—it is distributed in the χ^2 form with $(k-1)$ degrees of freedom. However, if the $p_{0i}(\mathbf{t})$ are " well-behaved " functions of $\mathbf{t}$, they will differ from the $p_{0i}(\theta)$ by the same order of magnitude as $\mathbf{t}$ does from $\boldsymbol{\theta}$. Then for all practical purposes (30.10) requires that

$$\mathbf{t} - \boldsymbol{\theta} = o(n^{-\frac{1}{2}}). \tag{30.11}$$

(30.11) is not customarily satisfied, since we usually have estimators with variances and covariances of order n^{-1} and then

$$\mathbf{t} - \boldsymbol{\theta} = O(n^{-\frac{1}{2}}). \tag{30.12}$$

In this "regular" case, therefore, our argument above does not hold. But it does hold in cases where estimators have variances of order n^{-2}, as we have found to be characteristic of estimators of parameters which locate the end-points of the range of a variable (cf. Exercises 14.8, 14.13 and 32.11). In such cases, therefore, we may use (30.8) with no new theory required. In the more common case where (30.12) holds, we must investigate further.

30.9 It will simplify our discussion if we first give Fisher's (1922a) alternative and revealing proof of the asymptotic distribution of (30.5) for the simple hypothesis case.

Suppose that we have k independent Poisson variates, the ith having parameter np_{0i}, where $n = \sum_{i=1}^{k} n_i$ and $\sum_i p_{0i} = 1$. The probability that the first takes the value n_1, the second n_2 and so on, is

$$P(n_1, n_2, \ldots, n_k, n) = \prod_{i=1}^{k} e^{-np_{0i}} (np_{0i})^{n_i}/n_i! = e^{-n} n^n \prod_{i=1}^{k} p_{0i}^{n_i}/n_i!; \tag{30.13}$$

n appears explicitly as a variate in (30.13), although of course the resulting $(k+1)$-variate distribution is singular since $n = \sum_i n_i$. Its marginal distribution $g(n)$ is easily found, since the sum of the k independent Poisson variables is itself (cf. Example 11.17, Vol. 1, 3rd edn) a Poisson variable with parameter equal to $\sum_{i=1}^{k} np_{0i} = n$. Thus

$$g(n) = e^{-n} n^n / n! \tag{30.14}$$

and

$$h(n_1, n_2, \ldots, n_k | n) = \frac{P(n_1, n_2, \ldots, n_k, n)}{g(n)}$$

$$= \frac{n!}{n_1! n_2! \ldots n_k!} p_{01}^{n_1} p_{02}^{n_2} \cdots p_{0k}^{n_k}. \tag{30.15}$$

We see at once that (30.15) is precisely the multinomial distribution of the n_i on which our test procedure is based. Thus, as an alternative to the proof of the asymptotic distribution of X^2 given in Example 15.3 (cf. **30.5**), we may obtain it by regarding the n_i as the values of k independent Poisson variables with parameters np_{0i}, conditional upon n being fixed. By Example 4.9, the standardized variable

$$x_i = \frac{n_i - np_{0i}}{(np_{0i})^{\frac{1}{2}}} \tag{30.16}$$

is asymptotically normal as $n \to \infty$. Hence, as $n \to \infty$,
$$X^2 = \sum_{i=1}^{k} x_i^2$$
is the sum of squares of k independent normal variates, subject to the single condition $\sum_i n_i = n$, which is equivalent to $\sum_i (np_{0i})^{\frac{1}{2}} x_i = 0$. By Example 11.6, X^2 therefore has a χ^2 distribution asymptotically, with $(k-1)$ degrees of freedom.

30.10 The utility of this alternative proof is that, in conjunction with Example 11.6, to which it refers, it shows clearly that if s further homogeneous linear conditions are imposed on the n_i, the only effect on the asymptotic distribution of X^2 will be to reduce the degrees of freedom from $(k-1)$ to $(k-s-1)$.

We now return to the composite hypothesis of **30.8** in the case when (30.12) holds. Suppose that we choose as our set of estimators $\mathbf{t}$ of $\boldsymbol{\theta}$ the Maximum Likelihood (or other asymptotically equivalent efficient) estimators, so that $\mathbf{t} = \hat{\boldsymbol{\theta}}$. Now the Likelihood Function L in this case is simply the multinomial (30.15) regarded as a function of the θ_j, on which the p_{0i} depend. Thus, generalizing (18.37) (where $s = 1$),
$$\frac{\partial \log L}{\partial \theta_j} = \sum_{i=1}^{k} n_i \frac{\partial p_{0i}}{\partial \theta_j} \frac{1}{p_{0i}}, \qquad j = 1, 2, \ldots, s, \tag{30.17}$$
and the ML estimators in this regular case are the roots of the s equations obtained by equating (30.17) to zero for each j. Clearly, each such equation is a homogeneous linear relationship among the n_i. We thus see that, in this regular case, we have s additional constraints imposed by the process of efficient estimation of $\boldsymbol{\theta}$ from the multinomial distribution, so that the statistic (30.8) is asymptotically distributed in the χ^2 form with $(k-s-1)$ degrees of freedom. A more rigorous and detailed proof is given by Cramér (1946)—see also Birch (1964b). We shall call $\hat{\boldsymbol{\theta}}$ the *multinomial* ML estimators.

The effect of estimation on the distribution of X^2

30.11 We may now, following Watson (1959), consider the general problem of the effect of estimating the unknown parameters on the asymptotic distribution of the X^2 statistic. We confine ourselves to the regular case, when (30.12) holds, and we write for any estimator $\mathbf{t}$ of $\boldsymbol{\theta}$
$$\mathbf{t} - \boldsymbol{\theta} = n^{-\frac{1}{2}} \mathbf{A} \mathbf{x} + o(n^{-\frac{1}{2}}). \tag{30.18}$$
where $\mathbf{A}$ is an arbitrary $(s \times k)$ matrix and $\mathbf{x}$ is the $(k \times 1)$ vector whose ith element is
$$x_i = \frac{n_i - np_{0i}(\boldsymbol{\theta})}{\{np_{0i}(\boldsymbol{\theta})\}^{\frac{1}{2}}}, \tag{30.19}$$
defined just as at (30.16) for the simple hypothesis case; we assume $\mathbf{A}$ to have been chosen so that
$$E(\mathbf{A}\mathbf{x}) = \mathbf{0}. \tag{30.20}$$
It follows at once from (30.18) and (30.20) that the dispersion matrix of $\mathbf{t}$, $\mathbf{V}(\mathbf{t})$, is of order n^{-1}. By a Taylor expansion applied to $\{p_{0i}(\mathbf{t}) - p_{0i}(\boldsymbol{\theta})\}$ in (30.9), we find that we may write
$$X^2 = \sum_{i=1}^{k} y_i^2$$

where, as $n \to \infty$,

$$y_i = x_i - n^{\frac{1}{2}} \sum_{j=1}^{s} (t_j - \theta_j) \frac{\partial p_{0i}(\boldsymbol{\theta})}{\partial \theta_j} \frac{1}{\{p_{0i}(\boldsymbol{\theta})\}^{\frac{1}{2}}} + o(1),$$

or, in matrix form,

$$\mathbf{y} = \mathbf{x} - n^{\frac{1}{2}} \mathbf{B}(\mathbf{t} - \boldsymbol{\theta}) + o(1), \tag{30.21}$$

where $\mathbf{B}$ is the $(k \times s)$ matrix whose (i,j)th element is

$$b_{ij} = \frac{\partial p_{0i}(\boldsymbol{\theta})}{\partial \theta_j} \frac{1}{\{p_{0i}(\boldsymbol{\theta})\}^{\frac{1}{2}}}. \tag{30.22}$$

Substituting (30.18) into (30.21), we find simply

$$\mathbf{y} = (\mathbf{I} - \mathbf{B}\mathbf{A})\mathbf{x} + o(1). \tag{30.23}$$

30.12 Now from equation (30.19) the x_i have zero means. As $n \to \infty$, they tend to multinormality, by the multivariate Central Limit theorem, and their dispersion matrix is, temporarily writing p_i for $p_{0i}(\boldsymbol{\theta})$,

$$\mathbf{V}(\mathbf{x}) = \begin{pmatrix} 1-p_1, & -(p_1 p_2)^{\frac{1}{2}}, & -(p_1 p_3)^{\frac{1}{2}}, & \ldots, & -(p_1 p_k)^{\frac{1}{2}} \\ -(p_2 p_1)^{\frac{1}{2}}, & 1-p_2, & -(p_2 p_3)^{\frac{1}{2}}, & \ldots, & -(p_2 p_k)^{\frac{1}{2}} \\ \vdots & & & & \\ -(p_k p_1)^{\frac{1}{2}}, & -(p_k p_2)^{\frac{1}{2}}, & -(p_k p_3)^{\frac{1}{2}}, & \ldots, & 1-p_k \end{pmatrix}$$
$$= \mathbf{I} - (\mathbf{p}^{\frac{1}{2}})(\mathbf{p}^{\frac{1}{2}})' \tag{30.24}$$

where $\mathbf{p}^{\frac{1}{2}}$ is the $(k \times 1)$ vector with ith element $\{p_{0i}(\boldsymbol{\theta})\}^{\frac{1}{2}}$. It follows at once from (30.23) and (30.24) that the y_i also are asymptotically normal with zero means and dispersion matrix

$$\mathbf{V}(\mathbf{y}) = (\mathbf{I} - \mathbf{B}\mathbf{A})\{\mathbf{I} - (\mathbf{p}^{\frac{1}{2}})(\mathbf{p}^{\frac{1}{2}})'\}(\mathbf{I} - \mathbf{B}\mathbf{A})'. \tag{30.25}$$

Thus $X^2 = \mathbf{y}'\mathbf{y}$ is asymptotically distributed as the sum of squares of k normal variates with zero means and dispersion matrix (30.25). If and only if $\mathbf{V}(\mathbf{y})$ is idempotent with r latent roots unity and $(k-r)$ zero, so that its trace is equal to r, we see as in **15.10-11** that the distribution of X^2 is of the χ^2 form with r degrees of freedom.

30.13 We now consider particular cases of (30.25). First, the case of a simple hypothesis, where no estimation is necessary, is formally obtainable by putting $\mathbf{A} \equiv \mathbf{0}$ in (30.18). (30.25) then becomes simply

$$\mathbf{V}(\mathbf{y}) = \mathbf{V}(\mathbf{x}) = \mathbf{I} - (\mathbf{p}^{\frac{1}{2}})(\mathbf{p}^{\frac{1}{2}})'. \tag{30.26}$$

Since $(\mathbf{p}^{\frac{1}{2}})'(\mathbf{p}^{\frac{1}{2}}) = \sum_{i=1}^{k} p_{0i}(\boldsymbol{\theta}) = 1$, (30.26) is seen on squaring it to be idempotent, and its trace is $(k-1)$. Thus X^2 is a χ^2_{k-1} variable in this case, as we already know from two different proofs.

30.14 The composite hypothesis case is not so straightforward. First, suppose as in **30.10** that the multinomial ML estimators $\hat{\boldsymbol{\theta}}$ are used. We seek the form of the matrix **A** in (30.8) when $\mathbf{t} = \hat{\boldsymbol{\theta}}$. Now we know from **18.26** that the elements of the reciprocal of the dispersion matrix of $\hat{\boldsymbol{\theta}}$ are asymptotically given by

$$\{\mathbf{V}(\hat{\boldsymbol{\theta}})\}_{jl}^{-1} = -E\left\{\frac{\partial^2 \log L}{\partial \theta_j \partial \theta_l}\right\}, \quad j, l = 1, 2, \ldots, s. \tag{30.27}$$

From (30.17), the multinomial ML equations give

$$\frac{\partial^2 \log L}{\partial \theta_j \partial \theta_l} = \sum_{i=1}^{k} \frac{n_i}{p_{0i}}\left(\frac{\partial^2 p_{0i}}{\partial \theta_j \partial \theta_l} - \frac{1}{p_{0i}}\frac{\partial p_{0i}}{\partial \theta_j}\frac{\partial p_{0i}}{\partial \theta_l}\right). \tag{30.28}$$

On taking expectations in (30.28), we find

$$-E\left\{\frac{\partial^2 \log L}{\partial \theta_j \partial \theta_l}\right\} = n\left\{\sum_{i=1}^{k} \frac{1}{p_{0i}}\frac{\partial p_{0i}}{\partial \theta_j}\frac{\partial p_{0i}}{\partial \theta_l} - \sum_{i=1}^{k}\frac{\partial^2 p_{0i}}{\partial \theta_j \partial \theta_l}\right\}. \tag{30.29}$$

The second term on the right of (30.29) is zero, since it is $\frac{\partial^2}{\partial \theta_j \partial \theta_l}\sum_{i=1}^{k} p_{0i}$. Thus, using (30.22),

$$-E\left\{\frac{\partial^2 \log L}{\partial \theta_j \partial \theta_l}\right\} = n\sum_{i=1}^{k} b_{ij}b_{il},$$

so that, from (30.27),

$$\{\mathbf{V}(\hat{\boldsymbol{\theta}})\}^{-1} = n\mathbf{B}'\mathbf{B}$$

or

$$\mathbf{C} = n\mathbf{V}(\hat{\boldsymbol{\theta}}) = (\mathbf{B}'\mathbf{B})^{-1}. \tag{30.30}$$

But from (30.18) and (30.24) we have

$$\mathbf{D} = n\mathbf{V}(\mathbf{t}) = \mathbf{A}\mathbf{V}(\mathbf{x})\mathbf{A}' = \mathbf{A}\{\mathbf{I} - (\mathbf{p}^{\frac{1}{2}})(\mathbf{p}^{\frac{1}{2}})'\}\mathbf{A}'. \tag{30.31}$$

Here (30.30) and (30.31) are alternative expressions for the same matrix.

We choose **A** to satisfy (30.30) by noting that

$$\mathbf{B}'(\mathbf{p}^{\frac{1}{2}}) = \mathbf{0} \tag{30.32}$$

(since the jth element of this $(s \times 1)$ vector is $\frac{\partial}{\partial \theta_j}\sum_{i=1}^{k} p_{0i} = 0$), and hence that if $\mathbf{A} = \mathbf{G}\mathbf{B}'$ where **G** is symmetric and non-singular, (30.31) gives $\mathbf{GB}'\mathbf{BG}'$. If this is to be equal to (30.30), we obviously have $\mathbf{G} = (\mathbf{B}'\mathbf{B})^{-1}$, so finally

$$\mathbf{A} = (\mathbf{B}'\mathbf{B})^{-1}\mathbf{B}' \tag{30.33}$$

in the case of multinomial ML estimation. (30.25) then becomes, using (30.32),

$$\mathbf{V}(\mathbf{y}) = \{\mathbf{I} - \mathbf{B}(\mathbf{B}'\mathbf{B})^{-1}\mathbf{B}'\}^2 - (\mathbf{p}^{\frac{1}{2}})(\mathbf{p}^{\frac{1}{2}})'$$
$$= \mathbf{I} - \mathbf{B}(\mathbf{B}'\mathbf{B})^{-1}\mathbf{B}' - (\mathbf{p}^{\frac{1}{2}})(\mathbf{p}^{\frac{1}{2}})'. \tag{30.34}$$

By squaring, this matrix is shown to be idempotent. Its rank is equal to its trace, which as in **19.9** is given by

$$\operatorname{tr}\mathbf{V}(\mathbf{y}) = \operatorname{tr}\{\mathbf{I} - (\mathbf{p}^{\frac{1}{2}})(\mathbf{p}^{\frac{1}{2}})'\} - \operatorname{tr}\mathbf{B}'.\mathbf{B}(\mathbf{B}'\mathbf{B})^{-1},$$

and using **30.13** this is

$$\operatorname{tr}\mathbf{V}(\mathbf{y}) = (k-1) - s.$$

Thus the distribution of X^2 is χ^2_{k-s-1} asymptotically, as we saw in **30.10**.

30.15 Our present approach enables us to inquire further: what happens to the asymptotic distribution of X^2 if some other estimators than $\hat{\theta}$ are used? This question was first considered in the simplest case by Fisher (1928b). Chernoff and Lehmann (1954) considered a case of particular interest, when the estimators used are the ML estimators based on the n individual observations and not the multinomial ML estimators $\hat{\theta}$, based on the k frequencies n_i, which we have so far discussed. If we have the values of the n observations, it is clearly an efficient procedure to utilize this knowledge in estimating θ, even though we are going to use the k-class frequencies alone in carrying out the test of fit. We shall find, however, that the X^2 statistic obtained in this way no longer has an asymptotic χ^2 distribution.

30.16 Let us return to the general expression (30.25) for the dispersion matrix. Multiplying it out, we rewrite it

$$\mathbf{V}(\mathbf{y}) = \{\mathbf{I} - (\mathbf{p}^{\frac{1}{2}})(\mathbf{p}^{\frac{1}{2}})'\} - \mathbf{BA}\{\mathbf{I} - (\mathbf{p}^{\frac{1}{2}})(\mathbf{p}^{\frac{1}{2}})'\}$$
$$- \{\mathbf{I} - (\mathbf{p}^{\frac{1}{2}})(\mathbf{p}^{\frac{1}{2}})'\}\mathbf{A}'\mathbf{B}' + \mathbf{BA}\{\mathbf{I} - (\mathbf{p}^{\frac{1}{2}})(\mathbf{p}^{\frac{1}{2}})'\}\mathbf{A}'\mathbf{B}'. \quad (30.35)$$

Rather than find the latent roots λ_i of $\mathbf{V}(\mathbf{y})$, we consider those of $\mathbf{I} - \mathbf{V}(\mathbf{y})$, which are $1 - \lambda_i$. We write this matrix in the form

$$\mathbf{I} - \mathbf{V}(\mathbf{y}) = (\mathbf{p}^{\frac{1}{2}})(\mathbf{p}^{\frac{1}{2}})' + \mathbf{B}[\mathbf{A}\{\mathbf{I} - (\mathbf{p}^{\frac{1}{2}})(\mathbf{p}^{\frac{1}{2}})'\} - \tfrac{1}{2}\mathbf{A}\{\mathbf{I} - (\mathbf{p}^{\frac{1}{2}})(\mathbf{p}^{\frac{1}{2}})'\}\mathbf{A}'\mathbf{B}']$$
$$+ [\{\mathbf{I} - (\mathbf{p}^{\frac{1}{2}})(\mathbf{p}^{\frac{1}{2}})'\}\mathbf{A}' - \tfrac{1}{2}\mathbf{BA}\{\mathbf{I} - (\mathbf{p}^{\frac{1}{2}})(\mathbf{p}^{\frac{1}{2}})'\}\mathbf{A}']\mathbf{B}'$$
$$= (\mathbf{p}^{\frac{1}{2}})(\mathbf{p}^{\frac{1}{2}})' + \mathbf{B}[\mathbf{A}\{\mathbf{I} - (\mathbf{p}^{\frac{1}{2}})(\mathbf{p}^{\frac{1}{2}})'\} - \tfrac{1}{2}\mathbf{DB}']$$
$$+ [\{\mathbf{I} - (\mathbf{p}^{\frac{1}{2}})(\mathbf{p}^{\frac{1}{2}})'\}\mathbf{A}' - \tfrac{1}{2}\mathbf{BD}]\mathbf{B}'. \quad (30.36)$$

On substituting (30.31), (30.36) may be written as the product of two partitioned matrices, giving

$$\mathbf{I} - \mathbf{V}(\mathbf{y}) = \begin{pmatrix} (\mathbf{p}^{\frac{1}{2}}) \\ \mathbf{B} \\ \{\mathbf{I} - (\mathbf{p}^{\frac{1}{2}})(\mathbf{p}^{\frac{1}{2}})'\}\mathbf{A}' - \tfrac{1}{2}\mathbf{BD} \end{pmatrix} \begin{pmatrix} (\mathbf{p}^{\frac{1}{2}})' \\ \mathbf{A}\{\mathbf{I} - (\mathbf{p}^{\frac{1}{2}})(\mathbf{p}^{\frac{1}{2}})'\} - \tfrac{1}{2}\mathbf{DB}' \\ \mathbf{B}' \end{pmatrix}'.$$

The matrices on the right may be transposed without affecting the non-zero latent roots. This converts their product from a $(k \times k)$ to a $(2s+1) \times (2s+1)$ matrix, which is reduced, on using (30.30) and (30.32), to

$$\begin{pmatrix} 1 & 0 & 0 \\ \hline 0 & & \\ & \mathbf{M} & \\ 0 & & \end{pmatrix} = \begin{pmatrix} 1 & 0_{1 \times s}, & 0_{1 \times s} \\ \hline 0_{s \times 1} & \mathbf{AB} - \tfrac{1}{2}\mathbf{DC}^{-1}, & \mathbf{D} - \tfrac{1}{2}\mathbf{ABD} - \tfrac{1}{2}\mathbf{DB}'\mathbf{A} + \tfrac{1}{4}\mathbf{DC}^{-1}\mathbf{D} \\ 0_{s \times 1} & \mathbf{C}^{-1}, & \mathbf{B}'\mathbf{A}' - \tfrac{1}{2}\mathbf{C}^{-1}\mathbf{D} \end{pmatrix}.$$

(30.37)

(30.37) has one latent root of unity and $2s$ others which are those of the matrix $\mathbf{M}$ partitioned off in its south-east corner. If $k \geqslant 2s+1$, which is almost invariably the case in applications, this implies that (30.36) has $(k-2s-1)$ zero latent roots, one of unity, and $2s$ others which are the roots of $\mathbf{M}$. Thus for $\mathbf{V}(\mathbf{y})$ itself, we have $(k-2s-1)$ latent roots of unity, one of zero and $2s$ which are the complements to unity of the latent roots of $\mathbf{M}$.

30.17 We now consider the problem introduced in **30.15**. Suppose that, to estimate θ, we use the ML estimators based on the n individual observations, which we shall call the "ordinary ML estimators" and denote by $\hat{\theta}^*$. We know from **18.26** that if f is the frequency function of the observations, we have asymptotically

$$\mathbf{D} = n\mathbf{V}(\hat{\theta}^*) = -\left\{E\left(\frac{\partial^2 \log f}{\partial \theta_j \partial \theta_i}\right)\right\}^{-1} \tag{30.38}$$

and that the elements of $\hat{\theta}^*$ are the roots of

$$\frac{\partial \log L}{\partial \theta_i} = 0, \quad i = 1, 2, \ldots, s,$$

where L is now the ordinary (not the multinomial) Likelihood Function. Thus if θ_0 is the true value, we have the Taylor expansion

$$0 = \left[\frac{\partial \log L}{\partial \theta_i}\right]_{\theta_j = \hat{\theta}_j^*} = \left[\frac{\partial \log L}{\partial \theta_i}\right]_{\theta_j = \theta_{0j}} + (\hat{\theta}_j^* - \theta_{0j})\left[\frac{\partial^2 \log L}{\partial \theta_i \partial \theta_j}\right]_{\theta_j = \theta_{0j} + \varepsilon} \quad i, j = 1, 2, \ldots, s,$$

and as in **18.26** this gives asymptotically, using (30.38),

$$n^{\frac{1}{2}}(\hat{\theta}^* - \theta) = n^{-\frac{1}{2}}\mathbf{D}\left(\frac{\partial \log \mathbf{L}}{\partial \theta}\right). \tag{30.39}$$

In this case, we evaluate $\mathbf{V}(\mathbf{y})$ directly from (30.21) with $\mathbf{t} = \hat{\theta}^*$, where we see that asymptotically

$$\mathbf{V}(\mathbf{y}) = E\{\mathbf{x} - n^{\frac{1}{2}}\mathbf{B}(\hat{\theta}^* - \theta)\}\{\mathbf{x} - n^{\frac{1}{2}}\mathbf{B}(\hat{\theta}^* - \theta)\}'$$
$$= \mathbf{V}(\mathbf{x}) + n\mathbf{B}\mathbf{V}(\hat{\theta}^*)\mathbf{B}' - (\mathbf{T}\mathbf{B}' + \mathbf{B}\mathbf{T}') \tag{30.40}$$

where

$$\mathbf{T} = E\{\mathbf{x}.n^{\frac{1}{2}}(\hat{\theta}^* - \theta)'\} = E\left\{\mathbf{x}.n^{-\frac{1}{2}}\left(\frac{\partial \log \mathbf{L}}{\partial \theta}\right)'\right\}\mathbf{D}$$

asymptotically by (30.39). We write this $\mathbf{T} = \mathbf{E}\mathbf{D}$.

If we write $\delta_{ir} = 1$ if the rth observation falls into the ith class, C_i, and $\delta_{ir} = 0$ otherwise, we can write (30.19) as

$$x_i = \sum_{r=1}^{n} (\delta_{ir} - p_{0i})/(np_{0i})^{\frac{1}{2}},$$

so that the characteristic term of $\mathbf{E}$ is

$$n^{-1}p_{0i}^{-\frac{1}{2}}E\left\{\sum_{r=1}^{n}(\delta_{ir} - p_{0i}).\frac{\partial \log L}{\partial \theta_j}\right\} = p_{0i}^{-\frac{1}{2}}E\left\{(\delta_{ir} - p_{0i})\frac{\partial \log L}{\partial \theta_j}\right\}$$
$$= p_{0i}^{-\frac{1}{2}}E\left\{\delta_{ir}\frac{\partial \log f(x_r|\theta)}{\partial \theta_j}\right\},$$

where we have used (17.18) and the independence of the observations. Thus we have

$$p_{0i}^{-\frac{1}{2}}\int_{C_i}\frac{\partial \log f}{\partial \theta_j}f\,dx = p_{0i}^{-\frac{1}{2}}\int_{C_i}\frac{\partial f}{\partial \theta_j}dx = p_{0i}^{-\frac{1}{2}}\frac{\partial}{\partial \theta_j}\int_{C_i}f\,dx = p_{0i}^{-\frac{1}{2}}\frac{\partial}{\partial \theta_j}p_{0i}.$$

This is precisely b_{ij} of (30.22). Thus

$$\mathbf{E} = \mathbf{B}, \quad \mathbf{T} = \mathbf{B}\mathbf{D}. \tag{30.41}$$

Using (30.24), (30.38) and (30.41) in (30.40), we obtain

$$\mathbf{V}(\mathbf{y}) = \mathbf{I} - (\mathbf{p}^{\frac{1}{2}})(\mathbf{p}^{\frac{1}{2}})' - \mathbf{B}\mathbf{D}\mathbf{B}'. \tag{30.42}$$

(30.42) is not idempotent, as may be seen by squaring it. Moreover, (30.41) shows that the non-negative dispersion matrix

$$\mathbf{V}\left\{\mathbf{B}'\mathbf{x} - n^{-\frac{1}{2}}\left(\frac{\partial \log \mathbf{L}}{\partial \theta}\right)\right\} = \mathbf{B}'\mathbf{B} + \mathbf{D}^{-1} - (\mathbf{B}'\mathbf{E} + \mathbf{E}'\mathbf{B}) = \mathbf{D}^{-1} - \mathbf{B}'\mathbf{B} = \mathbf{D}^{-1} - \mathbf{C}^{-1},$$

using (30.30). Thus $\mathbf{C}-\mathbf{D}$ is non-negative and (30.42) may be written in the form
$$\mathbf{V}(\mathbf{y}) = \mathbf{I} - (\mathbf{p}^{\frac{1}{2}})(\mathbf{p}^{\frac{1}{2}})' - \mathbf{BCB}' + \mathbf{B}(\mathbf{C}-\mathbf{D})\mathbf{B}'. \tag{30.43}$$
The first two terms on the right of (30.43) are what we got at (30.26) for the case when no estimation takes place, when $\mathbf{V}(\mathbf{y})$ has $(k-1)$ latent roots unity, and one of zero. The first three terms are (30.34), when, with multinomial ML estimation, $\mathbf{V}(\mathbf{y})$ has $(k-s-1)$ latent roots unity and $(s+1)$ of zero. Because of the non-negative definiteness of all the terms, reduction of (30.43) to canonical form shows that the latent roots of (30.43) are bounded by the corresponding latent roots of (30.26) and (30.34). Thus (30.43) has $(k-s-1)$ latent roots of unity, one of zero, and s between zero and unity, as established by Chernoff and Lehmann (1954).

In general, the values of the s latent roots depend upon $\boldsymbol{\theta}$, but if $\boldsymbol{\theta}$ contains only a location (and possibly also a scale) parameter, this is not so—cf. Exercise 30.21. In any case, it follows from the fact that the two sets of ML estimators $\hat{\boldsymbol{\theta}}$ and $\hat{\boldsymbol{\theta}}^*$ draw closer together as k increases, so that $\mathbf{D} \to \mathbf{C}$, that the last s latent roots tend to zero as $k \to \infty$.

30.19 What we have found, therefore, is that X^2 does not have an asymptotic χ^2 distribution when fully efficient (ordinary ML) estimators are used in estimating parameters—there is a partial recovery of the s degrees of freedom lost by the multinomial ML estimators. However, the distribution of X^2 is bounded between a χ^2_{k-1} and a χ^2_{k-s-1} variable, and as k becomes large these are so close together that the difference can be ignored—this is another way of expressing the final sentence in **30.18**. But for k small, the effect of using the χ^2_{k-s-1} distribution for test purposes may lead to serious error; for the probability of exceeding any given value will be greater than we suppose. s is rarely more than 1 or 2, but it is as well to be sure, when ordinary ML estimation is being used, that the critical values of χ^2_{k-s-1} and χ^2_{k-1} are both exceeded by X^2. The tables of χ^2 show that, for a test of size $\alpha = 0.05$, the critical value for $(k-1)$ degrees of freedom exceeds that for $(k-s-1)$ degrees of freedom, if s is small, by Cs, approximately, where C declines from about 1·5 at $(k-s-1) = 5$ to about 1·2 when $(k-s-1) = 30$. For $\alpha = 0.01$, the corresponding values of C are about 1·7 and 1·3.

The choice of classes for the X^2 test

30.20 The whole of the asymptotic theory of the X^2 test, which we have discussed so far is valid however we determine the k classes into which the observations are grouped, so long as they are determined *without reference to the observations*. The italicized condition is essential, for we have made no provision for the class boundaries themselves being random variables. However, it is common practice to determine the class boundaries, and sometimes even to fix k itself, after reference to the general picture presented by the observations. We must therefore discuss the formation of classes, and then consider how far it affects the theory we have developed.

We first consider the determination of class boundaries, leaving the choice of k until later. If there is a natural discreteness imposed by the problem (as in Example 30.1 where there are four natural groups) or if we have a sample of observations from a discrete distribution, the class-boundary problem arises only in the sense that we may decide (in order to reduce k, or in order to improve the accuracy of the asymptotic distribution of X^2 as we shall see in **30.30** below) to amalgamate some of the hypothetical frequencies at the discrete points. Indeed, if a discrete distribution has infinite range, like the Poisson, we are forced to some amalgamation of hypothetical frequencies if k is not to be infinite with most of the hypothetical frequencies very small indeed. But the class-boundary problem arises in its most acute form only when we are sampling from a continuous distribution. There are now no natural hypothetical frequencies at all. If we suppose k to be determined in advance in some way, how are the boundaries to be determined?

In practice, arithmetical convenience is usually allowed to dictate the solution: the classes are taken to cover equal ranges of the variate, except at an extreme where the range of the variate is infinite. The range of a class is roughly determined by the dispersion of the distribution, while the location of the distribution helps to determine where the central class should fall. Thus, if we wished to form $k = 10$ classes for a sample to be tested for normality, we might roughly estimate (perhaps by eye) the mean $\bar{x}$ and the standard deviation s of the sample and take the class-boundaries as $\bar{x} \pm \frac{1}{2}sj$, $j = 0, 1, 2, 3, 4$. The classes would then be

$(-\infty, \bar{x}-2s)$, $(\bar{x}-2s, \bar{x}-1\cdot5s)$, $(\bar{x}-1\cdot5s, \bar{x}-s)$, $(\bar{x}-s, \bar{x}-0\cdot5s)$, $(\bar{x}-0\cdot5s, \bar{x})$,
$(\bar{x}, \bar{x}+0\cdot5s)$, $(\bar{x}+0\cdot5s, \bar{x}+s)$, $(\bar{x}+s, \bar{x}+1\cdot5s)$, $(\bar{x}+1\cdot5s, \bar{x}+2s)$, $(\bar{x}+2s, \infty)$.

30.21 Although this procedure is not very precise, it clearly makes the class-boundaries random variables, and it is not obvious that the asymptotic distribution of X^2, calculated for classes formed in this way, is the same as when the classes are fixed. However, intuition suggests that since the asymptotic theory holds for *any* set of k fixed classes, it should hold also when the class-boundaries are determined from the sample. That this is so when the class-boundaries are determined by consistent estimation of parameters in the regular case was shown for the normal distribution by Watson (1957b) and for continuous distributions in general by A. R. Roy (1956), Watson (1958, 1959) and Chibisov (1971)—cf. also Dahiya and Gurland (1972). D. S. Moore (1971) extends the results to the multivariate case.

We may thus neglect the random variations of the class-boundaries so far as the asymptotic distribution of X^2, when H_0 holds, is concerned. Small-sample distributions, of course, will be affected, but nothing is yet known of the precise effects. (We discuss small-sample distributions of X^2 in the fixed-boundaries case in **30.30** below.)

The equal-probabilities method of constructing classes

30.22 We may now directly face the question of how class-boundaries should be determined, in the light of the assurance of the last paragraph of **30.21**. If we now seek an optimum method of boundary determination, it must be in terms of the power of the test; we should choose that set of boundaries which maximizes power for a test of given size. Unfortunately, there is as yet no method available for doing this, although it is to be hoped that the recent re-awakening of interest in the theory of X^2 tests will stimulate research in this field. We must therefore seek some means of avoiding the unpleasant fact that there is a multiplicity of possible sets of classes, any one of which will in general give a different result for the same data; we require a rule which is plausible and practical.

One such rule has been suggested by Mann and Wald (1942) and by Gumbel (1943); given k, choose the classes so that the hypothetical probabilities p_{0i} are all equal to $1/k$. This procedure is perfectly definite and unique. It varies arithmetically from the usual method, described in **30.20** (in which the classes are variate-intervals of equal width) in that we have to use tables to ensure that the p_{0i} are equal. This requires for exactness that the data should be available ungrouped. The procedure is illustrated in Example 30.2.

Example 30.2

Quenouille (1959) gives, apart from a change in location, 1000 random deviates from the distribution
$$dF = \exp(-x)\,dx, \qquad 0 \leqslant x \leqslant \infty.$$
The first 50 of these, arranged in order of variate-value, are:
0·01, 0·01, 0·04, 0·17, 0·18, 0·22, 0·22, 0·25, 0·25, 0·29, 0·42, 0·46, 0·47, 0·47, 0·56, 0·59, 0·67, 0·68, 0·70, 0·72, 0·76, 0·78, 0·83, 0·85, 0·87, 0·93, 1·00, 1·01, 1·01, 1·02, 1·03, 1·05, 1·32, 1·34, 1·37, 1·47, 1·50, 1·52, 1·54, 1·59, 1·71, 1·90, 2·10, 2·35, 2·46, 2·46, 2·50, 3·73, 4·07, 6·03.

Suppose that we wished to form four classes for a X^2 test. A natural grouping with equal-width intervals would be

Variate-values	Observed frequency	Hypothetical frequency
0–0·50	14	19·7
0·51–1·00	13	11·9
1·01–1·50	10	7·2
1·51 and over	13	11·2
	50	50·0

The hypothetical frequencies are obtained from the *Biometrika Tables* distribution function of a χ^2 variable with 2 degrees of freedom, which is just twice a variable with the distribution above. We find $X^2 = 3·1$ with 3 degrees of freedom, a value which would not reject the hypothetical parent distribution for any test of size less than $\alpha = 0·37$; the agreement of observation and hypothesis is therefore very satisfactory.

Let us now consider how the same data would be treated by the method of **30.22**. We first determine the values of the hypothetical variable, dividing it into four equal-probability classes—these are, of course, the quartiles. The *Biometrika Tables* give the values 0·288, 0·693, 1·386. We now form the table:

Variate-values	Observed frequency	Hypothetical frequency
0–0·28	9	12·5
0·29–0·69	9	12·5
0·70–1·38	17	12·5
1·39 and over	15	12·5
	50	50·0

X^2 is now easier to calculate, since (30.8) reduces to

$$X^2 = \frac{k}{n} \sum_{i=1}^{k} n_i^2 - n \qquad (30.44)$$

since all hypothetical probabilities $p_{0i} = 1/k$. We find here that $X^2 = 3\cdot 9$, which would not lead to rejection unless the test size exceeded $0\cdot 27$. The result is still very satisfactory, but the equal-probabilities test seems rather more critical of the hypothesis than the other test was.

It will be seen that there is little extra arithmetical work involved in the equal-probabilities method of carrying out the X^2 test. Instead of a regular class-width, with hypothetical frequencies to be looked up in a table (or, if necessary, to be calculated) we have irregular class-widths determined from the tables so that the hypothetical frequencies are equal.

We have had no parameters to estimate in this example. If s parameters must be estimated, we can only equalize the *estimated* hypothetical probabilities.

30.23 Apart from the virtue of removing the class-boundary decision from uncertainty, the equal-probabilities method of forming classes for the X^2 test will not necessarily increase the power of the test, for one would suspect that a " goodness-of-fit " hypothesis is likely to be most vulnerable at the extremes of the range of the variable, and the equal-probabilities method may well result in a loss of sensitivity at the extremes unless k is rather large. This brings us to the question of how k should be chosen, and in order to discuss this question we must consider the power of the X^2 test. First, we investigate the moments of the X^2 statistic.

The moments of the X^2 test statistic

30.24 We suppose, as before, that we have hypothetical probabilities p_{0i} when H_0 holds, so that our test statistic is, as at (30.8),

$$X^2 = \frac{1}{n} \sum_{i=1}^{k} \frac{n_i^2}{p_{0i}} - n.$$

We confine ourselves to the simple hypothesis. Suppose now that the true probabilities are p_{1i}, $i = 1, 2, \ldots, k$. The expected value of the test statistic is then

$$E(X^2) = \frac{1}{n} \sum_{i=1}^{k} \frac{1}{p_{0i}} E(n_i^2) - n.$$

From the moments of the multinomial distribution at (5.80),

$$E(n_i^2) = n p_{1i}(1-p_{1i}) + n^2 p_{1i}^2, \qquad (30.45)$$

whence

$$E(X^2) = \sum_{i=1}^{k} \frac{p_{1i}(1-p_{1i})}{p_{0i}} + n \left\{ \sum_{i=1}^{k} \frac{p_{1i}^2}{p_{0i}} - 1 \right\}. \qquad (30.46)$$

When H_0 holds, this reduces to

$$E(X^2 | H_0) = k - 1. \qquad (30.47)$$

This exact result is already known to hold asymptotically, since X^2 is then a χ_{k-1}^2

variate. If we differentiate (30.46) with respect to the p_{1i}, subject to $\sum_i p_{1i} = 1$, we find that, as $n \to \infty$, (30.46) has its minimum value when $p_{1i} = p_{0i}$. For any hypothesis H_1 specifying a set of probabilities $p_{1i} \neq p_{0i}$, we therefore have asymptotically

$$E(X^2 | H_1) > k-1. \tag{30.48}$$

(30.48), like the asymptotic argument based on the LR statistic in **30.6**, indicates that the critical region for the X^2 test consists of the upper tail, although this indication is not conclusive since the asymptotic distribution of X^2 is not of the χ^2 form when H_1 holds. This alternative distribution is, in fact, a non-central χ^2 under the conditions given in **30.27** below.

Even the variance of X^2 is a relatively complicated function of the p_{0i} and p_{1i} (cf. Exercise 30.5). However, in the equal-probabilities case $\left(p_{0i} = \dfrac{1}{k}\right)$, the asymptotic variance simplifies considerably and we find (the proof is left to the reader as Exercise 30.3)

$$\left. \begin{array}{l} \text{var}(X^2 | H_0) \sim 2(k-1), \\ \text{var}(X^2 | H_1) \sim 4(n-1)k^2 \{\sum_i p_{1i}^3 - (\sum_i p_{1i}^2)^2\}. \end{array} \right\} \tag{30.49}$$

From (30.46) we also have in the equal-probabilities case

$$E(X^2 | H_1) = (k-1) + (n-1)(k \sum_i p_{1i}^2 - 1). \tag{30.50}$$

(30.50) is always greater than $(k-1)$ for any n.

Consistency and unbiassedness of the X^2 test

30.25 Equations (30.49) and (30.50) are sufficient to demonstrate the consistency of the equal-probabilities X^2 test. For the test consists in comparing the value of X^2 with a fixed critical value, say c_α, in the upper tail of its distribution. Now when H_1 holds, the mean value and variance of X^2 are each of order n. By Tchebycheff's inequality (3.95),

$$P\{|X^2 - E(X^2)| \geq \lambda [\text{var}(X^2)]^{\frac{1}{2}}\} \leq \frac{1}{\lambda^2}. \tag{30.51}$$

Since c_α is fixed, it differs from $E(X^2)$ by a quantity of order n, so that if we require the probability that X^2 differs from its mean sufficiently to fall below c_α, the multiplier λ on the left of (30.51) must be of order $n^{\frac{1}{2}}$, and the right-hand side is therefore of order n^{-1}. Thus

$$\lim_{n \to \infty} P\{X^2 < c_\alpha\} = 0,$$

and the test is consistent for any H_1 specifying unequal class-probabilities.

The general X^2 test, with unequal p_{0i}, is also consistent against alternatives specifying at least one $p_{1i} \neq p_{0i}$, as is intuitively reasonable. A proof is given by Neyman (1949).

30.26 Although the X^2 test is consistent (and therefore asymptotically unbiassed), one would not expect it to be unbiassed in general against very close alternatives for small n. However, Mann and Wald (1942) have shown that the equal-probabilities test is locally unbiassed. Write P for the power of the test and expand P in a Taylor series with the k values $\theta_i = p_{1i} - (1/k)$ as arguments. We then have

$$P(\theta_1, \theta_2, \ldots, \theta_k) = P(0, 0, \ldots, 0) + \Sigma \theta_i \frac{\partial P}{\partial \theta_i}$$
$$+ \tfrac{1}{2} \left\{ \sum_i \theta_i^2 \frac{\partial^2 P}{\partial \theta_i^2} + \sum\sum_{i \neq j} \theta_i \theta_j \frac{\partial^2 P}{\partial \theta_i \partial \theta_j} \right\}, \qquad (30.52)$$

all derivatives being taken at the H_0 point $(0, 0, \ldots, 0)$. For a size-α test,
$$P(0, 0, \ldots, 0) = \alpha.$$

Further, since P is a symmetric function of the θ_i, all the $\frac{\partial P}{\partial \theta_i}$ are equal at $(0, 0, \ldots, 0)$, and similarly for the $\frac{\partial^2 P}{\partial \theta_i^2}$ and the $\frac{\partial^2 P}{\partial \theta_i \partial \theta_j}$. We may therefore rewrite (30.52) as

$$P = \alpha + \frac{\partial P}{\partial \theta_1} \sum_i \theta_i + \tfrac{1}{2} \left\{ \frac{\partial^2 P}{\partial \theta_1^2} \sum_i \theta_i^2 + \frac{\partial^2 P}{\partial \theta_1 \partial \theta_2} \sum\sum_{i \neq j} \theta_i \theta_j \right\}. \qquad (30.53)$$

Now
$$0 = \sum_i \theta_i = (\sum_i \theta_i)^2 = \sum_i \theta_i^2 + \sum\sum_{i \neq j} \theta_i \theta_j.$$

Thus (30.53) becomes simply

$$P = \alpha + \tfrac{1}{2} \left(\frac{\partial^2 P}{\partial \theta_1^2} - \frac{\partial^2 P}{\partial \theta_1 \partial \theta_2} \right) \sum_i \theta_i^2 + \ldots \qquad (30.54)$$

We may evaluate the second-order derivatives in (30.54) directly from the exact expression for the power

$$P = \sum_{X^2 > c_\alpha} \frac{n!}{n_1! n_2! \ldots n_k!} p_{11}^{n_1} p_{12}^{n_2} \ldots p_{1k}^{n_k}. \qquad (30.55)$$

Since $\theta_i = p_{1i} - \frac{1}{k}$, we find from (30.55)

$$\frac{\partial^2 P}{\partial \theta_1^2} - \frac{\partial^2 P}{\partial \theta_1 \partial \theta_2} = \Sigma \{n_1(n_1 - 1) - n_1 n_2\} \frac{n!}{n_1! n_2! \ldots n_k!} \left(\frac{1}{k}\right)^{n-2}$$
$$= k^2 \Sigma \{n_1^2 - n_1 - n_1 n_2\} f_n, \qquad (30.56)$$

where $f_n = \frac{n!}{n_1! n_2! \ldots n_k!} k^{-n}$ and all unlabelled summations are now over the critical region $X^2 \geqslant c_\alpha$, which, from the form of X^2 given at (30.44), is equivalent to

$$\sum_{i=1}^{k} n_i^2 \geqslant b_\alpha.$$

Now $\frac{1}{\alpha} \Sigma n_1^2 f_n$ is the mean of n_1^2 in the critical region, and this must exceed its overall mean, which by (30.45) is $\frac{n}{k}\left(1 - \frac{1}{k} + \frac{n}{k}\right)$. Thus

$$\Sigma n_1^2 f_n = \alpha \frac{n}{k}\left(1 - \frac{1}{k} + \frac{n}{k}\right) + d, \qquad (30.57)$$

where $d > 0$. By exactly the same argument, since n_1 is positive, we have $\frac{1}{\alpha}\Sigma n_1 f_n > \frac{n}{k}$ or

$$\Sigma n_1 f_n = \alpha(n/k) + e \tag{30.58}$$

where $e > 0$. Moreover, we obviously have in (30.57) and (30.58)

$$d > e. \tag{30.59}$$

From symmetry, we have

$$\Sigma n_1 n_2 f_n = \Sigma \left\{ \frac{1}{k(k-1)} (n^2 - \sum_i n_i^2) \right\} f_n$$

$$= \frac{\alpha n^2}{k(k-1)} - \frac{1}{k-1} \Sigma n_1^2 f_n. \tag{30.60}$$

Using (30.57)–(30.60), (30.56) becomes

$$\frac{1}{k^2}\left[\frac{\partial^2 P}{\partial \theta_1^2} - \frac{\partial^2 P}{\partial \theta_1 \partial \theta_2}\right] > \frac{k}{k-1}\cdot\alpha\frac{n}{k}\left(1 - \frac{1}{k} + \frac{n}{k}\right) - \alpha\frac{n}{k} - \alpha\frac{n^2}{k(k-1)} = 0. \tag{30.61}$$

Thus, in (30.54), the second term on the right is positive. The higher-order terms neglected in (30.54) involve third and higher powers of the θ_i and will therefore be of smaller modulus than the second-order term near H_0. Thus, $P \geqslant \alpha$ near H_0 and the equal-probabilities test is locally unbiassed, which is a recommendation of this class-formation procedure, since no such result is known to hold for the X^2 test in general.

The limiting power function

30.27 Suppose that, as in our discussion of ARE in Chapter 25, we allow H_1 to approach H_0 as n increases, at a rate sufficient to keep the power bounded away from 1. In fact, let $p_{1i} - p_{0i} = c_i n^{-\frac{1}{2}}$ where the c_i are fixed. Then the distribution of X^2 is asymptotically a non-central χ^2 with degrees of freedom $k-s-1$ (where s parameters are estimated by the multinomial ML estimators) and non-central parameter

$$\lambda = \sum_{i=1}^{k} \frac{c_i^2}{p_{0i}} = n \sum_{i=1}^{k} \frac{(p_{1i} - p_{0i})^2}{p_{0i}}. \tag{30.62}$$

This result, first announced by Eisenhart (1938), follows at once from the representation of **30.9–10**; its proof is left to the reader as Exercise 30.4. It now follows by using the Neyman–Pearson lemma (22.6) on (24.18) that the best critical region for testing $\lambda = 0$ consists of the upper tail of the distribution of X^2.

The approximation to the non-central χ^2 distribution in **24.5** enables us to evaluate the approximate power of the X^2 test. In fact, this is given precisely by the integral (24.30). For $\alpha = 0.05$, the exact tables by Patnaik described in **24.5** may be used.

Example 30.3

We may illustrate the use of the limiting power function by returning to the problem of Example 30.2 and examining the effect on the power of the equal-probabilities procedure of doubling k. To facilitate use of the *Biometrika Tables*, we actually take four classes with slightly unequal probabilities:

Values	p_{0i}	p_{1i}	$(p_{1i}-p_{0i})^2$	$(p_{1i}-p_{0i})^2/p_{0i}$
0–0·3	0·259	0·104	0·0240	0·0927
0·3–0·7	0·244	0·190	0·0029	0·0119
0·7–1·4	0·250	0·282	0·0010	0·0040
1·4 and over	0·247	0·424	0·0313	0·1267
				0·2353 $= \dfrac{\lambda}{n}$.

In the table, the p_{0i} are obtained from the Gamma distribution with parameter 1, as before, and the p_{1i} from the Gamma distribution with parameter 1·5. For these 4 classes, and $n = 50$ as in Example 30.2, we evaluate the non-central parameter of (30.62) as $\lambda = 0·2353 \times 50 = 11·8$. With 3 degrees of freedom for X^2, this gives a power when $\alpha = 0·05$ of 0·83, from Patnaik's table.

Suppose now that we form eight classes by splitting each of the above classes into two, with the new p_{0i} as equal as is convenient for use of the Tables. We find:

Values	p_{0i}	p_{1i}	$(p_{1i}-p_{0i})^2$	$(p_{1i}-p_{0i})^2/p_{0i}$
0–0·15	0·139	0·040	0·0098	0·0705
0·15–0·3	0·120	0·064	0·0031	0·0258
0·3 –0·45	0·103	0·071	0·0010	0·0097
0·45–0·7	0·141	0·119	0·0005	0·0035
0·7 –1·0	0·129	0·134	0·0000	0·0002
1·0 –1·4	0·121	0·148	0·0007	0·0058
1·4 –2·1	0·125	0·183	0·0034	0·0272
2·1 and over	0·122	0·241	0·0142	0·1163
				0·2590 $= \dfrac{\lambda}{n}$.

For $n = 50$, we now have $\lambda = 13·0$ with 7 degrees of freedom. The approximate power for $\alpha = 0·05$ is now about 0·75 from Patnaik's table. The doubling of k has increased λ, but only slightly. The power is actually reduced, because for given λ the central and non-central χ^2 distribution draw closer together as degrees of freedom increase (cf. Exercise 24.3) and here this effect is stronger than the increase in λ. However, n is too small here for us to place any exact reliance on the values of the power obtained from the limiting power function, and we should perhaps conclude that the doubling of k has affected the power very little.

The choice of k with equal probabilities

30.28 With the aid of the asymptotic power function of **30.27**, we can get a heuristic indication of how to choose k in the equal-probabilities case. The non-central parameter (30.62) is then, using θ_i as in **30.26**,

$$\lambda = nk \sum_{i=1}^{k} \theta_i^2. \qquad (30.63)$$

We now assume that $|\theta_i| \leq \dfrac{1}{k}$, all i, and consider what happens as k becomes large.

$\theta = \sum_{i=1}^{k} \theta_i^2$, as a function of k, will then be of the same order of magnitude as a sum of squares in the interval $\left(-\frac{1}{k}, \frac{1}{k}\right)$, i.e.

$$\theta \sim a \int_{-1/k}^{1/k} u^2 \, du = 2a \int_0^{1/k} u^2 \, du. \tag{30.64}$$

The asymptotic power of the test is a function $P\{k, \lambda\}$ which therefore is $P\{k, \lambda(k)\}$; it is a monotone increasing function of λ, and has its stationary values when $\dfrac{d\lambda(k)}{dk} = 0$. We thus put, using (30.63) and (30.64),

$$0 = \frac{1}{n}\frac{d\lambda}{dk} \sim \theta + k \cdot 2a\left(\frac{1}{k}\right)^2\left(-\frac{1}{k^2}\right) = \theta - 2a k^{-3}$$

giving

$$k^{-3} \sim \theta/(2a). \tag{30.65}$$

We cannot let $k \to \infty$ without restriction since all θ_i then $\to 0$, but we assume k large enough so that both the H_0 and H_1 distribution of X^2 are near normality, and the approximate power function of the test is (cf. (25.53)) therefore

$$P = G\left\{\frac{\left[\frac{\partial}{\partial \theta} E(X^2 | H_1)\right]_{\theta=0}}{[\operatorname{var}(X^2 | H_0)]^{\frac{1}{2}}} \cdot \theta - \lambda_\alpha \right\}, \tag{30.66}$$

where

$$G(-\lambda_\alpha) = \alpha \tag{30.67}$$

determines the size of the test. From (30.49) and (30.50)

$$\left[\frac{d}{d\theta} E(X^2 | H_1)\right]_{\theta=0} = (n-1)k, \tag{30.68}$$

$$\operatorname{var}(X^2 | H_0) = 2(k-1), \tag{30.69}$$

and if we insert these values and also (30.65) into (30.66), we obtain approximately

$$P = G\{2^{\frac{1}{2}} a(n-1) k^{-5/2} - \lambda_\alpha\}. \tag{30.70}$$

This is the approximate power function at the point where power is maximized for choice of k. If we choose a value P_0 at which we wish the maximization to occur, we have, on inverting (30.70),

$$G^{-1}\{P_0\} = 2^{\frac{1}{2}} a(n-1) k^{-5/2} - \lambda_\alpha, \tag{30.71}$$

or

$$k = b\left\{\frac{2^{\frac{1}{2}}(n-1)}{\lambda_\alpha + G^{-1}\{P_0\}}\right\}^{2/5}, \tag{30.72}$$

where $b = a^{2/5}$.

30.29 In the special case $P_0 = \frac{1}{2}$ (where we wish to choose k to maximize power when it is 0·5), $G^{-1}(0 \cdot 5) = 0$ and (30.72) simplifies. In this case, Mann and Wald (1942) obtained (30.72) by a much more sophisticated and rigorous argument—they found $b = 4$ in the case of the simple hypothesis. Our own derivation suggests that the same essential argument applies for the composite hypothesis, but b may be different in this case.

We conclude that k should be increased in the equal-probabilities case in proportion to $n^{2/5}$, and that k should be smaller if we are interested in the region of high power (when $G^{-1}\{P_0\}$ is large) than if we are interested in the "neighbouring" region of low power (when $G^{-1}\{P_0\}$ approaches $-\lambda_\alpha$, from above since the test is locally unbiassed).

With $b = 4$ and $P_0 = \frac{1}{2}$, (30.72) leads to much larger values of k than are commonly used. k will be doubled when n increases by a factor of $4\sqrt{2}$. When $n = 200$, $k = 31$ for $\alpha = 0.05$ and $k = 27$ for $\alpha = 0.01$—these are about the lowest values of k for which the approximate normality assumed in our argument (and also in Mann and Wald's) is at all accurate. In this case, Mann and Wald recommend the use of (30.72) when $n \geqslant 450$ for $\alpha = 0.05$ and $n \geqslant 300$ for $\alpha = 0.01$. It will be seen that n/k, the hypothetical expectation in each class, increases as $n^{3/5}$, and is equal to about 6 and 8 respectively when $n = 200$, $\alpha = 0.05$ and 0.01.

C. A. Williams (1950) reports that k can be halved from the Mann–Wald optimum without serious loss of power at the 0.50 point. But it should be remembered that n and k must be substantial before (30.72) produces good results. Example 30.4 illustrates the point which is also borne out by calculations made by Hamdan (1963) for tests of a normal mean. Hamdan (1968) verifies in the bivariate normal case that the use of (30.72) can result in loss of power compared with the use of equal-width classes.

Example 30.4

Consider again the problem of Example 30.3. We there found that we were at around the 0.8 value for power. From a table of the normal distribution $G^{-1}(0.8) = 0.84$. With $b = 4$, $\alpha = 0.05$, $\lambda_\alpha = 1.64$, (30.72) gives for the optimum k around this point

$$k = 4\left\{\frac{2^{\frac{1}{2}}(n-1)}{2\cdot 48}\right\}^{2/5} = 3\cdot 2\,(n-1)^{2/5}.$$

For $n = 50$, this gives $k = 15$ approximately.

Suppose now that we use the *Biometrika Tables* to construct a 15-class grouping with probabilities p_{0i} as nearly equal as is convenient. We find

Values	p_{0i}	p_{1i}	$(p_{1i}-p_{0i})^2/p_{0i}$
0–0.05	0.049	0.008	0.034
0.05–0.15	0.090	0.032	0.037
0.15–0.20	0.042	0.020	0.012
0.20–0.30	0.078	0.044	0.015
0.30–0.40	0.071	0.047	0.008
0.40–0.50	0.063	0.048	0.004
0.50–0.65	0.085	0.072	0.002
0.65–0.75	0.050	0.047	0.000
0.75–0.90	0.065	0.067	0.000
0.90–1.1	0.074	0.083	0.000
1.1–1.3	0.060	0.075	0.004
1.3–1.6	0.071	0.095	0.008
1.6–2.0	0.067	0.101	0.018
2.0–2.7	0.068	0.116	0.034
2.7 and over	0.067	0.145	0.098
			$\overline{0.274} = \lambda/n$.

Here $\lambda = 13.7$ and Patnaik's table gives a power of 0.64 for 14 degrees of freedom.

λ has again been increased, but power reduced because of the increase in k. We are not at the optimum here. With large k (and hence large n), the effect of increasing degrees of freedom would not offset the increase of λ in this way.

30.30 We must not make k too large, since the multinormal approximation to the multinomial distribution cannot be expected to be satisfactory if the np_{0i} are very small. A rough rule which is commonly used, although it seems to have no general theoretical basis, is that no expected frequency (np_{0i}) should be less than 5. Yarnold (1970), generalizing earlier recommendations by Cochran (1952, 1954), concludes from a detailed theoretical investigation of the simple H_0 case that the minimum expected frequency may be as small as $5p$, where p is the proportion of the k classes with expectations less than 5. If the H_0 distribution is unimodal, and equal-width classes are used in the conventional manner, small expected frequencies will occur only at the tails.

In the equal-probabilities case, all the expected frequencies will be equal. Slakter (1966, 1968) shows that even for fractional equal expected frequencies, the approximation remains good, but that power is approximately 80 per cent. of its nominal value—see also Kempthorne (1967). The Mann–Wald procedure of **30.29** leads to expected frequencies always greater than 5 for $n \geqslant 200$. It is interesting to note that in Examples 30.3–4, the application of this limit would have ruled out the 15-class procedure, and that the more powerful 8-class procedure, with expected frequencies ranging from 5 to 7, would have been acceptable.

> Hoeffding (1965) shows that for testing a simple H_0 (the composite case is less clear), if k is held fixed while $\alpha \to 0$ suitably as $n \to \infty$, the LR test is more powerful than the X^2 test. This result does not hold if k increases with n, e.g. in the equal-probabilities case as at (30.72).

Finally, we remark that the large-sample nature of the distribution theory of X^2 is not a disadvantage in practice, for we do not usually wish to test goodness-of-fit except in large samples.

Recommendations for the X^2 test

30.31 We summarize the above discussion with a few practical recommendations:
  (1) If the distribution being tested has been tabulated, use classes with equal, or nearly equal, probabilities.
  (2) Determine the number of classes when n exceeds 200 approximately by (30.72) with b between 2 and 4.
  (3) If parameters are to be estimated, use the ordinary ML estimators in the interests of efficiency, but recall that there is partial recovery of degrees of freedom (**30.19**) so that critical values should be adjusted upwards; if the multinomial ML estimators are used, no such adjustment is necessary.

None of the theory above will hold if the *form* (instead of the parameters alone) of $F_0(x)$ is estimated from the data used to test goodness of fit.

30.32 Apart from the difficulties we have already discussed in connexion with X^2 tests, which are not very serious, they have been criticized on two counts. In each case, the criticism is of the power of the test. Firstly, the fact that the essential underlying device is the reduction of the problem to a multinomial distribution problem itself implies the necessity for grouping the observations into classes. In a broad general sense, we must lose information by grouping in this way, and we suspect that the loss will be greatest when we are testing the fit of a continuous distribution. Secondly, the fact that the X^2 statistic is based on the *squares* of the deviations of observed from hypothetical frequencies implies that the X^2 test will be insensitive to the patterns of signs of these deviations, which is clearly informative. The first of these criticisms is the more radical, since it must clearly lead to the search for other test statistics to replace X^2, and we postpone discussion of such tests until after we have discussed the second criticism.

The signs of deviations

30.33 Let us consider how we should expect the pattern of deviations (of observed from hypothetical frequencies) to behave in some simple cases. Suppose that a simple hypothesis specifies a continuous unimodal distribution with location and scale parameters, say equal to mean and standard deviation; and suppose that the hypothetical mean is too high. For any set of k classes, the p_{0i} will be too small for low values of the variate, and too high thereafter. Since in large samples the observed proportions will converge stochastically to the true probabilities, the pattern of signs of observed deviations will be a series of positive deviations followed by a series of negative deviations. If the hypothetical mean is too low this pattern is reversed

Suppose now that the hypothetical value of the scale parameter is too low. The pattern of deviations in large samples is now a series of positives, followed by a series of negatives, followed by positives again. If the hypothetical scale parameter is too high, all these signs are reversed.

Now of course we do not knowingly use the X^2 test for changes in location and scale alone, since we can then find more powerful test statistics. However, when there is error in both location and scale parameters, the situation is essentially unchanged; we shall still have three (or in more complicated cases, somewhat more) " runs " of signs of deviations. More generally, whenever the parameters have true values differing from their hypothetical values, or when the true distributional form is one differing " smoothly " from the hypothetical form, we expect the signs of deviations to cluster in this way instead of being distributed randomly, as they should be if the hypothetical frequencies were the true ones.

30.34 This observation suggests that we supplement the X^2 test with a test of the number of runs of signs among the deviations, small numbers forming the critical region. The elementary theory of runs necessary for this purpose is given as Exercise 30.8. Before we can use it in any precise way, however, we must investigate the relationship between the " runs " test and the X^2 test. F. N. David (1947), Seal (1948) and Fraser (1950) showed that when H_0 holds the tests are asymptotically independent (cf. Exercise 30.7) and that for testing the simple hypothesis all patterns of signs are equiprobable, so that the distribution theory of Exercise 30.8 can be combined with the X^2 test as indicated in Exercise 30.9.

The supplementation by the " runs " test is likely to be valuable in increasing sensitivity when testing a simple hypothesis, as in the illustrative discussion above. For the composite hypothesis of particular interest where tests of fit are concerned, when all parameters are to be estimated from the sample, it is of no practical value, since the patterns of signs of deviations, although independent of X^2, are not equiprobable as in the simple hypothesis case, and the distribution theory of Exercise 30.8 is therefore of no use (cf. Fraser, 1950).

Other tests of fit

30.35 We now turn to the discussion of alternative tests of fit. Since these have striven to avoid the loss of information due to grouping suffered by the X^2 test, they cannot avail themselves of multinomial simplicities, and we must expect their theory to be more difficult. Before we discuss the more important tests individually, we remark on a feature they have in common.

It will have been noticed that, when using X^2 to test a simple hypothesis, its distribution is asymptotically χ^2_{k-1} *whatever the simple hypothesis may be*, although its exact distribution does depend on the hypothetical distribution specified. It is clear that this result is achieved because of the intervention of the multinomial distribution and its tendency to joint normality. Moreover, the same is true of the composite hypothesis situation if multinomial ML estimators are used—in this case $X^2 \to \chi^2_{k-s-1}$ *whatever the composite hypothesis may be*, though its exact distribution is even more clearly seen to be dependent on the composite hypothesis concerned. When other estimators are used (even when fully efficient ordinary ML estimators are used) these pleasant asymptotic properties do not hold: even the asymptotic distribution of X^2 now depends on the latent roots of the matrix (30.37), which are in general functions both of the hypothetical distribution and of the values of the parameters θ.

We express these results by saying that, in the first two instances above, the distribution of X^2 is asymptotically *distribution-free* (i.e. free of the influence of the hypothetical distribution's form and parameters), whereas in the third instance it is not asymptotically distribution-free or even *parameter-free* (i.e. free of the influence of the parameters of F_0 without being distribution-free).

30.36 We shall see that the most important alternative tests of fit all make use, directly or indirectly, of the *probability-integral transformation*, which we have encountered on various occasions (e.g. **1.27, 24.11**) as a means of transforming any known continuous distribution to the rectangular distribution on the interval (0, 1). In our present notation, if we have a simple hypothesis of fit specifying a d.f. $F_0(x)$, to which a f.f. $f_0(x)$ corresponds, then the variable $y = \int_{-\infty}^{x} f_0(u)\, du = F_0(x)$ is rectangularly distributed on (0, 1). Thus if we have a set of n observations x_i and transform them to a new set y_i by the probability-integral transformation for a known $F_0(x)$, and use a function of the y_i to test the departure of the y_i from rectangularity, the distribution of the test statistic will be distribution-free, not merely asymptotically but for any n.

When the hypothetical distribution is composite, say $F_0(x \mid \theta_1, \theta_2, \ldots, \theta_s)$ with the s parameters θ to be estimated, we must select s functions $t_1, \ldots, t_s$ of the x_i for this

purpose. The transformed variables are now
$$y_i = \int_{-\infty}^{x_i} f_0(u \mid t_1, t_2, \ldots, t_s) \, du,$$
but they are neither independent nor rectangularly distributed, and their distribution will depend in general both on the hypothetical distribution F_0 and on the true values of its parameters, as F. N. David and Johnson (1948) showed in detail. However (cf. Exercise 30.10), if F has only parameters of location and scale, suitably invariantly estimated, the distribution of the y_i will depend on the form of F but not on its parameters. It follows that for finite n, no test statistic based on the y_i can be distribution-free for a composite hypothesis of fit (although it may be parameter-free if only location and scale parameters are involved). Of course, such a test statistic may still be asymptotically distribution-free.

The Neyman-Barton " smooth " tests

30.37 The first of the tests of fit, alternative to X^2, which we shall discuss are the so-called " smooth " tests first developed by Neyman (1937a), who treated only the simple hypothesis, as we do now. Given $H_0 : F(x) = F_0(x)$, we transform the n observations x_i as in **30.36** by the probability integral transformation
$$y_i = \int_{-\infty}^{x_i} f_0(u) \, du = F_0(x_i), \qquad i = 1, 2, \ldots, n, \qquad (30.73)$$
and obtain n independent observations rectangularly distributed on the interval $(0, 1)$ when H_0 holds. We specify alternatives to H_0 as departures from rectangularity of the y_i, which nevertheless remain independent on $(0, 1)$. Neyman set up a system of distributions designed to allow the alternatives to vary smoothly from the H_0 (rectangular) distribution in terms of a few parameters. (It is this " smoothness " of the alternatives which has been transferred, by hypallage, to become a description of the tests.) In fact, Neyman specified for the frequency function of any y_i the alternatives
$$f(y \mid H_k) = c(\theta_1, \theta_2, \ldots, \theta_k) \exp\left\{1 + \sum_{r=1}^{k} \theta_r \pi_r(y)\right\}, \quad 0 \leqslant y \leqslant 1, \; k = 1, 2, 3, \ldots,$$
$$(30.74)$$
where c is a constant which ensures that (30.74) integrates to 1 and the $\pi_r(y)$ are Legendre polynomials transformed linearly so that they are orthonormal on the interval $(0, 1)$.(*) If we write $z = y - \tfrac{1}{2}$, the polynomials are, to the fourth order,

---

(*) The Legendre polynomials, say $L_r(z)$, are usually defined by
$$L_r(z) = (r! \, 2^r)^{-1} \frac{d^r}{dz^r} \{(z^2 - 1)^r\},$$
and satisfy the orthogonality conditions
$$\int_{-1}^{1} L_r(z) L_s(z) \, dz = \begin{cases} 0, & r \neq s, \\ \dfrac{2}{2r+1}, & r = s. \end{cases}$$
To render them orthonormal on $(-\tfrac{1}{2}, \tfrac{1}{2})$, therefore, we define polynomials $\pi_r(z)$ by
$$\pi_r(z) = (2r+1)^{\frac{1}{2}} L_r(2z)$$
We could now transfer to the interval $(0, 1)$ by writing $y = z + \tfrac{1}{2}$. It is more convenient, as in the text, to work in terms of $z = y - \tfrac{1}{2}$.

$$\left.\begin{array}{l}\pi_0(z) \equiv 1 \\ \pi_1(z) = 3^{\frac{1}{2}}.2z, \\ \pi_2(z) = 5^{\frac{1}{2}}.(6z^2-\tfrac{1}{2}), \\ \pi_3(z) = 7^{\frac{1}{2}}.(20z^3-3z), \\ \pi_4(z) = 3.(70z^4-15z^2+\tfrac{3}{8}).\end{array}\right\} \qquad (30.75)$$

30.38 The problem now is to find a test statistic for H_0 against H_k. We can see that if we rewrite (30.74) as

$$f(y \mid H_k) = c(\theta) \exp\left\{\sum_{r=0}^{k} \theta_r \pi_r(y)\right\}, \qquad 0 \leq y \leq 1, \quad k = 0, 1, 2, \ldots, \qquad (30.76)$$

defining $\theta_0 \equiv 1$, this includes H_0 also. We wish to test the simple

$$H_0 : \theta_1 = \theta_2 = \ldots = \theta_k = 0, \qquad (30.77)$$

or equivalently

$$H_0 : \sum_{r=1}^{k} \theta_r^2 = 0, \qquad (30.78)$$

against its composite negation. It will be seen that (30.76) is an alternative of the exponential family, linear in the θ_r and π_r. The Likelihood Function for n independent observations is

$$L(y \mid \theta) = \{c(\theta)\}^n \exp\left\{\sum_{r=0}^{k} \theta_r \sum_{i=1}^{n} \pi_r(y_i)\right\}. \qquad (30.79)$$

(30.79) clearly factorizes into k parts, and each statistic $t_r = \sum_{i=1}^{n} \pi_r(y_i)$ is sufficient for θ_r, and we therefore may confine ourselves to functions of the t_r in our search for a test statistic. When dealing with linear functions of the θ_r in **23.27–32**, we saw that the equivalent function of the t_r gives a UMPU test. Here we are interested in the sum of squares of the parameters, and it seems reasonable to use the corresponding function of the t_r, i.e. $\sum_{r=1}^{k} t_r^2$, as our test statistic, although we cannot expect it to have this strong optimum property. This was, in fact, apart from a constant, the statistic proposed by Neyman (1937a), who used a large-sample argument to justify its choice. E. S. Pearson (1938) showed that in large samples the statistic is equivalent to the LR test of (30.78). We write $u_r = n^{-\frac{1}{2}} t_r$; the test statistic is then[*]

$$p_k^2 = \sum_{r=1}^{k} u_r^2 = \frac{1}{n} \sum_{r=1}^{k} \left\{\sum_{i=1}^{n} \pi_r(y_i)\right\}^2. \qquad (30.80)$$

---

[*] The statistic is usually written ψ_k^2; we abandon this notation in accordance with our convention regarding Roman letters for statistics and Greek for parameters.

30.39 Since $u_r = n^{-\frac{1}{2}} \sum_{i=1}^{n} \pi_r(y_i)$, the u_r are asymptotically normally distributed by the Central Limit theorem, with mean and variance obtained from (30.79) as

$$E(u_r) = n^{\frac{1}{2}} E\{\pi_r(y)\} = n^{\frac{1}{2}} \theta_r, \qquad (30.81)$$
$$\text{var}(u_r) = \text{var}\{\pi_r(y)\} = 1, \qquad (30.82)$$

and they are uncorrelated since the π_r are orthogonal. Thus the test statistic (30.80) is asymptotically a sum of squares of k independent normal variables with unit variances and means all zero on H_0, but not otherwise. p_k^2 is therefore distributed asymptotically in the non-central χ^2 form with k degrees of freedom and non-central parameter, from (30.81),

$$\lambda = n \sum_{r=1}^{k} \theta_r^2. \qquad (30.83)$$

It follows at once that p_k^2 is a consistent (and, by **24.17**, asymptotically unbiassed) test, as Neyman (1937a) showed. F. N. David (1939) found that, when H_0 holds, the simplest test statistics p_1^2 and p_2^2 are adequately approximated by the (central) χ^2 distributions with 1 and 2 degrees of freedom respectively for $n \geq 20$.

30.40 The choice of k, the order of the system of alternatives (and the number of parameters by which the departure from H_0 is expressed) has to be made before a test is obtained. Clearly, we want no more parameters than are necessary for the alternative of interest, since they will "dilute" the test. Unfortunately, one frequently has no very precise alternative in mind when testing fit. This is a very real difficulty, and may be compared with the choice of number of classes in the X^2 test. In the latter case, we found that the choice could be based on sample size and test size alone; in our present uncertainty, there is no very clear guidance yet available.

30.41 In the first of a series of papers, Barton (1953-6), on whose work the following sections are based, has considered a slightly different general system of alternatives. He defines, instead of (30.76),

$$f(y|H_k) = \sum_{r=0}^{k} \theta_r \pi_r(y), \qquad 0 \leq y \leq 1, \; k = 0, 1, 2, \ldots, \qquad (30.84)$$

with $\theta_0 \equiv 1$ as before. No constant $c(\boldsymbol{\theta})$ is now required, since

$$\int_0^1 \left\{ \sum_{r=0}^{k} \theta_r \pi_r(y) \right\} dy = 1 + \sum_{r=1}^{k} \theta_r \int_0^1 \pi_r(y) \, dy = 1, \qquad (30.85)$$

since $\pi_0(y) \equiv 1$ and the π_r are orthogonal. However, we now must ensure that (30.84) is non-negative over the interval (0, 1), and this involves restriction of possible values of the θ_r. Thus, for example, with $k = 1$ the value of π_1 given in (30.75) indicates that we must restrict θ_1 by $|\theta_1| \leq 3^{-\frac{1}{2}}$.

Now if we write $\theta_r = n^{-\frac{1}{2}} \lambda_r, r \geq 1$, we see that we have a set of alternatives approaching H_0 as $n \to \infty$. What is more, as $n \to \infty$ we have

$$1 + n^{-\frac{1}{2}} \sum_r \lambda_r \pi_r(y) \sim \exp\{n^{-\frac{1}{2}} \sum_r \lambda_r \pi_r(y)\},$$

so that the asymptotic distribution of p_k^2 for the alternatives (30.76) will apply under

(30.84) with $\theta_r = n^{-\frac{1}{2}}\lambda_r$. In order to obtain the asymptotic non-central χ^2 distribution of p_k^2, in which the non-central parameter is now $\lambda = \sum_{r=1}^{k} \lambda_r^2$, we have had to let H_k tend to H_0 as $n \to \infty$. This is exactly what we did to obtain the corresponding result for the X^2 test in **30.27**.

30.42 If we do have a particular alternative distribution $g(y)$ in mind, we can express it in terms of a member of the class (30.84) as follows. Let us choose parameters θ_r in (30.84) to minimize the integral

$$Q^2 = \int_0^1 [g(y) - f(y|H_k)]^2 \, dy = \int_0^1 \left[g(y) - \left\{ 1 + \sum_{r=1}^{k} \theta_r \pi_r(y) \right\} \right]^2 dy. \tag{30.86}$$

Differentiating with respect to the θ_r, we find the necessary conditions for a minimum

$$\int_0^1 \pi_r(y) \left[g(y) - \left\{ 1 + \sum_{r=1}^{k} \theta_r \pi_r(y) \right\} \right] dy = 0, \quad \text{all } r,$$

and using the orthogonality of the $\pi_r(y)$, this becomes

$$E\{\pi_r(y) | H_k\} = \theta_r, \quad \text{all } r. \tag{30.87}$$

The minimum value of (30.86) is, as in ordinary Least Squares theory,

$$Q_{\min}^2 = \int_0^1 \{g(y) - 1\}^2 \, dy - \int_0^1 \left\{ \sum_{r=1}^{k} \theta_r \pi_r(y) \right\}^2 dy$$

$$= \int_0^1 g^2(y) \, dy - 1 - \sum_{r=1}^{k} \theta_r^2, \tag{30.88}$$

using the orthogonality again.

$Q_{\min}^2$ is non-negative by definition. Regarded as a function of k, it is seen to be non-increasing, and, since only $\sum_{r=1}^{k} \theta_r^2$ depends on k, $Q_{\min}^2 \to 0$ as $k \to \infty$. In fitting the model (30.84) to $g(y)$, therefore, we essentially have to judge the approximation of $\lambda = \sum_{r=1}^{k} \theta_r^2$ to $\int_0^1 g^2(y) \, dy - 1$, which bounds it above. The integral is in terms of the probability-integral-transformed variable—it is often more convenient to evaluate it in terms of the alternative distribution of x, untransformed. Call this $h(x)$. We then have, since $g(y) \, dy = h(x) \, dx$,

$$\int_0^1 g^2(y) \, dy = \int_{-\infty}^{\infty} h^2(x) \frac{dx}{dy} \, dx = \int_{-\infty}^{\infty} \left\{ \frac{h^2(x)}{h(x|H_0)} \right\} dx. \tag{30.89}$$

Example 30.5

Consider the normal distribution

$$h(x|\mu) = (2\pi)^{-\frac{1}{2}} \exp\{-\tfrac{1}{2}(x-\mu)^2\}, \quad -\infty \leq x \leq \infty,$$

with $H_0: \mu = 0$. Using (30.89),

$$\int_0^1 g^2(y) \, dy = \int_{-\infty}^{\infty} \left\{ \frac{h^2(x|\mu)}{h(x|0)} \right\} dx$$

$$= (2\pi)^{-\frac{1}{2}} \int_{-\infty}^{\infty} \exp\{-(x-\mu)^2 + \tfrac{1}{2}x^2\} \, dx$$

$$= \exp(\mu^2).(2\pi)^{-\frac{1}{2}} \int_{-\infty}^{\infty} \exp\{-\tfrac{1}{2}(x-2\mu)^2\} dx$$
$$= \exp(\mu^2).$$

Thus we must compare $\lambda = \sum_{r=1}^{k} \theta_r^2$ with

$$\exp(\mu^2) - 1 = \mu^2 + \frac{\mu^4}{2!} + \ldots \tag{30.90}$$

From (30.87),
$$\theta_r = \int_0^1 \pi_r(y) g(y) dy.$$

Because the $\pi_r(y-\tfrac{1}{2})$ are even functions for even r (cf. (30.75)) we have, since $g(y)$ is also even about the value $\tfrac{1}{2}$ for this symmetrical alternative,
$$\theta_{2r} = 0.$$

For odd r, we must evaluate individual terms. We find, using (30.75),

$$\theta_1 = \int_{-\frac{1}{2}}^{\frac{1}{2}} 3^{\frac{1}{2}}.2z g(z) dz = 3^{\frac{1}{2}}.2 \int_{-\infty}^{\infty} z h(x) dx$$
$$= 3^{\frac{1}{2}}.2 \int_{-\infty}^{\infty} \left\{ \int_{-\infty}^{x} h(u|H_0) du - \tfrac{1}{2} \right\} h(x|H_0) \exp\{-\tfrac{1}{2}(\mu^2 - 2\mu x)\} dx$$

and

$$\left[\frac{d\theta_1}{d\mu}\right]_{\mu=0} = 3^{\frac{1}{2}}.2 \int_{-\infty}^{\infty} x \left\{ \int_{-\infty}^{x} h(u|H_0) du - \tfrac{1}{2} \right\} h(x|H_0) dx$$
$$= 3^{\frac{1}{2}}.\tfrac{1}{2}\Delta_1,$$

where Δ_1 is Gini's mean difference (cf. Exercise 2.9), equal to $2/\pi^{\frac{1}{2}}$ in this normal case (cf. **10.14**), so that

$$\left[\frac{d\theta_1}{d\mu}\right]_{\mu=0} = \left(\frac{3}{\pi}\right)^{\frac{1}{2}}.$$

Thus for small variations $d\mu$ in μ, θ_1 alone will vary by $(3/\pi)^{\frac{1}{2}} d\mu \doteqdot 0.98 d\mu$, and if we use the p_1^2 test with $\theta_1 = (3/\pi)^{\frac{1}{2}} \mu$ we see from (30.90) that $\theta_1^2 \doteqdot \dfrac{3}{\pi}\mu^2$ will be very little less than the right-hand side, and we lose little efficiency in testing for a change in the mean μ of a normal distribution from zero. This is easily confirmed. The large-sample distribution of **30.41** is the non-central χ^2 with 1 degree of freedom and non-central parameter $n(3/\pi)\mu^2$, equivalent to a standardized normal deviate of $(3n/\pi)^{\frac{1}{2}}\mu$. The best test, based on the sample, uses a normal deviate of $n^{\frac{1}{2}}\mu$. The factor

$$(3/\pi)^{\frac{1}{2}} = 0.98$$

will make little difference to the power in large samples.

The advantage of the p_k^2 tests displayed here is that, given the alternative, we can choose k to give higher power than with the X^2 test.

30.43 Since **30.37**, we have confined ourselves to the simple hypothesis and ungrouped observations. We now turn to discussion of the grouping of data for the p_k^2 tests (which is in practice very necessary in view of the need for carrying out the probability integral transformation on every observation) and the extension of the tests to composite hypotheses, which is perhaps even more important. These subjects form the substance of the second and third of Barton's (1953–6) papers. The remarkable fact is that, once grouping has been carried out, the p_k^2 tests move into intimate relationship with the X^2 test.

Suppose that the range of the variate x is grouped into k classes, and let ξ_i be the median of the ith group from below. An obvious analogue of (30.73) is then

$$y_i' = \int_{-\infty}^{\xi_i} f_0(u)\,du = \sum_{j=1}^{i-1} p_{0j} + \tfrac{1}{2} p_{0i}, \qquad (30.91)$$

where the p_{0i} are hypothetical probabilities as before. We take all the y_i in a class to be replaced by the value y_i', and write $z_i' = y_i' - \tfrac{1}{2}$ as before. We now require a set of orthogonal polynomials $P_r(y')$ which will play the same role for grouped data as the standardized Legendre polynomials did in the ungrouped case. In view of the fact that the alternative hypothesis may now be formulated in terms of the variable x as

$$f(x\,|\,H_s) = \left\{\sum_{r=0}^{s} \theta_r P_r(y')\right\} f(x\,|\,H_0),$$

we have after grouping into k classes the alternative hypothetical probabilities expressed in terms of those tested by

$$p_{1i}\,|\,H_s = \left\{\sum_{r=0}^{s} \theta_r P_r(y')\right\} p_{0i}. \qquad (30.92)$$

It is therefore natural to specify the $P_r(y')$ by

$$P_0(y') \equiv 1, \qquad (30.93)$$

$$\left. \begin{array}{l} \sum_{i=1}^{k} p_{0i} P_r(y_i') P_t(y_i') = 1, \quad r = t, \\[4pt] \phantom{\sum_{i=1}^{k} p_{0i} P_r(y_i') P_t(y_i')} = 0, \quad r \ne t, \end{array} \right\} \quad r,t = 0,1,2,\ldots,k-1. \qquad (30.94)$$

Then (30.80) (with $k-1$ replacing k) becomes in grouped form, on using (30.91),

$$p_{k-1}^2 = \sum_{r=1}^{k-1} u_r^2 = \frac{1}{n}\sum_{r=1}^{k-1}\left\{\sum_{i=1}^{n} P_r(y_i)\right\}^2 = \frac{1}{n}\sum_{r=1}^{k-1}\left\{\sum_{i=1}^{k} n_i P_r(y_i')\right\}^2,$$

which by (30.93) is

$$= \frac{1}{n}\sum_{r=0}^{k-1}\left\{\sum_{i=1}^{k} n_i P_r(y_i')\right\}^2 - n = \sum_{r=0}^{k-1}\left\{\sum_{i=1}^{k} \frac{n_i}{(np_{0i})^{\frac{1}{2}}}\cdot p_{0i}^{\frac{1}{2}} P_r(y_i')\right\}^2 - n.$$

The summation is of the squares of weighted sums of the $p_{0i}^{\frac{1}{2}} P_r(y_i')$ (which are orthogonal by (30.94)), with weights $n_i/(np_{0i})^{\frac{1}{2}}$. We therefore have

$$p_{k-1}^2 + n = \sum_{i=1}^{k} \frac{n_i^2}{np_{0i}}$$

or, in virtue of (30.8),

$$p_{k-1}^2 = X^2 \qquad (30.95)$$

exactly.

30.44 Just as p_{k-1}^2 is identical with X^2 by (30.95), the lower-order tests p_r^2 ($r = 1, 2, \ldots, k-2$) can now be seen to be components or *partitions* of X^2, particular functions of the asymptotically standardized normal variates

$$x_i = \frac{n_i - np_{0i}}{(np_{0i})^{\frac{1}{2}}}$$

which we now discuss. If we write
$$u_r = \sum_{i=1}^{k} l_{ri} x_i, \qquad r = 1, 2, \ldots, k, \tag{30.96}$$
and choose the last row of the matrix $\mathbf{L} = \{l_{ri}\}$ to be
$$l_{ki} = p_{0i}^{\frac{1}{2}},$$
we have identically
$$u_k = 0,$$
and if we choose the other elements of $\mathbf{L}$ so that it is orthogonal, i.e.
$$\sum_{i=1}^{k} l_{ri} l_{si} = 1, \qquad r = s,$$
$$= 0, \qquad r \neq s,$$
the u_r will also be asymptotically standardized normal variates and, since they are orthogonal, asymptotically independent. Thus the sum of squares of any m ($m = 1, 2, \ldots, k-1$) of the u_r ($r = 1, 2, \ldots, k-1$) will be distributed like X^2 with m degrees of freedom, independently of any other sum based on different u_r. We shall return to X^2 partitioning problems in a particular context in Chapter 33.

From this point of view, the virtue of the $p_1^2, p_2^2, \ldots$ tests, where the data are grouped, is that they select the appropriate functions of the y_i for the test in hand to have maximum power; they isolate the important components of X^2.

30.45 With the remarks of **30.44** in mind, it is not surprising that when we come to consider the composite hypothesis, the theory of the grouped p_k^2 tests closely resembles that of the X^2 test already discussed in **30.11–21**. All the principal results carry over, as Barton showed (cf. Watson, 1959): if multinomial ML estimators are used, degrees of freedom are reduced by one for each parameter estimated; if the grouping is determined from the observations, this makes no difference (under regularity conditions) to the asymptotic distributions.

The main problem in the application of the p_k^2 tests to the composite hypothesis is that of choosing k. As we remarked in the simple hypothesis case, one often has no very precise alternative in mind in making a test of fit—otherwise one would, if possible, use a more specific test. In view of the fact that large samples are frequently used for tests of fit, so that grouping of observations is a practical necessity, the identity of the grouped p_{k-1}^2 test with the X^2 test means that, apart from partitioning problems, which are common to both types of test, there is no competition between them.

Tests of fit based on the sample distribution function

30.46 The remaining general tests of fit are all functions of the cumulative distribution of the sample, or *sample distribution function*, defined by
$$S_n(x) = \begin{cases} 0, & x < x_{(1)}, \\ \dfrac{r}{n}, & x_{(r)} \leq x < x_{(r+1)}, \\ 1, & x_{(n)} \leq x. \end{cases} \tag{30.97}$$
The $x_{(r)}$ are the order-statistics, i.e. the observations arranged so that
$$x_{(1)} \leq x_{(2)} \leq \ldots \leq x_{(n)}.$$

$S_n(x)$ is simply the proportion of the observations not exceeding x. If $F_0(x)$ is the true d.f., fully specified, from which the observations come, we have, for each value of x, from the Strong Law of Large Numbers,

$$\lim_{n \to \infty} P\{S_n(x) = F_0(x)\} = 1, \tag{30.98}$$

and in fact stronger results are available concerning the convergence of the sample d.f. to the true d.f.

In a sense, (30.98) is the fundamental relationship on which all statistical theory is based. If something like it did not hold, there would be no point in random sampling. In our present context, it is clear that a test of fit can be based on any measure of divergence of $S_n(x)$ and $F_0(x)$. We now suppose $F_0(x)$ to be continuous. Consider the test statistic

$$W^2 = \int_{-\infty}^{\infty} \{S_n(x) - F_0(x)\}^2 \, dF_0(x), \tag{30.99}$$

which was proposed by Smirnov (1936) after earlier suggestions by H. Cramér and R. von Mises. Now, from binomial theory (Example 3.2) with $p = F_0(x)$,

$$E\{S_n(x) - F_0(x)\}^2 = F_0(x)\{1 - F_0(x)\}/n. \tag{30.100}$$

Thus we have from (30.99) and (30.100)

$$E(W^2) = \frac{1}{n} \int_0^1 F_0(1 - F_0) \, dF_0 = \frac{1}{n}\left(\frac{1}{2} - \frac{1}{3}\right) = \frac{1}{6n}, \tag{30.101}$$

and Exercise 30.14 asks the reader to show similarly that

$$\operatorname{var}(W^2) = E(W^4) - E^2(W^2) = (4n - 3)/180n^3. \tag{30.102}$$

30.47 It will be noticed that the mean and variance of W^2 do not depend on F_0. In fact, the distribution of W^2 as a whole does not depend on F_0: the test is completely distribution-free for any n. This is easily seen directly, for if we apply the probability integral transformation (30.73) to x, we reduce (30.99) to

$$W^2 = \int_0^1 \{S_n(y) - y\}^2 \, dy, \tag{30.103}$$

i.e. we have reduced the problem of fit to testing whether, in a sample from the rectangular distribution on (0, 1), the sample departs too far from the d.f. of that distribution, $F(y) = y$.

From (30.101) and (30.102), it will be clear that the limiting distribution which must be sought is that of nW^2 (rather than the multiple $n^{\frac{1}{2}}$ which is commonly necessary because of the Central Limit theorem), which will have mean and variance asymptotically of order zero in n. The asymptotic theory of nW^2 is difficult, and the exact theory for finite n is unknown. Smirnov (1936) showed that its limiting c.f. is

$$\phi(t) = \lim_{n \to \infty} E\{\exp(itnW^2)\} = \left\{\frac{(2it)^{\frac{1}{2}}}{\sin[(2it)^{\frac{1}{2}}]}\right\}^{\frac{1}{2}}. \tag{30.104}$$

Anderson and Darling (1952) inverted $\phi(t)$ into a form suitable for numerical calculation, and tabulated the limiting distribution of nW^2 in inverse form, giving the values exceeded with probabilities 0·001, 0·01 (0·01) 0·99. Conventionally, the most important of their values for test purposes are:

Test size α	Critical value of nW^2
0·10	0·347
0·05	0·461
0·01	0·743
0·001	1·168

Large values of nW^2 form the critical region, as is evident from the motivation of the test.

Marshall (1958) showed that the asymptotic distribution of nW^2 is reached remarkably rapidly, the asymptotic critical values given above being adequate for n as low as 3.

E. S. Pearson and Stephens (1962) fit Johnson Type S_B distributions (cf. **6.27-34**) for $n = 5, 10, \infty$, to obtain critical values. See also Stephens and Maag (1968).

30.48 For the W^2 test, as for the ungrouped p_k^2 tests discussed in **30.37-42**, one needs to calculate $F_0(x)$ for each individual observation. Exercise 30.15 asks the reader to show that we may express the statistic as

$$nW^2 = \frac{1}{12n} + \sum_{r=1}^{n} \left\{ F_0(x_{(r)}) - \frac{2r-1}{2n} \right\}^2. \qquad (30.105)$$

The W^2 test has been investigated for the composite hypothesis, with one parameter unspecified, by Darling (1955). The test statistic is now no longer distribution-free in general, as we should expect from the discussion of **30.36**; the exception is when the parameter can be estimated with variance of order less than n^{-1}, when the limiting distribution is just as in the simple hypothesis case (cf. **30.8**, where we met the same phenomenon for X^2). If the parameter is of location or scale, estimated with variance of order n^{-1}, the limiting distribution is parameter-free (cf. **30.36**).

Anderson and Darling (1952, 1954) investigated an alternative test statistic for the simple hypothesis. It is simply (30.99) with the factor $[F_0(x)\{1-F_0(x)\}]^{-1}$ inserted in the integrand. They tabulate critical values of its asymptotic distribution for $\alpha = 0·10, 0·05, 0·01$. P. A. W. Lewis (1961) gives an exact tabulation of the d.f. for $n = 1$ and $n \to \infty$ and estimates based on sampling experiments for $n = 2$ (1) 8; the convergence to the asymptotic d.f. is extremely rapid.

Watson (1961) shows that if we modify the nW^2 statistic to

$$U^2 = n \int_0^1 \left[S_n(x) - F_0(x) - \int_0^1 \{S_n(x) - F_0(x)\} dF_0(x) \right]^2 dF_0(x),$$

the asymptotic distribution of $\pi^2 U^2$ is exactly that of nD_n^2 given at (30.132) below. E. S. Pearson and Stephens (1962) and Stephens (1963, 1964) give theoretical and empirical results on the distribution of U^2. Tiku (1965b) fits χ^2 approximations for U^2 and also for W^2.

The Kolmogorov statistic

30.49 We now come to the most important of the general tests of fit alternative to X^2. Like W^2, defined at (30.99), it is based on deviations of the sample d.f. $S_n(x)$ from the completely specified continuous hypothetical d.f. $F_0(x)$. The measure of deviation used, however, is very much simpler, being the maximum absolute difference between $S_n(x)$ and $F_0(x)$. Thus we define

$$D_n = \sup_x |S_n(x) - F_0(x)|. \tag{30.106}$$

The appearance of the modulus in the definition (30.106) might lead us to expect difficulties in the investigation of the distribution of D_n, but remarkably enough, the asymptotic distribution was obtained by Kolmogorov (1933) when he first proposed the statistic. The derivation which follows is due to Feller (1948).

30.50 We first note that the distribution of D_n is completely distribution-free when H_0 holds. We may see this very directly in this case, for if $S_n(x)$ and $F_0(x)$ are plotted as ordinates against x as abscissa, D_n is simply the value of the largest vertical difference between them. Clearly, if we make any one-to-one transformation of x, this will not affect the vertical difference at any point and, in particular, the value of D_n will be unaffected.

30.51 Now consider the values $x_{10}, x_{20}, \ldots, x_{n-1,0}$ defined by

$$F_0(x_{k0}) = k/n. \tag{30.107}$$

(If, for some k, (30.107) holds within an interval, we take x_{k0} to be the lower end-point of the interval.) Let c be a positive integer. If, for some value x,

$$S_n(x) - F_0(x) > c/n, \tag{30.108}$$

the inequality (30.108) will hold for all values of x in some interval at whose upper end-point x' it becomes an equality, i.e.

$$S_n(x') - F_0(x') = c/n. \tag{30.109}$$

Since $S_n(x)$ is by definition a step-function taking values which are multiples of $1/n$, and c is an integer, it follows from (30.109) that $F_0(x')$ is a multiple of $1/n$ and thus, from (30.107), $x' = x_{k0}$ for some k, so that (30.109) becomes

$$S_n(x_{k0}) - F_0(x_{k0}) = c/n,$$

i.e. from (30.107),

$$S_n(x_{k0}) = (k+c)/n. \tag{30.110}$$

From the definition of $S_n(x)$ at (30.97), this means that exactly $(k+c)$ of the observed values of x are less than x_{k0}, the hypothetical value below which k of them should fall. Conversely, if $x_{(k+c)} \leq x_{k0} < x_{(k+c+1)}$, (30.108) will follow immediately. We have therefore established the preliminary result that the inequality

$$S_n(x) - F_0(x) > c/n$$

holds for some x if and only if for some k

$$x_{(k+c)} \leq x_{k0} < x_{(k+c+1)}. \tag{30.111}$$

Thus we may confine ourselves to consideration of the probability that (30.111) occurs.

30.52 We denote the event (30.111) by $A_k(c)$. From (30.106), we see that the statistic D_n will exceed c/n if and only if at least one of the $2n$ events

$$A_1(c), A_1(-c), A_2(c), A_2(-c), \ldots, A_n(c), A_n(-c) \tag{30.112}$$

occurs. We now define the $2n$ mutually exclusive events U_r and V_r. U_r occurs if $A_r(c)$ is the first event in the sequence (30.112) to occur, and V_r occurs if $A_r(-c)$ is the first. Evidently

$$P\left\{D_n > \frac{c}{n}\right\} = \sum_{r=1}^{n} [P\{U_r\} + P\{V_r\}]. \qquad (30.113)$$

We have, from the definitions of $A_k(c)$ and U_r, V_r, the relations

$$\left.\begin{aligned}P\{A_k(c)\} &= \sum_{r=1}^{k} [P\{U_r\}P\{A_k(c)|A_r(c)\} + P\{V_r\}P\{A_k(c)|A_r(-c)\}], \\ P\{A_k(-c)\} &= \sum_{r=1}^{k} [P\{U_r\}P\{A_k(-c)|A_r(c)\} + P\{V_r\}P\{A_k(-c)|A_r(-c)\}].\end{aligned}\right\} \qquad (30.114)$$

From (30.111) and (30.107), we see that $P\{A_k(c)\}$ is the probability that exactly $(k+c)$ "successes" occur in n binomial trials with probability k/n, i.e.,

$$P\{A_k(c)\} = \binom{n}{k+c}\left(\frac{k}{n}\right)^{k+c}\left(1-\frac{k}{n}\right)^{n-(k+c)} \qquad (30.115)$$

Similarly, for $r \leqslant k$,

$$\left.\begin{aligned}P\{A_k(c)|A_r(c)\} &= \binom{n-(r+c)}{k-r}\left(\frac{k-r}{n-r}\right)^{k-r}\left(1-\frac{k-r}{n-r}\right)^{n-(k+c)} \\ P\{A_k(c)|A_r(-c)\} &= \binom{n-(r-c)}{k-r+2c}\left(\frac{k-r}{n-r}\right)^{k-r+2c}\left(1-\frac{k-r}{n-r}\right)^{n-(k+c)}\end{aligned}\right\} \qquad (30.116)$$

(30.115) and (30.116) hold for negative as well as positive c. Using them, we see that (30.114) is a set of $2n$ linear equations for the $2n$ unknowns $P\{U_r\}$, $P\{V_r\}$. If we solved these, and substituted into (30.113), we should obtain $P\left\{D_n > \frac{c}{n}\right\}$ for any c.

30.53 If we now write

$$p_k(c) = e^{-k}\frac{k^{k+c}}{(k+c)!}, \qquad (30.117)$$

we have

$$\left.\begin{aligned}P\{A_k(c)\} &= p_k(c)p_{n-k}(-c)/p_n(0), \\ P\{A_k(c)|A_r(c)\} &= p_{k-r}(0)p_{n-k}(-c)/p_{n-r}(-c), \\ P\{A_k(c)|A_r(-c)\} &= p_{k-r}(2c)p_{n-k}(-c)/p_{n-r}(c),\end{aligned}\right\} \qquad (30.118)$$

so that if we define

$$u_r = P\{U_r\}\frac{p_n(0)}{p_{n-r}(-c)}, \quad v_r = P\{V_r\}\frac{p_n(0)}{p_{n-r}(c)}, \qquad (30.119)$$

and substitute (30.115–19) into (30.114), the latter becomes simply

$$\left.\begin{aligned}p_k(c) &= \sum_{r=1}^{k} [u_r p_{k-r}(0) + v_r p_{k-r}(2c)], \\ p_k(-c) &= \sum_{r=1}^{k} [u_r p_{k-r}(-2c) + v_r p_{k-r}(0)].\end{aligned}\right\} \qquad (30.120)$$

TESTS OF FIT

The system (30.120) is to be solved for

$$\sum_{r=1}^{n}[P\{U_r\}+P\{V_r\}] = \frac{1}{p_n(0)}\sum_{r=1}^{n}[p_{n-r}(-c)u_r+p_{n-r}(c)v_r]. \qquad (30.121)$$

We therefore define

$$p_k = \frac{1}{p_n(0)}\sum_{r=1}^{k}p_{k-r}(-c)u_r, \quad q_k = \frac{1}{p_n(0)}\sum_{r=1}^{k}p_{k-r}(c)v_r, \qquad (30.122)$$

so that, from (30.121),

$$\sum_{r=1}^{n}[P\{U_r\}+P\{V_r\}] = p_n+q_n. \qquad (30.123)$$

We now set up generating functions for the p_k and q_k, namely

$$G_p(t) = \sum_{k=1}^{\infty}p_k t^k, \quad G_q(t) = \sum_{k=1}^{\infty}q_k t^k.$$

If we also define generating functions for the u_k, v_k and (for convenience) $n^{-\frac{1}{2}}p_k(c)$, namely

$$G_u(t) = \sum_{k=1}^{\infty}u_k t^k, \quad G_v(t) = \sum_{k=1}^{\infty}v_k t^k,$$

and

$$G(t,c) = n^{-\frac{1}{2}}\sum_{k=1}^{\infty}p_k(c)t^k,$$

we have from (30.122), the relationships

$$\left.\begin{array}{l}G_p(t) = G_u(t)\,G(t,-c)\,n^{\frac{1}{2}}/p_n(0),\\ G_q(t) = G_v(t)\,G(t,c)\,n^{\frac{1}{2}}/p_n(0).\end{array}\right\} \qquad (30.124)$$

30.54 We now consider the limiting form of (30.124). We put

$$c = z n^{\frac{1}{2}}$$

and let $n \to \infty$ and $c \to \infty$ with it so that z remains fixed.

We see from (30.117) that $p_k(c)$ is simply the probability of the value $(k+c)$ for a Poisson variate with parameter k, i.e. the probability of its being $c/k^{\frac{1}{2}}$ standard deviations above its mean. If k/n tends to some fixed value m, then as the Poisson variate tends to normality

$$p_k(c) \to (2\pi k)^{-\frac{1}{2}}\exp\left(-\tfrac{1}{2}\frac{c^2}{k}\right)$$

or, putting $k = mn$, $c = zn^{\frac{1}{2}}$,

$$n^{\frac{1}{2}}p_k(zn^{\frac{1}{2}}) \to (2\pi m)^{-\frac{1}{2}}\exp\left(-\tfrac{1}{2}\frac{z^2}{m}\right). \qquad (30.125)$$

Now since $G(t,c)$ is a generating function for the $n^{-\frac{1}{2}}p_k(c)$, we have

$$G(e^{-t/n}, zn^{\frac{1}{2}}) = n^{-\frac{1}{2}}\sum_{k=1}^{\infty}p_k(zn^{\frac{1}{2}})e^{-tk/n}$$

and under our limiting process this tends by (30.125) to

$$\lim_{n\to\infty} G(e^{-t/n}, zn^{\frac{1}{2}}) = (2\pi)^{-\frac{1}{2}}\int_0^{\infty}m^{-\frac{1}{2}}\exp\left(-tm-\tfrac{1}{2}\frac{z^2}{m}\right)dm. \qquad (30.126)$$

If we differentiate the integral I on the right of (30.126) with respect to $\frac{1}{2}z^2$, we find the simple differential equation

$$\frac{\partial I}{\partial(\frac{1}{2}z^2)} = -\left(\frac{t}{\frac{1}{2}z^2}\right)^{\frac{1}{2}} I$$

whose solution is

$$I = \left(\frac{\pi}{t}\right)^{\frac{1}{2}} \exp\{-(2tz^2)^{\frac{1}{2}}\}.$$

Thus

$$\lim_{n\to\infty} G(e^{-t/n}, zn^{\frac{1}{2}}) = (2t)^{-\frac{1}{2}} \exp\{-(2tz^2)^{\frac{1}{2}}\}. \tag{30.127}$$

(30.127) is an even function of z, and therefore of c.

Since, from (30.120),

$$\left.\begin{array}{l} G(t,c) = G_u(t)\,G(t,0) + G_v(t)\,G(t,2c), \\ G(t,-c) = G_u(t)\,G(t,-2c) + G_v(t)\,G(t,0), \end{array}\right\} \tag{30.128}$$

this evenness of (30.127) in c gives us

$$\lim_{n\to\infty} G_u(e^{-t/n}) = \lim_{n\to\infty} G_v(e^{-t/n})$$

$$= \frac{\lim G(e^{-t/n}, zn^{\frac{1}{2}})}{\lim G(e^{-t/n}, 0) + \lim G(e^{-t/n}, 2zn^{\frac{1}{2}})}$$

$$= \frac{\exp\{-(2tz^2)^{\frac{1}{2}}\}}{1 + \exp\{-(8tz^2)^{\frac{1}{2}}\}}, \tag{30.129}$$

by (30.127). Thus, in (30.124), remembering that

$$p_n(0) \sim (2\pi n)^{-\frac{1}{2}},$$

(30.127) and (30.129) give

$$\lim_{n\to\infty} n^{-1} G_p(e^{-t/n}) = \lim_{n\to\infty} n^{-1} G_q(e^{-t/n}) = \left(\frac{2\pi}{2t}\right)^{\frac{1}{2}} \frac{\exp\{-(8tz^2)^{\frac{1}{2}}\}}{1 + \exp\{-(8tz^2)^{\frac{1}{2}}\}} = L(t).$$

This may be expanded into geometric series as

$$L(t) = \left(\frac{2\pi}{2t}\right)^{\frac{1}{2}} \sum_{r=1}^{\infty} (-1)^{r-1} \exp\{-(8tr^2z^2)^{\frac{1}{2}}\}. \tag{30.130}$$

By the same integration as at (30.126), $L(t)$ is seen to be the one-sided Laplace transform $\int_0^\infty e^{-mt} f(m)\,dm$ of the function

$$f(m) = \sum_{r=1}^{\infty} (-1)^{r-1} \exp\{-2r^2 z^2/m\}. \tag{30.131}$$

(30.131) is thus the result of inverting either of the limiting generating functions of the p_k or q_k, of which the first is

$$\lim n^{-1} G_p(e^{-t/n}) = \lim n^{-1} \sum_{k=1}^{\infty} p_k e^{-tk/n} = \int_0^\infty (\lim p_k)\, e^{-tm}\,dm.$$

From (30.113) and (30.123), we require only the value $(p_n + q_n)$. We thus put $k = n$, i.e. $m = 1$, in (30.131) and after multiplying by two, obtain our final result

$$\lim_{n\to\infty} P\{D_n > zn^{-\frac{1}{2}}\} = 2 \sum_{r=1}^{\infty} (-1)^{r-1} \exp\{-2r^2 z^2\}. \tag{30.132}$$

Smirnov (1948) tabulates (30.132) (actually its complement) for $z = 0 \cdot 28 \, (0 \cdot 01) \, 2 \cdot 50 \, (0 \cdot 05) \, 3 \cdot 00$ to 6 d.p. or more. This is the whole effective range of the limiting distribution.

30.55 As well as deriving the limiting result (30.132), Kolmogorov (1933) gave recurrence relations for finite n, which have since been used to tabulate the distribution of D_n. Z. W. Birnbaum (1952) gives tables of $P\{D_n < c/n\}$ to 5 d.p., for $n = 1\,(1)\,100$ and $c = 1\,(1)\,15$, and inverse tables of the values of D_n for which this probability is 0·95 for $n = 2\,(1)\,5\,(5)\,30\,(10)\,100$ and for which the probability is 0·99 for $n = 2\,(1)\,5\,(5)\,30\,(10)\,80$. L. H. Miller (1956) gives inverse tables for $n = 1\,(1)\,100$ and probabilities 0·90, 0·95, 0·98, 0·99. Massey (1950a, 1951a) had previously given $P\{D_n < c/n\}$ for $n = 5\,(5)\,80$ and selected values of $c \leqslant 9$, and also inverse tables for $n = 1\,(1)\,20\,(5)\,35$ and probabilities 0·80, 0·85, 0·90, 0·95, 0·99.

It emerges that the critical values of the asymptotic distribution are:

Test size	Critical value of D_n
0·05	$1 \cdot 3581 \, n^{-\frac{1}{2}}$,
0·01	$1 \cdot 6276 \, n^{-\frac{1}{2}}$,

and that these are always greater than the exact values for finite n. The approximation for these values of α is satisfactory at $n = 80$.

Confidence limits for distribution functions

30.56 Because the distribution of D_n is distribution-free and adequately known for all n, and because it uses as its measure of divergence the maximum absolute deviation between $S_n(x)$ and $F_0(x)$, we may reverse the procedure of testing for fit and use D_n to set confidence limits for a (continuous) distribution function *as a whole*. For, whatever the true $F(x)$, we have, if d_α is the critical value of D_n for test size α,

$$P\{D_n = \sup_x |S_n(x) - F(x)| > d_\alpha\} = \alpha.$$

Thus we may invert this into the confidence statement

$$P\{S_n(x) - d_\alpha \leqslant F(x) \leqslant S_n(x) + d_\alpha, \text{ all } x\} = 1 - \alpha. \tag{30.133}$$

Thus we simply set up a band of width $\pm d_\alpha$ around the sample d.f. $S_n(x)$, and there is probability $1 - \alpha$ that the true $F(x)$ lies *entirely* within this band. This is a remarkably simple and direct method of estimating a distribution function. No other test of fit permits this inversion of test into confidence limits since none uses so direct and simply interpretable a measure of divergence as D_n.

One can draw useful conclusions from this confidence limits technique as to the sample size necessary to approximate a d.f. closely. For example, from the critical values given at the end of **30.55**, it follows that a sample of 100 observations would have probability 0·95 of having its sample d.f. everywhere within 0·13581 of the true d.f. To be within 0·05 of the true d.f. everywhere, with probability 0·99, would require a sample size of $(1 \cdot 6276/0 \cdot 05)^2$, i.e. more than 1000.

Noether (1963) shows that the left side of (30.133) holds with probability $\geqslant 1 - \alpha$ for discrete distributions. Thus the D_n test is then also conservative.

30.57 Because it is a modular quantity, D_n does not permit us to set one-sided confidence limits for $F(x)$, but we may consider positive deviations only and define

$$D_n^+ = \sup_x \{S_n(x) - F_0(x)\} \tag{30.134}$$

as was done by Wald and Wolfowitz (1939) and Smirnov (1939a).

To obtain the limiting distribution of D_n^+, we retrace the argument of **30.51-54**. We now consider only events $A_k(c)$ with $c > 0$ in (30.112). U_r is defined as before, but V_r is not considered. (30.114) is replaced by

$$P\{A_k(c)\} = \sum_{r=1}^{k} P\{U_r\} P\{A_k(c) \mid A_r(c)\}$$

and (30.128) by

$$G(t, c) = G_u(t) G(t, 0). \tag{30.135}$$

Instead of (30.129), we therefore have, using (30.127) and (30.135),

$$\lim_{n \to \infty} G_u(e^{-t/n}) = \exp\{-(2tz^2)^{\frac{1}{2}}\}.$$

The first equation in (30.124) holds, and we get, in the same way as before,

$$\lim_{n \to \infty} n^{-1} G_p(e^{-t/n}) = \left(\frac{2\pi}{2t}\right)^{\frac{1}{2}} \exp\{-(8tz^2)^{\frac{1}{2}}\}. \tag{30.136}$$

Again from (30.127), (30.136) is seen to be the one-sided Laplace transform of

$$f(m) = m^{-\frac{1}{2}} \exp(-2z^2/m)$$

and substitution of $m = 1$ as before gives

$$\lim_{n \to \infty} P\{D_n^+ > z n^{-\frac{1}{2}}\} = \exp(-2z^2), \tag{30.137}$$

which is Smirnov's (1939a) result. (30.137) may be rewritten

$$\lim_{n \to \infty} P\{2n(D_n^+)^2 \leqslant 2z^2\} = 1 - \exp(-2z^2). \tag{30.138}$$

Differentiation of (30.138) with respect to $(2z^2)$ shows that the variable $y = 2n(D_n^+)^2$ is asymptotically distributed in the negative exponential form

$$dF(y) = \exp(-y) dy, \quad 0 \leqslant y \leqslant \infty.$$

Alternatively, we may express this by saying that $2y = 4n(D_n^+)^2$ is asymptotically a χ^2 variate with 2 degrees of freedom. Evidently, exactly the same theory will hold if we consider only negative deviations.

30.58 Z. W. Birnbaum and Tingey (1951) give an expression for the exact distribution of D_n^+, and tabulate the values it exceeds with probabilities 0·10, 0·05, 0·01, 0·001, for $n = 5, 8, 10, 20, 40, 50$. As for D_n, the asymptotic values exceed the exact values, and the differences are small for $n = 50$.

We may evidently use D_n^+ to obtain one-sided confidence regions of the form $P\{S_n(x) - d_\alpha^+ \leqslant F(x)\} = 1 - \alpha$, where d_α^+ is the critical value of D_n^+.

Comparison of Kolmogorov's statistic with X^2

30.59 The D_n statistic will clearly not remain distribution-free in testing a composite hypothesis of fit (cf. **30.36**), and this represents a substantial disadvantage compared with the X^2 test. However, it permits the setting of exact confidence limits for the parent d.f., given only that it is continuous. Exercise 30.19 shows how to set somewhat analogous conservative confidence intervals, using X^2, for all the theoretical probabilities p_{0i} simultaneously.

Because of the strong convergence of $S_n(x)$ to the true d.f. $F(x)$ (cf. (30.98)), the D_n test is consistent against any alternative $G(x) \neq F(x)$. However, Massey (1950b, 1952) has given an example in which it is biassed (cf. Exercise 30.16). He also established a lower bound to the power of the test in large samples as follows.

30.60 Write $F_1(x)$ for the d.f. under the alternative hypothesis H_1, $F_0(x)$ for the d.f. being tested as before; and

$$\Delta = \sup_x |F_1(x) - F_0(x)|. \tag{30.139}$$

If d_α is the critical value of D_n as before, the power we require is

$$P = P\{\sup_x |S_n(x) - F_0(x)| > d_\alpha | H_1\}.$$

This is the probability of an inequality arising for *some* x. Clearly this is no less than the probability that it occurs at any particular value of x. Let us choose a particular value, x_Δ, at which F_0 and F_1 are at their farthest apart, i.e.

$$\Delta = F_1(x_\Delta) - F_0(x_\Delta). \tag{30.140}$$

Thus we have

$$P \geqslant P\{|S_n(x_\Delta) - F_0(x_\Delta)| > d_\alpha | H_1\}$$

or

$$P \geqslant 1 - P\{F_0(x_\Delta) - d_\alpha \leqslant S_n(x_\Delta) \leqslant F_0(x_\Delta) + d_\alpha | H_1\}. \tag{30.141}$$

Now, $S_n(x_\Delta)$ is binomially distributed with probability $F_1(x_\Delta)$ of falling below x_Δ. Thus we may approximate the right-hand side of (30.141) using the normal approximation to the binomial distribution, i.e. asymptotically

$$P \geqslant 1 - (2\pi)^{-\frac{1}{2}} \int_{\frac{F_0 - F_1 - d_\alpha}{\{F_1(1-F_1)/n\}^{\frac{1}{2}}}}^{\frac{F_0 - F_1 + d_\alpha}{\{F_1(1-F_1)/n\}^{\frac{1}{2}}}} \exp(-\tfrac{1}{2}u^2) \, du, \tag{30.142}$$

F_0 and F_1 being evaluated at x_Δ in (30.142) and hereafter. If F_1 is specified, (30.142) is the required lower bound for the power. Clearly, as $n \to \infty$ both limits of integration increase. If

$$d_\alpha < |F_0 - F_1| = \Delta, \tag{30.143}$$

they will both tend to $+\infty$ if $F_0 > F_1$ and to $-\infty$ if $F_0 < F_1$. Thus the integral will tend to zero and the power to 1. As n increases, d_α declines, so (30.143) is always ultimately satisfied. Hence the power $\to 1$ and the test is consistent.

If F_1 is not completely specified, we may still obtain a (worse) lower bound to the power from (30.142). Since $F_1(1 - F_1) \leqslant \tfrac{1}{4}$, we have, for large enough n,

$$P \geqslant 1 - (2\pi)^{-\frac{1}{2}} \int_{2n^{\frac{1}{2}}(F_0 - F_1 - d_\alpha)}^{2n^{\frac{1}{2}}(F_0 - F_1 + d_\alpha)} \exp(-\tfrac{1}{2}u^2) \, du$$

which, using the symmetry of the normal distribution, if $F_0 < F_1$, we may write as

$$P \geq 1 - (2\pi)^{-\frac{1}{2}} \int_{2n^{\frac{1}{2}}(\Delta - d_\alpha)}^{2n^{\frac{1}{2}}(\Delta + d_\alpha)} \exp(-\tfrac{1}{2}u^2)\,du. \qquad (30.144)$$

The bound (30.144) is in terms of the maximum deviation Δ alone.

> Z. W. Birnbaum (1953) obtained sharp upper and lower bounds for the power of D_n^+ in terms of Δ. Knott (1970) obtains small-sample power results for D_n^+ against shifts in location in the normal case.

30.61 Using (30.144) and calculations made by Williams (1950), Massey (1951a) compared the values of Δ for which the large-sample powers of the X^2 and the D_n tests are at least 0·5. For test size $\alpha = 0.05$, the D_n test can detect with power 0·5 a Δ about half the magnitude of that which the X^2 test can detect with this power; even with $n = 200$, the ratio of Δ's is 0·6, and it declines steadily in favour of D_n as n increases. For $\alpha = 0.01$ the relative performances are very similar. Since this comparison is based on the poor lower bound (30.144) to the power of D_n, we must conclude that D_n is a very much more sensitive test for the fit of a continuous distribution.

Kac *et al.* (1955) point out that if the Mann–Wald equal-probabilities procedure of **30.28–9** is used, the X^2 test requires Δ to be of order $n^{-2/5}$ to attain power $\tfrac{1}{2}$, whereas D_n requires Δ to be of order $n^{-\frac{1}{2}}$. Thus D_n asymptotically requires sample size to be of order $n^{4/5}$ compared to n for the X^2 test, and is asymptotically very much more efficient—in fact the relative efficiency of X^2 will tend to zero as n increases.

A detailed review of the theory of the W^2, D_n and related tests is given by Darling (1957).

Computation of D_n

30.62 If we are setting confidence limits for the unknown $F(x)$, no computations are required beyond the simple calculation of $S_n(x)$ and the setting of bounds distant $\pm d_\alpha$ from it. In using D_n for testing, however, we have to face the possibility of calculating $F_0(x)$ for every observed value of x, a procedure which is tedious even when $F_0(x)$ is well tabulated. However, because the test criterion is the maximum deviation between $S_n(x)$ and $F_0(x)$, it is often possible by preliminary examination of the data to locate the intervals in which the deviations are likely to be large. If initial calculations are made only for these values, computations may be stopped as soon as a single deviation exceeding d_α is found. (This abbreviation of the calculations is not possible for statistics like W^2, which depend on *all* deviations.)

A further considerable saving of labour may be effected as in the following example, due to Z. W. Birnbaum (1952).

Example 30.6

A sample of 40 observations is to hand, where values are arranged in order:

0·0475, 0·2153, 0·2287, 0·2824, 0·3743, 0·3868, 0·4421, 0·5033, 0·5945, 0·6004, 0·6255, 0·6331, 0·6478, 0·7867, 0·8878, 0·8930, 0·9335, 0·9602, 1·0448, 1·0556, 1·0894, 1·0999, 1·1765, 1·2036, 1·2344, 1·2543, 1·2712, 1·3507, 1·3515, 1·3528, 1·3774, 1·4209, 1·4304, 1·5137, 1·5288, 1·5291, 1·5677, 1·7238, 1·7919, 1·8794.

We wish to test, with $\alpha = 0.05$, whether the parent $F_0(x)$ is normal with mean 1 and variance 1/6. From Z. W. Birnbaum's (1952) tables we find for $n=40$, $\alpha=0.05$ that $d_\alpha = 0.2101$. Consider the smallest observation, $x_{(1)}$. To be acceptable, $F_0(x_{(1)})$ should lie between 0 and d_α, i.e. in the interval (0, 0·2101). The observed value of $x_{(1)}$ is 0·0475, and from tables of the normal d.f. we find $F_0(x_{(1)}) = 0.0098$, within the above interval, so the hypothesis is not rejected by this observation. Further, it cannot possibly be rejected by the next higher observations until we reach an $x_{(i)}$ for which *either* (a) $i/40 - 0.2101 > 0.0098$, i.e. $i > 8.796$, or (b) $F_0(x_{(i)}) > 0.2101 + 1/40$, i.e. $x_{(i)} > 0.7052$ (from the tables again). The 1/40 is added on the right of (b) because we know that $S_n(x_{(i)}) \geq 1/40$ for $i > 1$. Now from the data, $x_{(i)} > 0.7052$ for $i \geq 14$. We next need, therefore, to examine $i = 9$ (from the inequality (a)). We find there the acceptance interval for $F_0(x_{(9)})$

$$(S_9(x) - d_\alpha, S_8(x) + d_\alpha) = (9/40 - 0.2101, 8/40 + 0.2101) = (0.0149, 0.4101).$$

We find from the tables $F_0(x_{(9)}) = F_0(0.5945) = 0.1603$, which is acceptable. To reject H_0, we now require *either*

$$i/40 - 0.2101 > 0.1603, \text{ i.e. } i > 14.82$$

or $\quad F_0(x_{(i)}) > 0.4101 + 1/40$, i.e. $x_{(i)} > 0.9052$, i.e. $i \geq 17$.

We therefore proceed to $i = 15$, and so on. The reader should verify that only the 6 values $i = 1, 9, 15, 21, 27, 34$ require computations in this case. The hypothesis is accepted because in every one of these six cases the value of F_0 lies in the confidence interval; it would have been rejected, and computations ceased, if any one value had lain outside the interval.

Tests of normality

30.63 To conclude this chapter, we refer briefly to the problem of testing normality, i.e. the problem of testing whether the parent d.f. is a member of the family of normal distributions, the parameters being unspecified. Of course, any general test of fit for the composite hypothesis may be employed to test normality, and to this extent no new discussion is necessary. However, it is common to test the observed moment-ratios $\sqrt{b_1}$ and b_2 against their distributions given the hypothesis of normality (cf. **12.18** and Exercises 12.9–10 and the percentage points given in the *Biometrika Tables*) and these are sometimes called "tests of normality," although they are better called tests of skewness and kurtosis respectively. See **32.24** below for another use of these coefficients.

D. S. Moore (1971) gives values of X^2 for testing normality with $\alpha = 0.001, 0.005, 0.01, 0.05, 0.10, 0.25$ and $k = 5, 7, 9, 11, 15, 21$ when fully efficient ML estimators are used to determine boundaries with estimated equal probabilities in classes.

Kac *et al.* (1955) discuss the distributions of D_n and W^2 in testing normality when the two parameters (μ, σ^2) are estimated from the sample by $(\bar{x}, s^2)$. The limiting distributions are parameter-free (because μ and σ are location and scale parameters—cf. **30.36**) but are not obtained explicitly. Lilliefors (1967) uses extensive sampling experiments to give critical values of D_n in testing normality with $\alpha = 0.01, 0.05 (0.05) 0.20$ and $n = 4 (1) 20 (5) 30$, with an approximation for larger n. These critical values are roughly two-thirds of the simple H_0 values in **30.55**—e.g., for $\alpha = 0.05$, the value is $0.886n^{-\frac{1}{2}}$ and for $\alpha = 0.01$ it is $1.031n^{-\frac{1}{2}}$.

Shapiro and Wilk (1965) give a new criterion for testing normality based on the regression of the order-statistics upon their expected values, using the theory of **19.18–20** and extensive sampling experiments to establish its distribution. It is defined by $W = \left(\sum_{i=1}^{n} a_i x_{(i)}\right)^2 \bigg/ \sum_{i=1}^{n} (x_i - \bar{x})^2$ where the a_i are tabulated coefficients. Small values of W are critical. Shapiro *et al.* (1968) show by extensive sampling experiments that W is generally superior to the other tests given in this chapter for testing normality, although $\sqrt{b_1}$ and b_2 together are sensitive to non-normality.

Lilliefors (1969) discusses D_n for the exponential distribution with unknown mean. See also Srinivasan (1970), corrected by Schafer *et al.* (1972), for a modification of the estimation procedure when using D_n for composite H_0.

EXERCISES

30.1 Show that if, in testing a composite hypothesis, an inconsistent set of estimators **t** is used, the statistic $X^2 \to \infty$ as $n \to \infty$.

(cf. Fisher, 1924c)

30.2 Using (30.33), show that the matrix **M** defined at (30.37) reduces, when the vector of multinomial ML estimators $\hat{\theta}$ is used, to

$$\mathbf{M} = \begin{pmatrix} \tfrac{1}{2}\mathbf{I}, & \tfrac{1}{4}\mathbf{C} \\ \mathbf{C}^{-1}, & \tfrac{1}{2}\mathbf{I} \end{pmatrix},$$

and that **M** is idempotent with tr $\mathbf{M} = s$. Hence confirm the result of **30.10** and **30.14** that X^2 is asymptotically distributed like χ^2_{k-s-1} when $\hat{\theta}$ is used.

(Watson, 1959)

30.3 Show from the limiting joint normality of the n_i that as $n \to \infty$, the variance of the simple-hypothesis X^2 statistic in the equal-probabilities case ($p_{0i} = 1/k$) is

$$\operatorname{var}(X^2) = \lim_{n\to\infty} 2k^2 \left\{ \sum_i p_{1i}^2 - \left(\sum_i p_{1i}^2\right)^2 \right\} + 4(n-1)k^2 \left\{ \sum_i p_{1i}^3 - \left(\sum_i p_{1i}^2\right)^2 \right\}$$

where p_{1i}, $i = 1, 2, \ldots, k$ are the true class-probabilities. Verify that this reduces to the correct value $2(k-1)$ when

$$p_{1i} = p_{0i} = 1/k.$$

(Mann and Wald, 1942)

30.4 Establish the non-central χ^2 result of **30.27** for the alternative hypothesis distribution of the X^2 test statistic.

(cf. Cochran, 1952)

30.5 Show from the moments of the multinomial distribution (cf. (5.80)) that the exact variance of the simple-hypothesis X^2 statistic is given by

$$n \operatorname{var}(X^2) = 2(n-1)\left\{ 2(n-2)\sum_i \frac{p_{1i}^3}{p_{0i}^2} - (2n-3)\left(\sum_i \frac{p_{1i}^2}{p_{0i}}\right)^2 - 2\left(\sum_i \frac{p_{1i}^2}{p_{0i}}\right)\left(\sum_i \frac{p_{1i}}{p_{0i}}\right) \right.$$
$$\left. + 3\sum_i \frac{p_{1i}^2}{p_{0i}^2} \right\} - \left(\sum_i \frac{p_{1i}}{p_{0i}}\right)^2 + \sum_i \frac{p_{1i}}{p_{0i}^2}.$$

(Patnaik, 1949)

30.6 For the same alternative hypothesis as in Example 30.3, namely the Gamma distribution with parameter 1·5, use the *Biometrika Tables* to obtain the p_{1i} for the unequal-probabilities four-class grouping in Example 30.2. Calculate the non-central parameter (30.62) for this case, and show by comparison with Example 30.3 that the unequal-probabilities grouping would require about a 25 per cent larger sample than the equal-probabilities grouping in order to attain the same power against this alternative.

30.7 k independent standardized normal variables x_j are subject to c homogeneous linear constraints. Show that $S = \sum_{j=1}^{k} x_j^2$ is distributed independently of the signs of the x_j. If $c = 1$, and the constraint is $\sum_{j=1}^{k} x_j = 0$, show that all sequences of signs are equiprobable (except all signs positive, or all signs negative, which cannot occur), but that this is not so generally for $c > 1$. Hence show that any test based on the sequence of signs of the deviations of observed from hypothetical frequencies $(n_i - np_{0i})$ is asymptotically independent of the X^2 test when H_0 holds.

(F. N. David, 1947; Seal, 1948; Fraser, 1950)

30.8 M elements of one kind and N of another are arranged in a sequence at random $(M, N > 0)$. A *run* is defined as a subsequence of elements of one kind immediately preceded and succeeded by elements of the other kind. Let R be the number of runs in the whole sequence $(2 \leq R \leq M+N)$. Show that

$$P\{R = 2s\} = 2 \binom{M-1}{s-1}\binom{N-1}{s-1} \Big/ \binom{M+N}{M},$$

$$P\{R = 2s-1\} = \left\{ \binom{M-1}{s-2}\binom{N-1}{s-1} + \binom{M-1}{s-1}\binom{N-1}{s-2} \right\} \Big/ \binom{M+N}{M},$$

and that

$$E(R) = 1 + \frac{2MN}{M+N}$$

$$\mathrm{var}\ R = \frac{2MN(2MN - M - N)}{(M+N)^2(M+N-1)}.$$

(Stevens (1939); Wald and Wolfowitz (1940). Swed and Eisenhart (1943) tabulate the distribution of R for $M \leq N \leq 20$.)

30.9 From Exercises 30.7 and 30.8, show that if there are M positive and N negative deviations $(n_i - np_{0i})$, we may use the runs test to supplement the X^2 test for the simple hypothesis. From Exercise 16.4, show that if P_1 is the probability of a value of X^2 not less than that observed and P_2 is the probability of a value of R not greater than that observed, then $U = -2(\log P_1 + \log P_2)$ is asymptotically distributed like χ^2 with 4 degrees of freedom, large values of U forming the critical region for the combined test.

(F. N. David, 1947)

30.10 $x_1, x_2, \ldots, x_n$ are independent random variables with the same distribution $f(x|\theta_1, \theta_2)$. θ_1 and θ_2 are estimated by statistics $t_1(x_1, x_2, \ldots, x_n)$, $t_2(x_1, x_2, \ldots, x_n)$. Show that the random variables

$$y_i = \int_{-\infty}^{x_i} f(u|t_1, t_2)\, du$$

are not independent and that they have a distribution depending in general on f, θ_1 and θ_2; but that if f is of form $\theta_2^{-1} f\{(x-\theta_1)/\theta_2\}$, $\theta_2 > 0$, and t_1, t_2 respectively satisfy $t_1(\mathbf{x}+\alpha) = t_1(\mathbf{x}) + \alpha$, $t_2(\beta \mathbf{x}) = \beta t_2(\mathbf{x})$, $\beta > 0$, the distribution of y_i is not dependent on θ_1 and θ_2, but on the form of f alone.

(F. N. David and Johnson, 1948)

30.11 Show that for testing a composite hypothesis the X^2 test statistic using multinomial ML estimators is asymptotically equivalent to the LR test statistic when H_0 holds.

30.12 Show that Neyman's goodness-of-fit statistic (30.80) is equivalent to the LR test of the simple hypothesis (30.78) in large samples.

(E. S. Pearson, 1938)

30.13 Verify the values of the mean and variance (30.81–2).

30.14 Prove formula (30.102) for the variance of W^2.

30.15 Verify that nW^2 may be expressed in the form (30.105) by integrating (30.103) separately over each interval for which $S_n(x)$ is constant.

30.16 In testing a simple hypothesis specifying a d.f. $F_0(x)$, show diagrammatically that for a simple alternative $F_1(x)$ satisfying

$$F_1(x) < F_0(x) \quad \text{when} \quad F_0(x) < d_\alpha,$$
$$F_1(x) = F_0(x) \quad \text{elsewhere},$$

the D_n test (with critical value d_α) may be biassed.

(Massey, 1950b, 1952)

30.17 A random sample of n observations u_r is taken from the rectangular distribution on the interval $(0, 1)$, dividing that interval into $(n+1)$ lengths c_j, where $c_j \geq 0$ and $\sum_{j=1}^{n+1} c_j = 1$. The c_j are ordered so that $c_{(1)} \leq c_{(2)} \leq \ldots \leq c_{(n+1)}$. Show that the non-negative variables

$$\begin{cases} g_1 = (n+1)c_{(1)}, \\ g_j = (n+2-j)(c_{(j)} - c_{(j-1)}), \quad j = 2, 3, \ldots n+1; \; \sum_{j=1}^{n+1} g_j = 1, \end{cases}$$

have the distribution

$$dF_g = n! \, dg_1 \ldots dg_n,$$

and that the unordered c_j also have this distribution, so that

$$dF_c = n! \, dc_1 \ldots dc_n$$

(the $(n+1)$th variable being omitted in each case to remove the singularity of the distribution). Hence show that the variables

$$w_r = \sum_{j=1}^{r} g_j, \quad r = 1, 2, \ldots, n,$$

are distributed exactly as the order-statistics of the original sample, $u_{(r)}$, $r = 1, 2, \ldots, n$. Thus any test of fit based on the probability-integral transformation may be applied to the w_r, as well as to the $u_{(r)}$ obtained from the transformation.

(Durbin (1961), who finds from sampling experiments that a one-sided Kolmogorov test (D_n^-) applied to the w_r has better power properties than the ordinary two-sided D_n test for detecting changes in distributional form.)

30.18 Let θ be the unspecified parameters in testing a composite hypothesis of fit for n observations $\mathbf{x}$. Suppose that $\mathbf{t}$, with less than n components, is minimal sufficient for θ, and that we can make a $1-1$ transformation from $\mathbf{x}$ to $(\mathbf{t}, \mathbf{u})$, where $\mathbf{u}$ is distributed independently of $\mathbf{t}$. Show that if the value of $\mathbf{t}$ is discarded, and is replaced by a random observation $\mathbf{t}'$ from its distribution with a known value of θ, then the set of observations $\mathbf{x}'$ obtained by the inverse transformation from $(\mathbf{t}', \mathbf{u})$ is distributed independently of θ, so that the hypothesis of fit becomes simple.

(Durbin, 1961)

30.19 If X^2 for the simple or composite H_0 is asymptotically distributed in the χ^2 form with f degrees of freedom, its $100(1-\alpha)$ percentile being $\chi^2_{f,\alpha}$, show that if $\max \left| \dfrac{n_i}{n} - p_{0i} \right| = \Delta \leqslant 0.5$, the minimum possible value of X^2 for a fixed Δ, whatever the n_i and the p_{0i}, is $4n\Delta^2$. Putting $n \geqslant n_0 = \chi^2_{f,\alpha}/(4\Delta^2)$, show that
$$\text{Prob}\{X^2 \leqslant 4n\Delta^2\} \geqslant 1-\alpha$$
for any set of p_{0i} whatever; and that for sufficiently large n, a conservative set of $100(1-\alpha)$ per cent. confidence intervals for all the true p_{0i} simultaneously is given by
$$\left\{ \dfrac{n_i}{n} \pm 0.5\, (\chi^2_{f,\alpha}/n)^{\frac{1}{2}} \right\}, \quad i = 1, 2, \ldots, k.$$

(Naddeo, 1968)

30.20 Show that in the equal-probabilities case, X^2 defined at (30.44) varies by multiples of $2k/n$ and hence, using the argument of **31.80**, that we may expect the χ^2 approximation to the distribution of X^2 to be improved by a continuity correction of $-k/n$. Show also that the minimum attainable value of X^2 is
$$\dfrac{n - k\left[\dfrac{n}{k}\right]}{n} \left\{ k\left(\left[\dfrac{n}{k}\right]+1\right) - n \right\},$$
where $[z]$ is the integral part of z, and hence is zero only when n is an integral multiple of k.

30.21 Show that if the distribution tested is of the form $\dfrac{1}{\theta_2} f\!\left(\dfrac{x-\theta_1}{\theta_2} \right)$, $\theta_2 > 0$, the matrix **BDB**$'$ in (30.42) does not depend upon θ_1, θ_2, so that its latent roots also do not.

(cf. Watson, 1958)

CHAPTER 31

ROBUST AND DISTRIBUTION-FREE PROCEDURES

31.1 In the course of our examination of the various aspects of statistical theory which we have so far encountered, we have found on many occasions that excellent progress can be made when the underlying parent populations are normal in form. The basic reason for this is the spherical symmetry which characterizes normality, but this is not our present concern. What we have now to discuss is the extent to which we are likely to be justified if we apply this so-called " normal theory " in circumstances where the underlying distributions are not in fact normal. For, in the light of the relative abundance of theoretical results in the normal case, there is undoubtedly a temptation to regard distributions as normal unless otherwise proven, and to use the standard normal theory wherever possible. The question is whether such optimistic assumptions of normality are likely to be seriously misleading.

We may formulate the problem more precisely for hypothesis-testing problems in the manner of our discussion of similar regions in **23.4**. There, it will be recalled, we were concerned to establish the size of a test at a value α, irrespective of the values of some nuisance parameters. Our present question is of essentially the same kind, but it relates to the form of the underlying distribution itself rather than to its unspecified parameters: is the test size α sensitive to changes in the distributional form?

A statistical procedure which is insensitive to departures from the assumptions which underlie it is called " robust," an apt term introduced by G. E. P. Box (1953) and now in general use. Studies of robustness have been carried out by many writers. A good deal of their work has been concerned with the Analysis of Variance, and we postpone discussion of this until Volume 3. At present, we confine ourself to the results relevant to the procedures we have already encountered. G. E. P. Box and Andersen (1955) survey the subject generally.

The robustness of the standard " normal theory " procedures

31.2 Beginning with early experimental studies, notably by E. S. Pearson, the examination of robustness was continued by means of theoretical investigations, among which those of Bartlett (1935a), Geary (1936, 1947) and Gayen (1949–1951) are essentially similar in form. The observations are taken to come from parent populations specified by Gram–Charlier or Edgeworth series expansions, and corrective terms, to be added to the normal theory, are obtained as functions of the standardized higher cumulants, particularly κ_3 and κ_4. Their results may broadly be summarized by the statement that whereas tests on population means (i.e. " Student's " t-tests for the mean of a normal population and for the difference between the means of two normal populations with the same variance) are rather insensitive to departures from normality, tests on variances (i.e. the χ^2 test for the variance of a normal population, the F-test for the ratio of two normal population variances, and the modified LR test for the equality of several normal variances in Examples 24.4, 24.6) are very sensitive to such departures. Tests on means are robust; by comparison, tests on variances can only be described as frail. We have not the space here for a detailed derivation of these results, but it is easy to explain them in general terms.

31.3 The crucial point in the derivation of "Student's" t-distribution is the independence of its numerator and denominator, which holds exactly only for normal parent populations. If we are sampling from non-normal populations, the Central Limit theorem nevertheless assures us that the sample mean and the unbiassed variance estimator $s^2 = k_2$ will be asymptotically normally distributed. What is more, we know from Rule 10 for the sampling cumulants of k-statistics in **12.14** that

$$\kappa(2\,1) = \kappa_3/n, \tag{31.1}$$
$$\kappa(2^r\,1^s) = O(n^{-(r+s-1)}). \tag{31.2}$$

Since

$$\kappa(1^2) = \kappa_2/n, \quad \kappa(2^2) = \frac{\kappa_4}{n} + \frac{2\kappa_2^2}{n-1},$$

we have from (31.1) for the asymptotic correlation between $\bar{x}$ and s^2

$$\rho = \kappa_3/\{\kappa_2(\kappa_4 + 2\kappa_2^2)\}^{\frac{1}{2}}. \tag{31.3}$$

If the non-normal population is symmetrical, κ_3 and ρ of (31.3) are zero, and $\bar{x}$ and s^2 are exactly uncorrelated and asymptotically independent, so that the normal theory will hold for n large enough. If $\kappa_3 \neq 0$, (31.3) will be smaller when κ_4 is large, but will remain non-zero. The situation is saved, however, by the fact that the exact Student's t-distribution itself approaches normality as $n \to \infty$, as also, by the Central Limit theorem, does the distribution of

$$t = (\bar{x} - \mu)/(s^2/n)^{\frac{1}{2}}, \tag{31.4}$$

since s^2 converges stochastically to σ^2. The two limiting distributions are the same.

Thus, whatever the parent distribution, the statistic (31.4) tends to normality, and hence to the limiting normal theory. If the parent is symmetrical we may expect the statistic to approach its normal theory distribution (Student's t) more rapidly. This is, in fact, what the detailed investigations have confirmed: for small samples the normal theory is less robust in the face of parent skewness than for departure from normal kurtosis.

Efron (1969) discusses the robustness of t under symmetry conditions.

31.4 Similarly for the two-sample Student's t-statistic. If the two samples come from the same non-normal population and we use the normal test statistic

$$t = \{(\bar{x}_1 - \bar{x}_2) - (\mu_1 - \mu_2)\} \Big/ \left\{ \left[\frac{(n_1-1)s_1^2 + (n_2-1)s_2^2}{n_1+n_2-2}\right]\left(\frac{1}{n_1}+\frac{1}{n_2}\right)\right\}^{\frac{1}{2}}, \tag{31.5}$$

we find that the covariance between $(\bar{x}_1 - \bar{x}_2)$ and the term in square brackets in the denominator, say s^2, is given by

$$\text{cov} = \kappa_3 \left\{\frac{n_1-1}{n_1+n_2-2} \cdot \frac{1}{n_1} - \frac{n_2-1}{n_1+n_2-2} \cdot \frac{1}{n_2}\right\} = \frac{\kappa_3}{n_1+n_2-2}\left(\frac{1}{n_2} - \frac{1}{n_1}\right),$$

while the variances corresponding to this are

$$\text{var}(\bar{x}_1 - \bar{x}_2) = \kappa_2\left(\frac{1}{n_1} + \frac{1}{n_2}\right),$$
$$\text{var}(s^2) \sim (\kappa_4 + 2\kappa_2^2)/(n_1 + n_2).$$

The correlation is therefore asymptotically

$$\rho = \frac{\kappa_3}{\{\kappa_2(\kappa_4+2\kappa_2^2)\}^{\frac{1}{2}}} \frac{(n_1 n_2)^{\frac{1}{2}}}{n_1+n_2-2}\left(\frac{1}{n_2}-\frac{1}{n_1}\right). \tag{31.6}$$

Again, if $\kappa_3 = \rho = 0$, the asymptotic normality carries asymptotic independence with it. We also see that $\rho = 0$ if $n_1 = n_2$. In any case, as n_1 and n_2 become large, the Central Limit theorem brings (31.5) to asymptotic normality and hence to agreement with the Student's t-distribution.

Once again, these are precisely the results found by Bartlett (1935) and Gayen (1949–1951): if sample sizes are equal, even skewness in the parent is of little effect in disturbing normal theory. If the parent is symmetrical, the test will be robust even for differing sample sizes.

31.5 Studies have also been made of the effects of more complicated departures from normality in " Student's " t-tests. Hyrenius (1950) considered sampling from a mixture of normal distributions, and other Swedish writers, of whom Zackrisson (1959) gives references to earlier work, have considered various forms of populations composed of normal sub-populations. Robbins (1948) obtains the distribution of t when the observations come from normal populations differing only in means. For the two-sample test, Geary (1947) and Gayen (1949–1951) permit the samples to come from different populations.

31.6 When we turn to tests on variances, the picture is very different. The crucial point for normal theory in all tests on variances is that the ratio $z = \sum_{i=1}^{n}(x_i - \bar{x})^2/\sigma^2$ is distributed like χ^2 with $(n-1)$ degrees of freedom. If we consider the sampling cumulants of $k_2 = \kappa_2 z/(n-1)$, we see from (12.35) that

$$\operatorname{var} z = \left(\frac{n-1}{\kappa_2}\right)^2 \kappa(2^2) = 2(n-1) + \frac{(n-1)^2}{n}\frac{\kappa_4}{\kappa_2^2}$$

$$\sim (n-1)\left(2 + \frac{\kappa_4}{\kappa_2^2}\right), \tag{31.7}$$

while from (12.36)

$$\mu_3(z) = \left(\frac{n-1}{\kappa_2}\right)^3 \kappa(2^3)$$

$$= \left(\frac{n-1}{\kappa_2}\right)^3 \left\{\frac{\kappa_6}{n^2} + \frac{12\kappa_4\kappa_2}{n(n-1)} + \frac{4(n-2)\kappa_3^2}{n(n-1)^2} + \frac{8\kappa_2^3}{(n-1)^2}\right\}$$

$$\sim (n-1)\left\{\frac{\kappa_6}{\kappa_2^3} + \frac{12\kappa_4}{\kappa_2^2} + \frac{4\kappa_3^2}{\kappa_2^3} + 8\right\},$$

and similarly for higher moments from (12.37–39). These expressions make it obvious that the distribution of z depends on all the (standardized) cumulant ratios κ_3^2/κ_2^3, κ_4/κ_2^2, etc., and that the terms involving these ratios are of the same order in n as the normal theory constant terms. If, and only if, all higher cumulants are zero, so that the parent distribution is normal, these additional terms will disappear. Otherwise, (31.7) shows that even though z is asymptotically normally distributed, the large-sample distribution of z will not approach the normal theory χ^2 distribution. The Central Limit theorem does not rescue us here because z tends to a *different* normal distribution from the one we want.

31.7 Because κ_4 appears in (31.7) but κ_3 does not, we should expect deviations from normal kurtosis to exercise the greater effect on the distribution, and this is precisely the result found after detailed calculations by Gayen (1949–1951) for the χ^2 and variance-ratio tests for variances. G. E. P. Box (1953) found that the discrepancies from asymptotic normal theory became larger as more variances were compared, and his argument is simple enough to reproduce here.

Suppose that k samples of sizes n_i $(i = 1, 2, \ldots, k)$ are drawn from populations each of which has the same variance κ_2 and the same kurtosis coefficient $\gamma_2 = \kappa_4/\kappa_2^2$.

From (31.7), we then have asymptotically for any one sample

$$\operatorname{var}(s_i^2) = 2\kappa_2^2(1 + \tfrac{1}{2}\gamma_2)/n_i, \tag{31.8}$$

where s_i^2 is the unbiassed estimator of κ_2. Now by the Central Limit theorem, s_i^2 is asymptotically normal with mean κ_2 and variance (31.8), and is therefore distributed as if it came from a normal population and were based on $N_i = n_i/(1 + \tfrac{1}{2}\gamma_2)$ observations instead of n_i. Thus the effect on the modified LR criterion for comparing k normal variances, given at (24.44), is that $-2\log l^*/(1 + \tfrac{1}{2}\gamma_2)$ and not $-2\log l^*$ itself is distributed asymptotically as χ^2 with $k-1$ degrees of freedom.

The effects of this correction on the normal theory distribution can be quite extreme. We give in the table below some of Box's (1953) computations:

True probability of exceeding the asymptotic normal theory critical value for $\alpha = 0.05$

γ_2 \ k	2	3	5	10	30
−1	0·0056	0·0025	0·0008	0·0001	0·0⁶1
0	0·05	0·05	0·05	0·05	0·05
1	0·110	0·136	0·176	0·257	0·498
2	0·166	0·224	0·315	0·489	0·849

As is obvious from the table, the discrepancy from the normal theory value of 0·05 increases with $|\gamma_2|$, and with k for any fixed $\gamma_2 \neq 0$.

31.8 Although the result of **31.7** is asymptotic, Box (1953) shows that similar discrepancies occur for small samples. The lack of robustness in the variance test is so striking, indeed, that he was led to consider the criterion l^* of (24.44) as a test statistic for kurtosis, and found its sensitivity to be of the same order as the generally-used tests mentioned in **30.63**.

31.9 Finally, we mention briefly that Gayen (1949–1951) has considered the robustness both of the sample correlation coefficient r, and of Fisher's z-transformation of r to departures from bivariate normality. When the population correlation coefficient $\rho = 0$, and in particular when the variables are independent, the distribution of r is robust, even for sample size as low as 11; but for large values of ρ the departures from normal theory are appreciable. The z-transformation remains asymptotically normally distributed under parental non-normality, but the approach is less rapid. The mean and variance of z are, to order n^{-1}, unaffected by skewness in the parental marginal distributions, but the effect of departures from mesokurtosis may be considerable; the variance of z, in particular, is sensitive to the parental form, even in large samples, although the mean of z slowly approaches its normal value as n increases.

Hotelling (1961) makes a quite different approach to problems of robustness and gives a useful list of references.

Huber (1964) makes a general investigation of the robustness and efficiency of estimators of a location parameter—cf. also Bickel (1965), Gastwirth (1966) and Crow and Siddiqui (1967), who make recommendations for classes of symmetrical unimodal distributions including the normal, double exponential, rectangular and Cauchy. Siddiqui and Raghunandanan (1967) extend the results. See also A. Birnbaum and Laska (1967a), Gastwirth and Rubin (1969), Gastwirth and Cohen (1970), A. Birnbaum and Miké (1970), Jaeckel (1971 a, b), A. Birnbaum et al. (1971) and Miké (1971).

Transformations to normality

31.10 The investigation of robustness has as its aim the recognition of the range of validity of the standard normal theory procedures. As we have seen, this range may be wide or extremely narrow, but it is often difficult in practice to decide whether the standard procedures are likely to be approximately valid or misleading. Two other approaches to the non-fulfilment of normality assumptions have been made, which we now discuss.

The first possibility is to seek a transformation which will bring the observations close to the normal form, so that normal theory may be applied to the transformed observations. This may take the form discussed in **6.25–26**, where we normalize by finding a polynomial transformation. Alternatively, we may be able to find a simple normalizing functional transformation—cf. **6.27–35** and Fisher's z-transformation of the correlation coefficient at (16.75). The difficulty in both cases is that we must have knowledge of the underlying distribution before we know which transformation is best applied, information which is likely to be obtainable in theoretical contexts like the investigation of the sampling distribution of a statistic, but is harder to come by when the distribution of interest is arising in experimental work.

Fortunately, transformations designed to stabilize a variance (i.e. to render it independent of some parameter of the population) often also serve to normalize the distribution to which they are applied—Fisher's z-transformation of r is an example of this. Exercise 16.18 shows how a knowledge of the relation between mean and variance in the underlying distribution permits a simple variance-stabilizing transformation to be carried out. Such transformations are most commonly used in the Analysis of Variance, and we postpone detailed discussion of them until we treat that subject in Volume 3.

Distribution-free procedures

31.11 The second of the alternative approaches mentioned at the beginning of **31.10** is a radical one. Instead of holding to the standard normal theory methods (either because they are robust and approximately valid in non-normal cases or by transforming the observations to make them approximately valid), we abandon them entirely for the moment and approach our problems afresh. Can we find statistical procedures which remain valid for a wide class of parent distributions, say for all continuous distributions? If we can, they will necessarily be valid for normal distributions, and our robustness will be precise and assured. Such procedures are called *distribution-free*, as we have already seen in **30.35**, because their validity does not depend on the form of the underlying distributions at all, provided that they are continuous.

The remainder of this chapter, and parts of the two immediately following chapters, will be devoted to distribution-free methods. First, we discuss the relationship of distribution-free methods to the parametric–non-parametric distinction that we made in **22.3**.

31.12 It is easy to see that if we are dealing with a parametric problem (e.g. testing a parametric hypothesis or estimating a parameter) the method we use may or may not be distribution-free—e.g. **32.2–7** below discuss the Sign test, which tests the value of θ in (17.1), and remains a valid test for the population median of *any* continuous distribution; on the other hand, the optimum test based on the sample mean discussed in Examples 22.2–3 is not even valid for finite n if the population is non-normal, although its validity (not its optimality) is asymptotically rescued by the Central Limit theorem as in **31.3** above. It is perhaps not at once so clear that even if the problem is non-parametric, the method also may or may not be distribution-free. For example, in Chapter 30 we discussed tests of fit of composite hypotheses, where the problem is non-parametric, and found that the test statistic is not even asymptotically distribution-free in general when the estimators are not the multinomial ML estimators. Again, if we use the sample moment-ratio $b_2 = m_4/m_2^2$ as a test of normality, the problem is non-parametric but the distribution of b_2 is heavily dependent on the form of the parent.

However, most distribution-free procedures were devised for non-parametric problems, such as testing whether two continuous distributions are identical, and there is therefore a fairly free interchangeability of meanings in the terms " non-parametric " and " distribution-free " as used in the literature. We shall always use them in the quite distinct senses which we have defined: " non-parametric " is a description of the problem and " distribution-free " of the method used to solve the problem.

Distribution-free methods for non-parametric problems

31.13 The main classes of non-parametric problems which can be solved by distribution-free methods are as follows:

(1) *The two-sample problem*
   The hypothesis to be tested is that two populations, from each of which we have a random sample of observations, are identical.

(2) *The k-sample problem*
   This is the generalization of (1) to $k > 2$ populations.

(3) *Randomness*
   A series of n observations on a single variable is ordered in some way (usually through time). The hypothesis to be tested is that each observation comes independently from the same distribution.

(4) *Independence in a bivariate population*
   The hypothesis to be tested is that a bivariate distribution factorizes into two independent marginal distributions.

These are all hypothesis-testing problems, and it is indeed the case that most distribution-free methods are concerned with testing rather than estimation. However, we can find distribution-free

(1a) *Confidence intervals for a difference in location between two otherwise identical continuous distributions,*
(5) *Confidence intervals and tests for quantiles,*
and (6) *Tolerance intervals for a continuous distribution.*

In Chapter 30, we have already discussed

(7) *Distribution-free tests of fit*
and (8) *Confidence limits for a continuous distribution function.*

The categories listed above contain the bulk of the work done on distribution-free methods so far, although they are not exhaustive, as we shall see.

A very full bibliography of the subject is given by Savage (1962).

31.14 The reader will probably have noticed that problems (1) to (3) in **31.13** are all of the same kind, being concerned with testing the identity of a number of univariate continuous distributions, and he may have wondered why problem (4) has been grouped with them. The reason is that problem (4) can be modified to give problems (1) to (3). We shall indicate the relationship here briefly, and leave the details until we come to particular tests later.

Suppose that in problem (3) we numerically label the ordering of the variable x and regard this labelling as the observations on a variable y. Problem (3) is then reduced to testing the independence of x and the label variable y, i.e. to a special case of problem (4). Again in problem (4), suppose that the range of the second variable, say z, is dichotomized, and that we score $y = 1$ or 2 according to which part of the dichotomy an observed z falls into. If we now test the independence of x and y, we have reduced problem (4) to problem (1), for if x is independent of the y-classification, the distributions of x for $y = 1$ and for $y = 2$ must be identical. Similarly, we reduce problem (4) to problem (2) by polytomizing the range of z into $k > 2$ classes, scoring $y = 1, 2, \ldots, k$, and testing the independence of x and y.

The construction of distribution-free tests

31.15 How can distribution-free tests be constructed for non-parametric problems? We have already encountered two methods in our discussion of tests of fit in Chapter 30: one was to use the probability integral transformation which for simple hypotheses yields a distribution-free test; the second was to reduce the problem to a multinomial distribution problem, as for the X^2 test—we shall see in the next chapter that this latter device in its simplest form serves to produce a test (the so-called Sign Test) for problem (5) of **31.13**. But important classes of distribution-free tests for problems (1) to (4) rest on a different foundation, which we now examine.

If we know nothing of the form of the parent distributions, save perhaps that they are continuous, we obviously cannot find similar regions in the sample space by the methods used for parametric problems in Chapter 23. However, progress can be made. First, we make the necessary slight adjustments in our definitions of sufficiency and completeness.

In the absence of a parametric formulation, we must make these definitions refer directly to the parent d.f.; whereas previously we called a statistic t sufficient for the parameter θ if the factorization (17.68) were possible, we now define a family C of distributions and let θ be simply a variable indexing the membership of that family. With this understanding, t is called sufficient for the family C if the factorization (17.68) holds for all θ. Similarly, the definitions of completeness and bounded completeness of a family of distributions in **23.9** hold good for non-parametric situations if θ is taken as an indexing variable for members of the family.

31.16 Now we have seen in Examples 23.5 and 23.6 that the set of order-statistics $t = (x_{(1)}, x_{(2)}, \ldots, x_{(n)})$ is a sufficient statistic in some parametric problems, though not necessarily a minimal sufficient statistic. It is intuitively obvious that t will always be a sufficient statistic when all the observations come from the same parent distribution, for then no information is lost by ordering the observations. (It is also obvious that it will be minimal sufficient if nothing at all is known about the form of the parent distribution.) Now if the parent is continuous, we have observed in **23.5** that similar regions can always be constructed by permutation of the co-ordinates of the sample space, for tests of size which is a multiple of $(n!)^{-1}$. Such permutation leaves the set of order-statistics constant. If nothing whatever is known of the form of the parent, it is clear that we cannot get similar regions in any other way. Thus the result of **23.19** implies that the set of order-statistics is boundedly complete for the family of all continuous d.f.'s.(*)

We therefore see that if we wish to construct similar tests for hypotheses like those of problems (1)–(4) of **31.13**, we must use *permutation tests* which rest essentially on the fact, proved in **11.4** and obvious by symmetry, that any ordering of a sample from a continuous d.f. has the same probability $(n!)^{-1}$. There still remains the question of which permutation test to use for a particular hypothesis.

The efficiency of distribution-free tests

31.17 The search for distribution-free procedures is motivated by the desire to broaden the range of validity of our inferences. We cannot expect to make great gains in generality without some loss of efficiency in particular circumstances; that is to say, we cannot expect a distribution-free test, chosen in ignorance of the form of the parent distribution, to be as efficient as the test we would have used had we known that parental form. But to use this as an argument against distribution-free procedures is manifestly mistaken: it is precisely the absence of information as to parental form that leads us to choose a distribution-free method. The only "fair" standard of efficiency for a distribution-free test is that provided by other distribution-free tests. We should naturally choose the most efficient such test available.

But in what sense are we to judge efficiency? Even in the parametric case, UMP tests are rare, and we cannot hope to find distribution-free tests which are most powerful against all possible alternatives. We are thus led to examine the power of distribution-free tests against parametric alternatives to the non-parametric hypothesis tested. Despite its paradoxical sound, there is nothing contradictory about this, and the procedure has one great practical virtue. If we examine power against the alter-

(*) That it is actually complete is proved directly, e.g. by Lehmann (1959); the result is due to Scheffé (1943b).

natives considered in *normal* distribution theory, we obtain a measure of how much we can lose by using a distribution-free test if the assumptions of normal theory really are valid (though, of course, we would not know this in practice). If this loss is small, we are encouraged to sacrifice the little extra efficiency of the standard normal theory methods for the extended range of validity attached to the use of the distribution-free test.

We may take this comparison of normal theory tests with distribution-free tests a stage further. In certain cases, it is possible to examine the relative efficiency of the two methods for a wide range of underlying parent distributions; and it should be particularly noted that we have no reason to expect the normal theory method to maintain its efficiency advantages over the distribution-free method when the parent distribution is not truly normal. In fact, we might hazard a guess that distribution-free methods should suffer less from the falsity of the normality assumption than do the normal theory methods which depend upon that assumption. Such few investigations as have been carried out seem on the whole to support this guess.

Tests of independence

31.18 We begin our detailed discussion of distribution-free tests for non-parametric hypotheses, which will illustrate the general points made in **31.15–17**, with problem (4) of **31.13**—the problem of independence.

Suppose that we have a sample of n pairs (x,y) from a continuous bivariate distribution function $F(x,y)$ with continuous marginal distribution functions $G(x)$, $H(y)$. We wish to test

$$H_0: F(x,y) = G(x)H(y), \quad \text{all } x,y. \tag{31.9}$$

Under H_0, every one of the $n!$ possible orderings of the x-values is equiprobable, and independently of x, so is every one of $n!$ y-orderings; we therefore have $(n!)^2$ equiprobable points in the sample space. Since, however, we are interested only in the relationship between x and y, we are concerned only with different pairings of the n x's with the n y's, and there are $n!$ distinct sets of pairings (obtained, e.g. by keeping the y's fixed and permuting the x's) with equal probabilities $(n!)^{-1}$. From **31.16**, all similar size-α tests of H_0 contain $\alpha n! = N$ of these pairings (N assumed a positive integer).

Each of the $n!$ sets of pairings contains n values of (x,y) (some, of course, may coincide). The question is now: what function of the values (x,y) shall we take as our test statistic? Consider the alternative hypothesis H_1 that x and y are bivariate normally distributed with non-zero correlation parameter ρ. We may then write the Likelihood Function, by (16.47) and (16.50), as

$$L(x|H_1) = \{2\pi\sigma_x\sigma_y(1-\rho^2)^{\frac{1}{2}}\}^{-n} \exp\left\{-\frac{n}{2(1-\rho^2)}\left[\left(\frac{\bar{x}-\mu_x}{\sigma_x}\right)^2 - 2\rho\left(\frac{\bar{x}-\mu_x}{\sigma_x}\right)\left(\frac{\bar{y}-\mu_y}{\sigma_y}\right)\right.\right.$$
$$\left.\left. + \left(\frac{\bar{y}-\mu_y}{\sigma_y}\right)^2 + \left(\frac{s_x^2}{\sigma_x^2} - \frac{2\rho r s_x s_y}{\sigma_x\sigma_y} + \frac{s_y^2}{\sigma_y^2}\right)\right]\right\}. \tag{31.10}$$

Now changes in the pairings of the x's and y's leave the observed means and variances $\bar{x}$, $\bar{y}$, s_x^2, s_y^2, unchanged. The sample correlation coefficient r, however, is affected by the pairings through the term $\sum_{i=1}^{n} x_i y_i$ in its numerator. Evidently, (31.10) will be largest for any $\rho > 0$ when r is as large as possible, and for any $\rho < 0$ when r is as small as possible. By the Neyman–Pearson lemma of **22.10**, we shall obtain the most powerful permutation test by choosing as our critical regions those sets of pairings which maximize (31.10), for when H_0 holds, all pairings are equiprobable. Thus consideration of normal alternatives leads to the following test, first proposed on intuitive grounds by Pitman (1937b): reject H_0 against alternatives of positive correlation if r is large, against alternatives of negative correlation if r is small, and against general alternatives of non-independence if $|r|$ is large. The critical value in each case is to be determined from the distribution of r over the $n!$ distinct sets of pairings equiprobable on H_0.

Although Pitman's correlation test gives the most powerful permutation test of independence against normal alternatives, it is, of course, a valid test (i.e. it is a strictly size-α test) against any alternatives, and one may suppose that it will be reasonably powerful for a wide range of alternatives approximating normality.

The permutation distribution of r

31.19 Since
$$r = \left(\frac{1}{n}\sum_{i=1}^{n} x_i y_i - \bar{x}\bar{y}\right) \Big/ s_x s_y, \tag{31.11}$$

only $\sum_i x_i y_i$ is a random variable under permutation. We can obtain its exact distribution, and hence that of r, by enumeration of the $n!$ possibilities, but this becomes too tedious in practice when n is at all large. Instead, we approximate the exact distribution by fitting a distribution to its moments. We keep the y's fixed and permute the x's, and find
$$E(\Sigma x_i y_i) = \Sigma y_i E(x_i) = \Sigma y_i \bar{x} = n \bar{x}\bar{y},$$
whence, from (31.11),
$$E(r) = 0. \tag{31.12}$$

r is invariant under location changes in x and y, so for convenience we now measure from the means $(\bar{x}, \bar{y})$. We have
$$\operatorname{var}\left(\sum_i x_i y_i\right) = \sum_i y_i^2 \operatorname{var} x_i + \sum\sum_{i \neq j} y_i y_j \operatorname{cov}(x_i, x_j)$$
$$= \sum_i y_i^2 s_x^2 + \sum\sum_{i \neq j} y_i y_j \cdot \frac{1}{n(n-1)} \sum\sum_{i \neq j} x_i x_j$$
$$= n s_y^2 s_x^2 + \{(\Sigma y_i)^2 - \Sigma y_i^2\} \frac{1}{n(n-1)} \{(\Sigma x_i)^2 - \Sigma x_i^2\}$$
$$= n s_y^2 s_x^2 + n s_y^2 s_x^2/(n-1)$$
$$= n^2 s_y^2 s_x^2/(n-1).$$

Thus (31.11) gives
$$\operatorname{var} r = (n^2 s_x^2 s_y^2)^{-1} \operatorname{var}(\Sigma xy) = 1/(n-1). \tag{31.13}$$

The first two moments of r, given by (31.12) and (31.13), are quite independent of the actual values of (x, y) observed. By similar expectational methods, it will be found that
$$\left. \begin{array}{l} E(r^3) = \dfrac{n-2}{n(n-1)^2} \left(\dfrac{k_3}{k_2^{3/2}}\right)\left(\dfrac{k_3'}{(k_2')^{3/2}}\right), \\[2ex] E(r^4) = \dfrac{3}{n^2-1}\left\{1 + \dfrac{(n-2)(n-3)}{3n(n-1)^2}\left(\dfrac{k_4}{k_2^2}\right)\left(\dfrac{k_4'}{(k_2')^2}\right)\right\}, \end{array} \right\} \tag{31.14}$$

where the k's are the k-statistics of the observed x's and the k''s the k-statistics of the y's. Neglecting the differences between k-statistics and sample cumulants, we may rewrite (31.14) as

$$\left. \begin{array}{l} E(r^3) \doteqdot \dfrac{(n-2)}{n(n-1)^2} g_1 g_1', \\[6pt] E(r^4) \doteqdot \dfrac{3}{n^2-1}\left\{1 + \dfrac{(n-2)(n-3)}{3n(n-1)^2} g_2 g_2'\right\}, \end{array} \right\} \quad (31.15)$$

where g_1, g_2 are the measures of skewness and kurtosis of the x's, and g_1', g_2' those of the y's. If these are fixed, (31.14) may be written

$$\left. \begin{array}{l} E(r^3) = O(n^{-2}), \\[4pt] E(r^4) = \dfrac{3}{n^2-1}\{1 + O(n^{-1})\}. \end{array} \right\} \quad (31.16)$$

Thus, as $n \to \infty$, we have approximately

$$\left. \begin{array}{l} E(r^3) = 0, \\[4pt] E(r^4) = \dfrac{3}{n^2-1}. \end{array} \right\} \quad (31.17)$$

The moments (31.12), (31.13) and (31.17) are precisely those of (16.62), the symmetrical exact distribution of r in samples from a bivariate normal distribution with $\rho = 0$, as may easily be verified by integration of r^2 and r^4 in (16.62). Thus, to a close approximation, the *permutation* distribution of r is also

$$dF = \frac{1}{B\{\frac{1}{2}(n-2), \frac{1}{2}\}} (1-r^2)^{\frac{1}{2}(n-4)} dr, \quad -1 \leqslant r \leqslant 1, \quad (31.18)$$

and we may therefore use (31.18), or equivalently the fact that $t = \{(n-2)r^2/(1-r^2)\}^{\frac{1}{2}}$ has a " Student's " distribution with $(n-2)$ degrees of freedom, to carry out our tests on r. (31.18) is in fact very accurate even for small n, as we might guess from the exact agreement of its first two moments with those of the permutation distribution.

The convergence of the permutation and normal-theory distributions to a common limiting normal distribution has been rigorously proved by Hoeffding (1952).

31.20 It may at first seem surprising that the distribution-free permutation distribution of r, which is used in testing the non-parametric hypothesis (31.9), should agree so closely with the exact distribution (16.62) which was derived on the hypothesis of the independence and normality of x and y. But the reader should observe that the adequacy of the approximation to the third and fourth moments of the permutation distribution of r depends on the values of the g's in (31.15) : these will tend to be small if $F(x,y)$ is near-normal. In fact, we are now observing from the other end, so to speak, the phenomenon mentioned in **31.9**, namely the robustness of the distribution of r when $\rho = 0$.

But if the close coincidence of the permutation distribution with the normal-theory distribution is not altogether surprising, it is certainly very convenient and satisfying, since we may continue to use the normal-theory tables (here of Student's t) for the distribution-free test of the non-parametric hypothesis of independence.

Rank tests of independence

31.21 A minor disadvantage of r as a test of independence, briefly mentioned below (31.11), is that its exact distribution is very tedious to enumerate. The reason for this is simply that the exact distribution of r depends on the actual values of (x, y) observed, and these are, of course, random variables. Despite the excellence of the approximation to the distribution of r by (31.18), it is interesting to inquire how this difficulty can be removed—it is also useful in other contexts, for the approximation to a permutation distribution is not always quite so good.

The most obvious means of removing the dependence of the permutation distribution upon the randomly varying observations is to replace the values of (x,y) by new values (X, Y) (with correlation coefficient R) so determined that the permutation distribution of R is the same for every sample (although of course R itself will vary from sample to sample). We thus seek a set of conventional numbers (X, Y) to replace the observed (x, y). How should these be chosen? (X, Y) must not depend upon the actual values of (x, y), but evidently must reflect the order relationships between the observed values of x and y, since we are interested in the interdependence of the variables. We are thus led to consider functions of the *ranks* of x and y. We define the rank of y_i as its position among the order statistics; i.e.

$$\operatorname{rank}\{y_{(i)}\} = i.$$

We are reinforced in our inclination to consider tests based on ranks (otherwise called " rank order tests " or simply " rank tests ") by the fact that the ranks are invariant under any monotone transformations of the variables. Any such transformation will also leave the hypothesis of independence (31.9) invariant, and the ranks are therefore natural quantities to use. We have still not settled which functions of the ranks are to be used as our numbers (X, Y); the simplest obvious procedure is to use the ranks themselves, i.e. to replace the observed values x by their ranks among the x's, and the observed y's by their ranks.

31.22 If we do this, we calculate the correlation coefficient R between n pairs (X, Y), where $(X_1, X_2, \ldots, X_n)$ is a permutation of the first n natural numbers, and $(Y_1, Y_2, \ldots, Y_n)$ is another such permutation. In obtaining the permutation distribution of R, we may hold the Y's fixed and permute the X's as before, since there are only $n!$ distinct and equiprobable sets of pairings of (X, Y). We may thus without loss of generality arrange the n pairs of any sample so that the ranks Y are in the natural order $1, 2, \ldots, n$. If the rank X corresponding to the value $Y = i$ is denoted by X_i, we therefore have for the *rank correlation coefficient*

$$R = \left[\frac{1}{n}\sum_{i=1}^{n} iX_i - \{\tfrac{1}{2}(n+1)\}^2\right] \Big/ \{\tfrac{1}{12}(n^2-1)\}, \tag{31.19}$$

for the mean of the first n natural numbers is $\tfrac{1}{2}(n+1)$ and their variance $\tfrac{1}{12}(n^2-1)$. R is usually called Spearman's rank correlation coefficient, after the eminent psychologist who first introduced it in 1906 as a substitute for ordinary product-moment correlation; it is usually given the symbol r_s, which we shall now use for it. Since

$$\sum_{i=1}^{n} iX_i \equiv \tfrac{1}{6}n(n+1)(2n+1) - \tfrac{1}{2}\sum_{i=1}^{n}(X_i-i)^2, \tag{31.20}$$

r_s may alternatively be defined by

$$r_s = 1 - \frac{6}{n(n^2-1)} \sum_{i=1}^{n} (X_i - i)^2, \qquad (31.21)$$

which is usually more convenient for calculation.

> K. Pearson, in his biography of Galton, says that the latter "dealt with the correlation of ranks before he even reached the correlation of variates," i.e. about 1875, but Galton apparently published nothing explicitly.

31.23 Since the formulae (31.12–14) for the exact moments of r hold for arbitrary x and y, they hold for r_s defined by (31.21) in particular. Moreover, the natural numbers have all odd moments about the mean equal to zero by symmetry. This implies that the exact distribution of r_s is symmetrical and hence its odd moments are zero. If we substitute also for k_4, k_4' in (31.14), we obtain for the exact moments

$$\left. \begin{array}{l} E(r_s) = E(r_s^3) = 0, \\[4pt] \operatorname{var} r_s = \dfrac{1}{n-1}, \\[4pt] E(r_s^4) = \dfrac{3}{n^2-1}\left\{1 + \dfrac{12(n-2)(n-3)}{25n(n-1)^2}\right\}. \end{array} \right\} \qquad (31.22)$$

However, as indicated by the introductory discussion in **31.21**, the exact distribution of r_s can actually be tabulated once for all. Kendall (1962) gives tables of the frequency function of $\sum_i (X_i - i)^2$, the random component of r_s in (31.21), for $n = 4(1)10$. (The "tail" entries in Kendall's tables are reproduced in the *Biometrika Tables*.) Beyond this point, the approximation by (31.18) is adequate for practical purposes, as is shown by the following table comparing exact and approximate critical values of r_s for test sizes $\alpha = 0.05, 0.01$ and $n = 10$.

Comparison of exact and approximate critical values of r_s for $n = 10$

Two-sided test	Exact critical values (from Kendall (1955))	Approximate critical values from (31.18)
$\alpha = 0.05$:	0·648	0·632
$\alpha = 0.01$:	0·794	0·765

31.24 We chose r_s from among the possible rank tests of independence on grounds of simplicity; clearly any reasonable measure of the correlation between x and y, based on their rank values, will give a test of independence. Daniels (1944) defined a class of correlation coefficients which includes the ordinary product-moment correlation as well as r_s and others, and went on (Daniels, 1948) to show that these are all essentially *coefficients of disarray*, in the sense that if a pair of values of y are interchanged to bring them into the same order as the corresponding values of x, the value of any coefficient of this class will increase—(31.21) makes this clear for r_s in particular. Let us consider the question of measuring disarray among the ranks of x and y.

Suppose, as in **31.22**, that the ranks of y (which are there called Y) are arrayed in the natural order $1, 2, \ldots, n$ and that the corresponding ranks of x are $X_1, X_2, \ldots, X_n$, a permutation of $1, 2, \ldots, n$. A natural method of measuring the disarray of the

x-ranks, i.e. the extent of their departure from the order $1, 2, \ldots, n$, is to count the number of inversions of order among them. For example, in the x-ranking 3214 for $n = 4$, there are 3 inversions of order, namely 3–2, 3–1, 2–1. The number of such inversions, which we shall call Q, may range from 0 to $\tfrac{1}{2}n(n-1)$, these limits being reached respectively if the x-ranking is $1, 2, \ldots, n$ and $n, (n-1), \ldots, 1$. We may therefore define a coefficient

$$t = 1 - \frac{4Q}{n(n-1)}, \qquad (31.23)$$

which is symmetrically distributed on the range $(-1, +1)$ over the $n!$ equiprobable permutations, and therefore has expectation 0 when (31.9) holds.

The coefficient (31.23) had been discussed by several early writers (Fechner, Lipps) around the year 1900 and subsequently by several other writers, notably Lindeberg, in the 1920's (historical details are given by Kruskal (1958)), but first became widely used after a series of papers by M. G. Kendall starting in 1938 and consolidated in a monograph (Kendall, 1962) to which reference should also be made on questions concerning the use of t and r_s as *measures* of correlation. Here we are concerned only with their properties as distribution-free tests of (31.9).

31.25 The distribution of t, or equivalently of the number of inversions Q, over the $n!$ equiprobable x-rankings is easily established by the use of frequency-generating functions. Let the frequency function of Q in samples of size n be $f(Q, n)/n!$. We may generate the $n!$ x-rankings for sample size n from the $(n-1)!$ for sample size $(n-1)$ by inserting the new rank "n" in every possible position relative to the existing $(n-1)$. (Thus, e.g., the 2! rankings for $n = 2$

$$\begin{array}{c} 12 \\ 21 \end{array}$$

become the 3! rankings for $n = 3$

$$\begin{array}{ccc} 312 & 132 & 123 \\ 321 & 231 & 213.) \end{array}$$

In any ranking, the addition to Q brought about by this process is exactly equal to the number of ranks to the right of the point at which "n" is inserted. Any value of Q in the n-ranking is thus built up as the sum of n terms, each of which had a different value of Q in the $(n-1)$-ranking. This gives the relationship

$$f(Q, n) = f(Q, n-1) + f(Q-1, n-1) + f(Q-2, n-1) + \ldots$$
$$+ f(Q-(n-1), n-1). \qquad (31.24)$$

Using (31.24), we may build up the frequencies for sample size n from those for $(n-1)$ as in the following table, where any entry in row n is the sum of the entry immediately above it and the $(n-1)$ entries to the left of the latter:—

n \ Q	0	1	2	3	4	5	6	7	8	9	10	Total
2	1	1										2!
3	1	2	2	1								3!
4	1	3	5	6	5	3	1					4!
5	1	4	9	15	20	22	20	15	9	4	1	5!
⋮												

Now, if $f(Q,n)$ is the coefficient of θ^Q in a frequency-generating function $G(\theta,n)$, (31.24) implies that

$$G(\theta,n) = G(\theta,n-1)+\theta G(\theta,n-1)+\theta^2 G(\theta,n-1)+\ldots+\theta^{n-1}G(\theta,n-1)$$
$$= \frac{\theta^n-1}{\theta-1}G(\theta,n-1). \tag{31.25}$$

Applying (31.25) repeatedly, we find

$$G(\theta,n) = \left(\frac{\theta^n-1}{\theta-1}\right)\left(\frac{\theta^{n-1}-1}{\theta-1}\right)\cdots\left(\frac{\theta^3-1}{\theta-1}\right)G(\theta,2), \tag{31.26}$$

and since we see directly that

$$G(\theta,2) = 1.\theta^0 + 1.\theta^1 = \frac{\theta^2-1}{\theta-1},$$

(31.26) may be written

$$G(\theta,n) = \prod_{s=1}^{n}\left(\frac{\theta^s-1}{\theta-1}\right). \tag{31.27}$$

We obtain the characteristic function of Q by inserting the factor $(n!)^{-1}$ and replacing θ by $\exp(i\theta)$ in (31.27), so that

$$\phi(\theta) = \{n!(e^{i\theta}-1)^n\}^{-1}\prod_{s=1}^{n}(e^{i\theta s}-1). \tag{31.28}$$

The c.g.f. of Q is therefore

$$\psi(\theta) = \sum_{s=1}^{n}\log(e^{i\theta s}-1)-n\log(e^{i\theta}-1)-\log(n!). \tag{31.29}$$

If we substitute

$$e^{i\theta s}-1 \equiv e^{i\theta s/2}2\sinh(\tfrac{1}{2}i\theta s)$$

everywhere in (31.29), we reduce it to

$$\psi(\theta) = \tfrac{1}{2}i\theta\left(\sum_{s=1}^{n}s-n\right)+\sum_{s=1}^{n}\log\sinh(\tfrac{1}{2}i\theta s)-n\log\sinh(\tfrac{1}{2}i\theta)-\log(n!)$$
$$= \tfrac{1}{4}n(n-1)i\theta+\sum_{s=1}^{n}\log\left(\frac{\sinh(\tfrac{1}{2}i\theta s)}{\tfrac{1}{2}i\theta s}\right)-n\log\left(\frac{\sinh(\tfrac{1}{2}i\theta)}{\tfrac{1}{2}i\theta}\right), \tag{31.30}$$

and, using (3.61), (31.30) becomes

$$\psi(\theta) = \tfrac{1}{4}n(n-1)i\theta+\sum_{j=1}^{\infty}\frac{B_{2j}(i\theta)^{2j}}{2j(2j)!}\sum_{s=1}^{n}(s^{2j}-1), \tag{31.31}$$

where the B_{2j} are the (non-zero) even-order Bernoulli numbers defined in **3.25**.

Picking out the coefficients of $(i\theta)^{2j}/(2j)!$ in (31.31) we have, for the cumulants of Q,

$$\left.\begin{array}{l}\kappa_1 = \tfrac{1}{4}n(n-1), \quad \kappa_{2j+1} = 0, \quad j\geqslant 1, \\ \kappa_{2j} = \dfrac{B_{2j}}{2j}\left(\sum_{s=1}^{n}s^{2j}-n\right).\end{array}\right\} \tag{31.32}$$

From (31.23), this gives for the cumulants of the rank correlation statistic t itself

$$\left.\begin{array}{l}\kappa_{2j+1} = 0, \quad j\geqslant 0, \\ \kappa_{2j} = \dfrac{2^{4j-1}B_{2j}}{j\{n(n-1)\}^{2j}}\left(\sum_{s=1}^{n}s^{2j}-n\right).\end{array}\right\} \tag{31.33}$$

Thus, t is symmetrically distributed about zero and

$$\operatorname{var} t = \frac{2^3 B_2}{\{n(n-1)\}^2} \{\tfrac{1}{6} n(n+1)(2n+1) - n\}$$
$$= \frac{2(2n+5)}{9n(n-1)}. \tag{31.34}$$

31.26 Further, (31.33) shows that κ_{2j} is of order $n^{-4j} \sum_{s=1}^{n} s^{2j}$ in n. Since the summation is of order n^{2j+1}, this means that

$$\kappa_{2j} = O(n^{1-2j})$$

and hence the standardized cumulants

$$\frac{\kappa_{2j}}{(\kappa_2)^j} = O(n^{1-j}).$$

Thus

$$\lim_{n \to \infty} \frac{\kappa_{2j}}{(\kappa_2)^j} = 0, \quad j > 1, \tag{31.35}$$

and hence the distribution of t tends to normality with mean zero and variance given by (31.34). The tendency to normality is extremely rapid. Kendall (1962) gives the exact distribution function (generated from (31.24)) for $n = 4(1)10$. Beyond this point, the asymptotic normal distribution may be used with little loss of accuracy.

31.27 In **31.24** we arrived at the coefficient t by way of the realization that the number of inversions Q is a natural measure of the disarray of the x-ranking. If one thinks a little further about this, it seems reasonable to weight inversions unequally; e.g. in the x-ranking 24351, one feels that the inversion 5–1 ought to carry more weight, because it is a more extreme departure from the natural order $1, 2, \ldots, n$, than the inversion 4–3. A simple weighting which suggests itself is the distance apart of the ranks inverted; in the immediately preceding instance, this would give weights of 4 and 1 respectively to the two inversions. Thus, if we define

$$h_{ij} = \begin{cases} +1 & \text{if } X_i > X_j, \\ 0 & \text{otherwise,} \end{cases} \tag{31.36}$$

we now seek to use the weighted sum of inversions

$$V = \sum_{i<j} \sum h_{ij}(j-i) \tag{31.37}$$

instead of our previous sum of inversions

$$Q = \sum_{i<j} \sum h_{ij}. \tag{31.38}$$

However, use of (31.37) leads us straight back to r_s. We leave it to the reader to prove in Exercise 31.5 that

$$V \equiv \tfrac{1}{2} \sum_{i=1}^{n} (X_i - i)^2, \tag{31.39}$$

so that, from (31.21),

$$r_s = 1 - \frac{12V}{n(n^2-1)}, \tag{31.40}$$

which is a definition of r_s analogous to (31.23) for t.

31.28 Despite the apparently very different methods they use of weighting inversions, it is a remarkable fact that Q and V of (31.37–38), and hence the statistics t and r_s also, are very highly correlated when the hypothesis of independence (31.9) holds—the reader is left to obtain the actual value of their correlation coefficient in Exercise 31.6. It declines from 1 at $n = 2$ (when t and r_s are equivalent) to its minimum value of 0·98 at $n = 5$, and then increases towards 1 as $n \to \infty$. Thus the tests are asymptotically equivalent when H_0 holds, and this, together with the result of **25.13**, implies that, from the standpoint of asymptotic relative efficiency, both tests possess the same properties. Daniels (1944) showed that the limiting joint distribution of t and r_s when H_0 holds is bivariate normal.

31.29 In samples from a bivariate normal population, the high correlation between t and r_s persists even when the parent correlation coefficient $\rho \neq 0$; S. T. David et al. (1951) show that as $n \to \infty$, t and r_s have a correlation which tends to a value $\geqslant 0\cdot 984$ if $|\rho| \leqslant 0\cdot 8$, and to 0·937 when $\rho = 0\cdot 9$.

Hoeffding (1948a) showed that t and r_s are quite generally asymptotically distributed in the bivariate normal form, but that their correlation coefficient depends strongly on the parent bivariate distribution and may indeed be zero.

The efficiencies of tests of independence

31.30 We now examine the asymptotic relative efficiencies (ARE) of the three tests of independence so far considered, relative to the ordinary sample correlation coefficient r, when the alternative hypothesis is that of bivariate normality as at (31.10). By the methods of **23.27–36**, we see that r gives a UMPU test of $\rho = 0$ against one-sided and two-sided alternatives—the reader is asked to verify this in Exercise 31.21) Since by **31.19** the permutation test based on r is asymptotically equivalent to the normal-theory r-test for independence, we see that its ARE will be 1 compared to that test.

31.31 We now derive the ARE of the test based on t defined at (31.23). From the definition at (31.36) we see that
$$h_{ij} = \tfrac{1}{2}\{1 - \operatorname{sgn}(x_i - x_j)\operatorname{sgn}(y_i - y_j)\},$$
and since there are $\tfrac{1}{2}n(n-1)$ terms in $Q = \underset{i<j}{\Sigma\Sigma} h_{ij}$, we have for their mean
$$E\left\{\frac{Q}{\tfrac{1}{2}n(n-1)}\right\} = E(h_{ij}) = \tfrac{1}{2}\{1 - E[\operatorname{sgn}(x_i - x_j)\operatorname{sgn}(y_i - y_j)]\}, \qquad (31.41)$$
which from (31.23) gives
$$E(t) = 1 - 2E\left\{\frac{Q}{\tfrac{1}{2}n(n-1)}\right\} = E[\operatorname{sgn}(x_i - x_j)\operatorname{sgn}(y_i - y_j)].$$
Now if the parent distribution F of x and y is bivariate normal with correlation parameter ρ, so is that of $w = (x_i - x_j)$ and $z = (y_i - y_j)$. Thus
$$E(t) = E[\operatorname{sgn}(x_i - x_j)\operatorname{sgn}(y_i - y_j)] = \int_{-\infty}^{\infty}\int_{-\infty}^{\infty} \operatorname{sgn} w \operatorname{sgn} z \, dF$$
which on applying (4.8) becomes
$$= \int_{-\infty}^{\infty}\int_{-\infty}^{\infty}\left\{\frac{1}{\pi^2}\int_{-\infty}^{\infty}\frac{\sin t_1 w}{t_1}dt_1 \int_{-\infty}^{\infty}\frac{\sin t_2 z}{t_2}dt_2\right\}dF,$$

which may be rewritten

$$= \frac{1}{\pi^2}\int_{-\infty}^{\infty}\int_{-\infty}^{\infty}\left\{\int_{-\infty}^{\infty}\int_{-\infty}^{\infty}\exp(it_1 w + it_2 z)\,dF\right\}\frac{dt_1}{it_1}\frac{dt_2}{it_2}. \qquad (31.42)$$

The inner double integral in (31.42) is the c.f. of F, which is

$$\phi(t_1, t_2) = \exp\{-\tfrac{1}{2}(t_1^2 + t_2^2 + 2\rho t_1 t_2)\}.$$

If we insert this and differentiate the remaining double integral with respect to ρ, we find

$$\frac{\partial}{\partial \rho}E(t) = \frac{1}{\pi^2}\int_{-\infty}^{\infty}\int_{-\infty}^{\infty}\phi(t_1, t_2)\,dt_1\,dt_2. \qquad (31.43)$$

But the double integral on the right of (31.43) is simply evaluated as

$$\int_{-\infty}^{\infty}\exp\{-\tfrac{1}{2}t_2^2(1-\rho^2)\}\left[\int_{-\infty}^{\infty}\exp\{-\tfrac{1}{2}(t_1+\rho t_2)^2\}\,dt_1\right]dt_2 = 2\pi/(1-\rho^2)^{\frac{1}{2}}.$$

Thus (31.43) becomes

$$\frac{\partial}{\partial \rho}E(t) = \frac{2}{\pi(1-\rho^2)^{\frac{1}{2}}},$$

so that

$$\left[\frac{\partial}{\partial \rho}E(t)\right]_{\rho=0} = \frac{2}{\pi}. \qquad (31.44)$$

Also, from (31.34),

$$\text{var}(t \mid H_0) \sim \frac{4}{9n}, \qquad (31.45)$$

while for the ordinary correlation coefficient r, from (26.31),

$$\left[\frac{\partial}{\partial \rho}E(r)\right]_{\rho=0} \to 1 \qquad (31.46)$$

and from **31.19**

$$\text{var}\,r \sim \frac{1}{n}. \qquad (31.47)$$

Using (25.27) with $m = 1$, $\delta = \tfrac{1}{2}$, the results (31.44–47) give, for the ARE of t compared to r,

$$A_{t,r} = 9/\pi^2 \doteqdot 0.91. \qquad (31.48)$$

By the remark of **31.28**, (31.48) will hold also for the ARE of r_s compared to r, a result due originally to Hotelling and Pabst (1936).

31.32 Apart from the results of **31.30–31** against bivariate normal alternatives, little work has been done on the efficiencies of tests of independence, largely due to the difficulty of specifying non-normal alternatives to independence. A notable exception is the paper by Konijn (1956), which considers a class of alternatives to independence generated by linear transformations of two independent variables. He finds, as above, that t and r_s are often asymptotically equivalent tests, each having ARE close to that of the test based on the sample correlation coefficient r, equal to it or even (in case of an underlying double-exponential distribution) exceeding it.

31.33 A defect of all the tests we have considered is that they will not be consistent tests against *any* departure from the hypothesis of independence (31.9). To see this we need only remark that each is essentially based on a correlation coefficient of some kind, whose distribution will be free of location and scale parameters but will depend on the population correlation coefficient ρ. For departures from independence implying $\rho \neq 0$, these tests will be consistent. But it is perfectly possible in non-normal cases to have non-independence accompanied by $\rho = 0$ (cf. **26.6**), and we cannot expect our tests to be consistent against such alternatives. With this in mind, Hoeffding (1948b) proposed another distribution-free test of (31.9) which is consistent against any continuous alternative bivariate distribution with continuous marginal distributions. Hoeffding tabulates the distribution of his statistic for $n = 5, 6, 7$ and obtains its limiting c.f. and its cumulants. (The limiting d.f. is given by Blum *et al.* (1961).) He also proves that against this class of alternatives no rank test of independence exists which is unbiassed for every test size $\alpha = M/n!$ However, if randomization is permitted in the test function, Lehmann (1951) shows that generally unbiassed rank tests of independence do exist.

Tests of randomness against trend alternatives

31.34 As we remarked in **31.14**, problem (3) of **31.13** which is to test

$$H_0: F_1(x) = F_2(x) = \ldots = F_n(x), \quad \text{all } x, \tag{31.49}$$

where we have an observation from each of n continuous distributions ordered according to the value of some variable y, is equivalent to testing the independence of the x's and the y's. Thus any of our tests of independence may be used as a test of randomness. However, since the y-variable is not usually a random variable but merely a labelling of the distributions (through time or otherwise), any monotone transformation of y would do as well as y itself. It is therefore natural to confine our attention to rank tests of randomness, since the ranks are invariant under monotone transformation, which leaves the hypothesis (31.49) unchanged.

Mann (1945) seems to have been the first to recognize that a rank correlation statistic could be used to test randomness as well as independence and proposed the use of t (although of course r_s could be used just as well) against the class of alternatives

$$H_1: F_1(x) < F_2(x) < \ldots < F_n(x), \quad \text{all } x, \tag{31.50}$$

where the observations x_i remain independent.

Since (31.50) states that the probability of an observation falling below any fixed value increases monotonically as we pass along the sequence of n observations, it may be described as a *downward trend* alternative. The critical region for a size-α test therefore consists of the 100α per cent largest values of Q, the number of inversions defined at (31.38).

31.35 (31.50) implies that for $i < j$

$$P\{h_{ij} = 1\} = P\{X_i > X_j\} = \tfrac{1}{2} + \varepsilon_{ij}, \quad 0 < \varepsilon_{ij} \leq \tfrac{1}{2}. \tag{31.51}$$

We thus have, from (31.38),

$$E(Q \mid H_1) = \tfrac{1}{4}n(n-1) + \sum\sum_{i<j}\varepsilon_{ij} = \tfrac{1}{4}n(n-1) + S_n, \tag{31.52}$$

where S_n is the sum of the $\tfrac{1}{2}n(n-1)$ values ε_{ij}.

Now consider the variance of Q.

$$\operatorname{var}(Q \mid H_1) = \operatorname{var}\{\sum_{i<j} h_{ij}\} = \sum_{i<j} \operatorname{var}(h_{ij}) + \sum_{i<j}\sum_{k<l} \operatorname{cov}(h_{ij}, h_{kl}). \tag{31.53}$$

The covariance terms in (31.53) are of two kinds. Those involving four distinct suffixes are all zero, since the variables are then independent, and there are $\binom{n}{4}$ such terms. The remaining terms are non-zero and involve three distinct suffixes only, (i, j) and (k, l), having one suffix in common. The number of such terms is of order $\binom{n}{3}$, the number of ways of selecting three suffixes from n. Since there are only $\binom{n}{2}$ terms in the first summation of (31.53), we may therefore write

$$\text{var}(Q \mid H_1) = O(n^3). \tag{31.54}$$

31.26 shows that Q is asymptotically normally distributed when H_0 holds, and thus the critical region of the test consists asymptotically of the values of Q exceeding the value

$$Q_0 = \tfrac{1}{4} n(n-1) + d_\alpha \{\tfrac{1}{72} n(n-1)(2n+5)\}^{\frac{1}{2}} \tag{31.55}$$

where the term in braces in (31.55) is the variance of Q (obtained from (31.34) and (31.23)) and d_α is the appropriate standardized normal deviate.

31.36 From (31.52) and (31.55), we see that

$$P\{Q > Q_0 \mid H_1\} \sim P\{Q - E(Q \mid H_1) > d_\alpha [\tfrac{1}{72} n(n-1)(2n+5)]^{\frac{1}{2}} - S_n \mid H_1\}. \tag{31.56}$$

Using (31.54), we may write (31.56) asymptotically as

$$P\{Q > Q_0 \mid H_1\} \sim P\{Q - E(Q \mid H_1) > [\text{var}(Q \mid H_1)]^{\frac{1}{2}} [d_\alpha - c n^{-3/2} S_n]\}, \tag{31.57}$$

where c is some constant. We now impose the condition that

$$n^{-3/2} S_n \to \infty \tag{31.58}$$

as $n \to \infty$. Then

$$\lambda = d_\alpha - c n^{-3/2} S_n \to -\infty \tag{31.59}$$

and λ will be negative when n is large enough. By Tchebycheff's inequality (3.94), we have *a fortiori* for negative λ and any random variable x,

$$P\{x - E(x) > \lambda (\text{var } x)^{\frac{1}{2}}\} \geq 1 - \frac{1}{\lambda^2}. \tag{31.60}$$

Thus, when (31.58) holds, (31.57), (31.59) and (31.60) give

$$\lim_{n \to \infty} P\{Q > Q_0 \mid H_1\} = 1.$$

Thus the test of randomness is consistent provided that (31.58) holds. This is a rather mild requirement, for there are $\tfrac{1}{2} n(n-1)$ terms in S_n. Thus if there is a fixed non-zero lower bound to the ε_{ij}, (31.58) certainly holds. Commonly, one wishes to consider alternatives for which ε_{ij} is a function of the distance $|i-j|$ only; if it is an increasing function of this distance, (31.58) certainly holds.

As well as deriving a more general version of this result, in which the ε_{ij} need not all have the same sign, Mann (1945) derived a condition for the unbiassedness of the test, which is essentially that given as Exercise 31.8.

31.37 We now consider a particular trend alternative to randomness, where the mean of the variable x_i is a linear function of i, and its distribution is normal about that mean with constant variance. This is the ordinary linear regression model with normal errors. We have

$$x_i = \beta_0 + \beta_1 i + \delta_i, \tag{31.61}$$

where the errors δ_i are independently normally distributed, and variance σ^2 for all i. The test of randomness is equivalent to testing

$$H_0 : \beta_1 = 0 \tag{31.62}$$

in (31.61). We proceed to find the asymptotic relative efficiency (ARE) of the test based on t (or, equivalently, on Q) compared with the standard test, based on the sample regression coefficient

$$b = \frac{\Sigma(x_i - \bar{x})(i - \bar{i})}{\Sigma(i - \bar{i})^2} = \frac{\Sigma x_i i - \tfrac{1}{2} n(n+1)\bar{x}}{\tfrac{1}{12} n(n^2 - 1)}, \tag{31.63}$$

which is the LR test for (31.62) and (since there is only one constraint imposed by H_0) is UMP for one-sided alternatives, say $\beta_1 < 0$, and UMPU for two-sided alternatives $\beta_1 \neq 0$ (cf. **24.27**). We put $\sigma^2 = 1$ without loss of generality. We have, from Least Squares theory (cf. Examples 19.3, 19.6)

$$E(b \mid H_1) = \beta_1,$$

$$\operatorname{var}(b \mid H_0) = \frac{1}{\Sigma(i - \bar{i})^2} = \frac{12}{n(n^2 - 1)},$$

so that the ratio

$$\frac{\left\{ \left[\dfrac{\partial E(b \mid H_1)}{\partial \beta_1} \right]_{\beta_1 = 0} \right\}^2}{\operatorname{var}(b \mid H_0)} = \frac{n(n^2 - 1)}{12} \sim \frac{n^3}{12}. \tag{31.64}$$

31.38 To obtain the equivalent of (31.64) for the test based on t, we require the derivative of

$$E(Q \mid H_1) = E\left\{ \sum_{i<j} h_{ij} \right\} = \sum_{i<j} E(h_{ij}). \tag{31.65}$$

Now $(x_i - x_j)$ is, from the model (31.61), normally distributed with mean $\beta_1(i-j)$ and variance 2. Hence

$$E(h_{ij}) = P\{h_{ij} = 1\} = P\{x_i > x_j\}$$

$$= \int_0^\infty \frac{1}{2\pi^{\frac{1}{2}}} \exp\left\{ -\tfrac{1}{4}[t - \beta_1(i-j)]^2 \right\} dt$$

$$= \int_{-\beta_1(i-j)/2^{1/2}}^\infty (2\pi)^{-\frac{1}{2}} \exp(-\tfrac{1}{2} u^2) \, du.$$

Thus

$$\left[\frac{\partial}{\partial \beta_1} E(h_{ij}) \right]_{\beta_1 = 0} = \frac{i-j}{2^{\frac{1}{2}}} \cdot \frac{1}{(2\pi)^{\frac{1}{2}}} = \frac{(i-j)}{2\pi^{\frac{1}{2}}}. \tag{31.66}$$

From (31.65) and (31.66)

$$\left[\frac{\partial}{\partial \beta_1} E(Q \mid H_1) \right]_{\beta_1 = 0} = -\frac{1}{2\pi^{\frac{1}{2}}} \sum_{i<j} (j - i)$$

$$= -\frac{1}{2\pi^{\frac{1}{2}}} \frac{n(n^2 - 1)}{6} = \frac{-n(n^2 - 1)}{12\pi^{\frac{1}{2}}}. \tag{31.67}$$

Also, from (31.34) and (31.23)
$$\operatorname{var}(Q \mid H_0) = \tfrac{1}{72} n(n-1)(2n+5). \tag{31.68}$$
From (31.67) and (31.68)
$$\frac{\left\{\left[\frac{\partial}{\partial \beta_1} E(Q \mid H_1)\right]_{\beta_1=0}\right\}^2}{\operatorname{var}(Q \mid H_0)} = \frac{n^2(n^2-1)^2}{144\pi} \cdot \frac{72}{n(n-1)(2n+5)} \sim \frac{n^3}{4\pi}. \tag{31.69}$$

Use of (31.64) and (31.69) in (25.27) with $m = 1$, $\delta = \tfrac{3}{2}$, gives for the ARE of Q compared to b

$$A_{Q,b} = A_{t,b} = \left(\frac{3}{\pi}\right)^{1/3} \doteqdot 0.98. \tag{31.70}$$

Just as before (cf. **31.28**), the same result holds for the alternative coefficient r_s (or equivalently V); the direct evaluation of the ARE of V is left to the reader as Exercise 31.9.

Optimum rank tests of independence and of randomness

31.39 It is worth remarking that the two rank correlation coefficients t and r_s are even more efficient as tests of randomness against normal alternatives than as tests of bivariate independence against normal alternatives, the values of ARE given by (31.70) and (31.48) being $(3/\pi)^{1/3}$ and $(3/\pi)^2$ respectively. But although both of these values are near 1, they are not equal to 1, and we are left with the question whether distribution-free tests exist for these problems which have ARE of 1 compared with the best test.

In order to answer this question, let us return to our discussion of **31.21**, where the choice of r_s from among all possible rank tests was made on grounds of simplicity. In effect, we decided to replace the observed variate-values x by their ranks. Now since the permutation test based on the variate-values themselves has ARE 1 against normal alternatives (cf. **31.30**), we should expect to retain optimum efficiency if we replace the variate-values by functions of their ranks which, asymptotically, are perfectly correlated with the variate-values. Suppose, then, that after ranking the x observations, we replace them by the expected values of the order statistics in a sample of size n from a standardized normal distribution. These are a perfectly definite set of conventional numbers, usually called the *normal scores*; the point in using them is that as $n \to \infty$, the correlation of these numbers with the variate-values will tend to 1, and we shall obtain optimum rank tests against normal alternatives. The test statistic is therefore

$$\frac{\frac{1}{n}\sum_{i=1}^{n} i E(X_i, n) - \tfrac{1}{2}(n+1) \cdot \frac{1}{n}\sum_{i=1}^{n} E(X_i, n)}{\left[\frac{1}{12}(n^2-1) \cdot \frac{1}{n}\sum_{i=1}^{n}\left\{E(X_i, n) - \frac{1}{n}\sum_{i=1}^{n} E(X_i, n)\right\}^2\right]^{\frac{1}{2}}}, \tag{31.71}$$

where X_i is now the x-value corresponding to the ith largest value of y and $E(s, n)$ is the expected value of $x_{(s)}$ in a sample of size n from a standardized normal distribution. Neglecting constants, (31.71) is equivalent to testing with the statistic

$$c = \sum_{i=1}^{n} i E(X_i, n), \tag{31.72}$$

which therefore has ARE of 1 in testing independence or randomness against normal alternatives. Bhuchongkul (1964) confirms this result in the course of investigating the use of *any* conventional numbers in the test statistic.

The use of the normal scores as conventional numbers was first suggested by R. A. Fisher and F. Yates in the Introduction to their *Statistical Tables*, first published in 1938. The locally optimum properties of the test statistic (31.72) were demonstrated by Hoeffding (1950) and Terry (1952). A direct proof of the asymptotically perfect correlation between the expected values of the order statistics and the variate-values they replace is obtained from Hoeffding's (1953) theorem to the effect that for any parent d.f. $F(x)$ with finite mean, and any real continuous function $g(x)$ bounded in value by an integrable convex function,

$$\lim_{n \to \infty} \frac{1}{n} \sum_{m=1}^{n} g\{E(m,n)\} = \int_{-\infty}^{\infty} g(x) \, dF. \tag{31.73}$$

Successive substitution of $g(x) = \cos xt$, $g(x) = \sin xt$ in (31.73) shows that the limiting c.f. of the $E(m,n)$ is the c.f. of the distribution $F(x)$, which is $E\{\cos xt + i \sin xt\}$.

Bell and Doksum (1965) show that if instead of the normal scores $E(s, n)$ we use simply the observed $x_{(s)}$ in a sample of n random normal deviates, we get the same asymptotic properties in all the contexts considered in this chapter. The advantages are that no special table is needed, and that exact size α can be attained for tests; the disadvantage is that the small-sample power of these tests seems to be lower than for normal scores tests—Jogdeo (1966) shows that these tests have some undesirable properties. Bell and Doksum (1967) discuss distribution-free tests of independence, especially rank tests, and their optimum properties. They also propose randomized tests of the type just mentioned.

Brillinger (1966) points out that a set of ordered values $x_{(s)}$ are maximally correlated with the values $z_{(s)} = a + bE(x_{(s)})$ for any sample size. This follows from the fact that the correlation coefficient between the $x_{(s)}$ and $z_{(s)}$ equals the correlation ratio of x on its ordered values (cf. (26.40) and (26.45)).

31.40 As well as seeking optimum rank tests against normal alternatives, as in **31.39**, we may also ask whether there are any alternatives for which any particular rank test is optimum among rank tests. We do not pursue this subject here, because the inquiry would be artificial from our present viewpoint (cf. **31.17**), which essentially regards distribution-free procedures as perfectly robust substitutes for the standard normal-theory procedures. Our interest is therefore confined to comparisons of efficiency between distribution-free and standard normal-theory methods. An account of rank tests in general is given by Lehmann (1959), by Fraser (1957) and by Hájek and Šidák (1967).

31.41 Before leaving tests of randomness, we should mention that a variety of such tests have been proposed in the literature, none of which is as efficient against normal alternatives as those we have discussed. However, some of them are considerably simpler to compute than r_s or t, and very little less efficient. They are discussed in Exercises 31.10–12. Other tests have their ARE evaluated by Stuart (1954b, 1956).

Two-sample tests

31.42 We now consider problem (1) of **31.13**. Given independent random samples of sizes n_1, n_2 respectively from continuous distribution functions $F_1(x)$, $F_2(x)$, we wish to test the hypothesis

$$H_0 : F_1(x) = F_2(x), \quad \text{all } x. \tag{31.74}$$

As we remarked in **31.14**, this is equivalent to testing the independence of the variable x and a dummy variable dichotomized so that only two distinct values y arise. There are $n_1 + n_2 = n$ observations on the pair (x, y).

Let us for a moment consider the n values of x as being arranged over the n positions labelled

$$1, 2, 3, \ldots, n_1; \quad n_1+1, n_1+2, \ldots, n. \tag{31.75}$$

Under H_0, each of the $n!$ possible orderings of the x-values is equiprobable; but irrespective of whether H_0 holds, the $n_1!$ permutations of the positions in the first sample, and the $n_2!$ permutations of the positions in the second sample, do not affect the allocation of the n values to the two samples. Thus there are $n!/(n_1! n_2!)$ distinct allocations to the two samples, corresponding to the $\binom{n}{n_1}$ ways of selecting the members of the first sample from the n values.

31.43 For the hypothesis (31.74), unlike the others we have so far considered at (31.9) and (31.49), we may consider a class of alternatives much more general than those of standard normal theory, namely

$$H_1 : F_2(x) = F_1(x - \theta), \quad \text{all } x. \tag{31.76}$$

(31.76) states that the only difference between the two parent distributions is one of location. In terms of (31.76), (31.74) becomes

$$H_0 : \theta = 0. \tag{31.77}$$

We shall refer to (31.76) as the location-shift alternative hypothesis. It should be noted that although a location constant θ occurs in (31.76), it is not a parameter by our definition of **22.3**, since the form of the parent distribution $F_1(x)$ is unspecified, and thus the hypothesis (31.77) is non-parametric.

31.44 To suggest a statistic for testing H_0, we return to the case of normal alternatives. Consider two normal distributions differing only in location. Without loss of generality, we assume their common variance σ^2 to be equal to 1, and that the mean of the first distribution is zero. The Likelihood Function is therefore

$$\begin{aligned} L(x \mid H_1) &= (2\pi)^{-\frac{1}{2}n} \exp\left\{ -\tfrac{1}{2} \sum_{i=1}^{n_1} x_{1i}^2 - \tfrac{1}{2} \sum_{i=1}^{n_2} (x_{2i} - \theta)^2 \right\} \\ &= (2\pi)^{-\frac{1}{2}n} \exp\left\{ -\tfrac{1}{2} \sum_{i=1}^{n} x_i^2 + \theta \sum_{i=1}^{n_2} x_{2i} - \tfrac{1}{2} n_2 \theta^2 \right\}. \end{aligned} \tag{31.78}$$

From (31.78) we see that for $\theta > 0$, $L(x \mid H_1)$ will be maximized when $\sum_{i=1}^{n_2} x_{2i}$ is as large as possible and similarly for $\theta < 0$ when $\sum_{i=1}^{n_2} x_{2i}$ is as small as possible. By the Neyman–Pearson lemma of **22.10**, the most powerful critical region will consist of those of the $\binom{n}{n_1}$ equiprobable points in the sample space which maximize $L(x \mid H_1)$. We are thus led to use the statistic $\sum_{i=1}^{n_2} x_{2i}$, or equivalently the mean of the second sample,

$\bar{x}_2 = \frac{1}{n_2} \sum_{i=1}^{n_2} x_{2i}$. Since $n_1 \bar{x}_1 + n_2 \bar{x}_2 = n\bar{x}$, and the overall mean $\bar{x}$ is invariant under permutations, $\bar{x}_2$ determines the value of $\bar{x}_1$ also, and we may equivalently consider $\bar{x}_1$ or $\bar{x}_1 - \bar{x}_2$. For the two-sided alternative $\theta \neq 0$, we are inclined to use the "equal-tails" two-sided test on $\bar{x}_1 - \bar{x}_2$ or equivalently a one-sided test on $(\bar{x}_1 - \bar{x}_2)^2$, large values forming the critical region. It was in this form that the test statistic was first investigated by Pitman (1937a).

The permutation distribution of w

31.45 The observed variance of the combined samples, say s^2, is invariant under permutations. For fixed s^2, the statistic $(\bar{x}_1 - \bar{x}_2)^2$ can take values ranging between zero and its maximum value, which occurs when every member of the first sample is equal to $\bar{x}_1$, and every member of the second sample equals $\bar{x}_2$. We then have

$$ns^2 = n_1(\bar{x}_1 - \bar{x})^2 + n_2(\bar{x}_2 - \bar{x})^2 = \frac{n_1 n_2}{n}(\bar{x}_1 - \bar{x}_2)^2.$$

We thus have
$$0 \leq (\bar{x}_1 - \bar{x}_2)^2 \leq \frac{n^2 s^2}{n_1 n_2}.$$

If we therefore define
$$w = \frac{n_1 n_2}{n^2 s^2}(\bar{x}_1 - \bar{x}_2)^2, \tag{31.79}$$
we have for all possible samples
$$0 \leq w \leq 1. \tag{31.80}$$

31.46 To obtain the permutation distribution of w given H_0, we write it identically as
$$w = \frac{n_1}{n_2 s^2}(\bar{x}_1 - \bar{x})^2 \tag{31.81}$$

a form in which only $\bar{x}_1$ varies under permutation of the observations. The exact distribution of $\bar{x}_1$ may be tabulated by enumeration, but as previously remarked in **31.19** the process becomes tedious as n increases. In the form (31.81), however, we may use already-developed results to obtain the moments of w, for it is a multiple of the squared deviation of the sample mean from the population mean in sampling n_1 members from a finite population of n members. We found the necessary expectations at (12.114) and (12.120), which we rewrite in our present notation as

$$E(\bar{x}_1 - \bar{x})^2 = \frac{(n-n_1)s^2}{(n-1)n_1} = \frac{n_2 s^2}{n_1(n-1)} \tag{31.82}$$

and
$$E(\bar{x}_1 - \bar{x})^4 = \frac{n_2}{n_1^3(n-1)(n-2)(n-3)} \\ \times \{3n(n_1-1)(n_2-1)s^4 + [n(n+1) - 6n_1 n_2]m_4\},$$

where m_4 is the observed fourth moment of the combined samples. Thus
$$E(w) = 1/(n-1), \tag{31.83}$$
$$E(w^2) = \frac{1}{n_1 n_2 (n-1)(n-2)(n-3)} \\ \times \{3n_1 n_2(n-6) + 6n + [n(n+1) - 6n_1 n_2]g_2\}, \tag{31.84}$$

where g_2 is the measure of kurtosis $(m_4/s^4)-3$. When either n_1 or n_2 becomes large, and n with it, (31.84) is asymptotically

$$E(w^2) \sim \frac{3}{n^2-1}\left\{1+O\left(\frac{g_2}{3n_j}\right)\right\}, \qquad (31.85)$$

where n_j is the sample size which is not large. If both n_1 and $n_2 \to \infty$ (31.84) is

$$E(w^2) \sim \frac{3}{n^2-1}\left\{1+O\left(\frac{g_2}{n}\right)\right\}. \qquad (31.86)$$

Thus, especially when g_2 is small, we have

$$E(w^2) \doteq \frac{3}{n^2-1}. \qquad (31.87)$$

31.47 (31.83) and (31.87) are the first two moments about the origin of the Beta distribution of the first kind

$$dF = \frac{1}{B\{\tfrac{1}{2}, \tfrac{1}{2}n-1\}} w^{-\tfrac{1}{2}}(1-w)^{\tfrac{1}{2}n-2}\, dw, \qquad 0 \leqslant w \leqslant 1, \qquad (31.88)$$

which we may therefore expect to approximate the permutation distribution of w. In fact, Pitman (1937a, b) showed that the third moments also agree closely, and that the approximation is very good.

Now consider the ordinary "Student's" t^2-statistic for testing the difference in location between two normal distributions. In our present notation, we write it

$$\frac{t^2}{n-2} = \frac{(\bar{x}_1-\bar{x}_2)^2 \dfrac{n_1 n_2}{n}}{n_1 s_1^2 + n_2 s_2^2}, \qquad (31.89)$$

where s_1^2, s_2^2 are the separate sample variances. Using the identity

$$ns^2 \equiv n_1 s_1^2 + n_2 s_2^2 + \frac{n_1 n_2}{n}(\bar{x}_1-\bar{x}_2)^2$$

in (31.79) and (31.89) shows that

$$w \equiv \frac{1}{1+\dfrac{n-2}{t^2}} \qquad (31.90)$$

exactly. Thus we have been dealing with a monotone increasing function of t^2. What is more, in the exact normal theory, the transformation (31.90) applied to the "Student's" distribution with $\nu = n-2$ gives precisely the distribution (31.88). (In fact, we carried out essentially this transformation in reducing "Student's" distribution function to the Incomplete Beta function in **16.11**, except that there we transformed to $(1-w)$ and obtained (31.88) with $1-w$ replacing w.)

We have therefore found, exactly as in **31.19**, that the approximation to the permutation distribution in testing a non-parametric hypothesis is precisely the normal-theory distribution. In this particular case, we may test w, from (31.90), by putting $(n-2)w/(1-w) = t^2$ with $(n-2)$ degrees of freedom. For the one-sided tests discussed at the outset, we simply use t in the appropriate tail rather than t^2.

31.48 The wide applicability of t^2 as an approximation of the normal theory in this instance must clearly be attributed to the operation of Central Limit effects, since we are dealing with a difference between means; cf. the related discussion of robustness in **31.4**.

Just as we remarked in **31.30** previously, the asymptotic equivalence of the permutation distribution to that of the optimum normal theory test implies that the former has ARE of 1 against normal alternatives, in this case of a location-shift.

Distribution-free confidence intervals for a shift in location

31.49 We may now use the test statistic w of (31.79) to obtain distribution-free confidence intervals for the location-shift θ in (31.76). For, whatever the value of θ, (31.76) implies that the n_1 values x_{1i} $(i = 1, 2, \ldots, n_1)$ and the n_2 values $(x_{2i}+\theta)$, $(i = 1, 2, \ldots, n_2)$ are two samples which come from the same distribution $F_1(x)$. The distribution of w given H_0 is therefore applicable to these two samples.

Let us denote by $w(\theta)$ the calculated value of w for the two samples, which is evidently a function of θ. Let w_α be the upper critical value of w for a test of size α, i.e.
$$P\{w(\theta) \leq w_\alpha\} = 1-\alpha. \tag{31.91}$$
Using (31.90), (31.91) is equivalent to
$$P\{t^2(\theta) \leq t_\alpha^2\} = 1-\alpha, \tag{31.92}$$
where t^2 is defined at (31.89). The denominator of t^2 is a function of the separate sample variances only, and is therefore not a function of θ. Using (31.89) in (31.92), we therefore have
$$P\{[\bar{x}_1-(\bar{x}_2+\theta)]^2 \leq k_\alpha^2\} = 1-\alpha, \tag{31.93}$$
where
$$k_\alpha^2 = \frac{n(n_1 s_1^2 + n_2 s_2^2)}{n_1 n_2 (n-2)} t_\alpha^2. \tag{31.94}$$
Thus from (31.93) we have, whatever the true value of θ,
$$P\{(\bar{x}_1-\bar{x}_2)-k_\alpha \leq \theta \leq (\bar{x}_1-\bar{x}_2)+k_\alpha\} = 1-\alpha, \tag{31.95}$$
and (31.95) is a confidence interval for θ.

If the sample sizes are large enough for the permutation distribution of t^2 to be closely approximated by the exact "Student's" distribution, we obtain t_α^2 from the tables of the latter; otherwise, the exact permutation distribution of w must be used, with (31.90). We are then, of course, limited in the values of α we may choose for our test or confidence interval to multiples of $\binom{n}{n_1}^{-1}$.

> Van Eeden (1970) gives a point estimator of θ whose efficiency is robust in the sense that the MVB (17.22) is always attained asymptotically within a very wide class of regular distributions.

Consistency of the w-test

31.50 Using the result of the last section, we may easily see that w is a consistent test statistic for the hypothesis (31.77) against the alternative (31.76) with $\theta \neq 0$, provided that $F_1(x)$ has a finite variance. In fact, even if $F_1(x)$ and $F_2(x)$ have different finite variances and different means, w remains consistent, as Pitman (1948) showed.

Consider k^2, defined at (31.94). If $n_1, n_2 \to \infty$ so that $n_1/n_2 \to \lambda$, $0 < \lambda < \infty$, k^2 converges in probability to $\frac{(\lambda \sigma_1^2 + \sigma_2^2)}{n_1} t_\alpha^2$, which is of order n_1^{-1}, while by the Law of Large Numbers, $(\bar{x}_1 - \bar{x}_2)$ converges to $E(x_1) - E(x_2)$, which is the true value of θ, say θ_0. (31.95) now shows that for any α the confidence interval

$$I = (\bar{x}_1 - \bar{x}_2 - k, \bar{x}_1 - \bar{x}_2 + k)$$

is an interval, with length of order $n_1^{-\frac{1}{2}}$, for θ. If we choose $\alpha < \varepsilon$, we see that for any $\theta_1 \neq \theta_0$

$$\lim_{n_1 \to \infty} P\{\theta_1 \in I\} < \varepsilon \qquad (31.96)$$

and this is merely a translation into confidence intervals terminology of the consistency statement to be proved, for ε may be arbitrarily near zero. Ultimately, as n_1 increases, the interval will exclude θ_1 with probability tending to 1, i.e. the test of $\theta = \theta_0$ will reject $\theta_1 \neq \theta_0$. This argument also makes it clear that (31.77) may be replaced by $H_0: \theta = \theta_0$ if we add an increment θ_0 to each observation in the second sample.

Rank tests for the two-sample problem

31.51 Just as in **31.21** during our discussion of tests of independence, so here in the two-sample problem we see that if we wish to be in a position to tabulate the exact permutation distribution of a test statistic for any n, we must remove the dependence of the test statistic upon the actual values of the observations, which are random variables, and we are led to the use of rank tests, which are particularly appropriate because of their invariance under monotone transformation of the underlying variables, which leave the hypothesis (31.74) invariant. Once again, the simplest procedure is simply to replace the observations x_1 by their rank values, i.e. to rank the $n_1 + n_2 = n$ observations in a single sequence and replace the value $x_{(i)}$ by its rank X_i. We then have a set of n values X_i which are a permutation of the first n natural numbers, of which n_1 belong to the first sample and n_2 to the second.

Since, as we pointed out in **31.44**, the statistic w is equivalent to using the mean $\bar{x}_1$ of the first sample, the rank test obtained from w by replacing the observations by their ranks is equivalent to using

$$S = \sum_{i=1}^{n_1} X_i, \qquad (31.97)$$

the sum of the ranks in the first sample, which is analogous to r_s of (31.21) since both arise from replacing observations by ranks.

31.52 Now suppose that we seek an analogue of t, defined by (31.23), i.e. essentially of Q as defined at (31.38). We should obviously expect, if the hypothesis (31.74) holds, that the observations from the first and second samples would be thoroughly "mixed up" with no tendency for the ranks in the first sample to cluster at either or both ends of the range from 1 to n. Define a statistic U which counts the number of times a member of the first sample exceeds a member of the second sample, i.e.

$$U = \sum_{i=1}^{n_1} \sum_{j=1}^{n_2} h_{ij}, \qquad (31.98)$$

where h_{ij} is defined at (31.36) as before. U ranges in value from 0 to $n_1 n_2$.

Whereas in the case of tests of independence there is a genuine choice between r_s and t as test statistics (although they are equivalent from the viewpoint of ARE, as

we saw), for the two-sample situation the statistics (31.97) and (31.98) are functionally related. In fact
$$U \equiv S - \tfrac{1}{2}n_1(n_1+1). \tag{31.99}$$
To prove the relationship (31.99) it is only necessary to note that
$$U = \sum_{i=1}^{n_1}\left(\sum_{j=1}^{n_2} h_{ij}\right) = \sum_{i=1}^{n_1}(X_i - i).$$
Thus we may use whichever of U and S is more convenient. From both the theoretical and the computational viewpoint, U is simpler.

> The statistic S has been proposed independently by a great many writers (historical details are given by Kruskal and Wallis (1952–1953) and by Kruskal (1957)) and is usually called Wilcoxon's test statistic after Wilcoxon (1945, 1947); some authors call it the Mann–Whitney test statistic after authors who studied U a little later (Mann and Whitney, 1947) and others call it the Rank Sum test statistic.

The distribution of Wilcoxon's test statistic

31.53 We proceed to find the distribution of U when the hypothesis (31.74) holds. We may obtain its c.f. directly from that of Q given at (31.28). For Q is based on all $\tfrac{1}{2}n(n-1)$ possible comparisons between n observations, while U is based on the $n_1 n_2$ comparisons between the first n_1 and the second n_2 of them, the $\tfrac{1}{2}n_1(n_1-1)$ "internal" comparisons of the first sample and the $\tfrac{1}{2}n_2(n_2-1)$ of the second sample being excluded. We may write this relationship symbolically as
$$Q_n = Q_{n_1} + Q_{n_2} + U. \tag{31.100}$$
Since, given H_0, the components on the right of (31.100) are independent of each other, we have the relation between c.f.'s (cf. **7.18**)
$$\phi_n(\theta) = \phi_{n_1}(\theta)\phi_{n_2}(\theta)\,\Phi_U(\theta),$$
where the first three c.f.'s are those of Q with sample size equal to the suffix of ϕ, and $\Phi_U(\theta)$ is the c.f. of U. Thus
$$\Phi_U(\theta) = \phi_n(\theta)/\{\phi_{n_1}(\theta)\phi_{n_2}(\theta)\}, \tag{31.101}$$
or equivalently, taking logarithms for the c.g.f.,
$$\Psi_U(\theta) = \psi_n(\theta) - \psi_{n_1}(\theta) - \psi_{n_2}(\theta), \tag{31.102}$$
where ψ is defined by (31.30). Substituting this on the right of (31.102), we find
$$\Psi_U(\theta) = \tfrac{1}{2}n_1 n_2 i\theta + \sum_{j=1}^{\infty}\frac{B_{2j}(i\theta)^{2j}}{2j(2j)!}\left\{\sum_{s=1}^{n} - \sum_{s=1}^{n_1} - \sum_{s=1}^{n_2}\right\}(s^{2j}-1)$$
$$= \tfrac{1}{2}n_1 n_2 i\theta + \sum_{j=1}^{\infty}\frac{B_{2j}(i\theta)^{2j}}{2j(2j)!}\sum_{s=1}^{n_1}\{(n_2+s)^{2j}-s^{2j}\}. \tag{31.103}$$
The cumulants of U are therefore
$$\left.\begin{array}{l}\kappa_1 = \tfrac{1}{2}n_1 n_2, \quad \kappa_{2j+1} = 0, \quad j \geq 1,\\ \kappa_{2j} = \dfrac{B_{2j}}{2j}\sum_{s=1}^{n_1}\{(n_2+s)^{2j}-s^{2j}\}.\end{array}\right\} \tag{31.104}$$

In particular, we find
$$\kappa_2 = \tfrac{1}{12} n_1 n_2 (n+1). \tag{31.105}$$
(31.104) shows that the distribution of U is symmetrical whatever the values of n_1 and n_2. By (31.99), the distribution of S has the same cumulants except that its mean is $\tfrac{1}{2} n_1(n+1)$.

31.54 The exact distribution of S or U can easily be generated from the relation between frequency-generating functions equivalent to (31.101), which is
$$G_U(\theta) = \binom{n}{n_1}^{-1} G(\theta, n) / \{ G(\theta, n_1) G(\theta, n_2) \},$$
where $G_U(\theta)$ is the f.g.f. of U, and $G(\theta, n)$ the f.g.f. of Q given in **31.24**. Substituting its value from (31.27), we find
$$G_U(\theta) = \binom{n}{n_1}^{-1} \frac{\prod_{s=1}^{n} (\theta^s - 1)}{\prod_{s=1}^{n_1} (\theta^s - 1) \prod_{s=1}^{n_2} (\theta^s - 1)} = \binom{n}{n_1}^{-1} \prod_{s=1}^{n_1} \left(\frac{\theta^{n_2+s} - 1}{\theta^s - 1} \right). \tag{31.106}$$

The coefficient of θ^U in the second factor on the right of (31.106) is the number of ways in which n_1 of the first n integers can be chosen so that their sum is exactly $\tfrac{1}{2} n_1(n_1+1) + U$. We denote this by $f(U, n_1, n_2)$ and its cumulative sum from below by
$$A(U, n_1, n_2) = \sum_{r = \tfrac{1}{2} n_1(n_1+1)}^{U} f(r, n_1, n_2). \tag{31.107}$$

The function $A(U, n_1, \infty)$ was tabulated by Euler over two centuries ago. Fix and Hodges (1955) give tables which permit the calculation of $A(U, n_1, n_2)$ for n_1 (which without loss of generality may be taken $\leq n_2$) ≤ 12 and any n_2. Previously, White (1952) had tabulated critical values of S for "equal-tails" test sizes $\alpha = 0.05, 0.01$ and 0.001 and $n_1 + n_2 \leq 30$. Other tables are listed by Kruskal and Wallis (1952-1953) and Fix and Hodges (1955). Verdooren (1963) and Jacobson (1963) give further tables and corrections to earlier tables. Milton (1964) tabulates critical values for $n_1 \leq 20$, $n_2 \leq 40$ and one-tailed test sizes $\alpha = 0.0005, 0.001, 0.0025, 0.005, 0.01, 0.025, 0.05$ and 0.1.

31.55 The asymptotic normality of U when H_0 holds follows immediately from (31.104). If $n_1, n_2 \to \infty$ so that $0 < \lim n_1/n_2 = \lambda < \infty$, we write N indifferently for n_1, n_2 and n and see that κ_{2j} is at most of order N^{2j+1}, so that (as, indeed we saw at (31.105)) κ_2 is of order N^3. Thus
$$\kappa_{2j}/(\kappa_2)^j = O(N^{2j+1-3j}) = O(N^{1-j}),$$
whence
$$\lim_{N \to \infty} \kappa_{2j}/(\kappa_2)^j = 0, \quad j > 1,$$
and the distribution of U tends to normality with mean and variance given by (31.104–105). The tendency to normality is rapid, being effective when n_1 and n_2 are about 8 or so.

Under H_0, S is exactly the sum of n_1 integers randomly chosen from the first n, and thus asymptotically the sum of n_1 identical independent rectangular variates. This rectangular sum approximation to the distribution of S is more accurate than the normal, to which it tends rapidly as n_1 increases (cf. Example 11.9, Vol. 1), and percentiles of S/n are given for $n_1 \leq 30$ by Buckle *et al.* (1969).

Using a general theorem due to Hoeffding (1948a), Lehmann (1951) shows U to be asymptotically normal if the two samples come from different continuous distributions F_1, F_2 and $\lim n_1/n_2 = \lambda$ is bounded as above.

Lehmann (1963) obtains expressions for the distribution-free confidence intervals obtained from the Wilcoxon test statistic by the method of **31.49**—cf. Exercise 31.24.

Consistency and unbiassedness of the Wilcoxon test

31.56 It follows from the proof of consistency given in **31.50** for the w-test, which reduces to Wilcoxon's test when ranks replace variate-values, that the Wilcoxon test is consistent against alternatives for which F_1 and F_2, the underlying parent distributions, generate different mean *ranks* in their samples. Clearly, this will happen if and only if the probability p of an observation from F_1 exceeding one from F_2 differs from $\frac{1}{2}$. This result, given by Pitman (1948) and independently by later writers, may also be shown directly as indicated in Exercise 31.16.

31.57 If we consider the one-sided alternative hypothesis that the second sample comes from a "stochastically larger" distribution, i.e.

$$H_1 : F_1(x) > F_2(x), \quad \text{all } x, \tag{31.108}$$

it is a simple matter to prove that both Pitman's and Wilcoxon's one-sided tests are unbiassed against (31.108). In fact, Lehmann (1951) showed the unbiassedness of *any* similar critical region for (31.74) against (31.108) which satisfies the intuitively desirable condition (C) that if any member of the second sample is increased in value, the sample point remains in the critical region. For any pair F_1, F_2, let us define a function $h(x)$ by the equation

$$F_2(h(x)) = F_1(x) \tag{31.109}$$

so that, from (31.108),

$$h(x) > x. \tag{31.110}$$

Now consider the two samples with the n_2 members of the second sample transformed from x_i to $h(x_i)$ and the first sample unchanged. We see from (31.109) that the hypothesis of identical populations holds for the values thus transformed. If a region in the transformed sample space has content P, equation (31.110) and condition (C) assumed above ensure that for the untransformed sample space its content is $\alpha \leqslant P$. Since α is the size of the test on the untransformed variables, and P its power against (31.108), we see that the test is unbiassed.

Condition (C) is obviously satisfied by the one-sided Pitman and Wilcoxon tests.

31.58 If we now consider the general two-sided alternative hypothesis

$$H_1 : F_1(x) \neq F_2(x), \tag{31.111}$$

or even the more limited location-shift alternative (31.76) with θ unrestricted in sign, the Wilcoxon test is no longer unbiassed in general. For location-shift alternatives, Van der Vaart (1950, 1953) showed that if $n_1 = n_2$ or the common frequency function is symmetric about some value (not necessarily θ), the first derivative of the power function at $\theta = 0$ is zero if it exists, but that even then the test need not be unbiassed.

The ARE of the Wilcoxon test

31.59 We now confine ourselves to the location-shift situation (31.76), and find the ARE of Wilcoxon's test compared to "Student's" t-test (which is the optimum test of location-shift when F_1 is a normal distribution) when F_1 is an arbitrary continuous d.f. with finite variance σ^2.

"Student's" t defined at (31.89) is known to be asymptotically equivalent to using the statistic $(\bar{x}_1 - \bar{x}_2)$, which, whatever the form of F_1, tends to normality with mean $-\theta$ and variance $\sigma^2 \left(\dfrac{1}{n_1} + \dfrac{1}{n_2} \right) = \dfrac{n\sigma^2}{n_1 n_2}$. Thus we have

$$\frac{\left\{ \left[\dfrac{\partial}{\partial \theta} E(\bar{x}_1 - \bar{x}_2) \right]_{\theta=0} \right\}^2}{\mathrm{var}\{(\bar{x}_1 - \bar{x}_2) | \theta = 0\}} = \frac{n_1 n_2}{n \sigma^2}. \tag{31.112}$$

We now have to evaluate the equivalent of (31.112) for the Wilcoxon statistic U. From the definition of U at (31.98),

$$E(U) = n_1 n_2 E(h_{ij}) = n_1 n_2 p,$$

where p is the probability that an observation x_1 from the first distribution, $F_1(x)$, exceeds one x_2 from the second distribution, $F_1(x-\theta)$. This is the probability that $x_2 - x_1 < 0$. Using the formula (11.68) for the d.f. of the sum of two random variables (with $-y$ here being replaced by y in the argument of F_1 and suffixes 1, 2 interchanged to give the d.f. of $x_2 - x_1$), we find

$$p = H(0) = \int_{-\infty}^{\infty} F_1(0 + x - \theta) f_1(x) \, dx$$

whence

$$\frac{\partial p}{\partial \theta} = - \int_{-\infty}^{\infty} f_1(x - \theta) f_1(x) \, dx$$

and

$$\left[\frac{\partial E(U)}{\partial \theta} \right]_{\theta = 0} = n_1 n_2 \int_{-\infty}^{\infty} \{f_1(x)\}^2 \, dx. \tag{31.113}$$

(31.113) and (31.105) give

$$\frac{\left\{ \left[\dfrac{\partial E(U)}{\partial \theta} \right]_{\theta=0} \right\}^2}{\mathrm{var}(U | \theta = 0)} = \frac{12 n_1 n_2}{n+1} \left[\int_{-\infty}^{\infty} \{f_1(x)\}^2 \, dx \right]^2. \tag{31.114}$$

Using (25.27), (31.112) and (31.114), we have for the ARE of Wilcoxon's U test compared to "Student's" t,

$$A_{U, t} = 12 \sigma^2 \left[\int_{-\infty}^{\infty} \{f_1(x)\}^2 \, dx \right]^2, \tag{31.115}$$

a result due to Pitman (1948).

31.60 To evaluate (31.115), we only require the value of the integral

$$\int_{-\infty}^{\infty} \{f_1(x)\}^2 \, dx = E\{f_1(x)\}. \tag{31.116}$$

(31.116) is easily evaluated for particular distributions. When F_1 is normal, we have
$$E\{f_1(x)\} = (2\pi\sigma^2)^{-\frac{1}{2}} E\left\{\exp\left[-\tfrac{1}{2}\left(\frac{x-\mu}{\sigma}\right)^2\right]\right\}$$
$$= (2\pi\sigma^2)^{-\frac{1}{2}}\left[E\left\{\exp\left[it\left(\frac{x-\mu}{\sigma}\right)^2\right]\right\}\right]_{it=-\frac{1}{2}}$$
$$= (2\pi\sigma^2)^{-\frac{1}{2}}\left[(1-2it)^{-\frac{1}{2}}\right]_{it=-\frac{1}{2}} = (4\pi\sigma^2)^{-\frac{1}{2}}.$$
Thus, from (31.115) in the normal case
$$A_{U,t} = 3/\pi \doteq 0{\cdot}95. \tag{31.117}$$
The result is scale-free, as it always is since $\sigma E\{f_1(x)\}$ is scale-free. We may thus always re-scale in (31.115) if this is convenient.

31.61 It is easy to see that (31.115) can take infinite values (cf. Exercise 31.18). We now inquire whether there is a non-zero minimum below which it cannot fall. We wish to minimize $E\{f(x)\}$ for fixed σ^2, which we may take equal to unity. We also take $E(x) = 0$ without loss of generality. We thus require to minimize $\int_{-\infty}^{\infty} f^2(x)\,dx$ subject to the conditions $\int_{-\infty}^{\infty} x^2 f(x)\,dx = 1$, $\int_{-\infty}^{\infty} f(x)\,dx = 1$. Using Lagrange undetermined multipliers λ, μ, this is equivalent to minimizing the integral
$$\int_{-\infty}^{\infty} \{f^2(x) - 2\lambda(\mu^2 - x^2)f(x)\}\,dx. \tag{31.118}$$
Since $f(x)$ is non-negative, (31.118) is minimized for choice of $f(x)$ when
$$f(x) = \begin{cases} \lambda(\mu^2 - x^2), & x^2 \leqslant \mu^2, \\ 0, & x^2 > \mu^2, \end{cases} \tag{31.119}$$
a simple parabolic distribution. If λ and μ are found from the conditions
$$\int f(x)\,dx = \int x^2 f(x)\,dx = 1,$$
we find
$$\mu^2 = 5, \quad \lambda = 3/(20\sqrt{5}), \tag{31.120}$$
whence
$$\int f^2(x)\,dx = \frac{16}{15}\lambda^2\mu^5 = \frac{3}{5\sqrt{5}}. \tag{31.121}$$
Thus, from (31.121) and (31.115), the ARE for the distribution (31.119–120) is
$$\inf A_{U,t} = 108/125 = 0{\cdot}864. \tag{31.122}$$

31.62 The high value (31.122) for the minimum ARE of Wilcoxon's test compared to "Student's" t, which was first obtained by Hodges and Lehmann (1956), is very reassuring. In practical terms it means that in large samples we *cannot* lose more than 13·6 per cent efficiency in using Wilcoxon's rather than "Student's" test for a location shift; on the other hand, we may gain a very great deal—cf. Exercise 31.18 where it should be confirmed that for a Gamma distribution with parameter $p = 1$, $A_{U,t} = 3$. If the distribution is actually normal, the loss of efficiency is only about 5 per cent, by (31.117).

Van der Vaart (1950) showed that in the normal case the derivatives of the power function differ very little for very small sample sizes from their asymptotic values relative to those for the "Student's" t-test. Sundrum (1953) computed approximate power functions in the normal and rectangular cases for $n_1 = n_2 = 10$ which bear out, particularly in the normal case, the small power loss involved in using Wilcoxon's rather than "Student's" test. Dixon (1954) and Hodges and Lehmann (1956) give some small-sample results for the normal case which confirm this. Witting (1960), by using an Edgeworth expansion to order n^{-2}, shows that the value (31.117) for the ARE in the normal case holds very closely for sample sizes ranging from 4 to 40.

Chanda (1963) finds high ARE even for discrete populations. Noether (1967) shows that the test and the confidence intervals based upon it are conservative for discrete distributions. Haynam and Govindarajulu (1966) give exact power computations in the exponential and the rectangular cases.

A test with uniformly better ARE than "Student's" test

31.63 Although Wilcoxon's test performs very well compared to "Student's" test, as we have seen in **31.59–62**, we can do even better. Reverting to our discussion of **31.39**, we may obviously obtain ARE of 1 against *normal* location-shift alternatives by using the test statistic w (or equivalently the mean of the first sample, $\bar{x}_1$) with the observations replaced by the expected values of the order statistics in normal samples, which we denote by $E(s,n)$ as in **31.39**. The test statistic is therefore equivalent to

$$c_1 = \frac{1}{n_1} \sum_{i=1}^{n_1} E(X_i, n), \tag{31.123}$$

where X_i is the rank among the n observations of the ith observation in the first sample. (31.123) is usually called the *normal scores* test statistic. If we define

$$z_s = \begin{cases} 1 & \text{if the sth observation is from the first sample,} \\ 0 & \text{otherwise,} \end{cases}$$

we may rewrite (31.123) as

$$c_1 = \frac{1}{n_1} \sum_{s=1}^{n} E(s,n) z_s. \tag{31.124}$$

Hoeffding (1952) first demonstrated that the normal scores test has ARE 1 against normal alternatives. Terry (1952) gives its exact distribution for $n_1 + n_2 \leq 10$, and Klotz (1964) gives critical values for $n_1 + n_2 \leq 20$, when (31.74) holds. The asymptotic normality of c_1 (and of a wide class of similar statistics), whatever the parent distributions F_1, F_2 when $\lim n_1/n_2$ is bounded away from 0 and ∞, was demonstrated by Chernoff and Savage (1958). Clearly

$$E(c_1 | H_0) = \frac{1}{n_1} \sum_{s=1}^{n} E(s,n) E(z_s) = \frac{1}{n_1} \cdot \frac{n_1}{n} \sum_{s=1}^{n} E(s,n) = 0,$$

while $\text{var}(c_1 | H_0)$ is given at (31.133) below.

31.64 An alternative definition of c_1 is more convenient for the purpose of calculating its ARE. Define

$$H(x) = \frac{n_1}{n}F_1(x) + \frac{n_2}{n}F_2(x), \tag{31.125}$$

the distribution function of the combined parent populations, and its sample analogue

$$H_n(x) = \frac{n_1}{n}S_{n_1}(x) + \frac{n_2}{n}S_{n_2}(x), \tag{31.126}$$

where S_{n_1}, S_{n_2} are the sample distribution functions as defined at (30.97). If we also define the function $J_n(x)$ by

$$E(s,n) = J_n\left(\frac{s}{n}\right), \tag{31.127}$$

then c_1 may be defined in the integral form

$$c_1 = \int_{-\infty}^{\infty} J_n\{H_n(x)\} dS_{n_1}(x). \tag{31.128}$$

As $n \to \infty$ we have, from (31.125–126), $H_n(x) \to H(x)$, while from (31.127), under mild conditions (cf. 31.73)), $J_n(x) \to \Phi^{-1}(x) = J(x)$, where Φ is the standardized normal distribution function. Thus, as $n \to \infty$ we have from (31.128), under regularity conditions,

$$E(c_1) = \int_{-\infty}^{\infty} J\{H(x)\} dF_1(x). \tag{31.129}$$

Van der Waerden (1952, 1953) proposed a two-sample test based on the inverse normal d.f. transformation of the sample values, which is asymptotically equivalent to the c_1 test—cf. (31.129).

31.65 If we now consider location-shift alternatives (31.76), and differentiate (31.129) with respect to the location-shift θ, we have

$$\frac{\partial E(c_1)}{\partial \theta} = \int_{-\infty}^{\infty} J'\{H(x)\} \left\{\frac{d}{d\theta} H(x)\right\} dF_1(x), \tag{31.130}$$

and since

$$H(x) = \frac{n_1}{n}F_1(x) + \frac{n_2}{n}F_1(x-\theta)$$

we have

$$\frac{d}{d\theta} H(x) = -\frac{n_2}{n} f_1(x-\theta).$$

Putting this into (31.130), we find

$$\frac{\partial E(c_1)}{\partial \theta} = \frac{-n_2}{n} \int_{-\infty}^{\infty} J'\{H(x)\} f_1(x-\theta) dF_1(x). \tag{31.131}$$

Now when $\theta = 0$, $H(x) = F_1(x)$, so (31.131) gives

$$\left[\frac{\partial E(c_1)}{\partial \theta}\right]_{\theta=0} = \frac{-n_2}{n} \int_{-\infty}^{\infty} J'\{F_1(x)\} \{f_1(x)\}^2 dx. \tag{31.132}$$

When H_0 holds, the variance of c_1 is simply, from (31.124),

$$\text{var}(c_1 | H_0) = \frac{1}{n_1^2} \text{var}\left\{\sum_{s=1}^{n} E(s,n) z_s\right\}.$$

Here, only the z_s are random variables; in fact, they are identical 0–1 variables with probability $\frac{n_1}{n}$ of taking the value 1, but they are not independent, any pair having correlation $-\frac{1}{n-1}$ because of the symmetry. Thus

$$\operatorname{var}(c_1|H_0) = \frac{1}{n_1^2}\left[\sum_{s=1}^{n}\{E(s,n)\}^2 \operatorname{var} z_s + \sum\sum_{s\neq r} E(s,n)E(r,n)\operatorname{cov}(z_s,z_r)\right]$$

$$= \frac{\operatorname{var} z}{n_1^2}\left\{\sum_{s=1}^{n}\{E(s,n)\}^2 - \frac{1}{n-1}\left[\left(\sum_{s=1}^{n}E(s,n)\right)^2 - \sum_{s=1}^{n}\{E(s,n)\}^2\right]\right\},$$

and since $\sum_{s=1}^{n} E(s,n) = 0$ by the symmetry of the normal distribution and $\operatorname{var} z = \frac{n_1}{n}\left(1 - \frac{n_1}{n}\right)$, this reduces to

$$\operatorname{var}(c_1|H_0) = \frac{n_2}{n(n-1)n_1}\sum_{s=1}^{n}\{E(s,n)\}^2. \qquad (31.133)$$

This is an exact result. When $n \to \infty$, by (31.73),

$$\frac{1}{n}\sum_{s=1}^{n}\{E(s,n)\}^2 \to \int x^2 d\Phi(x) = 1$$

and

$$\operatorname{var}(c_1|H_0) \to \frac{n_2}{(n-1)n_1}. \qquad (31.134)$$

Thus, from (31.132) and (31.134)

$$\frac{\left\{\left[\frac{\partial}{\partial\theta}E(c_1)\right]_{\theta=0}\right\}^2}{\operatorname{var}(c_1|\theta=0)} \to \frac{n_1 n_2}{n}\left[\int_{-\infty}^{\infty}J'\{F_1(x)\}\{f_1(x)\}^2 dx\right]^2. \qquad (31.135)$$

Using (25.27), (31.135) and (31.112) give for the ARE of the normal scores test compared to Student's t

$$A_{c_1,t} = \sigma^2\left[\int_{-\infty}^{\infty}J'\{F_1(x)\}\{f_1(x)\}^2 dx\right]^2. \qquad (31.136)$$

In (31.136) we may put $\sigma^2 = 1$ by standardizing $F_1(x)$, since the result is scale-free, like (31.115)—cf. **31.60**. We now seek the minimum value of (31.136).

31.66 It will be remembered that $J(x)$ in (31.136) is defined as $\Phi^{-1}(x)$, the inverse of the standardized normal d.f. Thus

$$\Phi\{J(x)\} = x. \qquad (31.137)$$

Differentiating both sides, and writing $\phi(x) = \Phi'(x)$ for the normal f.f., we have

$$\frac{d}{dx}[\Phi\{J(x)\}] = \phi\{J(x)\}J'(x) = 1,$$

so that

$$J'(x) = \frac{1}{\phi\{J(x)\}}.$$

From (31.137), $\Phi\{J[F_1(x)]\} = F_1(x),$

so we have the equation in differentials

$$dF_1(x) = d\Phi\{J[F_1(x)]\} = \phi\{J[F_1(x)]\}dJ[F_1(x)]. \qquad (31.138)$$

Since
$$J'\{F_1(x)\} = \frac{1}{\phi\{J[F_1(x)]\}}, \tag{31.139}$$
the integral in square brackets in (31.136) becomes
$$I = \int_{-\infty}^{\infty} \frac{f_1 \, dF_1}{\phi\{J(F_1)\}}. \tag{31.140}$$

To minimize (31.136), we require to minimize (31.140) for choice of F_1, subject to the standardization conditions
$$\int x \, dF_1 = 0, \qquad \int x^2 \, dF_1 = 1.$$

31.67 Since ϕ is the standardized normal f.f., these conditions are obviously satisfied when $f_1 = \phi$, i.e.
$$J\{F_1(x)\} = x \tag{31.141}$$
and (31.140) is then equal to 1. Thus we have verified our statement of **31.63** that in the normal case the normal scores test has ARE 1.

Chernoff and Savage (1958) show that (31.141) is actually the unique solution of our minimization problem, i.e. that $I > 1$ for any non-normal F_1, so that the normal scores test has minimum ARE of 1 compared to the t-test. The following is a simplified proof.

We first recall from Exercise 9.13, Volume 1, 3rd edn, that for a positive random variable y, $E(y)E(1/y) \geq 1$. The integrand of (31.140) is such a variable, and thus
$$I \int_{-\infty}^{\infty} \frac{\phi\{J(F_1)\}}{f_1} dF_1 = I \int_{-\infty}^{\infty} \phi\{J(F_1)\} dx \geq 1. \tag{31.142}$$
Now, integrating by parts,
$$\int_{-\infty}^{\infty} \phi\{J(F_1)\} dx = [x\phi\{J(F_1)\}]_{-\infty}^{\infty} - \int_{-\infty}^{\infty} x \frac{d}{dx} \phi\{J(F_1)\} dx.$$
As $x \to \pm\infty$, $F_1 \to 1, 0$ and $J(F_1) \to \pm\infty$, so that the first term on the right-hand side vanishes since $x\phi(x)$ does so at its extremes. To evaluate the remaining integral on the right, we recall that since $\phi(x) \propto e^{-\frac{1}{2}x^2}$, $\phi'(x) = -x\phi(x)$. Thus
$$\frac{d}{dx}\phi\{J(F_1)\} = -J(F_1)\phi\{J(F_1)\} \cdot \frac{dJ(F_1)}{dx} = -J(F_1)\frac{dF_1}{dx},$$
using (31.138). We therefore have simply $\int_{-\infty}^{\infty} \phi\{J(F_1)\} dx = \int_{-\infty}^{\infty} x J(F_1) dF_1$, and (31.142) becomes $I \int_{-\infty}^{\infty} x J(F_1) dF_1 \geq 1$. We now use the Cauchy–Schwarz inequality on the integral, and obtain
$$I \left\{ \int_{-\infty}^{\infty} x^2 dF_1 \int_{-\infty}^{\infty} J^2(F_1) dF_1 \right\}^{\frac{1}{2}} \geq 1.$$
Each of these two integrals is unity, being the variance of a standardized distribution. Thus, finally,
$$I^2 = A_{c_1,t} \geq 1. \tag{31.143}$$
The equality in (31.143) holds only when (31.141) does, for only then do the two inequalities that we have used reduce to equalities.

Mikulski (1963) has shown under regularity conditions that no other distribution than the normal has this remarkable property that the efficiency of the best rank-order test compared to the best location-shift test always exceeds unity when the underlying distribution differs from that assumed in deriving the tests.

D. R. Cox (1964) considers similar uses of the exponential scores derived in Exercise 19.11.

31.68 The implication of the result of **31.63–67** is far-reaching. The normal scores test is distribution-free (completely robust) and has minimum ARE of 1 compared to the standard normal-theory test based on Student's t, which is only fairly robust. It is therefore difficult to make a case for the customary routine use of Student's test in testing for a location-shift between two samples when sample numbers are reasonably large. The t-test has no appreciable advantage (even in the normal case for which it is optimum) for sample sizes of 4 or 5 with α near 0·05—cf. Klotz (1964).

The labour of computing the normal scores test is very light. It consists of referring

$$\frac{c_1 - E(c_1 | H_0)}{\{\operatorname{var}(c_1 | H_0)\}^{\frac{1}{2}}} = \left\{\frac{n(n-1)}{n_1 n_2}\right\}^{\frac{1}{2}} \frac{\sum_{i=1}^{n_1} E\{X_i, n\}}{\left[\sum_{s=1}^{n} \{E(s, n)\}^2\right]^{\frac{1}{2}}}$$

to a table of the standardized normal distribution. Fisher and Yates' Tables give $T = \sum_{s=1}^{n} \{E(s, n)\}^2$ for $n = 1$ (1) 50 to 4 d.p. and the individual $E(s, n)$ to 2 d.p. Harter (1961a) gives all the $E(s, n)$ to 5 d.p. for $n = 2$ (1) 100 (25) 250 (50) 400. His table is reproduced by F. N. David et al. (1968) who also give T and all fourth-order sums of products to 5 d.p.

For $n = 50$, $\left[\frac{1}{n}\sum_{s=1}^{n}\{E(s, n)\}^2\right]^{\frac{1}{2}} = 0.97$, and thereafter tends to 1 as we saw below (31.133), so that this factor may be dropped, reducing the standardized test statistic to

$$\left(\frac{n-1}{n_1 n_2}\right)^{\frac{1}{2}} \sum_{i=1}^{n_1} E(X_i, n).$$

Hodges and Lehmann (1963) show that robust estimators of θ may be obtained from the Wilcoxon and the normal-scores tests which have the same efficiency properties as the tests have ARE—cf. also Høyland (1965) and Hodges (1968). Ramachandramurty (1966) gives forms of these estimators which are robust to scale-shift, and therefore applicable to the general problem of two means. See also A. Birnbaum and Laska (1967b).

Pratt (1964) studies the effect of an unknown scale difference on the size of the Wilcoxon, normal-scores, "Student's" t and other location tests. Asymptotically, the t-test is appreciably more robust only if n_1/n_2 is very near 1, and even then the gain is small if the scale multiple does not exceed 2.

31.69 Other tests for the two-sample problem, now rather overshadowed by the Wilcoxon and the normal scores tests, have been proposed. That of Wald and Wolfowitz (1940)—cf. Exercise 30.8—based on the number of runs, has the advantage of being consistent against the general alternative (31.111) if $\lim n_1/n_2$ is bounded away from 0 and ∞. Smirnov (1939b) proposed a test based on the maximum absolute difference d between the two sample distribution functions, and showed that $d \bigg/ \left(\frac{1}{n_1} + \frac{1}{n_2}\right)^{\frac{1}{2}}$ has the same limiting distribution as that of $D_n n^{\frac{1}{2}}$ given at (30.132). The convergence of the sample d.f.'s to the parent d.f.'s ensures that the test is consistent against (31.111). A lower bound for its power may be obtained just as for D_n in **30.60**, but the test may be biassed—cf. Exercise 30.16 for the D_n test.

Lehmann (1951) proposed a test which he showed to be always unbiassed.

Although rather little is known of the power of these tests, it is clear that they are less efficient against normal shift alternatives. In fact, Mood (1954) shows the Wald-Wolfowitz test to have ARE of 0 against normal location-shift or normal scale-shift alternatives.

The scale-shift alternative hypothesis

$$F_2(x) = F_1\left(\frac{x}{\theta}\right), \quad \theta > 0 \qquad (31.144)$$

in which we test $H_0 : \theta = 1$, is equivalent to a location-shift alternative for the logarithms of the variables if they are non-negative. Generally, however, we must seek new tests against (31.144). Mood (1954) proposed $W = \sum_{i=1}^{n_1} \{X_i - \frac{1}{2}(n+1)\}^2$, which is distribution-free, and showed that in the normal case its ARE compared to the optimum variance-ratio test is $15/(2\pi^2) \doteq 0.76$. For other parents, its ARE ranges from 0 to ∞. If (31.144) is replaced by the more general

$$F_2(x-\mu) = F_1\{(x-\nu)/\theta\}, \qquad (31.145)$$

where the unknown μ, ν are (nuisance) location parameters, Mood's test remains *asymptotically* distribution-free if μ and ν are estimated by the sample medians (cf. Crouse, 1964).

We may obtain ARE of 1 against normal alternatives by applying the variance-ratio test to the normal scores $E(X_i, n)$—cf. Capon (1961) who used the asymptotically equivalent statistic given by (31.123) with $E(X_i, n)$ replaced by its square. Klotz (1962) used the square of the inverse normal d.f. transformation, which is also asymptotically equivalent (cf. Van der Waerden's test in **31.64**) and tabulates critical values of this statistic for $8 \leq n_1 + n_2 \leq 20$. Raghavachari (1965a, b) confirms that the ARE of Klotz's test ranges from 0 to ∞ as a function of F in (31.144), and shows that its asymptotic properties against (31.144) persist against (31.145) when μ, ν are suitably consistently estimated from the samples, provided that F is symmetric and sufficiently regular.

Siegel and Tukey (1960) proposed a scale-shift test which has the same H_0 distribution as the Wilcoxon test (because it uses the set of sample rank values; these are allocated to the observations according to their distance from the extremes of the ordering) but an ARE of only $6/\pi^2 = 0.61$ in the normal case, as Klotz (1962) showed.

Van Eeden (1964) gives conditions for these and other scale-shift tests to be consistent. Hollander (1968) gives conditions under which scale-shift tests are uncorrelated with location-shift (e.g. the Wilcoxon and normal scores) tests; in particular, this is so when $F_2 \equiv F_1$ and is symmetric. Tamura (1963) showed that against normal scale-shift the ARE of Mood's test may be raised to 0.92 by using 5th powers instead of squares in its definition, and that for a test equivalent to the Siegel-Tukey test, the ARE may be raised to 0.96 by using tiny fractional powers of the ranks instead of the ranks themselves.

Moses (1963) discusses scale-shift tests generally. See also G. R. Shorack (1969) for ratios of scale parameters.

Mardia (1967, 1968, 1969) compares tests (including a generalization of the Wilcoxon test) for location-shifts in two samples from bivariate distributions. See also Fryer (1970).

k-sample tests

31.70 The generalization of two-sample tests to the k-sample problem (Problem (2) of **31.13**) is straightforward. The hypothesis to be tested is that $k \geq 2$ continuous distribution functions are identical, i.e.

$$H_0 : F_1(x) = F_2(x) = \ldots = F_k(x), \quad \text{all } x, \qquad (31.146)$$

on the basis of independent samples of n_p observations ($p = 1, 2, \ldots, k$) where $\sum_{p=1}^{k} n_p = n$. In the parametric theory, all of the F_p are assumed normal with common unknown variance σ^2 and different means θ_p, so that

$$H_1 : F_p(x) = F(x - \theta_p), \quad p = 1, 2, \ldots, k, \tag{31.147}$$

with not all the θ_p equal. (31.146) then becomes

$$H_0 : \theta_p = 0, \quad \text{all } p. \tag{31.148}$$

The LR test of (31.148) in the normal case is based on the statistic

$$F = \frac{1}{k-1} \sum_{p=1}^{k} n_p (\bar{x}_p - \bar{x})^2 \Big/ \frac{1}{n-k} \sum_{p=1}^{k} \sum_{j=1}^{n_t} (x_{pj} - \bar{x}_p)^2$$

(where $\bar{x}_p$ is the pth sample mean and $\bar{x}$ the overall sample mean), which is distributed in the variance-ratio (F) distribution with $(k-1, n-k)$ degrees of freedom. This follows immediately from the general LR theory of Chapter 24, and is the simplest case of the Analysis of Variance, which we shall discuss in Volume 3. As $(n-k) \to \infty$ it follows (cf. **16.22**(7)) that

$$S = \sum_{p=1}^{k} n_p (\bar{x}_p - \bar{x})^2 / \sigma^2 \tag{31.149}$$

is asymptotically a χ^2 variate with $k-1$ degrees of freedom.

This test has been shown by Gayen (1949–1951) to be remarkably robust to departures from normality, and we shall be discussing the robustness of Analysis of Variance procedures in general in Volume 3. Here we consider only distribution-free substitutes for the normal theory test.

31.71 Since (31.147) is clearly a generalization of the location-shift alternative (31.76) in the two-sample case, it is natural to seek generalizations of the two-sample tests to k samples. We consider two different approaches to the problem. First suppose that we simply replace the observations x by their ranks X. The statistic (31.149) then becomes

$$S = \sum_{p=1}^{k} n_p \{\bar{X}_p - \tfrac{1}{2}(n+1)\}^2 / \{\tfrac{1}{12}(n^2 - 1)\}, \tag{31.150}$$

reducing to the Wilcoxon test when $k = 2$. Kruskal and Wallis (1952–1953) proposed the statistic $H = (n-1)S/n$, large values of H forming the critical region of the test. They demonstrated its asymptotic χ^2_{k-1} distribution as in the parametric case. For $k = 3$, $n_p \leq 5$, they tabulate its exact distribution in the neighbourhood of its critical values for $\alpha = 0.10, 0.05, 0.01$. Kruskal (1952) showed that the H-test is consistent against any alternative hypothesis for which an observation from one of the parent distributions has probability $\neq \tfrac{1}{2}$ of exceeding a random observation from the k parents taken together. This is a generalization of the consistency condition for the Wilcoxon test in **31.56**.

> Puri (1964) considers the statistic obtained by replacing the observations by *any* set of conventional numbers, gives conditions for its asymptotic normality and obtains its ARE, which is independent of k. So far as the H-test is concerned, this means that its ARE is given by the Wilcoxon expression (31.115) (cf. Andrews (1954)) and the

analysis of **31.60-1** is applicable without change to the H-test. If normal scores are used instead of ranks, we obtain the generalization of the test of (31.124), the ARE is given by (31.136), and by **31.66-7** this is at least 1 as before. P. K. Sen (1967) gives a general theory of k-sample permutation tests.

31.72 In the two-sample case, we saw in **31.52** that it made no difference to the test statistic derived whether we replaced observations by ranks or counted the number of inversions of order between the samples. In the present k-sample case, it does matter. We now proceed to the inversions approach, and shall find that it leads to a different statistic from H of **31.71**.

Suppose that the statistic U of (31.98) is calculated for every pair of samples, there being $\frac{1}{2}k(k-1)$ pairs in all; we write U_{pq} for the value obtained from the pth and qth samples ($p, q = 1, 2, \ldots, k$; $p \neq q$) and

$$U = \sum_{p=1}^{k} \sum_{q=p+1}^{k} U_{pq}. \tag{31.151}$$

We may now very easily generalize the theory of **31.53**. (31.100) is replaced by $Q_n = \sum_{p=1}^{k} Q_{n_p} + U$, which leads to the c.f. relationship

$$\Phi_U(\theta) = \phi_n(\theta) \Big/ \prod_{p=1}^{k} \phi_{n_p}(\theta), \tag{31.152}$$

which is the analogue of (31.101). We find, corresponding to (31.103), for the c.g.f. of U

$$\Psi_U(\theta) = \tfrac{1}{4} i\theta \left(n^2 - \sum_{p=1}^{k} n_p^2 \right) + \sum_{j=1}^{\infty} \frac{B_{2j}(i\theta)^{2j}}{2j(2j)!} \left\{ \sum_{s=1}^{n} - \sum_{p=1}^{k} \sum_{s=1}^{n_p} \right\} s^{2j}. \tag{31.153}$$

The cumulants of U are therefore

$$\left. \begin{array}{l} \kappa_1 = \tfrac{1}{4}\left(n^2 - \sum_{p=1}^{k} n_p^2\right), \quad \kappa_{2j+1} = 0, \quad j > 0, \\[2mm] \kappa_{2j} = \dfrac{B_{2j}}{2j} \left\{ \sum_{s=1}^{n} - \sum_{p=1}^{k} \sum_{s=1}^{n_p} \right\} s^{2j}. \end{array} \right\} \tag{31.154}$$

In particular

$$\kappa_2 = \frac{1}{72}\left\{ n^2(2n+3) - \sum_{p=1}^{k} n_p^2 (2n_p + 3) \right\}. \tag{31.155}$$

31.73 The limit distribution of U also follows as in **31.55**. If the $n_p \to \infty$ so that n_p/n remains bounded for all p, we write N for any n_p or n and see that κ_{2j} is at most of order $(2j+1)$ in N, with κ_2 of order N^3. Thus κ_{2j}/κ_2^j is of order N^{1-j} at most and tends to zero for all $j > 1$, so that U tends to normality with mean and variance given by (31.154-155). Jonckheere (1954) shows that if only two of the k-sample sizes tend to infinity so that n_p/n, n_q/n remain bounded, the distribution of U still tends to normality—this may be seen from the consideration that U is the sum of $\tfrac{1}{2}k(k-1)$ (non-independent) Wilcoxon statistics U_{pq}. If r sample sizes, $r \geqslant 2$, tend to infinity, $\tfrac{1}{2}r(r-1)$ of the U_{pq} will tend to normality and will dominate U.

Jonckheere (1954) tabulated the exact distribution of U for samples of equal sizes m. His table covers $k = 3$, $m = 2(1)5$; $k = 4$, $m = 2(1)4$; $k = 5$, $m = 2, 3$; and $k = 6$, $m = 2$. Beyond this point, the normal approximation is adequate for equal sample sizes. Even for very unequal sample sizes, the normal approximation seems adequate for practical purposes when $n \geqslant 12$.

Terpstra (1952), who originally proposed the k-sample U-test and derived the c.f. and limiting distribution given above, gave necessary and sufficient conditions for the consistency of the test. If the probability that an observation from the pth distribution exceeds one from the qth is (cf. (31.51))

$$P\{x_{pi} > x_{qj}\} = \tfrac{1}{2} + \varepsilon_{pq}, \quad p \neq q,$$

and the weighted sum of the ε_{pq} is

$$S_n = \underset{p<q}{\sum \sum} n_p n_q \varepsilon_{pq}$$

then the conditions for consistency (as $n \to \infty$ with n_p/n bounded for all p) of the test using large values of U as critical region are (1) $S_n > 0$, (2) $(kn)^{-3/2} S_n \to \infty$. These are direct generalizations of (31.58), the condition for consistency of Q in testing randomness against downward trend.

31.74 So far as we know, the efficiencies of the two k-sample test statistics H and U have not been compared. The difficulty is that the forms of their limiting distributions are different; for fixed k, H has a non-central χ^2_{k-1} distribution, and U a normal distribution asymptotically when H_0 holds and presumably also under general alternatives. It seems likely that the U-test will be at its best when the alternatives are of the form (31.147) with $\theta_1 < \theta_2 < \ldots < \theta_k$, or in the more general situation when (31.147) is replaced by

$$F_1(x) < F_2(x) < \ldots < F_k(x), \quad \text{all } x. \tag{31.156}$$

(31.156) may be referred to as an *ordered* alternative hypothesis. Bartholomew (1961) (cf. Exercise 35.15, Volume 3) shows that U is asymptotically very efficient when the θ_i are equally-spaced and the n_i are all equal. The H-test, on the other hand, is likely to be more efficient against broader, more general, classes of alternatives.

> Hogg (1962) gives a method of constructing distribution-free k-sample tests from two-sample tests.

Tests of symmetry

31.75 In all of the hypotheses discussed in this chapter, we have been fundamentally concerned with n independent observations (usually on a single variate x but, in the case of testing independence, on a vector (x,y)). Our hypotheses have specified that certain of these observations are identically distributed, and proceeded to test some hypotheses concerning their parent distribution functions. We found (cf. **31.16**) that, to obtain similar tests of our hypotheses, we must restrict ourselves to permutation tests, the distribution theory of which assigns equal probability to each of the $n!$ orderings of the sample observations.

An implication of this procedure is that the tests we have derived remain valid if the hypotheses we have considered are replaced by the direct hypothesis that the joint distribution function of the observations is invariant under permutation of its arguments. For example, consider a two-sample test of the hypothesis

$$H_0 : F_1(x) = F_2(x), \quad \text{all } x, \tag{31.157}$$

where n_1, n_2 are the respective sizes of random samples from the two distributions and $n = n_1+n_2$. Write G for the joint distribution function of the n observations. Replace H_0 by the *hypothesis of symmetry*

$$H'_0 : G(x_1, x_2, \ldots, x_n) \equiv G(z_1, z_2, \ldots, z_n), \tag{31.158}$$

where the z's are any permutation of the x's. Then any similar test which is valid for (31.157) will be so for (31.158) also. This is not to say that the optimum properties of a test will remain the same for both hypotheses—a discussion of this point is given by Lehmann and Stein (1949). However, it does imply that any test of (31.157) cannot be consistent against the alternative hypothesis (31.158).

Practical situations are common in which a hypothesis of symmetry is appropriate. Since we have not discussed this problem so far even in the parametric case, we shall begin by a brief consideration of the latter in the simplest case.

The paired t-test

31.76 Suppose that variates x_1 and x_2 are jointly normally distributed with means and variances $(\mu_1, \sigma_1^2), (\mu_2, \sigma_2^2)$ respectively and correlation parameter ρ. We wish to test the composite hypothesis

$$H_0 : \Delta \equiv \mu_1 - \mu_2 = 0 \tag{31.159}$$

on the basis of m independent observations on (x_1, x_2). Consider the variable $y = x_1 - x_2$. It is normally distributed, with mean Δ and variance $\sigma^2 = \sigma_1^2 + \sigma_2^2 - 2\rho\sigma_1\sigma_2$. We have m observations on y available and may therefore test H_0 by the usual "Student's" t-test for the mean applied to the differences $(x_{1i}-x_{2i})$, $i = 1, 2, \ldots, m$. The procedure holds good when $\rho = 0$, when x_1 and x_2 are independent normal variates, and in this particular situation the test is a special case of that given at (21.51) with $n_1 = n_2 = m$.

31.77 Next simplify the example in **31.76** by putting $\sigma_1^2 = \sigma_2^2$. The joint distribution $F(x_1, x_2)$ is now symmetric in x_1 and x_2 save possibly for their means. When H_0 holds, we have complete symmetry. We may therefore write (31.159) as

$$H_0 : F(x_1, x_2) = F(x_2, x_1), \quad \text{all } x_1, x_2. \tag{31.160}$$

This is a typical symmetry hypothesis, which may formally be put into the general form (31.158) by writing G as the product of m factors (one for each observation on (x_1, x_2)).

We now abandon the normal theory of **31.76** and seek distribution-free methods of testing the non-parametric hypothesis (31.160) for arbitrary continuous F. If we take differences $y = x_1 - x_2$ as before, we see that H_0 implies the symmetry of the distribution of y about the point 0 or, if G is its d.f.,

$$H_0 : G(y) = 1 - G(-y), \quad \text{all } y. \tag{31.161}$$

We have thus reduced the hypothesis (31.160) of symmetry of a bivariate d.f. in its arguments to the hypothesis (31.161) of symmetry of a univariate distribution about a particular value, zero. This hypothesis is clearly of interest in its own right (i.e. we may simply be interested in the symmetry of a single variate), and we proceed to treat the problem in this form.

31.78 The hypothesis (31.161) implies that any observed absolute value of y, $|y_i|$, has equal probability of arising from a positive and a negative value of y_i. Thus there are 2^n equiprobable samples generated by associating every observed $|y_i|$ with a positive and a negative sign alternately, and combining all the $|y_i|$ in every possible way. We have formed the basis for a permutation test by permuting the sign attached to each $|y_i|$, and now have to select the test statistic. If we consider the alternative that the observations are normal with mean $\theta \neq 0$, we find as in **31.18** and **31.44** that the most powerful permutation test is based on the sum of the observed y_i (with their true signs, of course). We defer consideration of this test (suggested by Fisher (1935a)) since it is a special case of a permutation test in the Analysis of Variance, which we shall discuss in **37.38-41**, Volume 3.

31.79 Fisher's test of symmetry is equivalent to using the sum of the $|y_i|$ over observed positive values (since the sum of the $|y_i|$ is the same for each of the 2^n permutations). For a rank test of symmetry, we may replace observations by ranks and obtain the Wilcoxon test of symmetry, based on the sum of the ranks of the $|y_i|$ over observed positive values; or we may use the equivalent of the normal scores test, based on the expected values of the order-statistics from the positive half of a normal distribution. These three symmetry tests have ARE of 1, $3/\pi$ and 1 respectively, compared with the t-test against normal location-shift alternatives, just as for the two-sample location tests. Small-sample efficiency is high against normal alternatives (Klotz (1963)) for both rank tests—Klotz (1965) also considers an alternative measure of efficiency. The Wilcoxon symmetry test does not retain its high power under some non-normal (long tail) alternatives, but is still better than Student's t, although worse than the Sign test (H. J. Arnold (1965)) to be discussed in **32.6-7**, which is itself obtainable from Fisher's test by replacing the original paired observations (x_1, x_2) by their ranks, as can be seen from **37.40** and Exercise 37.14, Vol. 3. R. Thompson *et al.* (1967) report extensive sampling experiments comparing all these tests against various shift alternatives. See also Kempthorne and Doerfler (1969).

McCornack (1965) gives extensive tables of critical values for the Wilcoxon symmetry test. Exercise 31.25 gives the associated confidence intervals for the location constant in a symmetric distribution. Klotz (1963) and R. Thompson *et al.* (1967) tabulate the normal scores symmetry test.

The effects of discontinuities: continuity corrections and ties

31.80 In various places, in this chapter as elsewhere, we have approximated discontinuous distributions (in the present context the permutation distributions of test statistics) by their asymptotic forms, which are continuous. The approximation is often, but not always—cf. Plackett (1964) and **33.27** below—improved if we apply a *continuity correction*, which amounts to the following simple rule: when successive discrete probabilities in the exact distribution occur at values z_1, z_2, z_3, the probability at z_2 is taken to refer to the interval $(\frac{1}{2}(z_1+z_2), \frac{1}{2}(z_2+z_3))$. Thus, when we wish to evaluate the d.f. at the point z_2 from a continuous approximation, we actually evaluate it at the point $\frac{1}{2}(z_2+z_3)$.

31.81 There is another question connected with continuity which we should discuss here. Our hypotheses have been concerned with observations from continuous d.f.'s, and this implies that the probability of any pair of observations being precisely equal (a so-called *tie*) is zero and that we may therefore neglect the possibility. Thus we have throughout this chapter assumed that observations could be ordered without ties, so that the rank-order statistics were uniquely defined. However, in practice, observations are always rounded off to a few significant figures, and ties will therefore sometimes occur. Similarly, if the true parent d.f.'s are not in fact continuous, but are adequately represented by continuous d.f.'s, ties will occur. How are we to resolve the difficulty of obtaining a ranking in the presence of ties?

Two methods of treating ties have been discussed in the literature. The first is to order tied observations at random. This has the merit of simplicity and needs no new theory, but obviously sacrifices information contained in the observations and may be expected to lead to loss of efficeincy compared with the second method, which is to attribute to each of the tied observations the average rank of those tied. Putter (1955) shows that the ARE of the Wilcoxon test is less for random tie-breaking than when average ranks are allotted—see also Bühler (1967). McNeil (1967) shows that little ARE is lost through the existence of ties, and random tie-breaking, for a number of the tests discussed in this chapter. Kruskal and Wallis (1952–1953) and Kruskal (1952) present a discussion of ties in the *H*-test.

Until further information is available, the average-rank method is likely to be the more commonly used. Unfortunately, it removes the feature of rank order tests which we have remarked, that their exact distributions can be tabulated once for all. For, if the average-rank method of tie-breaking is used, the sum of a set of ranks is unaffected but, e.g., their variance is changed. The permutation distribution for small sample sizes now becomes a function of the number and extent of the ties observed, and this makes tabulation difficult. Kendall (1962) gives full details of and references to the necessary adjustments for the rank correlation coefficients and related statistics (which include the Wilcoxon test statistic), and other discussions of adjustments have been mentioned above.

31.82 Finally, we have seen in **30.56** and **31.62**, and will see again in **32.9** and **32.13**, that if the parent distribution is discrete, methods based on the continuity assumption prove to be conservative.

EXERCISES

31.1 By use of tables of the χ^2 distribution, verify the values of the true probabilities in the table in **31.7**.

31.2 Verify that the distribution (31.18) has moments (31.12), (31.13) and (31.17).

31.3 If r is the correlation coefficient defined at (31.11), if we transform the observed x- and y-values by $X = t_1(x)$, $Y = t_2(y)$, and calculate R, the correlation between the transformed values (X, Y), then every one of the equiprobable $n!$ permutations yields values r, R. Show that the correlation between r and R over the $n!$ permutations is given by
$$C(r, R) = C_1(x, X) C_2(y, Y),$$
i.e. that the correlation coefficient of the joint permutation distribution of the observed correlation coefficient and the correlation coefficient of the transformed observations is simply the product of the correlation between x and its transform with the correlation between y and its transform. (Daniels, 1944)

31.4 Derive the fourth moment of the rank correlation coefficient given at (31.22) from the general expression in (31.14).

31.5 Show that (31.21) and (31.40) are alternative definitions of (31.19) by proving the identities (31.20) and (31.39).

31.6 Using the definitions (31.37), (31.38), show that in the joint distribution of t and r_s over the $n!$ equiprobable permutations in the case of independence, their correlation coefficient is $2(n+1)/\{2n(2n+5)\}^{\frac{1}{2}}$. (cf. Daniels (1944))

31.7 A sample of n pairs (x, y) is obtained from a continuous bivariate distribution. Let $\tilde{x}, \tilde{y}$ be the sample medians of x and y and define the statistic
$$u = \sum_{i=1}^{n} \operatorname{sgn}(x_i - \tilde{x}) \operatorname{sgn}(y_i - \tilde{y}).$$
Show how u may be used to test the hypothesis of independence of x and y, and that its ARE compared to the sample correlation coefficient against bivariate normal alternatives is $4/\pi^2$. (This is called the *medial correlation test*.) (Blomqvist, 1950)

31.8 In **31.36**, use the result of Exercise 2.15, and the symmetry of the distribution of Q, to show that a sufficient condition for a size-α test of randomness to be strictly unbiassed against the alternatives (31.51) is that
$$S_n \geq \tfrac{1}{4}n(n-1) - (1-\alpha)\{\tfrac{1}{2}n(n-1) - Q_0\},$$
where Q_0 is the critical value of Q defined at (31.55).

(cf. Mann (1945))

31.9 Show that (31.69) holds for the statistic V defined by (31.37) as well as for Q, and hence that r_s as a test of randomness has the same ARE as the other rank correlation coefficient t.

31.10 In testing the hypothesis (31.49) of randomness against the normal regression alternatives (31.61), consider the class of test statistics
$$S = \Sigma w_{ij} h_{ij},$$

where the summation contains $\frac{1}{2}n$ terms (n a multiple of 6), the suffixes i, j each taking $\frac{1}{2}n$ different values and all suffixes being distinct, while the w_{ij} are weights. Thus S involves $\frac{1}{2}n$ comparisons between *independent* pairs of observations.

Show that the S-statistic with maximum ARE compared with b at (31.63–64) is

$$S_1 = \sum_{k=1}^{\frac{1}{2}n} (n-2k+1) h_{k, n-k+1}$$

with ARE

$$A_{S_1, b} = (2/\pi)^{\frac{1}{2}} \doteqdot 0 \cdot 86.$$

(D. R. Cox and Stuart, 1955)

31.11 In Exercise 31.10, show that if instead of S_1 we use the equally-weighted form

$$S_2 = \sum_{k=1}^{\frac{1}{2}n} h_{k, n-k+1},$$

the ARE is reduced to $\left(\dfrac{3}{2\pi}\right)^{\frac{1}{2}} \doteqdot 0 \cdot 78$, but that the maximum ARE attainable by an S-statistic with all weights 1 or 0 is $(16/9\pi)^{\frac{1}{2}} \doteqdot 0 \cdot 83$, which is the ARE of

$$S_3 = \sum_{k=1}^{\frac{1}{3}n} h_{k, \frac{2}{3}n+k}.$$

S_3 involves only $\frac{1}{3}n$ comparisons, between the " earliest " and " latest " observations.

(D. R. Cox and Stuart, 1955)

31.12 Define the statistic for testing randomness

$$B = \sum_{i=1}^{\frac{1}{2}n} \operatorname{sgn}(x_i - \tilde{x})$$

where $\tilde{x}$ is the median of the sample of size n (n even). Show that its ARE against normal alternatives is exactly that of S_2 in Exercise 31.11.

(D. R. Cox and Stuart (1955); cf. G. W. Brown and Mood (1951))

31.13 N samples, each of size n, are drawn independently from a continuous distribution with mean μ and variance σ^2, and the observations ranked from 1 to n in each sample. For the Nn combined observations, the correlation coefficient between the variate-values and the corresponding ranks is calculated. Show that as $N \to \infty$ this tends to

$$C_n = \left\{ \frac{12(n-1)}{\sigma^2(n+1)} \right\}^{\frac{1}{2}} \int_{-\infty}^{\infty} x \{F(x) - \tfrac{1}{2}\} dF(x)$$

$$= \left(\frac{n-1}{n+1}\right)^{\frac{1}{2}} \frac{3^{\frac{1}{2}} \Delta}{2\sigma},$$

where Δ is Gini's coefficient of mean difference defined by (2.24) and Exercise 2.9. Hence show that $\Delta \leqslant 2\sigma/3^{\frac{1}{2}}$.

In particular, show that for a normal distribution

$$C_n = \left\{ \left(\frac{n-1}{n+1}\right) \frac{3}{\pi} \right\}^{\frac{1}{2}},$$

so that

$$C = \lim_{n \to \infty} C_n = (3/\pi)^{\frac{1}{2}} \doteqdot 0 \cdot 98. \qquad \text{(Stuart, 1954c)}$$

31.14 Use (a) the theorem that the correlation between an efficient estimator and another estimator is the square root of the estimating efficiency of the latter (cf. (17.61)), (b) the relation between estimating efficiency and ARE given in **25.13,** (c) Daniels theorem of Exercise 31.3, and (d) the last result of Exercise 31.13 to establish the results for the ARE of the rank correlation coefficient r_s (and hence also t) as a test of independence (31.48) and as a test of randomness (31.70); and also to establish the ARE of Wilcoxon's rank-sum test against normal alternatives (31.117).

(Stuart, 1954c)

31.15 Obtain the variance (31.105) of Wilcoxon's test statistic by considering it as the sum of a sample of n_1 integers drawn from the finite population formed by the first n positive integers.

(Kruskal and Wallis, 1952–1953)

31.16 In **31.56** show for the two-sample Wilcoxon test statistic U that whatever the parent distribution F_1, F_2,
$$E(U) = n_1 n_2 p,$$
$$\operatorname{var} U = O(N^3),$$
where N stands indifferently for n_1, n_2, n. Hence, as $n_1, n_2 \to \infty$ with n_1/n_2 fixed, show that the test is consistent if $p \ne \tfrac{1}{2}$.

(Pitman, 1948)

31.17 Show that for the logistic distribution of Exercise 17.5, the ARE of the Wilcoxon test compared to "Student's" t-test for a location shift is $\pi^2/9$.

(This with the results of Exercise 17.5 and **25.13** implies that the Wilcoxon test is asymptotically best. Capon (1961) shows that this is generally so for locally most powerful rank tests, which the Wilcoxon is in this case.)

31.18 For the distribution
$$dF = \exp(-x) x^{p-1} dx / \Gamma(p), \qquad 0 \le x \le \infty, \qquad p > \tfrac{1}{2},$$
show that the ARE of the Wilcoxon test compared to the "Student's" t-test for a shift in location between two samples is
$$A_{U,t} = \frac{3p}{2^{4(p-1)} \{(2p-1) B(p,p)\}^2},$$
a monotone decreasing function of p. Verify that $A_{U,t} > 1 \cdot 25$ for $p \le 3$. Show that as $p \to \tfrac{1}{2}$, $A_{U,t} \to \infty$, and that as $p \to \infty$, $A_{U,t} \to 3/\pi$, agreeing with (31.117).

31.19. Show that the H-test of **31.71** reduces when $k = 2$ to the Wilcoxon test with critical region equally shared between the tails of the test statistic.

31.20 Using the result of **25.15** concerning the ARE of two test statistics with limiting non-central χ^2 distributions with equal degrees of freedom and only the non-central parameter a function of the distance from H_0, establish that the k-sample H-test of **31.71** has ARE, compared to the standard F-test in the normal case, equal to (31.115).

(cf. Andrews, 1954)

31.21 Show that in testing $\rho = 0$ for a bivariate normal population, the sample correlation coefficient r gives UMPU tests against one- and two-sided alternatives.

31.22 Show that Wilcoxon's test has ARE of 1 compared with "Student's" t-test against a location-shift for rectangular alternatives.

(Pitman, 1948)

31.23 Using (31.115), (31.136) and (31.140), show that the ARE of the Wilcoxon test compared to the normal scores test for location-shift alternatives is

$$A_{U, c_1} = 12 \left[\int \{f_1(x)\}^2 dx \Big/ \int \frac{\{f_1(x)\}^2 dx}{\phi\{J[F_1(x)]\}} \right]^2$$

and hence that

$$0 \leqslant A_{U, c_1} < \frac{6}{\pi}$$

(Hodges and Lehmann (1961) show that both equalities can be attained. See also Gastwirth (1970).)

31.24 Show that the intervals (31.95) based on the two-sample Wilcoxon statistic U at (31.98) are $\{D(U_\alpha), D(n_1 n_2 + 1 - U_\alpha)\}$, where Prob $\{U < U_\alpha \mid H_0\} = \tfrac{1}{2}\alpha$ and $D(r)$ is the rth smallest of the $n_1 n_2$ differences $(x_{1i} - x_{2j})$.   (Lehmann, 1963)

31.25 Show that the Wilcoxon test of symmetry, discussed in **31.79**, yields distribution-free confidence intervals for the location constant of a symmetric distribution $\{A(W_\alpha), A(\tfrac{1}{2}n(n-1)+1-W_\alpha)\}$ where W_α is the lower critical value of the "equal-tails" test and $A(r)$ is the rth smallest of the $\tfrac{1}{2}n(n-1)$ averages $\tfrac{1}{2}(x_i + x_j)$, $i < j = 1, 2, \ldots, n$.   (Lehmann, 1963)

31.26 In Exercise 20.19, show that $H_0 : \sigma_1 = \sigma_2$ is logically equivalent to $H_0 : (x+y)$ is independent of $(x-y)$, and hence that Exercise 31.21 provides UMPU tests for H_0 against $H_1 : \sigma_1 = \theta \sigma_2$.

CHAPTER 32

SOME USES OF ORDER-STATISTICS

32.1 In Chapter 31 we found that simple and remarkably efficient permutation tests of certain non-parametric hypotheses are obtained by the use of ranks, reflecting the order-relationships among the observations. In this chapter we first discuss the uses to which the order-statistics themselves can be put in providing distribution-free procedures for the non-parametric problems (5) and (6) listed in **31.13**. We then go on to consider uses of order-statistics in other (parametric) situations. The reader is reminded that the general distribution theory of order-statistics was discussed in Chapter 14, and that the theory of minimum-variance unbiassed estimation of location and scale parameters by linear functions of the order statistics was given in Chapter 19. A valuable general review of the literature of order-statistics was given by Wilks (1948), whose extensive bibliography is supplemented by the later one of F. N. David and Johnson (1956). Expositions of many branches of the theory, with extensive tables, are given in Sarhan and Greenberg (1962). See also the monograph by H. A. David (1970). Uses of order-statistics in testing and estimation are discussed and tabulated in the two volumes by Harter (1969). A review of the theory of "spacings" (differences between successive order-statistics) is given by Pyke (1965).

Sign test for quantiles

32.2 The so-called Sign test for the value of a quantile of a continuous distribution seems to have been the first distribution-free test ever used,[*] but the modern interest in it dates from the work of Cochran (1937).

Suppose that the parent d.f. is $F(x)$ and that
$$F(X_p) = p \qquad (32.1)$$
so that X_p is the p-quantile of the distribution, i.e. the value below which $100p$ per cent of the distribution lies. For any p, $0 < p < 1$, the value X_p is a location constant of the distribution. We wish to test the hypothesis
$$H_0 : X_p = x_0, \qquad (32.2)$$
where x_0 is some specified value. (If we take x_0 as our origin of measurement for convenience, we wish to test whether X_p is zero.)

32.3 If we have a sample of n observations, we know that the sample distribution function will converge in probability to the parent d.f. Let us, then, observe the relationship between the order-statistics $x_{(1)}, x_{(2)}, \ldots, x_{(n)}$ and the hypothetical value of

---

[*] Todhunter (1865) refers to its use in simple form by John Arbuthnot (Physician to Queen Anne, and formerly a mathematics teacher) to support *An Argument for Divine Providence taken from the constant Regularity observ'd in the Births of both Sexes* (1710–1712); Arbuthnot was a well-known wit and the author of the satire *The Art of Political Lying*.

X_p to be tested. We simply count how many of the sample observations fall below x_0, i.e. the statistic

$$S = n S_n(x_0) \qquad (32.3)$$

where $S_n(x)$ is the sample distribution function defined at (30.97). S counts the number of positive signs among the differences $(x_0 - x_i)$, and hence the test based on S is called the Sign test.(*) The distribution of S is at once seen to be binomial, for S is the sum of n independent observations on a 0–1 variable $h(x_0 - x)$ defined as at (31.36) with

$$P\{h(x_0 - x) = 1\} = P\{x < x_0\} = P,$$

say. The hypothesis (32.2) reduces to

$$H_0 : P = p, \qquad (32.4)$$

and we are simply testing the value of the binomial parameter P. We may wish to consider either one- or two-sided alternatives to (32.4).

If we specify nothing further about the parent d.f. $F(x)$, it is obvious intuitively that we cannot improve on S as a test statistic, and we find from binomial theory (cf. Exercise 22.2 and **23.31**) that for the one-sided alternative $H_1 : P > p$, the critical region consisting of large values of S is UMP, while for the two-sided alternative $H_2 : P \neq p$) a two-tailed critical region is UMPU.

In the most important case in practice, when $p = \frac{1}{2}$ and we are testing the *median* of the parent distribution, we have a symmetrical binomial distribution for S, and the UMPU critical region against H_2 is the equal-tails one.

A formal proof of these results is given by Lehmann (1959).

32.4 For small sample size n, therefore, tables of the binomial distribution are sufficient both to determine the size of the Sign test and to determine its power against any particular alternative value of P, and thus its power function for alternatives H_1 or H_2. As n increases, the tendency of the binomial distribution to normality enables us to say that $(S - nP)/\{nP(1-P)\}^{\frac{1}{2}}$ has a standardized normal distribution. If we use a continuity correction as in **31.80** for the discreteness of S, this amounts to replacing $|S - nP|$ by $|S - nP| - \frac{1}{2}$ in carrying out the test.

In the case of the median, when we are testing $P = \frac{1}{2}$, the tendency to normality is so rapid that special tables are hardly required at all, since we need only compare the value of

$$(|S - \tfrac{1}{2}n| - \tfrac{1}{2})/(\tfrac{1}{2}n^{\frac{1}{2}}) \qquad (32.5)$$

with the appropriate standardized normal deviate. Cochran (1937) gives exact critical values for $n \leqslant 50$ and test size $\alpha = 0{\cdot}05$, Dixon and Mood (1946) for $n \leqslant 100$ and $\alpha = 0{\cdot}25$, $0{\cdot}1$, $0{\cdot}05$ and $0{\cdot}01$, and MacKinnon (1964) for $n = 1\,(1)\,1000$ and $\alpha = 0{\cdot}50, 0{\cdot}10, 0{\cdot}05, 0{\cdot}02, 0{\cdot}01$ and $0{\cdot}001$.

(*) Because of the continuity of the parent d.f., the event $x_i = x_0$ can only occur with probability zero. If such "ties" occur in practice, the most powerful procedure is to ignore these observations for the purposes of the test, as was shown by Hemelrijk (1952)—cf. **31.81**.

Power of the Sign test for the median

32.5 The approximate power of the Sign test is also easily ascertained by use of the normal approximation. Neglecting the continuity correction of **32.4**, since this is small in large samples, we see that the critical region for the one-tailed test of $P = \frac{1}{2}$ against $P > \frac{1}{2}$ is

$$S \geqslant \tfrac{1}{2}n + d_\alpha \tfrac{1}{2} n^{\frac{1}{2}}$$

where d_α is the appropriate normal deviate for a test of size α. The power function is therefore approximately

$$Q_1(P) = \int_{\frac{1}{2}n+\frac{1}{2}d_\alpha n^{1/2}}^{\infty} \{2\pi n P(1-P)\}^{-\frac{1}{2}} \exp\left\{-\tfrac{1}{2}\frac{(u-nP)^2}{nP(1-P)}\right\} du$$

$$= \int_{\frac{n^{1/2}(\frac{1}{2}-P)+\frac{1}{2}d_\alpha}{\{P(1-P)\}^{1/2}}}^{\infty} (2\pi)^{-\frac{1}{2}} \exp(-\tfrac{1}{2}t^2)\, dt$$

$$= G\left\{\frac{n^{\frac{1}{2}}(P-\tfrac{1}{2}) - \tfrac{1}{2}d_\alpha}{[P(1-P)]^{\frac{1}{2}}}\right\}, \qquad (32.6)$$

where $G\{x\}$ is the normal d.f. From (32.6), it is immediate that as $n \to \infty$ the power $\to 1$ for any $P > \tfrac{1}{2}$, so that the test is consistent. The power function of the two-sided "equal-tails" test with critical region

$$|S - \tfrac{1}{2}n| \geqslant d_{\frac{1}{2}\alpha}\tfrac{1}{2}n^{\frac{1}{2}}$$

is similarly seen to be

$$Q_2(P) = G\left\{\frac{n^{\frac{1}{2}}(P-\tfrac{1}{2}) - \tfrac{1}{2}d_{\frac{1}{2}\alpha}}{[P(1-P)]^{\frac{1}{2}}}\right\} + G\left\{\frac{n^{\frac{1}{2}}(\tfrac{1}{2}-P) - \tfrac{1}{2}d_{\frac{1}{2}\alpha}}{[P(1-P)]^{\frac{1}{2}}}\right\}, \qquad (32.7)$$

which tends to 1 for any $P \neq \tfrac{1}{2}$ as $n \to \infty$. This establishes the consistency of the two-sided test against general alternatives.

Dixon (1953b) tabulates the power of the two-sided Sign test for test sizes $\alpha \leqslant 0.05$, $\alpha \leqslant 0.01$, n ranging from 5 to 100 and $P = 0.05\,(0.05)\,0.95$. MacStewart (1941) gives a table of the minimum sample size required to attain given power against given values of P. Gibbons (1964) examines the effect of non-normality on the power of the one-sided Sign test.

The Sign test in the symmetrical case

32.6 The power functions (32.6) and (32.7) are expressed in terms of the alternative hypothesis value of P. If we now wish to consider the efficiency of the Sign test in particular situations, we must particularize the distribution further. If we return to the original formulation (32.2) of the hypothesis, and restrict ourselves to the case of the median $X_{0.5} = M$, say, we wish to test

$$H_0 : M = M_0. \qquad (32.8)$$

If the parent distribution function is $F(x)$ as before and the f.f. is $f(x)$, we have for the value of P

$$P = F(M_0) = \int_{-\infty}^{M_0} f(x)\, dx. \qquad (32.9)$$

Suppose that we are interested in the relative efficiency of the Sign test where the parent F is known to be symmetrical, so that its mean and median M coincide. We may test the hypothesis (32.8) in this situation using as test statistic $\bar{x}$, the sample mean. If F has finite variance σ^2, $\bar{x}$ is asymptotically normal with mean M and variance σ^2/n, and in large samples it is equivalent to the " Student's " statistic

$$t = \frac{\bar{x} - M}{s/(n-1)^{\frac{1}{2}}}$$

where s^2 is the sample variance. For $\bar{x}$, we have

$$\frac{\partial}{\partial M} E(\bar{x} | M) = 1,$$

$$\text{var}(\bar{x} | M) = \sigma^2/n,$$

so that
$$\frac{\left\{\frac{\partial}{\partial M} E(\bar{x} | M)\right\}^2}{\text{var}(\bar{x} | M)} = \frac{n}{\sigma^2}. \tag{32.10}$$

For the Sign test statistic, we find it convenient first to measure from M as origin, so that

$$E(S | M) = nP = n \int_{-\infty}^{M_0 - M} f(x) \, dx,$$

and thus
$$\left\{\frac{\partial E(S | M)}{\partial M}\right\}_{M = M_0} = -nf(0).$$

Also,
$$\text{var}(S | M_0) = \tfrac{1}{4} n.$$

Thus, transferring back to the natural origin,

$$\frac{\left\{\frac{\partial E(S | M)}{\partial M}\right\}^2_{M = M_0}}{\text{var}(S | M_0)} = 4n\{f(M)\}^2. \tag{32.11}$$

From (32.10), (32.11) and (25.27) we find for the efficiency of the Sign test

$$A_{S, \bar{x}} = 4\sigma^2 \{f(M)\}^2, \tag{32.12}$$

a result due to Pitman.

32.7 There is clearly no non-zero lower bound to (32.12), as there was to the ARE for the Wilcoxon and normal scores tests in Chapter 31, since we may have the median ordinate $f(M) = 0$. In the normal case, $f(M) = (2\pi\sigma^2)^{-\frac{1}{2}}$, so (32.12) takes the value $2/\pi$. Since we are here testing symmetry about M_0, we may use the Wilcoxon test, as indicated in **31.79**, with ARE $3/\pi$ in the normal case and always exceeding 0·864. There is thus little except simplicity to recommend the use of the Sign test as a test of symmetry about a specified median : it is more efficient to test the sample mean in such a situation. The Sign test is useful when we wish to test for the median without the symmetry assumption.

> Dixon (1953b) tabulates the power efficiency of the two-sided Sign test in the normal case (and gives references to earlier work, notably by J. Walsh). He shows that the relative efficiency (i.e. the reciprocal of the ratio of sample sizes required by the Sign test and the " Student's " t-test to attain equal power for tests of equal size and against the same alternative—cf. **25.2**) decreases as any one of the sample size, test size, or the distance of P from $\tfrac{1}{2}$ increases.

s

Witting (1960) uses an Edgeworth expansion to order n^{-2} and shows that this second-order approximation gives results for the ARE little different from (32.12).

We can obtain a test of symmetry about an *unknown* mean by replacing M_0 in (32.8) by its estimator $\bar{x}$. Gastwirth (1971) shows that this modified Sign test statistic is not distribution-free and is far from robust.

Distribution-free confidence intervals for quantiles

32.8 The joint distribution of the order-statistics depends very directly upon the parent d.f. (cf., e.g., (14.1) and (14.2)) and therefore point estimation of the parent quantiles by order statistics is not distribution-free. Remarkably enough, however, pairs of order-statistics may be used to set distribution-free confidence intervals for any parent quantile.

Consider the pair of order-statistics $x_{(r)}$ and $x_{(s)}$, $r < s$, in a sample of n observations from the continuous d.f. $F(x)$. (14.2) gives the joint distribution of $F_r = F(x_{(r)})$ and $F_s = F(x_{(s)})$ as

$$dG_{r,s} = \frac{F_r^{r-1}(F_s - F_r)^{s-r-1}(1-F_s)^{n-s} dF_r dF_s}{B(r, s-r) B(s, n-s+1)}. \tag{32.13}$$

X_p, the p-quantile of $F(x)$, is defined by (32.1). For the probability that the interval $(x_{(r)}, x_{(s)})$ covers X_p, we have

$$P\{x_{(r)} \leqslant X_p \leqslant x_{(s)}\} = P\{x_{(r)} \leqslant X_p\} - P\{x_{(s)} < X_p | x_{(r)} \leqslant X_p\}. \tag{32.14}$$

Now since $x_{(r)} \leqslant x_{(s)}$, we can *only* have $x_{(s)} < X_p$ if $x_{(r)} < X_p$. Thus

$$P\{x_{(s)} < X_p | x_{(r)} \leqslant X_p\} = P\{x_{(s)} \leqslant X_p\}, \tag{32.15}$$

the equality on the right having zero probability because of the continuity of F. (32.14–15) give

$$P\{x_{(r)} \leqslant X_p \leqslant x_{(s)}\} = P\{x_{(r)} \leqslant X_p\} - P\{x_{(s)} \leqslant X_p\}. \tag{32.16}$$

Since $F(x)$ is monotone non-decreasing in x, the event $x_{(j)} \leqslant X_p$ is equivalent to $F(x_{(j)}) \leqslant F(X_p) = p$ by (32.1). Thus (32.16) is equivalent to

$$P\{x_{(r)} \leqslant X_p \leqslant x_{(s)}\} = P\{F(x_{(r)}) \leqslant p\} - P\{F(x_{(s)}) \leqslant p\}. \tag{32.17}$$

By (14.1), the variate $F(x_{(j)})$ has a Beta distribution with parameters j and $(n-j+1)$, whose distribution function at p is the Incomplete Beta Function $I_p(j, n-j+1)$. Thus finally (32.17) becomes

$$P\{x_{(r)} \leqslant X_p \leqslant x_{(s)}\} = I_p(r, n-r+1) - I_p(s, n-s+1) = 1-\alpha, \tag{32.18}$$

say, independent of the form of the continuous parent distribution F.

(32.18) is the interval analogue of the two-tailed Sign test for quantiles in **32.3**, for if that size-α test rejects H_0 at (32.4) when $S - np > a_\alpha$ or $< b_\alpha$, and we write $a_\alpha + np = r$, $b_\alpha + np = s - 1$, we have, using (32.3),

$$P\{r \leqslant nS_n(x_0) \leqslant s-1 | H_0\} = 1-\alpha,$$

which is another form of (32.18).

32.9 We see from (32.18) that the interval $(x_{(r)}, x_{(s)})$ covers the quantile X_p with a confidence coefficient which does not depend on $F(x)$ at all, and we thus have a distribution-free confidence interval for X_p. Since $I_p(a, b) = 1 - I_{1-p}(b, a)$, we may also write the confidence coefficient as

$$1 - \alpha = I_{1-p}(n-s+1, s) - I_{1-p}(n-r+1, r). \tag{32.19}$$

By the Incomplete Beta relation with the binomial expansion given in **5.7**, (32.19) may be expressed as

$$1 - \alpha = \left\{ \sum_{i=0}^{s-1} - \sum_{i=0}^{r-1} \right\} \binom{n}{i} p^i q^{n-i} = \sum_{i=r}^{s-1} \binom{n}{i} p^i q^{n-i}, \tag{32.20}$$

where $q = 1-p$. The confidence coefficient is therefore the sum of the terms in the binomial $(q+p)^n$ from the $(r+1)$th to the sth inclusive.

If we choose a pair of symmetrically placed order-statistics we have $s = n-r+1$, and find in (32.18–20)

$$1 - \alpha = I_p(r, n-r+1) - I_p(n-r+1, r)$$
$$= 1 - \{I_{1-p}(n-r+1, r) + I_p(n-r+1, r)\}, \tag{32.21}$$

$$= \sum_{i=r}^{n-r} \binom{n}{i} p^i q^{n-i}, \tag{32.22}$$

so that the confidence coefficient is the sum of the central $(n-2r+1)$ terms of the binomial, r terms at each end being omitted.

For any values of r and n, the confidence coefficient attaching to the interval $(x_{(r)}, x_{(n-r+1)})$ may be calculated from (32.21–2), if necessary using the *Tables of the Incomplete Beta Function*. The tables of the binomial distribution listed in **5.7** may also be used. Exercise 32.4 gives the reader an opportunity to practise the computation.

> Tukey and Scheffé (1945) show that if the parent distribution is discrete, the confidence intervals above cover X_p with probability $\geq 1-\alpha$.

32.10 In the special case of the parent median $X_{0\cdot 5}$, (32.21–22) reduce to

$$1 - \alpha = 1 - 2I_{0\cdot 5}(n-r+1, r) = 2^{-n} \sum_{i=r}^{n-r} \binom{n}{i}, \tag{32.23}$$

a particularly simple form. This confidence interval procedure for the median was first proposed by W. R. Thompson (1936). (32.23) is the interval analogue of the equal-tails Sign test for the median in **32.3-4**.

MacKinnon (1964) gives tables of r for $n = 1\,(1)\,1000$ and α as little as possible below 0·50, 0·10, 0·05, 0·02, 0·01 and 0·001. Van der Parren (1970) gives similar tables for $n = 3\,(1)\,150$ and $\alpha = 0\cdot 3,\ 0\cdot 2,\ 0\cdot 1,\ 0\cdot 05,\ 0\cdot 02$ and $0\cdot 01$, together with the exact value of $\alpha/2$ in each case.

Distribution-free tolerance intervals

32.11 In **20.37** we discussed the problem of finding tolerance intervals for a normal d.f. Suppose now that we require such intervals without making assumptions beyond continuity on the underlying distributional form. We require to calculate a randomly varying interval (l, u) such that

$$P\left\{\int_l^u f(x)\,dx \geq \gamma\right\} = \beta, \tag{32.24}$$

where $f(x)$ is the unknown continuous frequency function. It is not obvious that such a distribution-free procedure is possible, but Wilks (1941, 1942) showed that the order-statistics $x_{(r)}, x_{(s)}$, provide distribution-free tolerance intervals, and Robbins (1944) showed that *only* the order-statistics do so.

If we write $l = x_{(r)}$, $u = x_{(s)}$ in (32.24), we may rewrite it

$$P[\{F(x_{(s)}) - F(x_{(r)})\} \geq \gamma] = \beta. \tag{32.25}$$

We may obtain the exact distribution of the random variable $F(x_{(s)}) - F(x_{(r)})$ from (32.13) by the transformation $y = F(x_{(s)}) - F(x_{(r)})$, $z = F(x_{(r)})$, with Jacobian 1. (32.13) becomes

$$dH_{y,z} = \frac{z^{r-1} y^{s-r-1} (1-y-z)^{n-s} dy\,dz}{B(r, s-r) B(s, n-s+1)}, \qquad 0 \leq y+z \leq 1. \tag{32.26}$$

In (32.26) we integrate out z over its range $(0, 1-y)$, obtaining for the marginal distribution of y

$$dG_{s-r} = \frac{y^{s-r-1}\,dy}{B(r,s-r) B(s,n-s+1)} \int_0^{1-y} z^{r-1}(1-y-z)^{n-s}\,dz. \tag{32.27}$$

We put $z = (1-y)t$, reducing (32.27) to

$$\begin{aligned}
dG_{s-r} &= \frac{y^{s-r-1}(1-y)^{n-s+r}\,dy}{B(r,s-r) B(s,n-s+1)} \int_0^1 t^{r-1}(1-t)^{n-s}\,dt \\
&= y^{s-r-1}(1-y)^{n-s+r}\,dy\,\frac{B(r, n-s+1)}{B(r,s-r) B(s,n-s+1)} \\
&= \frac{y^{s-r-1}(1-y)^{n-s+r}\,dy}{B(s-r, n-s+r+1)}, \qquad 0 \leq y \leq 1.
\end{aligned} \tag{32.28}$$

Thus $y = F(x_{(s)}) - F(x_{(r)})$ is distributed as a Beta variate of the first kind. If we put $r = 0$ in (32.28) and interpret $F(x_{(0)})$ as zero (so that $x_{(0)} = -\infty$), (32.28) reduces to (14.1), with s written for r.

32.12 From (32.28), we see that (32.25) becomes

$$P\{y \geq \gamma\} = \int_\gamma^1 \frac{y^{s-r-1}(1-y)^{n-s+r}\,dy}{B(s-r, n-s+r+1)} = \beta, \tag{32.29}$$

which we may rewrite in terms of the Incomplete Beta Function as

$$P\{F(x_{(s)}) - F(x_{(r)}) \geq \gamma\} = 1 - I_\gamma(s-r, n-s+r+1) = \beta. \tag{32.30}$$

The relationship (32.30) for the distribution-free tolerance interval $(x_{(r)}, x_{(s)})$ contains five quantities: γ (the minimum proportion of $F(x)$ it is desired to cover), β (the probability with which we desire to do this), the sample size n, and the order-statistics' positions in the sample, r and s. Given any four of these, we can solve (32.30) for the fifth. In practice, β and γ are usually fixed at levels required by the problem, and r and s symmetrically chosen, so that $s = n-r+1$. (32.30) then reduces to

$$I_\gamma(n-2r+1, 2r) = 1-\beta. \tag{32.31}$$

The left-hand side of (32.31) is a monotone increasing function of n, and for any fixed β, γ, r we can choose n large enough so that (32.31) is satisfied. In practice, we must choose n as the nearest integer above the solution of (32.31). If $r = 1$, so that the extreme values in the sample are being used, (32.31) reduces to

$$I_\gamma(n-1, 2) = 1-\beta, \tag{32.32}$$

which gives the probability β with which the range of the sample of n observations covers at least a proportion γ of the parent d.f.

> The solution of (32.30) (and of its special cases (32.31–32)) has to be carried out numerically with the aid of the *Tables of the Incomplete Beta Function*, or equivalently (cf. 5.7) of the binomial d.f. Murphy (1948) gives graphs of γ as a function of n for $\beta = 0.90, 0.95$ and 0.99 and $r + (n-s+1) = 1\,(1)\,6\,(2)\,10\,(5)\,30\,(10)\,60\,(20)\,100$; these are exact for $n \leqslant 100$, and approximate up to $n = 500$.

Example 32.1

We consider the numerical solution of (32.32) for n. It may be rewritten

$$1-\beta = \frac{1}{B(n-1, 2)} \int_0^\gamma y^{n-2}(1-y)\,dy = n(n-1)\left\{\frac{\gamma^{n-1}}{n-1} - \frac{\gamma^n}{n}\right\}$$

$$= n\gamma^{n-1} - (n-1)\gamma^n. \tag{32.33}$$

For the values of β, γ which are required in practice (0.90 or larger, usually), n is so large that we may write (32.33) approximately as

$$1-\beta = n\gamma^{n-1}(1-\gamma),$$

$$\gamma = \left\{\left(\frac{1-\beta}{1-\gamma}\right)\frac{1}{n}\right\}^{1/(n-1)} \tag{32.34}$$

or

$$\log n + (n-1)\log \gamma = \log\{(1-\beta)/(1-\gamma)\}. \tag{32.35}$$

The derivative of the left-hand side of (32.35) with respect to n is $(1/n) + \log \gamma$ and for large n the left-hand side of (32.35) is a monotone decreasing function of n. Thus we may guess a trial value of n, compare the left with the (fixed) right-hand side of (32.35), and increase (decrease) n if the left (right) is greater. The value of n satisfying the approximation (32.35) will be somewhat too large to satisfy the exact relationship (32.33), since a positive term γ^n was dropped from the right of the latter, and we may safely use (32.35) unadjusted. Alternatively, we may put the solution of (32.35) into (32.33) and adjust to obtain the correct value.

Example 32.2

We illustrate Example 32.1 with a particular computation. Let us put $\beta = \gamma = 0.99$. (32.35) is then

$$\log n + (n-1)\log 0.99 = 0,$$

the right-hand side, of course, being zero whenever $\beta = \gamma$. We may use logs to base 10, since the adjustment to natural logs cancels through (32.35). Thus we have to solve

$$\log_{10} n - 0{\cdot}00436\,(n-1) = 0.$$

We first guess $n = 1000$. This makes the left-hand side negative, so we reduce n to 500, which makes it positive. We then progress iteratively as follows:

n	$\log_{10} n$	$0{\cdot}00436\,(n-1)$
1000	3	4·36
500	2·6990	2·18
700	2·8451	3·05
650	2·8129	2·83
600	2·7782	2·61
640	2·8062	2·79
645	2·8096	2·81

We now put the value $n = 645$ into the exact (32.33). Its right-hand side is

$$645\,(0{\cdot}99)^{644} - 644\,(0{\cdot}99)^{645} = 1{\cdot}004 - 0{\cdot}992 = 0{\cdot}012.$$

Its left-hand side is $1 - \beta = 0{\cdot}01$, so the agreement is good and we may for all practical purposes take $n = 645$ in order to get a 99 per cent tolerance interval for 99 per cent of the parent d.f.

32.13 We have discussed only the simplest case of setting distribution-free tolerance intervals for a univariate continuous distribution. Extensions to multivariate tolerance regions, including the discontinuous case, have been made by Wald (1943b), Scheffé and Tukey (1945), Tukey (1947, 1948), Fraser and Wormleighton (1951), Fraser (1951, 1953), and Kemperman (1956). Walsh (1962) considers symmetrical continuous distributions. Goodman and Madansky (1962) consider parameter-free and distribution-free tolerance limits for the exponential distribution. Guttman (1970) reviews the subject as a whole.

Scheffé and Tukey (1945) and Tukey (1948) show that if the parent distribution is discrete, the above tolerance intervals and regions have probability $\geqslant 1 - \alpha$.

Point estimation using order-statistics

32.14 As we remarked at the beginning of **31.8**, we cannot make distribution-free point estimates using the order-statistics because their joint distribution depends heavily upon the parent d.f. $F(x)$. We are now, therefore, re-entering the field of parametric problems, and we ask what uses can be made of the order-statistics in estimating parameters. These are two essentially different contexts in which the order-statistics may be considered:

(1) We may deliberately use functions of the order-statistics to estimate parameters, even though we know these estimating procedures are inefficient, because of the simplicity and rapidity of the computational procedures. (We discussed essentially this point in Example 17.13 in another connexion.) In 14.6–7 we gave some numerical values concerning the efficiencies of multiples of the sample median and mid-range as estimators of the mean of a normal population, and also of the sample interquartile range as an estimator of the normal population standard deviation. These three estimators are examples of easily computed inefficient statistics.

(2) For some reason, not all the sample members may be available for estimation

purposes, and we must perforce use an estimator which is a function of only some of them. The distinction between (1) and (2) thus essentially concerns the background of the problem. Formally, however, we may subsume (1) under (2) as the extreme case when the number of sample members not available is equal to zero.

Truncation and censoring

32.15 Before proceeding to any detail, we briefly discuss the circumstances in which sample members are not available. Suppose first that the underlying variate x simply cannot be observed in part or parts of its range. For example, if x is the distance from the centre of a vertical circular target of fixed radius R on a shooting range, we can only observe x for shots actually hitting the target. If we have no knowledge of how many shots were fired at the target (say, n) we simply have to accept the m values of x observed on the target as coming from a distribution ranging from 0 to R. We then say that the distribution of x is *truncated on the right* at R. Similarly, if we define y in this example as the distance of a shot from the vertical line through the centre of the target, y may range from $-R$ to $+R$ and its distribution is *doubly truncated*. Similarly, we may have a variate *truncated on the left* (e.g. if observations below a certain value are not recorded). Generally, a variate may be multiply truncated in several parts of its range simultaneously. A truncated variate differs in no essential way from any other but it is treated separately because its distribution is generated by an underlying untruncated variable, which may be of familiar form. Thus, in Exercise 17.27, we considered a Poisson distribution truncated on the left to exclude the zero frequency.

Tukey (1949) and W. L. Smith (1957) have shown that truncation at fixed points does not alter any properties of sufficiency and completeness possessed by a statistic.

32.16 On the other hand, consider our target example of **32.15** again, but now suppose that we know how many shots were fired at the target. We still only observe m values of x, all between O and R inclusive, but we know that $n-m = r$ further values of x exist, and that these will exceed R. In other words, we have observed the first m order-statistics $x_{(1)}, \ldots, x_{(m)}$ in a sample of size n. The sample of x is now said to be *censored on the right* at R. (Censoring is a property of the sample whereas truncation is a property of the distribution.) Similarly, we may have *censoring on the left* (e.g. in measuring the response to a certain stimulus, a certain minimum response may be necessary in order that measurement is possible at all) and *double censoring*, where the lowest r_1 and the highest r_2 of a sample of size n are not observed, only the other $m = n-(r_1+r_2)$ being available for estimation purposes.

There is a further distinction to be made in censored samples. In the examples we have mentioned, the censoring arose because the variate-values occurred outside some observable range; the censoring took place at certain fixed points. This is called Type I censoring. Type II censoring is said to occur when *a fixed proportion of the sample size n* is censored at the lower and/or upper ends of the range of x. In practice, Type II censoring often occurs when x, the variate under observation, is a time-period (e.g., the period to failure of a piece of equipment undergoing testing) and the experimental time available is limited. It may then be decided to stop when the first m of the n observations are to hand. It follows that Type II censoring is usually on the right of the variable.

From the theoretical point of view, the prime distinction between Type I and Type II censoring is that in the former case m (the number of observations) is a random variable, while in the latter case it is fixed in advance. The theory of Type II censoring is correspondingly simpler.

Of course, single truncation or censoring is merely a special case of double truncation or censoring, where one terminal of the distribution is unrestricted, while an "ordinary" situation is, so to speak, the doubly extreme case when there is no restriction of any kind.

32.17 There is by now an extensive literature on problems of truncation and censoring. To give a detailed account of the subject would take too much space. We shall therefore summarize the results in sections **32.17-22**, leaving the reader who is interested in the subject to follow up the references. We classify estimation problems into three main groups.

(A) Maximum likelihood estimators

A solution to any of the problems may be obtained by ML estimation; the likelihood equations are usually soluble only by iterative methods. For example if a continuous variate with frequency function $f(x|\theta)$ is doubly truncated at known points a, b, with $a < b$, the LF if n observations are made is

$$L_T(x|\theta) = \prod_{i=1}^{n} f(x_i|\theta) \Big/ \left\{ \int_a^b f(x|\theta)\,dx \right\}^n, \qquad (32.36)$$

the denominator in (32.36) arising because the truncated variate has f.f.

$$f(x|\theta) \Big/ \int_a^b f(x|\theta)\,dx.$$

(32.36) can be maximized by the usual methods.

Consider now the same variate, doubly censored at the fixed points a, b, with r_1 small and r_2 large sample members unobserved. For this Type I censoring, the LF is

$$L_I(x|\theta) \propto \left\{ \int_{-\infty}^{a} f(x|\theta)\,dx \right\}^{r_1} \prod_{i=1}^{n-r_1-r_2} f(x_i|\theta) \left\{ \int_b^{\infty} f(x|\theta)\,dx \right\}^{r_2}, \qquad (32.37)$$

and r_1 and r_2 are, of course, random variables.

On the other hand, if the censoring is of Type II, with r_1 and r_2 fixed, the LF is

$$L_{II}(x|\theta) \propto \left\{ \int_{-\infty}^{x_{(r_1+1)}} f(x|\theta)\,dx \right\}^{r_1} \prod_{i=r_1+1}^{n-r_2} f(x_{(i)}|\theta) \left\{ \int_{x_{(n-r_2)}}^{\infty} f(x|\theta)\,dx \right\}^{r_2}. \qquad (32.38)$$

(32.37) and (32.38) are of exactly the same form. They differ in that the limits of integration are random variables in (32.38) but not in (32.37), and that r_1, r_2 are random variables in (32.37) but not in (32.38). Given a set of observations, however, the formal similarity permits the same methods of iteration to be used in obtaining the ML solutions. Moreover, as $n \to \infty$, the two types of censoring are asymptotically equivalent.

B. R. Rao (1958a) showed under regularity conditions that censoring always results in a loss of estimation efficiency, but that truncation need not do so. This is also true (Swamy (1962a)) if the observations are grouped. Cf. Exercise 32.26.

Halperin (1952a) showed, under regularity conditions similar to those of **18.16** and **18.26**, that the ML estimators of parameters from Type II censored samples are consistent, asymptotically normally distributed, and efficient—cf. Exercise 32.15.

Hartley (1958) gives a general method for iterative solution of likelihood equations for incomplete data (covering both truncation and censoring) from discrete distributions.

(B) Minimum variance unbiassed linear estimators

A second approach is to seek the linear function of the available order statistics which is unbiassed with minimum variance in estimating the parameter of interest. To do this, we use the method of LS applied to the ordered observations. We have already considered the theory when all observations are available in **19.18–21**, and this may be applied directly to truncated situations, provided that the expectation vector and dispersion matrix of the order-statistics are calculated for the truncated distribution itself and not for the underlying distribution upon which the truncation took place. The practical difficulty here is that this dispersion matrix is a function of the truncation points a, b, so that the MV unbiassed linear function will differ as a and b vary. There has been little or no work done in this field, presumably because of this difficulty.

When we come to censored samples, a difficulty persists for Type I censoring, since we do not know how many order-statistics will fall within the censoring limits (a, b). Thus an estimator must be defined separately for every value of r_1 and r_2 and its expectation and variance should be calculated over all possible values of r_1 and r_2 with the appropriate probability for each combination. Again, we know of no case where this has been done. However, for Type II censoring, the problem does not arise, since r_1 and r_2 are fixed in advance, and we always know which $(n-r_1-r_2)$ order-statistics will be available for estimation purposes. Given their expectations and dispersion matrix, we may apply the LS theory of **19.18–21** directly. Moreover, the expectations and the dispersion matrix of all n order-statistics need be calculated only once for each n. For each r_1, r_2 we may then select the $(n-r_1-r_2)$ expectations of the available observations and the submatrix which is their dispersion matrix.

Chernoff *et al.* (1967) prove general formulae for linearly combining functions of the order-statistics to estimate location and scale parameters fully efficiently in censored or uncensored samples. Cf. also Siddiqui and Butler (1969) for the multivariate case.

A number of authors have suggested simpler procedures to avoid the computational complexities of the ML and LS approaches. The most general results have been obtained by Blom (1958), who derived " nearly " unbiassed " nearly " efficient linear estimators, as did Plackett (1958), who showed that the ML estimators of location and scale parameters are asymptotically linear, and that the MV linear unbiassed estimators are asymptotically normally distributed and efficient. Thus, asymptotically at least, the two approaches draw together.

Gastwirth (1966) examines the relation between MV linear unbiassed estimators and the corresponding asymptotically most powerful rank tests.

32.18 We now briefly give an account of the results available for each of the principal distributions which have been studied from the standpoint of truncation and censoring; the numerical details are too extensive to be reproduced here.

The normal distribution

Swamy (1962b) shows that truncation always reduces efficiency when both mean and variance are estimated and (1963) usually also does so when the observations are grouped —cf. Grundy (1952). For single and double truncation, ML estimation has been discussed by Cohen (1950a, 1957) who gives graphs to aid iterative solution of ML equations; Cohen and Woodward (1953) give tables, while Hald (1949) and Halperin (1952b) give graphs, for ML estimation with single truncation. Iterative ML procedures for singly and doubly Type II censored samples are given by Harter and Moore (1966a) who review earlier work. The ML estimators tend to be somewhat more precise, especially when censoring is strongly asymmetric, than the MV unbiassed linear estimators studied by Sarhan and Greenberg (1956, 1958) whose book (1962) gives tables of the coefficients of these estimators for all combinations of tail censoring numbers r_1, r_2 when $n = 1(1)20$. The linearized ML estimators proposed by Plackett (1958) never have efficiency less than 99·98 per cent for $n = 10$. Dixon (1957) shows that for estimating the mean of the population the very simple "trimmed" estimator

$$t = \frac{1}{n-2} \sum_{i=2}^{n-1} x_{(i)}$$

never has efficiency less than 99 per cent for $n = 3(1)20$, and presumably for $n > 20$ also, while the mean of the "best two" observations (i.e. those whose mean is unbiassed with minimum variance) has efficiency falling slowly from 86·7 per cent at $n = 5$ to its asymptotic value of 81 per cent. The "best two" observations are approximately $x_{(0·27n)}$ and $x_{(0·73n)}$ (cf. Exercise 32.14). Similar simple estimates of the population standard deviation σ are given by unbiassed multiples of the statistic

$$u = \sum_i \{x_{(n-i+1)} - x_{(i)}\},$$

the summation containing 1, 2, 3, or 4 values of i. The best statistic of this type never has efficiency less than 96 per cent in estimating σ.

Dixon (1960) shows that if i observations are censored in each tail, the "Winsorized" estimator of the mean

$$m_w = \frac{1}{n} \left[\sum_{s=i+2}^{n-i-1} x_{(s)} + (i+1)(x_{(i+1)} + x_{(n-i)}) \right]$$

has at least 99·9 per cent efficiency compared with the MV unbiassed linear estimator, and that for single censoring of i observations (say, to the right) the similar estimator

$$m_a = \frac{1}{n+a-1} \left[\sum_{s=2}^{n-i-1} x_{(s)} + (i+1) x_{(n-i)} + a x_{(1)} \right],$$

with a chosen to make m_a unbiassed, is at least 96 per cent efficient. See also Tiku (1967).

> For some general results on the efficiency of trimmed and Winsorized estimators of the mean, for symmetric and symmetric unimodal distributions, cf. the papers cited in **31.9**.

Walsh (1950a) shows that estimation of a percentage point of a normal distribution by the appropriate order-statistic is very efficient (although the estimation procedure is actually valid for any continuous d.f.) for Type II single censoring when the great majority of the sample is censored.

Saw (1959) has shown that in singly Type II censored samples, the population mean can be estimated with asymptotic efficiency at least 94 per cent by a properly weighted combination of the observation nearest the censoring point (x_c) and the simple mean of the other observations, and the population standard deviation estimated with asymptotic efficiency 100 per cent by using the sum and the sum of squares of the other observations about x_c. Saw gives tables of the appropriate weights for $n \leqslant 20$. For Type I censored samples, Saw (1961) proposes simple linear estimators of high efficiency.

32.19 *The exponential distribution*

The distribution $f(x) = \exp\{-(x-\mu)/\sigma\}/\sigma$, $\mu \leqslant x \leqslant \infty$, has been studied very fully from the standpoint of truncation and censoring, the reason being its importance in studies of the durability of certain products, particularly electrical and electronic components. A very full bibliography of this field of *life testing* is given by Mendenhall (1958), supplemented by Govindarajulu (1964).

ML estimation of σ (with μ known) for single truncation or Type I censoring on the right is considered by Deemer and Votaw (1955)—cf. Exercise 32.16. Their results are generalized to censored samples from mixtures of several exponential distributions by Mendenhall and Hader (1958). For Type II censoring on the right, the ML estimator of σ is given by Epstein and Sobel (1953), and the estimator shown to be also the MV unbiassed linear estimator by Sarhan (1955)—cf. Exercises 32.17–18.

Sarhan and Greenberg (1957) give tables, for sample sizes up to 10, of the coefficients of the MV unbiassed linear estimators of σ alone, and of (μ, σ) jointly, for all combinations of Type II censoring in the tails. MV unbiassed estimators based on one or two order-statistics are given by Harter (1961b), Sarhan *et al.* (1963) and Siddiqui (1963); and those based on 3, 4 or 5 order-statistics by Kulldorff (1963) who gives some general theory. See also Laurent (1963). Saleh (1966) derives estimators based on k order-statistics.

32.20 *The Poisson distribution*

Cohen (1954) gives ML estimators and their asymptotic variances for singly and doubly truncated and (Type I) censored Poisson distributions, and discusses earlier, less general, work on this distribution. Cohen (1960b) gives tables and a chart for ML estimation when zero values are truncated. Tate and Goen (1958) obtain the MV unbiassed estimator when truncation is on the left, and, in the particular case when only zero values are truncated, compare it with the (biassed) ML estimator and a simple unbiassed estimator suggested by Plackett (1953)—cf. Exercises 32.20, 32.22–4.

Cohen (1960a) discusses ML estimation of the Poisson parameter and a parameter θ, when a proportion θ of the values "1" observed are misclassified as "0," and the same author (1960c) gives the ML estimation procedure when the zero values and (erroneously) some of the "1" values have been truncated.

32.21 *Other distributions*

For the Gamma distribution with three parameters

$$dF = \frac{1}{\Gamma(\beta)} \exp\{-\alpha(x-\mu)\}\{\alpha(x-\mu)\}^{\beta-1} d\{\alpha(x-\mu)\},$$

Chapman (1956) considers truncation on the right, and proposes simplified estimators of (α, β) with μ known and of (α, β, μ) jointly. Cohen (1950b) had considered estimation by the method of moments in the truncated case. Raj (1953) and Den Broeder (1955) considered censored and truncated situations, the latter paper being concerned with the estimation of α alone with restriction in either tail of the distribution. Wilk *et al.* (1962) give ML estimators of (α, β) with μ known, and of (α, β, μ) jointly, for Type II censoring on the right. Harter and Moore (1965, 1967a) consider ML estimation for doubly censored samples from the Gamma and from the Weibull distribution (14.54) (Volume 1), and (1967b) from the logistic distribution of Exercises 18.29 and 32.12. For logistic estimation using order-statistics, see Gupta and Gnanadesikan (1966), Gupta *et al.* (1967) and Raghunandanan and Srinivasan (1970).

Sarhan and Greenberg (1959) consider MV unbiassed linear estimation for rectangular distributions—cf. Exercise 32.25. Downton (1966a, b) and Harter and Moore (1968) consider the extreme-value distribution treated in Exercise 18.6. Govindarajulu (1966) does the same for the symmetrically censored double exponential f.f. Finney (1949a), Rider (1955), Sampford (1955), Wilkinson (1961) and S. M. Shah (1961) discuss singly truncated binomial and negative binomial distributions. S. M. Shah (1966) discusses the doubly truncated binomial.

Harter and Moore (1966b) consider local ML estimation for censored samples from the 3-parameter lognormal distribution (with unknown starting-point); the strict ML estimator is infinite—cf. Exercise 18.23. See also Tiku (1968).

Tests of hypotheses in censored samples

32.22 In distinction from the substantial body of work on estimation discussed in **32.17–21**, much less work has so far been done on hypothesis-testing problems for truncated and censored situations. Epstein and Sobel (1953) and Epstein (1954) discuss tests for censored exponential distributions; F. N. David and Johnson (1954, 1956) give various simple tests for censored normal samples based on sample medians and quantiles. Halperin (1961a) gives simple confidence intervals for singly censored exponential and normal samples.

Gehan (1965a, b) (cf. also Halperin (1960)) has extended the Wilcoxon test to Type I censored samples, and A. P. Basu (1967a) investigates a similar extension by M. Sobel for Type II censoring. See also Hettmansperger (1968). Gastwirth (1965) derives asymptotically most powerful rank tests for censored samples. A. P. Basu (1967b) generalizes the k-sample tests based on (31.150) and (31.151) for Type II censoring on the right. See also R. A. Shorack (1968) and Breslow (1970b).

Outlying observations

32.23 In the final sections of this chapter, we shall briefly discuss a problem which, at some time or other, faces every practical statistician, and perhaps, indeed, most practical scientists. The problem is to decide whether one or more of a set of observations has come from a different population from that generating the other observations; it is distinguished from the ordinary two-sample problem by the fact that we do not know in advance *which* of the set of observations may be from the discrepant population—if we did, of course, we could apply two-sample techniques which we have discussed in earlier chapters. In fact, we are concerned with whether " contamination " has taken place.

The setting in which the problem usually arises is that of a suspected instrumental or recording error; the scientist examines his data in a general way, and suspects that some (usually only one) of the observations are too extreme (high, or low, or both) to be consistent with the assumption that they have all been generated by the same parent. What is required is some objective method of deciding whether this suspicion is well-founded.

Because the scientist's suspicion is produced by the behaviour in the tails of his observed distribution, the " natural " test criteria which suggest themselves are based on the behaviour of the extreme order-statistics, and in particular on their deviation from some measure of location for the unsuspected observations; or (especially in the case where " high " and " low " errors are suspected) the sample range itself may be used as a test statistic. Thus, for example, Irwin (1925) investigated the distribution of $(x_{(p)} - x_{(p-1)})/\sigma$ in samples from a normal population (see also E. S. Pearson (1926) and Sillitto (1951)), and " Student " (1927) recommended the use of the range for testing outlying observations.

Since these very early discussions of the problem, a good deal of work has been done along the same lines, practically all of which considers only the case of a normal parent. Now it is clear that the distribution of extreme observations is sensitive to the parental distributional form (cf. Chapter 14), so that these procedures are very unlikely to be robust to departures from normality, but it is difficult in general to do other than take a normal parent—the same objection on grounds of non-robustness would lie for any other parent.

32.24 Ferguson (1961), following Dixon (1950), sets up two general alternative hypotheses. *Model A* (the location-shifts hypothesis) is that n independent normal observations x_i have common variance σ^2 and means $E(x_i) = \mu + \sigma \Delta a_{\nu_i}$ where the a's are known constants, not all equal, and $(\nu_1, \nu_2, \ldots, \nu_n)$ is an unknown permutation of the integers 1 to n. *Model B* (the scale-shifts hypothesis) is that the x_i are independent and normal with common mean μ and variances $V(x_i) = \sigma^2 \exp(\Delta a_{\nu_i})$. We test $H_0 : \Delta = 0$ in either model. Considering only tests invariant under location and scale changes, Ferguson (1961) shows that in Model A, with μ unknown, the locally most powerful test of H_0 against the one-sided $H_1 : \Delta > 0$ is based on $\sqrt{b_1}$, the skewness coefficient of the x's. Large values of $\sqrt{b_1}$ are critical if $k_3(a)$, the third k-statistic of the a's, is positive, and small values of $\sqrt{b_1}$ are critical if $k_3(a) < 0$. If (as occurs in problems of outliers), $(n-l)$ of the a's are zero and $l/n < \tfrac{1}{2}$, then $k_3(a) > 0$, so that large values of $\sqrt{b_1}$ are always critical if less than half the observations are outliers. For the two-sided $H_1' : \Delta \neq 0$, the test based on b_2, the kurtosis coefficient of the x's, is locally most powerful unbiassed, large or small values being critical according as $k_4(a) > 0$ or < 0.

If $(n-l)$ of the a's are zero, $k_4(a) > 0$ if $l/n \leqslant \cdot 21$, so large values of b_2 are always critical if less than 21 per cent of observations are outliers. For Model B, where only the one-sided $H_1 : \Delta > 0$ is relevant, small scale-shifts in outlier problems are always upward, and the locally most powerful test is based on large values of b_2 whatever the a's may be (so that any number of outliers is permissible here).

However, " locally " most powerful means " near $\Delta = 0$ ", so evidence is still required of the efficiency of these tests for large shifts. Moreover, these tests have a formidable rival, since Paulson (1952) and Kudo (1956) have shown that for Model A, with at most one outlier, the probability of correctly rejecting the outlier is maximized if we use as our criterion the value of the *studentized extreme deviate* (cf. W. R. Thompson (1935), E. S. Pearson and Chandra Sekar (1936))

$$t_n = \frac{x_{(n)} - \bar{x}}{s} \quad \left(\text{or } t_1 = \frac{\bar{x} - x_{(1)}}{s}\right) \tag{32.39}$$

(where s^2 is the pooled estimate of σ^2 from all available observations) for one-sided alternatives $\Delta > 0$ or $\Delta < 0$ respectively. The same property holds for the *studentized maximum absolute deviate*

$$t_{\max} = \max \{t_n, t_1\} \tag{32.40}$$

for two-sided alternatives $\Delta \neq 0$ in Model A and also (Ferguson, 1961) for $\Delta > 0$ in Model B.

Ferguson (1961) carried out sampling experiments on Model A based on 25,000 random normal deviates successively assembled into samples of size $n = 5(5)25$, with one outlier differing from the other observations by $\sigma(\sigma)15\sigma$. Of the one-sided tests, t_n behaved slightly better than $\sqrt{b_1}$, while b_2 and $t_{\max}$ differed little as two-sided tests.

The Biometrika Tables give $\cdot 95$ and $\cdot 99$ quantiles for $\sqrt{b_1}$ for $n \geqslant 25$ and of b_2 for $n \geqslant 200$. Ferguson (1961) estimates their quantiles by sampling experiments for $n = 5(5)25$. Quesenberry and David (1961) tabulate $t_1(t_n)$ and $t_{\max}$.

32.25 Dixon (1950, 1951) considered ratios of form $\frac{x_{(r)} - x_{(1)}}{x_{(n-s)} - x_{(1)}}$ as rejection criteria and conducted sampling experiments to examine their powers. When σ is known he found that the *standardized extreme deviate*

$$u_n = \frac{x_{(n)} - \bar{x}}{\sigma} \quad \left(\text{or } u_1 = \frac{\bar{x} - x_{(1)}}{\sigma}\right) \tag{32.41}$$

and the *standardized range*

$$w = \frac{x_{(n)} - x_{(1)}}{\sigma} \tag{32.42}$$

were about equally powerful. H. A. David *et al.* (1954) tabulate the percentiles of the *studentized range*

$$q = \frac{x_{(n)} - x_{(1)}}{s}. \tag{32.43}$$

(32.43) is also tabulated, in the simplified case where its numerator and denominator are derived from independent samples, in *The Biometrika Tables*, by Pachares (1959) and by Harter (1960). (32.39) is tabulated with the same simplification in *The Biometrika Tables*, by Nair (1948, 1952), H. A. David (1956), Pillai (1959) and by Pillai and Tienzo (1959)—see also H. A. David and Paulson (1965). (32.40) is similarly tabulated by Halperin *et al.* (1955). Anscombe (1960) investigated the effect of rejecting outliers on subsequent estimation, mainly when σ is known. Bliss *et al.* (1956) gave a range criterion for rejecting a single outlier among k normal samples of size n, with tables. Other criteria are discussed by Grubbs (1950) and Dixon (1950, 1951, 1953a).

Dixon (1962, Chapter 10H of Sarhan and Greenberg (1962)) gives an extensive review of the subject, including many of the tables referred to above, and some others.

An obvious way of increasing the robustness of tests and estimators to the presence of outliers is to base them upon the "central" part of the sample—e.g., the "trimmed" and "Winsorized" estimators of **32.18**, and the tests in **32.22**.

Non-normal situations

32.26 One of the few general methods of handling the problem of outlying observations is due to Darling (1952), who obtains an integral form for the c.f. of the distribution of

$$z_n = \frac{\sum_{i=1}^{n} x_i}{x_{(n)}} \qquad (32.44)$$

where the n observations x_i are identical independent *positive-valued* variates with a fully specified distribution. In particular cases, this c.f. may be inverted. Darling goes on to consider the case of χ^2 variates in detail. Here, we consider only the simpler case of rectangular variates, where Darling's result may be derived directly.

32.27 Suppose we have observations $x_1, x_2, \ldots, x_n$ rectangularly distributed on the interval $(0, \theta)$. Then we know from **17.40** that the largest observation $x_{(n)}$ is sufficient for θ, and from **23.12** that $x_{(n)}$ is a complete sufficient statistic. By the result of Exercise 23.7, therefore, any statistic whose distribution does not depend upon θ will be distributed independently of $x_{(n)}$. Now clearly z_n as defined at (32.44) is of degree zero in θ. Thus z_n is distributed independently of $x_{(n)}$, and the conditional distribution of z_n given $x_{(n)}$ is the same as its unconditional (marginal) distribution. But, given $x_{(n)}$, any $x_{(i)}$ ($i < n$) is uniformly distributed on the range $(0, x_{(n)})$. Thus $x_{(i)}/x_{(n)}$, given $x_{(n)}$, is uniformly distributed on the range $(0, 1)$ and we see from (32.44) that z_n is distributed exactly like the sum of $(n-1)$ independent rectangular variates on $(0, 1)$ plus the constant 1 ($= x_{(n)}/x_{(n)}$).

Since we have seen in Example 11.9 that the sum of n independent rectangular variates tends to normality (and is actually close to normality even for $n = 3$), it follows that z_n is asymptotically normally distributed with mean and variance exactly given by

$$E(z_n) = (n-1)\tfrac{1}{2} + 1 = \tfrac{1}{2}(n+1), \qquad \text{var } z_n = (n-1)\tfrac{1}{12}. \qquad (32.45)$$

Small values of z_n (corresponding to large values of $x_{(n)}$) form the critical region for the hypothesis that all n observations are identically distributed against the alternative that the largest of them comes from an "outlying" distribution.

A. P. Basu (1965) discusses outliers for the exponential distribution.

32.28 We must now refer to the possibility of using distribution-free methods to solve the "outlier" problem without specific distributional assumptions. It is clear that, if the extreme observations are under suspicion, this would automatically stultify any attempt to use an ordinary two-sample test based on rank order, such as were discussed in Chapter 31, for this problem. However, if we are prepared to make the assumption of symmetry in the (continuous) parent distribution, we are in a position to do something, for we may then compare the behaviour of the observations in the suspected "tail" of the observed distribution with the behaviour in the other "tail" which is supposed to be well behaved. Thus, for large n, we may consider the absolute deviations from the sample mean (or median) of the k largest and k smallest observations, rank these $2k$ values, and use a symmetry test to decide whether they may be regarded as homogeneous. The test will be approximate, since the centre of symmetry is unknown and we estimate it by the sample mean or median, but otherwise this is simply an application of a test of symmetry (cf. **31.78–9**) to the tails of the distribution. If n is reasonably large, and k large enough to give a reasonable choice of test size α, the procedure should be sensitive enough for practical purposes.

Essentially similar, but more complicated, distribution-free tests of whether a group of 4 or more observations are to be regarded as "outliers" have been proposed by Walsh (1950b).

32.29 Finally, we observe that if distribution-free methods of testing and estimation are used, they will generally be less affected by the presence of outliers than are methods based upon distributional assumptions, for they use order properties, rather than metric properties, of the observations, as we saw in Chapter 31.

EXERCISES

32.1 For a frequency function $f(x)$ which takes the largest value at the median M, show that the ARE of the Sign test compared to "Student's" t-test, given at (32.12), is never less than $\frac{1}{3}$, and attains this value when $f(x)$ is rectangular.

(Hodges and Lehmann, 1956)

32.2 Show that if n independent observations come from the same continuous distribution $F(x)$, any symmetric function of them is distributed independently of any function of their rank order. Hence show that the Sign test and a rank correlation test may be used in combination to test the hypothesis that $F(x)$ has median M_0 against the alternative that *either* the observations are identically distributed with median $\neq M_0$ *or* the median trends upwards (or downwards) for each succeeding observation.   (Savage, 1957)

32.3 Obtain the result (32.12) for the ARE of the Sign test for symmetry from the efficiency of the sample median relative to the sample mean in *estimating* the mean of a symmetrical distribution (cf. **25.13**).

32.4 In setting confidence intervals for the median of a continuous distribution using the symmetrically spaced order-statistics $x_{(r)}$ and $x_{(n-r+1)}$, show from (32.23) that for $n = 30$, the values of r shown below give the confidence coefficients shown:

r	$1-\alpha$	r	$1-\alpha$
8	0·995	12	0·80
9	0·98	13	0·64
10	0·96	14	0·42
11	0·90	15	0·14

32.5 Show that in testing the hypothesis $H_0 : \theta = \theta_0$ against the alternative $H_1 : \theta = \theta_1 > \theta_0$ for a sample of n observations from
$$dF = \tfrac{1}{2}\exp\{-|x-\theta|\}dx, \quad -\infty \leqslant x \leqslant \infty,$$
the test based on $\left[\dfrac{\partial \log L}{\partial \theta}\right]_{\theta=\theta_0}$ is equivalent to the Sign test, which is therefore a locally most powerful one-sided test by **22.18**.

(cf. Lehmann, 1959)

32.6 Show from Example 32.1 that the range of a sample of size $n = 100$ from a continuous distribution has probability exceeding 0·95 per cent covering at least 95 per cent of the parent d.f., but that if we wish to cover at least 99 per cent with probability 0·95, n must be about 475 or more.

32.7 Show from Example 32.1 that if we wish to find a distribution-free tolerance interval for a proportion γ of a continuous d.f., with probability β near 1, a small increase in β from β_1 to β_2 requires an increase in sample size from n_1 to $n_2 = n_1 \left(\dfrac{1-\beta_2}{1-\beta_1}\right)$ approximately.

32.8 $F(x|\theta)$ is any continuous d.f., $F(a|\theta) = 0$, $F(b|\theta) = 1$, and values $\lambda_0, \lambda_1, \ldots, \lambda_{n+1}$ are defined by $F(\lambda_i|\theta_0) = \dfrac{i}{n+1}$, where θ_0 is the true value of θ. Consider a sample of n observations, and divide the sample space into $(n+1)^n$ parts, with probabilities $P_r, r = 1, 2, \ldots, (n+1)^n$, by the planes $x_i = \lambda_j, i = 1, 2, \ldots n; j = 0, 1, \ldots,$

$n+1$. If t is any asymptotically unbiassed estimator of θ, and t_r its value at some arbitrarily chosen point in the rth of the $(n+1)^n$ parts of the sample space, show that, as $n \to \infty$,

$$\operatorname{var} t \sim \sum_r (t_r - \theta)^2 P_r \geq \left[n^2 \sum_{i=0}^{n} \left\{ \left(\frac{\partial F}{\partial \theta} \right)_{\lambda_{i+1}} - \left(\frac{\partial F}{\partial \theta} \right)_{\lambda_i} \right\}^2 \right]^{-1}.$$

(Blom, 1958)

32.9 Show that if $\partial F/\partial \theta$ is a continuous function of x within the range of x and tends to zero at its extremities, and the integral $E\left(\dfrac{\partial \log f}{\partial \theta}\right)^2 > 0$ exists, where f is the f.f. $\dfrac{\partial F(x|\theta)}{\partial x}$, then the asymptotic lower bound to the variance of an unbiassed estimator given in Exercise 32.8 reduces to the MVB (17.24).

32.10 Show that in estimating the mean of a rectangular distribution of known range, the result of Exercise 32.8 gives an asymptotic bound $(2n^2)^{-1}$ for the variance, and that this is attained by the sample midrange $t = \frac{1}{2}(x_{(1)} + x_{(n)})$.

32.11 Show that if $\partial F/\partial \theta$, considered as a function of x, has a denumerable number of discontinuities, and $E\left(\dfrac{\partial \log f}{\partial \theta}\right)^2$ exists, the asymptotic variance bound of Exercise 32.8 is of order n^{-2}.

(Blom, 1958)

32.12 For samples of size n from the logistic distribution
$$dF = e^{-x} dx/(1 + e^{-x})^2, \quad -\infty \leq x \leq \infty,$$
show that the c.f. of the rth order-statistic $x_{(r)}$ is

$$\phi_{r,n}(t) = \frac{\Gamma(r+it)\Gamma(n-r+1-it)}{\Gamma(r)\Gamma(n-r+1)},$$

so that, from (16.29), $\frac{1}{2}\left(x_{(r)} - \log \dfrac{n-r+1}{r} \right)$ is distributed in Fisher's z form (16.26) with $\nu_1 = 2r$, $\nu_2 = 2(n-r+1)$; and hence that the cumulants of $x_{(r)}$ are, for $r > \frac{1}{2}(n+1)$,

$$\kappa_1 = \sum_{s=n-r+1}^{r-1} \frac{1}{s}, \quad \kappa_2 = \frac{\pi^2}{3} - \left(\sum_{s=1}^{r-1} \frac{1}{s^2} + \sum_{s=1}^{n-r} \frac{1}{s^2} \right),$$

$$\kappa_3 = 2 \sum_{s=n-r+1}^{r-1} \frac{1}{s^3}, \quad \kappa_4 = \frac{2\pi^4}{15} - 6\left(\sum_{s=1}^{r-1} \frac{1}{s^4} + \sum_{s=1}^{n-r} \frac{1}{s^4} \right).$$

Show that $\phi_{r+1, n+1}(t) = \left(1 + \dfrac{it}{r} \right) \phi_{r,n}(t)$, so that $\mu_{k,r,n} = E(x_{(r)}^k | n)$ satisfies $\mu_{k, r+1, n+1} = \mu_{k,r,n} + \dfrac{k}{r} \mu_{k-1, r, n}$.

(Plackett, 1958; cf. B. K. Shah, 1970. A. Birnbaum and Dudman (1963) table the means and standard deviations for $n = 1\,(1)\,10\,(5)\,20, 50, 100$; S. S. Gupta and Shah (1965) table the first four moments, together with selected quantiles of the distributions, for all the order-statistics when $n = 1\,(1)\,10$, and quantiles only for the extreme and central order-statistics for $n = 11 (1) 25$. Tarter and Clark (1965) give formulae for the covariances of logistic order-statistics, which are tabulated for $n \leqslant 10$ by B. K. Shah (1966).)

32.13 A continuous distribution has d.f. $F(x)$ and ξ is defined by $F(\xi) = p$. Show by expanding x in a Taylor series about the value $E\{x_{(r)}\}$ that in samples of size n, as $n \to \infty$ with $r = [np]$,

$$E(x_{(r)}) = p + O(n^{-1})$$

and that

$$F[E\{x_{(r+1)}\}] - F[E\{x_{(r)}\}] = \frac{1}{n} + O(n^{-2}). \qquad \text{(Plackett, 1958)}$$

32.14 Show that in samples from a normal population, if we estimate the mean μ by

$$t = \tfrac{1}{2}(x_{(i)} + x_{(n-i+1)}),$$

we obtain minimum variance asymptotically by choosing $i = 0 \cdot 2703 n$. Similarly, if we estimate σ by

$$s = \frac{1}{k}(x_{(n-i+1)} - x_{(i)}),$$

show that we obtain minimum variance when $i = 0 \cdot 0692 n$.

(Benson, 1949; the results were first given by Karl Pearson in 1920—cf. also Moore (1956))

32.15 Show (cf. (32.37–38)) that censoring has the effect of attaching a discrete probability to the original parent distribution $f(x|\theta)$ in each range where censoring takes place. Hence show that, e.g. for Type I censoring above a fixed point x_0, the asymptotic variance of the ML estimator of θ is

$$\operatorname{var} \hat{\theta} \sim 1 \bigg/ \bigg[n \bigg\{ \int_{-\infty}^{x_0} \left(\frac{\partial \log f}{\partial \theta} \right)^2 f \, dx + P \left(\frac{\partial \log P}{\partial \theta} \right)^2 \bigg\} \bigg],$$

where

$$P = \int_{x_0}^{\infty} f \, dx. \qquad \text{(cf. Halperin (1952a))}$$

32.16 Show that for a sample of $(n+m)$ observations from

$$dF = \theta e^{-\theta x}, \qquad 0 \leqslant x \leqslant \infty;\ \theta > 0,$$

n of which are measured between 0 and a fixed value x_0, and m of which are Type I censored, having values greater than x_0, the ML estimator of θ is

$$\hat{\theta} = n \bigg/ \bigg\{ \sum_{i=1}^{n} x_{(i)} + m x_0 \bigg\}$$

and that

$$\operatorname{var} \hat{\theta} \sim \frac{\theta^2}{n} \bigg/ \{1 - \exp(-\theta x_0)\},$$

so that the asymptotic variance of $\hat{\theta}$ is a monotonic decreasing function of x_0.

(Deemer and Votaw, 1955)

32.17 In Exercise 32.16, show that if the censoring is Type II, with a fixed number m of the sample order-statistics censored on the right, show that the ML estimator of $\lambda = 1/\theta$ is

$$\hat{\lambda} = \frac{1}{n} \bigg\{ \sum_{i=1}^{n} x_{(i)} + m x_{(n)} \bigg\}$$

and that its asymptotic variance is

$$\operatorname{var} \hat{\lambda} \sim \lambda^2/n. \qquad \text{(Epstein and Sobel, 1953)}$$

32.18 In Exercise 32.17, show that the variance of $\hat{\lambda}$ is exact for any value of n and that $\hat{\lambda}$ is the MV unbiassed linear estimator of λ. (Sarhan, 1955)

32.19 Show that if a Poisson distribution with parameter θ is truncated on the right, there exists no unbiassed estimator of θ. (Tate and Goen, 1958)

32.20 If values $\leq a$ are removed by truncation, show that the parameter θ of a Poisson distribution may be estimated unbiassedly by

$$t = \frac{1}{n} \sum_{r=a+2}^{\infty} r n_r,$$

where n_r is the frequency observed at the value r. Show that

$$\operatorname{var} t = \frac{\theta}{n}\left\{1 + \frac{\theta^{a+1}}{a!\left(e^\theta - \sum_{r=0}^{a} \frac{\theta^r}{r!}\right)}\right\}$$

and that an unbiassed estimator of this variance is

$$\widehat{\operatorname{var} t} = \frac{1}{n}\left\{t + \frac{(a+1)(a+2)n_{a+2}}{n}\right\}$$

(Subrahmaniam (1965) who shows that t has efficiency ≥ 90 per cent compared to the ML estimator when $\theta \leq 1$. If $a = 0$, t is Plackett's (1953) estimator and has efficiency ≥ 95 per cent for all θ—cf. also J. Roy and Mitra (1957).)

32.21 For a sample of n observations, x_r, show that

$$\sum_{r=1}^{n} x_r^2 \equiv n\bar{x}^2 + \sum_{r=1}^{n-1} \frac{z_r^2}{r(r+1)},$$

where

$$z_r = r x_{(r+1)} - \sum_{s=1}^{r} x_{(s)}, \qquad r = 1, 2, \ldots, n-1,$$

and hence, applying the Helmert transformation of Example 11.3 to the n order-statistics in the form

$$y_r = z_r / \{r(r+1)\}^{\frac{1}{2}}, \qquad r = 1, 2, \ldots, n-1, \qquad y_n = n^{\frac{1}{2}} \bar{x},$$

show that in samples from a standardized normal population the joint frequency function of the z_r is

$$n^{\frac{1}{2}} (2\pi)^{-\frac{1}{2}(n-1)} \exp\left\{-\frac{1}{2} \sum_{r=1}^{n-1} z_r^2 / [r(r+1)]\right\}, \quad 0 \leq z_1 \leq z_2 \leq \ldots \leq z_{n-1} = n(x_{(n)} - \bar{x}).$$

Defining the functions

$$G_r(x) = \int_0^x \exp\left\{-\frac{1}{2}\frac{t^2}{r(r+1)}\right\} G_{r-1}(t)\, dt, \qquad G_0(x) \equiv 1,$$

show that the distribution function of $u = (x_n - \bar{x})$, say $P_n(u)$, satisfies the relationship

$$P_n(u) = n^{\frac{1}{2}} (2\pi)^{-\frac{1}{2}(n-1)} G_{n-1}(nu).$$

(Nair (1948); McKay (1935) had obtained an equivalent result)

32.22 In Exercise 32.20, show that the ML estimator is a root of

$$\theta / (1 - e^{-\theta}) = \bar{x}$$

where $\bar{x}$ is the observed mean, and that

$$\operatorname{var} \hat{\theta} \sim \frac{\theta}{n} \cdot (1 - e^{-\theta})^2 / \{1 - (\theta+1) e^{-\theta}\}.$$

Hence show that

$$\lim_{\theta \to 0} n \operatorname{var} \hat{\theta} = 2, \qquad \lim_{\theta \to \infty} n \operatorname{var} \hat{\theta} = 1,$$

and that $n \operatorname{var} \hat{\theta}$ never lies outside these limits. (cf. Cohen, 1960b)

32.23 Show that the ML estimator in Exercise 32.22 may be expressed as
$$\hat{\theta} = \bar{x} + \sum_{r=1}^{\infty} \frac{(-1)^r \bar{x}^r}{r!} \frac{d^{r-1}}{d\bar{x}^{r-1}} e^{-r\bar{x}} = \bar{x} - \sum_{r=1}^{\infty} \frac{r^{r-1}}{r!} (\bar{x} e^{-\bar{x}})^r.$$
(Irwin, 1959)

32.24 In Exercise 32.20, show that an estimate of the number of observations *including the missing values* is given by the ratio $(\Sigma x)^2 / \Sigma x(x-1)$, where the summations are over values from 1 to ∞. Hence show how θ may be estimated iteratively.
(Irwin, 1959; the method is due to A. G. McKendrick)

32.25 In Type II censored samples from the rectangular distribution $dF = dx/\sigma$, $\mu - \tfrac{1}{2}\sigma \leq x \leq \mu + \tfrac{1}{2}\sigma$, show from (32.38) that the extreme observed values $x_{(r_1+1)}$, $x_{(n-r_2)}$ are a pair of sufficient statistics for μ and σ. Hence show that the MV unbiassed linear estimators of the parameters are
$$m = \frac{(n-2r_2-1)x_{(r+1)1} + (n-2r_1-1)x_{(n-r_2)}}{2(n-r_1-r_2-1)},$$
$$s = \frac{n+1}{n-r_1-r_2-1}(x_{(n-r_2)} - x_{(r_1+1)}),$$
with variances
$$\operatorname{var} m = \frac{\sigma^2 \{(r_1+1)(n-2r_2-1) + (n-2r_1-1)(r_2+1)\}}{4(n+1)(n+2)(n-r_1-r_2-1)},$$
$$\operatorname{var} s = \frac{\sigma^2 (r_1+r_2+2)}{(n+2)(n-r_1-r_2-1)},$$
thus generalizing Example 19.10.
(Sarhan and Greenberg, 1959)

32.26 Referring to **18.16**, show that if the parent distribution $f(x \mid \theta)$ is truncated at points a below, b above, the reciprocal of the asymptotic variance of $\hat{\theta}$ becomes
$$R_{a,b}^2(\theta) = n \left[\left\{ \int_a^b \frac{1}{f} \left(\frac{\partial f}{\partial \theta} \right)^2 dx \Big/ \int_a^b f \, dx \right\} - \left\{ \left(\int_a^b \frac{\partial f}{\partial \theta} dx \right)^2 \Big/ \left(\int_a^b f \, dx \right)^2 \right\} \right],$$
tending to the previous definition (18.30) as $a \to -\infty$, $b \to +\infty$. Show that
$$R_{-\infty,\infty}^2(\theta) - R_{a,b}^2(\theta) \text{ is not necessarily positive, but that } R_{-\infty,\infty}^2(\theta) - \left(\int_a^b f \, dx \right) R_{a,b}(\theta) > 0$$
always, so that we may expect some increase in the asymptotic variance of the ML estimator unless truncation is very severe.
(cf. B. R. Rao (1958a))

CHAPTER 33

CATEGORIZED DATA

33.1 For most of this book, we have been concerned with the analysis of measured observations, but in Chapter 31 we investigated the use of rank order statistics in testing hypotheses; such statistics may be constructed from measured observations or, alternatively, may be obtained from the ranks of the observations if no measurements are available or even possible. In the present chapter, we shall discuss *categorized data*; by this we mean data which are presented in the form of frequencies falling into certain categories or classes. We have, of course, discussed the problems of grouping variate-values into classes, e.g. in connexion with the calculation of moments. There, however, the grouping was undertaken as a matter of computational convenience or necessity, and we were in any case largely concerned with univariate situations; here, we specifically confine ourselves to problems arising in connexion with the statistical relationships (whether of dependence or interdependence) between two or more " variables " expressed in categorized form. We have put quotes on the word " variables " because it is to be interpreted in the most general sense.

A categorized " variable " may simply be a convenient classification of a measurable variable into groups, in the manner already familiar to us. On the other hand, it may not be expressible in terms of an underlying measurable variable at all. For example, we may classify men by (a) their height, (b) their hair colour, (c) their favourite film actor; (a) is a categorization of a measurable variable, but (b) and (c) are not. There is a further distinction between (b) and (c), for hair colour itself may be expressed on an ordered scale, according to pigmentation, from light to dark; this is not so for (c). Although, of course, one could impose various types of classifications upon the film actors named, the actors are intrinsically not ordered in any way. We refer to (b) as an *ordered* classification or categorization, and (c) as an *unordered* one. As an extreme case of an unordered classification, we may consider a classification which is simply a labelling of different samples (which we wish to compare in respect of some other variable).

33.2 There is a further point to be borne in mind: on occasion, the two variables being investigated may simply be the same variable observed on two different occasions (e.g. before and after some event) or on two related samples (e.g. father and son, husband and wife, etc.). We shall refer to such a situation as one with *identical categorizations*. Identical categorizations may, of course, be of any of the types (a), (b) or (c) in **33.1**.

Association in 2 × 2 tables

33.3 Historically, a very large part of the literature on categorized variables has been concerned with the problems of measurement and testing of the interdependence of two such variables, or, as it is generally known, the problem of *association*. We leave aside entirely the problem of estimating interdependence in the case where the form of underlying measurable variables is known or assumed—this has been discussed for the bivariate normal case in **26.27–33**. In other words, we confine ourselves to non-parametric problems.

33.4 Consider in the first place a population classified according to the presence or absence of an attribute A. The simplest kind of problem in interdependence arises when there are two attributes A, B, and if we denote the absence of A by α and the absence of B by β, the numbers falling into the four possible sub-groups may, in an obvious notation, be represented by

	B	not-B	Totals
A	(AB)	$(A\beta)$	(A)
not-A	(αB)	$(\alpha\beta)$	(α)
Totals	(B)	(β)	n

(33.1)

We shall often write this 2×2 *table* (sometimes called a *fourfold table*) in a form which has already occurred at (26.58):

$$\begin{array}{cc|c} a & b & a+b \\ c & d & c+d \\ \hline a+c & b+d & n \end{array} \qquad (33.2)$$

If there is no association between A and B, that is to say, if the possession of A is irrelevant to the possession of B, there must be the same proportion of A's among the B's as among the not-B's. Thus, by definition, the attributes are *independent*(*) in this set of n observations if

$$\frac{a}{a+c} = \frac{b}{b+d} = \frac{a+b}{n}. \qquad (33.3)$$

It follows that

$$\left. \begin{array}{l} \dfrac{c}{a+c} = \dfrac{d}{b+d} = \dfrac{c+d}{n}, \\[6pt] \dfrac{a}{a+b} = \dfrac{c}{c+d} = \dfrac{a+c}{n}, \\[6pt] \dfrac{b}{a+b} = \dfrac{d}{c+d} = \dfrac{b+d}{n}. \end{array} \right\} \qquad (33.4)$$

(*) It would perhaps be better to use a neutral word like "unassociated" rather than "independent" to describe the relationship (33.3), since it does not imply (though it is implied by) the stochastic independence of numerical variables which may have generated the 2×2 table. The distinction we are making is precisely analogous to that between lack of correlation and independence—cf. **26.10**. However, historical usage is against us here, and in any case there is some danger of confusion between "unassociated" and "dissociated," to be defined at the end of **33.4**. We shall therefore continue to use "independence," as applied to categorized variables, to mean "lack of association."

(33.3) may be rewritten

$$a = \frac{(a+b)(a+c)}{n}. \quad (33.5)$$

If now, in any given table,

$$a > \frac{(a+b)(a+c)}{n}, \quad (33.6)$$

there are relatively more A's among the B's than among the not-B's, and we shall say that A and B are *positively associated*, or simply *associated*. *Per contra*, if

$$a < \frac{(a+b)(a+c)}{n} \quad (33.7)$$

we shall say that A and B are *negatively associated* or *dissociated*.

Example 33.1

The following table (Greenwood and Yule, 1915, *Proc. Roy. Soc. Medicine*, **8**, 113) shows 818 cases classified according to inoculation against cholera (attribute A) and freedom from attack (attribute B).

	Not attacked	Attacked	TOTALS
Inoculated	276	3	279
Not-inoculated	473	66	539
TOTALS	749	69	818

If the attributes were independent, the frequency in the inoculated-not-attacked class would be $\frac{279 \times 749}{818} = 255$. The observed frequency is greater than this and hence inoculation is positively associated with exemption from attack.

Measures of association

33.5 If we are to sum up the strength of association between two attributes in a single coefficient, it is natural to require that the limits of variation of the coefficient should be known, and that it shall take the central value or the lowest value of its range when there is no association ("independence"). We may always make location and scale changes in any coefficient to bring it within the range $(-1, +1)$; independence should then correspond to a value of zero for the coefficient. This convention has the advantage of agreeing with the properties of the product-moment correlation coefficient (cf. **26.9**). Alternatively, the range $(0, 1)$ may be preferred, zero being the value taken in the case of independence.

Another obvious desideratum in a measure of association is that it should increase as the relationship proceeds from dissociation to association. Consider the difference between observed and "independence" frequencies in the cell corresponding to (AB),

$$D = a - \frac{(a+b)(a+c)}{n} = \frac{ad - bc}{n}. \quad (33.8)$$

For constant marginal frequencies, it is evident that the difference in any cell between observed and "independence" frequencies is $\pm D$ and thus D determines uniquely the departure from independence. We thus require that our coefficient should increase with D.

33.6 Following Yule (1900, 1912) we define a coefficient of association, Q, by the equation

$$Q = \frac{ad-bc}{ad+bc} = \frac{nD}{ad+bc}. \qquad (33.9)$$

It is zero if the attributes are independent, for then $D = 0$. It can equal $+1$ only if $bc = 0$, in which case there is *complete association* (either all A's are B's or all B's are A's), and -1 only if $ad = 0$, in which case there is *complete dissociation*. Furthermore, Q increases with D, for if we write $e = bc/(ad)$, we have

$$Q = (1-e)/(1+e) = 2/(1+e) - 1,$$

so that

$$\frac{dQ}{de} = -\frac{2}{(1+e)^2} < 0,$$

and as $\frac{dD}{de}$ is also negative, $\frac{dQ}{dD}$ is positive.

Yule also proposed a so-called *coefficient of colligation*

$$Y = \frac{1 - \left(\frac{bc}{ad}\right)^{\frac{1}{2}}}{1 + \left(\frac{bc}{ad}\right)^{\frac{1}{2}}} = \frac{(ad)^{\frac{1}{2}} - (bc)^{\frac{1}{2}}}{(ad)^{\frac{1}{2}} + (bc)^{\frac{1}{2}}}, \qquad (33.10)$$

but it is easy to show that

$$Q = \frac{2Y}{1 + Y^2} \qquad (33.11)$$

and nothing much seems to be gained by the use of Y. It is easily seen to satisfy our conditions.

Yet a third coefficient, to which we shall return in **33.17** below, is

$$V = \frac{(ad - bc)}{\{(a+b)(a+c)(b+d)(c+d)\}^{\frac{1}{2}}}. \qquad (33.12)$$

This is evidently zero when $D = 0$ and increases with D. If $V^2 = 1$, we have

$$(a+b)(a+c)(b+d)(c+d) = (ad-bc)^2,$$

giving

$$4abcd + a^2(bc+bd+cd) + b^2(ac+ad+cd) + c^2(ab+ad+bd) + d^2(ac+ab+bc) = 0.$$

Since no frequency can be negative, this can only hold if at least two of a, b, c, d are zero. If the frequencies in the same row and column vanish, the case is purely nugatory. We have then only to consider $a = 0$, $d = 0$ or $b = 0$, $c = 0$. In the first case $V = -1$, in the second $V = +1$. It cannot lie outside these limits.

33.7 It will be observed that whereas $|V| = 1$ only if two frequencies in the 2×2 table vanish, $|Q|$ and $|Y|$ are unity if only one frequency vanishes. This raises a point in connexion with the definition of *complete* association. We shall say that association is complete if all A's are B's, notwithstanding that all B's are not A's. If all dumb men are deaf, there is complete association between dumbness and deafness, however many deaf men there are who are not dumb.(*) The coefficient V is unity only if all A's are B's and all B's are A's, a condition which we could, if so desired, describe as *absolute* association.

We must point out in this connexion that statistical association is different from association in the colloquial sense. In current speech we say that A and B are associated if they occur together fairly often; but in statistics they are associated only if A occurs relatively more frequently among the B's than among the not-B's. If 90 per cent of smokers have poor digestions, we cannot say that smoking and poor digestion are associated until it is shown that less than 90 per cent of non-smokers have poor digestions.

Standard errors of the coefficients

33.8 We now consider the 2×2 table as a sample, and derive the standard errors of the coefficients of **33.6** on the hypothesis of independence. We have, writing ∂ for the differential to avoid confusion with d,

$$\frac{\partial e}{e} = \frac{\partial b}{b} + \frac{\partial c}{c} - \frac{\partial a}{a} - \frac{\partial d}{d},$$

whence

$$\frac{\operatorname{var} e}{e^2} = \sum_u \frac{\operatorname{var} u}{u^2} + 2 \sum_{u,v} \left\{ \pm \frac{\operatorname{cov}(u,v)}{uv} \right\}, \qquad (33.13)$$

where u and v are to be summed over a, b, c, d. Using multinomial results typified by

$$\operatorname{var} a = \frac{a(n-a)}{n},$$

$$\operatorname{cov}(a,b) = -\frac{ab}{n},$$

we find, on substitution in (33.13),

$$\operatorname{var} e = e^2 \left(\frac{1}{a} + \frac{1}{b} + \frac{1}{c} + \frac{1}{d} \right). \qquad (33.14)$$

It is then easy to derive

$$\operatorname{var} Q = \frac{1}{4}(1-Q^2)^2 \left\{ \frac{1}{a} + \frac{1}{b} + \frac{1}{c} + \frac{1}{d} \right\}, \qquad (33.15)$$

$$\operatorname{var} Y = \frac{1}{16}(1-Y^2)^2 \left\{ \frac{1}{a} + \frac{1}{b} + \frac{1}{c} + \frac{1}{d} \right\}. \qquad (33.16)$$

(*) If this asymmetrical convention is followed, complete association between A and B is not, in general, the same as complete association between B and A.

The sampling variance of V may be found similarly, but involves rather more lengthy algebra. We have

$$\operatorname{var} V = \frac{1}{n}\left[1 - V^2 + (V + \tfrac{1}{2}V^3)\frac{(a-d)^2 - (b-c)^2}{\{(a+b)(a+c)(b+d)(c+d)\}^{\frac{1}{2}}} \right. $$
$$\left. - \frac{3}{4}V^2\left\{\frac{(a+b-c-d)^2}{(a+b)(c+d)} - \frac{(a+c-b-d)^2}{(a+c)(b+d)}\right\}\right]. \tag{33.17}$$

These formulae assume, as usual in large-sample theory, that the observed frequencies may be used instead of their expectations in the sampling variances.

Partial association

33.9 The coefficients described above measure the interdependence of two attributes in the statistical sense, but in order to help us to decide whether such dependence has any causal significance it is often necessary, just as in **27.1** for correlations, to consider association in sub-populations. Suppose, for example, that a positive association is noticed between inoculation and freedom from attack. It is tempting to infer that the inoculation confers exemption, but this is not necessarily so. It might be that the people who are inoculated are drawn largely from the richer classes, who live in better hygienic conditions and are therefore better equipped to resist attack or less exposed to risk. In other words, the association of A and B might be due to the association of both with a third attribute C (wealth). We therefore consider the association of A and B conditional upon C being fixed.

Associations in sub-populations are called *partial associations*. Analogously to (33.6), A and B are said to be positively associated in the population of C's if

$$(ABC) > \frac{(AC)(BC)}{(C)} \tag{33.18}$$

where (ABC) represents the number of members bearing the attributes A, B and C; and so on. We may also define coefficients of partial association, colligation, etc., such as

$$Q_{AB.C} = \frac{(ABC)(\alpha\beta C) - (A\beta C)(\alpha B C)}{(ABC)(\alpha\beta C) + (A\beta C)(\alpha B C)}, \tag{33.19}$$

which is derived from (33.9) by adding C to all the symbols representing the frequencies.

Example 33.2

The following example, though not in line with modern genetical thought, has a historical interest as showing some early attempts at discussing heredity in a quantitative way.

Galton's *Natural Inheritance* gives particulars, for 78 families containing not less than six brothers or sisters, of eye-colour in parent and child. Denoting a light-eyed child by A, a light-eyed parent by B and a light-eyed grandparent by C, we trace every possible line of descent and record whether a light-eyed child has light-eyed parent and grandparent, the number of such being denoted by (ABC) and so on. The symbol $(A\beta\gamma)$, for example, denotes the number of light-eyed children whose parents and grandparents have not-light eyes. The eight possible classes are

$(ABC) = 1928$	$(\alpha BC) = 303$
$(AB\gamma) = 596$	$(\alpha B\gamma) = 225$
$(A\beta C) = 552$	$(\alpha\beta C) = 395$
$(A\beta\gamma) = 508$	$(\alpha\beta\gamma) = 501.$

The first question we discuss is: does there exist any association between parent and offspring with regard to eye-colour? We consider both the grandparent–parent group (association of B's and C's) and the parent–child group (association of A's and B's).

The proportion of light-eyed among children of light-eyed parents is $(BC)/(C)$ = 2231/3178 = 70·2 per cent. That of light-eyed among children of not-light-eyed parents, $(B\gamma)/(\gamma)$, is 821/1830 = 44·9 per cent. Likewise $(AB)/(B)$ = 82·7 per cent and $(A\beta)/(\beta)$ = 54·2 per cent. Evidently there is some positive association in this set of observations between parent and offspring in regard to eye-colour.

Consider now the relationship between eye-colours of grandparents and grandchildren. The proportion of light-eyed among grand-children of light-eyed grandparents is $\frac{(AC)}{(C)} = \frac{2480}{3178}$ = 78·0 per cent. That among grand-children of not-light-eyed grandparents, $(A\gamma)/(\gamma)$ is 1104/1830 = 60·3 per cent.

Thus the association between eye-colour in grandparents and grand-children is also positive. In tabular form, the data are:

Attributes	A	α	Totals	Attributes	C	γ	Totals	Attributes	C	γ	Totals
B	2524	528	3052	B	2231	821	3052	A	2480	1104	3584
β	1060	896	1956	β	947	1009	1956	α	698	726	1424
Totals	3584	1424	5008		3178	1830	5008		3178	1830	5008

The coefficients of association and colligation Q and Y are

	Q	Y
Grandparents–parents	0·487	0·260
Parents–children	0·603	0·336
Grandparents–grand-children	0·401	0·209

Now the question arises: is the resemblance between grandparent and grandchild due merely to that between grandparent and parent, parent and child? To investigate this, we consider the associations of grandparent and grand-child in the sub-populations "parents light-eyed" and "parents not-light-eyed"; that is, the associations of A and C in B and β.

Among light-eyed parents, the proportion of light-eyed amongst grand-children of light-eyed grandparents = $\frac{(ABC)}{(BC)} = \frac{1938}{2231}$ = 86·4 per cent, while the proportion of light-eyed amongst grand-children of not-light-eyed grandparents = $\frac{(AB\gamma)}{(B\gamma)} = \frac{596}{821}$ = 72·6 per cent.

Among not-light-eyed parents, the proportion of light-eyed amongst the grand-children of light-eyed grandparents = $\frac{(A\beta C)}{(\beta C)} = \frac{552}{947}$ = 58·3 per cent, and the pro-

portion of light-eyed amongst the grand-children of not-light-eyed grandparents
$$= \frac{(A\beta\gamma)}{(\beta\gamma)} = \frac{508}{1009} = 50\cdot 3 \text{ per cent.}$$

In both cases, the partial association is well marked and positive. The association between grandparents and grand-children cannot, then, be due wholly to the associations between grandparents and parents, parents and children. This was interpreted to indicate the existence of ancestral heredity, as it is called, as well as parental heredity. The relevant tables are:

Table 33.1

Parents light-eyed

	Grandparents		
Grand-children	BC	$B\gamma$	TOTALS
AB	1928	596	2524
αB	303	225	428
TOTALS	2231	821	3052

Parents not-light-eyed

	Grandparents		
Grand-children	βC	$\beta\gamma$	TOTALS
$A\beta$	552	508	1060
$\alpha\beta$	395	501	896
TOTALS	947	1009	1956

The coefficients of association and colligation are:
$$Q_{AC.B} = 0\cdot 412, \qquad Q_{AC.\beta} = 0\cdot 159,$$
$$Y_{AC.B} = 0\cdot 216, \qquad Y_{AC.\beta} = 0\cdot 080.$$

33.10 If there are p different attributes under consideration, the number of partial associations can become very large, even for moderate p. For example, we can choose two in $\binom{p}{2}$ ways and consider their associations in all the possible sub-populations of the other $(p-2)$, which are seen to be 3^{p-2} in number. Thus there are $\binom{p}{2} 3^{p-2}$ associations.

One of the principal difficulties, in fact, in discussing data subject to multiple dichotomy (and, even more, multiple polytomy) is the sheer volume of the large number of tables which results.

One result in this connexion is worth noticing. We have, generalizing D in equation (33.8),

$$\begin{aligned}
D_{AB.C} + D_{AB.\gamma} &= \left\{(ABC) - \frac{(AC)(BC)}{(C)}\right\} + \left\{(AB\gamma) - \frac{(A\gamma)(B\gamma)}{(\gamma)}\right\} \\
&= (AB) - \frac{(A)(B)}{n} - \frac{n}{(C)(\gamma)} \left\{(AC)(BC) - \frac{(A)(C)(BC)}{n} \right. \\
&\quad \left. - \frac{(B)(C)(AC)}{n} + \frac{(A)(B)(C)^2}{n^2}\right\} \\
&= D_{AB} - \frac{n}{(C)(\gamma)} \left\{(AC) - \frac{(A)(C)}{n}\right\}\left\{(BC) - \frac{(B)(C)}{n}\right\} \\
&= D_{AB} - \frac{n}{(C)(\gamma)} D_{AC} D_{BC}.
\end{aligned} \qquad (33.20)$$

If, then, A and B are independent in both (C) and (γ), $D_{AB.C} = D_{AB.\gamma} = 0$ and (33.20) gives

$$D_{AB} = \frac{n}{(C)(\gamma)} D_{AC} D_{BC}, \qquad (33.21)$$

i.e., A and B are not independent in the population as a whole unless C is independent of A or B or both in that population. Compare (27.67) for partial correlation, where it follows that $\rho_{12\cdot 3} = 0$ only implies $\rho_{12} = 0$ if ρ_{13} or $\rho_{23} = 0$.

This result indicates that illusory associations may arise when two populations (C) and (γ) are amalgamated, or that real associations may be masked. If A and C, B and C, are associated, we have, from (33.20)

$$D_{AB} = \frac{n}{(C)(\gamma)} D_{AC} D_{BC} + D_{AB.C} + D_{AB.\gamma}, \qquad (33.22)$$

so that if A and B are associated positively in (C) and negatively in (γ), D_{AB} may be zero, that is to say, A and B may appear independent in the whole population.

Example 33.3

Consider the case in which some patients are treated for a disease and others not. If A denotes recovery and B denotes treatment, suppose that the frequencies in the 2×2 table are:

	B	β	Totals
A	100	200	300
α	50	100	150
Totals	150	300	450

Here $(AB) = 100 = \dfrac{(A)(B)}{n}$, so that the attributes are independent. So far as can be seen, treatment exerts no effect on recovery.

Denoting male sex by C and female sex by γ, suppose the frequencies among males and females are:

Males

	BC	βC	Totals
AC	80	100	180
αC	40	80	120
Totals	120	180	300

Females

	$B\gamma$	$\beta\gamma$	Totals
$A\gamma$	20	100	120
$\alpha\gamma$	10	20	30
Totals	30	120	150

In the male group we now have

$$Q_{AB.C} = \frac{(80 \times 80) - (100 \times 40)}{(80 \times 80) + (100 \times 40)} = 0.231,$$

and in the female group

$$Q_{AB.\gamma} = -0.429.$$

Thus treatment was positively associated with recovery among the males and negatively associated with it among the females. The apparent independence in the combined table is due to the cancelling of these associations.

More paradoxically, two tables may have associations of the same sign and yet, when merged, form a table with association of the opposite sign—the reader may like to experiment numerically to produce such a result.

Probabilistic interpretations of measures of association

33.11 The measures of association discussed in **33.5–10** were developed from 1900 onwards, and set out to summarize the strength of association in a single comprehensive coefficient. But, just as we saw in **26.10** for the correlation coefficient, it is not always reasonable to suppose that any single coefficient can do this adequately. Goodman and Kruskal (1954, 1959), in addition to giving a detailed discussion of the history of measures of association and a very full bibliography, make a powerful plea for choosing a measure which is interpretable for the purpose in hand. Most of their discussion is couched in terms of polytomized tables, i.e. tables with two or more categories in the row and column classifications, and we shall refer to their work in **33.35, 33.40** below when considering polytomies. Here, however, we remark that in a 2×2 table the coefficients Q and Y defined at (33.10–11) can be given operational interpretations, under certain conditions.

33.12 Consider the selection at random of two individuals from a population of n individuals classified into a table (33.1), so that each individual will fall into one of the four categories in the body of the table. Let us score for $i = 1, 2$,

$$a_i \begin{cases} = +1 \text{ if an individual possesses } A, \\ = 0 \text{ otherwise,} \end{cases}$$

$$b_i \begin{cases} = +1 \text{ if an individual possesses } B, \\ = 0 \text{ otherwise,} \end{cases}$$

and $a_0 = a_1 - a_2$, $b_0 = b_1 - b_2$.

Define the probabilities

$$\pi_s = P\{a_0 b_0 = 1\}, \qquad \pi_d = P\{a_0 b_0 = -1\}, \qquad \pi_t = P\{a_0 b_0 = 0\}.$$

Then the coefficient

$$\gamma = \frac{\pi_s - \pi_d}{1 - \pi_t} \tag{33.23}$$

is the probability $\pi_s/(1-\pi_t)$ that the two individuals selected from the population have their A- and B-categories different and in the same order (if we sense the table (33.1) from left to right and top to bottom) minus the probability $\pi_d/(1-\pi_t)$ that they have different A- and B-categories in opposite orders. Clearly, using the notation (33.2),

$$\gamma = (ad-bc)/(ad+bc) \equiv Q$$

as defined at (33.9). Q therefore has a direct probabilistic interpretation as above.

33.13 Similarly, consider choosing a single individual at random from the population, and suppose that we are asked (without prior knowledge) to guess whether it is A or not-A. The best estimate we can make is to guess that the individual comes from the larger of the frequencies (A), (α) in (33.1); the probability that this estimate is correct is, from (33.2),

$$\rho_{m.} = \frac{1}{n}\max(a+b, c+d).$$

Similarly, if we have to guess whether the individual is B or not-B, we guess the larger of (B), (β), with probability of success

$$\rho_{.m} = \frac{1}{n}\max(a+c, b+d).$$

Thus if we are asked to guess the A-category half of the time and the B-category the other half of the time, the probability of success is $\frac{1}{2}(\rho_{m.}+\rho_{.m})$ and the probability of error is

$$\pi_0 = 1 - \tfrac{1}{2}(\rho_{m.}+\rho_{.m}). \tag{33.24}$$

Suppose now that we *know* the individual's B-category and are asked to guess the A-category. The best guess is now the larger category *in the appropriate column* of the table, with probability of success equal to $\dfrac{\max(a,c)}{a+c}$ or $\dfrac{\max(b,d)}{b+d}$ in the respective columns. Since these columns will occur in random sampling with probabilities $(a+c)/n$, $(b+d)/n$ respectively, the overall probability of success in guessing the A-category given the B-category is

$$\pi_A = \frac{1}{n}\{\max(a,c) + \max(b,d)\}. \tag{33.25}$$

The overall probability of success in guessing the B-category given the A-category will similarly be

$$\pi_B = \frac{1}{n}\{\max(a,b) + \max(c,d)\}. \tag{33.26}$$

If, as in the no-information situation above, we had to guess the categories alternately, the probability of success would be the mean of (33.25) and (33.26) and the probability of error therefore would be

$$\pi_1 = 1 - \frac{1}{2n}\{\max(a,b) + \max(a,c) + \max(b,d) + \max(c,d)\}. \tag{33.27}$$

33.14 We now define the coefficient, from (33.24) and (33.27),

$$\lambda = \frac{\pi_0 - \pi_1}{\pi_0}, \tag{33.28}$$

which is the relative reduction in error-probability produced by knowledge of one category in predicting the other. Clearly, $\pi_0 \geqslant \pi_1$, so we have

$$0 \leqslant \lambda \leqslant 1. \tag{33.29}$$

Now, Yule (1912) suggested that the value of a "reasonable" measure of association should not be affected if each row and each column of the 2×2 table is separately multiplied through by an arbitrary positive constant. Using this invariance principle,

CATEGORIZED DATA

we multiply the table (33.2) by the constants:

First row: $(cd)^{\frac{1}{2}}/n$,　　First column: $(bd)^{\frac{1}{2}}/n$,
Second row: $(ab)^{\frac{1}{2}}/n$,　　Second column: $(ac)^{\frac{1}{2}}/n$,

and transform it (neglecting constants common to all four frequencies) to

$$\begin{array}{cc|c} (ad)^{\frac{1}{2}} & (bc)^{\frac{1}{2}} & \frac{1}{2}m \\ (bc)^{\frac{1}{2}} & (ad)^{\frac{1}{2}} & \frac{1}{2}m \\ \hline \frac{1}{2}m & \frac{1}{2}m & m \end{array} \qquad (33.30)$$

Further obvious multiplications show that any coefficient of association satisfying the invariance principle must be a function of the "cross-ratio" ad/bc alone. Q and Y of (33.9–10) obviously satisfy this condition, but V of (33.12) does not. Edwards (1963) derives this condition from other propositions.

Yule's invariance principle enables us to relate λ at (33.28) to Y. For the moment, suppose $ad > bc$. From (33.28), we then have for the transformed table (33.30),

$$\lambda = \frac{\frac{1}{2} - \left[1 - \frac{2}{m}(ad)^{\frac{1}{2}} \right]}{\frac{1}{2}}$$

$$= \frac{2(ad)^{\frac{1}{2}} - \frac{1}{2}m}{\frac{1}{2}m},$$

which since $(ad)^{\frac{1}{2}} + (bc)^{\frac{1}{2}} = \frac{1}{2}m$ becomes

$$\lambda = \frac{(ad)^{\frac{1}{2}} - (bc)^{\frac{1}{2}}}{(ad)^{\frac{1}{2}} + (bc)^{\frac{1}{2}}}. \qquad (33.31)$$

(33.31) is identical with the definition of Y at (33.10), but we have chosen its sign arbitrarily by taking $ad > bc$. Thus, generally,

$$\lambda \equiv |Y|, \qquad (33.32)$$

conferring a probabilistic interpretation upon the magnitude of Y.

Large-sample tests of independence in a 2 × 2 table

33.15 We now consider the observed frequencies in a 2×2 table to be a sample, and we suppose that in the parent population the true probabilities corresponding to the frequencies a, b, c, d are $p_{11}, p_{12}, p_{21}, p_{22}$ respectively. We write the probabilities

$$\begin{array}{cc|c} p_{11} & p_{12} & p_{1.} \\ p_{21} & p_{22} & p_{2.} \\ \hline p_{.1} & p_{.2} & 1 \end{array} \qquad (33.33)$$

with $p_1 = p_{11} + p_{12}$, and so forth. We suppose the observations drawn with replacement from the population (or, equivalently that the parent population is infinite).

We also rewrite the table (33.2) in the notationally symmetric form

$$\begin{array}{cc|c} n_{11} & n_{12} & n_{1.} \\ n_{21} & n_{22} & n_{2.} \\ \hline n_{.1} & n_{.2} & n \end{array}$$

The distribution of the sample frequencies is given by the multinomial whose general term is

$$L = \frac{n!}{n_{11}!\, n_{12}!\, n_{21}!\, n_{22}!} p_{11}^{n_{11}} p_{12}^{n_{12}} p_{21}^{n_{21}} p_{22}^{n_{22}}. \tag{33.34}$$

To estimate the p_{ij}, we find the Maximum Likelihood solutions for variations in the p_{ij} subject to $\Sigma p_{ij} = 1$. If λ is a Lagrange multiplier, this leads to

$$\frac{n_{11}}{p_{11}} - \lambda = 0 \quad \text{or} \quad n_{11} = \lambda p_{11}$$

and three similar equations. Summing these, we find $\lambda = n$ and the proportions p_{ij} are simply estimated by

$$\hat{p}_{11} = n_{11}/n \tag{33.35}$$

and three similar equations. This is as we should expect. The estimators are unbiassed.

We know, and have already used the fact in **33.8**, that the variances of the n_{ij} are typified by

$$\operatorname{var} n_{11} = n p_{11}(1 - p_{11})$$

and the covariances by

$$\operatorname{cov}(n_{11}, n_{12}) = -n p_{11} p_{12}.$$

These are exact results, and we also know (cf. Example 15.3) that in the limit the joint distribution of the n_{ij} tends to the multinormal with these variances and covariances. We may now also observe that the asymptotic multinormality follows from the fact that these are ML estimators and satisfy the conditions of **18.26**.

33.16 Now suppose we wish to test the hypothesis of independence in the 2×2 table, which is

$$H_0 : p_{11} p_{22} = p_{12} p_{21}. \tag{33.36}$$

This hypothesis is, of course, composite, imposing one constraint, and having two degrees of freedom. We allow p_{11} and p_{12} to vary and express p_{21} and p_{22} by

$$p_{21} = \frac{p_{11}(1 - p_{11} - p_{12})}{p_{11} + p_{12}}, \quad p_{22} = \frac{p_{12}(1 - p_{11} - p_{12})}{p_{11} + p_{12}}. \tag{33.37}$$

The logarithm of the Likelihood Function is therefore, neglecting constants,

$$\log L = n_{11} \log p_{11} + n_{12} \log p_{12} + n_{21} \log p_{21} + n_{22} \log p_{22}$$
$$= n_{11} \log p_{11} + n_{12} \log p_{12} + n_{21} \{\log p_{11} + \log(1 - p_{11} - p_{12}) - \log(p_{11} + p_{12})\}$$
$$\quad + n_{22} \{\log p_{12} + \log(1 - p_{11} - p_{12}) - \log(p_{11} + p_{12})\}$$
$$= n_{.1} \log p_{11} + n_{.2} \log p_{12} + n_{2.} \{\log(1 - p_{1.}) - \log p_{1.}\}.$$

To estimate the parameters, we put

$$0 = \frac{\partial \log L}{\partial p_{11}} = \frac{n_{.1}}{p_{11}} - n_{2.} \left\{ \frac{1}{1 - p_{1.}} + \frac{1}{p_{1.}} \right\} = \frac{n_{.1}}{p_{11}} - \frac{n_{2.}}{p_{1.}(1 - p_{1.})}, \tag{33.38}$$

$$0 = \frac{\partial \log L}{\partial p_{12}} = \frac{n_{.2}}{p_{12}} - \frac{n_{2.}}{p_{1.}(1 - p_{1.})}, \tag{33.39}$$

giving for the ML estimators under H_0

$$\hat{p}_{11} = \frac{n_{1.}}{n}\frac{n_{.1}}{n}, \quad \hat{p}_{12} = \frac{n_{1.}}{n}\frac{n_{.2}}{n}. \tag{33.40}$$

(33.37) gives analogous expressions for $\hat{p}_{21}$ and $\hat{p}_{22}$. Thus we estimate the cell probabilities from the products of the proportional marginal frequencies. This justifies the definition of association by comparison with those products in **33.4–5**.

Substituting these ML estimators into the LF, we have

$$L(n_{ij}|H_0, \hat{p}_{ij}) \propto (n_{1.}n_{.1})^{n_{11}}(n_{1.}n_{.2})^{n_{12}}(n_{2.}n_{.1})^{n_{21}}(n_{2.}n_{.2})^{n_{22}}/n^{2n}, \tag{33.41}$$

while the unconditional maximum of the LF is obtained by inserting the estimators (33.35) to obtain

$$L(n_{ij}|\hat{p}_{ij}) \propto n_{11}^{n_{11}} n_{12}^{n_{12}} n_{21}^{n_{21}} n_{22}^{n_{22}}/n^n. \tag{33.42}$$

(33.41–2) give for the LR test statistic

$$l = \left(\frac{n_{1.}n_{.1}}{n\,n_{11}}\right)^{n_{11}} \left(\frac{n_{1.}n_{.2}}{n\,n_{12}}\right)^{n_{12}} \left(\frac{n_{2.}n_{.1}}{n\,n_{21}}\right)^{n_{21}} \left(\frac{n_{2.}n_{.2}}{n\,n_{22}}\right)^{n_{22}}. \tag{33.43}$$

Writing $n\hat{p}_{ij} = n_{i.}n_{.j}/n = e_{ij}$, this becomes

$$l = \left(\frac{e_{11}}{n_{11}}\right)^{n_{11}} \left(\frac{e_{12}}{n_{12}}\right)^{n_{12}} \left(\frac{e_{21}}{n_{21}}\right)^{n_{21}} \left(\frac{e_{22}}{n_{22}}\right)^{n_{22}}. \tag{33.44}$$

33.17 The general result of **24.7** now shows that $-2\log l$ is asymptotically distributed as χ^2 with one degree of freedom. This is easily seen as follows. Writing $D_{ij} = n_{ij} - e_{ij}$ (cf. (33.8)), and expanding as far as $D^2 (= D_{ij}^2$, all i, j), we have

$$-2\log l = 2\sum_{i=1}^{2}\sum_{j=1}^{2} e_{ij}\left(1 + \frac{D_{ij}}{e_{ij}}\right)\left(\frac{D_{ij}}{e_{ij}} - \tfrac{1}{2}\frac{D^2}{e_{ij}^2}\right)$$

$$= D^2 \sum_i \sum_j \frac{1}{e_{ij}}. \tag{33.45}$$

(33.45) may be rewritten

$$-2\log l = \sum_i \sum_j \frac{(n_{ij}-e_{ij})^2}{e_{ij}} \equiv X^2. \tag{33.46}$$

We have thus demonstrated in a particular case the asymptotic equivalence of the LR and X^2 goodness-of-fit tests which we remarked in **30.5**. (33.46) could have been derived directly by observing that the composite H_0 implies a set of hypothetical frequencies e_{ij}, and that the test of independence amounts to testing the goodness-of-fit of the observations to these hypothetical frequencies. As in **30.10**, the number of degrees of freedom is the number of classes (4) minus 1 minus the number of parameters estimated (2), i.e. one.

We leave the reader to show that the X^2 statistic at (33.46) is identically equal to nV^2, where V is the measure of association defined at (33.12), and is therefore not a function of the cross-ratio. Fienberg and Gilbert (1970) demonstrate this geometrically.

Exact test of independence: models for the 2 × 2 table

33.18 The tests of independence derived in **33.15–17** are asymptotic in n, the sample size. Before we can devise exact tests of independence in 2×2 tables, we must consider some distinctions first made by Barnard (1947a, b) and E. S. Pearson (1947).

It will be recalled that the expected values in the cells of the 2×2 table on the hypothesis of independence of the two categorized variables are

$$e_{ij} = \frac{n_{i.} \, n_{.j}}{n}, \qquad i,j = 1,2, \tag{33.47}$$

depending only on the four marginal frequencies and upon the sample size, n. Since we are now concerned with exact arguments, we must explicitly take account of the manner in which the table was formed, and in particular of the manner in which the marginal frequencies arose. Even with n fixed, we still have three distinct possibilities in respect of the marginal frequencies. Both sets of marginal frequencies may be random variables, as in the case where a sample of size n is taken from a bivariate distribution and subsequently classified into a *double dichotomy*. Alternatively, one set of marginal frequencies may be fixed, because that classification is merely a labelling of two samples (say, Men and Women) which are to be compared in respect of the other classification (say, numbers infected and not-infected by a particular disease). If the numbers in the two samples are fixed in advance (e.g. if it is decided to examine fixed numbers of Men and of Women for the disease), we have one fixed set of marginal frequencies and one set variable. When we are thus comparing two (or more) samples in respect of a characteristic, we often refer to it as a test of *homogeneity* in two (or k) samples.

Finally, we have the third possibility, in which both sets of marginal frequencies are fixed in advance. This is much rarer in practice than the other two cases, and the reader may like to try to construct a situation to which this applies before reading on. The classical example of such a situation (cf. Fisher (1935a)) concerns a psycho-physical experiment: a human subject is tested n times to verify his power of recognition of two objects (e.g. the taste of butter and of margarine). Each object is presented a certain number of times (not necessarily the same number for the two objects) *and the subject is informed of these numbers*. The subject, if rational, then makes the marginal frequencies of his assertions (" butter " or " margarine ") coincide with the known frequency with which they have been presented to him.

Example 33.4

To make the distinction of **33.18** clearer, let us discuss some actual examples. The table in Example 33.1 above is certainly not of our last type, with both sets of marginal frequencies fixed, but it is not clear, without further information, which of the other types it belongs to. Possibly 818 persons were examined and then classified into the 2×2 table. Alternatively, two samples of 279 inoculated and 539 not-inoculated persons were separately examined and each classified into " attacked " and " not-attacked." It is also possible that two samples of 69 attacked and 749 not-attacked persons were classified into " inoculated " and " not-inoculated." There are thus three ways in which the table might have been formed, one of the double-dichotomy type and two of the homogeneity type. Reference to the actual process by which the observations were collected would be necessary to resolve the choice.

CATEGORIZED DATA

To illustrate the last type in **33.18**, we give a fictitious table referring to the butter-margarine tasting experiment there described:

		Identification made by subject		
		Butter	Margarine	
Object actually presented	Butter	4	11	15
	Margarine	11	14	25
		15	25	40

33.19 We have no right to expect the same method of analysis to remain appropriate to the three different real situations discussed in **33.18** (although we shall see in **33.24** below that, so far as tests of independence are concerned, the Case I test turns out to be optimum in the other two situations). We therefore now make probabilistic formulations of the three different situations. We begin with the both-margins-fixed situation, since this is the simplest.

Case I: Both margins fixed

On the hypothesis, which we write

$$H_0 : \frac{p_{11}}{p_{1.}} = \frac{p_{21}}{p_{2.}}, \qquad (33.48)$$

the probability of observing the table

n_{11}	n_{12}	$n_{1.}$
n_{21}	n_{22}	$n_{2.}$
$n_{.1}$	$n_{.2}$	n

(33.49)

when all marginal frequencies are fixed is

$$P_I = P\{n_{ij}|n, n_{1.}, n_{.1}\} = P\{n_{ij}|n, n_{1.}\}/P\{n_{.1}|n, n_{1.}\}$$

$$= \binom{n_{1.}}{n_{11}}\binom{n_{2.}}{n_{21}} \Big/ \binom{n}{n_{.1}}$$

$$= \frac{n_{1.}!\, n_{.1}!\, n_{2.}!\, n_{.2}!}{n!\, n_{11}!\, n_{12}!\, n_{21}!\, n_{22}!}. \qquad (33.50)$$

(33.50) is symmetrical in the frequencies n_{ij} and in the marginal frequencies, as it must be from the symmetry of the situation. Since all marginal frequencies are fixed, only one of the n_{ij} may vary independently, and we may take this to be n_{11} without loss of generality. Regarding (33.50) as the distribution of n_{11}, we see that it is a hypergeometric distribution (cf. **5.18**). In fact, (33.50) is simply the hypergeometric f.f. (5.48) with the substitutions

$$N \equiv n, \quad n \equiv n_{1.}, \quad Np \equiv n_{.1}, \quad Nq \equiv n_{.2}, \quad N-n \equiv n_{2.},$$
$$j \equiv n_{11}, \quad n-j \equiv n_{12}, \quad Np-j \equiv n_{21}, \quad Nq-(n-j) \equiv n_{22}.$$

The mean and variance of n_{11} are therefore, from (5.53) and (5.55),

$$\left. \begin{array}{l} E(n_{11}) = n_{1.}\, n_{.1}/n, \\[4pt] \operatorname{var} n_{11} = \dfrac{n_{1.}\, n_{.1}\, n_{2.}\, n_{.2}}{n^2(n-1)}, \end{array} \right\} \qquad (33.51)$$

and n_{11} is asymptotically normal with these moments. Thus

$$t = \frac{n_{11} - n_{1.} \, n_{.1}/n}{\left\{\dfrac{n_{1.} \, n_{.1} n_{2.} \, n_{.2}}{n^2(n-1)}\right\}^{\frac{1}{2}}} \tag{33.52}$$

is asymptotically a standardized normal variate. Replacing $(n-1)$ by n, we see that (33.52) is equivalent to $n^{\frac{1}{2}}V$, where V is defined at (33.12) and hence (cf. **33.17**) t^2 is equivalent to the X^2 statistic defined at (33.46). This confirms that the general large-sample test of **33.17** applies in this situation.

33.20 We may use (33.50) to evaluate the exact probability of any given configuration of frequencies. If we sum these probabilities over the " tail " of the distribution of n_{11}, we may construct a critical region for an exact test, first proposed by R. A. Fisher. The procedure is illustrated in the following example.

Example 33.5 (Data from Yates, 1934, quoting M. Hellman)

The following table shows 42 children according to the nature of their teeth and type of feeding.

	Normal teeth	Mal-occluded teeth	TOTALS
Breast-fed	4	16	20
Bottle-fed	1	21	22
TOTALS	5	37	42

These data evidently do not leave both margins fixed, but for the present we use them illustratively and we shall see later (**33.24**) that this is justified.

We choose as n_{11} a frequency with the smallest range of variation, i.e. one of the two frequencies having the smallest marginal frequencies. In this particular case, given the fixed marginal frequency $n_{.1} = 5$, the range of variation of n_{11} is from 0 to 5.

The probability that $n_{11} = 0$ is, from (33.50),

$$\frac{5! \, 37! \, 20! \, 22!}{42! \, 20! \, 0! \, 5! \, 17!} = 0 \cdot 030,96.$$

The probabilities for $n_{11} = 1, 2, \ldots$ are obtained most easily by multiplying by

$$\frac{5 \times 20}{1 \times 18}, \quad \frac{4 \times 19}{2 \times 19}, \quad \frac{3 \times 18}{3 \times 20}, \text{ etc.,}$$

and are as follows:

Number of normal breast-fed children (n_{11})	Probability	Probabilities cumulated upwards
0	0·0310	1·0001
1	0·1720	0·9691
2	0·3440	0·7971
3	0·3096	0·4531
4	0·1253	0·1435
5	0·0182	0·0182
	1·0001	

To test independence against the alternative that normal teeth are positively associated with breast-feeding, we use a critical region consisting of large values of n_{11} (the number of normal breast-fed children). We have a choice of two " reasonable " values for the size of the exact test. For $\alpha = 0.0182$, only $n_{11} = 5$ would lead to rejection of the hypothesis; for $\alpha = 0.1435$, $n_{11} = 4$ or 5 leads to rejection. Probably, the former critical region would be used by most statisticians, leading in this particular case ($n_{11} = 4$) to acceptance of the hypothesis of independence.

33.21 Tables for use in the exact test based on (33.50) have been computed. Finney (1948) gives the values of n_{21} (his b) required to reject the hypothesis of independence for values of $n_{1.}$, $n_{2.}$ (or $n_{.1}$, $n_{.2}$) and n_{11} up to 15 and single-tailed tests of sizes $\alpha \leqslant 0.05, 0.025, 0.01, 0.005$, together with the exact size in each case. Finney's table is reproduced in the *Biometrika Tables*. These tables are extended up to $n_{1.}, n_{2.} = 30$ in Finney *et al.* (1963, 1966) and up to 50 for $\alpha \leqslant 0.05, 0.01$. Armsen (1955) gives tables for one- and two-tailed tests of sizes $\alpha \leqslant 0.05, 0.01$ and n ranging to 50. Bross and Kasten (1957) give charts for one-sided test sizes $\alpha = 0.05, 0.025, 0.01, 0.005$ or two-sided tests of size 2α, and minimum marginal frequency (say $n_{.1}) \leqslant 50$, based on the approximation of the hypergeometric (33.50) by the binomial $\binom{n_{1.}}{n_{11}} \left(\frac{n_{.1}}{n}\right)^{n_{11}} \left(\frac{n_{.2}}{n}\right)^{n_{21}}$.

The critical values in the charts are conservative unless $n_{.1}/n$ is small.

Case II: One margin fixed; homogeneity

33.22 We write the hypothesis (which is now one of equality of probabilities in two populations) in the form (33.48) as before, but $n_{1.}$ and $n_{2.}$ are fixed and n_{11}, n_{21} are independent random variables, so that $n_{.1}$ (and hence its complement $n_{.2}$) is a random variable. We test the hypothesis (33.48) by considering the corresponding difference of proportions

$$u = \frac{n_{11}}{n_{1.}} - \frac{n_{21}}{n_{2.}}. \qquad (33.53)$$

On the hypothesis, this is asymptotically normal with mean zero and variance

$$\operatorname{var} u = p(1-p)\left(\frac{1}{n_{1.}} + \frac{1}{n_{2.}}\right),$$

where p is the hypothetical common value of $\frac{p_{11}}{p_{1.}}, \frac{p_{21}}{p_{2.}}$.

We estimate $p(1-p)$ unbiassedly by the pooled estimator

$$\frac{n}{n-1} \cdot \left(\frac{n_{.1}}{n}\right)\left(\frac{n_{.2}}{n}\right),$$

so the estimated variance of u is

$$\widehat{\operatorname{var}}\, u = \frac{n_{.1} n_{.2}}{n(n-1)} \left(\frac{1}{n_{1.}} + \frac{1}{n_{2.}}\right) = \frac{n_{.1} n_{.2}}{(n-1) n_{1.} n_{2.}}.$$

Thus we have an asymptotic standardized normal variate

$$t = \frac{u - E(u)}{(\widehat{\operatorname{var}}\, u)^{\frac{1}{2}}} = \frac{\dfrac{n_{11}}{n_{1.}} - \dfrac{n_{21}}{n_{2.}}}{\left\{\dfrac{n_{.1} n_{.2}}{(n-1) n_{1.} n_{2.}}\right\}^{\frac{1}{2}}}, \qquad (33.54)$$

and this is identical with (33.52). The large-sample tests are therefore identical in the two cases.

But for small samples, the test is different. On the hypothesis, we now have

$$P_{II} = P\{n_{11}|p, n_{1\cdot}\} P\{n_{21}|p, n_{2\cdot}\}$$
$$= \binom{n_{1\cdot}}{n_{11}} p^{n_{11}}(1-p)^{n_{12}} \binom{n_{2\cdot}}{n_{21}} p^{n_{21}}(1-p)^{n_{22}}, \quad (33.55)$$

that is to say, to P_I defined at (33.50) multiplied by a binomial factor

$$\binom{n}{n_{\cdot 1}} p^{n_{\cdot 1}}(1-p)^{n_{\cdot 2}}.$$

This must evidently be so, for we have the original probability for fixed $n_{\cdot 1}$ now multiplied by the probability of $n_{\cdot 1}$ itself. Unlike (33.50), (33.55) depends on an unknown parameter, p, and cannot be evaluated.

Case III: No margin fixed; double dichotomy

33.23 We now turn to the case where n is fixed, but none of the marginal totals. The hypothesis is now genuinely one of bivariate independence. We have already derived the large-sample test in this context in **33.15–17**. The exact probability of n_{11} is now

$$P_{III} = P\{n_{11}|n_{1\cdot}, n_{\cdot 1}, n\} P\{n_{\cdot 1}|p, n\} P\{n_{1\cdot}|p', n\}, \quad (33.56)$$

where p' is the hypothetical common value of $p_{11}/p_{\cdot 1}$, $p_{12}/p_{\cdot 2}$. The first two factors on the right of (33.56) are equivalent to (33.55), and the third is

$$P\{n_{1\cdot}|p', n\} = \binom{n}{n_{1\cdot}} (p')^{n_{1\cdot}}(1-p')^{n_{2\cdot}}. \quad (33.57)$$

Thus (33.56) depends on two unknown parameters (p, p') and cannot be evaluated.

The optimum exact test for 2 × 2 tables

33.24 We may now demonstrate the remarkable result, first given by Tocher (1950), that the exact test based on the Case I probabilities (33.50) actually gives UMPU tests for Cases II and III. The argument can be made very simple. (33.55), the Case II distribution of n_{11} given H_0, contains a single nuisance parameter p, the hypothetical common value of $p_{11}/p_{1\cdot}$, $p_{21}/p_{2\cdot}$. It is easily verified that when H_0 holds, the pooled estimator $\hat{p} = \frac{n_{11}+n_{21}}{n_{1\cdot}+n_{2\cdot}} = \frac{n_{\cdot 1}}{n}$ is sufficient for p. By **23.10**, it is complete and distributed in the linearized exponential form (23.17). Thus one- and two-sided UMPU tests of H_0 will, by **23.30–1**, be based on the conditional distribution of n_{11} given $n_{\cdot 1}$, i.e. (since $n_{1\cdot}$ is already fixed) upon (33.50). Cf. Exercise 23.22.

Similarly, in Case III, we have two nuisance parameters (p, p') in (33.56) for which $\left(\frac{n_{\cdot 1}}{n}, \frac{n_{1\cdot}}{n}\right)$ are jointly sufficient and complete when H_0 holds. Thus, again from **23.10** and **23.30–1**, UMPU tests of H_0 will be based on the conditional distribution of n_{11} given $(n_{\cdot 1}, n_{1\cdot})$, i.e. upon (33.50).

Thus the conditional Case I distribution (all marginal frequencies fixed) provides UMPU tests for both the homogeneity and double-dichotomy situations.

CATEGORIZED DATA

It should be remarked that these results only hold strictly if randomization is permitted in order to obtain tests of any size α; the discreteness of the distributions in a 2×2 table limits our choice of test size—cf. Example 33.5 and also **20.22**. Unless the frequencies are very small, however, there is usually at least one "reasonable" value of α available for the conditional exact test based on (33.50), so that the difficulty is theoretical rather than practical.

33.25 Although the same test is valid in the three situations, its power function will differ, since the alternative to independence must obviously be different in the three situations. For Case II (homogeneity), Bennett and Hsu (1960) give charts of the power function using the Finney *et al.* (1963, 1966) tables (cf. **33.21**). Patnaik (1948) gave approximations adequate for larger samples—see also J. Hannan and Harkness (1963). E. S. Pearson and Merrington (1948) carried out sampling experiments on the power functions of the exact and asymptotic tests in Case I (both margins fixed). Harkness and Katz (1964) compare the power functions in the three cases—see also Harkness (1965).

33.26 Berger (1961) gives large-sample χ^2 tests for the equality of a measure of association in two separate 2×2 tables. Three measures of association are considered: (a) the ratio p_{11}/p_{12} in (33.33); (b) Q defined by (33.9); and (c) one equivalent to V defined at (33.12). Goodman (1963a) generalizes to the case of k separate 2×2 tables and (1964a) gives other methods based on the cross-ratio ad/bc. See also **33.62** below.

Continuity correction in the large-sample X^2 test

33.27 A continuity correction of the type mentioned in **31.80** was found by Yates (1934) to improve the fit of the continuous approximation (33.52) to the discrete exact distribution (33.50) in Case I. The correction requires (using (33.12)) that

$$X^2 \equiv nV^2 = \frac{n(ad-bc)^2}{(a+b)(a+c)(b+d)(c+d)} \tag{33.58}$$

should have the term $(ad-bc)$ in its numerator replaced by $|ad-bc| - \tfrac{1}{2}n$, which is the same as increasing (if $ad > bc$) b and c by $\tfrac{1}{2}$, and reducing a and d by $\tfrac{1}{2}$. Thus the corrected test statistic is

$$X_c^2 = \frac{n\{|ad-bc|-\tfrac{1}{2}n\}^2}{(a+b)(a+c)(b+d)(c+d)}. \tag{33.59}$$

The effect is illustrated in Example 33.6.

Example 33.6

In Example 33.5, we found the probability that $n_{11} \geq 4$ to be 0·1435. Let us compare this with the result obtained by using the asymptotic χ^2 distribution with one degree of freedom. From the table of Example 33.5 we have, for (33.58),

$$X^2 = 42 \frac{\{4(21) - 16(1)\}^2}{20 \cdot 22 \cdot 5 \cdot 37} = 2 \cdot 386.$$

From Appendix Table 4b, $P\{X^2 \geq 2 \cdot 386\} = 0 \cdot 122$, a more exact value being 0·1224. This, however, is the probability associated with a two-tailed test, because X^2 is the square of a normal deviate. For comparison with the exact test, we have to halve this, obtaining 0·0612. The approximation to the exact value of 0·1435 is very poor.

If we apply a continuity correction, the corrected value X_c^2 is then, by (33.59),
$$\frac{42(68-21)^2}{20.22.5.37} = 1\cdot140.$$
The corresponding probability from the χ^2 table is 0·2854, one half of which is 0·1427, this time in excellent agreement with the exact value of 0·1435.

> Cochran (1954) recommends that the continuity-corrected X^2 be used as an adequate approximation to the exact test for $n \geqslant 40$; and if no hypothetical frequency is less than 5, for $n \geqslant 20$ also.
>
> In Cases II and III, a continuity correction does not generally improve the fit of (33.52) to (33.55) or (33.56)—cf. Plackett (1964).
>
> Lancaster (1949a) examined the effect of continuity corrections in cases where a number of X^2 values from different tables are added together. In such circumstances, each X^2 should not be corrected for continuity, or a serious bias may result. Lancaster shows that where the original tables cannot be pooled, the best procedure is to add the uncorrected X^2 values. Similar results for one-tailed tests are given by Yates (1955).

The general $r \times c$ table : measurement of association

33.28 We now consider the more general situation in which two variables are classified into two or more categories. We extend our notation to write the $r \times c$ table in the form:

$$\begin{array}{cccc|c} n_{11} & n_{12} & \ldots & n_{1c} & n_{1.} \\ n_{21} & n_{22} & \ldots & n_{2c} & n_{2.} \\ \cdot & \cdot & & \cdot & \cdot \\ \cdot & \cdot & & \cdot & \cdot \\ \cdot & \cdot & & \cdot & \cdot \\ n_{r1} & n_{r2} & \ldots & n_{rc} & n_{r.} \\ \hline n_{.1} & n_{.2} & \ldots & n_{.c} & n \end{array} \qquad (33.60)$$

In the older literature, (33.60) is called a *contingency table*. The discussion of **33.1–2** applies to this general two-variable categorization.

The problem of measuring association in such a table presents severe difficulties which are, in a sense, inherent. In Chapter 26 we found in the case of measured variates that it may be impossible to express a complicated pattern of interdependence in terms of a single coefficient, and this holds similarly in the present situation. The most successful attempts to do so have been based on more or less latent assumptions about the nature of underlying variate-distributions.

33.29 In (33.60), if the two variables were independent, the frequency in the ith row and jth column would be $n_{i.}n_{.j}/n$. The deviation from independence in that particular cell of the table is therefore measured by
$$D_{ij} = n_{ij} - n_{i.}n_{.j}/n, \qquad (33.61)$$
the generalization of (33.8). We may define a coefficient of association in terms of the so-called *square contingency* $\sum_{i,j} D_{ij}^2/(n_{i.}n_{.j})$ and shall write
$$X^2 = \sum_{i,j} \frac{D_{ij}^2}{n_{i.}n_{.j}/n} \equiv n\left\{\sum_{i,j}\frac{n_{ij}^2}{n_{i.}n_{.j}} - 1\right\}, \qquad (33.62)$$

CATEGORIZED DATA

the generalization of (33.46). On the hypothesis of independence, X^2 is asymptotically distributed in the χ^2 form, as is easy to see from the goodness-of-fit standpoint mentioned in **33.17** above; the degrees of freedom are given by

$$(rc-1)-(r-1)-(c-1) = (r-1)(c-1),$$

the number of classes minus 1 minus the number of parameters fitted.

33.30 X^2 itself is not a convenient measure of association, since its upper limit is infinite as n increases. Following Karl Pearson (1904), we put

$$P = \left(\frac{X^2}{n+X^2}\right)^{\frac{1}{2}}, \tag{33.63}$$

and call P Pearson's *coefficient of contingency*. It was proposed because it may be shown that if a bivariate normal distribution with correlation parameter ρ is classified into a contingency table, then $P^2 \to \rho^2$ as the number of categories in the table increases. For finite r and c, however, the coefficient P has limitations. It vanishes, as it should, when there is complete independence; and conversely, if $P = 0$, we have $X^2 = 0$ so that every deviation D_{ij} is zero. Clearly $0 \leq P \leq 1$. But in general, P cannot attain the same upper limit, and therefore fails to satisfy a desideratum mentioned in **33.5**. Consider, for example, a " square " table with $r = c$, in which only the leading diagonal frequencies n_{ii} are non-zero. Then $n_{i.} = n_{.i} = n_{ii}$, all i, and by (33.62)

$$X^2 = n(r-1),$$

so that, from (33.63),
$$P = \left(\frac{r-1}{r}\right)^{\frac{1}{2}}.$$

Thus even in such a case of *complete association*—cf. **33.7**—the value of P, its maximum, depends on the number of rows and columns in the table.

To remedy this, Tschuprow proposed the alternative function of X^2

$$T = \left\{\frac{X^2}{n[(r-1)(c-1)]^{\frac{1}{2}}}\right\}^{\frac{1}{2}}, \tag{33.64}$$

which attains $+1$ when $r = c$ in a case of complete association as above, but cannot do so if $r \neq c$. In fact, it is easy to see, just as above, that the maximum attainable value for X^2 is $n \times \min(r-1, c-1)$ (attained when all the frequencies lie in a longest diagonal of the table) and thus the attainable upper bound for P is

$$\left\{\frac{\min(r-1, c-1)}{1+\min(r-1, c-1)}\right\}^{\frac{1}{2}},$$

while that for T is

$$\left\{\frac{\min(r-1, c-1)}{[(r-1)(c-1)]^{\frac{1}{2}}}\right\}^{\frac{1}{2}} = \left\{\frac{\min(r-1, c-1)}{\max(r-1, c-1)}\right\}^{\frac{1}{2}}.$$

Following Cramér (1946), we may define a further modification, which can always attain $+1$, by

$$C = \left\{\frac{X^2}{n \min(r-1, c-1)}\right\}^{\frac{1}{2}} = T\left\{\frac{\max(r-1, c-1)}{\min(r-1, c-1)}\right\}^{\frac{1}{4}}. \tag{33.65}$$

Evidently $C = T$ when the table is square but $C > T$ otherwise, although the difference

will not be very large unless r and c are very different. We also see that
$$\frac{P^2}{T^2} = \frac{[(r-1)(c-1)]^{\frac{1}{2}}}{1+(X^2/n)},$$
so that as n increases we expect to have $P > T$ if independence holds, when X^2 has expectation $(r-1)(c-1)$. The difference $P-T$ is often substantial. Cf. Exercise 33.4.

An interpretation of C^2 is given in **33.51** below.

Example 33.7

The table (from W. H. Gilby, *Biometrika*, 8, 94) shows the distribution of 1725 school children who were classified (1) according to their standard of clothing, and (2) according to their intelligence, the standards in the latter case being A = mentally deficient, B = slow and dull, C = dull, D = slow but intelligent, E = fairly intelligent, F = distinctly capable, G = very able.

Table 33.2

Standard of clothing	A and B	C	D	E	F	G	TOTALS
Very well clad	33	48	113	209	194	39	636
Well clad	41	100	202	255	138	15	751
Poor but passable	39	58	70	61	33	4	265
Very badly clad	17	13	22	10	10	1	73
TOTALS	130	219	407	535	375	59	1725

We investigate the association between standard of clothing and intelligence. We first work out the "independence" frequencies $n_{i.} n_{.j}/n$. For example, $n_{1.} n_{.1}/n$ is $636 \times 130/1725 = 47 \cdot 930$. The term $nD_{11}^2/n_{1.} n_{.1}$ in (33.61) is then
$$(33-47 \cdot 93)^2/47 \cdot 93 = 4 \cdot 65.$$

The sum of the 24 such terms in the table will be found to be $X^2 = 174 \cdot 82$.

It is quicker to calculate X^2 from the extreme right-hand side of (33.62), i.e. to calculate
$$X^2 = n \left\{ \sum_{i,j} \frac{n_{ij}^2}{n_{i.} n_{.j}} - 1 \right\}, \tag{33.66}$$
and with a calculating machine this is expeditiously evaluated by first dividing the square of every frequency in the ith row by its row total, and then dividing all the resulting quotients in the jth column by that column's original total frequency. We should, in this case, first have

$$\frac{33^2}{636} \quad \frac{48^2}{636} \quad \frac{113^2}{636} \quad \frac{209^2}{636} \quad \frac{194^2}{636} \quad \frac{39^2}{636},$$

$$\frac{41^2}{752} \quad \frac{100^2}{751} \quad \frac{202^2}{751} \quad \frac{255^2}{751} \quad \frac{138^2}{751} \quad \frac{15^2}{751},$$

and two further rows, and then divide the columns of this array by 130, 219, etc. We then have only to subtract 1 from the total of the 24 entries in the final array, and multiply by 1725 ($= n$) to have (33.66). The reader should check the computation by both methods.

With $X^2 = 174 \cdot 82$, we now have, from (33.63–5), the coefficients
$$P = \left(\frac{174 \cdot 82}{1725 + 174 \cdot 82}\right)^{\frac{1}{2}} = 0 \cdot 30,$$
$$T = \left\{\frac{174 \cdot 82}{1725 \, (3 \times 5)^{\frac{1}{2}}}\right\}^{\frac{1}{2}} = 0 \cdot 16,$$
$$C = \left\{\frac{174 \cdot 82}{1725 \times 3}\right\}^{\frac{1}{2}} = 0 \cdot 18.$$

The relationship between the three coefficients is as we should expect from the remarks at the end of **33.30**, with C little larger than T, but P nearly twice as large, even though its attainable upper bound here is $\left(\frac{3}{4}\right)^{\frac{1}{2}} = 0 \cdot 866$ against $\left(\frac{3}{5}\right)^{\frac{1}{4}} = 0 \cdot 880$ for T (and, of course, 1 for C). Thus it happens that the attainable upper bounds for P and T are almost the same in this particular case, but P gives the impression of a considerably stronger association between the variables. Exercise 33.3 finds similar results for another set of data.

All three coefficients are monotone functions of X^2, and we may therefore test independence directly by X^2, which we have seen to be 174·82. As there are $(r-1)(c-1) = 15$ degrees of freedom, this far exceeds any critical value likely to be used in practice—a test of size $\alpha = 0 \cdot 001$ would use a critical value of 37·697.

Models for the $r \times c$ table

33.31 In discussing measures of association in the $r \times c$ table, distinctions as to the underlying model, similar to those made for the 2×2 table in **33.18–23**, are necessary. S. N. Roy and Mitra (1956) have explicitly extended that discussion of the three types of table (both sets of marginal frequencies fixed, one set fixed, neither set fixed) to the general $r \times c$ table; no new point arises. Roy and Mitra go on to show (as we did for the 2×2 table) that, on the hypothesis of independence, X^2 defined at (33.62) is asymptotically distributed in the X^2 form with $(r-1)(c-1)$ degrees of freedom under all three models. It is intuitively obvious that the differences between the models, given the hypothesis of independence, vanish asymptotically, since any marginal frequencies which are random variables will converge in probability to their expectations.

Exact treatment of the $r \times c$ table on the lines of **33.20** is necessarily a tedious piece of enumeration. G. H. Freeman and Halton (1951) give details of the method.

> Geometrical interpretations of independence, dependence and of fixed marginal frequencies are given by Fienberg (1968). Goodman (1968) gives general methods of analysing independence in tables which may have zero frequencies in some cells.

Standard errors of the coefficients and of X^2

33.32 The coefficients of association (33.63–5) are all monotone increasing function of X^2, and their standard errors can therefore be deduced from the standard error of X^2 by the use of (10.14). For example, the standard error of $(X^2)^{\frac{1}{2}}$, to which both (33.64) and (33.65) reduce apart from constants, is, to order n^{-1},

$$\text{var}\{(X^2)^{\frac{1}{2}}\} = \{\tfrac{1}{2}(\mathbf{X}^2)^{-\frac{1}{2}}\}^2 \, \text{var}\, X^2 = \frac{1}{4\mathbf{X}^2} \text{var}\, X^2, \qquad (33.67)$$

where $\mathbf{X}^2$ is the population value of X^2, i.e. (33.62) calculated from the population frequencies. Clearly, the first approximation (33.67) is only valid if $\mathbf{X}^2 \neq 0$, i.e. it does not hold in the case where the two categorized variables are independent in the population. This is because the sample X^2 converges in probability to the population $\mathbf{X}^2$, and since $X^2 \geqslant 0$ its distribution has a singularity when $\mathbf{X}^2 = 0$ and its variance is of order n^{-2}—cf. the analogous situation for the squared multiple correlation coefficient in **27.33**. Fortunately, we need not pursue the case of independence, since we may then test with X^2 itself. If we wish to estimate non-zero population coefficients $\mathbf{T}$ and $\mathbf{C}$ defined in terms of $\mathbf{X}^2$ by the population analogues of (33.64–5), and set standard errors to the estimates, we have

$$\left. \begin{array}{l} \operatorname{var} T = \dfrac{1}{4n^2[(r-1)(c-1)]\mathbf{T}^2} \operatorname{var} X^2, \quad \mathbf{T} \neq 0, \\[2mm] \operatorname{var} C = \dfrac{1}{4n^2[\min(r-1, c-1)]^2 \mathbf{C}^2} \operatorname{var} X^2, \quad \mathbf{C} \neq 0. \end{array} \right\} \qquad (33.68)$$

For Pearson's coefficient (33.63), the same difficulty arises in the case of independence, since

$$\left(\frac{\partial P}{\partial (X^2)} \right)^2 = \frac{n^2}{4X^2(n+X^2)^3},$$

so that we may only write

$$\operatorname{var} P = \frac{n^2}{4\mathbf{X}^2(n+\mathbf{X}^2)^3} \operatorname{var} X^2 \quad \text{if} \quad \mathbf{X}^2 \neq 0. \qquad (33.69)$$

This, unlike the expressions (33.68), cannot be written in terms of a parent coefficient $\mathbf{P}$ alone.

33.33 From **33.32**, we see that we need only calculate the variance of X^2 to obtain the standard errors we require. The variance of X^2 in the case of non-independence is complicated, but was worked out by K. Pearson (1915) for a table with fixed marginal frequencies, and more accurately by Young and Pearson (1915, 1919) who gave the variance to order $1/n^3$. Kondo (1929) gives the mean and variance to order $1/n^2$ when the marginal frequencies are random variables, and shows explicitly that the variance is of order $1/n^2$ in the case of independence. (The exact variance of X^2 in the independence case is given by Haldane (1940)—a specialization of his result is given as Exercise 33.9.) The formulae are lengthy, and we shall quote only the first approximation given by K. Pearson (1915):

$$\text{estimated var } X^2 = 4n \left\{ \frac{1}{n} \sum_i \sum_j \frac{(n_{ij} - n_{i.}n_{.j}/n)^2}{(n_{i.}n_{.j}/n)^2} + \frac{X^2}{n} - \left(\frac{X^2}{n} \right)^2 \right\}. \qquad (33.70)$$

If (33.70) is substituted into (33.68–9) and X^2, T, C, written for $\mathbf{X}^2$, $\mathbf{T}$, $\mathbf{C}$, we have the required standard errors.

33.34 The summation on the right of (33.70) differs from the definition of X^2 at (33.62) only in that the denominator term is squared. If the marginal frequencies all increase with n, this implies that the dominating term on the right of (33.70) will be the middle one in the braces, so that asymptotically we may estimate the variance of X^2 by $4nX^2/n = 4X^2$. This may also be seen directly. Under the conditions of **30.27**, we have that X^2 will be distributed asymptotically in the non-central χ^2 form, in this context with $\nu = (r-1)(c-1)$ degrees of freedom and non-central parameter (30.62), where the p_{0i} here are the "independence" frequencies. By Exercise 24.1, the variance of such a distribution is

$$2(\nu+2\lambda) = 2\left\{(r-1)(c-1)+2n\sum_i\sum_j\frac{(p_{ij}-p_{i.}p_{.j})^2}{p_{i.}p_{.j}}\right\}.$$

For large n, the first term on the right is negligible,(*) and the second is estimated by

$$4n\sum_i\sum_j\frac{(n_{ij}/n - n_{i.}n_{.j}/n^2)^2}{n_{i.}n_{.j}/n^2} = 4X^2.$$

Thus the leading term in the standard error of X^2 derives from its asymptotic non-central χ^2 distribution. It is worth noting that the non-central parameter λ in the distribution of X^2 is estimated by X^2 itself, so that the use of X^2 and its standard error is equivalent to setting approximate limits for λ. Bulmer (1958) discusses confidence limits for $\lambda^{\frac{1}{2}}$, which is a natural "distance" parameter.

Substitution of var $X^2 = 4X^2$ into (33.68) gives, after simplifying, the approximations

$$\text{var } T = 1/\{n[(r-1)(c-1)]^{\frac{1}{2}}\}, \quad \text{var } C = 1/\{n \min(r-1, c-1)\}.$$

Other measures of association

33.35 It cannot be pretended that any of the coefficients (33.63–5) based on the X^2 statistic has been shown to be a satisfactory measure of association, principally because their values have no simple probabilistic interpretation—cf. Blalock (1958). Altham (1970a) gives propositions which, as in **33.14** above, imply that association should be measured by a function of the $(r-1)(c-1)$ cross-ratios in the table—even in the 2×2 case, we have seen that $X^2 \equiv nV^2$ is not such a function. A number of more readily interpretable measures were proposed by Goodman and Kruskal (1954, 1959). We have already seen in **33.11–14** that two of their principal suggestions reduce in the case of a 2×2 table to two of the conventional measures of association. For the general $r \times c$ table, this is not so, and the Goodman-Kruskal coefficients are not functions of the X^2 statistic, unlike the measures so far discussed. For example, generalizing the approach of **33.13–14**, they define a population coefficient which is the relative decrease in the probability of incorrectly predicting one variable when the second variable is known, and a symmetrized coefficient which takes both predictive directions into account. For ordered classifications, the approach of **33.12** generalizes similarly—we refer to it in our discussion of ordered tables in **33.40** below. Goodman and Kruskal (1963) develop formulae for the standard errors of some of their coefficients under various sampling models. See also Goodman (1964b).

Lancaster and Hamdan (1964) generalize the tetrachoric method of **26.27–9** to a polychoric method for estimation of the correlation coefficient in the general $r \times c$ table. The method, which uses sets of $(r-1)$ and $(c-1)$ orthonormal functions as in **33.44–5**, requires electronic computing facilities, but gives much better results than P at (33.63), which essentially (cf. **33.34**) equates X^2 to its asymptotic value in the bivariate normal case, $n\rho^2(1-\rho^2)^{-1}$, and often gives unduly low estimates.

(*) Except in the case of independence, when the second term is zero—this is the singularity referred to in **33.32** above.

A LR statistic may be defined which is asymptotically equivalent to X^2, as in **30.5** and Exercise 30.11. Gabriel (1966) shows how this may be used to test subsets of the categories, and of the populations, compared—cf. the discussion of simultaneous test procedures in **35.54**, **35.63** (Vol. 3). Nathan (1969) extends the LR analysis to data sampled from a population stratified into sub-populations.

Ordered tables : rank measures of association

33.36 If there is a natural ordering (cf. **33.1**) of row- and of column-categories in a $r \times c$ table, we are presented with a new situation, which was not distinguished in the 2×2 table case because with only two categories the two possible orders of the categories can only change the sign of any measure of association. With three or more categories, the knowledge that there is an order between the categories conveys new statistical information which we may use in measuring association. Generally, we are unable to assume any underlying metric for the categories; we know that the categories proceed, say, from "high" to "low" values of an underlying variable, but we can attach no numerical values to them. In such a case, we may make what is perhaps a slightly unexpected application of the rank-order statistic t discussed in **31.24**. For we may regard an $r \times c$ table with a grand total of n observations as a way of displaying the rankings of n objects according to two variables, for one of which only r separate ranks are distinguished and for the other of which only c separate ranks are distinguished. From this point of view, the marginal frequencies in the table are the numbers of observations "tied" (cf. **31.81**) at the different rank values distinguished. The case where there are no "ties" corresponds to a table of all whose marginal frequencies are unity.

33.37 The measurement of association is now seen to be simply the problem of measuring the correlation between the two rankings. Either of the coefficients t and r_s defined at (31.23), (31.40) may be used, but we shall discuss only the former.

Some slight problems are produced by the fact that we are interested here in rankings with many ties. In the first place, we can no longer define the rank correlation coefficient in terms of a simple 0–1 scoring system as in the definition of h_{ij} at (31.36), for we now have three possibilities instead of two. We therefore define

$$a_{ij} = \begin{cases} +1 & \text{if } x_i < x_j, \\ 0 & \text{if } x_i = x_j, \\ -1 & \text{if } x_i > x_j, \end{cases} \qquad b_{ij} = \begin{cases} +1 & \text{if } y_i < y_j, \\ 0 & \text{if } y_i = y_j, \\ -1 & \text{if } y_i > y_j. \end{cases}$$

Our measure of rank correlation is now to be based on the sum

$$S = \sum_{i,j} a_{ij} b_{ij}, \qquad i,j = 1, 2, \ldots, n\,;\ i \neq j. \tag{33.71}$$

If we wish to standardize S to lie in the range $(-1, +1)$ and attain its endpoints in the extreme cases of complete dissociation and complete association, thus satisfying the desideratum of **33.5**, we have a choice of several possibilities:

(1) If there were no ties, no a_{ij} or b_{ij} could be zero, and (33.71) would vary between $\pm n(n-1)$ inclusive. The measure of association would then be $S/\{n(n-1)\}$. The reader may satisfy himself that this is identical with t of (31.23), from the definitions

of h_{ij} and a_{ij}, b_{ij}. If some scores a_{ij}, b_{ij} are zero, this measure, which we shall now write

$$t_a = \frac{S}{n(n-1)}, \qquad (33.72)$$

can no longer attain ± 1; its actual limits of variation depend on the number of zero scores.

(2) If we rewrite the denominator (33.72) for the case of no ties as

$$n(n-1) = \{\sum_{i,j} a_{ij}^2 \sum_{i,j} b_{ij}^2\}^{\frac{1}{2}},$$

which makes clear that t is a correlation coefficient between the two sets of scores (cf. Daniels (1944)), we may define

$$t_b = \frac{\sum_{i,j} a_{ij} b_{ij}}{\{\sum_{i,j} a_{ij}^2 \sum_{i,j} b_{ij}^2\}^{\frac{1}{2}}}. \qquad (33.73)$$

t_a and t_b are identical when there are no zero scores, but otherwise the denominator of (33.73) is smaller than that of (33.72) and thus $t_b > t_a$. Even so, t_b cannot attain ± 1 generally, for the Cauchy inequality

$$(\sum a_{ij} b_{ij})^2 \leqslant \sum a_{ij}^2 \sum b_{ij}^2$$

only becomes an equality when the sets of scores a_{ij}, b_{ij} are proportional, which here means that all the observations must be concentrated in a positive or negative leading diagonal of the table (i.e. north-west to south-east or north-east to south-west). If no marginal frequency is to be zero, this means that only for a square table (i.e. an $r \times r$ table) can t_b attain ± 1.

(3) For a non-square $r \times c$ table $(r \neq c)$, $|\sum_{i,j} a_{ij} b_{ij}|$ attains its maximum when all the observations lie in cells of a longest diagonal of the table (i.e. a diagonal containing $m = \min(r, c)$ cells) and are as equally as possible divided between these cells. If n is a multiple of m (as we may suppose here, since n is usually large and m a small integer), the reader may satisfy himself that this maximum is $n^2(m-1)/m$, and thus a third measure is

$$t_c = \frac{m \sum_{i,j} a_{ij} b_{ij}}{n^2(m-1)}. \qquad (33.74)$$

t_c can attain ± 1 for any $r \times c$ table, apart from the slight effect produced by n not being a multiple of m. For large n, (33.72) and (33.74) show that t_c is nearly $m t_a/(m-1)$.

33.38 The coefficients t_b and t_c do not differ much in value if each margin contains approximately equal frequencies. For

$$\sum_{i,j} a_{ij}^2 = n(n-1) - \sum_{p=1}^{c} n_{.p}(n_{.p}-1)$$

$$= n^2 - \sum_{p=1}^{c} n_{.p}^2,$$

and similarly

$$\sum_{i,j} b_{ij}^2 = n^2 - \sum_{p=1}^{r} n_{p.}^2.$$

584   THE ADVANCED THEORY OF STATISTICS

Thus the denominator of t_b is

$$(\Sigma a_{ij}^2 \Sigma b_{ij}^2)^{\frac{1}{2}} = \left\{\left(n^2 - \sum_{p=1}^{c} n_{.p}^2\right)\left(n^2 - \sum_{p=1}^{r} n_{p.}^2\right)\right\}^{\frac{1}{2}} \quad (33.75)$$

$$= n^2 \left\{\left(1 - \frac{\sum_{p=1}^{c} n_{.p}^2}{n^2}\right)\left(1 - \frac{\sum_{p=1}^{r} n_{p.}^2}{n^2}\right)\right\}^{\frac{1}{2}}, \quad (33.76)$$

while that of t_c is

$$n^2 \frac{(m-1)}{m} = n^2 \left(1 - \frac{1}{m}\right). \quad (33.77)$$

If all the marginal column frequencies $n_{.p}$ are equal, and all the marginal row frequencies $n_{p.}$ are equal, (33.76) reduces to

$$n^2 \left\{\left(1 - \frac{1}{c}\right)\left(1 - \frac{1}{r}\right)\right\}^{\frac{1}{2}} \quad (33.78)$$

approximately. (33.78) is the same as (33.77) if the table is square ($r = c = m$); otherwise (33.78) is the larger and thus t_b the smaller. This tends to be more than offset by the fact that if the marginal frequencies are not precisely equal, the sums of squares will be increased, and (33.76), the denominator of t_b, therefore decreases.

The following example (cf. also Kendall (1962)) illustrates the computations of the coefficients.

Example 33.8

In the table below, we are interested in the association between distance vision in right and left eye.

Table 33.3—3242 men aged 30–39 employed in U.K. Royal Ordnance factories 1943–6 : unaided distance vision

Right eye \ Left eye	Highest grade	Second grade	Third grade	Lowest grade	TOTALS
Highest grade	821	112	85	35	1053
Second grade	116	494	145	27	782
Third grade	72	151	583	87	893
Lowest grade	43	34	106	331	514
TOTALS	1052	791	919	480	3242

The numerator of all forms of t is calculated by taking each cell in the table in turn and multiplying its frequency positively by all frequencies to its south-east and negatively by all frequencies to its south-west. Cells in the same row and column are always ignored. (There is no need to apply the process to the last row of the table, which has nothing below it.) $\sum_{i,j} a_{ij} b_{ij}$ is twice the sum of all these terms, because we may have $i < j$ or $i > j$. For this particular table, we have

$$821 (494 + 145 + 27 + 151 + 583 + 87 + 34 + 106 + 331)$$
$$+ 112 (145 + 27 + 583 + 87 + 106 + 331 - 116 - 72 - 43),$$

and so on. As we proceed down the table, fewer terms enter the brackets. The reader should verify that we find, on summing inside the brackets,

$$821\,(1958) + 112\,(1048) + 85\,(-465) + 35\,(-1744)$$
$$+ 116\,(1292) + 494\,(992) + 145\,(118) + 27\,(-989)$$
$$+ 72\,(471) + 151\,(394) + 583\,(254) + 87\,(-183)$$
$$= 2{,}480{,}223.$$

Thus the numerator is
$$\Sigma a_{ij} b_{ij} = 4{,}960{,}446.$$
From (33.75), the denominator of t_b is
$$[\{3242^2 - (1052^2 + 791^2 + 919^2 + 480^2)\}\{3242^2 - (1053^2 + 782^2 + 893^2 + 514^2)\}]^{\frac{1}{2}}$$
$$= [7{,}728{,}586 \times 7{,}703{,}218]^{\frac{1}{2}}.$$
Thus
$$t_b = \frac{4{,}960{,}446}{[7{,}728{,}586 \times 7{,}703{,}218]^{\frac{1}{2}}} = 0\cdot 643.$$
From (33.74), on the other hand,
$$t_c = \frac{4 \times 4{,}960{,}446}{3242^2 \times 3} = 0\cdot 629.$$

We therefore find t_b a trifle larger in this case, where both sets of marginal frequencies vary by about a factor of 2 from largest to smallest. A similar result is found in Exercise 33.10, where the range of variation of marginal frequencies is about threefold.

33.39 Apart from the question of attaining the limits ± 1 discussed in **33.37** above, the main difference between the forms t_b and t_c is that an upper bound (see (33.81) below) can be set for the standard error of t_c in sampling n observations, the marginal frequencies not being fixed; in such a situation, t_b is a ratio of random variables and its standard error is not known. If the marginal frequencies are fixed, t_b is no longer a ratio of random variables, but its distribution has only been investigated on the hypothesis of independence of the two variables categorized in the table—of course, if we wish only to test independence, we need concern ourselves only with the common numerator $\Sigma a_{ij} b_{ij}$—the details of the test are given by Kendall (1962). Stuart (1953) showed how the upper bound for the variance of t_c may be used to test the difference between two values found for different tables. This is fairly obvious, and we omit it here.

33.40 Goodman and Kruskal (1954) proposed a measure of association for ordered tables which is closely related to the t coefficients we have discussed. It has the same numerator, but yet another different denominator, and is

$$G = \frac{\Sigma a_{ij} b_{ij}}{n^2 - \sum_{p=1}^{c} n_{.p}^2 - \sum_{p=1}^{r} n_{p.}^2 + \sum_{p=1}^{r} \sum_{q=1}^{c} n_{pq}^2}. \tag{33.79}$$

If we compare the denominator of G with that of t_b at (33.75), which is identically equal to
$$\{[n^2 - \tfrac{1}{2}(\textstyle\sum_p n_{.p}^2 + \sum_p n_{p.}^2)]^2 - \tfrac{1}{4}(\textstyle\sum_p n_{.p}^2 - \sum_p n_{p.}^2)^2\}^{\frac{1}{2}}.$$
and is thus very nearly $n^2 - \tfrac{1}{2}(\sum_p n_{.p}^2 + \sum_p n_{p.}^2)$, it will be seen that the denominator of G

is in practice likely to be smaller always. What is more, it is easily seen that G can attain its limits ± 1 if all the observations lie in a longest diagonal of the table. Thus G is rather similar to t_c. Goodman and Kruskal (1963) give the standard error of G, a method of computing it, and a simple upper bound for it which is estimated from

$$\operatorname{var} G \leqslant \frac{2}{n}\frac{(1-G^2)}{D_G/n^2}, \qquad (33.80)$$

where D_G is the denominator of G at (33.79). This compares with the upper bound for the variance of t_c

$$\operatorname{var} t_c \leqslant \frac{2}{n}\left\{\left(\frac{m}{m-1}\right)^2 - t_c^2\right\}. \qquad (33.81)$$

Goodman and Kruskal (1963) show in two worked examples that G tends to be larger than t_c, but that the upper bound for its standard error is considerably smaller; the details are given in Exercise 33.11. If this is shown to be true in general, this fact, together with the direct interpretability of G in terms of order-relationships in random sampling (it gives the probability of the orders of x and y agreeing minus that of their disagreeing, conditional upon there being no ties—cf. **33.12** for the 2×2 case) would make it likely to become the standard measure of association for the ordered case.

Small-sample experiments by Rosenthal (1966) verify the applicability of the asymptotic theory for G.

Scoring methods with pre-assigned scores

33.41 Returning to our general discussion of $r \times c$ tables, we now consider the possibilities of imposing a metric on the categories in the table. If we assign numerical scores to the categories for each variable, we bring the problems of measuring interdependence and dependence back to the ordinary (grouped) bivariate table, which we discussed at length in Chapter 26. Thus, we may calculate correlation and regression coefficients in the ordinary way. The difficulty is to decide on the appropriate scoring system to use. We have discussed this from the standpoint of rank tests in **31.21–4**, where we saw that different tests resulted from different sets of "conventional numbers" (i.e. "scores" in our present terminology). Here, the difficulty is more acute, as we are seeking a measure, and not merely a test of independence.

The simplest scoring system uses the sets of natural numbers $1, 2, \ldots, r$ and $1, 2, \ldots, c$ for row and column categories respectively. Alternatively, we could use the sets of normal scores $E(s, r)$ and $E(s, c)$ discussed in **31.39**. Example 33.9 illustrates the procedures.

Example 33.9

Let us calculate the correlation coefficients, using the scoring systems of **33.41**, for the data of Example 33.8. For the natural numbers scoring, we assign scores 1, 2, 3, 4 to the categories from "highest" to "lowest" for left eye (x) and (because the table here happens to be square) similarly for right eye (y). We find, with $n = 3242$,

$$\Sigma x = (1052 \times 1) + (791 \times 2) + (919 \times 3) + (480 \times 4) = 7311,$$
$$\Sigma y = (1053 \times 1) + (782 \times 2) + (893 \times 3) + (514 \times 4) = 7352,$$
$$\Sigma x^2 = (1052 \times 1^2) + \ldots = 20{,}167,$$
$$\Sigma y^2 = (1053 \times 1^2) + \ldots = 20{,}442,$$
$$\Sigma xy = (821 \times 1 \times 1) + (112 \times 1 \times 2) + \ldots = 19{,}159.$$

Thus the correlation coefficient is, for natural number scores,

$$r_1 = \frac{19{,}159 - (7311)(7352)/3242}{[\{20{,}167 - (7311)^2/3242\}\{20{,}442 - (7352)^2/3242\}]^{\frac{1}{2}}}$$

$$= \frac{2579}{(3677 \times 3772)^{\frac{1}{2}}} = 0 \cdot 69.$$

This is not very different from the values of the ranking measures $t_b = 0 \cdot 64$, $t_c = 0 \cdot 63$ found in Example 33.8, and we should expect this since the "natural numbers" scoring system is closely related to the rank correlation coefficients, as we saw in Chapter 31.

Suppose now that we use the normal scores

$$E(1,4) = -1 \cdot 029,$$
$$E(2,4) = -0 \cdot 297,$$
$$E(3,4) = +0 \cdot 297,$$
$$E(4,4) = +1 \cdot 029,$$

obtained from the *Biometrika Tables*, Table 28.

We now simplify the computations into the form

$$\Sigma x = 1 \cdot 029(480 - 1052) + 0 \cdot 297(919 - 791) = -550 \cdot 6,$$
$$\Sigma y = 1 \cdot 029(514 - 1053) + 0 \cdot 297(893 - 782) = -521 \cdot 7,$$
$$\Sigma x^2 = (1 \cdot 029)^2(480 + 1 \cdot 052) + (0 \cdot 297)^2(919 + 791) = 1773,$$
$$\Sigma y^2 = (1 \cdot 029)^2(514 + 1053) + (0 \cdot 297)^2(893 + 782) = 1807,$$
$$\Sigma xy = (1 \cdot 029)^2(821 + 331 - 35 - 43) + (0 \cdot 297)^2(494 + 583 - 145 - 151)$$
$$\quad + (1 \cdot 029)(0 \cdot 297)(116 + 112 + 106 + 87 - 72 - 34 - 85 - 27)$$
$$= 1268.$$

Thus the correlation coefficient for normal scores is

$$r_2 = \frac{1268 - (550 \cdot 6)(521 \cdot 7)/3242}{[\{1773 - (550 \cdot 6)^2/3242\}\{1807 - (521 \cdot 7)^2/3242\}]^{\frac{1}{2}}}$$

$$= \frac{1179}{(1680 \times 1723)^{\frac{1}{2}}} = 0 \cdot 69,$$

exactly the same to two decimal places as we found for natural number scores. It hardly seems worth the extra trouble of computation to use the normal scores, at least when the number of categories is as small as 4.

33.42 If one were strictly trying to impose a normal metric upon the $r \times c$ table, a more reasonable system would be to assign scores to the categories which correspond to the proportions observed in sampling from a normal distribution. Thus, in Example 33.9, we should calculate the "cutting points" of a standardized normal distribution which give relative frequencies

$$\frac{1052}{3242}, \quad \frac{791}{3242}, \quad \frac{919}{3242}, \quad \frac{480}{3242},$$

and use as the "left eye" scores the means within these four sections of the normal distribution.

We need not make the calculation for the moment, but it is clear that the set of scores obtained will differ from the crude normal scores used in Example 33.9. We return to this scoring system in **33.50** below.

We do not further pursue the study of scoring methods with pre-assigned scoring systems, because it is clear that by putting "information" into the table in this way, we are making distributional assumptions which may lead us astray if they are incorrect. On the whole, we should generally prefer to avoid this by using the rank order methods of **33.36–40**. Yates (1948) first proposed the natural numbers scoring system of **33.41**, and E. J. Williams (1952) surveyed scoring methods generally.

Least Squares analysis in a $r \times c$ table

33.43 If we are investigating the dependence of one categorization (say, the rows) upon the other, we can avoid imposing a metric upon the latter. Define a score x_i ($i = 1, 2, \ldots, c-1$) for all columns but the last, such that $x_i = +1$ for each observation in the ith column, and $x_i = 0$ otherwise. The set of $(c-1)$ x-scores then identifies the column for every observation, the cth column being indicated by all $(c-1)$ scores being zero. Now, if the row-categorization has (or has imposed upon it) a metric, an ordinary LS analysis of its dependence upon the x-scores may be carried out by the methods of Chapter 19, in which there will be a parameter θ_i for each x_i. Linear models for the $2 \times c$ table are further discussed in **33.55** below.

The choice of "optimum" scores: canonical analysis

33.44 However, we may approach the problem of scoring the categories in a (not necessarily ordered) $r \times c$ table from quite another viewpoint. We may ask: what scores should be allotted to the categories in order to maximize the correlation coefficient between the two variables? Surprisingly enough, it emerges that these "optimum" scores are closely connected with the transformation of the frequencies in the table to bivariate normal frequencies. We first prove a theorem, due to Lancaster (1957), for ungrouped observations.

Let x and y be distributed in the bivariate normal form with correlation ρ. Let $x' = x'(x)$ and $y' = y'(x)$ be new variables, functions respectively of x alone and y alone, with $E\{(x')^2\}$ and $E\{(y')^2\}$ both finite. Then we may validly write

$$x' = a_0 + a_1 H_1(x) + a_2 H_2(x) + \ldots, \tag{33.82}$$

where the H_r are the Tchebycheff–Hermite polynomials defined by (6.21), standardized so that

$$\int_{-\infty}^{\infty} H_r^2(x) \alpha(x) \, dx = 1.$$

$\sum_{i=1}^{\infty} a_i^2$ will be convergent. The correlation is unaffected by changes of origin or scale, so we may write $a_0 = 0$, and hence

$$x' = \sum_{i=1}^{\infty} a_i H_i, \quad \sum_{i=1}^{\infty} a_i^2 = 1,$$

and similarly we may write

$$y' = \sum_{i=1}^{\infty} b_i H_i, \quad \sum_{i=1}^{\infty} b_i^2 = 1.$$

Now $H_r(x)$ is, by **6.14**, the coefficient of $t^r/r!$ in $\exp(tx - \tfrac{1}{2}t^2)$. Since the expectation of $\exp(tx - \tfrac{1}{2}t^2 + uy - \tfrac{1}{2}u^2)$ equals $\exp(\rho tu)$, we have

$$\int_{-\infty}^{\infty}\int_{-\infty}^{\infty} H_r(x) H_s(y) f\, dx\, dy = \begin{cases} \rho^r, & r = s, \\ 0, & r \neq s, \end{cases} \qquad (33.83)$$

where f is the bivariate normal frequency.

The variances of x' and y' are unity in virtue of the orthogonality of the H_r, and hence their correlation is

$$\operatorname{cov}(x', y') = \sum_{i=1}^{\infty} a_i b_i \rho^i. \qquad (33.84)$$

Now this is less than $|\rho|$ unless $a_1^2 = b_1^2 = 1$. The other a's and b's must then vanish. Hence the maximum correlation between x' and y' is $|\rho|$ and we have Lancaster's theorem: if a bivariate distribution of (x, y) can be obtained from the bivariate normal by separate transformations on x and y, the correlation in the transformed distribution cannot in absolute value exceed ρ, that in the bivariate normal distribution.

33.45 Suppose now that we seek a second pair of such transforms of x and y separately, say x'' and y''. If we require these to be standardized and uncorrelated with the first pair (x', y'), the Tchebycheff–Hermite representation

$$x'' = \sum_{i=1}^{\infty} c_i H_i(x), \quad y'' = \sum_{i=1}^{\infty} d_i H_i(y),$$

together with the orthogonality laid down requires at once that $c_1 = d_1 = 0$. Thus we obtain

$$x'' = \sum_{i=2}^{\infty} c_i H_i, \quad y'' = \sum_{i=2}^{\infty} d_i H_i,$$

and, as at (33.84),

$$\operatorname{cov}(x'', y'') = \sum_{i=2}^{\infty} c_i d_i \rho^i, \qquad (33.85)$$

which is maximized in absolute value only if $c_2^2 = d_2^2 = 1$ and all other $c_r, d_r = 0$, when the correlation is ρ^2. We may proceed similarly to further pairs of variables (x''', y'''), $(x^{(iv)}, y^{(iv)})$, etc., obtaining maximized correlations $|\rho|^3, \rho^4$, etc.

The transformed pairs of variables are known as the *canonical variables*. What we have shown is that the rth pair of canonical variables has *canonical correlation* $|\rho|^r$. Evidently, from our proof, the canonical variables themselves are simply the Tchebycheff–Hermite forms in the bivariate normal variables (x, y), i.e.

$$(x^{(r)}, y^{(r)}) = (H_r(x), H_r(y)). \qquad (33.86)$$

Lancaster (1958) further extends this type of analysis.

33.46 The results of **33.44–5** apply to ungrouped bivariate population values. In practice, when we have a sample in an ordered $r \times c$ table, there is no difficulty in making separate transformations of the variables to achieve marginal univariate normal distributions for them—this is essentially what we discussed in **33.42**—but we should be fortunate if we found these separate transformations to result in a bivariate normal distribution of the frequencies in the body of the table. However, the theoretical implication of the result is clear: if we seek separate scoring systems for the two categorized variables such as to maximize their correlation, we are basically trying to produce a bivariate normal distribution by operation upon the margins of the table.

33.47 Suppose, then, that we allot scores x_i, y_j ($i = 1, 2, \ldots, r$; $j = 1, 2, \ldots, c$) to the categories of an $r \times c$ table. Without loss of generality we may take them to be in standard measure (zero mean, unit variance). Then we have

$$\operatorname{var} x = \sum_i n_{i.} x_i^2/n = \operatorname{var} y = \sum_j n_{.j} y_j^2/n = 1, \tag{33.87}$$

$$\operatorname{cov}(x, y) = \operatorname{corr}(x, y) = \sum_i \sum_j n_{ij} x_i y_j/n. \tag{33.88}$$

We require to maximize (33.88) for variation in x and y subject to (33.87). If λ, μ are Lagrange undetermined multipliers, this leads to the equations

$$\sum_j n_{ij} y_j - \lambda n_{i.} x_i = 0, \qquad i = 1, 2, \ldots, r, \tag{33.89}$$

$$\sum_i n_{ij} x_i - \mu n_{.j} y_j = 0, \qquad j = 1, 2, \ldots, c. \tag{33.90}$$

Multiplying (33.89) by x_i and summing over i, we have

$$\lambda = R,$$

where R is the correlation we are seeking. Similarly we find $\mu = R$, and hence, from (33.89–90),

$$\left.\begin{array}{l}\sum_j n_{ij} y_j = R n_{i.} x_i, \qquad i = 1, 2, \ldots, r, \\ \sum_i n_{ij} x_i = R n_{.j} y_j, \qquad j = 1, 2, \ldots, c.\end{array}\right\} \tag{33.91}$$

Eliminating x and y, we have a determinantal equation which may be written symbolically

$$\begin{vmatrix} R n_{i.} & n_{ij} \\ n_{ij} & R n_{.j} \end{vmatrix} = 0. \tag{33.92}$$

We shall study this in considering the theory of canonical correlations in Volume 3. It is enough here to note that R can be expressed in terms of the cell frequencies. In fact, (33.92) is an equation in R^2 with a number of roots. There are, in general, $m = \min(r, c)$ non-zero roots, one of which is identically unity; we are interested only in the $m-1$ others. These are the *canonical correlations*. That there can be only m non-zero canonical correlations follows from the fact that the rank of the array of frequencies $\{n_{ij}\}$ is at most m. We require the largest root, R_1. The others, apart from sampling effects, are powers of this largest, as we have seen in **33.45**.

33.48 It follows from (33.92) that if the canonical correlations, the roots of (33.92), are $R_1, R_2, \ldots, R_{m-1}$, then

$$n_{ij} = \frac{n_{i.} n_{.j}}{n} \left\{ 1 + \sum_{s=1}^{m-1} R_s x_s y_s \right\}. \tag{33.93}$$

In the limit, as the categories of the $r \times c$ table become finer and $m \to \infty$, this reduces because of **33.44–5** to the tetrachoric series

$$f = (2\pi)^{-1} \exp\{-\tfrac{1}{2}(x^2 + y^2)\} \left\{ 1 + \sum_{j=1}^{\infty} H_j(x) H_j(y) \rho^j \right\}, \tag{33.94}$$

where f is the bivariate normal frequency. (33.94) is simply another form of (26.66), which differs only in the factor $j!$ in its denominator, since the H_j were not there standardized.

CATEGORIZED DATA

33.49 When the largest canonical correlation R_1 has been determined from (33.92), we can immediately calculate the "optimum" sets of scores giving this correlation. This is perhaps most easily done by returning to (33.91). If the second equation there is multiplied by $n_{pj}/\{(n_{p.}n_{i.})^{\frac{1}{2}}n_{.j}\}$ and summed over j, it becomes

$$\sum_i (\sum_j n_{ij} x_i) \frac{n_{pj}}{(n_{p.}n_{i.})^{\frac{1}{2}}n_{.j}} = \frac{R\sum_j n_{pj} y}{(n_{p.}n_{i.})^{\frac{1}{2}}} = \left(\frac{n_{p.}}{n_{i.}}\right)^{\frac{1}{2}} R^2 x_p. \qquad (33.95)$$

(33.95) may be rewritten

$$\sum_i (x_i n_{i.}^{\frac{1}{2}}) \sum_j \frac{n_{ij} n_{pj}}{(n_{p.}n_{i.})^{\frac{1}{2}}n_{.j}} = R^2 (x_p n_{p.}^{\frac{1}{2}}), \qquad (33.96)$$

which makes it clear that the squared canonical correlations are the latent roots of the $(r \times r)$ matrix $\mathbf{NN}'$, where $\mathbf{N}$ is the $(r \times c)$ matrix whose elements are $\frac{n_{ij}}{(n_{i.}n_{.j})^{\frac{1}{2}}}$, for (33.96) in matrix terms is

$$\mathbf{NN}'\mathbf{u} = R^2 \mathbf{u}, \qquad (33.97)$$

where $\mathbf{u}$ is the $(r \times 1)$ vector with elements $x_i n_{i.}^{\frac{1}{2}}$.

Since the rank of $\mathbf{N}$, and hence of $\mathbf{NN}'$, is at most $m = \min(r, c)$, $\mathbf{NN}'$ will have m non-zero latent roots in general as stated in **33.47**. It is easily verified that $R^2 = 1$ is always a root, the latent vector $\mathbf{u}$ then having elements $(n_{p.})^{\frac{1}{2}}$, i.e. $x_p \equiv 1$ for this root. Leaving aside this root, which is irrelevant to the problem of association, we see that once the largest latent root R_1^2 of (33.97) is determined, the corresponding latent vector $\mathbf{u}_1$ gives the vector of scores for the first canonical x-variable. Similarly, the scores for the first canonical y-variable are given by the latent vector $\mathbf{v}_1$ in

$$\mathbf{N}'\mathbf{N}\mathbf{v} = R^2 \mathbf{v}, \qquad (33.98)$$

where $\mathbf{v}$ is the $(c \times 1)$ vector with elements $y_j n_{.j}^{\frac{1}{2}}$. The non-zero latent roots of the $(c \times c)$ matrix $\mathbf{N}'\mathbf{N}$ are, of course, the same as those of $\mathbf{NN}'$, namely the squared canonical correlations. However, there is no need to solve both (33.97) and (33.98); we need only solve one (it is naturally easier to choose the one corresponding to the smaller of r and c, i.e. $\mathbf{NN}'$ if $r \leq c$) and then obtain the other set of scores from (33.91), which we rewrite

$$\left. \begin{array}{l} y_j = \dfrac{1}{R n_{.j}} \sum_i n_{ij} x_i, \\[2mm] x_i = \dfrac{1}{R n_{i.}} \sum_j n_{ij} y_j. \end{array} \right\} \qquad (33.99)$$

Example 33.10

Let us make a canonical analysis of the data of Example 33.8. We first rewrite the table with the marginal frequencies replaced by their square roots:

821	112	85	35	32·449,961,5
116	494	145	27	27·964,262,9
72	151	583	87	29·883,105,6
43	34	106	331	22·671,568,1
32·434,549,5	28·124,722,2	30·315,012,8	21·908,902,3	

We now construct the matrix **N** by dividing the n_{ij} in the table by the product of the corresponding marginal square roots, e.g. $821/(32\cdot 434{,}549{,}5 \times 32\cdot 449{,}961{,}5)$. We obtain:

$$\mathbf{N} = \begin{pmatrix} 0\cdot780047593 & 0\cdot122720070 & 0\cdot086406614 & 0\cdot049230386 \\ 0\cdot127892990 & 0\cdot628109454 & 0\cdot171043616 & 0\cdot044069667 \\ 0\cdot074284618 & 0\cdot179664791 & 0\cdot643554112 & 0\cdot132884065 \\ 0\cdot058476184 & 0\cdot053322330 & 0\cdot154229180 & 0\cdot666385924 \end{pmatrix} \quad (33.100)$$

Here we have $r = c$, so it is immaterial which of $\mathbf{NN'}$ and $\mathbf{N'N}$ we work with. We shall compute

$$\mathbf{NN'} = \begin{pmatrix} 0\cdot633424197 & 0\cdot193793122 & 0\cdot142143279 & 0\cdot098290784 \\ 0\cdot193793122 & 0\cdot442076157 & 0\cdot238281615 & 0\cdot096718276 \\ 0\cdot142143279 & 0\cdot238281615 & 0\cdot469617711 & 0\cdot201730920 \\ 0\cdot098290784 & 0\cdot096718276 & 0\cdot201730920 & 0\cdot474119575 \end{pmatrix}. \quad (33.101)$$

We recall that the sum of the latent roots is equal to the trace of the matrix, so that if the trace of the matrix does not much exceed 1, the largest canonical correlation must be small—this is a useful preliminary check. Here, the trace exceeds 2, so that R_1^2 could be as large as 1.

We now obtain the latent roots. We must solve the characteristic equation

$$|\mathbf{NN'} - \lambda \mathbf{I}| = 0. \quad (33.102)$$

If we subtract λ from each diagonal element in $\mathbf{NN'}$, and expand the determinant of the resulting matrix, we find that it reduces to the quartic equation

$$\lambda^4 - 2\cdot019237640\,\lambda^3 + 1\cdot343416989\,\lambda^2 - 0\cdot355747132\,\lambda + 0\cdot031567594 = 0.$$

Since one root of this equation must be unity, the left-hand side has a factor $(\lambda - 1)$, and we may write it as

$$(\lambda - 1)(\lambda^3 - 1\cdot019237640\,\lambda^2 + 0\cdot324179349\,\lambda - 0\cdot031567783) = 0.$$

We are thus left with a cubic equation, which is solved by standard methods. The roots are the squared canonical correlations

$$\left.\begin{array}{l} R_1^2 = \lambda_1 = 0\cdot48516, \\ R_2^2 = \lambda_2 = 0\cdot34604, \\ R_3^2 = \lambda_3 = 0\cdot18803. \end{array}\right\} \quad (33.103)$$

It will be noticed that $R_1 = 0\cdot697$ is not very much larger than the correlations of $0\cdot69$ obtained with natural number and normal scores in Example 33.9.

We now require the latent vectors corresponding to λ_1. We first solve the set of equations for the elements of $\mathbf{u}_1$

$$\mathbf{NN'u}_1 = 0\cdot48516\mathbf{u}_1,$$

and find on dividing the elements u_i of $\mathbf{u}_1$ by $n_i^{\frac{1}{2}}$ that the canonical scores for the row-categories are

$$\mathbf{x} = \begin{pmatrix} u_1/n_{1.}^{\frac{1}{2}} \\ u_2/n_{2.}^{\frac{1}{2}} \\ u_3/n_{3.}^{\frac{1}{2}} \\ u_4/n_{4.}^{\frac{1}{2}} \end{pmatrix} = \begin{pmatrix} -1 \cdot 307 \\ +0 \cdot 021 \\ +0 \cdot 739 \\ +1 \cdot 362 \end{pmatrix}. \tag{33.104}$$

The canonical y-scores are obtained from (33.99). Thus, for example,

$$y_1 = \{(821 \times -1 \cdot 307) + (116 \times +0 \cdot 021) + (72 \times +0 \cdot 739) + (43 \times +1 \cdot 362)\}/\{1052(0 \cdot 48516)^{\frac{1}{2}}\}.$$

The set of scores is

$$\mathbf{y} = \begin{pmatrix} -1 \cdot 309 \\ +0 \cdot 040 \\ +0 \cdot 730 \\ +1 \cdot 406 \end{pmatrix}. \tag{33.105}$$

The sets of scores (33.104) and (33.105), when weighted by the row or column frequencies, have zero means but have had to be adjusted so that their variances are unity, since latent vectors have an arbitrary scale constant.

33.50 We now recall the implication of Lancaster's theorem discussed in **33.46**: the choice of row- and column-scores to maximize the correlation between the categorized variables is essentially equivalent to transforming the margins of the table to univariate normality, with the intention of rendering the body of the table bivariate normal. Let us, therefore, apply to the data of Example 33.10 the normal scoring system outlined in **33.42**. We shall then be able to see whether the resulting scores for the categories agree well with the canonical scoring in Example 33.10.

Example 33.11

The two sets of proportional marginal frequencies in Example 33.8, together with the corresponding ranges of a standardized univariate normal distribution (obtained from the *Biometrika Tables*) are:

$p_i = \left(\dfrac{n_{i.}}{n}\right)$	Corresponding normal range (a_i, b_i)	$p_i = \left(\dfrac{n_{.i}}{n}\right)$	Corresponding normal range (a_i, b_i)
0·3245	$(-\infty, \quad -0 \cdot 4551)$	0·3248	$(-\infty, \quad -0 \cdot 4543)$
0·2440	$(-0 \cdot 4551, \ +0 \cdot 1726)$	0·2412	$(-0 \cdot 4543, \ +0 \cdot 1662)$
0·2835	$(+0 \cdot 1726, \ +1 \cdot 0450)$	0·2755	$(+0 \cdot 1662, \ +1 \cdot 0006)$
0·1480	$(+1 \cdot 0450, \quad \infty)$	0·1585	$(+1 \cdot 0006, \quad \infty)$
1·0000		1·0000	

(33.106)

The mean value within a range (a_i, b_i) of a standardized normal distribution containing a fraction p_i of the distribution is

$$\mu_i = \frac{1}{p_i} \int_{a_i}^{b_i} (2\pi)^{-\frac{1}{2}} t e^{-\frac{1}{2}t^2} dt \propto \frac{1}{p_i} (e^{-\frac{1}{2}a_i^2} - e^{-\frac{1}{2}b_i^2}). \tag{33.107}$$

We neglect the factor $(2\pi)^{-\frac{1}{2}}$, since we are interested only in correlations, and scale changes in scores do not affect these. The values of the scores are found, using 4-figure logarithms, to be:

Row scores	Column scores
−2·778	−2·777
−0·343	−0·350
+1·432	+1·380
+3·914	+3·825

Unlike the scores in Example 33.10, these are not exactly standardized, even if we restore the neglected factor $(2\pi)^{-\frac{1}{2}}$, for the use of means within ranges of the standardized normal distribution introduces a grouping approximation. We find, for the sums and sums of squares of these scores (weighted, of course, by row and column marginal frequencies respectively):

	Row scores	Column scores
Sum	+98	−94
Sum of squares	17,923	16,984
Mean	0·030	−0·029
Standard deviation	2·351	2·289

Adjusting the scores by subtracting the mean and dividing by the standard deviation we obtain:

Standardized row scores (x)	Standardized column scores (y)	
−1·194	−1·205	
−0·159	−0·140	
+0·596	+0·616	(33.108)
+1·652	+1·684	

The scores (33.108) agree only rather roughly with those in (33.104–5). We may only regard the method of the present example as giving a crude approximation to the canonical scores.

Partitions of X^2: canonical components

33.51 The canonical correlations discussed in **33.44–9** have a close relationship with the X^2 statistic (33.62). Consider again the matrix $\mathbf{NN}'$ defined in **33.49**. Its diagonal elements are

$$(\mathbf{NN}')_{ii} = \sum_j \frac{n_{ij}^2}{n_{i.}\, n_{.j}},$$

and thus we have

$$\operatorname{tr}\mathbf{NN}' = \sum_i \sum_j \frac{n_{ij}^2}{n_{i.}\, n_{.j}} = \frac{X^2}{n} + 1, \qquad (33.109)$$

by (33.62). Remembering that the trace of a matrix is the sum of the latent roots, and that the latent roots of $\mathbf{NN}'$ are $1, R_1^2, \ldots, R_{m-1}^2$, we therefore have from (33.109)

$$X^2 = n(R_1^2 + R_2^2 + \ldots + R_{m-1}^2). \qquad (33.110)$$

We thus display the squared canonical correlations, multiplied by n, as *components* or *partitions* of X^2.

It is tempting to suppose that the components in (33.110) are themselves asymptotically independent χ^2 variables on the hypothesis of independence, the degrees of freedom of nR_s^2 being $(r-s)(c-s)-(r-s+1)(c-s+1) = r+c-2s-1$. However, Lancaster (1963) shows that this is not so.

From (33.110) it follows at once that
$$\frac{1}{m-1} \sum_{j=1}^{m-1} R_j^2 = C^2,$$
defined at (33.65). The square of Cramér's association measure is the mean squared canonical correlation.

Example 33.12

In Example 33.10, we have $n = 3242$, and the squared canonical correlations are given by (33.103). Thus, by (33.110),
$$X^2 = 3242(0\cdot 48516 + 0\cdot 34604 + 0\cdot 18803) = 3,304,$$
as may be verified by direct calculation in Example 33.8.

33.52 There are many (indeed an infinite number of) other ways in which X^2 may be partitioned; the formal structure of such partitions was given in **30.44**. Whether a particular partitioning has statistical interest depends on the purpose of the analysis. As a preliminary, it should be noticed that X^2 itself is, in fact, a component of a larger such quantity, which we shall denote by X_T^2.

We no longer restrict ourselves to ordered tables, but consider only the independence case, when $p_{ij} = p_{i.}p_{.j}$ for all i, j. The probability of the observed frequencies n_{ij} is then

$$P\{n_{ij}|p_{ij}, n\} = \frac{n!}{\prod_{i,j} n_{ij}!} \prod_{i,j} (p_{ij})^{n_{ij}} = \frac{n!}{\prod_{i,j} n_{ij}!} \prod_i p_{i.}^{n_{i.}} \prod_{i,j} \left(\frac{p_{ij}}{p_{i.}}\right)^{n_{ij}} \qquad (33.111)$$

$$= \frac{n! \prod_i p_{i.}^{n_{i.}}}{\prod_i n_{i.}!} \cdot \frac{n! \prod_j p_{.j}^{n_{.j}}}{\prod_j n_{.j}!} \cdot \frac{\prod_i n_{i.}! \prod_j n_{.j}!}{\prod_{i,j} n_{ij}! \, n!} \qquad (33.112)$$

just as in **33.23** for the 2×2 table. The left-hand side of (33.112) and each of the three factors on its right can be approximated by χ^2 distributions. Writing
$$n_{ij} = np_{ij} + e_{ij},$$
we find on using Stirling's series that the left-hand side of (33.122) is asymptotically
$$\log P\{n_{ij}|p_{ij}, n\} = \text{constant} - \tfrac{1}{2} \sum_{i,j} \{e_{ij}^2/(np_{ij})\}.$$

Thus
$$X_T^2 = \sum_{i,j} \frac{(n_{ij} - np_{i.}p_{.j})^2}{np_{i.}p_{.j}} \qquad (33.113)$$

is asymptotically distributed like the sum of squares of rc standardized normal variates subject to one linear constraint $(\sum_i \sum_j n_{ij} = n)$ and is therefore a χ^2 variate with $rc-1$ degrees of freedom.

Similarly,
$$X_R^2 = \sum_i \frac{(n_{i.} - np_{i.})^2}{np_{i.}} \qquad (33.114)$$

is asymptotically a χ^2_{r-1},

$$X_C^2 = \sum_j \frac{(n_{.j} - np_{.j})^2}{np_{.j}} \qquad (33.115)$$

is asymptotically a χ^2_{c-1}, and we already know that the " ordinary " X^2, which we now write

$$X_{RC}^2 = \sum_i \sum_j \frac{(n_{ij} - n_{i.}n_{.j}/n)^2}{n_{i.}n_{.j}/n}, \qquad (33.116)$$

is asymptotically a χ^2 with $(r-1)(c-1)$ degrees of freedom. (33.113–16) give the asymptotic partitioning

$$X_T^2 = X_R^2 + X_C^2 + X_{RC}^2 \qquad (33.117)$$

which is, in fact, a way of reflecting the factorization (33.112). Degrees of freedom on the right of (33.117) add to the degrees of freedom on its left, i.e.

$$rc - 1 \equiv (r-1) + (c-1) + (r-1)(c-1). \qquad (33.118)$$

Thus we see (as we did in **33.29**) that the degrees of freedom $(r-1)+(c-1)$ on the right of (33.118) are lost to the " ordinary " X^2 (which is X_{RC}^2 in our present notation) because we have to estimate row- and column-probabilities from the table—if these were known *a priori*, (33.113) could be used instead. This is merely another instance of the loss of degrees of freedom due to estimation of parameters, which we remarked in **19.9**.

Example 33.13 (Lancaster, 1949b)

A sampling experiment was (in effect) conducted nine times according to variations of factor A (threefold) and factor B (threefold). The frequencies (which were occurrences in sampling from Poisson populations) were as follows:

3,009	2,832	3,008	8,849
3,047	3,051	2,997	9,095
2,974	3,038	3,018	9,030
9,030	8,921	9,023	26,974

This is one of the relatively infrequent cases where we have prior marginal probabilities. Here $p_{ij} = \frac{1}{9}(i, j = 1, 2, 3)$ and $p_{i.} = p_{.j} = \frac{1}{3}$. Using equation (33.117) we find:

Component	Value	Degrees of freedom	Critical value $\alpha = 0.05$
X_R^2	3·615	2	5·991
X_C^2	0·828	2	5·991
X_T^2	11·864	8	15·507
$X_T^2 - X_R^2 - X_C^2$	7·421	4	9·488

None of the three values X_R^2, X_C^2, X_T^2 exceeds its random sampling limit given in the last column. The conditions of experimentation seem to have been about constant.

CATEGORIZED DATA

Had we not possessed information about the marginal probabilities, we should have had to estimate them. We then find $X_{RC}^2 = 7·547$ with 4 d.f. This is not quite the same as the value of 7·421 for $X_T^2 - X_R^2 - X_C^2$ in the above table, but the difference is trivial. It is, of course, due to the fact that the partition (33.117) is strictly an asymptotic one.

33.53 By essentially the method of **33.52**, Lancaster (1949b) partitions X^2 for an $r \times c$ table into $(r-1)(c-1)$ components, each having a single degree of freedom. Each degree of freedom corresponds to X^2 for a particular 2×2 classification of the table. We shall not give the details here, but the method is easily understood from two examples. The 2×3 table

n_{11}	n_{12}	n_{13}	$n_{1.}$
n_{21}	n_{22}	n_{23}	$n_{2.}$
$n_{.1}$	$n_{.2}$	$n_{.3}$	n

for which X^2 has 2 degrees of freedom, has the 2×2 component tables:

n_{11}	n_{12}	$n_{11}+n_{12}$		$(n_{11}+n_{12})$	n_{13}	$n_{1.}$
n_{21}	n_{22}	$n_{21}+n_{22}$	and	$(n_{21}+n_{22})$	n_{23}	$n_{2.}$
$n_{.1}$	$n_{.2}$	$n_{.1}+n_{.2}$		$(n_{.1}+n_{.2})$	$n_{.3}$	n

(33.119)

If X^2 is calculated for each of these 2×2 tables in the ordinary way, their sum will approximately be the X^2 of the original 2×3 table. Similarly, for a 3×3 table, with 4 degrees of freedom, the four component 2×2 tables are:

n_{11}	n_{12}	$n_{11}+n_{12}$	$(n_{11}+n_{12})$	n_{13}	$n_{1.}$
n_{21}	n_{22}	$n_{21}+n_{22}$	$(n_{21}+n_{22})$	n_{23}	$n_{2.}$
$(n_{11}+n_{21})(n_{12}+n_{22})$	$(n_{11}+n_{12}+n_{21}+n_{22})$	$(n_{11}+n_{12}+n_{21}+n_{22})(n_{13}+n_{23})$	$n_{1.}+n_{2.}$
$(n_{11}+n_{21})(n_{12}+n_{22})$	$(n_{11}+n_{12}+n_{21}+n_{22})$	$(n_{11}+n_{12}+n_{21}+n_{22})(n_{13}+n_{23})$	$n_{1.}+n_{2.}$
n_{31}	n_{32}	$(n_{31}+n_{32})$	$(n_{31}+n_{32})$	n_{33}	$n_{3.}$
$n_{.1}$	$n_{.2}$	$n_{.1}+n_{.2}$	$(n_{.1}+n_{.2})$	$n_{.3}$	n

(33.120)

The procedure is quite general, but must be used with care as the partitioning is not unique (since rows and columns may in general be permuted). The components are only additive asymptotically, as in Example 33.13.

> Lancaster (1949b, 1950) and Irwin (1949) give a method of partitioning X^2 *exactly* into $(r-1)(c-1)$ components corresponding to 2×2 tables, but the approximate partition is good enough for most practical purposes—cf. Exercise 33.15. A. W. Kimball (1954) simplifies the computations for the exact partitioning.

33.54 Other types of partitioning of X^2 are discussed by Cochran (1954) in a review of methods for $r \times c$ tables (and, indeed, also for goodness-of-fit tests) which includes a discussion of the problems of handling tables with small hypothetical (independence) frequencies in some or many cells without destroying the χ^2 approximation to the distribution of X^2. His recommendations are that if only 1 cell out of 5 or more, or 2 cells out of 10 or more, have hypothetical frequencies smaller than 5, a minimum hypothetical frequency of 1 is allowable. If there are more such cells, a minimum hypothetical frequency of 2 is usually adequate if there are fewer than 30 degrees of freedom. For more than 30 d. of f., the exact mean and variance of X^2 given by Haldane (1939) should be used and X^2 taken to be asymptotically normal with these moments.

For ordered tables, Quenouille (1948) in an unpublished paper gives partitions of X^2 which extract linear, quadratic, etc., components.

$2 \times c$ tables : the binomial homogeneity test

33.55 A particular case of the $r \times c$ table which is of special interest is the $2 \times c$ table, where we are comparing c samples in respect of the possession or non-possession of an attribute. The general formula (33.62) for X^2 reduces here to

$$X^2 = \sum_{i=1}^{2} \sum_{j=1}^{c} \frac{(n_{ij} - n_{i.} \, n_{.j}/n)^2}{n_{i.} \, n_{.j}/n}, \qquad (33.121)$$

where $n_{.j}$ is the jth sample size and $n = \sum_j n_{.j}$ as before. Useful exact and approximate methods of calculating (33.121) are given in Exercise 33.21. If we write

$$\hat{p} = n_{1.}/n$$

for the ML estimate from the table of the probability of observing a "success" (i.e. an entry in the first row of the table), (33.121) may be expressed as

$$\begin{aligned} X^2 &= \sum_{j=1}^{c} \left[\frac{(n_{1j} - n_{.j}\hat{p})^2}{n_{.j}\hat{p}} + \frac{\{(n_{.j} - n_{1j}) - n_{.j}(1-\hat{p})\}^2}{n_{.j}(1-\hat{p})} \right] \\ &= \sum_{j=1}^{c} \frac{(n_{1j} - n_{.j}\hat{p})^2}{n_{.j}\hat{p}(1-\hat{p})}, \end{aligned} \qquad (33.122)$$

distributed asymptotically as χ^2 with $c-1$ degrees of freedom. The test of the homogeneity of the c binomial samples based on (33.122) is essentially based on the sum of squares of c independent binomial variables each measured from its expectation (estimated on the hypothesis of homogeneity) and divided by its estimated standard error $\{n_{.j}\hat{p}(1-\hat{p})\}^{\frac{1}{2}}$. There are $c-1$ degrees of freedom because we estimate the expectation linearly from the data—if it were given independently of the observations as p, we would replace $\hat{p}$ by p in (33.122) and have the full c degrees of freedom for X^2.

Armitage (1955) gives an expository account of tests for trend in the probabilities underlying an ordered $2 \times c$ table, which are essentially applications in this simpler situation of the rank-order and scoring methods which are discussed generally earlier in this chapter.

> An alternative method of analysing a $2 \times c$ table is to consider the *logistic transform*, λ_i, defined in Exercise 17.25, of the probability θ_i that an observation in the ith column falls in the first row. We thus obtain the *linear logistic model* $\lambda = X\beta$, treated in the monograph by D. R. Cox (1970)—the general linear model, $\theta = E(y) = X\beta$, here suffers from the problems of heteroscedasticity (since var $y_i = \theta_i(1-\theta_i)$) and may in any case give estimates $\hat{\theta}_i$ not in the interval (0, 1), which can only be adjusted with computational difficulty. Gart (1971) reviews methods of comparing proportions based on this model.

The Poisson homogeneity test

33.56 Now consider what happens to the statistic (33.122) when the hypothetical underlying binomial distribution tends to the Poisson distributions in the classical manner of **5.8**, so that

$$n_{.j} \to \infty, \quad p \to 0, \quad n_{.j}p \to \lambda.$$

We then have c independent observations on this Poisson variable, namely the n_{1j}. The statistic (33.122) with $\hat{p}$ replaced by p reduces to

$$X^2 = \sum_{j=1}^{c} \frac{(n_{1j}-\lambda)^2}{\lambda}, \tag{33.123}$$

which is asymptotically a χ^2 variable with c degrees of freedom as in **33.55**. If, as is usual, λ must be estimated, we use the complete sufficient unbiassed estimator $n_{1.}/c = \bar{n}_1$, which is the mean of the c observations. Thus

$$X^2 = \sum_{j=1}^{c} \frac{(n_{1j}-\bar{n}_1)^2}{\bar{n}_1} \tag{33.124}$$

has $(c-1)$ degrees of freedom as a test of homogeneity of c Poisson frequencies, a degree of freedom having been lost by the estimation process just as for (33.122).

The tests of this and the last section, which are due to R. A. Fisher, are sometimes called the *dispersion tests* of the binomial and Poisson distributions. This is because each is the sum of a number of c terms, each term being the ratio of a variate squared about its estimated expectation to an estimate of its variance—in the case of (33.124), the Poisson population mean and variance are equal, so that $\bar{n}_1$ estimates both.

33.57 Cochran (1954) gives a detailed account and bibliography of the binomial and Poisson dispersion tests, and especially of the partitioning of degrees of freedom from X^2 in each case.

It appears, in particular, that the dispersion statistic (33.124) often gives a more powerful test of the hypothesis that a sample originated from a Poisson distribution than does the X^2 goodness-of-fit test based on grouping the observations into the frequencies with which the values 0, 1, 2, ... are observed. The basic reason for this is that for Poisson distributions with small values of the parameter λ, the observed frequencies fall off sharply after a certain value, which is as low as 4 or 5 if λ is 1 or less (cf. Table 5.3 and Example 19.11). Thus, unless n is extremely large, a goodness-of-fit test can only have a few degrees of freedom (about 5) since the values in the upper tail must be pooled into a single class to obtain a sufficiently large hypothetical frequency for the test to be valid (cf. **30.30**). This does not apply to the dispersion test, where the number of degrees of freedom is equal to $c-1$, one less than the number of observations, no grouping being necessary—this point is perhaps obscured by our derivation of the test through the $2 \times c$ table, but is clear from (33.124) directly. Thus, for "reasonable" sample sizes, the dispersion test may be expected to be more powerful.

> Potthoff and Whittinghill (1966a, b) give other statistics for testing binomial, multinomial and Poisson homogeneity. See also Buhler *et al.* (1965) and Wisniewski (1968).

Armitage (1966) considers the moments of the X^2 dispersion tests when the data are sampled from a population stratified into sub-populations.

Multi-way tables

33.58 Except in **33.9–10**, we have been considering the relationships between two categorized variables; our $r \times c$ tables have been two-way tables. It is natural to generalize the problem to $p \geqslant 3$ variables categorized in *multi-way tables*, or, as they are sometimes called, *complex contingency tables*. This was first done by K. Pearson (1904, 1916) for an underlying multinormal distribution. If the pth variable is polytomized into r_p categories, we have a $r_1 \times r_2 \times \ldots \times r_p$ table, which can only be physically represented in p dimensions. In the simplest case when $p = 3$, we can represent the $r_1 \times r_2 \times r_3$ table as a solid with cells arrayed in rows, columns and "layers," and to avoid subscripts we shall use the initial letters as in the two-way case and call this a $r \times c \times l$ table. In point of fact, the three-way table is the only multivariate one which has received more than formal attention in the literature, since no new theoretical points arise when $p > 3$; but we shall see that the generalization from two to three dimensions does introduce new considerations.

B. N. Lewis (1962) gives an extended review of the subject.

33.59 Let us first consider the approach of **33.52**, where we partitioned the two-way $r \times c$ table in the case of independence. If we write n_{ijk} for the observed frequency and p_{ijk} for the probability in the cell in the ith row, jth column and kth layer, the hypothesis of complete independence is

$$H_0 : p_{ijk} = p_{i..} p_{.j.} p_{..k}, \qquad (33.125)$$

where a dot denotes summation over that subscript as before. In the two-way case, we had the partition (33.117–18) into "rows," "columns" and "rows × columns" components, with $(r-1)$, $(c-1)$ and $(r-1)(c-1)$ degrees of freedom respectively. In the present three-way case, we have the asymptotically additive components:

	Component	Degrees of freedom
Rows	X_R^2	$r-1$
Columns	X_C^2	$c-1$
Layers	X_L^2	$l-1$
Rows × columns	X_{RC}^2	$(r-1)(c-1)$
Rows × layers	X_{RL}^2	$(r-1)(l-1)$
Columns × layers	X_{CL}^2	$(c-1)(l-1)$
Rows × columns × layers	X_{RCL}^2	$(r-1)(c-1)(l-1)$
TOTALS	X_T^2	$rcl-1$

(33.126)

In the $2 \times 2 \times 2$ table, each component in (33.126) has 1 degree of freedom.

If we regard the $r \times c \times l$ table as a parallelepiped, the variation is thus expressed first of all in terms of edges, secondly in terms of faces, and finally in terms of the main body of the table.

The individual components in (33.126) are easily calculated. If there are hypothetical probabilities for any or all of the $p_{i..}, p_{.j.}, p_{..k}$, the corresponding components (X_R^2, X_C^2 and X_L^2) are simply the goodness-of-fit X^2 values for the row, column and layer marginal distributions, taken separately. If there are no hypothetical probabilities in any case, the corresponding one of these components is identically zero. We now compute the "ordinary" X^2 for testing independence in each of the three two-way tables. From the $(r \times c)$ table X^2, we subtract $(X_R^2 + X_C^2)$; from the $(r \times l)$ table X^2, we subtract $(X_R^2 + X_L^2)$; and from the $(c \times l)$ table X^2, we subtract $(X_C^2 + X_L^2)$. The results are X_{RC}^2, X_{RL}^2 and X_{CL}^2 respectively. Finally, we compute the X^2 for testing independence in the $(r \times c \times l)$ table. This is X_T^2, and X_{RCL}^2 is obtained by differencing.

Example 33.14 (Lancaster (1951), quoting data of Roberts *et al.*)

The following show some data for rats in a $2 \times 2 \times 2$ table classified according to whether they do or do not possess attributes, A, B, D. As before, we use α, β, δ to denote absence of the attributes.

The basic frequencies are:

$(ABD) = 475$	$(\alpha BD) = 467$
$(AB\delta) = 460$	$(\alpha B\delta) = 440$
$(A\beta D) = 462$	$(\alpha\beta D) = 494$
$(A\beta\delta) = 509$	$(\alpha\beta\delta) = 427$

We arrange these in three 2×2 tables thus:

Attributes	α	A	Totals	Attributes	δ	D	Totals	Attributes	δ	D	Totals
β	921	971	1892	β	936	956	1892	α	867	961	1828
B	907	935	1842	B	900	942	1842	A	969	937	1906
Totals	1828	1906	3734	Totals	1836	1898	3734	Totals	1836	1898	3734

The hypothetical probabilities of all the attributes are $\frac{1}{2}$. Thus for A we have

$$X_A^2 = \frac{(1828 - 3734/2)^2}{1867} + \frac{(1906 - 3734/2)^2}{1867} = 1 \cdot 6294. \tag{33.127}$$

Similarly, we find for the other components the values in the third column of the following table:

Component	Degrees of freedom	X^2 (prior hypothetical probabilities)	X^2 (parameters estimated)	
A	1	1·6294	0	
B	1	0·6695	0	
D	1	1·0295	0	
AB	1	0·1296	0·1176	(33.128)
AD	1	4·2517	4·3426	
BD	1	0·1296	0·1397	
ABD	1	2·7863	2·6904	
Totals	7	10·6256	7·2904	

For one degree of freedom, the 95 per cent point of χ^2 is 3·84 and the 97½ per cent point is 5·02. The only component near these values is the AD term, which lies between them. If there is any connexion between the factors at all, therefore, one would look for it between A and D, but the hypothesis of independence is not very strongly suspect. In fact, we find $V_{AD} = \left(\dfrac{X_{AD}^2}{n}\right)^{\frac{1}{2}} = 0.03$. In any case, the component ABD is within sampling limits, and if A and D were connected, we should expect the ABD component to be large. Furthermore, we must bear in mind, as always when partitioning X^2, that the separation of a single test into a number (here 7) increases the probability of some component falling outside its random sampling limits. On the whole, therefore, the conclusion seems to be that all three factors are independent (or so weakly dependent that there is no decisive indication of interdependence).

Had we not had prior probabilities but estimated them by marginal frequencies, we should have obtained the values of X^2 in the last column of (33.128). The values are very close to the previous ones, as they should be, and the same conclusion is reached.

It may be noted that we might have had prior information about some of the probabilities but not of others. In such a case, we should estimate those unknown and proceed as before.

33.60 The nature of the general multi-way table makes it possible to consider a large number of hypotheses other than that of complete independence, stated at (33.125). S. N. Roy and Mitra (1956), who make similar distinctions regarding the structure of multi-way tables as we did for 2×2 and $r \times c$ tables in **33.18** and **33.31** above, develop large-sample X^2 tests for a number of these. For example, we may wish to test

$$H_1 : \frac{p_{ijk}}{p_{..k}} = \frac{p_{i.k}}{p_{..k}} \frac{p_{.jk}}{p_{..k}}, \qquad (33.129)$$

which states that in a layer of a three-way table, the row and column variables are independent. This is the analogue of a zero partial correlation between rows and columns with layers fixed. Or we may wish to test

$$H_2 : p_{ijk} = p_{ij.} \, p_{..k}, \qquad (33.130)$$

which asserts the independence of the row-column classification, considered as a bivariate distribution, from the layers. This is the analogue of a zero multiple correlation of layers upon rows and columns.

By summing the two sides of (33.130) first over j and then over i, we see that it implies both

$$p_{i.k} = p_{i..} \, p_{..k} \qquad (33.131)$$

and

$$p_{.jk} = p_{.j.} \, p_{..k}. \qquad (33.132)$$

However (33.131–2) do not alone imply (33.130). S. N. Roy and Kastenbaum (1956) have investigated what additional hypothesis was necessary to ensure that (33.131–2) lead to (33.130). They rejected because of its mathematical intractability the natural

$$H_3 : p_{ijk} = \frac{p_{ij.}\, p_{.jk}\, p_{i.k}}{p_{.j.}\, p_{..k}\, p_{i..}}, \qquad (33.133)$$

and instead suggested

$$H_4 : p_{ijk} = a_{ij} a_{jk} a_{ik}, \qquad (33.134)$$

where the a's are arbitrary positive numbers. They show that (33.134) and (33.131–2) imply (33.130). (33.133) may be otherwise expressed as in Exercise 33.31.

33.61 In accordance with the terminology of the Analysis of Variance (Volume 3), (33.134) is the hypothesis that the *second-order interaction* in the table is zero. This problem was first considered for the $2 \times 2 \times 2$ table by Bartlett (1935b) and for the $2 \times 2 \times l$ table by Norton (1945). Lancaster (1951) proposed an alternative method based on the component X^2_{RCL} in (33.126), whose interpretation as a test of second-order interaction is critically discussed by Plackett (1962)—see also an illuminating discussion of the analogies with Analysis of Variance by Darroch (1962). Plackett proposed another test which is simplified by Goodman (1963b), who generalizes (1964c) the method to interactions of any order, discusses (1963a) the $2 \times 2 \times l$ case, gives (1964a) other methods based on cross-ratios, and gives (1964d) simple methods of testing and obtaining confidence intervals for second-order interactions. A general discussion of the estimation and testing of interactions is given by Goodman (1970, 1971). Goodman (1969, 1971) gives a method of partitioning the LR statistic mentioned in **33.35** into components relating to the second-order interaction in a three-way table, to partial association between two of the variables, and to the associations between the other two pairs; degrees of freedom are split up just as in the last four rows of (33.126). Altham (1970b) extends her work on cross-ratios (cf. **33.35**) to the three-way table.

Lindley (1964) gives a related Bayesian analysis of contingency tables—see also Bloch and Watson (1967).

33.62 Birch (1963) considers ML estimation of parameters in multi-way tables. He also (1964c) discusses a test for the existence of partial association in a $2 \times 2 \times l$ table, due essentially to Cochran (1954), and based on the statistic $n_{11.}$, approximately normal with mean and variance obtained by summing those in (33.51) over the l layers.

If $\theta_k = \log \frac{p_{11k} p_{22k}}{p_{12k} p_{21k}}$, H_0 is that all $\theta_k = 0$, and the test is UMPU against H_1: all θ_k equal and positive. Estimation of the assumed common value of θ_k is also discussed. Testing that the θ_k are *equal* is equivalent to testing the second-order interaction—cf. **33.61**. The theory is generalized to $r \times c \times l$ tables by Birch (1965) and Zelen (1971).

Lancaster (1960) has extended the ideas of canonical analysis to the multi-way table.

EXERCISES

33.1 Show that the coefficient of association Q is greater in absolute value than the coefficient of colligation Y, except when both are zero or unity in absolute value.

33.2 Derive the standard error of the coefficient V given at (33.17).

33.3 For the 3×4 table (from Ammon, *Zur Anthropologie der Badener*)

Eye colour group	B_1	B_2	B_3	B_4	Totals
A_1	1768	807	189	47	2811
A_2	946	1387	746	53	3132
A_3	115	438	288	16	857
Totals	2829	2632	1223	116	$6800 = n$

show that $X^2 = 1075 \cdot 2$, and hence that the coefficients (33.63–5) are

$$P = 0 \cdot 36,$$
$$T = 0 \cdot 25,$$
$$C = 0 \cdot 28.$$

33.4 Show that for an $r \times c$ table the Pearson coefficient of contingency P is equal to the Tschuprow coefficient T for two values of X^2/n, one of which is zero; that for X^2/n between these values $P > T$, and for X^2/n greater than the higher value $T > P$.

33.5 In experiments on the immunization of cattle from tuberculosis, the following results were secured:—

Table 33.4—Data from Report on the Spahlinger Experiments in Northern Ireland, 1931–1934

(H.M. Stationery Office, 1935)

	Died of Tuberculosis or very seriously affected	Unaffected or only slightly affected	Totals
Inoculated with vaccine	6	13	19
Not inoculated or inoculated with control media	8	3	11
Totals	14	16	30

Show that for this table, on the hypothesis that inoculation and susceptibility to tuberculosis are independent, $X^2 = 4 \cdot 75$, so that the hypothesis is rejected for $\alpha \geqslant 0 \cdot 029$; that with a correction for continuity the corresponding value of α is $0 \cdot 072$; and that by the exact method of **33.19–20**, $\alpha = 0 \cdot 070$.

33.6 Show that if two rows or two columns of an $r \times c$ table are amalgamated, X^2 for testing independence in the new table cannot be greater than X^2 for the original table, and in general will be less.

33.7 Show that if f is a standardized p-variate normal distribution with dispersion matrix $\mathbf{V}$ and marginal distributions $f_1, f_2, \ldots, f_p$,

$$\phi^2 = \frac{X^2}{n} = \int \cdots \int \frac{(f - f_1 f_2 \cdots f_p)^2}{f_1 f_2 \cdots f_p} dx_1 \cdots dx_p = \frac{1}{(|\mathbf{V}||\mathbf{W}|)^{\frac{1}{2}}} - 1,$$

where $\mathbf{W} = 2\mathbf{I} - \mathbf{V}$.

(K. Pearson, 1904)

33.8 In a multi-way table based on classification of a standardized p-variate normal distribution according to variates with correlations ρ_{ij}, show that

$$\log(1 + \phi^2) = -\tfrac{1}{2} \log|\mathbf{I} + \mathbf{P}| - \tfrac{1}{2} \log|\mathbf{I} - \mathbf{P}|,$$

where ϕ^2 is defined in Exercise 33.7 and $\mathbf{P}$ is the matrix with elements ρ_{ij}, $i \neq j$ and 0, $i = j$. Hence, by expanding, show that

$$\phi^2 > \tfrac{1}{2} \operatorname{tr} \mathbf{P}^2 = \sum_{i<j} \rho_{ij}^2.$$

(Lancaster, 1957)

33.9 For the $r \times c$ table with both sets of marginal frequencies fixed, consider the statistic

$$\frac{H}{n^2} = \sum_{i=1}^{r} \sum_{j=1}^{c} \left\{ \left(n_{ij} - \frac{n_{i.} n_{.j}}{n} \right)^2 \Big/ n_{.j} \right\} = \sum_{i=1}^{r} \frac{n_{i.}}{n} \sum_{j=1}^{c} \frac{D_{ij}^2}{n_{i.} n_{.j}/n},$$

which is a weighted sum of the contributions of the rows of the table to X^2 at (33.62), the weights being the proportional row frequencies $n_{i.}/n$. Show that on the hypothesis of independence

$$E(H) = (c-1) n (n^2 - \sum_i n_{i.}^2)/(n-1),$$

$$\operatorname{var} H = \frac{4\{(\sum_i n_{i.}^2)^2 - n \sum_i n_{i.}^3\}}{(n-1)(n-2)} \left\{ \left(c^2 - n \sum_j \frac{1}{n_{.j}}\right)(n-2) - 4(c-1)(n-c) \right\}$$

$$+ \left[\frac{2n^2 (n^2 - \sum_i n_{i.}^2)(\sum_i n_{i.}^2 - n)}{(n-1)^2 (n-2)(n-3)} + \frac{4(n+6)\{(\sum_i n_{i.}^2)^2 - n \sum_i n_{i.}^3\}}{(n-1)(n-2)(n-3)} \right]$$

$$\times \left\{ n(n-1)\left(c - 1 + \frac{1}{n} - \sum_j \frac{1}{n_{.j}}\right) + (c-1)^2 \right\}.$$

If all row marginal totals are equal, so that $n_{i.} = n/r$, show that $H = \frac{n^2}{r} X^2$, and that

$$E(H) = \frac{n^3}{(n-1)r}(r-1)(c-1)$$

$$\operatorname{var} H = \frac{2n^5 (n-r)(r-1)}{(n-1)^2 (n-2)(n-3) r^2} \left\{ n(n-1)\left((c-1) + \frac{1}{n} - \sum_j \frac{1}{n_{.j}}\right) + (c-1)^2 \right\},$$

so that

$$E(X^2) \to (r-1)(c-1),$$
$$\operatorname{var}(X^2) \to 2(r-1)(c-1),$$

as they must since its limiting distribution is of the χ^2 form with $(r-1)(c-1)$ degrees of freedom.

(C. A. B. Smith (1951–2); cf. also Haldane (1940))

33.10 Show as in Example 33.8 that for Table 33.5 $\sum_{i,j} a_{ij} b_{ij} = 2 \times 13{,}264{,}256$ and hence show that

$$t_b = 0\cdot 658,$$
$$t_c = 0\cdot 633.$$

Table 33.5—7477 women aged 30–39 employed in U.K. Royal Ordnance factories 1943–6 : unaided distance vision

Right eye \ Left eye	Highest grade	Second grade	Third grade	Lowest grade	TOTALS
Highest grade	1520	266	124	66	1976
Second grade	234	1512	432	78	2256
Third grade	117	362	1772	205	2456
Lowest grade	36	82	179	492	789
TOTALS	1907	2222	2507	841	7477

33.11 In the 4 × 4 tables of Example 33.8 and Exercise 33.10, show that the coefficient G defined by (33.79) takes the values 0·776, 0·798 respectively, with maximum standard errors given by (33.80) as 0·022, 0·014 respectively. Show also that the maximum standard errors obtained for the t_c values of 0·629, 0·633 are 0·029, 0·019 respectively.

(Goodman and Kruskal, 1963)

33.12 The following data, due to D. Chapman, relate the conditions under which homework was carried out (rated from the best, A_1, to the worst, A_5) and the teacher's assessment of the quality of the work (from best, B_1, to worst, B_3) :

	A_1	A_2	A_3	A_4	A_5	TOTALS
B_1	141	67	114	79	39	440
B_2	131	66	143	72	35	447
B_3	36	14	38	28	16	132
TOTALS	308	147	295	179	90	1019

Show by assigning natural-number scores to the categories that the regression coefficient of homework quality rating upon homework conditions (0·025 in these units) is within ordinary sampling fluctuation limits of zero, its standard error being 0·016.

(Yates, 1948)

33.13 The table below, due to A. R. Treloar, relates the periodontal condition of 135 women to their average daily calcium intake :

		Average grams of calcium per day			
		0–0·40	0·40–0·55	0·55–0·70	over 0·70
Periodontal condition	A	5	3	10	11
	B	4	5	8	6
	C	26	11	3	6
	D	23	11	1	2

CATEGORIZED DATA

Show that the canonical correlations are

$$R_1 = 0.56273, \qquad \begin{cases} nR_1^2 = 42.74984, \\ nR_2^2 = 1.59497, \\ nR_3^2 = 0.00003, \end{cases}$$
$$R_2 = 0.10869, \quad \text{so that}$$
$$R_3 = 0.00045,$$
$$\overline{X^2 = 44.34484,}$$

only the first of these components of X^2 exceeding conventional critical values for χ^2 distributions. Show that the canonical scores corresponding to R_1 are:—

Periodontal condition	Calcium intake
A: -1.3880	0–0.40 : 0.8397
B: -1.0571	0.40–0.55 : 0.4819
C: 0.6016	0.55–0.70 : -1.5779
D: 0.9971	over 0.70 : -1.1378

In particular, note the change of trend in the calcium intake scores at 0.70, which confirms the impression from the data that there seems to be a limit above which increased calcium intake does not further improve periodontal condition.

(E. J. Williams, 1952)

33.14 In Example 33.10, show that

$$\mathbf{N'N} = \begin{pmatrix} 0.633768533 & 0.192522708 & 0.146101456 & 0.092877193 \\ 0.192522708 & 0.444704410 & 0.241885812 & 0.093129969 \\ 0.146101456 & 0.241885812 & 0.474670557 & 0.200085907 \\ 0.092877193 & 0.093129969 & 0.200085907 & 0.466094141 \end{pmatrix}$$

and show that its four latent roots are given by unity and (33.103). Use (33.98) to evaluate the vector of y-scores (33.105) directly, and thence obtain the x-scores (33.104) by use of (33.99).

33.15 For the data of Example 33.13, show that the four 2×2 tables (33.120) yield components

$$2.860, \quad 2.180,$$
$$2.526, \quad 0.005,$$

totalling 7.571, as against $X^2 = 7.547$ for the original 3×3 table.

(Lancaster, 1949b)

33.16 In **33.53**, use the method of **33.52** to demonstrate the asymptotic partition of X^2 for the 2×3 table into single (2×2) components (33.119), and show that the argument may be extended to the general $r \times c$ table.

(Lancaster, 1949b)

33.17 In the multinomial $(p_1+p_2+ \ldots +p_r)^n$, define

$$w_i = \frac{n_i - np_i}{\{np_i(1-p_i)\}^{\frac{1}{2}}}, \qquad i = 1, 2, \ldots, r.$$

Show that, in the notation of partial correlations and regressions,

$$\rho_{ij} = -\{p_i p_j/(1-p_i)(1-p_j)\}^{\frac{1}{2}},$$
$$\rho_{ij.kl\ldots m} = -(p_i p_j/[\{(1-p_i)-(p_k+p_l+ \ldots +p_m)\}\{(1-p_j)-(p_k+p_l+ \ldots +p_m)\}])^{\frac{1}{2}},$$
$$\sigma_2^2 = 1, \quad \sigma_{i.j}^2 = \{1-(p_i+p_j)\}/\{(1-p_i)(1-p_j)\},$$
$$\beta_{ij.kl\ldots m} = -\{p_i p_j(1-p_j)/(1-p-i)\}^{\frac{1}{2}}/\{1-p_j(p_k+p_l+ \ldots +p_m)\}.$$

Show that w_1/σ_1, $w_{2.1}/\sigma_{2.1}$, $w_{3.21}/\sigma_{3.21}$, etc., are asymptotically normally distributed with zero means, unit variances and zero correlations. Show further that $w_{k.12..(k-1)} = 0$ and hence that the first $(k-1)$ w's provide a partition of X^2 into $(k-1)$ independent components.

(Lancaster, 1949b)

33.18 Let **C** and **D** be orthogonal matrices of rank r and c respectively and let the $(rc \times rc)$ matrix **K** be their direct product $\mathbf{C} \times \mathbf{D}$ and the variables
$$x_{ij} = (n_{ij} - np_{i.}p_{.j})/(np_{i.}p_{.j})^{\frac{1}{2}}$$
be arranged as a column vector
$$x \equiv \{x_{11}, x_{12}, \ldots, x_{1c}, x_{21}, x_{22}, \ldots, x_{2c}, \ldots, x_{r1}, x_{r2}, \ldots, x_{rc}\}'.$$
Show that by a suitable choice of the elements of **C** and **D**, the matrix $\mathbf{Y} = \mathbf{Kx}$ gives the components of X^2 for the $r \times c$ contingency table, y_{11} referring to the total, $y_{1k} (k \neq 1)$ the column totals, $y_{k1} (k \neq 1)$ the row totals and the remaining terms the other $(r-1)(c-1)$ degrees of freedom (cf. **33.52**).

(Lancaster, 1951)

33.19 In the particular case of a 3×3 table in Exercise 33.18, take
$$\mathbf{C} = \mathbf{D} = \begin{pmatrix} \frac{1}{\sqrt{3}} & \frac{1}{\sqrt{3}} & \frac{1}{\sqrt{3}} \\ \frac{1}{\sqrt{2}} & \frac{-1}{\sqrt{2}} & 0 \\ \frac{1}{\sqrt{6}} & \frac{1}{\sqrt{6}} & \frac{-2}{\sqrt{6}} \end{pmatrix}.$$
For the data of Example 33.13, obtain the matrix of x-values
$$\begin{pmatrix} 0 & -0.812829 & -0.409012 \\ -1.834459 & 1.653094 & 1.471167 \\ -0.499424 & -1.587173 & -0.070020 \end{pmatrix}$$
and hence verify the table of that example.

33.20 Show that for a $2 \times c$ table with $n_{1.} = n_{2.}$, X^2 reduces to the form
$$X^2 = \sum_{j=1}^{c} \frac{(n_{1j} - n_{2j})^2}{n_{1j} + n_{2j}}.$$
This is the test statistic for testing the homogeneity of two equal-sized samples polytomized into c categories.

33.21 In the $2 \times c$ table, suppose that $n_{1.} > n_{2.}$, and choose small integers k, h with $k > h \geqslant 1$ such that k/h approximates $n_{1.}/n_{2.}$ and hence $(hn_{1.} - kn_{2.})/n$ is small. Show that for the table, (33.121) may be written exactly as
$$X^2 = \frac{n^2}{(h+k)^2 n_{1.} n_{2.}} \left\{ \sum_{j=1}^{c} (hn_{1j} - kn_{2j})^2/n_{.j} - (hn_{1.} - kn_{2.})^2/n \right\},$$
and approximately as
$$X^2 \doteqdot \frac{1}{hk} \left\{ \sum_{j=1}^{c} (hn_{1j} - kn_{2j})^2/n_{.j} - (hn_{1.} - kn_{2.})^2/n \right\}$$
with an error of a factor less than $1 - \frac{(hn_{1.} - kn_{2.})}{n}$.

(Haldane, 1955b)

33.22 For the 2 × 11 table,

25	80	38	52	9	21	33	24	30	51	56	419
1	15	12	8	0	7	6	2	7	7	3	68
26	95	50	60	9	28	39	26	37	58	59	487

use the results of the last exercise with $h = 1$, $k = 6$ to show that $X^2 = 16.709$ by the exact formula, and 16.393 by the approximate one. This is an approximation error of 1.9 per cent, against the 2.3 per cent maximum allowed.

(Haldane, 1955b)

33.23 Show (cf. Exercise 33.20) that in a $r \times r$ contingency table, we may test complete symmetry (i.e. $H_0 : p_{ij} = p_{ji}$, $i,j = 1, 2, \ldots, r$, where p_{ij} is the probability of an observation occurring in the ith row, jth column) by the statistic

$$u = \sum_{i<j} \sum \frac{(n_{ij} - n_{ji})^2}{n_{ij} + n_{ji}}$$

asymptotically distributed as χ^2 with $\tfrac{1}{2}r(r-1)$ degrees of freedom.

(Bowker, 1948—see also Ireland *et al.* (1969))

33.24 Show that in a $r \times r$ table with identical categorizations (cf. **33.2**) we may test the homogeneity of the two underlying marginal distributions (i.e. $H_0 : p_{i.} = p_{.i}$, $i = 1, 2, \ldots, r$) by the quadratic form in the $(r-1)$ asymptotically normal variables $d_i = (n_{i.} - n_{.i})$, $i = 1, 2, \ldots, r-1$,

$$Q = \mathbf{d}' \mathbf{V}^{-1} \mathbf{d} = \sum_{i=1}^{r-1} \sum_{j=1}^{r-1} V^{ij} d_i d_j,$$

where V^{ij} is the (i,j)th element of the inverse of the $(r-1) \times (r-1)$ estimated dispersion matrix $\mathbf{V}$ of the d_i, whose elements are

$$V_{ii} = n_{i.} + n_{.i} - 2n_{ii} - (n_{i.} - n_{.i})^2/n, \quad V_{ij} = -\{n_{ij} + n_{ji} + (n_{i.} - n_{.i})(n_{j.} - n_{.j})/n\}, \quad i \neq j.$$

Show that Q is asymptotically distributed as χ^2 with $(r-1)$ degrees of freedom.

(cf. Stuart, 1955b; Madansky (1963) investigates the LR Test for marginal homogeneity in a multi-way table—cf. also Ireland *et al.* (1969), Kullback (1971) and Fryer (1971).)

33.25 Show that in a $r \times r$ table, the hypothesis of common proportionality of the diagonal cells, i.e.

$$H_0 : \frac{p_{ii}}{p_{i.}p_{.i}} = \frac{p_{jj}}{p_{j.}p_{.j}}, \quad i,j = 1, 2, \ldots, r,$$

may be tested by the statistic

$$V = \sum_{i=1}^{r} n_{i.} n_{.i} \frac{\left\{ \dfrac{n_{ii}}{n_{i.} n_{.i}} - \dfrac{\sum_{i} n_{ii}}{\sum_{i} n_{i.} n_{.i}} \right\}^2}{\sum_{i} n_{ii} / \sum_{i} n_{i.} n_{.i}},$$

asymptotically distributed as χ^2 with $(r-1)$ degrees of freedom.

(J. Durbin; cf. Glass (1954), p. 234)

33.26 Verify the values of the two sets of components of X^2 in (33.128), using the method given below (33.127).

33.27 Re-arrange the frequencies of Example 33.13, except 3018, in the $2 \times 2 \times 2$ table

$$\begin{array}{cc|cc} 3009 & 2832 & 3008 & 2974 \\ 3047 & 3051 & 2997 & 3038 \end{array},$$

the rows and columns being as laid out and the figures on the right being a "layer" above those on the left. Obtain the following partitions of X^2, the hypothetical probabilities all being $\frac{1}{2}$:

	X^2 (hypothetical probabilities)	X^2 (parameters estimated from data)
R	4·0115	0
C	1·1503	0
L	0·2540	0
RC	2·7357	2·7824
RL	1·7372	1·7547
LC	1·3524	1·3607
RCL	0·4690	0·5107
	11·7101	6·4085

(Lancaster, 1951)

33.28 In a $2 \times 2 \times 2$ table, let

$$T = (p_{1..})^{\frac{1}{2}}, \quad U = (p_{.1.})^{\frac{1}{2}}, \quad V = (p_{..1})^{\frac{1}{2}},$$
$$t = (p_{2..})^{\frac{1}{2}}, \quad u = (p_{.2.})^{\frac{1}{2}}, \quad v = (p_{..2})^{\frac{1}{2}},$$

and the (8×8) matrix $\mathbf{M}$ be the direct product of the matrices

$$\begin{pmatrix} V & v \\ -v & V \end{pmatrix} \times \begin{pmatrix} U & u \\ -u & U \end{pmatrix} \times \begin{pmatrix} T & t \\ -t & T \end{pmatrix}.$$

Let
$$x_{ijk} = (n_{ijk} - np_{i..}p_{.j.}p_{..k})/(np_{i..}p_{.j.}p_{..k})^{\frac{1}{2}},$$
and x represent the column vector

$$\mathbf{x} \equiv (x_{111} \; x_{211} \; x_{121} \; x_{221} \; x_{112} \; x_{212} \; x_{122} \; x_{222})'.$$

Show that the elements of $\mathbf{Mx}$ are the components of (asymptotically independent) χ variables for, in this order (cf. (33.126)),

$$T, R, C, RC, L, RL, CL, RCL.$$

(Lancaster, 1951)

33.29 The marginal probabilities of a $r \times c$ table, namely the $p_{i.}, p_{.j}$, are known, and it is required to estimate the cell probabilities p_{ij}. A sample of n observations is taken from the multinomial distribution with probabilities p_{ij}. Show that the ML estimators $\hat{p}_{ij}$ of the p_{ij} are the solutions of the $(r-1)(c-1)$ equations

$$\frac{n_{ij}}{p_{ij}} - \frac{n_{ic}}{p_{ic}} - \frac{n_{rj}}{p_{rj}} + \frac{n_{rc}}{p_{rc}} = 0, \quad i = 1, 2, \ldots, r-1; \; j = 1, 2, \ldots, c-1.$$

Show also that these are the modified MV unbiassed linear estimators of the p_{ij} obtained by applying (19.59) to the n_{ij} ($i = 1, 2, \ldots, r-1$: $j = 1, 2, \ldots, c-1$), their exact dispersion matrix $\mathbf{V}$ having been modified by replacing p_{ij} by $\hat{p}_{ij}$ throughout.

(El-Badry and Stephan (1955); cf. also J. H. Smith (1947))

33.30 On c separate occasions, the same set of n individuals are observed in respect of the possession (scored 1) or absence (scored 0) of an attribute, and the results put in the form of a $2 \times 2 \times \ldots \times 2 = 2^c$ table. Let T_j ($j = 1, 2, \ldots, c$) be the total number of 1's on the jth occasion and u_i ($i = 1, 2, \ldots, 2^c$) be the number of 1's among the c coordinates of the ith cell in the table. Show that if the probability of a " 1 " is identical on all c occasions, the statistic

$$Q = c(c-1) \sum_{j=1}^{c} (T_j - \bar{T})^2 / (c \sum_i u_i - \sum_i u_i^2)$$

(where the summations in the denominator are over all non-empty cells) is asymptotically distributed as χ^2 with $(c-1)$ degrees of freedom.

(Cochran, 1950; Madansky (1963) generalizes to the r^c table)

33.31 Defining the ratio $R_{ij.} = p_{ij.}/(p_{i..}\,p_{.j.})$, and $R_{i.k}$, $R_{.jk}$ similarly, show that (33.133) may be written

$$H_3 : R_{ij.} = (p_{ijk}/p_{..k})/\{(p_{i.k}/p_{..k})(p_{.jk}/p_{..k})\},$$

with similar expressions for H_3 in terms of $R_{i.k}$ and $R_{j.k}$, and show that H_3 is the hypothesis that each of these three ratios is invariant under variation in its suppressed categorized variable.

33.32 In Exercise 33.29, replace the ML estimation procedure by the following: use $n_{ij}/n = p_{ij}^{(0)}$ as an initial estimate of p_{ij}. Then alternately adjust it proportionally to the known row and column marginal probabilities, using

$$p_{ij}^{(2m-1)} = p_{ij}^{(2m-2)} p_{i.}/p_{i.}^{(2m-2)},$$
$$p_{ij}^{(2m)} = p_{ij}^{(2m-1)} p_{.j}/p_{.j}^{(2m-1)},$$

for $m = 1, 2, \ldots$. Show that this iterative proportional fitting procedure (IPFP) preserves the $(r-1)(c-1)$ cross-ratios in the originally observed table.

(Mosteller (1968). The IPFP is due to Deming and Stephan (1940). Fienberg (1970) shows that it converges to the true p_{ij}.)

CHAPTER 34

SEQUENTIAL METHODS

34.1 When considering sampling problems in the foregoing chapters we have usually assumed that the sample size n was fixed. This may be because we chose it beforehand; or it may be because n was not at our choice, as for example when we are presented with the results of a finished experiment; or it may be due to the fact that the sample size was determined by some other criterion, as when we decide to observe for a given period of time. We have made our inferences in domains for which n is a constant. For example, in setting a standard error to an estimate, we were effectively making probability statements within a field of samples all of size n. If n is determined in some way that is independent of the values of the observations, such a conditional argument is reasonable, but it is as well to realize that we are treating n as an ancillary statistic in the sense defined in **23.37**, and using the Conditionality Principle discussed there, which we saw did not necessarily lead to optimum procedures. This does not imply general acceptance of the Conditionality Principle, however: we hope we will not be thought quite cynical if we add that, in our view, the reason why so many statistical arguments are made conditionally upon an observed value of n is that this procedure is very convenient mathematically—cf. **22.32** above.

Sequential procedures

34.2 Occasionally, however, the sample size is a random variable explicitly dependent upon the values of the observations. One of the simplest cases is one we have already touched upon in Example 9.13 (Vol. 1, 3rd edn, p. 225). Suppose we are sampling human beings one by one to discover what proportion belong to a rare bloodgroup. Instead of sampling, say, 1000 individuals and counting the number of members of that blood-group we may prefer to go on sampling until 20 such members have occurred. We shall see later why this may be a preferable procedure; for the moment we take for granted that it is worth considering. In successive trials of such an inquiry we should doubtless find that for a fixed number of successes, say 20, the number n required to achieve them varied considerably. It must be at least 20 but it might be infinite (although the probability of going on indefinitely is zero, so that we are almost certain to stop sooner or later).

34.3 Procedures like this are called *sequential*. Their typical feature is a *sampling scheme*, which lays down a *stopping rule* under which we decide at each stage of the drawing whether to stop or to continue sampling. In our present example the rule is very simple: if we draw a failure, continue; if we draw a success, continue also unless 19 successes have previously occurred, in which event, stop. The decision at any point is, in general, dependent on the observations made up to that point. Thus, for a sequence of values $x_1, x_2, \ldots, x_n$, the sample size at which we stop is not independent of the x's. It is this fact which gives sequential analysis its characteristic features.

SEQUENTIAL METHODS

Sequential methods were first developed during the Second World War, principally by Wald (whose work is summarized in his book, Wald (1947)) in U.S.A., and simultaneously in England by G. A. Barnard (1946).

34.4 The ordinary case where we fix a sample number beforehand can be regarded as a very special case of a sequential scheme. The sampling procedure is then: go on until you have obtained n members, irrespective of what actual values arise. This, however, is a stopping rule of such a degenerate kind that it is of no interest here.

If the probability is unity that the procedure will terminate, the scheme is said to be *closed*. If there is a non-zero probability that sampling can continue indefinitely the scheme is called *open*. We shall not seriously consider open schemes in this chapter. They are obviously of little practical use compared to closed schemes, and we usually have to reduce them to closed form by putting an upper limit to the size of the sample. Such truncation often makes their properties difficult to determine exactly.

Usage in this matter is not entirely uniform in the literature of the subject. "Closed" sometimes means "truncated," that is to say, applies to the case where the stopping rule puts an upper limit to the sample size. Correspondingly, "open" sometimes means "non-truncated."

Example 34.1

As an example of a fairly simple sequential scheme let us consider sampling from a (large) population with proportion ϖ of successes. We will proceed until m successes are observed and then stop. It scarcely needs proof that such a scheme is closed. The probability that in an infinite sequence we do not observe m successes is zero.

The probability of $m-1$ successes in the first $n-1$ trials together with a success at the nth trial is (cf. **5.14–15**)

$$\binom{n-1}{m-1} \varpi^m \chi^{n-m}, \qquad n = m, \, m+1, \ldots, \tag{34.1}$$

where $\chi = 1 - \varpi$. This gives us the distribution of n. The frequency-generating function of n (with the origin at zero) is given by

$$\{\varpi t/(1-\chi t)\}^m. \tag{34.2}$$

Thus for the cumulant-generating function we have

$$\psi(t) = \log\left(\frac{\varpi e^t}{1-\chi e^t}\right)^m = m \log\left(\frac{\varpi}{e^{-t}-\chi}\right).$$

Expanding this as far as the coefficient of t^2 we find

$$\kappa_1 = m/\varpi, \tag{34.3}$$
$$\kappa_2 = m(1-\varpi)/\varpi^2 = m\chi/\varpi^2. \tag{34.4}$$

Thus the mean value of the sample size n is m/ϖ. It does not follow that m/n is an unbiassed estimator of ϖ. Such an unbiassed estimator is, in fact, given by

$$p = (m-1)/(n-1), \tag{34.5}$$

for if $m > 1$,
$$E\left(\frac{m-1}{n-1}\right) = \sum_{n=m}^{\infty} \frac{m-1}{n-1} \binom{n-1}{m-1} \varpi^m \chi^{n-m}$$
$$= \varpi \sum_{n=m}^{\infty} \binom{n-2}{m-2} \varpi^{m-1} \chi^{n-m} = \varpi, \tag{34.6}$$

while if $m = 1$, $p = 0$ for $n > 1$ and $p = 1$ for $n = 1$ remains unbiassed.

The variance of this estimator is not expressible in a very concise form. We have

$$E\left(\frac{m-1}{n-1}\right)^2 = (m-1)\varpi^{m-1}\chi^{1-m} \sum_{n=m}^{\infty} \binom{n-2}{m-2} \frac{\chi^{n-1}}{n-1}$$

$$= (m-1)\varpi^m \chi^{1-m} \int_0^{\chi} \sum_{n=m}^{\infty} \binom{n-2}{m-2} t^{n-2} dt$$

$$= (m-1)\varpi^m \chi^{1-m} \int_0^{\chi} t^{m-2}(1-t)^{1-m} dt. \qquad (34.7)$$

Putting $u = \varpi t/\{\chi(1-t)\}$ we find

$$E\left(\frac{m-1}{n-1}\right)^2 = (m-1)\varpi^2 \int_0^1 \frac{u^{m-2} du}{\varpi + \chi u}$$

$$= (m-1)\varpi^2 \int_0^1 u^{m-2} \left\{\sum_{j=0}^{\infty} \chi^j (1-u)^j\right\} du$$

$$= (m-1)\varpi^2 \sum_{j=0}^{\infty} \chi^j B(m-1, j+1)$$

$$= \varpi^2 \left[1 + \frac{\chi}{m} + \frac{2!\chi^2}{m(m+1)} + \frac{3!\chi^3}{m(m+1)(m+2)} + \ldots\right]. \qquad (34.8)$$

Hence, subtracting ϖ^2, we have

$$\text{var } p = \frac{\varpi^2 \chi}{m} \left[1 + \frac{2\chi}{m+1} + \frac{6\chi^2}{(m+1)(m+2)} + \ldots\right]. \qquad (34.9)$$

We can obtain an unbiassed estimator of var p in a simple closed form. In the same manner that we arrived at (34.6) we have

$$E\frac{(m-1)(m-2)}{(n-1)(n-2)} = \varpi^2.$$

Hence
$$E\left\{\left(\frac{m-1}{n-1}\right)^2 - \frac{(m-1)(m-2)}{(n-1)(n-2)}\right\} = E\left(\frac{m-1}{n-1}\right)^2 - \varpi^2.$$

Thus
$$\text{Est. var } p = \left(\frac{m-1}{n-1}\right)^2 - \frac{(m-1)(m-2)}{(n-1)(n-2)}$$

$$= \frac{(m-1)(n-m)}{(n-1)^2(n-2)}$$

$$= \frac{p(1-p)}{n-2} = \frac{p^2 q}{m-q-2p}. \qquad (34.10)$$

We note that for large n this is asymptotically equal to the corresponding result for fixed sample size n.

An estimator of the coefficient of variation of p for this negative binomial distribution is given by

$$\left(\frac{p^2 q}{m-q-2p}\right)^{\frac{1}{2}}/p,$$

and for small p this becomes approximately $(m-1)^{-\frac{1}{2}}$. Thus for the sequential process the *relative* sampling variation of p is approximately constant.

SEQUENTIAL METHODS

This sequential scheme is often called *inverse* binomial sampling—it was first discussed by Haldane (1945) and Finney (1949b). W. Knight (1965) unifies its theory for binomial, Poisson, hypergeometric and exponential distributions.

34.5 The sampling of attributes plays such a large part in sequential analysis that we may, before proceeding to more general considerations, discuss a useful diagrammatic method of representing the process.

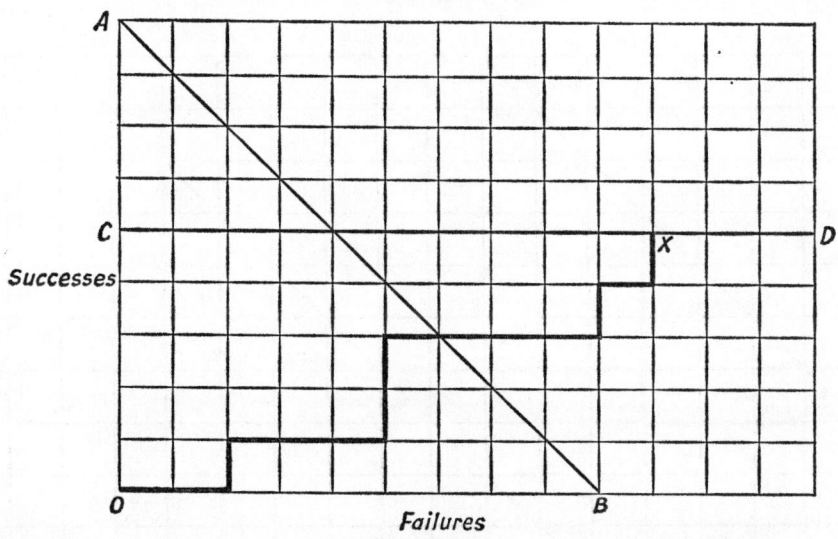

Fig. 34.1

Take a grid such as that of Fig. 34.1 and measure number of failures along the abscissa, number of successes along the ordinate. The sequential drawing of a sample may be represented on this grid by a path from the origin, moving one step to the right for a failure F and one step upwards for a success S. The path OX corresponds, for example, to the sequence $FFSFFFFSSFFFFSFS$. A stopping rule is equivalent to some sort of barrier on the diagram. For example, the line AB is such that $S+F = 9$ and thus corresponds to the case of fixed sample size $n = 9$. The line CD corresponds to $S = 5$ and is thus of the type we considered in Exercise 34.1 with $m = 5$. The path OX, involving a sample of 15, is then one sample which would terminate at X. If X is the point whose co-ordinates are (x, y) the number of different paths from O to X is the number of ways in which x can be selected from $(x+y)$. The probability of arriving at X is this number times the probability of x S's and y F's, namely

$$\binom{x+y}{x} \varpi^x \chi^y.$$

Example 34.2. Gambler's Ruin

One of the oldest problems in the theory of probability concerns a sequential process. Consider two players, A and B, playing a series of games at each of which A's chance of success is ϖ and B's is $1-\varpi$. The loser at each game pays the winner one unit. If A starts with a units and B with b units what are their chances of ruin (a player being ruined when he has lost his last unit)?

A series of games like this is a sequential set representable on a diagram like Fig. 34.1. We may take A's winning as a success. The game continues so long as A or B has any money left but stops when A has $a+b$ (when B has lost all his initial stake) or when B has $a+b$ (when A has lost his initial stake). The boundaries of the scheme are therefore the lines $y-x = -a$ and $y-x = b$.

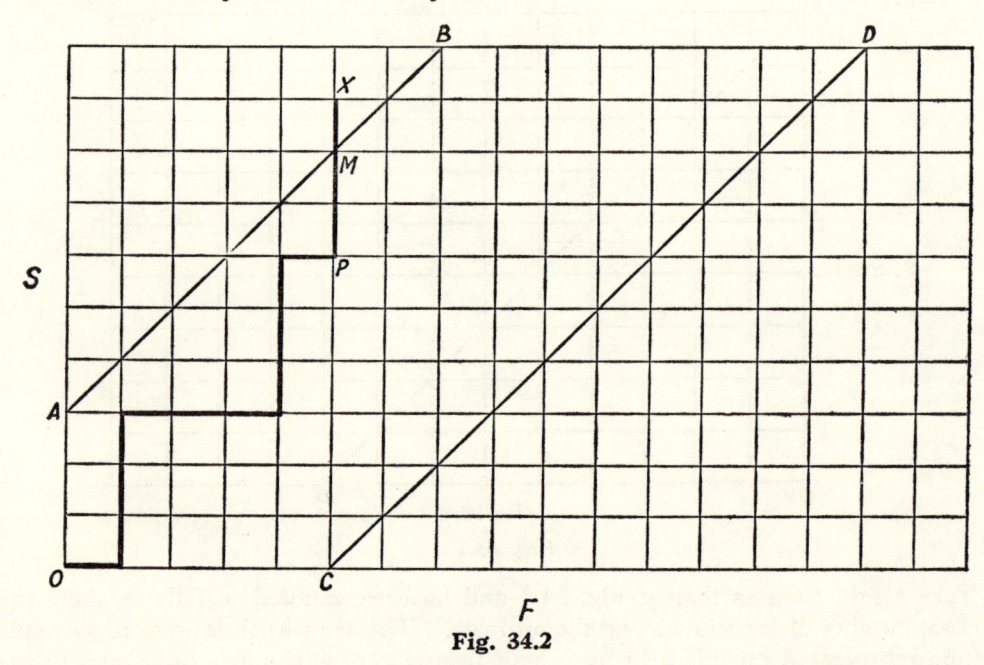

Fig. 34.2

Fig. 34.2 shows the situation for the case $a = 5$, $b = 3$. The lines AB, CD are at 45° to the axes and go through $F = 0$, $S = 3$ and $F = 5$, $S = 0$ respectively. For any point between these lines $S-F$ is less than 3 and $F-S$ is less than 5. On AB, $S-F$ is 3, and if a path arrives at that line B has lost three more games than A and is ruined; similarly, if the path arrives at CD, B is ruined. The sequential scheme is, then: if the point lies between the lines, continue sampling; if it reaches AB, stop with the ruin of B; if it reaches CD, stop with the ruin of A.

The actual probabilities are easily obtained. Let u_x be the probability that A will be ruined when he possesses x units. By considering a further game we see that

$$u_x = \varpi u_{x+1} + \chi u_{x-1}, \qquad (34.11)$$

with boundary conditions

$$u_0 = 1, \quad u_{a+b} = 0. \qquad (34.12)$$

The general solution of (34.11) is

$$u_x = A t_1^x + B t_2^x$$

where t_1 and t_2 are the roots of

$$\varpi t^2 - t + \chi = 0,$$

namely $\qquad t = 1 \quad \text{and} \quad t = \chi/\varpi.$

Provided that $\varpi \neq \chi$, the solution is then found to be, on using (34.12),

$$u_x = \frac{\left(\frac{\chi}{\varpi}\right)^{a+b} - \left(\frac{\chi}{\varpi}\right)^x}{\left(\frac{\chi}{\varpi}\right)^{a+b} - 1}, \qquad \varpi \neq \tfrac{1}{2}. \qquad (34.13)$$

If, however, $\varpi = \tfrac{1}{2}$, the solution is

$$u_x = \frac{a+b-x}{a+b}. \qquad (34.14)$$

In particular, at the start of the game, for $\varpi = \tfrac{1}{2}$, $x = a$,

$$u_a = \frac{b}{a+b}. \qquad (34.15)$$

34.6 We can obviously generalize this kind of situation in many ways and, in particular, can set up various types of boundary. A closed scheme is one for which it is virtually certain that the boundary will be reached.

Suppose, in particular, that the scheme specifies that if A loses he pays one unit but if B loses he pays k units. The path on Fig. 34.2 representing a series then consists of steps of unity parallel to the abscissa and k units parallel to the ordinate. And this enables us to emphasize a point which is constantly bedevilling the mathematics of sequential schemes: a path may not end exactly on a boundary, but may cross it. For example, with $k=3$ such a path might be OX in Fig. 34.2. After two successes and five failures we arrive at P. Another success would take us to X, crossing the boundary at M. We stop, of course, at this stage, whether the boundary is reached or crossed. The point of the example is that the exact probability of reaching the boundary at M is zero—in fact, this point is inaccessible. As we shall see, such discontinuities sometimes make it difficult to put forward exact and concise statements about the probabilities of what we are doing. We refer to such situations as "end-effects." In most practical circumstances they can be neglected.

Sequential tests of hypotheses

34.7 Let us apply the ideas of sequential analysis to testing hypotheses and, in the first instance, to choosing between H_0 and H_1. We suppose that these hypotheses concern a parameter θ which may take values θ_0 and θ_1 respectively; i.e. H_0 and H_1 are simple. We seek a sampling scheme which divides the sample space into three mutually exclusive domains: (a) domain ω_a, such that if the sample point falls within it we accept H_0 (and reject H_1); (b) domain ω_r, such that if the sample point falls within it we accept H_1 (and reject H_0); (c) the remainder of the sampling space, ω_c— if a point falls here we continue sampling. In Example 34.2, taking A's ruin as H_0, B's ruin as H_1, the region ω_a is the region to the right of CD, including the line itself; ω_r is the region above AB, including the line itself; ω_c is the region between the lines.

Operating characteristic

34.8 The probability of accepting H_0 when H_1 is true is a function of θ_1 which we shall denote by $K(\theta_1)$. If the scheme is closed the probability of rejecting H_0 when

H_1 is true is then $1-K(\theta_1)$. Considered as a function of θ_1 for different values of θ_1 this is simply the power function. As in our previous work we could, of course, work in terms of power; but in sequential analysis it has become customary to work with $K(\theta_1)$ itself.

$K(\theta)$ considered as a function of θ is called the "Operating Characteristic" (OC) of the scheme. Graphed as ordinate against θ as abscissa it gives us the "OC curve," the complement (to unity) of the Power Function.

Average sample number

34.9 A second function which is used to describe the performance of a sequential test is the "Average Sample Number" (ASN). This is the mean value of the sample size n required to reach a decision to accept H_0 or H_1 and therefore to discontinue sampling. The OC for H_0 and H_1 does not depend on the sample number, but only on constants determined initially by the sampling scheme. The ASN is the expected amount of sampling that we have to do to implement the scheme.

Example 34.3

Consider sampling from a (large) population of attributes of which proportion ϖ are successes, and let ϖ be small. We are interested in the possibility that ϖ is less than some given value ϖ_0. This is, for example, a frequently arising situation where a manufacturer of some item wishes to guarantee that the proportion of rejects in a batch of articles is below some declared figure. Consider first of all the alternative $\varpi_1 > \varpi_0$.

We will take a very simple scheme. If no success appears we proceed to sample until a pre-assigned sample number n_0 has appeared and accept ϖ_0. If, however, a success appears we accept ϖ_1 and stop sampling.

If the true probability of success is ϖ, the probability that we accept the hypothesis is then $(1-\varpi)^{n_0} = \chi^{n_0}$. This is the OC. It is a J-shaped curve decreasing monotonically from $\varpi = 0$ to $\varpi = 1$. For two particular values we merely take the ordinates at ϖ_0 and ϖ_1.

The common sense of the situation requires that we should accept the smaller of ϖ_0 and ϖ_1 if no success appears, and the larger if a success does appear. Let ϖ_0 be the smaller; then the probability of a Type I error α equals $1-\chi_0^{n_0}$ and that of an error of Type II, β, equals $\chi_1^{n_0}$. If we were to interchange ϖ_0 and ϖ_1, the α-error would be $1-\chi_1^{n_0}$ and the β-error $\chi_0^{n_0}$, both of which are greater than in the former case.

We can use the OC in this particular case to provide a test of the composite hypothesis $H_0 : \varpi \leqslant \varpi_0$ against $H_1 : \varpi > \varpi_0$. In fact, if $\varpi < \varpi_0$ the chance of an α-error is less than $1-\chi_0^{n_0}$ and the chance of a β-error is less than $\chi_0^{n_0}$.

The ASN is found by ascertaining the mean value of m, the sample number at which we terminate. For any given ϖ this is clearly

$$\sum_{m=1}^{n_0-1} m\varpi(1-\varpi)^{m-1} + n_0(1-\varpi)^{n_0-1}$$

$$= -\varpi\frac{\partial}{\partial\varpi}\sum_{0}^{n_0-1}(1-\varpi)^m + n_0(1-\varpi)^{n_0-1}$$

$$= \frac{1-(1-\varpi)^{n_0}}{\varpi}. \tag{34.16}$$

The ASN in this case is also a decreasing function of ϖ since it equals
$$\frac{1-\chi^{n_0}}{1-\chi} = 1+\chi+\chi^2+ \ldots +\chi^{n_0-1}.$$

We observe that the ASN will differ according to whether ϖ_0 or ϖ_1 is the true value.

A comparison of the results of the sequential procedure with those of an ordinary fixed sample-size is not easy to make for discontinuous distributions, especially as we have to compare two kinds of error. Consider, however, $\varpi_0 = 0.1$ and $n = 30$. From tables of the binomial (e.g. *Biometrika Tables*, Table 37) we see that the probability of 5 successes or more is about 0.18. Thus on a fixed sample-size basis we may reject $\varpi = 0.1$ in a sample of 30 with a Type I error of 0.18. For the alternative $\varpi = 0.2$ the probability of 4 or fewer successes is 0.26, which is then the Type II error.

With the sequential test, for a sample of n_0 the Type I error is $1-\chi_0^{n_0}$ and the Type II error is $\chi_1^{n_0}$. For a sample of 2 the Type I error is 0.19 and the Type II error 0.64. For a sample of 6 the errors are 0.47 and 0.26 respectively. We clearly cannot make both types of errors correspond in this simple case, but it is evident that samples of smaller size are needed in the sequential case to fix either type of error at a given level. With more flexible sequential schemes, both types of error can be fixed at given levels with smaller ASN than the fixed-size sample number. In fact, their economy in sample number is one of their principal recommendations—cf. Example 34.10.

Wald's probability-ratio test

34.10 Suppose we take a sample of m values in succession, $x_1, x_2, \ldots, x_m$, from a population $f(x, \theta)$. At any stage the ratio of the probabilities of the sample on hypotheses $H_0(\theta = \theta_0)$ and $H_1(\theta = \theta_1)$ is

$$L_m = \prod_{i=1}^{m} f(x_i, \theta_1) \bigg/ \prod_{i=1}^{m} f(x_i, \theta_0), \qquad (34.17)$$

We select two numbers A and B, related to the desired α- and β-errors in a manner to be described later, and set up a sequential test as follows: so long as $B < L_m < A$ we continue sampling; at the first occasion when $L_m \geqslant A$ we accept H_1; at the first occasion when $L_m \leqslant B$ we accept H_0.

An equivalent but more convenient form for computation is the logarithm of L_m, the critical inequality then being

$$\log B < \sum_{i=1}^{m} \log f(x_i, \theta_1) - \sum_{i=1}^{m} \log f(x_i, \theta_0) < \log A. \qquad (34.18)$$

This family of tests we shall refer to as " sequential probability-ratio tests " (SPR tests).

34.11 We shall often find it convenient to write

$$z_i = \log \{f(x_i, \theta_1)/f(x_i, \theta_0)\}, \qquad (34.19)$$

and the critical inequality (34.18) is then equivalent to a statement concerning the cumulative sums of z_i's. Let us first of all prove that a SPR test terminates with probability unity, i.e. is closed.

The sampling terminates if either
$$\Sigma z_i \geqslant \log A$$
or
$$\Sigma z_i \leqslant \log B.$$

The z_i's are independent random variables with variance, say $\sigma^2 > 0$. $\sum_{i=1}^{m} z_i$ then has a variance $m\sigma^2$. As m increases, the dispersion becomes greater and the probability that a value of Σz_i remains within the finite limits $\log B$ and $\log A$ tends to zero. More precisely, the mean $\bar{z}$ tends under the central limit effect to a (normal) distribution with variance σ^2/m, and hence the probability that it falls between $(\log B)/m$ and $(\log A)/m$ tends to zero.

> It was shown by Stein (1946) that $E(e^{mt})$ exists for any complex number t whose real part is less than some $t_0 > 0$. It follows that the random variable m has moments of all orders.

Example 34.4

Consider again the binomial distribution, the probability of success being ϖ. If there are k successes in the first m trials the SPR criterion is given by

$$\log L_m = k \log \frac{\varpi_1}{\varpi_0} + (m-k) \log \frac{1-\varpi_1}{1-\varpi_0}. \tag{34.20}$$

This quantity is computed as we go along, the sampling continuing until we reach the boundary values $\log B$ or $\log A$. How we decide upon A and B will appear in a moment.

34.12 It is a remarkable fact that the numbers A and B can be derived very simply (at least to an acceptable degree of approximation) from the probabilities of errors of the first and second kinds, α and β, without knowledge of the parent population. There are thus no distributional problems to be solved. This does not mean that the sequential process is distribution-free. All that is happening is that our knowledge of the frequency distribution is put into the criterion L_m of (34.17) and we work with this ratio of likelihoods directly. It will not, then, come as a surprise to find that SPR tests have certain optimum properties; for they use all the available information, including the order in which the sample values occur.

Consider a sample for which L_m lies between A and B for the first $n-1$ trials and then becomes $\geqslant A$ at the nth trial so that we accept H_1 (and reject H_0). By definition, the probability of getting such a sample is at least A times as large under H_1 as under H_0. This, being true for any one sample, is true for all and for the aggregate of all possible samples resulting in the acceptance of H_1. The probability of accepting H_1 when H_0 is true is α, and that of accepting H_1 when H_1 is true is $1-\beta$. Hence
$$1-\beta \geqslant A\alpha$$
or
$$A \leqslant \frac{1-\beta}{\alpha}. \tag{34.21}$$

In like manner we see from the cases in which we accept H_0 that
$$\beta \leqslant B(1-\alpha),$$
or
$$B \geqslant \frac{\beta}{1-\alpha}. \tag{34.22}$$

34.13 If our boundaries were such that A and B were exactly attained when attained at all, i.e. if there were no end-effects, we could write
$$A = \frac{1-\beta}{\alpha}, \quad B = \frac{\beta}{1-\alpha}. \tag{34.23}$$
In point of fact, Wald (1947) showed that for all practical purposes these equalities could be assumed to hold. Suppose that we have exactly
$$a = \frac{1-\beta}{\alpha}, \quad b = \frac{\beta}{1-\alpha} \tag{34.24}$$
and that the true errors of first and second kind for the limits a and b are α', β'. We then have, from (34.21),
$$\frac{\alpha'}{1-\beta'} \leqslant \frac{1}{a} = \frac{\alpha}{1-\beta}, \tag{34.25}$$
and from (34.22)
$$\frac{\beta'}{1-\alpha'} \leqslant b = \frac{\beta}{1-\alpha}. \tag{34.26}$$
Hence
$$\alpha' \leqslant \frac{\alpha(1-\beta')}{1-\beta} \leqslant \frac{\alpha}{1-\beta}, \tag{34.27}$$
$$\beta' \leqslant \frac{\beta(1-\alpha')}{1-\alpha} \leqslant \frac{\beta}{1-\alpha}. \tag{34.28}$$
Furthermore,
$$\alpha'(1-\beta) + \beta'(1-\alpha) \leqslant \alpha(1-\beta') + \beta(1-\alpha')$$
or
$$\alpha' + \beta' \leqslant \alpha + \beta. \tag{34.29}$$

Now in practice α and β are small, often conventionally 0·01 or 0·05. It follows from (34.27) and (34.28) that the amount by which α' can exceed α, or β' exceed β, is negligible. Moreover, from (34.29) we see that either $\alpha' \leqslant \alpha$ or $\beta' \leqslant \beta$. Hence, by using a and b in place of A and B, the worst we can do is to increase one of the errors, and then only by a very small amount. Such a procedure, then, will always be on the safe side in the sense that for all practical purposes it will not increase the errors of wrong decision. To avoid tedious repetition we shall henceforward use the equalities (34.23) except where the contrary is specified.

The inequalities for A and B were also derived for the critical value $1/k_\alpha$ of the fixed sample size probability-ratio test in Exercise 22.14. The fact that A and B practically attain their limits implies that the two sequential critical values enclose the single fixed-n critical value, as is intuitively acceptable.

Example 34.5

Consider again the binomial of Example 34.4 with $\alpha = 0·01$, $\beta = 0·10$, $\varpi_0 = 0·01$

and $\varpi_1 = 0\cdot 03$. We have, for k successes and $n-k$ failures (taking logarithms to base 10),

$$\log\frac{\beta}{1-\alpha} \leqslant (n-k)\log\frac{1-\varpi_1}{1-\varpi_0} + k\log\frac{\varpi_1}{\varpi_0} \leqslant \log\frac{1-\beta}{\alpha}$$

or

$$\log\frac{10}{99} \leqslant (n-k)\log\frac{97}{99} + k\log 3 \leqslant \log 90$$

or

$$-0\cdot 995{,}653 \leqslant -0\cdot 008{,}863{,}5\,(n-k) + (0\cdot 477{,}121)\,k \leqslant 1\cdot 954{,}243.$$

Dividing through by $0\cdot 008{,}863{,}5$ we find, to the nearest integer,

$$-112 \leqslant 54k - n \leqslant 220. \tag{34.30}$$

For a test of this kind, for example, if no success occurred in the first 112 drawings we should accept ϖ_0. If one occurred at the 100th drawing and another at the 200th, we could not accept before the 220th (i.e., $112 + (2 \times 54)$) drawing. And if, by the 200th drawing, there had occurred 6 successes, say at the 50th, 100th, 125th, 150th, 175th, 200th, we could not reject, $54k - n$ being 124 at the 200th drawing; but if that experience was then repeated, the quantity $54k - n$ would exceed 220 and we should accept ϖ_1.

The OC of the SPR test

34.14 Consider the function

$$L^h = \left\{\frac{f(x,\theta_1)}{f(x,\theta_0)}\right\}^h, \tag{34.31}$$

where h is a function of θ. $L^h f(x,\theta)$, say $g(x,\theta)$, is a frequency function for any value of θ provided that

$$E(L^h) = \int\left\{\frac{f(x,\theta_1)}{f(x,\theta_0)}\right\}^h f(x,\theta)\,dx = 1 \tag{34.32}$$

It may be shown (cf. Exercise 34.4) that there is at most one non-zero value of h satisfying this equation. Consider the rule: accept H_0, continue sampling, or accept H_1 according to the inequality

$$B^h \leqslant \frac{\Pi\left\{L^h f(x,\theta)\right\}}{\Pi\left\{f(x,\theta)\right\}} \leqslant A^h. \tag{34.33}$$

This is evidently equal to the ordinary rule of (34.18) provided that $h > 0$. Consider testing H: that the true distribution is $f(x,\theta)$, against G: that the true distribution is $g(x,\theta)$. If α', β' are the two errors, the likelihood ratio is the one appearing in (34.33), and we then have

$$A^h = \frac{1-\beta'}{\alpha'}, \quad B^h = \frac{\beta'}{1-\alpha'}, \tag{34.34}$$

and hence

$$\alpha' = \frac{1-B^h}{A^h - B^h},$$

and since α' is the power function when H_1 holds, its complement, the OC, is given by

$$1-\alpha' = K(\theta) = \frac{A^h - 1}{A^h - B^h}. \tag{34.35}$$

The same formula holds if $h < 0$.

We can now find the OC of the test. When $h(\theta) = 1$ we have the performance at $\theta = \theta_0$. When $h(\theta) = -1$ we have the performance at $\theta = \theta_1$. For other value we have, in effect, to solve (34.32) for θ and then substitute in (34.35). But this is, in fact, not necessary in order to plot the OC curve of $K(\theta)$ against θ. We can take $h(\theta)$ itself as a parameter and plot (34.35) against it.

Example 34.6

Consider once again the binomial of previous Examples. We may write for the discrete values 1 (success) and 0 (failure)

$$f(1, \varpi) = \varpi,$$
$$f(0, \varpi) = 1 - \varpi.$$

Then (34.32) becomes

$$\varpi \left(\frac{\varpi_1}{\varpi_0}\right)^h + (1-\varpi)\left(\frac{1-\varpi_1}{1-\varpi_0}\right)^h = 1,$$

or

$$\varpi = \frac{1 - \left(\frac{1-\varpi_0}{1-\varpi_1}\right)^h}{\left(\frac{\varpi_1}{\varpi_0}\right)^h - \left(\frac{1-\varpi_1}{1-\varpi_0}\right)^h}, \tag{34.36}$$

For $A = (1-\beta)/\alpha$, $B = \beta/(1-\alpha)$ we then have from (34.35)

$$K(\varpi) = \frac{\left(\frac{1-\beta}{\alpha}\right)^h - 1}{\left(\frac{1-\beta}{\alpha}\right)^h - \left(\frac{\beta}{1-\alpha}\right)^h}. \tag{34.37}$$

We can now plot $K(\varpi)$ against ϖ by using (34.36) and (34.37) as parametric equations in h.

The ASN of the SPR test

34.15 Consider a sequence of n random variables z_i. If n were a fixed number we should have

$$E\left(\sum_{i=1}^{n} z_i\right) = nE(z).$$

This is not true for sequential sampling, but we have instead the result

$$E\left(\sum_{i=1}^{n} z_i\right) = E(n)E(z), \tag{34.38}$$

which is not quite as obvious as it looks. The result is due to Wald and to Blackwell (1946), the following proof being due to Johnson (1959b).

Let each z_i have mean value μ, $E|z_i| \leq C < \infty$, and let the probability that n takes the value k be $P_k (k = 1, 2, \ldots)$. Consider the "marker" variable y_i which is unity if z_i is observed (i.e. if $n \geq i$) and zero in the opposite case. Then

$$P(y_i = 1) = P(n \geq i) = \sum_{j=i}^{\infty} P_j. \tag{34.39}$$

Now let $Z_n = \sum_{i=1}^{n} z_i$. Then

$$Z_n = \sum_{i=1}^{\infty} y_i z_i,$$

$$E(Z_n) = E \sum_{i=1}^{\infty} (y_i z_i) = \sum_{i=1}^{\infty} E(y_i z_i), \tag{34.40}$$

which will be finite if $E(n)$ is finite. Furthermore, since y_i depends only on $z_1, z_2, \ldots, z_{i-1}$ and not on z_i, we have

$$E(y_i z_i) = E(y_i) E(z_i). \tag{34.41}$$

Hence
$$E(Z_n) = \Sigma E(y_i) E(z_i) = \mu \Sigma E(y_i)$$
$$= \mu \sum_{i=1}^{\infty} \{P_i + P_{i+1} + \ldots\}$$
$$= \mu \sum_{i=1}^{\infty} i P_i$$
$$= \mu E(n),$$

whence (34.38) follows.

We then have

$$E(n) = \frac{E(Z_n)}{E(z)}. \tag{34.42}$$

But, to our usual approximation, Z_n can take only two values for the sampling to terminate, $\log A$ with probability $1 - K(\theta)$ and $\log B$ with probability $K(\theta)$. Thus

$$E(n) = \frac{K \log B + (1-K) \log A}{E(z)}, \tag{34.43}$$

which is the approximate formula for the average sample number.

Example 34.7

For the binomial we find

$$E(z) = E \log \left(\frac{f(x, \varpi_1)}{f(x, \varpi_0)} \right)$$
$$= \varpi \log \frac{\varpi_1}{\varpi_0} + (1 - \varpi) \log \frac{1 - \varpi_1}{1 - \varpi_0}. \tag{34.44}$$

The ASN can then be calculated from (34.43) when ϖ_1, ϖ_2, A and B (or α and β) are given. It is, of course, a function of ϖ.

34.16 For practical application, sequential testing for attributes is often expressed in such a way that the calculations are in terms of integers. Equation (34.30) is a case in point. We may rewrite it as

$$332 \geqslant 220 + (n-k) - 53k \geqslant 0$$

We may imagine a game in which we start with a score of 220. If a failure occurs we add one to the score; if a success occurs we lose 53 units. The game stops as soon as the score falls to zero or rises to 332, corresponding to acceptance of the values ϖ_0 and ϖ_1 respectively.

34.17 On such a scheme, suppose that we start with a score S_2. For every failure we gain one unit, but for every success we lose b units. If the score rises by S_1 so as to reach $S_1 + S_2 \ (= 2S, \text{ say})$ we accept one hypothesis; if it falls to zero we accept the other. Let the score at any point be x and the probability be u_x that it will ultimately reach $2S$ without in the meantime falling to zero. Consider the outcome of the next trial. A failure increases the score by unity to $x+1$, a success diminishes it by b to $x-b$. Thus

$$u_x = (1-\varpi) u_{x+1} + \varpi u_{x-b}, \qquad (34.45)$$

with initial conditions

$$u_0 = u_{-1} = u_{-2} = \ldots = u_{-b+1} = 0, \qquad (34.46)$$
$$u_{2S} = 1. \qquad (34.47)$$

For $b = 1$ this equation is easy to solve, as in Example 34.2. For $b > 1$ (and we shall suppose it integral) the solution is more cumbrous. We quote without proof the solution obtained by Burman (1946)

$$u_x = \frac{F(x)}{F(2S)} \qquad (34.48)$$

where

$$\left.\begin{array}{l} F(x) = \chi^{-x}\left\{1 - \binom{x-b-1}{1}\varpi\chi^b + \binom{x-2b-1}{2}(\varpi\chi^b)^2 \right. \\ \left. \qquad\qquad + \binom{x-3b-1}{3}(\varpi\chi^b)^3 + \ldots\right\}, \qquad x > 0, \\ = 0, \qquad x \leqslant 0. \end{array}\right\} \qquad (34.49)$$

Here the series continues as long as $x - kb - 1$ is positive. Burman also gave expressions for the ASN and the variance of the sample number.

34.18 Anscombe (1949a) tabulated functions of this kind. Putting

$$R_1 = \frac{S_1}{b+1}, \quad R_2 = \frac{S_2}{b+1}, \qquad (34.50)$$

Anscombe tabulates R_1, R_2 for certain values of the errors α, β (actually $1-\alpha$ and β) and the ratio S_1/S_2, the values for $\varpi(b+1)$ being also provided.

Given $\varpi_0, \varpi_1, \alpha, \beta$ we can find R_1 and R_2. There remains an element of choice according to how we fix the ratio S_1/S_2.

Thus, for $\varpi_0 = 0.01$, $\varpi_1 = 0.03$, $S_2 = 2S_1$, $\alpha = 0.01$, $\beta = 0.10$ we find $R_2 = 4$, $R_1 = 2$ approximately. Also $\varpi(b+1) = 0.571$ or $b = 56$. We then find, from (34.50), $S_1 = 114$, $S_2 = 228$. The agreement with Example 34.5 ($S_1 = 112$, $S_2 = 220$, $b = 53$) is very fair. The ASN for $\varpi = 0.01$ is 253 and that for $\varpi = 0.03$ is 306.

34.19 It is instructive to consider what happens in the limit when the units 1 and b are small compared to the total score 211. We can imagine this, on the diagram of Fig. 34.1, as a shrinkage of the mesh so that the routes approach a continuous random path of a particle subject to infinitesimal disturbances in two perpendicular directions. From this viewpoint the subject links up with the theory of Brownian motion and diffusion. If Δ is the difference operator defined by

$$\Delta u_x = u_{x+1} - u_x$$

we may write equation (34.45) in the form

$$\{(1-\varpi)(1+\Delta) + \varpi(1+\Delta)^{-b} - 1\} u_x = 0. \tag{34.51}$$

For small b this is nearly equivalent to

$$\{(1-\varpi) + \varpi - 1 + (1-\varpi)\Delta - b\varpi\Delta + \tfrac{1}{2}b(b+1)\varpi\Delta^2\} u_x = 0,$$

namely, to

$$\{(1-\varpi - b\varpi)\Delta + \tfrac{1}{2}b(b+1)\varpi\Delta^2\} u_x = 0. \tag{34.52}$$

In the limit this becomes

$$\frac{d^2 u}{dx^2} + \lambda \frac{du}{dx} = 0, \tag{34.53}$$

where

$$\lambda = \frac{2(1-\varpi - b\varpi)}{\varpi b(b+1)}. \tag{34.54}$$

The general solution of (34.53) is

$$u_x = k_1 + k_2 e^{-\lambda x},$$

and since the boundary conditions are

$$u_{2S} = 1, \quad u_0 = 0,$$

we have

$$u_x = \frac{1 - \exp(-\lambda x)}{1 - \exp\{-\lambda(S_1 + S_2)\}}. \tag{34.55}$$

Thus for $x = S_2$ the probability of acceptance is

$$u_{S_2} = \frac{\exp(\lambda S_2) - 1}{\exp(\lambda S_2) - \exp(-\lambda S_1)}. \tag{34.56}$$

34.20 As before, write

$$R_1(b+1) = S_1, \quad R_2(b+1) = S_2,$$

and let ϖ tend to zero so that $\varpi(b+1) = \gamma$, say, remains finite. From (34.54) we see that λ tends to zero, but

$$b\lambda \to \frac{2\{1 - \varpi(b+1)\}}{\varpi(b+1)} = \frac{2(1-\gamma)}{\gamma} = \delta, \text{ say.} \tag{34.57}$$

Then λS_1 tends to $S_1 \delta/b$, i.e. to $R_1 \delta$, provided that $\gamma \neq 1$, and (34.56) becomes

$$u_{S_2} = \frac{\exp(R_2 \delta) - 1}{\exp(R_2 \delta) - \exp(-R_1 \delta)}, \qquad \gamma \neq 1. \tag{34.58}$$

If δ tends to zero we find

$$u_{S_2} = \frac{R_2}{R_2 + R_1}, \qquad \gamma = 1. \tag{34.59}$$

(34.58) may be compared to (34.37) which, for small h, can be written

$$u_{S_2} = \frac{\exp\left(\frac{1-\beta}{\alpha} h\right) - 1}{\exp\left(\frac{1-\beta}{\alpha} h\right) - \exp\left(-\frac{1-\alpha}{\beta} h\right)}. \tag{34.60}$$

34.21 The use of sequential methods in the control of the quality of manufactured products has led to considerable developments of the kind of results mentioned in **34.17** and **34.18**. We shall not have the space here to discuss the subject in detail and the reader who is interested is referred to some of the textbooks on quality control. We will merely mention some of the extensions of the foregoing theory by way of illustrating the scope of the subject.

(a) *Stopping rules*. Even for a closed scheme it may be desirable to call a halt in the sampling at some stage. For instance, circumstances may prevent sampling beyond a certain point of time; or, in clinical trials, medical etiquette may require a change of treatment to a new drug which looks promising even before its value is fully established. Sequential schemes may be truncated in various ways, the simplest being to require stopping either after a given sample size has been reached or when a given time has elapsed. In such cases our general notions about performance characteristics and average sample numbers remain unchanged, but the actual mathematics and arithmetic are usually far more troublesome. Armitage (1957) considers sequential sampling under various restrictions.

(b) *Rectifying inspection*. In the schemes we have considered the hypotheses were that the batch or population under inquiry should be accepted or rejected as having a specified proportion of an attribute. If the attribute is " defectiveness " we may prefer not to reject a batch *in toto* but to inspect every member of it and to replace the defective ones. This does not of itself affect the general character of the scheme—the decision to reject is replaced by a decision to rectify—but it does, of course, affect the proportion of rejects in the whole aggregate of batches to which the sampling plan is applied—what is known as the average outgoing quality level (AOQL); and hence it affects the values of the parameters which we put into the plan. The theory was examined by Bartky (1943).

(c) *Double sampling*. As an extension of this idea, we may find it economical to proceed in stages. For example, we may decide to have four possible decisions: to accept outright; to reject outright; to continue sampling; to suspend judgment but to inspect fully. There is evidently a wide variety of possible choice here. An excellent example is the double sampling procedure of Dodge and Romig (1944). We shall encounter the idea again later in the chapter (**34.36**).

Example 34.8

Consider the testing, for the mean of a normal distribution with unit variance, of $H_0(\mu = \mu_0)$ against $H_1(\mu = \mu_1)$. With z defined as at (34.19) we have

$$z_i = -\tfrac{1}{2}(x_i - \mu_1)^2 + \tfrac{1}{2}(x_i - \mu_0)^2$$
$$= (\mu_1 - \mu_0) x_i - \tfrac{1}{2}(\mu_1^2 - \mu_0^2).$$
$$Z_m = \sum_{i=1}^{m} z_i = m(\mu_1 - \mu_0)\bar{x} - \tfrac{1}{2}m(\mu_1^2 - \mu_0^2). \tag{34.61}$$

We accept H_0 and H_1 according as this quality $\leq \log B$ or $\geq \log A$. For the appropriate OC curve we have, from (34.35),

$$K(\mu) = \frac{A^h - 1}{A^h - B^h}, \tag{34.62}$$

where h is given by

$$\frac{1}{\sqrt{(2\pi)}} \int_{-\infty}^{\infty} \exp[h\{(\mu_1 - \mu_0)x - \tfrac{1}{2}(\mu_1^2 - \mu_0^2)\}] \exp\{-\tfrac{1}{2}(x - \mu)^2\} dx = 1, \tag{34.63}$$

which is easily seen to be equivalent to

$$\exp\{\mu^2 - h\mu_1^2 + h\mu_0^2 - (\mu - h\mu_1 + h\mu_0)^2\} = 1$$

or to

$$h = \frac{\mu_1 + \mu_0 - 2\mu}{\mu_1 - \mu_0}, \qquad \mu_1 \neq \mu_0, \quad h \neq 0. \tag{34.64}$$

We can then draw the OC curve for a range of values of μ by calculating h from (34.64) and substituting in (34.62).

Likewise for the ASN we have

$$E(z) = \frac{1}{\sqrt{(2\pi)}} \int_{-\infty}^{\infty} \exp\{-\tfrac{1}{2}(x - \mu)^2\} \{(\mu_1 - \mu_0)x - \tfrac{1}{2}(\mu_1^2 - \mu_0^2)\} dx$$
$$= (\mu_1 - \mu_0)\mu - \tfrac{1}{2}(\mu_1^2 - \mu_0^2). \tag{34.65}$$

Again, for a range of μ the ASN can be determined from this equation in conjunction with (34.62) and

$$E(n) = \frac{K \log B + (1 - K) \log A}{E(z)}. \tag{34.66}$$

Manly (1970a) gives charts that permit choice of A and B for specified α, β, and also give the value of $E(n)$.

Example 34.9

Suppose that the mean of a normal distribution is known to be μ. To test a hypothesis, concerning its variance, $H_0: \sigma^2 = \sigma_0^2$ against $H_1: \sigma^2 = \sigma_1^2$, we have

$$Z_m = \Sigma z_i = -m \log \sigma_1 - \frac{1}{2\sigma_1^2} \Sigma(x - \mu)^2 + m \log \sigma_0 + \frac{1}{2\sigma_0^2} \Sigma(x - \mu)^2. \tag{34.67}$$

This lies between the limits $\log\{\beta/(1-\alpha)\}$ and $\log\{(1-\beta)/\alpha\}$ if

$$\log \frac{\beta}{1-\alpha} < -m \log \frac{\sigma_1}{\sigma_0} - \tfrac{1}{2}\left(\frac{1}{\sigma_1^2} - \frac{1}{\sigma_0^2}\right) \Sigma(x - \mu)^2 < \log \frac{1-\beta}{\alpha}. \tag{34.68}$$

With some rearrangement we find that this is equivalent to

$$\frac{2\log\frac{1-\beta}{\alpha}+m\log\frac{\sigma_1^2}{\sigma_0^2}}{\frac{1}{\sigma_0^2}-\frac{1}{\sigma_1^2}} \leqslant \Sigma(x-\mu)^2 \leqslant \frac{2\log\frac{\beta}{1-\alpha}+m\log\frac{\sigma_1^2}{\sigma_0^2}}{\frac{1}{\sigma_0^2}-\frac{1}{\sigma_1^2}}. \quad (34.69)$$

The OC and ASN are given in Exercises 34.18 and 34.19.

If the mean is not known the test remains the same except that the test statistic $\Sigma(x-\mu)^2$ is replaced by $\Sigma(x-\bar{x})^2$ and the value m in the inequality (34.69) is replaced by $(m-1)$.

The efficiency of a sequential test

34.22 In general, many different tests may be derived for given α and β, θ_0 and θ_1. There is no point in comparing their power for given sample numbers because they are arranged so as to have the same β-error. We may, however, define efficiency in terms of sample size or ASN. The test with the smaller ASN may reasonably be said to be the more efficient. Following Wald (1947) we shall prove that when end-effects are negligible the SPR test is a most efficient test. More precisely, if S' is a SPR test and S is some other test based on the sum of logarithms of identically distributed variables,

$$E_i(n \mid S) \geqslant E_i(n \mid S'), \quad i = 0, 1, \quad (34.70)$$

where E_i denotes the expected value of n on hypothesis H_i.

Note first of all that if u is any random variable, $u - E(u)$ is the value measured from the mean, and

$$\exp\{u - E(u)\} \geqslant 1 + \{u - E(u)\}.$$

On taking expectations we have

$$E[\exp\{u - E(u)\}] \geqslant 1, \quad (34.71)$$

which gives

$$E(\exp u) \geqslant \exp\{E(u)\}. \quad (34.72)$$

We also have, from (34.42), for *any* closed sequential test based on the sums of type Z_n

$$E_i(n \mid S) = \frac{E_i(\log L_n \mid S)}{E_i(z)}. \quad (34.73)$$

If E^* denotes the conditional expectation when H_0 is true, and E^{**} the conditional expectation when H_1 is true, we have, as at (34.22), neglecting end-effects,

$$E^*(L_n \mid S) = \frac{\beta}{1-\alpha}, \quad (34.74)$$

and similarly, as at (34.21),

$$E^{**}(L_n \mid S) = \frac{1-\beta}{\alpha}. \quad (34.75)$$

Hence

$$E_0(n \mid S) = \frac{1}{E_0(z)}\{(1-\alpha)E^*(\log L_n \mid S) + \alpha E^{**}(\log L_n \mid S)\}. \quad (34.76)$$

In virtue of (34.72), (34.74) and (34.75) we then have $E_0(z) \leq 0$ and

$$E_0(n|S) \geq \frac{1}{E_0(z)}\left\{(1-\alpha)\log\frac{\beta}{1-\alpha}+\alpha\log\frac{1-\beta}{\alpha}\right\}, \quad (34.77)$$

and interchanging H_0 and H_1, α and β in (34.77) gives

$$E_1(n|S) \geq \frac{1}{E_1(z)}\left\{\beta\log\frac{\beta}{1-\alpha}+(1-\beta)\log\frac{1-\beta}{\alpha}\right\}. \quad (34.78)$$

When $S = S'$ these inequalities, as at (34.43), are replaced (neglecting end-effects) by equalities. Hence (34.70).

Example 34.10

One of the recommendations of the sequential method, as we have remarked, is that for a given (α, β), it requires a smaller sample on the average than the method employing a fixed sample size. General formulae comparing the two would be difficult to derive, but we may illustrate the point on the testing of a mean in normal variation (Example 34.8).

For fixed n and α the test consists of finding a deviate d such that

$$\text{Prob}\{\mu_0-d \leq \bar{x} \leq \mu_0+d|H_0\} = 1-\alpha,$$
$$\text{Prob}\{\mu_0-d \leq \bar{x} \leq \mu_0+d|H_1\} = \beta,$$

and, putting

$$\lambda_0 = \sqrt{n}(d-\mu_0),$$
$$\lambda_1 = \sqrt{n}(d-\mu_1),$$

we have

$$n = \frac{(\lambda_1-\lambda_0)^2}{(\mu_0-\mu_1)^2}. \quad (34.79)$$

Given α, β, μ_0 and μ_1, n is determinable. Let us compare it with the ASN of a SPR test. Taking the approximate formula (34.43), which is

$$E_i(n) = \frac{1}{E_i(z)}[K(\mu)\log B + \{1-K(\mu)\}\log A], \quad (34.80)$$

we find, since

$$E_1(z) = \tfrac{1}{2}(\mu_0-\mu_1)^2$$

and

$$E_0(z) = -\tfrac{1}{2}(\mu_0-\mu_1)^2,$$

$$\frac{E_1(n)}{n} = \frac{2}{(\lambda_1-\lambda_0)^2}\{\beta\log B+(1-\beta)\log A\}. \quad (34.81)$$

Likewise we find

$$\frac{E_0(n)}{n} = -\frac{2}{(\lambda_1-\lambda_0)^2}\{(1-\alpha)\log B+\alpha\log A\}. \quad (34.82)$$

Thus, for $\alpha = 0.01, \beta = 0.03, A = 97, B = 3/99$ and we find $\lambda_0 = 2.5758, \lambda_1 = -1.8808$. The ratio $E_0(n)/n$ is then 0.43 and $E_1(n)/n = 0.55$. We thus require in the sequential case, on the average, either 43 or 55 per cent of the fixed sample size needed to attain the same performance.

> It should be emphasized that a reduced ASN will only be obtained by using a SPR test when one of H_0 or H_1 is true, and not necessarily for other parameter values, not

even for intermediate values. Kiefer and Weiss (1957) provided a theoretical basis for improving upon the SPR test in this respect. The normal case was studied by T. W. Anderson (1960), by Weiss (1962) and by D. Freeman and Weiss (1964); the latter two papers also solve the binomial case explicitly—see also Alling (1966) and Breslow (1970a). D. G. Hoel and Mazumdar (1969) treat the scale parameter of an exponential distribution.

Composite hypotheses

34.23 Although we have considered the test of a simple H_0 against a simple H_1, the OC and ASN functions are, in effect, calculated against a range of alternatives and therefore give us the performance of the test for a simple H_0 against a composite H_1. We now consider the case of a composite H_0. Suppose that θ may vary in some domain Ω. We require to test that it lies in some sub-domain ω_a against the alternatives that it lies either in a rejection sub-domain ω_r, or in a region of indifference $\Omega - \omega_a - \omega_r$ (which may be empty). We shall require of the errors two things: the probability that an error of the first kind, $\alpha(\theta)$, which in general varies with θ, shall not exceed some fixed number α for all θ in ω_a; and the probability that an error of the second kind, $\beta(\theta)$, shall not exceed β for all θ in ω_r. Wherever our parameter point θ really lies, then, we shall have upper limits to the errors, given by α and β.

34.24 Such a requirement, however, hardly constitutes a very effective criterion. We are always on the safe side, but may be so far on the safe side in particular cases as to lose a good deal of efficiency. Wald (1947) suggested that it might be better to consider the average of $\alpha(\theta)$ over ω_a and of $\beta(\theta)$ over ω_r as reasonable criteria. This raises the question as to what sort of average should be used. Wald defines two weighting functions, $w_a(\theta)$ and $w_r(\theta)$ such that

$$\int_{\omega_a} w_a(\theta)\,d\theta = 1, \qquad \int_{\omega_r} w_r(\theta)\,d\theta = 1, \tag{34.83}$$

and we then define

$$\int_{\omega_a} w_a(\theta)\alpha(\theta)\,d\theta = \alpha, \tag{34.84}$$

$$\int_{\omega_r} w_r(\theta)\beta(\theta)\,d\theta = \beta. \tag{34.85}$$

By these means we reduce the problem to one of testing simple hypotheses. In fact, if we let

$$L_{0m} = \int_{\omega_a} f(x_1,\theta)f(x_2,\theta)\ldots f(x_m,\theta)\,\alpha(\theta)\,d\theta, \tag{34.86}$$

$$L_{1m} = \int_{\omega_r} f(x_1,\theta)f(x_2,\theta)\ldots f(x_m,\theta)\,\beta(\theta)\,d\theta, \tag{34.87}$$

the likelihood ratio L_{0m}/L_{1m} can be used in the ordinary way with errors α and β. We may, if we like, regard (34.86) and (34.87) as the posterior probabilities of the sample when θ itself has prior probabilities $w_a(\theta)$ and $w_r(\theta)$.

34.25 This procedure, of course, throws the problem into the form of finding or choosing the weight functions $w_a(\theta)$ and $w_r(\theta)$. We are in the same position as the

632   THE ADVANCED THEORY OF STATISTICS

probabilist wishing to apply Bayes' theorem. We may resolve it by some form of Bayes' postulate, e.g. by assuming that $w_a(\theta) = 1$ everywhere in ω_a. Another possibility is to choose $w_a(\theta)$ and $w_r(\theta)$ so as to optimize some properties of the test.

For example, the choice of the test is made when we select α and β (or, to our approximation, A and B) and the weight functions. Among all such tests there will be maximum values of $\alpha(\theta)$ and $\beta(\theta)$. If we choose the weight functions so as to minimize (max α, max β), we have a test which, for given A and B, has the lowest possible bound to the average errors. If it is not possible to minimize the maxima of α and β simultaneously we may, perhaps, minimize the maximum of some function such as their sum.

A sequential *t*-test

34.26 A test proposed by Wald (1947) and, in a modified form, by other writers sets out to test the mean μ of a normal distribution when the variance is unknown. It is known as the sequential *t*-test because it deals with the same problem as "Student's" *t* in the fixed-sample case; but it does not free itself from the scale parameter σ in the same way and the name is, perhaps, somewhat misleading.

Specifically, we wish to test that, compared to some value μ_0, the deviation $(\mu-\mu_0)/\sigma$ is small, say $< \delta$. The three sub-domains of **34.23** are then as follows:

ω_a consists of (μ_0, σ) for all σ;

ω_r consists of values for which $|\mu-\mu_0| \geqslant \sigma\delta$, for all σ;

$\Omega - \omega_a - \omega_r$ consists of values for which $0 \leqslant |\mu-\mu_0| < \sigma\delta$, for all σ.

We define weight functions for σ as follows:

$$v_{ac} = \frac{1}{c}, \quad 0 \leqslant \sigma \leqslant c, \\ = 0 \text{ elsewhere.} \qquad (34.88)$$

$$v_{rc} = \frac{1}{2c}, \quad 0 \leqslant \sigma \leqslant c, \quad \mu = \mu_0 + \delta\sigma \\ = 0 \text{ elsewhere.} \qquad (34.89)$$

Then

$$L_{1m} = \int_0^c v_{rc} \frac{1}{(2\pi)^{\frac{1}{2}m} \sigma^m} \exp\left\{-\frac{1}{2\sigma^2} \Sigma(x_i-\mu)^2\right\} d\sigma$$

$$= \frac{1}{(2\pi)^{\frac{1}{2}m}} \cdot \frac{1}{2c} \int_0^c \left[\frac{1}{\sigma^m} \exp\left\{-\frac{1}{2\sigma^2}\Sigma(x_i-\mu_0-\delta\sigma)^2\right\} \right. $$
$$\left. +\frac{1}{\sigma^m}\exp\left\{-\frac{1}{2\sigma^2}\Sigma(x_i-\mu_0+\delta\sigma)^2\right\}\right] d\sigma. \qquad (34.90)$$

$$L_{0m} = \frac{1}{(2\pi)^{\frac{1}{2}m}} \frac{1}{c} \int_0^c \frac{1}{\sigma^m} \exp\left\{-\frac{1}{2\sigma^2}\Sigma(x_i-\mu_0)^2\right\} d\sigma. \qquad (34.91)$$

The limit of the ratio L_{1m}/L_{0m} as c tends to infinity then becomes

$$\lim L_{1m}/L_{0m} = \frac{\frac{1}{2}\int_0^\infty \frac{d\sigma}{\sigma^m}\left[\exp\left\{-\frac{1}{2\sigma^2}\Sigma(x_i-\mu_0-\delta\sigma)^2\right\}+\exp\left\{-\frac{1}{2\sigma^2}\Sigma(x_i-\mu_0+\delta\sigma)^2\right\}\right]}{\int_0^\infty \frac{d\sigma}{\sigma^m}\exp\left\{-\frac{1}{2\sigma^2}\Sigma(x_i-\mu_0)^2\right\}}.$$

$$(34.92)$$

This depends on the x's, which are observed, and on μ_0 and δ, which are given, but not on σ, which we have integrated out of the problem by the weight functions (34.88) and (34.89). If we can evaluate the integrals in (34.92) we can apply this ratio to give a sequential test.

34.27 The rather arbitrary-looking weight functions are, in fact, such as to optimize the test. To prove this we first of all establish (a) that $\alpha(\mu, \sigma)$ is constant in ω_a; (b) that $\beta(\mu, \sigma)$ is a function of $|(\mu-\mu_0)/\sigma|$ alone : and (c) that $\beta(\mu, \sigma)$ is monotonically decreasing in $|(\mu-\mu_0)/\sigma|$.

If $\bar{x}$ is the sample mean and S^2 is the sum $\Sigma(x_i-\bar{x})^2$, the distribution of the ratio $(\bar{x}-\mu_0)/S$ depends only on $(\mu-\mu_0)/\sigma$. If then we can show that (34.92) is a single-valued function of $(\bar{x}-\mu_0)/S$, the properties (a) and (b) will follow, for $(\mu-\mu_0)/\sigma$ is zero in ω_a and $\beta(\mu, \sigma)$ depends only on the distribution of (34.92). Now the numerator and denominator of (34.92) are both homogeneous functions of $(x_i-\mu_0)$ of degree $m-1$, as may be verified by putting $x_i = \lambda x_i$, $\mu_0 = \lambda \mu_0$, $\sigma = \lambda \sigma$. Thus the ratio of (34.92) is of degree zero. Further, it is a function of $\Sigma(x-\mu_0)^2$ and $\Sigma(x-\mu_0)$ only, and hence we do not change it by putting $(x_i-\mu_0)/\sqrt{\Sigma(x_i-\mu_0)^2}$ for $x_i-\mu_0$. The ratio is, then, a function only of $\Sigma(x_i-\mu_0)/\sqrt{\Sigma(x_i-\mu_0)^2}$, and is, in actual fact, a function of the square of that quantity, namely of

$$\frac{(\bar{x}-\mu_0)^2}{\Sigma(x_i-\mu_0)^2} = \frac{(\bar{x}-\mu_0)^2}{n(\bar{x}-\mu_0)^2 + S^2}.$$

It is therefore a single-valued function of $(\bar{x}-\mu_0)/S$.

To show that $\beta(\mu, \sigma)$ is monotonically decreasing in $|(\mu-\mu_0)/\sigma|$ it is sufficient to show that the ratio (34.92) is a strictly increasing function of $|(\bar{x}-\mu_0)/S|$, or equivalently of $(\bar{x}-\mu_0)^2/\Sigma(x_i-\mu_0)^2$. Now for fixed $\Sigma(x_i-\mu_0)^2$ the denominator of (34.92) is fixed and the numerator is an increasing function of $(\bar{x}-\mu_0)^2$. Thus the whole ratio is increasing in $(\bar{x}-\mu_0)^2$ for fixed $\Sigma(x_i-\mu_0)^2$ and the required result follows.

34.28 Under these conditions we can prove that the sequential t-test is optimal. In fact, any test is optimal if (i) $\alpha(\theta)$ is constant in ω_a; (ii) $\beta(\theta)$ is constant over the boundary of ω_r; and (iii) if $\beta(\theta)$ does not exceed its boundary value for any θ inside ω_r.

To prove this, let v_a and v_r be two weight functions obeying these conditions and w_a, w_r two other weight functions. Let α, β be the errors for the first set, α^*, β^* those for the second. Then we have

$$\frac{1-\beta}{\alpha} = A, \quad \frac{\beta}{1-\alpha} = B,$$

and hence

$$\int_{\omega_a} \alpha^*(\theta) w_a(\theta) d\theta = \alpha = \frac{1-B}{A-B}, \qquad (34.93)$$

$$\int_{\omega_r} \beta^*(\theta) w_r(\theta) d\theta = \beta = \frac{B(A-1)}{A-B}. \qquad (34.94)$$

Thus, in ω_a the maximum of $\alpha^*(\theta)$ is greater than $(1-B)/(A-B)$, for the integral of $w_a(\theta)$ over that region is unity. But if v_a has constant α over that domain, its maximum

is equal to $(1-B)/(A-B)$. Hence
$$\max \alpha^*(\theta) \geqslant \max \alpha(\theta) \text{ in } \omega_a.$$
Likewise, in ω_r the maximum of $\beta^*(\theta)$ is attained somewhere outside ω_r and cannot exceed $B(A-1)/(A-B)$, whereas for $\beta(\theta)$ the maximum value must be attained somewhere. Hence
$$\max \beta^*(\theta) \geqslant \max \beta(\theta) \text{ in } \omega_r.$$
The result follows. The conditions we have considered are sufficient but by no means necessary.

34.29 Some tables for the use of this test were provided by Armitage (1947). The integrals occurring in (34.92) are, in fact, expressible in terms of the confluent hypergeometric function and, in turn, in terms of the distribution of non-central t. *Tables to facilitate sequential t-Tests* (U.S.A. National Bureau of Standards, Applied Mathematics Series 7, 1951) is a co-operative work with an introduction by K. J. Arnold. See also later work by Armitage and others, most recently Myers *et al.* (1966).

An alternative method of attack is given by D. R. Cox (1952a, b) for the case where the distribution of a set of sufficient statistics factorizes as at (23.118). A SPR test of θ_l can then be based on T_s, free of the nuisance (often scale) parameters θ_k, in the ordinary way. This approach also requires an invariance condition—cf. Hall *et al.* (1965). Hajnal (1961) develops a two-sample sequential t-test along these lines.

D. R. Cox (1963) gives a large-sample sequential test for any composite hypothesis, based on ML theory—cf. Exercise 34.21.

Sequential estimation; the moments and distribution of n

34.30 In testing hypotheses we usually fix the errors in advance and proceed with the sampling until acceptance or rejection is reached. We may also use the sequential process for estimation, but our problems are then more difficult to solve and may even be difficult to formulate. We draw a sequence of observations with the object of estimating some parameter of the parent; but in general it is not easy to decide what is the appropriate estimator, what biases are present, what are the sampling errors or what should be the rules determining the ending of the sampling process. The basic difficulty is that the sample size has a complicated distribution. A secondary nuisance is the end-effect to which we have already referred.(*)

34.31 We derived at (34.42) the mean of n, and now rewrite (34.42) in the form
$$E\{Z_n - nE(z)\} = 0. \qquad (34.95)$$
Let us assume that absolute second moments exist, that the variance of each z_i is equal to σ^2, and that $E(n^2)$ exists. The variance of $\{Z_n - nE(z)\}$ may then be derived (the proof is left to the reader as Exercise 34.5) as
$$E\{Z_n - nE(z)\}^2 = \sigma^2 E(n) \qquad (34.96)$$
and it follows that, with $E(z) = \mu$ as before,
$$\mu^2 E(n^2) = \sigma^2 E(n) + 2\mu E(nZ_n) - E(Z_n^2).$$

(*) Lehmann and Stein (1950) considered the concept of completeness (cf. 23.9) in the sequential case, but general criteria are not easy to apply even in attribute sampling—cf. de Groot (1959).

SEQUENTIAL METHODS

If Z_n and n are uncorrelated, this simplifies to
$$\mu^2 \operatorname{var} n = \sigma^2 E(n) - \operatorname{var} Z_n. \tag{34.97}$$
Similar results for higher moments have been obtained by Wolfowitz (1947)—cf. Exercise 34.6. See also Chow *et al.* (1965).

Approximate expressions for the first four moments of n are obtained by B. K. Ghosh (1969) by differentiating its c.f., given in Exercise 34.11 below. He applies his results to the cases of the normal mean, normal variance, the binomial, the Poisson and the exponential scale parameter. The distribution of n is also discussed—Exercise 34.13 below treats the case of the normal mean. See also Manly (1970b) and Chanda (1971).

34.32 Now let
$$Y_n = \sum_{i=1}^{n} \frac{\partial}{\partial \theta} \log f(x_i, \theta). \tag{34.98}$$
Then, under regularity conditions, $E(\partial \log f / \partial \theta) = 0$ as at (17.18) and we have
$$E(Y_n) = 0, \tag{34.99}$$
and
$$\operatorname{var} Y_n = E\left(\Sigma \frac{\partial \log f}{\partial \theta}\right)^2 = E(n) E\left(\frac{\partial \log f}{\partial \theta}\right)^2. \tag{34.100}$$
as at (34.95). If t is an estimator of θ with bias $b(\theta)$, i.e. is such that
$$E(t) = \theta + b(\theta)$$
we have, differentiating this equation,
$$\operatorname{cov}\left(t, \Sigma \frac{\partial \log f}{\partial \theta}\right) = E\left(t \Sigma \frac{\partial \log f}{\partial \theta}\right) = 1 + b'(\theta). \tag{34.101}$$
Then, by the Cauchy–Schwarz inequality
$$\operatorname{var} t \, E\left(\Sigma \frac{\partial \log f}{\partial \theta}\right)^2 \geqslant \{1 + b'(\theta)\}^2,$$
and hence, by (34.100),
$$\operatorname{var} t \geqslant \frac{\{1 + b'(\theta)\}^2}{E(n) E\left(\frac{\partial \log f}{\partial \theta}\right)^2}, \tag{34.102}$$
which is Wolfowitz's form of the lower bound to the variance in sequential estimation. It consists simply of putting $E(n)$ for n in the MVB (17.22). Wolfowitz (1947) also gives an extension of the result to the simultaneous estimation of several parameters.

Example 34.11

Consider the binomial with unit index
$$f(x, \varpi) = \varpi^x (1 - \varpi)^{1-x}, \quad x = 0, 1.$$
We have
$$\frac{\partial \log f}{\partial \varpi} = \frac{x}{\varpi} - \frac{1-x}{1-\varpi}, \quad E\left(\frac{\partial \log f}{\partial \varpi}\right)^2 = \frac{1}{\varpi(1-\varpi)}.$$
If p is an unbiassed estimator of ϖ in a sample from this distribution, we then have
$$\operatorname{var} p \geqslant \frac{\varpi(1-\varpi)}{E(n)}. \tag{34.103}$$

34.33 If n is large the theory of sequential estimation is very much simplified in virtue of a general result due to Anscombe (1949b, 1952, 1953). Simply stated, this amounts to saying that for statistics where a Central Limit effect is present, the formulae for standard errors are the same for sequential samples as for samples of fixed size. We might argue this heuristically from (34.102). n varies about its mean n_0 with standard deviation of order $n_0^{-\frac{1}{2}}$ and thus formulae accurate to order n^{-1} remain accurate to that order if we use n_0 instead of n. More formally:

Let $\{Y_n\}$, $n = 1, 2, \ldots$ be a sequence of random variables. Let there exist a real number θ, a sequence of positive numbers $\{w_n\}$, and a distribution function $F(x)$ such that

(a) Y_n converges to θ in the scale of w_n, namely

$$P\left\{\frac{Y_n - \theta}{w_n} \leqslant x\right\} \to F(x) \text{ as } n \to \infty; \tag{34.104}$$

(b) $\{Y_n\}$ is uniformly continuous in probability, namely given (small) positive ε and η,

$$P\left\{\left|\frac{Y_{n'} - Y_n}{w_n}\right| < \varepsilon \text{ for all } n, n' \text{ such that } |n' - n| < \varepsilon n\right\} > 1 - \eta. \tag{34.105}$$

Let $\{n_r\}$ be an increasing sequence of positive integers tending to infinity and $\{N_r\}$ be a sequence of random variables taking positive integral values such that $N_r/n_r \to 1$ in probability as $r \to \infty$. Then

$$P\left\{\frac{Y_{N_r} - \theta}{w_{N_r}} \leqslant x\right\} \to F(x) \text{ as } r \to \infty \tag{34.106}$$

in all continuity points of $F(x)$.

The complexity of the enunciation and the proof are due to the features we have already noticed: end-effects (represented by the relation between N_r and n_r) and the variation in n_r.

In fact, let (34.105) be satisfied with ν large enough so that for any $n_r > \nu$

$$P\{|N_r - n_r| < cn_r\} > 1 - \eta. \tag{34.107}$$

Consider the event E: $|N_r - n_r| < cn_r$ and $|Y_{N_r} - Y_{n_r}| < \varepsilon w_{N_r}$,

and the events A: $|Y_{n'} - Y_n| < \varepsilon w_n$, all n' such that $|n' - n| < \varepsilon n$,

B: $|N_r - n_r| < cn_r$.

Then
$$P(E) \geqslant P\{A \text{ and } B\} = P(A) - P\{A \text{ and not-}B\}$$
$$\geqslant P(A) - P(\text{not-}B)$$
$$\geqslant 1 - 2\eta. \tag{34.108}$$

Also
$$P\{Y_{N_r} - \theta \leqslant xw_{N_r}\} = P\{Y_{N_r} - \theta \leqslant xw_{n_r} \text{ and } E\}$$
$$+ P\{Y_{N_r} - \theta \leqslant xw_{n_r} \text{ and not-}E\}.$$

Thus, in virtue of the definition of E we find

$$P\{Y_{n_r} - \theta \leqslant (x - \varepsilon)w_{n_r}\} - 2\eta < P\{Y_{n_r} - \theta \leqslant xw_{n_r}\}$$
$$< P\{Y_{n_r} - \theta \leqslant (x + \varepsilon)w_{n_r}\} + 2\eta,$$

and (34.106) follows. It is to be noted that the proof does not assume N_r and Y_n to be independent.

34.34 To apply this result to sequential estimation, let $x_1, x_2, \ldots$ be a sequence of observations and Y_n an estimator of a parameter θ, D_n an estimator of the scale w_n of Y_n. The sampling rule is: given some constant k, sample until the first occurring $D_n \leqslant k$ and then calculate Y_n. We show that Y_n is an estimator of θ with scale asymptotically equal to k if k is small.

Let conditions (34.104) and (34.105) be satisfied and $\{k_r\}$ be a sequence of positive numbers tending to zero. Let $\{N_r\}$ be the sequence of random variables such that N_r is the least integer n for which $D_n \leqslant k_r$; and let $\{n_r\}$ be the sequence such that n_r is the least n for which $w_n \leqslant k_r$. We require two further conditions:

(c) $\{w_n\}$ converges monotonically to zero and $w_n/w_{n+1} \to 1$ as $n \to \infty$;
(d) N_r is a random variable for all r and $N_r/n_r \to 1$ in probability as $r \to \infty$.

Condition (c) implies that $w_{n_r}/k_r \to 1$ as $n \to \infty$. It then follows from our previous result that

$$P\left\{\frac{Y_{N_r}-\theta}{k_r} \leqslant x\right\} \to F(x) \text{ as } r \to \infty. \tag{34.109}$$

34.35 It may also be shown that if the x's are independently and identically distributed, the conditions (a) and (c)—which are easily verifiable—together imply condition (b) and the distribution of their sum tends to a distribution function. In particular, these conditions are satisfied for Maximum Likelihood estimators, for estimators based on means of some functions of the observations, and for quantiles.

Example 34.12

Consider the estimation of the mean μ of a normal distribution with unknown variance σ^2. We require of the estimator a (small) variance k^2.

The obvious statistic is $Y_n = \bar{x}_n$. For fixed n this has variance σ^2/n estimated as

$$D_n^2 = \frac{1}{n(n-1)} \Sigma (x_i - \bar{x})^2. \tag{34.110}$$

Conditions (a) and (c) are obviously satisfied and in virtue of the result quoted in **34.35** this entails the satisfaction of condition (b). To show that (d) holds, transform by Helmert's transformation

$$\xi_i = \left(x_{i+1} - \frac{1}{i}\sum_{j=1}^{i} x_j\right)\sqrt{\frac{i}{i+1}}.$$

Then

$$D_n^2 = \frac{1}{n(n-1)} \sum_{i=1}^{n-1} \xi_i^2.$$

By the Strong Law of Large Numbers, given ε, η, there is a ν such that

$$P\left\{\left|\frac{1}{n-1}\sum_{i=1}^{n-1}\xi_i^2 - \sigma^2\right| < \varepsilon \text{ for all } n > \nu\right\} > 1 - \eta. \tag{34.111}$$

If k is small enough, the probability exceeds $1-\eta$ that $D_n \leqslant k$ for any n in the range $2 \leqslant n \leqslant \nu$. Thus, given $N > \nu$, (34.111) implies that

$$\left| \frac{N}{\sigma^2/k^2} - 1 \right| < \frac{\varepsilon}{\sigma^2}$$

with probability exceeding $1-\eta$. Hence, as k tends to zero, condition (d) holds.

The rule is, then, that we select k and proceed until $D_n \leqslant k$. The mean $\bar{x}$ then has variance approximately equal to k^2.

Example 34.13

Consider the Poisson distribution with parameter equal to λ. If we proceed until the variance of the mean, estimated as $\bar{x}/n$, is less than k^2, we have an estimator $\bar{x}$ of λ with variance k^2. This is equivalent to proceeding until the number of successes falls below $k^2 n^2$. But we should not use this result for small n.

On the other hand, suppose we wanted to specify in advance not the variance but the coefficient of variation, say l. The method would then fail. It would propose that we proceed until $\bar{x}/\sqrt{(\bar{x}/n)}$ is less than l, i.e. until $n\bar{x} \leqslant l^2$ or the sum of observations falls below l^2. But the sum must ultimately exceed any finite number. This is related to the result noted in Example 34.1 where we saw that for sequential sampling of rare attributes the coefficient of variation is approximately constant.

Stein's double-sampling method

34.36 At the end of Example 23.7, we observed that for fixed n no similar test of the mean μ of a normal population with unknown variance σ^2 could have power independent of σ^2. This implies (cf. **23.26**) that no confidence interval of pre-assigned length can be found for μ. However, if we use a sequential method, these statements are no longer true, as Stein (1945) pointed out.

34.37 We consider a normal population with mean μ and variance σ^2 and require to estimate μ with confidence coefficient $1-\alpha$, the length of the confidence-interval being l. We choose first of all a sample of fixed size n_0, and then a further sample $n-n_0$ where n now depends on the observations in the first sample.

Take a "Student's" t-variable with n_0-1 degrees of freedom, and let the probability that it lies in the range $-t_\alpha$ to t_α be $1-\alpha$. Define

$$\sqrt{z} = \frac{l}{2t_\alpha}. \tag{34.112}$$

Let s^2 be the estimated variance of the sample of n_0 values, i.e.,

$$s^2 = \frac{1}{n_0-1} \sum_{i=1}^{n_0} (x_i - \bar{x})^2. \tag{34.113}$$

We determine n by

$$n = \max\{n_0, 1 + [s^2/z]\}, \tag{34.114}$$

where $[s^2/z]$ means the greatest integer less than s^2/z.

Consider the n observations altogether, and let them have mean Y_n. Then Y_n is distributed independently of s and consequently $(Y_n-\mu)\sqrt{n}$ is independent of s; and hence $(Y_n-\mu)\sqrt{n}/s$ is distributed as t with n_0-1 d.f. Hence

$$P\left\{\left|\frac{(Y_n-\mu)\sqrt{n}}{s}\right| < t_\alpha\right\} = 1-\alpha,$$

or
$$P\left\{Y_n - \frac{st_\alpha}{\sqrt{n}} \leq \mu \leq Y_n + \frac{st_\alpha}{\sqrt{n}}\right\} = 1-\alpha,$$

or
$$P\{Y_n - \tfrac{1}{2}l \leq \mu \leq Y_n + \tfrac{1}{2}l\} \geq 1-\alpha. \tag{34.115}$$

The appearance of the inequality in (34.115) is due to the end-effect that s^2/z may not be integral, which in general is small, so that the limits given by $Y_n \pm \tfrac{1}{2}l$ are close to the exact limits for confidence coefficient $1-\alpha$. In point of fact we can, by a device suggested by Stein, obtain exact limits, though the procedure entails rejecting observations and is probably not worth while in practice.

Seelbinder (1953) and Moshman (1958) discuss the optimum choice of first sample size in Stein's method. Bhattacharjee (1965) shows that Stein's procedure is more sensitive to non-normality than "Student's" t-test (cf. **31.3**), and that, as we should expect, non-normality re-introduces the dependence of the interval length (and corresponding test power) upon σ^2.

34.38 Chapman (1950) extended Stein's method to testing the ratio of the means of two normal variables, the test being independent of both variances. It depends, however, on the distribution of the difference of two t-variables, for which Chapman provides some tables. D. R. Cox (1952c) considered the problem of estimation in double sampling, obtaining a number of asymptotic results. He also considered corrections to the single and double sampling results to improve the approximations of asymptotic theory. A. Birnbaum and Healy (1960) discuss a general class of double sampling procedures to attain prescribed variance. Graybill and Connell (1964) give a double sampling procedure for estimating a normal variance within a fixed interval. Goldman and Zeigler (1966) compare different methods in estimating a normal mean or variance: Stein's is best for the mean. S. Banerjee (1967) generalizes Stein's method to obtain pre-assigned length intervals for the problem of two means (cf. **21.11-25**).

Distribution-free tests

34.39 By the use of order-statistics we can reduce many procedures to the binomial case. Consider, for example, the testing of the hypothesis that the mean of a normal distribution is greater than μ_0 (a one-sided test). Replace the mean by the median and variate values by a score of, say, $+$ if the sample value falls above it and $-$ in the opposite case. On the hypothesis $H_0: \mu = \mu_0$ these signs will be distributed binomially with $\varpi = \tfrac{1}{2}$. On the hypothesis $H_1: \mu = \mu_0 + k\sigma$ the probability of a positive sign is

$$\varpi_1 = \frac{1}{\sqrt{(2\pi)}} \int_{-k}^{\infty} \exp(-\tfrac{1}{2}x^2)\,dx. \tag{34.116}$$

We may then set up a SPR test of ϖ_0 against ϖ_1 in the usual manner. This will have a type I error α and a type II error β of accepting H_0 when H_1 is true; and this type II error will be $\leq \beta$ when $\mu - \mu_0 > k\sigma$. This is, in fact, a sequential form of the Sign test of **32.2-7**.

Tests of this kind are often remarkably efficient, and the sacrifice of efficiency may be well worth while for the simplicity of application. Armitage (1947) compared this particular test with Wald's t-test and came to the conclusion that, as judged by sample number, the optimum test is not markedly superior to the Sign test.

34.40 Jackson (1960) has provided a bibliography on sequential analysis, classified by topic. Johnson (1961) gives a useful review of the subject.

Decision functions

34.41 In closing this chapter, we may refer briefly to a development of Wald's ideas on sequential procedures towards a general theory of decisions. A situation is envisaged in which, at some stage of the sampling at least, one has to take a decision, e.g. to accept a hypothesis, or to continue sampling. The consequences of these decisions are assumed to be known, and it is further assumed that they can be evaluated numerically. The problem is then to decide on optimum decision rules. Various possible principles can be adopted, e.g. to act so as to maximize expected gain or to minimize expected loss. Some writers have gone so far as to argue that all estimation and hypothesis-testing are, in fact, decision-making operations. We emphatically disagree, both that all statistical inquiry emerges in decision and that the consequences of many decisions can be evaluated numerically. And even in cases where both points may be conceded, it appears to us questionable whether some of the principles that have been proposed are such as a reasonable person would use in practice. That statistics is solely the science of decision-making seems to us a patent exaggeration. But, like some questions in probability, this is a matter on which each individual has to make up his own mind—with such aid from the theory of decision functions as he can get.

The leading expositions of this theory are the pioneer work by Wald himself (1950) and that by Blackwell and Girshick (1954).

EXERCISES

34.1 In Example 34.1, show by use of Exercise 9.13 that (34.3) implies the biasedness of m/n for ϖ.

34.2 Referring to Example 34.6, sketch the OC curve for a binomial with $\alpha = 0.01$, $\beta = 0.03$, $\varpi_1 = 0.1$, $\varpi_2 = 0.2$. (The curve is half a bell-shaped curve with a maximum at $\varpi = 0$ and zero at $\varpi = 1$. Six points are enough to give its general shape.) Similarly, sketch the ASN curve for the same binomial.

34.3 Two samples, each of size n, are drawn from populations, P_1 and P_2, with proportions ϖ_1 and ϖ_2 of an attribute. They are paired off in order of occurrence. t_1 is the number of pairs in which there is a success from P_1 and a failure from P_2; t_2 is the number of pairs in which there is a failure from P_1 and a success from P_2. Show that in the (conditional) set of such pairs the probability of a member of t_1 is

$$\varpi = (1-\varpi_1)\varpi_2 / \{\varpi_1(1-\varpi_2) + \varpi_2(1-\varpi_1)\}.$$

Considering this as an ordinary binomial in the set of $t = t_1 + t_2$ values, show how to test the hypothesis that $\varpi_1 \geqslant \varpi_2$ by testing $\varpi = \tfrac{1}{2}$. Hence derive a sequential test for $\varpi_1 \geqslant \varpi_2$.

If
$$u = \frac{\varpi_2(1-\varpi_1)}{\varpi_1(1-\varpi_2)},$$

show that $\varpi = u/(1+u)$ and hence derive the following acceptance and rejection numbers:

$$a_t = \frac{\log \dfrac{\beta}{1-\alpha}}{\log u_1 - \log u_0} + t \frac{\log \dfrac{1+u_1}{1+u_0}}{\log u_1 - \log u_0},$$

$$r_t = \frac{\log \dfrac{1-\beta}{\alpha}}{\log u_1 - \log u_0} + t \frac{\log \dfrac{1+u_1}{1+u_0}}{\log u_1 - \log u_0},$$

where u_i is the value of u corresponding to H_i ($i = 0, 1$).

(Wald, 1947)

34.4 Referring to the function $h \neq 0$ of **34.14** show that if z is a random variable such that $E(z)$ exists and is not zero; if there exists a positive δ such that $P(e^z < 1-\delta) > 0$ and $P(e^z > 1+\delta) > 0$; and if for any real h, $E(\exp hz) = g(h)$ exists, then

$$\lim_{h \to \infty} g(h) = \infty = \lim_{h \to -\infty} g(h)$$

and that $g''(h) > 0$ for all real values of h. Hence show that $g(h)$ is strictly decreasing over the interval $(-\infty, h^*)$ and strictly increasing over (h^*, ∞), where h^* is the value for which $g(h)$ is a minimum. Hence show that there exists at most one $h \neq 0$ for which $E(\exp hz) = 1$.

(Wald, 1947)

34.5 In **34.31**, deduce the expressions (34.96-7).

(cf. Johnson, 1959b)

34.6 In Exercise 34.5, show that the third moment of $Z_n - n\mu$ is

$$E(Z_n - n\mu)^3 = \mu_3 E(n) - 3\sigma^2 E\{n(Z_n - n\mu)\},$$

where μ_3 is the third moment of z.

(Wolfowitz, 1947)

34.7 If z is defined as at (34.19), let t be a complex variable such that $E(\exp zt) = \phi(t)$ exists in a certain part of the complex plane. Show that
$$E[\{\exp(tZ_n)\}\{\phi(t)\}^{-n}] = 1$$
for any point where $|\phi(t)| \geqslant 1$.

(Wald, 1947)

34.8 Putting $t = h$ in the foregoing exercise show that, if E_b refers to expectation under the restriction that $Z_n \leqslant -b$ and E_a to the restriction $Z_n \geqslant a$, then
$$K(h)E_b \exp(hZ_n) + \{1-K(h)\}E_a \exp(hZ_n) = 1,$$
where K is the OC. Hence, neglecting end-effects, show that
$$K(h) = \frac{e^{h(a+b)} - e^{hb}}{e^{h(a+b)} - 1}, \quad h \neq 0,$$
$$= \frac{a}{a+b}, \quad h = 0.$$

(Girshick, 1946)

34.9 Differentiating the identity of Exercise 34.7 with respect to t and putting $t = 0$, show that
$$E(n) = \frac{a\{1-K(h)\} - bK(h)}{E(z)}$$
and hence derive equation (34.43)

(Girshick, 1946)

34.10 Assuming, as in the previous exercise, that the identity is differentiable, derive the results of Exercises 34.7 and 34.8.

34.11 In the identity of Exercise 34.7, put
$$-\log \phi(t) = \tau$$
where τ is purely imaginary. Show that if $\phi(t)$ is not singular at $t = 0$ and $t = h$, this equation has two roots $t_1(\tau)$ and $t_2(\tau)$ for sufficiently small values of τ. In the manner of Exercise 34.8, show that the characteristic function of n is given asymptotically by
$$E(e^{n\tau}) = \frac{A^{t_2} - A^{t_1} + B^{t_1} - B^{t_2}}{B^{t_1}A^{t_2} - A^{t_1}B^{t_2}}.$$

(Wald, 1947)

34.12 In the case when z is normal with mean μ and variance σ^2, show that t_1 and t_2 in Exercise 34.11 are
$$t_1 = -\frac{\mu}{\sigma^2} + \frac{1}{\sigma^2}(\mu^2 - 2\sigma^2\tau)^{\frac{1}{2}},$$
$$t_2 = -\frac{\mu}{\sigma^2} - \frac{1}{\sigma^2}(\mu^2 - 2\sigma^2\tau)^{\frac{1}{2}},$$
where the sign of the radical is determined so that the real part of $\mu^2 - 2\sigma^2\tau$ is positive.

In the limiting case $B = 0$, A finite (when of necessity $E(z) > 0$ if $E(n)$ is to exist), show that the c.f. is
$$A^{-t_1}$$
and in the case B finite, $A = 0$ (when $E(z) < 0$), show that the c.f. is
$$B^{-t_1}.$$

(Wald, 1947)

34.13 In the first of the two limiting cases of the previous exercise, show that the distribution of $m = \mu^2 n / 2\sigma^2$ is given by
$$dF(m) = \frac{c}{2\Gamma(\tfrac{1}{2}) m^{3/2}} \exp\left(-\frac{c^2}{4m} - m + c\right) dm, \quad 0 \leq m \leq \infty,$$
where $c = \mu \log A / \sigma^2$.

For large c show that $2m/c$ is approximately normal with unit mean and variance $1/c$.

(Wald, 1947, who also shows that when A, B are finite the distribution of n is the weighted sum of a number of variables of the above type.)

34.14 Values of u are observed from the exponential distribution
$$dF = e^{-\lambda u} \lambda\, du, \quad 0 \leq u \leq \infty.$$
Show that a sequential test of $\lambda = \lambda_0$ against $\lambda = \lambda_1$ is given by
$$k_1 + (\lambda_1 - \lambda_0) \sum_{j=1}^n u_j \leq n \log(\lambda_1/\lambda_0) \leq k_2 + (\lambda_1 - \lambda_0) \sum_{j=1}^n u_j,$$
where k_1 and k_2 are constants.

Compare this with the test of Exercise 34.3 in the limiting case when ϖ_1 and ϖ_2 tend to zero so that $\varpi_1 t = \lambda_0$ and $\varpi_2 t = \lambda_1$ remain finite.

(Anscombe and Page, 1954)

34.15 It is required to estimate a parameter θ with a small variance $a(\theta)/\lambda$ when λ tends to infinity. If t_m is an unbiassed estimator in samples of fixed size m with variance $v(\theta)/m$; if $\gamma_1(t_m) = O(m^{-\frac{1}{2}})$ and $\gamma_2(t_m) = O(m^{-1})$; and if $a(t_m)$ and $b(t_m)$ can be expanded in series to give asymptotic means and standard errors, consider the double sampling rule:

(a) Take a sample of size $N\lambda$ and let t_1 be the estimate of θ from it.
(b) Take a second sample of size $\max\{0, [\{n_0(t_1) - N\}\lambda]\}$ where $n_0(t_1) = v(t_1)/a(t_1)$. Let t_2 be the estimate of θ from the second sample.
(c) Let $t = \dfrac{Nt_1 + \{n_0(t_1) - N\} t_2}{n_0(t_1)}$ if $n_0(t_1) \geq N$.
(d) Assume that $N < n_0(\theta)$ and the distribution of $m_0(t_1) = 1/n_0(t_1)$ is such that the event $n_0(t_1) < N$ may be ignored.

Show that under this rule
$$E(t) = \theta + O(\lambda^{-1}),$$
$$\operatorname{var} t = a(\theta) \lambda^{-1} \{1 + O(\lambda^{-1})\}.$$

(D. R. Cox, 1952c)

34.16 In the previous exercise, take the same procedure except that $n_0(t_1)$ is replaced by
$$n(t_1) = n_0(t_1)\left\{1 + \frac{b(t_1)}{\lambda}\right\}.$$
Show that
$$E(t) = \theta + m_0'(\theta) v(\theta) \lambda^{-1} + O(\lambda^{-2}).$$
Put
$$t' = t - m_0'(t) v(t) \lambda^{-1} \quad \text{if} \quad N \leq n(t_1)$$
$$= 0 \text{ otherwise,}$$
and hence show that t' has bias $O(\lambda^{-2})$.

Show further that if we put
$$b(\theta) = n_0(\theta) v(\theta) \{2m_0(\theta) m_0'(\theta) \gamma_1(\theta) v^{-\frac{1}{2}}(\theta) + m_0'^2(\theta) + 2m_0(\theta) m_0''(\theta) + m_0''(\theta)/(2N)\},$$
then
$$\operatorname{var} t' = a(\theta) \lambda^{-1} + O(\lambda^{-3}).$$
(D. R. Cox, 1952c)

34.17 Applying Exercise 34.15 to the binomial distribution, with
$$a(\varpi) = a\varpi^2, \quad v(\varpi) = \varpi(1-\varpi), \quad \gamma_1(\varpi) = \frac{(1-2\varpi)}{\{\varpi(1-\varpi)\}^{\frac{1}{2}}},$$
show that the total sample size is
$$n(t_1) = \frac{1-t_1}{at_1} + \frac{3}{t_1(1-t_1)} + \frac{1}{aNt_1}$$
and the estimator $t' = t - \dfrac{at}{1-t}$.

Thus N should be chosen as large as possible, provided that it does not exceed $(1-\varpi)/(a\varpi)$.
(D. R. Cox, 1952c)

34.18 Referring to Example 34.9, show that
$$K(\sigma) = \left\{ \left(\frac{1-\beta}{\alpha}\right)^h - 1 \right\} \bigg/ \left\{ \left(\frac{1-\beta}{\alpha}\right)^h - \left(\frac{\beta}{1-\alpha}\right)^h \right\}$$
where h is given by
$$\sigma \left(\frac{\sigma_1}{\sigma_0}\right)^h = \left\{ \frac{h}{\sigma_1^2} - \frac{h}{\sigma_0^2} + \frac{1}{\sigma^2} \right\}^{-\frac{1}{2}},$$
provided that the expression in brackets on the right is positive. Hence show how to draw the OC curve.
(Wald, 1947)

34.19 In the previous exercise derive the expression for the ASN
$$\frac{K(\sigma)\{h_0 - h_1\} + h_1}{\sigma^2 - \gamma} \quad \text{where} \quad \gamma = \log(\sigma_1^2/\sigma_0^2) \bigg/ \left(\frac{1}{\sigma_0^2} - \frac{1}{\sigma_1^2}\right).$$
(Wald, 1947)

34.20 Justify the statement in the last sentence of Example 34.9, giving a test of normal variances when the parent mean is unknown.
(Girshick, 1946)

34.21 In **34.10**, $f(x, \theta)$ is replaced by $f(x, \theta, \phi)$ where ϕ is a nuisance parameter, so that H_0 and H_1 are composite. $(\hat{\theta}_n, \hat{\phi}_n)$ is the ML estimator of (θ, ϕ) after n observations. If θ_0 and θ_1 differ from the true value of θ by amounts of order $n^{-\frac{1}{2}}$, show that a SPR test based on
$$t_n(\hat{\phi}) = \log L_n(x, \theta_1, \hat{\phi}) - \log L_n(x, \theta_0, \hat{\phi})$$
is asymptotically equivalent to (34.18) for the simple hypothesis when ϕ is known, if and only if
$$\frac{1}{n} \frac{\partial^2 \log L_n(x, \theta, \phi)}{\partial \theta \, \partial \phi} \to 0$$
in probability, i.e. if $\hat{\theta}$ and $\hat{\phi}$ are asymptotically independent. In any case, show that
$$t_n(\hat{\phi}) \sim (\theta_1 - \theta_0) \{\hat{\theta} - \tfrac{1}{2}(\theta_0 + \theta_1)\} E\left\{ -\frac{\partial^2 \log L_n}{\partial \theta^2} \right\},$$
so that $T_n = n\{\hat{\theta} - \tfrac{1}{2}(\theta_0 + \theta_1)\}$ may be used as a test statistic with mean and variance at once obtained from those of $\hat{\theta}$.
(D. R. Cox, 1963)

34.22 For a sample from the distribution $f(x \mid \theta) = g(x)/h(\theta)$, $a \leq x \leq \theta$, show that the SPR test of $H_0: \theta = \theta_0$ against $H_1: \theta = \theta_1$ $(0 < \theta_0 < \theta_1)$ has $\alpha = 0$, and has the form:— accept H_1 if $x_n > \theta_0$, $n = 1, 2, \ldots$; accept H_0 if $x_n \leq \theta_0$ and $(\theta_0/\theta_1)^n \leq \beta$; continue sampling otherwise.

Show directly that when H_0 holds, the sample size is constant at $n_0 = \log \beta / \log (\theta_0/\theta_1)$, and that when H_1 holds the ASN is exactly $\left(1 - \beta\right) \bigg/ \left(1 - \dfrac{\theta_0}{\theta_1}\right)$ if n_0 is an integer. Verify these formulae for the ASN from (34.43).

APPENDIX TABLES

1. The frequency function of the normal distribution
2. The distribution function of the normal distribution
3. Quantiles of the d.f. of χ^2
4a. The distribution function of χ^2 for one degree of freedom, $0 \leqslant \chi^2 \leqslant 1$
4b. The distribution function of χ^2 for one degree of freedom, $1 \leqslant \chi^2 \leqslant 10$
5. Quantiles of the d.f. of t
6. 5 per cent. points of z
7. 5 per cent. points of F
8. 1 per cent. points of z
9. 1 per cent. points of F

Appendix Table 1 Frequency function of the normal distribution $y = \dfrac{1}{\sqrt{(2\pi)}} e^{-\frac{1}{2}x^2}$ with first and second differences

x	y	$\Delta^1\,(-)$	Δ^2	x	y	$\Delta^1\,(-)$	Δ^2
0·0	0·39894	199	−392	2·5	0·01753	395	+79
0·1	0·39695	591	−374	2·6	0·01358	316	+66
0·2	0·39104	965	−347	2·7	0·01042	250	+53
0·3	0·38139	1312	−308	2·8	0·00792	197	+45
0·4	0·36827	1620	−265	2·9	0·00595	152	+36
0·5	0·35207	1885	−212	3·0	0·00443	116	+27
0·6	0·33322	2097	−159	3·1	0·00327	89	+23
0·7	0·31225	2256	−104	3·2	0·00238	66	+17
0·8	0·28969	2360	− 52	3·3	0·00172	49	+13
0·9	0·26609	2412	0	3·4	0·00123	36	+10
1·0	0·24197	2412	+ 46	3·5	0·00087	26	+ 7
1·1	0·21785	2366	+ 84	3·6	0·00061	19	+ 6
1·2	0·19419	2282	+118	3·7	0·00042	13	+ 4
1·3	0·17137	2164	+143	3·8	0·00029	9	+ 2
1·4	0·14973	2021	+161	3·9	0·00020	7	+ 3
1·5	0·12952	1860	+173	4·0	0·00013	4	—
1·6	0·11092	1687	+177	4·1	0·00009	3	—
1·7	0·09405	1510	+177	4·2	0·00006	2	—
1·8	0·07895	1333	+170	4·3	0·00004	2	—
1·9	0·06562	1163	+162	4·4	0·00002	—	—
2·0	0·05399	1001	+150	4·5	0·00002	—	—
2·1	0·04398	851	+137	4·6	0·00001	—	—
2·2	0·03547	714	+120	4·7	0·00001	—	—
2·3	0·02833	594	+108	4·8	0·00000	—	—
2·4	0·02239	486	+ 91				

Appendix Table 2 Distribution function of the normal distribution

The table shows the area under the curve $y = (2\pi)^{-\frac{1}{2}} e^{-\frac{1}{2}x^2}$ lying to the left of specified deviates x; e.g. the area corresponding to a deviate 1·86 (= 1·5 + 0·36) is 0·9686.

Deviate	0·0 +	0·5 +	1·0 +	1·5 +	2·0 +	2·5 +	3·0 +	3·5 +
0·00	5000	6915	8413	9332	9772	$9^2$379	$9^2$865	$9^3$77
0·01	5040	6950	8438	9345	9778	$9^2$396	$9^2$869	$9^3$78
0·02	5080	6985	8461	9357	9783	$9^2$413	$9^2$874	$9^3$78
0·03	5120	7019	8485	9370	9788	$9^2$430	$9^2$878	$9^3$79
0·04	5160	7054	8508	9382	9793	$9^2$446	$9^2$882	$9^3$80
0·05	5199	7088	8531	9394	9798	$9^2$461	$9^2$886	$9^3$81
0·06	5239	7123	8554	9406	9803	$9^2$477	$9^2$889	$9^3$81
0·07	5279	7157	8577	9418	9808	$9^2$492	$9^2$893	$9^3$82
0·08	5319	7190	8599	9429	9812	$9^2$506	$9^2$897	$9^3$83
0·09	5359	7224	8621	9441	9817	$9^2$520	$9^2$900	$9^3$83
0·10	5398	7257	8643	9452	9821	$9^2$534	$9^3$03	$9^3$84
0·11	5438	7291	8665	9463	9826	$9^2$547	$9^3$06	$9^3$85
0·12	5478	7324	8686	9474	9830	$9^2$560	$9^3$10	$9^3$85
0·13	5517	7357	8708	9484	9834	$9^2$573	$9^3$13	$9^3$86
0·14	5557	7389	8729	9495	9838	$9^2$585	$9^3$16	$9^3$86
0·15	5596	7422	8749	9505	9842	$9^2$598	$9^3$18	$9^3$87
0·16	5636	7454	8770	9515	9846	$9^2$609	$9^3$21	$9^3$87
0·17	5675	7486	8790	9525	9850	$9^2$621	$9^3$24	$9^3$88
0·18	5714	7517	8810	9535	9854	$9^2$632	$9^3$26	$9^3$88
0·19	5753	7549	8830	9545	9857	$9^2$643	$9^3$29	$9^3$89
0·20	5793	7580	8849	9554	9861	$9^2$653	$9^3$31	$9^3$89
0·21	5832	7611	8869	9564	9864	$9^2$664	$9^3$34	$9^3$90
0·22	5871	7642	8888	9573	9868	$9^2$674	$9^3$36	$9^3$90
0·23	5910	7673	8907	9582	9871	$9^2$683	$9^3$38	$9^4$04
0·24	5948	7704	8925	9591	9875	$9^2$693	$9^3$40	$9^4$08
0·25	5987	7738	8944	9599	9878	$9^2$702	$9^3$42	$9^4$12
0·26	6026	7764	8962	9608	9881	$9^2$711	$9^3$44	$9^4$15
0·27	6064	7794	8980	9616	9884	$9^2$720	$9^3$46	$9^4$18
0·28	6103	7823	8997	9625	9887	$9^2$728	$9^3$48	$9^4$22
0·29	6141	7852	9015	9633	9890	$9^2$736	$9^3$50	$9^4$25
0·30	6179	7881	9032	9641	9893	$9^2$744	$9^3$52	$9^4$28
0·31	6217	7910	9049	9649	9896	$9^2$752	$9^3$53	$9^4$31
0·32	6255	7939	9066	9656	9898	$9^2$760	$9^3$55	$9^4$33
0·33	6293	7967	9082	9664	9901	$9^2$767	$9^3$57	$9^4$36
0·34	6331	7995	9099	9671	9904	$9^2$774	$9^3$58	$9^4$39
0·35	6368	8023	9115	9678	9906	$9^2$781	$9^3$60	$9^4$41
0·36	6406	8051	9131	9686	9909	$9^2$788	$9^3$61	$9^4$43
0·37	6443	8078	9147	9693	9911	$9^2$795	$9^3$62	$9^4$46
0·38	6480	8106	9162	9699	9913	$9^2$801	$9^3$64	$9^4$48
0·39	6517	8133	9177	9706	9916	$9^2$807	$9^3$65	$9^4$50
0·40	6554	8159	9192	9713	9918	$9^2$813	$9^3$66	$9^4$52
0·41	6591	8186	9207	9719	9920	$9^2$819	$9^3$68	$9^4$54
0·42	6628	8212	9222	9726	9922	$9^2$825	$9^3$69	$9^4$56
0·43	6664	8238	9236	9732	9925	$9^2$831	$9^3$70	$9^4$58
0·44	6700	8264	9251	9738	9927	$9^2$836	$9^3$71	$9^4$59
0·45	6736	8289	9265	9744	9929	$9^2$841	$9^3$72	$9^4$61
0·46	6772	8315	9279	9750	9931	$9^2$846	$9^3$73	$9^4$63
0·47	6808	8340	9292	9756	9932	$9^2$851	$9^3$74	$9^4$64
0·48	6844	8365	9306	9761	9934	$9^2$856	$9^3$75	$9^4$66
0·49	6879	8389	9319	9767	9936	$9^2$861	$9^3$76	$9^4$67

Note—Decimal points in the body of the table are omitted. Repeated 9's are indicated by powers, e.g. $9^3$71 stands for 0·99971.

Appendix Table 3 Quantiles of the d.f. of χ^2

(Reproduced from Table III of Sir Ronald Fisher's *Statistical Methods for Research Workers*, Oliver and Boyd Ltd., Edinburgh, by kind permission of the author and publishers)

$P = 1-F$	0·99	0·98	0·95	0·90	0·80	0·70	0·50	0·30	0·20	0·10	0·05	0·02	0·01
$v=1$	0·0³157	0·0³628	0·0²393	0·0158	0·0642	0·148	0·455	1·074	1·642	2·706	3·841	5·412	6·635
2	0·0201	0·0404	0·103	0·211	0·446	0·713	1·386	2·408	3·219	4·605	5·991	7·824	9·210
3	0·115	0·185	0·352	0·584	1·005	1·424	2·366	3·665	4·642	6·251	7·815	9·837	11·345
4	0·297	0·429	0·711	1·064	1·649	2·195	3·357	4·878	5·989	7·779	9·488	11·668	13·277
5	0·554	0·752	1·145	1·610	2·343	3·000	4·351	6·064	7·289	9·236	11·070	13·388	15·086
6	0·872	1·134	1·635	2·204	3·070	3·828	5·348	7·231	8·558	10·645	12·592	15·033	16·812
7	1·239	1·564	2·167	2·833	3·822	4·671	6·346	8·383	9·803	12·017	14·067	16·622	18·475
8	1·646	2·032	2·733	3·490	4·594	5·527	7·344	9·524	11·030	13·362	15·507	18·168	20·090
9	2·088	2·532	3·325	4·168	5·380	6·393	8·343	10·656	12·242	14·684	16·919	19·679	21·666
10	2·558	3·059	3·940	4·865	6·179	7·267	9·342	11·781	13·442	15·987	18·307	21·161	23·209
11	3·053	3·609	4·575	5·578	6·989	8·148	10·341	12·899	14·631	17·275	19·675	22·618	24·725
12	3·571	4·178	5·226	6·304	7·807	9·034	11·340	14·011	15·812	18·549	21·026	24·054	26·217
13	4·107	4·765	5·892	7·042	8·634	9·926	12·340	15·119	16·985	19·812	22·362	25·472	27·688
14	4·660	5·368	6·571	7·790	9·467	10·821	13·339	16·222	18·151	21·064	23·685	26·873	29·141
15	5·229	5·985	7·261	8·547	10·307	11·721	14·339	17·322	19·311	22·307	24·996	28·259	30·578
16	5·812	6·614	7·962	9·312	11·152	12·624	15·338	18·418	20·465	23·542	26·296	29·633	32·000
17	6·408	7·255	8·672	10·085	12·002	13·531	16·338	19·511	21·615	24·769	27·587	30·995	33·409
18	7·015	7·906	9·390	10·865	12·857	14·440	17·338	20·601	22·760	25·989	28·869	32·346	34·805
19	7·633	8·567	10·117	11·651	13·716	15·352	18·338	21·689	23·900	27·204	30·144	33·687	36·191
20	8·260	9·237	10·851	12·443	14·578	16·266	19·337	22·775	25·038	28·412	31·410	35·020	37·566
21	8·897	9·915	11·591	13·240	15·445	17·182	20·337	23·858	26·171	29·615	32·671	36·343	38·932
22	9·542	10·600	12·338	14·041	16·314	18·101	21·337	24·939	27·301	30·813	33·924	37·659	40·289
23	10·196	11·293	13·091	14·848	17·187	19·021	22·337	26·018	28·429	32·007	35·172	38·968	41·638
24	10·856	11·992	13·848	15·659	18·062	19·943	23·337	27·096	29·553	33·196	36·415	40·270	42·980
25	11·524	12·697	14·611	16·473	18·940	20·867	24·337	28·172	30·675	34·382	37·652	41·566	44·314
26	12·198	13·409	15·379	17·292	19·820	21·792	25·336	29·246	31·795	35·563	38·885	42·856	45·642
27	12·879	14·125	16·151	18·114	20·703	22·719	26·336	30·319	32·912	36·741	40·113	44·140	46·963
28	13·565	14·847	16·928	18·939	21·588	23·647	27·336	31·391	34·027	37·916	41·337	45·419	48·278
29	14·256	15·574	17·708	19·768	22·475	24·577	28·336	32·461	35·139	39·087	42·557	46·693	49·588
30	14·953	16·306	18·493	20·599	23·364	25·508	29·336	33·530	36·250	40·256	43·773	47·962	50·892

Note.—For values of v greater than 30 the quantity $\sqrt{(2\chi^2)}$ may be taken to be distributed normally about mean $\sqrt{(2v-1)}$ with unit variance.

Appendix Table 4a Distribution function of χ^2 for one degree of freedom for values $\chi^2 = 0$ to $\chi^2 = 1$ by steps of $0\cdot 01$

χ^2	$P = 1 - F$	Δ	χ^2	$P = 1 - F$	Δ
0	1·00000	7966	0·50	0·47950	436
0·01	0·92034	3280	0·51	0·47514	430
0·02	0·88754	2505	0·52	0·47084	423
0·03	0·86249	2101	0·53	0·46661	418
0·04	0·84148	1842	0·54	0·46243	411
0·05	0·82306	1656	0·55	0·45832	406
0·06	0·80650	1516	0·56	0·45426	400
0·07	0·79134	1404	0·57	0·45026	395
0·08	0·77730	1312	0·58	0·44631	389
0·09	0·76418	1235	0·59	0·44242	384
0·10	0·75183	1169	0·60	0·43858	379
0·11	0·74014	1111	0·61	0·43479	374
0·12	0·72903	1060	0·62	0·43105	369
0·13	0·71843	1015	0·63	0·42736	365
0·14	0·70828	974	0·64	0·42371	360
0·15	0·69854	938	0·65	0·42011	355
0·16	0·68916	905	0·66	0·41656	351
0·17	0·68011	874	0·67	0·41305	346
0·18	0·67137	845	0·68	0·40959	343
0·19	0·66292	820	0·69	0·40616	338
0·20	0·65472	795	0·70	0·40278	334
0·21	0·64677	773	0·71	0·39944	330
0·22	0·63904	752	0·72	0·39614	326
0·23	0·63152	731	0·73	0·39288	322
0·24	0·62421	713	0·74	0·38966	318
0·25	0·61708	696	0·75	0·38648	315
0·26	0·61012	679	0·76	0·38333	311
0·27	0·60333	663	0·77	0·38022	308
0·28	0·59670	648	0·78	0·37714	304
0·29	0·59022	634	0·79	0·37410	301
0·30	0·58388	620	0·80	0·37109	297
0·31	0·57768	607	0·81	0·36812	294
0·32	0·57161	595	0·82	0·36518	291
0·33	0·56566	583	0·83	0·36227	287
0·34	0·55983	572	0·84	0·35940	285
0·35	0·55411	560	0·85	0·35655	281
0·36	0·54851	551	0·86	0·35374	278
0·37	0·54300	540	0·87	0·35096	276
0·38	0·53760	530	0·88	0·34820	272
0·39	0·53230	521	0·89	0·34548	270
0·40	0·52709	512	0·90	0·34278	267
0·41	0·52197	503	0·91	0·34011	264
0·42	0·51694	495	0·92	0·33747	261
0·43	0·51199	487	0·93	0·33486	258
0·44	0·50712	479	0·94	0·33228	256
0·45	0·50233	471	0·95	0·32972	253
0·46	0·49762	463	0·96	0·32719	251
0·47	0·49299	457	0·97	0·32468	248
0·48	0·48842	449	0·98	0·32220	246
0·49	0·48393	443	0·99	0·31974	243
0·50	0·47950	436	1·00	0·31731	241

Appendix Table 4b Distribution function of χ^2 for one degree of freedom for values of χ^2 from 1 to 10 by steps of 0·1

χ^2	$P = 1 - F$	Δ	χ^2	$P = 1 - F$	Δ
1·0	0·31731	2304	5·5	0·01902	106
1·1	0·29427	2095	5·6	0·01796	99
1·2	0·27332	1911	5·7	0·01697	94
1·3	0·25421	1749	5·8	0·01603	89
1·4	0·23672	1605	5·9	0·01514	83
1·5	0·22067	1477	6·0	0·01431	79
1·6	0·20590	1361	6·1	0·01352	74
1·7	0·19229	1258	6·2	0·01278	71
1·8	0·17971	1163	6·3	0·01207	66
1·9	0·16808	1078	6·4	0·01141	62
2·0	0·15730	1000	6·5	0·01079	59
2·1	0·14730	929	6·6	0·01020	56
2·2	0·13801	864	6·7	0·00964	52
2·3	0·12937	803	6·8	0·00912	50
2·4	0·12134	749	6·9	0·00862	47
2·5	0·11385	699	7·0	0·00815	44
2·6	0·10686	651	7·1	0·00771	42
2·7	0·10035	609	7·2	0·00729	39
2·8	0·09426	568	7·3	0·00690	38
2·9	0·08858	532	7·4	0·00652	35
3·0	0·08326	497	7·5	0·00617	33
3·1	0·07829	465	7·6	0·00584	32
3·2	0·07364	436	7·7	0·00552	30
3·3	0·06928	408	7·8	0·00522	28
3·4	0·06520	383	7·9	0·00494	26
3·5	0·06137	359	8·0	0·00468	25
3·6	0·05778	337	8·1	0·00443	24
3·7	0·05441	316	8·2	0·00419	23
3·8	0·05125	296	8·3	0·00396	21
3·9	0·04829	279	8·4	0·00375	20
4·0	0·04550	262	8·5	0·00355	19
4·1	0·04288	246	8·6	0·00336	18
4·2	0·04042	231	8·7	0·00318	17
4·3	0·03811	217	8·8	0·00301	16
4·4	0·03594	205	8·9	0·00285	15
4·5	0·03389	192	9·0	0·00270	14
4·6	0·03197	181	9·1	0·00256	14
4·7	0·03016	170	9·2	0·00242	13
4·8	0·02846	160	9·3	0·00229	12
4·9	0·02686	151	9·4	0·00217	12
5·0	0·02535	142	9·5	0·00205	10
5·1	0·02393	134	9·6	0·00195	11
5·2	0·02259	126	9·7	0·00184	10
5·3	0·02133	119	9·8	0·00174	9
5·4	0·02014	112	9·9	0·00165	8
5·5	0·01902	106	10·0	0·00157	8

Appendix Table 5 Quantiles of the d.f. of t

(Reproduced from Sir Ronald Fisher and Dr F. Yates: *Statistical Tables for Biological, Medical and Agricultural Research*, Oliver and Boyd Ltd., Edinburgh, by kind permission of the authors and publishers)

$P=2(1-F)$	0·9	0·8	0·7	0·6	0·5	0·4	0·3	0·2	0·1	0·05	0·02	0·01	0·001
$\nu=1$	0·158	0·325	0·510	0·727	1·000	1·376	1·963	3·078	6·314	12·706	31·821	63·657	636·619
2	0·142	0·289	0·445	0·617	0·816	1·061	1·386	1·886	2·920	4·303	6·965	9·925	31·598
3	0·137	0·277	0·424	0·584	0·765	0·978	1·250	1·638	2·353	3·182	4·541	5·841	12·924
4	0·134	0·271	0·414	0·569	0·741	0·941	1·190	1·533	2·132	2·776	3·747	4·604	8·610
5	0·132	0·267	0·408	0·559	0·727	0·920	1·156	1·476	2·015	2·571	3·365	4·032	6·869
6	0·131	0·265	0·404	0·553	0·718	0·906	1·134	1·440	1·943	2·447	3·143	3·707	5·959
7	0·130	0·263	0·402	0·549	0·711	0·896	1·119	1·415	1·895	2·365	2·998	3·499	5·408
8	0·130	0·262	0·399	0·546	0·706	0·889	1·108	1·397	1·860	2·306	2·896	3·355	5·041
9	0·129	0·261	0·398	0·543	0·703	0·883	1·100	1·383	1·833	2·262	2·821	3·250	4·781
10	0·129	0·260	0·397	0·542	0·700	0·879	1·093	1·372	1·812	2·228	2·764	3·169	4·587
11	0·129	0·260	0·396	0·540	0·697	0·876	1·088	1·363	1·796	2·201	2·718	3·106	4·437
12	0·128	0·259	0·395	0·539	0·695	0·873	1·083	1·356	1·782	2·179	2·681	3·055	4·318
13	0·128	0·259	0·394	0·538	0·694	0·870	1·079	1·350	1·771	2·160	2·650	3·012	4·221
14	0·128	0·258	0·393	0·537	0·692	0·868	1·076	1·345	1·761	2·145	2·624	2·977	4·140
15	0·128	0·258	0·393	0·536	0·691	0·866	1·074	1·341	1·753	2·131	2·602	2·947	4·073
16	0·128	0·258	0·392	0·535	0·690	0·865	1·071	1·337	1·746	2·120	2·583	2·921	4·015
17	0·128	0·257	0·392	0·534	0·689	0·863	1·069	1·333	1·740	2·110	2·567	2·898	3·965
18	0·127	0·257	0·392	0·534	0·688	0·862	1·067	1·330	1·734	2·101	2·552	2·878	3·922
19	0·127	0·257	0·391	0·533	0·688	0·861	1·066	1·328	1·729	2·093	2·539	2·861	3·883
20	0·127	0·257	0·391	0·533	0·687	0·860	1·064	1·325	1·725	2·086	2·528	2·845	3·850
21	0·127	0·257	0·391	0·532	0·686	0·859	1·063	1·323	1·721	2·080	2·518	2·831	3·819
22	0·127	0·256	0·390	0·532	0·686	0·858	1·061	1·321	1·717	2·074	2·508	2·819	3·792
23	0·127	0·256	0·390	0·532	0·685	0·858	1·060	1·319	1·714	2·069	2·500	2·807	3·767
24	0·127	0·256	0·390	0·531	0·685	0·857	1·059	1·318	1·711	2·064	2·492	2·797	3·745
25	0·127	0·256	0·390	0·531	0·684	0·856	1·058	1·316	1·708	2·060	2·485	2·787	3·725
26	0·127	0·256	0·390	0·531	0·684	0·856	1·058	1·315	1·706	2·056	2·479	2·779	3·707
27	0·127	0·256	0·389	0·531	0·684	0·855	1·057	1·314	1·703	2·052	2·473	2·771	3·690
28	0·127	0·256	0·389	0·530	0·683	0·855	1·056	1·313	1·701	2·048	2·467	2·763	3·674
29	0·127	0·256	0·389	0·530	0·683	0·854	1·055	1·311	1·699	2·045	2·462	2·756	3·659
30	0·127	0·256	0·389	0·530	0·683	0·854	1·055	1·310	1·697	2·042	2·457	2·750	3·646
40	0·126	0·255	0·388	0·529	0·681	0·851	1·050	1·303	1·684	2·021	2·423	2·704	3·551
60	0·126	0·254	0·387	0·527	0·679	0·848	1·046	1·296	1·671	2·000	2·390	2·660	3·460
120	0·126	0·254	0·386	0·526	0·677	0·845	1·041	1·289	1·658	1·980	2·358	2·617	3·373
∞	0·126	0·253	0·385	0·524	0·674	0·842	1·036	1·282	1·645	1·960	2·326	2·576	3·291

Appendix Table 6 5 per cent. points of the distribution of z
(values at which the d.f. = 0.95)

(Reprinted from Table VI of Sir Ronald Fisher's *Statistical Methods for Research Workers*, Oliver and Boyd Ltd., Edinburgh, by kind permission of the author and publishers)

	\multicolumn{10}{c	}{Values of v_1}
v_2	1	2	3	4	5	6	8	12	24	∞
1	2·5421	2·6479	2·6870	2·7071	2·7194	2·7276	2·7380	2·7484	2·7588	2·7693
2	1·4592	1·4722	1·4765	1·4787	1·4800	1·4808	1·4819	1·4830	1·4840	1·4851
3	1·1577	1·1284	1·1137	1·1051	1·0994	1·0953	1·0899	1·0842	1·0781	1·0716
4	1·0212	0·9690	0·9429	0·9272	0·9168	0·9093	0·8993	0·8885	0·8767	0·8639
5	0·9441	0·8777	0·8441	0·8236	0·8097	0·7997	0·7862	0·7714	0·7550	0·7368
6	0·8948	0·8188	0·7798	0·7558	0·7394	0·7274	0·7112	0·6931	0·6729	0·6499
7	0·8606	0·7777	0·7347	0·7080	0·6896	0·6761	0·6576	0·6369	0·6134	0·5862
8	0·8355	0·7475	0·7014	0·6725	0·6525	0·6378	0·6175	0·5945	0·5682	0·5371
9	0·8163	0·7242	0·6757	0·6450	0·6238	0·6080	0·5862	0·5613	0·5324	0·4979
10	0·8012	0·7058	0·6553	0·6232	0·6009	0·5843	0·5611	0·5346	0·5035	0·4657
11	0·7889	0·6909	0·6387	0·6055	0·5822	0·5648	0·5406	0·5126	0·4795	0·4387
12	0·7788	0·6786	0·6250	0·5907	0·5666	0·5487	0·5234	0·4941	0·4592	0·4156
13	0·7703	0·6682	0·6134	0·5783	0·5535	0·5350	0·5089	0·4785	0·4419	0·3957
14	0·7630	0·6594	0·6036	0·5677	0·5423	0·5233	0·4964	0·4649	0·4269	0·3782
15	0·7568	0·6518	0·5950	0·5585	0·5326	0·5131	0·4855	0·4532	0·4138	0·3628
16	0·7514	0·6451	0·5876	0·5505	0·5241	0·5042	0·4760	0·4428	0·4022	0·3490
17	0·7466	0·6393	0·5811	0·5434	0·5166	0·4964	0·4676	0·4337	0·3919	0·3366
18	0·7424	0·6341	0·5753	0·5371	0·5099	0·4894	0·4602	0·4255	0·3827	0·3253
19	0·7386	0·6295	0·5701	0·5315	0·5040	0·4832	0·4535	0·4182	0·3743	0·3151
20	0·7352	0·6254	0·5654	0·5265	0·4986	0·4776	0·4474	0·4116	0·3668	0·3057
21	0·7322	0·6216	0·5612	0·5219	0·4938	0·4725	0·4420	0·4055	0·3599	0·2971
22	0·7294	0·6182	0·5574	0·5178	0·4894	0·4679	0·4370	0·4001	0·3536	0·2892
23	0·7269	0·6151	0·5540	0·5140	0·4854	0·4636	0·4325	0·3950	0·3478	0·2818
24	0·7246	0·6123	0·5508	0·5106	0·4817	0·4598	0·4283	0·3904	0·3425	0·2749
25	0·7225	0·6097	0·5478	0·5074	0·4783	0·4562	0·4244	0·3862	0·3376	0·2685
26	0·7205	0·6073	0·5451	0·5045	0·4752	0·4529	0·4209	0·3823	0·3330	0·2625
27	0·7187	0·6051	0·5427	0·5017	0·4723	0·4499	0·4176	0·3786	0·3287	0·2569
28	0·7171	0·6030	0·5403	0·4992	0·4696	0·4471	0·4146	0·3752	0·3248	0·2516
29	0·7155	0·6011	0·5382	0·4969	0·4671	0·4444	0·4117	0·3720	0·3211	0·2466
30	0·7141	0·5994	0·5362	0·4947	0·4648	0·4420	0·4090	0·3691	0·3176	0·2419
60	0·6933	0·5738	0·5073	0·4632	0·4311	0·4064	0·3702	0·3255	0·2654	0·1644
∞	0·6729	0·5486	0·4787	0·4319	0·3974	0·3706	0·3309	0·2804	0·2085	0

Appendix Table 7 5 per cent. points of the variance ratio F
(values at which the d.f. = 0·95)

(Reproduced from Sir Ronald Fisher and Dr F. Yates: *Statistical Tables for Biological, Medical and Agricultural Research*, Oliver and Boyd Ltd., Edinburgh, by kind permission of the authors and publishers)

v_2 \ v_1	1	2	3	4	5	6	8	12	24	∞
1	161·40	199·50	215·70	224·60	230·20	234·00	238·90	243·90	249·00	254·30
2	18·51	19·00	19·16	19·25	19·30	19·33	19·37	19·41	19·45	19·50
3	10·13	9·55	9·28	9·12	9·01	8·94	8·84	8·74	8·64	8·53
4	7·71	6·94	6·59	6·39	6·26	6·16	6·04	5·91	5·77	5·63
5	6·61	5·79	5·41	5·19	5·05	4·95	4·82	4·68	4·53	4·36
6	5·99	5·14	4·76	4·53	4·39	4·28	4·15	4·00	3·84	3·67
7	5·59	4·74	4·35	4·12	3·97	3·87	3·73	3·57	3·41	3·23
8	5·32	4·46	4·07	3·84	3·69	3·58	3·44	3·28	3·12	2·93
9	5·12	4·26	3·86	3·63	3·48	3·37	3·23	3·07	2·90	2·71
10	4·96	4·10	3·71	3·48	3·33	3·22	3·07	2·91	2·74	2·54
11	4·84	3·98	3·59	3·36	3·20	3·09	2·95	2·79	2·61	2·40
12	4·75	3·88	3·49	3·26	3·11	3·00	2·85	2·69	2·50	2·30
13	4·67	3·80	3·41	3·18	3·02	2·92	2·77	2·60	2·42	2·21
14	4·60	3·74	3·34	3·11	2·96	2·85	2·70	2·53	2·35	2·13
15	4·54	3·68	3·29	3·06	2·90	2·79	2·64	2·48	2·29	2·07
16	4·49	3·63	3·24	3·01	2·85	2·74	2·59	2·42	2·24	2·01
17	4·45	3·59	3·20	2·96	2·81	2·70	2·55	2·38	2·19	1·96
18	4·41	3·55	3·16	2·93	2·77	2·66	2·51	2·34	2·15	1·92
19	4·38	3·52	3·13	2·90	2·74	2·63	2·48	2·31	2·11	1·88
20	4·35	3·49	3·10	2·87	2·71	2·60	2·45	2·28	2·08	1·84
21	4·32	3·47	3·07	2·84	2·68	2·57	2·42	2·25	2·05	1·81
22	4·30	3·44	3·05	2·82	2·66	2·55	2·40	2·23	2·03	1·78
23	4·28	3·42	3·03	2·80	2·64	2·53	2·38	2·20	2·00	1·76
24	4·26	3·40	3·01	2·78	2·62	2·51	2·36	2·18	1·98	1·73
25	4·24	3·38	2·99	2·76	2·60	2·49	2·34	2·16	1·96	1·71
26	4·22	3·37	2·98	2·74	2·59	2·47	2·32	2·15	1·95	1·69
27	4·21	3·35	2·96	2·73	2·57	2·46	2·30	2·13	1·93	1·67
28	4·20	3·34	2·95	2·71	2·56	2·44	2·29	2·12	1·91	1·65
29	4·18	3·33	2·93	2·70	2·54	2·43	2·28	2·10	1·90	1·64
30	4·17	3·32	2·92	2·69	2·53	2·42	2·27	2·09	1·89	1·62
40	4·08	3·23	2·84	2·61	2·45	2·34	2·18	2·00	1·79	1·51
60	4·00	3·15	2·76	2·52	2·37	2·25	2·10	1·92	1·70	1·39
120	3·92	3·07	2·68	2·45	2·29	2·17	2·02	1·83	1·61	1·25
∞	3·84	2·99	2·60	2·37	2·21	2·09	1·94	1·75	1·52	1·00

Lower 5 per cent. points are found by interchange of v_1 and v_2, i.e. v_1 must always correspond to the greater mean square.

Appendix Table 8 1 per cent. points of the distribution of z
(values at which the d.f. = 0·99)

(Reprinted from Table VI of Sir Ronald Fisher's *Statistical Methods for Research Workers*, Oliver and Boyd Ltd., Edinburgh, by kind permission of the author and publishers)

	\multicolumn{10}{c}{Values of v_1}
v_2	1	2	3	4	5	6	8	12	24	∞
1	4·1535	4·2585	4·2974	4·3175	4·3297	4·3379	4·3482	4·3585	4·3689	4·3794
2	2·2950	2·2976	2·2984	2·2988	2·2991	2·2992	2·2994	2·2997	2·2999	2·3001
3	1·7649	1·7140	1·6915	1·6786	1·6703	1·6645	1·6569	1·6489	1·6404	1·6314
4	1·5270	1·4452	1·4075	1·3856	1·3711	1·3609	1·3473	1·3327	1·3170	1·3000
5	1·3943	1·2929	1·2449	1·2164	1·1974	1·1838	1·1656	1·1457	1·1239	1·0997
6	1·3103	1·1955	1·1401	1·1068	1·0843	1·0680	1·0460	1·0218	0·9948	0·9643
7	1·2526	1·1281	1·0672	1·0300	1·0048	0·9864	0·9614	0·9335	0·9020	0·8658
8	1·2106	1·0787	1·0135	0·9734	0·9459	0·9259	0·8983	0·8673	0·8319	0·7904
9	1·1786	1·0411	0·9724	0·9299	0·9006	0·8791	0·8494	0·8157	0·7769	0·7305
10	1·1535	1·0114	0·9399	0·8954	0·8646	0·8419	0·8104	0·7744	0·7324	0·6816
11	1·1333	0·9874	0·9136	0·8674	0·8354	0·8116	0·7785	0·7405	0·6958	0·6408
12	1·1166	0·9677	0·8919	0·8443	0·8111	0·7864	0·7520	0·7122	0·6649	0·6061
13	1·1027	0·9511	0·8737	0·8248	0·7907	0·7652	0·7295	0·6882	0·6386	0·5761
14	1·0909	0·9370	0·8581	0·8082	0·7732	0·7471	0·7103	0·6675	0·6159	0·5500
15	1·0807	0·9249	0·8448	0·7939	0·7582	0·7314	0·6937	0·6496	0·5961	0·5269
16	1·0719	0·9144	0·8331	0·7814	0·7450	0·7177	0·6791	0·6339	0·5786	0·5064
17	1·0641	0·9051	0·8229	0·7705	0·7335	0·7057	0·6663	0·6199	0·5630	0·4879
18	1·0572	0·8970	0·8138	0·7607	0·7232	0·6950	0·6549	0·6075	0·5491	0·4712
19	1·0511	0·8897	0·8057	0·7521	0·7140	0·6854	0·6447	0·5964	0·5366	0·4560
20	1·0457	0·8831	0·7985	0·7443	0·7058	0·6768	0·6355	0·5864	0·5253	0·4421
21	1·0408	0·8772	0·7920	0·7372	0·6984	0·6690	0·6272	0·5773	0·5150	0·4294
22	1·0363	0·8719	0·7860	0·7309	0·6916	0·6620	0·6196	0·5691	0·5056	0·4176
23	1·0322	0·8670	0·7806	0·7251	0·6855	0·6555	0·6127	0·5615	0·4969	0·4068
24	1·0285	0·8626	0·7757	0·7197	0·6799	0·6496	0·6064	0·5545	0·4890	0·3967
25	1·0251	0·8585	0·7712	0·7148	0·6747	0·6442	0·6006	0·5481	0·4816	0·3872
26	1·0220	0·8548	0·7670	0·7103	0·6699	0·6392	0·5952	0·5422	0·4748	0·3784
27	1·0191	0·8513	0·7631	0·7062	0·6655	0·6346	0·5902	0·5367	0·4685	0·3701
28	1·0164	0·8481	0·7595	0·7023	0·6614	0·6303	0·5856	0·5316	0·4626	0·3624
29	1·0139	0·8451	0·7562	0·6987	0·6576	0·6263	0·5813	0·5269	0·4570	0·3550
30	1·0116	0·8423	0·7531	0·6954	0·6540	0·6226	0·5773	0·5224	0·4519	0·3481
60	0·9784	0·8025	0·7086	0·6472	0·6028	0·5687	0·5189	0·4574	0·3746	0·2352
∞	0·9462	0·7636	0·6651	0·5999	0·5522	0·5152	0·4604	0·3908	0·2913	0

Appendix Table 9 1 per cent. points of the variance ratio F
(values at which the d.f. = 0·99)

(Reproduced from Sir Ronald Fisher and Dr F. Yates: *Statistical Tables for Biological, Medical and Agricultural Research*, Oliver and Boyd Ltd., Edinburgh, by kind permission of the authors and publishers)

v_2 \ v_1	1	2	3	4	5	6	8	12	24	∞
1	4052	4999	5403	5625	5764	5859	5981	6106	6234	6366
2	98·49	99·00	99·17	99·25	99·30	99·33	99·36	99·42	99·46	99·50
3	34·12	30·81	29·46	28·71	28·24	27·91	27·49	27·05	26·60	26·12
4	21·20	18·00	16·69	15·98	15·52	15·21	14·80	14·37	13·93	13·46
5	16·26	13·27	12·06	11·39	10·97	10·67	10·27	9·89	9·47	9·02
6	13·74	10·92	9·78	9·15	8·75	8·47	8·10	7·72	7·31	6·88
7	12·25	9·55	8·45	7·85	7·46	7·19	6·84	6·47	6·07	5·65
8	11·26	8·65	7·59	7·01	6·63	6·37	6·03	5·67	5·28	4·86
9	10·56	8·02	6·99	6·42	6·06	5·80	5·47	5·11	4·73	4·31
10	10·04	7·56	6·55	5·99	5·64	5·39	5·06	4·71	4·33	3·91
11	9·65	7·20	6·22	5·67	5·32	5·07	4·74	4·40	4·02	3·60
12	9·33	6·93	5·95	5·41	5·06	4·82	4·50	4·16	3·78	3·36
13	9·07	6·70	5·74	5·20	4·86	4·62	4·30	3·96	3·59	3·16
14	8·86	6·51	5·56	5·03	4·69	4·46	4·14	3·80	3·43	3·00
15	8·68	6·36	5·42	4·89	4·56	4·32	4·00	3·67	3·29	2·87
16	8·53	6·23	5·29	4·77	4·44	4·20	3·89	3·55	3·18	2·75
17	8·40	6·11	5·18	4·67	4·34	4·10	3·79	3·45	3·08	2·65
18	8·28	6·01	5·09	4·58	4·25	4·01	3·71	3·37	3·00	2·57
19	8·18	5·93	5·01	4·50	4·17	3·94	3·63	3·30	2·92	2·49
20	8·10	5·85	4·94	4·43	4·10	3·87	3·56	3·23	2·86	2·42
21	8·02	5·78	4·87	4·37	4·04	3·81	3·51	3·17	2·80	2·36
22	7·94	5·72	4·82	4·31	3·99	3·76	3·45	3·12	2·75	2·31
23	7·88	5·66	4·76	4·26	3·94	3·71	3·41	3·07	2·70	2·26
24	7·82	5·61	4·72	4·22	3·90	3·67	3·36	3·03	2·66	2·21
25	7·77	5·57	4·68	4·18	3·86	3·63	3·32	2·99	2·62	2·17
26	7·72	5·53	4·64	4·14	3·82	3·59	3·29	2·96	2·58	2·13
27	7·68	5·49	4·60	4·11	3·78	3·56	3·26	2·93	2·55	2·10
28	7·64	5·45	4·57	4·07	3·75	3·53	3·23	2·90	2·52	2·06
29	7·60	5·42	4·54	4·04	3·73	3·50	3·20	2·87	2·49	2·03
30	7·56	5·39	4·51	4·02	3·70	3·47	3·17	2·84	2·47	2·01
40	7·31	5·18	4·31	3·83	3·51	3·29	2·99	2·66	2·29	1·80
60	7·08	4·98	4·13	3·65	3·34	3·12	2·82	2·50	2·12	1·60
120	6·85	4·79	3·95	3·48	3·17	2·96	2·66	2·34	1·95	1·38
∞	6·64	4·60	3·78	3·32	3·02	2·80	2·51	2·18	1·79	1·00

Lower 1 per cent. points are found by interchange of v_1 and v_2, i.e. v_1 must always correspond to the greater mean square.

REFERENCES

Note. References to R. A. Fisher, J. Neyman, E. S. Pearson, K. Pearson, "Student", A. Wald, S. S. Wilks and G. U. Yule that are marked with an asterisk are reproduced in the following collections:

R. A. Fisher, *Contributions to Mathematical Statistics*. Wiley, New York, 1950.
A Selection of Early Statistical Papers of J. Neyman. Cambridge Univ. Press, 1967.
Joint Statistical Papers of J. Neyman and E. S. Pearson. Cambridge Univ. Press, 1967.
The Selected Papers of E. S. Pearson. Cambridge Univ. Press, 1966.
Karl Pearson's Early Statistical Papers. Cambridge Univ. Press, London, 1948.
" Student's " Collected Papers. Biometrika Office, University College London, 1942.
Selected Papers in Statistics and Probability by Abraham Wald. McGraw-Hill, New York, 1955; reprinted by Stanford Univ. Press, 1957.
S. S. Wilks: Collected Papers. Contributions to Mathematical Statistics. Wiley, New York, 1967.
Statistical Papers of George Udny Yule. Griffin, London, 1971.

ADAMS, J. E., GRAY, H. L. and WATKINS, T. A. (1971). On asymptotic characterization of bias reduction by jackknifing. *Ann. Math. Statist.*, 48, 1606.
AITKEN, A. C. (1933). On the graduation of data by the orthogonal polynomials of least squares. *Proc. Roy. Soc. Edin.*, A, 53, 54.
 On fitting polynomials to weighted data by least squares. *Ibid.*, 54, 1.
 On fitting polynomials to data with weighted and correlated errors. *Ibid.*, 54, 12.
AITKEN, A. C. (1935). On least squares and linear combination of observations. *Proc. Roy. Soc. Edin.*, A, 55, 42.
AITKEN, A. C. (1948). On the estimation of many statistical parameters. *Proc. Roy. Soc. Edin.*, A, 62, 369.
AITKEN, A. C. and SILVERSTONE, H. (1942). On the estimation of statistical parameters. *Proc. Roy. Soc. Edin.*, A, 61, 186.
ALLAN, F. E. (1930). The general form of the orthogonal polynomials for simple series with proofs of their simple properties. *Proc. Roy. Soc. Edin.*, A, 50, 310.
ALLEN, H. V. (1938). A theorem concerning the linearity of regression. *Statist. Res. Mem.*, 2, 60.
ALLING, D. W. (1966). Closed sequential tests for binomial probabilities. *Biometrika*, 53, 73.
ALTHAM, P. M. E. (1970a). The measurement of association of rows and columns for an $r \times s$ contingency table. *J.R. Statist. Soc.*, B, 32, 63.
ALTHAM, P. M. E. (1970b). The measurement of association in a contingency table: three extensions of the cross-ratios and metrics methods. *J.R. Statist. Soc.*, B, 32, 395.
AMOS, D. E. (1964). Representations of the central and non-central t distributions. *Biometrika*, 51, 451.
ANDERSEN, E. B. (1970). Sufficiency and exponential families for discrete sample spaces. *J. Amer. Statist. Ass.*, 65, 1248.
ANDERSON, T. W. (1955). Some statistical problems in relating experimental data to predicting performance of a production process. *J. Amer. Statist. Ass.*, 50, 163.
ANDERSON, T. W. (1960). A modification of the sequential probability ratio test to reduce sample size. *Ann. Math. Statist.*, 31, 165.
ANDERSON, T. W. (1962). Least squares and best unbiased estimates. *Ann. Math. Statist.*, 33, 266.

REFERENCES

ANDERSON, T. W. and DARLING, D. A. (1952). Asymptotic theory of certain "goodness of fit" criteria based on stochastic processes. *Ann. Math. Statist.*, **23**, 193.

ANDERSON, T. W. and DARLING, D. A. (1954). A test of goodness of fit. *J. Amer. Statist. Ass.*, **49**, 765.

ANDREWS, F. C. (1954). Asymptotic behaviour of some rank tests for analysis of variance. *Ann. Math. Statist.*, **25**, 724.

ANSCOMBE, F. J. (1949a). Tables of sequential inspection schemes to control fraction defective. *J.R. Statist. Soc.*, **A, 112**, 180.

ANSCOMBE, F. J. (1949b). Large-sample theory of sequential estimation. *Biometrika*, **36**, 455.

ANSCOMBE, F. J. (1950). Sampling theory of the negative binomial and logarithmic series distributions. *Biometrika*, **37**, 358.

ANSCOMBE, F. J. (1952). Large-sample theory of sequential estimation. *Proc. Camb. Phil. Soc.*, **48**, 600.

ANSCOMBE, F. J. (1953). Sequential estimation. *J.R. Statist. Soc.*, **B, 15**, 1.

ANSCOMBE, F. J. (1960). Rejection of outliers. *Technometrics*, **2**, 123.

ANSCOMBE, F. J. (1967). Topics in the investigation of linear relations fitted by the method of least squares. *J.R. Statist. Soc.*, **B, 29**, 1.

ANSCOMBE, F. J. and PAGE, E. S. (1954). Sequential tests for binomial and exponential populations. *Biometrika*, **41**, 252.

ANTLE, C., KLIMKO, L. and HARKNESS, W. (1970). Confidence intervals for the parameters of the logistic distribution. *Biometrika*, **57**, 397.

ARMITAGE, P. (1947). Some sequential tests of Student's hypothesis. *J.R. Statist. Soc.*, **B, 9**, 250.

ARMITAGE, P. (1955). Tests for linear trends in proportions and frequencies. *Biometrics*, **11**, 375.

ARMITAGE, P. (1957). Restricted sequential procedures. *Biometrika*, **44**, 9.

ARMITAGE, P. (1966). The chi-square test for heterogeneity of proportions, after adjustment for stratification. *J.R. Statist. Soc.*, **B, 28**, 150 (Addendum: **29**, 197).

ARMSEN, P. (1955). Tables for significance tests of 2×2 contingency tables. *Biometrika*, **42**, 494.

ARNOLD, H. J. (1965). Small sample power of the one sample Wilcoxon tests for non-normal shift alternatives. *Ann. Math. Statist.*, **36**, 1767.

ARVESEN, J. N. (1969). Jackknifing U-statistics. *Ann. Math. Statist.*, **40**, 2076. (Corrections **41**, 1375.)

ASKOVITZ, S. I. (1957). A short-cut graphic method for fitting the best straight line to a series of points according to the criterion of least squares. *J. Amer. Statist. Ass.*, **52**, 13.

ASPIN, A. A. (1948). An examination and further development of a formula arising in the problem of comparing two mean values. *Biometrika*, **35**, 88.

ASPIN, A. A. (1949). Tables for use in comparisons whose accuracy involves two variances, separately estimated. *Biometrika*, **36**, 290.

ATKINSON, A. C. (1970). A method of discriminating between models. *J.R. Statist. Soc.*, **B, 32**, 323.

BAHADUR, R. R. (1964). On Fisher's bound for asymptotic variances. *Ann. Math. Statist.*, **35**, 1545.

BAHADUR, R. R. (1967). Rates of convergence of estimates and test statistics. *Ann. Math. Statist.*, **38**, 303.

BAIN, L. J. (1967). Reducing a random sample to a smaller set, with applications. *J. Amer. Statist. Ass.*, **62**, 510.

BAIN, L. J. (1969). The moments of a noncentral t and noncentral F-distribution. *Amer. Statistician*, **23(4)**, 33.

BANERJEE, S. (1967). Confidence interval of preassigned length for the Behrens-Fisher problem. *Ann. Math. Statist.*, **38**, 1175.

BARANKIN, E. W. (1949). Locally best unbiased estimates. *Ann. Math. Statist.*, **20**, 477.

REFERENCES

BARANKIN, E. W. (1960a). Application to exponential families of the solution of the minimal dimensionality problem for sufficient statistics. *Bull. Int. Statist. Inst.*, **38** (4), 141.

BARANKIN, E. W. (1960b). Sufficient parameters: solution of the minimal dimensionality problem. *Ann. Inst. Statist. Math.*, **12**, 91.

BARANKIN, E. W. (1961). A note on functional minimality of sufficient statistics. *Sankhyā*, A, **23**, 401.

BARANKIN, E. W. and KATZ, M., Jr. (1959). Sufficient statistics of minimal dimension. *Sankhyā*, **21**, 217.

BARANKIN, E. W. and MAITRA, A. P. (1963). Generalization of the Fisher-Darmois-Koopman-Pitman theorem on sufficient statistics. *Sankhyā*, A, **25**, 217.

BARNARD, G. A. (1946). Sequential tests in industrial statistics. *Suppl. J.R. Statist. Soc.*, **8**, 1.

BARNARD, G. A. (1947a). Significance tests for 2×2 tables. *Biometrika*, **34**, 123.

BARNARD, G. A. (1947b). 2×2 tables. A note on E. S. Pearson's paper. *Biometrika*, **34**, 168.

BARNARD, G. A. (1950). On the Fisher-Behrens test. *Biometrika*, **37**, 203.

BARNARD, G. A. (1963). The logic of least squares. *J.R. Statist. Soc.*, B, **25**, 124.

BARNETT, V. D. (1966a). Evaluation of the maximum-likelihood estimator where the likelihood equation has multiple roots. *Biometrika*, **53**, 151.

BARNETT, V. D. (1966b). Order statistics estimators of the location of the Cauchy distribution. *J. Amer. Statist. Ass.*, **61**, 1205.

BARR, D. R. (1966). On testing the equality of uniform and related distributions. *J. Amer. Statist. Ass.*, **61**, 856.

BARTHOLOMEW, D. J. (1961). Ordered tests in the analysis of variance. *Biometrika*, **48**, 325.

BARTHOLOMEW, D. J. (1967). Hypothesis testing when the sample size is treated as a random variable. *J.R. Statist. Soc.*, B, **29**, 53.

BARTKY, W. (1943). Multiple sampling with constant probability. *Ann. Math. Statist.*, **14**, 363.

BARTLETT, M. S. (1935a). The effect of non-normality on the t-distribution. *Proc. Camb. Phil. Soc.*, **31**, 223.

BARTLETT, M. S. (1935b). Contingency table interactions. *Suppl. J.R. Statist. Soc.*, **2**, 248.

BARTLETT, M. S. (1936). The information available in small samples. *Proc. Camb. Phil. Soc.*, **32**, 560.

BARTLETT, M. S. (1937). Properties of sufficiency and statistical tests. *Proc. Roy. Soc.*, A, **160**, 268.

BARTLETT, M. S. (1938). The characteristic function of a conditional statistic. *J. Lond. Math. Soc.*, **13**, 62.

BARTLETT, M. S. (1939). A note on the interpretation of quasi-sufficiency. *Biometrika*, **31**, 391.

BARTLETT, M. S. (1949). Fitting a straight line when both variables are subject to error. *Biometrics*, **5**, 207.

BARTLETT, M. S. (1951). An inverse matrix adjustment arising in discriminant analysis. *Ann. Math. Statist.*, **22**, 107.

BARTLETT, M. S. (1953, 1955). Approximate confidence intervals. *Biometrika*, **40**, 12, 306 and **42**, 201.

BARTON, D. E. (1953). On Neyman's smooth test of goodness of fit and its power with respect to a particular system of alternatives. *Skand. Aktuartidskr.*, **36**, 24.

BARTON, D. E. (1955). A form of Neyman's ψ_k^2 test of goodness of fit applicable to grouped and discrete data. *Skand. Aktuartidskr.*, **38**, 1.

BARTON, D. E. (1956). Neyman's ψ_k^2 test of goodness of fit when the null hypothesis is composite. *Skand. Aktuartidskr.*, **39**, 216.

BASU, A. P. (1965). On some tests of hypotheses relating to the exponential distribution when some outliers are present. *J. Amer. Statist. Ass.*, **60**, 548.

BASU, A. P. (1967a). On the large sample properties of a generalized Wilcoxon-Mann-Whitney statistic. *Ann. Math. Statist.*, **38**, 905.

BASU, A. P. (1967b). On two K-sample rank tests for censored data. *Ann. Math. Statist.*, **38**, 1520.

REFERENCES

Basu, A. P. (1969). On some tests for several linear relations. *J.R. Statist. Soc.*, B, **31**, 65.

Basu, D. (1955). On statistics independent of a complete sufficient statistic. *Sankhyā*, **15**, 377.

Bateman, G. I. (1949). The characteristic function of a weighted sum of non-central squares of normal variables subject to s linear restraints. *Biometrika*, **36**, 460.

Beale, E. M. L. (1960). Confidence regions in non-linear estimation. *J.R. Statist. Soc.*, B, **22**, 41.

Beale, E. M. L., Kendall, M. G. and Mann, D. W. (1967). The discarding of variables in multivariate analysis. *Biometrika*, **54**, 357.

Bell, C. B. and Doksum, K. A. (1965). Some new distribution-free statistics. *Ann. Math. Statist.*, **36**, 203.

Bell, C. B. and Doksum, K. A. (1967). Distribution-free tests of independence. *Ann. Math. Statist.*, **38**, 429.

Bellman, R. and Roth, R. (1969). Curve fitting by segmented straight lines. *J. Amer. Statist. Ass.*, **64**, 1079.

Bement, T. R. and Williams, J. S. (1969). Variance of weighted regression estimators when sampling errors are independent and heteroscedastic. *J. Amer. Statist. Ass.*, **64**, 1369.

Bennett, B. M. and Hsu, P. (1960). On the power function of the exact test for the 2×2 contingency table. *Biometrika*, **47**, 393.

Benson, F. (1949). A note on the estimation of mean and standard deviation from quantiles. *J.R. Statist. Soc.*, B, **11**, 91.

Berger, A. (1961). On comparing intensities of association between two binary characteristics in two different populations. *J. Amer. Statist. Ass.*, **56**, 889.

Berkson, J. (1938). Some difficulties of interpretation encountered in the application of the chi-square test. *J. Amer. Statist Ass.*, **33**, 526.

Berkson, J. (1950). Are there two regressions? *J. Amer. Statist Ass.*, **45**, 164.

Berkson, J. (1955). Maximum likelihood and minimum χ^2 estimates of the logistic function. *J. Amer. Statist. Ass.*, **50**, 130.

Berkson, J. (1956). Estimation by least squares and by maximum likelihood. *Proc. 3rd Berkeley Symp. Math. Statist. and Prob.*, **1**, 1.

Bernstein, S. (1928). Fondements géométriques de la théorie des corrélations. *Metron*, **7**, (2) 3.

Bhattacharjee, G. P. (1965). Effect of non-normality on Stein's two sample test. *Ann. Math. Statist.*, **36**, 651.

Bhattacharyya, A. (1943). On some sets of sufficient conditions leading to the normal bivariate distribution. *Sankhyā*, **6**, 399.

Bhattacharyya, A. (1946-7-8). On some analogues of the amount of information and their use in statistical estimation. *Sankhyā*, **8**, 1, 201, 315.

Bhuchongkul, S. (1964). A class of nonparametric tests for independence in bivariate populations. *Ann. Math. Statist.*, **35**, 138.

Bickel, P. J. (1965). On some robust estimates of location. *Ann. Math. Statist.*, **36**, 847.

Birch, M. W. (1963). Maximum likelihood in three-way contingency tables. *J.R. Statist. Soc.*, B, **25**, 220.

Birch, M. W. (1964a). A note on the maximum likelihood estimation of a linear structural relationship. *J. Amer. Statist. Ass.*, **59**, 1175.

Birch, M. W. (1964b). A new proof of the Pearson–Fisher theorem. *Ann. Math. Statist.*, **35**, 817.

Birch, M. W. (1964c). The detection of partial association, I: The 2×2 case. *J.R. Statist. Soc.*, B, **26**, 313.

Birch, M. W. (1965). The detection of partial association, II: The general case. *J.R. Statist. Soc.*, B, **27**, 111.

Birnbaum, A. (1962). On the foundations of statistical inference. *J. Amer. Statist. Ass.*, **57**, 269.

Birnbaum, A. (1970). On Durbin's modified principle of conditionality. *J. Amer. Statist. Ass.*, **65**, 402.

REFERENCES

BIRNBAUM, A. and DUDMAN, J. (1963). Logistic order statistics. *Ann. Math. Statist.*, **34**, 658.

BIRNBAUM, A. and HEALY, W. C., Jr. (1960). Estimates with prescribed variance based on two-stage sampling. *Ann. Math. Statist.*, **31**, 662.

BIRNBAUM, A. and LASKA, E. (1967a). Optimal robustness: a general method, with applications to linear estimators of location. *J. Amer. Statist. Ass.*, **62**, 1230.

BIRNBAUM, A. and LASKA, E. (1967b). Efficiency robust two-sample tests. *J. Amer. Statist. Ass.*, **62**, 1241.

BIRNBAUM, A. and MIKÉ, V. (1970). Asymptotically robust estimators of location. *J. Amer. Statist. Ass.*, **65**, 1265.

BIRNBAUM, A., LASKA, E. and MEISNER, M. (1971). Optimally robust linear estimators of location. *J. Amer. Statist. Ass.*, **66**, 302.

BIRNBAUM, Z. W. (1952). Numerical tabulation of the distribution of Kolmogorov's statistic for finite sample size. *J. Amer. Statist. Ass.*, **47**, 425.

BIRNBAUM, Z. W. (1953). On the power of a one-sided test of fit for continuous probability functions. *Ann. Math. Statist.*, **24**, 484.

BIRNBAUM, Z. W. and TINGEY, F. H. (1951). One-sided confidence contours for probability distribution functions. *Ann. Math. Statist.*, **22**, 592.

BLACKWELL, D. (1946). On an equation of Wald. *Ann. Math. Statist.*, **17**, 84.

BLACKWELL, D. (1947). Conditional expectation and unbiased sequential estimation. *Ann. Math. Statist.*, **18**, 105.

BLACKWELL, D. and GIRSHICK, M. A. (1954). *Theory of Games and Statistical Decisions*. Wiley, New York.

BLALOCK, H. M., Jr. (1958). Probabilistic interpretations for the mean square contingency. *J. Amer. Statist. Ass.*, **53**, 102.

BLISCHKE, W. R., TRUELOVE, A. J. and MUNDLE, P. B. (1969). On non-regular estimation. I. Variance bounds for estimators of location parameters. *J. Amer. Statist. Ass.*, **64**, 1056.

BLISS, C. I., COCHRAN, W. G. and TUKEY, J. W. (1956). A rejection criterion based upon the range. *Biometrika*, **43**, 448.

BLOCH, D. A. (1966). A note on the estimation of the location parameter of the Cauchy distribution. *J. Amer. Statist. Ass.*, **61**, 852.

BLOCH, D. A. and WATSON, G. S. (1967). A Bayesian study of the multinomial distribution. *Ann. Math. Statist.*, **38**, 1423.

BLOM, G. (1958). *Statistical estimates and transformed Beta-variables*. Almqvist and Wiksell, Stockholm; Wiley, New York.

BLOMQVIST, N. (1950). On a measure of dependence between two random variables. *Ann. Math. Statist.*, **21**, 593.

BLUM, J. R., KIEFER, J. and ROSENBLATT, M. (1961). Distribution free tests of independence based on the sample distribution function. *Ann. Math. Statist.*, **32**, 485.

BLYTH, C. R. and HUTCHINSON, D. W. (1960). Table of Neyman-shortest unbiased confidence intervals for the binomial parameter. *Biometrika*, **47**, 381.

BLYTH, C. R. and HUTCHINSON, D. W. (1961). Table of Neyman-shortest unbiased confidence intervals for the Poisson parameter. *Biometrika*, **48**, 191.

BOHRER, R. and FRANCIS, G. K. (1972). Sharp one-sided confidence bounds for linear regression over intervals. *Biometrika*, **59**, 99.

BOWDEN, D. C. (1970). Simultaneous confidence bands for linear regression models. *J. Amer. Statist. Ass.*, **65**, 413.

BOWKER, A. H. (1946). Computation of factors for tolerance limits on a normal distribution when the sample is large. *Ann. Math. Statist.*, **17**, 238.

BOWKER, A. H. (1947). Tolerance limits for normal distributions. *Selected Techniques of Statistical Analysis*. McGraw-Hill, New York.

BOWKER, A. H. (1948). A test for symmetry in contingency tables. *J. Amer. Statist. Ass.*, **43**, 572.

BOWMAN, K. O. (1972). Tables of the sample size requirement. *Biometrika*, **59**, 234.

BOWMAN, K. O. and SHENTON, L. R. (1969). Remarks on maximum likelihood estimators for the Gamma distribution. *1st Internat. Conf. Quality Control, Tokyo, 1969*, 519.
Box, G. E. P. (1949). A general distribution theory for a class of likelihood criteria. *Biometrika*, **36**, 317.
Box, G. E. P. (1953). Non-normality and tests on variances. *Biometrika*, **40**, 318.
Box, G. E. P. and ANDERSEN, S. L. (1955). Permutation theory in the derivation of robust criteria and the study of departures from assumption. *J.R. Statist Soc.*, B, **17**, 1.
Box, M. J. (1971). Bias in non-linear estimation. *J.R. Statist. Soc.*, B, **33**, 171.
BRADU, D. and MUNDLAK, Y. (1970). Estimation in lognormal linear models. *J. Amer. Statist. Ass.*, **65**, 198.
BRANDNER, F. A. (1933). A test of the significance of the difference of the correlation coefficients in normal bivariate samples. *Biometrika*, **25**, 102.
BRESLOW, N. (1970a). Sequential modification of the UMP test for binomial probabilities. *J. Amer. Statist. Ass.*, **65**, 639.
BRESLOW, N. (1970b). A generalized Kruskal-Wallis test for comparing K samples subject to unequal patterns of censorship. *Biometrika*, **57**, 579.
BRILLINGER, D. R. (1962). Examples bearing on the definition of fiducial probability with a bibliography. *Ann. Math. Statist.*, **33**, 1349.
BRILLINGER, D. R. (1964). The asymptotic behaviour of Tukey's general method of setting approximate confidence limits (the jackknife) when applied to maximum likelihood estimates. *Rev. Int. Statist. Inst.*, **32**, 202.
BRILLINGER, D. R. (1966). An extremal property of the conditional expectation. *Biometrika*, **53**, 594.
BROSS, I. D. J. and KASTEN, E. L. (1957). Rapid analysis of 2×2 tables. *J. Amer. Statist. Ass.*, **52**, 18.
BROWN, G. W. and MOOD, A. M. (1951). On median tests for linear hypotheses. *Proc. 2nd Berkeley Symp. Math. Statist. and Prob.*, 159.
BROWN, L. D. (1964). Sufficient statistics in the case of independent random variables. *Ann. Math. Statist.*, **35**, 1456.
BROWN, L. D. (1971). Non-local asymptotic optimality of appropriate likelihood ratio tests. *Ann. Math. Statist.*, **42**, 1206.
BROWN, R. L. (1957). Bivariate structural relation. *Biometrika*, **44**, 84.
BROWN, R. L. and FEREDAY, F. (1958). Multivariate linear structural relations. *Biometrika*, **45**, 136.
BUCKLE, N., KRAFT, C. and VAN EEDEN, C. (1969). An approximation to the Wilcoxon-Mann-Whitney distribution. *J. Amer. Statist. Ass.*, **64**, 591.
BÜHLER, W. J. (1967). The treatment of ties in the Wilcoxon test. *Ann. Math. Statist.*, **38**, 519.
BUHLER, W., FEIN, H., GOLDSMITH, D., NEYMAN, J. and PURI, P. S. (1965). Locally optimal test for homogeneity with respect to very rare events. *Proc. Nat. Acad. Sci. U.S.A.*, **54**, 673.
BULGREN, W. G. (1971). On representations of the doubly non-central F distribution. *J. Amer. Statist. Ass.*, **66**, 184.
BULGREN, W. G. and AMOS, D. E. (1968). A note on representations of the doubly non-central t distribution. *J. Amer. Statist. Ass.*, **63**, 1013.
BULMER, M. G. (1958). Confidence limits for distance in the analysis of variance. *Biometrika*, **45**, 360.
BURMAN J. P. (1946). Sequential sampling formulae for a binomial population. *J.R. Statist. Soc.*, B, **8**, 98.

CAPON, J. (1961). Asymptotic efficiency of certain locally most powerful rank tests. *Ann. Math. Statist.*, **32**, 88.
CHAN, L. K. (1970). Linear estimation of the location and scale parameters of the Cauchy distribution based on sample quantiles. *J. Amer. Statist. Ass.*, **65**, 851.

REFERENCES

CHANDA, K. C. (1963). On the efficiency of two-sample Mann–Whitney test for discrete populations. *Ann. Math. Statist.*, **34**, 612.

CHANDA, K. C. (1971). Asymptotic distribution of the sample size for a sequential probability ratio test. *J. Amer. Statist. Ass.*, **66**, 178.

CHANDLER, K. N. (1950). On a theorem concerning the secondary subscripts of deviations in multivariate correlation using Yule's notation. *Biometrika*, **37**, 451.

CHAPMAN, D. G. (1950). Some two sample tests. *Ann. Math. Statist.*, **21**, 601.

CHAPMAN, D. G. (1956). Estimating the parameters of a truncated Gamma distribution. *Ann. Math. Statist.*, **27**, 498.

CHAPMAN, D. G. and ROBBINS, H. (1951). Minimum variance estimation without regularity assumptions. *Ann. Math. Statist.*, **22**, 581.

CHERNOFF, H. (1949). Asymptotic studentisation in testing of hypotheses. *Ann. Math. Statist.*, **20**, 268.

CHERNOFF, H. (1951). A property of some Type A regions. *Ann. Math. Statist.*, **22**, 472.

CHERNOFF, H. (1952). A measure of asymptotic efficiency for tests of a hypothesis based on the sum of observations. *Ann. Math. Statist.*, **23**, 493.

CHERNOFF, H. (1954). On the distribution of the likelihood ratio. *Ann. Math. Statist.*, **25**, 573.

CHERNOFF, H. and LEHMANN, E. L. (1954). The use of maximum likelihood estimates in χ^2 tests for goodness of fit. *Ann. Math. Statist.*, **25**, 579.

CHERNOFF, H. and SAVAGE, I. R. (1958). Asymptotic normality and efficiency of certain non-parametric test statistics. *Ann. Math. Statist.*, **29**, 972.

CHERNOFF, H., GASTWIRTH, J. L. and JOHNS, M. V., Jr. (1967). Asymptotic distribution of linear combinations of functions of order statistics with applications to estimation. *Ann. Math. Statist.*, **38**, 52.

CHEW, V. (1970). Covariance matrix estimation in linear models. *J. Amer. Statist. Ass.*, **65**, 173.

CHIBISOV, D. M. (1971). Certain chi-square type tests for continuous distributions. *Theory Prob. Applic.*, **16**, 1.

CHIPMAN, J. S. (1964). On least squares with insufficient observations. *J. Amer. Statist. Ass.*, **59**, 1078.

CHOI, S. C. and WETTE, R. (1969). Maximum likelihood estimation of the parameters of the gamma distribution and their bias. *Technometrics*, **11**, 683.

CHOW, Y. S., ROBBINS, H. and TEICHER, H. (1965). Moments of randomly stopped sums. *Ann. Math. Statist.*, **36**, 789.

CLARK, R. E. (1953). Percentage points of the incomplete beta function. *J. Amer. Statist. Ass.*, **48**, 831.

CLOPPER, C. J. and PEARSON, E. S. (1934)*. The use of confidence or fiducial limits illustrated in the case of the binomial. *Biometrika*, **26**, 404.

COCHRAN, W. G. (1937). The efficiencies of the binomial series tests of significance of a mean and of a correlation coefficient. *J.R. Statist. Soc.*, **100**, 69.

COCHRAN, W. G. (1938). The omission or addition of an independent variate in multiple linear regression. *Suppl. J.R. Statist. Soc.*, **5**, 171.

COCHRAN, W. G. (1950). The comparison of percentages in matched samples. *Biometrika*, **37**, 256.

COCHRAN, W. G. (1952). The χ^2 test of goodness of fit. *Ann. Math. Statist.*, **23**, 315.

COCHRAN, W. G. (1954). Some methods for strengthening the common χ^2 tests. *Biometrics*, **10**, 417.

COHEN, A. and STRAWDERMAN, W. E. (1971). Unbiasedness of tests for homogeneity of variances. *Ann. Math. Statist.* **42**, 355.

COHEN, A. C., Jr. (1950a). Estimating the mean and variance of normal populations from singly truncated and doubly truncated samples. *Ann. Math. Statist.*, **21**, 557.

COHEN, A. C., Jr. (1950b). Estimating parameters of Pearson Type III populations from truncated samples. *J. Amer. Statist. Ass.*, **45**, 411.

COHEN, A. C., Jr. (1954). Estimation of the Poisson parameter from truncated samples and from censored samples. *J. Amer. Statist. Ass.*, **49**, 158.
COHEN, A. C., Jr. (1957). On the solution of estimating equations for truncated and censored samples from normal populations. *Biometrika*, **44**, 225.
COHEN, A. C., Jr. (1960a). Estimating the parameter of a modified Poisson distribution. *J. Amer. Statist. Ass.*, **55**, 139.
COHEN, A. C., Jr. (1960b). Estimating the parameter in a conditional Poisson distribution. *Biometrics*, **16**, 203.
COHEN, A. C., Jr. (1960c). Estimation in truncated Poisson distribution when zeros and some ones are missing. *J. Amer. Statist. Ass.*, **55**, 342.
COHEN, A. C., Jr. and WOODWARD, J. (1953). Tables of Pearson–Lee–Fisher functions of singly truncated normal distributions. *Biometrics*, **9**, 489.
COX, C. P. (1958). A concise derivation of general orthogonal polynomials. *J.R. Statist. Soc.*, B, **20**, 406.
COX, D. R. (1952a). Sequential tests for composite hypotheses. *Proc. Camb. Phil. Soc.*, **48**, 290.
COX, D. R. (1952b). A note on the sequential estimation of means. *Proc. Camb. Phil. Soc.*, **48**, 447.
COX, D. R. (1952c). Estimation by double sampling. *Biometrika*, **39**, 217.
COX, D. R. (1956). A note on the theory of quick tests. *Biometrika*, **43**, 478.
COX, D. R. (1958a). The regression analysis of binary sequences. *J.R. Statist. Soc.*, B, **20**, 215.
COX, D. R. (1958b). Some problems connected with statistical inference. *Ann. Math. Statist.*, **29**, 357.
COX, D. R. (1961). Tests of separate families of hypotheses. *Proc. 4th Berkeley Symp. Math. Statist. and Prob.*, **1**, 105.
COX, D. R. (1962). Further results on tests of separate families of hypotheses. *J.R. Statist. Soc.*, B, **24**, 406.
COX, D. R. (1963). Large sample sequential tests for composite hypotheses. *Sankhyā*, A, **25**, 5.
COX, D. R. (1964). Some applications of exponential order scores. *J.R. Statist. Soc.*, B, **26**, 103.
COX, D. R. (1967). Fieller's theorem and a generalization. *Biometrika*, **54**, 567.
COX, D. R. (1970). *Analysis of Binary Data*. Methuen, London.
COX, D. R. and HINKLEY, D. V. (1968). A note on the efficiency of least-squares estimates. *J.R. Statist. Soc.*, B, **30**, 284.
COX, D. R. and STUART, A. (1955). Some quick sign tests for trend in location and dispersion. *Biometrika*, **42**, 80.
CRAMÉR, H. (1946). *Mathematical Methods of Statistics*. Princeton Univ. Press.
CREASY, M. A. (1956). Confidence limits for the gradient in the linear functional relationship. *J.R. Statist. Soc.*, B, **18**, 65.
CROUSE, C. F. (1964). Note on Mood's test. *Ann. Math. Statist.*, **35**, 1825.
CROW, E. L. (1956). Confidence intervals for a proportion. *Biometrika*, **43**, 423.
CROW, E. L. and GARDNER, R. S. (1959). Confidence intervals for the expectation of a Poisson variable. *Biometrika*, **46**, 441.
CROW, E. L. and SIDDIQUI, M. M. (1967). Robust estimation of location. *J. Amer. Statist. Ass.*, **62**, 353.

DAHIYA, R. C. and GURLAND, J. (1972). Pearson chi-squared test of fit with random intervals. *Biometrika*, **59**, 147.
DANIELS, H. E. (1944). The relation between measures of correlation in the universe of sample permutations. *Biometrika*, **33**, 129.
DANIELS, H. E. (1948). A property of rank correlations. *Biometrika*, **35**, 416.

DANIELS, H. E. (1951–2). The theory of position finding. *J.R. Statist. Soc.*, B, 13, 186 and 14, 246.
DANIELS, H. E. (1961). The asymptotic efficiency of a maximum likelihood estimator. *Proc. 4th Berkeley Symp. Math. Statist. and Prob.*, 1, 151.
DANIELS, H. E. and KENDALL, M. G. (1958). Short proof of Miss Harley's theorem on the correlation coefficient. *Biometrika*, 45, 571.
DANTZIG, G. B. (1940). On the non-existence of tests of " Student's " hypothesis having power functions independent of σ. *Ann. Math. Statist.*, 11, 186.
DAR, S. N. (1962). On the comparison of the sensitivities of experiments. *J.R. Statist. Soc.*, B, 24, 447.
DARLING, D. A. (1952). On a test for homogeneity and extreme values. *Ann. Math. Statist.*, 23, 450.
DARLING, D. A. (1955). The Cramér–Smirnov test in the parametric case. *Ann. Math. Statist.*, 26, 1.
DARLING, D. A. (1957). The Kolmogorov–Smirnov, Cramér–von Mises tests. *Ann. Math. Statist.*, 28, 823.
DARMOIS, G. (1935). Sur les lois de probabilité à estimation exhaustive. *C.R. Acad. Sci., Paris*, 200, 1265.
DARROCH, J. N. (1962). Interactions in multi-factor contingency tables. *J.R. Statist. Soc.*, B, 24, 251.
DASGUPTA, P. (1968). Tables of the non-centrality parameter of F-test as a function of power. *Sankhyā*, B, 30, 73.
DAVID, F. N. (1937). A note on unbiased limits for the correlation coefficient. *Biometrika*, 29, 157.
DAVID, F. N. (1938). *Tables of the Correlation Coefficient*. Cambridge Univ. Press.
DAVID, F. N. (1939). On Neyman's " smooth " test for goodness of fit. *Biometrika*, 31, 191.
DAVID, F. N. (1947). A χ^2 " smooth " test for goodness of fit. *Biometrika*, 34, 299.
DAVID, F. N. (1950). An alternative form of χ^2. *Biometrika*, 37, 448.
DAVID, F. N. and JOHNSON, N. L. (1948). The probability integral transformation when parameters are estimated from the sample. *Biometrika*, 35, 182.
DAVID, F. N. and JOHNSON, N. L. (1954). Statistical treatment of censored data. Part I. Fundamental formulae. *Biometrika*, 41, 228.
DAVID, F. N. and JOHNSON, N. L. (1956). Some tests of significance with ordered variables. *J.R. Statist. Soc.*, B, 18, 1.
DAVID, F. N., BARTON, D. E., GANESHALINGAM, S., HARTER, H. L., KIM, P. J., MERRINGTON, M. and WALLEY, D. (1968). *Normal Centroids, Medians and Scores for Ordinal Data*. Cambridge Univ. Press.
DAVID, H. A. (1956). Revised upper percentage points of the extreme studentized deviate from the sample mean. *Biometrika*, 43, 449.
DAVID, H. A. (1970). *Order Statistics*. Wiley, New York.
DAVID, H. A. and PAULSON, A. S. (1965). The performance of several tests for outliers. *Biometrika*, 52, 429.
DAVID, H. A., HARTLEY, H. O. and PEARSON, E. S. (1954). The distribution of the ratio, in a single normal sample, of range to standard deviation. *Biometrika*, 41, 482.
DAVID, S. T., KENDALL, M. G. and STUART, A. (1951). Some questions of distribution in the theory of rank correlation. *Biometrika*, 38, 131.
DAVIS, A. W. and SCOTT, A. J. (1971). On the k-sample Behrens–Fisher distribution. (Div. of Math. Statist. Tech. Paper No. 33, C.S.I.R.O., Australia.)
DAVIS, R. C. (1951). On minimum variance in nonregular estimation. *Ann. Math. Statist.*, 22, 43.
DEEMER, W. L., Jr. and VOTAW, D. F., Jr. (1955). Estimation of parameters of truncated or censored exponential distributions. *Ann. Math. Statist.*, 26, 498.
DE GROOT, M. H. (1959). Unbiased sequential estimation for binomial populations. *Ann. Math. Statist.*, 30, 80.

REFERENCES

DEMING, W. E. and STEPHAN, F. F. (1940). On a least squares adjustment of a sampled frequency table when the expected marginal totals are known. *Ann. Math. Statist.*, **11**, 427.

DEMPSTER, A. P. (1963). Further examples of inconsistencies in the fiducial argument. *Ann. Math. Statist.*, **34**, 884.

DEMPSTER, A. P. (1964). On the difficulties inherent in Fisher's fiducial argument. *J. Amer. Statist. Ass.*, **59**, 56.

DEMPSTER, A. P. and SCHATZOFF, M. (1965). Expected significance level as a sensitivity index for test statistics. *J. Amer. Statist. Ass.*, **60**, 420.

DEN BROEDER, G. G., Jr. (1955). On parameter estimation for truncated Pearson Type III distributions. *Ann. Math. Statist.*, **26**, 659.

DENNY, J. L. (1967). Sufficient conditions for a family of probabilities to be exponential. *Proc. Nat. Acad. Sci., U.S.A.*, **57**, 1184.

DIXON, W. J. (1950). Analysis of extreme values. *Ann. Math. Statist.*, **21**, 488.

DIXON, W. J. (1951). Ratios involving extreme values. *Ann. Math. Statist.*, **22**, 68.

DIXON, W. J. (1953a). Processing data for outliers. *Biometrics*, **9**, 74.

DIXON, W. J. (1953b). Power functions of the Sign Test and power efficiency for normal alternatives. *Ann. Math. Statist.*, **24**, 467.

DIXON, W. J. (1954). Power under normality of several nonparametric tests. *Ann. Math. Statist.*, **25**, 610.

DIXON, W. J. (1957). Estimates of the mean and standard deviation of a normal population. *Ann. Math. Statist.*, **28**, 806.

DIXON, W. J. (1960). Simplified estimation from censored normal samples. *Ann. Math. Statist.*, **31**, 385.

DIXON, W. J. and MOOD, A. M. (1946). The statistical sign test. *J. Amer. Statist. Ass.*, **41**, 557.

DODGE, H. F. and ROMIG, H. G. (1944). *Sampling Inspection Tables*. Wiley, New York.

DOLBY, G. R. and LIPTON, S. (1972). Maximum likelihood estimation of the general nonlinear functional relationship with replicated observations and correlated errors. *Biometrika*, **59**, 121.

DORFF, M. and GURLAND, J. (1961a). Estimation of the parameters of a linear functional relation. *J.R. Statist. Soc.*, B, **23**, 160.

DORFF, M. and GURLAND, J. (1961b). Small sample behaviour of slope estimators in a linear functional relation. *Biometrics*, **17**, 283.

DOWNTON, F. (1953). A note on ordered least-squares estimation. *Biometrika*, **40**, 457.

DOWNTON, F. (1966a). Linear estimates with polynomial coefficients. *Biometrika*, **53**, 129.

DOWNTON, F. (1966b). Linear estimates of parameters in the extreme value distribution. *Technometrics*, **8**, 3.

DRAPER, N. R., GUTTMAN I. and KANEMASU, H. (1971). The distribution of certain regression statistics. *Biometrika*, **58**, 295.

DUNCAN, A. J. (1957). Charts of the 10% and 50% points of the operating characteristic curves for fixed effects analysis of variance F-tests, $\alpha = 0\cdot 10$ and $0\cdot 05$. *J. Amer. Statist. Ass.*, **52**, 345.

DUNN, O. J. (1968). A note on confidence bands for a regression line over a finite range. *J. Amer. Statist. Ass.*, **63**, 1028.

DUNN, O. J. and CLARK, V. (1969, 1971). Correlation coefficients measured on the same individuals; *and* Comparisons of tests of the equality of dependent correlation coefficients. *J. Amer. Statist. Ass.*, **64**, 366 and **66**, 904.

DURBIN, J. (1953). A note on regression when there is extraneous information about one of the coefficients. *J. Amer. Statist. Ass.*, **48**, 799.

DURBIN, J. (1954). Errors in variables. *Rev. Int. Statist. Inst.*, **22**, 23.

DURBIN, J. (1960). Estimation of parameters in time-series regression models. *J.R. Statist. Soc.*, B, **22**, 139.

DURBIN, J. (1961). Some methods of constructing exact tests. *Biometrika*, **48**, 41.

DURBIN, J. (1970). On Birnbaum's theorem on the relation between sufficiency, conditionality and likelihood. *J. Amer. Statist. Ass.*, **65**, 395.

DURBIN, J. and KENDALL, M. G. (1951). The geometry of estimation. *Biometrika*, **38**, 150.

DYNKIN, E. B. (1951). Necessary and sufficient statistics for a family of probability distributions. (Russian.) *Usp. Mat. Nauk (N.S.)*, **6**, No. 1 (41), 68.

EDWARDS, A. W. F. (1963). The measure of association in a 2×2 table. *J.R. Statist. Soc.*, A, **126**, 109.

EFRON, B. (1967). The power of the likelihood ratio test. *Ann. Math. Statist.*, **38**, 802.

EFRON, B. (1969). Student's t-test under symmetry conditions. *J. Amer. Statist. Ass.*, **64**, 1278.

EISENHART, C. (1938). The power function of the χ^2 test. *Bull. Amer. Math. Soc.*, **44**, 32.

EL-BADRY, M. A. and STEPHAN, F. F. (1955). On adjusting sample tabulations to census counts. *J. Amer. Statist. Ass.*, **50**, 738.

ELLISON, B. E. (1964). On two-sided tolerance intervals for a normal distribution. *Ann. Math. Statist.*, **35**, 762.

EPSTEIN, B. (1954). Truncated life tests in the exponential case. *Ann. Math. Statist.*, **25**, 555.

EPSTEIN, B. and SOBEL, M. (1953). Life testing. *J. Amer. Statist. Ass.*, **48**, 486.

EPSTEIN, B. and SOBEL, M. (1954). Some theorems relevant to life testing from an exponential distribution. *Ann. Math. Statist.*, **25**, 373.

EZEKIEL, M. J. B. and FOX, K. A. (1959). *Methods of Correlation and Regression Analysis: Linear and Curvilinear*. Wiley, New York.

FEDER, P. I. (1968). On the distribution of the log likelihood ratio test statistic when the true parameter is "near" the boundaries of the hypothesis regions. *Ann. Math. Statist.*, **39**, 2044.

FELLER, W. (1938). Note on regions similar to the sample space. *Statist. Res. Mem.*, **2**, 117.

FELLER, W. (1948). On the Kolmogorov–Smirnov limit theorems for empirical distributions. *Ann. Math. Statist.*, **19**, 177.

FEND, A. V. (1959). On the attainment of Cramér–Rao and Bhattacharyya bounds for the variance of an estimate. *Ann. Math. Statist.*, **30**, 381.

FERGUSON, T. S. (1961). On the rejection of outliers. *Proc. 4th Berkeley Symp. Math. Statist. and Prob.*, **1**, 253.

FÉRON, R. and FOURGEAUD, C. (1952). Quelques propriétés caractéristiques de la loi de Laplace–Gauss. *Publ. Inst. Statist. Paris*, **1**, 44.

FIELLER, E. C. (1940). The biological standardisation of insulin. *Suppl. J.R. Statist. Soc.*, **7**, 1.

FIELLER, E. C. (1954). Some problems in interval estimation. *J.R. Statist. Soc.*, B, **16**, 175.

FIENBERG, S. E. (1968). The geometry of an $r \times c$ contingency table. *Ann. Math. Statist.*, **39**, 1186.

FIENBERG, S. E. (1970). An iterative procedure for estimation in contingency tables. *Ann. Math. Statist.*, **41**, 907.

FIENBERG, S. E. and GILBERT, J. P. (1970). The geometry of a two by two contingency table. *J. Amer. Statist. Ass.*, **65**, 694.

FINNEY, D. J. (1941). On the distribution of a variate whose logarithm is normally distributed. *Suppl. J.R. Statist. Soc.*, **7**, 155.

FINNEY, D. J. (1948). The Fisher–Yates test of significance in 2×2 contingency tables. *Biometrika*, **35**, 145.

FINNEY, D. J. (1949a). The truncated binomial distribution. *Ann. Eugen.*, **14**, 319.

FINNEY, D. J. (1949b). On a method of estimating frequencies. *Biometrika*, **36**, 233.

FINNEY, D. J., LATSCHA, R., BENNETT, B. M., HSU, P. and HORST, C. (1963, 1966). *Tables for Testing Significance in a 2×2 Contingency Table*, with *Supplement*. Cambridge Univ. Press.

REFERENCES

FISHER, R. A. (1921a).* On the mathematical foundations of theoretical statistics. *Phil. Trans.*, A, **222**, 309.

FISHER, R. A. (1921b).* Studies in crop variation. I. An examination of the yield of dressed grain from Broadbalk. *J. Agric. Sci.*, **11**, 107.

FISHER, R. A. (1921c). On the " probable error " of a coefficient of correlation deduced from a small sample. *Metron*, **1**, (4), 3.

FISHER, R. A. (1922a).* On the interpretation of chi-square from contingency tables, and the calculation of P. *J.R. Statist. Soc.*, **58**, 87.

FISHER, R. A. (1922b).* The goodness of fit of regression formulae and the distribution of regression coefficients. *J.R. Statist. Soc.*, **85**, 597.

FISHER, R. A. (1924a). The distribution of the partial correlation coefficient. *Metron*, **3**, 329.

FISHER, R. A. (1924b). The influence of rainfall on the yield of wheat at Rothamsted. *Phil. Trans.*, B, **213**, 89.

FISHER, R. A. (1924c).* The conditions under which χ^2 measures the discrepancy between observation and hypothesis. *J.R. Statist. Soc.*, **87**, 442.

FISHER, R. A. (1925–). *Statistical Methods for Research Workers*. Oliver and Boyd, Edinburgh.

FISHER, R. A. (1925).* Theory of statistical estimation. *Proc. Camb. Phil. Soc.*, **22**, 700.

FISHER, R. A. (1928a).* The general sampling distribution of the multiple correlation coefficient. *Proc. Roy. Soc.*, A, **121**, 654.

FISHER, R. A. (1928b).* On a property connecting the χ^2 measure of discrepancy with the method of maximum likelihood. *Atti Congr. Int. Mat.*, Bologna, **6**, 94.

FISHER, R. A. (1935a). *The Design of Experiments*. Oliver and Boyd, Edinburgh.

FISHER, R. A. (1935b).* The fiducial argument in statistical inference. *Ann. Eugen.*, **6**, 391.

FISHER, R. A. (1939).* The comparison of samples with possibly unequal variances. *Ann. Eugen.*, **9**, 174.

FISHER, R. A. (1941).* The negative binomial distribution. *Ann. Eugen.*, **11**, 182.

FISHER, R. A. (1956). *Statistical methods and scientific inference*. Oliver and Boyd, Edinburgh.

FIX, E. (1949a). Distributions which lead to linear regressions. *Proc. 1st Berkeley Symp. Math. Statist. and Prob.*, 79.

FIX, E. (1949b). Tables of noncentral χ^2. *Univ. Calif. Publ. Statist.*, **1**, 15.

FIX, E. and HODGES, J. L., Jr. (1955). Significance probabilities of the Wilcoxon test. *Ann. Math. Statist.*, **26**, 301.

FOLKS, J. L. and ANTLE, C. E. (1967). Straight line confidence regions for linear models. *J. Amer. Statist. Ass.*, **62**, 1365.

FOX, L. (1950). Practical methods for the solution of linear equations and the inversion of matrices. *J.R. Statist. Soc.*, B, **12**, 120.

FOX, L. and HAYES, J. G. (1951). More practical methods for the inversion of matrices. *J.R. Statist. Soc.*, B, **13**, 83.

FOX, M. (1956). Charts of the power of the F-test. *Ann. Math. Statist.*, **27**, 484.

FRASER, D. A. S. (1950). Note on the χ^2 smooth test. *Biometrika*, **37**, 447.

FRASER, D. A. S. (1951). Sequentially determined statistically equivalent blocks. *Ann. Math. Statist.*, **22**, 372.

FRASER, D. A. S. (1953). Nonparametric tolerance regions. *Ann. Math. Statist.*, **24**, 44.

FRASER, D. A. S. (1957). *Nonparametric Methods in Statistics*. Wiley, New York.

FRASER, D. A. S. (1961a). On fiducial inference. *Ann. Math. Statist.*, **32**, 661.

FRASER, D. A. S. (1961b). The fiducial method and invariance. *Biometrika*, **48**, 261.

FRASER, D. A. S. (1962). On the consistency of the fiducial method. *J.R. Statist. Soc.*, B, **24**, 425.

FRASER, D. A. S. (1963). On sufficiency and the exponential family. *J.R. Statist. Soc.*, B, **25**, 115.

FRASER, D. A. S. (1964). Fiducial inference for location and scale parameters. *Biometrika*, **51**, 17.

FRASER, D. A. S. and GUTTMAN, I. (1956). Tolerance regions. *Ann. Math. Statist.*, **27**, 162.

FRASER, D. A. S. and WORMLEIGHTON, R. (1951). Non-parametric estimation IV. *Ann. Math. Statist.*, **22**, 294.

FREEMAN, D. and WEISS, L. (1964). Sampling plans which approximately minimize the maximum expected sample size. *J. Amer. Statist. Ass.*, **59**, 67.

FREEMAN, G. H. and HALTON, J. H. (1951). Note on an exact treatment of contingency, goodness of fit and other problems of significance. *Biometrika*, **38**, 141.

FREUND, R. J., VAIL, R. W. and CLUNIES-ROSS, C. W. (1961). Residual analysis. *J. Amer. Statist. Ass.*, **56**, 98 (corrigenda: 1005).

FRYER, J. G. (1970). On the nonparametric tests of David and Fix for the bivariate two-sample location problem. *J. Amer. Statist. Ass.*, **65**, 1297.

FRYER, J. G. (1971). On the homogeneity of the marginal distributions of a multidimensional contingency table. *J.R. Statist. Soc.*, A, **134**, 368.

GABRIEL, K. R. (1966). Simultaneous test procedures for multiple comparisons on categorical data. *J. Amer. Statist. Ass.*, **61**, 1081.

GAFARIAN, A. V. (1964). Confidence bands in straight line regression. *J. Amer. Statist. Ass.* **59**, 182.

GART, J. J. (1971). The comparison of proportions: a review of significance tests, confidence intervals and adjustments for stratification. *Rev. Int. Statist. Inst.*, **39**, 148.

GART, J. J. and PETTIGREW, H. M. (1970). On the conditional moments of the k-statistics for the Poisson distribution. *Biometrika*, **57**, 661.

GARWOOD, F. (1936). Fiducial limits for the Poisson distribution. *Biometrika*, **28**, 437.

GASTWIRTH, J. L. (1965). Asymptotically most powerful rank tests for the two-sample problem with censored data. *Ann. Math. Statist.*, **36**, 1243.

GASTWIRTH, J. L. (1966). On robust procedures. *J. Amer. Statist. Ass.*, **61**, 929.

GASTWIRTH, J. L. (1970). On asymptotic relative efficiencies of a class of rank tests. *J.R. Statist. Soc.*, B, **32**, 227.

GASTWIRTH, J. L. (1971). On the sign test for symmetry. *J. Amer. Statist. Ass.*, **66**, 821.

GASTWIRTH, J. L. and COHEN, M. L. (1970). Small sample behaviour of some robust linear estimators of location. *J. Amer. Statist. Ass.*, **65**, 946.

GASTWIRTH, J. L. and RUBIN, H. (1969). On robust linear estimators. *Ann. Math. Statist.*, **40**, 24.

GAYEN, A. K. (1949). The distribution of "Student's" t in random samples of any size drawn from non-normal universes. *Biometrika*, **36**, 353.

GAYEN, A. K. (1950). The distribution of the variance ratio in random samples of any size drawn from non-normal universes. Significance of difference between the means of two non-normal samples. *Biometrika*, **37**, 236, 399.

GAYEN, A. K. (1951). The frequency distribution of the product-moment correlation coefficient in random samples of any size drawn from non-normal universes. *Biometrika*, **38**, 219.

GEARY, R. C. (1936). The distribution of "Student's" ratio for non-normal samples. *Suppl. J.R. Statist. Soc.*, **3**, 178.

GEARY, R. C. (1942a). The estimation of many parameters. *J.R. Statist. Soc.*, **105**, 213.

GEARY, R. C. (1942b). Inherent relations between random variables. *Proc. R. Irish Acad.*, A, **47**, 63.

GEARY, R. C. (1943). Relations between statistics: the general and the sampling problem when the samples are large. *Proc. R. Irish Acad.*, A, **49**, 177.

GEARY, R. C. (1944). Comparison of the concepts of efficiency and closeness for consistent estimates of a parameter. *Biometrika*, **33**, 123.

GEARY, R. C. (1949). Determination of linear relations between systematic parts of variables with errors of observation the variances of which are unknown. *Econometrica*, **17**, 30.

GEARY, R. C. (1953). Non-linear functional relationship between two variables when one variable is controlled. *J. Amer. Statist. Ass.*, **48**, 94.

GEHAN, E. A. (1965a). A generalized Wilcoxon test for comparing arbitrarily singly-censored samples. *Biometrika*, **52**, 203.

REFERENCES

GEHAN, E. A. (1965b). A generalized two-sample Wilcoxon test for doubly censored data. *Biometrika*, **52**, 650.

GEISSER, S. and CORNFIELD, J. (1963). Posterior distributions for multivariate normal parameters. *J.R. Statist. Soc.*, **B**, **25**, 368.

GHOSH, B. K. (1969). Moments of the distribution of sample size in a SPRT. *J. Amer. Statist. Ass.*, **64**, 1560.

GHOSH, J. K. and SINGH, R. (1966). Unbiased estimation of location and scale parameters. *Ann. Math. Statist.*, **37**, 1671.

GIBBONS, J. D. (1964). Effect of non-normality on the power function of the sign test. *J. Amer. Statist. Ass.*, **59**, 142.

GIBSON, W. M. and JOWETT, G. H. (1957). Three-group regression analysis. Part I, Simple Regression Analysis. *Applied Statist.*, **6**, 114.

GIRSHICK, M. A. (1946). Contributions to the theory of sequential analysis. *Ann. Math. Statist.*, **17**, 123 and 282.

GJEDDEBAEK, N. F. (1949–61). Contribution to the study of grouped observations. I–VI. *Skand. Aktuartidskr.*, **32**, 135; **39**, 154; **40**, 20; *Biometrics*, **15**, 433; *Skand. Aktuartidskr.*, **42**, 194; **44**, 55.

GLASS, D. V., ed. (1954). *Social Mobility in Britain*. Routledge and Kegan Paul, London.

GODAMBE, V. P. (1960). An optimum property of regular maximum likelihood estimation. *Ann. Math. Statist.*, **31**, 1208.

GOHEEN, H. W. and KAVRUCK, S. (1948). A worksheet for tetrachoric r and standard error of tetrachoric r using Hayes' diagrams and tables. *Psychometrika*, **13**, 279.

GOLDBERGER, A. S. (1961). Stepwise least squares: residual analysis and specification error. *J. Amer. Statist. Ass.*, **56**, 998.

GOLDBERGER, A. S. and JOCHEMS, D. B. (1961). Note on stepwise least squares. *J. Amer. Statist. Ass.*, **56**, 105.

GOLDMAN, A. S. and ZEIGLER, R. K. (1966). Comparisons of some two stage sampling methods. *Ann. Math. Statist.*, **37**, 891.

GOOD, I. J., GOVER, T. N. and MITCHELL, G. J. (1970). Exact distributions for X^2 and for the likelihood-ratio statistic for the equiprobable multinomial distribution. *J. Amer. Statist. Ass.*, **65**, 267.

GOODMAN, L. A. (1963a). On methods for comparing contingency tables. *J.R. Statist. Soc.*, **A**, **126**, 94.

GOODMAN, L. A. (1963b). On Plackett's test for contingency table interactions. *J.R. Statist. Soc.*, **B**, **25**, 179.

GOODMAN, L. A. (1964a). Simultaneous confidence limits for cross-product ratios in contingency tables. *J.R. Statist. Soc.*, **B**, **26**, 86.

GOODMAN, L. A. (1964b). Simultaneous confidence intervals for contrasts among multinomial populations. *Ann. Math. Statist.*, **35**, 716.

GOODMAN, L. A. (1964c). Interactions in multidimensional contingency tables. *Ann. Math. Statist.*, **35**, 632.

GOODMAN, L. A. (1964d). Simple methods for analysing three-factor interaction in contingency tables. *J. Amer. Statist. Ass.*, **59**, 319.

GOODMAN, L. A. (1968). The analysis of cross-classified data: independence, quasi-independence and interactions in contingency tables with or without missing entries. *J. Amer. Statist. Ass.*, **63**, 1091.

GOODMAN, L. A. (1969). On partitioning χ^2 and detecting partial association in three-way contingency tables. *J.R. Statist. Soc.*, **B**, **31**, 486.

GOODMAN, L. A. (1970). The multivariate analysis of qualitative data: interactions among multiple classifications. *J. Amer. Statist. Ass.*, **65**, 226.

GOODMAN, L. A. (1971). Partitioning of chi-square, analysis of marginal contingency tables, and estimation of expected frequencies in multidimensional contingency tables. *J. Amer. Statist. Ass.*, **66**, 339.

GOODMAN, L. A. and KRUSKAL, W. H. (1954, 1959, 1963). Measures of association for cross classifications. Parts I, II and III. *J. Amer. Statist. Ass.*, **49**, 732; **54**, 123; and **58**, 310.

GOODMAN, L. A. and MADANSKY, A. (1962). Parameter-free and nonparametric tolerance limits: the exponential case. *Technometrics*, **4**, 75.

GOVINDARAJULU, Z. (1964). A supplement to Mendenhall's bibliography on life testing and related topics. *J. Amer. Statist. Ass.*, **59**, 1231.

GOVINDARAJULU, Z. (1966). Best linear estimates under symmetric censoring of the parameters of a double exponential population. *J. Amer. Statist. Ass.*, **61**, 248.

GRAY, H. L., WATKINS, T. A. and ADAMS, J. E. (1972). On the jackknife statistic; its extensions, and its relation to e_n-transformations. *Ann. Math. Statist.*, **43**, 1.

GRAYBILL, F. A. and BOWDEN, D. C. (1967). Linear segment confidence bands for simple linear models. *J. Amer. Statist. Ass.*, **62**, 403.

GRAYBILL, F. A. and CONNELL, T. L. (1964). Sample size required for estimating the variance within d units of the true value. *Ann. Math. Statist.*, **35**, 438.

GRAYBILL, F. A. and MARSAGLIA, G. (1957). Idempotent matrices and quadratic forms in the general linear hypothesis. *Ann. Math. Statist.*, **28**, 678.

GRUBBS, F. E. (1950). Sample criteria for testing outlying observations. *Ann. Math. Statist.*, **21**, 27.

GRUNDY, P. M. (1952). The fitting of group truncated and grouped censored normal distributions. *Biometrika*, **39**, 252.

GUENTHER, W. C. (1964). Another derivation of the non-central chi-square distribution. *J. Amer. Statist. Ass.*, **59**, 957.

GUENTHER, W. C. and WHITCOMB, M. G. (1966). Critical regions for tests of interval hypotheses about the variance. *J. Amer. Statist. Ass.*, **61**, 204.

GUEST, P. G. (1954). Grouping methods in the fitting of polynomials to equally spaced observations. *Biometrika*, **41**, 62.

GUEST, P. G. (1956). Grouping methods in the fitting of polynomials to unequally spaced observations. *Biometrika*, **43**, 149.

GUILFORD, J. P. and LYONS, T. C. (1942). On determining the reliability and significance of a tetrachoric coefficient of correlation. *Psychometrika*, **7**, 243.

GUMBEL, E. J. (1943). On the reliability of the classical chi-square test. *Ann. Math. Statist.*, **14**, 253.

GUPTA, S. S. and GNANADESIKAN, M. (1966). Estimation of the parameters of the logistic distribution. *Biometrika*, **53**, 565.

GUPTA, S. S., QUREISHI, A. S. and SHAH, B. K. (1967). Best linear unbiased estimators of the parameters of the logistic distribution using order statistics. *Technometrics*, **9**, 43.

GUPTA, S. S. and SHAH, B. K. (1965). Exact moments and percentage points of the order statistics and the distribution of the range from the logistic distribution. *Ann. Math Statist.*, **36**, 907.

GURLAND, J. (1968). A relatively simple form of the distribution of the multiple correlation coefficient. *J.R. Statist. Soc.*, **B**, **30**, 276.

GURLAND, J. and MILTON, R. (1970). Further consideration of the distribution of the multiple correlation coefficient. *J.R. Statist. Soc.*, **B**, **32**, 381.

GUTTMAN, I. (1957). On the power of optimum tolerance regions when sampling from normal distributions. *Ann. Math. Statist.*, **28**, 773.

GUTTMAN, I. (1970). *Statistical Tolerance Regions: Classical and Bayesian*. Griffin, London.

HAAS, G., BAIN, L. and ANTLE, C. (1970). Inferences for the Cauchy distribution based on maximum likelihood estimators. *Biometrika*, **57**, 403.

HÁJEK, J. and ŠIDÁK, Z. (1967). *Theory of Rank Tests*. Academia, Prague.

HAJNAL, J. (1961). A two-sample sequential t-test. *Biometrika*, **48**, 65.

HALD, A. (1949). Maximum likelihood estimation of the parameters of a normal distribution which is truncated at a known point. *Skand. Aktuartidskr.*, **32**, 119.

HALDANE, J. B. S. (1940). The mean and variance of χ^2 when used as a test of homogeneity, when expectations are small. *Biometrika*, **31**, 346.
HALDANE, J. B. S. (1945). On a method of estimating frequencies. *Biometrika*, **33**, 222.
HALDANE, J. B. S. (1955a). Substitutes for χ^2. *Biometrika*, **42**, 265.
HALDANE, J. B. S. (1955b). The rapid calculation of χ^2 as a test of homogeneity from a $2 \times n$ table. *Biometrika*, **42**, 519.
HALDANE, J. B. S. and SMITH, S. M. (1956). The sampling distribution of a maximum-likelihood estimate. *Biometrika*, **43**, 96.
HALL, W. J., WIJSMAN, R. A. and GHOSH, J. K. (1965). The relationship between sufficiency and invariance with applications in sequential analysis. *Ann. Math. Statist.*, **36**, 575.
HALMOS, P. R. and SAVAGE, L. J. (1949). Application of the Radon–Nikodym theorem to the theory of sufficient statistics. *Ann. Math. Statist.*, **20**, 225.
HALPERIN, M. (1952a). Maximum likelihood estimation in truncated samples. *Ann. Math. Statist.*, **23**, 226.
HALPERIN, M. (1952b). Estimation in the truncated normal distribution. *J. Amer. Statist. Ass.*, **47**, 457.
HALPERIN, M. (1960). Extension of the Wilcoxon–Mann–Whitney test to samples censored at the same fixed point. *J. Amer. Statist. Ass.*, **55**, 125.
HALPERIN, M. (1961a). Confidence intervals from censored samples. *Ann. Math. Statist.*, **32**, 828.
HALPERIN, M. (1961b). Fitting of straight lines and prediction when both variables are subject to error. *J. Amer. Statist. Ass.*, **56**, 657.
HALPERIN, M. (1963). Confidence interval estimation in non-linear regression. *J.R. Statist. Soc.*, B, **25**, 330.
HALPERIN, M. (1964). Interval estimation of non-linear parametric functions, II. *J. Amer. Statist. Ass.*, **59**, 168.
HALPERIN, M. and GURIAN, J. (1968). Confidence bands in linear regression with constraints on the independent variables. *J. Amer. Statist. Ass.*, **63**, 1020.
HALPERIN, M. and MANTEL, N. (1963). Interval estimation of non-linear parametric functions. *J. Amer. Statist. Ass.*, **58**, 611.
HALPERIN, M., GREENHOUSE, S. W., CORNFIELD, J. and ZALOKAR, J. (1955). Tables of percentage points for the studentized maximum absolute deviate in normal samples. *J. Amer. Statist. Ass.*, **50**, 185.
HALPERIN, M., RASTOGI, S. C., HO, I. and YANG, Y. Y. (1967). Shorter confidence bands in linear regression. *J. Amer. Statist. Ass.*, **62**, 1050.
HAMDAN, M. A. (1963). The number and width of classes in the chi-square test. *J. Amer. Statist. Ass.*, **58**, 678.
HAMDAN, M. A. (1968). Optimum choice of classes for contingency tables. *J. Amer. Statist. Ass.*, **63**, 291.
HAMDAN, M. A. (1970). The equivalence of tetrachoric and maximum likelihood estimates of ρ in 2×2 tables. *Biometrika*, **57**, 212.
HAMILTON, M. (1948). Nomogram for the tetrachoric correlation coefficient. *Psychometrika*, **13**, 259.
HAMMERSLEY, J. M. (1950). On estimating restricted parameters. *J.R. Statist. Soc.*, B, **12**, 192.
HANNAN, E. J. (1956). The asymptotic powers of certain tests based on multiple correlations. *J.R. Statist. Soc.*, B, **18**, 227.
HANNAN, J. and HARKNESS, W. (1963). Normal approximation to the distribution of two independent binomials, conditional on fixed sum. *Ann. Math. Statist.*, **34**, 1593.
HANNAN, J. F. and TATE, R. F. (1965). Estimation of the parameters for a multivariate normal distribution when one variable is dichotomized. *Biometrika*, **52**, 664.
HARKNESS, W. L. (1965). Properties of the extended hypergeometric distribution. *Ann. Math. Statist.*, **36**, 938.
HARKNESS, W. L. and KATZ, L. (1964). Comparison of the power functions for the test of independence in 2×2 contingency tables. *Ann. Math. Statist.*, **35**, 1115.

REFERENCES

HARLEY, B. I. (1956). Some properties of an angular transformation for the correlation coefficient. *Biometrika*, **43**, 219.

HARLEY, B. I. (1957). Further properties of an angular transformation of the correlation coefficient. *Biometrika*, **44**, 273.

HARTER, H. L. (1960). Tables of range and studentized range. *Ann. Math. Statist.*, **31**, 1122.

HARTER, H. L. (1961a). Expected values of normal order statistics. *Biometrika*, **48**, 151, 476.

HARTER, H. L. (1961b). Estimating the parameters of negative exponential populations from one or two order statistics. *Ann. Math. Statist.*, **32**, 1078.

HARTER, H. L. (1963). Percentage points of the ratio of two ranges and power of the associated test. *Biometrika*, **50**, 187.

HARTER, H. L. (1964). Criteria for best substitute interval estimators, with an application to the normal distribution. *J. Amer. Statist. Ass.*, **59**, 1133.

HARTER, H. L. (1969). *Order Statistics and their Use in Testing and Estimation*. Vol. 1, *Tests Based on Range and Studentized Range of Samples from a Normal Population*; Vol. 2, *Estimates Based on Order Statistics of Samples from Various Populations*. Aerospace Research Laboratories, U.S. Air Force; U.S. Govt Printing Office, Washington.

HARTER, H. L. and MOORE, A. H. (1965). Maximum-likelihood estimation of the parameters of gamma and Weibull populations from complete and from censored samples. *Technometrics*, **7**, 639.

HARTER, H. L. and MOORE, A. H. (1966a). Iterative maximum-likelihood estimation of the parameters of normal populations from singly and doubly censored samples. *Biometrika*, **53**, 205.

HARTER, H. L. and MOORE, A. H. (1966b). Local maximum-likelihood estimation of the parameters of three-parameter lognormal populations from complete and censored samples. *J. Amer. Statist. Ass.*, **61**, 842. (Corrections, **63**, 1549.)

HARTER, H. L. and MOORE, A. H. (1967a). Asymptotic variances and covariances of maximum-likelihood estimators, from censored samples, of the parameters of Weibull and gamma populations. *Ann. Math. Statist.*, **38**, 557.

HARTER, H. L. and MOORE, A. H. (1967b). Maximum-likelihood estimation, from censored samples, of the parameters of a logistic distribution. *J. Amer. Statist. Ass.*, **62**, 675.

HARTER, H. L. and MOORE, A. H. (1968). Maximum-likelihood estimation, from doubly censored samples, of the parameters of the first asymptotic distribution of extreme values. *J. Amer. Statist. Ass.*, **63**, 889.

HARTLEY, H. O. (1958). Maximum likelihood estimation from incomplete data. *Biometrics*, **14**, 174.

HARTLEY, H. O. (1964). Exact confidence regions for the parameters in non-linear regression laws. *Biometrika*, **51**, 347.

HARTLEY, H. O. and BOOKER, A. (1965). Nonlinear least squares estimation. *Ann. Math. Statist.*, **36**, 638.

HAYES, S. P., Jr. (1943). Tables of the standard error of the tetrachoric correlation coefficient. *Psychometrika*, **8**, 193.

HAYES, S. P., Jr. (1946). Diagrams for computing tetrachoric correlation coefficients from percentage differences. *Psychometrika*, **11**, 163.

HAYNAM, G. E. and GOVINDARAJULU, Z. (1966). Exact power of Mann–Whitney test for exponential and rectangular alternatives. *Ann. Math. Statist.*, **37**, 945.

HEALY, M. J. R. (1955). A significance test for the difference in efficiency between two predictors. *J.R. Statist. Soc.*, **B**, **17**, 266.

HEDAYAT, A. and ROBSON, D. S. (1970). Independent stepwise residuals for testing homoscedasticity. *J. Amer. Statist. Ass.*, **65**, 1573.

HEMELRIJK, J. (1952). A theorem on the Sign Test when ties are present. *Proc. Kon. Ned. Akad. Wetensch.*, **A**, **55**, 322.

HETTMANSPERGER, T. P. (1968). On the trimmed Mann–Whitney statistic. *Ann. Math. Statist.*, **39**, 1610.

REFERENCES

HILL, B. M. (1963a). The three-parameter lognormal distribution and Bayesian analysis of a point-source epidemic. *J. Amer. Statist. Ass.*, **58**, 72.

HILL, B. M. (1963b). Information for estimating the proportions in mixtures of exponential and normal distributions. *J. Amer. Statist. Ass.*, **58**, 918.

HINKLEY, D. V. (1969). Inference about the intersection in two-phase regression. *Biometrika*, **56**, 495.

HODGES, J. L., Jr. (1967). Efficiency in normal samples and tolerance of extreme values for some estimates of location. *Proc. 5th Berkeley Symp. Math. Statist. and Prob.*, **1**, 163.

HODGES, J. L., Jr. and LEHMANN, E. L. (1956). The efficiency of some nonparametric competitors of the t-test. *Ann. Math. Statist.*, **27**, 324.

HODGES, J. L., Jr. and LEHMANN, E. L. (1961). Comparison of the normal scores and Wilcoxon tests. *Proc. 4th Berkeley Symp. Math. Statist. and Prob.*, **1**, 307.

HODGES, J. L., Jr. and LEHMANN, E. L. (1963). Estimates of location based on rank tests. *Ann. Math. Statist.*, **34**, 598.

HODGES, J. L., Jr. and LEHMANN, E. L. (1967). Moments of chi and power of t. *Proc. 5th Berkeley Symp. Math. Statist. and Prob.*, **1**, 187.

HODGES, J. L., Jr. and LEHMANN, E. L. (1968). A compact table for power of the t-test. *Ann. Math. Statist.*, **39**, 1629.

HODGES, J. L., Jr. and LEHMANN, E. L. (1970). Deficiency. *Ann. Math. Statist.*, **41**, 783.

HOEFFDING, W. (1942). A class of statistics with asymptotically normal distribution. *Ann. Math. Statist.*, **19**, 293.

HOEFFDING, W. (1948). A non-parametric test of independence. *Ann. Math. Statist.*, **19**, 546.

HOEFFDING, W. (1950). "Optimum" non-parametric tests. *Proc. 2nd Berkeley Symp. Math. Statist. and Prob.*, 83.

HOEFFDING, W. (1952). The large-sample power of tests based on permutations of observations. *Ann. Math. Statist.*, **23**, 169.

HOEFFDING, W. (1953). On the distribution of the expected values of the order statistics. *Ann. Math. Statist.*, **24**, 93.

HOEFFDING, W. (1965). Asymptotically optimal tests for multinomial distributions. *Ann. Math. Statist.*, **36**, 369.

HOEL, D. G. and MAZUMDAR, M. (1969). A class of sequential tests for an exponential parameter. *J. Amer. Statist. Ass.*, **64**, 1549.

HOEL, P. G. (1951). Confidence regions for linear regression. *Proc. 2nd Berkeley Symp. Math. Statist. and Prob.*, 75.

HOGBEN, D., PINKHAM, R. S. and WILK, M. B. (1961). The moments of the non-central t-distribution. *Biometrika*, **48**, 465.

HOGG, R. V. (1956). On the distribution of the likelihood ratio. *Ann. Math. Statist.*, **27**, 529.

HOGG, R. V. (1961). On the resolution of statistical hypotheses. *J. Amer. Statist. Ass.*, **56**, 978.

HOGG, R. V. (1962). Iterated tests of the equality of several distributions. *J. Amer. Statist. Ass.*, **57**, 579.

HOGG, R. V. and CRAIG, A. T. (1956). Sufficient statistics in elementary distribution theory. *Sankhyā*, **17**, 209.

HOGG, R. V. and TANIS, E. A. (1963). An iterated procedure for testing the equality of several exponential distributions. *J. Amer. Statist. Ass.*, **58**, 435.

HOLLANDER, M. (1968). Certain uncorrelated nonparametric test statistics. *J. Amer. Statist. Ass.*, **63**, 707.

HOOKER, R. H. (1907). Correlation of the weather and crops. *J.R. Statist. Soc.*, **70**, 1.

HORA, R. B. and BUEHLER, R. J. (1966). Fiducial theory and invariant estimation. *Ann. Math. Statist.*, **37**, 643.

HORA, R. B. and BUEHLER, R. J. (1967). Fiducial theory and invariant prediction. *Ann. Math. Statist.*, **38**, 795.

HOTELLING, H. (1940). The selection of variates for use in prediction, with some comments on the general problem of nuisance parameters. *Ann. Math. Statist.*, **11**, 271.

REFERENCES

HOTELLING, H. (1953). New light on the correlation coefficient and its transforms. *J.R. Statist. Soc.*, B, **15**, 193.

HOTELLING, H. (1961). The behavior of some standard statistical tests under nonstandard conditions. *Proc. 4th Berkeley Symp. Math. Statist. and Prob.*, **1**, 319.

HOTELLING, H. and PABST, M. R. (1936). Rank correlation and tests of significance involving no assumptions of normality. *Ann. Math. Statist.*, **7**, 529.

HOWE, W. G. (1969). Two-sided tolerance limits for normal populations—some improvements. *J. Amer. Statist. Ass.*, **64**, 610.

HOYLE, M. H. (1968). The estimation of variances after using a Gaussianating transformation. *Ann. Math. Statist.*, **39**, 1125.

HØYLAND, A. (1965). Robustness of the Hodges–Lehmann estimates for shift. *Ann. Math. Statist.*, **36**, 174.

HSU, P. L. (1941). Analysis of variance from the power function standpoint. *Biometrika*, **32**, 62.

HUBER, P. J. (1964). Robust estimation of a location parameter. *Ann. Math. Statist.*, **35**, 73.

HUBER, P. J. (1967). The behaviour of maximum likelihood estimates under nonstandard conditions. *Proc. 5th Berkeley Symp. Math. Statist. and Prob.*, **1**, 221.

HUDSON, D. J. (1966). Fitting segmented curves whose join points have to be estimated. *J. Amer. Statist. Ass.*, **61**, 1097.

HUDSON, D. J. (1969). Least-squares fitting of a polynomial constrained to be either non-negative, non-decreasing or convex. *J.R. Statist. Soc.*, B, **31**, 113.

HUZURBAZAR, V. S. (1948). The likelihood equation, consistency and the maxima of the likelihood function. *Ann. Eugen.*, **14**, 185.

HUZURBAZAR, V. S. (1949). On a property of distributions admitting sufficient statistics. *Biometrika*, **36**, 71.

HUZURBAZAR, V. S. (1955). Confidence intervals for the parameter of a distribution admitting a sufficient statistic when the range depends on the parameter. *J.R. Statist. Soc.*, B, **17**, 86.

HYRENIUS, H. (1950). Distribution of "Student"–Fisher's t in samples from compound normal functions. *Biometrika*, **37**, 429.

IRELAND, C. T., KU, H. H. and KULLBACK, S. (1969). Symmetry and marginal homogeneity of an $r \times r$ contingency table. *J. Amer. Statist. Ass.*, **64**, 1323.

IRWIN, J. O. (1925). The further theory of Francis Galton's individual difference problem. On a criterion for the rejection of outlying observations. *Biometrika*, **17**, 100 and 238.

IRWIN, J. O. (1949). A note on the sub-division of χ^2 into components. *Biometrika*, **36**, 130.

IRWIN, J. O. (1959). On the estimation of the mean of a Poisson distribution from a sample with the zero class missing. *Biometrics*, **15**, 329.

JACKSON, J. E. (1960). Bibliography on sequential analysis. *J. Amer. Statist. Ass.*, **55**, 516.

JACOBSON, J. E. (1963). The Wilcoxon two-sample statistic: tables and bibliography. *J. Amer. Statist. Ass.*, **58**, 1086.

JAECKEL, L. A. (1971a, b). Robust estimates of location: symmetry and asymmetric contamination; *and* Some flexible estimates of location. *Ann. Math. Statist.*, **42**, 1020, 1540.

JEFFREYS, H. (1948). *Theory of Probability*. 2nd edn, Oxford Univ. Press.

JENKINS, W. L. (1955). An improved method for tetrachoric r, *and* Note by J. A. Fishman (1956). *Psychometrika*, **20**, 253 and **21**, 305.

JOGDEO, K. (1966). On randomized rank score procedures of Bell and Doksum. *Ann. Math. Statist.*, **37**, 1697.

JOHNSON, N. L. (1950). On the comparison of estimators. *Biometrika*, **37**, 281.

JOHNSON, N. L. (1959a). On an extension of the connexion between Poisson and χ^2 distributions. *Biometrika*, **46**, 352.

JOHNSON, N. L. (1959b). A proof of Wald's theorem on cumulative sums. *Ann. Math. Statist.*, **30**, 1245.

JOHNSON, N. L. (1961). Sequential analysis: a survey. *J.R. Statist. Soc.*, A, **124**, 372.
JOHNSON, N. L. and PEARSON, E. S. (1969). Tables of percentage points of non-central χ. *Biometrika*, **56**, 255.
JOHNSON, N. L. and WELCH, B. L. (1939). Applications of the non-central t distribution. *Biometrika*, **31**, 362.
JOINER, B. L. (1969). The median significance level and other small sample measures of test efficacy. *J. Amer. Statist. Ass.*, **64**, 971.
JONCKHEERE, A. R. (1954). A distribution-free k-sample test against ordered alternatives. *Biometrika*, **41**, 133.

KABE, D. G. (1963). Extension of Cochran's formulae for addition or omission of a variate in multiple regression analysis. *J. Amer. Statist. Ass.*, **58**, 527.
KAC, M., KIEFER, J. and WOLFOWITZ, J. (1955). On tests of normality and other tests of goodness of fit based on distance methods. *Ann. Math. Statist.*, **26**, 189.
KALE, B. K. (1961). On the solution of the likelihood equation by iteration processes. *Biometrika*, **48**, 452.
KALE, B. K. (1962). On the solution of likelihood equations by iteration processes. The multiparametric case. *Biometrika*, **49**, 479.
KASTENBAUM, M. A., HOEL, D. G. and BOWMAN, K. O. (1970a). Sample size requirements: one-way analysis of variance. *Biometrika*, **57**, 421.
KASTENBAUM, M. A., HOEL, D. G. and BOWMAN, K. O. (1970b). Sample size requirements: randomized block designs. *Biometrika*, **57**, 573.
KATTI, S. K. and GURLAND, J. (1962). Efficiency of certain methods of estimation for the negative binomial and the Neyman type A distributions. *Biometrika*, **49**, 215.
KEMPERMAN, J. H. B. (1956). Generalised tolerance limits. *Ann. Math. Statist.*, **27**, 180.
KEMPTHORNE, O. (1952). *The Design and Analysis of Experiments*. Wiley, New York.
KEMPTHORNE, O. (1967). The classical problem of inference—goodness of fit. *Proc. 5th Berkeley Symp. Math. Statist. and Prob.*, **1**, 235.
KEMPTHORNE, O. and DOERFLER, T. E. (1969). The behaviour of some significance tests under experimental randomization. *Biometrika*, **56**, 231.
KENDALL, M. G. (1951). Regression, structure and functional relationship. Part I. *Biometrika*, **38**, 11.
KENDALL, M. G. (1952). Regression, structure and functional relationship. Part II. *Biometrika*, **39**, 96.
KENDALL, M. G. (1948). *Rank Correlation Methods*. 4th edn, 1971, Griffin, London.
KENNEDY, W. J. and BANCROFT, T. A. (1971). Model building for prediction in regression based upon repeated significance tests. *Ann. Math. Statist.*, **42**, 1273.
KEYNES, J. M. (1911). The principal averages and the laws of error which lead to them. *J.R. Statist. Soc.*, **74**, 322.
KIEFER, J. (1952). On minimum variance estimators. *Ann. Math. Statist.*, **23**, 627.
KIEFER, J. and WEISS, L. (1957). Some properties of generalized sequential probability ratio tests. *Ann. Math. Statist.*, **28**, 57.
KIEFER, J. and WOLFOWITZ, J. (1956). Consistency of the maximum likelihood estimator in the presence of infinitely many incidental parameters. *Ann. Math. Statist.*, **27**, 887.
KIMBALL, A. W. (1954). Short-cut formulas for the exact partition of χ^2 in contingency tables. *Biometrics*, **10**, 452.
KIMBALL, B. F. (1946). Sufficient statistical estimation functions for the parameters of the distribution of maximum values. *Ann. Math. Statist.*, **17**, 299.
KLOTZ, J. H. (1962). Nonparametric tests for scale. *Ann. Math. Statist.*, **33**, 498.
KLOTZ, J. H. (1963). Small sample power and efficiency for the one sample Wilcoxon and normal scores tests. *Ann. Math. Statist.*, **34**, 624.
KLOTZ, J. H. (1964). On the normal scores two-sample rank test. *J. Amer. Statist. Ass.*, **59**, 652.
KLOTZ, J. H. (1965). Alternative efficiencies for signed rank tests. *Ann. Math. Statist.*, **36**, 1759.

REFERENCES

KNIGHT, W. (1965). A method of sequential estimation applicable to the hypergeometric, binomial, Poisson, and exponential distributions. *Ann. Math. Statist.*, **36**, 1494.

KNOTT, M. (1970). The small-sample power of one-sided Kolmogorov tests for a shift in location of the normal distribution. *J. Amer. Statist. Ass.*, **65**, 1384.

KOLMOGOROV, A. (1933). Sulla determinazione empirica di una legge di distribuzione. *G. Ist. Ital. Attuari*, **4**, 83.

KOŁODZIECZYK, S. (1935). On an important class of statistical hypotheses. *Biometrika*, **27**, 161.

KONDO, T. (1929). On the standard error of the mean square contingency. *Biometrika*, **21**, 376.

KONIJN, H. S. (1956, 1958). On the power of certain tests for independence in bivariate populations. *Ann. Math. Statist.*, **27**, 300 and **29**, 935.

KOOPMAN, B. O. (1936). On distributions admitting a sufficient statistic. *Trans. Amer. Math. Soc.*, **39**, 399.

KRAMER, K. H. (1963). Tables for constructing confidence limits on the multiple correlation coefficient. *J. Amer. Statist. Ass.*, **58**, 1082.

KRISHNAN, M. (1966). Locally unbiased type M test. *J.R. Statist. Soc.*, **B, 28**, 298.

KRISHNAN, M. (1967). The moments of a doubly non-central t distribution. *J. Amer. Statist. Ass.*, **62**, 278.

KRISHNAN, M. (1968). Series representations of the doubly noncentral t-distribution. *J. Amer. Statist. Ass.*, **63**, 1004.

KRUSKAL, W. H. (1952). A nonparametric test for the several sample problem. *Ann. Math. Statist.*, **23**, 525.

KRUSKAL, W. H. (1957). Historical notes on the Wilcoxon unpaired two-sample test. *J. Amer. Statist. Ass.*, **52**, 356.

KRUSKAL, W. H. (1958). Ordinal measures of association. *J. Amer. Statist. Ass.*, **53**, 814.

KRUSKAL, W. (1961). The coordinate-free approach to Gauss–Markov estimation, and its application to missing and extra observations. *Proc. 4th Berkeley Symp. Math. Statist. and Prob.*, **1**, 435.

KRUSKAL, W. H. and WALLIS, W. A. (1952–3). Use of ranks in one-criterion variance analysis. *J. Amer. Statist. Ass.*, **47**, 583 and **48**, 907.

KUDO, A. (1956). On the testing of outlying observations. *Sankhyā*, **17**, 67.

KULLBACK, S. (1971). Marginal homogeneity of multidimensional contingency tables. *Ann. Math. Statist.*, **42**, 594.

KULLDORFF, G. (1958). Maximum likelihood estimation of the mean/standard deviation of a normal random variable when the sample is grouped. *Skand. Aktuartidskr.*, **41**, 1 and 18.

KULLDORFF, G. (1963). Estimation of one or two parameters of the exponential distribution on the basis of suitably chosen order statistics. *Ann. Math. Statist.*, **34**, 1419.

LAHA, R. G. (1956). On a characterization of the stable law with finite expectation. *Ann. Math. Statist.*, **27**, 187.

LANCASTER, H. O. (1949a). The combination of probabilities arising from data in discrete distributions. *Biometrika*, **36**, 370.

LANCASTER, H. O. (1949b). The derivation and partition of χ^2 in certain discrete distributions. *Biometrika*, **36**, 117.

LANCASTER, H. O. (1950). The exact partition of χ^2 and its application to the problem of the pooling of small expectations. *Biometrika*, **37**, 267.

LANCASTER, H. O. (1951). Complex contingency tables treated by the partition of χ^2. *J.R. Statist. Soc.*, **B, 13**, 242.

LANCASTER, H. O. (1957). Some properties of the bivariate normal distribution considered in the form of a contingency table. *Biometrika*, **44**, 289.

LANCASTER, H. O. (1958). The structure of bivariate distributions. *Ann. Math. Statist.*, **29**, 719.

LANCASTER, H. O. (1960). On tests of independence in several dimensions. *J. Aust. Math. Soc.*, **1**, 241.

REFERENCES

LANCASTER, H. O. (1963). Canonical correlations and partitions of χ^2. *Quart. J. Math.*, Oxford, **(2) 14**, 220.

LANCASTER, H. O. and HAMDAN, M. A. (1964). Estimation of the correlation coefficient in contingency tables with possibly nonmetrical characters. *Psychometrika*, **29**, 383.

LARSON, H. J. and BANCROFT, T. A. (1963a). Sequential model building for prediction in regression analysis, I. *Ann. Math. Statist.* **34**, 462.

LARSON, H. J. and BANCROFT, T. A. (1963b). Biases in prediction by regression for certain incompletely specified models. *Biometrika*, **50**, 391.

LAURENT, A. G. (1963). Conditional distribution of order statistics and distribution of the reduced ith order statistic of the exponential model. *Ann. Math. Statist.*, **34**, 652.

LAWLEY, D. N. (1956). A general method for approximating to the distribution of likelihood ratio criteria. *Biometrika*, **43**, 295.

LAWTON, W. H. (1965). Some inequalities for central and non-central distributions. *Ann. Math. Statist.*, **36**, 1521.

LECAM, L. (1953). On some asymptotic properties of maximum likelihood estimates and related Bayes' estimates. *Univ. Calif. Publ. Statist.*, **1**, 277.

LECAM, L. (1970). On the assumptions used to prove asymptotic normality of maximum likelihood estimates. *Ann. Math. Statist.*, **41**, 802.

LEE, Y. S. (1971). Some results on the sampling distribution of the multiple correlation coefficient. *J.R. Statist. Soc.*, **B, 33**, 117.

LEE, Y. S. (1972). Tables of upper percentage points of the multiple correlation coefficient. *Biometrika*, **59**, 175.

LEHMANN, E. L. (1947). On optimum tests of composite hypotheses with one constraint. *Ann. Math. Statist.*, **18**, 473.

LEHMANN, E. L. (1949). Some comments on large sample tests. *Proc. 1st Berkeley Symp. Math. Statist. and Prob.*, 451.

LEHMANN, E. L. (1950). Some principles of the theory of testing hypotheses. *Ann. Math. Statist.*, **21**, 1.

LEHMANN, E. L. (1951). Consistency and unbiassedness of certain non-parametric tests. *Ann. Math. Statist.*, **22**, 165.

LEHMANN, E. L. (1959). *Testing Statistical Hypotheses.* Wiley, New York.

LEHMANN, E. L. (1963). Nonparametric confidence intervals for a shift parameter. *Ann. Math. Statist.*, **34**, 1507.

LEHMANN, E. L. and SCHEFFÉ, H. (1950, 1955). Completeness, similar regions and unbiased estimation. *Sankhyā*, **10**, 305 and **15**, 219.

LEHMANN, E. L. and STEIN, C. (1948). Most powerful tests of composite hypotheses. I. Normal distributions. *Ann. Math. Statist.*, **19**, 495.

LEHMANN, E. L. and STEIN, C. (1949). On the theory of some non-parametric hypotheses. *Ann. Math. Statist.*, **20**, 28.

LEHMANN, E. L. and STEIN, C. (1950). Completeness in the sequential case. *Ann. Math. Statist.*, **21**, 376.

LEHMER, E. (1944). Inverse tables of probabilities of errors of the second kind. *Ann. Math. Statist.*, **15**, 388.

LEWIS, B. N. (1962). On the analysis of interaction in multidimensional contingency tables. *J.R. Statist. Soc.*, **A, 125**, 88.

LEWIS, P. A. W. (1961). Distribution of the Anderson–Darling statistic. *Ann. Math. Statist.*, **32**, 1118.

LEWIS, T. O. and ODELL, P. L. (1966). A generalization of the Gauss–Markov theorem. *J. Amer. Statist. Ass.*, **61**, 1063.

LEWONTIN, R. C. and PROUT, T. (1956). Estimation of the number of different classes in a population. *Biometrics*, **12**, 211.

LILLIEFORS, H. W. (1967). On the Kolmogorov–Smirnov test for normality with mean and variance unknown. *J. Amer. Statist. Ass.*, **62**, 399.

LILLIEFORS, H. W. (1969). On the Kolmogorov–Smirnov test for the exponential distribution with mean unknown. *J. Amer. Statist. Ass.*, **64**, 387.
LINDLEY, D. V. (1947). Regression lines and the linear functional relationship. *Suppl. J.R. Statist. Soc.*, **9**, 218.
LINDLEY, D. V. (1950). Grouping corrections and Maximum Likelihood equations. *Proc. Camb. Phil. Soc.*, **46**, 106.
LINDLEY, D. V. (1953a). Statistical inference. *J.R. Statist. Soc.*, **B, 15**, 30.
LINDLEY, D. V. (1953b). Estimation of a functional relationship. *Biometrika*, **40**, 47.
LINDLEY, D. V. (1958a). Fiducial distributions and Bayes' theorem. *J.R. Statist. Soc.*, **B, 20**, 102.
LINDLEY, D. V. (1958b). Discussion of Cox (1958a).
LINDLEY, D. V. (1964). The Bayesian analysis of contingency tables. *Ann. Math. Statist.*, **35**, 1622.
LINDLEY, D. V., EAST, D. A. and HAMILTON, P. A. (1960). Tables for making inferences about the variance of a normal distribution. *Biometrika*, **47**, 433.
LINNIK, YU. V. (1964). On the Behrens–Fisher problem. *Bull. Int. Statist. Inst.*, **40(2)**, 833.
LINNIK, YU. V. (1967). On the elimination of nuisance parameters in statistical problems. *Proc. 5th Berkeley Symp. Math. Statist. and Prob.*, **1**, 267.
LLOYD, E. H. (1952). Least-squares estimation of location and scale parameters using order statistics. *Biometrika*, **39**, 88.
LOCKS, M. O., ALEXANDER, M. J. and BYARS, B. J. (1963). New tables of the noncentral t-distribution. *Aeronaut. Res. Lab. (Ohio)*, no. ARL 63–19.

McCORNACK, R. L. (1965). Extended tables of the Wilcoxon matched pair signed rank statistic. *J. Amer. Statist. Ass.*, **60**, 864.
McELROY, F. W. (1967). A necessary and sufficient condition that ordinary least-squares estimators be best linear unbiased. *J. Amer. Statist. Ass.*, **62**, 1302.
McKAY, A. T. (1935). The distribution of the difference between the extreme observation and the sample mean in samples of n from a normal universe. *Biometrika*, **27**, 466.
MACKINNON, W. J. (1964). Table for both the sign test and distribution-free confidence intervals of the median for sample sizes to 1,000. *J. Amer. Statist. Ass.*, **59**, 935.
McNEIL, D. R. (1967). Efficiency loss due to grouping in distribution-free tests. *J. Amer. Statist. Ass.*, **62**, 954.
McNOLTY, F. (1962). A contour-integral derivation of the non-central chi-square distribution. *Ann. Math. Statist.*, **33**, 796.
MACSTEWART, W. (1941). A note on the power of the sign test. *Ann. Math. Statist.*, **12**, 236.
MADANSKY, A. (1959). The fitting of straight lines when both variables are subject to error. *J. Amer. Statist. Ass.*, **54**, 173.
MADANSKY, A. (1962). More on length of confidence intervals. *J. Amer. Statist. Ass.*, **57**, 586.
MADANSKY, A. (1963). Tests of homogeneity for correlated samples. *J. Amer. Statist. Ass.*, **58**, 97.
MANLY, B. F. J. (1970a). The choice of a Wald test on the mean of a normal population. *Biometrika*, **57**, 91.
MANLY, B. F. J. (1970b). On the distribution of the decisive sample number of certain sequential tests. *Biometrika*, **57**, 367.
MANN, H. B. (1945). Non-parametric tests against trend. *Econometrica*, **13**, 245.
MANN, H. B. (1949). *Analysis and Design of Experiments: Analysis of Variance and Analysis of Variance Designs*. Dover, New York.
MANN, H. B. and WALD, A. (1942). On the choice of the number of intervals in the application of the chi-square test. *Ann. Math. Statist.*, **13**, 306.
MANN, H. B. and WHITNEY, D. R. (1947). On a test of whether one of two random variables is stochastically larger than the other. *Ann. Math. Statist.*, **18**, 50.
MARDIA, K. V. (1967, 1968, 1969). A non-parametric test for the bivariate two-sample location problem; Small sample power of a non-parametric test for the bivariate two-sample

location problem in the normal case; *and* On the null distribution of a non-parametric test for the bivariate two-sample problem. *J.R. Statist. Soc.*, B, **29**, 320; **30**, 83; and **31**, 98.

MARITZ, J. S. (1953). Estimation of the correlation coefficient in the case of a bivariate normal population when one of the variables is dichotomised. *Psychometrika*, **18**, 97.

MARSAGLIA, G. (1964). Conditional means and covariances of normal variables with singular covariance matrix. *J. Amer. Statist. Ass.*, **59**, 1203.

MASSEY, F. J., Jr. (1950a). A note on the estimation of a distribution function by confidence limits. *Ann. Math. Statist.*, **21**, 116.

MASSEY, F. J., Jr. (1950b, 1952). A note on the power of a non-parametric test. *Ann. Math. Statist.*, **21**, 440 and **23**, 637.

MASSEY, F. J., Jr. (1951). The Kolmogorov-Smirnov test of goodness of fit. *J. Amer. Statist. Ass.*, **46**, 68.

MAULDON, J. G. (1955). Pivotal quantities for Wishart's and related distributions, and a paradox in fiducial theory. *J.R. Statist. Soc.*, B, **17**, 79.

MEHTA, J. S. and SRINIVASAN, R. (1970). On the Behrens-Fisher problem. *Biometrika*, **57**, 649.

MENDENHALL, W. (1958). A bibliography on life testing and related topics. *Biometrika*, **45**, 521.

MENDENHALL, W. and HADER, R. J. (1958). Estimation of parameters of mixed exponentially distributed failure time distributions from censored life test data. *Biometrika*, **45**, 504.

MICKEY, M. R. and BROWN, M. B. (1966). Bounds on the distribution functions of the Behrens-Fisher statistic. *Ann. Math. Statist.*, **37**, 639.

MIKÉ, V. (1971). Efficiency-robust systematic linear estimators of location. *J. Amer. Statist. Ass.*, **66**, 594.

MIKULSKI, P. W. (1963). On the efficiency of optimal nonparametric procedures in the two sample case. *Ann. Math. Statist.*, **34**, 22.

MILLER, L. H. (1956). Table of percentage points of Kolmogorov statistics. *J. Amer. Statist. Ass.*, **51**, 111.

MILLER, R. G., Jr. (1964). A trustworthy jackknife. *Ann. Math. Statist.*, **35**, 1549.

MILTON, R. C. (1964). An extended table of critical values for the Mann-Whitney (Wilcoxon) two-sample statistic. *J. Amer. Statist. Ass.*, **59**, 925.

MITRA, S. K. and RAO, C. R. (1969). Conditions for optimality and validity of simple least squares theory. *Ann. Math. Statist.*, **40**, 1617.

MOOD, A. M. (1954). On the asymptotic efficiency of certain nonparametric two-sample tests. *Ann. Math. Statist.*, **25**, 514.

MOORE, D. S. (1971). A chi-square statistic with random cell boundaries. *Ann. Math. Statist.*, **42**, 147.

MOORE, P. G. (1956). The estimation of the mean of a censored normal distribution by ordered variables. *Biometrika*, **43**, 482.

MORAN, P. A. P. (1950). The distribution of the multiple correlation coefficient. *Proc. Camb. Phil. Soc.*, **46**, 521.

MOSES, L. E. (1963). Rank tests of dispersion. *Ann. Math. Statist.*, **34**, 973.

MOSHMAN, J. (1958). A method for selecting the size of the initial sample in Stein's two sample procedure. *Ann. Math. Statist.*, **29**, 1271.

MOSTELLER, F. (1968). Association and estimation in contingency tables. *J. Amer. Statist. Ass.*, **63**, 1.

MURPHY, R. B. (1948). Non-parametric tolerance limits. *Ann. Math. Statist.*, **19**, 581.

MYERS, M. H., SCHNEIDERMAN, M. A. and ARMITAGE, P. (1966). Boundaries for closed (wedge) sequential t test plans. *Biometrika*, **53**, 431.

NADDEO, A. (1968). Confidence intervals for the frequency function and the cumulative frequency function of a sample drawn from a discrete random variable. *Rev. Int. Statist. Inst.*, **36**, 313.

NAIR, K. R. (1948). The distribution of the extreme deviate from the sample mean and its studentized form. *Biometrika*, **35**, 118.

NAIR, K. R. (1952). Tables of percentage points of the "Studentized" extreme deviate from the sample mean. *Biometrika*, **39**, 189.
NAIR, K. R. and BANERJEE, K. S. (1942). A note on fitting of straight lines if both variables are subject to error. *Sankhyā*, **6**, 331.
NAIR, K. R. and SHRIVASTAVA, M. P. (1942). On a simple method of curve fitting. *Sankhyā*, **6**, 121.
NATHAN, G. (1969). Tests of independence in contingency tables from stratified samples. *New Developments in Survey Sampling*, ed. N. L. Johnson and H. Smith, Jr., Wiley-Interscience, New York.
NEYMAN, J. (1935).* Sur la vérification des hypothèses statistiques composées. *Bull. Soc. Math. France*, **63**, 1.
NEYMAN, J. (1937a).* "Smooth test" for goodness of fit. *Skand. Aktuartidskr.*, **20**, 149.
NEYMAN, J. (1937b).* Outline of a theory of statistical estimation based on the classical theory of probability. *Phil. Trans.*, A, **236**, 333.
NEYMAN, J. (1938a)*. On statistics the distribution of which is independent of the parameters involved in the original probability law of the observed variables. *Statist. Res. Mem.*, **2**, 58.
NEYMAN, J. (1938b). Tests of statistical hypothèses which are unbiassed in the limit. *Ann. Math. Statist.*, **9**, 69.
NEYMAN, J. (1949)*. Contribution to the theory of the χ^2 test. *Proc. 1st Berkeley Symp. Math. Statist. and Prob.*, 239.
NEYMAN, J. (1951). Existence of consistent estimates of the directional parameter in a linear structural relation between two variables. *Ann. Math. Statist.*, **22**, 497.
NEYMAN, J. and PEARSON, E. S. (1928)*. On the use and interpretation of certain test criteria for the purposes of statistical inference. *Biometrika*, **20A**, 175 and 263.
NEYMAN, J. and PEARSON, E. S. (1931)*. On the problem of *k* samples. *Bull. Acad. Polon. Sci.*, **3**, 460.
NEYMAN, J. and PEARSON, E. S. (1933a)*. On the testing of statistical hypotheses in relation to probabilities *a priori*. *Proc. Camb. Phil. Soc.*, **29**, 492.
NEYMAN, J. and PEARSON, E. S. (1933b)*. On the problem of the most efficient tests of statistical hypotheses. *Phil. Trans.*, A, **231**, 289.
NEYMAN, J. and PEARSON, E. S. (1936a)*. Sufficient statistics and uniformly most powerful tests of statistical hypotheses. *Statist. Res. Mem.*, **1**, 113.
NEYMAN, J. and PEARSON, E. S. (1936b)*. Unbiassed critical regions of Type A and Type A_1. *Statist. Res. Mem.*, **1**, 1.
NEYMAN, J. and PEARSON, E. S. (1938)*. Certain theorems on unbiassed critical regions of Type A, *and* Unbiassed tests of simple statistical hypotheses specifying the values of more than one unknown parameter. *Statist. Res. Mem.*, **2**, 25.
NEYMAN, J. and SCOTT, E. L. (1948). Consistent estimates based on partially consistent observations. *Econometrica*, **16**, 1.
NEYMAN, J. and SCOTT, E. L. (1951). On certain methods of estimating the linear structural relation. *Ann. Math. Statist.*, **22**, 352.
NEYMAN, J. and TOKARSKA, B. (1936)*. Errors of the second kind in testing "Student's" hypothesis. *J. Amer. Statist. Ass.*, **31**, 318.
NEYMAN, J., IWASKIEWICZ, K. and KOŁODZIECZYK, S. (1935)*. Statistical problems in agricultural experimentation. *Suppl. J.R. Statist. Soc.*, **2**, 107.
NOETHER, G. E. (1955). On a theorem of Pitman. *Ann. Math. Statist.*, **26**, 64.
NOETHER, G. E. (1957). Two confidence intervals for the ratio of two probabilities, and some measures of effectiveness. *J. Amer. Statist. Ass.*, **52**, 36.
NOETHER, G. E. (1963). Note on the Kolmogorov statistic in the discrete case. *Metrika*, **7**, 115.
NOETHER, G. E. (1967). Wilcoxon confidence intervals for location parameters in the discrete case. *J. Amer. Statist. Ass.*, **62**, 184.
NORTON, H. W. (1945). Calculation of chi-square for complex contingency tables. *J. Amer. Statist. Ass.*, **40**, 125.

REFERENCES

OGBURN, W. F. (1935). Factors in the variation of crime among cities. *J. Amer. Statist. Ass.*, **30**, 12.

OLKIN, I. and PRATT, J. W. (1958). Unbiased estimation of certain correlation coefficients. *Ann. Math. Statist.*, **29**, 201.

OWEN, D. B. (1963). Factors for one-sided tolerance limits and for variables sampling plans. *Sandia Corp. Monogr.* (Off. Techn. Serv., Dept. Commerce, Washington), SCR–607.

OWEN, D. B. (1965). The power of Student's t-test. *J. Amer. Statist. Ass.*, **60**, 320.

PACHARES, J. (1959). Table of the upper 10% points of the studentized range. *Biometrika*, **46**, 461.

PACHARES, J. (1960). Tables of confidence limits for the binomial distribution. *J. Amer. Statist. Ass.*, **55**, 521.

PACHARES, J. (1961). Tables for unbiased tests on the variance of a normal population. *Ann. Math. Statist.*, **32**, 84.

PATIL, G. P. and SHORROCK, R. (1965). On certain properties of the exponential-type families. *J.R. Statist. Soc.*, B, **27**, 94.

PATNAIK, P. B. (1948). The power function of the test for the difference between two proportions in a 2×2 table. *Biometrika*, **35**, 157.

PATNAIK, P. B. (1949). The non-central χ^2- and F-distributions and their applications. *Biometrika*, **36**, 202.

PAULSON, E. (1941). On certain likelihood-ratio tests associated with the exponential distribution. *Ann. Math. Statist.*, **12**, 301.

PAULSON, E. (1952). An optimum solution to the k-sample slippage problem for the normal distribution. *Ann. Math. Statist.*, **23**, 610.

PEARSON, E. S. (1926). A further note on the distribution of range in samples taken from a normal population. *Biometrika*, **18**, 173.

PEARSON, E. S. (1938)*. The probability integral transformation for testing goodness of fit and combining independent tests of significance. *Biometrika*, **30**, 134.

PEARSON, E. S. (1947)*. The choice of statistical tests illustrated in the interpretation of data classed in a 2×2 table. *Biometrika*, **34**, 139.

PEARSON, E. S. (1959). Note on an approximation to the distribution of non-central χ^2. *Biometrika*, **46**, 364.

PEARSON, E. S. and CHANDRA SEKAR, C. (1936)*. The efficiency of statistical tools and a criterion for the rejection of outlying observations. *Biometrika*, **28**, 308.

PEARSON, E. S. and HARTLEY, H. O. (1951). Charts of the power function for analysis of variance tests derived from the non-central F-distribution. *Biometrika*, **38**, 112.

PEARSON, E. S. and MERRINGTON, M. (1948)*. 2×2 tables: the power function of the test on a randomised experiment. *Biometrika*, **35**, 331.

PEARSON, E. S. and STEPHENS, M. A. (1962). The goodness-of-fit tests based on W_N^2 and U_N^2. *Biometrika*, **49**, 397.

PEARSON, E. S. and TIKU, M. L. (1970). Some notes on the relationship between the distributions of central and non-central F. *Biometrika*, **57**, 175.

PEARSON, K. (1897). On a form of spurious correlation which may arise when indices are used in the measurement of organs. *Proc. Roy. Soc.*, **60**, 489.

PEARSON, K. (1900).* On a criterion that a given system of deviations from the probable in the case of a correlated system of variables is such that it can be reasonably supposed to have arisen in random sampling. *Phil. Mag.*, (5), **50**, 157.

PEARSON, K. (1904).* On the theory of contingency and its relation to association and normal correlation. *Drapers' Co. Memoirs, Biometric Series*, **No. 1**, London.

PEARSON, K. (1909). On a new method for determining correlation between a measured character A and a character B, of which only the percentage of cases wherein B exceeds (or falls short of) a given intensity is recorded for each grade of A. *Biometrika*, **7**, 96.

PEARSON, K. (1913). On the probable error of a correlation coefficient as found from a fourfold table. *Biometrika*, **9**, 22.

REFERENCES

PEARSON, K. (1915). On the probable error of a coefficient of mean square contingency. *Biometrika*, **10**, 570.

PEARSON, K. (1916). On the general theory of multiple contingency with special reference to partial contingency. *Biometrika*, **11**, 145.

PEARSON, K. (1917). On the probable error of biserial η. *Biometrika*, **11**, 292.

PEERS, H. W. (1965). On confidence points and Bayesian probability points in the case of several parameters. *J.R. Statist. Soc.*, B, **27**, 9.

PILLAI, K. C. S. (1959). Upper percentage points of the extreme studentized deviate from the sample mean. *Biometrika*, **46**, 473.

PILLAI, K. C. S. and TIENZO, B. P. (1959). On the distribution of the extreme studentized deviate from the sample mean. *Biometrika*, **46**, 467.

PITMAN, E. J. G. (1936). Sufficient statistics and intrinsic accuracy. *Proc. Camb. Phil. Soc.*, **32**, 567.

PITMAN, E. J. G. (1937a). Significance tests which may be applied to samples from any population. *Suppl. J.R. Statist. Soc.*, **4**, 119.

PITMAN, E. J. G. (1937b). Significance tests which may be applied to samples from any populations. II. The correlation coefficient test. *Suppl. J.R. Statist. Soc.*, **4**, 225.

PITMAN, E. J. G. (1937c). The "closest" estimates of statistical parameters. *Proc. Camb. Phil. Soc.*, **33**, 212.

PITMAN, E. J. G. (1938). The estimation of the location and scale parameters of a continuous population of any given form. *Biometrika*, **30**, 391.

PITMAN, E. J. G. (1939a). A note on normal correlation. *Biometrika*, **31**, 9.

PITMAN, E. J. G. (1939b). Tests of hypotheses concerning location and scale parameters. *Biometrika*, **31**, 200.

PITMAN, E. J. G. (1948). *Non-Parametric Statistical Inference*. University of North Carolina Institute of Statistics (mimeographed lecture notes).

PITMAN, E. J. G. (1957). Statistics and science. *J. Amer. Statist. Ass.*, **52**, 322.

PLACKETT, R. L. (1949). A historical note on the method of least squares. *Biometrika*, **36**, 458.

PLACKETT, R. L. (1950). Some theorems in least squares. *Biometrika*, **37**, 149.

PLACKETT, R. L. (1953). The truncated Poisson distribution. *Biometrics*, **9**, 485.

PLACKETT, R. L. (1958). Linear estimation from censored data. *Ann. Math. Statist.*, **29**, 131.

PLACKETT, R. L. (1962). A note on interactions in contingency tables. *J.R. Statist. Soc.*, B, **24**, 162.

PLACKETT, R. L. (1964). The continuity correction in 2×2 tables. *Biometrika*, **51**, 327.

POTTHOFF, R. F. and WHITTINGHILL, M. (1966a). Testing for homogeneity. I. The binomial and multinomial distributions. *Biometrika*, **53**, 167.

POTTHOFF, R. F. and WHITTINGHILL, M. (1966b). Testing for homogeneity. II. The Poisson distribution. *Biometrika*, **53**, 183.

PRATT, J. W. (1961). Length of confidence intervals. *J. Amer. Statist. Ass.*, **56**, 549.

PRATT, J. W. (1963). Shorter confidence intervals for the mean of a normal distribution with known variance. *Ann. Math. Statist.*, **34**, 574.

PRATT, J. W. (1964). Robustness of some procedures for the two-sample location problem. *J. Amer. Statist. Ass.*, **59**, 665.

PRESS, S. J. (1966). A confidence interval comparison of two test procedures proposed for the Behrens–Fisher problem. *J. Amer. Statist. Ass.*, **61**, 454.

PRICE, R. (1964). Some non-central F-distributions expressed in closed form. *Biometrika*, **51**, 107.

PRZYBOROWSKI, J. and WILÉNSKI, M. (1935). Statistical principles of routine work in testing clover seed for dodder. *Biometrika*, **27**, 273.

PURI, M. L. (1964). Asymptotic efficiency of a class of c-sample tests. *Ann. Math. Statist.*, **35**, 102.

PUTTER, J. (1955). The treatment of ties in non-parametric tests. *Ann. Math. Statist.*, **26**, 368.

PYKE, R. (1965). Spacings. *J.R. Statist. Soc.*, B, **27**, 395.

QUENOUILLE, M. H. (1948). Partitioning chi-squared. *Unpublished.*
QUENOUILLE, M. H. (1956). Notes on bias in estimation. *Biometrika*, **43**, 353.
QUENOUILLE, M. H. (1958). *Fundamentals of Statistical Reasoning.* Griffin, London.
QUENOUILLE, M. H. (1959). Tables of random observations from standard distributions. *Biometrika*, **46**, 178.
QUESENBERRY, C. P. and DAVID, H. A. (1961). Some tests for outliers. *Biometrika*, **48**, 379.

RAGHAVACHARI, M. (1965a). The two-sample scale problem when locations are unknown. *Ann. Math. Statist.*, **36**, 1236.
RAGHAVACHARI, M. (1965b). On the efficiency of the normal scores test relative to the F-test. *Ann. Math. Statist.*, **36**, 1306.
RAGHUNANDANAN, K. and SRINIVASAN, R. (1970). Simplified estimation of parameters in a logistic distribution. *Biometrika*, **57**, 677.
RAJ, D. (1953). Estimation of the parameters of Type III populations from truncated samples. *J. Amer. Statist. Ass.*, **48**, 366.
RAMACHANDRAN, K. V. (1958). A test of variances. *J. Amer. Statist. Ass.*, **53**, 741.
RAMACHANDRAMURTY, P. V. (1966). On some nonparametric estimates for shift in the Behrens–Fisher situation. *Ann. Math. Statist.*, **37**, 593.
RAO, B. R. (1958a). On the relative efficiencies of ban estimates based on doubly truncated and censored samples. *Proc. Nat. Inst. Sci. India*, A, **24**, 366.
RAO, B. R. (1958b). On an analogue of Cramér–Rao's inequality. *Skand. Aktuartidskr.*, **41**, 57.
RAO, C. R. (1945). Information and accuracy attainable in the estimation of statistical parameters. *Bull. Calcutta Math. Soc.*, **37**, 81.
RAO, C. R. (1947). Minimum variance and the estimation of several parameters. *Proc. Camb. Phil. Soc.*, **43**, 280.
RAO, C. R. (1952). *Advanced Statistical Methods in Biometric Research.* Wiley, New York.
RAO, C. R. (1957). Theory of the method of estimation by minimum chi-square. *Bull. Int. Statist. Inst.*, **35 (2)**, 25.
RAO, C. R. (1961). Asymptotic efficiency and limiting information. *Proc. 4th Berkeley Symp. Math. Statist. and Prob.*, **1**, 531.
RAO, C. R. (1962a). Efficient estimates and optimum inference procedures in large samples. *J.R. Statist. Soc.*, B, **24**, 46.
RAO, C. R. (1962b). Apparent anomalies and irregularities in maximum likelihood estimation. *Sankhyā*, A, **24**, 73.
RAO, C. R. (1967). Least squares theory using an estimated dispersion matrix and its application to measurement of signals. *Proc. 5th Berkeley Symp. Math. Statist. and Prob.*, **1**, 355.
RAO, C. R. (1970). Estimation of heteroscedastic variances in linear models. *J. Amer. Statist. Ass.*, **65**, 161.
RESNIKOFF, G. J. (1962). Tables to facilitate the computation of percentage points of the non-central t-distribution. *Ann. Math. Statist.*, **33**, 580.
RESNIKOFF, G. J. and LIEBERMAN, G. J. (1957). *Tables of the non-central* t-*distribution.* Stanford Univ. Press.
RICHARDSON, D. H. and WU, D. M. (1970). Least squares and grouping method estimators in the errors in variables model. *J. Amer. Statist. Ass.*, **65**, 724.
RIDER, P. R. (1955). Truncated binomial and negative binomial distributions. *J. Amer. Statist. Ass.*, **50**, 877.
ROBBINS, H. (1944). On distribution-free tolerance limits in random sampling. *Ann. Math. Statist.*, **15**, 214.
ROBBINS, H. (1948). The distribution of Student's t when the population means are unequal. *Ann. Math. Statist.*, **19**, 406.
ROBISON, D. E. (1964). Estimates for the points of intersection of two polynomial regressions. *J. Amer. Statist. Ass.*, **59**, 214.

ROBSON, D. S. (1959). A simple method for constructing orthogonal polynomials when the independent variable is unequally spaced. *Biometrics*, **15**, 187.

ROBSON, D. S. and WHITLOCK, J. H. (1964). Estimation of a truncation point. *Biometrika*, **51**, 33.

ROSENTHAL, I. (1966). Distribution of the sample version of the measure of association, Gamma. *J. Amer. Statist. Ass.*, **61**, 440.

ROTHENBERG, T. J., FISHER, F. M. and TILANUS, C. B. (1964). A note on estimation from a Cauchy sample. *J. Amer. Statist. Ass.*, **59**, 460.

ROY, A. R. (1956). On χ^2-statistics with variable intervals. Technical Report, Stanford University, Statistics Department.

ROY, J. and MITRA, S. K. (1957). Unbiassed minimum variance estimation in a class of discrete distributions. *Sankhyā*, **18**, 371.

ROY, K. P. (1957). A note on the asymptotic distribution of likelihood ratio. *Bull. Calcutta Statist. Ass.*, **7**, 73.

ROY, S. N. (1954). Some further results in simultaneous confidence interval estimation. *Ann. Math. Statist.*, **25**, 752.

ROY, S. N. and BOSE, R. C. (1953). Simultaneous confidence interval estimation. *Ann. Math. Statist.*, **24**, 513.

ROY, S. N. and KASTENBAUM, M. A. (1956). On the hypothesis of no " interaction " in a multiway contingency table. *Ann. Math. Statist.*, **27**, 749.

ROY, S. N. and MITRA, S. K. (1956). An introduction to some non-parametric generalisations of analysis of variance and multivariate analysis. *Biometrika*, **43**, 361.

RUSHTON, S. (1951). On least squares fitting of orthonormal polynomials using the Choleski method. *J.R. Statist. Soc.*, **B, 13**, 92.

SALEH, A. K. M. E. (1966). Estimation of the parameters of the exponential distribution based on optimum order statistics in censored samples. *Ann. Math. Statist.*, **37**, 1717.

SAMPFORD, M. R. (1955). The truncated negative binomial distribution. *Biometrika*, **42**, 58.

SANKARAN, M. (1964). On an analogue of Bhattacharya bound. *Biometrika*, **51**, 268.

SARHAN, A. E. (1954). Estimation of the mean and standard deviation by order statistics. *Ann. Math. Statist.*, **25**, 317.

SARHAN, A. E. (1955). Estimation of the mean and standard deviation by order statistics, II, III. *Ann. Math. Statist.*, **26**, 505 and 576.

SARHAN, A. E. and GREENBERG, B. G. (1956). Estimation of location and scale parameters by order statistics from singly and doubly censored samples. I. The normal distribution up to samples of size 10. *Ann. Math. Statist.*, **27**, 427.

SARHAN, A. E. and GREENBERG, B. G. (1957). Tables for best linear estimates by order statistics of the parameters of single exponential distributions from singly and doubly censored samples. *J. Amer. Statist. Ass.*, **52**, 58.

SARHAN, A. E. and GREENBERG, B. G. (1958). Estimation of location and scale parameters by order statistics from singly and doubly censored samples. II. Tables for the normal distribution for samples of sizes $11 \leqslant n \leqslant 15$. *Ann. Math. Statist.*, **29**, 79.

SARHAN, A. E. and GREENBERG, B. G. (1959). Estimation of location and scale parameters for the rectangular population from censored samples. *J.R. Statist. Soc.*, **B, 21**, 356.

SARHAN, A. E. and GREENBERG, B. G. (editors) (1962). *Contributions to Order Statistics*. Wiley, New York.

SARHAN, A. E., GREENBERG, B. G. and OGAWA, J. (1963). Simplified estimates for the exponential distribution. *Ann. Math. Statist.*, **34**, 102.

SAVAGE, I. R. (1957). On the independence of tests of randomness and other hypotheses. *J. Amer. Statist. Ass.*, **52**, 53.

SAVAGE, I. R. (1962). *Bibliography of non-parametric statistics*. Harvard Univ. Press.

SAVAGE, L. J. (1970). Comments on a weakened principle of conditionality. *J. Amer. Statist. Ass.*, **65**, 399.

SAW, J. G. (1959). Estimation of the normal population parameters given a singly censored sample. *Biometrika*, **46**, 150.

SAW, J. G. (1961). Estimation of the normal population parameters given a type I censored sample. *Biometrika*, **48**, 367.

SCHAFER, R. E., FINKELSTEIN, J. M. and COLLINS, J. (1972). On a goodness-of-fit test for the exponential distribution with mean unknown. *Biometrika*, **59**, 222.

SCHEFFÉ, H. (1942a). On the theory of testing composite hypotheses with one constraint. *Ann. Math. Statist.*, **13**, 280.

SCHEFFÉ, H. (1942b). On the ratio of the variances of two normal populations. *Ann. Math. Statist.*, **13**, 371.

SCHEFFÉ, H. (1943a). On solutions of the Behrens–Fisher problem based on the t-distribution. *Ann. Math. Statist.*, **14**, 35.

SCHEFFÉ, H. (1943b). On a measure problem arising in the theory of non-parametric tests. *Ann. Math. Statist.*, **14**, 227.

SCHEFFÉ, H. (1944). A note on the Behrens–Fisher problem. *Ann. Math. Statist.*, **15**, 430.

SCHEFFÉ, H. (1958). Fitting straight lines when one variable is controlled. *J. Amer. Statist. Ass.*, **53**, 106.

SCHEFFÉ, H. (1970a). Multiple testing versus multiple estimation. Improper confidence sets. Estimation of directions and ratios. *Ann. Math. Statist.*, **41**, 1.

SCHEFFÉ, H. (1970b). Practical solutions of the Behrens–Fisher problem. *J. Amer. Statist. Ass.*, **65**, 1501.

SCHEFFÉ, H. and TUKEY, J. W. (1945). Non-parametric estimation: I. Validation of order statistics. *Ann. Math. Statist.*, **16**, 187.

SCHEUER, E. M. and SPURGEON, R. A. (1963). Some percentage points of the noncentral t-distribution. *J. Amer. Statist. Ass.*, **58**, 176.

SCHUCANY, W. R., GRAY, H. L. and OWEN, D. B. (1971). On bias reduction in estimation. *J. Amer. Statist. Ass.*, **66**, 524.

SEAL, H. L. (1948). A note on the χ^2 smooth test. *Biometrika*, **35**, 202.

SEAL, H. L. (1967). The historical development of the Gauss linear model. *Biometrika*, **54**, 1.

SEELBINDER, B. M. (1953). On Stein's two-stage sampling scheme. *Ann. Math. Statist.*, **24**, 640.

SEN, P. K. (1967). On some multisample permutation tests based on a class of U-statistics. *J. Amer. Statist. Ass.*, **62**, 1201.

SHAH, B. K. (1966). On the bivariate moments of order statistics from a logistic distribution. *Ann. Math. Statist.*, **37**, 1002.

SHAH, B. K. (1970). Note on moments of a logistic order statistics. *Ann. Math. Statist.*, **41**, 2150.

SHAH, S. M. (1961). The asymptotic variances of method of moments estimates of the parameters of the truncated binomial and negative binomial distributions. *J. Amer. Statist. Ass.*, **56**, 990.

SHAH, S. M. (1966). On estimating the parameter of a doubly truncated binomial distribution. *J. Amer. Statist. Ass.*, **61**, 259.

SHAPIRO, S. S. and WILK, M. B. (1965). An analysis of variance test for normality (complete samples). *Biometrika*, **52**, 591.

SHAPIRO, S. S., WILK, M. B. and CHEN, H. J. (1968). A comparative study of various tests for normality. *J. Amer. Statist. Ass.*, **63**, 1343.

SHARPE, K. (1970). Robustness of normal tolerance intervals. *Biometrika*, **57**, 71.

SHENTON, L. R. (1949). On the efficiency of the method of moments and Neyman's Type A contagious distribution. *Biometrika*, **36**, 450.

SHENTON, L. R. (1950). Maximum likelihood and the efficiency of the method of moments. *Biometrika*, **37**, 111.

SHENTON, L. R. (1951). Efficiency of the method of moments and the Gram–Charlier Type A distribution. *Biometrika*, **38**, 58.

REFERENCES

SHENTON, L. R. and BOWMAN, K. (1963). Higher moments of a maximum-likelihood estimate. *J.R. Statist. Soc.*, B, **25**, 305.

SHORACK, G. R. (1969). Testing and estimating ratios of scale parameters. *J. Amer. Statist. Ass.*, **64**, 999.

SHORACK, R. A. (1968). Recursive generation of the distribution of several non-parametric test statistics under censoring. *J. Amer. Statist. Ass.*, **63**, 353.

SICHEL, H. S. (1951-2). New methods in the statistical evaluation of mine sampling data. *Trans. Inst. Mining Metallurgy*, **61**, 261.

SIDDIQUI, M. M. (1963). Optimum estimators of the parameters of negative exponential distributions from one or two order statistics. *Ann. Math. Statist.*, **34**, 117.

SIDDIQUI, M. M. and BUTLER, C. (1969). Asymptotic joint distribution of linear systematic statistics from multivariate distributions. *J. Amer. Statist. Ass.*, **64**, 300.

SIDDIQUI, M. M. and RAGHUNANDANAN, K. (1967). Asymptotically robust estimators of location. *J. Amer. Statist. Ass.*, **62**, 950.

SIEGEL, S. and TUKEY, J. W. (1960). A nonparametric sum of ranks procedure for relative spread in unpaired samples. *J. Amer. Statist. Ass.*, **55**, 429.

SIEVERS, G. L. (1969). On the probability of large deviations and exact slopes. *Ann. Math. Statist.*, **40**, 1908.

SILLITTO, G. P. (1951). Interrelations between certain linear systematic statistics of samples from any continuous population. *Biometrika*, **38**, 377.

SILVERSTONE, H. (1957). Estimating the logistic curve. *J. Amer. Statist. Ass.*, **52**, 567.

SILVEY, S. D. (1969). Multicollinearity and imprecise estimation. *J.R. Statist. Soc.*, B, **31**, 539.

SLAKTER, M. J. (1966). Comparative validity of the chi-square and two modified chi-square goodness-of-fit tests for small but equal expected frequencies. *Biometrika*, **53**, 619.

SLAKTER, M. J. (1968). Accuracy of an approximation to the power of the chi-square goodness of fit test with small but equal expected frequencies. *J. Amer. Statist. Ass.*, **63**, 912.

SMIRNOV, N. V. (1936). Sur la distribution de ω^2. *C.R. Acad. Sci., Paris*, **202**, 449.

SMIRNOV, N. V. (1939a). Sur les écarts de la courbe de distribution empirique. *Rec. Math. (Matemat. Sbornik)*, N.S., **6** (48), 3.

SMIRNOV, N. V. (1939b). On the estimation of the discrepancy between empirical curves of distribution for two independent samples. *Bull. Math. Univ. Moscou, Série Int.*, **2**, No. 2, 3.

SMIRNOV, N. V. (1948). Table for estimating the goodness of fit of empirical distributions. *Ann. Math. Statist.*, **19**, 279.

SMITH, C. A. B. (1951). A test for heterogeneity of proportions. *Ann. Eugen.*, **16**, 16.

SMITH, C. A. B. (1952). A simplified heterogeneity test. *Ann. Eugen.*, **17**, 35.

SMITH, J. H. (1947). Estimation of linear functions of cell proportions. *Ann. Math. Statist.*, **18**, 231.

SMITH, K. (1916). On the "best" values of the constants in frequency distributions. *Biometrika*, **11**, 262.

SMITH, W. L. (1957). A note on truncation and sufficient statistics. *Ann. Math. Statist.*, **28**, 247.

SOLARI, M. E. (1969). The "maximum likelihood solution" of the problem of estimating a linear functional relationship. *J.R. Statist. Soc.*, B, **31**, 372.

SOPER, H. E. (1914). On the probable error of the biserial expression for the correlation coefficient. *Biometrika*, **10**, 384.

SPJØTVOLL, E. (1968). Most powerful tests for some non-exponential families. *Ann. Math. Statist*, **39**, 772.

SPRENT, P. (1966). A generalized least-squares approach to linear functional relationships. *J.R. Statist. Soc.*, B, **28**, 278.

SPROTT, D. A. (1960). Necessary restrictions for distributions *a posteriori*. *J.R. Statist. Soc.*, B, **22**, 312.

SPROTT, D. A. (1961). An example of an ancillary statistic and the combination of two samples by Bayes' theorem. *Ann. Math. Statist.*, **32**, 616.

SRINIVASAN, R. (1970). An approach to testing the goodness of fit of incompletely specified distributions. *Biometrika*, **57**, 605.

STEIN, C. (1945). A two-sample test for a linear hypothesis whose power is independent of the variance. *Ann. Math. Statist.*, **16**, 243.

STEIN, C. (1946). A note on cumulative sums. *Ann. Math. Statist.*, **17**, 498.

STEPHENS, M. A. (1963). The distribution of the goodness-of-fit statistic U_N^2. I. *Biometrika*, **50**, 303.

STEPHENS, M. A. (1964). The distribution of the goodness-of-fit statistic, U_N^2. II. *Biometrika*, **51**, 393.

STEPHENS, M. A. and MAAG, U. R. (1968). Further percentage points for W_N^2. *Biometrika*, **55**, 428.

STERNE, T. E. (1954). Some remarks on confidence or fiducial limits. *Biometrika*, **41**, 275.

STEVENS, W. L. (1939). Distribution of groups in a sequence of alternatives. *Ann. Eugen.*, **9**, 10.

STEVENS, W. L. (1950). Fiducial limits of the parameter of a discontinuous distribution. *Biometrika*, **37**, 117.

STUART, A. (1953). The estimation and comparison of strengths of association in contingency tables. *Biometrika*, **40**, 105.

STUART, A. (1954a). Too good to be true? *Applied Statist.*, **3**, 29.

STUART, A. (1954b). Asymptotic relative efficiencies of distribution-free tests of randomness against normal alternatives. *J. Amer. Statist. Ass.*, **49**, 147.

STUART, A. (1954c). The correlation between variate-values and ranks in samples from a continuous distribution. *Brit. J. Statist. Psychol.*, **7**, 37.

STUART, A. (1955a). A paradox in statistical estimation. *Biometrika*, **42**, 527.

STUART, A. (1955b). A test for homogeneity of the marginal distributions in a two-way classification. *Biometrika*, **42**, 412.

STUART, A. (1956). The efficiencies of tests of randomness against normal regression. *J. Amer. Statist. Ass.*, **51**, 285.

STUART, A. (1958). Equally correlated variates and the multinormal integral. *J.R. Statist. Soc.*, B, **20**, 273.

STUART, A. (1967). The average critical value method and the asymptotic relative efficiency of tests. *Biometrika*, **54**, 308.

"STUDENT" (1927)*. Errors of routine analysis. *Biometrika*, **19**, 151.

SUBRAHMANIAM, K. (1965). A note on estimation in the truncated Poisson. *Biometrika*, **52**, 279.

SUKHATME, P. V. (1936). On the analysis of k samples from exponential populations with special reference to the problem of random intervals. *Statist. Res. Mem.*, **1**, 94.

SUNDRUM, R. M. (1953). The power of Wilcoxon's 2-sample test. *J.R. Statist. Soc.*, B, **15**, 246.

SUNDRUM, R. M. (1954). On the relation between estimating efficiency and the power of tests. *Biometrika*, **41**, 542.

SWAMY, P. S. (1962a). On the amount of information supplied by censored samples of grouped observations in the estimation of statistical parameters. *Biometrika*, **49**, 245.

SWAMY, P. S. (1962b). On the joint efficiency of the estimates of the parameters of normal populations based on singly and doubly truncated samples. *J. Amer. Statist. Ass.*, **57**, 46.

SWAMY, P. S. (1963). On the amount of information supplied by truncated samples of grouped observations in the estimation of the parameters of normal populations. *Biometrika*, **50**, 207.

SWED, F. S. and EISENHART, C. (1943). Tables for testing randomness of grouping in a sequence of alternatives. *Ann. Math. Statist.*, **14**, 66.

SWINDEL, B. F. (1968). On the bias of some least-squares estimators of variance in a general linear model. *Biometrika*, **55**, 313.

REFERENCES

TAGUTI, G. (1958). Tables of tolerance coefficients for normal populations. *Rep. Statist. Appl. Res. (JUSE)*, **5**, 73.

TAKEUCHI, K. (1969). A note on the test for the location parameter of an exponential distribution. *Ann. Math. Statist.*, **40**, 1838.

TAMURA, R. (1963). On a modification of certain rank tests. *Ann. Math. Statist.*, **34**, 1101.

TANG, P. C. (1938). The power function of the analysis of variance tests with tables and illustrations of their use. *Statist. Res. Mem.*, **2**, 126.

TARTER, M. E. and CLARK, V. A. (1965). Properties of the median and other order statistics of logistic variates. *Ann. Math. Statist.*, **36**, 1779.

TATE, R. F. (1953). On a double inequality of the normal distribution. *Ann. Math. Statist.*, **24**, 132.

TATE, R. F. (1954). Correlation between a discrete and a continuous variable. Point-biserial correlation. *Ann. Math. Statist.*, **25**, 603.

TATE, R. F. (1955). The theory of correlation between two continuous variables when one is dichotomised. *Biometrika*, **48**, 205.

TATE, R. F. (1959). Unbiased estimation: functions of location and scale parameters. *Ann. Math. Statist.*, **30**, 341.

TATE, R. F. and GOEN, R. L. (1958). Minimum variance unbiassed estimation for the truncated Poisson distribution. *Ann. Math. Statist.*, **29**, 755.

TATE, R. F. and KLETT, G. W. (1959). Optimal confidence intervals for the variance of a normal distribution. *J. Amer. Statist. Ass.*, **54**, 674.

TEICHER, H. (1961). Maximum likelihood characterization of distributions. *Ann. Math. Statist.*, **32**, 1214.

TERPSTRA, T. J. (1952). The asymptotic normality and consistency of Kendall's test against trend, when ties are present in one ranking. *Proc. Kon. Ned. Akad. Wetensch.*, A, **55**, 327.

TERRY, M. E. (1952). Some rank order tests which are most powerful against specific parametric alternatives. *Ann. Math. Statist.*, **23**, 346.

THATCHER, A. R. (1964). Relationships between Bayesian and confidence limits for predictions. *J.R. Statist. Soc.*, B, **26**, 176.

THEIL, H. (1950). A rank-invariant method of linear and polynomial regression analysis. *Indag. Math.*, **12**, 85 and 173.

THEIL, H. and VAN YZEREN, J. (1956). On the efficiency of Wald's method of fitting straight lines. *Rev. Int. Statist. Inst.*, **24**, 17.

THOMPSON, J. R. (1968). Some shrinkage techniques for estimating the mean. *J. Amer. Statist. Ass.*, **63**, 113.

THOMPSON, R., GOVINDARAJULU, Z. and DOKSUM, K. A. (1967). Distribution and power of the absolute normal scores test. *J. Amer. Statist. Ass.*, **62**, 966.

THOMPSON, W. R. (1935). On a criterion for the rejection of observations and the distribution of the ratio of deviation to sample standard deviation. *Ann. Math. Statist.*, **6**, 214.

THOMPSON, W. R. (1936). On confidence ranges for the median and other expectation distributions for populations of unknown distribution form. *Ann. Math. Statist.*, **7**, 122.

TIKU, M. L. (1965a). Laguerre series forms of non-central χ^2 and F distributions. *Biometrika*, **52**, 415.

TIKU, M. L. (1965b). Chi-square approximations for the distributions of goodness-of-fit statistics U_N^2 and W_N^2. *Biometrika*, **52**, 630.

TIKU, M. L. (1966). A note on approximating the non-central F distribution. *Biometrika*, **53**, 606.

TIKU, M. L. (1967a). Tables of the power of the F-test. *J. Amer. Statist. Ass.*, **62**, 525. (Corrections, **63**, 1551.)

TIKU, M. L. (1967b). Estimating the mean and standard deviation from a censored normal sample. *Biometrika*, **54**, 155.

TIKU, M. L. (1968). Estimating the parameters of log-normal distribution from censored samples. *J. Amer. Statist. Ass.*, **63**, 134.

TOCHER, K. D. (1950). Extension of the Neyman–Pearson theory of tests to discontinuous variates. *Biometrika*, **37**, 130.
TODHUNTER, I. (1865). *A History of the Mathematical Theory of Probability from the time of Pascal to that of Laplace.* Macmillan, London.
TRICKETT, W. H., WELCH, B. L. and JAMES, G. S. (1956). Further critical values for the two-means problem. *Biometrika*, **43**, 203.
TUKEY, J. W. (1947). Non-parametric estimation, II. Statistically equivalent blocks and tolerance regions—the continuous case. *Ann. Math. Statist.*, **18**, 529.
TUKEY, J. W. (1948). Non-parametric estimation, III. Statistically equivalent blocks and multivariate tolerance regions—the discontinuous case. *Ann. Math. Statist.*, **19**, 30.
TUKEY, J. W. (1949). Sufficiency, truncation and selection. *Ann. Math. Statist.*, **20**, 309.
TUKEY, J. W. (1957). Some examples with fiducial relevance. *Ann. Math. Statist.*, **28**, 687.

VAN DER PARREN, J. L. (1970). Tables for distribution-free confidence limits for the median. *Biometrika*, **57**, 613.
VAN DER VAART, H. R. (1950). Some remarks on the power function of Wilcoxon's test for the problem of two samples, I, II. *Proc. Kon. Ned. Akad. Wetensch.*, A, **53**, 494 and 507.
VAN DER VAART, H. R. (1953). An investigation on the power function of Wilcoxon's two sample test if the underlying distributions are not normal. *Proc. Kon. Ned. Akad. Wetensch.*, A, **56**, 438.
VAN DER WAERDEN, B. L. (1952). Order tests for the two-sample problem and their power. *Proc. Kon. Ned. Akad. Wetensch.*, A, **55**, 453.
VAN DER WAERDEN, B. L. (1953). Order tests for the two-sample problem. *Proc. Kon. Ned. Akad. Wetensch.*, A, **56**, 303 and 311.
VAN EEDEN, C. (1963). The relation between Pitman's asymptotic relative efficiency of two tests and the correlation coefficient between their test statistics. *Ann. Math. Statist.*, **34**, 1442.
VAN EEDEN, C. (1964). Note on the consistency of some distribution-free tests for dispersion. *J. Amer. Statist. Ass.*, **59**, 105.
VAN EEDEN, C. (1970). Efficiency-robust estimation of location. *Ann. Math. Statist.*, **41**, 172.
VERDOOREN, L. R. (1963). Extended tables of critical values for Wilcoxon's test statistic. *Biometrika*, **50**, 177.
VILLEGAS, C. (1961). Maximum likelihood estimation of a linear functional relationship. *Ann. Math. Statist.*, **32**, 1048.
VILLEGAS, C. (1964). Confidence region for a linear relation. *Ann. Math. Statist.*, **35**, 780.
VILLEGAS, C. (1969). On the least squares estimation of non-linear relations. *Ann. Math. Statist.*, **40**, 462.

WALD, A. (1940).* The fitting of straight lines if both variables are subject to error. *Ann. Math. Statist.*, **11**, 284.
WALD, A. (1941).* Asymptotically most powerful tests of statistical hypotheses. *Ann. Math. Statist.*, **12**, 1.
WALD, A. (1942).* On the power function of the analysis of variance test. *Ann. Math. Statist.*, **13**, 434.
WALD, A. (1943a).* Tests of statistical hypotheses concerning several parameters when the number of observations is large. *Trans. Amer. Math. Soc.*, **54**, 426.
WALD, F. (1943b).* An extension of Wilks' method for setting tolerance limits. *Ann. Math. Statist.*, **14**, 45.
WALD, A. (1947). *Sequential Analysis.* Wiley, New York.
WALD, A. (1949).* Note on the consistency of the maximum likelihood estimate. *Ann. Math. Statist.*, **20**, 595.
WALD, A. (1950). *Statistical Decision Functions.* Wiley, New York.

REFERENCES

WALD, A. (1955).* Testing the difference between the means of two normal populations with unknown standard deviations. *Selected Papers in Statistics and Probability by Abraham Wald.* McGraw-Hill, New York. (Reprint by Stanford Univ. Press.)

WALD, A. and WOLFOWITZ, J. (1939).* Confidence limits for continuous distribution functions. *Ann. Math. Statist.*, **10**, 105.

WALD, A. and WOLFOWITZ, J. (1940).* On a test whether two samples are from the same population. *Ann. Math. Statist.*, **11**, 147.

WALD, A. and WOLFOWITZ, J. (1946).* Tolerance limits for a normal distribution. *Ann. Math. Statist.*, **17**, 208.

WALKER, A. M. (1963). A note on the asymptotic efficiency of an asymptotically normal estimator sequence. *J.R. Statist. Soc.*, B, **25**, 195.

WALLACE, D. L. (1958). Asymptotic approximations to distributions. *Ann. Math. Statist.*, **29**, 635.

WALLACE, T. D. (1964). Efficiencies for stepwise regressions. *J. Amer. Statist. Ass.*, **59**, 1179.

WALLACE, T. D. and TORO-VIZCARRONDO, C. E. (1969). Tables for the mean square error test for exact linear restrictions in regression. *J. Amer. Statist. Ass.*, **64**, 1649.

WALLIS, W. A. (1951). Tolerance intervals for linear regression. *Proc. 2nd Berkeley Symp. Math. Statist. and Prob.*, 43.

WALSH, J. E. (1946). On the power function of the sign test for slippage of means. *Ann. Math. Statist.*, **17**, 358.

WALSH, J. E. (1947). Concerning the effect of intraclass correlation on certain significance tests. *Ann. Math. Statist.*, **18**, 88.

WALSH, J. E. (1950a). Some estimates and tests based on the r smallest values in a sample. *Ann. Math. Statist.*, **21**, 386.

WALSH, J. E. (1950b). Some non-parametric tests of whether the largest observations of a set are too large or too small. *Ann. Math. Statist.*, **21**, 583.

WALSH, J. E. (1962). Distribution-free tolerance intervals for continuous symmetrical populations. *Ann. Math. Statist.*, **33**, 1167.

WALTON, G. S. (1970). A note on nonrandomized Neyman-shortest unbiased confidence intervals for the binomial and Poisson parameters. *Biometrika*, **57**, 223.

WATSON, G. S. (1957a). Sufficient statistics, similar regions and distribution-free tests. *J.R. Statist. Soc.*, B, **19**, 262.

WATSON, G. S. (1957b). The χ^2 goodness-of-fit test for normal distributions. *Biometrika*, **44**, 336.

WATSON, G. S. (1958). On chi-square goodness-of-fit tests for continuous distributions. *J.R. Statist. Soc.*, B, **20**, 44.

WATSON, G. S. (1959). Some recent results in chi-square goodness-of-fit tests. *Biometrics*, **15**, 440.

WATSON, G. S. (1961). Goodness-of-fit tests on a circle. *Biometrika*, **48**, 109.

WATSON, G. S. (1967). Linear least squares regression. *Ann. Math. Statist.*, **38**, 1679.

WEISS, L. (1962). On sequential tests which minimize the maximum expected sample size. *J. Amer. Statist. Ass.*, **57**, 551.

WEISSBERG, A. and BEATTY, G. H. (1960). Tables of tolerance-limit factors for normal distributions. *Technometrics*, **2**, 483.

WELCH, B. L. (1938). The significance of the difference between two means when the population variances are unequal. *Biometrika*, **29**, 350.

WELCH, B. L. (1939). On confidence limits and sufficiency, with particular reference to parameters of location. *Ann. Math. Statist.*, **10**, 58.

WELCH, B. L. (1947). The generalisation of "Student's" problem when several different population variances are involved. *Biometrika*, **34**, 28.

WELCH, B. L. (1965). On comparisons between confidence point procedures in the case of a single parameter. *J.R. Statist. Soc.*, B, **27**, 1.

WELCH, B. L. and PEERS, H. W. (1963). On formulae for confidence points based on integrals of weighted likelihoods. *J.R. Statist. Soc.*, B, **25**, 318.

REFERENCES

WHITAKER, L. (1914). On Poisson's law of small numbers. *Biometrika*, **10**, 36.

WHITE, C. (1952). The use of ranks in a test of significance for comparing two treatments. *Biometrics*, **8**, 33.

WICKSELL, S. D. (1917). The correlation function of Type A. *Medd. Lunds Astr. Obs.*, Series 2, No. 17.

WICKSELL, S. D. (1934). Analytical theory of regression. *Medd. Lunds Astr. Obs.*, Series 2, No. 69.

WILCOXON, F. (1945). Individual comparisons by ranking methods. *Biometrics Bull.*, **1**, 80.

WILCOXON, F. (1947). Probability tables for individual comparisons by ranking methods. *Biometrics*, **3**, 119.

WILK, M. B., GNANADESIKAN, R. and HUYETT, M. J. (1962). Estimation of parameters of the gamma distribution using order statistics. *Biometrika*, **49**, 525.

WILKINSON, G. N. (1961). Estimation of proportion from zero-truncated binomial data. *Biometrics*, **17**, 153.

WILKS, S. S. (1938a)*. The large-sample distribution of the likelihood ratio for testing composite hypotheses. *Ann. Math. Statist.*, **9**, 60.

WILKS, S. S. (1938b)*. Shortest average confidence intervals from large samples. *Ann. Math. Statist.*, **9**, 166.

WILKS, S. S. (1938c)*. Fiducial distributions in fiducial inference. *Ann. Math. Statist.*, **9**, 272.

WILKS, S. S. (1941)*. Determination of sample sizes for setting tolerance limits. *Ann. Math. Statist.*, **12**, 91.

WILKS, S. S. (1942)*. Statistical prediction with special reference to the problem of tolerance limits. *Ann. Math. Statist.*, **13**, 400.

WILKS, S. S. (1948)*. Order statistics. *Bull. Amer. Math. Soc.*, **54**, 6.

WILKS, S. S. and DALY, J. F. (1939)*. An optimum property of confidence regions associated with the likelihood function. *Ann. Math. Statist.*, **10**, 225.

WILLIAMS, C. A., Jr. (1950). On the choice of the number and width of classes for the chi-square test of goodness of fit. *J. Amer. Statist. Ass.*, **45**, 77.

WILLIAMS, E. J. (1952). Use of scores for the analysis of association in contingency tables. *Biometrika*, **39**, 274.

WILLIAMS, E. J. (1959). The comparison of regression variables. *J.R. Statist. Soc.*, B, **21**, 396.

WISE, M. E. (1963). Multinomial probabilities and the χ^2 and X^2 distributions. *Biometrika*, **50**, 145.

WISE, M. E. (1964). A complete multinomial distribution compared with the X^2 approximation and an improvement to it. *Biometrika*, **51**, 277.

WISHART, J. (1931). The mean and second moment coefficient of the multiple correlation coefficient, in samples from a normal population. *Biometrika*, **22**, 353.

WISHART, J. (1932). A note on the distribution of the correlation ratio. *Biometrika*, **24**, 441.

WISNIEWSKI, T. K. M. (1968). Testing for homogeneity of a binomial series. *Biometrika* **55**, 426.

WITTING, H. (1960). A generalized Pitman efficiency for nonparametric tests. *Ann. Math. Statist.*, **31**, 405.

WOLFOWITZ, J. (1947). The efficiency of sequential estimates and Wald's equation for sequential processes. *Ann. Math. Statist.*, **18**, 215.

WOLFOWITZ, J. (1949). The power of the classical tests associated with the normal distribution. *Ann. Math. Statist.*, **20**, 540.

WOODCOCK, E. R. and EAMES, A. R. (1970). *Confidence limits for numbers from 0 to 1200 based on the Poisson distribution.* Authority Health and Safety Branch, AHSB(S) R.179, H.M.S.O., London.

WORKING, H. and HOTELLING, H. (1929). The application of the theory of error to the interpretation of trends. *J. Amer. Statist. Ass.*, **24** (Suppl.), 73.

REFERENCES

WYNN, H. P. and BLOOMFIELD, P. (1971). Simultaneous confidence bands in regression analysis. *J.R. Statist. Soc.*, B, **33**, 202.

YARNOLD, J. K. (1970). The minimum expectation in X^2 goodness of fit tests and the accuracy of approximations for the null distribution. *J. Amer. Statist. Ass.*, **65**, 864.

YATES, F. (1934). Contingency tables involving small numbers and the χ^2 test. *Suppl. J.R. Statist. Soc.*, **1**, 217.

YATES, F. (1939a). An apparent inconsistency arising from tests of significance based on fiducial distributions of unknown parameters. *Proc. Camb. Phil. Soc.*, **35**, 579.

YATES, F. (1939b). Tests of significance of the differences between regression coefficients derived from two sets of correlated variates. *Proc. Roy. Soc. Edin.*, A, **59**, 184.

YATES, F. (1948). The analysis of contingency tables with groupings based on quantitative characters. *Biometrika*, **35**, 176.

YATES, F. (1955). A note on the application of the combination of probabilities test to a set of 2 × 2 tables. *Biometrika*, **42**, 404.

YOUNG, A. W. and PEARSON, K. (1915, 1919). On the probable error of a coefficient of contingency without approximation; and *Peccavimus! Biometrika*, **11**, 215 and **12**, 259.

YULE, G. U. (1900)*. On the association of attributes in statistics. *Phil. Trans.*, A, **194**, 257.

YULE, G. U. (1907)*. On the theory of correlation for any number of variables treated by a new system of notation. *Proc. Roy. Soc.*, A, **79**, 182.

YULE, G. U. (1912)*. On the methods of measuring association between two attributes. *J.R. Statist. Soc.*, **75**, 579.

YULE, G. U. (1926)*. Why do we sometimes get nonsense-correlations between time-series?—A study in sampling and the nature of time-series. *J.R. Statist. Soc.*, **89**, 1.

ZACKRISSON, U. (1959). The distribution of "Student's" t in samples from individual non-normal populations. *Publ. Statist. Inst., Univ. Gothenburg*, No. 6.

ZACKS, S. (1970). Uniformly most accurate upper tolerance limits for monotone likelihood ratio families of discrete distributions. *J. Amer. Statist. Ass.*, **65**, 307.

ZAHN, D. A. and ROBERTS, G. C. (1971). Exact χ^2 criterion tables with cell expectations one: an application to Coleman's measure of consensus. *J. Amer. Statist. Ass.*, **66**, 145.

ZELEN, M. (1971). The analysis of several 2×2 tables. *Biometrika*, **58**, 129.

ZYSKIND, G. (1963). A note on residual analysis. *J. Amer. Statist. Ass.*, **58**, 1125.

ZYSKIND, G. (1969). Parametric augmentations and error structures under which certain simple least squares and analysis of variance procedures are also best. *J. Amer. Statist. Ass.*, **64**, 1353.

INDEX

(References are to pages)

Adams, J. E., jackknifing, 6.
Aitken, A. C., MVB, 9; minimization of generalized variance, 85; more general linear model, (Exercises 19.2, 19.5) 100; orthogonal polynomials, 375.
Alexander, M. J., tables of non-central t, 265.
Allan, F. E., orthogonal polynomials, 374–5.
Allen, H. V., linearity of regression, 432.
Alling, D. W., binomial sequential tests, 631.
Alternative hypothesis (H_1), see Hypotheses.
Altham, P. M. E., cross-ratios in contingency tables, 581, 603.
Ammon, O., data on eye- and hair-colour, (Exercise 33.3) 604.
Amos, D. E., non-central t' and t'', 265.
Amount of information, 10.
Ancillary statistics, 226.
Andersen, E. B., sufficiency in discrete case, 27.
Andersen, S. L., robustness, 483.
Anderson, R. L., tables of orthogonal polynomials, 375.
Anderson, T. W., LS estimators and MV, 84; controlled variables, 425; tests of fit, 468; efficiency of SPR tests, 631.
Andrews, F. C., ARE of Kruskal–Wallis test, 522, (Exercise 31.20) 530.
Anscombe, F. J., ML estimation in negative binomial, (Exercises 18.26–7) 75; stepwise regression, 349; outlying observations, 548; tables for sequential sampling of attributes, 625; sequential estimation, 636; sequential test in the exponential distribution, (Exercise 34.14) 643.
Antle, C., ML estimation in Cauchy, 52; ML estimation of logistic parameters, (Exercise 18.29) 76; confidence regions for linear models, 385.
Arbuthnot, J., early use of Sign test, 532 footnote.
ARE, see Asymptotic relative efficiency.
Armitage, P., tests for trend in probabilities, 598; X^2 dispersion tests for stratified populations, 600; sequential tests, 627, 634, 639.
Armsen, P., tables of exact test of independence in 2×2 table, 573.

Arnold, H. J., power of Wilcoxon symmetry test, 526.
Arnold, K. J., sequential t-test tables, 634.
Array, 293.
Arvesen, J. N., jackknifing, 6.
Askovitz, S. I., graphical fitting of regression line, (Exercise 28.19) 389.
ASN, average sample number, 618.
Aspin, A. A., tables for problem of two means, 155.
Association, in 2×2 tables, 556–61; and independence, 557 footnote; in $r \times c$ tables, 576–94; partial, 561–5, 600–3; see Categorized data.
Asymptotic local efficiency, 279.
Atkinson, A. C., discriminating between linear models, 257.
Asymptotic relative efficiency (ARE), 276–87; definition, 277; and derivatives of power function, 278–82; tests which cannot be compared by, 278, 280–1; and maximum power loss, 283–4; and estimating efficiency, 284–5; and correlation, 285, (Exercise 25.9) 289; non-normal cases, 285–6.
Attributes, sampling of, see Binomial distribution.
Average sample number (ASN), 618.

Bahadur, R. R., efficiency, 44; comparison of tests, 287.
Bain, L. J., ML estimation in Cauchy, 52; problems of two means, 151; moments of non-central distributions, (Exercises 24.21–2) 272.
BAN (Best asymptotically normal) estimators, 95–9.
Bancroft, T. A., bias in linear models under selection, 379.
Banerjee, K. S., estimation of linear relation, 421.
Banerjee, S., double sampling for problem of two means, 639.
Barankin, E. W., bounds for variance, 16; sufficiency, 28; minimal sufficiency, 203 footnote.

INDEX

Barnard, G. A., optimum LS properties, 83 footnote; frequency justification of fiducial inference, 165; models for the 2×2 table, 570; sequential methods, 613.

Barnett, V. D., iterative ML estimation, 51, 52; Cauchy location estimation, 52.

Barr, D. R., testing non-regular distributions, 250.

Bartholomew, D. J., tests with n random, 192; efficiency of k-sample test, 524.

Bartky, W., rectifying inspection, 627.

Bartlett, M. S., approximate confidence intervals, 117, 133; conditional distribution in problem of two means, (Exercise 21.10) 167; sufficiency and similar regions, 198; quasi-sufficiency, 226; approximations to distribution of LR statistic, 243; modification of LR statistic, (Example 24.4) 245; conditional c.f., 332; adjustment in LS for extra observation, (Exercise 28.20) 389; estimation of functional relation, 420-1, (Exercise 29.11) 434; robustness, 483, 485; interaction in multi-way tables, 603.

Barton, D. E., " smooth " tests of fit, 460-66; tables of normal scores, 520.

Basu, A. P., estimating several linear relations, 410; tests in censored samples, 546; outliers for exponential distributions, 549.

Basu, D., sufficiency, completeness and independence, (Exercise 23.7) 229-30.

Bateman, G. I., non-central normal variates subject to constraints, (Exercise 24.2) 268.

Bayesian intervals, 157-9; critical discussion, 159-61; and fiducial theory, 161-5; and confidence intervals, 165.

BCR, best critical region, 173; see Tests of hypotheses.

Beale, E. M. L., non-linear estimation, 133; discarding variables, 350.

Beatty, G. H., tables of normal tolerance limits, 136.

Behrens, W. V., fiducial solution to problem of two means, 156.

Bell, C. B., tests using random normal deviates, 505.

Bellman, R., LS fitting, 371.

Bement, T. R., LS estimation, 91.

Bennett, B. M., test of independence in 2×2 table, 573, 575.

Benson, F., estimation using best two order-statistics, (Exercise 32.14) 553.

Berger, A., comparisons of 2×2 tables, 575.

Berkson, J., sampling experiments on BAN estimators, 99; choice of test size, 190; controlled variables, 424.

Bernstein, S., characterization of bivariate normality, 368 footnote.

Best asymptotically normal (BAN) estimators, 95-9.

Bhattacharjee, G. P., non-normality and Stein's test, 639.

Bhattacharyya, A., lower bound for variance, 12; covariance between MVB estimator and unbiassed estimators of its cumulants, (Exercise 17.4) 32; characterizations of bivariate normality, (Exercises 28.7-11) 386-7.

Bhuchongkul, S., tests of independence, 504.

Bias in estimation, 4-5; corrections for, 5-7, (Exercises 17.13, 17.17-18) 34-5; see Unbiassed estimation.

Bias in tests, 209-15; see Unbiassed tests, Tests of hypotheses.

Bickel, P. J., robustness and efficiency of location estimators, 487.

Bienaymé-Tchebycheff inequality, and consistent estimation, (Example 17.2) 3-4.

Binomial distribution, unbiassed estimation of square of parameter, θ, (Example 17.4) 6; MVB for θ, (Example 17.9) 11; estimation of $\theta(1-\theta)$, (Example 17.11) 15; unbiassed estimation of functions of θ, (Exercise 17.12) 34; estimation of linear relation between functions of parameters of independent binomials, (Exercise 17.25) 36; ML estimator biassed in sequential sampling, (Exercise 18.18) 73; equivalence of ML and MCS estimators, (Exercise 19.15) 101; confidence intervals, (Example 20.2) 108, 123-5, (Exercise 20.8) 137-8; tables and charts of confidence intervals, 123; tolerance limits, 136; confidence intervals for the ratio of two binomial parameters, (Exercise 20.9) 138; prediction by Bayesian and confidence methods, 164; fiducial intervals, (Exercise 21.3) 166; testing simple H_0 for θ, (Exercise 22.2) 193, 221; minimal sufficiency, (Exercise 23.11) 230; truncated, 546; homogeneity (dispersion) tests, 598-600, (Exercises 33.21-2) 608-9; (inverse) sequential sampling, 5, (Examples 34.1-7) 613-27, (Exercises 34.1-3) 641; MVB in sequential estimation, (Example 34.11) 635; double sampling, (Exercise 34.17) 644; see Sign test.

Birch, M. W., ML estimation of structural relation, 397, (Exercise 29.18) 435; limiting distribution of X^2, 442; multi-way tables, 603.

Birnbaum, A., Conditionality and Likelihood Principles, 227; robustness, 487, 520; logistic order statistics, (Exercise 32.12) 552; double sampling, 639.

Birnbaum, Z. W., tabulation of Kolmogorov test, 473; one-sided test of fit, 474, 476; computation of Kolmogorov statistic, (Example 30.6) 476–7.

Biserial correlation, 319–23, (Exercises 26.5, 26.10–12) 326–7.

Bivariate normal distribution, ML estimation of correlation parameter, ρ, alone, (Example 18.3) 40–1; indeterminate ML estimator of a function of ρ, (Example 18.4) 44; asymptotic variance of $\hat{\rho}$, (Example 18.6) 47; ML estimation of various combinations of parameters, (Example 18.14) 59, (Example 18.15) 61, (Exercises 18.11–14) 72; charts of confidence intervals for ρ, 123; estimation of ratio of means, (Example 20.7) 130–2; intervals for ratio of variances, (Exercise 20.19) 139; power of tests for ρ, (Example 22.7) 180; testing ratio of means, (Exercises 23.32) 233; joint c.f. of squares of variates, (Example 26.1) 295; linear regressions of squares of variates, (Example 26.3) 297, (Example 26.5) 298; estimation of ρ, 305–7; confidence intervals and tests for ρ, 307–8; tests of independence and regression tests, 308; correlation ratios, (Example 26.8) 310; estimation of ρ when parent dichotomized, (Exercises 26.3–4) 325; distribution of regression coefficient, (Exercise 26.9) 325–6; ML estimation in biserial situation, (Exercises 26.10–12) 326–7; LR test for ρ, (Exercise 26.15) 327; ML estimation and LR tests for common correlation parameter of two distributions, (Exercises 26.19–22) 328; linear regressions, (Example 28.1) 364; joint distribution of sums of squares, (Example 28.2) 364–5; characterization, 367–8, (Exercises 28.7–11) 386–7; robustness of tests for ρ, 486; efficiencies of tests of independence, 491–2, 499–500; efficiencies of tests of regression, 502–4; UMPU tests of $\rho = 0$ and $\sigma_1 = \sigma_2$, (Exercises 31.21, 31.26) 530–1; interpretation of contingency coefficient, 577; transformations and canonical correlations, 588–9.

Blackwell, D., sufficiency and MV estimation, 25; sequential sampling, 623; decision functions, 640.

Blalock, H. M., Jr., interpretation of X^2 measures, 581.

Blischke, W. R., variance bounds, 16.

Bliss, C. I., outlying observations, 549.

Bloch, D., Cauchy location estimation, 52; Bayesian contingency table analysis, 603.

Blom, G., simple estimation in censored samples, 543; asymptotic MVB, (Exercises 32.8–11) 551–2.

Blomqvist, N., ARE of tests, 279; medial correlation test, (Exercise 31.7) 528.

Bloomfield, P., confidence bands in regression, 385.

Blum, J. R., distribution of Hoeffding's test, 501.

Blyth, C. R., tables of binomial and Poisson confidence intervals, 125.

Bohrer, R., confidence bounds in linear regression, 385.

Booker, A., non-linear LS, 79.

Bose, R. C., simultaneous confidence intervals, 133.

Bounded completeness, see Completeness.

Bowden, D. C., confidence bounds in linear regression, 385.

Bowker, A. H., tables of tolerance limits, 135; approximations for tolerance limits, (Exercises, 20.20–2) 139–40; test of complete symmetry in $r \times r$ table, (Exercise 33.23) 609.

Bowman, K., cumulants of ML estimators, 49; ML estimators for Gamma, 69; tables of power of LR test, 263.

Box, G. E. P., approximations to distribution of LR statistic, 244; robustness, 483, 486.

Box, M. J., bias in non-linear estimation, (Exercise 18.15) 72, 79.

Bradu, D., lognormal linear models, 84.

Brandner, F. A., ML estimation and LR tests for bivariate normal correlation parameters, (Exercises 26.19–22) 328.

Brillinger, D. R., removal of ML bias, 44; fiducial paradoxes, 162; maximal correlation with order-statistics scores, 505.

Breslow, N., censored Kruskal–Wallis test, 546; sequential binomial test, 631.

Bross, I. D. J., charts for test of independence in 2×2 table, 573.

INDEX

Brown, G. W., median test of randomness, (Exercise 31.12) 529.

Brown, L., sufficient statistics, 27; LR tests, 287.

Brown, M. B., problem of two means, 154.

Brown, R. L., data on functional relation, (Example 29.1) 404; confidence intervals for functional relations, 407; linear functional relations in several variables, 410, (Exercises 29.6–8) 433–4.

Buckle, N., rectangular sum approximation to Wilcoxon test, 512.

Buehler, R. J., fiducial inference, 162.

Buhler, W., homogeneity tests, 599.

Bühler, W. J., ties in Wilcoxon test, 527.

Bulgren, W. G., doubly non-central distributions, 262, 265.

Bulmer, M. G., confidence limits for distance parameter, 581.

Burman, J. P., sequential sampling, 625.

Butler, C., distributions of multivariate order-statistics, 543.

Byars, B. J., tables of non-central t, 265.

c_1 (normal scores) test, 516–21.

Canonical analysis of $r \times c$ tables, 588–95.

Capon, J., scale-shift tests, 521; locally most powerful rank tests, (Exercise 31.17) 530.

Categorized data, 556–611; measures of association, 556–61, 576–88, (Exercises 33.1–4, 33.6) 604–5; partial association, 561–5; probabilistic interpretations, 565–567, 581; large-sample tests of independence, 567–9, 579–81; exact test of independence, different models, 569–75, 579; continuity correction, 575–6, (Exercise 33.5) 604; ordered tables, rank measures, 582–6, (Exercises 33.10–11) 606; ordered tables, scoring methods and canonical analysis, 586–95, (Exercises 33.12–14) 606–7; use of LS, 588; partitions of X^2, canonical components, 594–8, (Exercise 33.13) 606–7; binomial and Poisson dispersion tests, 598–600; multi-way tables, 600–3; test of complete symmetry, (Exercise 33.23) 609; test of identical margins, (Exercise 33.24) 609; estimation of cell probabilities, (Exercises 33.29, 33.32) 610–11; test of identical margins in 2^c table, (Exercise 33.30) 611.

Cauchy distribution, uselessness of sample mean as estimator of median, 2–3; sample median consistent, (Example 17.5) 8; MVB for location, (Example 17.7) 11; ML and order-statistics estimators, (Example 18.9) 52; testing simple H_0 for median, (Example 22.4) 176–7, (Exercise 22.4) 193; completeness, 199, (Exercise 23.5) 229.

Censored samples, 541–6; *see also* under names of distributions.

Central and non-central confidence intervals, 107–8.

Centre of location, 66.

Chan, L. K., Cauchy estimation, 52.

Chanda, K. C., Wilcoxon test in discrete cases, 516; sequential tests, 635.

Chandler, K. N., relations between residuals, (Exercise 27.5) 357.

Chandra Sekar, C., outlying observations, 548.

Chapman, D. G., bound for variance of estimator, 16; estimation of normal standard deviation, (Exercise 17.6) 33; truncated Gamma distribution, 546; double sampling, 639.

Characteristic functions, and completeness, 199; conditional, 332–3.

Chen, H. J., testing normality, 478.

Chernoff, H., asymptotic expansions in problem of two means, 155; reduction of size of critical regions, 185 footnote; LR test, 241; measure of test efficiency, 287, (Exercises 25.5–6) 288; distribution of X^2 test of fit, 445, 447; ARE of normal scores test, 516, 519; efficient estimation from order-statistics, 543.

Chew, V., LS estimation, 91.

Chibisov, D. M., tests of fit, 448.

Chipman, J. S., singular LS estimation, 89.

Choi, S. C., ML bias in Gamma distribution, (Exercise 18.15) 72.

Chow, Y. S., moments in sequential estimation, 635.

Clark, R. E., tables of confidence limits for the binomial parameter, 123.

Clark, V., dependent correlation coefficients, 308, (Exercise 28.22) 390.

Clark, V. A., logistic order-statistics, (Exercise 32.12) 553.

Clitic curve, 362.

Clopper, C. J., charts of confidence intervals for the binomial parameter, 123.

Closeness in estimation, 8.

Clunies-Ross, C. W., two-stage LS, (Exercise 28.24) 390.

Cochran, W. G., Sign test, (Exercise 25.1) 288, 532–3; adjustment of regression analysis, 371; critical region for X^2 test, 439; choice of classes for X^2 test, 457; limiting power function of X^2 test, (Exercise 30.4) 479; outlying observations, 549; X^2 in 2×2 tables, 576; partitions of X^2, 597; dispersion tests, 599; test for partial association, 603; test of identical margins in 2^c table, (Exercise 33.30) 611.

Cohen, A., unbiassed tests for variances, 255.

Cohen, A. C., Jr., truncated normal distributions, 544; truncated and censored Poisson distributions, 545, (Exercise 32.22) 554; truncated Gamma distribution, 546.

Cohen, M. L., robust estimation of location, 487.

Colligation coefficient, 559.

Collins, J., test of fit for exponential, 478.

Combination of tests, (Exercise 30.9) 480.

Completeness, 199; and unique estimation, 199; of sufficient statistics, 199–203; and similar regions, 205; in non-parametric problems, 490; sequential, 634 footnote.

Complex contingency tables, see Multi-way tables.

Components of X^2, 465–6; in $r \times c$ tables, 594–8.

Composite hypotheses, 195–233; see Hypotheses, Tests of hypotheses.

Conditional and unconditional inference in regression, 371.

Conditional tests, 225–8, 371.

Conditionality Principle, 226–7.

Confidence belt, coefficient, 105; see Confidence intervals.

Confidence distribution, 107, 162.

Confidence intervals, 103–40; distributions with one parameter, 103–25; limits, 104; belt, coefficient, 105; graphical representation, 106, 110, 113, 114, 116, 124, 126, 130; central and non-central, 107–8; discontinuous distributions, 108–10, 123–5; conservative, 108–10; for large samples, 110–113, 117–19; nestedness, 112–17; difficult cases, 113–17, (Example 20.7) 130–2, (Exercise 28.21) 389; shortest, 119–23; minimum average length in large samples, 120–1; most selective, 122, 128, 215; and tests, 122–3, 215; unbiassed, 122, 215; tables and charts, 123, 125; distributions with several parameters, 125–8; choice of statistic, 128; studentization, 128–32; simultaneous intervals for several parameters, 132–3; when range depends on parameter, (Exercises 20.13–16) 138–9; problem of two means, 146–55, (Exercises 21.9, 21.12) 167–8; critical discussion, 159–61, 165.

Confidence limits, 104; see Confidence intervals.

Confidence regions, 133, (Exercise 20.5) 137; for a regression, 380–5.

Connell, T. L., double sampling for normal variance, 639.

Consistency in estimation, 3–4, 96 footnote, 273; of ML estimators, 41–4, 46, 57.

Consistency in tests, 250; of LR tests, 250–1.

Constraints, imposed by a hypothesis, 171, 195.

Contingency, tables, 576; coefficient, 577; see Categorized data.

Continuity corrections, in distribution-free tests, 526–7; in 2×2 tables, 575–6.

Controlled variables, 424–5, 429.

Convergence in probability, 3; see Consistency in estimation.

Cornfield, J., fiducial inference, 164; tables of studentized maximum absolute deviate, 548.

Corrections, for bias in estimation, 5–7, (Exercises 17.17–18) 34–5; to ML estimators for grouping, (Exercises 18.24–5) 74–5.

Correlation, between estimators, 17–18; generally, 290–360; and interdependence, 290–291, 300–1; and causation, 291–2; coefficient, 299–300; historical note, 300; computation of coefficient, (Examples 26.6–7) 301–4; scatter diagram, 304; standard error, 304–5; estimation and testing in normal samples, 305–8; ratios, 308–10; linearity of regression, 308–10, (Exercise 26.24) 329; LR tests of coefficient, ratio and linearity of regression, 311–13; intra-class, 314–16, (Exercise 26.14) 327; tetrachoric, 316–19; biserial, 319–23, (Exercises 26.10–12) 326–327; point-biserial, 323–4; coefficient increased by Sheppard's grouping corrections, (Exercise 26.16) 327; attenuation, (Exercise 26.17) 327; spurious, (Exercise 26.18) 327–8; matrix, 330; determinant, 331; see also Multiple correlation, Partial correlation, Regression.

Covariance, 295; see Correlation, Regression.

Cox, C. P., orthogonal polynomials, 375.

INDEX

Cox, D. R., estimation of linear relation between functions of parameters of independent binomials, (Exercise 17.25) 36; efficiency of LS estimators, 84; confidence distribution, 107; conditional tests, 227; testing ratio of normal means, (Exercise 23.32) 233; LR statistics, 257; ARE and maximum power loss, 284; regression of efficient on inefficient test statistic, (Exercise 25.7) 288; exponential scores, 520; ARE of simple tests of randomness, (Exercises 31.10–12) 528–9; linear logistic model, 598; sequential procedures for composite H_0, 634, (Exercise 34.21) 644; double sampling, 639, (Exercises 34.15–17) 643–4.

Craig, A. T., completeness of sufficient statistics, 200; completeness and independence, (Exercises 23.8–9) 230; LR test for rectangular distribution, (Exercise 24.9) 269; sufficiency in exponential family with range a function of parameters, (Exercise 24.17) 271.

Cramér, H., MVB, 9; efficiency and asymptotic normality of ML estimators, 46; distribution of X^2 test of fit, 442; test of fit, 467; coefficient of association, 577, 595.

Cramér–Rao inequality, see Minimum Variance Bound.

Creasy, M. A., confidence intervals for linear functional relation, 404.

Crime in cities, Ogburn's data, (Example 27.2) 344–5.

Critical region, see Tests of hypotheses.

Cross-ratio, in 2×2 table, 567, 575, 581, 603.

Crouse, C. F., Mood's test, 521.

Crow, E. L., tables of confidence limits for proportions and Poisson parameters, 123; robustness, 487.

Dahiya, R. C., X^2 test, 448.

Daly, J. F., smallest confidence regions, 133.

Daniels, H. E., asymptotic normality and efficiency of ML estimators, 45; non-unique ML estimators, (Exercise 18.33) 76–7; minimization of generalized variance by LS estimators, 85; estimation of a function of the normal correlation parameter, 306; coefficients of correlation and disarray, 495, 583; joint distribution of rank correlations, 499, (Exercise 31.6) 528; theorem on correlations, (Exercise 31.3) 528.

Dantzig, G. B., no test of normal mean with power independent of unknown variance, 206.

Dar, S. N., ratio of non-central F variables, 264.

Darling, D. A., tests of fit, 468, 476; distribution of ratio of sum to extreme value, 549.

Darmois, G., sufficiency, 26, 28.

Darroch, J. N., interactions in multi-way tables, 603.

Dasgupta, P., tables of power of LR test, 263.

David, F. N., modified MCS estimators, (Exercise 19.14) 101; charts for the bivariate normal correlation parameter, 123, 307; bias in testing normal correlation parameter, 307; runs test to supplement X^2 test of fit, 458, (Exercises 30.7, 30.9) 479–80; probability integral transformation with parameters estimated, 460, (Exercise 30.10) 480; " smooth " tests of fit, 462; tables of normal scores, 520; bibliography of order-statistics, 532; tests on censored samples, 546.

David, H. A., order-statistics, 532; tables of studentized deviates and of studentized range, 548.

David, S. T., joint distribution of rank correlations, 481.

Davis, A. W., Behrens–Fisher distribution, 157.

Davis, R. C., sufficiency with terminals dependent on the parameter, 30.

Decision functions, 640.

Deemer, W. L., Jr., truncation and censoring for the exponential distribution, 545, (Exercise 32.16) 553.

Deficiency, 20.

Degrees of freedom of a hypothesis, 171, 195.

De Groot, M. H., sequential sampling for attributes, 634 footnote.

Deming, W. E., iterative estimation in $r \times c$ table, (Exercise 33.32) 611.

Dempster, A. P., fiducial inference, 161, 162; comparison of tests, 287.

Den Broeder, G. G., Jr., truncated and censored Gamma variates, 546.

Denny, J. L., sufficiency and exponential family, 27.

Dependence and interdependence, 290–1.

Disarray, 495.

Discontinuities, and confidence intervals, 108–110, 123–5; and tests, 174–5; effect on distribution-free tests, 526–7; correction in 2×2 tables, 575–6.

Dispersion tests, binomial and Poisson, 598–600.

Distribution-free procedures, tests of fit, 459–460, 467–8; confidence limits for a continuous d.f., 473–4; in general, 487–8; and non-parametric problems, 488; classification of, 488–9; construction of tests, 489–90; efficiency of tests, 490–1; tests of independence, 491–501, 504–5; tests of randomness, 501–5; two-sample tests, 505–21; confidence intervals for a location shift, 509, (Exercise 31.17) 530; k-sample tests, 521–4; tests of symmetry, 524–6; tests for quantiles, 532–6; confidence intervals for quantiles, 536–7; tolerance intervals, 537–40; tests for outliers, 549–50; categorized data, 556–603; sequential tests, 639–40.

Distribution function, sample, 466–7; confidence limits for, 473–4.

Dixon, W. J., efficiency of tests, 276; power of Wilcoxon test, 516; Sign test, 533, 534, 535; censored normal samples, 544; outlying observations, 547, 548, 549.

Dodge, H. F., double sampling, 627.

Doerfler, T. E., comparison of tests, 526.

Doksum, K. A., tests using random normal deviates, 505; symmetry tests, 526.

Dolby, G. R., non-linear functional relationship, 429.

Dorff, M., estimation of functional and structural relations, 399, 424.

Double exponential (Laplace) distribution, ML estimation of mean, (Example 18.7) 47–8, (Exercise 18.1) 70; testing against normal form, (Example 22.5) 177; censored samples, 546; Sign test asymptotically most powerful for location, (Exercise 32.5) 551.

Double sampling, 623, 638–9, (Exercises 34.15–17) 643–4.

Doubly non-central distributions, F'', 262; t'', 265.

Downton, F., ordered LS estimation, 91; extreme-value distribution, 546.

Draper, N. R., regression with selected variables, 379.

Dudman, J., logistic order statistics, (Exercise 32.12) 552.

Duncan, A. J., charts of power function for linear hypothesis, 263.

Dunn, O. J., dependent correlation coefficients, 308, (Exercise 28.22) 390; confidence bounds in regression, 385.

Durbin, J., generalization of MVB, (Exercise 18.32) 76; geometry of LS theory, 84; Conditionality and Likelihood Principles, 227; supplementary information in regression, (Exercises 28.17–18) 388–9; instrumental variables, 415, (Exercise 29.17) 435; tests of fit and distribution of rectangular order-statistics, (Exercise 30.17) 481; removing nuisance parameter by discarding sufficient statistic, (Exercise 30.18) 481; test of diagonal proportionality in $r \times r$ table, (Exercise 33.25) 609.

Dynkin, E. B., minimal sufficiency, 203.

Eames, A. R., Poisson confidence limits, 123.

East, D. A., tables of confidence intervals for a normal variance, 123.

Edgeworth, F. Y., correlation coefficient, 300.

Edwards, A. W. F., association and cross-ratio, 567.

Efficiency, in estimation, and correlation between estimators, 17–18; definition, 19; measurement, 19–21; partition of error in inefficient estimator, (Exercise 17.11) 34; of ML estimators, 44–9, 57–8; of method of moments, 67–9, (Exercises 18.9, 18.16, 18.20, 18.27–8) 71–5; and power of tests, 179–80, (Exercise 22.7) 193; and ARE of tests, 284–5.

Efficiency of tests, 273–6; see Asymptotic Relative Efficiency.

Efron, B., power of tests, (Exercise 22.15) 194; robustness of Student's t-test, 484.

Eisenhart, C., limiting power function of X^2 test of fit, 453; tables of runs test, (Exercise 30.8) 490.

El-Badry, M. A., ML estimation of cell probabilities in $r \times c$ table, (Exercise 33.29) 610.

Ellison, B. E., tolerance intervals, 135.

Epstein, B., an independence property of the exponential distribution, (Exercise 23.10) 230; censoring in exponential samples, 545–6, (Exercise 32.17) 553.

Errors, in LS model, 79; from regression, 335 footnote; rule for cancelling subscripts, (Exercise 27.5) 357; identical, 366; of

INDEX

observation and functional relations, 391–393; of observation in regression, 430–432.

Estimates and estimators, 2.

Estimation, point, 1–102; interval, 102–168; and completeness, 199; *see* Efficiency in estimation.

Estimators, and estimates, 2.

Exhaustive, 202 footnote.

Exponential distribution, m.s.e. of $\bar{x}$, 22, 26; sufficiency of smallest observation for lower terminal, (Example 17.19) 30; sufficiency when upper terminal a function of scale parameter, (Example 17.20) 30; MV unbiased estimation, (Exercise 17.24) 36; ML estimation of scale parameter by sample mean, (Exercise 18.2) 70; ML estimation and sufficiency when both terminals depend on parameters, (Exercise 18.5) 70; grouping correction to ML estimator of scale parameter, (Exercise 18.25) 75; ordered LS estimation of location and scale parameters, (Exercises 19.11–12) 101; confidence limits for location parameter, (Exercise 20.16) 139; UMP tests for location parameter, (Example 22.6) 178; satisfies condition for two-sided BCR, 182; UMP test without single sufficient statistic, (Examples 22.9–10) 184–6, (Exercise 22.11) 194; UMP similar one-sided tests of composite H_0 for scale parameter, (Example 23.9) 207–8; independence of two statistics, (Exercise 23.10) 230; non-completeness and similar regions, (Exercise 23.15) 231; UMPU test of composite H_0 for scale parameter, (Exercise 23.24) 232; UMP similar test of composite H_0 for location parameter, (Exercise 23.25) 232; LR tests, 255, (Exercises 24.11–13, 24.16, 24.18) 270–1; test of fit on random deviates, (Examples 30.2–4) 449–57, (Exercise 30.6) 479; power of Wilcoxon test, 516; use of scores, 520; truncation and censoring, 545–6, (Exercises 32.16–18) 533–4; inverse sampling, 615; sequential test for scale parameter, (Exercise 34.14) 643.

Exponential family of distributions, 12; attainability of MVB, 15; as characteristic form of distribution admitting sufficient statistics, 26, 28; sufficient statistics distributed in same form, (Exercise 17.14) 34; and completeness, 199–200; UMPU tests for, 216–26; with range dependent on parameters, (Exercise 24.17) 271.

Extreme-value distribution, ML estimation in, (Exercise 18.6) 70; censored samples, 546.

Ezekiel, M. J. B., confidence intervals for multinormal multiple correlation, 356.

F distribution, non-central, 262–4.

Fechner, G. T., work on rank correlation, 496.

Feder, P. I., LR test, 241, 257.

Fein, H., homogeneity tests, 599.

Feller, W., similar regions, 197, (Exercises 23.1–2) 229; distribution of Kolmogorov test statistic, 469.

Fend, A. V., variance bounds, 14; ML estimator and MV, (Exercise 18.37) 77.

Fereday, F., linear functional relations in several variables, 410, (Exercises 29.6–8) 433–4.

Ferguson, T. S., rejection of outliers, 547–8.

Féron, R., characterization of bivariate normality, 368 footnote.

Fiducial inference, generally, 141–5; in "Student's" distribution, 145–6, (Exercise 21.7) 167; in problem of two means, 155–157; critical discussion, 159–65; paradoxes, 161–2; concordance with Bayesian inference, 162–5.

Fiducial intervals, probability, *see* Fiducial inference.

Fieller, E. C., difficult confidence intervals, 132.

Fienberg, S. E., contingency tables, 569, 579, (Exercise 33.32) 611.

Finkelstein, J. M., test of fit for exponential, 478.

Finney, D. J., ML estimation in lognormal distribution, (Exercises 18.7–9, 18.19–20) 71, 73; truncated binomial distribution, 546; tables of exact test of independence in 2×2 table, 573, 575; inverse sampling, 615.

Fisher, F. M., Cauchy location estimation, 52.

Fisher, R. A. (Sir Ronald), definition of LF; 8 footnote; sufficiency, 22, 202 footnote, partition of error in inefficient estimator, (Exercise 17.11) 34; ML principle, 37; successive approximation to ML estimator, 51, (Example 18.10) 52; use of LF, 64; ML estimation of location and scale parameters, 64; efficiency of method of moments, 68, (Exercise 18.16) 72–3,

Fisher, R. A. (Sir Ronald)—*contd.*
(Exercise 18.27) 75; consistency, 96 footnote, 273; fiducial inference, 141; fiducial solution to problem of two means, 156; fiducial paradoxes, 162; fiducial intervals for future observations, (Exercise 21.7) 167; exhaustiveness and sufficiency, 202 footnote; ancillary statistics, 226; non-central χ^2 and F distributions, 239, 263; transformation of correlation coefficient, 307; tests of correlation and regression, 311; distribution of intraclass correlation, 316, (Exercise 26.14) 327; distribution of partial correlations, 346; distributions of multiple correlations, 351–4, (Exercises 27.13–16) 358; orthogonal polynomials, 374–5; test of difference between regressions, (Exercise 28.15) 387; distribution of X^2 test of fit, 441, 445, (Exercise 30.1) 479; tests using normal scores, 505; test of symmetry, 526; exact treatment of 2×2 table, 570, 572; dispersion tests, 599.

Fit, *see* Tests of fit.

Fix, E., power of LR tests, 241; linearity of regression, 432; tables of Wilcoxon test, 512.

Folks, J. L., confidence regions for linear models, 385.

Fourfold table, *see* 2×2 table.

Fourgeaud, C., characterization of bivariate normality, 368 footnote.

Fox, K. A., confidence intervals for multinormal multiple correlation, 356.

Fox, L., matrix inversion, 82.

Fox, M., charts of power function for linear hypothesis, 263.

Francis, G. K., confidence bounds in linear regression, 385.

Fraser, D. A. S., tolerance intervals for a normal distribution, 136; fiducial inference, 162, 164, 165; minimal sufficiency, 203 footnote; runs test to supplement X^2 test of fit, 458, (Exercise 30.7) 479–80; rank tests, 505; tolerance regions, 540.

Fréchet, M., linear regressions and correlation ratios, (Exercise 26.24) 329.

Freeman, D., sequential tests, 631.

Freeman, G. H., exact test of independence in $r \times c$ tables, 579.

Freund, R. J., two-stage LS, (Exercise 28.24) 390.

Fryer, J. G., bivariate location tests, 521; homogeneity in contingency tables, (Exercise 33.24) 609.

Functional and structural relations, 290; generally, 391–435; notation, 391–2; linear, 392–3; and regression, 392–3, 396, 402; ML estimation, 395–404, 426–427; geometrical interpretation, 400–1, 426–7; confidence intervals and tests, 404–8; several variables, 408–10; use of product-cumulants, 411–14, 427–8; instrumental variables, 414–24, 428–9; use of ranks, 422–4; controlled variables, 424–5, 429; non-linear relations, 426–30.

Gabriel, K. R., multiple comparisons in $r \times c$ tables, 582.

Gafarian, A. V., confidence regions for regression lines, 385.

Galton, F., regression and correlation, 300; rank correlation, 495; data on eye-colour, (Example 33.2) 561.

Gambler's ruin, (Example 34.2) 561.

Gamma distribution, MVB estimation of scale parameter, (Exercise 17.1) 32; sufficiency properties, (Exercise 17.9) 33; estimation of lower terminal, (Exercise 17.22) 35; ML estimation of location and scale parameters, (Example 18.17) 66–7; efficiency of method of moments, (Example 18.18) 68–9; confidence intervals for scale parameter, (Exercise 20.1) 137; fiducial intervals for scale parameter, (Example 21.2) 144–5; distribution of linear function of Gamma variates, (Exercise 21.8) 167; non-existence of similar regions, (Exercise 23.2) 229; completeness and a characterization, (Exercise 23.27) 233; connexions with rectangular distribution, 246–7; ARE of Wilcoxon test, (Exercise 31.18) 530; truncation and censoring, 546.

Ganeshalingam, S., tables of normal scores, 520.

Gardner, R. S., tables of limits for a Poisson parameter, 123.

Gart, J. J., conditional moments of Poisson k-statistics, (Exercise 17.19) 35; comparison of proportions, 598.

Garwood, F., tables of confidence intervals for the Poisson parameter, 123.

Gastwirth, J. L., robustness, 487; ARE of rank tests, (Exercise 31.23) 531; modified Sign test, 536; efficient estimation from order-statistics, 543; MV estimators and

INDEX

rank tests, 543 ; rank tests for censored samples, 546.

Gauss, C., sample mean as ML estimator, (Exercise 18.2) 70 ; originator of LS theory, 83, 85.

Gayen, A. K., studies of robustness, 483, 485, 486, 522.

Geary, R. C., " close " estimators, 8 ; asymptotic minimization of generalized variance by ML estimators, 58 ; average critical value, (Exercise 25.10) 289 ; functional and structural relations, 411, 426, 429, (Exercises 29.2, 29.15) 433, 435 ; robustness, 483, 485.

Gehan, E. A., Wilcoxon test for censored samples, 546.

Geisser, S., fiducial inference, 164.

General linear hypothesis, see Linear model.

Generalized variance, 58 ; minimized asymptotically by ML estimators, 58 ; minimized in linear model by LS estimators, 85.

Ghosh, B. K., moments of n in SPRT, 635.

Ghosh, J. K., completeness, 199 ; invariance and sufficiency, 267, 634.

Gibbons, J. D., non-normality and Sign test, 515.

Gibson, W. M., estimation of functional relation, 421.

Gilbert, J. P., 2×2 table, 569.

Gilby, W. H., data on clothing and intelligence, (Example 33.7) 578.

Girshick, M. A., decision functions, 640 ; sequential analysis, (Exercises 34.8, 34.9, 34.20) 642, 644.

Gjeddebaek, N. F., grouping and ML, (Exercise 18.25) 74–5.

Glass, D. V., Durbin's test of diagonal proportionality in $r \times r$ table, (Exercise 33.25) 609.

Gnanadesikan, M., censored samples, 546 ; logistic estimators, 546.

Godambe, V. P., generalization of MVB, (Exercise 18.32) 76.

Goen, R. L., truncated Poisson distribution, 545, (Exercise 32.19) 554.

Goheen, H. W., tetrachoric correlation, 319.

Goldberger, A. S., two-stage LS (Exercise 28.24) 390.

Goldman, A. S., double sampling, 639.

Goldsmith, D., homogeneity tests, 599.

Good, I. J., distribution of X^2, 438.

Goodman, L. A., tolerance intervals, 540 ; association in categorized data, 565, 579, 581, 585–6, (Exercise 33.11) 606 ; 2×2 tables, 575 ; interactions in multi-way tables, 603.

Goodness-of-fit, 436 ; see Tests of fit.

Gover, T. N., distribution of X^2, 438.

Govindarajulu, Z., power of Wilcoxon test, 516 ; symmetry tests, 526 ; bibliography of life-testing, 545 ; censored double exponential, 546.

Gram–Charlier Series, Type A, efficiency of method of moments, (Exercise 18.28) 75.

Gray, H. L., jackknifing, 6.

Graybill, F. A., quadratic forms in non-central normal variates, 240 ; confidence bands in regression, 385 ; double sampling for normal variance, 639.

Greenberg, B. G., order-statistics, 532 ; censored samples, 544, 545, 546, 549, (Exercise 32.25) 555.

Greenhouse, S. W., tables of studentized maximum absolute deviate, 548.

Greenwood, M., data on inoculation, (Example 33.1) 558.

Grouping, corrections to ML estimators, (Exercises 18.24–5) 74–5 ; for instrumental variables, 415–23, 428–9 ; truncated and censored data, 543, 544.

Grubbs, F. E., tests for outliers, 549.

Grundy, P. M., grouped truncated and censored normal data, 544.

Guenther, W. C., unbiassed tests for a normal variance, 222 ; non-central χ^2, 238.

Guest, P. G., orthogonal polynomials, 375.

Guilford, J. P., tetrachoric correlation, 319.

Gumbel, E. J., choice of classes for X^2 test of fit, 449.

Gupta, S. S., logistic estimators, 546 ; logistic tables, (Exercise 33.12) 552.

Gurian, J., confidence bands in linear regression, 385.

Gurland, J., estimation in negative binomial and Neyman Type A distributions, (Exercise 18.28) 75 ; multiple R^2, (Exercises 27.23–5) 359–60 ; estimation of functional and structural relations, 399, 424 ; X^2 test, 448.

Guttman, I., tolerance intervals, 136, 540 ; regression with selected variables, 379.

H_0, hypothesis tested, null hypothesis, 171.
H_1, alternative hypothesis, 171.
Haas, G., ML estimation in Cauchy, 52.

INDEX

Hader, R. J., censored exponential samples, 545.
Hájek, J., problem of two means, 154; rank tests, 505.
Hajnal, J., sequential two-sample t-test, 634.
Hald, A., truncated normal distribution, 544.
Haldane, J. B. S., cumulants of a ML estimator, 49; mean of modified MCS estimator, (Exercise 19.14) 101; standard error of X^2 in $r \times c$ tables, 580, (Exercise 33.9) 605; approximation to X^2 for $2 \times c$ table, (Exercises 33.21–2) 608–9; inverse sampling, 615.
Hall, W. J., invariance and sufficiency, 267, 634.
Halmos, P. R., sufficiency, 23.
Halperin, M., confidence intervals for non-linear functions, 133, (Exercise 28.3) 386; confidence bands in regression, 385; linear functional relation, 421; ML estimation in censored samples, 543, (Exercise 32.15) 553; truncated normal distribution, 544; confidence intervals for censored samples, 546; Wilcoxon test for censored samples, 546; tables of studentized maximum absolute deviate, 548.
Halton, J. H., exact test of independence in $r \times c$ tables, 579.
Hamdan, M. A., tetrachoric correlation, 319; choice of classes for X^2 test, 456; polychoric estimation, 581.
Hamilton, M., nomogram for tetrachoric correlation, 319.
Hamilton, P. A., tables of confidence intervals for a normal variance, 123.
Hammersley, J. M., estimation of restricted parameters, (Exercises 18.21–2) 73–4.
Hannan, E. J., ARE for non-central χ^2 variates, 286; regressors, 361.
Hannan, J. F., multinormal biserial methods, (Exercise 26.12) 327; power of test in 2×2 tables, 575.
Harkness, W. L., ML estimation of logistic parameters, (Exercise 18.29) 76; power of test in 2×2 tables, 575.
Harley, B. I., estimation of a function of normal correlation parameter, 306.
Harter, H. L., shortest confidence intervals, 122; ratio of normal ranges, (Exercise 23.14) 231; tables of normal scores, 520; order-statistics, 532; censored samples, 544, 545, 546; tables of studentized range, 548.
Hartley, H. O., non-linear LS, 79; charts of power function for linear hypothesis, 263; confidence regions for non-linear regression, (Exercise 28.3) 386; iterative solution of ML equations for incomplete data, 543; tables of studentized range, 548.
Hayes, J. G., matrix inversion, 82.
Hayes, S. P., Jr., nomogram and tables for tetrachoric correlation, 319.
Hayman, G. E., power of Wilcoxon test, 516.
Healy, M. J. R., comparison of predictors, (Exercise 28.22) 389–90.
Healy, W. C., Jr., double sampling, 639.
Hedayat, A., independent LS residuals, (Exercise 19.18) 102.
Hemelrijk, J., power of Sign test, 533 footnote.
Hettmansperger, T. P., trimmed Wilcoxon statistic, 546.
Hill, B. M., estimating proportions in mixtures of distributions, (Exercise 17.2) 32; ML estimation in lognormal distribution, (Exercise 18.23) 74.
Hinkley, D. V., efficiency of LS estimators, 84; LS fitting, 371.
Ho, I., confidence bands in linear regression, 385.
Hodges, J. L., Jr., variance of median, 7; deficiency, 20, (Exercise 17.8) 33; power of t-test, 265; efficiency of tests, 276; tables of Wilcoxon test, 512; ARE of Wilcoxon test, 515; power of Wilcoxon test, 516; robust estimation, 520; Wilcoxon and normal scores tests ARE, (Exercise 31.23) 531; minimum ARE of Sign test, (Exercise 32.1) 551.
Hoeffding, W., efficiency of LR and X^2 tests in multinomial samples, 287, 457; permutation and normal-theory distributions, 493; joint distribution of rank correlations, 499; test of independence, 501; optimum properties of tests using normal scores, 505, 516; asymptotic distribution of expected values of order-statistics, 505; distribution of Wilcoxon test, 513.
Hoel, D. G., tables of power of LR tests, 263; sequential test, 631.
Hoel, P. G., confidence regions for regression lines, 380, 383, 385, (Exercise 28.13) 387.
Hogben, D., moments of non-central t, 265.
Hogg, R. V., completeness of sufficient statistics, 200; completeness, sufficiency and independence, (Exercises 23.8–9) 230; tests in k samples, (Exercise 24.6) 269, (Exercise 24.13) 270, 524; LR tests when range depends on parameter, 246, (Exer-

cises 24.8–9) 269 ; sufficiency in exponential family when range a function of parameters, (Exercise 24.17) 271.

Hollander, M., location- and scale-shift tests, 521.

Homogeneity tests, binomial and Poisson, 598–600.

Hooker, R. H., data on weather and crops, (Example 27.1) 342.

Hora, R. B., fiducial inference, 162.

Horst, C., tables of exact test of independence in 2×2 tables, 573, 575.

Hotelling, H., estimation of functions of normal correlation parameter, 306 ; variance of multiple correlation coefficient, 355 ; confidence region for regression line, 380, 383–5, (Exercise 28.13) 387 ; comparison of predictors, (Exercise 28.22) 389–90 ; robustness, 487 ; efficiency of rank correlation test, 500.

Houseman, E. E., tables of orthogonal polynomials, 375.

Howe, W. G., normal tolerance limits, 136.

Hoyle, M. H., estimating polynomials in normal parameters, (Exercise 17.10) 33–4.

Høyland, A., robust estimation, 520.

Hsu, P., test of independence in 2×2 tables, 573, 575.

Hsu, P. L., optimum property of LR test for linear hypothesis, 266.

Huber, P. J., ML estimators, 43, 46 ; robustness of location estimators, 487.

Hudson, D. J., fitting by LS, 371.

Hutchinson, D. W., tables of binomial and Poisson confidence intervals, 125.

Huyett, M. J., censored Gamma samples, 546.

Huzurbazar, V. S., uniqueness of ML estimators, 39, 55 ; consistency of ML estimators, 43 ; confidence intervals when range depends on parameter, (Exercises 20.13–16) 138–9.

Hypergeometric distribution, inverse sampling, 615.

Hypotheses, statistical 169, 195 ; parametric and non-parametric, 170 ; simple and composite, 170–1, 195 ; degrees of freedom, constraints, 171 ; critical regions and alternative hypotheses, 171–2 ; null hypothesis, 171 footnote ; see Tests of hypotheses.

Hyrenius, H., Student's t in mixture samples, 485.

Identical categorizations, 556.

Identical errors, 366.

Identifiability, (Example 18.4), 44 ; in structural relations, 396, 398–9, 408, 412.

Incidental parameters, 399.

Independence, proofs using sufficiency and completeness, (Exercises 23.6–10) 229–30 ; and correlation, 295–6 ; tests of, 491–501, 504–5 ; and association, 557 footnote ; frequencies, 558–9.

Information, amount of, 10 ; matrix, 28.

Instrumental variables, 414–24, 428–9.

Interactions, in multi-way tables, 603.

Interdependence, 290–1.

Intersection, of polynomial regressions, 380.

Interval estimation, 103–68.

Intra-class correlation, 314–16, (Exercise 26.14) 327 ; in multinormal distribution, (Exercise 27.17) 359.

Invariant estimators, 22 ; tests, 252–5, 266–7 ; estimators in tests of fit, 460, 468, 477, (Exercise 30.10) 480 ; and rank tests, 494, 501, 510.

Inverse sampling, 615.

Inversions of order, 496.

Ireland, C. T., contingency tables, (Exercises 35.23–4) 609.

Irwin, J. O., truncated Poisson distribution, (Exercises 32.23–4) 555 ; outlying observations, 547 ; components of X^2, 597.

Iwaskiewicz, K., tables of power function of Student's t test, 265.

Jackson, J. E., bibliography of sequential analysis, 640.

Jacobson, J. E., tables of Wilcoxon test, 512.

Jaeckel, L. A., robust estimation, 487.

James, G. S., tables for problem of two means, 155.

Jeffreys, Sir Harold, Bayesian intervals, 157–61.

Jenkins, W. L., nomogram for tetrachoric correlation, 319.

Jochems, D. B., two stage LS, (Exercise 28.24) 390.

Jogdeo, K., tests using random normal deviates, 505.

Johns, M. V., Jr., efficient estimation from order statistics, 543.

Johnson, N. L., comparison of estimators, 8, 22 ; non-central χ^2, 239 ; non-central t-distribution, 265 ; non-central χ^2 and Poisson distributions, (Exercise 24.19) 271–2 ; probability integral transforma-

Johnson, N. L.—*contd.*
    tion with parameters estimated, 460, (Exercise 30.10) 480; bibliography of order-statistics, 532; tests on censored samples, 546; SPR tests, 623, (Exercise 34.5) 641; review of sequential analysis, 640.

Joiner, B. L., test efficiency, 287.
Jonckheere, A. R., *k*-sample test, 523.
Jowett, G. H., estimation of functional relation, 421.

k-sample tests, (Examples 24.4, 24.6) 244, 254; 247–9; (Exercises 24.4–7) 268–9; (Exercises 24.11–13) 270; 521–5.
Kabe, D. G., adjustment of regression analysis, 371.
Kac, M., comparison of X^2 and Kolmogorov tests of fit, 476; tests of normality, 477.
Kale, B. K., iterative ML estimation, 51, 62.
Kanemasu, H., regression with selected variables, 379.
Kasten, E. L., charts for test of independence in 2×2 tables, 573.
Kastenbaum, M. A., tables of power of LR test, 263; hypotheses in multi-way tables, 603.
Katti, S. K., estimation in negative binomial and Neyman Type A distributions, (Exercise 18.28) 75.
Katz, L., power of test in 2×2 tables, 575.
Katz, M., Jr., minimal sufficiency, 203 footnote.
Kavruck, S., tetrachoric correlation, 319.
Kemperman, J. H. B., tolerance regions, 540.
Kempthorne, O., Tang's tables, 263; X^2 test, 457; comparison of tests, 526.
Kendall, M. G., geometrical interpretation of LS theory, 84; estimation of a function of normal correlation parameter, 306; discarding variables, 350; linear regression with identical observation errors, 431, (Exercise 29.13) 434; distributions of rank correlations, 495–9; ties in rank correlation, 527; rank measure of association, 584–5.
Kennedy, W. J., bias in linear models under selection, 379.
Keynes, J. M., characterizations of distributions by forms of ML estimators, (Exercises 18.2–3) 70.
Kiefer, J., bounds for estimation variance, 16; non-existence of a ML estimator, (Exercise 18.34) 77; consistency of ML estimators, 402; comparison of X^2 and Kolmogorov tests of fit, 476; tests of normality, 477; distribution of Hoeffding's test, 501; sequential tests, 631.
Kim, P. J., tables of normal scores, 520.
Kimball, A. W., computation of X^2 partitions, 597.
Kimball, B. F., ML estimation in extreme-value distribution, (Exercise 18.6) 70.
Klett, G. W., tables of confidence intervals for a normal variance, 123; shortness of these intervals, (Exercise 20.10) 138.
Klimko, L., ML estimation of logistic parameters, (Exercise 18.29) 76.
Klotz, J., normal scores test, 516, 520; test for scale-shift, 521; tests of symmetry, 526.
Knight, W., inverse sampling, 615.
Knott, M., power of tests of fit, 476.
Kolmogorov, A., test of fit, 468–77.
Kolodzieczyk, S., general linear hypothesis, 259; tables of power function of Student's *t*-test, 265.
Kondo, T., standard error of X^2 in $r \times c$ table, 580.
Konijn, H. S., ARE of tests, 278; linear transformations of independent variates, 500.
Koopman, B. O., distributions with sufficient statistics, 26, 28.
Kraft, C., rectangular sum approximation to Wilcoxon test, 512.
Kramer, K. H., confidence intervals for multinormal multiple correlation, 356.
Krishnan, M., Type M unbiassed tests, 211; doubly non-central *t*, 265.
Kruskal, W. H., geometry of LS, 85; linear regressions, correlation coefficients and ratios, (Exercise 26.24) 329; history of rank correlation, 496; Wilcoxon test, 511, 512, (Exercise 31.15) 530; *k*-sample test, 522; ties, 527; association in categorized data, 565, 581, 585–6, (Exercise 33.11) 606.
Ku, H. H., contingency tables, (Exercises 33.23–4) 609.
Kudo, A., rejection of outliers, 548.
Kullback, S., contingency tables, (Exercises 33.23–4) 609.
Kulldorff, G., grouping and ML, (Exercise 18.25) 75; censored exponential samples, 545.
Kurtic curve, 362.

Laha, R. G., linearity of regression, 432.
Lambda criterion, 234 footnote.

INDEX

Lancaster, H. O., continuity corrections when pooling X^2 values, 576; polychoric estimation, 581; canonical correlations and transformation to bivariate normality, 588–9; components of X^2, 595, (Example 33.13) 596, 597, (Exercises 33.15–18) 607–8; multi-way tables, (Example 33.14) 601, 603, (Exercise 33.8) 605, (Exercise 33.27–8) 610.

Laplace distribution, *see* Double exponential.

Laplace transform, 199 footnote.

Larson, H. J., bias in linear models under selection, 379.

Laska, E., robustness, 487, 520.

Latscha, R., tables of exact test of independence in 2×2 tables, 573, 575.

Laurent, A. G., censored exponential samples, 545.

Lawley, D. N., approximations to distribution of LR statistic, 243.

Lawton, W. H., problem of two means, 154.

Least Squares (LS) estimators, 78–95, (Exercises 19.1–12) 100–1; and ML equivalent in normal case, 78–9, 84; LS principle, 79; in linear model, 79–91, (Exercises 19.1–9) 100; unbiased, 81; dispersion matrix, 82; MV property, 83; m.s.e., 84; geometrical interpretation, 84–5; minimization of generalized variance, 85; estimation of error variance, 85–6; irrelevance of normality assumption to estimation properties in linear model, 86–87; singular case, 88–91; with linear constraints, 91, (Exercise 19.17) 102; ordered estimation of location and scale parameters, 91–5, (Exercises 19.10–12) 100–1; inefficient in Poisson case, (Exercise 19.16) 101; uncorrelated residuals, (Exercise 19.18) 102; general linear hypothesis, 257–61; approximate linear regression, 298–9, 338–9; in linear regression model, 369–85; use of supplementary information, (Exercises 28.17–18) 388–9; graphical, (Exercise 28.19) 389; adjustment for an extra observation, (Exercise 28.20) 389; comparison of predictors, (Exercise 28.22) 389–90; two-stage, (Exercise 28.24) 390; in $r \times c$ tables, 588; *see* Linear model, Regression.

Lecam, L., ML estimators, 46; super-efficiency, 46.

Lee, Y. S., multiple R^2, 354, 355, (Exercise 27.23) 359; tables of R, 356.

Legendre polynomials, 460 footnote.

Lehmann, E. L., example of absurd unbiased estimator, 5, (Exercise 17.26) 36; deficiency, 20, (Exercise 17.8) 33; testing hypotheses, 169; optimum test property of sufficient statistics, 196; completeness, 199; completeness of linear exponential family, 200; minimal sufficiency, 203; completeness and similar regions, 205; problem of two means, (Example 23.10) 208; non-similar tests of composite H_0, 209; unbiased tests, 215; UMPU tests for the exponential family, 216; non-completeness of Cauchy family, (Exercise 23.5) 229; minimal sufficiency for binomial and rectangular distributions, (Exercises 23.11–12) 230; a UMPU test for normal distribution, (Exercise 23.16) 231; UMPU tests for Poisson and binomial distributions, (Exercises 23.21–2) 232; UMP tests for exponential and rectangular distributions, (Exercises 23.24–6) 232–3; a useless LR test, (Example 24.7) 256; power of t-test, 265; optimum properties of LR tests, 265; asymptotically UMP tests for a normal mean, (Example 25.1) 274; efficiency of tests, 276; UMP invariant tests for correlation and multiple correlation coefficients, 307, 356; confidence intervals in regression, (Exercise 28.21) 389; distribution of X^2 test of fit, 445, 447; completeness of order-statistics, 490 footnote; unbiased rank tests of independence, 501; rank tests, 505; distribution of Wilcoxon test, 513; confidence intervals based on Wilcoxon tests, 513, (Exercises 31.24–5) 531; consistency of Wilcoxon test, 513; ARE of Wilcoxon test, 515; power of Wilcoxon test, 516; robust estimation, 520; unbiased two-sample test, 521; tests of symmetry, 525; Wilcoxon and normal scores tests ARE, (Exercise 31.23) 531; optimum properties of Sign test, 533; minimum ARE of Sign test, (Exercise 32.1) 551; Sign test asymptotically most powerful location test for double exponential, (Exercise 32.5) 551; sequential completeness, 634 footnote.

Lehmer, E., tables of power function for linear hypothesis, 263.

Lewis, B. N., multi-way tables, 600.

Lewis, P. A. W., tables of test of fit, 468.
Lewis, T. O., singular LS estimation, 89.
Lewontin, R. C., ML estimation of number of classes in a multinomial, (Exercise 18.10) 71–2.
LF, *see* Likelihood function.
Lieberman, G. J., tables of non-central t-distribution, 265.
Life testing, 545.
Likelihood equation, 38 ; *see* Maximum Likelihood.
Likelihood function (LF), 8 ; use of, 64, *see* Maximum Likelihood.
Likelihood Principle, 226.
Likelihood Ratio (LR) tests, normal distribution, (Example 22.8) 184 ; generally, 234–257 ; and ML, 234–5, 250–1 ; not necessarily similar, 235–6 ; approximations to distributions, 237, 240–4 ; asymptotic distribution, 240–1 ; asymptotic power and tables, 241–2, 263–4 ; when range depends on parameter, 246–50, (Exercises 24.8–9) 269 ; properties, 250–7 ; consistency, 241–2, 250–1 ; unbiassedness, 251–2, 255–6, (Exercises 24.14–18) 271 ; other properties, 255–6 ; a useless test, (Example 24.7) 256 ; for linear hypotheses, 259–61 ; power function, 263–4 ; optimum properties, 265–7 ; efficient in multinomial samples, 287 ; in normal correlation and regression, 311–313, 351 ; of fit, 437–40, (Exercise 30.11) 480 ; of independence in 2×2 tables, 567–9 ; in $r\times c$ tables, 582.
Lilliefors, H. W., tests of fit, 477, 478.
Lindeberg, J. W., early work on rank correlation, 496.
Lindley, D. V., grouping corrections to ML estimators, (Exercises 18.24–5) 74–5 ; tables of confidence intervals for a normal variance, 123 ; concordance of fiducial and Bayesian inference, 162–4, (Exercise 21.11) 167–8 ; choice of test size, 191 ; conditional tests, 227 ; inconsistency of ML estimators in functional relation problem, 403 ; controlled variables, 424 ; observational errors and linearity of regressions, 430, 432, (Exercises 29.12, 29.14) 434–5 ; Bayesian contingency analysis, 603.
Linear model, 79–91 ; and normality assumption, 86–7 ; lognormal, 84 ; more general, 87, (Exercises 19.2–3, 19.5) 100 ; general linear hypothesis, 257 ; canonical form, 258 ; LR statistic in, 260 ; and LS theory, 260 ; power function of LR test, 263–4 ; optimum properties of LR test, 265–7 ; and regression, 369–85 ; meaning of " linear," 370–1 ; confidence intervals and tests for parameters, 377–80, (Exercises 28.15–22) 388–90 ; confidence regions for a regression line, 380–5 ; supplementary information in regression, (Exercises 28.17–18) 388 ; adjustment for an extra observation, (Exercise 28.20) 389 ; linear logistic model, 598 ; *see* Least Squares, Regression.
Linear regression, *see* Regression.
Linnik, Yu V., problem of two means, 209.
Lipps, G. F., early work on rank correlation, 496.
Lipton, S., non-linear functional relationship, 429.
Lloyd, E. H., ordered LS estimation of location and scale parameters, 91, (Exercise 19.10) 100–1.
Local efficiency, asymptotic, 279.
Location, centre of, 66.
Location and scale parameters, ML estimation of, 64–7, (Exercises 18.2, 18.4) 70 ; ML estimators asymptotically independent for symmetrical parent, 65–6 ; LS estimation by order-statistics, 91–5, (Exercises 19.10–12) 100–1 ; ordered LS estimators uncorrelated for symmetrical parent, 93–4 ; completeness, 199, 200 ; no test for location with power independent of scale parameter, (Exercise 23.30) 233 ; unbiassed invariant tests for, 252–5, (Exercise 24.15) 271 ; estimation of, in testing fit, 447, 459–60, 468, 477–8, (Exercises 30.10, 30.21) 480–2.
Location-shift, 506 ; confidence intervals for, 509, (Exercises 31.24–5) 531.
Locks, M. O., tables of non-central t, 265.
Logistic distribution, MVB, (Exercise 17.5) 33 ; ML estimation in, (Exercise 18.29) 76 ; ARE of Wilcoxon test, (Exercise 31.17) 530 ; censored samples, 546 ; cf. and cumulants of order-statistics, (Exercise 32.12) 552–3 ; linear logistic models, 598.
Lognormal distribution, ML estimation in, (Exercises 18.7–9, 18.19–20, 18.23), 71–4 ; in linear models, 84 ; censored samples, 546.
LR, *see* Likelihood Ratio.

LS, *see* Least Squares.
Lyons, T. C., tetrachoric correlation, 319.

Maag, U. R., test of fit, 468.
McCornack, R. L., tables of Wilcoxon symmetry test, 526.
McElroy, F. W., LS estimators, 91.
McKay, A. T., distribution of extreme deviate from mean, (Exercise 32.21) 554.
McKendrick, A. G., estimation in censored samples, (Exercise 32.24) 555.
MacKinnon, W. J., tables of Sign test and confidence intervals for median, 533, 537.
McNeil, D. R., ties in distribution-free tests, 527.
McNolty, F., non-central χ^2, 238.
MacStewart, W., power of Sign test, 534.
Madansky, A., shortest confidence intervals, 122; functional and structural relations, 399, 412, 421, (Exercise 29.16) 435; tolerance intervals, 540; multi-way tables, (Exercises 33.24, 33.30) 609, 611.
Maitra, A. P., sufficiency, 28.
Manly, B. F. J., sequential tests, 628, 635.
Mann, D. W., discarding variables, 350.
Mann, H. B., Tang's tables, 263; choice of classes for X^2 test of fit, 449, 455–7; unbiassedness of X^2 test, 452; variance of X^2, (Exercise 30.3) 479; rank correlation test for trend, 501–2, (Exercise 31.8) 528; Wilcoxon test, 511.
Mantel, N., confidence intervals for non-linear functions, 133.
Mardia, K. V., bivariate two-sample test, 521.
Maritz, J. S., biserial correlation, 323.
Marsaglia, G., quadratic forms in non-central normal variates, 240; multinormal regression, (Exercise 27.21) 359.
Massey, F. J., Jr., tables of Kolmogorov's test of fit, 473; power of Kolmogorov test, 475–6, (Exercise 30.16) 481.
Mauldon, J. G., fiducial paradox, 162.
Maximum Likelihood (ML) estimators, in general, 37–77; ML principle, 37; and MVB estimation, 38–9, (Exercise 18.32) 76; and sufficiency, 38, 39, (Example 18.5) 44–5, 54–5; uniqueness in presence of sufficient statistics, 38–9, 55–6; not generally unbiassed, (Example 18.1) 40, 44, (Example 18.11) 56, (Exercise 18.37) 77; large-sample optimum properties, 40; consistency and inconsistency, 41–4, 57, (Example 18.16), 402–4; cases of indeterminacy, (Example 18.4) 44, (Exercises 18.17, 18.23, 18.33–4) 73–7, 400; efficiency and asymptotic normality, 44–8, 57–9; asymptotic variance equal to MVB, 45–6; simplification of asymptotic variance and dispersion matrix in presence of sufficiency, 48, 58; m.s.e., 48–9; cumulants of, 49–50; successive approximations to, 51–4, (Exercises 18.35–6) 77; of several parameters, 54–62; asymptotic minimization of generalized variance, 58; non-identical parents, 62–3; of location and scale parameters, 63–7; characterization of parents having ML estimators of given form, (Exercises 18.2–3) 70; of parameters restricted to integer values, (Exercises 18.21–2) 73–4; corrections for grouping, (Exercises 18.24–5) 74–5; and LS equivalent in normal case, 78–9, 84; in structural and functional relations, 395–404, 426–7; choice of, in testing fit, 445–7; for truncation and censoring problems, 542–3; in sequential tests, (Exercise 34.21) 644.
Mazumdar, M., sequential test, 631.
MCS, *see* Minimum Chi-Square.
Mean-square-error (m.s.e.), estimation by minimizing, 21–2, 26, (Exercise 17.16) 348–9.
Medial correlation, (Exercise 31.7) 528.
Median, Sign test for, 532–6, (Exercises 32.1, 32.5) 551; confidence intervals for, 537, (Exercise 32.4) 551.
Mehta, J. S., problem of two means, 155.
Meisner, M., robustness, 487.
Mendelian pea data, (Example 30.1) 439.
Mendenhall, W., censored exponential samples, 545; bibliography of life testing, 545.
Merrington, M., tables of normal scores, 520; power functions of tests of independence in 2×2 tables, 575.
Method of moments, *see* Moments.
Mickey, M. R., problem of two means, 154.
Miké, V., robustness, 487.
Mikulski, P. W., efficiency of rank-order tests, 520.
Miller, L. H., tables of Kolmogorov test of fit, 473.
Miller, R. G., Jr., jackknifing, 6.
Milton, R., multiple R^2, (Exercises 27.24–5) 360; tables of Wilcoxon test, 512.
Minimal sufficiency, *see* Sufficiency.

INDEX

Minimum Chi-Square estimators, 96–9 ; modified, 97 ; asymptotically equivalent to ML estimators, (Exercise 19.13) 101 ; mean and variance, (Exercise 19.14) 101.

Minimum mean-square-error, *see* Mean-square-error.

Minimum Variance (MV) estimators, 8–18 ; and MVB estimators, 12 ; unique, 17 ; correlation with, 17–18 ; uncorrelated with others having zero mean, 18 ; and sufficient statistics, 25–6, (Exercise 17.24) 36 ; among unbiassed linear combinations, 83 ; and ML estimation, 38–9, (Exercise 18.37) 77 ; uniqueness and completeness, 199 ; in truncation and censoring problems, 543–4.

Minimum Variance Bound (MVB), 9 ; condition for attainment, 10 ; for particular problems, (Examples 17.6–10) 10–12 ; MVB estimation and MV estimation, 12 ; improvements to, 12–17 ; asymptotically attained for any function of the parameter, 16 ; and sufficiency, 24 ; for several parameters, 28 ; smaller than variance bound when several parameters unknown, (Exercise 17.20) 35 ; relaxation of regularity conditions for, (Exercises 17.21–2) 35 ; analogues, (Exercise 17.23) 35 ; and ML estimation, 38, 46, (Exercise 18.4) 70 ; generalization, (Exercise 18.32) 76 ; robustly attainable, 509 ; asymptotic improvements in non-regular cases, (Exercises 32.8–11) 551–2 ; in sequential sampling, 634–5.

Mises, R. von, test of fit, 467.

Mitchell, G. J., distribution of X^2, 438.

Mitra, S. K., LS estimators, 91 ; truncated Poisson, (Exercise 32.20) 554 ; models for the $r \times c$ table, 579 ; hypotheses in multi-way tables, 602.

Mixtures of distributions, MVB for proportions, (Exercise 17.2) 32.

ML, *see* Maximum Likelihood.

Moments, method of, efficiency, 67–9, (Exercises 18.9, 18.16, 18.20, 18.27–8) 71–75.

Mood, A. M., two-sample tests, 521 ; median test of randomness, (Exercise 31.12) 529 ; critical values for Sign test, 533.

Moore, A. H., censored samples, 544, 546.

Moore, D. S., X^2 test, 448, 477.

Moore, P. G., estimation using best two order-statistics, (Exercise 32.14) 553.

Moran, P. A. P., distribution of multiple correlation coefficient, 352.

Moses, L. E., scale-shift tests, 521.

Moshman, J., double sampling, 639.

Mosteller, F., iterative estimation in contingency tables and cross-ratios, (Exercise 33.32) 611.

m.s.e., *see* Mean-square-error.

Multicollinearity, 84.

Multinomial distribution, for ML estimation, 49 ; successive approximation to a ML estimator, (Example 18.10) 52–4 ; ML estimation of number of classes when all probabilities equal, (Exercise 18.10) 71–2 ; efficiency of LR tests, 287 ; as basis of tests of fit, 437, 442 ; tests of fit on pea data, (Example 30.1) 439–40 ; homogeneity test, 599.

Multinormal distribution, variance of a quadratic form, (Exercise 19.3) 100 ; case where single sufficient statistic exists without UMP test, (Example 22.11) 186 ; single sufficient statistic for two parameters, (Example 23.4) 202 ; LR tests of independence, 255 ; biserial methods, (Exercise 26.12) 327 ; partial correlation and regression, 330–8, 345–6, (Exercises 27.4, 27.6, 27.21) 357–9 ; invariance of independence under orthogonal transformation, 346, (Exercise 27.7) 357 ; multiple correlation, 351–6 ; with intra-class correlation matrix, (Exercises 27.17–19) 359 ; limiting value of X^2, (Exercises 33.7–8) 605.

Multiple comparisons, in $r \times c$ tables, 582.

Multiple correlation, 347–56 ; coefficient, 347–8 ; geometrical interpretation, 348–9 ; discarding of variables, 349–50 ; conditional sampling distribution in normal case, 350–1, (Exercise 27.13–15) 358 ; unconditional distribution in multinormal case, 351–6, (Exercises 27.14–16, 27.23–5) 358–60 ; unbiassed estimation in multinormal case, 356 ; with uncorrelated regressors (Exercise 27.20) 359.

Multi-way tables, 600–3.

Mundlak, Y., lognormal linear models, 84.

Murphy, R. B., charts for tolerance intervals, 539.

MV, *see* Minimum Variance.

MVB, *see* Minimum Variance Bound.

Myers, M. H., sequential t-tests, 634.

Naddeo, A., confidence intervals from X^2, (Exercise 30.19) 482.

INDEX

Nair, K. R., estimation of functional relation, 420–1; distribution of extreme deviate from sample mean, (Exercise 32.21) 554; tables of studentized extreme deviate, 548.

Nathan, G., stratified contingency tables, 582.

Negative binomial distribution, ML estimation, (Exercises 18.26–7) 75; tolerance limits, 136; truncated, 546; sequential sampling for attributes, (Example 34.1) 613–15.

Neyman, J., consistency of ML estimators, 63; BAN estimators, 95–7; confidence intervals, 103, 116 footnote; most selective intervals, 122; intervals for upper terminal of a rectangular distribution, (Exercises 20.3–4) 137; tests of hypotheses, 169; maximizing power, 173; testing simple H_0 against simple H_1, 174; BCR in tests for normal parameters, (Example 22.8) 183; UMP tests and sufficient statistics, 185, (Exercise 22.11) 194; sufficiency and similar regions, 198; bias in test for normal variance, (Example 23.12) 212; unbiassed tests, 215; sufficiency, similar regions and independence, (Exercises 23.3–4) 229; LR method, 234; tables of power function of t-test, 265; LR tests in k normal samples, (Exercises 24.4–6) 268–9; incidental and structural parameters, 399; consistent estimation of structural relation, 402, 416; consistency of X^2 test of fit, 451; "smooth" test of fit, 460–2; homogeneity tests, 599. See Type A.

Neyman–Pearson lemma, 173–7; extension, 217–18.

Noether, G. E., confidence intervals for ratio of binomial parameters, (Exercise 20.9) 138; ARE, 276; conservativeness of distribution-free procedures in discrete cases, 473, 516.

Non-central confidence intervals, 107.

Non-central F-distribution, 262–4, (Exercise 24.21) 272.

Non-central t-distribution, 264–5, (Exercise 24.22) 272.

Non-central χ^2 distribution, 237–41, 263, (Exercises 24.1–3) 268, (Exercises 24.19–20) 271–2; and ARE, 285–6.

Non-parametric hypotheses, 170; and distribution-free methods, 488.

Normal distribution, estimation of mean, 2, 7; MVB for mean, (Example 17.6) 10; MVB for variance, (Example 17.10) 11; estimation efficiency of sample median, (Example 17.12) 20; estimation efficiency of sample mean deviation, (Example 17.13) 20; sufficiency in estimating mean and variance, (Example 17.15) 24, (Example 17.17) 28; estimation of standard deviation, (Exercise 17.6) 33; MV unbiassed estimation of square of mean, (Exercise 17.7) 33; deficiency in estimating σ^2, (Exercise 17.8) 33; minimum mean-square-error estimation of σ^p, (Exercise 17.16) 34; estimation efficiency of mean difference, (Exercise 17.27) 36; ML estimator of mean, (Example 18.2) 40, (Exercise 18.2) 67; ML estimator of standard deviation, (Example 18.8) 48, (Exercise 18.36) 77; ML estimation of mean and variance, (Examples 18.11, 18.13) 56, 58; estimation of mean restricted to integer values, (Exercises 18.21–2) 73–4; grouping corrections to ML estimators of mean and variance, (Exercise 18.25) 74–5; ML estimation of common mean of populations with different variances, (Exercise 18.30) 76; ML estimation of mean functionally related to variance, (Exercise 18.31) 76; non-existence of a ML estimator, (Exercise 18.34) 77; LS and ML equivalent, 78–9; confidence intervals for mean, (Examples 20.1, 20.3, 20.5) 105, 111, 115; confidence intervals for variance, (Example 20.6) 118–9, (Exercise 20.6) 137, (Exercise 20.10) 138; tables of confidence intervals for variance and for ratio of two variances, 123; confidence intervals for mean with variance unknown, 128–30; point and interval estimation of ratio of two means, (Example 20.7) 130–2; tolerance intervals, 134–6, (Exercises 20.20–2) 140; confidence regions for mean and variance, (Exercise 20.5) 137; confidence intervals for ratio of two variances, (Exercise 20.7) 137, (Exercise 23.18) 232; fiducial intervals for mean, (Example 21.1) 143–4; fiducial intervals for mean with variance unknown, 145–6; confidence and fiducial intervals for problem of two means, 146–57, (Exercises 21.4–5, 21.9–10) 167, (Example 23.10) 208–9, (Example 23.16) 224, (Example 24.2) 236; Bayesian intervals, 157–9; testing simple H_0 for mean, (Examples 22.1–3) 172–6, 180–1, (Examples 22.12–14) 188–92, (Exercise 22.12)

Normal distribution—*contd.*
194, (Example 23.11) 210, 211, 221, (Examples 25.1–5) 274–85; testing normal against double exponential form, (Example 22.5) 177; testing various hypotheses for mean and variance, (Example 22.8) 182–4; testing simple H_0 for variance, (Exercises 22.3, 22.5) 193, 221; non-existence of similar regions (Example 23.1) 197–8; testing composite H_0 for mean, (Example 23.7) 205–6, 215, (Example 23.14) 222, (Example 23.17) 227, (Example 24.1) 235; testing composite H_0 for difference between two means, variances equal, (Example 23.8) 206–7, (Example 23.15) 223; testing composite H_0 for variance, (Examples 23.12–14) 211–15, 221, (Exercise 23.13) 231, (Examples 24.3, 24.5), 242, 251; testing composite H_0 for linear functions of differing means, and for common variance, (Example 23.15) 222–3, (Exercise 23.19) 232; testing composite H_0 for weighted sum of reciprocals of differing variances, (Example 23.16) 223–4; testing composite H_0 for variance-ratio, (Example 23.16) 223–4, (Exercises 23.14, 23.17–18) 231–2; proofs of independence properties using completeness, (Exercises 23.8–9) 230; "peculiar" UMPU tests, (Exercises 23.16, 23.21) 231–2; minimality and single sufficiency, (Exercise 23.31) 233; testing ratio of means, (Exercise 23.32) 233; testing equality of several variances, (Examples 24.4, 24.6) 244, 254, (Exercises 24.4, 24.7) 268–9; in general linear hypothesis, 258–9; tables of power function of t-test, 265; LR tests for k samples, (Exercises 24.4–6) 268–9; asymptotically most powerful tests for the mean, (Example 25.1) 274; ARE of sample median, (Examples 25.2, 25.4–5) 278–85; failure of product-cumulant method in estimating functional relation, 412–14, 427; testing normality, 436, 477–8, 486; choice of classes for X^2 test, 448, 456; "smooth" test for mean, (Example 30.5) 463; robustness of tests, 483–7; use of normal scores in tests, 504–5; choice of two-sample test, 515, 520, 521; paired t-test, 525; truncation and censoring, 544–6; criteria for rejecting outlying observations, 547–9; estimation using best two order-statistics, (Exercise 32.14) 553; distribution of extreme deviate from sample mean, (Exercise 32.21) 554; sequential test of simple H_0 for mean, (Examples 34.8, 34.10) 628, 630; sequential tests for variance, (Example 34.9) 628, (Exercises 34.18–20) 644; sequential t-test, 632–4; sequential estimation of mean, (Example 34.12) 637; double sampling for mean, variance and two means, 638–9; distribution of sample size in sequential sampling, (Exercises 34.12–13) 642–3; *see also* Bivariate normal, Multinormal.

Normal scores, 504–5; in two-sample tests, 516–20, 521, (Exercise 31.23) 531.

Norton, H. W., interaction in multi-way tables, 603.

Nuisance parameter, 195, 217; removal of, (Exercise 30.18) 481; *see also* Studentization.

Null hypothesis, 171 footnote.

OC, operating characteristic, 617–18.

Odell, P. L., singular LS estimation, 89.

Ogawa, J., censored exponential samples, 545.

Ogburn, W. F., data on crime in cities, (Example 27.2) 344.

Olkin, I., unbiassed estimation of normal correlation parameter, 306–7; unbiassed estimation of multiple correlation coefficient, 356.

Operating characteristic, 617–18.

Order-statistics, for Cauchy location, (Example 18.9) 52; in LS estimation of location and scale parameters, 91–5; and similar regions, 197; in tests of fit, (Exercise 30.17) 481; completeness, 490; use of normal scores, 504–5; asymptotic distribution of expected values, 505; Sign test for quantiles, 532–6; confidence intervals for quantiles, 536–7; tolerance intervals, 537–40; in point estimation, 540–1; truncation and censoring, 541–6; outlying observations, 546–50; estimation of quantiles, (Exercise 32.13) 553.

Ordered alternative, for k-sample test, 524.

Ordered categorization, 556.

Ordered $r \times c$ tables, 582–8.

Orthogonal, regression analysis, 371; polynomials in regression, 372–6, (Exercise 28.23) 390; Legendre polynomials, 460 footnote.

INDEX

Outlying observations, 546–50, (Exercise 32.21) 554.
Owen, D. B., jackknifing, 6 ; tables of power of Student's t-test, 265.

Pabst, M. R., efficiency of rank correlation test, 500.
Pachares, J., tables of confidence limits for the binomial parameter and normal variances, 123 ; tables of the studentized range, 548.
Page, E. S., sequential test in the exponential distribution, (Exercise 34.14) 643.
Paired t-test, 525.
Parameter, 1, 170, 506 ; nuisance, 195, 217, (Exercise 30.18) 481.
Parameter-free tests of fit, 459–60, 468, 477, (Exercise 30.10) 480.
Parametric hypotheses, 170.
Partial association, 561–5, 600–3.
Partial correlation and regression, 330–47 ; partial correlation, 330 ; linear partial regression, 334 ; relations between quantities of different orders, 335–8, (Exercises 27.1–3, 27.5) 357 ; approximate linear partial regression, 338–9 ; estimation of population coefficients, 339–40 ; geometrical interpretation, 340–2 ; computations, (Examples 27.1–2) 342–5 ; sampling distributions, 345–6.
Partitions of X^2, 465–6 ; in $r \times c$ tables, 594–8.
Patil, G. P., attainability of MVB, 15.
Patnaik, P. B., non-central χ^2 and F distributions, 239, 241, 263–4, (Exercise 24.2) 268 ; variance of X^2, (Exercise 30.5) 479 ; power function of test of independence in 2×2 tables, 575.
Paulson, A. S., tests for outliers, 548.
Paulson, E., LR tests for exponential distributions, 255, (Exercises 24.16, 24.18) 271 ; rejection of outliers, 548.
Pearson distributions, efficient estimation in, 67–9, (Exercise 18.16) 72.
Pearson, E. S., charts of confidence intervals for the binomial parameter, 123 ; tests of hypotheses, 169 ; maximizing power, 173 ; testing simple H_0 against simple H_1, 174 ; BCR in tests for normal parameters, (Example 22.8) 183 ; UMP tests and sufficient statistics, 185, (Exercise 22.11) 194 ; choice of test size, 191 ; bias in test for normal variance, (Example 23.12) 212 ; unbiased tests, 215 ; LR method, 234 ; non-central χ^2, 239 ; charts of power function for linear hypothesis, 263 ; non-central F, 264, (Exercise 24.21) 272 ; LR tests in k normal samples, (Exercises 24.4–6) 268–9 ; tests of fit, 461, 468, (Exercise 30.12) 481 ; studies of robustness, 483 ; rejection of outlying observations, 547, 548 ; tables of studentized range, 548 ; models for 2×2 tables, 570 ; power functions of tests in 2×2 tables, 575.
Pearson, K., development of correlation theory, 291, 495 ; tetrachoric correlation, 318 ; biserial correlation, (Table 26.7) 319, 321, 322 ; " spurious " correlation, (Exercise 26.18) 328 ; X^2 test of fit, 438 ; estimation using best two order-statistics, (Exercise 32.14) 553 ; coefficient of contingency, 577 ; standard error of X^2 in $r \times c$ tables, 580 ; multi-way tables, 600, (Exercise 33.7) 605.
Peers, H. W., Bayesian and confidence intervals, 164.
Permutation tests, 490.
Pettigrew, H. M. conditional moments of Poisson k-statistics, (Exercise 17.19) 35.
Pillai, K. C. S. tables of studentized extreme deviate, 548.
Pinkham, R. S., moments of non-central t, 265.
Pitman, E. J. G., " close " estimation, 8 ; minimum m.s.e. invariant estimators, 22 ; distributions possessing sufficient statistics, 26, 28, 30 ; confidence intervals for ratio of variances in bivariate normal distribution, (Exercise 20.19) 139 ; sufficiency of LR on simple H_0, (Exercise 22.13) 194 ; unbiased invariant tests for location and scale parameters, 252 ; ARE, 276 ; ARE of Sign test, (Exercise 25.2) 288, 535 ; ARE of Wilcoxon test, (Exercises 25.3–4) 288, 514, (Exercise 31.22) 530 ; test of independence, 492 ; two-sample test, 507–8 ; consistency of Wilcoxon test, 513, (Exercise 31.16) 530.
Plackett, R. L., on origins of LS theory, 83 ; LS theory in the singular case, 88, (Exercises 19.7–8) 100 ; general linear model, (Exercise 19.2) 100 ; continuity corrections, 526, 576 ; simplified estimation in censored samples, 543 ; censored normal samples, 544 ; truncated Poisson distribution, 545, (Exercise 32.20) 554 ; c.f. and cumulants of logistic order-statistics, (Exercise 32.12) 552 ; estimation of quantiles,

Plackett, R. L.—*contd.*
(Exercise 32.13) 553 ; interaction in multiway tables, 603.

Point-biserial correlation, 323–4, (Exercise 26.5) 325.

Point estimation, 1–102.

Poisson distribution, MVB for parameter, (Example 17.8) 11 ; conditional moments of k-statistics, (Exercise 17.19) 35 ; absurd unbiased estimator in truncated case, (Exercise 17.26) 36 ; MCS estimation, (Example 19.11) 97–9 ; in linear model, (Exercise 19.16) 101 ; confidence intervals, (Example 20.4) 111–13 ; tables of confidence intervals, 123 ; tolerance limits, 136 ; testing simple H_0, (Exercise 22.1) 193, 221 ; UMPU tests for difference between two Poisson parameters, (Exercise 23.21) 232 ; and non-central χ^2 distribution, (Exercise 24.19) 271 ; truncation and censoring, 545, (Exercises 32.19–20, 32.22–4) 554–5 ; dispersion test, 599–600 ; inverse sampling, 615 ; sequential estimation, (Example 34.13) 638.

Polychoric estimation of normal correlation, 581.

Polynomial regression, 370–1, 372–6.

Polytomy, 565.

Potthoff, R. F., testing homogeneity, 599.

Power, of a test, 172–3 ; function, 187–8 ; *see* Tests of hypotheses.

Pratt, J. W., expected length of confidence intervals, (Exercises 20.17–18) 139 ; unbiased estimation of normal correlation parameter, 306–7 ; unbiased estimation of multinormal multiple correlation, 356 ; robustness of two-sample tests, 520.

Prediction intervals, 378.

Press, S. J., confidence interval comparisons for problem of two means, 155.

Price, R., non-central F distributions, 263.

Probability integral transformation, in tests of fit, 459, 467, (Exercise 30.10) 480.

Prout, T., *see* Lewontin, R.C.

Przyborowski, J., confidence limits for the Poisson parameter, 123.

Pseudo-inverse of singular matrix, (Exercise 27.21) 359.

Puri, M. L., k-sample tests, 522.

Puri, P. S., homogeneity tests, 599.

Putter, J., treatment of ties, 527.

Pyke, R., spacings, 532.

Quadratic forms, mean and variance, (Exercises 19.3, 19.9) 100 ; non-central χ^2 distribution of, 239–40, (Exercise 24.10) 270.

Quantiles, tests and confidence intervals for, 532–7.

Quenouille, M. H., jackknifing, 5, (Exercises 17.17–18) 34–5 ; fiducial paradoxes, 162 ; random exponential deviates, (Example 30.2) 449 ; partitions of X^2, 598.

Quesenberry, C. P., tests for outliers, 548.

Qureishi, A. S., estimation of logistic parameters, 546.

Raghavachari, M., scale-shift tests, 521.

Raghunandanan, K., robustness, 487 ; logistic estimation, 546.

Raj, D., truncated and censored Gamma distributions, 546.

Ramachandramurty, P. V., robust estimation, 520.

Ramachandran, K. V., tables of confidence limits for normal variance and ratio, 123 ; 204, (Exercise 23.17) 231.

Randomness, tests of, 501–5.

Rank tests, 494 ; using rank correlation coefficients, 494–504 ; optimum, 504–5 ; for two- and k-sample problems, 510, 522–3 ; independence of symmetric functions, (Exercise 32.2) 551.

Ranks, as instrumental variables, 422–4, (Exercises 29.9–10) 434.

Rao, B. R., analogue of MVB, (Exercise 17.23) 35 ; censoring, truncation and efficiency, 543, (Exercise 32.26) 535.

Rao, C. R., MVB, 9, (Exercise 17.20) 35 ; sufficiency and MV estimation, 25, 28 ; efficiency and correlation, 46 ; LS estimators, 91 ; MCS estimation, 97 ; second-order estimating efficiency, 99 ; power of tests, (Exercise 22.15) 194.

Rastogi, S. C., confidence bands in linear regression, 385.

Ratio, confidence intervals for, *see* particular distributions.

Rectangular distribution, sufficiency of largest observation for upper terminal, (Example 17.16) 24 ; sufficiency when both terminals depend on parameter, (Examples 17.18, 17.21–3) 30–1 ; MV unbiased estimation, (Exercise 17.24) 36 ; ML estimator of terminal, (Example 18.1) 40, 48–9 ; a ML estimator a function of only one of a pair of sufficient statistics, (Example 18.5)

44–5 ; ML estimation of both terminals, (Example 18.12) 57 ; non-unique ML estimator of location, (Exercise 18.17) 73 ; ordered LS estimation of location and scale, (Example 19.10) 94 ; confidence intervals for upper terminal, (Exercises 20.2–4, 20.15) 137–9 ; UMP one-sided tests of simple H_0 for location parameter, (Exercise 22.8) 193 ; minimal sufficiency, (Exercise 23.12) 230 ; power of conditional test, (Exercise 23.23) 232 ; UMP test of composite H_0 for location parameter, (Exercise 23.26) 233 ; connexions with χ^2 distribution, 246–7 ; LR test for location parameter, (Exercise 24.9) 269, (Exercise 24.16) 271 ; distribution of order-statistics, (Exercise 30.17) 481 ; efficiency of Wilcoxon two-sample test, 516, (Exercise 31.22) 530 ; asymptotic variance bound for mean, (Exercise 32.10) 552 ; estimation from censored samples, (Exercise 32.25) 555.

Regression, and dependence, 290–1 ; curves, 293 ; and covariance, 295–6 ; linear, 296–9 ; coefficients, 297 ; equations, 297 ; historical note, 300 ; computation, (Examples 26.6–7) 301–4 ; scatter diagram, 304 ; standard errors, 304–5 ; tests and independence tests, 308 ; linearity and correlation coefficient and ratios, 311–13 ; LR test of linearity, 313 ; discarding variables, 349–51 ; stepwise, 349–50 ; generally, 361–90 ; analytical theory, 361–9, (Exercise 28.4) 386 ; criteria for linearity, 365–7 ; general linear model, 369–85 ; adjustment, 371, (Exercise 28.20) 389 ; segmented curves, 371 ; conditional and unconditional inference, 371 ; orthogonal, 371–6, 379, (Exercise 28.23) 390 ; confidence intervals and tests for parameters, 377–80, (Exercises 28.14–18, 21–2) 387–90 ; confidence region for a regression line, 380–5, (Exercise 28.13) 387 ; tests of difference between coefficients, (Exercises 28.14–15) 387 ; use of supplementary information, (Exercises 28.17–18) 388 ; graphical fitting, (Exercise 28.19) 389 ; and functional relations, 392–3, 394–5, 396, 402 ; and controlled variables, 424–5, effect of observational errors, 430–2 ; *see also* Correlation, Least Squares, Linear model, Multiple correlation, Partial correlation and regression.

Regressor, 361.

Residuals, 85, 335 footnote; rule for cancellation of subscripts, (Exercise 27.5) 357.

Resnikoff, G. J., tables of non-central t-distribution, 265.

Restricted parameters, ML estimation of, (Exercises 18.21–2) 73–4.

Richardson, D. H., estimation of functional relation, 421.

Rider, P. R., truncated distributions, 546.

Roberts, G. C., distribution of X^2, 438.

Robbins, H., variance bounds for estimators, 16 ; estimation of normal standard deviation, (Exercise 17.6) 33 ; distribution of Student's t when means of observations differ, 485 ; tolerance intervals and order-statistics, 538 ; moments in sequential estimation, 635.

Robison, D. E., intersection of polynomial regressions, 380.

Robson, D. S., jackknife estimation of terminal, 6, (Exercise 17.13) 34 ; uncorrelated LS residuals, (Exercise 19.18) 102 ; orthonormal polynomials, (Exercise 28.23) 390.

Robustness, 483–7, 509, 520.

Romig, H. G., double sampling, 627.

Rosenblatt, M., distribution of Hoeffding's test, 501.

Rosenthal, I., sampling experiments on G in ordered tables, 586.

Roth, R., LS fitting, 371.

Rothenberg, T. J., Cauchy location estimation, 52.

Roy, A. R., choice of classes for X^2 test, 448.

Roy, J., truncated Poisson, (Exercise 32.20) 554.

Roy, K. P., asymptotic distribution of LR statistic, 241.

Roy, S. N., simultaneous confidence intervals, 133 ; models for the $r \times c$ table, 579 ; hypotheses in multi-way tables, 602–3.

Rubin, H., robust linear estimators, 487.

Runs test, (Exercises 30.7–9) 479–80, 520.

Rushton, S., orthogonal polynomials, 375.

Saleh, A. K. M. E., censored exponential samples, 545.

Sampford, M. R., truncated negative binomial, 546.

Sample d.f., 466–7.

Sampling variance, as criterion of estimation, 7 ; *see* Minimum Variance estimators.

718 INDEX

Sankaran, M., variance bound, (Exercise 17.23) 35.

Sarhan, A. E., ordered LS estimation of location and scale parameters, (Exercises 19.11–12) 101, (Exercise 32.18) 554 ; order-statistics, 532 ; censored samples, 544, 545, 546, 549, (Exercise 32.25) 555.

Savage, I. R., bibliography of non-parametric statistics, 489 ; normal scores two-sample test, 516, 519 ; independence of symmetric and rank statistics, (Exercise 32.2) 551.

Savage, L. J., sufficiency, 23 ; Conditionality and Likelihood Principles, 227.

Saw, J. G., censored normal samples, 545.

Scale parameters, *see* Location and scale parameters.

Scale-shift alternative, 521.

Scatter diagram, 304.

Scedastic curve, 362.

Schafer, R. E., test of fit for exponential, 478.

Schatzoff, M., comparison of tests, 287.

Scheffé, H., ratio of normal means, (Example 20.7) 132 ; problem of two means, 148–53, 155, (Example 23.10) 208 ; linear function of χ^2 variates, (Exercise 21.8) 167 ; completeness, 199 ; completeness of the linear exponential family, 200 ; minimal sufficiency, 203 ; completeness and similar regions, 205 ; unbiased tests, 215 ; UMPU tests for the exponential family, 216 ; non-completeness of Cauchy family, (Exercise 23.5) 229 ; minimal sufficiency for binomial and rectangular distributions, (Exercises 23.11–12) 230 ; a UMPU test for normal distribution, (Exercise 23.16) 231 ; unbiased confidence interval for ratio of normal variances, (Exercise 23.18) 232 ; UMPU tests for Poisson and binomial distributions, (Exercises 23.21–2) 232 ; controlled variables, 425, 429 ; completeness of order-statistics, 490 footnote ; confidence intervals for quantiles, 537 ; tolerance regions, 540.

Scheuer, E. M., tables of non-central t, 265.

Schneiderman, M. A., sequential t-tests, 634.

Schucany, W. R., jackknifing, 6.

Scoring for parameters, in ML estimation, 51.

Scott, A. J., Behrens–Fisher distribution, 157.

Scott, E. L., consistency of ML estimators, 63, 402, 416 ; incidental and structural parameters, 399.

Seal, H. L., on origins of LS theory, 83 ; runs test as supplement to X^2 test of fit, 458, (Exercise 30.7) 480.

Seelbinder, B. M., double sampling, 639.

Segmented curves, fitting by LS, 371.

Sen, P. K., k-sample permutation tests, 523.

Sequential methods, 612–45 ; for attributes, (Exercise 18.18) 73, 612–27 ; closed, open, and truncated schemes, 613 ; tests of hypotheses, 617 ; OC, 617–18 ; ASN, 618–19 ; SPR tests, 619–31, (Exercises 34.4, 34.7–11, 34.21–2) 641–5 ; stopping rules, rectifying inspection and double sampling in applications, 627 ; efficiency, 629–31 ; composite hypotheses, 631–4, (Exercise 34.21) 644 ; sequential t-test, 632–4 ; estimation, 634–8, (Exercises 34.5–6) 641 ; double sampling, 627, 638–9, (Exercises 34.15–16) 643 ; distribution-free, 639–40.

Shah, B. K., logistic distribution, (Exercise 32.12) 552–3.

Shah, S. M., truncated binomials, 546.

Shapiro, S. S., testing normality, 478.

Sharpe, K., tolerance intervals, 136.

Shaw, G. B., on correlation and causality, 291.

Shenton, L. R., cumulants of ML estimators, 50 ; ML estimators for Gamma, 69 ; efficiency of method of moments, (Exercise 18.23) 75.

Sheppard's corrections, and ML grouping corrections, (Exercise 18.25) 74–5.

Shorak, G. R., tests for ratios of scale parameters, 521.

Shorak, R. A., tests in censored samples, 546.

Shorrock, R., attainability of MVB, 15.

Shrivastava, M. P., estimation of functional relation, 420.

Sichel, H. S., ML estimation in lognormal distribution, (Exercises 18.7–9) 71.

Sidák, Z., rank tests, 505.

Siddiqui, M. M., robustness, 487 ; multivariate order-statistics, 543 ; censored exponential samples, 545.

Siegel, S., two-sample test against scale-shift, 521.

Sievers, G. L., test efficiency, 287.

Sign test, ARE, (Exercises 25.1–2) 288, 532–536, (Exercises 32.1–3, 32.5) 551 ; sequential, 639.

Significance level, 171 footnote.

Significance tests, *see* Tests of hypotheses.

Sillitto, G. P., order-statistics, 547.

Silverstone, H., MVB, 9 ; comparison of BAN estimators, 99.

Silvey, S. D., multicollinearity in LS, 84.

Similar regions, similar tests, 196–8, 205 ; see Tests of hypotheses.

Simple hypotheses, 169–94 ; see Hypotheses, Tests of hypotheses.

Singh, R., completeness, 199.

Singular matrix, pseudo-inverse of, (Exercise 27.21) 359.

Size of a test, 171 ; choice of, 190–2.

Slakter, M. J., approximation of X^2, 457.

Smirnov, N. V., test of fit, 467 ; tables of Kolmogorov test of fit, 473 ; one-sided test of fit, 474 ; two-sample test, 520.

Smith, C. A. B., moments of X^2 in a $r \times c$ table, (Exercise 33.9) 605.

Smith, J. H., ML estimation of cell probabilities in a $r \times c$ table, (Exercise 33.29) 610.

Smith, K., MCS estimation in the Poisson distribution, (Example 19.11) 98–9.

Smith, S. M., cumulants of a ML estimator, 49.

Smith, W. L., truncation and sufficiency, 541.

" Smooth " tests of fit, 460–6.

Sobel, M., an independence property of the exponential distribution, (Exercise 23.10) 230 ; censored samples, 545, 546, (Exercise 32.17) 553.

Solari, M. E., linear functional relationship, 400.

Soper, H. E., standard error of biserial correlation coefficient, 322–3.

Spjøtvoll, E., tests for non-exponential families, 226.

SPR tests, sequential probability ratio tests, 619–31, (Exercises 34.4, 34.7–11, 34.21–2) 641–5.

Sprent, P., linear functional relations, 410.

Sprott, D. A., fiducial and Bayesian methods, 164.

Spurgeon, R. A., tables of non-central t, 265.

"Spurious " correlation, (Exercise 26.18) 327–8.

Srinivasan, R., problem of two means, 155 ; tests of fit, 478 ; logistic estimation, 546.

Statistic, 1.

Statistical relationship, 290.

Stein, C., estimation of restricted parameters, (Exercise 18.22) 74 ; non-similar tests of composite H_0, 209 ; a useless LR test, (Example 24.7) 256 ; invariance and sufficiency, 267 ; tests of symmetry, 525 ; SPR tests, 620 ; sequential completeness, 634 footnote ; double sampling, 638–9.

Stephan, F. F., estimation of cell probabilities in a $r \times c$ table, (Exercises 33.29, 33.32) 610–11.

Stephens, M. A., tests of fit, 468.

Stepwise regression, 349–50, 379.

Sterne, T. E., confidence intervals for a proportion, 123.

Stevens, W. L., confidence intervals in the discontinuous case, 124 ; fiducial intervals for binomial parameter, (Exercise 21.3) 166 ; runs test, (Exercise 30.8) 480.

Stochastic convergence, 3 ; see Consistency in estimation.

Stochastically larger, 495.

Stopping rule, 592.

Strawderman, W. E., unbiassed tests for variances, 255.

Structural parameters, 393.

Structural relationship, see Functional and structural relations.

Stuart, A., estimation of normal correlation parameter, (Exercise 18.12) 72 ; ARE and maximum power loss, 284 ; average critical value and ARE, (Exercise 25.10) 289 ; intra-class correlated multinormal distribution, (Exercises 27.18–19) 359 ; critical region for X^2 test of fit, 439 ; joint distribution of rank correlations, 499 ; ARE of tests of randomness, 505, (Exercises 31.10–12) 528–9 ; correlation between ranks and variate-values (Exercises 31.13–14) 529–30 ; testing difference in strengths of association, 585 ; test of identical margins in $r \times r$ tables, (Exercise 33.24) 609.

" Student " (W. S. Gosset), studentization, 124 ; outlying observations, 547.

Studentization, 128–30, (Example 20.7) 30–2.

Studentized extreme deviate, 548.

Studentized maximum absolute deviate, 548.

Studentized range, 548.

Subrahmaniam, K., truncated Poisson, (Exercise 32.20) 554.

Sufficiency, generally, 22–31 ; definition, 22–4 ; factorization criterion, 23 ; and MVB, 24 ; functional relationship between sufficient statistics, 25 ; and MV estimation, 25–6 ; distributions possessing sufficient statistics, 26–7 ; for several parameters, 27–8 ; when range of parent depends on parameter, 29–31, (Exercise 24.17) 271 ; single and joint, 27 ; distribution of sufficient statistics for exponential

Sufficiency—*contd.*
　family, (Exercise 17.14) 34, (Exercise 24.17) 271 ; and ML estimation, 38–9, (Example 18.5) 44, 54–7 ; and BCR for tests, 177–8, 185–7, (Exercises 22.11, 22.13) 194 ; optimum test property of sufficient statistics, 196 ; and similar regions, 198, 205 ; minimal, 202–4, (Exercise 18.13) 72, (Exercise 23.31) 233 ; ancillary statistics and quasi-sufficiency, 226 ; independence and completeness, (Exercises 23.6–7) 229–30 ; and LR tests, 255 ; and invariance, 267 ; and nuisance parameters, (Exercise 30.18) 481 ; and truncation, 541.

Sufficient statistics, *see* Sufficiency.

Sukhatme, P. V., LR tests for k exponential distributions, (Exercises 24.11–13) 270.

Sundrum, R. M., estimation efficiency and power, 179, (Exercise 22.7) 193 ; power of Wilcoxon test, 516.

Superefficiency, 46.

Supplementary information, in regression, (Exercises 28.17–18) 388 ; instrumental variables, 414–24.

Swamy, P. S., efficiency under truncation and censoring, 543, 544.

Swed, F. S., tables of runs test, (Exercise 30.8) 480.

Swindel, B. F., LS estimation of variance, 91.

Symmetrical distributions, ML estimators of location and scale parameters asymptotically independent, 65–6 ; ordered LS estimators of location and scale parameters uncorrelated, 93–4 ; condition for LS estimator of location parameter to be sample mean, (Exercise 19.10) 100 ; ARE of Sign test, 534–5, (Exercise 32.1) 552 ; tests of symmetry, 524–6.

Symmetry tests, 524–6, (Exercise 31.25) 531.

t distribution, non-central, 264–5 ; tables of power of t-test, 265.

Taguti, G., tables of normal tolerance limits, 136.

Takeuchi, K., exponential location test, (Exercise 23.25) 232.

Tamura, R., scale-shift tests, 521.

Tang, P. C., non-central F distribution and tables of power function for linear hypothesis, 263–4 ; non-central χ^2 distribution, (Exercise 24.1) 268.

Tanis, E. A., testing exponential distributions, (Exercise 24.13) 270.

Tarter, M. E., logistic order-statistics, (Exercise 32.12) 553.

Tate, R. F., MV unbiassed estimation, (Exercise 17.24) 36 ; tables of confidence intervals for a normal variance, 123 ; shortness of these intervals, (Exercise 20.10) 138 ; biserial and point-biserial correlation, 323–4, (Exercises 26.10–12) 326–7 ; truncated Poisson distribution, 545, (Exercise 32.19) 554.

Tchebycheff inequality, and consistent estimation, (Example 17.2) 3–4.

Teicher, H., characterization by form of ML estimator (Exercise 18.2) 70 ; moments in sequential estimation, 635.

Terminal, estimation of, (Exercise 17.13) 34.

Terpstra, T. J., k-sample test, 524.

Terry, M. E., tests using normal scores, 505, 516.

Tests of fit, 436–82 ; LR and Pearson tests for simple H_0, 437–40 ; X^2 notation, 438 footnote ; composite H_0, 440–7 ; choice of classes for X^2 test, 447–50, 454–7 ; moments of X^2 statistic, 450–1, (Exercises 30.3, 30.5) 479 ; consistency and unbiassedness of X^2 test, 451–3 ; limiting power of X^2 test, 453–4 ; recommendations for X^2, 457 ; use of signs of deviations, 458–9, (Exercises 30.7–9) 479–80 ; other tests than X^2, 459–78 ; " smooth " tests, 460–6 ; connexion between X^2 and " smooth " tests, 465 ; components of X^2, 465–6 ; tests based on sample d.f., 466–478 ; Smirnov test, 467–8 ; Kolmogorov test, 468–78 ; comparison of X^2 and Kolmogorov tests, 474–6 ; tests of normality, 477–8.

Tests of hypotheses, 169–289 ; and confidence intervals, 122–3, 128, 215 ; size, 171, 190–2 ; power, 172–3, 187–8 ; BCR, 173 ; simple H_0 against simple H_1, 174 ; randomization in discontinuous case, 174–5 ; BCR and sufficient statistics, 177–8, 185–7 ; power and estimation efficiency, 179, (Exercise 22.7) 193 ; simple H_0 against composite H_1, 180–2 ; UMP tests, 180 ; no UMP test generally against two-sided H_1, 181–2 ; UMP tests with more than one parameter, 182–5, (Exercise 22.11) 194 ; one- and two-sided tests, 188–90 ; composite H_0, 195 ;

optimum property of sufficient statistics, 196; similar regions and tests, 196–9, 205, 214–15; existence of similar regions, 197–8, (Exercises 23.1–3) 229, (Exercise 23.29) 233; similar regions, sufficiency and bounded completeness, 198, 205–9; most powerful similar regions, 205; non-similar tests of composite H_0, 209; bias in, 209–15; unbiassed tests and similar tests, 214; older terminology, 215; UMPU tests for the exponential family, 216–26; ancillary statistics and conditional tests, 226–8; LR tests, 234–57; unbiassed invariant tests for location and scale parameters, 252–5, (Exercise 24.15) 271; general linear hypothesis, 257–67; comparison of, 273–89; *see also* Asymptotic relative efficiency, Hypotheses, Likelihood Ratio tests, Sequential methods.

Tetrachoric, correlation, 316–19, (Exercises 26.5–6) 325; series and canonical correlations, 590.

Thatcher, A. R. binomial prediction, 164.

Theil, H., estimation of functional relation, 421, 422, 429, (Exercises 29.9–10) 434.

Thompson, J. R., minimum m.s.e. estimation, (Example 17.14) 22.

Thompson, R., symmetry tests, 526.

Thompson, W. R., confidence intervals for median, 518; outlying observations, 548.

Tienzo, B. P., studentized extreme deviate, 548.

Ties, and distribution-free tests, 526–7.

Tiku, M. L., non-central χ^2, 239; non-central F, 263, 264, (Exercise 24.21) 272; tables of power of LR test, 263; tests of fit, 468; censored samples, 544, 546.

Tilanus, C. B., Cauchy location estimation, 52.

Tocher, K. D., optimum exact test for 2×2 table, 574.

Todhunter, I., Arbuthnot's use of Sign test, 532 footnote.

Tokarska, B., tables of power function of Student's t-test, 265.

Tolerance intervals, for a normal distribution, 133–6, (Exercises 20.20–2) 139–40; distribution-free, 537–40, (Exercises 32.6–7) 551.

Tolerance regions, 541.

Toro-Vizcarrondo, C. E., non-central F, 263.

Transformations, of functional relations to linearity, 429–30; to normality, 487.

Trend, 501.

Trickett, W. H., tables for problem of two means, 155.

Trimmed estimators, 544, 549.

Truelove, A. J., variance bounds, 16.

Truncated distributions, 522–7; estimation of truncation point, (Exercise 17.13) 34.

Truncated estimators, and m.s.e., 21, (Exercise 17.15) 34.

Truncated sequential schemes, 613.

Tschuprow, A. A., coefficient of association, 577.

Tukey, J. W., fiducial paradox, 162; two-sample test against scale-shift, 521; confidence intervals for quantiles, 537; tolerance regions, 540; truncation and sufficiency, 541; outlying observations, 549.

Two means problem, *see* Normal distribution.

Two-sample tests, 505–21.

2×2 tables, association in, 556–61, (Exercises 33.1–2) 604; partial association, 561–5; probabilistic interpretation of measures, 565–7; large-sample tests of independence, 567–9; exact test of independence for different models, 569–75; continuity correction, 575–6, (Exercise 33.5) 604; components of X^2 in $2 \times 2 \times 2$ tables, 600, (Exercises 33.27–8) 610; interaction in $2 \times 2 \times 2$ tables, 603; test of identical margins in 2^c tables, (Exercise 33.30) 611.

Type A, A_1, B, B_1, C tests, 215.

Type A contagious distribution, ML and moments estimators, (Exercise 18.28), 75.

Type M unbiassed tests, 211.

Type I, Type II censoring, 541–2.

Type IV (Pearson) distribution, centre of location and efficiency of method of moments, (Exercise 18.16) 72–3.

UMP, uniformly most powerful, 180.

UMPU, uniformly most powerful unbiassed, 211.

Unbiassed estimation, 4–5; *see* Bias in estimation.

Unbiassed tests, 211; *see* Tests of hypotheses.

Uniform distribution, *see* Rectangular distribution.

Uniformly most powerful (UMP), 180; *see* Tests of hypotheses.

Uniformly most powerful unbiassed (UMPU), 211; *see* Tests of hypotheses.

Vail, R. W., two-stage LS, (Exercise 28.24) 390.

Van der Parren, tables of confidence limits for median, 537.
Van der Reyden, D., tables of orthogonal polynomials, 375.
Van der Vaart, H. R., Wilcoxon two-sample test, 513, 516.
Van der Waerden, B. L., two-sample test, 517.
Van Eeden, C., ARE and correlation, (Exercise 25.9) 289; robust estimation, 509; rectangular sum approximation to Wilcoxon test, 512; scale-shift tests, 521.
Van Yzeren, J., estimation of functional relation, 421.
Variance, generalized, see Generalized variance.
Variance, lower bounds to, see Minimum Variance Bound.
Verdooren, L. R., tables of Wilcoxon test, 512.
Villegas, C., non-linear LS, 79, 429; linear functional relations, 410.
Votaw, D. F., Jr., truncation and censoring for the exponential distribution, 545, (Exercise 32.16) 553.

Wald, A., consistency of ML estimators, 43; asymptotic normality of ML estimators, 57; tolerance intervals for a normal distribution, 134–5; problem of two means, 155; asymptotic distribution of LR statistic, 241; test consistency, 250, 273; asymptotic properties of LR tests, 256; optimum property of LR test of linear hypothesis, 266; asymptotically most powerful tests, 274; estimation of functional relation, 416–17, (Exercises 29.4–5) 433; choice of classes for X^2 test of fit, 449, 455–7; unbiassedness of X^2 test of fit, 452; one-sided test of fit, 474; variance of X^2 statistic, (Exercise 30.3) 479; runs test, (Exercise 30.8) 480, 520; tolerance regions, 540; sequential methods, 613, 619–23, 629, 631–2, (Exercises 34.3–4, 34.7, 34.11–13, 34.18–19) 641–4.
Walker, A. M., efficiency of estimators, 46.
Wallace, D. L., asymptotic expansions in problem of two means, 155.
Wallace, T. D., non-central F, 263; two-stage LS, (Exercise 28.24) 390.
Walley, D., tables of normal scores, 520.
Wallis, W. A., tolerance limits, 135; Wilcoxon test, 511, 512, (Exercise 31.15) 530; k-sample test, 522; treatment of ties, 527.

Walsh, J. E., measure of test efficiency, 286; effect of intra-class correlation on Student's t-test, (Exercise 27.17) 359; power of Sign test, 535; tolerance intervals, 540; censored normal samples, 545; outlying observations, 550.
Walton, G. S., binomial and Poisson confidence intervals, 123.
Watkins, T. A., jackknifing, 6.
Watson, G. S., LS theory, 91; problem of two means, (Example 23.10) 209; non-completeness and similar regions, (Exercise 23.15) 231; distribution of X^2 test of fit, 442, 448, 466, (Exercises 30.2, 30.21) 479–82; modified Smirnov test, 468; Bayesian contingency table analysis, 603.
Weather and crops, Hooker's data, (Example 27.1) 342–4.
Weibull distribution, censored samples, 546.
Weiss, L., sequential tests, 631.
Weissberg, A., tables of normal tolerance limits, 136.
Welch, B. L., problem of two means, 153–5; Bayesian and confidence intervals, 164; power of conditional tests, 226, 228, (Exercise 23.23) 232; non-central t-distribution, 265.
Wette, R., ML bias in Gamma distribution, (Exercise 18.15) 72.
Whitaker, L., data on deaths of women, 98.
Whitcomb, M. G., unbiassed tests for a normal variance, 222.
White, C., tables of Wilcoxon test, 512.
Whitlock, J. H., jackknife estimation of terminal, 6, (Exercise 17.13) 34.
Whitney, D. R., Wilcoxon test, 511.
Whittinghill, M., testing homogeneity, 599.
Wicksell, S. D., analytical theory of regression, 363, (Exercise 28.4) 386; criterion for linearity of regression, 366.
Widder, D. V., Laplace transform, 199 footnote.
Wiener, N., completeness for location parameter, 199.
Wijsman, R. A., invariance and sufficiency, 267, 634.
Wilcoxon, F., two-sample test, 511.
Wilcoxon test, 510–16, 520, 521; ARE, (Exercises 25.3–4) 288, 514–16, (Exercises 31.14–19, 31.22–24) 529–31; for symmetry, 526, (Exercise 31.25) 531; extension to censored samples, 546.
Wilénski, H., confidence limits for the Poisson distribution, 123.

INDEX

Wilk, M. B., moments of non-central t, 265; testing normality, 478; censored Gamma samples, 546.

Wilkinson, G. N., truncated binomial, 546.

Wilks, S. S., shortest confidence intervals, 120; smallest confidence regions, 133; confidence intervals for upper terminals of rectangular distribution, (Exercise 20.2) 137; asymptotic distribution of LR statistic, 241; review of literature of order-statistics, 532; tolerance intervals, 538.

Williams, C. A., Jr., X^2 test of fit, 456, 476.

Williams, E. J., comparison of predictors, (Exercise 28.22) 389–90; scoring methods in $r \times c$ tables, 585; canonical analysis, (Exercise 33.13) 607.

Williams, J. S., LS estimation, 91.

Winsorized estimators, 544, 549.

Wise, M. E., approximations to X^2, 438.

Wishart, J., non-central χ^2 and F distributions, 239, 263, (Exercise 24.1) 268; moments of multiple correlation coefficient, 354, (Exercises 27.10–11) 358.

Wisniewski, T. K. M., binomial homogeneity, 599.

Witting, H., ARE of Wilcoxon test, 516; ARE of Sign test, 536.

Wolfowitz, J., non-existence of ML estimator, (Exercise 18.34) 77; tolerance intervals for a normal distribution, 134–5; test consistency, 250, 273; optimum properties of LR test for linear hypothesis, 265; consistency of ML estimators, 402; one-sided test of fit, 474; comparison of X^2 and Kolmogorov tests, 476; tests of normality, 477; runs test, (Exercise 30.8) 480, 520; sequential estimation, 635, (Exercise 34.6) 641.

Women's Measurements and Sizes, data from, (Tables 26.1–2) 292–3.

Woodcock, E. R., Poisson confidence limits, 123.

Woodward, J., truncated normal distribution, 544.

Working, H., confidence region for a regression line, 380, 383–5, (Exercise 28.13) 387.

Wormleighton, R., tolerance regions, 540.

Wu, D. M., estimation of functional relation, 421.

Wynn, H. P., confidence bands in regression, 385.

χ^2 distribution, *see* Gamma distribution, Non-central χ^2 distribution.

X^2 test of fit, 437–59, 465–6, 474–6, 477, (Exercises 30.1–6, 30.9, 30.19–21) 479–82; partitions of, 465–6; in 2×2 tables, 569, 575; in $r \times c$ tables, 576–9, (Exercise 33.9) 605; partitions in $r \times c$ tables, 594–8; dispersion tests, 598–600, (Exercises 33.20–22) 608–9; in multi-way tables, 600–3, (Exercises 33.7–8) 605; *see* Tests of fit.

Yang, Y. Y., confidence bands in linear regression, 385.

Yarnold, J. K., X^2 tests, 457.

Yates, F., fiducial inference, 161; testing difference between regressions, (Exercise 28.14) 387; tests using normal scores, 505, 520; data on mal-occluded teeth, (Example 33.5) 572; continuity correction in 2×2 tables, 575; scoring in $r \times c$ tables, 588, (Exercise 33.12) 606.

Yields of wheat and potatoes, data on, (Table 26.3) 303.

Young, A. W., standard error of X^2 in $r \times c$ tables, 580.

Yule, G. U., development of correlation theory, 291; partial correlation notation, 330; large-sample distributions of partial coefficients, 346; relations between coefficients of different orders, (Exercise 27.1) 357; data on inoculation, (Example 33.1) 558; coefficients of association and colligation, 559; invariance principle for measures of association, 566–7.

Zackrisson, U., distribution of Student's t in mixed normal samples, 485.

Zacks, S., tolerance limits, 136.

Zahn, D. A., distribution of X^2, 438.

Zalokar, J., tables of studentized maximum absolute deviate, 548.

Zeigler, R. K., double sampling, 639.

Zelen, M., several 2×2 tables, 603.

Zyskind, G., two-stage LS, (Exercise 28.24) 390.